Common Groups in Organic Chemistry

Organic Groups

Abbreviation	Meaning	Structure
Ac	acetyl	$CH_3-\overset{\overset{O}{\|\|}}{C}-R$
	allyl	$H_2C{=}CH-CH_2-R$
Boc	*t*-butyloxycarbonyl	$(CH_3)_3C-O-\overset{\overset{O}{\|\|}}{C}-R$
Bn	benzyl	$Ph-CH_2-R$
n-Bu	*n*-butyl	$CH_3-CH_2-CH_2-CH_2-R$
i-Bu	isobutyl	$(CH_3)_2CH-CH_2-R$
s-Bu	*sec*-butyl	$CH_3-CH_2-\underset{\underset{CH_3}{\|}}{C}H-R$
t-Bu	*tert*-butyl	$(CH_3)_3C-R$
Bz	benzoyl	$Ph-\overset{\overset{O}{\|\|}}{C}-R$
Cbz (or Z)	benzyloxycarbonyl	$Ph-CH_2-O-\overset{\overset{O}{\|\|}}{C}-R$
Et	ethyl	CH_3-CH_2-R
c-Hx	cyclohexyl	(cyclohexyl)$-R$
Me	methyl	CH_3-R
Ph	phenyl	(phenyl)$-R$
Pr	propyl	$CH_3-CH_2-CH_2-R$
i-Pr	isopropyl	$(CH_3)_2CH-R$
Sia	*secondary* isoamyl	$(CH_3)_2CH-\underset{\underset{CH_3}{\|}}{C}H-R$
THP	tetrahydropyranyl	(tetrahydropyranyl)$-R$
Ts	*para*-toluenesulfonyl, "tosyl"	$CH_3-\text{(aryl)}-\overset{\overset{O}{\|\|}}{\underset{\underset{O}{\|\|}}{S}}-R$
	vinyl	$H_2C{=}\overset{\overset{H}{\|}}{\underset{\underset{R}{}}{C}}$

Not all of these abbreviations are used in this text, but they are provided for reference.

Common Reagents and Solvents

Abbreviation		Structure
Ac₂O	acetic anhydride	$CH_3-\overset{\overset{O}{\|\|}}{C}-O-\overset{\overset{O}{\|\|}}{C}-CH_3$
DCC	dicyclohexylcarbodiimide	(cyclohexyl)$-N{=}C{=}N-$(cyclohexyl)
DIBAL or DIBAH	diisobutylalumium hydride	$[(CH_3)_2CHCH_2]_2AlH$
DME, "glyme"	1,2-dimethoxyethane	$CH_3-O-CH_2CH_2-O-CH_3$
diglyme	bis(2-methoxyethyl) ether	$(CH_3-O-CH_2CH_2)_2O$
DMF	dimethylformamide	$H-\overset{\overset{O}{\|\|}}{C}-N\,(CH_3)_2$
DMSO	dimethyl sulfoxide	$CH_3-\overset{\overset{O}{\|\|}}{S}-CH_3$
EtOH	ethanol	CH_3CH_2OH
EtO⁻	ethoxide ion	$CH_3CH_2-O^-$
Et₂O	diethyl ether	$CH_3CH_2-O-CH_2CH_3$
HMPA, HMPT	hexamethylphosphoric triamide	$[(CH_3)_2N]_3\,P{=}O$
LAH	lithium aluminum hydride	$LiAlH_4$
LDA	lithium diisopropylamide	$[(CH_3)_2CH]_2N^-\ \ Li^+$
MCPBA	*meta*-chloroperoxybenzoic acid	(Cl-aryl)$-\overset{\overset{O}{\|\|}}{C}-O-O-H$
MeOH	methanol	CH_3OH
MeO⁻	methoxide ion	CH_3-O^-
MVK	methyl vinyl ketone	$CH_3-\overset{\overset{O}{\|\|}}{C}-CH{=}CH_2$
NBS	*N*-bromosuccinimide	(succinimidyl)$-Br$
PCC	pyridinium chlorochromate	$pyr \cdot CrO_3 \cdot HCl$
Pyr	pyridine	(pyridine)
t-BuOH	*tertiary* butyl alcohol	$(CH_3)_3C-OH$
t-BuOK	potassium *tertiary*-butoxide	$(CH_3)_3C-O^-\ K^+$
THF	tetrahydrofuran	(tetrahydrofuran)
TMS	tetramethylsilane	$(CH_3)_4Si$

ORGANIC CHEMISTRY

ORGANIC CHEMISTRY

FIFTH EDITION

L. G. WADE, JR.

WHITMAN COLLEGE

PEARSON EDUCATION, INC.
UPPER SADDLE RIVER, NEW JERSEY 07458

Library of Congress Cataloging-in-Publication Data

Wade, L. G.
    Organic chemistry / L. G. Wade , Jr—5th ed.
        p. cm.
    Includes bibliographical references and index.
    ISBN 0-13-033832-X (alk. paper)
    1. Chemistry, Organic. I. Title.

QD251.3.W33 2003
547—dc21   2002070086

Senior Editor: Nicole Folchetti
Editor in Chief, Development: Ray Mullaney
Editor in Chief, Physical Sciences: John Challice
Development Editor, Text: John Murdzek
Development Editor, Media: Carol Pritchard-Martinez
Media Editor: Paul Draper
Art Director: Jonathan Boylan
Assistant Art Director: John Christiana
Executive Managing Editor: Kathleen Schiaparelli
Assistant Managing Editor, Science Media: Nicole Bush
Assistant Managing Editor, Science Supplements: Dinah Thong
Project Manager: Kristen Kaiser
Senior Marketing Manager: Steve Sartori
Media Production Editor: Elizabeth Wright
Supplements Production Editor: Natasha Wolfe
Editorial Assistants: Nancy Bauer/Eliana Ortiz
Photo Editor: Debbie Latronica
Creative Director: Carole Anson
Director, Creative Services: Paul Belfanti
Vice President ESM Production and Manufacturing: David W. Riccardi
Manufacturing Manager: Trudy Pisciotti
Assistant Manufacturing Manager: Michael Bell
Photo Researcher: Julie Tesser
Art Studio: Artworks: Managing Editor, AV Production & Management: Patty Burns; Production Manager: Ronda Whitson;
    Manager, Production Technologies: Matthew Haas; AV Art Editor: Connie Long;
    Illustrators: Kathryn Anderson, Royce Copenheaver, Jay McElroy, Daniel Knopsnyder, Stacy Smith;
    Quality Assurance: Pamela Taylor, Timothy Nguyen
Contributing Art Studio: Preparé, Inc.
Interior & Cover Design: Thomas Nery
Cover Photographs: Strawberry Poison Dart Frog, Kennan Ward Photography; Pharmaceutical Bottles, Getty Images/PhotoDisc;
    Red Roses, Getty Images; Red Anthias, Getty Images/PhotoDisc
Production Services/Composition: Preparé, Inc.

© 2003, 1999, 1995, 1991, 1987 by Pearson Education, Inc.
Pearson Education, Inc.
Upper Saddle River, NJ 07458

Printed in the United States of America
10 9 8 7 6 5 4 3 2 1

ISBN 0-13-033832-X

Pearson Education Ltd., *London*
Pearson Education Australia Pty. Ltd., *Sydney*
Pearson Education Singapore, Pte. Ltd.
Pearson Education North Asia Ltd., *Hong Kong*
Pearson Education Canada, Inc., *Toronto*
Pearson Educacíon de Mexico, S.A. de C.V.
Pearson Education—Japan, *Tokyo*
Pearson Education Malaysia, Pte. Ltd.

**To my students and colleagues
at Whitman College**

BRIEF CONTENTS

CONTENTS

electrostatic potential map
of chloromethane

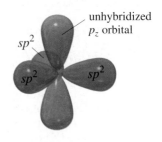

unhybridized
p_z orbital

sp^2

sp^2 sp^2

sp^2 hybrid carbon atom
(viewed from the side)

Chapter 3
Structure and Stereochemistry of Alkanes 80

Chapter 4
The Study of Chemical Reactions 124

Chapter 5
Stereochemistry 167

Chapter 6
Alkyl Halides: Nucleophilic Substitution and Elimination 212

ethyl bromide (1°)
attack is easy

Chapter 7
Structure and Synthesis of Alkenes 272

Chapter 8
Reactions of Alkenes 314

Chapter 9
Alkynes 370

Chapter 10
Structure and Synthesis of Alcohols 405

electrostatic potential
map of methyllithium

Chapter 11
Reactions of Alcohols 445

Chapter 12
Infrared Spectroscopy and Mass Spectrometry 490

Chapter 13
Nuclear Magnetic Resonance Spectroscopy 539

Chapter 14
Ethers, Epoxides, and Sulfides 600

18-crown-6
with K⁺ solvated

Chapter 15
Conjugated Systems, Orbital Symmetry,
and Ultraviolet Spectroscopy 638

Chapter 16
Aromatic Compounds 679

Chapter 17
Reactions of Aromatic Compounds 722

Chapter 18
Ketones and Aldehydes 774

Chapter 19
Amines 836

electrostatic potential
map of trimethylamine

Chapter 20
Carboxylic Acids 900

Chapter 21
Carboxylic Acid Derivatives 940

Chapter 22
Alpha Substitutions and Condensations of Enols and Enolate Ions 1003

Chapter 23
Carbohydrates and Nucleic Acids 1057

Chapter 24
Amino Acids, Peptides, and Proteins 1114

Chapter 25
Lipids 1162

Chapter 26
Synthetic Polymers 1182

Appendices

Mechanism Boxes

PREFACE

To the Student

As you begin your study of organic chemistry, you might feel overwhelmed by the number of compounds, names, reactions, and mechanisms that confront you. You may even wonder whether you can learn all this material in a single year. The most important function of a textbook is to organize the material to show that most of organic chemistry consists of a few basic principles and many extensions and applications of these principles. Relatively little memorization is required if you grasp the major concepts and develop flexibility in applying those concepts. Frankly, I have a poor memory, and I hate memorizing lists of information. I don't remember the specifics of most of the reactions and mechanisms in this book, but I can work them out by remembering a few basic principles, like "alcohol dehydrations usually go by E1 mechanisms."

Still, you'll have to learn some facts and fundamental principles to serve as the working "vocabulary" of each chapter. As a student, I learned this the hard way when I made a D on my second organic chemistry exam. I thought organic would be like general chemistry, where I could memorize a couple of equations and fake my way through the exams. For example, in the ideal gas chapter, I would memorize $pv = nrT$, and I was good to go. When I tried the same approach in organic, I got a D. We learn by making mistakes, and I learned a lot in organic chemistry.

In writing this book, I've tried to point out a small number of important facts and principles that should be learned to prepare for solving problems. For example, of the hundreds of reaction mechanisms shown in this book, about 20 are the fundamental mechanistic steps that combine into the longer, more complicated mechanisms. I've highlighted these fundamental mechanisms in *Key Mechanism* boxes to alert you to their importance. Spectroscopy is another area where a student might feel pressured to memorize hundreds of facts, such as NMR chemical shifts and infrared vibration frequencies. I couldn't do that, so I've always gotten by with knowing about a dozen NMR chemical shifts and about a dozen IR vibration frequencies, and knowing how they are affected by other influences. I've listed those important infrared frequencies in Table 12-2 and the important NMR chemical shifts in Table 13-3.

Don't try to memorize your way through this course. It doesn't work; you have to know what's going on so you can apply the material. Also, don't think (like I did) that you can get by without memorizing *anything*. Read the chapter, listen carefully to the lectures, and *work the problems*. The problems will tell you whether or not you know the material. If you can do the problems, you should do well on the exams. If you can't do the problems, you probably won't be able to do the exams, either. If you keep having to look up something to do the problems, that item is a good one to learn.

Here are some hints I give my students at the beginning of the course:

1. Read the material in the book before the lecture (expect 13–15 pages per lecture). Knowing what to expect and what is in the book, you can take fewer notes and spend more time listening and understanding the lecture.
2. After the lecture, review you notes and the book, and do the in-chapter problems. Also, read the material for the next lecture.
3. If you are confused about something, visit your instructor during office hours immediately, before you fall behind. Bring your attempted solutions to problems with you to show your instructor where you are having trouble.
4. To study for the exam, begin by reviewing each chapter and your notes, then concentrate on the end-of-chapter problems. Also use old exams for practice, if available.

Remember the two "golden rules" of organic chemistry.

1. **DON'T GET BEHIND!** The course moves too fast, and it's hard to catch up.
2. **WORK LOTS OF PROBLEMS.** Everyone needs the practice, and problems show where you need more work.

Study Aids

Several kinds of study aids are provided to emphasize and review the most important points.

Summary Tables   Whenever a large amount of material lends itself to a concise summary, a summary table is provided to compare and contrast this material. For example, the following summary table compares the factors affecting S_N1 and S_N2 reactions:

SUMMARY   Nucleophilic Substitutions

	S_N1	S_N2
promoting factors		
nucleophile	weak nucleophiles are OK	strong nucleophile needed
substrate (RX)	$3° > 2°$	$CH_3X > 1° > 2°$
solvent	good ionizing solvent needed	wide variety of solvents
leaving group	good one required	good one required
other	$AgNO_3$ forces ionization	
characteristics		
kinetics	first order, $k_r[RX]$	second order, $k_r[RX][Nuc:^-]$
stereochemistry	mixture of inversion and retention	complete inversion
rearrangements	common	impossible

Reaction Summaries   At the conclusion of each section on syntheses or reactions of a functional group ("Reactions of Alkenes," for example), a summary table is provided for efficient review. Each summary, highlighted by a blue background, includes cross-references to reactions that are discussed elsewhere.

SUMMARY   Reactions of Alkenes

1. *Electrophilic Additions*

   a. *Addition of hydrogen halides* (Section 8-3)

(HX = HCl, HBr, or HI)   Markovnikov orientation
(anti-Markovnikov with HBr and peroxides)

Example

Mechanism Boxes   In this edition, I've added a new design element to help you find the important mechanisms easily when you review a chapter. These mechanism boxes (about 100 total) have large blue headings that make them easy to see as you thumb through the chapter. Please let me know if you think of ways they might be made more helpful. In choosing which mechanisms to put in boxes, I've tried to include most of the standard mechanisms that most students can work through and understand after they've finished studying the chapter.

MECHANISM   Formation of Halohydrins

Step 1: Formation of a halonium ion.

(X = Cl, Br, or I)   halonium ion

Step 2: Water opens the halonium ion; deprotonation gives the halohydrin.

back-side attack

halohydrin
Markovnikov orientation
anti stereochemistry

The most important of these mechanisms (about 20) are in *Key Mechanism* boxes. These are the fundamental mechanistic principles that make organic chemistry (and biochemistry) work. They are the pieces that compose most of the other, longer mechanisms given in the book. Thirty years from now, even if you don't remember anything else from your organic chemistry course, I hope you will still understand these fundamental ways that compounds react.

KEY MECHANISM   Electrophilic Addition to Alkenes

Step 1: Attack of the pi bond on the electrophile forms a carbocation.

+ on the more substituted carbon

Step 2: Attack by a nucleophile gives the product of addition.

Problems   The in-chapter problems appear right after the relevant sections of the text. These problems provide immediate review and reinforcement of the material as you learn it, helping to make sure you understand each section well enough before moving on to the next. Later, end-of-chapter problems promote additional review and practice. Your instructor may choose to assign specific problems that reflect the emphasis of the lectures. Problems with red stars (*) are more difficult problems that require extra thought and perhaps some extension of the material presented in the chapter.

Solved Problems   Where appropriate, solved problems (highlighted by a beige background) are provided to show how you might approach a particular type of problem and what kind of answer is expected. For example, a solved problem might work through a mechanism to show how it is broken down into individual steps and how red curved arrows show movement of electrons.

SOLVED PROBLEM 7-5

Propose a mechanism for the sulfuric acid–catalyzed dehydration of *t*-butyl alcohol.

SOLUTION

The first step is protonstion of the hydroxyl group, which converts it to a good leaving group.

The second step is ionization of the protonated alcohol to give a carbocation.

Abstraction of a proton completes the mechanism.

PROBLEM-SOLVING **HINT**

Alcohol dehydrations usually go through E1 elimination of the protonated alcohol. Reactivity is:

$$3° > 2° \gg 1°$$

Rearrangements are common.

PROBLEM 7-25

Propose mechanisms for the following reactions:

(a)

cyclopentanol   cyclopentene

Solved problems are often followed by another problem to give students an immediate opportunity to practice the principles covered in the solved problems.

Glossaries   Each chapter ends with a glossary that defines and explains technical terms introduced in that chapter. New terms defined in the glossary are printed in boldface the first time they appear in the chapter. Don't think of the glossaries as simply dictionaries for looking up words. The Index works better for that. The real purpose of the glossaries is as study aids for reviewing the material. Read through them after you read each chapter, and they will help to jog your memory as you go over the definitions and make sure you understand and can use all the new terms.

Chapter 8
Glossary

addition A reaction involving an increase in the number of groups attached to the alkene and a decrease in the number of elements of unsaturation. (p. 314)

   anti addition: An addition in which two groups add to opposite faces of the double bond (as in addition of Br_2). (p. 326)

   electrophilic addition: An addition in which the electrophile bonds to one of the double-bonded carbons first, followed by the nucleophile. (p. 315)

   syn addition: An addition in which two groups add to the same face of the double bond (as in osmium tetroxide hydroxylation). (p. 326)

addition polymer (chain-growth polymer) A polymer that results from rapid addition of one molecule at a time to a growing polymer chain, usually with a reactive intermediate (cation, radical, or anion) at the growing end of the chain. (p. 352)

alkoxymercuration The addition of mercuric acetate to an alkene in an alcohol solution, forming an alkoxymercurial intermediate. Demercuration gives an ether. (p. 326)

Problem-Solving Strategies   The problem-solving strategies (highlighted by a green background) suggest methods for approaching complicated problems, such as those that require proposing mechanisms and developing multistep syntheses. Students often have trouble seeing how to approach problem solving, and these strategies are meant to help you break problems down into simpler pieces. Although organic chemistry cannot be reduced to a rote process that guarantees an answer, experienced chemists instinctively approach problems in ways that are more likely to lead to solutions. The suggestions in the problem-solving discussions approximate what an experienced chemist is likely to do in approaching these problems. They serve as a starting point, not a guaranteed route to the answers.

PROBLEM-SOLVING STRATEGY
Organic Synthesis

Alkyl halides and alkenes are readily made from other compounds, and they are easily converted to other functional groups. This flexibility makes them useful as reagents and intermediates for organic synthesis. Alkenes are particularly important for industrial syntheses because they are inexpensive and available in large quantities from cracking and dehydrogenation of petroleum fractions.

   Organic synthesis is the preparation of desired compounds from readily available materials. Synthesis is one of the major areas of organic chemistry, and nearly every chapter of this book involves organic synthesis in some way. A synthesis may be a simple one-step reaction, or it may involve many steps and incorporate a subtle strategy for assembling the correct carbon skeleton with all the functional groups in the right positions.

   Many of the problems in this book are synthesis problems. In some synthesis problems, you are asked to show how to convert a given starting material to the desired product. There are obvious one-step answers to some of these problems, while others may require several steps and there may be many correct answers. In solving multistep synthetic problems, it is often helpful to analyze the problem backward: Begin with the desired product (called the *target compound*) and see how it might be mentally changed or broken down to give the starting materials. This backward approach to synthesis is called a **retrosynthetic analysis**.

PROBLEM-SOLVING HINT

To move a proton (as in a tautomerism) under acidic conditions, try adding a proton in the new position, then removing it from the old position.

Problem-Solving Hints   These suggestions (green headings in the marginal column at the side of the page) are provided to remind you of facts or principles that are likely to be useful for solving common types of problems. These are the tips I give my own students when I help them work problems and review for exams. These hints highlight material that is sometimes overlooked but plays an important role in solving problems.

Essential Problem-Solving Skills   This list is provided at the end of each chapter to remind you of the kinds of skills needed to solve typical problems associated with the material in that chapter. When you finish a chapter, this list can point out concepts you might need to review, or it might suggest types of problems and solutions you have not considered. Reviewing the problem-solving skills is often a good prelude to doing the end-of-chapter problems.

Essential Problem-Solving Skills in Chapter 8

1. Predict the products of additions, oxidations, reductions, and cleavages of alkenes, including
   (a) Orientation of reaction (regiochemistry)
   (b) Stereochemistry
2. Propose logical mechanisms to explain the observed products of alkene reactions, including regiochemistry and stereochemistry.
3. Use retrosynthetic analysis to solve multistep synthesis problems with alkenes as reagents, intermediates, or products.
4. When more than one method is usable for a chemical transformation, choose the better method and explain its advantages.
5. Use clues provided by products of reactions, such as ozonolysis, to determine the structure of an unknown alkene.

Use of Color to Help Organize Material   Color is used to highlight major features for easy location: the blue backgrounds of summary tables and beige backgrounds of solved problems shown above are examples.

1. Problem-Solving features; background color of problem-solving strategies and skills, and headings for problem-solving hints, are shown in green.
2. Key definitions and rules are in blue type.

   MARKOVNIKOV'S RULE The addition of a proton acid to the double bond of an alkene results in a product with the acid proton bonded to the carbon atom that already holds the greater number of hydrogen atoms.

3. Curved red arrows are used throughout for "electron pushing," to show the flow of electrons through the course of a reaction.

cyclopentene oxide

trans-1,2-cyclopentanediol
(racemic)

4. A blue pointing hand indicates the introduction of an important general type of reaction. In most cases, specific examples and variations of that type of reaction will follow the general reaction.

$$E2 \text{ rate} = k_r[R\!-\!X][B^-] \qquad \text{transition state}$$

The variety of available colors makes it possible to highlight and distinguish key aspects of reactions, structures, and molecular drawings, and to distinguish atoms and bonds within molecules and transition states.

You will occasionally come across notes in the margins that highlight the many biological, medical, industrial, and societal uses of organic chemistry. These margin notes were added to provide a direct connection between organic chemistry concepts and real-life applications of those concepts.

Phenol was initially named carbolic acid because it was isolated from the acidic fraction of coal tar. It is highly corrosive and readily absorbed through the skin so that it must be handled with care.

pyridine

*sp*² hybrid

ethyl bromide (1°)
attack is easy

isopropyl bromide (2°)
attack is possible

t-butyl bromide (3°)
attack is impossible

I am always interested to hear from students using this book. If you have any suggestions about how the book might be made better, or if you've found an error, please let me know. (L. G. Wade, Whitman College, Walla Walla, WA, 99362: E-mail wadelg@whitman.edu). I take students' suggestions seriously, and hundreds of them now appear in this book. For example, Whitman student Brian Lian suggested Figure 21-9, and University of Minnesota student (and racing driver) Jim Coleman gave me the facts on methanol use at Indianapolis.

Good luck with your study of organic chemistry. I'm certain you will enjoy this course, especially if you let yourself relax and develop an interest in how organic compounds influence our lives. My goal in writing this book has been to make the process a little easier: to build the concepts logically on top of each other, so they flow naturally from one to the next. The hints and suggestions for problem-solving have helped my students in the past, and I hope some of them will help you to learn and use the material. Even if your memory is worse than mine (highly unlikely), you should be able to do well in organic chemistry. I hope this will be a good learning experience for all of us.

Student Supplements

Solutions Manual (0-13-060028-8)   Brief answers to many of the in-chapter problems are given at the back of this book. These answers are sufficient for a student on the right track, but they are of limited use to one who is having difficulty working the problems. The *Solutions Manual*, prepared by Jan W. Simek of California Polytechnic State University, contains complete solutions to all the problems. The Solutions Manual also gives helpful hints on how to approach each kind of problem. This supplement is a useful aid for any student, and it is particularly valuable for students who feel they understand the material but need more help with problem solving. Appendix 1 of the *Solutions Manual* summarizes the IUPAC system of nomenclature. Appendix 2 reviews and demonstrates how acidity varies with structure in organic molecules, and how one can predict the direction of an acid–base equilibrium.

Wade Companion Web site www.prenhall.com/wade   Students and instructors can use this student resource free of charge. Instructors can also use of the Syllabus Manager tool on the Companion Web site to create a syllabus, accessible anywhere and anytime by students, with assignments from on-line and print materials. Many of the resources on the Instructor Resource CD are also available to students and instructors on this Web site, as well as practice exercises and links to current readings and external Web sites.

The Companion Web site is a valuable self-study resource for students. Several sets of *practice exercises* in multiple-choice format test their knowledge of the chapter. Hints for each question are provided, and once students submit a quiz for grading, answer-specific feedback is immediately provided.

Several galleries feature interactive elements that help explain difficult concepts in the chapter: ***Student Tutorials*** show students how to think about certain crucial topics in organic chemistry; the ***Molecule Gallery*** features hundreds of 3D molecular models showing alternate representations; the ***Animation Gallery*** has over forty animations of key mechanisms.

Other features of the Web site include a ***Current Topics*** module that highlights recent research and developments in topics related to each chapter, and **Web Destinations** with links to additional resources on the Web. For students who are using ChemOffice Ltd., a workbook featuring ***ChemOffice Exercises*** is also available on the site. This site has been thoroughly updated and enhanced from the previous edition.

Internet Guide for Chemistry (ISBN: 0-13-045763-9)   The Internet Guide for Chemistry is a quick-start guide for chemistry students who are new to the Internet and the World Wide Web. It goes over the basics of what the World Wide Web is and how it has changed the world, how to choose an Internet provider, how to select Internet software, and some browser fundamentals. It delves into tools that are more specially tuned to chemistry, and then turns to a primer on doing Web-based research, including a new section on using search engines effectively. With the Internet Guide, any chemistry student should feel more comfortable using the Internet, including the many Web-based resources that are part of the educational package from Prentice Hall.

ChemOffice Ltd. Software (ISBN: 0-13-082932-3)   Issued on CD, this software package includes student versions of Cambridge Software's popular molecular modeling software, ChemDraw and Chem3D. The package is functionally similar to ChemOfficePro v5.5. A workbook featuring chapter-specific exercises written especially for Wade 5/e is available for download on the Companion Web site for Wade.

Molecular Model Kits

Model kits allow students to construct their own molecules and thus illuminate concepts related to electron structure, stereochemistry, isomerism, ring strain, conformations of cyclic functional groups, nomenclature, and more.

Prentice Hall Molecular Model Kit (ISBN: 0-205-08136-3)   Every organic chemistry student needs a set of molecular models. These models are used to demonstrate a multitude of principles, including stereochemistry, ring strain, conformations of cyclic and acyclic systems, and many others. These principles are ideally presented with this durable model kit. The kit allows students to build space-filling and ball-and-stick models of organic molecules.

Brumlik Framework Molecular Model Kit (ISBN: 0-13-330076-5)   Models constructed with this kit allow students to see the relationship between atoms in organic molecules, including precise interatomic distances and bond angles. The flexible bonds can form strained systems, with the amount of bend in the bonds giving a qualitative idea of the amount of strain.

Brumlik Universal Molecular Model Kit (ISBN: 0-13-931700-7)   A scientifically accurate molecular model set that demonstrates the framework of a molecule, the space-filling capacity of a molecule, and molecular orbitals. This kit features color-coded atomic valence spheres and connectors. Its parts are fully interchangeable with the Brumlik Framework Molecular Model Kit.

Molecular Modeling Workbook (0-13-100828-5)   Featuring **SpartanView** and **SpartanBuild** software, this workbook includes a software tutorial and numerous challenging exercises students can tackle to solve problems involving structure building and analysis, using the tools included in the two pieces of Spartan software. Available free when packaged with the text; please ask your Prentice Hall representative for details, or e-mail chemistryservice@prenhall.com.

New York Times Themes of the Times   This newspaper supplement features recent articles on such topics as chemistry and health, environmental problems, and advances in chemistry as reported in the *New York Times*. It is available at no charge to users of Wade's *Organic Chemistry*.

To the Instructor

In writing the first edition of this text, my goal was to produce a modern, readable text that uses the most effective techniques of presentation and review. Subsequent editions extended and refined that goal, with substantial rewriting and reorganization and with the addition of several new features. This fifth edition incorporates even more refinements than the fourth, with revisions in the organization, writing, and graphics.

1. ***Mechanism Boxes.*** About 100 of the most important mechanisms have been organized into mechanism boxes, with large blue headings for easy review. I've tried to choose most of the standard mechanisms that nearly everyone teaches; yet, in some cases, it seems that other mechanisms would be good candidates. If there are additional mechanisms that should be boxed, or some that should not be boxed, please let me know what you think.

   In choosing the Key Mechanisms, I've used two major criteria. If the mechanism is one of the fundamental mechanisms that make up the longer, more complex mechanisms, then it must be a Key Mechanism. Examples are S_N1, S_N2, E1, E2, nucleophilic acyl substitution, electrophilic aromatic substitution, etc. The other criterion is more subjective. If the mechanism is one of the ones I routinely expect students to do on exams, then it is a Key Mechanism. Examples are formation of imines and acetals, aldol and Claisen condensations, etc. If you feel I have left one out or included one that should not be a Key Mechanism, please let me know.

2. ***Updated Coverage.*** Several new sections have been included to cover new material or material of current interest.

   Chapter 4: A new section on free-radical inhibitors has been added to show students how some of the common inhibitors break the free-radical chain reaction, and their importance in chemistry and biochemistry.

   Chapter 5: Using the Mislow and Siegel definition (*J. Am. Chem. Soc.* **1984**, *106*, 3319), I have introduced the term *stereocenter* and explained the differences between this term and the IUPAC terms *chirality center* and *asymmetric carbon atom* (or *chiral carbon atom*). The term *stereocenter* is much broader than the more precise term *asymmetric carbon atom*, and it assumes that one already knows the stereochemical properties of the molecule (to know which bonds will give rise to stereoisomers upon their interchange). Therefore, I have continued to encourage students to identify the (immediately apparent) asymmetric carbon atoms to use as tools in examining a molecule to determine its stereochemistry.

   Chapter 8: The Nobel prize-winning asymmetric reduction work by Noyori and Knowles is discussed, together with its implications for enantioselective drug synthesis.

   Chapter 14: The Nobel prize-winning Sharpless asymmetric epoxidation is discussed, together with the factors that selectively enhance the formation of one enantiomer of the product.

   Chapter 12: All the infrared spectra have been printed in a larger format, allowing the positions of absorptions to be determined more precisely.

   Chapter 13: The NMR spectra have been converted to high-field (300 MHz) spectra from the excellent Aldrich collection.

   Chapter 16: A section has been added discussing the aromaticity of Fullerenes and their relationship to other allotropes of carbon.

   Chapter 24: I've added a new section to discuss *prions*: proteins which are thought to be infectious because of misfolding, resulting in clumping and formation of plaques.

3. ***Electrostatic Potential Maps.*** I have introduced electrostatic potential maps in cases where they might help students to visualize the charge distribution of a species in a way that helps to explain the electrophilic or nucleophilic nature of a compound. In introducing EPMs, I've emphasized their qualitative nature without stressing their mathematical derivation. As a result, I've explained and used EPMs much like they are introduced in the general chemistry textbooks, and I've simply built on their use in general chemistry.

The entire book has been edited, with many extensive sections rewritten to enhance clarity. As in the first edition, each new topic is introduced carefully and explained thoroughly. Many introductory sections have been rewritten to update them and make them more approachable for students. Whenever possible, illustrations have been added or modified to help students visualize the physical concepts.

The emphasis continues to be on *chemical reactivity*. Chemical reactions are introduced as soon as possible, and each functional group is considered in view of its reactivity toward electrophiles, nucleophiles, oxidants, reductants, and other reagents. "Electron-pushing" mechanisms are stressed throughout as a means of explaining and predicting this reactivity. Structural concepts such as stereochemistry and spectroscopy are thoroughly treated as useful techniques that enhance the fundamental study of chemical reactivity.

Organization

This book maintains the traditional organization that concentrates on one functional group at a time while comparing and contrasting the reactivity of different functional groups. Reactions are emphasized, beginning with Lewis acid–base reactions in Chapter 1, continuing with thermodynamics and kinetics in Chapter 4, and covering most of the important substitution, addition, and elimination reactions in the three chapters following stereochemistry.

Spectroscopic techniques (IR, MS, and NMR) are covered in Chapters 12 and 13, so that they can be included in the first semester. This early coverage is needed to allow effective use of spectroscopy in the laboratory. Still, a large amount of organic chemistry has been covered before this digression into structure determination. The principles of spectroscopy are practiced and reinforced in later chapters, where the characteristic spectral features of each functional group are summarized and reinforced by practice problems.

Key Features

Flexibility of Coverage

No two instructors teach organic chemistry exactly the same way. This book covers all the fundamental topics in detail, building each new concept on those that come before. Many topics may be given more or less emphasis at the discretion of the instructor. Examples of these topics are ^{13}C NMR spectroscopy, ultraviolet spectroscopy, conservation of orbital symmetry, nucleic acids, and the special topics chapters: lipids and synthetic polymers.

Another area of flexibility is in the problems. The wide-ranging problem sets review the material from several viewpoints, and more study problems are provided than most students are able to complete. This large variety allows the instructor to select the most appropriate problems for the individual course.

Up-to-Date Treatment

In addition to the classical reactions, this book covers many techniques and reactions that have more recently gained wide use among practicing chemists. Molecular-orbital theory is introduced early and used to explain electronic effects in conjugated and aromatic systems, pericyclic reactions, and ultraviolet spectroscopy. Carbon-13 NMR spectroscopy is treated as the routine tool it has become in most research laboratories. Many of the newer synthetic techniques are also included, such as asymmetric hydrogenation and epoxidation, use of LDA, Birch reduction. Swern oxidations, alkylation of 1,3-dithianes, and oxidations using pyridinium chlorochromate.

Reaction Mechanisms

Reaction mechanisms are important in all areas of organic chemistry, but they are difficult for many students. Students fall into the trap of memorizing a mechanism while not understanding why it proceeds as it does. This book stresses the principles used to predict mechanisms. Problem-solving sections develop basic techniques for approaching mechanism problems, and they work to minimize rote memorization. These techniques emphasize deciding whether the reaction is acidic, basic, or free radical in nature, then breaking it down into Lewis acid–base interactions and using "electron pushing arrows" to illustrate these individual steps. Important mechanisms are highlighted by placing them in the *Mechanism* and *Key Mechanism* boxes.

Introduction to Mechanisms Using Free-Radical Halogenation

The advantages and disadvantages of using free-radical halogenation to introduce reaction mechanisms have been debated for many years. The principal objection to free-radical halogenation is that it is not a useful synthetic reaction. But useful reactions such as nucleophilic substitution and additions to alkenes are complicated by participation of the solvent and other effects. Gas-phase free-radical halogenation allows a clearer treatment of kinetics and thermodynamics, as long as its disadvantages as a synthetic reaction are carefully discussed and the student is aware of the limitations.

Organic Synthesis

Organic synthesis is stressed throughout this book, with progressive discussions of the process involved in developing a synthesis. *Retrosynthetic analysis* is emphasized, and the student learns to work backward from the target compound and forward from the starting materials to find a common intermediate.

Typical yields have been provided for many synthetic reactions, although I hope students will not misuse these numbers. Too often students consider the yield of a reaction to be a fixed characteristic just as the melting point of a compound is fixed. In practice, many factors affect product yields, and literature values for apparently similar reactions often differ by a factor of 2 or more. The yields given in this book are *typical* yields that a good student with excellent technique might obtain.

Spectroscopy

Spectroscopy is one of the most important tools of the organic chemist. This book develops the theory for each type of spectroscopy and then discusses the characteristic spectral features. The most useful and dependable characteristics are summarized into a small number of rules of thumb that allow the student to interpret most spectra without looking up or memorizing large tables of data. For reference use, extensive tables of NMR and IR data and a more complete version of the Woodward–Fieser rules for UV are provided as appendices.

This approach is particularly effective with IR and NMR spectroscopy, and with mass spectrometry. Practical rules are given to help students see what information is available in the spectrum and what spectral characteristics usually correspond to what structural features. Sample problems show how the information from various spectra is combined to propose a structure. The emphasis is on helping students develop an intuitive feel for using spectroscopy to solve structural problems.

Nomenclature

IUPAC nomenclature is stressed throughout the book, but common nomenclature is also discussed and used to develop students' familiarity. Teaching only the IUPAC nomenclature might be justifiable in theory, but such an approach would handicap students in their further study and use of the literature. Much of the literature of chemistry, biology, and medicine uses common names such as methyl ethyl ketone, isovaleric acid, methyl *t*-butyl ether, γ-aminobutyric acid, and ε-caprolactam. This book emphasizes why systemic nomenclature is often preferred, yet it encourages familiarity with common names as well.

I've enjoyed working on this new edition, and I hope that it is an improved fine-tuning of the fourth edition. I've tried to make this book as error-free as possible, but I'm sure some errors have slipped by. If you find errors, or have suggestions about how the book might be made better, please let me know (L. G. Wade, Whitman College, Walla Walla, WA, 99362; E-mail wadelg@whitman.edu). Errors can be fixed quickly in the next printing. I've already started a file of possible changes and improvements for the sixth edition, and I hope many of the current users will contribute suggestions to this file. I hope this book makes your job easier and helps more of your students to succeed. That's the most important reason why I wrote it.

Instructor Supplements

Instructor's Resource CD-ROM (0-13-0604496)   An instructor CD that contains almost all the art from the text, 42 animations, and 135 tutorials. Using the included MediaPortfolio software, instructors can browse for figures and other media elements by thumbnail and description, as well as search by key word or title. In addition, all of the Student Tutorials available on the Companion Website are available on the IRCD as well. The images and videos can be cut and pasted, or dragged into your MS PowerPoint® lecture presentation or other documents. The set also contains two different prebuilt PowerPoint® Presentations for every chapter, all images in .PDF format, as well as all the responsive media elements specifically developed for Organic Chemistry, Fifth Edition.

Transparency Pack (0-13-033836-2)   This package comprises 250 four-color acetates of the most useful images, computer art, and line drawings from the text. The Transparency Pack is available at no charge to adopters of Organic Chemistry, Fifth Edition.

Test Item File (0-13-033833-8)   by Gary Hollis, Roanoke College, is a printed version of all questions found on the TestGen-EQ software. The new Fifth Edition contains over 1800 questions, including 20% new questions with varied levels of difficulty.

TestGen-EQ (0-13-033834-6)   This computerized version of the Test Item File included electronic versions of all 1800+ test questions. TestGen-EQ allows you to create and tailor exams to your own needs and includes tools for course management, algorithmic question generation, and administering tests over a local area network.

Course Management Options   Combine the assets of the Companion Website and the Test Item File for Wade 5e with the power of full customization and the functionality of a robust course management system. Course management systems include an online gradebook, discussion forums and bulletin boards, document upload and download areas, and other features. Prentice Hall offers full courses, matched to the book, suitable for a variety of platforms:

- **WebCT**
- **BlackBoard**
- **CourseCompass** Powered by Blackboard, CourseCompass is a nationally hosted alternative to a university-licensed course management platform. CourseCompass is designed to be accessible to the novice, but also contains even more power tools than are available in a standard BlackBoard course, such as full integration with TestGen test creation software.

Acknowledgments

I am pleased to thank the many talented people who helped with this revision. Particular thanks are due to Ray Mullaney and John Murdzek, who made thousands of useful suggestions throughout the writing and revision process, and who helped to shape this new edition. Special thanks are also due to Jan W. Simek, author of the *Solutions Manual*, who made a multitude of useful and perceptive suggestions.

I would like to thank the reviewers for their valuable insight and commentary. Although I did not adopt all their suggestions, most of them were helpful and contributed to the quality of the final product.

Fourth Edition Reviewers

Mohammed Ali	Southeast Missouri State University
Paul T. Buonora	University of Scranton
Jeff Charonnat	California State University, Northridge
Arnold Craig	Montana State University
Rhoda E. R. Craig	Kalamazoo College
William Dailey	University of Pennsylvania
S. Todd Deal	Georgia Southern University
Roger D. Frampton	Tidewater Community College
Catherine Franklin	SUNY, University at Albany
Philip Hampton	University of New Mexico
Catherine Hagen Howard	Texarkana College
Norman R. Hunter	University of Manitoba
T. G. Jackson	University of South Alabama
Francis M. Klein	Creighton University
Eugene A. Kline	Tennessee Technological University
N. Dale Ledford	University of South Alabama
Clifford C. Leznoff	York University
R. Daniel Libby	Moravian College
James W. Long	University of Oregon
James G. Macmillan	University of Nothern Iowa
Donald S. Matteson	Washington State University
Robert McClelland	University of Toronto
James C. McKenna	Oklahoma City Community College
Gary W. Morrow	University of Dayton
Richard Narske	Augustana College
Nicholas R. Natale	University of Idaho
Bjorn Olesen	Southeast Missouri State University
Michael Pena	Arizona State University
Bryan W. Roberts	University of Pennsylvania
Joseph M. Ross	
Melvin L. Rueppel	Rueppel Consulting
William N. Setzer	University of Alabama in Hunstville
Warren V. Sherman	Chicago State University
Tami Spector	University of San Francisco
Kenneth W. Stagliano	Illinois Institute of Technology
David H. Thompson	Purdue University
Maria Vogt	Bloomfield College.

Critique Reviewers

Simon Bott	University of Houston
Weldon Burnham	Richland College
Robert S. Coleman	Ohio State University
Sergio J. Cortes	University of Texas at Dallas
Steven Todd Deal	Georgia Southern University
Gary Earl	Augustana College
K. R. Fountain	Truman State University
Christopher M. Hadad	Ohio State University
Christopher Allen Hansen	Washington State University
Eric Miller	San Juan College
Nicholas R. Natale	University of Idaho
Tony E. Nicolas	New York City Technical College

John G. O'Brien	Truman State University
Michael W. Pelter	Purdue University Calumet
William R. Pool	Westminster College
Erik E. Simanek	Texas A&M University
Donald G. Slavin	Community College of Philadelphia
Greg Slough	Kalamazoo College
Kenneth W. Stagliano	Illinois Institute of Technology
Maria Vogt	Bloomfield College

Draft Reviewers

Neil T. Allison	University of Arkansas
Jeff Charonnat	California State University, Northridge
William Dailey	University of Pennsylvania
John DiCesare	University of Tulsa
John Farrar	University of St. Francis
John F. Helling	University of Florida
Thomas G. Jackson	University of Alabama
Madeline Jouille	University of Pennsylvania
Stephen Kass	University of Minnesota
Eugene A. Kline	Tennessee Technological University
William Mancini	Paradise Valley Community College
Jerrold Meinwold	Cornell University
Nicholas R. Natale	University of Idaho
Snorri T. Sigurdsson	University of Washington
Maria Vogt	Bloomfield College
Stephen A. Woski	University of Alabama
Catherine Woytowicz	George Washington University

Accuracy Reviewers

Donnie Byers	Johnson County Community College
Jeff Charonnat	California State University, Northridge
Sergio J. Cortes	University of Texas at Dallas
Giovanna Ghirlanda	University of Pennsylvania
Christopher Hadad	Ohio State University
Catherine Woytowicz	George Washington University

Application Notes Contributor

Christian P. Whitman	The University of Texas at Austin

Finally, I want to thank the people at Prentice Hall, whose dedication and flexibility contributed to the completion of this project. As acquisitions editor, Nicole Folchetti kept the project moving, ensured the needed resources were available, and made many useful comments and suggestions. Production editor Fran Daniele kept the production process organized, on track, and on schedule. It has been a pleasure working with all these thoroughly professional and competent people.

L. G. Wade, Jr.

ABOUT THE AUTHOR

L. G. "Skip" Wade decided to become a chemistry major during his sophomore year at Rice University, while taking organic chemistry from Professor Ronald M. Magid. After receiving his B.A. from Rice in 1969, Wade went on to Harvard University, where he did research with Professor James D. White. While at Harvard, he served as the Head Teaching Fellow for the organic laboratories and was strongly influenced by the teaching methods of two master educators, Professors Leonard K. Nash and Frank H. Westheimer.

After completing his Ph.D. at Harvard in 1974, Dr. Wade joined the chemistry faculty at Colorado State University. Over the course of fifteen years at Colorado State, Dr. Wade taught organic chemistry to thousands of students working toward careers in all areas of biology, chemistry, human medicine, veterinary medicine, and environmental studies. He also authored research papers in organic synthesis and in chemical education, as well as eleven books reviewing current research in organic synthesis. Since 1989, Dr. Wade has been a chemistry professor at Whitman College, where he teaches organic chemistry and pursues research interests in organic synthesis and forensic chemistry. Dr. Wade received the A. E. Lange Award for Distinguished Science Teaching at Whitman in 1993.

Dr. Wade's interest in forensic science has led him to testify as an expert witness in court cases involving drugs and firearms, and he has worked as a police firearms instructor, drug consultant, and boating safety officer. He also enjoys repairing and restoring old violins and bows, which he has done professionally for many years.

CHAPTER 1

Introduction and Review

The modern definition of **organic chemistry** is *the chemistry of carbon compounds*. What is so special about carbon that a whole branch of chemistry is devoted to its compounds? Unlike most other elements, carbon forms strong bonds to other carbon atoms and to a wide variety of other elements. Chains and rings of carbon atoms can be built up to form an endless variety of molecules. It is this diversity of carbon compounds that provides the basis for life on Earth. Living creatures are composed largely of complex organic compounds that serve structural, chemical, or genetic functions.

The term **organic** literally means "derived from living organisms." Originally, the science of organic chemistry was the study of compounds extracted from living organisms and their natural products. Compounds such as sugar, urea, starch, waxes, and plant oils were considered "organic," and people accepted **Vitalism**: the belief that natural products needed a "vital force" to create them. Organic chemistry, then, was the study of compounds having the vital force. Inorganic chemistry was the study of gases, rocks, and minerals and the compounds that could be made from them.

In the nineteenth century, experiments showed that organic compounds could be synthesized from inorganic compounds. In 1828, the German chemist Friedrich Wöhler converted ammonium cyanate, made from ammonia and cyanic acid, to urea simply by heating it in the absence of oxygen.

1-1
The Origins of Organic Chemistry

$$\underset{\substack{\text{ammonium cyanate}\\ \text{(inorganic)}}}{NH_4^+ \ ^-OCN} \xrightarrow{\text{heat}} \underset{\substack{\text{urea}\\ \text{(organic)}}}{H_2N-\overset{\overset{\textstyle O}{\|}}{C}-NH_2}$$

Urea had always come from living organisms and was presumed to contain the vital force, yet ammonium cyanate is inorganic and thus lacks the vital force. Some chemists claimed that a trace of vital force from Wöhler's hands must have contaminated the reaction, but most recognized the possibility of synthesizing organic compounds from inorganics. Many other syntheses were carried out, and the vital force theory was eventually discarded.

Since Vitalism was disproved in the early nineteenth century, you'd think it would be extinct by now. And you'd be wrong! Vitalism lives on today in the minds of those who believe that "natural" (plant-derived) vitamins, flavor compounds, etc. are somehow different and more healthful than the identical "artificial" (synthesized) compounds.

The Jarvik 7 artificial heart, composed largely of synthetic organic materials.

As chemists, we know that plant-derived compounds and the synthesized compounds are identical. Assuming they are pure, the only way to tell them apart is through ^{14}C dating: Compounds synthesized from petrochemicals have a lower content of radioactive ^{14}C and appear old because their ^{14}C has decayed over time. Plant-derived compounds are recently synthesized from CO_2 in the air. They have a higher content of radioactive ^{14}C. Some large chemical suppliers provide isotope ratio analyses to show that their "naturals" have high ^{14}C content and are plant-derived. Such a sophisticated analysis lends a high-tech flavor to this twenty-first-century form of Vitalism.

Even though organic compounds do not need a vital force, they are still distinguished from inorganic compounds. The distinctive feature of organic compounds is that they *all* contain one or more carbon atoms. Still, not all carbon compounds are organic; substances such as diamond, graphite, carbon dioxide, ammonium cyanate, and sodium carbonate are derived from minerals and have typical inorganic properties. Most of the millions of carbon compounds are classified as organic, however.

We ourselves are composed largely of organic molecules, and we are nourished by the organic compounds in our food. The proteins in our skin, the lipids in our cell membranes, the glycogen in our livers, and the DNA in the nuclei of our cells are all organic compounds. Our bodies are also regulated and defended by complex organic compounds.

Chemists have learned to synthesize or simulate many of these complex molecules. The synthetic products serve as drugs, medicines, plastics, pesticides, paints, and fibers. Many of the most important advances in medicine are actually advances in organic chemistry. New synthetic drugs are developed to combat disease, and new polymers are molded to replace failing organs. Organic chemistry has gone full circle. It began as the study of compounds derived from "organs," and now it gives us the drugs and materials we need to save or replace those organs.

One of nicotine's effects is to increase the concentration of a chemical in the brain's reward system. Release of this chemical makes smokers feel good and reinforces the need to smoke.

One of the reasons chemists synthesize derivatives of complex organic compounds like morphine is to discover new drugs that retain the good properties (potent pain-relieving) but not the bad properties (highly addictive).

nicotine   vitamin C   carmine   morphine

Four examples of organic compounds in living organisms. Tobacco contains nicotine, an addictive alkaloid. Rose hips contain vitamin C, essential for preventing scurvy. The red dye carmine comes from cochineal insects, shown on prickly pear cactus. Opium poppies contain morphine, a pain-relieving, addictive alkaloid.

Before we begin our study of organic chemistry, we must review some basic principles. Many of these concepts of atomic and molecular structure are crucial to your understanding of the structure and bonding of organic compounds.

1-2A　Structure of the Atom

Atoms are made up of protons, neutrons, and electrons. Protons are positively charged and are found together with (uncharged) neutrons in the nucleus. Electrons, which have a negative charge that is equal in magnitude to the positive charge on the proton, occupy the space surrounding the nucleus (Figure 1-1). Protons and neutrons have similar masses, about 1800 times the mass of an electron. Almost all the atom's mass is in the nucleus, but it is the electrons that take part in chemical bonding and reactions.

Each element is distinguished by the number of protons in the nucleus (the atomic number). The number of neutrons is usually similar to the number of protons, although the number of neutrons may vary. Atoms with the same number of protons but different numbers of neutrons are called **isotopes**. For example, the most common kind of carbon atom has six protons and six neutrons in its nucleus. Its mass number (the sum of the protons and neutrons) is 12, and we write its symbol as ^{12}C. About 1% of carbon atoms have seven neutrons; the mass number is 13, written ^{13}C. A very small fraction of carbon atoms have eight neutrons and a mass number of 14. The ^{14}C isotope is radioactive, with a half-life (the time it takes for half of the nuclei to decay) of 5730 years. The predictable decay of ^{14}C is used to determine the age of organic materials up to about 50,000 years old.

1-2B　Electronic Structure of the Atom

An element's chemical properties are determined by the number of protons in the nucleus and the corresponding number of electrons around the nucleus. The electrons form bonds and determine the structure of the resulting molecules. Because they are small and light, electrons show properties of both particles and waves; in many ways, the electrons in atoms and molecules behave more like waves than like particles.

Electrons that are bound to nuclei are found in **orbitals**. The Heisenberg uncertainty principle states that we can never determine exactly where the electron is; nevertheless, we can determine the **electron density**, the probability of finding the electron in a particular part of the orbital. An orbital, then, is an allowed energy state for an electron, with an associated probability function that defines the distribution of electron density in space.

Atomic orbitals are grouped into different "shells" at different distances from the nucleus. Each shell is identified by a principal quantum number n, with $n = 1$ for the lowest-energy shell closest to the nucleus. As n increases, the shells are farther from the nucleus, higher in energy, and can hold more electrons. Most of the common elements in organic compounds are found in the first two rows of the periodic table, indicating that their electrons are found in the first two electron shells. The first shell ($n = 1$) can hold two electrons, and the second shell ($n = 2$) can hold eight.

The first electron shell contains just the $1s$ orbital. All s orbitals are spherically symmetrical, meaning that they are nondirectional. The electron density is only a function of the distance from the nucleus. The electron density of the $1s$ orbital is graphed in Figure 1-2. Notice how the electron density is highest *at* the nucleus and falls off exponentially with increasing distance from the nucleus. The $1s$ orbital might be imagined as a cotton boll, with the cottonseed at the middle representing the nucleus. The density of the cotton is highest nearest the seed, and it becomes less dense at greater distances from this "nucleus."

The second electron shell consists of the $2s$ and $2p$ orbitals. The $2s$ orbital is spherically symmetrical like the $1s$ orbital, but its electron density is not a simple exponential function. The $2s$ orbital has a smaller amount of electron density close to the nucleus. Most of the electron density is farther away, beyond a region of zero electron density called

1-2
Principles of Atomic Structure

cloud of electrons

nucleus
(protons and neutrons)

▲ **Figure 1-1**
An atom has a dense, positively charged nucleus surrounded by a cloud of electrons.

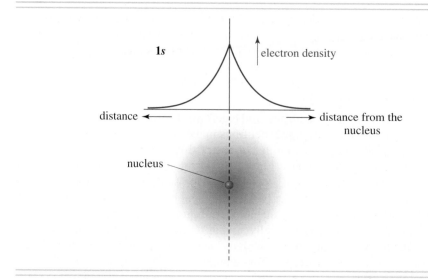

▶ **Figure 1-2**
Graph and diagram of the $1s$ atomic orbital. The electron density is highest at the nucleus and drops off exponentially with increasing distance from the nucleus in any direction.

a **node**. Because most of the $2s$ electron density is farther from the nucleus than that of the $1s$, the $2s$ orbital is higher in energy. Figure 1-3 shows a graph of the $2s$ orbital.

In addition to the $2s$ orbital, the second shell also contains three $2p$ atomic orbitals, one oriented in each of the three spatial directions. These orbitals are called the $2p_x$, the $2p_y$, and the $2p_z$, according to their direction along the x, y, or z axis. The $2p$ orbitals are slightly higher in energy than the $2s$, because the average location of the electron in a $2p$ orbital is farther from the nucleus. Each p orbital consists of two lobes, one on either side of the nucleus, with a **nodal plane** at the nucleus. The nodal plane is a flat (planar) region of space, including the nucleus, with zero electron density. The three $2p$ orbitals differ only in their spatial orientation, so they have identical energies. Orbitals with identical energies are called **degenerate orbitals**. Figure 1-4 shows the shapes of the three degenerate $2p$ atomic orbitals.

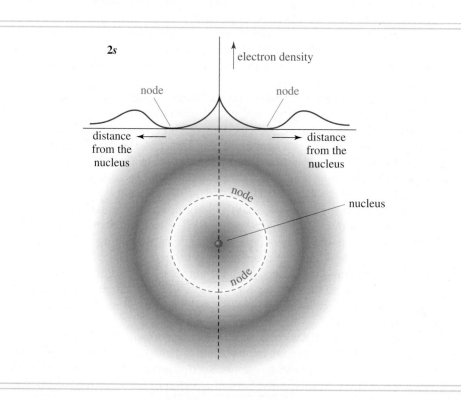

▶ **Figure 1-3**
The $2s$ orbital has a small region of high electron density close to the nucleus, but most of the electron density is farther from the nucleus, beyond a node, or region of zero electron density.

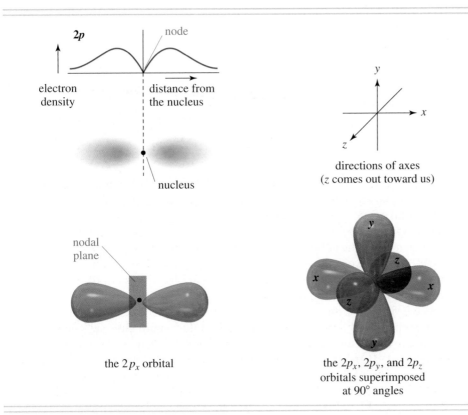

◀ **Figure 1-4**
The $2p$ orbitals. There are three $2p$ orbitals, oriented at right angles to each other. Each is labeled according to its orientation along the x, y, or z axis.

The *Pauli exclusion principle* tells us that each orbital can hold a maximum of two electrons, provided that their spins are paired. The first shell (one $1s$ orbital) can accommodate two electrons. The second shell (one $2s$ orbital and three $2p$ orbitals) can accommodate eight electrons, and the third shell (one $3s$ orbital, three $3p$ orbitals, and five $3d$ orbitals) can accommodate 18 electrons.

1-2C Electronic Configurations of Atoms

Aufbau means "building up" in German, and the *aufbau principle* tells us how to build up the electronic configuration of an atom's ground (most stable) state. Starting with the lowest-energy orbital, we fill the orbitals in order until we have added the proper number of electrons. Table 1-1 shows the ground-state electronic configurations of the elements in the first two rows of the periodic table.

TABLE 1-1 Electronic Configurations of the Elements of the First and Second Rows

Element	Configuration	Valence Electrons
H	$1s^1$	1
He	$1s^2$	2
Li	$1s^2 2s^1$	1
Be	$1s^2 2s^2$	2
B	$1s^2 2s^2 2p_x^1$	3
C	$1s^2 2s^2 2p_x^1 2p_y^1$	4
N	$1s^2 2s^2 2p_x^1 2p_y^1 2p_z^1$	5
O	$1s^2 2s^2 2p_x^2 2p_y^1 2p_z^1$	6
F	$1s^2 2s^2 2p_x^2 2p_y^2 2p_z^1$	7
Ne	$1s^2 2s^2 2p_x^2 2p_y^2 2p_z^2$	8

Relative orbital energies

$-2p_x \quad -2p_y \quad -2p_z$

energy $\quad -2s$

$-1s$

▶ **Figure 1-5**

First three rows of the periodic table. The organization of the periodic table results from the filling of atomic orbitals in order of increasing energy. For these representative elements, the number of the column corresponds to the number of valence electrons.

Partial periodic table

IA							noble gases (VIII)
H	IIA	IIIA	IVA	VA	VIA	VIIA	He
Li	Be	B	C	N	O	F	Ne
Na	Mg	Al	Si	P	S	Cl	Ar

Lithium carbonate, a salt of lithium, is a mood-stabilizing agent used to treat the psychiatric disorder known as mania. Mania is characterized by behaviors such as elated mood, feelings of greatness, racing thoughts, and an inability to sleep. We don't know how lithium carbonate helps to stabilize these patients' moods.

Two additional concepts are illustrated in Table 1-1. The **valence electrons** are those electrons that are in the outermost shell. Carbon has four valence electrons, nitrogen has five, and oxygen has six. Helium has two valence electrons, and neon has eight, corresponding to a filled first shell and second shell, respectively. In general (for the representative elements), the column or group number of the periodic table corresponds to the number of valence electrons (Figure 1-5). Hydrogen and lithium have one valence electron, and they are both in the first column (group IA) of the periodic table. Carbon has four valence electrons, and it is in group IVA of the periodic table.

Notice in Table 1-1 that carbon's third and fourth valence electrons are not paired; they occupy separate orbitals. Although the Pauli exclusion principle says that two electrons can occupy the same orbital, the electrons repel each other, and pairing requires additional energy. **Hund's rule** states that when there are two or more orbitals of the same energy, electrons will go into *different* orbitals rather than pair up in the same orbital. The first $2p$ electron (boron) goes into one $2p$ orbital, the second (carbon) goes into a different orbital, and the third (nitrogen) occupies the last $2p$ orbital. The fourth, fifth, and sixth $2p$ electrons must pair up with the first three electrons.

> **PROBLEM 1-1**
>
> Write the electronic configurations of the third-row elements shown in the partial periodic table in Figure 1-5.

1-3
Bond Formation: The Octet Rule

In 1915, G. N. Lewis proposed several new theories describing how atoms bond together to form molecules. One of these theories states that a filled shell of electrons is especially stable, and *atoms transfer or share electrons in such a way as to attain a filled shell of electrons*. A filled shell of electrons is simply the electron configuration of a noble gas such as He, Ne, or Ar. This principle has come to be called the **octet rule** because a filled shell implies eight valence electrons for the elements in the second row of the periodic table.

1-3A Ionic Bonding

There are two ways that atoms can interact to attain noble-gas configurations. Sometimes atoms attain noble-gas configurations by transferring electrons from one atom to another. For example, lithium has one electron more than the helium configuration, and fluorine has one electron less than the neon configuration. Lithium easily loses its valence electron, and fluorine easily gains one:

$$Li \overset{\frown}{} \ddot{\underset{\displaystyle ..}{F}}: \longrightarrow Li^+ + :\ddot{\underset{\displaystyle ..}{F}}:^- \longrightarrow Li^+ :\ddot{\underset{\displaystyle ..}{F}}:^-$$

electron transfer   He configuration   Ne configuration   ionic bond

A transfer of one electron gives each element a noble-gas configuration. The resulting ions have opposite charges, and they attract each other to form an **ionic bond**. Ionic bonding usually results in the formation of a large crystal lattice rather than individual molecules. Ionic bonding is common in inorganic compounds but relatively uncommon in organic compounds.

1-3B Covalent Bonding

Covalent bonding, in which electrons are shared rather than transferred, is the most common type of bonding in organic compounds. Hydrogen, for example, needs a second electron to achieve the noble-gas configuration of helium. If two hydrogen atoms come together and form a bond, they "share" their two electrons, and each atom has two electrons in its valence shell.

$$H \cdot \ + \ H \cdot \ \longrightarrow \ H:H \quad \text{each H shares two electrons}$$
$$\text{(He configuration)}$$

We will study covalent bonding in more detail in Chapter 2.

1-4 Lewis Structures

One way to symbolize the bonding in a covalent molecule is to use **Lewis structures**. In a Lewis structure, each valence electron is symbolized by a dot. A bonding pair of electrons is symbolized by a pair of dots or by a dash (—). We try to arrange all the atoms so they have their appropriate noble-gas configurations: two electrons for hydrogen, and octets for the second-row elements.

Consider the Lewis structure of methane (CH_4).

$$
\begin{array}{ccc}
\text{H} & & \text{H} \\
\ddot{} & & | \\
\text{H} : \overset{\text{H}}{\underset{\ddot{\text{H}}}{\text{C}}} : \text{H} & \text{or} & \text{H}-\text{C}-\text{H} \\
& & | \\
& & \text{H}
\end{array}
$$

methane

Carbon contributes four valence electrons, and each hydrogen contributes one, to give a total of eight electrons. All eight electrons surround carbon to give it an octet, and each hydrogen atom shares two of the electrons.

The Lewis structure for ethane (C_2H_6) is more complex.

$$
\begin{array}{ccc}
\text{H H} & & \text{H H} \\
\ddot{}\ \ddot{} & & |\ \ | \\
\text{H} : \ddot{\text{C}} : \ddot{\text{C}} : \text{H} & \text{or} & \text{H}-\text{C}-\text{C}-\text{H} \\
\ddot{}\ \ddot{} & & |\ \ | \\
\text{H H} & & \text{H H}
\end{array}
$$

ethane

Once again, we have computed the total number of valence electrons (14) and distributed them so that each carbon atom is surrounded by eight and each hydrogen by two. The only possible structure for ethane is the one shown, with the two carbon atoms sharing a pair of electrons and each hydrogen atom sharing a pair with one of the carbons. The ethane structure shows the most important characteristic of carbon—its ability to form strong carbon–carbon bonds.

Valence-shell electrons that are *not* shared between two atoms are called **nonbonding electrons**. A pair of nonbonding electrons is often called a **lone pair**. Oxygen atoms, nitrogen atoms, and the halogens (F, Cl, Br, I) usually have nonbonding electrons in their stable compounds. These lone pairs of nonbonding electrons help to determine the reactivity of their parent compounds. The following Lewis structures show one lone pair of electrons on the nitrogen atom of methylamine, and two lone pairs on the oxygen atom of ethanol. Halogen atoms usually have three lone pairs, as shown in the structure of chloromethane.

methylamine   ethanol   chloromethane

A correct Lewis structure should show any lone pairs. Organic chemists often draw structures that omit most or all of the lone pairs. These are not true Lewis structures because you must imagine the correct number of nonbonding electrons.

PROBLEM 1-2

Draw Lewis structures for the following compounds.
(a) ammonia, NH_3 (b) water, H_2O
(c) hydronium ion, H_3O^+ (d) propane, C_3H_8
(e) ethylamine, $CH_3CH_2NH_2$ (f) dimethyl ether, CH_3OCH_3
(g) fluoroethane, CH_3CH_2F (h) 2-propanol, $CH_3CH(OH)CH_3$
(i) borane, BH_3 (j) boron trifluoride, BF_3
Explain what is unusual about the bonding in compounds in parts (i) and (j).

1-5
Multiple Bonding

In drawing Lewis structures in Section 1-4, we placed just one pair of electrons between any two atoms. The sharing of one pair between two atoms is called a **single bond**. Many molecules have adjacent atoms sharing two or even three electron pairs. The sharing of two pairs is called a **double bond**, and the sharing of three pairs is called a **triple bond**. Ethylene (C_2H_4) is an organic compound with a double bond. When we draw a Lewis structure for ethylene, the only way to show both carbon atoms with octets is to draw them sharing two pairs of electrons. The following examples show organic compounds with double bonds. In each case, four electrons (two pairs) are shared between two atoms to give them octets. A double dash ($=$) symbolizes a double bond.

ethylene   formaldehyde   formaldimine

Acetylene (C_2H_2) has a triple bond. Its Lewis structure shows three pairs of electrons between the carbon atoms to give them octets. The following examples show organic compounds with triple bonds. A triple dash ($\equiv$) symbolizes a triple bond.

$$H:C:::C:H \qquad H:\overset{..}{\underset{..}{C}}:C:::C:\overset{..}{\underset{..}{C}}:H \qquad H:\overset{..}{\underset{..}{C}}:C:::N:$$
$$\qquad\qquad H\qquad\qquad\qquad H \qquad H \qquad\qquad H$$

or   or   or

$$H-C\equiv C-H \qquad H-\underset{\underset{H}{|}}{\overset{\overset{H}{|}}{C}}-C\equiv C-\underset{\underset{H}{|}}{\overset{\overset{H}{|}}{C}}-H \qquad H-\underset{\underset{H}{|}}{\overset{\overset{H}{|}}{C}}-C\equiv N:$$

acetylene   dimethylacetylene   acetonitrile

All these Lewis structures show that carbon normally forms four bonds in neutral organic compounds. Nitrogen generally forms three bonds, and oxygen usually forms two. Hydrogen and the halogens usually form only one bond. The number of bonds an atom usually forms is called its **valence**. Carbon is tetravalent, nitrogen is trivalent, oxygen is divalent, and hydrogen and the halogens are monovalent. By remembering the usual number of bonds for these common elements, we can write organic structures more easily. If we draw a structure with each atom having its usual number of bonds, a correct Lewis structure usually results.

SUMMARY **Common Bonding Patterns (Uncharged)**

$$-\overset{|}{\underset{|}{C}}- \qquad -\overset{..}{N}- \qquad -\overset{..}{\underset{..}{O}}- \qquad -H \qquad -\overset{..}{\underset{..}{Cl}}:$$

	carbon	nitrogen	oxygen	hydrogen	halogens
valence:	4	3	2	1	1
lone pairs:	0	1	2	0	3

PROBLEM-SOLVING **HINT**

These "usual numbers of bonds" might be single bonds, or they might be combined into double and triple bonds. For example, three bonds to nitrogen might be three single bonds, one single bond and one double bond, or one triple bond ($:N\equiv N:$). In working problems, consider all possibilities.

PROBLEM 1-3

Write a Lewis structure for each of the following molecular formulas.
(a) N_2
(b) HCN
(c) HONO
(d) CO_2
(e) H_2CNH
(f) HCO_2H
(g) C_2H_3Cl
(h) HNNH
(i) C_3H_6
(j) C_3H_4 (two double bonds)
(k) C_3H_4 (one triple bond)

PROBLEM 1-4

Circle any lone pairs (pairs of nonbonding electrons) in the structures you drew for Problem 1-3.

A bond with the electrons shared equally between the two atoms is called a **nonpolar covalent bond**. The bond in H_2 and the C—C bond in ethane are nonpolar covalent bonds. In most bonds between two different elements, the bonding electrons are attracted more strongly to one of the two nuclei. An unequally shared pair of bonding electrons is called a **polar covalent bond**.

1-6
Electronegativity and Bond Polarity

nonpolar   polar   ionic bond
covalent bond   covalent bond

▶ **Figure 1-6**

▶ **Figure 1-6**
Chloromethane contains a polar carbon–chlorine bond with a partial negative charge on chlorine and a partial positive charge on carbon. The electrostatic potential map shows a red region (electron-rich) around the partial negative charge and a blue region (electron-poor) around the partial positive charge. Other colors show intermediate values of electrostatic potential.

chloromethane   chloromethane

When carbon is bonded to chlorine, for example, the bonding electrons are attracted more strongly to the chlorine atom. The carbon atom bears a small partial positive charge, and the chlorine atom bears an equal amount of negative charge. Figure 1-6 shows the polar carbon–chlorine bond in chloromethane. We symbolize the bond polarity by an arrow with its head at the negative end of the polar bond and a plus sign at the positive end. The bond polarity is measured by its **dipole moment** (μ), defined to be the amount of charge separation (δ^+ and δ^-) multiplied by the bond length. The symbol δ^+ means "a small amount of positive charge"; δ^- means "a small amount of negative charge."

Figure 1-6 also shows an **electrostatic potential map (EPM)** for chloromethane, using color to represent the calculated charge distribution in a molecule. Red shows electron-rich regions, and blue shows electron-poor regions. Orange, yellow, and green show intermediate levels of electrostatic potential. In chloromethane, the red region shows the partial negative charge on chlorine, and the blue region shows the partial positive charges on carbon and the hydrogen atoms.

We often use **electronegativities** as a guide in predicting whether a given bond will be polar and the direction of its dipole moment. The Pauling electronegativity scale, most commonly used by organic chemists, is based on bonding properties, and it is useful for predicting the polarity of covalent bonds. Elements with higher electronegativities generally have more attraction for the bonding electrons. Therefore, in a bond between two different atoms, the atom with the higher electronegativity is the negative end of the dipole. Figure 1-7 shows Pauling electronegativities for some of the important elements in organic compounds.

Notice that the electronegativities increase from left to right across the periodic table. Nitrogen, oxygen, and the halogens are all more electronegative than carbon; sodium, lithium, and magnesium are less electronegative. Hydrogen's electronegativity is similar to that of carbon, so we usually consider C—H bonds to be nonpolar. We will consider the polarity of bonds and molecules in more detail in Section 2-9.

PROBLEM 1-5

Use electronegativities to predict the direction of the dipole moments of the following bonds.
(a) C—Cl   (b) C—O   (c) C—N   (d) C—S   (e) C—B
(f) N—Cl   (g) N—O   (h) N—S   (i) N—B   (j) B—Cl

H						
2.2						
Li	Be	B	C	N	O	F
1.0	1.6	1.8	2.5	3.0	3.4	4.0
Na	Mg	Al	Si	P	S	Cl
0.9	1.3	1.6	1.9	2.2	2.6	3.2
K						Br
0.8						3.0
						I
						2.7

▶ **Figure 1-7**
The electronegativities of some of the elements found in organic compounds.

In polar bonds, the partial charges (δ^+ and δ^-) on the bonded atoms are *real*. **Formal charges** provide a method for keeping track of electrons, but they may or may not correspond to real charges. In most cases, if the Lewis structure shows that an atom has a formal charge, it actually bears at least part of that charge. The concept of formal charge helps us determine which atoms bear most of the charge in a charged molecule, and it also helps us to see charged atoms in molecules that are neutral overall.

To calculate formal charges, count how many electrons contribute to the charge of each atom and compare that number with the number of valence electrons in the free, neutral atom (given by the group number in the periodic table). The electrons that contribute to an atom's charge are:

1. *All* its unshared (nonbonding) electrons; plus
2. *Half* the (bonding) electrons it shares with other atoms, or one electron of each bonding pair.

The formal charge of a given atom can be calculated by the formula

$$\text{formal charge (FC)} = [\text{group number}] - [\text{nonbonding electrons}] - \tfrac{1}{2}[\text{shared electrons}]$$

1-7
Formal Charges

SOLVED PROBLEM 1-1

Compute the formal charge (FC) on each atom in the following structures.
(a) Methane (CH_4)

$$\begin{array}{c} H \\ H : \overset{..}{\underset{..}{C}} : H \\ H \end{array}$$

SOLUTION

Each of the hydrogen atoms in methane has one bonding pair of electrons (two shared electrons). Half of two shared electrons is one electron, and one valence electron is what hydrogen needs to be neutral. Hydrogen atoms with one bond are formally neutral: $FC = 1 - 0 - 1 = 0$.

The carbon atom has four bonding pairs of electrons (eight electrons). Half of eight shared electrons is four electrons, and four electrons are what carbon (group IVA) needs to be neutral. Carbon is formally neutral whenever it has four bonds: $FC = 4 - 0 - \tfrac{1}{2}(8) = 0$.

(b) The hydronium ion, H_3O^+

$$H : \overset{\overset{\displaystyle ..}{O}}{\underset{\displaystyle H}{\underset{..}{}}} : H$$

- two nonbonding electrons
- three bonds, six bonding electrons

SOLUTION

In drawing the Lewis structure for this ion, we use eight electrons: six from oxygen plus three from the hydrogens, minus one because the ion has a positive charge. Each hydrogen has one bond and is formally neutral. Oxygen is surrounded by an octet, with six bonding electrons and two nonbonding electrons. Half the bonding electrons plus all the nonbonding electrons contribute to its charge: $6/2 + 2 = 5$; but oxygen (group VIA) needs six valence electrons to be neutral. Consequently, the oxygen atom has a formal charge of $+1$: $FC = 6 - 2 - \tfrac{1}{2}(6) = +1$.

(c) $H_3N - BH_3$

- Boron has four bonds, eight bonding electrons
- Nitrogen has four bonds, eight bonding electrons

SOLUTION

This is a neutral compound where the individual atoms are formally charged. The Lewis structure shows that both nitrogen and boron have four shared bonding pairs of electrons. Both boron and nitrogen have $8/2 = 4$ electrons contributing to their charges. Nitrogen (group V) needs five valence electrons to be neutral, so it bears a formal charge of $+1$. Boron (group III) needs only three valence electrons to be neutral, so it bears a formal charge of -1.

$$\text{Nitrogen:} \quad FC = 5 - 0 - \tfrac{1}{2}(8) = +1$$

$$\text{Boron:} \quad FC = 3 - 0 - \tfrac{1}{2}(8) = -1$$

(d) $[H_2CNH_2]^+$

$$\underset{H}{\overset{H}{>}}C=\overset{+}{N}\underset{H}{\overset{H}{<}}$$

SOLUTION

In this structure, both carbon and nitrogen have four shared pairs of bonding electrons. With four bonds, carbon is formally neutral; however, nitrogen is in group V, and it bears a formal positive charge: $FC = 5 - 0 - 4 = +1$.

This compound might also be drawn with the following Lewis structure:

$$\underset{H}{\overset{H}{>}}\overset{+}{C}-\overset{\cdot\cdot}{N}\underset{H}{\overset{H}{<}}$$

In this structure, the carbon atom has three bonds with six bonding electrons. We calculate that $6/2 = 3$ electrons, so carbon is one short of the four needed to be formally neutral: $FC = 4 - 0 - \tfrac{1}{2}(6) = +1$.

Nitrogen has six bonding electrons and two nonbonding electrons. We calculate that $6/2 + 2 = 5$, so the nitrogen is uncharged in this second structure:

$$FC = 5 - 2 - \tfrac{1}{2}(6) = 0$$

The significance of these two Lewis structures is discussed in Section 1–9.

Most organic compounds contain only a few common elements, usually with complete octets of electrons. The summary table on the facing page shows the most commonly occurring bonding structures, using dashes to represent bonding pairs of electrons. Use the rules for calculating formal charges to verify the charges shown on these structures. A good understanding of the structures shown here will help you to draw organic compounds and their ions quickly and correctly.

1-8
Ionic Structures

Some organic compounds contain ionic bonds. For example, the structure of methylammonium chloride (CH_3NH_3Cl) cannot be drawn using just covalent bonds. That would require nitrogen to have five bonds, implying ten electrons in its valence shell. The correct structure shows the chloride ion ionically bonded to the rest of the structure.

$$H-\overset{\overset{\displaystyle H}{|}}{\underset{\underset{\displaystyle H}{|}}{C}}-\overset{\overset{\displaystyle H}{|}}{\underset{\underset{\displaystyle H}{|}}{\overset{+}{N}}}-H \quad :\overset{\cdot\cdot}{\underset{\cdot\cdot}{Cl}}:^-$$

methylammonium chloride

$$H-\overset{\overset{\displaystyle H}{|}}{\underset{\underset{\displaystyle H}{|}}{C}}-\overset{\overset{\displaystyle H \quad H}{|}}{\underset{\underset{\displaystyle H \quad H}{|}}{N}}{\overset{H}{\diagdown}}\quad \overset{\cdot\cdot}{\underset{\cdot\cdot}{Cl}}:$$

too many electrons around nitrogen

cannot be drawn covalently

SUMMARY Common Bonding Patterns in Organic Compounds and Ions

Atom	Valence Electrons	Positively Charged	Neutral	Negatively Charged
B	3		$-\overset{\textstyle\mid}{\underset{\textstyle\mid}{B}}-$ (no octet)	$-\overset{\textstyle\mid}{\underset{\textstyle\mid}{B}}-$
C	4	$-\overset{+}{\underset{\textstyle\mid}{C}}-$ (no octet)	$-\overset{\textstyle\mid}{\underset{\textstyle\mid}{C}}-$	$-\overset{..}{\underset{\textstyle\mid}{C}}-$
N	5	$-\overset{\textstyle\mid}{\underset{\textstyle\mid}{N}}{}^{+}-$	$-\overset{..}{\underset{\textstyle\mid}{N}}-$	$-\overset{..}{\underset{..}{N}}=$
O	6	$-\overset{..}{\underset{\textstyle\mid}{O}}{}^{+}-$	$-\overset{..}{\underset{..}{O}}-$	$-\overset{..}{\underset{..}{O}}:^{-}$
halogen	7	$-\overset{..}{\underset{..}{Cl}}{}^{+}-$	$-\overset{..}{\underset{..}{Cl}}:$	$:\overset{..}{\underset{..}{Cl}}:^{-}$

PROBLEM-SOLVING HINT

This is a very important table. Work enough problems to become familiar with these bonding patterns so you can recognize other patterns as being either unusual or wrong.

Some molecules can be drawn either covalently or ionically. For example, sodium acetate ($NaOCOCH_3$) may be drawn with either a covalent bond or an ionic bond between sodium and oxygen. Because sodium generally forms ionic bonds with oxygen (as in NaOH), the ionically bonded structure is usually preferred. In general, bonds between atoms with very large electronegativity differences (about 2 or more) are usually drawn as ionic.

$$Na^+ \quad {}^-:\overset{..}{\underset{..}{O}}-\overset{\displaystyle\overset{..}{O}}{\underset{\displaystyle\underset{H}{|}}{C}}-\overset{\displaystyle\overset{H}{|}}{\underset{\displaystyle\underset{H}{|}}{C}}-H \qquad Na-\overset{..}{\underset{..}{O}}-\overset{\displaystyle\overset{..}{O}}{\underset{\displaystyle\underset{H}{|}}{C}}-\overset{\displaystyle\overset{H}{|}}{\underset{\displaystyle\underset{H}{|}}{C}}-H$$

more common   less common

PROBLEM 1-6

Draw Lewis structures for the following compounds and ions, showing appropriate formal charges.

(a) $[CH_3OH_2]^+$

(b) NH_4Cl

(c) $(CH_3)_2NH_2Cl$

(d) $NaOCH_3$

(e) $^+CH_3$

(f) $^-CH_3$

(g) $NaBH_4$

(h) $NaBH_3CN$

(i) $(CH_3)_2O-BF_3$

(j) $[HONH_3]^+$

(k) $KOC(CH_3)_3$

(l) $[H_2C=OH]^+$

1-9A Resonance Hybrids

Some compounds' structures are not adequately represented by a single Lewis structure. When two or more valence-bond structures are possible, differing only in the placement of electrons, the molecule will usually show characteristics of both structures. The different structures are called **resonance structures** or **resonance forms** because they are not different compounds, just different ways of drawing the same compound.

1-9
Resonance

The actual molecule is said to be a **resonance hybrid** of its resonance forms. In Solved Problem 1-1(d) we saw that the ion $[H_2CNH_2]^+$ might be represented by either of the following resonance forms:

resonance forms of a resonance hybrid   combined representation

The actual structure of this ion is a resonance hybrid of the two structures. In the actual molecule, the positive charge is **delocalized** (spread out) over both the carbon atom and the nitrogen atom. In the left resonance form, the positive charge is on carbon, but carbon does not have an octet. Nitrogen's nonbonding electrons can move into the bond (as indicated by the red arrow) to give the second structure with a double bond a positive charge on nitrogen and an octet on carbon. The combined representation attempts to combine the two resonance forms into a single picture with the charge shared by carbon and nitrogen.

Spreading the positive charge over two atoms makes the ion more stable than it would be if the entire charge were localized only on the carbon or only on the nitrogen. We call this a **resonance-stabilized** cation. Resonance is most important when it allows a charge to be delocalized over two or more atoms, as in this example.

Resonance stabilization plays a crucial role in organic chemistry, especially in the chemistry of compounds having double bonds. We will use the concept of resonance frequently throughout this course. For example, the acidity of acetic acid (below) is enhanced by resonance effects. When acetic acid loses a proton, the resulting acetate ion has a negative charge delocalized over both of the oxygen atoms. Each oxygen atom bears half of the negative charge, and this delocalization stabilizes the ion. Each of the carbon–oxygen bonds is halfway between a single bond and a double bond, and they are said to have a *bond order* of $1\frac{1}{2}$.

acetic acid   acetate ion

We use a single double-headed arrow between resonance forms (and often enclose them in brackets) to indicate that the actual structure is a hybrid of the Lewis structures we have drawn. By contrast, an equilibrium is represented by two arrows in opposite directions.

Some uncharged molecules actually have resonance-stabilized structures with equal positive and negative formal charges. We can draw two Lewis structures for nitromethane (CH_3NO_2), but both of them have a formal positive charge on nitrogen and a negative charge on one of the oxygens. Thus, nitromethane has a positive charge on the nitrogen atom and a negative charge spread equally over the two oxygen atoms. The N—O bonds are midway between single and double bonds, as indicated in the combined representation:

resonance forms   combined representation

Remember that individual resonance forms do not exist. The molecule does not "resonate" between these structures. It is a hybrid with some characteristics of both. An analogy is a mule, which is a hybrid of a horse and a donkey. The mule does not "resonate" between looking like a horse and looking like a donkey; it looks like a mule all the time, with the broad back of the horse and the long ears of the donkey.

1-9B   Major and Minor Resonance Contributors

Two or more correct Lewis structures for the same compound may or may not represent electron distributions of equal energy. Although separate resonance forms do not exist, we can estimate their relative energies as if they did exist. More stable resonance forms are closer representations of the real molecule than less stable ones. The two resonance forms shown earlier for the acetate ion have similar bonding, and they are of identical energy. The same is true for the two resonance forms of nitromethane. The following resonance forms are bonded differently, however.

major contributor   minor contributor

These structures are not equal in estimated energy. The first structure has the positive charge on nitrogen. The second has the positive charge on carbon, and the carbon atom does not have an octet. The first structure is more stable because it has an additional bond and all the atoms have octets. Many stable ions have a positive charge on a nitrogen atom with four bonds (see Summary Table, page 13). We call the more stable resonance form the **major contributor**, and the less stable form is the **minor contributor**. The structure of the actual compound resembles the major contributor more than it does the minor contributor.

Many organic molecules have major and minor resonance contributors. Formaldehyde ($H_2C{=}O$) can be written with a negative charge on oxygen, balanced by a positive charge on carbon. This polar resonance form is higher in estimated energy than the double-bonded structure because it has charge separation, fewer bonds, and a positively charged carbon atom without an octet. The charge-separated structure is only a minor contributor, but it helps to explain why the formaldehyde $C{=}O$ bond is very polar, with a partial positive charge on carbon and a partial negative charge on oxygen. The electrostatic potential map (EPM) also shows an electron-rich region (red) around oxygen and an electron-poor region (blue) around carbon in formaldehyde.

all octets   no octet on C   dipole moment   EPM of formaldehyde
no charge separation   charge separation
(major contributor)   (minor contributor)

In drawing resonance forms, we try to draw structures that are as low in energy as possible. The best candidates are those that have the maximum number of octets and the maximum number of bonds. Also, we look for structures with the minimum amount of charge separation.

Only electrons can be delocalized. Unlike electrons, nuclei cannot be delocalized. They must remain in the same places, with the same bond distances and angles, in all the resonance contributors. The following general rules will help us to draw realistic resonance structures:

1. All the resonance structures must be valid Lewis structures for the compound.

2. Only the placement of the electrons may be shifted from one structure to another. (Electrons in double bonds and lone pairs are the ones that are most commonly shifted.) Nuclei cannot be moved, and the bond angles must remain the same.

3. The number of unpaired electrons (if any) must remain the same. Most stable compounds have no unpaired electrons, and all the electrons must remain paired in all the resonance structures.

4. The major resonance contributor is the one with the lowest energy.
   Good contributors generally have all octets satisfied, as many bonds as possible, and as little charge separation as possible. Negative charges are more stable on the more electronegative atoms.

5. Resonance stabilization is most important when it serves to delocalize a charge over two or more atoms.

SOLVED PROBLEM 1-2

For each of the following compounds, draw the important resonance forms. Indicate which structures are major and minor contributors or whether they would have the same energy.

(a) $[CH_3OCH_2]^+$

SOLUTION

minor contributor   major contributor

The first (minor) structure has a carbon atom with only six electrons around it. The second (major) structure has octets on all atoms and an additional bond.

(b)

SOLUTION

minor contributor   major contributor

Both of these structures have octets on oxygen and both carbon atoms, and they have the same number of bonds. The first structure has the negative charge on carbon; the second has it on oxygen. Oxygen is the more electronegative element, so the second structure is the major contributor.

(c) H_2SO_4

SOLUTION

The first structure, with more bonds and less charge separation, is possible because sulfur is a third-row element with accessible *d* orbitals, giving it an expandable valence. For example, SF_6 is a stable compound with 12 electrons around sulfur. Theoretical calculations suggest that the last structure, with octets on all atoms, may be the major resonance contributor, however. We cannot always predict the major contributor of a resonance hybrid.

PROBLEM 1-7

Draw the important resonance forms for the following molecules and ions.

(a) CO_3^{2-} **(b)** NO_3^- **(c)** NO_2^- **(d)** $H_2C{=}CH{-}CH_2^+$

(e) $H_2C{=}CH{-}CH_2^-$ **(f)** SO_4^{2-} **(g)** $[CH_3C(OCH_3)_2]^+$

PROBLEM 1-8

For each of the following compounds, draw the important resonance forms. Indicate which structures are major and minor contributors or whether they have the same energy.

(a) $[H_2CNO_2]^-$ **(b)** $H_2C{=}CH{-}NO_2$ **(c)** $[H_2COH]^+$

(d) H_2CNN **(e)** $[H_2CCN]^-$ **(f)** $H_2N{-}\overset{+}{C}H{-}CH{=}CH{-}NH_2$

(g) $H{-}\overset{\overset{\displaystyle O}{\|}}{C}{-}\overset{-}{C}H{-}\overset{\overset{\displaystyle O}{\|}}{C}{-}H$ **(h)** $H{-}\overset{\overset{\displaystyle O}{\|}}{C}{-}NH_2$

PROBLEM-SOLVING HINT

In drawing resonance forms for ions, see how you can delocalize the charge over several atoms. Try to spread a negative charge over electronegative elements like oxygen and nitrogen. Try to spread a positive charge over as many carbons as possible, but especially over any atoms that can bear the positive charge and still have an octet; for example, oxygen (with three bonds) or nitrogen (with four bonds).

Several kinds of formulas are used by organic chemists to represent organic compounds. Some of these formulas involve a shorthand notation that requires some explanation. **Structural formulas** actually show which atoms are bonded to which. There are two types of structural formulas, complete Lewis structures and condensed structural formulas. In addition, there are several ways of drawing condensed structural formulas. As we have seen, a Lewis structure symbolizes a bonding pair of electrons as a pair of dots or as a dash ($-$). Lone pairs of electrons are shown as pairs of dots.

1-10 Structural Formulas

1-10A Condensed Structural Formulas

Condensed structural formulas (Table 1-2) are written without showing all the individual bonds. In a condensed structure, each central atom is shown together with the atoms that are bonded to it. The atoms bonded to a central atom are often listed after the central atom (as in CH_3CH_3 rather than $H_3C{-}CH_3$) even if that is not their actual bonding order. In many cases, if there are two or more identical groups, parentheses and a subscript may be used to represent all the identical groups. Nonbonding electrons are rarely shown in condensed structural formulas.

TABLE 1-2 Examples of Condensed Structural Formulas

Compound	Lewis Structure	Condensed Structural Formula
ethane		CH_3CH_3
isobutane		$(CH_3)_3CH$
n-hexane		$CH_3(CH_2)_4CH_3$

(continues on next page)

TABLE 1-2   *(continued)*

Compound	Lewis Structure	Condensed Structural Formula
diethyl ether		$CH_3CH_2OCH_2CH_3$ or CH_3CH_2—O—CH_2CH_3 or $(CH_3CH_2)_2O$
ethanol		CH_3CH_2OH
isopropyl alcohol		$(CH_3)_2CHOH$
dimethylamine		$(CH_3)_2NH$

When a condensed structural formula is written for a compound containing double or triple bonds, the multiple bonds are often drawn as they would be in a Lewis structure. Table 1-3 shows examples of condensed structural formulas containing multiple bonds. Notice that the —CHO group of an aldehyde and the —COOH group of a carboxylic acid are actually bonded differently from what the condensed notation suggests.

As you can see from Tables 1-2 and 1-3, the distinction between a complete Lewis structural formula and a condensed structural formula can be blurry. Chemists often draw

TABLE 1-3   Condensed Structural Formulas for Double and Triple Bonds

Compound	Lewis Structure	Condensed Structural Formula
2-butene		$CH_3CHCHCH_3$ or $CH_3CH{=}CHCH_3$
acetonitrile		CH_3CN or $CH_3C{\equiv}N$
acetaldehyde		CH_3CHO or $CH_3\overset{\text{O}}{\overset{\|}{C}}H$
acetone		CH_3COCH_3 or $CH_3\overset{\text{O}}{\overset{\|}{C}}CH_3$
acetic acid		CH_3COOH or $CH_3\overset{\text{O}}{\overset{\|}{C}}{-}OH$ or CH_3CO_2H

formulas with some parts condensed and other parts completely drawn out. You should work with these different types of formulas so that you understand what all of them mean.

PROBLEM 1-9

Draw complete Lewis structures for the following condensed structural formulas.
(a) $CH_3(CH_2)_3CH(CH_3)_2$ **(b)** $(CH_3)_2CHCH_2Cl$ **(c)** $CH_3CH_2COCHCH_2$
(d) CH_3CH_2CHO **(e)** CH_3COCN **(f)** $(CH_3)_3CCOOH$ **(g)** $(CH_3CH_2)_2CO$

1-10B   Line–Angle Formulas

Another kind of shorthand used for organic structures is the **line–angle formula**, sometimes called a **skeletal structure** or a **stick figure**. Line–angle formulas are often used for cyclic compounds and occasionally for noncyclic ones. In a stick figure, bonds are represented by lines, and carbon atoms are assumed to be present wherever two lines meet or a line begins or ends. Nitrogen, oxygen, and halogen atoms are shown, but hydrogen atoms are not usually drawn unless they are bonded to a drawn atom. Each carbon atom is assumed to have enough hydrogen atoms to give it a total of four bonds. Nonbonding electrons are rarely shown. Table 1-4 shows some examples of line–angle drawings.

TABLE 1-4   Examples of Line–Angle Drawings

Compound	Condensed Structure	Line–Angle Formula
hexane	$CH_3(CH_2)_4CH_3$	
2-hexene	$CH_3CH=CHCH_2CH_2CH_3$	
3-hexanol	$CH_3CH_2CH(OH)CH_2CH_2CH_3$	
2-cyclohexenone		
2-methylcyclohexanol		
nicotinic acid (a vitamin, also called niacin)		

PROBLEM 1-10

Give Lewis structures corresponding to the following line–angle structures.

(a)   **(b)**   **(c)**   **(d)**

(e)   **(f)**   **(g)**   **(h)**

PROBLEM 1-11

Draw condensed structural formulas corresponding to the following line–angle structures.

1-11
Molecular Formulas and Empirical Formulas

Before we can write possible structural formulas for a compound, we need to know its molecular formula. The **molecular formula** simply gives the number of atoms of each element in one molecule of the compound. For example, the molecular formula for 1-butanol is $C_4H_{10}O$.

$$CH_3CH_2CH_2CH_2OH$$

1-butanol, molecular formula $C_4H_{10}O$

Calculation of the Empirical Formula   Molecular formulas can be determined by a two-step process. The first step is the determination of an **empirical formula**, simply the relative ratios of the elements present. Suppose, for example, that an unknown compound was found by quantitative elemental analysis to contain 40.0% carbon and 6.67% hydrogen. The remainder of the weight (53.3%) is assumed to be oxygen. To convert these numbers to an empirical formula, we can follow a simple procedure:

1. Assume the sample contains 100 g, so the percent value gives the number of grams of each element. Divide the number of grams of each element by the atomic weight to get the number of moles of that atom in the 100-g sample.
2. Divide each of these numbers of moles by the smallest one. This step should give recognizable ratios.

For the unknown compound, we do the following computations:

$$\frac{40.0 \text{ g C}}{12.0 \text{ g/mol}} = 3.33 \text{ mol C}; \quad \frac{3.33 \text{ mol}}{3.33 \text{ mol}} = 1$$

$$\frac{6.67 \text{ g H}}{1.01 \text{ g/mol}} = 6.60 \text{ mol H}; \quad \frac{6.60 \text{ mol}}{3.33 \text{ mol}} = 1.98 \cong 2$$

$$\frac{53.3 \text{ g O}}{16.0 \text{ g/mol}} = 3.33 \text{ mol O}; \quad \frac{3.33 \text{ mol}}{3.33 \text{ mol}} = 1$$

The first computation divides the number of grams of carbon by 12, the number of grams of hydrogen by 1, and the number of grams of oxygen by 16. We compare these numbers by dividing them by the smallest number, 3.33. The final result is a ratio of one carbon to two hydrogens to one oxygen. This result gives the empirical formula $C_1H_2O_1$ or CH_2O, which simply shows the ratios of the elements. The molecular formula can be any multiple of this empirical formula, because any multiple also has the same ratio of elements. Possible molecular formulas are CH_2O, $C_2H_4O_2$, $C_3H_6O_3$, $C_4H_8O_4$, etc.

Calculation of the Molecular Formula   How do we know the correct molecular formula? We can choose the right multiple of the empirical formula if we know the molecular weight. Molecular weights can be determined by methods that relate the freezing-point depression or boiling-point elevation of a solvent to the molal concentration of the unknown. If the compound is volatile, we can convert it to a gas and use its volume to determine the number of moles according to the gas law. Newer methods include *mass spectrometry*, which we will cover in Chapter 11.

For our example (empirical formula CH_2O), let's assume that the molecular weight is determined to be about 60. The weight of one CH_2O unit is 30, so our unknown compound must contain twice this many atoms. The molecular formula must be $C_2H_4O_2$. The compound might be acetic acid.

$$\underset{\text{acetic acid, } C_2H_4O_2}{CH_3-\overset{\overset{\textstyle O}{\|}}{C}-OH}$$

In Chapters 12, 13, and 15 we will use spectroscopic techniques to determine the complete structure for a compound once we know its molecular formula.

PROBLEM 1-12

Compute the empirical and molecular formulas for each of the following elemental analyses. In each case, propose at least one structure that fits the molecular formula.

	C	H	N	Cl	MW
(a)	40.0%	6.67%	0	0	90
(b)	32.0%	6.67%	18.7%	0	75
(c)	37.2%	7.75%	0	55.0%	64
(d)	38.4%	4.80%	0	56.8%	125

PROBLEM-SOLVING HINT

If an elemental analysis does not add up to 100%, the missing percentage is assumed to be oxygen.

1-12
Arrhenius Acids and Bases

The properties and reactions of acids and bases are central to our study of organic chemistry. We need to consider exactly what is meant by the terms **acid** and **base**. Most people would agree that H_2SO_4 is an acid and NaOH is a base. Is BF_3 an acid or a base? Is ethylene ($H_2C{=}CH_2$) an acid or a base? To answer these questions, we need to understand the three different definitions of acids and bases: the Arrhenius definition, the Brønsted–Lowry definition, and the Lewis definition.

Acidic compounds were first classified on the basis of their sour taste. The Latin terms *acidus* (sour) and *acetum* (vinegar) gave rise to our modern terms *acid* and *acetic acid*. Alkaline compounds (bases) were substances that neutralize acids, such as limestone and plant ashes (*al kalai* in Arabic).

The *Arrhenius* theory, developed at the end of the nineteenth century, defined acids as substances that dissociate in water to give H_3O^+ ions. The stronger acids, such as sulfuric acid (H_2SO_4), were assumed to dissociate to a greater degree than weaker acids, such as acetic acid (CH_3COOH).

$$\underset{\text{sulfuric acid}}{H_2SO_4} \;+\; H_2O \;\rightleftharpoons\; H_3O^+ \;+\; HSO_4^-$$

$$\underset{\text{acetic acid}}{CH_3-\overset{\overset{\textstyle O}{\|}}{C}-OH} \;+\; H_2O \;\rightleftharpoons\; H_3O^+ \;+\; CH_3-\overset{\overset{\textstyle O}{\|}}{C}-O^-$$

According to the Arrhenius definition, bases are substances that dissociate in water to give hydroxide ions. Strong bases, such as NaOH, were assumed to dissociate more completely than weaker, sparingly soluble bases such as $Mg(OH)_2$.

$$NaOH \;\rightleftharpoons\; Na^+ \;+\; {}^-OH$$

$$Mg(OH)_2 \;\rightleftharpoons\; Mg^{2+} \;+\; 2\,{}^-OH$$

The acidity or basicity of an aqueous (water) solution is measured by the concentration of H_3O^+. This value also implies the concentration of ^-OH because these two concentrations are related by the water ion-product constant:

$$K_w \;=\; [H_3O^+][^-OH] \;=\; 1.00 \times 10^{-14} \quad \text{(at 24°C)}$$

In a neutral solution, the concentrations of H_3O^+ and ^-OH are equal.

$$[H_3O^+] \;=\; [^-OH] \;=\; 1.0 \times 10^{-7}\,M \quad \text{in a neutral solution}$$

Acidic and basic solutions are defined by an excess of H_3O^+ or ^-OH.

$$\text{acidic:} \quad [H_3O^+] \;>\; 10^{-7}\,M \quad \text{and} \quad [^-OH] \;<\; 10^{-7}\,M$$

$$\text{basic:} \quad [H_3O^+] \;<\; 10^{-7}\,M \quad \text{and} \quad [^-OH] \;>\; 10^{-7}\,M$$

Because these concentrations can span a wide range of values, the acidity or basicity of a solution is usually measured on a logarithmic scale. The **pH** is defined as the negative logarithm (base 10) of the H_3O^+ concentration.

$$pH \;=\; -\log_{10}[H_3O^+]$$

A neutral solution has a pH of 7, an acidic solution has a pH less than 7, and a basic solution has a pH greater than 7.

> **PROBLEM 1-13**
>
> Calculate the pH of the following solutions.
> (a) 5.00 g of HBr in 100 mL of aqueous solution
> (b) 1.50 g of NaOH in 50 mL of aqueous solution

The Arrhenius definition was an important contribution to understanding many acids and bases, but it does not explain why a compound such as ammonia (NH_3) neutralizes acids, even though it has no hydroxide ion in its molecular formula. In Section 1-13 we discuss a more versatile theory of acids and bases that will include ammonia and a wider variety of organic acids and bases.

1-13
Brønsted–Lowry Acids and Bases

In 1923, Brønsted and Lowry defined acids and bases on the basis of the transfer of protons. *A* **Brønsted–Lowry acid** is *any species that can donate a proton*, and a **Brønsted–Lowry base** is *any species that can accept a proton*. These definitions also include all the Arrhenius acids and bases because compounds that dissociate to give H_3O^+ are proton donors, and compounds that dissociate to give ^-OH are proton acceptors. (Hydroxide ion accepts a proton to form H_2O.)

In addition to Arrhenius acids and bases, the Brønsted–Lowry definition includes bases that have no hydroxide ions, yet can accept protons. Consider the following examples of acids donating protons to bases. NaOH is a base under either the Arrhenius or Brønsted–Lowry definition. The other three are Brønsted–Lowry bases but not Arrhenius bases because they have no hydroxide ions.

$$
\begin{array}{ccccccc}
\text{HCl} & + & \text{NaOH} & \rightleftharpoons & \text{NaCl} & + & \text{HOH} \\
\text{proton} & & \text{proton} & & & & \\
\text{donor} & & \text{acceptor} & & & &
\end{array}
$$

$$
\begin{array}{ccccccc}
H_2SO_4 & + & :NH_3 & \rightleftharpoons & HSO_4^- & + & H{-}\overset{+}{N}H_3 \\
\text{proton} & & \text{proton} & & & & \\
\text{donor} & & \text{acceptor} & & & &
\end{array}
$$

HCl + [H, CH₃ / C=C / H, H structure] ⇌ Cl⁻ + [H—C—C⁺ with CH₃, H, H structure]

proton donor   proton acceptor

HNO₃ + [H / C=N: / H, H structure] ⇌ NO₃⁻ + [H / C=N⁺ / H, H structure]

proton donor   proton acceptor

When a base accepts a proton, it becomes an acid capable of returning that proton. When an acid donates its proton, it becomes a base capable of accepting that proton back. One of the most important principles of the Brønsted–Lowry definition is this concept of **conjugate acids** and **bases**. For example, NH_4^+ and NH_3 are a conjugate acid–base pair. NH_3 is the base; when it accepts a proton, it is transformed into its conjugate acid, NH_4^+. Many compounds (water, for instance) can react either as an acid or as a base. Here are some additional examples of conjugate acid–base pairs.

H_2SO_4 + H_2O ⇌ HSO_4^- + H_3O^+
acid   base   conjugate base   conjugate acid

H_2O + $:NH_3$ ⇌ ^-OH + NH_4^+
acid   base   conjugate base   conjugate acid

[H—C(=O)—OH] + CH₃—Ö:⁻ ⇌ [H—C(=O)—Ö:⁻] + CH₃—Ö—H
acid   base   conjugate base   conjugate acid

1-13A   Acid Strength

The strength of a Brønsted–Lowry acid is expressed as it is in the Arrhenius definition, by the extent of its ionization in water. The general reaction of an acid (HA) with water is the following:

$$HA + H_2O \underset{}{\overset{K_a}{\rightleftharpoons}} H_3O^+ + A^- \qquad K_a = \frac{[H_3O^+][A^-]}{[HA]}$$

acid   base

↑———— conjugate acid–base pair ————↑

K_a is called the *acid-dissociation constant*, and its size indicates the relative strength of the acid. The stronger the acid is, the more it dissociates, giving a larger value of K_a. Acid dissociation constants vary over a wide range. Strong acids are almost completely ionized in water, and their dissociation constants are greater than 1. Most organic acids are weak acids, with values of K_a less than 10^{-4}. Many organic compounds are extremely weak acids; for example, methane and ethane are essentially nonacidic, with K_a values less than 10^{-40}.

Because they span such a wide range, acid-dissociation constants are often expressed on a logarithmic scale. The pK_a of an acid is defined just like the pH of a solution: as the negative logarithm (base 10) of K_a.

$$pK_a = -\log_{10} K_a$$

SOLVED PROBLEM 1-3

Calculate K_a and pK_a for water.

SOLUTION

The equilibrium that defines K_a for water is

$$H_2O \quad + \quad H_2O \quad \xrightleftharpoons{\quad K_a \quad} \quad H_3O^+ \quad + \quad ^-OH$$

acid (HA)   solvent   conjugate base (A⁻)

Water serves as both the acid and the solvent in this dissociation. The equilibrium expression is

$$K_a = \frac{[H_3O^+][A^-]}{[HA]} = \frac{[H_3O^+][^-OH]}{[H_2O]}$$

We already know that $[H_3O^+][^-OH] = 1.00 \times 10^{-14}$, the ion-product constant for water.
The concentration of H_2O in water is simply the number of moles of water in 1 L (about 1 kg).

$$\frac{1000 \text{ g/L}}{18 \text{ g/mol}} = 55.6 \text{ mol/L}$$

Substitution gives

$$K_a = \frac{[H_3O^+][^-OH]}{[H_2O]} = \frac{1.00 \times 10^{-14}}{55.6} = 1.8 \times 10^{-16} \text{ M}$$

The logarithm of 1.8×10^{-16} is -15.7, so the pK_a of water is 15.7.

PROBLEM-SOLVING **HINT**

In most cases, the pK_a of an acid corresponds to the pH where the acid is about half dissociated. At a lower (more acidic) pH, the acid is mostly undissociated; at a higher (more basic) pH, the acid is mostly dissociated.

Strong acids generally have pK_a values around 0, and weak acids, such as most organic acids, have pK_a values that are greater than 4. *Weaker acids have larger values of pK_a.* Table 1-5 lists values of K_a and pK_a for some common inorganic and organic compounds. Notice that the pK_a values increase as the K_a values decrease.

PROBLEM 1-14

Ammonia appears in Table 1-5 both as an acid and as a conjugate base.
(a) Explain how ammonia can act as both an acid and a base. Which of these roles does it commonly fill in aqueous solutions?
(b) Show how water can serve as both an acid and a base.
(c) Show how methanol (CH_3OH) can serve as both an acid and a base. Write an equation for the reaction of methanol with sulfuric acid.

1-13B   Base Strength

The strength of an acid is inversely related to the strength of its conjugate base. For an acid (HA) to be strong, its conjugate base (A⁻) must be stable in its anionic form; otherwise, HA would not easily lose its proton. Therefore, the conjugate base of a strong acid must be a weak base. On the other hand, if an acid is weak, its conjugate is a strong base.

$$HCl \quad + \quad H_2O \quad \rightleftharpoons \quad H_3O^+ \quad + \quad Cl^-$$

strong acid   weak base

$$CH_3-\overset{\cdot\cdot}{\underset{\cdot\cdot}{O}}H \quad + \quad H_2O \quad \rightleftharpoons \quad H_3O^+ \quad + \quad CH_3\overset{\cdot\cdot}{\underset{\cdot\cdot}{O}}{:}^-$$

weak acid   strong base

In the reaction of an acid with a base, the equilibrium generally favors the *weaker* acid and base. For example, in the preceding reactions, H_3O^+ is a weaker acid than HCl

TABLE 1-5   Relative Strength of Some Common Organic and Inorganic Acids and Their Conjugate Bases

	Acid				Conjugate Base		K_a	pK_a
strong acids	HCl hydrochloric acid	$+ H_2O$	$\rightleftharpoons$	$H_3O^+ +$	Cl^- chloride ion		1.6×10^2	-2.2
	HF hydrofluoric acid	$+ H_2O$	$\rightleftharpoons$	$H_3O^+ +$	F^- fluoride ion		6.8×10^{-4}	3.17
	H—C(=O)—OH formic acid	$+ H_2O$	$\rightleftharpoons$	$H_3O^+ +$	H—C(=O)—O$^-$ formate ion		1.7×10^{-4}	3.76
	CH_3—C(=O)—OH acetic acid	$+ H_2O$	$\rightleftharpoons$	$H_3O^+ +$	CH_3—C(=O)—O$^-$ acetate ion		1.8×10^{-5}	4.74
weak acids	$H-C\equiv N:$ hydrocyanic acid	$+ H_2O$	$\rightleftharpoons$	$H_3O^+ +$	$^-:C\equiv N:$ cyanide ion		6.0×10^{-10}	9.22
	$^+NH_4$ ammonium ion	$+ H_2O$	$\rightleftharpoons$	$H_3O^+ +$	$:NH_3$ ammonia		5.8×10^{-10}	9.24
	CH_3—OH methyl alcohol	$+ H_2O$	$\rightleftharpoons$	$H_3O^+ +$	CH_3O^- methoxide ion		3.2×10^{-16}	15.5
	H_2O water	$+ H_2O$	$\rightleftharpoons$	$H_3O^+ +$	HO^- hydroxide ion		1.8×10^{-16}	15.7
very weak	NH_3 ammonia	$+ H_2O$	$\rightleftharpoons$	$H_3O^+ +$	$^-:\ddot{N}H_2$ amide ion		10^{-33}	33
not acidic	CH_4 methane	$+ H_2O$	$\rightleftharpoons$	$H_3O^+ +$	$^-:CH_3$ methyl anion		$<10^{-40}$	>40

stronger → weaker (acid direction); weaker → stronger bases (conjugate base direction)

but a stronger acid than CH_3OH. It also follows that H_2O is a stronger base than Cl^- but a weaker base than CH_3O^-.

The strength of a base is measured much like the strength of an acid, by using the equilibrium constant of the hydrolysis reaction.

$$A^- \quad + \quad H_2O \quad \xrightarrow{K_b} \quad HA \quad + \quad {}^-OH$$

conjugate base   conjugate acid

The equilibrium constant (K_b) for this reaction is called the *base-dissociation constant* for the base A^-. Because this constant spans a wide range of values, it is often given in logarithmic form. The negative logarithm (base 10) of K_b is defined as pK_b.

$$K_b = \frac{[HA][^-OH]}{[A^-]} \qquad pK_b = -\log_{10} K_b$$

The acid-base properties of many natural products are important for their isolation, distribution in the body, and therapeutic effects. For example, morphine (page 2) is isolated from the opium poppy and crosses into the brain as the free base, where the nitrogen is not charged. However, it exerts its pain-relieving effects as the charged species.

When we multiply K_a by K_b, we can see how the acidity of an acid is related to the basicity of its conjugate base.

$$(K_a)(K_b) = \frac{[H_3O^+][A^-]}{[HA]} \frac{[HA][^-OH]}{[A^-]} = [H_3O^+][^-OH] = 1.0 \times 10^{-14}$$
$$\text{water ion-product constant}$$

$$(K_a)(K_b) = 10^{-14}$$

Logarithmically,

$$pK_a + pK_b = -\log 10^{-14} = 14$$

The product of K_a and K_b must always equal the ion-product constant of water, 10^{-14}. If the value of K_a is large, the value of K_b must be small; that is, the stronger an acid, the weaker its conjugate base. Similarly, a small value of K_a (weak acid) implies a large value of K_b (strong base).

The stronger an acid, the weaker its conjugate base.

The weaker an acid, the stronger its conjugate base.

Acid–base reactions favor the weaker acid and the weaker base.

PROBLEM-SOLVING HINT

An acid will donate a proton to the conjugate base of any weaker acid (smaller K_a or higher pK_a).

PROBLEM 1-15 (Partially Solved)

Write equations for the following acid–base reactions. Use the information in Table 1-5 to predict whether the equilibrium will favor the reactants or the products.
(a) HCOOH + $^-$CN
(b) CH_3COO^- + CH_3OH
(c) CH_3OH + $NaNH_2$
(d) $NaOCH_3$ + HCN
(e) HCl + H_2O
(f) H_3O^+ + CH_3O^-

Solution to (a): Cyanide is the conjugate base of HCN. It can accept a proton from formic acid:

Reading from Table 1-5, formic acid (pK_a = 3.76) is a stronger acid than HCN (pK_a = 9.22), and cyanide is a stronger base than formate. The products (weaker acid and base) are favored.

SOLVED PROBLEM 1-4

Each of the following compounds can act as an acid. Show the reaction of each compound with a general base (A^-), and show the Lewis structure of the conjugate base that results.
(a) CH_3CH_2OH
(b) CH_3NH_2
(c) CH_3COOH

SOLUTION
(a) Ethanol (CH_3CH_2OH) can lose the O—H proton to give a conjugate base that is an organic analogue of hydroxide ion.

(C—H protons are much less acidic than O—H protons because carbon is less electronegative than oxygen, and the negative charge is therefore less stable on carbon.)

(b) Methylamine (CH_3NH_2) is a very weak acid. A very strong base can abstract a proton to give a powerful conjugate base.

$$CH_3-\overset{\overset{\displaystyle H}{|}}{\underset{\cdot\cdot}{N}}-H \quad + \quad A:^- \quad \rightleftharpoons \quad CH_3-\overset{\cdot\cdot}{\underset{\cdot\cdot}{N}}-H \quad + \quad HA$$

methylamine   very strong   (powerful base)
(very weak acid)   base

(c) Acetic acid (CH_3COOH) is a moderately strong acid, giving the resonance-stabilized acetate ion as its conjugate base.

$$CH_3-\overset{\overset{\displaystyle \cdot\cdot}{\overset{\displaystyle O}{\|}}}{C}-\overset{\cdot\cdot}{\underset{\cdot\cdot}{O}}-H \quad + \quad A:^- \quad \rightleftharpoons$$

acetic acid
(moderate acid)

$$\left[CH_3-\overset{\overset{\displaystyle \cdot\cdot}{\overset{\displaystyle O}{\|}}}{C}-\overset{\cdot\cdot}{\underset{\cdot\cdot}{O}}:^- \quad \longleftrightarrow \quad CH_3-\overset{\overset{\displaystyle :\overset{\cdot\cdot}{O}:^-}{|}}{C}=\overset{\cdot\cdot}{\underset{\cdot\cdot}{O}} \right] \quad + \quad HA$$

acetate ion
(moderate base)

SOLVED PROBLEM 1-5

Each of the compounds in Solved Problem 1-4 can also react as a base. Show the reaction of each compound with a general acid (HA), and show the Lewis structure of the conjugate acid that results.

SOLUTION

(a) Ethanol can undergo protonation on its oxygen atom. Notice that one of the lone pairs of the oxygen forms the new O—H bond.

$$CH_3CH_2-\overset{\cdot\cdot}{\underset{\cdot\cdot}{O}}-H \quad + \quad HA \quad \rightleftharpoons \quad CH_3CH_2-\overset{\overset{\displaystyle H}{|}}{\underset{\cdot\cdot}{O}}{}^+\!\!-H \quad + \quad A:^-$$

ethanol   acid   (strong acid)
(weak base)

(b) The nitrogen atom of methylamine has a pair of electrons that can bond to a proton.

$$CH_3-\overset{\cdot\cdot}{N}H_2 \quad + \quad HA \quad \rightleftharpoons \quad CH_3-\overset{\overset{\displaystyle H}{|}}{\underset{+}{N}}H_2 \quad + \quad A:^-$$

methylamine   acid   (moderate acid)
(moderate base)

(c) Acetic acid has nonbonding electrons on both of its oxygen atoms. Either of these oxygen atoms might become protonated, but protonation of the double-bonded oxygen is favored because protonation of this oxygen gives a symmetrical, resonance-stabilized conjugate acid.

$$CH_3-\overset{\overset{\displaystyle \cdot\cdot}{\overset{\displaystyle O}{\|}}}{C}-\overset{\cdot\cdot}{\underset{\cdot\cdot}{O}}-H \quad + \quad HA \quad \rightleftharpoons$$

acetic acid
(very weak base)

$$\left[CH_3-\overset{\overset{\displaystyle +\overset{\cdot\cdot}{O}-H}{\|}}{C}-\overset{\cdot\cdot}{\underset{\cdot\cdot}{O}}-H \quad \longleftrightarrow \quad CH_3-\overset{\overset{\displaystyle :\overset{\cdot\cdot}{O}-H}{|}}{\underset{+}{C}}-\overset{\cdot\cdot}{\underset{\cdot\cdot}{O}}-H \quad \longleftrightarrow \quad CH_3-\overset{\overset{\displaystyle :\overset{\cdot\cdot}{O}-H}{|}}{C}=\overset{\cdot\cdot}{\underset{+}{O}}-H \right] \quad + \quad A:^-$$

conjugate acid of acetic acid
(very strong acid)

PROBLEM 1-16

Solved Problem 1-5(c) showed protonation of the double-bonded oxygen in acetic acid. Show the product of protonation on the other (—OH) oxygen. Explain why protonation of the double-bonded oxygen is favored.

PROBLEM 1-17

(a) Rank ethanol, methylamine, and acetic acid in decreasing order of acidity.
(b) Rank ethanol, methylamine (pK_b 3.36), and ethoxide ion ($CH_3CH_2O^-$) in decreasing order of basicity. In each case, explain your ranking.

1-13C   Structural Effects on Acidity

How can we look at a structure and predict whether a compound will be a strong acid, a weak acid, or not an acid at all? To be a Brønsted–Lowry acid (HA), a compound must contain a hydrogen atom that can be lost as a proton. A strong acid must have a stable conjugate base ($A\colon^-$) after losing the proton.

The stability of the conjugate base is a good guide to acidity. More-stable anions tend to be weaker bases, and their conjugate acids tend to be stronger acids. Some of the factors that affect the stability of conjugate bases are electronegativity, size, and resonance.

Electronegativity   A more electronegative element bears a negative charge more easily, giving a more stable conjugate base and a stronger acid. Electronegativities increase from left to right in the periodic table:

Size   The negative charge of an anion is more stable if it is spread over a larger region o f space. Within a column of the periodic table, acidity increases down the column, as the size of the elements increases.

Resonance Stabilization   The negative charge of a conjugate base may be delocalized over two or more atoms by resonance. Depending on how electronegative those
atoms are, and how many share the charge, resonance delocalization is often the dom-
inant effect helping to stabilize an anion. Consider the following conjugate bases.

Conjugate Base	Acid	pK_a
$CH_3CH_2-\ddot{O}:^-$ ethoxide ion	CH_3CH_2-OH ethanol	15.9 (weak acid)
$\left[CH_3-\overset{\overset{..}{O}}{\underset{}{C}}-\ddot{O}:^- \longleftrightarrow CH_3-\overset{:\ddot{O}:^-}{\underset{}{C}}=\ddot{O} \right]$ acetate ion	$CH_3-\overset{O}{\underset{}{C}}-OH$ acetic acid	4.74 (moderate acid)
$\left[CH_3-\overset{\overset{..}{O}}{\underset{\underset{..}{.\ddot{O}.}}{S}}-\ddot{O}:^- \longleftrightarrow CH_3-\overset{:\ddot{O}:^-}{\underset{\underset{..}{.\ddot{O}.}}{S}}=\ddot{O} \longleftrightarrow CH_3-\overset{\overset{..}{O}}{\underset{\underset{..}{:\ddot{O}:^-}}{S}}=\ddot{O} \right]$ methanesulfonate ion	$CH_3-\overset{O}{\underset{O}{S}}-OH$ methanesulfonic acid	−1.2 (strong acid)

Ethoxide ion is the strongest of these three bases. Ethoxide has a negative charge
localized on one oxygen atom; acetate ion has the negative charge shared by two oxygen atoms; and the methanesulfonate ion has the negative charge spread over three oxygen
atoms. The pK_a values of the conjugate acids of these anions show that acids are stronger
if they deprotonate to give resonance-stabilized conjugate bases.

PROBLEM 1-18

Write equations for the following acid–base reactions. Label the conjugate acids and bases, and
show any resonance stabilization. Predict whether the equilibrium favors the reactants or products.
(a) $CH_3CH_2OH + CH_3NH^-$ **(b)** $CH_3CH_2COOH + CH_3NHCH_3$
(c) $CH_3OH + H_2SO_4$ **(d)** $NaOH + H_2S$
(e) $CH_3NH_3^+ + CH_3O^-$ **(f)** $CH_3O^- + CH_3COOH$
(g) $CH_3SO_3^- + CH_3COOH$

The Brønsted–Lowry definition of acids and bases depends on the transfer of a proton from
the acid to the base. The base uses a pair of nonbonding electrons to form a bond to the
proton. G. N. Lewis reasoned that this kind of reaction does not need a proton. Instead, a
base could use its lone pair of electrons to bond to some other electron-deficient atom. In
effect, we can look at an acid–base reaction from the viewpoint of the *bonds* that are
formed and broken rather than a proton that is transferred. The following reaction shows
the proton transfer with emphasis on the bonds being broken and formed. Organic chemists
routinely use curved arrows to show the movement of the participating electrons.

1-14
Lewis Acids and Bases

$$B:^- \curvearrowright H:A \rightleftharpoons B:H + ^-:A$$

Lewis bases are species with nonbonding electrons that can be donated to form
new bonds. **Lewis acids** are species that can accept these electron pairs to form new
bonds. Since a Lewis acid *accepts* a pair of electrons, it is called an **electrophile**, from
the Greek words meaning "lover of electrons." A Lewis base is called a **nucleophile**,
or "lover of nuclei," because it donates electrons to a nucleus with an empty (or easily
vacated) orbital. In this book, we sometimes use colored type for emphasis: blue for
nucleophiles, green for electrophiles, and occasionally red for acidic protons.

The Lewis acid–base definitions include reactions having nothing to do with protons. Below are some examples of Lewis acid–base reactions. Notice that the common Brønsted–Lowry acids and bases also fall under the Lewis definition, with a proton serving as the electrophile. Curved arrows (red) are used to show the movement of electrons, generally from the nucleophile to the electrophile.

$$B\!:\!\!^{-} \qquad H^{+} \qquad \longrightarrow \qquad B\!-\!H$$

nucleophile   electrophile   bond formed

nucleophile   electrophile   bond formed

nucleophile   electrophile   bond formed

Some of the terms associated with acids and bases have evolved specific meanings in organic chemistry. When organic chemists use the term *base*, they usually mean a proton acceptor (a Brønsted–Lowry base). Similarly, the term *acid* usually means a proton acid (a Brønsted–Lowry acid). When the acid–base reaction involves formation of a bond to some other element (especially carbon), organic chemists refer to the electron donor as a *nucleophile* (Lewis base) and the electron acceptor as an *electrophile* (Lewis acid).

The **curved-arrow formalism** is used to show the flow of an electron pair *from the electron donor to the electron acceptor*. The movement of each pair of electrons involved in making or breaking bonds is indicated by its own separate arrow, as shown in the preceding set of reactions. In this book, these curved arrows are always printed in red. In the reaction of CH_3O^- with CH_3Cl above, one curved arrow shows the lone pair on oxygen forming a bond to carbon. Another curved arrow shows that the $C-Cl$ bonding pair detaches from carbon and becomes a lone pair on the Cl^- product.

PROBLEM-SOLVING HINT

Use one curved arrow for each pair of electrons participating in the reaction.

nucleophile   electrophile

The curved-arrow formalism is universally used for keeping track of the flow of electrons in reactions. We have also used this device (in Section 1-9, for example) to keep track of electrons in resonance structures as we imagined their "flow" in going from one resonance structure to another. Remember that electrons do not "flow" in resonance structures: They are simply delocalized. Still, the curved-arrow formalism helps our *minds* flow from one resonance structure to another. We will find ourselves constantly using these (red) curved arrows to keep track of electrons, both as reactants change to products and as we imagine additional resonance structures of a hybrid.

PROBLEM 1-19 (Partially Solved)

In the following acid–base reactions
(1) Determine which species are acting as acids and which are acting as bases.
(2) Use the curved-arrow formalism to show the movement of electron pairs in these reactions, as well as the imaginary movement in the resonance hybrids of the products.
(3) Indicate which reactions are best termed Brønsted–Lowry acid–base reactions.

(a)

$$CH_3-\overset{\overset{\textstyle O}{\|}}{C}-H \; + \; HCl \quad \longrightarrow \quad CH_3-\overset{\overset{\textstyle \overset{+}{O}-H}{\|}}{C}-H \; + \; Cl^-$$

acetaldehyde

This reaction is a proton transfer from HCl to the $C=O$ group of acetaldehyde. Therefore, it is a Brønsted–Lowry acid–base reaction, with HCl acting as the acid (proton donor) and acetaldehyde acting as the base (proton acceptor). Before drawing any curved arrows, remember that arrows must show the movement of electrons *from* the electron-pair donor (the base) *to* the electron-pair acceptor (the acid). An arrow must go *from* the electrons on acetaldehyde that form the bond *to* the hydrogen atom, and the bond to chlorine must break, with the chloride ion taking these electrons. Drawing these arrows is easier once we draw good Lewis structures for all the reactants and products.

The resonance forms of the product show that a pair of electrons can be moved between the oxygen atom and the $C=O$ pi bond. The positive charge is delocalized over the carbon and oxygen atoms, with most of the positive charge on oxygen because all octets are satisfied in that resonance structure.

(b)

$$CH_3-\overset{\overset{\textstyle O}{\|}}{C}-H \; + \; CH_3-O^- \quad \longrightarrow \quad CH_3-\overset{\overset{\textstyle O^-}{|}}{\underset{\underset{\textstyle O-CH_3}{|}}{C}}-H$$

acetaldehyde

In this case, no proton has been transferred, so this is not a Brønsted–Lowry acid–base reaction. Instead, a bond has formed between the $C=O$ carbon atom and the oxygen of the CH_3-O^- group. Drawing the Lewis structures helps to show that the CH_3-O^- group (the nucleophile in this reaction) donates the electrons to form the new bond to acetaldehyde (the electrophile). This result agrees with our intuition that a negatively charged ion is likely to be electron-rich and therefore an electron donor.

electrophile   nucleophile

Notice that acetaldehyde acts as the nucleophile (base) in part (a) and the electrophile (acid) in part (b). Like most organic compounds, acetaldehyde is both acidic and basic. It acts as a base if we add a strong enough acid to make it donate electrons or accept a proton. It acts as an acid if the base we add is strong enough to donate an electron pair or abstract a proton.

(c) $BH_3 \; + \; CH_3-O-CH_3 \quad \longrightarrow \quad CH_3-\overset{\overset{\textstyle {}^-BH_3}{|}}{\underset{+}{O}}-CH_3$

(d) $CH_3-\overset{\overset{\textstyle O}{\|}}{C}-H \; + \; {}^-OH \quad \longrightarrow \quad CH_3-\overset{\overset{\textstyle O^-}{|}}{\underset{\underset{\textstyle OH}{|}}{C}}-H$

(e) $CH_3-\overset{\overset{\textstyle O}{\|}}{C}-H \; + \; {}^-OH \quad \longrightarrow \quad \left[H-\overset{\overset{\textstyle H}{|}}{\underset{-}{C}}-\overset{\overset{\textstyle O}{\|}}{C}-H \quad \longleftrightarrow \quad H-\overset{\overset{\textstyle H}{|}}{C}=\overset{\overset{\textstyle O^-}{|}}{C}-H \right] + H_2O$

(f) $CH_3-NH_2 \; + \; CH_3-Cl \quad \longrightarrow \quad CH_3-\overset{+}{N}H_2-CH_3 \; + \; Cl^-$

PROBLEM-SOLVING **HINT**

The curved arrows we use in mechanisms show the *flow of electrons* and not the movement of atoms. We will use these curved arrows constantly throughout this course.

Chapter 1 Glossary

Each chapter ends with a glossary that summarizes the most important new terms in the chapter. These glossaries are more than just a dictionary to look up unfamiliar terms as you encounter them (the index serves that purpose). The glossary is one of the tools for reviewing the chapter. You can read carefully through the glossary to see if you understand and remember all the terms and associated chemistry mentioned there. Anything that seems unfamiliar should be reviewed by turning to the page number given in the glossary listing.

acids and **bases** (pp. 21–31)

    (**Arrhenius** definitions)   **acid:** dissociates in water to give H_3O^+
                                  base: dissociates in water to give ^-OH

    (**Brønsted–Lowry** definitions)   **acid:** proton donor
                                  base: proton acceptor

    (**Lewis** definitions)   **acid:** electron-pair acceptor (electrophile)
                                  base: electron-pair donor (nucleophile)

conjugate acid The acid that results from protonation of a base. (p. 23)

conjugate base The base that results from loss of a proton from an acid. (p. 23)

covalent bonding Bonding that occurs by the sharing of electrons in the region between two nuclei. (p. 7)

    single bond: A covalent bond that involves the sharing of one pair of electrons. (p. 8)

    double bond: A covalent bond that involves the sharing of two pairs of electrons. (p. 8)

    triple bond: A covalent bond that involves the sharing of three pairs of electrons. (p. 8)

curved-arrow formalism A method of drawing curved arrows to keep track of electron movement from nucleophile to electrophile (or within a molecule) during the course of a reaction. (p. 30)

degenerate orbitals Orbitals with identical energies. (p. 4)

dipole moment (μ) A measure of the polarity of a bond (or a molecule), proportional to the product of the charge separation times the bond length. (p. 10)

electron density The relative probability of finding an electron in a certain region of space. (p. 3)

electronegativity A measure of an element's ability to attract electrons. Elements with higher electronegativities attract electrons more strongly. (p. 10)

electrophile An electron-pair acceptor (Lewis acid). (p. 29)

electrostatic potential map (EPM) A computer-calculated molecular representation that uses colors to show the charge distribution in a molecule. In most cases, the EPM uses red to show electron-rich regions (most negative electrostatic potential) and blue to show electron-poor regions (most positive electrostatic potential). The intermediate colors orange, yellow, and green show regions with intermediate electrostatic potentials. (p. 10)

empirical formula The ratios of atoms in a compound. (p. 20) See also **molecular formula**.

formal charges A method for keeping track of charges, showing what charge would be on an atom in a particular Lewis structure. (p. 11)

Hund's rule When there are two or more unfilled orbitals of the same energy (degenerate orbitals), the lowest-energy configuration places the electrons in different orbitals (with parallel spins) rather than paired in the same orbital. (p. 6)

ionic bonding Bonding that occurs by the attraction of oppositely charged ions. Ionic bonding usually results in the formation of a large three-dimensional crystal lattice. (p. 7)

isotopes Atoms with the same number of protons but different numbers of neutrons; atoms of the same element but with different atomic masses. (p. 3)

Lewis acid, Lewis base See **acids** and **bases**.

Lewis structure A structural formula that shows all valence electrons, with the bonds symbolized by dashes (—) or by pairs of dots, and nonbonding electrons symbolized by dots. (p. 7)

line–angle formula (skeletal structure, stick figure) A shorthand structural formula with bonds represented by lines and carbon atoms wherever two lines meet or a line begins or bends. Nitrogen, oxygen, and halogen atoms are shown, but hydrogen atoms are not. Each carbon atom is assumed to have enough hydrogens to give it four bonds. (p. 19)

Lewis structure of 2-cyclohexenol

2-cyclohexenol
equivalent line–angle formula

lone pair A pair of nonbonding electrons. (p. 7)

molecular formula The number of atoms of each element in one molecule of a compound. The **empirical formula** simply gives the ratios of atoms of the different elements. For example, the molecular formula of glucose is $C_6H_{12}O_6$. Its empirical formula is CH_2O. Neither the molecular formula nor the empirical formula gives structural information. (p. 20)

node In an orbital, a region with zero electron density. (p. 4)

nodal plane A flat (planar) region of space with zero electron density. (p. 4)

nonbonding electrons Valence electrons that are not used for bonding. A pair of nonbonding electrons is often called a **lone pair**. (p. 7)

nucleophile An electron-pair donor (Lewis base). (p. 29)

octet rule Atoms generally form bonding arrangements that give them filled shells of electrons (noble-gas configurations). For the second-row elements, this configuration has eight valence electrons. (p. 6)

orbital An allowed energy state for an electron bound to a nucleus; the probability function that defines the distribution of electron density in space. The *Pauli exclusion principle* states that up to two electrons can occupy each orbital if their spins are paired. (p. 3)

organic chemistry (new definition): The chemistry of carbon compounds. (old definition): The study of compounds derived from living organisms and their natural products. (p. 1)

pH A measure of the acidity of a solution, defined as the negative logarithm (base 10) of the H_3O^+ concentration. $pH = -\log_{10}[H_3O^+]$ (p. 22)

polar covalent bond A covalent bond in which electrons are shared unequally. A bond with equal sharing of electrons is called a **nonpolar covalent bond**. (p. 9)

resonance hybrid A molecule or ion for which two or more valid Lewis structures can be drawn, differing only in the placement of the valence electrons. These Lewis structures are called **resonance forms** or **resonance structures**. Individual resonance forms do not exist, but we can estimate their relative energies. The more important (lower-energy) structures are called **major contributors**, and the less important (higher-energy) structures are called **minor contributors**. When a charge is spread over two or more atoms by resonance, it is said to be **delocalized**, and the molecule is said to be **resonance stabilized**. (pp. 13–16)

structural formulas A **complete structural formula** (such as a Lewis structure) shows all the atoms and bonds in the molecule. A **condensed structural formula** shows each central atom along with the atoms bonded to it. A **line–angle formula** assumes that there is a carbon atom wherever two lines meet or a line begins or ends. See Section 1-10 for examples. (p. 17)

valence The number of bonds an atom usually forms. (p. 9)

valence electrons Those electrons that are in the outermost shell. (p. 6)

Vitalism The belief that syntheses of organic compounds require the presence of a "vital force." (p. 1)

Essential Problem-Solving Skills in Chapter I

1. Draw and interpret Lewis, condensed, and line–angle structural formulas. Show which atoms bear formal charges.

2. Draw resonance forms, and use them to predict stabilities.

3. Calculate empirical and molecular formulas from elemental compositions.

4. Predict relative acidities and basicities based on structure, bonding, and resonance of conjugate acid–base pairs.

5. Calculate, use, and interpret values of K_a and pK_a.

6. Identify nucleophiles (Lewis bases) and electrophiles (Lewis acids), and write equations for Lewis acid–base reactions using curved arrows to show the flow of electrons.

Study Problems

It's easy to fool yourself into thinking you understand organic chemistry when you actually do not. As you read through this book, all the facts and ideas may make sense, yet you have not learned to combine and use those facts and ideas. An examination is a painful time to learn that you do not really understand the material.

The best way to learn organic chemistry is to use it. You will certainly need to read and reread all the material in the chapter, but this level of understanding is just the beginning. Problems are provided so you can work with the ideas, applying them to new compounds and new reactions that you have never seen before. By working problems, you force yourself to use the material and fill in the gaps in your understanding. You also increase your level of self-confidence and your ability to do well on exams.

Several kinds of problems are included in each chapter. There are problems within the chapters, giving examples and drill over the material as it is covered. Work these problems as you read through the chapter to ensure your understanding as you go along. Answers to many of these in-chapter problems are found at the back of this book. Study Problems at the end of each chapter give you additional experience using the material, and they force you to think in depth about the ideas. Some of the study problems have short answers in the back of this book, and all of them have detailed answers in the accompanying Solutions Manual.

Taking organic chemistry without working the problems is like skydiving without a parachute. Initially there is a breezy sense of freedom and daring. But then, there is the inevitable jolt that comes at the end for those who went unprepared.

1-20   Define and give an example for each term.
   (a) isotopes   (b) orbital   (c) node
   (d) degenerate orbitals   (e) valence electrons   (f) ionic bonding
   (g) covalent bonding   (h) Lewis structure   (i) nonbonding electrons
   (j) single bond   (k) double bond   (l) triple bond
   (m) polar bond   (n) formal charges   (o) resonance forms
   (p) molecular formula   (q) empirical formula   (r) Arrhenius acid and base
   (s) Bronsted–Lowry acid and base   (t) Lewis acid and base   (u) electrophile
   (v) nucleophile

1-21   Name the element that corresponds to each electronic configuration.
   (a) $1s^2 2s^2 2p^2$ (b) $1s^2 2s^2 2p^4$ (c) $1s^2 2s^2 2p^6 3s^2 3p^3$ (d) $1s^2 2s^2 2p^6 3s^2 3p^5$

1-22   There is a small portion of the periodic table that you must know to do organic chemistry. Construct this part from memory, using the following steps.
   (a) From memory, make a list of the elements in the first two rows of the periodic table, together with their numbers of valence electrons.
   (b) Use this list to construct the first two rows of the periodic table.
   (c) Organic compounds often contain sulfur, phosphorus, chlorine, bromine, and iodine. Add these elements to your periodic table.

1-23   For each compound, state whether its bonding is covalent, ionic, or a mixture of covalent and ionic.
   (a) NaCl   (b) NaOH   (c) CH_3Li (d) CH_2Cl_2 (e) $NaOCH_3$ (f) HCO_2Na (g) CF_4

1-24   (a) Both PCl_3 and PCl_5 are stable compounds. Draw Lewis structures for these two compounds.
   (b) NCl_3 is a known compound, but all attempts to synthesize NCl_5 have failed. Draw Lewis structures for NCl_3 and a hypothetical NCl_5, and explain why NCl_5 is an unlikely structure.

1-25   Draw a Lewis structure for each species.
   (a) N_2H_4 (b) N_2H_2 (c) $(CH_3)_4NCl$ (d) CH_3CN (e) CH_3CHO (f) $CH_3S(O)CH_3$
   (g) H_2SO_4 (h) CH_3NCO (i) $CH_3OSO_2OCH_3$ (j) $CH_3C(NH)CH_3$ (k) $(CH_3)_3CNO$

1-26   Draw a Lewis structure for each compound. Include all nonbonding pairs of electrons.
   (a) $CH_3CHCHCH_2CHCHCOOH$ (b) $NCCH_2COCH_2CHO$
   (c) $CH_2CHCH(OH)CH_2CO_2H$ (d) $CH_3CH(CH_3)CH_2C(CH_2CH_3)_2CHO$

1-27   Draw a line–angle formula for each compound in Problem 1-26.

1-28   Draw Lewis structures for
   (a) two compounds of formula C_4H_{10} (b) two compounds of formula C_2H_7N
   (c) three compounds of formula $C_3H_8O_2$ (d) two compounds of formula C_2H_4O

1-29   Draw a complete structural formula and a condensed structural formula for
   (a) three compounds of formula C_3H_8O (b) five compounds of formula C_3H_6O

1-30   Some of the following molecular formulas correspond to stable compounds. When possible, draw a stable structure for each formula.

$$CH_2 \quad CH_3 \quad CH_4 \quad CH_5$$
$$C_2H_2 \quad C_2H_3 \quad C_2H_4 \quad C_2H_5 \quad C_2H_6 \quad C_2H_7$$
$$C_3H_3 \quad C_3H_4 \quad C_3H_5 \quad C_3H_6 \quad C_3H_7 \quad C_3H_8 \quad C_3H_9$$

Can you propose a general rule for the numbers of hydrogen atoms in stable hydrocarbons?

1-31   Draw complete Lewis structures, including lone pairs, for the following compounds.

(a) pyridine

(b) pyrrolidine

(c) morpholine

(d) NH₂⟍⟋⟍COOH
γ-aminobutyric acid
(a neurotransmitter)

(e)

(f) ⟍CHO

(g) ⟨⟩—SO₃H

(h)

1-32   Give the molecular formula of each compound shown in Problem 1-31.

1-33   Compound X, isolated from lanolin (sheep's wool fat), has the pungent aroma of dirty sweatsocks. A careful analysis showed that compound X contains 62.0% carbon and 10.4% hydrogen. No nitrogen or halogen was found.
   (a) Compute an empirical formula for compound X.
   (b) A molecular weight determination showed that compound X has a molecular weight of approximately 117. Find the molecular formula of compound X.
   (c) There are many possible structures that have this molecular formula. Draw complete structural formulas for four of them.

1-34   For each of the following structures,
   (1) Draw a Lewis structure; fill in any nonbonding electrons.
   (2) Calculate the formal charge on each atom other than hydrogen. All are electrically neutral except as noted.

(a) $\left[\begin{array}{c} H \\ \diagdown \\ C=N=N \\ \diagup \\ H \end{array} \right.$ ⟷ $\left. \begin{array}{c} H \\ \diagdown \\ C-N\equiv N \\ \diagup \\ H \end{array} \right]$ (b) $(CH_3)_3NO$
(trimethylamine oxide)

(c) $[CH_2=CH-CH_2]^+$ (d) CH_3NO_2 (e) $[(CH_3)_3O]^+$

1-35   (1) From what you remember of electronegativities, show the direction of the dipole moments of the following bonds.
   (2) In each case, predict whether the dipole moment is relatively large or small.
   (a) $C-Cl$ (b) $C-H$ (c) $C-Li$ (d) $C-N$ (e) $C-O$
   (f) $C-B$ (g) $C-Mg$ (h) $N-H$ (i) $O-H$ (j) $C-Br$

1-36   Determine whether the following pairs of structures are actually different compounds or simply resonance forms of the same compounds.

(a)   and

(b)   and

(c)   and

(d) $CH_2=\overset{\overset{\displaystyle O^-}{|}}{C}-H$ and $\overset{-}{C}H_2-\overset{\overset{\displaystyle O}{||}}{C}-H$

(e) $H-\overset{\overset{\displaystyle O}{||}}{C}-CH_3$ and $H-\overset{\overset{\displaystyle O-H}{|}}{C}=CH_2$

(f) $H-\overset{\overset{\displaystyle O-H}{|}}{\underset{+}{C}}-H$ and $H-\overset{\overset{\displaystyle {}^+O-H}{||}}{C}-H$

(g) $H-\overset{\overset{\displaystyle O}{||}}{C}-NH_2$ and $H-\overset{\overset{\displaystyle O^-}{|}}{C}=\overset{+}{N}H_2$

(h) $CH_2=C=O$ and $H-C\equiv C-OH$

(i) $CH_2=CH-\overset{+}{C}H_2$ and $\overset{+}{C}H_2-CH=CH_2$ (j) $CH_3-\overset{\overset{\displaystyle O}{||}}{C}-CH=CH_2$ and $CH_3-\overset{\overset{\displaystyle O^-}{|}}{C}=CH-\overset{+}{C}H_2$

1-37   Draw the important resonance forms to show the delocalization of charges in the following ions.

(a) $CH_3-\overset{\overset{\displaystyle O}{||}}{C}-\overset{-}{C}H_2$ (b) $H-\overset{\overset{\displaystyle O}{||}}{C}-CH=CH-\overset{-}{C}H_2$ (c) ⟨⟩—$\overset{+}{C}H_2$ (d)

(e) **(f)** **(g)** **(h)**

(i) $CH_3-CH=CH-CH=CH-\overset{+}{CH}-CH_3$ **(j)** $CH_3-CH=CH-CH=CH-CH_2-\overset{+}{CH_2}$

1-38 **(a)** Draw the resonance forms for SO_2 (bonded $O-S-O$).
(b) Draw the resonance forms for ozone (bonded $O-O-O$).
(c) Sulfur dioxide has one more resonance form than ozone. Explain why this structure is not possible for ozone.

*1-39 The following compound can become protonated on any of the three nitrogen atoms. One of these nitrogens is much more basic than the others, however.
(a) Draw the important resonance forms of the products of protonation on each of the three nitrogen atoms.
(b) Determine which nitrogen atom is the most basic.

$$CH_3-NH-C\overset{\displaystyle NH}{\underset{\displaystyle NH_2}{}}$$

1-40 In the following sets of resonance forms, label the major and minor contributors and state which structures would be of equal energy. Add any missing resonance forms.

(a) $\left[CH_3-\ddot{\overset{..}{C}H}-C\equiv N\colon \longleftrightarrow CH_3-CH=C=\ddot{N}\colon^- \right]$

(b) $\left[CH_3-\overset{\displaystyle O^-}{C}=CH-\overset{+}{CH}-CH_3 \longleftrightarrow CH_3-\overset{\displaystyle O^-}{C}-CH=CH-CH_3 \right]$

(c) $\left[CH_3-\overset{\displaystyle O}{\overset{\|}{C}}-\overset{-}{C}H-\overset{\displaystyle O}{\overset{\|}{C}}-CH_3 \longleftrightarrow CH_3-\overset{\displaystyle O^-}{C}=CH-\overset{\displaystyle O}{\overset{\|}{C}}-CH_3 \right]$

(d) $\left[CH_3-\overset{-}{C}H-CH=CH-NO_2 \longleftrightarrow CH_3-CH=CH-\overset{-}{C}H-NO_2 \right]$

(e) $\left[CH_3-CH_2-\overset{\displaystyle NH_2}{\underset{+}{C}}-NH_2 \longleftrightarrow CH_3-CH_2-\overset{\displaystyle NH_2}{C}=\overset{+}{N}H_2 \right]$

1-41 For each pair of ions, determine which ion is more stable. Use resonance forms to explain your answers.

(a) $CH_3-\overset{+}{CH}-CH_3$ or $CH_3-\overset{+}{CH}-OCH_3$

(b) $CH_2=CH-\overset{+}{CH}-CH_3$ or $CH_2=CH-CH_2-\overset{+}{CH_2}$

(c) $\overset{-}{C}H_2-CH_3$ or $\overset{-}{C}H_2-C\equiv N\colon$

(d) or

(e) $CH_3-\overset{\displaystyle CH_3}{\underset{\displaystyle \overset{+}{C}-CH_3}{N}}$ or $CH_3-\overset{\displaystyle CH_3}{\underset{\displaystyle \overset{+}{C}-CH_3}{CH}}$

1-42 Rank the following species in order of increasing acidity. Explain your reasons for ordering them as you do.
$$NH_3 \quad H_2SO_4 \quad CH_3OH \quad CH_3COOH$$

1-43 Rank the following species in order of increasing basicity. Explain your reasons for ordering them as you do.
$$CH_3O^- \quad CH_3COOH \quad CH_3COO^- \quad NaOH \quad NH_2^- \quad HSO_4^-$$

1-44 The K_a of phenylacetic acid is 5.2×10^{-5}, and the pK_a of propionic acid is 4.87.

phenylacetic acid, $K_a = 5.2 \times 10^{-5}$ propionic acid, $pK_a = 4.87$

(a) Calculate the pK_a of phenylacetic acid and the K_a of propionic acid.

(b) Which of these is the stronger acid? Calculate how much stronger an acid it is.

(c) Predict whether the following equilibrium will favor the reactants or the products.

$$\text{C}_6\text{H}_5\text{—CH}_2\text{COO}^- + \text{CH}_3\text{CH}_2\text{COOH} \rightleftharpoons \text{C}_6\text{H}_5\text{—CH}_2\text{COOH} + \text{CH}_3\text{CH}_2\text{COO}^-$$

1-45   Label the reactants in these acid–base reactions as Lewis acids (electrophiles) or Lewis bases (nucleophiles). Use curved arrows to show the movement of electron pairs in the reactions.

(a) $\text{CH}_3\overset{\cdot\cdot}{\underset{\cdot\cdot}{\text{O}}}:^- + \text{CH}_3\text{—}\overset{\cdot\cdot}{\underset{\cdot\cdot}{\text{Cl}}}: \longrightarrow \text{CH}_3\text{—}\overset{\cdot\cdot}{\underset{\cdot\cdot}{\text{O}}}\text{—CH}_3 + :\overset{\cdot\cdot}{\underset{\cdot\cdot}{\text{Cl}}}:^-$

(b) $\text{CH}_3\text{—}\overset{+}{\underset{\underset{\text{CH}_3}{|}}{\overset{\cdot\cdot}{\text{O}}}}\text{—CH}_3 + :\underset{\underset{\text{H}}{|}}{\overset{\cdot\cdot}{\text{O}}}\text{—H} \longrightarrow \text{CH}_3\text{—}\underset{\underset{\text{CH}_3}{|}}{\overset{\cdot\cdot}{\text{O}}}: + \text{CH}_3\text{—}\overset{+}{\underset{\underset{\text{H}}{|}}{\overset{\cdot\cdot}{\text{O}}}}\text{—H}$

(c) $\text{H}\text{—}\overset{\overset{\text{O}}{\|}}{\text{C}}\text{—H} + :\text{NH}_3 \longrightarrow \text{H}\text{—}\underset{\underset{^+\text{NH}_3}{|}}{\overset{\overset{:\overset{\cdot\cdot}{\text{O}}:^-}{|}}{\text{C}}}\text{—H}$

(d) $\text{CH}_3\text{—}\overset{\cdot\cdot}{\text{NH}}_2 + \text{CH}_3\text{—CH}_2\text{—}\overset{\cdot\cdot}{\underset{\cdot\cdot}{\text{Cl}}}: \longrightarrow \text{CH}_3\text{—}\overset{+}{\text{NH}}_2\text{—CH}_2\text{CH}_3 + :\overset{\cdot\cdot}{\underset{\cdot\cdot}{\text{Cl}}}:^-$

(e) $\text{CH}_3\text{—}\overset{\overset{\cdot\cdot\overset{\cdot}{\text{O}}}{\|}}{\text{C}}\text{—CH}_3 + \text{H}_2\text{SO}_4 \longrightarrow \text{CH}_3\text{—}\overset{\overset{\cdot\cdot\overset{+}{\text{O}}\text{—H}}{\|}}{\text{C}}\text{—CH}_3 + \text{HSO}_4^-$

(f) $(\text{CH}_3)_3\text{CCl} + \text{AlCl}_3 \longrightarrow (\text{CH}_3)_3\text{C}^+ + {}^-\text{AlCl}_4$

(g) $\text{CH}_3\text{—}\overset{\overset{\cdot\cdot\overset{\cdot}{\text{O}}}{\|}}{\text{C}}\text{—CH}_3 + {}^-:\overset{\cdot\cdot}{\underset{\cdot\cdot}{\text{O}}}\text{—H} \longrightarrow \text{CH}_3\text{—}\overset{\overset{:\overset{\cdot\cdot}{\text{O}}:^-}{|}}{\text{C}}\text{=CH}_2 + \text{H}\text{—}\overset{\cdot\cdot}{\underset{\cdot\cdot}{\text{O}}}\text{—H}$

(h) $\text{CH}_2\text{=CH}_2 + \text{BF}_3 \longrightarrow {}^-\text{BF}_3\text{—CH}_2\text{—}\overset{+}{\text{CH}}_2$

(i) ${}^-\text{BF}_3\text{—CH}_2\text{—}\overset{+}{\text{CH}}_2 + \text{CH}_2\text{=CH}_2 \longrightarrow {}^-\text{BF}_3\text{—CH}_2\text{—CH}_2\text{—CH}_2\text{—}\overset{+}{\text{CH}}_2$

1-46   Predict the products of the following acid_base reactions.

(a) $\text{H}_2\text{SO}_4 + \text{CH}_3\text{COO}^- \rightleftharpoons$

(b) $\text{CH}_3\text{COOH} + (\text{CH}_3)_3\text{N}: \rightleftharpoons$

(c) $\text{C}_6\text{H}_5\text{—}\overset{\overset{\text{O}}{\|}}{\text{C}}\text{—OH} + {}^-\text{OH} \rightleftharpoons$

(d) $(\text{CH}_3)_3\overset{+}{\text{NH}} + {}^-\text{OH} \rightleftharpoons$

(e) $\text{HO}\text{—}\overset{\overset{\text{O}}{\|}}{\text{C}}\text{—OH} + 2\,{}^-\text{OH} \rightleftharpoons$

(f) $\text{H}_2\text{O} + \text{NH}_3 \rightleftharpoons$

(g) $\text{HCOOH} + \text{CH}_3\text{O}^- \rightleftharpoons$

*\ **1-47**   Methyllithium (CH_3Li) is often used as a base in organic reactions.

(a) Predict the products of the following acid–base reaction.

$$\text{CH}_3\text{CH}_2\text{—OH} + \text{CH}_3\text{—Li} \longrightarrow$$

(b) What is the conjugate acid of CH_3Li? Would you expect CH_3Li to be a strong base or a weak base?

*\ **1-48**   In 1934, Edward A. Doisy of Washington University extracted 3000 lb of hog ovaries to isolate a few milligrams of pure estradiol, a potent female hormone. Doisy burned 5.00 mg of this precious sample in oxygen and found that 14.54 mg of CO_2 and 3.97 mg of H_2O were generated.

(a) Determine the empirical formula of estradiol.

(b) The molecular weight of estradiol was later determined to be 272. Determine the molecular formula of estradiol.

CHAPTER 2

Structure and Properties of Organic Molecules

In Chapter 1, we considered how atoms bond together to gain noble-gas configurations, forming molecules in the process. Using the octet rule, we drew Lewis structures for organic molecules and used these diagrams to determine which bonds are single bonds, double bonds, and triple bonds. We discussed various ways of drawing organic structures, and we saw how resonance structures represent molecules whose actual bonding cannot be described by a single Lewis structure.

The material in Chapter 1 explains little about the actual shapes and properties of organic molecules. These aspects of molecular structure can be explained by considering how the atomic orbitals on an atom mix to form hybrid atomic orbitals and how orbitals on different atoms combine to form molecular orbitals. We will use these orbitals to account for the shapes and properties we observe in organic molecules.

2-1
Wave Properties of Electrons in Orbitals

We like to picture the atom as a miniature solar system, with the electrons orbiting around the nucleus. This solar system picture satisfies our intuition, but it does not correspond with what we know about the atom. About 1923, Louis de Broglie suggested that the properties of electrons in atoms are better explained by treating the electrons as waves rather than as particles.

There are two general kinds of waves, *traveling waves* and *standing waves*. Examples of traveling waves are the sound waves that carry a thunderclap and the water waves that form the wake of a boat. Standing waves vibrate in a fixed location. Standing waves are found inside an organ pipe, where the rush of air creates a vibrating air column, and in the wave pattern of a guitar string when it is plucked. An electron in an atomic orbital is like a stationary, bound vibration: a standing wave.

We can understand the features of an orbital (a three-dimensional standing wave) more easily by using a guitar string as a one-dimensional analogy (see Figure 2-1). If you pluck a guitar string at its middle, a standing wave results. This vibration has all of the string displaced upward for a fraction of a second, then downward for an equal time. If we draw an instantaneous picture of the waveform, it shows the string displaced in a smooth curve either upward or downward, depending on the exact instant of the picture.

A 1s orbital is like this guitar string, except that it is three-dimensional. The orbital can be described by its **wave function** ψ, which is the mathematical description of the shape of the wave as it vibrates. All of the wave is positive in sign for a brief instant; then it is negative in sign. The electron density at any point is given by ψ^2, the square of the wave function at that point. *Notice that the plus sign and the minus sign of these wave functions are not charges. The plus or minus sign is the instantaneous sign of the constantly changing wave function.* The 1s orbital is spherically symmetrical, and it is often represented by a circle (representing a sphere) with a nucleus in the center and with a plus or minus sign to indicate the instantaneous sign of the wave function (Figure 2-2).

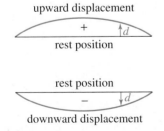

▲ Figure 2-1
A standing wave. The fundamental frequency of a guitar string is a standing wave with the string alternately displaced upward and downward.

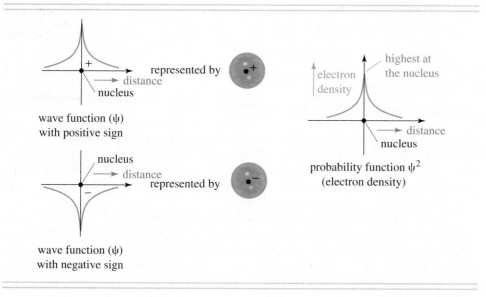

The $1s$ orbital is similar to the fundamental vibration of a guitar string. The wave function is instantaneously all positive or all negative. The square of the wave function gives the electron density. A circle with a nucleus is used to represent the spherically symmetrical s orbital.

If we gently place a finger at the center of a guitar string while plucking the string, the finger keeps the midpoint of the string from moving. The position at the midpoint is always zero, so this point is a **node**. The string vibrates in two parts, with the two halves vibrating in opposite directions. We say that the two halves of the string are *out of phase*: When one is displaced upward, the other is displaced downward. Figure 2-3 shows this first harmonic of the guitar string.

The first harmonic of the guitar string resembles the $2p$ orbital (Figure 2-4). We have drawn the $2p$ orbital as two "lobes," separated by a node (a nodal plane). The two lobes of the p orbital are out of phase with each other. Whenever the wave function has a plus sign in one lobe, it has a minus sign in the other lobe.

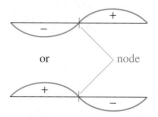

▲ **Figure 2-3**
First harmonic of a guitar string. The two halves of the string are separated by a node, a point with zero displacement. The two halves vibrate out of phase with each other.

2-1A   Linear Combination of Atomic Orbitals

Atomic orbitals can combine and overlap to give more complex standing waves. We can add and subtract their wave functions to give the wave functions of new orbitals. This process is called the **linear combination of atomic orbitals** (LCAO). The number of new orbitals generated always equals the number of orbitals we started with.

1. When orbitals on *different* atoms interact, they produce **molecular orbitals** (MOs) that lead to bonding (or antibonding).

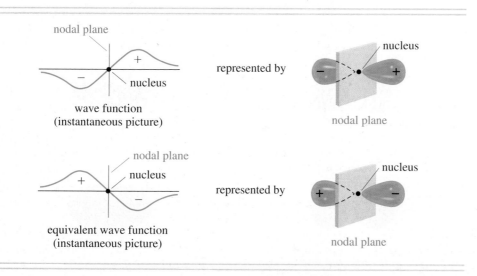

◄ **Figure 2-4**
The $2p$ orbital has two lobes, separated by a nodal plane. The two lobes are out of phase with each other. When one has a plus sign, the other has a minus sign.

2. When orbitals on the *same* atom interact, they give **hybrid atomic orbitals** that define the geometry of the bonds.

We begin by looking at how atomic orbitals on different atoms interact to give molecular orbitals. Then we consider how atomic orbitals on the same atom can interact to give hybrid atomic orbitals.

2-2
Molecular Orbitals

The stability of a covalent bond results from a large amount of electron density in the bonding region, the space between the two nuclei (Figure 2-5). In the bonding region, the electrons are close to both nuclei, resulting in a lowering of the overall energy. The bonding electrons also mask the positive charges of the nuclei, so they do not repel each other as much as they would otherwise.

There is always an optimum distance for the two bonded nuclei. If they are too far apart, their attraction for the bonding electrons is diminished. If they are too close together, their electrostatic repulsion pushes them apart. The internuclear distance where attraction and repulsion are balanced, which also gives the minimum energy (the strongest bond), is called the *bond length*.

2-2A   The Hydrogen Molecule; Sigma Bonding

The hydrogen molecule is the simplest example of covalent bonding. As two hydrogen atoms approach each other, their 1*s* wave functions can add *constructively* so that they reinforce each other, or *destructively* so that they cancel out where they overlap. Figure 2-6

▶ **Figure 2-5**
A bonding molecular orbital places a large amount of electron density in the bonding region, the space between the two nuclei.

▶ **Figure 2-6**
Formation of a σ-bonding MO. When the 1*s* orbitals of two hydrogen atoms overlap in phase, they interact constructively to form a bonding MO. The electron density in the bonding region (between the nuclei) is increased. The result is a cylindrically symmetrical bond, or sigma bond.

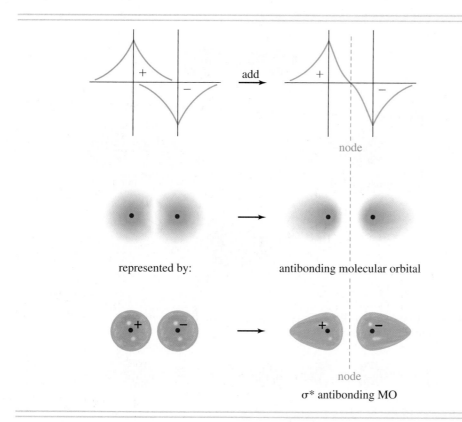

add

node

represented by: antibonding molecular orbital

σ^* antibonding MO

node

◀ **Figure 2-7**
Formation of a σ^* antibonding MO. When two $1s$ orbitals overlap out of phase, they interact destructively to form an antibonding MO. The positive and negative values of the wave functions tend to cancel out in the region between the nuclei, and a node separates the nuclei. We use an asterisk (*) to designate antibonding orbitals such as this sigma antibonding orbital, σ^*.

shows how the wave functions interact constructively when they are in phase and have the same sign in the region between the nuclei. The wave functions reinforce each other and increase the electron density in this bonding region. The result is a **bonding molecular orbital** (bonding MO).

The bonding MO depicted in Figure 2-6 has most of its electron density centered *along* the line connecting the nuclei. This type of bond is called a *cylindrically symmetrical bond* or a **sigma bond** (σ bond). Sigma bonds are the most common bonds in organic compounds. All single bonds in organic compounds are sigma bonds, and every double or triple bond contains one sigma bond.

When two hydrogen $1s$ orbitals overlap out of phase with each other, an **antibonding molecular orbital** results (Figure 2-7). The two $1s$ wave functions have opposite signs, so they tend to cancel out where they overlap. The result is a node (actually a nodal plane) separating the two atoms. The presence of a node separating the two nuclei usually indicates that the orbital is antibonding.

Figure 2-8 shows the relative energies of the atomic orbitals and the molecular orbitals of the H_2 system. When the $1s$ orbitals are in phase, the resulting molecular orbital is a sigma bonding MO, with lower energy than that of a $1s$ atomic orbital. When two $1s$ orbitals overlap out of phase, they form an antibonding (σ^*) orbital with higher energy than that of a $1s$ atomic orbital. The two electrons in the H_2 system are found with paired spins in the sigma bonding MO, giving a stable H_2 molecule. In stable molecules, the antibonding orbitals (such as σ^*) are usually vacant.

2-2B   Sigma Overlap Involving p Orbitals

When two p orbitals overlap along the line between the nuclei, a bonding orbital and an antibonding orbital result. Once again, most of the electron density is centered along the line between the nuclei. This linear overlap is another type of sigma bonding MO.

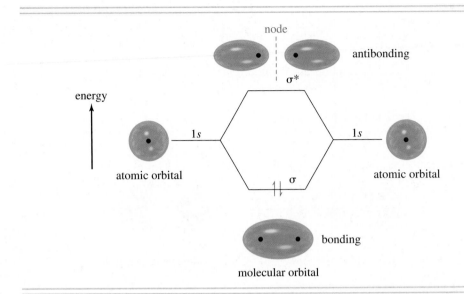

▶ Figure 2-8
When the two hydrogen 1s orbitals overlap, a sigma bonding MO and a sigma antibonding MO result. Two electrons (represented by arrows) go into the bonding MO with opposite spins, forming a stable H_2 molecule.

The constructive overlap of two p orbitals along the line joining the nuclei forms a σ bond represented as follows:

SOLVED PROBLEM 2-1

Draw the σ^* antibonding orbital that results from destructive overlap of the two p_x orbitals shown above.

SOLUTION

This orbital results from the destructive overlap of lobes of the two p orbitals with opposite phases. If the signs are reversed on one of the orbitals, adding the two orbitals gives an antibonding orbital with a node separating the two nuclei:

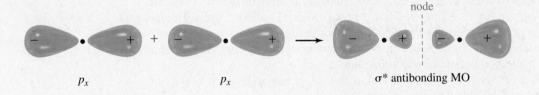

Overlap of an s orbital with a p orbital also gives a bonding MO and an antibonding MO, as shown below. Constructive overlap of the s orbital with the p_x orbital gives a sigma bonding MO with its electron density centered along the line between the nuclei. Destructive overlap gives a σ^* antibonding orbital with a node separating the nuclei.

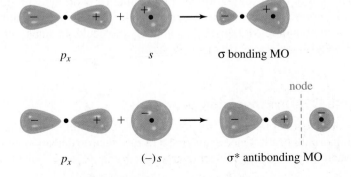

A **pi bond** (π **bond**) results from overlap between two *p* orbitals oriented perpendicular to the line connecting the nuclei (Figure 2-9). These parallel orbitals overlap sideways, with most of the electron density centered *above and below* the line connecting the nuclei. This overlap is parallel, not linear (as in a sigma bond), so a pi molecular orbital is *not* cylindrically symmetrical. Figure 2-9 shows a pi bonding MO and the corresponding π* antibonding MO.

2-3
Pi Bonding

2-3A   Single and Double Bonds

A **double bond** requires the presence of four electrons in the bonding region between the nuclei. The first pair of electrons goes into the sigma bonding MO, forming a strong sigma bond. The second pair of electrons cannot go into the same orbital or the same space. It goes into a pi bonding MO, with its electron density centered above and below the sigma bond.

The combination of a sigma bond and a pi bond is the normal structure of a double bond. Figure 2-10 shows the structure of ethylene, an organic molecule containing a carbon–carbon double bond.

Thus far, we have discussed only bonds involving overlap of simple *s* and *p* atomic orbitals. Although these simple bonds are occasionally seen in organic compounds, they are not as common as bonds formed using **hybrid atomic orbitals**. Hybrid atomic orbitals result from the mixing of orbitals on the *same* atom. The geometry of these hybrid orbitals helps us to account for the actual structures and bond angles observed in organic compounds.

destructive (antibonding) interaction   π* antibonding MO

constructive (bonding) interaction   π bonding MO

◀ **Figure 2-9**
The sideways overlap of two *p* orbitals leads to a pi bonding MO and a pi antibonding MO. A pi bond is not as strong as most sigma bonds.

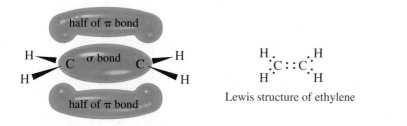

Lewis structure of ethylene

◀ **Figure 2-10**
The second bond of a double bond is a pi bond. The pi bond has its electron density centered in two lobes, above and below the sigma bond. Together, the two lobes of the pi bonding molecular orbital constitute one bond.

2-4
Hybridization and Molecular Shapes

If we predict the bond angles of organic molecules using just the simple *s* and *p* orbitals, we expect bond angles of about 90°. The *s* orbitals are nondirectional, and the *p* orbitals are oriented at 90° to one another (see Figure 1-3). Experimental evidence shows, however, that bond angles in organic compounds are usually close to 109°, 120°, or 180° (Figure 2-11). A common way of accounting for these bond angles is the *valence-shell electron pair repulsion theory* (VSEPR theory): Electron pairs repel each other, and the bonds and lone pairs around a central atom generally are separated by the largest possible angles. An angle of 109.5° is the largest possible separation for four pairs of electrons; 120° is the largest separation for three pairs; and 180° is the largest separation for two pairs. All the structures in Figure 2-11 have bond angles that separate their bonds about as far apart as possible.

The shapes of these molecules cannot result from bonding between simple *s* and *p* atomic orbitals. Although these orbitals have the lowest energies for isolated atoms in space, they are not the best for forming bonds. To explain the shapes of common organic molecules, we assume that the *s* and *p* orbitals combine to form hybrid atomic orbitals that separate the electron pairs more widely in space and place more electron density in the bonding region between the nuclei.

2-4A *sp* Hybrid Orbitals

Orbitals can interact to form new orbitals. We have used this principle to form molecular orbitals by adding and subtracting atomic orbitals on *different* atoms, but we can also add and subtract orbitals on the *same* atom. Consider the result when we combine a *p* orbital and an *s* orbital on the same atom (Figure 2-12).

▶ **Figure 2-11**
The angles between the *p* orbitals are all 90°, but few organic compounds have bond angles of 90°. Their bond angles are usually close to 109°, 120°, or 180°.

methane, 109.5°   ethylene, close to 120°   acetylene, 180°

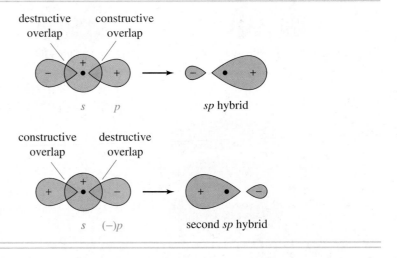

▶ **Figure 2-12**
Addition of an *s* orbital to a *p* orbital gives an *sp* hybrid atomic orbital, with most of its electron density on one side of the nucleus. Adding the *p* orbital with opposite phase gives the other *sp* hybrid atomic orbital, with most of its electron density on the opposite side of the nucleus from the first hybrid.

The resulting orbital is called an *sp* **hybrid orbital**. Its electron density is concentrated toward one side of the atom. We started with two orbitals (*s* and *p*), so we must finish with two *sp* hybrid orbitals. The second *sp* hybrid orbital results if we add the *p* orbital with the opposite phase (Fig. 2-12).

The result of this hybridization is a pair of directional *sp* hybrid orbitals pointed in opposite directions. These hybridized orbitals provide enhanced electron density in the bonding region for a sigma bond toward the left of the atom and for another sigma bond toward the right. In addition, these hybrid orbitals give a bond angle of 180°, separating the bonding electrons as much as possible. In general, *sp* hybridization results in this **linear** bonding arrangement.

SOLVED PROBLEM 2-2

Draw the Lewis structure for beryllium hydride, BeH_2. Draw the orbitals that overlap in the bonding of BeH_2, and label the hybridization of each orbital. Predict the H—Be—H bond angle.

SOLUTION

First, we draw a Lewis structure for BeH_2.

$$H\!:\!Be\!:\!H$$

There are only four valence electrons in BeH_2 (two from Be and one from each H), so the Be atom cannot have an octet of electrons. The bonding must involve orbitals on Be that give the strongest bonds (the most electron density in the bonding region) and also allow the two pairs of electrons to be separated as far as possible.

Hybrid orbitals concentrate the electron density in the bonding region, and *sp* hybrids give 180° separation for two pairs of electrons. Hydrogen cannot use hybridized orbitals, since the closest available *p* orbitals are the 2*p*'s, and they are much higher in energy than the 1*s*. The bonding in BeH_2 results from overlap of *sp* hybrid orbitals on Be with the 1*s* orbitals on hydrogen. Figure 2-13 shows how this occurs.

2-4B   *sp*² Hybrid Orbitals

To orient three bonds as far apart as possible, bond angles of 120° are required. When an *s* orbital combines with two *p* orbitals, the resulting three hybrid orbitals are oriented at 120° angles to each other (Figure 2-14). These orbitals are called **sp² hybrid orbitals** because they are composed of one *s* and two *p* orbitals. The 120° arrangement is called **trigonal** geometry, in contrast to the linear geometry associated with *sp* hybrid orbitals. There is still an unhybridized *p* orbital (p_z) perpendicular to the plane of the three *sp*² hybrid orbitals.

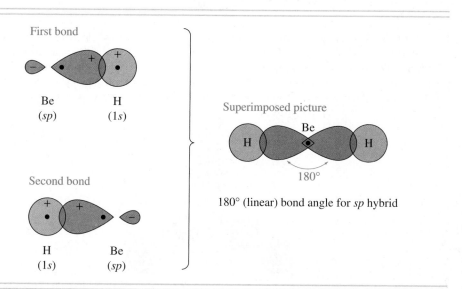

◄ Figure 2-13

The bonding of BeH_2. To form two sigma bonds, the two *sp* hybridized atomic orbitals on Be overlap with the 1*s* orbitals of hydrogen. The bond angle is 180° (linear).

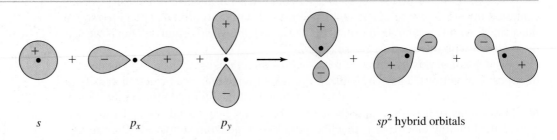

s p_x p_y sp^2 hybrid orbitals

▶ **Figure 2-14**
Hybridization of an s orbital with two
p orbitals gives a set of three sp^2
hybrid orbitals. The bond angles
associated with this trigonal structure
are about 120°. The remaining p
orbital is perpendicular to the plane
of the three hybrid orbitals.

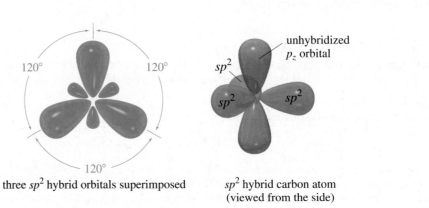

three sp^2 hybrid orbitals superimposed sp^2 hybrid carbon atom
(viewed from the side)

SOLVED PROBLEM 2-3

Borane (BH_3) is not stable under normal conditions, but it has been detected at low pressure.
(a) Draw the Lewis structure for borane.
(b) Draw a diagram of the bonding in this molecule, and label the hybridization of each orbital.
(c) Predict the H—B—H bond angle.

SOLUTION

There are only six valence electrons in borane. Boron has a single bond to each of the three
hydrogen atoms.

           The best bonding orbitals are those that provide the greatest electron density in the
bonding region while keeping the three pairs of bonding electrons as far apart as possible.
Hybridization of an s orbital with two p orbitals gives three sp^2 hybrid orbitals directed 120°
apart. Overlap of these orbitals with the hydrogen 1s orbitals gives a planar, trigonal molecule. (Note that the small back lobes of the hybrid orbitals have been omitted.)

PROBLEM-SOLVING HINT
The number of hybrid orbitals formed
is always the same as the total
number of s and p orbitals hybridized.

number of orbitals	hybrid	angle
2	sp	180°
3	sp^2	120°
4	sp^3	109.5°

Hybridization of an s orbital with all three p orbitals gives four sp^3 hybrid orbitals with tetrahedral geometry and 109.5° bond angles.

2-4C sp^3 Hybrid Orbitals

Many organic compounds contain carbon atoms that are bonded to four other atoms. When four bonds are oriented as far apart as possible, they form a regular tetrahedron (109.5° bond angles), as pictured in Figure 2-15. This **tetrahedral** arrangement can be explained by combining the s orbital with all three p orbitals. The resulting four orbitals are called sp^3 **hybrid orbitals** because they are composed of one s and three p orbitals.

Methane (CH_4) is the simplest example of sp^3 hybridization (Figure 2-16). The Lewis structure for methane has eight valence electrons (four from carbon and one from each hydrogen), corresponding to four C—H single bonds. Tetrahedral geometry separates these bonds by the largest possible angle, 109.5°.

Methanotrophs are bacteria that grow using methane as their only food source. It may be possible to harness this capability and produce various chemicals (like methanol) from methane or to reduce the methane levels in the atmosphere. High levels of methane are believed to contribute to global warming.

◀ **Figure 2-16**
Methane has tetrahedral geometry, using four sp^3 hybrid orbitals to form sigma bonds to the four hydrogen atoms.

Figures 2-15 and 2-16 were more difficult to draw than the earlier figures because they depict three-dimensional objects on a two-dimensional piece of paper. The p_z orbital should look like it points in and out of the page, and the tetrahedron should look three-dimensional. These drawings use perspective and the viewer's imagination to add the third dimension.

The use of perspective is difficult when a molecule is large and complicated. Organic chemists have developed a shorthand notation to simplify three-dimensional drawings. Dashed lines indicate bonds that go backward, away from the reader. Wedge-shaped bonds depict bonds that come forward, toward the reader. Straight lines are bonds in the plane of the page. Dashed lines and wedges show perspective in the second drawing of methane in Figure 2-16.

2-5
Drawing Three-Dimensional Molecules

The three-dimensional structure of ethane, C_2H_6, has the shape of two tetrahedra joined together. Each carbon atom is sp^3 hybridized, with four sigma bonds formed by the four sp^3 hybrid orbitals. Dashed lines represent bonds that go away from the viewer, wedges represent bonds that come out toward the viewer, and other bond lines are in the plane of the page. All the bond angles are close to $109.5°$.

PROBLEM-SOLVING HINT

When showing perspective, do not draw another bond between the two bonds in the plane of the paper. Such a drawing shows an incorrect shape.

ethane   ethane   ethane

PROBLEM 2-1

(a) Use your molecular models to make ethane, and compare the model with the structures above.

(b) Make a model of propane (C_3H_8), and draw this model using dashed lines and wedges to represent bonds going back and coming forward.

2-6
General Rules of Hybridization and Geometry

At this point, we can consider some general rules for determining the hybridization of orbitals and the bond angles of atoms in organic molecules. After stating these rules, we solve some problems to show how the rules are used.

Rule 1:   Both sigma bonding electrons and lone pairs occupy hybrid orbitals. The number of hybrid orbitals on an atom is computed by adding the number of sigma bonds and the number of lone pairs of electrons on that atom.

Because the first bond to another atom is always a sigma bond, the number of hybrid orbitals may be computed by adding the number of lone pairs to the number of atoms bonded to the central atom.

Rule 2:   Use the hybridization and geometry that give the widest possible separation of the calculated number of bonds and lone pairs.

Summary of Hybridization and Geometry

Hybrid Orbitals	Hybridization	Geometry	Approximate Bond Angles
2	$s + p = sp$	linear	180°
3	$s + p + p = sp^2$	trigonal	120°
4	$s + p + p + p = sp^3$	tetrahedral	109.5°

The number of hybrid orbitals obtained equals the number of atomic orbitals combined. Lone pairs of electrons take up more space than bonding pairs of electrons, thus they compress the bond angles.

Rule 3:   If two or three pairs of electrons form a multiple bond between two atoms, the first bond is a sigma bond formed by a hybrid orbital. The second bond is a pi bond, consisting of two lobes above and below the sigma bond, formed by two *p* orbitals. The third bond of a triple bond is another pi bond, perpendicular to the first pi bond (see Figure 2-18).

Solved Problems 2-4 through 2-8 show how to use these rules to predict the hybridization and bond angles in organic compounds.

SOLVED PROBLEM 2-4

Predict the hybridization of the nitrogen atom in ammonia, NH_3. Draw a picture of the three-dimensional structure of ammonia, and predict the bond angles.

SOLUTION

The hybridization depends on the number of sigma bonds plus lone pairs. A Lewis structure provides this information.

$$
\begin{array}{cc}
\quad\ \ \text{H} & \quad\ \ \ \text{H} \\
\quad\ \ \ | & \quad\ \ \ | \\
\text{H}:\!\overset{..}{\underset{..}{\text{N}}}\!: \quad\text{or}\quad & \text{H}\!-\!\overset{}{\underset{|}{\text{N}}}\!:\!\diagdown \\
\quad\ \ \ \text{H} & \quad\ \ \ \text{H}\qquad\text{lone pair}
\end{array}
$$

In this structure, there are three sigma bonds and one pair of nonbonding electrons. Four hybrid orbitals are required, implying sp^3 hybridization and tetrahedral geometry around the nitrogen atom, with bond angles of about 109.5°. The resulting structure is much like that of methane, except that one of the sp^3 hybrid orbitals is occupied by a lone pair of electrons.

$$
\text{H}\cdots\!\overset{\overset{\displaystyle(\cdot\cdot)}{}}{\underset{\underset{\text{H}\quad 107.3°}{}}{\text{N}}}\!\!\diagup\text{H}
$$

Notice that the bond angles in ammonia (107.3°) are slightly smaller than the ideal tetrahedral angle, 109.5°. The nonbonding electrons are more diffuse than a bonding pair of electrons, and they take up more space. The lone pair repels the electrons in the N—H bonds, compressing the bond angle.

PROBLEM 2-2

(a) Predict the hybridization of the oxygen atom in water, H_2O. Draw a picture of its three-dimensional structure, and explain why its bond angle is 104.5°.

(b) The electrostatic potential maps for ammonia and water are shown below. The structure of ammonia is shown within its EPM. Note how the lone pair creates a region of high electron potential (red), and the hydrogens are in regions of low electron potential (blue). Show how your three-dimensional structure of water corresponds with its EPM.

NH_3

H_2O

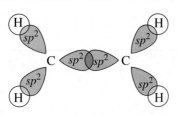

σ bond framework
(viewed from above the plane)

π bond
(viewed from alongside the plane)

ethylene

▲ **Figure 2-17**
The carbon atoms in ethylene are sp^2 hybridized, with trigonal bond angles of about 120°. All the carbon and hydrogen atoms lie in the same plane.

SOLVED PROBLEM 2-5

Predict the hybridization, geometry, and bond angles for ethylene (C_2H_4).

SOLUTION

The Lewis structure of ethylene is

Each carbon atom has an octet, and there is a double bond between the carbon atoms. Each carbon is bonded to three other atoms (three sigma bonds), and there are no lone pairs. The carbon atoms are sp^2 hybridized, and the bond angles are trigonal: about 120°. The double bond is composed of a sigma bond formed by overlap of two sp^2 hybridized orbitals, plus a pi bond formed by overlap of the unhybridized p orbitals remaining on the carbon atoms. Because the pi bond requires parallel alignment of its two p orbitals, the ethylene molecule must be planar (Fig. 2-17).

PROBLEM 2-3

Predict the hybridization, geometry, and bond angles for the central atoms in
(a) 2-butene, $CH_3CH{=}CHCH_3$ **(b)** $CH_2{=}NCH_3$

SOLVED PROBLEM 2-6

Predict the hybridization, geometry, and bond angles for the carbon atoms in acetylene, C_2H_2.

SOLUTION

The Lewis structure of acetylene is

$$H{:}C{:::}C{:}H \quad \text{or} \quad H{-}C{\equiv}C{-}H$$

Both carbon atoms have octets, but each carbon is bonded to just two other atoms, requiring two sigma bonds. There are no lone pairs. Each carbon atom is sp hybridized and linear (180° bond angles). The sp hybrid orbitals are generated from the s orbital and the p_x orbital (the p orbital directed along the line joining the nuclei). The p_y orbitals and the p_z orbitals are unhybridized.

The **triple bond** is composed of one sigma bond, formed by overlap of sp hybrid orbitals, plus two pi bonds. One pi bond results from overlap of the two p_y orbitals and another from overlap of the two p_z orbitals (Figure 2-18).

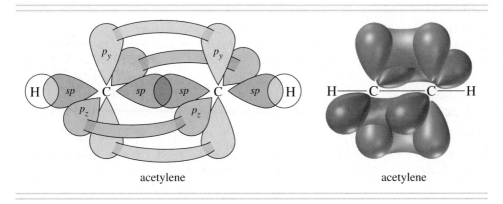

The carbon atoms in acetylene are *sp* hybridized, with linear (180°) bond angles. The triple bond contains one sigma bond and two perpendicular pi bonds.

PROBLEM 2-4

Predict the hybridization, geometry, and bond angles for the carbon and nitrogen atoms in acetonitrile ($CH_3 — C \equiv N$:).

SOLVED PROBLEM 2-7

Predict the hybridization, geometry, and bond angles for the carbon and oxygen atoms in acetaldehyde (CH_3CHO).

SOLUTION

The Lewis structure for acetaldehyde is

The oxygen atom and both carbon atoms have octets of electrons. The CH_3 carbon atom is sigma bonded to four atoms, so it is sp^3 hybridized (and tetrahedral). The $C = O$ carbon is bonded to three atoms (no lone pairs), so it is sp^2 hybridized and its bond angles are about 120°.

The oxygen atom is probably sp^2 hybridized because it is bonded to one atom (carbon) and has two lone pairs, requiring a total of three hybrid orbitals. We cannot experimentally measure the angles of the lone pairs on oxygen, however, so it is impossible to confirm whether the oxygen atom is really sp^2 hybridized.

The double bond between carbon and oxygen looks just like the double bond in ethylene. There is a sigma bond formed by overlap of sp^2 hybrid orbitals and a pi bond formed by overlap of the unhybridized *p* orbitals on carbon and oxygen (Fig. 2-19).

PROBLEM-SOLVING **HINT**

Begin with a good Lewis structure, and use hybrid orbitals for the sigma bonds and lone pairs. Use pi bonds between unhybridized *p* orbitals for the second and third bonds of double and triple bonds.

▲ **Figure 2-19**
The CH_3 carbon in acetaldehyde is sp^3 hybridized, with tetrahedral bond angles of about 109.5°. The carbonyl ($C = O$) carbon is sp^2 hybridized, with bond angles of about 120°. The oxygen atom is probably sp^2 hybridized, but we cannot measure any bond angles to verify this prediction.

PROBLEM 2-5

(1) Draw a Lewis structure for each of the following compounds.
(2) Label the hybridization, geometry, and bond angles around each atom other than hydrogen.
(3) Draw a three-dimensional representation (using wedges and dashed lines) of the structure.

(a) CO_2 (b) $CH_3—O—CH_3$ (c) $(CH_3)_3N$
(d) HCOOH   (e) HCN   (f) $CH_3CH=CH_2$
(g) ozone (O_3), bonded OOO

PROBLEM 2-6

Allene, $CH_2=C=CH_2$, has the structure shown below. Explain how the bonding in allene requires the two $=CH_2$ groups at its ends to be at right angles to each other.

H_⁗C=C=C with H above and H below

allene

SOLVED PROBLEM 2-8

In Chapter 1, we considered the electronic structure of $[CH_2NH_2]^+$. Predict its hybridization, geometry, and bond angles.

SOLUTION

This is a tricky question. This ion has two important resonance forms:

resonance forms   combined representation

When resonance is involved, different resonance forms may suggest different hybridization and bond angles. Only electrons can be delocalized, however. The molecule can have only one set of bond angles, which must be compatible with all the important resonance forms. Looking at either resonance form for $[CH_2NH_2]^+$, we would predict sp^2 hybridization (120° bond angles) for the carbon atom; however, the first resonance form suggests sp^3 hybridization for nitrogen (109° bond angles), while the second suggests sp^2 hybridization (120° bond angles). Which is correct?

Experiments show that the bond angles on both carbon and nitrogen are about 120°, implying sp^2 hybridization. Nitrogen cannot be sp^3 hybridized because there must be an unhybridized p orbital available to form the pi bond in the second resonance form. In the first resonance form below, we picture the lone pair residing in this unhybridized p orbital.

In general, resonance-stabilized structures have bond angles appropriate for the largest number of pi bonds needed at each atom; that is, with unhybridized p orbitals available for all the pi bonds shown in any important resonance form.

PROBLEM-SOLVING HINT

To predict the hybridization and geometry of an atom in a resonance hybrid, consider the resonance form with the most pi bonds to that atom.

PROBLEM 2-7

(1) Draw the important resonance forms for each compound.
(2) Label the hybridization and bond angles around each atom other than hydrogen.
(3) Use a three-dimensional drawing to show where the electrons are pictured to be in each resonance form.

(a) $[H_2COH]^+$ (b) $\left[\begin{array}{c} O \\ \| \\ H-C-CH_2 \end{array}\right]^-$ (c) $[H_2CCN]^-$ (d) $B(OH)_3$

Some bonds rotate easily, but others do not. When we look at a structure, we must recognize which bonds rotate and which do not. If a bond rotates easily, each molecule can rotate through the different angular arrangements of atoms. If a bond cannot rotate, however, different angular arrangements may be distinct compounds (isomers) with different properties.

2-7
Bond Rotation

2-7A   Rotation of Single Bonds

In ethane (CH_3—CH_3), both carbon atoms are sp^3 hybridized and tetrahedral. Ethane looks like two methane molecules that have each had a hydrogen plucked off (to form a methyl group) and are joined by overlap of their sp^3 orbitals (Fig. 2-20).

We can draw many structures for ethane, differing only in how one methyl group is twisted in relation to the other one. Such structures, differing only in rotations about a single bond, are called *conformations*. Two of the infinite number of conformations of ethane are shown in Figure 2-20. Construct a molecular model of ethane, and twist the model into these two conformations.

Which of these structures for ethane is the "right" one? Are the two methyl groups lined up so that their C—H bonds are parallel (*eclipsed*), or are they *staggered*, as in the drawing on the right? The answer is that both structures, and all the possible structures in between, are correct structures for ethane, and a real ethane molecule rotates through all these conformations. The two carbon atoms are bonded by overlap of their sp^3 orbitals to form a sigma bond along the line between the carbons. This sigma bond can be twisted without destroying the linear overlap of the two sp^3 orbitals. No matter how you turn one of the methyl groups, its sp^3 orbital still overlaps with the sp^3 orbital of the other carbon atom.

PROBLEM-SOLVING HINT

Drawings that differ only by rotations of single bonds usually represent the same compound. For example, the following drawings all represent *n*-butane:

$CH_3CH_2CH_2CH_3$

n-butane

$$H_3C-\underset{\underset{H}{|}}{\overset{\overset{H}{|}}{C}}-\underset{\underset{H}{|}}{\overset{\overset{CH_3}{|}}{C}}-H$$

n-butane

$$H-\underset{\underset{H}{|}}{\overset{\overset{CH_3}{|}}{C}}-\underset{\underset{H}{|}}{\overset{\overset{CH_3}{|}}{C}}-H$$

n-butane

$$H-\underset{\underset{CH_3}{|}}{\overset{\overset{H}{|}}{C}}-\underset{\underset{H}{|}}{\overset{\overset{CH_3}{|}}{C}}-H$$

n-butane

methyl group   methyl group   ethane

eclipsed   staggered

◀ **Figure 2-20**
Ethane is composed of two methyl groups bonded by overlap of their sp^3 hybrid orbitals. These methyl groups may rotate with respect to each other.

2-7B Rigidity of Double Bonds

Not all bonds allow free rotation; ethylene, for example is quite rigid. In ethylene, the double bond between the two CH_2 groups consists of a sigma bond and a pi bond. When we twist one of the two CH_2 groups, the sigma bond is unaffected but the pi bond loses its overlap. The two *p* orbitals cannot overlap when the two ends of the molecule are at right angles, and the pi bond is effectively broken.

We can make the following generalization:

Rotation about single bonds is allowed, but double bonds are rigid and cannot be twisted.

Because double bonds are rigid, we can separate and isolate compounds that differ only in how their substituents are arranged on a double bond. For example, the double bond in 2-butene (CH_3—CH=CH—CH_3) prevents the two ends of the molecule from rotating. Two different compounds are possible, and they have different physical properties:

cis-2-butene
bp = 3.7°C

trans-2-butene
bp = 0.9°C

The molecule with the methyl groups on the same side of the double bond is called *cis*-2-butene, and the one with the methyl groups on opposite sides is called *trans*-2-butene. These kinds of molecules are discussed further in Section 2-8B.

PROBLEM 2-8

Two compounds are known with the formula CH_3—CH=N—CH_3.
(a) Draw a Lewis structure for this molecule, and label the hybridization of each carbon and nitrogen atom.
(b) What two compounds have this formula?
(c) Explain why there is only one compound known with the formula $(CH_3)_2CNCH_3$.

2-8
Isomerism

Isomers are different compounds with the same molecular formula. There are several types of isomerism in organic compounds, and we will cover them in detail in Chapter 5 (Stereochemistry). For now, we need to recognize the two large classes of isomers: constitutional isomers and stereoisomers.

2-8A Constitutional Isomerism

Constitutional isomers (or **structural isomers**) are isomers that differ in their bonding sequence; that is, their atoms are connected differently. Let's use butane as an example.

If you were asked to draw a structural formula for C_4H_{10}, either of the following structures would be correct:

$$CH_3{-}CH_2{-}CH_2{-}CH_3 \qquad CH_3{-}\overset{\overset{\displaystyle CH_3}{|}}{CH}{-}CH_3$$

n-butane   isobutane

These two compounds are isomers because they have the same molecular formula. They are constitutional isomers because their atoms are connected differently. The first compound (*n*-butane for "normal" butane) has its carbon atoms in a straight chain four carbons long. The second compound ("isobutane" for "an isomer of butane") has a branched structure with a longest chain of three carbon atoms and a methyl side chain.

There are three constitutional isomers of pentane (C_5H_{12}), whose common names are *n*-pentane, isopentane, and neopentane. The number of isomers increases rapidly as the number of carbon atoms increases.

PROBLEM-SOLVING HINT

Constitutional isomers (structural isomers) differ in the order in which their atoms are bonded.

$$CH_3{-}CH_2{-}CH_2{-}CH_2{-}CH_3 \qquad CH_3{-}\overset{\overset{\displaystyle CH_3}{|}}{CH}{-}CH_2{-}CH_3 \qquad CH_3{-}\overset{\overset{\displaystyle CH_3}{|}}{\underset{\underset{\displaystyle CH_3}{|}}{C}}{-}CH_3$$

n-pentane   isopentane   neopentane

Constitutional isomers may differ in ways other than the branching of their carbon chain. They may differ in the position of a double bond or other group or by having a ring or some other feature. Notice how the following constitutional isomers all differ in which atoms are bonded to which other atoms. (Check the number of hydrogens bonded to each carbon.) These compounds are not isomers of the pentanes shown above, however, because these have a different molecular formula (C_5H_{10}).

$$H_2C{=}CH{-}CH_2CH_2CH_3 \qquad CH_3{-}CH{=}CH{-}CH_2CH_3$$

1-pentene   2-pentene   cyclopentane   methylcyclobutane

2-8B   Stereoisomers

Stereoisomers are isomers that differ only in how their atoms are oriented in space. Their atoms are bonded in the same order, however. For example, *cis*- and *trans*-2-butene have the same connections of bonds, so they are not constitutional isomers. They are stereoisomers because they differ only in the spatial orientation of the groups attached to the double bond. The cis isomer has the two methyl groups on the same side of the double bond, and the trans isomer has them on opposite sides. In contrast, 1-butene is a constitutional isomer of *cis*- and *trans*-2-butene.

cis-2-butene   *trans*-2-butene   1-butene

Cis and trans isomers are only one type of stereoisomerism. The study of the structure and chemistry of stereoisomers is called **stereochemistry**. We will encounter stereochemistry throughout our study of organic chemistry, and Chapter 5 is devoted entirely to this field.

PROBLEM-SOLVING HINT

Similar groups on the same side of the double bond: *cis*.
Similar groups on opposite sides of the double bond: *trans*.

Cis-trans isomers are also called **geometric isomers** because they differ in the geometry of the groups on a double bond. The cis isomer is always the one with similar groups on the same side of the double bond, and the trans isomer has similar groups on opposite sides of the double bond.

To have cis-trans isomerism, there must be two different groups on each end of the double bond; otherwise, there can be no cis or trans isomers. For example, 1-butene has two identical hydrogens on one end of the double bond. Reversing their positions does not give a different compound. Similarly, 2-methyl-2-butene has two identical methyl groups on one end of the double bond. Reversing the methyl groups does not give a different compound.

PROBLEM-SOLVING HINT

Identical groups on one of the double-bonded carbons implies no cis-trans isomerism.

identical

1-butene
no cis or trans

identical

2-methyl-2-butene
no cis or trans

PROBLEM 2-9

Which of the following compounds show cis-trans isomerism? Draw the cis and trans isomers of those that do.
(a) $CHF=CHF$
(b) $F_2C=CH_2$
(c) $CH_2=CH-CH_2-CH_3$

(d) ⬡$=CHCH_3$
(e) ⬡$-CHCHCH_3$
(f) ⬡$=CHCH_3$

PROBLEM 2-10

Give the relationship between the following pairs of structures. The possible relationships are the following:

    same compound   constitutional isomers (structural isomers)
    cis-trans isomers   not isomers (different molecular formula)

(a) Br, H C=C H, Br   and   Br, Br C=C H, H
(b) $CH_3CH_2CHCH_2CH_3$ (with CH_2CH_3)   and $CH_3CH_2CHCH_2CH_2CH_3$ (with CH_3)

(c) Br, H C=C H, Br   and   Br, H C=C Br, H
(d) Br, H C=C H, Br   and   H, Br C=C Br, H

(e) H—C(H)(Cl)—C(Cl)(H)—H   and   H—C(Cl)(H)—C(Cl)(H)—H
(f) H—C(H)(CH$_3$)—C(CH$_3$)(H)—H   and   H—C(CH$_3$)(H)—C(H)(H)—CH$_3$

(g) $CH_3-CH_2-CH_2-CH_3$ and $CH_3-CH=CH-CH_3$
(h) $CH_2=CH-CH_2CH_2CH_3$ and $CH_3-CH=CH-CH_2CH_3$
(i) $CH_2=CHCH_2CH_2CH_3$ and $CH_3CH_2CH_2CH=CH_2$

(j) [cyclohexane with two CH_3 groups]   and   [cyclohexane with two CH_3 groups]

(k) [cyclohexene with CH_3]   and   [cyclohexene with CH_3]

In Section 1-6, we reviewed the concept of polar covalent bonds between atoms with different electronegativities. Now we are ready to combine this concept with molecular geometry to study the polarity of entire molecules.

2-9A   Bond Dipole Moments

Bond polarities can range from nonpolar covalent, through polar covalent, to totally ionic. In the following examples, ethane has a nonpolar covalent C—C bond. Methylamine, methanol, and chloromethane have increasingly polar (C—N, C—O, and C—Cl) covalent bonds. Methylammonium chloride $CH_3NH_3^+$ Cl^-) has an ionic bond between the methylammonium ion and the chloride ion.

$$H_3C\text{—}CH_3 \qquad H_3C\text{—}NH_2 \qquad H_3C\text{—}OH \qquad H_3C\text{—}Cl \qquad H_3CNH_3^+ \ Cl^-$$

$$\text{ethane} \qquad\quad \text{methylamine} \qquad \text{methanol} \qquad \text{chloromethane} \qquad \text{methylammonium}$$
$$\text{chloride}$$

nonpolar $\qquad\qquad$ increasing polarity $\qquad\qquad\qquad\qquad$ ionic

The polarity of an individual bond is measured as its **bond dipole moment**, μ, defined as

$$\mu = \delta \times d$$

where δ is the amount of charge at either end of the dipole and d is the distance between the charges.

Dipole moments are expressed in units of the **debye** (D), where 1 debye = 3.34×10^{-30} coulomb meters. If a proton and an electron (charge 1.60×10^{-19} coulomb) were 1 Å apart (distance 10^{-10} meter), the dipole moment would be

$$\mu = (1.60 \times 10^{-19} \text{ coulomb}) \times (10^{-10} \text{ meter}) = 1.60 \times 10^{-29} \text{ coulomb meter}$$

Expressed in debyes,

$$\mu = \frac{1.60 \times 10^{-29} \text{ C} \cdot \text{m}}{3.34 \times 10^{-30} \text{ C} \cdot \text{m/D}} = 4.8 \text{ D}$$

A simple rule of thumb, using common units, is that

$$\mu \text{ (in debyes)} = 4.8 \times \delta \text{ (electron charge)} \times d \text{ (in angstroms)}$$

Dipole moments are measured experimentally, and they can be used to calculate other information such as bond lengths and charge separations.

SOLVED PROBLEM 2-9

Calculate the charge separation for a typical C—O single bond, with a bond length of 1.43 Å and a dipole moment of 0.86 D.

SOLUTION

$$\mu = 0.86 \text{ D}$$

$$\overset{\delta^+}{C} \underset{1.43 \text{ Å}}{\text{—}} \overset{\delta^-}{O}$$

Using the formula for the dipole moment, we have

$$0.86\,D = 4.8 \times \delta \times 1.43\,\text{Å}$$

$$\delta = 0.125\,e$$

The amount δ of charge separation is about 0.125 electronic charge, so the carbon atom has about an eighth of a positive charge, and the oxygen atom has about an eighth of a negative charge.

PROBLEM 2-11

A typical C—Cl bond has a dipole moment of about 1.56 D and a bond length of about 1.81 Å. Calculate the amount of charge separation in such a bond.

PROBLEM 2-12

The C=O double bond has a dipole moment of about 2.4 D and a bond length of about 1.21 Å.
(a) Calculate the amount of charge separation in this bond.
(b) Use this information to evaluate the relative importance of the following two resonance contributors.

Bond dipole moments in organic compounds range from zero in symmetrical bonds to about 3.6 D for the strongly polar C≡N: triple bond. Table 2-1 shows typical dipole moments for some of the bonds common in organic molecules. Recall that the positive end of the crossed arrow corresponds to the less electronegative (partial positive charge) end of the dipole.

2-9B   Molecular Dipole Moments

A **molecular dipole moment** is the dipole moment of the molecule taken as a whole. It is a good indicator of a molecule's overall polarity. Molecular dipole moments can be measured directly, in contrast to bond dipole moments, which must be estimated by comparing various compounds. The value of the molecular dipole moment is equal to the *vector* sum of the individual bond dipole moments. This vector sum reflects both the magnitude and the direction of each individual bond dipole moment.

For example, formaldehyde has one strongly polar C=O bond, while carbon dioxide has two. We might expect CO_2 to have the larger dipole moment, but its dipole moment is actually zero. The symmetry of the carbon dioxide molecule explains this surprising result. The structures of formaldehyde and carbon dioxide are shown on the next page, together with their electrostatic potential maps.

TABLE 2-1	Bond Moments (Debye) for Some Common Covalent Bonds		
Bond	**Dipole Moment, μ**	**Bond**	**Dipole Moment, μ**
C—N	0.22 D	H—C	0.3 D
C—O	0.86 D	H—N	1.31 D
C—F	1.51 D	H—O	1.53 D
C—Cl	1.56 D	C=O	2.4 D
C—Br	1.48 D	C≡N	3.6 D
C—I	1.29 D		

formaldehyde EPM	$\mu = 2.3$ D formaldehyde	$\mu = 0$ carbon dioxide	carbon dioxide EPM

The electrostatic potential maps of formaldehyde and CO_2 show the bond dipole moments, with red at the negative ends and blue at the positive ends of the dipoles. In carbon dioxide, the bond dipole moments are oriented in opposite directions, and they cancel each other.

Figure 2-21 shows some examples of molecular dipole moments. Notice that the dipole moment of C—H bonds is small, so we often treat C—H bonds as nearly nonpolar. Also note that the tetrahedral symmetry of CCl_4 positions the four C—Cl dipole moments in directions so that they cancel. A partial canceling of the bond dipole moments explains why $CHCl_3$, with three C—Cl bonds, has a smaller molecular dipole moment than CH_3Cl, with only one.

Lone pairs of electrons contribute to the dipole moments of bonds and molecules. Each lone pair corresponds to a charge separation, with the nucleus having a partial positive charge balanced by the negative charge of the lone pair. Figure 2-22 shows four molecules with lone pairs and large dipole moments. Notice how the lone pairs

$\mu = 1.9$ D chloromethane	$\mu = 1.0$ D chloroform	$\mu = 0$ carbon tetrachloride	$\mu = 1.9$ D cis-1,2-dibromoethene

◀ **Figure 2-21**
A molecular dipole moment is the vector sum of the individual bond dipole moments.

$\mu = 1.5$ D ammonia	$\mu = 1.9$ D water	$\mu = 2.9$ D acetone	$\mu = 3.9$ D acetonitrile

◀ **Figure 2-22**
The presence of lone pairs may have a large effect on the molecular dipole moment.

contribute to the large dipole moments, especially in the $C=O$ and $C\equiv N$ bonds. Also notice how the electrostatic potential maps show red areas (high negative potential) in the electron-rich regions of the lone pairs.

PROBLEM 2-13

The $N-F$ bond is more polar than the $N-H$ bond, but NF_3 has a *smaller* dipole moment than NH_3. Explain this curious result.

$$NH_3 \qquad\qquad NF_3$$
$$\mu = 1.5\,D \qquad\qquad \mu = 0.2\,D$$

PROBLEM 2-14

For each of the following compounds
(1) Draw the structure.
(2) Show how the bond dipole moments (and those of any nonbonding pairs of electrons) contribute to the molecular dipole moment.
(3) Predict whether the compound has a large (>1 D), small, or zero dipole moment.

(a) CH_2Cl_2	(b) CH_3F	(c) CF_4	(d) CH_3OH
(e) O_3	(f) HCN	(g) CH_3CHO	(h) $H_2C=NH$
(i) $(CH_3)_3N$	(j) $CH_2=CHCl$	(k) BF_3	(l) $BeCl_2$
(m) NH_4^+			

PROBLEM 2-15

Two isomers of 1,2-dichloroethene are known. One has a dipole moment of 2.4 D; the other has zero dipole moment. Draw the two isomers and explain why one has zero dipole moment.

$$CHCl=CHCl$$

1,2-dichloroethene

2-10
Intermolecular Attractions and Repulsions

When two molecules approach, they attract or repel each other. This interaction can be described fairly simply in the case of atoms (like the noble gases) or simple molecules such as H_2 or Cl_2. In general, the forces are attractive until the molecules come so close that they infringe on each other's atomic radius. When this happens, the small attractive force quickly becomes a large repulsive force, and the molecules "bounce" off each other. With complicated organic molecules, these attractive and repulsive forces are more difficult to predict. We can still describe the nature of the forces, however, and we can show how they affect the physical properties of organic compounds.

Attractions between molecules are particularly important in solids and liquids. In these "condensed" phases, the molecules are continuously in contact with each other. The melting points, boiling points, and solubilities of organic compounds show the effects of these forces. Three major kinds of attractive forces cause molecules to associate into solids and liquids: the dipole–dipole forces of polar molecules, the London forces that affect all molecules, and the "hydrogen bonds" that link molecules having $-OH$ or $-NH$ groups.

2-10A   Dipole–Dipole Forces

Most molecules have permanent dipole moments as a result of their polar bonds. Each molecular dipole moment has a positive end and a negative end. The most stable arrangement has the positive end of one dipole close to the negative end of another. When two negative ends or two positive ends approach each other, they repel. They may turn and orient themselves in the more stable positive-to-negative arrangement. **Dipole–dipole forces**, therefore, are generally attractive intermolecular forces resulting from the attraction of the positive and negative ends of the dipole moments of polar molecules. Figure 2-23 shows the attractive and repulsive orientations of polar molecules, using chloromethane as the example.

attraction (common)

symbolized by

repulsion (uncommon)

or

◀ **Figure 2-23**
Dipole–dipole interactions result from the approach of two polar molecules. If their positive and negative ends approach, the interaction is an attractive one. If two negative ends or two positive ends approach, the interaction is repulsive. In a liquid or a solid, the molecules are mostly oriented with the positive and negative ends together, and the net force is attractive.

Polar molecules are mostly oriented in the lower-energy positive-to-negative arrangement, and the net force is attractive. This attraction must be overcome when the liquid vaporizes, resulting in larger heats of vaporization and higher boiling points for strongly polar compounds.

2-10B   The London Dispersion Force

Carbon tetrachloride (CCl_4) has zero dipole moment, yet its boiling point is higher than that of chloroform ($\mu = 1.0$ D). Clearly, there must be some kind of force other than dipole–dipole forces holding the molecules of carbon tetrachloride together.

$\mu = 0$
carbon tetrachloride, bp = 77°C

$\mu = 1.0$ D
chloroform, bp = 62°C

In nonpolar molecules such as carbon tetrachloride, the principal attractive force is the **London dispersion force**, one of the **van der Waals forces** (Fig. 2-24). The London

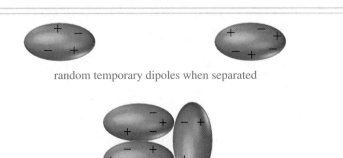

random temporary dipoles when separated

correlated temporary dipoles when in contact

◀ **Figure 2-24**
London dispersion forces result from the attraction of correlated temporary dipoles.

force arises from temporary dipole moments that are induced in a molecule by other near-by molecules. Even though carbon tetrachloride has no permanent dipole moment, the electrons are not always evenly distributed. A small temporary dipole moment is induced when one molecule approaches another molecule in which the electrons are slightly displaced from a symmetrical arrangement. The electrons in the approaching molecule are displaced slightly so that an attractive dipole–dipole interaction results.

These temporary dipoles last only a fraction of a second, and they change continuously; yet they are correlated so their net force is attractive. This attractive force depends on close surface contact of two molecules, so it is roughly proportional to the molecular surface area. Carbon tetrachloride has a larger surface area than chloroform (a chlorine atom is much larger than a hydrogen atom), so the intermolecular van der Waals attractions between carbon tetrachloride molecules are stronger than they are between chloroform molecules.

We can see the effects of London forces in the boiling points of simple hydrocarbons. If we compare the boiling points of several different isomers, the isomers with larger surface areas (and greater potential for London force attraction) have higher boiling points. The boiling points of three isomers of molecular formula C_5H_{12} are given below. The long-chain isomer (*n*-pentane) has the greatest surface area and the highest boiling point. As the amount of chain branching increases, the molecule becomes more spherical and its surface area decreases. The most highly branched isomer (neopentane) has the smallest surface area and the lowest boiling point.

$$CH_3-CH_2-CH_2-CH_2-CH_3 \qquad CH_3-\overset{\overset{\displaystyle CH_3}{|}}{C}H-CH_2-CH_3 \qquad CH_3-\overset{\overset{\displaystyle CH_3}{|}}{\underset{\underset{\displaystyle CH_3}{|}}{C}}-CH_3$$

n-pentane, bp = 36°C   isopentane, bp = 28°C   neopentane, bp = 10°C

2-10C   Hydrogen Bonding

A **hydrogen bond** is not a true bond but a particularly strong dipole–dipole attraction. A hydrogen atom can participate in hydrogen bonding if it is bonded to oxygen, nitrogen, or fluorine. Organic compounds do not contain H—F bonds, so we consider only N—H and O—H hydrogens to be hydrogen bonded (Figure 2-25).

The O—H and N—H bonds are strongly polarized, leaving the hydrogen atom with a partial positive charge. This electrophilic hydrogen has a strong affinity for nonbonding electrons, and it forms intermolecular attachments with the nonbonding electrons on oxygen or nitrogen atoms.

Although hydrogen bonding is a strong form of intermolecular attraction, it is much weaker than a normal C—H, N—H, or O—H covalent bond. Breaking a hydrogen

▶ **Figure 2-25**
Hydrogen bonding is a strong intermolecular attraction between an electrophilic O—H or N—H hydrogen atom and a pair of nonbonding electrons.

methanol, $\overset{\delta^+}{H}—\overset{\delta^-}{\underset{..}{\overset{..}{O}}}—CH_3$ methylamine, $\overset{\delta^+}{H}—\overset{\delta^-}{\underset{..}{N}}\overset{\displaystyle H}{\underset{\displaystyle CH_3}{}}$

bond requires about 5 kcal/mol (20 kJ/mol), compared with about 100 kcal/mol (about 400 kJ/mol) required to break a C—H, N—H, or O—H bond.

Hydrogen bonding has a large effect on the physical properties of organic compounds, as shown by the boiling points of ethanol (ethyl alcohol) and dimethyl ether, two isomers of molecular formula C_2H_6O:

$$CH_3—CH_2—OH \qquad\qquad CH_3—O—CH_3$$

ethanol, bp 78°C   dimethyl ether, bp −25°C

These two isomers have the same size and the same molecular weight. Alcohols like ethanol have O—H hydrogens, however, so they are extensively hydrogen bonded. Dimethyl ether has no O—H hydrogen, so it cannot form hydrogen bonds. As a result of its hydrogen bonding, ethanol has a boiling point more than 100°C higher than that of dimethyl ether.

The effect of N—H hydrogen bonding on boiling points can be seen in the isomers of formula C_3H_9N shown below. Trimethylamine has no N—H hydrogens, so it is not hydrogen bonded. Ethylmethylamine has one N—H hydrogen atom, and the resulting hydrogen bonding raises its boiling point about 34°C above that of trimethylamine. Propylamine, with two N—H hydrogens, is more extensively hydrogen bonded and has the highest boiling point of these three isomers.

trimethylamine, bp 3.5°C   ethylmethylamine, bp 37°C   propylamine, bp 49°C

Alcohols form stronger hydrogen bonds than amines, probably because oxygen is more electronegative than nitrogen. Thus, the O—H bond is more strongly polarized than the N—H bond. This effect is seen in the boiling points above, with more than 100° difference in the boiling points of ethanol and dimethyl ether, compared with a 34° difference for ethylmethylamine and trimethylamine.

> Hydrogen bonding is also essential for the structural integrity of many biological molecules. For example, the double helix structure of DNA is maintained, in part, by hydrogen bonds between the bases: adenine pairs with thymine and guanine pairs with cytosine. Figure 23-25 shows this base-pairing.

PROBLEM 2-16

Draw the hydrogen bonding that takes place between
(a) two molecules of ethanol.
(b) two molecules of propylamine.

SOLVED PROBLEM 2-10

Rank the following compounds in order of increasing boiling points. Explain the reasons for your chosen order.

neopentane   *n*-hexane   2,3-dimethylbutane

1-pentanol   2-methyl-2-butanol

> **PROBLEM-SOLVING HINT**
> To predict relative boiling points, look for differences in
> 1. hydrogen bonding,
> 2. molecular weight and surface area, and
> 3. dipole moments.

SOLUTION

To predict relative boiling points, we should look for differences in (1) hydrogen bonding, (2) molecular weight and surface area, and (3) dipole moments. Except for neopentane, these compounds have similar molecular weights. Neopentane is the lightest, and it is a compact spherical structure that minimizes van der Waals attractions. Neopentane is the lowest-boiling compound.

Neither *n*-hexane nor 2,3-dimethylbutane is hydrogen bonded, so they will be next higher in boiling points. Because 2,3-dimethylbutane is more highly branched (and has a smaller surface area) than *n*-hexane, 2,3-dimethylbutane will be lower boiling than *n*-hexane. So far, we have

neopentane $<$ 2,3-dimethylbutane $<$ *n*-hexane $<$ the others

The two remaining compounds are both hydrogen bonded, and 1-pentanol has more area for van der Waals forces. Therefore, 1-pentanol should be the highest-boiling compound. We predict the following order:

neopentane $<$ 2,3-dimethylbutane $<$ *n*-hexane $<$ 2-methyl-2-butanol $<$ 1-pentanol

10°C   58°C   69°C   102°C   138°C

The actual boiling points are given here to show that our prediction is correct.

PROBLEM 2-17

For each pair of compounds, circle the compound you expect to have the higher boiling point. Explain your reasoning.

(a) $(CH_3)_3C-C(CH_3)_3$ and $(CH_3)_2CH-CH_2CH_2-CH(CH_3)_2$
(b) $CH_3(CH_2)_6CH_3$ and $CH_3(CH_2)_5CH_2OH$
(c) $HOCH_2-(CH_2)_4-CH_2OH$ and $(CH_3)_3CCH(OH)CH_3$
(d) $(CH_3CH_2CH_2)_2NH$ and $(CH_3CH_2)_3N$

(e) ⬡NH   and   ⬡—NH_2

2-11
Polarity Effects on Solubilities

In addition to affecting boiling points and melting points, intermolecular forces determine the solubility properties of organic compounds. The general rule is that "*like dissolves like*." Polar substances dissolve in polar solvents, and nonpolar substances dissolve in nonpolar solvents. We discuss the reasons for this rule now, then apply the rule in later chapters as we discuss the solvent properties of organic compounds.

We should consider four different cases: (1) a polar solute with a polar solvent, (2) a polar solute with a nonpolar solvent, (3) a nonpolar solute with a nonpolar solvent, and (4) a nonpolar solute with a polar solvent. We will use sodium chloride and water as examples of polar solutes and solvents, and paraffin "wax" and gasoline as examples of nonpolar solutes and solvents.

Polar Solute in a Polar Solvent (Dissolves)   When you think about sodium chloride dissolving in water, it seems remarkable that the oppositely charged ions can be separated from each other. A great deal of energy is required to separate these ions. A polar solvent (such as water) can separate the ions because it *solvates* them (Fig. 2-26). If water is the solvent, the solvation process is called *hydration*. As the salt dissolves, water molecules surround each ion, with the appropriate end of the water dipole moment next to the ion. In the case of the positive (sodium) ion, the oxygen atom of the water molecule approaches. The negative ions (chloride) are approached by the hydrogen atoms of the water molecules.

Because water molecules are strongly polar, a large amount of energy is released when the sodium and chloride ions are hydrated. This energy is nearly sufficient to overcome the lattice energy of the crystal. The salt dissolves, partly because of strong solvation by water molecules and partly because of the increase in entropy (randomness or freedom of movement) when it dissolves.

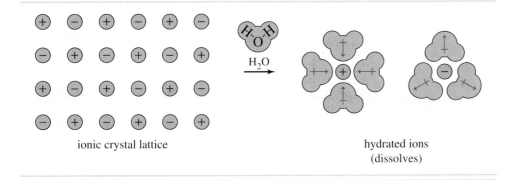

◀ **Figure 2-26**
The hydration of sodium and chloride ions by water molecules overcomes the lattice energy of sodium chloride. The salt dissolves.

Polar Solute in a Nonpolar Solvent (Does Not Dissolve)   If you stir sodium chloride with a nonpolar solvent such as turpentine or gasoline, you will find that the salt does not dissolve (Fig. 2-27). The nonpolar molecules of these solvents do not solvate ions very strongly, and they cannot overcome the large lattice energy of the salt crystal. This is a case where the attractions of the ions in the solid for each other are much greater than their attractions for the solvent.

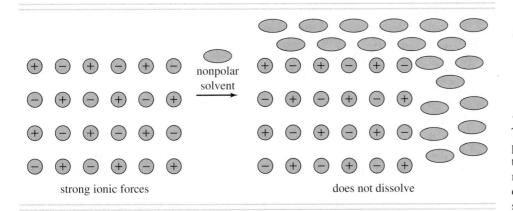

◀ **Figure 2-27**
The intermolecular attractions of polar substances are stronger than their attractions for nonpolar solvent molecules. Thus, a polar substance does not dissolve in a nonpolar solvent.

Nonpolar Solute in a Nonpolar Solvent (Dissolves)   Paraffin "wax" dissolves in gasoline. Both paraffin and gasoline are mixtures of nonpolar hydrocarbons (Fig. 2-28). The molecules of a nonpolar substance (paraffin) are weakly attracted to each other, and these van der Waals attractions are easily overcome by van der Waals attractions with the solvent. Although there is little change in energy when the nonpolar substance dissolves in a nonpolar solvent, there is a large increase in entropy.

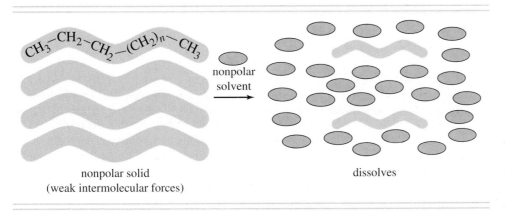

◀ **Figure 2-28**
The weak intermolecular attractions of a nonpolar substance are overcome by the weak attractions for a nonpolar solvent. The nonpolar substance dissolves.

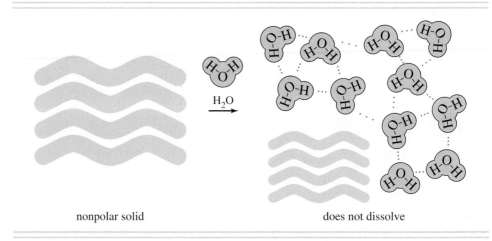

nonpolar solid   does not dissolve

▶ **Figure 2-29**
If a nonpolar molecule were to dissolve in water, it would break up the hydrogen bonds between the water molecules. Therefore, nonpolar substances do not dissolve in water.

Motor oil and water do not mix because the nonpolar oil molecules cannot displace the strong intermolecular attractions between water molecules.

Nonpolar Solute in a Polar Solvent (Does Not Dissolve) Anyone who does home canning knows that a nonpolar solid such as paraffin does not dissolve in a polar solvent such as water. Why not? The nonpolar molecules are only weakly attracted to each other, and little energy is required to separate them. The problem is that the water molecules are strongly attracted to each other by their hydrogen bonding. If a nonpolar paraffin molecule were to dissolve, it would have to displace some of these hydrogen bonds, yet there is almost no energy released from solvation of the nonpolar molecule. In effect, the hydrogen-bonded network of water molecules *excludes* the paraffin molecules (Fig. 2-29).

Figures 2-26 through 2-29 show why the saying "like dissolves like" is generally true. Polar substances dissolve in polar solvents, and nonpolar substances dissolve in nonpolar solvents. This general rule also applies to the mixing of liquids. Everyone knows that water and gasoline (or oil) do not mix. Gasoline and oil are both nonpolar hydrocarbons, however, and they mix freely with each other. They do not dissolve in water because they would have to break up the hydrogen bonds of the water molecules.

Ethanol is a polar molecule, and it is miscible with water; that is, it mixes freely with water in all proportions. Ethanol has an O—H group that forms hydrogen bonds with water molecules. When ethanol dissolves in water, it forms new ethanol–water hydrogen bonds to replace the water–water hydrogen bonds that are broken:

$$CH_3-CH_2-O\overset{H}{\underset{}{}} \quad + \quad \overset{}{\underset{H \quad H}{O}} \quad \longrightarrow \quad CH_3-CH_2-O\overset{}{\underset{H}{}}$$

In Section 2-12, we will see many kinds of organic compounds with a wide variety of "functional groups." As you encounter these new compounds, you should look to see whether the molecules are polar or nonpolar and whether they can engage in hydrogen bonding.

Most vitamins consist of charged groups, making them water-soluble. As a result, they are rapidly eliminated and generally nontoxic. Vitamins A and D, however, are nonpolar and are stored in the fat tissue of the body, which is also nonpolar. Hence, these two vitamins are potentially toxic in large doses.

PROBLEM 2-18

Circle the member of each pair that is more soluble in water.
(a) $CH_3CH_2OCH_2CH_3$ or $CH_3CH_2CH_2CH_2CH_3$
(b) $CH_3CH_2NHCH_3$ or $CH_3CH_2CH_2CH_3$

(c) CH_3CH_2OH or $CH_3CH_2CH_2CH_2OH$

(d)

or

In future chapters, we will study many different types of organic compounds. The various kinds of compounds are briefly described here so that you can recognize them as you encounter them. For the purpose of this brief survey, we divide organic compounds into three classes: (1) hydrocarbons, (2) compounds containing oxygen, and (3) compounds containing nitrogen.

2-12 Hydrocarbons

   The hydrocarbons are compounds composed entirely of carbon and hydrogen. The major classes of hydrocarbons are alkanes, alkenes, alkynes, and aromatic hydrocarbons.

2-12A Alkanes

Alkanes are hydrocarbons that contain only single bonds. Alkane names generally have the *-ane* suffix, while the first part of the name gives the number of carbon atoms. Table 2-2 shows how the prefixes in the names correspond with the number of carbon atoms.

   The **cycloalkanes** are a special class of alkanes that are in the form of a ring. Figure 2-30 shows some cycloalkanes.

TABLE 2-2   Correspondence of Prefixes and Numbers of Carbon Atoms

Alkane Name	Number of Carbons	Alkane Name	Number of Carbons
*meth*ane	1	*hex*ane	6
*eth*ane	2	*hept*ane	7
*prop*ane	3	*oct*ane	8
*but*ane	4	*non*ane	9
*pent*ane	5	*dec*ane	10

CH_4 $CH_3—CH_3$ $CH_3—CH_2—CH_3$ $CH_3—CH_2—CH_2—CH_3$ $CH_3—\overset{\displaystyle CH_3}{\overset{|}{CH}}—CH_3$

methane   ethane   propane   butane   isobutane

cyclopentane

or

cyclopentane

cyclohexane

or

cyclohexane

◀ **Figure 2-30**
Cycloalkanes are alkanes in the form of a ring.

▲ **Figure 2-31**
Alkyl groups are named like the alkanes they are derived from, with a -*yl* suffix.

Alkanes are the major components of heating gases (natural gas and liquefied petroleum gas), gasoline, motor oil, fuel oil, and paraffin "wax." Other than combustion, alkanes undergo few reactions. In fact, when a molecule contains an alkane portion and a nonalkane portion, we often ignore the presence of the alkane portion because it is relatively unreactive.

Alkanes undergo few reactions because they have no **functional group**, where reactions usually occur. Functional groups are reactive parts such as double bonds, hydroxyl groups, or halogen atoms. Most organic compounds are characterized and classified by the functional groups they contain.

A molecule with a functional group may be represented as the functional group with **alkyl groups** attached. An **alkyl group** is an alkane portion of a molecule, with one hydrogen atom removed to allow bonding to the functional group. Figure 2-31 shows how an alkyl group is named, using a substituted cycloalkane as the example.

We are often concerned with the structure of only the most important part of a molecule. In these cases, we often use the symbol R as a substituent. The R group is simply an alkyl group. We presume that the exact nature of the R group is not important.

2-12B   Alkenes

Alkenes are hydrocarbons that contain carbon–carbon double bonds. A carbon–carbon double bond is the most reactive part of an alkene, so we say that the double bond is the *functional group* of the alkene. Alkene names end in the -*ene* suffix, as shown by the following examples:

$$CH_2{=}CH_2 \qquad\qquad CH_2{=}CH{-}CH_3$$
ethene (ethylene)   propene (propylene)

$$CH_2{=}CH{-}CH_2{-}CH_3 \qquad CH_3{-}CH{=}CH{-}CH_3$$
1-butene   2-butene

Carbon–carbon double bonds cannot rotate, and many alkenes show geometric (cis-trans) isomerism (Sections 2-7B, 2-8B). The following are the cis-trans isomers of some simple alkenes:

cis-2-butene   *trans*-2-butene   *cis*-2-hexene   *trans*-2-hexene

Cycloalkenes are also common. Unless the rings are very large, cycloalkenes are always the cis isomers, and the term *cis* is omitted from the names. In a large ring, a trans double bond may occur, giving a *trans*-cycloalkene.

cyclopentene   cyclohexene   *trans*-cyclodecene

2-12C   Alkynes

Alkynes are hydrocarbons with carbon–carbon triple bonds as their functional group. Alkyne names generally have the *-yne* suffix, although some of their common names (*acetylene*, for example) do not conform to this rule. The triple bond is linear, so there is no possibility of geometric (cis-trans) isomerism in alkynes.

$$H-C\equiv C-H \qquad\qquad H-C\equiv C-CH_3$$

ethyne (acetylene)   propyne (methylacetylene)

$$H-C\equiv C-CH_2-CH_3 \qquad\qquad CH_3-C\equiv C-CH_3$$

1-butyne   2-butyne

In an alkyne, four atoms must be in a straight line. These four collinear atoms are not easily bent into a ring, so cycloalkynes are rare. Cycloalkynes are stable only if the ring is large.

2-12D   Aromatic Hydrocarbons

The compounds below may look like cycloalkenes, but their properties are different from those of simple alkenes. These **aromatic hydrocarbons** (also called **arenes**) are all derivatives of *benzene*, represented by a six-membered ring with three double bonds. This bonding arrangement is particularly stable, for reasons that are explained in Chapter 16.

benzene   ethylbenzene   an alkylbenzene

> The presence of a methyl or ethyl group can sometimes make a big difference. For example, benzene is quite toxic and causes leukemia, while methyl benzene (and ethyl benzene) are much less toxic because enzymes can oxidize the methyl group.

Just as a generic alkyl group substituent is represented by R, a generic aryl group is represented by Ar. When a benzene ring serves as a substituent, it is called a phenyl group, abbreviated Ph.

an arylcyclopentane   might be   phenylcyclopentane   or other compounds

PROBLEM 2-19

Classify the following hydrocarbons, and draw a Lewis structure for each one. A compound may fit into more than one of the following classifications:

alkane   cycloalkane   aromatic hydrocarbon
alkene   cycloalkene
alkyne   cycloalkyne

(a) $CH_3(CH_2)_3CH_3$ **(b)** $CH_3CHCHCH_2CH_3$ **(c)** $CH_3CCCH_2CH_2CH_3$

(d) CH_2—C≡C—CH_2 **(e)**   **(f)**
CH_2—CH_2—CH_2—CH_2

(g)   **(h)**   **(i)**

2-13
Organic Compounds Containing Oxygen

Many organic compounds contain oxygen atoms bonded to alkyl groups. The major classes of oxygen-containing compounds are alcohols, ethers, ketones, aldehydes, carboxylic acids, and acid derivatives.

2-13A   Alcohols

Alcohols are organic compounds that contain the **hydroxyl group** (—OH) as their functional group. The general formula for an alcohol is R—OH. Alcohols are among the most polar organic compounds because the hydroxyl group is strongly polar and can participate in hydrogen bonding. Some of the simple alcohols like ethanol and methanol are miscible (soluble in all proportions) with water. Names of alcohols end in the *-ol* suffix from the word "alcohol," as shown for the following common alcohols.

R—OH CH_3—OH CH_3CH_2—OH CH_3—CH_2—CH_2—OH CH_3—$\overset{\displaystyle OH}{\underset{}{CH}}$—$CH_3$
an alcohol   methanol   ethanol   1-propanol   2-propanol
          (methyl alcohol)   (ethyl alcohol)   (*n*-propyl alcohol)   (isopropyl alcohol)

Alcohols are some of the most common organic compounds. Methyl alcohol (methanol), also known as "wood alcohol," is used as an industrial solvent and as an automobile racing fuel. Ethyl alcohol (ethanol) is sometimes called "grain alcohol" because it is produced by the fermentation of grain or almost any other organic material. "Isopropyl alcohol" is the common name for 2-propanol, used as "rubbing alcohol."

2-13B   Ethers

Ethers are composed of two alkyl groups bonded to an oxygen atom. The general formula for an ether is R—O—R′. (The symbol R′ represents another alkyl group, either the same as or different from the first.) Like alcohols, ethers are much more polar than hydrocarbons. Ethers have no O—H hydrogens, however, so they cannot hydrogen bond with themselves. Ether names are often formed from the names of the alkyl groups and the word "ether." Diethyl ether is the common "ether" used for starting engines in cold weather and once used for surgical anesthesia.

R—O—R′ CH_3—O—CH_3 CH_3CH_2—O—CH_2CH_3 —O—CH_3
ROR′, an ether   dimethyl ether   diethyl ether

cyclohexyl methyl ether

2-13C Aldehydes and Ketones

The **carbonyl group**, C=O, is the functional group for both aldehydes and ketones. A **ketone** has two alkyl groups bonded to the carbonyl group; an **aldehyde** has one alkyl group and a hydrogen atom bonded to the carbonyl group. Ketone names generally have the *-one* suffix; aldehyde names use either the *-al* suffix or the *-aldehyde* suffix.

The carbonyl group is strongly polar, and most ketones and aldehydes are somewhat soluble in water. Both acetone and acetaldehyde are miscible with water. Acetone, often used as nail polish remover, is a common solvent with low toxicity.

RCOR′, a ketone   2-propanone (acetone)   2-butanone (methyl ethyl ketone)   cyclohexanone

or RCHO   or CH$_3$CHO   or CH$_3$CH$_2$CHO   or CH$_3$CH$_2$CH$_2$CHO
an aldehyde   ethanal (acetaldehyde)   propanal (propionaldehyde)   butanal (butyraldehyde)

2-13D Carboxylic Acids

Carboxylic acids contain the **carboxyl group**, —COOH, as their functional group. The general formula for a carboxylic acid is **R**—COOH (*or* RCO$_2$H). The carboxyl group is a combination of a carbonyl group and a hydroxyl group, but this combination has different properties from those of ketones and alcohols. Carboxylic acids owe their acidity (pK_a of about 5) to the resonance-stabilized *carboxylate anions* formed by deprotonation. The following reaction shows the dissociation of a carboxylic acid:

carboxylic acid   carboxylate anion

Systematic names for carboxylic acids use the *-oic acid* suffix, but historical names are commonly used. Formic acid was first isolated from ants, genus *Formica*. Acetic acid, found in vinegar, gets its name from the Latin word for "sour" *(acetum)*. Propionic acid gives the tangy flavor to sharp cheeses, and butyric acid provides the pungent aroma of rancid butter.

methanoic acid   ethanoic acid   propanoic acid   butanoic acid
(formic acid)   (acetic acid)   (propionic acid)   (butyric acid)

Carboxylic acids are strongly polar, like ketones, aldehydes, and alcohols. They are relatively soluble in water; in fact, all four of the carboxylic acids shown above are miscible (soluble in all proportions) with water.

PROBLEM 2-20

Draw a Lewis structure, and classify each of the following compounds. The possible classifications are as follows:

    alcohol   ketone   carboxylic acid
    ether   aldehyde

(a) CH_3CH_2CHO
(d) $CH_3CH_2OCH_2CH_3$

(b) $CH_3CH_2CH(OH)CH_3$
(c) $CH_3COCH_2CH_3$

(e) [cyclohexane ring with —COOH]
(f) [tetrahydropyran ring with O]

(g) [cyclopentane ring with =O]
(h) [cyclopentane ring with —CHO]
(i) [cyclopentane ring with —CH$_2$OH]

2-13E   Carboxylic Acid Derivatives

Several related functional groups can be formed from carboxylic acids. Each contains the carbonyl group bonded to an oxygen or other electron-withdrawing element. Among these functional groups are **acid chlorides, esters**, and **amides**. All of these groups can be converted back to carboxylic acids by acidic or basic hydrolysis.

$R—C(=O)—OH$	$R—C(=O)—Cl$	$R—C(=O)—O—R'$	$R—C(=O)—NH_2$
or $R—COOH$	or $R—COCl$	or $R—COOR'$	or $R—CONH_2$
carboxylic acid	acid chloride	ester	amide

$CH_3—C(=O)—OH$	$CH_3—C(=O)—Cl$	$CH_3—C(=O)—O—CH_2CH_3$	$CH_3—C(=O)—NH_2$
or CH_3COOH	or CH_3COCl	or $CH_3COOCH_2CH_3$	or CH_3CONH_2
acetic acid	acetyl chloride	ethyl acetate	acetamide

2-14
Organic Compounds Containing Nitrogen

Nitrogen is another element often found in the functional groups of organic compounds. The most common "nitrogenous" organic compounds are amines, amides, and nitriles.

2-14A   Amines

Amines are alkylated derivatives of ammonia. Like ammonia, amines are basic.

$$R—\ddot{N}H_2 + H_2O \rightleftharpoons R—\overset{+}{N}H_3 \quad {}^-OH \qquad K_b \cong 10^{-3}$$

Because of their basicity ("alkalinity"), naturally occurring amines are often called *alkaloids*. Simple amines are named by naming the alkyl groups bonded to nitrogen and adding the word "amine." The structures of some simple amines are shown below, together with the structure of nicotine, a toxic alkaloid found in tobacco leaves.

$$R—\ddot{N}H_2 \quad \text{or} \quad R—\ddot{N}H—R' \quad \text{or} \quad R—\underset{..}{\overset{R'}{N}}—R''$$

amines

$$CH_3—\ddot{N}H_2 \qquad CH_3—\ddot{N}H—CH_2CH_3 \qquad (CH_3CH_2)_3N\colon$$

methylamine   ethylmethylamine   triethylamine   piperidine   nicotine

2-14B   Amides

Amides are acid derivatives that result from a combination of an acid with ammonia or an amine. Proteins have the structure of long-chain, complex amides.

$$R-\overset{\overset{\textstyle O}{\|}}{C}-NH_2 \quad \text{or} \quad R-\overset{\overset{\textstyle O}{\|}}{C}-NHR' \quad \text{or} \quad R-\overset{\overset{\textstyle O}{\|}}{C}-NR_2'$$

amides

$$CH_3-\overset{\overset{\textstyle O}{\|}}{C}-NH_2 \qquad CH_3-\overset{\overset{\textstyle O}{\|}}{C}-NH-CH_3$$

acetamide   *N*-methylacetamide   *N,N*-dimethylbenzamide

2-14C   Nitriles

A **nitrile** is a compound containing the **cyano group**, $-C\equiv N$. The cyano group was introduced in Section 2-6 as an example of *sp* hybridized bonding.

$$R-C\equiv N\colon \quad CH_3-C\equiv N\colon \quad CH_3CH_2-C\equiv N\colon$$

a nitrile   acetonitrile   propionitrile   benzonitrile

All of these classes of compounds are summarized in the table of **Common Organic Compounds and Functional Groups**, given on the front inside cover for convenient reference.

PROBLEM 2-21

Draw a Lewis structure, and classify each of the following compounds.
(a) $CH_3CH_2CONHCH_3$ **(b)** $(CH_3CH_2)_2NH$ **(c)** $(CH_3)_2CHCOOCH_3$
(d) $(CH_3CH_2)_3CCH_2COCl$ **(e)** $(CH_3CH_2)_2O$ **(f)** $CH_3CH_2CH_2CN$

(g) $(CH_3)_3CCH_2CH_2COOH$ **(h)**   **(i)**

(j)   **(k)**   **(l)**

(m)   **(n)**   **(o)**   **(p)**

PROBLEM 2-22

Circle the functional groups in the following structures. State to which class (or classes) of compounds the structure belongs.
(a) $CH_2{=}CHCH_2CH_3$ **(b)** CH_3-O-CH_3 **(c)** CH_3CHO
(d) $HCONH_2$ **(e)** CH_3NHCH_3 **(f)** $R-COOH$

(g) (structure: benzene ring with CH$_2$OH)

(h) (structure: cyclohexene ring with CN)

(i) (structure: cyclopropane with =O)

(j) hydrocortisone

(k) vitamin E

Chapter 2 Glossary

acid chloride An acid derivative with a chlorine atom in place of the hydroxyl group. (p. 72)

$$R-\overset{\overset{\displaystyle O}{\|}}{C}-Cl$$

alcohol A compound that contains a hydroxyl group; R—OH. (p. 70)

aldehyde A carbonyl group with one alkyl group and one hydrogen; $R-\overset{\overset{\displaystyle O}{\|}}{C}-H$. (p. 71)

alkanes Hydrocarbons containing only single bonds. (p. 67)

alkenes Hydrocarbons containing C=C double bonds. (p. 68)

alkynes Hydrocarbons containing C≡C triple bonds. (p. 69)

alkyl group A hydrocarbon group with only single bonds; an alkane with one hydrogen removed, to allow bonding to another group; symbolized by R. (p. 68)

amide An acid derivative that contains an amine instead of the hydroxyl group of the acid. (p. 73)

$$R-\overset{\overset{\displaystyle O}{\|}}{C}-NH_2 \qquad R-\overset{\overset{\displaystyle O}{\|}}{C}-NHR' \qquad R-\overset{\overset{\displaystyle O}{\|}}{C}-NR'_2$$

amine An alkylated analogue of ammonia; R—NH$_2$, R$_2$NH, or R$_3$N. (p. 72)

aromatic hydrocarbons (arenes) Hydrocarbons containing a *benzene ring*, a six-membered ring with three double bonds. (p. 69)

bond dipole moment A measure of the polarity of an individual bond in a molecule, defined as $\mu = (4.8 \times d \times \delta)$. μ is the dipole moment in **debyes** (10^{-10} esu-Å), d is the bond length in angstroms, and δ is the effective amount of charge separated, in units of the electronic charge. (p. 57)

carbonyl group The $\overset{\diagdown}{}C$=O functional group, as in a ketone or aldehyde. (p. 71)

carboxyl group The —COOH functional group, as in a carboxylic acid. (p. 71)

carboxylic acid A compound that contains the carboxyl group; $R-\overset{\overset{\displaystyle O}{\|}}{C}-OH$. (p. 71)

cis-trans isomers (geometric isomers) Stereoisomers that differ in their cis-trans arrangement on a ring or a double bond. [The cis isomer has similar groups on the same side, while the trans isomer has similar groups on opposite sides.] (p. 56)

constitutional isomers (structural isomers) Isomers whose atoms are connected differently; they differ in their bonding sequence. (p. 54)

cyano group The —C≡N functional group, as in a nitrile. (p. 73)

dipole–dipole forces Attractive intermolecular forces resulting from the attraction of the positive and negative ends of the molecular dipole moments of polar molecules. (p. 60)

dipole moment See **bond dipole moment** and **molecular dipole moment.**

double bond A bond containing four electrons between two nuclei. One pair of electrons forms a sigma bond, and the other pair forms a pi bond. (p. 43)

ester An acid derivative with an alkyl group replacing the acid proton; $R-\overset{\overset{\displaystyle O}{\|}}{C}-OR'$. (p. 72)

ether A compound with an oxygen bonded between two alkyl groups; R—O—R′. (p. 70)

functional group The reactive, nonalkane part of an organic molecule. (p. 68)

geometric isomers See **cis-trans isomers**. (p. 56)

hybrid atomic orbital A directional orbital formed from a combination of s and p orbitals on the same atom. Two orbitals are formed by sp hybridization, three orbitals by sp^2 hybridization, and four orbitals by sp^3 hybridization. (p. 44)

> sp **hybrid orbitals** give a bond angle of 180° (**linear** geometry).
>
> sp^2 **hybrid orbitals** give bond angles of 120° (**trigonal** geometry).
>
> sp^3 **hybrid orbitals** give bond angles of 109.5° (**tetrahedral** geometry).

hydrocarbons Compounds composed exclusively of carbon and hydrogen.

> **alkanes:** Hydrocarbons containing only single bonds. (p. 67)
>
> **alkenes:** Hydrocarbons containing $C{=}C$ double bonds. (p. 68)
>
> **alkynes:** Hydrocarbons containing $C{\equiv}C$ triple bonds. (p. 69)
>
> **cycloalkanes, cycloalkenes, cycloalkynes:** Alkanes, alkenes, and alkynes in the form of a ring. (p. 68)
>
> **aromatic hydrocarbons:** Hydrocarbons containing a *benzene ring*, a six-membered ring with three double bonds. (p. 69)

benzene

hydrogen bond A particularly strong intermolecular attraction between a nonbonding pair of electrons and an electrophilic O—H or N—H hydrogen. Hydrogen bonds have bond energies of about 5 kcal/mol (21 kJ/mol), compared with about 100 kcal/mol (about 400 kJ/mol) for typical C—H bonds. (p. 62)

hydroxyl group The —OH functional group, as in an alcohol. (p. 70)

isomers Different compounds with the same molecular formula. (p. 54)

> **constitutional isomers (structural isomers)** are connected differently; they differ in their bonding sequence.
>
> **stereoisomers** differ only in how their atoms are oriented in space.
>
> **cis-trans isomers (geometric isomers)** are stereoisomers that differ in their cis-trans arrangement on a ring or a double bond.
>
> **stereochemistry** is the study of the structure and chemistry of stereoisomers.

ketone A carbonyl group with two alkyl groups attached; R—C—R′. (p. 71)

linear combination of atomic orbitals (LCAO) Wave functions can add to each other to produce the wave functions of new orbitals. The number of new orbitals generated equals the original number of orbitals. (p. 39)

London dispersion forces Intermolecular forces resulting from the attraction of correlated temporary dipole moments induced in adjacent molecules. (p. 61)

molecular dipole moment The vector sum of the bond dipole moments (and any nonbonding pairs of electrons) in a molecule; a measure of the polarity of a molecule. (p. 58)

molecular orbital (MO) An orbital formed by overlap of atomic orbitals on different atoms. MOs can be either bonding or antibonding, but only the bonding MOs are filled in most stable molecules. (p. 40)

> A **bonding molecular orbital** places a large amount of electron density in the bonding region between the nuclei. The energy of an electron in a bonding MO is lower than it is in an atomic orbital.
>
> An **antibonding molecular orbital** places most of the electron density outside the bonding region. The energy of an electron in an antibonding MO is higher than it is in an atomic orbital.

nitrile A compound containing a cyano group, —C≡N. (p. 73)

node In an orbital, a region of space with zero electron density. (p. 39)

pi bond (π bond) A bond formed by sideways overlap of two p orbitals. A pi bond has its electron density in two lobes, one above and one below the line joining the nuclei. (p. 43)

sigma bond (σ bond) A bond with most of its electron density centered along the line joining the nuclei; a cylindrically symmetrical bond. Single bonds are normally sigma bonds. (p. 41)

stereochemistry The study of the structure and chemistry of stereoisomers. (p. 55)

stereoisomers Isomers that differ only in how their atoms are oriented in space. (p. 55)

structural isomers (IUPAC term: **constitutional isomers**) Isomers whose atoms are connected differently; they differ in their bonding sequence. (p. 54)

triple bond A bond containing six electrons between two nuclei. One pair of electrons forms a sigma bond and the other two pairs form two pi bonds at right angles to each other. (p. 50)

van der Waals forces The attractive forces between neutral molecules, including dipole–dipole forces and London forces. (p. 61)

> **dipole–dipole forces:** The forces between polar molecules resulting from attraction of their permanent dipole moments.
>
> **London forces:** Intermolecular forces resulting from the attraction of correlated temporary dipole moments induced in adjacent molecules.

wave function (ψ) The mathematical description of an orbital. The square of the wave function (ψ^2) is proportional to the electron density. (p. 38)

Essential Problem-Solving Skills in Chapter 2

1. Draw the structure of a single bond, a double bond, and a triple bond.
2. Predict the hybridization and geometry of the atoms in a molecule.
3. Draw a good three-dimensional representation of a given molecule.
4. Identify constitutional isomers and stereoisomers.
5. Identify polar and nonpolar molecules, and predict which ones can engage in hydrogen bonding.
6. Predict general trends in the boiling points and solubilities of compounds, based on their size, polarity, and hydrogen-bonding ability.
7. Identify the general classes of hydrocarbons, and draw structural formulas for examples.
8. Identify the classes of compounds containing oxygen or nitrogen, and draw structural formulas for examples.

Study Problems

2-23 Define and give examples of the following terms.

(a) bonding MO	(b) antibonding MO	(c) hybrid atomic orbital
(d) sigma bond	(e) pi bond	(f) double bond
(g) triple bond	(h) constitutional isomers	(i) cis-trans isomers
(j) stereoisomers	(k) bond dipole moment	(l) molecular dipole moment
(m) dipole–dipole forces	(n) London forces	(o) hydrogen bonding
(p) miscible liquids	(q) hydrocarbons	(r) alkyl group
(s) functional group		

2-24 Give a definition and an example for each class of organic compounds.

(a) alkane	(b) alkene	(c) alkyne
(d) alcohol	(e) ether	(f) ketone
(g) aldehyde	(h) aromatic hydrocarbon	(i) carboxylic acid
(j) ester	(k) amine	(l) amide
(m) nitrile		

2-25 If the carbon atom in CH_2Cl_2 were flat, there would be two stereoisomers. The carbon atom in CH_2Cl_2 is actually tetrahedral. Make a model of this compound, and determine whether there are any stereoisomers of CH_2Cl_2.

$$H\!-\!\overset{\overset{\displaystyle H}{|}}{\underset{\underset{\displaystyle Cl}{|}}{C}}\!-\!Cl \qquad Cl\!-\!\overset{\overset{\displaystyle H}{|}}{\underset{\underset{\displaystyle H}{|}}{C}}\!-\!Cl$$

2-26 Cyclopropane (C_3H_6, a three-membered ring) is more reactive than most other cycloalkanes.
(a) Draw a Lewis structure for cyclopropane.
(b) Compare the bond angles of the carbon atoms in cyclopropane with those in an acyclic (noncyclic) alkane.
(c) Suggest why cyclopropane is so reactive.

2-27 For each of the following compounds
(1) Give the hybridization and approximate bond angles around each atom except hydrogen.
(2) Draw a three-dimensional diagram, including any lone pairs of electrons.
(a) H_3O^+ **(b)** ^-OH **(c)** NH_2NH_2 **(d)** $(CH_3)_3N$ **(e)** $(CH_3)_4N^+$
(f) CH_3COOH **(g)** $CH_3CH{=}NH$ **(h)** CH_3OH **(i)** CH_2O

2-28 For each of the following compounds
(1) Draw a Lewis structure.
(2) Show the kinds of orbitals that overlap to form each bond.
(3) Give approximate bond angles around each atom except hydrogen.
(a) $(CH_3)_2NH$ **(b)** CH_3CH_2OH **(c)** $CH_3{-}CH{=}N{-}CH_3$
(d) $CH_3{-}CH{=}C(CH_3)_2$ **(e)** $CH_3{-}C{\equiv}C{-}CHO$ **(f)** $H_2N{-}CH_2{-}CN$

(g) $CH_3{-}\overset{\overset{\displaystyle O}{\|}}{C}{-}CH_3$ **(h)**   **(i)**

2-29 In most amines, the nitrogen atom is sp^3 hybridized, with a pyramidal structure and bond angles close to $109°$. In formamide, the nitrogen atom is found to be planar, with bond angles close to $120°$. Explain this surprising finding. (*Hint*: Consider resonance forms and the overlap needed in them.)

$$H{-}\overset{\overset{\displaystyle O}{\|}}{C}{-}\overset{\displaystyle ..}{N}H_2$$
formamide

2-30 Predict the hybridization and geometry of the carbon and nitrogen atoms in the following ions. (*Hint*: Resonance)

(a) $CH_3{-}\overset{\overset{\displaystyle O}{\|}}{C}{-}\overset{\displaystyle -}{\overset{..}{C}}H_2$ **(b)** $H_2N{-}CH{=}CH{-}\overset{+}{C}H_2$

2-31 Draw orbital pictures of the pi bonding in the following compounds.
(a) CH_3COCH_3 **(b)** HCN **(c)** $CH_2{=}CH{-}CH{=}CH{-}CN$ **(d)** $CH_3{-}C{\equiv}C{-}CHO$

2-32 **(a)** Draw the structure of *cis*-$CH_3{-}CH{=}CH{-}CH_2CH_3$ showing the pi bond with its proper geometry.
(b) Circle the six coplanar atoms in this compound.
(c) Draw the trans isomer, and circle the coplanar atoms. Are there still six?
(d) Circle the coplanar atoms in the following structure.

2-33 In 2-pentyne ($CH_3CCCH_2CH_3$) there are four atoms in a straight line. Use dashed lines and wedges to draw a three-dimensional representation of this molecule, and circle the four atoms that are in a straight line.

2-34 Which of the following compounds show cis-trans isomerism? Draw the cis and trans isomers of the ones that do.
(a) $CH_2{=}C(CH_3)_2$ **(b)** $CH_3CH{=}CHCH_3$ **(c)** $CH_3{-}C{\equiv}C{-}CH_3$
(d) cyclopentene,   **(e)** $CH_3{-}CH{=}\underset{\underset{\displaystyle CH_2CH_2CH_3}{|}}{C}{-}CH_2{-}CH_3$

2-35 Give the relationships between the following pairs of structures. The possible relationships are: same compound, cis-trans isomers, constitutional (structural) isomers, not isomers (different molecular formula).
(a) $CH_3CH_2CH_2CH_3$ and $(CH_3)_3CH$ **(b)** $CH_2{=}CH{-}CH_2Cl$ and $CHCl{=}CH{-}CH_3$
(c)   and   **(d)**   and
(e)   and   **(f)**   and

(g) and   **(h)** and

2-36   Sulfur dioxide has a dipole moment of 1.60 D. Carbon dioxide has a dipole moment of zero, even though C—O bonds are more polar than S—O bonds. Explain this apparent contradiction.

2-37   For each of the following compounds
   (1) Draw the Lewis structure.
   (2) Show how the bond dipole moments (and those of any nonbonding pairs of electrons) contribute to the molecular dipole moment.
   (3) Predict whether the compound will have a large (>1 D), small, or zero dipole moment.
   (a) CH_3—CH=N—CH_3 **(b)** CH_3—CN **(c)** CBr_4

   (d) CH_3—$\overset{\overset{\displaystyle O}{\|}}{C}$—$CH_3$ **(e)**   **(f)**   **(g)**

2-38   Diethyl ether and 1-butanol are isomers, and they have similar solubilities in water. Their boiling points are very different, however. Explain why these two compounds have similar solubility properties but dramatically different boiling points.

$$CH_3CH_2-O-CH_2CH_3 \qquad CH_3CH_2CH_2CH_2-OH$$
diethyl ether, bp 35°C   1-butanol, bp 118°C
8.4 mL dissolves in 100 mL H_2O 9.1 mL dissolves in 100 mL H_2O

2-39   *N*-methylpyrrolidine has a boiling point of 81°C, and piperidine has a boiling point of 106°C.
   (a) Explain this large difference (25°C) in boiling point for these two isomers.
   (b) Tetrahydropyran has a boiling point of 88°C, and cyclopentanol has a boiling point of 141°C. These two isomers have a boiling point difference of 53°C. Explain why the two oxygen-containing isomers have a much larger boiling point difference than the two amine isomers.

N-methylpyrrolidine, bp 81°C   piperidine, bp 106°C

tetrahydropyran, bp 88°C   cyclopentanol, bp 141°C

2-40   Which of the following pure compounds can form hydrogen bonds? Which can form hydrogen bonds with water?
   (a) $(CH_3CH_2)_2NH$ **(b)** $(CH_3CH_2)_3N$ **(c)** $CH_3CH_2CH_2OH$
   (d) $(CH_3CH_2CH_2)_2O$ **(e)** $CH_3(CH_2)_3CH_3$ **(f)** CH_2=CH—CH_2CH_3
   (g) CH_3COCH_3 **(h)** CH_3CH_2COOH **(i)** CH_3CH_2CHO

   (j)   **(k)**   **(l)** CH_3—$\overset{\overset{\displaystyle O}{\|}}{C}$—$NH_2$

2-41   Predict which compound in each pair has a higher boiling point. Explain your prediction.
   (a) $CH_3CH_2OCH_3$ or $CH_3CH(OH)CH_3$
   (b) $CH_3CH_2CH_2CH_3$ or $CH_3CH_2CH_2CH_2CH_3$
   (c) $CH_3CH_2CH_2CH_2CH_3$ or $(CH_3)_2CHCH_2CH_3$
   (d) $CH_3CH_2CH_2CH_2CH_3$ or $CH_3CH_2CH_2CH_2CH_2Cl$

2-42   Circle the functional groups in the following structures. State to which class (or classes) of compounds the structure belongs.

   (a)   **(b)** COOH   **(c)** CHO   **(d)**

(e)

(f)

(g) CH₂OCH₃ ... CN

(h) $CH_3-CH-COOCH_3$ with NH₂

2-43   Dimethyl sulfoxide (DMSO) has been used as an anti-inflammatory rub for racehorses. DMSO and acetone seem to have similar structures, but the C=O carbon atom in acetone is planar, while the S=O sulfur atom in DMSO is pyramidal. Draw correct Lewis structures for DMSO and acetone, predict the hybridizations, and explain these observations.

$$CH_3-\overset{O}{\underset{||}{S}}-CH_3 \qquad CH_3-\overset{O}{\underset{||}{C}}-CH_3$$
DMSO   acetone

2-44   Many naturally occurring compounds contain more than one functional group. Identify the functional groups in the following compounds.
(a) Penicillin G is a naturally occurring antibiotic.
(b) Dopamine is the neurotransmitter that is deficient in Parkinson's disease.
(c) Thyroxine is the principal thyroid hormone.
(d) Testosterone is a male sex hormone.

penicillin G

dopamine

thyroxine

testosterone

CHAPTER 3

Structure and Stereochemistry of Alkanes

Whenever possible, we will study organic chemistry using families of compounds to organize the material. The properties and reactions of the compounds in a family are similar, just as their structures are similar. By considering how the structural features of a class of compounds determine their properties, we can predict the properties and reactions of similar new compounds. This organization elevates organic chemistry from a catalog of many individual compounds to a systematic study of a few types of compounds. Organic molecules are classified according to their reactive parts, called *functional groups*. We considered some of the common functional groups in Sections 2-12 through 2-14.

An **alkane** is a hydrocarbon that contains only single bonds. The alkanes are the simplest and least reactive class of organic compounds because they contain only carbon and hydrogen and they have no reactive functional groups. Although alkanes undergo reactions such as cracking and combustion at high temperatures, they are much less reactive than other classes of compounds having functional groups.

3-1
Classification of Hydrocarbons (Review)

We classify hydrocarbons according to their bonding (Section 2-12), as shown in Table 3-1. A hydrocarbon with a carbon–carbon double bond (such as ethylene) is called an *alkene*. If a hydrocarbon has a carbon–carbon triple bond (like acetylene), it is called an *alkyne*. Hydrocarbons with aromatic rings (resembling benzene) are called *aromatic hydrocarbons*.

If a hydrocarbon has no double or triple bonds, it is said to be **saturated** because it has the maximum number of bonded hydrogens. Another way to describe *alkanes*, then, is as the class of **saturated hydrocarbons**.

TABLE 3-1	Summary of Hydrocarbon Classification	
Compound Type	**Functional Group**	**Example**
alkanes	none (no double or triple bonds)	$CH_3-CH_2-CH_3$, propane
alkenes	$C=C$ double bond	$CH_2=CH-CH_3$, propene
alkynes	$-C\equiv C-$ triple bond	$H-C\equiv C-CH_3$, propyne
aromatics	benzene ring	ethylbenzene

Table 3-2 shows the structures and formulas of the first 20 unbranched alkanes. Any isomers of these compounds have the same molecular formulas even though their structures are different. Notice how the molecular formulas increase by two hydrogen atoms each time a carbon atom is added.

The structures of the alkanes in Table 3-2 are purposely written as chains of $-CH_2-$ groups (**methylene groups**), terminated at each end by a hydrogen atom. This is the general formula for the unbranched (straight-chain) alkanes. These alkanes differ only by the number of methylene groups in the chain. If the molecule contains n carbon atoms, it must contain $(2n + 2)$ hydrogen atoms. Figure 3-1 shows this representation of alkane structure and how it leads to formulas of the form C_nH_{2n+2}.

3-2
Molecular Formulas of Alkanes

TABLE 3-2   Formulas and Physical Properties of the Unbranched Alkanes, Called the *n*-Alkanes

Alkane	Number of Carbons	Structure	Formula	Boiling Point (°C)	Melting Point (°C)	Density[a]
methane	1	$H-CH_2-H$	CH_4	−164	−183	0.55
ethane	2	$H-(CH_2)_2-H$	C_2H_6	−89	−183	0.51
propane	3	$H-(CH_2)_3-H$	C_3H_8	−42	−189	0.50
butane	4	$H-(CH_2)_4-H$	C_4H_{10}	0	−138	0.58
pentane	5	$H-(CH_2)_5-H$	C_5H_{12}	36	−130	0.63
hexane	6	$H-(CH_2)_6-H$	C_6H_{14}	69	−95	0.66
heptane	7	$H-(CH_2)_7-H$	C_7H_{16}	98	−91	0.68
octane	8	$H-(CH_2)_8-H$	C_8H_{18}	126	−57	0.70
nonane	9	$H-(CH_2)_9-H$	C_9H_{20}	151	−51	0.72
decane	10	$H-(CH_2)_{10}-H$	$C_{10}H_{22}$	174	−30	0.73
undecane	11	$H-(CH_2)_{11}-H$	$C_{11}H_{24}$	196	−26	0.74
dodecane	12	$H-(CH_2)_{12}-H$	$C_{12}H_{26}$	216	−10	0.75
tridecane	13	$H-(CH_2)_{13}-H$	$C_{13}H_{28}$	235	−5	0.76
tetradecane	14	$H-(CH_2)_{14}-H$	$C_{14}H_{30}$	254	6	0.76
pentadecane	15	$H-(CH_2)_{15}-H$	$C_{15}H_{32}$	271	10	0.77
hexadecane	16	$H-(CH_2)_{16}-H$	$C_{16}H_{34}$	287	18	0.77
heptadecane	17	$H-(CH_2)_{17}-H$	$C_{17}H_{36}$	303	23	0.76
octadecane	18	$H-(CH_2)_{18}-H$	$C_{18}H_{38}$	317	28	0.76
nonadecane	19	$H-(CH_2)_{19}-H$	$C_{19}H_{40}$	330	32	0.78
eicosane	20	$H-(CH_2)_{20}-H$	$C_{20}H_{42}$	343	37	0.79
triacontane	30	$H-(CH_2)_{30}-H$	$C_{30}H_{62}$	>450	66	0.81

[a]Densities are given in g/mL at 20°C, except for methane and ethane, whose densities are given at their boiling points.

▲ **Figure 3-1**
Examples of the general alkane molecular formula, C_nH_{2n+2}.

A series of compounds, like the *n*-alkanes, that differ only by the number of —CH$_2$— groups, is called a homologous series, and the individual members of the series are called **homologs**. For example, butane is a homolog of propane, and both of these are homologs of hexane and decane.

Although we have derived the C$_n$H$_{2n+2}$ formula using the *n*-alkanes, its use is not limited to unbranched molecules. Any isomer of one of these *n*-alkanes has the same molecular formula. Just as butane and pentane follow the C$_n$H$_{2n+2}$ rule, their branched isomers isobutane, isopentane, and neopentane also follow the rule.

PROBLEM 3-1

Using the general molecular formula for alkanes,
(a) Predict the molecular formula of the C$_{25}$ straight-chain alkane.
(b) Predict the molecular formula of 4,6-diethyl-12-(3,5-dimethyloctyl)triacontane, an alkane containing 44 carbon atoms.

3-3
Nomenclature of Alkanes

The names *methane, ethane, propane,* and *butane* have historical roots. From pentane on, alkanes are named using the Greek word for the number of carbon atoms, plus the suffix -*ane* to identify the molecule as an alkane. Table 3-2 gives the names and physical properties of the *n*-alkanes up to 20 carbon atoms.

3-3A   Common Names

If all alkanes had unbranched (straight-chain) structures, their nomenclature would be simple. Most alkanes have structural isomers, however, and we need a way of naming all the different isomers. For example, there are two isomers of formula C$_4$H$_{10}$. The unbranched isomer is simply called *butane* (or *n-butane*, meaning "normal" butane), and the branched isomer is called *isobutane*, meaning an "isomer of butane."

$$CH_3{-}CH_2{-}CH_2{-}CH_3 \qquad CH_3{-}\overset{\overset{\displaystyle CH_3}{|}}{CH}{-}CH_3$$

butane (*n*-butane)   isobutane

The three isomers of C$_5$H$_{12}$ are called *pentane* (or *n-pentane*), *isopentane*, and *neopentane*.

$$CH_3{-}CH_2{-}CH_2{-}CH_2{-}CH_3 \qquad CH_3{-}\overset{\overset{\displaystyle CH_3}{|}}{CH}{-}CH_2{-}CH_3 \qquad CH_3{-}\overset{\overset{\displaystyle CH_3}{|}}{\underset{\underset{\displaystyle CH_3}{|}}{C}}{-}CH_3$$

pentane (*n*-pentane)   isopentane   neopentane

Isobutane, isopentane, and *neopentane* are **common names** or **trivial names,** meaning historical names arising from common usage. Common names cannot easily describe the larger, more complicated molecules having many isomers, however. The number of isomers for any molecular formula grows rapidly as the number of carbon atoms increases. For example, there are five structural isomers of hexane, 18 isomers of octane, and 75 isomers of decane! We need a system of nomenclature that enables us to name complicated molecules without having to memorize hundreds of these historical common names.

3-3B   IUPAC or Systematic Names

A group of chemists representing the countries of the world met in 1892 to devise a system for naming compounds that would be simple to use, require a minimum of memorization, and yet be flexible enough to name even the most complicated organic compounds. This was the first meeting of the group that came to be known as the International Union of Pure and Applied Chemistry, abbreviated **IUPAC**. This international group has developed a detailed system of nomenclature that we call the **IUPAC rules**. The IUPAC rules are accepted throughout the world as the standard method for naming organic compounds. The names that are generated using this system are called **IUPAC names** or **systematic names**.

   The IUPAC system works consistently to name many different families of compounds. We will consider the naming of alkanes in detail, and later extend these rules to other kinds of compounds as we encounter them. The IUPAC system uses the longest chain of carbon atoms as the main chain, which is numbered to give the locations of side chains. We will use four rules to organize this process.

Rule 1: The Main Chain   The first rule of nomenclature gives the base name of the compound:

> Find the longest continuous chain of carbon atoms, and use the name of this chain as the base name of the compound.

For example, the longest chain of carbon atoms in the following compound contains six carbons, so the compound is named as a *hexane* derivative. The longest chain is rarely drawn in a straight line; look carefully to find it.

3-methyl*hexane*

   The groups attached to the main chain are called **substituents** because they are substituted (in place of a hydrogen atom) on the main chain. When there are two longest chains of equal length, use the chain with the greater number of substituents as the main chain. The following compound contains two different seven-carbon chains and is named as a *heptane*. We choose the chain on the right as the main chain because it has more substituents (in red) attached to the chain.

wrong
seven-carbon chain, but only three substituents

correct
seven-carbon chain, four substituents

PROBLEM-SOLVING HINT

When looking for the longest continuous chain (to give the base name), look to find all the different chains of that length. Often, the longest chain with the most substituents is not obvious.

Rule 2: Numbering the Main Chain   To give the locations of the substituents, assign a number to each carbon atom on the main chain.

> Number the longest chain, beginning with the end of the chain nearest a substituent.

We start the numbering from the end nearest a branch so the numbers of the substituted carbons will be as low as possible. In the heptane structure on the right on page 83, numbering from top to bottom gives the first branch at C3 (carbon atom 3), but numbering from bottom to top gives the first branch at C2. Numbering from bottom to top is correct. (If each end had a substituent the same distance in, we would start at the end nearer the second branch point.)

incorrect

correct

3-ethyl-2,4,5-trimethylheptane

Rule 3: Naming Alkyl Groups Next, name the substituent groups.

Name the substituent groups attached to the longest chain as **alkyl groups**. Give the location of each alkyl group by the number of the main-chain carbon atom to which it is attached.

Alkyl groups are named by replacing the -*ane* suffix of the alkane name with -*yl*. *Methane* becomes *methyl*; *ethane* becomes *ethyl*.

CH_4, methane $\qquad\qquad$ CH_3—, methyl group

CH_3—CH_3, ethane $\qquad\qquad$ CH_3—CH_2—, ethyl group

CH_3—CH_2—CH_3, propane $\quad$ CH_3—CH_2—CH_2—, propyl group

The following alkanes show the use of alkyl group nomenclature.

3-methylhexane

3-ethyl-6-methylnonane

Figure 3-2 gives the names of the most common alkyl groups, those having up to four carbon atoms. The *propyl* and *butyl* groups are simply unbranched three- and four-carbon alkyl groups. These groups are often named as "*n*-propyl" and "*n*-butyl" groups, however, to eliminate any question about which kind of propyl group or butyl group is meant.

The simple branched alkyl groups are usually known by common names. The isopropyl and isobutyl groups have a characteristic "iso" $(CH_3)_2CH$ grouping, just as in isobutane.

isopropyl group $\qquad$ isobutyl group $\qquad$ isobutane

The names of the *secondary*-butyl (*sec*-butyl) and *tertiary*-butyl (*tert*-butyl or *t*-butyl) groups are based on the **degree of alkyl substitution** of the carbon atom attached to the main chain. In the *sec*-butyl group, the carbon atom bonded to the main chain is

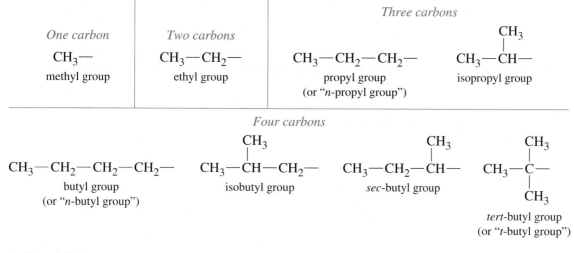

▲ **Figure 3-2**
Some common alkyl groups.

secondary (2°), or bonded to two other carbon atoms. In the *t*-butyl group, it is **tertiary** (3°), or bonded to three other carbon atoms. In both the *n*-butyl group and the isobutyl group, the carbon atoms bonded to the main chain are **primary** (1°), bonded to only one other carbon atom.

$$
\begin{array}{ccc}
\overset{\displaystyle H}{\underset{\displaystyle H}{R-C-}} & \overset{\displaystyle R}{\underset{\displaystyle H}{R-C-}} & \overset{\displaystyle R}{\underset{\displaystyle R}{R-C-}} \\
\text{a } \textit{primary } (1°) \text{ carbon} & \text{a } \textit{secondary } (2°) \text{ carbon} & \text{a } \textit{tertiary } (3°) \text{ carbon}
\end{array}
$$

$$
\begin{array}{ccc}
\overset{\displaystyle H}{\underset{\displaystyle H}{CH_3CH_2CH_2-C-}} & \overset{\displaystyle CH_3}{\underset{\displaystyle H}{CH_3CH_2-C-}} & \overset{\displaystyle CH_3}{\underset{\displaystyle CH_3}{CH_3-C-}} \\
\textit{n}\text{-butyl group } (1°) & \textit{sec}\text{-butyl group } (2°) & \textit{t}\text{-butyl group } (3°)
\end{array}
$$

SOLVED PROBLEM 3-1

Give the structures of 4-isopropyloctane and 5-*t*-butyldecane.

SOLUTION

4-Isopropyloctane has a chain of eight carbons, with an isopropyl group on the fourth carbon.
5-*t*-Butyldecane has a chain of ten carbons, with a *t*-butyl group on the fifth.

$$
\begin{array}{c}
CH_3-CH-CH_3 \\
| \\
CH_3-CH_2-CH_2-CH-CH_2-CH_2-CH_2-CH_3 \\
\text{4-isopropyloctane}
\end{array}
$$

$$
\begin{array}{c}
CH_3 \\
| \\
CH_3-C-CH_3 \\
| \\
CH_3-CH_2-CH_2-CH_2-CH-CH_2-CH_2-CH_2-CH_2-CH_3 \\
\text{5-}\textit{t}\text{-butyldecane}
\end{array}
$$

Haloalkanes can be named just like alkanes, with the the halogen atom treated as a substituent. Halogen substituents are named *fluoro-*, *chloro-*, *bromo-*, and *iodo-*.

$$CH_3-\overset{\overset{\displaystyle Br}{|}}{CH}-CH_2CH_3$$
2-bromobutane

$$CH_3-\overset{\overset{\displaystyle CH_3}{|}}{CH}-\overset{\overset{\displaystyle Cl}{|}}{CH}-CH_2CH_3$$
3-chloro-2-methylpentane

$$CH_3-\overset{\overset{\displaystyle F}{|}}{CH}-CH_2F$$
1,2-difluoropropane

PROBLEM 3-2

Name the following alkanes and haloalkanes.

(a) $CH_3-\overset{\overset{\displaystyle CH_2-CH_3}{|}}{CH}-CH_2-CH_3$

(b) $CH_3-\overset{\overset{\displaystyle Br}{|}}{CH}-\overset{\overset{\displaystyle CH_2CH_3}{|}}{CH}-CH_3$

(c) $CH_3-CH_2-\overset{\overset{\displaystyle CH_3-CH_2}{|}}{CH}-\overset{\overset{\displaystyle CH_2CH(CH_3)_2}{|}}{CH}-CH_2-CH_2-CH_3$

(d) $CH_3-CH_2-CH_2-CH_2-CH_2-CH_2-\overset{\overset{\displaystyle CH_3-CH-CH_3}{\overset{\displaystyle |}{\underset{\displaystyle |}{CH_2}}}}{CH}-\overset{\overset{\displaystyle CH_3}{\underset{\displaystyle |}{CH}}}{CH}-CH_3$

Rule 4: Organizing Multiple Groups   The final rule tells how to organize the names of compounds with more than one substituent.

> When two or more substituents are present, list them in alphabetical order. When two or more of the *same* alkyl substituent are present, use the prefixes *di-*, *tri-*, *tetra-*, etc. to avoid having to name the alkyl group twice.

di- means 2	*penta-* means 5
tri- means 3	*hexa-* means 6
tetra- means 4	

Using this rule, we can construct names for some complicated structures. Let's finish naming the heptane on p. 83. This compound has an ethyl group on C3 and three methyl groups on C2, C4, and C5. The ethyl group is listed alphabetically before the methyl groups.

3-ethyl-2,4,5-trimethylheptane

SOLVED PROBLEM 3-2

Give a systematic (IUPAC) name for the following compound.

SOLUTION

The longest carbon chain contains eight carbon atoms, so this compound is named as an octane. Numbering from left to right gives the first branch on C2; numbering from right to left gives the first branch on C3, so we number from left to right.

There are four methyl groups: two on C2, one on C3, and one on C6. These four groups will be listed as "2,2,3,6-tetramethyl. ..." There is an isopropyl group on C4. Listing the isopropyl group and the methyl groups alphabetically, we have

4-isopropyl-2,2,3,6-tetramethyloctane

PROBLEM-SOLVING HINT

When substituents are alphabetized, *iso-* is used as part of the alkyl group name, but the hyphenated prefixes are not. Thus *isobutyl* is alphabetized with *i*, but *n*-butyl, *t*-butyl, and *sec*-butyl are alphabetized with *b*. The number prefixes *di-, tri-, tetra-*, etc. are ignored in alphabetizing.

SUMMARY   Rules for Naming Alkanes

To name an alkane, we follow four rules:
1. Find the longest continuous chain of carbon atoms, and use this chain as the base name.
2. Number the longest chain, beginning with the end nearest a branch.
3. Name the substituents on the longest chain (as alkyl groups). Give the location of each substituent by the number of the main-chain carbon atom to which it is attached.
4. When two or more substituents are present, list them in alphabetical order. When two or more of the *same* alkyl substituent are present, use the prefixes *di-, tri-, tetra-*, and so on (ignored in alphabetizing) to avoid having to name the alkyl group twice.

PROBLEM 3-3

Write structures for the following compounds.
(a) 3-ethyl-3-methylpentane
(b) 3-methyl-5-propylnonane
(c) 4-*t*-butyl-2-methylheptane
(d) 5-isopropyl-3,3,4-trimethyloctane

PROBLEM-SOLVING HINT

Always compare the total number of carbon atoms in the name with the number in the structure to make sure they match. For example, an *isopropyldimethyloctane* should have 3 + 2 + 8 carbon atoms.

PROBLEM 3-4

Provide IUPAC names for the following compounds.
(a) $(CH_3)_2CHCH_2CH_3$
(b) $CH_3\text{—}C(CH_3)_2\text{—}CH_3$

(c) $CH_3CH_2CH_2\overset{\overset{\displaystyle CH_2CH_3}{|}}{CH}\text{—}CH(CH_3)_2$
(d) $CH_3\text{—}\overset{\overset{\displaystyle CH_3}{|}}{CH}\text{—}CH_2\text{—}\overset{\overset{\displaystyle CH_2CH_3}{|}}{CH}\text{—}CH_3$

(e) $CH_3CH_2\overset{\overset{\displaystyle C(CH_3)_3}{|}}{CH}CHCH_3$
$\quad\quad\quad\quad\underset{\underset{\displaystyle CH(CH_3)_2}{|}}{}$
(f) $(CH_3)_3C\text{—}\overset{\overset{\displaystyle CH_3\text{—}CHCH_2CH_3}{|}}{CH}\text{—}CH_2CH_2CH_3$

PROBLEM 3-5

Give structures and names for
(a) the five isomers of C_6H_{14}
(b) the nine isomers of C_7H_{16}

Complex Substituents   Complex alkyl groups are named by a systematic method using the longest alkyl chain as the base alkyl group. The base alkyl group is numbered beginning with the carbon atom (the "head carbon") bonded to the main chain. The substituents on the base alkyl group are listed with appropriate numbers, and parentheses are used to set off the name of the complex alkyl group. The following examples illustrate the systematic method for naming complex alkyl groups.

$$-\overset{1}{C}H-\overset{2}{C}H-\overset{3}{C}H_3$$

a (1-ethyl-2-methylpropyl) group

$$-\overset{1}{C}-\overset{2}{C}H_2-\overset{3}{C}H-\overset{4}{C}H_3$$

a (1,1,3-trimethylbutyl) group

3-ethyl-5-(1-ethyl-2-methylpropyl)nonane

1,1-dimethyl-3-(1,1,3-trimethylbutyl)cyclooctane

PROBLEM 3-6

Draw the structures of the following groups, and give their more common names.
(a) the (1-methylethyl) group   **(b)** the (2-methylpropyl) group
(c) the (1-methylpropyl) group   **(d)** the (1,1-dimethylethyl) group

PROBLEM 3-7

Draw the structures of the following compounds.
(a) 4-(1-methylethyl)heptane   **(b)** 5-(1,2,2-trimethylpropyl)nonane

PROBLEM 3-8

Without looking at the structures, give a molecular formula for each compound in Problem 3-7. Use the names of the groups to determine the number of carbon atoms, then use the $(2n + 2)$ rule.

3-4
Physical Properties of Alkanes

Oil floats on water. Note how the oil slick (from the leaking Exxon Valdez) spreads across the top of the water. Oil recovery booms, containing nonpolar fibers, are used to soak up and contain the spilled oil. Note how most of the oil slick ends at the oil recovery booms.

Alkanes are used primarily as fuels, solvents, and lubricants. Natural gas, gasoline, kerosene, heating oil, lubricating oil, and paraffin "wax" are all composed primarily of alkanes, with different physical properties resulting from different ranges of molecular weights.

3-4A   Solubilities and Densities of Alkanes

Alkanes are nonpolar, so they dissolve in nonpolar or weakly polar organic solvents. Alkanes are said to be **hydrophobic** ("water hating") because they do not dissolve in water. Alkanes are good lubricants and preservatives for metals because they keep water from reaching the metal surface and causing corrosion.

Densities of the *n*-alkanes are listed in Table 3-2 (page 81). Alkanes have densities around 0.7 g/mL, compared with a density of 1.0 g/mL for water. Because alkanes are less dense than water and insoluble in water, a mixture of an alkane (such as gasoline or oil) and water quickly separates into two phases, with the alkane on top.

3-4B   Boiling Points of Alkanes

Table 3-2 also gives the boiling points and melting points of the unbranched alkanes. The boiling points increase smoothly with increasing numbers of carbon atoms and increasing molecular weights. Larger molecules have larger surface areas, resulting in increased intermolecular van der Waals attractions. These increased attractions must be overcome for vaporization and boiling to occur. Thus, a larger molecule, with greater surface area and greater van der Waals attractions, boils at a higher temperature.

Alkane boiling points. The boiling points of the unbranched alkanes (blue) are compared with those of some branched alkanes (green). Because of their smaller surface areas, branched alkanes have lower boiling points than unbranched alkanes.

A graph of n-alkane boiling points versus the number of carbon atoms (the blue line in Figure 3-3) shows the increase in boiling points with increasing molecular weight. Each additional CH_2 group increases the boiling point by about 30°C up to about ten carbons, and by about 20°C in higher alkanes.

The green line in Figure 3-3 represents the boiling points of some branched alkanes. In general, a branched alkane boils at a lower temperature than the n-alkane with the same number of carbon atoms. This difference in boiling points arises because branched alkanes are more compact, with less surface area for London force interactions.

3-4C   Melting Points of Alkanes

The blue line in Figure 3-4 is a graph of the melting points of the n-alkanes. Like their boiling points, the melting points increase with increasing molecular weight. The melting point graph is not smooth, however. Alkanes with even numbers of carbon atoms pack better into a solid structure, so that higher temperatures are needed to melt them. Alkanes with odd numbers of carbon atoms do not pack as well, and they melt at lower temperatures. The sawtooth-shaped graph of melting points is smoothed by drawing separate lines (green and red) for the alkanes with even and odd numbers of carbon atoms.

Branching of the chain also affects an alkane's melting point. A branched alkane generally melts at a *higher* temperature than the n-alkane with the same number of carbon atoms. Branching of an alkane gives it a more compact three-dimensional structure, which packs more easily into a solid structure and increases the melting point. The boiling points and melting points of three isomers of formula C_6H_{14} show that the boiling

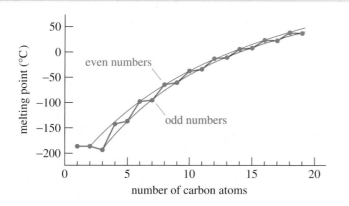

Alkane melting points. The melting point curve for n-alkanes with even numbers of carbon atoms is slightly higher than the curve for alkanes with odd numbers of carbons.

points decrease and the melting points increase as the shape of the molecule becomes more highly branched and compact.

$$CH_3-CH-CH_2-CH_2-CH_3 \quad (CH_3)_2CH-CH(CH_3)_2 \quad CH_3-C(CH_3)_2-CH_2-CH_3$$

| bp 60°C | bp 58°C | bp 50°C |
| mp −154°C | mp −135°C | mp −98°C |

PROBLEM 3-9

List each set of compounds in order of increasing boiling point.
(a) hexane, octane, and decane
(b) octane, $(CH_3)_3C-C(CH_3)_3$, and $CH_3CH_2C(CH_3)_2CH_2CH_2CH_3$

PROBLEM 3-10

Repeat Problem 3-9, listing the compounds in order of increasing melting point.

3-5
Uses and Sources of Alkanes

Clean-burning propane-powered vehicles help to reduce air pollution in urban areas.

Distillation separates alkanes into fractions with similar boiling points. These fractions are suited for different uses based on their physical properties such as volatility and viscosity.

3-5A   Major Uses of Alkanes

C_1–C_2   The first four alkanes (methane, ethane, propane, and butane) are gases at room temperature and atmospheric pressure. Methane and ethane are difficult to liquefy, so they are usually handled as compressed gases. Upon cooling to cryogenic (very low) temperatures, however, methane and ethane become liquids. *Liquefied natural gas*, mostly methane, can be transported in special refrigerated tankers more easily than it can be transported as a compressed gas.

C_3–C_4   Propane and butane are easily liquefied at room temperature under modest pressure. These gases, often obtained along with liquid petroleum, are stored in low-pressure cylinders of *liquefied petroleum gas (LPG)*. Propane and butane are good fuels, both for heating and for internal combustion engines. They burn cleanly, and pollution-control equipment is rarely necessary. In many agricultural areas, propane and butane are more cost-effective tractor fuels than gasoline and diesel fuel. Propane and butane have largely replaced Freons® (see Section 6-3D) as propellants in aerosol cans. Unlike alkanes, the chlorofluorocarbon Freon® propellants are suspected of damaging the earth's protective ozone layer.

C_5–C_8   The next four alkanes are free-flowing, volatile liquids. Isomers of pentane, hexane, heptane, and octane are the primary constituents of gasoline. Their volatility is crucial for this use because the injection system simply squirts a stream of gasoline into the intake air as it rushes through. If gasoline did not evaporate easily, it would reach the cylinder in the form of droplets. Droplets cannot burn as efficiently as a vapor, so the engine would smoke and give low mileage.

In addition to being volatile, gasoline must resist the potentially damaging explosive combustion known as *knocking*. The antiknock properties of gasoline are rated by an **octane number** that is assigned by comparing the gasoline to a mixture of *n*-heptane (which knocks badly) and isooctane (2,2,4-trimethylpentane, which is not prone to knocking). The gasoline being tested is used in a test engine with variable compression ratio. Higher compression ratios induce knocking, so the compression ratio is increased until knocking begins. Tables are available that show the percentage of isooctane in an isooctane/heptane blend that begins to knock at any given compression ratio. The octane number assigned to the gasoline is simply the percentage of isooctane in an isooctane/heptane mixture that begins to knock at that same compression ratio.

$$CH_3CH_2CH_2CH_2CH_2CH_2CH_3$$

n-heptane (0 octane)
prone to knocking

2,2,4-trimethylpentane (100 octane)
"isooctane," resists knocking

C$_9$–C$_{16}$ The nonanes (C$_9$) through about the hexadecanes (C$_{16}$) are higher-boiling liquids that are somewhat viscous. These alkanes are used in kerosene, jet fuel, and diesel fuel. **Kerosene**, the lowest boiling of these fuels, was once widely available but is now harder to find. It is less volatile than gasoline and less prone to forming explosive mixtures. Kerosene was used in kerosene lamps and heaters, which use wicks to allow this heavier fuel to burn. Jet fuel is similar to kerosene, but more highly refined and less odorous.

Diesel fuel is not very volatile, so it does not evaporate in the intake air. In a diesel engine, the fuel is sprayed directly into the cylinder right at the top of the compression stroke. The hot, highly compressed air in the cylinder causes the fuel to burn quickly, swirling and vaporizing as it burns. Some of the alkanes in diesel fuel have fairly high freezing points, and they may solidify in cold weather. This partial solidification causes the diesel fuel to turn into a waxy, semisolid mass. Owners of diesel engines in cold climates often mix a small amount of gasoline with their diesel fuel in the winter. The added gasoline dissolves the frozen alkanes, diluting the slush and allowing it to be pumped to the cylinders.

C$_{16}$ and Up Alkanes with more than 16 carbon atoms are most often used as lubricating and heating oils. These are sometimes called "mineral" oils because they come from petroleum, which was once considered a mineral.

Paraffin "wax" is not a true wax, but a purified mixture of high-molecular-weight alkanes with melting points well above room temperature. The true waxes are long-chain esters, discussed in Chapter 25.

3-5B Alkane Sources; Petroleum Refining

Alkanes are derived mostly from petroleum and petroleum by-products. *Petroleum*, often called *crude oil*, is pumped from wells that reach into pockets containing the remains of prehistoric plants. The principal constituents of crude oil are alkanes, some aromatics, and some undesirable compounds containing sulfur and nitrogen. The composition of petroleum and the amounts of contaminants vary from one source to another, and a refinery must be carefully adjusted to process a particular type of crude oil. Because of their different qualities, different prices are paid for light Arabian crude, West Texas crude, and other classes of crude petroleum.

The first step in refining petroleum is a careful fractional distillation. The products of that distillation are not pure alkanes but mixtures of alkanes with useful ranges of boiling points. Table 3-3 shows the major fractions obtained from distillation of crude petroleum.

The large distillation tower at left is used to separate petroleum into fractions based on their boiling points. The "cat cracker" at right uses catalysts and high temperatures to crack large molecules into smaller ones.

TABLE 3-3	Major Fractions Obtained from Distillation of Crude Petroleum		
Boiling Range (°C)	**Number of Carbons**	**Fraction**	**Use**
under 30°	2–4	petroleum gas	LP gas for heating
30°–180°	4–9	gasoline	motor fuel
160°–230°	8–16	kerosene	heat, jet fuel
200°–320°	10–18	diesel	motor fuel
300°–450°	16–30	heavy oil	heating, lubrication
>300° (vacuum)	>25	petroleum "jelly," paraffin "wax"	
residue	>35	asphalt	

After distillation, **catalytic cracking** converts some of the less valuable fractions to more valuable products. Catalytic cracking involves heating alkanes in the presence of materials that catalyze the cleavage of large molecules into smaller ones. Cracking is often used to convert higher-boiling fractions into mixtures that can be blended with gasoline. When cracking is done in the presence of hydrogen (**hydrocracking**), the product is a mixture of alkanes, free of sulfur and nitrogen impurities. The following reaction shows the catalytic hydrocracking of a molecule of tetradecane into two molecules of heptane.

$$CH_3 - (CH_2)_{12} - CH_3 + H_2 \xrightarrow[\text{SiO}_2 \text{ or Al}_2\text{O}_3 \text{ catalyst}]{\text{heat}} 2 \, CH_3 - (CH_2)_5 - CH_3$$

3-5C   Natural Gas

Natural gas was once treated as a waste product of petroleum production and destroyed by flaring it off. Now natural gas is an equally valuable natural resource, pumped and stored throughout the world. Natural gas is about 70% methane, 10% ethane, and 15% propane, depending on the source of the gas. Small amounts of other hydrocarbons and contaminants are also present. Natural gas is often found above pockets of petroleum, although it is also found in places where there is little, if any, recoverable petroleum. Natural gas is used primarily as a fuel to heat buildings and to generate electricity. It is also important as a starting material for the production of fertilizers.

3-6
Reactions of Alkanes

Alkanes are the least reactive class of organic compounds. Their low reactivity is reflected in another term for alkanes: **paraffins**. The name *paraffin* comes from two Latin terms, *parum*, meaning "too little," and *affinis*, meaning "affinity." Chemists found that alkanes do not react with strong acids or bases or with most other reagents. They attributed this low reactivity to a lack of affinity for other reagents, so they coined the name "paraffins."

Most useful reactions of alkanes take place under energetic or high-temperature conditions. These conditions are inconvenient in a laboratory because they require specialized equipment, and the rate of the reaction is difficult to control. Alkane reactions often form mixtures of products that are difficult to separate. These mixtures may be of commercial importance for an industry, however, where the products may be separated and sold separately. Newer methods of selective functionalization may eventually change this picture. For now, however, the following alkane reactions are rarely seen in laboratory applications, but they are widely used in the chemical industry and even in your home and car.

3-6A   Combustion

Combustion is a rapid oxidation that takes place at high temperatures, converting alkanes to carbon dioxide and water. Little control over the reaction is possible, except for moderating the temperature and controlling the fuel/air ratio to achieve efficient burning.

$$\text{☞} \quad C_nH_{(2n+2)} \quad + \quad \text{excess O}_2 \quad \xrightarrow{\text{heat}} \quad n \, CO_2 \quad + \quad (n+1) \, H_2O$$

Example
$$CH_3CH_2CH_3 \quad + \quad 5 \, O_2 \quad \xrightarrow{\text{heat}} \quad 3 \, CO_2 \quad + \quad 4 \, H_2O$$

Unfortunately, the burning of gasoline and fuel oil pollutes the air and depletes the petroleum resources needed for lubricants and chemical feedstocks. Solar and nuclear heat sources cause less pollution, and they do not deplete these important natural resources. Facilities that use these more environment-friendly heat sources are more expensive than those that rely on the combustion of alkanes, however.

3-6B   Cracking and Hydrocracking

As discussed in Section 3-5B, catalytic **cracking** of large hydrocarbons at high temperatures produces smaller hydrocarbons. The cracking process usually operates under conditions

Combustion is the most common reaction of alkanes. Lightning initiated this fire in a tank containing 3 million gallons of gasoline at the Shell Oil storage facility in Woodbridge, NJ (June 11, 1996).

that give the maximum yields of gasoline. In **hydrocracking**, hydrogen is added to give saturated hydrocarbons; cracking without hydrogen gives mixtures of alkanes and alkenes.

Catalytic hydrocracking

$C_{12}H_{26}$
long-chain alkane

$\xrightarrow[\text{catalyst}]{H_2, \text{ heat}}$

C_5H_{12}

C_7H_{16}
shorter-chain alkanes

Catalytic cracking

$C_{12}H_{26}$
long-chain alkane

$\xrightarrow[\text{catalyst}]{\text{heat}}$

C_5H_{10}

C_7H_{16}
shorter-chain alkanes and alkenes

Incomplete combustion of gasoline and other motor fuels releases significant quantities of volatile organic compounds (VOCs) into the atmosphere. VOCs are composed of short-chained alkanes, alkenes, aromatic compounds, and a variety of other hydrocarbons. VOCs are components of air pollution and contribute to cardiac and respiratory diseases.

3-6C Halogenation

Alkanes can react with halogens (F_2, Cl_2, Br_2, I_2) to form alkyl halides. For example, methane reacts with chlorine (Cl_2) to form chloromethane (methyl chloride), dichloromethane (methylene chloride), trichloromethane (chloroform), and tetrachloromethane (carbon tetrachloride).

$$CH_4 + Cl_2 \xrightarrow{\text{heat or light}} CH_3Cl + CH_2Cl_2 + CHCl_3 + CCl_4 + HCl$$

Heat or light is usually needed to initiate this **halogenation**. Reactions of alkanes with chlorine and bromine proceed at moderate rates and are easily controlled. Reactions with fluorine are often too fast to control, however. Iodine reacts very slowly or not at all. We discuss the halogenation of alkanes in Chapter 4.

Although alkanes are not as reactive as other classes of organic compounds, they have many of the same structural characteristics. We will use simple alkanes as examples to study some of the properties of organic compounds, including the structure of sp^3 hybridized carbon atoms and properties of C—C and C—H single bonds.

3-7
Structure and Conformations of Alkanes

3-7A Structure of Methane

The simplest alkane is *methane*, CH_4. Methane is perfectly tetrahedral, with the 109.5° bond angles predicted for an sp^3 hybrid carbon. Four hydrogen atoms are covalently bonded to the central carbon atom, with bond lengths of 1.09 Å.

3-7B   Conformations of Ethane

Ethane, the two-carbon alkane, is composed of two methyl groups with overlapping sp^3 hybrid orbitals forming a sigma bond between them.

ethane   ethane   ethane

The two methyl groups are not fixed in a single position but are relatively free to rotate about the sigma bond connecting the two carbon atoms. The bond maintains its linear bonding overlap as the carbon atoms turn. The different arrangements formed by rotations about a single bond are called **conformations**, and a specific conformation is called a **conformer** ("**confor**mational **iso**mer"). Pure conformers cannot be isolated in most cases, because the molecules are constantly rotating through all the possible conformations.

linear overlap of sigma bond   overlap maintained

In drawing conformations, we often use **Newman projections**, a way of drawing a molecule looking straight down the bond connecting two carbon atoms (Fig. 3-5). The front carbon atom is represented by three lines (three bonds) coming together in a

▶ **Figure 3-5**
The Newman projection looks straight down the carbon–carbon bond.

viewed from the end   perspective drawing   Newman projection

Newman projections: $\theta = 0°$

Sawhorse structures:

eclipsed, $\theta = 0°$ staggered, $\theta = 60°$ skew, $\theta = $ anything else

◀ **Figure 3-6**
Ethane conformations. The eclipsed conformation has a dihedral angle $\theta = 0°$, and the staggered conformation has $\theta = 60°$. Any other conformation is called a *skew* conformation.

Y shape. The back carbon is represented by a circle with three bonds pointing out from it. Until you become familiar with Newman projections, you should make models and compare your models with the drawings.

An infinite number of conformations are possible for ethane, because the angle between the hydrogen atoms on the front and back carbons can take on an infinite number of values. Figure 3-6 uses Newman projections and sawhorse structures to illustrate some of these ethane conformations. **Sawhorse structures** picture the molecule looking down at an angle toward the carbon–carbon bond. Sawhorse structures can be misleading, depending on how the eye sees them. We will generally use perspective or Newman projections to draw molecular conformations.

Any conformation can be specified by its **dihedral angle** (θ), the angle between the C—H bonds on the front carbon atom and the C—H bonds on the back carbon in the Newman projection. Two of the conformations have special names. The conformation with $\theta = 0°$ is called the **eclipsed conformation** because the Newman projection shows the hydrogen atoms on the back carbon to be hidden (eclipsed) by those on the front carbon. The **staggered conformation**, with $\theta = 60°$, has the hydrogen atoms on the back carbon staggered halfway between the hydrogens on the front carbon. Any other intermediate conformation is called a **skew conformation**.

In a sample of ethane gas at room temperature, the ethane molecules are rotating and their conformations are constantly changing. These conformations are not all equally favored, however. The lowest-energy conformation is the staggered conformation, with the electron clouds in the C—H bonds separated as much as possible. The eclipsed conformation places the C—H electron clouds closer together; it is about 3.0 kcal/mol (12.6 kJ/mol) higher in energy than the staggered conformation. Three kilocalories is not a large amount of energy, and at room temperature, most molecules have enough kinetic energy to overcome this small rotational barrier.

Figure 3-7 shows how the potential energy of ethane changes as the carbon–carbon bond rotates. The *y* axis shows the potential energy relative to the most stable (staggered) conformation. The *x* axis shows the dihedral angle as it increases from 0° (eclipsed) through 60° (staggered) and on through additional eclipsed and staggered conformations as θ continues to increase. As ethane rotates toward an eclipsed conformation, its potential energy increases, and there is resistance to the rotation. This resistance to twisting (torsion) is called **torsional strain**, and the 3.0 kcal/mol (12.6 kJ/mol) of energy required is called **torsional energy**.

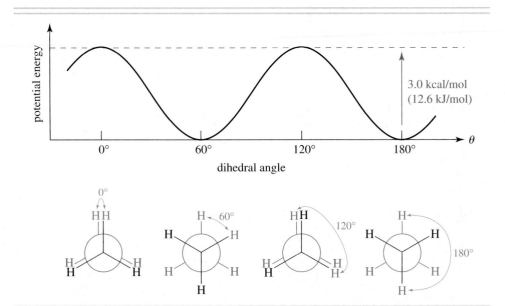

▶ **Figure 3-7**
The torsional energy of ethane is lowest in the staggered conformation. The eclipsed conformation is about 3.0 kcal/mol (12.6 kJ/mol) higher in energy. At room temperature, this barrier is easily overcome and the molecules rotate constantly.

Conformational analysis is the study of the energetics of different conformations. Many reactions depend on a molecule's ability to twist into a particular conformation; conformational analysis can help to predict which conformations are favored, and which reactions are more likely to take place. We will apply conformational analysis to propane and butane first, and later to some interesting cycloalkanes.

3-7C   Conformations of Propane

Propane is the three-carbon alkane, with formula C_3H_8. Figure 3-8 shows a three-dimensional representation of propane and a Newman projection looking down one of the carbon–carbon bonds.

Figure 3-9 shows a graph of the torsional energy of propane as one of the carbon–carbon bonds rotates. The torsional energy of the eclipsed conformation is about 3.3 kcal/mol (13.8 kJ/mol), only 0.3 kcal (1.2 kJ) more than that required for ethane. Apparently, the torsional strain resulting from eclipsing a carbon–hydrogen bond with a carbon–methyl bond is only 0.3 kcal (1.2 kJ) more than the strain of eclipsing two carbon–hydrogen bonds.

PROBLEM-SOLVING **HINT**

A C—H bond eclipsed with another C—H bond contributes 1.0 kcal/mol torsional energy (one-third of eclipsed ethane). A C—H bond eclipsed with a C—CH₃ bond contributes 1.3 kcal/mol.

PROBLEM 3-11

Draw a graph, similar to Figure 3-9, of the torsional strain of 2-methylpropane as it rotates about the bond between C1 and C2. Show the dihedral angle and draw a Newman projection for each staggered and eclipsed conformation.

▶ **Figure 3-8**
Propane is shown here as a perspective drawing and as a Newman projection looking down one of the carbon–carbon bonds.

viewed from the end   perspective drawing   Newman projection

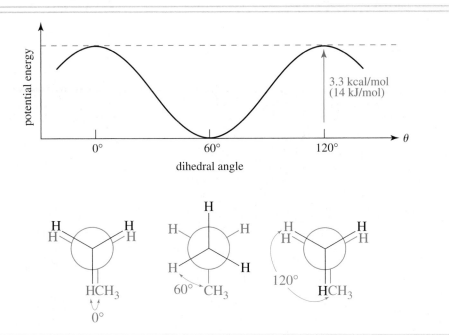

◀ Figure 3-9
Torsional energy of propane. When a C—C bond of propane rotates, the torsional energy varies much like it does in ethane, but with 0.3 kcal/mol (1.2 kJ/mol) of additional torsional energy in the eclipsed conformation.

Butane is the four-carbon alkane, with molecular formula C_4H_{10}. We refer to *n*-butane as a straight-chain alkane, but the chain of carbon atoms is not really straight. The angles between the carbon atoms are close to the tetrahedral angle, about 109.5°. Rotations about any of the carbon–carbon bonds are possible; Figure 3-10 shows Newman projections, looking along the central C2—C3 bond, for four conformations of butane. Construct butane with your molecular models, and sight down the C2—C3 bond. Notice that we have defined the dihedral angle θ as the angle between the two end methyl groups.

3-8
Conformations of Butane

totally eclipsed (0°) gauche (60°) eclipsed (120°) anti (180°)

▲ Figure 3-10
Butane conformations. Rotations about the center bond in butane give different molecular shapes. Three of these conformations have specific names.

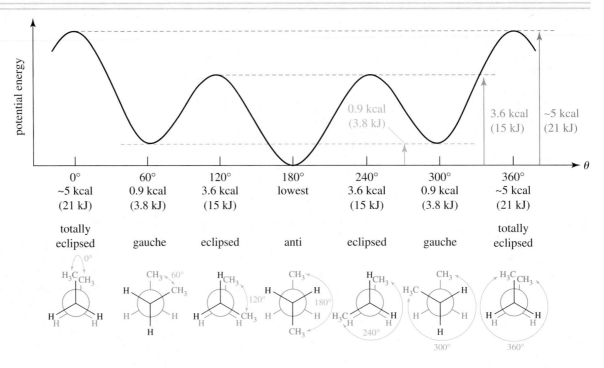

▲ **Figure 3-11**

Torsional energy of butane. The anti conformation is lowest in energy, and the totally eclipsed conformation is highest in energy.

Three of the conformations shown in Figure 3-10 are given special names. When the methyl groups are pointed in the same direction ($\theta = 0°$), they eclipse each other. This conformation is called **totally eclipsed**, to distinguish it from the other eclipsed conformations like the one at $\theta = 120°$. At $\theta = 60°$, the butane molecule is staggered and the methyl groups are toward the left and right of each other. This 60° conformation is called **gauche** (pronounced gōsh), a French word meaning "left" or "awkward."

Another staggered conformation occurs at $\theta = 180°$, with the methyl groups pointing in opposite directions. This conformation is called **anti** because the methyl groups are "opposed."

3-8A   Torsional Energy of Butane

A graph of the relative torsional energies of the butane conformations is shown in Figure 3-11. All the staggered conformations (anti and gauche) are lower in energy than any of the eclipsed conformations. The anti conformation is lowest in energy because it places the bulky methyl groups as far apart as possible. The gauche conformations, with the methyl groups separated by just 60°, are 0.9 kcal (3.8 kJ) higher in energy than the anti conformation because the methyl groups are close enough that their electron clouds begin to repel each other. Use your molecular models to compare the crowding of the methyl groups in these conformations.

3-8B   Steric Hindrance

The totally eclipsed conformation is about 1.4 kcal (5.9 kJ) higher in energy than the other eclipsed conformations, because it forces the two end methyl groups so close together that their electron clouds experience a strong repulsion. This kind of interference between two bulky groups is called **steric strain** or **steric hindrance**. The following structure shows the interference between the methyl groups in the totally eclipsed conformation.

Totally eclipsed conformation of butane

Rotating the totally eclipsed conformation 60° to a gauche conformation releases most, but not all, of this steric strain. The gauche conformation is still 0.9 kcal/mol (3.8 kJ/mol) higher in energy than the most stable anti conformation.

What we have learned about the conformations of butane can be applied to other alkanes. We can predict that carbon–carbon single bonds will assume staggered conformations whenever possible to avoid eclipsing of the groups attached to them. Among the staggered conformations, the anti conformation is preferred because it has the lowest torsional energy. We must remember, however, that there is enough thermal energy present at room temperature for the molecules to rotate rapidly among all the different conformations. The relative stabilities are important because more molecules will be found in the more stable conformations than in the less stable ones.

PROBLEM-SOLVING **HINT**

A $C\!-\!CH_3$ bond eclipsed with another $C\!-\!CH_3$ bond contributes about 3 kcal/mol torsional energy. (Totally eclipsed butane is about 5 kcal/mol, with 1.0 for each of the two interactions between $C\!-\!H$ bonds, leaving 3 kcal for the methyl–methyl eclipsing.)

PROBLEM 3-12

Draw a graph, similar to Figure 3-11, of the torsional energy of 2-methylbutane as it rotates about the C2—C3 bond.

3-9
Conformations of Higher Alkanes

The higher alkanes resemble butane in their preference for anti and gauche conformations about the carbon–carbon bonds. The lowest-energy conformation for any straight-chain alkane is the one with all the internal carbon–carbon bonds in its anti conformations. These anti conformations give the chain a zigzag shape. At room temperature, the internal carbon–carbon bonds undergo rotation, and many molecules contain gauche conformations. Gauche conformations make kinks in the zigzag structure. Nevertheless, we frequently draw alkane chains in a zigzag structure to represent the most stable arrangement.

anti conformation

gauche conformation

octane, all anti conformation

PROBLEM 3-13

Draw a perspective representation of the most stable conformation of 3-methylhexane.

3-10
Cycloalkanes

Many organic compounds are **cyclic**: They contain rings of atoms. The carbohydrates we eat are cyclic, the nucleotides that make up our DNA and RNA are cyclic, and the antibiotics we use to treat diseases are cyclic. In this chapter, we use the cycloalkanes as examples to discuss the properties and stability of cyclic compounds.

Cycloalkanes are alkanes that contain rings of carbon atoms. Simple cycloalkanes are named like acyclic (noncyclic) alkanes, with the prefix *cyclo-* indicating the presence of a ring. For example, the cycloalkane with four carbon atoms in a ring is called *cyclobutane*. The cycloalkane with seven carbon atoms in a ring is *cycloheptane*. Line–angle formulas are often used for drawing the rings of cycloalkanes (Figure 3-12).

3-10A General Molecular Formulas of Cycloalkanes

Simple cycloalkanes are rings of CH_2 groups (methylene groups). Each one has exactly twice as many hydrogen atoms as carbon atoms, giving the general molecular formula C_nH_{2n}. This general formula has two fewer hydrogen atoms than the $(2n + 2)$ formula for an acyclic alkane because a ring has no ends, and no hydrogens are needed to cap off the ends of the chain.

3-10B Physical Properties of Cycloalkanes

Most cycloalkanes resemble the acyclic (noncyclic), open-chain alkanes in their physical properties and in their chemistry. They are nonpolar, relatively inert compounds with boiling points and melting points that depend on their molecular weights. The cycloalkanes are held in a more compact cyclic shape, so their physical properties are similar to those of the compact, branched alkanes. The physical properties of some common cycloalkanes are listed in Table 3-4.

Cyclopropane was once used as general anesthetic because its vapors, like those of other simple alkanes and cycloalkanes, cause sleepiness and a loss of consciousness. After inhalation into the lungs, cyclopropane goes into the blood. Due to its nonpolar nature, it rapidly leaves the blood and passes through the nonpolar membranes surrounding the central nervous system, where it produces anesthesia. Cyclopropane is no longer used as an anesthetic because it is highly flammable (like ether) and can cause explosions when mixed with air.

TABLE 3-4 Physical Properties of Some Simple Cycloalkanes

Cycloalkane	Formula	Boiling Point (°C)	Melting Point (°C)	Density
cyclopropane	C_3H_6	−33	−128	0.72
cyclobutane	C_4H_8	−12	−50	0.75
cyclopentane	C_5H_{10}	49	−94	0.75
cyclohexane	C_6H_{12}	81	7	0.78
cycloheptane	C_7H_{14}	118	−12	0.81
cyclooctane	C_8H_{16}	148	14	0.83

▲ Figure 3-12
Structures of some cycloalkanes.

3-10C   Nomenclature of Cycloalkanes

Cycloalkanes are named much like acyclic alkanes. Substituted cycloalkanes use the cycloalkane for the base name, with the alkyl groups named as substituents. If there is just one substituent, no numbering is needed.

methylcyclopentane   *t*-butylcycloheptane   (1,2-dimethylpropyl)cyclohexane

If there are two or more substituents on the ring, the ring carbons are numbered to give the lowest possible numbers for the substituted carbons. The numbering begins with one of the substituted ring carbons and continues in the direction that gives the lowest possible numbers to the other substituents. In the name, the substituents are listed in alphabetical order. When the numbering could begin with either of two alkyl groups (as in a disubstituted cycloalkane), begin with the one that is alphabetically first.

1-ethyl-2-methylcyclobutane   1,1,3-trimethylcyclopentane   1,1-diethyl-4-isopropylcyclohexane

When the acyclic portion of the molecule contains more carbon atoms than the cyclic portion (or when it contains an important functional group), the cyclic portion is sometimes named as a cycloalkyl substituent.

4-cyclopropyl-3-methyloctane   5-cyclobutyl-1-pentyne   cyclopentylcyclohexane

PROBLEM 3-14

Give IUPAC names for the following compounds.

(a)   (b)   (c)

PROBLEM 3-15

Draw the structure and give the molecular formula for each of the following compounds.
(a) cyclododecane   (b) propylcyclohexane
(c) cyclopropylcyclopentane   (d) 3-ethyl-1,1-dimethylcyclohexane

3-11
cis-trans Isomerism in Cycloalkanes

Open-chain alkanes undergo rotations about their carbon–carbon single bonds, so they are free to assume any of an infinite number of conformations. Alkenes have rigid double bonds that prevent rotation, giving rise to cis and trans isomers with different orientations of the groups on the double bond (Section 2-8). Cycloalkanes are similar to alkenes in this respect. A cycloalkane has two distinct faces. If two substituents point toward the same face, they are **cis**. If they point toward opposite faces, they are **trans**. These **geometric isomers** cannot interconvert without breaking and re-forming bonds.

Figure 3-13 compares the cis-trans isomers of 2-butene with those of 1,2-dimethyl-cyclopentane. Make models of these compounds to convince yourself that *cis*- and *trans*-1,2-dimethylcyclopentane cannot interconvert by simple rotations about the bonds.

PROBLEM 3-16

Which of the following cycloalkanes are capable of geometric (cis-trans) isomerism? Draw the cis and trans isomers.
(a) 3-ethyl-1,1-dimethylcyclohexane   (b) 1,4-dimethylcyclohexane
(c) 1-ethyl-3-methylcyclopentane   (d) 1-cyclopropyl-2-methylcyclohexane

PROBLEM 3-17

Give IUPAC names for the following cycloalkanes.

▶ **Figure 3-13**
Cis-trans isomerism in cycloalkanes. Like alkenes, cycloalkane rings are restricted from free rotation. Two substituents on a cycloalkane must be either on the same side (cis) or on opposite sides (trans) of the ring.

3-12
Stabilities of Cycloalkanes; Ring Strain

Although all the simple cycloalkanes (up to about C_{20}) have been synthesized, the most common rings contain five or six carbon atoms. We will study the stabilities and conformations of these rings in detail because they help to determine the properties of many important organic compounds.

Why are five-membered and six-membered rings more common than the other sizes? Adolf von Baeyer first attempted to explain the relative stabilities of cyclic molecules in the late nineteenth century, and he was awarded a Nobel Prize for this work in 1905. Baeyer reasoned that the carbon atoms in acyclic alkanes have bond angles of 109.5°. (We now explain this bond angle by the tetrahedral geometry of the sp^3 hybridized carbon atoms.)

If a cycloalkane requires bond angles other than 109.5°, the orbitals of its carbon–carbon bonds cannot achieve optimum overlap, and the cycloalkane must have

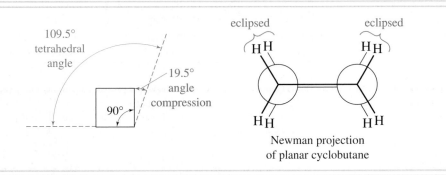

The ring strain of a planar cyclobutane results from two factors: angle strain from the compressing of the bond angles to 90° rather than the tetrahedral angle of 109.5°, and torsional strain from eclipsing of the bonds.

some **angle strain** (sometimes called **Baeyer strain**) associated with it. Figure 3-14 shows that a planar cyclobutane, with 90° bond angles, is expected to have significant angle strain.

In addition to this angle strain, the Newman projection in Figure 3-14 shows that the bonds are eclipsed, resembling the *totally eclipsed* conformation of butane (Section 3-7). This eclipsing of bonds gives rise to torsional strain. Together, the angle strain and the torsional strain add to give what we call the **ring strain** of the cyclic compound. The amount of ring strain depends primarily on the size of the ring.

Before we discuss the ring strain of different cycloalkanes, we need to consider how ring strain is measured. In theory, we should measure the total amount of energy in the cyclic compound and subtract the amount of energy in a similar, strain-free reference compound. The difference should be the amount of extra energy due to ring strain in the cyclic compound. These measurements are commonly made using heats of combustion.

3-12A   Heats of Combustion

The **heat of combustion** is the amount of heat released when a compound is burned with an excess of oxygen in a sealed container called a *bomb calorimeter*. If the compound has extra energy as a result of ring strain, that extra energy is released in the combustion. The heat of combustion is usually measured by the temperature rise in the water bath surrounding the "bomb."

A cycloalkane can be represented by the molecular formula $(CH_2)_n$, so the general reaction in the bomb calorimeter is:

$$\text{cycloalkane, } (CH_2)_n + \tfrac{3}{2}nO_2 \longrightarrow nCO_2 + nH_2O + n(\text{energy per } CH_2)$$

heat of combustion

The molar heat of combustion of cyclohexane is nearly twice that of cyclopropane, simply because cyclohexane contains twice as many methylene (CH_2) groups per mole. To compare the relative stabilities of cycloalkanes, we divide the heat of combustion by the number of methylene (CH_2) groups. The result is the energy per CH_2 group. These normalized energies allow us to compare the relative amounts of ring strain (per methylene group) in the cycloalkanes.

Table 3-5 shows the heats of combustion for some simple cycloalkanes. The reference value of 157.4 kcal (659 kJ) per mole of CH_2 groups comes from an unstrained long-chain alkane. The values show large amounts of ring strain in cyclopropane and cyclobutane. Cyclopentane, cycloheptane, and cyclooctane have much smaller amounts of ring strain, and cyclohexane has no ring strain at all. We will discuss several of these rings in detail to explain this pattern of ring strain.

TABLE 3-5   Heats of Combustion (per Mole) for Some Simple Cycloalkanes

Ring Size	Cycloalkane	Molar Heat of Combustion	Heat of Combustion per CH_2 Group	Ring Strain per CH_2 Group	Total Ring Strain
3	cyclopropane	499.8 kcal	166.6 kcal	9.2 kcal	27.6 kcal (115 kJ)
4	cyclobutane	655.9 kcal	164.0 kcal	6.6 kcal	26.4 kcal (110 kJ)
5	cyclopentane	793.5 kcal	158.7 kcal	1.3 kcal	6.5 kcal (27 kJ)
6	cyclohexane	944.5 kcal	157.4 kcal	0.0 kcal	0.0 kcal (0.0 kJ)
7	cycloheptane	1108.3 kcal	158.3 kcal	0.9 kcal	6.3 kcal (26 kJ)
8	cyclooctane	1268.9 kcal	158.6 kcal	1.2 kcal	9.6 kcal (40 kJ)
reference: long-chain alkane			157.4 kcal	0.0 kcal	0.0 kcal (0.0 kJ)

3-12B   Cyclopropane

The information in Table 3-5 shows that cyclopropane bears more ring strain per methylene group than any of the other cycloalkanes. Two factors contribute to this large ring strain. First is the angle strain required to compress the bond angles from the tetrahedral angle of 109.5° to the 60° angles of cyclopropane. The bonding overlap of the carbon–carbon sp^3 orbitals is weakened when the bond angles differ so much from the tetrahedral angle. The sp^3 orbitals cannot point directly toward each other, and they overlap at an angle to form weaker "bent bonds" (Fig. 3-15).

Torsional strain is the second factor in cyclopropane's large ring strain. The three-membered ring is planar, and all the bonds are eclipsed. A Newman projection of one of the carbon–carbon bonds (Fig. 3-16) shows that the conformation resembles the

▶ **Figure 3-15**
Angle strain in cyclopropane. The bond angles are compressed to 60° from the usual 109.5° bond angle of sp^3 hybridized carbon atoms. This severe angle strain leads to nonlinear overlap of the sp^3 orbitals and "bent bonds."

▲ **Figure 3-16**
Torsional strain in cyclopropane. All the carbon–carbon bonds are eclipsed, generating torsional strain that contributes to the total ring strain.

totally eclipsed conformation of butane. The torsional strain in cyclopropane is not as great as its angle strain, but it helps to account for the large total ring strain.

Cyclopropane is generally more reactive than other alkanes. Reactions that open the cyclopropane ring release 27.6 kcal (115 kJ) of ring strain, which provides an additional driving force for these reactions.

PROBLEM 3-18

The heat of combustion of *cis*-1,2-dimethylcyclopropane is larger than that of the trans isomer. Which isomer is more stable? Use drawings to explain this difference in stability.

3-12C Cyclobutane

The total ring strain in cyclobutane is almost as great as that in cyclopropane, but is distributed over four carbon atoms. If cyclobutane were perfectly planar and square, it would have 90° bond angles. A planar geometry requires eclipsing of all the bonds, however, as in cyclopropane. To reduce this torsional strain, cyclobutane actually assumes a slightly folded form with bond angles of 88°. These smaller bond angles require slightly more angle strain than 90° angles, but the relief of some of the torsional strain appears to compensate for a small increase in angle strain (Figure 3-17).

PROBLEM 3-19

trans-1,2-Dimethylcyclobutane is more stable than *cis*-1,2-dimethylcyclobutane, but *cis*-1,3-dimethylcyclobutane is more stable than *trans*-1,3-dimethylcyclobutane. Use drawings to explain these observations.

slightly folded conformation

Newman projection of one bond

▲ **Figure 3-17**
The conformation of cyclobutane is slightly folded. Folding gives partial relief from the eclipsing of bonds, as shown in the Newman projection. Compare this actual structure with the hypothetical planar structure in Figure 3-14.

3-12D Cyclopentane

If cyclopentane had the shape of a planar, regular pentagon, its bond angles would be 108°, close to the tetrahedral angle of 109.5°. A planar structure would require all the bonds to be eclipsed, however. Cyclopentane actually assumes a slightly puckered "envelope" conformation that reduces the eclipsing and lowers the torsional strain (Figure 3-18). This puckered shape is not fixed, but undulates by the thermal up-and-down motion of the five methylene groups. The "flap" of the envelope seems to move around the ring as the molecule undulates.

The conformation of cyclopentane is important because ribose and deoxyribose, the sugar components of RNA and DNA, respectively, assume cyclopentane-like ring conformations. These conformations are crucial to the properties and reactions of RNA and DNA.

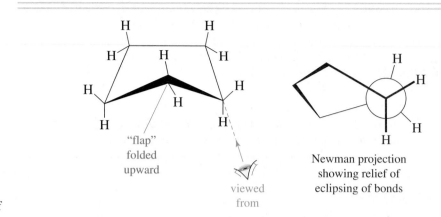

▶ **Figure 3-18**
The conformation of cyclopentane is slightly folded, like the shape of an envelope. This puckered conformation reduces the eclipsing of adjacent CH_2 groups.

3-13
Cyclohexane Conformations

We will cover the conformations of cyclohexane in more detail than other cycloalkanes because cyclohexane ring systems are particularly common. Carbohydrates, steroids, plant products, pesticides, and many other important compounds contain cyclohexane-like rings whose conformations and stereochemistry are critically important to their reactivity. The abundance of cyclohexane rings in nature is probably due to both their stability and the selectivity offered by their predictable conformations.

The combustion data (Table 3-5) show that cyclohexane has *no* ring strain. Cyclohexane must have bond angles that are near the tetrahedral angle (no angle strain) and also have no eclipsing of bonds (no torsional strain). A planar, regular hexagon would have bond angles of 120° rather than 109.5°, implying some angle strain. A planar ring would also have torsional strain because the bonds on adjacent CH_2 groups would be eclipsed. Therefore, the cyclohexane ring cannot be planar.

3-13A   Chair and Boat Conformations

Cyclohexane achieves tetrahedral bond angles and staggered conformations by assuming a puckered conformation. The most stable conformation is the **chair conformation** shown in Figure 3-19. Build a molecular model of cyclohexane, and compare its shape with the drawings in Figure 3-19. In the chair conformation, the angles between

chair conformation   viewed along the "seat" bonds

Newman projection

▶ **Figure 3-19**
The chair conformation of cyclohexane has one methylene group puckered upward and another puckered downward. Viewed from the Newman projection, the chair has no eclipsing of the carbon–carbon bonds. The bond angles are 109.5°.

boat conformation   symmetrical boat

Newman projection   "twist" boat

◀ **Figure 3-20**
In the symmetrical boat conformation of cyclohexane, eclipsing of bonds results in torsional strain. In the actual molecule, the boat is skewed to give the twist boat, a conformation with less eclipsing of bonds and less interference between the two flagpole hydrogens.

the carbon–carbon bonds are all 109.5°. The Newman projection looking down the "seat" bonds shows all the bonds in staggered conformations.

The **boat conformation** of cyclohexane (Figure 3-20) also has bond angles of 109.5° and avoids angle strain. The boat conformation resembles the chair conformation except that the "footrest" methylene group is folded upward. The boat conformation suffers from torsional strain, however, because there is eclipsing of bonds.

This eclipsing forces two of the hydrogens on the ends of the "boat" to interfere with each other. These hydrogens are called **flagpole hydrogens** because they point upward from the ends of the boat like two flagpoles. Figure 3-20 uses a Newman projection of the carbon–carbon bonds along the sides of the boat to show this eclipsing.

A cyclohexane molecule in the boat conformation actually exists as a slightly skewed **twist boat conformation**, also shown in Figure 3-20. If you assemble your molecular model in the boat conformation and twist it slightly, the flagpole hydrogens move away from each other and the eclipsing of the bonds is reduced. Even though the twist boat is lower in energy than the symmetrical boat, it is still about 5.5 kcal/mol (23 kJ/mol) higher in energy than the chair conformation. When someone refers to the "boat conformation," the twist boat is often intended.

At any instant, most of the molecules in a cyclohexane sample are in chair conformations. The energy barrier between the boat and chair is sufficiently low, however, that the conformations interconvert many times each second. The interconversion from the chair to the boat takes place by the footrest of the chair flipping upward and forming the boat. The highest-energy point in this process is the conformation where the footrest is planar with the sides of the molecule. This unstable arrangement is called the **half-chair conformation**. Figure 3-21 shows how the energy of cyclohexane varies as it interconverts between the boat and chair forms.

The conformations of biological molecules are critical for their activities. For example, steroids will fit into their receptors in only one conformation. The correct fit activates the receptor, resulting in a biological activity.

3-13B   Axial and Equatorial Positions

If we could freeze cyclohexane in the chair conformation, we would see that there are two different kinds of carbon–hydrogen bonds. Six of the bonds (one on each carbon atom) are directed up and down, parallel to the axis of the ring. These are called **axial bonds**. The other six bonds point out from the ring, along the "equator" of the ring.

▶ **Figure 3-21**
Conformational energy of cyclohexane. The chair conformation is most stable, followed by the twist boat. To convert between these two conformations, the molecule must pass through the unstable half-chair conformation.

These are called **equatorial bonds**. The axial bonds and hydrogens are shown in red in Figure 3-22, and the equatorial bonds and hydrogens are shown in green.

Each carbon atom in cyclohexane is bonded to two hydrogen atoms, one directed upward and one downward. As the carbon atoms are numbered in Figure 3-22, C1 has an axial bond upward and an equatorial bond downward. C2 has an equatorial bond upward and an axial bond downward. The pattern alternates. The odd-numbered carbon atoms have axial bonds up and equatorial bonds down, like C1. The even-numbered carbons have equatorial bonds up and axial bonds down, like C2. This pattern of alternating axial and equatorial bonds is helpful for predicting the conformations of substituted cyclohexanes, as we see in Sections 3-13 and 3-14.

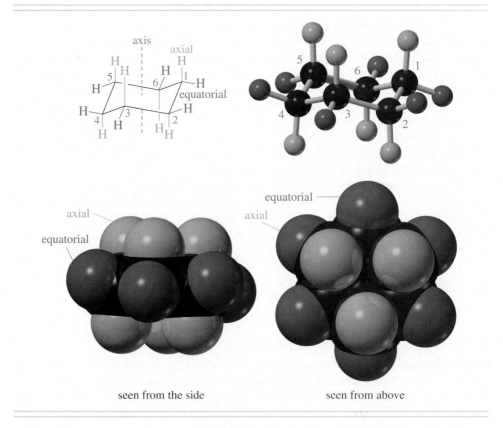

seen from the side   seen from above

▶ **Figure 3-22**
Axial bonds are directed vertically, parallel to the axis of the ring. Equatorial bonds are directed outward, toward the equator of the ring. As they are numbered here, the odd-numbered carbons have their *upward* bonds axial and their *downward* bonds equatorial. The even-numbered carbons have their *downward* bonds axial and their *upward* bonds equatorial.

PROBLEM-SOLVING STRATEGY
Drawing Chair Conformations

Drawing realistic pictures of cyclohexane conformations is not difficult, but certain rules should be followed to show the actual positions and angles of the substituents on the ring. Make a cyclohexane ring with your models, put it in a chair conformation, and use it to follow along with this discussion. When you hold your model at the angle that corresponds to a drawing, the angles of the bonds in the model should correspond to the angles in the drawing.

To draw the carbon–carbon bond framework, first draw two parallel lines, slightly slanted and slightly offset. The atoms at the ends of these bonds lie in a plane, and they define what will be the "armrests" of our chair.

Draw the headrest and footrest carbons, and draw the lines connecting them to the armrests. The two lines connecting the headrest carbon should be parallel to the two lines connecting the footrest.

Notice that the carbon–carbon bond framework uses lines with only three different slopes, labeled *a, b,* and *c*. Compare this drawing with your model, and notice the pairs of carbon–carbon bonds with three distinct slopes.

We can draw the chair with the headrest to the left and the footrest to the right, or vice versa. Practice drawing it both ways.

Now fill in the axial and equatorial bonds. The axial bonds are drawn vertically, either up or down. When a vertex of the chair points upward, its axial bond also points upward. If the vertex points downward, its axial bond points downward. C1 is a downward-pointing vertex, and its axial bond also points downward. C2 points upward, and its axial bond points upward.

The equatorial bonds take more thought. Each carbon atom is represented by a vertex formed by two lines (bonds), having two of the possible slopes *a, b,* and *c*. Each equatorial bond should have the third slope: the slope that is *not* represented by the two lines forming the vertex.

Look at your model as you add the equatorial bonds. The vertex C1 is formed by lines of slopes b and c, so its equatorial bond should have slope a. The equatorial bond at C2 should have slope b, and so on. Notice the W- and M-shaped patterns that result when these bonds are drawn correctly.

PROBLEM 3-20

The cyclohexane chair drawn above has the headrest to the left and the footrest to the right. Draw a cyclohexane chair with its axial and equatorial bonds, having the headrest to the right and the footrest to the left.

PROBLEM 3-21

Draw 1,2,3,4,5,6-hexamethylcyclohexane with all the methyl groups
(a) in axial positions   **(b)** in equatorial positions

If your cyclohexane rings look awkward or slanted when using the analytical approach shown above, then try the artistic approach:* Draw a wide M, and draw a wide W below it, displaced about half a bond length to one side or the other. Connect the second atoms and the fourth atoms to give the cyclohexane ring with four equatorial bonds.

W displaced to the right   W displaced to the left

The other two equatorial bonds are drawn parallel to the ring connections. The axial bonds are then drawn vertically.

3-14
Conformations of Monosubstituted Cyclohexanes

A substituent on a cyclohexane ring (in the chair conformation) can occupy either an axial or an equatorial position. In many cases, the reactivity of the substituent depends on whether its position is axial or equatorial. The two possible chair conformations for methylcyclohexane are shown in Figure 3-23. These conformations are in equilibrium because they interconvert at room temperature. The twist boat serves as an intermediate in this **chair–chair interconversion**, sometimes called a "ring-flip." Place different-colored atoms in the axial and equatorial positions of your cyclohexane model, and notice that the chair–chair interconversion changes axial to equatorial and equatorial to axial.

The two chair conformations of methylcyclohexane interconvert at room temperature, so the one that is lower in energy predominates. Careful measurements have shown that the chair with the methyl group in an equatorial position is the most stable conformation. It is about 1.7 kcal/mol (7.1 kJ/mol) lower in energy than the conformation with the methyl group in an axial position. Both of these chair conformations are lower in energy than any boat conformation. We can show how the 1.7 kcal energy difference between the axial and equatorial positions arises by examining molecular models and Newman projections of the two conformations. First, make a model of methylcyclohexane and use it to follow this discussion.

*See V. Dragojlovic, *J. Chem. Educ.* 2001, *78*, 923.

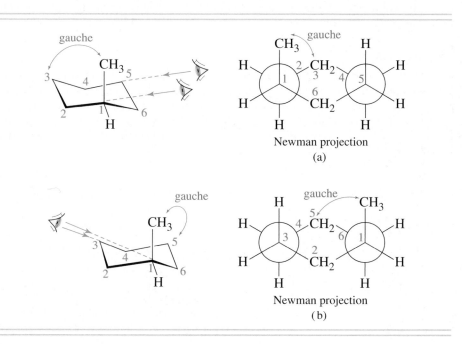

▲ Figure 3-23
Chair–chair interconversion of methylcyclohexane. The methyl group is axial in one conformation, and equatorial in the other.

Consider a Newman projection looking along the armrest bonds of the conformation with the methyl group axial (Figure 3-24a): The methyl group is on C1, and we are looking from C1 toward C2. There is a 60° angle between the bond to the methyl group and the bond from C2 to C3, and the methyl substituent and C3 are in a gauche relationship. In our analysis of torsional strain in butane, we saw that a gauche interaction raises the energy of a conformation by 0.9 kcal/mol (3.8 kJ/mol) relative to the anti conformation. This axial methyl group is also gauche to C5, as you will see if you look along the C1—C6 bond in your model. Figure 3-24b shows this second gauche relationship.

Newman projection
(a)

Newman projection
(b)

◀ Figure 3-24
(a) When the methyl substituent is in an axial position on C1, it is gauche to C3. (b) The axial methyl group on C1 is also gauche to C5 of the ring.

▶ **Figure 3-25**
Looking down the C1—C2 bond of the equatorial conformation. Notice that the methyl group is anti to C3.

The Newman projection for the conformation with the methyl group equatorial shows that the methyl group has an anti relationship to both C3 and C5. Figure 3-25 shows the Newman projection along the C1—C2 bond, with the anti relationship of the methyl group to C3.

PROBLEM 3-22

Draw a Newman projection, similar to Figure 3-25, down the C1—C6 bond in the equatorial conformation of methylcyclohexane. Show that the equatorial methyl group is also anti to C5. (Using your models will help.)

The axial methylcyclohexane conformation has two gauche interactions, each representing about 0.9 kcal (3.8 kJ) of additional energy. The equatorial methyl group has no gauche interactions. Therefore, we predict that the axial conformation is higher in energy by 1.8 kcal (7.6 kJ) per mole, in good agreement with the experimental value of 1.7 kcal (7.1 kJ) per mole. Figure 3-26 shows that the gauche relationship of the axial methyl group with C3 and C5 places the methyl hydrogens close to the axial hydrogens on these carbons, and their electron clouds begin to interfere. This form of steric hindrance is called a **1,3-diaxial interaction** because it involves substituents on the carbon atom that would be numbered C3 if the carbon bearing the methyl group was numbered C1. These 1,3-diaxial interactions are not present in the equatorial conformation.

A larger group usually has a larger energy difference between the axial and equatorial positions, because the 1,3-diaxial interaction shown in Figure 3-26 is stronger for larger groups. Table 3-6 shows the energy differences between the axial and equatorial positions for several alkyl groups and functional groups. The axial position is higher in energy in each case.

PROBLEM 3-23

Table 3-6 shows that the axial–equatorial energy difference for methyl, ethyl, and isopropyl groups increases gradually: 1.7, 1.8, and 2.1 kcal/mol (7.1, 7.5, and 8.8 kJ/mol). The *t*-butyl group jumps to an energy difference of 5.4 kcal/mol (23 kJ/mol), over twice the value for the isopropyl group. Draw pictures of the axial conformations of isopropylcyclohexane and *t*-butylcyclohexane, and explain why the *t*-butyl substituent experiences such a large increase in axial energy over the isopropyl group.

▶ **Figure 3-26**
The axial substituent interferes with the axial hydrogens on C3 and C5. This interference is called a 1,3-diaxial interaction.

TABLE 3-6 Energy Differences Between the Axial and Equatorial Conformations of Monosubstituted Cyclohexanes

	X	E(axial)—E(equatorial)	
		(kcal/mol)	(kJ/mol)
	—F	0.2	0.8
	—CN	0.2	0.8
	—Cl	0.5	2.1
	—Br	0.6	2.5
	—OH	1.0	4.1
	—COOH	1.4	5.9
	—CH_3	1.7	7.1
	—CH_2CH_3	1.8	7.5
	—$CH(CH_3)_2$	2.1	8.8
	—$C(CH_3)_3$	5.4	23

axial ⇌ equatorial

PROBLEM 3-24

Draw the most stable conformation of
(a) ethylcyclohexane   (b) isopropylcyclohexane   (c) *t*-butylcyclohexane

The steric interference between substituents in axial positions is particularly severe when there are large substituents on two carbon atoms that bear a 1,3-diaxial relationship (*cis* on C1 and C3, or C1 and C5). Figure 3-27 shows the large 1,3-diaxial interaction between the two methyl groups in the unfavorable conformation of *cis*-1,3-dimethylcyclohexane. The 1,3-diaxial interference is relieved when the molecule flips to the diequatorial conformation. Use your models to compare the diaxial and diequatorial forms of *cis*-1,3-dimethylcyclohexane.

Either of the chair conformations of *trans*-1,3-dimethylcyclohexane has one methyl group in an axial position and one in an equatorial position. These conformations have equal energies, and they are present in equal amounts.

3-15
Conformations of Disubstituted Cyclohexanes

Chair conformations of trans-*1,3-dimethylcyclohexane*

same energy

Now we can compare the relative stabilities of the cis and trans isomers of 1,3-dimethylcyclohexane. The most stable conformation of the cis isomer has both methyl groups in equatorial positions. Either conformation of the trans isomer places one

1,3-diaxial interaction

diaxial—very unfavorable   diequatorial—much more stable

◀ **Figure 3-27**

Two chair conformations are possible for *cis*-1,3-dimethylcyclohexane. One has both methyl groups in axial positions, with a 1,3-diaxial interaction between them. The more stable conformation has both methyl groups in equatorial positions.

methyl group in an axial position. The trans isomer is therefore higher in energy than the cis isomer by about 1.7 kcal/mol, the energy difference between axial and equatorial methyl groups. Remember that the cis and trans isomers cannot interconvert, and there is no equilibrium between these isomers.

PROBLEM-SOLVING HINT

Ring-flips change the axial or equatorial positioning of groups, but they cannot change their cis-trans relationships. Converting cis into trans would require breaking and re-forming bonds.

PROBLEM-SOLVING HINT

If you number the carbons in a cyclohexane, the odd-numbered carbons are similar, as are the even-numbered carbons. If the odd-numbered carbons all have their *up* bond axial and their *down* bond equatorial, the even-numbered carbons will all have their *down* bond axial and their *up* bond equatorial. For example, *cis*-1,3 (both up, both odd) will be both axial or both equatorial. *cis*-1,2 (both up, one odd, one even) will be one axial, one equatorial. This tip allows you to predict the answers before you draw them.

SOLVED PROBLEM 3-3

(a) Draw both chair conformations of *cis*-1,2-dimethylcyclohexane, and determine which conformer is more stable.
(b) Repeat for the trans isomer.
(c) Predict which isomer (cis or trans) is more stable.

SOLUTION

(a) There are two chair conformations possible for the cis isomer, and these two conformations interconvert at room temperature. Each of these conformations places one methyl group axial and one equatorial, giving them the same energy.

same energy

(b) There are two chair conformations of the trans isomer that interconvert at room temperature. One of these has both methyl groups axial, and the other has both equatorial. The diequatorial conformation is more stable because neither methyl group occupies the more hindered axial position.

higher energy (diaxial)   lower energy (diequatorial)

(c) The trans isomer is more stable. The most stable conformation of the trans isomer is diequatorial and therefore about 1.7 kcal/mol (7.1 kJ/mol) lower in energy than either conformation of the cis isomer, each having one methyl axial and one equatorial. Remember that cis and trans are distinct isomers and cannot interconvert.

PROBLEM 3-25

(a) Draw both chair conformations of *cis*-1,4-dimethylcyclohexane, and determine which conformer is more stable.
(b) Repeat for the trans isomer.
(c) Predict which isomer (cis or trans) is more stable.

PROBLEM 3-26

Use your results from Problem 3-25 to complete the following table. Each entry shows the positions of two groups arranged as shown. For example, two groups that are trans on adjacent carbons (*trans*-1,2) must be both equatorial (e,e) or both axial (a,a).

Positions	cis	trans
1,2	(e,a) or (a,e)	(e,e) or (a,a)
1,3		
1,4		

3-15A Substituents of Different Sizes

In many substituted cyclohexanes, the substituents are of different sizes. As shown in Table 3-6 (p. 113), the energy difference between the axial and equatorial positions for a larger group is greater than that for a smaller group. In general, if both groups cannot be equatorial, the most stable conformation has the larger group equatorial and the smaller group axial.

SOLVED PROBLEM 3-4

Draw the most stable conformation of *trans*-1-ethyl-3-methylcyclohexane.

SOLUTION

First, we draw the two conformations.

Both of these conformations require one group to be axial while the other is equatorial. The ethyl group is bulkier than the methyl group, so the conformation with the ethyl group equatorial is more stable. These chair conformations are in equilibrium at room temperature, and the one with the equatorial ethyl group predominates.

PROBLEM 3-27

Draw the two chair conformations of each of the following substituted cyclohexanes. In each case, label the more stable conformation.
(a) *cis*-1-ethyl-2-methylcyclohexane (b) *trans*-1-ethyl-2-methylcyclohexane
(c) *cis*-1-ethyl-4-isopropylcyclohexane (d) *trans*-1-ethyl-4-methylcyclohexane

PROBLEM-SOLVING STRATEGY
Recognizing cis and trans Isomers

Some students find it difficult to look at a chair conformation and tell whether a disubstituted cyclo-hexane is the cis isomer or the trans isomer. In the following drawing, the two methyl groups appear to be oriented in similar directions. They are actually trans but are often mistaken for cis.

trans-1,2-dimethylcyclohexane

This ambiguity is eliminated by recognizing that each of the ring carbons has two avail-able bonds, one upward and one downward. In this drawing, the methyl group on C1 is on the downward bond, and the methyl on C2 is on the upward bond. Because one is down and one is up, their relationship is trans. A cis relationship would require both groups to be upward or both to be downward.

PROBLEM 3-28

Name the following compounds. Remember that two *up* bonds are cis; two *down* bonds are cis; one *up* bond and one *down* bond are trans.

3-15B   Extremely Bulky Groups

Some groups, such as *tertiary*-butyl groups, are so bulky that they are extremely hindered in axial positions. Regardless of the other groups present, cyclohexanes with *t*-butyl substituents are most stable when the *t*-butyl group is in an equatorial position. The following figure shows the severe steric interactions in a chair conformation with a *t*-butyl group axial.

If two *t*-butyl groups are attached to the ring, both of them are much less hindered in equatorial positions. When neither chair conformation allows both bulky groups to be equatorial, they may force the ring into a twist boat conformation. For example, *cis*-1,4-di-*t*-butylcyclohexane (Figure 3-28) is most stable in the twist boat.

PROBLEM 3-29

Draw the most stable conformation of
(a) *cis*-1-*t*-butyl-3-ethylcyclohexane
(b) *trans*-1-*t*-butyl-2-methylcyclohexane
(c) *trans*-1-*t*-butyl-3-(1,1-dimethylpropyl)cyclohexane

▶ **Figure 3-28**
The most stable conformation of *cis*-1,4-di-*t*-butylcyclohexane is a twist boat. Either of the chair conformations requires one of the bulky *t*-butyl groups to occupy an axial position.

t-butyl group moves out of the axial position   twist boat

Two or more rings can be joined into *bicyclic* or *polycyclic* systems. There are three ways that two rings may be joined. **Fused rings** are most common, sharing two adjacent carbon atoms and the bond between them. **Bridged rings** are also common, sharing two nonadjacent carbon atoms (the **bridgehead carbons**) and one or more carbon atoms (the **bridge**) between them. **Spirocyclic compounds**, in which the two rings share only one carbon atom, are relatively rare.

3-16
Bicyclic Molecules

fused bicyclic

bicyclo[4.4.0]decane
(decalin)

bridged bicyclic

bicyclo[2.2.1]heptane
(norbornane)

spirocyclic

spiro[4.4]nonane

3-16A   Nomenclature of Bicyclic Alkanes

The name of a bicyclic compound is based on the name of the alkane having the same number of carbons as there are in the ring system. This name follows the prefix *bicyclo* and a set of brackets enclosing three numbers. The following examples contain eight carbon atoms and are named bicyclo[4.2.0]octane and bicyclo[3.2.1]octane, respectively.

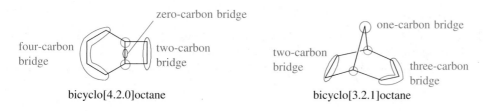

zero-carbon bridge

four-carbon bridge

two-carbon bridge

bicyclo[4.2.0]octane

one-carbon bridge

two-carbon bridge

three-carbon bridge

bicyclo[3.2.1]octane

Bicyclic molecules are found in many natural product structures. A derivative of bicyclo[3.2.1]octane, in which nitrogen replaces the carbon at the one-carbon bridge, is found in cocaine, an addictive drug that produces euphoria.

All fused and bridged bicyclic systems have three bridges connecting the two bridgehead atoms (red circles) where the rings connect. The numbers in the brackets give the number of carbon atoms in each of the three bridges connecting the bridgehead carbons, in order of decreasing size.

PROBLEM 3-30

Name the following compounds.

(a)   **(b)**   **(c)**   **(d)**

3-16B   *cis-* and *trans-*Decalin

Decalin (bicyclo[4.4.0]decane) is the most common example of a fused-ring system. Two geometric isomers of decalin exist, as shown in Figure 3-29. One has the rings fused using two cis bonds, while the other is fused using two trans bonds. You should make a model of decalin to follow this discussion.

If we consider the left ring in the drawing of *cis*-decalin, the bonds to the right ring are both directed downward (and the attached hydrogens are directed upward). These bonds are therefore cis, and this is a cis ring fusion. In *trans*-decalin, one of the bonds to the right ring is directed upward and the other downward. These bonds are trans, and this is a trans ring fusion. The six-membered rings in both isomers assume chair conformations, as shown in Figure 3-29.

cis-decalin   *trans*-decalin

cis-decalin   *trans*-decalin

▶ **Figure 3-29**
cis-Decalin has a ring fusion where the second ring is attached by two cis bonds. *trans*-Decalin is fused using two trans bonds. (The other hydrogens are omitted for clarity.)

The conformation of *cis*-decalin is somewhat flexible, while the trans isomer is quite rigid. If one of the rings in the trans isomer did a chair–chair interconversion, the bonds to the second ring would both become axial and would be directed 180° apart. This is an impossible conformation, and it prevents any chair–chair interconversion in *trans*-decalin.

> **PROBLEM 3-31**
>
> Use your models to do a chair–chair interconversion on each ring of the conformation of *cis*-decalin shown in Figure 3-29. Draw the conformation that results.

Chapter 3 Glossary

acyclic Not cyclic. (p. 100)

alkane A hydrocarbon having only single bonds; a **saturated hydrocarbon**; general formula: C_2H_{2n+2}. (p. 80)

alkyl group The group of atoms remaining after a hydrogen atom is removed from an alkane; an alkane-like substituent. Symbolized by R. (p. 84)

angle strain or Baeyer strain The strain associated with compressing bond angles to smaller (or larger) angles. (p. 103)

anti conformation A conformation with a 180° dihedral angle between the largest groups. Usually the lowest-energy conformation. (p. 98)

aromatic hydrocarbon A hydrocarbon having a benzene-like aromatic ring. (p. 69)

axial bond One of six bonds (three up and three down) on the chair conformation of the cyclohexane ring that are parallel to the "axis" of the ring. (p. 107)

bridged bicyclic compound A compound containing two rings joined at nonadjacent carbon atoms. (p. 117)

bridged bicyclic systems (bridgeheads circled)

bridgehead carbons The carbon atoms shared by two or more rings. Three chains of carbon atoms (bridges) connect the bridgeheads. (p. 117)

chair–chair interconversion (ring-flip) The process of one chair conformation of a cyclohexane flipping into another one, with all the axial and equatorial positions reversed. The boat conformation is an intermediate for the chair–chair interconversion. (p. 110) chair (methyls axial)   boat   chair (methyls equatorial)

cis-trans isomers (geometric isomers) Stereoisomers that differ only with respect to their cis or trans arrangement on a ring or double bond. (p. 102)

  cis: Having two similar groups directed toward the same face of a ring or double bond.
  trans: Having two similar groups directed toward opposite faces of a ring or double bond.

cis-2-butene   *trans*-2-butene   *cis*-1,2-dimethylcyclopentane   *trans*-1,2-dimethylcyclopentane

combustion A rapid oxidation at high temperatures in the presence of air or oxygen. (p. 92)

common names The names that have developed historically, generally with a specific name for each compound; also called **trivial names**. (p. 82)

conformational analysis The study of the energetics of different conformations. (p. 96)

conformations or conformers Structures that are related by rotations about single bonds. In most cases, conformations interconvert at room temperature, and they are not true isomers. (p. 94)

totally eclipsed conformation   gauche conformation   anti conformation

conformations of cyclohexanes (p. 106)

chair   half-chair   boat   twist boat

  chair conformation: The most stable conformation of cyclohexane, with one part puckered upward and another part puckered downward.
  boat conformation: The less stable puckered conformation of cyclohexane, with both parts puckered upward. The most stable boat is actually the **twist boat** (or simply **twist**) conformation. Twisting minimizes torsional strain and steric strain.
  flagpole hydrogens: Two hydrogens (blue) in the boat conformation point upward like flagpoles. The twist boat reduces the steric repulsion of the flagpole hydrogens.
  half-chair conformation: The unstable conformation halfway between the chair conformation and the boat conformation. Part of the ring is flat in the half-chair conformation.

constitutional isomers (structural isomers) Isomers whose atoms are connected differently; they differ in their bonding sequence. (p. 54)

cracking Heating large alkanes to cleave them into smaller molecules. (p. 92)

  catalytic cracking: Cracking in the presence of a catalyst.
  hydrocraking: Catalytic cracking in the presence of hydrogen to give mixtures of alkanes.

cyclic Containing a ring of atoms. (p. 100)

cycloalkane An alkane containing a ring of carbon atoms; general formula: C_nH_{2n}. (p. 100)

degree of alkyl substitution The number of alkyl groups bonded to a carbon atom in a compound or in an alkyl group. (p. 84)

| primary (1°) | secondary (2°) | tertiary (3°) | quaternary (4°) |
| carbon atom | carbon atom | carbon atom | carbon atom |

1,3-diaxial interaction The strong steric hindrance between two axial groups on cyclohexane carbons with one carbon between them. (p. 112)

dihedral angle (θ) (see also **conformations**) The angle between two specified groups in a Newman projection. (p. 95)

eclipsed conformation Any conformation with bonds directly lined up with each other, one behind the other, in the Newman projection. The conformation with $\theta = 0°$ is an eclipsed conformation. See also **staggered conformation**. (p. 95)

equatorial bond One of the six bonds (three down and three up) on the cyclohexane ring that are directed out toward the "equator" of the ring. The equatorial positions are shown in green in the following drawing. (p. 108)

axis

axial bonds in red;
equatorial bonds in green

fused-ring system A molecule in which two or more rings share two adjacent carbon atoms. (p. 117)

fused-ring systems

gauche conformation A conformation with a 60° dihedral angle between the largest groups. (p. 98)

geometric isomers See **cis-trans isomers**, the IUPAC term. (p. 102)

halogenation The reaction of alkanes with halogens, in the presence of heat or light, to give products with halogen atoms substituted for hydrogen atoms. (p. 93)

$$R{-}H + X_2 \xrightarrow{\text{heat or light}} R{-}X + HX \qquad X = F, Cl, Br$$

heat of combustion The heat given off when a mole of a compound is burned with excess oxygen to give CO_2 and H_2O in a *bomb calorimeter*. A measure of the energy content of a molecule. (p. 103)

homologs Two compounds that differ only by one or more $-CH_2-$ groups. (p. 82)

hydrophilic Attracted to water, soluble in water.

hydrophobic Repelled by water; insoluble in water. (p. 88)

IUPAC names The systematic names that follow the rules adopted by the International Union of Pure and Applied Chemistry. (p. 83)

kerosene A thin, volatile oil distilled from petroleum, with a boiling range higher than that of gasoline and lower than that of diesel fuel. Kerosene was once used in lanterns and heaters, but now most of this petroleum fraction is further refined for use as jet fuel. (p. 91)

methine group The $-CH-$ group.

methylene group The $-CH_2-$ group. (p. 81)

methyl group The $-CH_3$ group. (p. 84)

***n*-alkane, normal alkane, or straight-chain alkane** An alkane with all its carbon atoms in a single chain, with no branching or alkyl substituents. (p. 82)

Newman projections A way of drawing the conformations of a molecule by looking straight down the bond connecting two carbon atoms. (p. 94)

a Newman projection of butane in the anti conformation

octane number A rating of the antiknock properties of a gasoline blend. Its octane number is the percentage of isooctane (2,2,4-trimethylpentane) in an isooctane/heptane blend that begins to knock at the same compression ratio as the gasoline being tested. (p. 90)

paraffins Another term for alkanes. (p. 92)

ring strain The extra strain associated with the cyclic structure of a compound, as compared with a similar acyclic compound. Composed of angle strain and torsional strain. (p. 103)

    angle strain or Baeyer strain: The strain associated with compressing bond angles to smaller (or larger) angles.

    torsional strain: The strain associated with eclipsing of bonds in the ring.

saturated Having no double or triple bonds. (p. 80)

sawhorse structures A way of picturing conformations by looking down at an angle toward the carbon–carbon bond. (p. 95)

skew conformation Any conformation that is not precisely staggered or eclipsed. (p. 95)

spirocyclic compounds Bicyclic compounds in which the two rings share only one carbon atom. (p. 117)

staggered conformation Any conformation with the bonds equally spaced in the Newman projection. The conformation with $\theta = 60°$ is a staggered conformation. (p. 95)

eclipsed conformation of ethane   staggered conformation of ethane

steric hindrance or **steric strain** The interference between two bulky groups that are so close together that their electron clouds experience a repulsion. (p. 98)

substituent A side chain or appendage on the main chain. (p. 83)

systematic names Same as IUPAC names, the names that follow the rules adopted by the International Union of Pure and Applied Chemistry. (p. 83)

torsional energy or **conformational energy** The energy required to twist a bond into a specific conformation. (p. 95)

torsional strain The resistance to twisting about a bond. (p. 95)

totally eclipsed conformation A conformation with a 0° dihedral angle between the largest groups. Usually the highest-energy conformation. (p. 98)

Essential Problem-Solving Skills in Chapter 3

1. Explain and predict trends in physical properties of alkanes.
2. Correctly name alkanes, cycloalkanes, and bicyclic alkanes.
3. Given the name of an alkane, draw the structure and give the molecular formula.
4. Compare the energies of alkane conformations and predict the most stable conformation.
5. Compare the energies of cycloalkanes, and explain ring strain.
6. Identify and draw cis and trans stereoisomers of cycloalkanes.
7. Draw accurate cyclohexane conformations, and predict the most stable conformations of substituted cyclohexanes.

Study Problems

3-32   Define and give an example for each term.

(a) alkane	**(b)** alkene	**(c)** alkyne
(d) saturated	**(e)** hydrophobic	**(f)** aromatic
(g) hydrophilic	**(h)** *n*-alkane	**(i)** methylene group
(j) methyl group	**(k)** common name	**(l)** systematic name
(m) conformers	**(n)** eclipsed	**(o)** Newman projection
(p) staggered	**(q)** gauche	**(r)** anti conformation
(s) an acyclic alkane	**(t)** cis-trans isomers on a ring	**(u)** chair conformation
(v) boat conformation	**(w)** twist boat	**(x)** half-chair conformation
(y) axial position	**(z)** equatorial position	**(A)** catalytic cracking
(B) chair–chair interconversion	**(C)** fused ring system	**(D)** bridged bicyclic compound
(E) bridgehead carbon atoms	**(F)** combustion	

3-33   Which of the following Lewis structures represent the same compound? Which ones represent different compounds?

3-34   Draw the structure that corresponds with each name.
  (a) 3-ethyloctane
  (b) 4-isopropyldecane
  (c) *sec*-butylcycloheptane
  (d) 2,3-dimethyl-4-propylnonane
  (e) 2,2,4,4-tetramethylhexane
  (f) *trans*-1,3-diethylcyclopentane
  (g) *cis*-1-ethyl-4-methylcyclohexane
  (h) isobutylcyclopentane
  (i) *t*-butylcyclohexane
  (j) pentylcyclohexane
  (k) cyclobutylcyclohexane
  (l) *cis*-1-bromo-3-chlorocyclohexane

3-35   Each of the following descriptions applies to more than one alkane. In each case, draw and name two structures that match the description.
  (a) a methylheptane
  (b) a diethyldecane
  (c) a *cis*-diethylcycloheptane
  (d) a *trans*-dimethylcyclopentane
  (e) a (2,3-dimethylpentyl)cycloalkane
  (f) a bicyclodecane

3-36   Write structures for a homologous series of alcohols (R—OH) having from one to six carbons.

3-37   Give the IUPAC names of the following alkanes.
  (a) $CH_3C(CH_3)_2CH(CH_2CH_3)CH_2CH_2CH(CH_3)_2$
  (b) CH_3CH_2—CH—CH_2CH_2—CH—CH_3
                     | |
               CH_3CHCH_3 CH_3CHCH_3

  (c)
  (d)
  (e)
  (f) CH_2CH_3 ... $CH_2CH_2CH_3$
  (g) $C(CH_2CH_3)_3$
  (h) CH_2CH_3 ... $CH(CH_3)_2$

3-38   Draw and name eight isomers of molecular formula C_8H_{18}.

3-39   The following names are all incorrect or incomplete, but they represent real structures. Draw each structure and name it correctly.
  (a) 2-ethylpentane
  (b) 3-isopropylhexane
  (c) 5-chloro-4-methylhexane
  (d) 2-dimethylbutane
  (e) 2-cyclohexylbutane
  (f) 2,3-diethylcyclopentane

3-40   In each pair of compounds, which compound has the higher boiling point? Explain your reasoning.
  (a) octane or 2,2,3-trimethylpentane
  (b) heptane or 2-methylnonane
  (c) 2,2,5-trimethylhexane or nonane

3-41   There are eight different five-carbon alkyl groups.
  (a) Draw them.
  (b) Give them systematic names.
  (c) In each case, label the degree of substitution (primary, secondary, or tertiary) of the head carbon atom, bonded to the main chain.

3-42   Use a Newman projection, about the indicated bond, to draw the most stable conformer for each compound.
  (a) 3-methylpentane about the C2—C3 bond
  (b) 3,3-dimethylhexane about the C3—C4 bond

3-43   **(a)** Draw the two chair conformations of *cis*-1,3-dimethylcyclohexane, and label all the positions as axial or equatorial.
  (b) Label the higher energy conformation and the lower energy conformation.
  (c) The energy difference in these two conformations has been measured to be about 5.4 kcal (23 kJ) per mole. How much of this energy difference is due to the torsional energy of gauche relationships?
  (d) How much energy is due to the additional steric strain of the 1,3-diaxial interaction?

3-44   Draw the two chair conformations of each compound, and label the substituents as axial and equatorial. In each case, determine which conformation is more stable.
  (a) *cis*-1-ethyl-2-isopropylcyclohexane
  (b) *trans*-1-ethyl-2-isopropylcyclohexane
  (c) *cis*-1-ethyl-3-methylcyclohexane
  (d) *trans*-1-ethyl-3-methylcyclohexane
  (e) *cis*-1-ethyl-4-methylcyclohexane

3-45   Using what you know about the conformational energetics of substituted cyclohexanes, predict which of the two decalin isomers is more stable. Explain your reasoning.

3-46   The most stable form of the common sugar glucose contains a six-membered ring in the chair conformation with all the substituents equatorial. Draw this most stable conformation of glucose.

HO O CH_2OH

HO OH

OH

glucose

The Study of Chemical Reactions

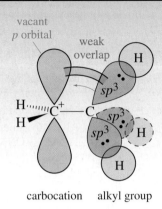

vacant
p orbital
weak overlap

carbocation   alkyl group

4-1
Introduction

The most interesting and useful aspect of organic chemistry is the study of reactions. We cannot remember thousands of specific organic reactions, but we can organize the reactions into logical groups based on how the reactions work and what intermediates are involved. We begin our study by considering the halogenation of alkanes, a relatively simple reaction that takes place in the gas phase, without a solvent to complicate the reaction. In practice, alkanes are so unreactive that they are rarely used as starting materials for most organic syntheses. We begin with them because we have already studied their structure and properties, and their reactions are relatively uncomplicated. Once we have used alkanes to develop the tools for studying reactions, we will apply those tools to a variety of more useful reactions.

A reaction equation, with the reactants on the left and the products on the right, is only the first step in our study of a reaction. If we truly want to understand a reaction, we must also know the **mechanism**, the step-by-step pathway from reactants to products. To know how well the reaction goes to products, we study its **thermodynamics**, the energetics of the reaction at equilibrium. The amounts of reactants and products present at equilibrium depend on their relative stabilities.

Even though the equilibrium may favor the formation of a product, the reaction may not take place at a useful rate. To use a reaction in a realistic time period (and to keep the reaction from becoming violent), we study its **kinetics**, the variation of reaction rates with different conditions and concentrations of reagents. Understanding the reaction's kinetics helps us to propose reaction mechanisms that are consistent with the properties we observe.

4-2
Chlorination of Methane

The reaction of methane with chlorine produces a mixture of chlorinated products, whose composition depends on the amount of chlorine added and also on the reaction conditions. Either light or heat is needed for the reaction to take place at a useful rate. The first reaction with chlorine is:

$$
\begin{array}{c}
\text{H} \\
| \\
\text{H}-\text{C}-\text{H} \\
| \\
\text{H}
\end{array}
\; + \; \text{Cl}-\text{Cl} \quad \xrightarrow{\text{heat or light}} \quad
\begin{array}{c}
\text{H} \\
| \\
\text{H}-\text{C}-\text{Cl} \\
| \\
\text{H}
\end{array}
\; + \; \text{H}-\text{Cl}
$$

methane   chlorine   chloromethane   hydrogen
                                                            (methyl chloride)   chloride

This reaction may continue; heat or light is needed for each step:

$$
\underset{\substack{\displaystyle |\\ H}}{\overset{\substack{\displaystyle H\\ |}}{H-C-Cl}} \xrightarrow{Cl_2} \underset{\substack{\displaystyle |\\ H}}{\overset{\substack{\displaystyle Cl\\ |}}{H-C-Cl}} \xrightarrow{Cl_2} \underset{\substack{\displaystyle |\\ H}}{\overset{\substack{\displaystyle Cl\\ |}}{Cl-C-Cl}} \xrightarrow{Cl_2} \underset{\substack{\displaystyle |\\ Cl}}{\overset{\substack{\displaystyle Cl\\ |}}{Cl-C-Cl}}
$$

$$+ \quad HCl \qquad\qquad + \quad HCl \qquad\qquad + \quad HCl$$

This sequence raises several questions about the chlorination of methane. Why is heat or light needed for the reaction to go? Why do we get a mixture of products? Is there any way to modify the reaction to get just one pure product? Are the observed products formed because they are the most stable products possible? Or are they favored because they are formed faster than any other products?

We will answer these questions, but first we must understand three aspects of the reaction: the mechanism, the thermodynamics, and the kinetics.

1. The **mechanism** is the complete, step-by-step description of exactly which bonds break and which bonds form in what order to give the observed products.
2. **Thermodynamics** is the study of the energy changes that accompany chemical and physical transformations. It allows us to compare the stability of reactants and products and predict which compounds are favored by the equilibrium.
3. **Kinetics** is the study of reaction rates, determining which products are formed fastest. Kinetics also helps to predict how the rate will change if we change the reaction conditions.

We will use the chlorination of methane to show how we study a reaction. Before we can propose a detailed mechanism for the chlorination, we must learn everything we can about how the reaction works and what factors affect the reaction rate and the product distribution.

A careful study of the chlorination of methane has established three important characteristics:

1. The chlorination does not occur at room temperature in the absence of light. The reaction begins when light falls on the mixture or when it is heated. Thus, we know this reaction requires some form of energy to *initiate* it.
2. The most effective wavelength of light is a blue color that is strongly absorbed by chlorine gas. This finding implies that light is absorbed by the chlorine molecule, activating chlorine so that it initiates the reaction with methane.
3. The light-initiated reaction has a high *quantum yield*. This means that many molecules of the product are formed for every photon of light absorbed. Our mechanism must explain how hundreds of individual reactions of methane with chlorine result from the absorption of a single photon by a single molecule of chlorine.

A **chain reaction** mechanism has been proposed to explain the chlorination of methane. A chain reaction consists of three kinds of steps:

1. The **initiation step**, which generates a reactive intermediate.
2. **Propagation steps**, in which the reactive intermediate reacts with a stable molecule to form another reactive intermediate, allowing the chain to continue until the supply of reactants is exhausted or the reactive intermediate is destroyed.
3. **Termination steps**, side reactions that destroy reactive intermediates and tend to slow or stop the reaction.

4-3
The Free-Radical Chain Reaction

In studying the chlorination of methane, we will consider just the first reaction to form chloromethane (common name *methyl chloride*). This reaction is a **substitution**: Chlorine does not add to methane, but a chlorine atom *substitutes* for one of the hydrogen atoms, which appears in the HCl by-product.

| methane | chlorine | chloromethane (methyl chloride) |

4-3A   The Initiation Step: Generation of Radicals

Blue light, absorbed by chlorine but not by methane, promotes this reaction. Therefore, initiation probably results from the absorption of light by a molecule of chlorine. Blue light has about the right energy to split a chlorine molecule (Cl_2) into two chlorine atoms (58 kcal/mol).* The splitting of a chlorine molecule by absorption of a photon is shown as follows:

$$:\ddot{Cl}:\ddot{Cl}: \;+\; \text{photon } (h\nu) \longrightarrow \;:\ddot{Cl}\cdot \;+\; \cdot\ddot{Cl}:$$

Notice the fishhook-shaped half-arrows used to show the movement of single unpaired electrons. Just as we use curved arrows to represent the movement of electron *pairs*, we use these curved half-arrows to represent the movement of single unpaired electrons. These half-arrows show that the two electrons in the Cl—Cl bond separate, and one leaves with each chlorine atom.

The splitting of a Cl_2 molecule is an initiation step that produces two highly reactive chlorine atoms. A chlorine atom is an example of a **reactive intermediate**: a short-lived species that is never present in high concentration because it reacts as quickly as it is formed. Each Cl· atom has an odd number of valence electrons (seven), one of which is unpaired. The unpaired electron is called the *odd electron* or the *radical electron*. Species with unpaired electrons are called **radicals** or **free radicals**. Radicals are electron-deficient because they lack an octet. The odd electron readily combines with an electron in another atom to complete an octet and form a bond.

Figure 4-1 shows the Lewis structures of some free radicals. Radicals are often represented by a structure with a single dot representing the unpaired odd electron.

> Free radicals may play a role in diseases and accelerate aging. In the course of everyday life, reactive oxygen species are encountered in the environment and produced in the body. These compounds break down into short-lived hydroxyl radicals, which can react with the body's proteins and DNA. The resulting damage accumulates and may result in heart disease, cancer, and premature aging.

PROBLEM 4-1

Draw Lewis structures for the following free radicals.
(a) the *n*-propyl radical, CH_3—CH_2—$\dot{C}H_2$ **(b)** the *t*-butyl radical, $(CH_3)_3C\cdot$
(c) the isopropyl radical   **(d)** the iodine atom

Lewis structures

$:\ddot{\underset{..}{Cl}}\cdot$ $:\ddot{\underset{.}{Br}}\cdot$ $H:\ddot{\underset{..}{O}}\cdot$ $H:\overset{H}{\underset{H}{C}}\cdot$ $H:\overset{H}{\underset{H}{C}}:\overset{H}{\underset{H}{C}}\cdot$

Written

| Cl· | Br· | HO· | $CH_3\cdot$ | $CH_3CH_2\cdot$ |
| chlorine atom | bromine atom | hydroxyl radical | methyl radical | ethyl radical |

▶ **Figure 4-1**
Free radicals are reactive species with odd numbers of electrons. The unpaired electron is represented by a dot in the formula.

*The energy of a photon of light is related to its frequency ν by the relationship $E = h\nu$, where h is Planck's constant. Blue light has an energy of about 60 kcal (250 kJ) per einstein (an einstein is a mole of photons).

4-3B Propagation Steps

When a chlorine radical collides with a methane molecule, it abstracts (removes) a hydrogen atom from methane. One of the electrons in the C—H bond remains on carbon, while the other combines with the odd electron on the chlorine atom to form the H—Cl bond.

First propagation step

$$
\underset{\text{methane}}{\text{H}-\overset{\displaystyle \text{H}}{\underset{\displaystyle \text{H}}{\text{C}}}-\text{H}} \quad + \quad \underset{\text{chlorine atom}}{\text{Cl}\cdot} \quad \longrightarrow \quad \underset{\text{methyl radical}}{\text{H}-\overset{\displaystyle \text{H}}{\underset{\displaystyle \text{H}}{\text{C}}}\cdot} \quad + \quad \underset{\text{hydrogen chloride}}{\text{H}-\text{Cl}}
$$

This step forms only one of the final products: the molecule of HCl. A later step must form chloromethane. Notice that the first propagation step begins with one free radical (the chlorine atom) and produces another free radical (the methyl radical). The regeneration of a free radical is characteristic of a propagation step of a chain reaction. The reaction can continue because another reactive intermediate is produced.

In the second propagation step, the methyl radical reacts with a molecule of chlorine to form chloromethane. The odd electron of the methyl radical combines with one of the two electrons in the Cl—Cl bond to give the Cl—CH_3 bond, and the chlorine atom is left with the odd electron.

Second propagation step

$$
\underset{\text{methyl radical}}{\text{H}-\overset{\displaystyle \text{H}}{\underset{\displaystyle \text{H}}{\text{C}}}\cdot} \quad + \quad \underset{\text{chlorine molecule}}{\text{Cl}-\text{Cl}} \quad \longrightarrow \quad \underset{\text{chloromethane}}{\text{H}-\overset{\displaystyle \text{H}}{\underset{\displaystyle \text{H}}{\text{C}}}-\text{Cl}} \quad + \quad \underset{\text{chlorine atom}}{\text{Cl}\cdot}
$$

In addition to forming chloromethane, the second propagation step produces another chlorine atom. The chlorine atom can react with another molecule of methane, giving HCl and a methyl radical, which reacts with Cl_2 to give chloromethane and regenerate yet another chlorine atom. In this way, the chain reaction continues until the supply of the reactants is exhausted or some other reaction consumes the radical intermediates. The chain reaction explains why many molecules of methyl chloride and HCl are formed by each photon of light that is absorbed. We can summarize the reaction mechanism as follows:

KEY MECHANISM   Free-Radical Halogenation

Initiation

$$\text{Cl}-\text{Cl} \quad + \quad h\nu \text{ (light)} \quad \longrightarrow \quad 2\ \text{Cl}\cdot$$

Propagation

continues
the chain

The overall reaction is simply the sum of the propagation steps.

$$
\underset{\displaystyle \overset{\textstyle H}{|}}{\underset{\displaystyle H}{H-C-H}} \;+\; Cl-Cl \;\longrightarrow\; \underset{\displaystyle \overset{\textstyle H}{|}}{\underset{\displaystyle H}{H-C-Cl}} \;+\; H-Cl
$$

PROBLEM 4-2

(a) Write the propagation steps leading to the formation of dichloromethane (CH_2Cl_2) from chloromethane.
(b) Explain why free-radical halogenation usually gives mixtures of products.
(c) How could an industrial plant control the proportions of reactants to favor production of CCl_4? To favor CH_3Cl?

4-3C   Termination Reactions

If anything happens to consume some of the free-radical intermediates without generating new ones, the chain reaction will slow or stop. Such a side reaction is called a **termination reaction**: a step that produces fewer reactive intermediates (free radicals) than it consumes. The following are some of the possible termination reactions in the chlorination of methane:

Some anti-cancer agents act by generating highly reactive hydroxyl radicals, which damage and degrade the DNA of the rapidly dividing tumor cells. As a result, the cells die and the tumor shrinks. One example of a radical generator is bleomycin, which is used for the treatment of testicular cancer.

Combination of any two free radicals is a termination step because it decreases the number of free radicals. Other termination steps involve reactions of free radicals with the walls of the vessel or other contaminants. Although the first of these termination steps gives chloromethane, one of the products, it consumes the free radicals that are necessary for the reaction to continue, thus breaking the chain. Its contribution to the amount of product obtained from the reaction is small compared with the contribution of the propagation steps.

While a chain reaction is in progress, the concentration of radicals is very low. The probability that two radicals will combine in a termination step is lower than the probability that each will encounter a molecule of reactant and give a propagation step. The termination steps become important toward the end of the reaction, when there are relatively few molecules of reactants available. At this point, the free radicals are less likely to encounter a molecule of reactant than they are to encounter each other (or the wall of the container). The chain reaction quickly stops.

PROBLEM 4-3

Each of the following proposed mechanisms for the free-radical chlorination of methane is wrong. Explain how the experimental evidence disproves each mechanism.
(a) $Cl_2 + hv \rightarrow Cl_2^*$ (an "activated" form of Cl_2)
  $Cl_2^* + CH_4 \rightarrow HCl + CH_3Cl$
(b) $CH_4 + hv \rightarrow CH_3\cdot + H\cdot$
  $CH_3\cdot + Cl_2 \rightarrow CH_3Cl + Cl\cdot$
  $Cl\cdot + H\cdot \rightarrow HCl$

PROBLEM 4-4

Free-radical chlorination of hexane gives very poor yields of 1-chlorohexane, while cyclohexane can be converted to chlorocyclohexane in good yield.
(a) How do you account for this difference?
(b) What ratio of reactants (cyclohexane and chlorine) would you use for the synthesis of chlorocyclohexane?

PROBLEM-SOLVING **HINT**

In a free-radical chain reaction, initiation steps generally create new free radicals. Propagation steps usually combine a free radical and a reactant to give a product and another free radical. Termination steps generally decrease the number of free radicals.

4-4
Equilibrium Constants and Free Energy

Now that we have determined a mechanism for the chlorination of methane, we can consider the energetics of the individual steps. Let's begin by reviewing some of the principles needed for this discussion.

Thermodynamics is the branch of chemistry that deals with the energy changes accompanying chemical and physical transformations. These energy changes are most useful for describing the properties of systems at equilibrium. Let's review how energy and entropy variables describe an equilibrium.

The equilibrium concentrations of reactants and products are governed by the **equilibrium constant** of the reaction. For example, if A and B react to give C and D, then the equilibrium constant K_{eq} is defined by the following equation:

$$A + B \rightleftharpoons C + D$$

$$K_{eq} = \frac{[\text{products}]}{[\text{reactants}]} = \frac{[C][D]}{[A][B]}$$

The value of K_{eq} tells us the position of the equilibrium: whether the products or the reactants are more stable, and therefore energetically favored. If K_{eq} is larger than 1, the reaction is favored as written from left to right. If K_{eq} is less than 1, the reverse reaction is favored (from right to left as written).

The chlorination of methane has a large equilibrium constant of about 1.1×10^{19}.

$$CH_4 + Cl_2 \rightleftharpoons CH_3Cl + HCl$$

$$K_{eq} = \frac{[CH_3Cl][HCl]}{[CH_4][Cl_2]} = 1.1 \times 10^{19}$$

The equilibrium constant for chlorination is so large that the remaining amounts of the reactants are close to zero at equilibrium. Such a reaction is said to *go to completion*, and the value of K_{eq} is a measure of the reaction's tendency to go to completion.

From the value of K_{eq} we can calculate the change in **free energy** (sometimes called **Gibbs free energy**) that accompanies the reaction. Free energy is represented by G, and the change (Δ) in free energy associated with a reaction is represented by ΔG, the difference between the free energy of the products and the free energy of the reactants. ΔG is a measure of the amount of energy available to do work.

$$\Delta G = (\text{free energy of products}) - (\text{free energy of reactants})$$

If the energy levels of the products are lower than the energy levels of the reactants (a "downhill" reaction), then the reaction is energetically favored; and this equation gives a negative value of ΔG, corresponding to a decrease in the energy of the system.

The **standard Gibbs free energy change** $\Delta G°$ is most commonly used. The symbol ° designates a reaction involving reactants and products in their standard states (pure substances in their most stable states at 25°C and 1 atm pressure). The relationship between $\Delta G°$ and K_{eq} is given by the expression

$$K_{eq} = e^{-\Delta G°/RT}$$

or conversely, by

$$\Delta G° = -RT(\ln K_{eq}) = -2.303RT(\log_{10} K_{eq})$$

where

$R = 1.987$ cal/kelvin-mol (8.314 J/kelvin-mol), the gas constant

$T = $ absolute temperature, in kelvins*

$e = 2.718$, the base of natural logarithms

The value of RT at 25°C is about 0.592 kcal/mol (2.48 kJ/mol).

The formula shows that a reaction is favored ($K_{eq} > 1$) if it has a *negative* value of $\Delta G°$ (energy is released). A reaction that has a positive value of $\Delta G°$ (energy must be added) is unfavorable. These predictions agree with our intuition that reactions should go from higher-energy states to lower-energy states, with a net decrease in free energy.

SOLVED PROBLEM 4-1

Calculate the value of $\Delta G°$ for the chlorination of methane.

SOLUTION

$$\Delta G° = -2.303RT(\log K_{eq})$$

K_{eq} for the chlorination is 1.1×10^{19}, and $\log K_{eq} = 19.04$

At 25°C (about 298°Kelvin), the value of RT is

$$RT = (1.987 \text{ cal/kelvin-mol})(298 \text{ kelvins}) = 592 \text{ cal/mol, or } 0.592 \text{ kcal/mol}$$

Substituting, we have:

$$\Delta G° = (-2.303)(0.592 \text{ kcal/mol})(19.04) = -25.9 \text{ kcal/mol } (-108.2 \text{ kJ/mol})$$

This is a large negative value for $\Delta G°$, showing that this chlorination has a large driving force that pushes it toward completion.

PROBLEM-SOLVING **HINT**

A reaction with a negative ΔG is favored.
A reaction with a positive ΔG is unfavorable.

In general, a reaction goes nearly to completion ($>99\%$) for values of $\Delta G°$ that are more negative than about -3.0 kcal. Table 4-1 shows what percentages of the starting materials are converted to products at equilibrium for reactions with various values of $\Delta G°$.

---

*Absolute temperatures (in kelvins) are correctly given without a degree sign, as in the equation 25°C = 298K. We will often include the degree sign, however, to distinguish absolute temperatures (K) from equilibrium constants (K) as in 25°C = 298°K.

TABLE 4-1 Product Composition as a Function of $\Delta G°$ at 25°C

$\Delta G°$ kcal/mol	$\Delta G°$ kJ/mol	K	Conversion to Products
+1.0	(+4.2)	0.18	15%
+0.5	(+2.1)	0.43	30%
0.0	(0.0)	1.0	50%
−0.5	(−2.1)	2.3	70%
−1.0	(−4.2)	5.4	84%
−2.0	(−8.4)	29	97%
−3.0	(−13)	159	99.40%
−4.0	(−17)	860	99.88%
−5.0	(−21)	4660	99.98%

PROBLEM 4-5

The following reaction has a value of $\Delta G° = -0.50$ kcal/mol (−2.1 kJ/mol).

$$CH_3Br + H_2S \rightleftharpoons CH_3SH + HBr$$

(a) Calculate K_{eq} at room temperature (25°C) for this reaction as written.
(b) Starting with a 1 M solution of CH_3Br and H_2S, calculate the final concentrations of all four species at equilibrium.

PROBLEM 4-6

At room temperature (25°C), the reaction of two molecules of acetone to form diacetone alcohol proceeds to an extent of about 5%. Determine the value of $\Delta G°$ for this reaction.

Two factors contribute to the change in free energy: the change in **enthalpy** and the change in **entropy** multiplied by the temperature.

$$\Delta G° = \Delta H° - T\Delta S°$$

$\Delta G°$ = (free energy of products) − (free energy of reactants)

$\Delta H°$ = (enthalpy of products) − (enthalpy of reactants)

$\Delta S°$ = (entropy of products) − (entropy of reactants)

At low temperatures, the enthalpy term ($\Delta H°$) is usually much larger than the entropy term ($-T\Delta S°$), and the entropy term is sometimes ignored.

4-5 Enthalpy and Entropy

4-5A Enthalpy

The **change in enthalpy** ($\Delta H°$) is the heat of reaction—the amount of heat evolved or consumed in the course of a reaction, usually given in kilocalories (or kilojoules) per mole. The enthalpy change is a measure of the relative strength of bonding in the products and reactants. Reactions tend to favor products with the lowest enthalpy (those with the strongest bonds).

If weaker bonds are broken and stronger bonds are formed, heat is evolved and the reaction is **exothermic** (negative value of $\Delta H°$). In an exothermic reaction, the enthalpy term makes a favorable negative contribution to $\Delta G°$. If stronger bonds are broken and weaker bonds are formed, then energy is consumed in the reaction, and the reaction is **endothermic** (positive value of $\Delta H°$). In an endothermic reaction, the enthalpy term makes an unfavorable positive contribution to $\Delta G°$.

The value of $\Delta H°$ for the chlorination of methane is about -25.0 kcal/mol (-104.5 kJ/mol). This is a highly exothermic reaction, with the decrease in enthalpy serving as the primary driving force.

4-5B   Entropy

Entropy is often described as randomness, or freedom of motion. Reactions tend to favor products with the greatest entropy. Notice the negative sign in the entropy term of the free-energy expression. A positive value of the entropy change ($\Delta S°$), indicating that the products have more freedom of motion than the reactants, makes a favorable (negative) contribution to $\Delta G°$.

In many cases, the enthalpy change ($\Delta H°$) is much larger than the entropy change ($\Delta S°$), and the enthalpy term dominates the equation for $\Delta G°$. Thus, a negative value of $\Delta S°$ does not necessarily mean that the reaction has an unfavorable value of $\Delta G°$. The formation of strong bonds (the change in enthalpy) is usually the most important component in the driving force for a reaction.

In the chlorination of methane, the value of $\Delta S°$ is $+2.9$ eu (entropy units or cal/ kelvin-mole). The $-T\Delta S°$ term in the free energy is

$$-T\Delta S° = -(298°K)(2.9 \text{ cal/kelvin-mol}) = -860 \text{ cal/mol}$$

$$= -0.86 \text{ kcal/mol } (-3.6 \text{ kJ/mol})$$

The value of $\Delta G° = -25.9$ kcal/mol is divided into enthalpy and entropy terms:

$$\Delta G° = \Delta H° - T\Delta S° = -25.0 \text{ kcal/mol} - 0.86 \text{ kcal/mol}$$

$$= -25.9 \text{ kcal/mol } (-108.2 \text{ kJ/mol})$$

The enthalpy change is the largest factor in the driving force for chlorination. This is the case in most organic reactions: The entropy term is often small in relation to the enthalpy term. When we discuss chemical reactions involving the breaking and forming of bonds, we can often use the values of the enthalpy changes ($\Delta H°$), under the assumption that $\Delta G° \cong \Delta H°$. We must be cautious in making this approximation, however, because some reactions have relatively small changes in enthalpy and larger changes in entropy.

SOLVED PROBLEM 4-2

Predict whether the value of $\Delta S°$ for the dissociation of Cl_2 is positive (favorable) or negative (unfavorable). What effect does the entropy term have on the sign of the value of $\Delta G°$ for this reaction?

$$Cl_2 \xrightarrow{hv} 2 \ Cl\cdot$$

SOLUTION

Two isolated chlorine atoms have much more freedom of motion than a single chlorine molecule. Therefore, the change in entropy is positive. This positive (favorable) value of $T\Delta S°$ is small, however, compared with the much larger, positive (unfavorable) value of $\Delta H°$. The chlorine molecule is much more stable than two chlorine atoms, showing that the positive enthalpy term predominates.

PROBLEM 4-7

When ethene is mixed with hydrogen in the presence of a platinum catalyst, hydrogen adds across the double bond to form ethane. At room temperature, the reaction goes to completion. Predict the signs of $\Delta H°$ and $\Delta S°$ for this reaction. Explain these signs in terms of bonding and freedom of motion.

PROBLEM-SOLVING HINT

In general, two smaller molecules (or fragments, such as radicals) have more freedom of motion (greater entropy) than one larger molecule.

ethene   ethane

PROBLEM 4-8

For each reaction, estimate whether $\Delta S°$ for the reaction is positive, negative, or impossible to predict.

(a) $C_{10}H_{22} \xrightarrow[\text{catalyst}]{\text{heat}} C_3H_6 + C_7H_{16}$

n-decane   propene   heptane

(b) The formation of diacetone alcohol:

(c) $CH_3-\overset{\overset{\displaystyle O}{\|}}{C}-OH + CH_3OH \underset{}{\overset{H^+}{\rightleftarrows}} CH_3-\overset{\overset{\displaystyle O}{\|}}{C}-OCH_3 + H_2O$

We can put known amounts of methane and chlorine into a bomb calorimeter and use a hot wire to initiate the reaction. The temperature rise in the calorimeter is used to calculate the precise value of the heat of reaction, $\Delta H°$. This measurement shows that 25 kcal (105 kJ) of heat is evolved (exothermic) for each mole of methane converted to chloromethane. Thus, $\Delta H°$ for the reaction is negative, and the heat of reaction is given as

$$\Delta H° = -25 \text{ kcal/mol (105 kJ/mol)}$$

In many cases, we want to predict whether a particular reaction will be endothermic or exothermic, without actually measuring the heat of reaction. We can calculate an approximate heat of reaction by adding and subtracting the energies involved in the breaking and forming of bonds. To do this calculation, we need to know the energies of the affected bonds.

The **bond-dissociation energy (BDE)** is the amount of energy required to break a particular bond **homolytically**, that is, in such a way that each bonded atom retains one of the bond's two electrons. In contrast, when a bond is broken **heterolytically**, one of the atoms retains both electrons.

4-6
Bond-Dissociation Energies

Homolytic cleavage (free radicals result)

$A\!:\!B \longrightarrow A\cdot + \cdot B \qquad \Delta H° = \text{bond-dissociation energy}$

$:\!\ddot{C}l\!:\!\ddot{C}l\!: \longrightarrow 2 :\!\ddot{C}l\cdot \qquad \Delta H° = 58 \text{ kcal/mol (242 kJ/mol)}$

Heterolytic cleavage (ions result)

$A\!:\!B \longrightarrow A^+ + \ ^-\!:\!B$

$(CH_3)_3C\!-\!\ddot{C}l\!: \longrightarrow (CH_3)_3C^+ + :\!\ddot{C}l\!:^- \qquad (\Delta H° \text{ varies with solvent})$

Homolytic cleavage (**radical cleavage**) forms free radicals, while **heterolytic cleavage** (**ionic cleavage**) forms ions. Enthalpies for heterolytic (ionic) cleavage depend strongly on the solvent's ability to solvate the ions that result. Homolytic cleavage is used to define bond-dissociation energies because the values do not vary so much with different solvents or with no solvent. Note that a curved arrow is used to show the movement of the electron pair in an ionic cleavage, and curved half-arrows to show the separation of individual electrons in a homolytic cleavage.

Energy is released when bonds are formed, and energy is consumed to break bonds. Therefore, bond-dissociation energies are always positive (endothermic). The overall enthalpy change for a reaction is the sum of the dissociation energies of the bonds broken minus the sum of the dissociation energies of the bonds formed.

$$\Delta H° = \Sigma(\text{BDE of bonds broken}) - \Sigma(\text{BDE of bonds formed})$$

By studying the heats of reaction for many different reactions, chemists have developed reliable tables of bond-dissociation energies. Table 4-2 gives the bond-dissociation energies for the homolysis of bonds in a variety of molecules.

TABLE 4-2   Bond-Dissociation Energies for Homolytic Cleavages
A:B $\longrightarrow$ A· + ·B

Bond	Bond-Dissociation Energy kcal/mol	kJ/mol	Bond	Bond-Dissociation Energy kcal/mol	kJ/mol
H—X bonds and X—X bonds			Bonds to secondary carbons		
H—H	104	435	$(CH_3)_2CH$—H	95	397
D—D	106	444	$(CH_3)_2CH$—F	106	444
F—F	38	159	$(CH_3)_2CH$—Cl	80	335
Cl—Cl	58	242	$(CH_3)_2CH$—Br	68	285
Br—Br	46	192	$(CH_3)_2CH$—I	53	222
I—I	36	151	$(CH_3)_2CH$—OH	91	381
H—F	136	569	Bonds to tertiary carbons		
H—Cl	103	431	$(CH_3)_3C$—H	91	381
H—Br	88	368	$(CH_3)_3C$—F	106	444
H—I	71	297	$(CH_3)_3C$—Cl	79	331
HO—H	119	498	$(CH_3)_3C$—Br	65	272
HO—OH	51	213	$(CH_3)_3C$—I	50	209
Methyl bonds			$(CH_3)_3C$—OH	91	381
CH_3—H	104	435	Other C—H bonds		
CH_3—F	109	456	$PhCH_2$—H (benzylic)	85	356
CH_3—Cl	84	351	CH_2=$CHCH_2$—H (allylic)	87	364
CH_3—Br	70	293	CH_2=CH—H (vinyl)	108	452
CH_3—I	56	234	Ph—H (aromatic)	110	460
CH_3—OH	91	381	C—C bonds		
Bonds to primary carbons			CH_3—CH_3	88	368
CH_3CH_2—H	98	410	CH_3CH_2—CH_3	85	356
CH_3CH_2—F	107	448	CH_3CH_2—CH_2CH_3	82	343
CH_3CH_2—Cl	81	339	$(CH_3)_2CH$—CH_3	84	351
CH_3CH_2—Br	68	285	$(CH_3)_3C$—CH_3	81	339
CH_3CH_2—I	53	222			
CH_3CH_2—OH	91	381			
$CH_3CH_2CH_2$—H	98	410			
$CH_3CH_2CH_2$—F	107	448			
$CH_3CH_2CH_2$—Cl	81	339			
$CH_3CH_2CH_2$—Br	68	285			
$CH_3CH_2CH_2$—I	53	222			
$CH_3CH_2CH_2$—OH	91	381			

We can use values from Table 4-2 to predict the heat of reaction for the chlorination of methane. This reaction involves the breaking (positive values) of a CH_3—H bond and a Cl—Cl bond, and the formation (negative values) of a CH_3—Cl bond and a H—Cl bond.

4-7
Enthalpy Changes in Chlorination

Overall reaction

$$CH_3-H \ + \ Cl-Cl \ \longrightarrow \ CH_3-Cl \ + \ H-Cl$$

Bonds broken	$\Delta H°$ (per mole)	Bonds formed	$\Delta H°$ (per mole)
Cl—Cl	+58 kcal (242 kJ)	H—Cl	−103 kcal (431 kJ)
CH_3—H	+104 kcal (435 kJ)	CH_3—Cl	−84 kcal (351 kJ)
total	+162 kcal (677 kJ)	total	−187 kcal (782 kJ)

$$\Delta H° = +162 \text{ kcal} + (-187) \text{ kcal} = -25 \text{ kcal/mol} (-105 \text{ kJ/mol})$$

The bond-dissociation energies also provide the heat of reaction for each individual step:

First propagation step

$$Cl\cdot \ + \ CH_4 \ \longrightarrow \ CH_3\cdot \ + \ HCl$$

Breaking a CH_3—H bond	+104 kcal/mol (+435 kJ/mol)
Forming an H—Cl bond	−103 kcal/mol (−431 kJ/mol)
Step total	+1 kcal/mol (+4 kJ/mol)

Second propagation step

$$CH_3\cdot \ + \ Cl_2 \ \longrightarrow \ CH_3Cl \ + \ Cl\cdot$$

Breaking a Cl—Cl bond	+58 kcal/mol (+243 kJ/mol)
Forming a CH_3—Cl bond	−84 kcal/mol (−352 kJ/mol)
Step total	−26 kcal/mol (−109 kJ/mol)

$$\text{Grand total} = +1 \text{ kcal/mol} + (-26 \text{ kcal/mol}) = -25 \text{ kcal/mol} (-105 \text{ kJ/mol})$$

The sum of the values of $\Delta H°$ for the individual propagation steps gives the overall enthalpy change for the reaction. The initiation step, $Cl_2 \rightarrow 2 \, Cl\cdot$ is not added to give the overall enthalpy change, because it is not necessary for each molecule of product formed. The first splitting of a chlorine molecule simply begins the chain reaction, which generates hundreds or thousands of molecules of chloromethane.

PROBLEM 4-9

(a) Propose a mechanism for the free-radical chlorination of ethane,

$$CH_3-CH_3 \ + \ Cl_2 \ \xrightarrow{hv} \ CH_3-CH_2Cl \ + \ HCl$$

(b) Calculate $\Delta H°$ for each step in this reaction.
(c) Calculate the overall value of $\Delta H°$ for this reaction.

PROBLEM-SOLVING HINT

Bond-dissociation energies are for breaking bonds, which costs energy. In calculating values of $\Delta H°$, use positive BDE values for bonds that are broken and negative values for bonds that are formed.

Alternative Mechanism The mechanism we have used is not the only one that might be proposed to explain the reaction of methane with chlorine. We know that the initiating step must be the splitting of a molecule of Cl_2, but there are other propagation steps that would form the correct products:

(a) $Cl\cdot + CH_3$—H $\longrightarrow$ CH_3—Cl + H· $\qquad \Delta H° = +104 \text{ kcal} - 84 \text{ kcal} = +20 \text{ kcal}$ (+84 kJ)

(b) H· + Cl—Cl $\longrightarrow$ H—Cl + Cl· $\qquad \underline{\Delta H° = + 58 \text{ kcal} - 103 \text{ kcal} = -45 \text{ kcal} (-189 \text{ kJ})}$

$\qquad\qquad\qquad\qquad\qquad\qquad\qquad\qquad\qquad$ Total $\qquad$ −25 kcal (−105 kJ)

This alternative mechanism seems plausible, but step (a) is endothermic by 20 kcal/mol (84 kJ/mol). The previous mechanism provides a lower energy alternative. When a chlorine atom collides with a methane molecule, it will not react to give methyl chloride and a hydrogen atom ($\Delta H° = +20 \text{ kcal} = +84 \text{ kJ}$); it will react to give HCl and a methyl radical ($\Delta H° = +1 \text{ kcal} = +4 \text{ kJ}$), the first propagation step of the correct mechanism.

> ### PROBLEM 4-10
>
> **(a)** Using bond-dissociation energies from Table 4-2 (page 134), calculate the heat of reaction for each of the steps in the free-radical bromination of methane.
>
> $$Br_2 + CH_4 \xrightarrow{\text{heat or light}} CH_3Br + HBr$$
>
> **(b)** Calculate the overall heat of reaction.

4-8
Kinetics and the Rate Equation

Kinetics is the study of reaction rates. How fast a reaction goes is just as important as the position of its equilibrium. Just because thermodynamics favors a reaction (negative $\Delta G°$) does not necessarily mean the reaction will actually occur. For example, a mixture of gasoline and oxygen does not react without a spark or a catalyst. Similarly, a mixture of methane and chlorine does not react if it is kept cold and dark.

The **rate of a reaction** is how fast the products appear and the reactants disappear. We can determine the rate by measuring the increase in the concentrations of the products with time, or the decrease in the concentrations of the reactants with time.

Reaction rates depend on the concentrations of the reactants. The greater the concentrations, the more often the reactants collide and the greater the chance of reaction. A **rate equation** (sometimes called a **rate law**) is the relationship between the concentrations of the reactants and the observed reaction rate. Each reaction has its own rate equation, *determined experimentally* by changing the concentrations of the reactants and measuring the change in the rate. For example, consider the general reaction

$$A + B \longrightarrow C + D$$

The reaction rate is usually proportional to the concentrations of the reactants ([A] and [B]) raised to some powers, a and b. We can use a general rate expression to represent this relationship as

$$\text{rate} = k_r[A]^a[B]^b$$

where k_r is the **rate constant**, and the values of the powers (a and b) must be determined experimentally. We *cannot* guess or calculate the rate equation from just the stoichiometry of the reaction. The rate equation depends on the mechanism of the reaction and on the rates of the individual steps in the mechanism.

In the general rate equation, the power a is called the **order** of the reaction with respect to reactant A, and b is the order of the reaction with respect to B. The sum of these powers ($a + b$) is called the **overall order** of the reaction.

The following reaction has a simple rate equation:

$$CH_3-Br + {}^-OH \xrightarrow{\text{H}_2\text{O/acetone}} CH_3-OH + Br^-$$

Experiments show that doubling the concentration of methyl bromide, $[CH_3Br]$, doubles the rate of reaction. Doubling the concentration of hydroxide ion, $[^-OH]$, also doubles the rate. Thus, the rate is proportional to both $[CH_3Br]$ and $[^-OH]$, so the rate equation has the following form:

$$\text{rate} = k_r[CH_3Br][^-OH]$$

This rate equation is *first order* in each of the two reagents because it is proportional to the first power of their concentrations. The rate equation is *second order overall* because the sum of the powers of the concentrations in the rate equation is 2; that is, (first order) + (first order) = second order overall.

Reactions of the same overall type do not necessarily have the same form of rate equation. For example, the following similar reaction has a different kinetic order:

$$(CH_3)_3C-Br + {}^-OH \xrightarrow{\text{H}_2\text{O/acetone}} (CH_3)_3C-OH + Br^-$$

Doubling the concentration of *t*-butyl bromide ($[(CH_3)_3C-Br]$) causes the rate to double, but doubling the concentration of hydroxide ion ($[^-OH]$) has no effect on the rate of this particular reaction. The rate equation is

$$\text{rate} = k_r[(CH_3)_3C-Br]$$

This reaction is first order in *t*-butyl bromide, and zeroth order in hydroxide ion (proportional to $[^-OH]$ to the zeroth power). It is first order overall.

The most important fact to remember is that *the rate equation must be determined experimentally*. We cannot predict the form of the rate equation from the stoichiometry of the reaction. We determine the rate equation experimentally, then use that information to propose consistent mechanisms.

SOLVED PROBLEM 4-3

Chloromethane reacts with dilute sodium cyanide ($Na^+ \ {}^-C{\equiv}N$) according to the following equation:

$$\underset{\text{chloromethane}}{CH_3-Cl} + \underset{\text{cyanide}}{{}^-C{\equiv}N} \longrightarrow \underset{\text{acetonitrile}}{CH_3-C{\equiv}N} + \underset{\text{chloride}}{Cl^-}$$

When the concentration of chloromethane is doubled, the rate is observed to double. When the concentration of cyanide ion is tripled, the rate is observed to triple.
(a) What is the kinetic order with respect to chloromethane?
(b) What is the kinetic order with respect to cyanide ion?
(c) What is the kinetic order overall?
(d) Write the rate equation for this reaction.

SOLUTION
(a) When $[CH_3Cl]$ is doubled, the rate doubles, which is two to the first power. The reaction is first order with respect to chloromethane.
(b) When $[^-CN]$ is tripled, the reaction rate triples, which is three to the first power. The reaction is first order with respect to cyanide ion.
(c) First order plus first order equals second order overall.
(d) rate $= k_r[CH_3Cl][^-CN]$

PROBLEM 4-11

The reaction of *t*-butyl chloride with methanol is found to follow the rate equation given below:

$$(CH_3)_3C—Cl \quad + \quad CH_3—OH \quad \longrightarrow \quad (CH_3)_3C—OCH_3 \quad + \quad HCl$$

t-butyl chloride   methanol   methyl *t*-butyl ether

$$rate = k_r[(CH_3)_3C—Cl]$$

(a) What is the kinetic order with respect to *t*-butyl chloride?
(b) What is the kinetic order with respect to methanol?
(c) What is the kinetic order overall?

PROBLEM 4-12

Under certain conditions, the bromination of cyclohexene follows an interesting rate law:

$$rate = k_r[cyclohexene][Br_2]^2$$

(a) What is the kinetic order with respect to cyclohexene?
(b) What is the kinetic order with respect to bromine?
(c) What is the overall kinetic order?

PROBLEM 4-13

When a small piece of platinum is added to a mixture of ethene and hydrogen, the following reaction occurs:

ethene   hydrogen   ethane

Doubling the concentration of hydrogen has no effect on the reaction rate. Doubling the concentration of ethene also has no effect.
(a) What is the kinetic order of this reaction with respect to ethene? Hydrogen? What is the overall order?
(b) Write the unusual rate equation for this reaction.
(c) Explain this strange rate equation, and suggest what one might do to accelerate the reaction.

4-9
Activation Energy and the Temperature Dependence of Rates

Each reaction has its own characteristic rate constant, k_r. Its value depends on the conditions of the reaction, especially the temperature. This temperature dependence is expressed by the Arrhenius equation,

$$k_r = Ae^{-E_a/RT}$$

where

A = a constant (the "frequency factor")

E_a = activation energy

R = the gas constant, 1.987 cal/kelvin-mole

T = the absolute temperature

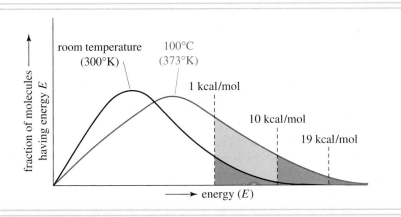

◀ **Figure 4-2**
Graph showing how the number of molecules having a given activation energy decreases as the activation energy increases. At a higher temperature (red curve), more collisions have the needed energy.

The **activation energy**, E_a, is the minimum kinetic energy the molecules must have to overcome the repulsions between their electron clouds when they collide. The exponential term $e^{-E_a/RT}$ corresponds to the fraction of collisions in which the particles have the minimum energy E_a needed to react. We can calculate E_a for a reaction by measuring how k_r varies with temperature, and substituting into the Arrhenius equation.

The *frequency factor A* accounts for the frequency of collisions and the fraction of collisions with the proper orientation for the reaction to occur. In most cases, only a small fraction of collisions occur between molecules with enough speed and with just the right orientation for reaction to occur. Far more collisions occur without enough kinetic energy or without the proper orientation, and the molecules simply bounce off each other.

The Arrhenius equation implies that the rate of a reaction depends on the fraction of molecules with kinetic energy of at least E_a. Figure 4-2 shows how the distribution of kinetic energies in a sample of a gas depends on the temperature. The black curved line shows the molecular energy distribution at room temperature, and the dashed lines show the energy needed to overcome barriers of 1 kcal (4 kJ), 10 kcal (42 kJ), and 19 kcal (79 kJ). The area under the curve to the right of each barrier corresponds to the number of molecules with enough energy to overcome that barrier. The red curve shows how the energy distribution is shifted at 100°C. At 100°C, many more molecules have the energy needed to overcome the energy barriers, especially the 19 kcal/mol barrier.

As the temperature increases, a larger fraction of molecular collisions have enough kinetic energy to result in reaction, and the reaction rate increases. Table 4-3 shows the dependence of reaction rates on the temperature by listing values of the relative rate constant k_{rel}, which is just the exponential term $e^{-E_a/RT}$ for some typical values of E_a and some convenient temperatures. With a typical activation energy of about 10 to 15 kcal/mol (40 to 60 kJ/mol), the reaction rate approximately doubles when the temperature is raised by 10°C, as from 27°C (near room temperature) to 37°C (body temperature).

Because the relative rate constant k_{rel} increases quickly when the temperature is raised, it might seem that raising the temperature would always be a good way to save

TABLE 4-3   Variation of the Rate with Temperature

E_a (per mole)	Values of $k_{rel} = e^{-E_a/RT}$ (units of 10^{-9})		
	27°C (300°K)	37°C (310°K)	100°C (373°K)
5 kcal (21 kJ)	240,000	320,000	1,200,000
10 kcal (42 kJ)	58	99	1,500
15 kcal (63 kJ)	0.014	0.031	1.9
20 kcal (84 kJ)	0.0000033	0.0000098	0.0023

time by making reactions go faster. The problem with raising the temperature is that *all* reactions are accelerated, including all the unwanted side reactions. We try to find a temperature that allows the desired reaction to go at a reasonable rate without producing unacceptable rates of side reactions.

4-10
Transition States

The activation energy E_a represents the energy difference between the reactants and the **transition state**, the highest-energy state in a molecular collision that leads to reaction. In effect, the activation energy is the barrier that must be overcome for the reaction to take place. The value of E_a is always positive, and its magnitude depends on the relative energy of the transition state. The term *transition state* implies that this configuration is the transition between the reactants and products, and the molecules can either go on to products or return to reactants.

Unlike the reactants or products, a transition state is unstable and cannot be isolated. It is not an intermediate, because an **intermediate** is a species that exists for some finite length of time, even if it is very short. An intermediate has at least some stability, but the transition state is a transient on the path from one intermediate to another. The transition state is often symbolized by a superscript double dagger ($\ddagger$), and the changes in variables such as free energy, enthalpy, and entropy involved in achieving the transition state are symbolized $\Delta G^\ddagger$, $\Delta H^\ddagger$, and $\Delta S^\ddagger$. $\Delta G^\ddagger$ is similar to E_a, and the symbol $\Delta G^\ddagger$ is often used in speaking of the activation energy.

Transition states have high energies because bonds must begin to break before other bonds can form. The following equation shows the reaction of a chlorine radical with methane. The transition state shows the C—H bond partially broken and the H—Cl bond partially formed. Transition states are often enclosed by brackets to emphasize their transient nature.

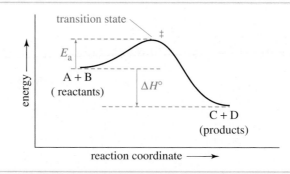

transition state

Reaction-Energy Diagrams   The concepts of transition state and activation energy are easier to understand graphically. Figure 4-3 shows a **reaction-energy diagram** for a one-step exothermic reaction. The vertical axis of the energy diagram represents the total potential energy of all the species involved in the reaction. The horizontal axis is called the **reaction coordinate**. The reaction coordinate symbolizes the progress of the reaction, going from the reactants on the left to the products on the right. The transition state is the highest point on the graph, and the activation energy is the energy difference between the reactants and the transition state. The heat of reaction ($\Delta H°$) is the difference in energy between the reactants and the products.

▶ **Figure 4-3**
Reaction-energy diagram for a one-step exothermic reaction. The reactants are toward the left, and the products are toward the right. The vertical axis represents the potential energy. The transition state is the highest point on the graph, and the activation energy is the energy difference between the reactants and the transition state.

If a **catalyst** were added to the reaction in Figure 4-3, it would create a transition state of lower energy, thereby lowering the activation energy. Addition of a catalyst would not change the energies of the reactants and products, however, so the heat of reaction would be unaffected.

SOLVED PROBLEM 4-4

Consider the following reaction:

$$CH_4 + Cl\cdot \longrightarrow \cdot CH_3 + HCl$$

This reaction has an activation energy (E_a) of +4 kcal/mol (+17 kJ/mol) and a $\Delta H°$ of +1 kcal/mol (+4 kJ/mol). Draw a reaction-energy diagram for this reaction.

SOLUTION

We draw a diagram that shows the products to be 1 kcal *higher* in energy than the reactants. The barrier is made to be 4 kcal higher in energy than the reactants.

> Enzymes serve as biological catalysts. They speed up reactions without changing the energies of the reactants (called substrates) and products. Without enzymes, most of the reactions in our cells would not go fast enough to keep us alive.

PROBLEM 4-14

(a) Draw the reaction-energy diagram for the reverse reaction:

$$CH_3\cdot + HCl \longrightarrow CH_4 + Cl\cdot$$

(b) What is the activation energy for this reverse reaction?
(c) What is the heat of reaction ($\Delta H°$) for this reverse reaction?

PROBLEM 4-15

(a) Draw a reaction-energy diagram for the following reaction:

$$CH_3\cdot + Cl_2 \longrightarrow CH_3Cl + Cl\cdot$$

The activation energy is 1 kcal/mol (4 kJ/mol), and the overall $\Delta H°$ for the reaction is −26 kcal/mol (−109 kJ/mol).
(b) Give the equation for the reverse reaction.
(c) What is the activation energy for the reverse reaction?

4-11
Rates of Multistep Reactions

Many reactions proceed by mechanisms involving several steps and several intermediates. The reaction of methane with chlorine, for example, goes through two propagation steps. The propagation steps are shown below, along with their heats of reaction and their activation energies. Just the propagation steps are shown because the rate of the initiation step is controlled by the amount of light or heat available to split chlorine molecules.

▶ **Figure 4-4**
Combined reaction-energy diagram for the chlorination of methane. The energy maxima are transition states, and the energy minima are intermediates.

Step	$\Delta H°$ (per mole)	E_a (per mole)
$CH_4 + Cl\cdot \longrightarrow CH_3\cdot + HCl$	+1 kcal (+4 kJ)	4 kcal (17 kJ)
$CH_3\cdot + Cl_2 \longrightarrow CH_3Cl + Cl\cdot$	−26 kcal (−109 kJ)	1 kcal (4 kJ)

In this reaction, $Cl\cdot$ and $CH_3\cdot$ are *reactive intermediates*. Unlike transition states, these intermediates are stable as long as they do not collide with other atoms or molecules. They are free radicals, however, and they are quite reactive toward other molecules. Figure 4-4 shows a single reaction-energy profile that includes both propagation steps of the chlorination. The energy maxima (high points) are the unstable transition states, and the energy minima (low points) are the intermediates. This complete energy profile provides most of the important information about the energetics of the reaction.

The Rate-Limiting Step   In a multistep reaction, each step has its own characteristic rate. There can be only one overall reaction rate, however, and it is controlled by the **rate-limiting step** (also called the **rate-determining step**). In general, the *highest-energy* step of a multistep reaction is the "bottleneck," and it determines the overall rate. How can we tell which step is rate limiting? If we have the reaction-energy diagram, it is simple: The highest point in the energy diagram is the transition state with the highest energy—the transition state for the rate-limiting step.

The highest point in the energy diagram of the chlorination of methane (Figure 4-4) is the transition state for the reaction of methane with a chlorine radical. This step must be rate limiting. If we calculate a rate for this slow step, it will be the rate for the overall reaction. The second, faster step will consume the products of the slow step as fast as they are formed.

4-12 Temperature Dependence of Halogenation

We now apply what we know about rates to the reaction of methane with halogens. The rate-limiting step for chlorination is the endothermic reaction of the chlorine atom with methane to form a methyl radical and a molecule of HCl.

Rate-limiting step

$$CH_4 + Cl\cdot \longrightarrow CH_3\cdot + HCl$$

The activation energy for this step is 4 kcal/mol (17 kJ/mol). At room temperature, the value of $e^{-E_a/RT}$ is 1300×10^{-6}. This value represents a rate that is fast but controllable.

In a free-radical chain reaction, every propagation step must occur quickly, or the free radicals will undergo unproductive collisions and become involved in termination steps. We can predict how quickly the various halogen atoms react with methane by using the measured activation energies of the slowest steps:

Reaction	E_a (per mole)	Relative rate ($e^{-E_a/RT} \times 10^6$) 27°C (300°K)	Relative rate ($e^{-E_a/RT} \times 10^6$) 227°C (500°K)
F· + CH$_4$ ⟶ HF + CH$_3$·	1.2 kcal (5 kJ)	140,000	300,000
Cl· + CH$_4$ ⟶ HCl + CH$_3$·	4 kcal (17 kJ)	1300	18,000
Br· + CH$_4$ ⟶ HBr + CH$_3$·	18 kcal (75 kJ)	9×10^{-8}	0.015
I· + CH$_4$ ⟶ HI + CH$_3$·	34 kcal (140 kJ)	2×10^{-19}	2×10^{-9}

Using these relative rates, we can make predictions about the reactions of methane with halogen radicals. The reaction with fluorine should be difficult to control because its rate is very high. Chlorine should react moderately at room temperature, but it may become difficult to control if the temperature rises much (the rate at 500°K is rather high). The reaction with bromine is very slow, but heating might give an observable rate. Iodination is probably out of the question because its rate is exceedingly slow, even at 500°K.

Laboratory halogenations show that our predictions are right. In fact, fluorine reacts explosively with methane, and chlorine reacts at a moderate rate. A mixture of bromine and methane must be heated to react, and iodine does not react at all.

PROBLEM 4-16

The bromination of methane proceeds through the following steps:

	$\Delta H°$ (per mole)	E_a (per mole)
Br$_2$ $\xrightarrow{h\nu}$ 2 Br·	+46 kcal (192 kJ)	46 kcal (192 kJ)
CH$_4$ + Br· → CH$_3$· + HBr	16 kcal (67 kJ)	18 kcal (75 kJ)
CH$_3$· + Br$_2$ → CH$_3$Br + Br·	−24 kcal (−101 kJ)	1 kcal (4 kJ)

(a) Draw a complete reaction-energy diagram for this reaction.
(b) Label the rate-limiting step.
(c) Draw the structure of each transition state.
(d) Compute the overall value of $\Delta H°$ for the bromination.

PROBLEM 4-17

(a) Using the BDEs in Table 4-2 (p. 134), compute the value of $\Delta H°$ for each step in the iodination of methane.
(b) Compute the overall value of $\Delta H°$ for iodination.
(c) Suggest *two* reasons why iodine does not react well with methane.

Up to now, our discussions of halogenation have used methane as the starting material. Using such a simple compound, we could concentrate on the thermodynamics and kinetics of the reaction. Now we consider halogenation of the "higher" alkanes, meaning those of higher molecular weight.

4-13
Halogenation of Higher Alkanes

4-13A Chlorination of Propane: Product Ratios

Halogenation is a substitution, where a halogen atom replaces a hydrogen.

$$R-H + X_2 \longrightarrow R-X + H-X$$

In methane, all four hydrogen atoms are identical, and it does not matter which hydrogen is replaced. In the higher alkanes, replacement of different hydrogen atoms leads to different products. In the chlorination of propane, for example, two mono-chlorinated (just one chlorine atom) products are possible. One has the chlorine atom on a primary carbon atom, and the other has the chlorine atom on the secondary carbon atom.

$$
\underset{\text{propane}}{\overset{\overset{\text{1° carbon} \qquad \text{2° carbon}}{\diagdown \qquad \diagup}}{CH_3-CH_2-CH_3}} + Cl_2 \xrightarrow{h\nu,\ 25°C} \underset{\substack{\text{1-chloropropane, 40\%} \\ (n\text{-propyl chloride})}}{\overset{\overset{\displaystyle Cl}{|}}{CH_2-CH_2-CH_3}} + \underset{\substack{\text{2-chloropropane, 60\%} \\ (\text{isopropyl chloride})}}{\overset{\overset{\displaystyle Cl}{|}}{CH_3-CH-CH_3}}
$$

The product ratio shows that replacement of hydrogen atoms by chlorine is not random. Propane has six primary hydrogens (hydrogens bonded to primary carbons) and only two secondary hydrogens (bonded to the secondary carbon), yet the major product results from substitution of a secondary hydrogen. We can calculate how reactive each kind of hydrogen is by dividing the amount of product observed by the number of hydrogens that can be replaced to give that product.

Figure 4-5 shows the definition of primary, secondary, and tertiary hydrogens and the calculation of their relative reactivity. Replacing either of the two secondary hydrogens accounts for 60% of the product, and replacing any of the six primary hydrogens accounts for 40% of the product. We calculate that each secondary hydrogen is 4.5 times as reactive as each primary hydrogen. To explain this preference for reaction at the secondary position, we must look carefully at the reaction mechanism (Figure 4-6).

When a chlorine atom reacts with propane, abstraction of a hydrogen atom can give either a primary radical or a secondary radical. The structure of the radical formed in this step determines the structure of the observed product, either 1-chloropropane or 2-chloropropane. The product ratio shows that the secondary radical is formed preferentially. This preference for reaction at the secondary position results from the greater stability of the secondary free radical and the transition state leading to it.

▶ Figure 4-5
There are six primary hydrogens in propane and only two secondary hydrogens, yet the major product results from replacement of a secondary hydrogen.

$$
\underset{\text{primary (1°) hydrogens}}{\overset{\overset{\displaystyle H}{|}}{R-\underset{\underset{\displaystyle H}{|}}{C}-H}} \qquad \underset{\text{secondary (2°) hydrogens}}{\overset{\overset{\displaystyle R}{|}}{R-\underset{\underset{\displaystyle H}{|}}{C}-H}} \qquad \underset{\text{tertiary (3°) hydrogen}}{\overset{\overset{\displaystyle R}{|}}{R-\underset{\underset{\displaystyle R}{|}}{C}-H}}
$$

Six primary (1°) hydrogens   *relative reactivity*

$$
\underset{}{H_2C\overset{\diagup CH_3}{\underset{\diagdown CH_3}{}}} \xrightarrow[\text{replacement}]{Cl_2,\ h\nu} \underset{\text{primary chloride}}{CH_3-CH_2-CH_2-Cl} \qquad \frac{40\%}{6\ \text{hydrogens}} = 6.67\%\ \text{per H}
$$

Two secondary (2°) hydrogens

$$
\underset{}{\overset{H}{\underset{H}{\diagdown}}C\overset{CH_3}{\underset{CH_3}{\diagup}}} \xrightarrow[\text{replacement}]{Cl_2,\ h\nu} \underset{\text{secondary chloride}}{\overset{\overset{\displaystyle Cl}{|}}{CH_3-CH-CH_3}} \qquad \frac{60\%}{2\ \text{hydrogens}} = 30.0\%\ \text{per H}
$$

The 2° hydrogens are $\dfrac{30.0}{6.67} = 4.5$ times as reactive as the 1° hydrogens.

Initiation: Splitting of the chlorine molecule

$$Cl_2 \;+\; h\nu \;\longrightarrow\; 2\,Cl\cdot$$

First propagation step: Abstraction (removal) of a primary or secondary hydrogen

$$CH_3{-}CH_2{-}CH_3 \;+\; Cl\cdot \;\longrightarrow\; \cdot CH_2{-}CH_2{-}CH_3 \quad \text{or} \quad CH_3{-}\overset{\cdot}{C}H{-}CH_3 \;+\; HCl$$

primary radical   secondary radical

Second propagation step: Reaction with chlorine to form the alkyl chloride

$$\cdot CH_2{-}CH_2{-}CH_3 \;+\; Cl_2 \;\longrightarrow\; Cl{-}CH_2{-}CH_2{-}CH_3 \;+\; Cl\cdot$$

primary radical   primary chloride
(1-chloropropane)

$$\text{or} \quad CH_3{-}\overset{\cdot}{C}H{-}CH_3 \;+\; Cl_2 \;\longrightarrow\; CH_3{-}\overset{\displaystyle Cl}{\overset{|}{C}H}{-}CH_3 \;+\; Cl\cdot$$

secondary radical   secondary chloride
(2-chloropropane)

▲ Figure 4-6

The mechanism for free-radical chlorination of propane. The first propagation step forms either a primary radical or a secondary radical. This radical determines whether the final product will be the primary chloride or the secondary chloride.

PROBLEM 4-18

What would be the product ratio in the chlorination of propane if all the hydrogens were abstracted at equal rates?

PROBLEM 4-19

Classify each hydrogen atom in the following compounds as primary (1°), secondary (2°), or tertiary (3°).

(a) butane   (b) isobutane   (c) 2-methylbutane
(d) cyclohexane   (e) norbornane (see page 117)

4-13B Free-Radical Stabilities

Figure 4-7 shows the energy required (the bond-dissociation energy) to form a free radical by breaking a bond between a hydrogen atom and a carbon atom. This energy is greatest for a methyl carbon, and it decreases for a primary carbon, a secondary carbon, and a tertiary carbon. The more highly substituted the carbon atom, the less energy is required to form the free radical.

From the information in Figure 4-7, we conclude that free radicals are more stable if they are more highly substituted. The following free radicals are listed in decreasing order of stability.

$$
\underset{3^\circ}{\overset{\displaystyle R}{\underset{\displaystyle R}{R{-}\overset{|}{\underset{|}{C}}\cdot}}} \;>\;
\underset{2^\circ}{\overset{\displaystyle R}{\underset{\displaystyle H}{R{-}\overset{|}{\underset{|}{C}}\cdot}}} \;>\;
\underset{1^\circ}{\overset{\displaystyle H}{\underset{\displaystyle H}{R{-}\overset{|}{\underset{|}{C}}\cdot}}} \;>\;
\underset{Me\cdot}{\overset{\displaystyle H}{\underset{\displaystyle H}{H{-}\overset{|}{\underset{|}{C}}\cdot}}}
$$

tertiary > secondary > primary > methyl

Formation of a methyl radical   *Bond-dissociation energy*

$$CH_4 \longrightarrow H\cdot + \cdot CH_3 \qquad \Delta H° = 104 \text{ kcal (435 kJ)}$$

Formation of a primary (1°) radical

$$CH_3-CH_2-CH_3 \longrightarrow H\cdot + CH_3-CH_2-\dot{C}H_2 \quad \Delta H° = 98 \text{ kcal (410 kJ)}$$

Formation of a secondary (2°) radical

$$CH_3-CH_2-CH_3 \longrightarrow H\cdot + CH_3-\dot{C}H-CH_3 \quad \Delta H° = 95 \text{ kcal (397 kJ)}$$

Formation of a tertiary (3°) radical

$$
\begin{array}{c}
CH_3 \\
| \\
CH_3-\underset{\underset{CH_3}{|}}{C}-H
\end{array}
\longrightarrow
H\cdot +
\begin{array}{c}
CH_3 \\
| \\
CH_3-\underset{\underset{CH_3}{|}}{C}\cdot
\end{array}
\qquad \Delta H° = 91 \text{ kcal (381 kJ)}
$$

▶ **Figure 4-7**
Bond-dissociation energies show that more highly substituted free radicals are more stable than less highly substituted ones.

In the chlorination of propane, the secondary hydrogen atom is abstracted more often because the secondary radical and the transition state leading to it are lower in energy than the primary radical and its transition state. Using the bond-dissociation energies in Table 4-2 (page 134), we can calculate $\Delta H°$ for each of the possible reaction steps. Abstraction of the secondary hydrogen is 3 kcal/mol (13 kJ/mol) more exothermic than abstraction of the primary hydrogen.

$$1° \text{ H:} \quad CH_3-CH_2-CH_3 + Cl\cdot \longrightarrow CH_3-CH_2-CH_2\cdot + H-Cl$$

Energy required to break the $CH_3CH_2CH_2\frac{2}{5}H$ bond	$+98$ kcal/mol ($+410$ kJ/mol)
Energy released in forming the $H\frac{2}{5}Cl$ bond	-103 kcal/mol (-431 kJ/mol)
Total energy for reaction at the primary position:	-5 kcal/mol (-21 kJ/mol)

$$2° \text{ H:} \quad CH_3-\underset{\underset{CH_3}{|}}{\overset{\overset{CH_3}{|}}{CH_2}} + Cl\cdot \longrightarrow CH_3-\underset{|}{\overset{\overset{CH_3}{|}}{CH}}\cdot + H-Cl$$

Energy required to break the $CH_3-CH\frac{2}{5}H$ bond	$+95$ kcal/mol ($+397$ kJ/mol)
Energy released in forming the $H\frac{2}{5}Cl$ bond	-103 kcal/mol (-431 kJ/mol)
Total energy for reaction at the secondary position:	-8 kcal/mol (-34 kJ/mol)

A reaction-energy diagram for this rate-limiting first propagation step appears in Figure 4-8. The activation energy to form the secondary radical is slightly lower, so the secondary radical is formed faster than the primary radical.

SOLVED PROBLEM 4-5

Tertiary hydrogen atoms react with Cl· about 5.5 times as fast as primary ones. Predict the product ratios for chlorination of isobutane.

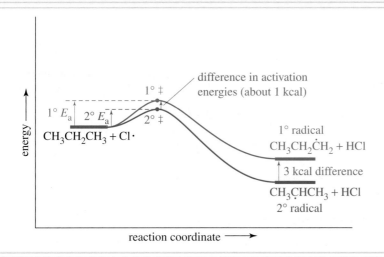

◀ **Figure 4-8**
Reaction-energy diagram for the first propagation step in the chlorination of propane. Formation of the secondary radical has a lower activation energy than does formation of the primary radical.

SOLUTION

There are nine primary hydrogens and one tertiary hydrogen in isobutane.

(9 primary hydrogens) × (reactivity 1.0) = 9.0 relative amount of reaction

(1 tertiary hydrogen) × (reactivity 5.5) = 5.5 relative amount of reaction

Even though the primary hydrogens are less reactive, there are so many of them that the primary product is the major product. The product ratio will be 9.0 : 5.5 or about 1.6 : 1.

$$\text{Fraction of primary} = \frac{9.0}{9.0 + 5.5} = 62\%$$

$$\text{Fraction of tertiary} = \frac{5.5}{9.0 + 5.5} = 38\%$$

major product
62%

minor product
38%

PROBLEM 4-20

Use the bond-dissociation energies in Table 4-2 (page 134) to calculate the heats of reaction for the two possible first propagation steps in the chlorination of isobutane. Use this information to draw a reaction-energy diagram like Figure 4-8, comparing the activation energies for formation of the two radicals.

PROBLEM 4-21

Predict the ratios of products that result when isopentane (2-methylbutane) is chlorinated.

PROBLEM 4-22

(a) When *n*-heptane burns in a gasoline engine, the combustion process takes place too quickly. The explosive detonation makes a noise called knocking. When 2,2,4-trimethylpentane (isooctane) is burned, combustion takes place in a slower, more controlled manner. Combustion is a free-radical chain reaction, and its rate depends on the reactivity of the free-radical intermediates. Explain why isooctane has less tendency to knock than does *n*-heptane.

(b) Alkoxy radicals (R—O·) are generally more stable than alkyl (R·) radicals. Write an equation showing an alkyl free radical (from burning gasoline) abstracting a hydrogen atom from *t*-butyl alcohol, $(CH_3)_3COH$. Explain why *t*-butyl alcohol works as an anti-knock additive for gasoline.

4-13C Bromination of Propane

Figure 4-9 shows the free-radical reaction of propane with bromine. Notice that this reaction is both heated to 125°C and irradiated with light to achieve a moderate rate. The secondary bromide (2-bromopropane) is favored by a 97:3 product ratio. From this product ratio, we calculate that the two secondary hydrogens are each 97 times as reactive as one of the primary hydrogens.

The 97:1 reactivity ratio for bromination is much larger than the 4.5:1 ratio for chlorination. We say that bromination is more *selective* than chlorination because the major reaction is favored by a larger amount. To explain this enhanced selectivity, we must consider the transition states and activation energies for the rate-limiting step.

As with chlorination, the rate-limiting step in bromination is the first propagation step: abstraction of a hydrogen atom by a bromine radical. The energetics of the two possible hydrogen abstractions are shown below. Compare these numbers with the energetics of the first propagation step of chlorination shown on page 146. The bond energies are taken from Table 4-2 (page 134).

$$1° \text{ H:} \quad CH_3—CH_2—CH_3 \ + \ Br· \ \longrightarrow \ CH_3—CH_2—CH_2· \ + \ H—Br$$

Energy required to break the $CH_3CH_2CH_2\!\!-\!\!H$ bond	+98 kcal/mol (+410 kJ/mol)
Energy released in forming the $H\!\!-\!\!Br$ bond	−88 kcal/mol (−368 kJ/mol)
Total energy for reaction at the primary position:	+10 kcal/mol (+42 kJ/mol)

$$CH_3—CH_2—CH_3 \ + \ Br_2 \ \xrightarrow{hv,\ 125°C} \ \overset{\overset{\displaystyle Br}{|}}{CH_3—CH_2—CH_2} \ + \ \overset{\overset{\displaystyle Br}{|}}{CH_3—CH—CH_3} \ + \ HBr$$

primary bromide, 3%   secondary bromide, 97%

Relative reactivity

six primary hydrogens $\dfrac{3\%}{6} = 0.5\%$ per H

two secondary hydrogens $\dfrac{97\%}{2} = 48.5\%$ per H

The 2° hydrogens are $\dfrac{48.5}{0.5} = 97$ times as reactive as the 1° hydrogens.

▲ **Figure 4-9**
This 97:3 ratio of products shows that bromine abstracts a secondary hydrogen 97 times as rapidly as a primary hydrogen. Bromination (reactivity ratio 97:1) is much more selective than chlorination (reactivity ratio 4.5:1).

◀ **Figure 4-10**
Reaction-energy diagram for the first
propagation step in the bromination
of propane. The energy difference in
the transition states is nearly as large
as the energy difference in the
products.

$$2° \text{ H: } CH_3-\overset{\overset{\displaystyle CH_3}{|}}{CH_2} + Br \cdot \longrightarrow CH_3-\overset{\overset{\displaystyle CH_3}{|}}{CH} \cdot + H-Br$$

Energy required to break the $CH_3-\overset{CH_3}{\underset{\vert}{CH}} \overset{2}{\underset{\overline{}}{}} H$ bond	+95 kcal/mol (+397 kJ/mol)
Energy released in forming the $H \overset{2}{\underline{\overline{}}} Br$ bond	−88 kcal/mol (−368 kJ/mol)
Total energy for reaction at the secondary position:	+7 kcal/mol (+29 kJ/mol)

The energy differences between chlorination and bromination result from the difference in the bond-dissociation energies of H—Cl (103 kcal) and H—Br (88 kcal). The HBr bond is weaker, and abstraction of a hydrogen atom by Br· is endothermic. This endothermic step explains why bromination is much slower than chlorination, but it still does not explain the enhanced selectivity observed with bromination.

Consider the reaction-energy diagram for the first propagation step in the bromination of propane (Fig. 4-10). Although the difference in values of $\Delta H°$ between abstraction of a primary hydrogen and a secondary hydrogen is still 3 kcal/mol (13 kJ/mol), the energy profile for bromination shows a much larger difference in activation energies for abstraction of the primary and secondary hydrogens than we saw for chlorination (Fig. 4-8).

4-14
The Hammond Postulate

Figure 4-11 summarizes the energy diagrams for bromination and chlorination of propane. Together, these energy diagrams explain enhanced selectivity observed in bromination.

When we compare the reaction-energy diagrams for the first propagation steps of chlorination and bromination, there are two important differences:

1. The first propagation step is endothermic for bromination but exothermic for chlorination.

2. The transition states forming the 1° and 2° radicals for the endothermic bromination have a larger energy difference than those for the exothermic chlorination, even though the energy difference of the products is the same (3 kcal, or 13 kJ) in both reactions.

(a) BROMINATION
endothermic
TS close to products
large difference in E_a

(b) CHLORINATION
exothermic
TS close to reactants
small difference in E_a

▲ **Figure 4-11**
(**a**) In the endothermic bromination, the transition states are closer to the products (the radicals) in energy and in structure. The difference in the 1° and 2° activation energies is about 2.5 kcal (10 kJ), nearly the entire energy difference of the radicals. (**b**) In the exothermic chlorination, the transition states are closer to the reactants in energy and in structure. The difference in activation energies for chlorination is about 1 kcal (4 kJ), only a third of the energy difference of the radicals.

In general, we will find that these differences are related:

In an endothermic reaction, the transition state is closer to the products in energy and in structure. In an exothermic reaction, the transition state is closer to the reactants in energy and in structure.

Figure 4-12 compares the transition states for bromination and chlorination. The product-like transition state for bromination has the C—H bond nearly broken and a great deal of radical character on the carbon atom. The energy of this transition state reflects most of the energy difference of the radical products. The reactant-like transition state for chlorination has the C—H bond just beginning to break, with little radical character on the carbon atom. This transition state reflects only a small part (about a third) of the energy difference of the radical products. Therefore, chlorination is less selective.

▶ **Figure 4-12**
In the endothermic bromination, the transition state resembles the products (the free radical and HBr). In the exothermic chlorination, the free radical has just begun to form in the transition state, so the transition state resembles the reactants.

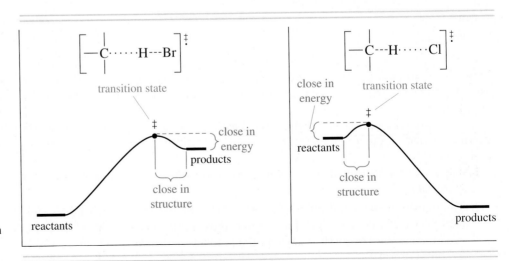

These reactions are examples of a more general principle, called the **Hammond postulate**.

> HAMMOND POSTULATE: Related species that are closer in energy are also closer in structure. The structure of a transition state resembles the structure of the closest stable species.

This general rule tells us something about the transition states in endothermic and exothermic reactions. The transition state is always the point of highest energy on the energy diagram. Its structure resembles either the reactants or the products, whichever ones are higher in energy. In an endothermic reaction, the products are higher in energy, and the transition state is product-like. In an exothermic reaction, the reactants are higher in energy, and the transition state is reactant-like. Thus, the Hammond postulate helps us understand why exothermic processes tend to be less selective than similar endothermic processes.

PROBLEM 4-23

(a) Compute the heats of reaction for abstraction of a primary hydrogen and a secondary hydrogen from propane by a fluorine radical.

$$CH_3-CH_2-CH_3 + F\cdot \longrightarrow CH_3-CH_2-\dot{C}H_2 + HF$$

$$CH_3-CH_2-CH_3 + F\cdot \longrightarrow CH_3-\dot{C}H-CH_3 + HF$$

(b) How selective do you expect free-radical fluorination to be?
(c) What product distribution would you expect to obtain from the free-radical fluorination of propane?

PROBLEM-SOLVING STRATEGY
Proposing Reaction Mechanisms

Throughout this course, we will propose mechanisms to explain reactions. Methods for dealing with different types of mechanisms will be discussed as we encounter them. These techniques for dealing with a variety of mechanisms are collected in Appendix 4, but at this point we consider only free-radical mechanisms like those in this chapter.

Free-Radical Reactions

General principles: Free-radical reactions generally proceed by chain-reaction mechanisms, using an initiator with an easily broken bond (such as chlorine, bromine, or a peroxide) to start the chain reaction. In drawing the mechanism, expect free-radical intermediates (especially highly substituted or resonance-stabilized intermediates). Watch for the most stable free radicals, and avoid any high-energy radicals such as hydrogen atoms.

1. **Draw a step that breaks the weak bond in the initiator.**
   A free-radical reaction usually begins with an initiation step in which the initiator undergoes homolytic (free-radical) cleavage to give two radicals.

2. **Draw a reaction of the initiator with one of the starting materials.**
   One of the initiator radicals reacts with one of the starting materials to give a free-radical version of the starting material. The initiator might abstract a hydrogen atom or add to a double bond, depending on what reaction leads to formation of the observed product. You might want to consider bond-dissociation energies to see which reaction is energetically favored.

3. **Draw a reaction of the free-radical version of the starting material with another starting material molecule to form a bond needed in the product and generate a new radical intermediate.**
   Check your intermediates to be sure that you have used the most stable radical intermediates. For a realistic chain reaction, no new initiation steps should be required; a radical should be regenerated in each propagation step.

4. Draw termination step(s).

The reaction ends with termination steps, which are side reactions rather than part of the product-forming mechanism. The reaction of any two free radicals to give a stable molecule is a termination step, as is a collision of a free radical with the container.

Before we illustrate this procedure, let's consider a few common mistakes. Avoiding these mistakes will help you to draw correct mechanisms throughout this course.

Common Mistakes to Avoid

1. Do not use condensed or line–angle formulas for reaction sites. Draw all the bonds and all the substituents of each carbon atom affected throughout the mechanism. Three-bonded carbon atoms in intermediates are most likely to be radicals in the free-radical reactions we have studied. If you draw condensed formulas or line–angle formulas, you will likely misplace a hydrogen atom and show a reactive species on the wrong carbon.
2. Do not show more than one step occurring at once, unless they really do occur at once.

PROBLEM-SOLVING **HINT**

Free-radical bromination is highly selective, chlorination is moderately selective, and fluorination is nearly nonselective.

SAMPLE PROBLEM

Draw a mechanism for the reaction of methylcyclopentane with bromine under irradiation with light. Predict the major product.

In every mechanism problem, we first draw what we know, showing all the bonds and all the substituents of each carbon atom that may be affected throughout the mechanism.

1. **Draw a step involving cleavage of the weak bond in the initiator.**

The use of light with bromine suggests a free-radical reaction, with light providing the energy for dissociation of Br_2. This homolytic cleavage initiates the chain reaction by generating two Br· radicals.

Initiation step

$$Br - Br \xrightarrow{hv} Br\cdot + \cdot Br$$

2. **Draw a reaction of the initiator with one of the starting materials.**

One of these initiator radicals should react with methylcyclopentane to give a free-radical version of methylcyclopentane. As we have seen, a bromine or chlorine radical can abstract a hydrogen atom from an alkane to generate an alkyl radical. The bromine radical is highly selective, and the most stable alkyl radical should result. Abstraction of the tertiary hydrogen atom gives a tertiary radical.

First propagation step

3. **Draw a reaction of the free-radical version of the starting material with another starting material molecule to form a bond needed in the product and to generate a new radical intermediate.**

The alkyl radical should react with another starting-material molecule, in another propagation step, to generate a product and another radical. Reaction of the alkyl radical with Br_2 gives 1-bromo-1-methylcyclopentane (the major product) and another bromine radical to continue the chain.

Second propagation step

major product

4. **Draw termination step(s).**

It is left to you to add some possible termination steps and summarize the mechanism developed above.

As practice in using a systematic approach to proposing mechanisms for free-radical reactions, work Problem 4-24 by going through the four steps outlined above.

PROBLEM 4-24

2,3-Dimethylbutane reacts with bromine in the presence of light to give a good yield of a monobrominated product. Further reaction gives a dibrominated product. Predict the structures of these products, and propose a mechanism for the formation of the monobrominated product.

PROBLEM 4-25

In the presence of a small amount of bromine, cyclohexene undergoes the following light-promoted reaction:

cyclohexene   3-bromocyclohexene

(a) Propose a mechanism for this reaction.
(b) Draw the structure of the rate-limiting transition state.
(c) Use the Hammond postulate to predict which intermediate most closely resembles this transition state.
(d) Explain why cyclohexene reacts with bromine much faster than cyclohexane, which must be heated to react.

4-15
Radical Inhibitors

We often want to prevent or retard free-radical reactions. For example, oxygen in the air oxidizes and spoils foods, solvents, and other compounds mostly by free-radical chain reactions. Chemical intermediates may decompose or polymerize by free-radical chain reactions. Even the cells in living systems are damaged by radical reactions, which can lead to aging, cancerous mutations, or cell death.

Radical inhibitors are often added to foods and chemicals to retard spoilage by radical chain reactions. Chain reactions depend on the individual steps being fast, so that each initiation step results in many molecules reacting, as in the reaction-energy diagram at the left of the figure. (Only the radicals are shown.)

chain reaction   with inhibitor I

The diagram at right in the figure shows how an inhibitor (I) can stop the chain by reacting with a radical intermediate in a fast, highly exothermic step to form an intermediate that is relatively stable. The next step in the chain becomes endothermic and very slow.

"Butylated hydroxyanisole" (BHA) is often added to foods as an antioxidant. It stops oxidation by reacting with radical intermediates to form a relatively stable free-radical intermediate (BHA radical). The BHA radical can react with a second free radical to form an even more stable quinone with all its electrons paired.

BHA   BHA radical   a quinone

vitamine E
(R = alkyl chain)

Radical inhibitors also help to protect the cells of living systems. Like BHA, vitamin E is also a phenol (an aromatic ring with an —OH group), and it is thought to react with radicals by losing the OH hydrogen atom as shown above for BHA. Ascorbic acid (vitamin C) is also thought to protect cells from free radicals, possibly by the following mechanism:

ascorbic acid
(vitamin C)

stabilized free radical

PROBLEM 4-26

Draw resonance forms to show how the BHA radical is stabilized by delocalization of the radical electron over other atoms in the molecule.

PROBLEM 4-27

Write an equation for the reaction of vitamin E with an oxidizing radical (RO·) to give ROH and a less reactive free radical.

4-16
Reactive
Intermediates

The free radicals we have studied are one class of reactive intermediates. **Reactive intermediates** are short-lived species that are never present in high concentrations because they react as quickly as they are formed. In most cases, reactive intermediates are fragments of molecules (like free radicals), often having atoms with unusual numbers of bonds. Some of the common reactive intermediates contain carbon atoms with only two or three bonds, compared with carbon's four bonds in its stable compounds. Such species react quickly with a variety of compounds to give more stable products with tetravalent carbon atoms.

Although reactive intermediates are not stable compounds, they are important to our study of organic chemistry. Most reaction mechanisms involve reactive intermediates. If you are to understand these mechanisms and propose mechanisms of your own, you need to know how reactive intermediates are formed and how they are likely to react. In this chapter, we consider their structure and stability, and in later chapters, we see how they are formed and ways they react to give stable compounds.

Species with trivalent (three-bonded) carbon are classified according to their charge, which depends on the number of nonbonding electrons. The *carbocations* have no nonbonding electrons and are positively charged. The *radicals* have one nonbonding electron and are neutral. The *carbanions* have a pair of nonbonding electrons and are negatively charged.

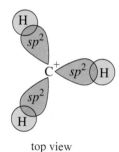

top view

side view

▲ **Figure 4-13**
The methyl cation is similar to BH_3. The carbon atom is sigma bonded to three hydrogen atoms by overlap of its sp^2 hybrid orbitals with the s orbitals of hydrogen. There is a vacant p orbital perpendicular to the plane of the three C—H bonds.

The most common intermediates with a divalent (two-bonded) carbon atom are the *carbenes*. A carbene has two nonbonding electrons on the divalent carbon atom, making it uncharged.

4-16A   Carbocations

A **carbocation** (also called a **carbonium ion** or a **carbenium ion**) is a species that contains a carbon atom bearing a positive charge. The positively charged carbon atom is bonded to three other atoms, and it has no nonbonding electrons, so it has only six electrons in its valence shell. It is sp^2 hybridized, with a planar structure and bond angles of about 120°. For example, the methyl cation (CH_3^+) is planar, with bond angles of exactly 120°. The unhybridized p orbital is vacant, and lies perpendicular to the plane of the C—H bonds (Figure 4-13). The structure of CH_3^+ is similar to the structure of BH_3, discussed in Chapter 2.

With only six electrons in the positive carbon's valence shell, a carbocation is a powerful electrophile (Lewis acid), and it may react with any nucleophile it encounters. Carbocations are proposed as intermediates in many organic reactions, some of which we will encounter in Chapter 6.

Like free radicals, carbocations are *electron-deficient* species: They have fewer than eight electrons in the valence shell. Also like free radicals, carbocations are stabilized by alkyl substituents. An alkyl group stabilizes an electron-deficient carbocation in two ways: (1) through an inductive effect, and (2) through the partial overlap of filled orbitals with empty ones. The **inductive effect** is a donation of electron density through the sigma bonds of the molecule. The positively charged carbon atom withdraws some electron density from alkyl groups bonded to it.

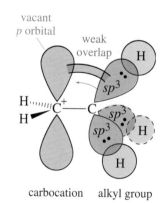

carbocation   alkyl group

▲ **Figure 4-14**
A carbocation is stabilized by overlap of filled orbitals on an adjacent alkyl group with the vacant p orbital of the carbocation. Overlap between a sigma bond and a p orbital is called hyperconjugation.

Alkyl substituents also have filled sp^3 orbitals that can overlap with the empty p orbital on the positively charged carbon atom, further stabilizing the carbocation (Figure 4-14). Even though the attached alkyl group rotates, one of its sigma bonds is always aligned with the empty p orbital on the carbocation. The pair of electrons in this sigma bond spreads out into the empty *p* orbital, stabilizing the electron-deficient carbon atom. This type of overlap between a *p* orbital and a sigma bond is called *hyperconjugation*. In general, more highly substituted carbocations are more stable.

Stability of carbocations

$$3° > 2° > 1° > methyl$$

Unsaturated carbocations are also stabilized by **resonance stabilization**. If a pi bond is adjacent to a carbocation, the filled *p* orbitals of the pi bond will overlap with the empty *p* orbital of the carbocation. The result is a delocalized ion, with the positive charge shared by two atoms. Resonance delocalization is particularly effective in stabilizing carbocations.

Carbocations are common intermediates in organic reactions. Highly substituted alkyl halides can ionize when they are heated in a polar solvent. The strongly electrophilic carbocation reacts with any available nucleophile, often the solvent.

PROBLEM 4-28

The triphenylmethyl cation is so stable that some of its salts can be stored for months. Explain why this cation is so stable.

triphenylmethyl cation

PROBLEM 4-29

Rank the following carbocations in decreasing order of stability. Classify each as primary, secondary, or tertiary.

(a) The isopentyl cation, $(CH_3)_2CHCH_2$—CH_2^+

(b) The 3-methyl-2-butyl cation, CH_3—$\overset{+}{CH}$—$CH(CH_3)_2$

(c) The 2-methyl-2-butyl cation, CH_3—$\overset{+}{C}(CH_3)CH_2CH_3$

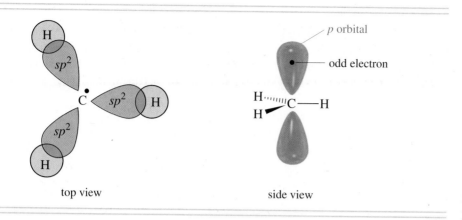

top view   side view

◀ **Figure 4-15**
The structure of the methyl radical is like that of the methyl cation (Fig. 4-13), except there is an additional electron. The odd electron is in the *p* orbital perpendicular to the plane of the three C—H bonds.

4-16B   Free Radicals

Like carbocations, **free radicals** are sp^2 hybridized and planar. Unlike carbocations, however, the *p* orbital perpendicular to the plane of the C—H bonds of the radical is not empty; it contains the odd electron. Figure 4-15 shows the structure of the methyl radical.

Radicals and carbocations are both electron deficient because they lack an octet around the carbon atom. Like carbocations, radicals are stabilized by the electron-donating effect of alkyl groups, making more highly substituted radicals more stable. This effect is confirmed by the bond-dissociation energies shown in Figure 4-7: Less energy is required to break a C—H bond to form a more highly substituted radical.

Vitamin B$_{12}$ is an essential dietary factor and a deficiency results in anemia and neurological damage. The vitamin assists two different enzymes in the production and the stabilization of methyl radicals. These methyl radicals are then used for the synthesis of important cellular components.

$$
\begin{array}{ccccccc}
 & R & & R & & H & & H \\
 & | & & | & & | & & | \\
R-&C\cdot& > & R-C\cdot & > & R-C\cdot & > & H-C\cdot \\
 & | & & | & & | & & | \\
 & R & & H & & H & & H \\
\end{array}
$$

most stable   least stable

Stability of radicals

$$3° \quad > \quad 2° \quad > \quad 1° \quad > \quad methyl$$

Like carbocations, radicals can be stabilized by resonance. Overlap with the *p* orbitals of a pi bond allows the odd electron to be delocalized over two carbon atoms. Resonance delocalization is particularly effective in stabilizing a radical.

PROBLEM 4-30

Rank the following radicals in decreasing order of stability. Classify each as primary, secondary, or tertiary.

(a) The isopentyl radical, $(CH_3)_2CHCH_2-\dot{C}H_2$

(b) The 3-methyl-2-butyl radical, $CH_3-\dot{C}H-CH(CH_3)_2$

(c) The 2-methyl-2-butyl radical, $CH_3-\dot{C}(CH_3)CH_2CH_3$

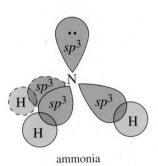

▲ **Figure 4-16**
Both the methyl anion and ammonia
have an sp^3 hybridized central atom,
with a nonbonding pair of electrons
occupying one of the tetrahedral
positions.

4-16C   Carbanions

A **carbanion** has a trivalent carbon atom that bears a negative charge. There are eight electrons around the carbon atom (three bonds and one lone pair), so it is not electron deficient; rather, it is electron rich and a strong nucleophile (Lewis base). A carbanion has the same electronic structure as an amine. Compare the structures of a methyl carbanion and ammonia:

methyl anion   ammonia

The hybridization and bond angles of a simple carbanion also resemble those of an amine. The carbon atom is sp^3 hybridized and tetrahedral. One of the tetrahedral positions is occupied by an unshared lone pair of electrons. Figure 4-16 compares the orbital structures and geometry of ammonia and the methyl anion.

Like amines, carbanions are nucleophilic and basic. A carbanion has a negative charge on its carbon atom, however, making it a more powerful base and a stronger nucleophile than an amine. For example, a carbanion is sufficiently basic to remove a proton from ammonia.

$$R_3C{:}^- \;+\; {:}NH_3 \;\longrightarrow\; R_3CH \;+\; {}^-{:}\ddot{N}H_2$$

The stability order of carbanions reflects their high electron density. Alkyl groups and other electron-donating groups actually destabilize a carbanion. The order of stability is the opposite of that for carbocations and free radicals.

least stable   most stable

Stability of carbanions

$$3° \;<\; 2° \;<\; 1° \;<\; \text{methyl}$$

Carbanions that occur as intermediates in organic reactions are almost always bonded to stabilizing groups. They can be stabilized either by inductive effects or by resonance. For example, halogen atoms are electron withdrawing, so they stabilize carbanions through the inductive withdrawal of electron density. Resonance also plays an important role in stabilizing carbanions. A carbonyl group (C=O) stabilizes an adjacent carbanion by overlap of its pi bond with the nonbonding electrons of the carbanion. The negative charge is delocalized onto the electronegative oxygen atom of the carbonyl group.

resonance-stabilized carbanion

This resonance-stabilized carbanion must be sp^2-hybridized and planar for effective delocalization of the negative charge onto oxygen (Section 2-6). Resonance-stabilized carbanions are the most common type of carbanions we will encounter in organic reactions.

PROBLEM 4-31

Acetylacetone (2,4-pentanedione) reacts with sodium hydroxide to give water and the sodium salt of a carbanion. Write a complete structural formula for the carbanion, and use resonance forms to show the stabilization of the carbanion.

$$H_3C-\overset{\overset{\displaystyle O}{\|}}{C}-CH_2-\overset{\overset{\displaystyle O}{\|}}{C}-CH_3$$

acetylacetone (2,4-pentanedione)

PROBLEM 4-32

Acetonitrile ($CH_3C{\equiv}N$) is deprotonated by very strong bases. Write resonance forms to show the stabilization of the carbanion that results.

4-16D   Carbenes

Carbenes are uncharged reactive intermediates containing a divalent carbon atom. The simplest carbene has the formula $:CH_2$ and is called *methylene*, just as a $-CH_2-$ group in a molecule is called a *methylene group*. One way of generating carbenes is to form a carbanion that can expel a halide ion. For example, a strong base can abstract a proton from tribromomethane ($CHBr_3$) to give an inductively stabilized carbanion. This carbanion expels bromide ion to give dibromocarbene.

$$\underset{\text{tribromomethane}}{Br-\overset{\overset{\displaystyle Br}{|}}{\underset{\underset{\displaystyle Br}{|}}{C}}-H \quad {}^-OH} \quad \rightleftharpoons \quad H_2O \;+\; \underset{\text{a carbanion}}{Br-\overset{\overset{\displaystyle Br}{|}}{\underset{\underset{\displaystyle Br}{|}}{C}}:^-} \quad \longrightarrow \quad \underset{\text{dibromocarbene}}{\overset{\displaystyle Br}{\underset{\displaystyle Br}{>}}C:} \;+\; Br^-$$

The electronic structure of dibromocarbene is shown below. The carbon atom has only six electrons in its valence shell. It is sp^2 hybridized, with trigonal geometry. An unshared pair of electrons occupies one of the sp^2 hybrid orbitals, and there is an empty p orbital extending above and below the plane of the atoms. A carbene has both a lone pair of electrons and an empty p orbital, so it can react as a nucleophile or as an electrophile.

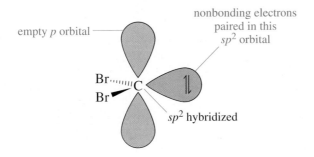

empty p orbital

nonbonding electrons paired in this sp^2 orbital

sp^2 hybridized

Methylene itself is formed when diazomethane (CH_2N_2) is heated or irradiated with light. The diazomethane molecule splits to form a stable nitrogen molecule and the very reactive carbene.

$$\underset{\text{diazomethane}}{\overset{\displaystyle H}{\underset{\displaystyle H}{>}}\overset{-}{C}{:}-\overset{+}{N}{\equiv}N:} \quad \xrightarrow{\text{heat or light}} \quad \underset{\text{methylene}}{\overset{\displaystyle H}{\underset{\displaystyle H}{>}}C:} \;+\; \underset{\text{nitrogen}}{:N{\equiv}N:}$$

The most common synthetic reaction of carbenes is their addition to double bonds to form cyclopropane rings. For example, dibromocarbene adds to cyclohexene to give an interesting bicyclic compound.

cyclohexene   dibromocarbene

Carbenes have never been purified or even made in a high concentration, because when two carbenes collide, they immediately dimerize to give an alkene.

Carbenes and carbenoids (carbene-like reagents) are useful both for the synthesis of other compounds and for the investigation of reaction mechanisms. The carbene intermediate is generated in the presence of its target compound, so that it can react immediately, and the concentration of the carbene is always low. Reactions using carbenes are discussed in Chapter 8.

PROBLEM 4-33

When it is strongly heated, ethyl diazoacetate decomposes to give nitrogen gas and a carbene. Draw a Lewis structure of the carbene.

$$:N\overset{+}{\equiv}N-\overset{-}{C}H-\overset{\overset{\displaystyle O}{\|}}{C}-O-CH_2CH_3$$

ethyl diazoacetate

SUMMARY   **Reactive Intermediates**

	Structure	Stability			Properties
carbocations	$-\overset{\|}{\underset{\|}{C}}{}^{+}$	$3° >$	$2° > 1° >$	${}^{+}CH_3$	electrophilic strong acids
radicals	$-\overset{\|}{\underset{\|}{C}}\cdot$	$3° >$	$2° > 1° >$	$\cdot CH_3$	electron deficient
carbanions	$-\overset{\|}{\underset{\|}{C}}:^{-}$	$^{-}:CH_3 >$	$1° > 2° >$	$3°$	nucleophilic strong bases
carbenes	$\diagup\!\!\diagdown C:$				both nucleophilic and electrophilic

---

Chapter 4 Glossary

activation energy (E_a) The energy difference between the reactants and the transition state; the minimum energy the reactants must have for the reaction to occur. (p. 139)
bond-dissociation energy **(BDE)** The amount of energy required to break a particular bond homolytically, to give radicals. (p. 133)

$$A{:}B \longrightarrow A\cdot + B\cdot \qquad \Delta H° = BDE$$

carbanion A strongly nucleophilic species with a negatively charged carbon atom having only three bonds. The carbon atom has a nonbonding pair of electrons. (p. 158)

carbene A highly reactive species with only two bonds to an uncharged carbon atom and with a nonbonding pair of electrons. The simplest carbene is methylene, $:CH_2$. (p. 159)

carbocation (carbonium ion, carbenium ion) A strongly electrophilic species with a positively charged carbon atom having only three bonds. (p. 155)

catalyst A substance that increases the rate of a reaction (by lowering E_a) without being consumed in the reaction. (p. 141)

chain reaction A multistep reaction where a reactive intermediate formed in one step brings about a second step that generates the intermediate needed for the following step. (p. 125)

    initiation step: The preliminary step in a chain reaction, where the reactive intermediate is first formed.

    propagation steps: The steps in a chain reaction that are repeated over and over to form the product. The sum of the propagation steps should give the net reaction.

    termination steps: Any steps where a reactive intermediate is consumed without another one being generated.

enthalpy (heat content; H) A measure of the heat energy in a system. In a reaction, the heat absorbed or evolved is called the *heat of reaction*, $\Delta H°$. A decrease in enthalpy (negative $\Delta H°$) is favorable for a reaction. (p. 131)

    endothermic: Consuming heat (having a positive $\Delta H°$).

    exothermic: Giving off heat (having a negative $\Delta H°$).

entropy (S) A measure of disorder or freedom of motion. An increase in entropy (positive $\Delta S°$) is favorable for a reaction. (p. 132)

equilibrium A state of a system such that no more change is taking place; the rate of the forward reaction equals the rate of the reverse reaction. (p. 129)

equilibrium constant A quantity calculated from the relative amounts of the products and reactants present at equilibrium. (p. 129) For the reaction

$$a\text{A} + b\text{B} \rightleftharpoons c\text{C} + d\text{D}$$

the equilibrium constant is

$$K_{eq} = \frac{[\text{C}]^c [\text{D}]^d}{[\text{A}]^a [\text{B}]^b}$$

free energy (**Gibbs free energy**; G) A measure of a reaction's tendency to go in the direction written. A decrease in free energy (negative ΔG) is favorable for a reaction. (p. 129)

    Free-energy change is defined: $\Delta G = \Delta H - T\Delta S$

standard Gibbs free energy change: ($\Delta G°$) The free-energy change corresponding to reactants and products in their standard states (pure substances in their most stable states) at 25°C and 1 atm pressure. $\Delta G°$ is related to K_{eq} by

$$K_{eq} = e^{-\Delta G°/RT} \qquad \text{(p. 130)}$$

Hammond postulate Related species (on a reaction-energy diagram) that are closer in energy are also closer in structure. In an exothermic reaction, the transition state is closer to the reactants in energy and in structure. In an endothermic reaction, the transition state is closer to the products in energy and in structure. (p. 149)

heterolytic cleavage (ionic cleavage) The breaking of a bond in such a way that one of the atoms retains both of the bond's electrons. A heterolytic cleavage forms two ions. (p. 133)

$$\text{A} \!:\! \text{B} \longrightarrow \text{A} \!:\!^- + \text{B}^+$$

homolytic cleavage (radical cleavage) The breaking of a bond in such a way that each atom retains one of the bond's two electrons. A homolytic cleavage produces two radicals. (p. 133)

$$\text{A} \!:\! \text{B} \longrightarrow \text{A} \cdot + \cdot \text{B}$$

inductive effect A donation (or withdrawal) of electron density through sigma bonds. (p. 155)

intermediate A molecule or a fragment of a molecule that is formed in a reaction and exists for a finite length of time before it reacts in the next step. An intermediate corresponds to a relative minimum (a low point) in the reaction-energy diagram. (p. 140)

reactive intermediate: A short-lived species that is never present in high concentration because it reacts as quickly as it is formed. (p. 155)

kinetics The study of reaction rates. (p. 136)

mechanism The step-by-step pathway from reactants to products showing which bonds break and which bonds form in what order. The mechanism should include the structures of all intermediates and arrows to show the movement of electrons. (p. 125)

potential-energy diagram See **reaction-energy diagram**. (p. 140)

radical (free radical) A highly reactive species in which one of the atoms has an odd number of electrons. Most commonly, a radical contains a carbon atom with three bonds and an "odd" (unpaired) electron. (p. 157)

radical inhibitor A compound added to prevent the propagation of free-radical chain reactions. In most cases, the inhibitor reacts to form a radical that is too stable to propagate the chain. (p. 153)

rate of a reaction The amount of product formed or reactant consumed per unit of time. (p. 136)

rate equation (rate law) The relationship between the concentrations of the reagents and the observed reaction rate. (p. 136) A general rate law for the reaction $A + B \longrightarrow C + D$ is

$$\text{rate} = k_r[A]^a[B]^b$$

kinetic order: The power of a concentration term in the rate equation. The rate equation above is ath order in [A], bth order in [B], and $(a + b)$th order overall.

rate constant: The constant k_r in the rate equation.

rate-limiting step (rate-determining step) The slowest step in a multistep sequence of reactions. In general, the rate-limiting step is the step with the highest-energy transition state. (p. 142)

reaction-energy diagram (potential-energy diagram) A plot of potential-energy changes as the reactants are converted to products. The vertical axis is potential energy (usually free energy, but occasionally enthalpy). The horizontal axis is the **reaction coordinate**, a measure of the progress of the reaction. (p. 140)

resonance stabilization Stabilization that takes place by delocalization of electrons in a pi-bonded system. Cations, radicals, and anions are often stabilized by resonance delocalization. (p. 156)

resonance stabilization of cations, radicals, and anions

substitution A reaction in which one atom replaces another, usually as a substituent on a carbon atom. (p. 126)

thermodynamics The study of the energy changes accompanying chemical transformations. Thermodynamics is generally concerned with systems at equilibrium. (p. 129)

transition state (activated complex) The state of highest energy between reactants and products. A relative maximum (high point) on the reaction-energy diagram. (p. 140)

Essential Problem-Solving Skills in Chapter 4

1. Explain the mechanism and energetics of the free-radical halogenation of alkanes.
2. Based on the selectivity of halogenation, predict the products of halogenation of an alkane.
3. Calculate free-energy changes from equilibrium constants.
4. Calculate enthalpy changes from bond-dissociation energies.
5. Determine the order of a reaction, and suggest a possible mechanism based on its rate equation.
6. Use energy diagrams to discuss transition states, activation energies, intermediates, and the rate-determining step of a multistep reaction.
7. Use the Hammond postulate to predict whether a transition state will be reactant-like or product-like.
8. Describe the structures of carbocations, carbanions, free radicals, and carbenes and the structural features that stabilize them. Explain which are electrophilic and which are nucleophilic.

Study Problems

4-34   Define and give an example for each term.

(a) homolytic cleavage	**(b)** heterolytic cleavage	**(c)** free radical
(d) carbocation	**(e)** carbanion	**(f)** carbene
(g) carbonium ion	**(h)** intermediate	**(i)** catalyst
(j) transition state	**(k)** rate equation	**(l)** equilibrium constant
(m) rate constant	**(n)** reaction mechanism	**(o)** chain reaction
(p) substitution reaction	**(q)** activation energy	**(r)** bond-dissociation energy
(s) rate-limiting step	**(t)** Hammond postulate	**(u)** resonance stabilization

4-35   Consider the following reaction-energy diagram.

(a) Label the reactants and the products. Label the activation energy for the first step and the second step.
(b) Is the overall reaction endothermic or exothermic? What is the sign of $\Delta H°$?
(c) Which points in the curve correspond to intermediates? Which correspond to transition states?
(d) Label the transition state of the rate-limiting step. Does its structure resemble the reactants, the products, or an intermediate?

4-36   Draw a reaction-diagram profile for a one-step exothermic reaction. Label the parts that represent the reactants, products, transition state, activation energy, and heat of reaction.

4-37   Draw a reaction-energy diagram for a two-step endothermic reaction with a rate-limiting second step.

4-38   Treatment of *t*-butyl alcohol with concentrated HCl gives *t*-butyl chloride.

$$\underset{\text{\textit{t}-butyl alcohol}}{CH_3-\overset{\overset{\displaystyle CH_3}{|}}{\underset{\underset{\displaystyle CH_3}{|}}{C}}-OH} \; + \; H^+ \; + \; Cl^- \longrightarrow \underset{\text{\textit{t}-butyl chloride}}{CH_3-\overset{\overset{\displaystyle CH_3}{|}}{\underset{\underset{\displaystyle CH_3}{|}}{C}}-Cl} \; + \; H_2O$$

When the concentration of H^+ is doubled, the reaction rate doubles. When the concentration of *t*-butyl alcohol is tripled, the reaction rate triples. When the chloride ion concentration is quadrupled, however, the reaction rate is unchanged. Write the rate equation for this reaction.

4-39   Label each hydrogen atom in the following compounds as primary (1°), secondary (2°), or tertiary (3°).
   (a) $CH_3CH_2CH(CH_3)_2$ **(b)** $(CH_3)_3CCH_2C(CH_3)_3$

(c)   **(d)**   **(e)**

4-40   Use bond-dissociation energies (Table 4-2, p. 134) to calculate values of $\Delta H°$ for the following reactions.
   (a) $CH_3-CH_3 + I_2 \longrightarrow CH_3CH_2I + HI$ **(b)** $CH_3CH_2Cl + HI \longrightarrow CH_3CH_2I + HCl$
   (c) $(CH_3)_3C-OH + HCl \longrightarrow (CH_3)_3C-Cl + H_2O$ **(d)** $CH_3CH_2CH_3 + H_2 \longrightarrow CH_3CH_3 + CH_4$
   (e) $CH_3CH_2OH + HBr \longrightarrow CH_3CH_2-Br + H_2O$

4-41   Use the information in Table 4-2 (p. 134) to rank the following radicals in decreasing order of stability.

$·CH_3$ $CH_3\overset{·}{C}H_2$ (benzyl) $\overset{·}{C}H_2$ $(CH_3)_3C·$ $(CH_3)_2\overset{·}{C}H$ $CH_2=CH-\overset{·}{C}H_2$

4-42   For each of the following alkanes,
   (1) Draw all the possible monochlorinated derivatives.
   (2) Determine whether free-radical chlorination would be a good way to make any of these monochlorinated derivatives.
   (a) cyclopentane   **(b)** methylcyclopentane
   (c) 2,3-dimethylbutane   **(d)** 2,2,3,3-tetramethylbutane

4-43   Write a mechanism for the light-initiated reaction of cyclohexane with chlorine to give chlorocyclohexane. Label the initiation and propagation steps.

$$\text{cyclohexane} + Cl_2 \xrightarrow{h\nu} \text{chlorocyclohexane} + HCl$$

4-44   Draw the important resonance structures of the following free radicals.

   (a) $CH_2=CH-\overset{·}{C}H_2$ **(b)** (cyclohexadienyl)$-\overset{·}{C}H_2$ **(c)** $CH_3-\overset{\overset{\displaystyle O}{||}}{C}-O·$ **(d)**   **(e)**

***4-45**   In the presence of a small amount of bromine, the following light-promoted reaction has been observed.

$$H_3C-\text{(cyclopentene)}-CH_3 + Br_2 \xrightarrow{h\nu} H_3C-\text{(cyclopentene)}\begin{smallmatrix}CH_3\\Br\end{smallmatrix} + H_3C-\text{(cyclopentene)}\begin{smallmatrix}CH_3\\H, Br\end{smallmatrix}$$

   (a) Write a mechanism for this reaction. Your mechanism should explain how both products are formed. (*Hint*: Notice which H atom has been lost in both products.)
   (b) Explain why only this hydrogen atom has been replaced, in preference to any of the other hydrogen atoms in the starting material.

4-46   For each compound, predict the major product of free-radical bromination. Remember that bromination is highly selective, and only the most stable radical will be formed.
   (a) cyclohexane   **(b)** methylcyclopentane   **(c)** 2,2,3-trimethylbutane
   (d) decalin   **(e)** 3-methyloctane   **(f)** hexane

   (g) (ethylbenzene $-CH_2CH_3$)   **(h)** (decalin)   (2 products)

   ethylbenzene

4-47   When exactly 1 mole of methane is mixed with exactly 1 mole of chlorine and light is shone on the mixture, a chlorination reaction occurs. The products are found to contain substantial amounts of di-, tri-, and tetrachloromethane.

   (a) Explain how a mixture is formed from this stoichiometric mixture of reactants, and propose mechanisms for the formation of these compounds from chloromethane.
   (b) How would you run this reaction to get a good conversion of methane to CH_3Cl? Methane to CCl_4?

4-48 The chlorination of pentane gives a mixture of three monochlorinated products.
   (a) Draw their structures.
   (b) Predict the ratios in which these monochlorination products will be formed, remembering that a chlorine atom abstracts a secondary hydrogen about 4.5 times as fast as it abstracts a primary hydrogen.

4-49 **(a)** Draw the structure of the transition state for the second propagation step in the chlorination of methane.

$$\cdot CH_3 + Cl_2 \longrightarrow CH_3Cl + Cl\cdot$$

   Be careful to show whether the transition state is product-like or reactant-like and which of the two partial bonds is stronger.
   (b) Repeat for the second propagation step in the bromination of methane.

4-50 Peroxides are often added to free-radical reactions as initiators because the oxygen–oxygen bond is homolytically cleaved rather easily. For example, the bond-dissociation energy of the O—O bond in hydrogen peroxide (H—O—O—H) is only 51 kcal/mol (213 kJ/mol). Give a mechanism for the hydrogen peroxide–initiated reaction of cyclopentane with chlorine.

4-51 When dichloromethane is treated with strong NaOH, an intermediate is generated that reacts like a carbene. Draw the structure of this reactive intermediate, and give a mechanism for its formation.

*__4-52__ When ethene is treated in a calorimeter with H_2 and a Pt catalyst, the heat of reaction is found to be -32.7 kcal/mol (-137 kJ/mol), and the reaction goes to completion. When the reaction takes place at 1400 K, the equilibrium is found to be evenly balanced, with $K_{eq} = 1$. Compute the value of ΔS for this reaction.

$$CH_2{=}CH_2 + H_2 \xrightarrow{\text{Pt catalyst}} CH_3{-}CH_3 \qquad \Delta H = -32.7 \text{ kcal/mol}$$

*__4-53__ When a small amount of iodine is added to a mixture of chlorine and methane, it prevents chlorination from occurring. Therefore, iodine is a *free-radical inhibitor* for this reaction. Calculate $\Delta H°$ values for the possible reactions of iodine with species present in the chlorination of methane, and use these values to explain why iodine inhibits the reaction. (The I—Cl bond-dissociation energy is 50 kcal/mol or 211 kJ/mol.)

*__4-54__ Tributyltin hydride (Bu_3SnH) is used synthetically to reduce alkyl halides, replacing a halogen atom with hydrogen. Free-radical initiators promote this reaction, and free-radical inhibitors are known to slow it or stop it. Your job is to develop a mechanism, using the following reaction as the example.

   The following bond-dissociation energies may be helpful:

Br—Br	46 kcal/mol
H—Br	88 kcal/mol
Bu_3Sn—H	74 kcal/mol
Bu_3Sn—Br	132 kcal/mol

   (a) Propose initiation and propagation steps to account for this reaction.
   (b) Calculate values of ΔH for your proposed steps to show that they are energetically feasible.
      (*Hint*: A trace of Br_2 and light suggests it's there only as an initiator, to create Br$\cdot$ radicals. Then decide which atom can be abstracted most favorably from the starting materials by the Br$\cdot$ radical. That should complete the initiation. Now decide what energetically favored propagation steps will accomplish the reaction.)

*__4-55__ When healthy, Earth's stratosphere contains a low concentration of ozone (O_3) that absorbs potentially harmful ultraviolet (UV) radiation by the cycle shown at right: Chlorofluorocarbon refrigerants, such as Freon 12® (CF_2Cl_2), are stable in the lower atmosphere, but in the stratosphere, they absorb high-energy UV radiation to generate chlorine radicals.

$$CF_2Cl_2 \xrightarrow{hv} \cdot CF_2Cl + Cl\cdot$$

The presence of a small number of chlorine radicals appears to lower ozone concentrations dramatically. The following reactions are all known to be exothermic (except the one requiring light) and to have high rate constants. Propose two mechanisms to explain how a small number of chlorine radicals can destroy large numbers of ozone molecules. Which of the two mechanisms is more likely when the concentration of chlorine atoms is very small?

$$Cl—O—O—Cl \xrightarrow{h\nu} O_2 + 2\,Cl\cdot \qquad Cl—O\cdot + O \longrightarrow O_2 + Cl\cdot$$

$$Cl\cdot + O_3 \longrightarrow Cl—O\cdot + O_2 \qquad 2\,Cl—O\cdot \longrightarrow Cl—O—O—Cl$$

*4-56   Deuterium (D) is the hydrogen isotope of mass number 2, with a proton and a neutron in its nucleus. The chemistry of deuterium is nearly identical to the chemistry of hydrogen, except that the C—D bond is slightly stronger than the C—H bond by 1.2 kcal/mol (5.0 kJ/mol). Reaction rates tend to be slower if a C—D bond (as opposed to a C—H bond) is broken in a rate-limiting step.

   This effect, called a *kinetic isotope effect*, is clearly seen in the chlorination of methane. Methane undergoes free-radical chlorination 12 times as fast as tetradeuteriomethane (CD_4).

$$\textit{Faster:} \quad CH_4 + Cl\cdot \longrightarrow CH_3Cl + HCl \qquad \text{relative rate} = 12$$

$$\textit{Slower:} \quad CD_4 + Cl\cdot \longrightarrow CH_3Cl + DCl \qquad \text{relative rate} = 1$$

(a) Draw the transition state for the rate-limiting step of each of these reactions, showing how a bond to hydrogen or deuterium is being broken in this step.

(b) Monochlorination of deuterioethane (C_2H_5D) leads to a mixture containing 93% C_2H_4DCl and 7% C_2H_5Cl. Calculate the relative rates of abstraction per hydrogen and deuterium in the chlorination of ethane.

(c) Consider the thermodynamics of the chlorination of methane and the chlorination of ethane, and use the Hammond postulate to explain why one of these reactions has a much larger isotope effect than the other.

CHAPTER 5

Stereochemistry

Stereochemistry is the study of the three-dimensional structure of molecules. No one can understand organic chemistry, biochemistry, or biology without using stereochemistry. Biological systems are exquisitely selective, and they often discriminate between molecules with subtle stereochemical differences. We have seen (Section 2-8) that isomers are grouped into two broad classes: constitutional isomers and stereoisomers. **Constitutional isomers** (structural isomers) differ in their bonding sequence; their atoms are connected differently. **Stereoisomers** have the same bonding sequence, but they differ in the orientation of their atoms in space.

Differences in spatial orientation might seem unimportant, but stereoisomers often have remarkably different physical, chemical, and biological properties. For example, the cis and trans isomers of butenedioic acid are a special type of stereoisomer called *cis-trans isomers* (or *geometric isomers*). Both compounds have the formula HOOC—CH=CH—COOH, but they differ in how these atoms are arranged in space. The cis isomer is called *maleic acid*, and the trans isomer is called *fumaric acid*. Fumaric acid is an essential metabolic intermediate in both plants and animals, but maleic acid is toxic and irritating to tissues.

5-1
Introduction

A double bond in rhodopsin, a visual pigment found in your eyes that enables you to see at night, is converted from the cis isomer to the trans isomer when light strikes the eye. As a result, a nerve impulse travels to the brain and you see the source of the light.

fumaric acid, mp 287°C
essential metabolite

maleic acid, mp 138°C
toxic irritant

rhodopsin

The discovery of stereochemistry was one of the most important breakthroughs in the structural theory of organic chemistry. Stereochemistry explained why several types of isomers exist, and it forced scientists to propose the tetrahedral carbon atom. In this chapter, we study the three-dimensional structures of molecules to understand their stereochemical relationships. We compare the various types of stereoisomers and study ways to differentiate among stereoisomers. In future chapters, we will see how stereochemistry plays a major role in the properties and reactions of organic compounds.

167

5-2
Chirality

What is the difference between your left hand and your right hand? They look similar, yet a left-handed glove does not fit the right hand. The same principle applies to your feet. They look almost identical, yet the left shoe fits painfully on the right foot. The relationship between your two hands or your two feet is that they are nonsuperimposable (nonidentical) mirror images of each other. Objects that have left-handed and right-handed forms are called **chiral** (*kī′ rel,* rhymes with "spiral"), the Greek word for "handed."

We can tell whether an object is chiral by looking at its mirror image (Figure 5-1). Every physical object (with the possible exception of a vampire) has a mirror image, but *a chiral object has a mirror image that is different from the original object.* For example, a chair and a spoon and a glass of water all look the same in a mirror. Such objects are called **achiral**, meaning "not chiral." A hand looks different in the mirror. If the original hand were the right hand, it would look like a left hand in the mirror.

right hand   left hand

▶ **Figure 5-1**
Use of a mirror to test for chirality. An object is chiral if its mirror image is different from the original object.

Besides shoes and gloves, we encounter many other chiral objects every day (Figure 5-2). What is the difference between an English car and an American car? The English car has the steering wheel on the right-hand side, while the American car has it on the left. To a first approximation, the English and American cars are nonsuperimposable mirror images. Most screws have right-hand threads and are turned clockwise to tighten. The mirror image of a right-handed screw is a left-handed screw, turned counterclockwise to tighten. Those of us who are left-handed realize that scissors are chiral. Most scissors are right-handed. If you use them in your left hand, they cut poorly, if at all. A left-handed person must go to a well-stocked store to find a pair of left-handed scissors, the mirror image of the "standard" right-handed scissors.

mirror

mirror   mirror

▶ **Figure 5-2**
Common chiral objects. Many objects come in "left-handed" and "right-handed" versions.

PROBLEM 5-1

Determine whether the following objects are chiral or achiral.

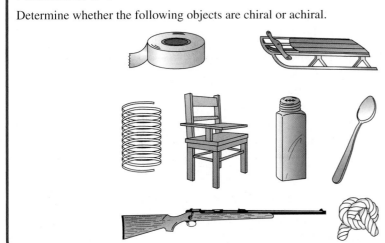

5-2A Chirality and Enantiomerism in Organic Molecules

Like other objects, molecules are either chiral or achiral. For example, consider the two geometric isomers of 1,2-dichlorocyclopentane (Figure 5-3). The cis isomer is achiral because its mirror image is superimposable on the original molecule. Two molecules are said to be **superimposable** if they can be placed on top of each other and the three-dimensional position of each atom of one molecule coincides with the equivalent atom of the other molecule. To draw the mirror image of a molecule, simply draw the same structure with left and right reversed. The up-and-down and front-and-back directions are unchanged. These two mirror-image structures are identical (superimposable), and *cis*-1,2-dichlorocyclopentane is achiral.

The mirror image of *trans*-1,2-dichlorocyclopentane is different from (nonsuper-imposable with) the original molecule. These are two different compounds, and we should expect to discover two mirror-image isomers of *trans*-1,2-dichlorocyclopentane. Make models of these isomers to convince yourself that they are different no matter how you twist and turn them. Nonsuperimposable mirror-image molecules are called **enantiomers**. A chiral compound always has an enantiomer (a nonsuperimposable mirror image). An achiral compound always has a mirror image that is the same as the original molecule. Let's review the definitions of these words.

enantiomers: mirror-image isomers; pairs of compounds that are nonsuperim-posable mirror images

chiral: ("handed") different from its mirror image; having an enantiomer

achiral: ("not handed") identical with its mirror image; not chiral

Any compound that is chiral must have an enantiomer. Any compound that is achiral cannot have an enantiomer.

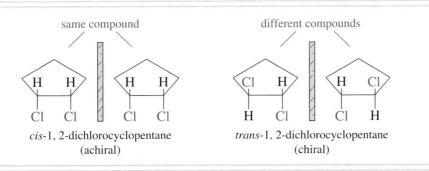

cis-1, 2-dichlorocyclopentane (achiral)

trans-1, 2-dichlorocyclopentane (chiral)

◀ **Figure 5-3**
Stereoisomers of 1,2-dichlorocyclopentane. The cis isomer has no enantiomers; it is achiral. The trans isomer is chiral; it can exist in either of two nonsuperimposable enantiomeric forms.

PROBLEM-SOLVING HINT

Stereochemistry is a difficult topic for many students. Use your models to help you see the relationships between structures. Once you have experience working with these three-dimensional relationships, you may (or may not) be able to visualize them without constructing models.

PROBLEM 5-2

Make a model and draw a three-dimensional structure for each compound. Then draw the mirror image of your original structure and determine whether the mirror image is the same compound. Label each structure as being chiral or achiral, and label pairs of enantiomers.

(a) *cis*-1,2-dimethylcyclobutane

(b) *trans*-1,2-dimethylcyclobutane

(c) *cis*- and *trans*-1,3-dimethylcyclobutane

(d) 2-bromobutane

(e)

(f)

5-2B Asymmetric Carbon Atoms, Chirality Centers, and Stereocenters

The three-dimensional drawing of 2-bromobutane in Figure 5-4 shows that 2-bromobutane cannot be superimposed on its mirror image. This simple molecule is chiral, with two distinct enantiomers. What is it about a molecule that makes it chiral? The most common feature (but not the only one) that lends chirality is a carbon atom that is bonded to four different groups. Such a carbon atom is called an **asymmetric carbon atom** or a **chiral carbon atom**, and is often designated by a *. Carbon atom 2 of 2-bromobutane is bonded to a hydrogen atom, a bromine atom, a methyl group, and an ethyl group. It is an asymmetric carbon atom, and it is responsible for the chirality of 2-bromobutane.

An asymmetric carbon atom is the most common example of a **chirality center**, the IUPAC term for any atom holding a set of ligands in a spatial arrangement that is not superimposable on its mirror image. Chirality centers belong to an even broader group called stereocenters. A **stereocenter** (or **stereogenic atom**) is any atom at which the interchange of two groups gives a stereoisomer.* Asymmetric carbons and the double-bonded carbon atoms in cis-trans isomers are the most common types of stereocenters. Figure 5-5 compares these successively broader definitions.

Make a model of an asymmetric carbon atom, bonded to four different-colored atoms. Also make its mirror image, and try to superimpose the two (Figure 5-6). No matter how you twist and turn the models, they never look exactly the same.

PROBLEM-SOLVING HINT

Every object has a mirror image. Is its mirror image the same or different? Different: The object is chiral. Same: The object is achiral.

▶ **Figure 5-4**
2-Bromobutane is chiral by virtue of an asymmetric carbon atom (chiral carbon atom), marked by a *.

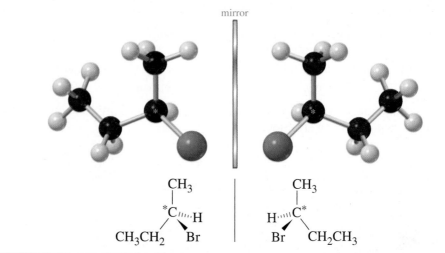

mirror

$$\overset{CH_3}{\underset{CH_3CH_2\qquad Br}{\overset{*}{C}}{}_{\!\!\!\!\!\!\textit{////}}H} \qquad \overset{CH_3}{\underset{Br\qquad CH_2CH_3}{H_{\textit{\tiny\textbackslash\textbackslash\textbackslash}}\overset{*}{C}}}$$

*The term *stereocenter* (*stereogenic atom*) is not consistently defined. The original (Mislow) definition is given above. Some sources simply define it as a synonym for an *asymmetric carbon* (*chiral carbon*) or for a *chirality center*.

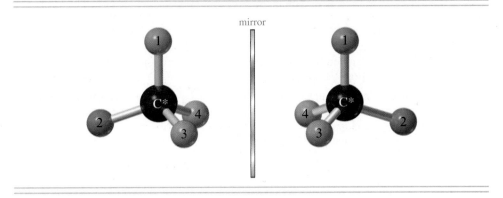

◀ **Figure 5-5**
Asymmetric carbon atoms are examples of chirality centers, which are examples of stereocenters.

◀ **Figure 5-6**
Enantiomers of an asymmetric carbon atom. These two mirror images are nonsuperimposable.

If two groups on a carbon atom are the same, however, the arrangement usually is not chiral. Figure 5-7 shows the mirror image of a tetrahedral structure with only three different groups; two of the four groups are the same. If the structure on the right is rotated 180°, it can be superimposed on the left structure.

We can generalize at this point, but keep in mind that the ultimate test for chirality is always whether the molecule's mirror image is the same or different.

1. If a compound has no asymmetric carbon atom, it is usually achiral.

2. If a compound has just one asymmetric carbon atom, it is chiral.

3. If a compound has more than one asymmetric carbon, it may or may not be chiral. (We will see examples in Section 5-12.)

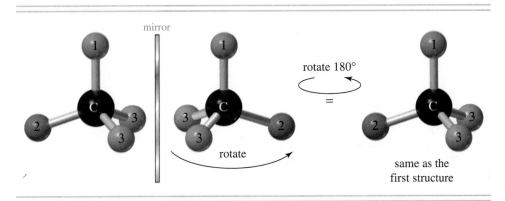

◀ **Figure 5-7**
A carbon atom bonded to just three different types of groups is not chiral.

SOLVED PROBLEM 5-1

Star (*) each asymmetric carbon atom in the following structure:

SOLUTION

There are three asymmetric carbons, starred in red.

(1) The (CHOH) carbon of the side chain is asymmetric. Its four substituents are the ring, a hydrogen atom, a hydroxyl group, and a methyl group.

(2) Carbon atom C1 of the ring is asymmetric. Its four substituents are the side chain, a hydrogen atom, the part of the ring closer to the chlorine atom ($-CH_2-CHCl-$), and the part of the ring farther from the chlorine atom ($-CH_2-CH_2-CH_2-CHCl-$).

(3) The ring carbon bearing the chlorine atom is asymmetric. Its four substituents are the chlorine atom, a hydrogen atom, the part of the ring closer to the side chain, and the part of the ring farther from the side chain.

Notice that different groups might be different in any manner. For example, the ring carbon bearing the chlorine atom is asymmetric even though two of its ring substituents initially appear to be $-CH_2-$ groups. These two parts of the ring are different because one is closer to the side chain and one is farther away. The *entire* structure of the group must be considered.

PROBLEM-SOLVING HINT

To draw the mirror image of a structure, keep up-and-down and front-and-back aspects as they are in the original structure, but reverse left and right.

PROBLEM 5-3

Draw a three-dimensional structure for each compound, and star all asymmetric carbon atoms. Draw the mirror image for each structure, and state whether you have drawn a pair of enantiomers or just the same molecule twice. Build molecular models of any of these examples that seem difficult to you.

(a) 1-bromobutane

(b) 1-pentanol

(c) 2-pentanol

(d) 3-pentanol

(e) chlorocyclohexane

(f) $CH_3-CH-COOH$ with NH_2

(g) *cis*-1,2-dichlorocyclobutane

(h) *trans*-1,2-dichlorocyclobutane

(i) *trans*-1,3-dichlorocyclobutane

(j)

(k)

PROBLEM-SOLVING HINT

To see whether a ring carbon is asymmetric, see if there is a difference in the path around the ring in each direction. If there is, then the two ring bonds are "different groups."

PROBLEM 5-4

For each of the stereocenters (circled) in Figure 5-5,
(a) Draw the compound with two of the groups on the stereocenter interchanged.
(b) Give the relationship of the new compound to the original compound.

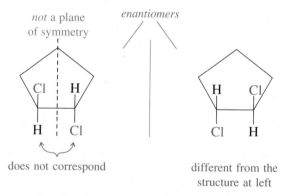

σ

internal mirror plane
of symmetry (σ)

◀ **Figure 5-8**
cis-1,2-Dichlorocyclopentane has a
mirror plane of symmetry. Any
compound with an internal mirror
plane of symmetry *cannot* be chiral.

5-2C Mirror Planes of Symmetry

In Figure 5-3 we saw that *cis*-1,2-dichlorocyclopentane is achiral. Its mirror image was found to be identical with the original molecule. Figure 5-8 shows a shortcut that often shows whether a molecule is chiral.

If we draw a line down the middle of *cis*-1,2-dichlorocyclopentane, bisecting a carbon atom and two hydrogen atoms, the part of the molecule that appears to the right of the line is the mirror image of the part on the left. This kind of symmetry is called an **internal mirror plane**, sometimes symbolized by the Greek lowercase letter sigma (σ). Since the right-hand side of the molecule is the reflection of the left-hand side, the molecule's mirror image is the same as the original molecule.

Notice below that the chiral trans isomer of 1,2-dichlorocyclopentane does not have a mirror plane of symmetry. The chlorine atoms do not reflect into each other across our hypothetical mirror plane. One of them is directed up, the other down.

not a plane
of symmetry

enantiomers

Cl   H   H   Cl

H   Cl   Cl   H

does not correspond

different from the
structure at left

We can generalize from these and other examples to state the following principle:

Any molecule that has an internal mirror plane of symmetry cannot be chiral, even though it may contain asymmetric carbon atoms.

The converse is not true, however. When we cannot find a mirror plane of symmetry, that does not necessarily mean that the molecule must be chiral. The following example has no internal mirror plane of symmetry, yet the mirror image is superimposable on the original molecule. You may need to make models to show that these mirror images are just two drawings of the same compound.

Cl ... Br   Br ... Cl

Br ' Cl   =   Cl ' Br

(hydrogens are omitted for clarity)

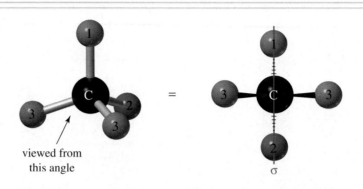

▶ **Figure 5-9**
A carbon atom with two identical substituents (only three different substituents) usually has an internal mirror plane of symmetry. The structure is not chiral.

Using what we know about mirror planes of symmetry, we can see why a chiral (asymmetric) carbon atom is special. Figure 5-4 showed that an asymmetric carbon has a mirror image that is nonsuperimposable on the original structure. If a carbon atom has only three different kinds of substituents, however, it has an internal mirror plane of symmetry (Figure 5-9). Therefore, it cannot contribute to chirality in a molecule.

> **PROBLEM 5-5**
>
> For each compound, determine whether the molecule has an internal mirror plane of symmetry. If it does, draw the mirror plane on a three-dimensional drawing of the molecule. If the molecule does not have an internal mirror plane, determine whether or not the structure is chiral.
> **(a)** methane
> **(b)** *cis*-1,2-dibromocyclobutane
> **(c)** *trans*-1,2-dibromocyclobutane
> **(d)** 1,2-dichloropropane
>
> **(e)** glyceraldehyde, $HOCH_2$—CH—CHO
> |
>   OH
>
> **(f)**
>
> **(g)** CH_3
>
>   CH
> CH_3 CH_3
>
> **(h)** CH_3—CH—COOH
> |
> NH_2
>   alanine

5-3
(*R*) and (*S*) Nomenclature of Asymmetric Carbon Atoms

Alanine, from Problem 5-5(h), is one of the amino acids found in common proteins. Alanine has an asymmetric carbon atom, and it exists in two enantiomeric forms.

natural alanine   unnatural alanine

These mirror images are different, and this difference is reflected in their biochemistry. Only the enantiomer on the left can be metabolized by the usual enzyme; the one on the right is not recognized as a useful amino acid. Both are named alanine, however, or 2-aminopropanoic acid in the IUPAC system. We need a simple way to distinguish between enantiomers and to give each of them a unique name.

The **Cahn–Ingold–Prelog convention** is the most widely accepted system for naming the configurations of chirality centers. Each asymmetric carbon atom is assigned a letter (R) or (S) based on its three-dimensional configuration. To determine the name, we follow a two-step procedure that assigns "priorities" to the four substituents and then assigns the name based on the positions of these substituents. Here is the procedure:

1. Assign a "priority" to each group bonded to the asymmetric carbon. We speak of group 1 as having the highest priority, group 2 second, group 3 third, and group 4 as having the lowest priority.

(a) Look at the first atom of the group—the atom bonded to the asymmetric carbon. Atoms with higher atomic numbers receive higher priorities. For example, if the four groups bonded to an asymmetric carbon atom were H, CH_3, NH_2, and F, the fluorine atom (atomic number 9) would have the highest priority, followed by the nitrogen atom of the NH_2 group (atomic number 7), then by the carbon atom of the methyl group (atomic number 6). Note that we look only at the atomic number of the atom directly attached to the asymmetric carbon, not the entire group. Hydrogen comes last.

With different isotopes of the same element, the heavier isotopes have higher priorities. For example, tritium (^3H) receives a higher priority than deuterium (^2H), followed by hydrogen (^1H).

Examples of priority for atoms bonded to an asymmetric carbon:

$$I > Br > Cl > S > F > O > N > {}^{13}C > {}^{12}C > Li > {}^{3}H > {}^{2}H > {}^{1}H$$

(b) In case of ties, use the next atoms along the chain as tiebreakers. For example, we assign a higher priority to isopropyl $-CH(CH_3)_2$ than to ethyl $-CH_2CH_3$. The first carbon atom in the ethyl group is bonded to two hydrogens and one carbon, while the first carbon in the isopropyl group is bonded to two carbons and one hydrogen. An ethyl group and a $-CH_2CH_2Br$ have identical first atoms and second atoms, but the bromine atom in the third position gives $-CH_2CH_2Br$ a higher priority than $-CH_2CH_3$. One high-priority atom takes priority over any number of lower-priority atoms.

Examples

(c) Treat double and triple bonds as if each were a bond to a separate atom. For this method, imagine that each pi bond is broken and the atoms at both ends duplicated. Note than when you break a bond, you always add *two* imaginary atoms.

R—C≡C—H becomes R—C—C—H
break and duplicate

OH   OH
| |
R—C=O becomes R—C—O
break and duplicate

2. Using a three-dimensional drawing or a model, put the fourth-priority group in back and view the molecule along the bond from the asymmetric carbon to the fourth-priority group. Draw an arrow from the first-priority group, through the second, to the third. If the arrow points clockwise, the asymmetric carbon atom is called (*R*) (Latin, *rectus*, "upright"). If the arrow points counterclockwise, the chiral carbon atom is called (*S*) (Latin, *sinister*, "left").

Alternatively, you can draw the arrow and imagine turning a car's steering wheel in that direction. If the car would go to the left, the asymmetric carbon atom is designated (*S*). If the car would go to the right, the asymmetric carbon atom is designated (*R*).

Let's use the enantiomers of alanine as an example. The naturally occurring enantiomer is the one on the left, determined to have the (*S*) configuration.

natural (*S*)-alanine   unnatural (*R*)-alanine

Of the four atoms attached to the asymmetric carbon in alanine, nitrogen has the largest atomic number, giving it the highest priority. Next is the —COOH carbon atom, since it is bonded to oxygen atoms. Third is the methyl group, followed by the hydrogen atom. When we position the natural enantiomer with its hydrogen atom pointing away from us, the arrow from —NH_2 to —COOH to —CH_3 points counterclockwise. Thus, the naturally occurring enantiomer of alanine has the (S) configuration. Make models of these enantiomers to illustrate how they are named (R) and (S).

SOLVED PROBLEM 5-2

Draw the enantiomers of 1,3-dibromobutane and label them as (R) and (S). (Making a model is particularly helpful for this type of problem.)

$$\overset{*}{CH_2-CH_2-CH-CH_3}$$
$$\underset{Br}{\mid}\qquad\qquad\underset{Br}{\mid}$$

SOLUTION

The third carbon atom in 1,3-dibromobutane is asymmetric. The bromine atom receives first priority, the (—CH_2CH_2Br) group second priority, the methyl group third, and the hydrogen fourth. The following mirror images are drawn with the hydrogen atom back, ready to assign (R) or (S) as shown.

SOLVED PROBLEM 5-3

The structure of one of the enantiomers of carvone is shown below. Find the asymmetric carbon atom, and determine whether it has the (R) or the (S) configuration.

SOLUTION

The asymmetric carbon atom is one of the ring carbons, as indicated by the asterisk in the structure below. Although there are two —CH_2— groups bonded to the carbon, they are different —CH_2— groups. One is a —CH_2—CO— group, and the other is a —CH_2—CH=C group. The groups are assigned priorities, and this is found to be the (S) enantiomer.

(S)-carvone

Scientists frequently use the isotopes of hydrogen to assign the configuration of the products of biological reactions. Ethanol, made chiral by the presence of a deuterium (2H) at C-2, is one of the early examples.

PROBLEM 5-6

Star (*) each asymmetric carbon atom in the following examples, and determine whether it has the (*R*) or the (*S*) configuration.

(a) HO⋯C with CH₃ (up), H and CH₂CH₃

(b) Br⋯C with H (up), H₃C and CH₂CH₃

(c) structure with H, H, CH₃, CH₃, CH₃

(d) cyclopentane with Cl, H, H, Cl

(e) cyclopentane with H, H, Cl, Cl

(f) norbornene structure

(g) norbornene with =O

(h) limonene structure

*(i) structure with CHO, C⋯CH=CH₂, (CH₃O)₂CH, CH(CH₃)₂

PROBLEM 5-7

In Problem 5-3, you drew the enantiomers for a number of chiral compounds. Now go back and designate each asymmetric carbon atom as either (*R*) or (*S*).

5-4
Optical Activity

Mirror-image molecules have nearly identical physical properties. Compare the following properties of (*R*)-2-bromobutane and (*S*)-2-bromobutane.

	(*R*)-2-Bromobutane	(*S*)-2-Bromobutane
boiling point (°C)	91.2	91.2
melting point (°C)	−112	−112
refractive index	1.436	1.436
density	1.253	1.253

Differences in enantiomers become apparent in their interactions with other chiral molecules, such as enzymes. Still, we need a simple method to distinguish between enantiomers and measure their purity in the laboratory. **Polarimetry** is a common method used to distinguish between enantiomers, based on their ability to rotate the plane of polarized light in opposite directions. For example, the two enantiomers of thyroid hormone are shown below. The (*S*) enantiomer has a powerful effect on the metabolic rate of all the cells in the body. The (*R*) enantiomer is useless. In the laboratory, we distinguish between the enantiomers by observing that the active one rotates the plane of polarized light to the left.

thyroid hormone (*S*)
rotates polarized light to the left

wrong enantiomer (*R*)
rotates polarized light to the right

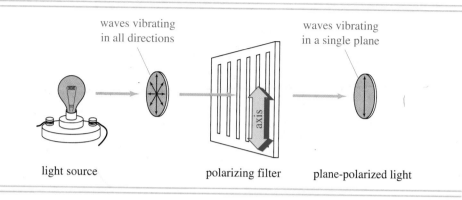

The waves of plane-polarized light vibrate primarily in a single plane.

5-4A Plane-Polarized Light

Most of what we see is unpolarized light, vibrating randomly in all directions. **Plane-polarized light** is composed of waves that vibrate in only one plane. Although there are other types of "polarized light," the term usually refers to plane-polarized light.

When unpolarized light passes through a polarizing filter, the randomly vibrating light waves are filtered so that most of the light passing through is vibrating in one direction (Figure 5-10). The direction of vibration is called the *axis* of the filter. Polarizing filters may be made from carefully cut calcite crystals or from specially treated plastic sheets. Plastic polarizing filters are often used as lenses in sunglasses, because the axis of the filters can be positioned to filter out reflected glare.

When light passes first through one polarizing filter and then through another, the amount of light emerging depends on the relationship between the axes of the two filters (Figure 5-11). If the axes of the two filters are lined up (parallel), then nearly all the light that passes through the first filter also passes through the second. If the axes of the two filters are perpendicular (*crossed poles*), however, all the polarized light that emerges

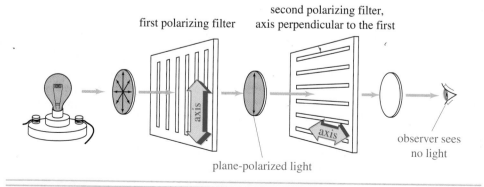

◀ **Figure 5-11**
When the axis of a second polarizing filter is parallel to the first, a maximum amount of light passes through. When the axes of the filters are perpendicular, no light passes through.

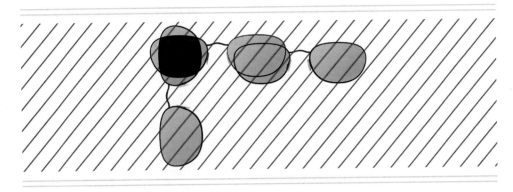

▶ **Figure 5-12**
Using sunglasses to demonstrate parallel axes of polarization and crossed poles. When the pairs of sunglasses are parallel, a maximum amount of light passes through. When they are perpendicular, very little light passes through.

from the first filter is stopped by the second. At intermediate angles of rotation, intermediate amounts of light pass through.

You can demonstrate this effect for yourself by wearing a pair of polarized sunglasses while looking at a light source through another pair (Figure 5-12). The second pair seems to be transparent, as long as its axis is lined up with the pair you are wearing. When the second pair is rotated to 90°, however, the lenses become opaque, as if they were covered with black ink.

5-4B   Rotation of Plane-Polarized Light

When polarized light passes through a solution containing a chiral compound, the chiral compound causes the plane of vibration to rotate. Rotation of the plane of polarized light is called **optical activity**, and substances that rotate the plane of polarized light are said to be **optically active**.

Before the relationship between chirality and optical activity was known, enantiomers were called **optical isomers** because they seemed identical except for their opposite optical activity. The term was loosely applied to more than one type of isomerism among optically active compounds, however, and this ambiguous term has been replaced by the well-defined term *enantiomers*.

Two enantiomers have identical physical properties, except for the direction they rotate the plane of polarized light.

> Enantiomeric compounds rotate the plane of polarized light by exactly the same amount but in opposite directions.

If the (R) isomer rotates the plane 30° clockwise, the (S) isomer will rotate it 30° counterclockwise. If the (R) enantiomer rotates the plane 5° counterclockwise, the (S) enantiomer will rotate it 5° clockwise.

> We cannot predict which direction a particular enantiomer [either (R) or (S)] will rotate the plane of polarized light.

(R) and (S) are simply names, while the direction and magnitude of rotation are physical properties that must be measured.

PROBLEM-SOLVING HINT

Don't confuse the process for *naming* a structure (R) or (S) with the process for *measuring* an optical rotation. Just because we use the terms *clockwise* and *counterclockwise* in naming (R) and (S) does not mean that light follows our naming rules.

5-4C   Polarimetry

A **polarimeter** measures the rotation of polarized light. It has a tubular cell filled with a solution of the optically active material and a system for passing polarized light through the solution and measuring the rotation as the light emerges (Figure 5-13). The light from a sodium lamp is filtered so that it consists of just one wavelength (one color), because most compounds rotate different wavelengths of light by different amounts. The wavelength of light most commonly used for polarimetry is a yellow emission line in the spectrum of sodium, called the *sodium D line*.

Monochromatic (one-color) light from the source passes through a polarizing filter, then through the sample cell containing a solution of the optically active compound. On

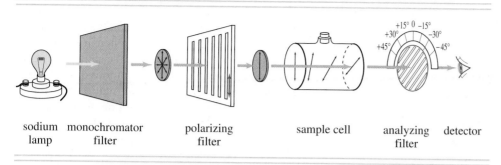

Schematic diagram of a polarimeter. The light originates at a source (usually a sodium lamp) and passes through a polarizing filter and the sample cell. An optically active solution rotates the plane of polarized light. The analyzing filter is another polarizing filter equipped with a protractor. It is turned until a maximum amount of light is observed, and the rotation is read from the protractor.

sodium lamp   monochromator filter   polarizing filter   sample cell   analyzing filter   detector

leaving the sample cell, the polarized light encounters another polarizing filter. This filter is movable, with a scale allowing the operator to read the angle between the axis of the second (analyzing) filter and the axis of the first (polarizing) filter. The operator rotates the analyzing filter until the maximum amount of light is transmitted, then the observed rotation is read from the protractor. The observed rotation is symbolized by α, the Greek letter alpha.

Compounds that rotate the plane of polarized light toward the right (clockwise) are called **dextrorotatory**, from the Greek word *dexios*, meaning "toward the right." Compounds that rotate the plane toward the left (counterclockwise) are called **levorotatory**, from the Latin word *laevus*, meaning "toward the left." These terms are sometimes abbreviated by a lowercase *d* or *l*. Using IUPAC notation, the direction of rotation is specified by the (+) or (−) sign of the rotation:

> Dextrorotatory (clockwise) rotations are (+).
> Levorotatory (counterclockwise) rotations are (−).

For example, the isomer of 2-butanol that rotates the plane of polarized light clockwise is named (+)-2-butanol or *d*-2-butanol. Its enantiomer, (−)-2-butanol or *l*-2-butanol, rotates the plane counterclockwise by exactly the same amount.

You can see the principle of polarimetry by using two pairs of polarized sunglasses, a beaker, and some corn syrup or sugar solution. Wear one pair of sunglasses, look down at a light, and hold another pair of sunglasses above the light. Notice that the most light is transmitted through the two pairs of sunglasses when their axes are parallel. Very little light is transmitted when their axes are perpendicular.

Put syrup into the beaker, and hold the beaker above the bottom pair of sunglasses so the light passes through one pair of sunglasses (the polarizing filter), then the beaker (the optically active sample), and then the other pair of sunglasses (the analyzing filter); see Figure 5-14.

Again, check the angles giving maximum and minimum light transmission. Is the syrup solution dextrorotatory or levorotatory? Did you notice the color variation as you rotated the filter? You can see why just one color of light should be used for accurate work.

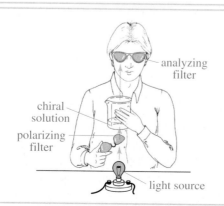

analyzing filter

chiral solution

polarizing filter

light source

◀ **Figure 5-14**
Apparatus for using a lightbulb and two pairs of polarized sunglasses as a simple polarimeter.

5-4D   Specific Rotation

The angular rotation of polarized light by a chiral compound is a characteristic physical property of that compound, just like the boiling point or the density. The rotation (α) observed in a polarimeter depends on the concentration of the sample solution and the length of the cell, as well as the optical activity of the compound. For example, twice as concentrated a solution would give twice the original rotation. Similarly, a 20-cm cell gives twice the rotation observed using a similar concentration in a 10-cm cell.

To use the rotation of polarized light as a characteristic property of a compound, we must standardize the conditions for measurement. We define a compound's **specific rotation** [α] as the rotation found using a 10-cm (1-dm) sample cell and a concentration of 1 g/mL. Other cell lengths and concentrations may be used, as long as the observed rotation is divided by the path length of the cell (l) and the concentration (c).

$$[\alpha] = \frac{\alpha(\text{observed})}{c \cdot l}$$

where

$$\alpha(\text{observed}) = \text{rotation observed in the polarimeter}$$

$$c = \text{concentration, g/mL}$$

$$l = \text{length of sample cell (path length), decimeters (dm)}$$

SOLVED PROBLEM 5-4

When one of the enantiomers of 2-butanol is placed in a polarimeter, the observed rotation is 4.05° counterclockwise. The solution was made by diluting 6 g of 2-butanol to a total of 40 mL, and the solution was placed into a 200-mm polarimeter tube for the measurement. Determine the specific rotation for this enantiomer of 2-butanol.

SOLUTION

Since it is levorotatory, this must be (−)-2-butanol. The concentration is 6 g per 40 mL = 0.15 g/ml, and the path length is 200 mm = 2 dm. The specific rotation is

$$[\alpha]_D^{25} = \frac{-4.05°}{(0.15)(2)} = -13.5°$$

A rotation depends on the wavelength of light used and also on the temperature, so these data are given together with the rotation. In Solved Problem 5-4, the "25" means that the measurement was made at 25°C, and the "D" means that the light used was the D line of the sodium spectrum.

Without even measuring it, we can predict that the specific rotation of the other enantiomer of 2-butanol will be

$$[\alpha]_D^{25} = +13.5°$$

where the (+) refers to the clockwise direction of the rotation. This enantiomer would be called (+)-2-butanol. We could refer to this pair of enantiomers as (+)-2-butanol and (−)-2-butanol or as (R)-2-butanol and (S)-2-butanol.

Does this mean that (R)-2-butanol is the dextrorotatory isomer because it is named (R), and (S)-2-butanol is levorotatory because it is named (S)? Not at all! The rotation of a compound, (+) or (−), is something that we measure in the polarimeter, depending on how the molecule interacts with light. The (R) and (S) nomenclature is our own artificial way of describing how the atoms are arranged in space.

In the laboratory, we can measure a rotation and see whether a particular substance is $(+)$ or $(-)$. On paper, we can determine whether a particular drawing is named (R) or (S). But it is difficult to predict whether a structure we call (R) will rotate polarized light clockwise or counterclockwise. Similarly, it is difficult to predict whether a dextrorotatory substance in a flask has the (R) or (S) configuration.

PROBLEM 5-8

A solution of 2.0 g of $(+)$-glyceraldehyde, $HOCH_2$—$CHOH$—CHO, in 10 mL of water was placed in a 100-mm cell. Using the sodium D line, a rotation of $+1.74°$ was found at 25°C. Determine the specific rotation of $(+)$-glyceraldehyde.

PROBLEM 5-9

A solution of 0.5 g of $(-)$-epinephrine (see Fig. 5-15) dissolved in 10 mL of dilute HCl was placed in a 20-cm polarimeter tube. Using the sodium D line, the rotation was found to be $-5.0°$ at 25°C. Determine the specific rotation of epinephrine.

PROBLEM 5-10

A chiral sample gives a rotation that is close to 180°. How can one tell whether this rotation is $+180°$ or $-180°$?

If the direction of rotation of polarized light were the only difference between enantiomers, one might ask whether the difference was important. Biological systems commonly distinguish between enantiomers, and two enantiomers may have totally different biological properties. In fact, any **chiral probe** can distinguish between enantiomers, and a polarimeter is only one example of a chiral probe. Another example is your hand. If you needed to sort a box of gloves into right-handed gloves and left-handed gloves, you could distinguish between them by checking to see which ones fit your right hand.

Enzymes in living systems are chiral, and they are capable of distinguishing between enantiomers. Usually, only one enantiomer of a pair fits properly into the chiral active site of an enzyme. For example, the levorotatory form of epinephrine is one of the principal hormones secreted by the adrenal medulla. When synthetic epinephrine is given to a patient, the $(-)$ form has the same stimulating effect as the natural hormone. The $(+)$ form lacks this effect and is mildly toxic. Figure 5-15 shows a simplified picture of how only the $(-)$ enantiomer fits into the enzyme's active site.

Biological systems are capable of distinguishing between the enantiomers of many different chiral compounds. In general, just one of the enantiomers produces the characteristic effect; the other either produces no effect or has a different effect. Even your nose is capable of distinguishing between some enantiomers. For example, $(-)$-carvone is the fragrance associated with spearmint oil, while $(+)$-carvone has the tangy odor of caraway seed. The receptor sites for the sense of smell must be chiral, therefore, just as the active sites in most enzymes are chiral. In general, enantiomers do not interact identically with other *chiral* molecules, whether or not they are of biological origin.

5-5
Biological Discrimination of Enantiomers

Enzymes can exist as two enantiomers, although only one enantiomer is found in nature. In 1992, Stephen Kent and co-workers reported the synthesis of both enantiomers of an enzyme that cuts peptide substrates, and showed for the first time that each enzyme acts only on the corresponding enantiomeric peptide substrate.

$(+)$-carvone (caraway seed) $(-)$-carvone (spearmint)

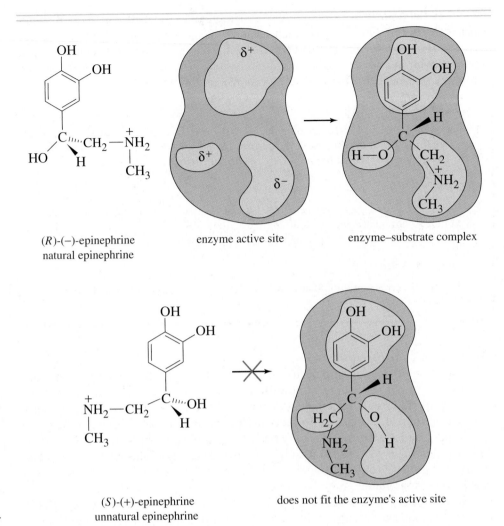

▶ **Figure 5-15**
Chiral recognition of epinephrine by an enzyme. Only the levorotatory enantiomer fits into the active site of the enzyme.

PROBLEM 5-11

If you had the two enantiomers of carvone in unmarked bottles, could you use just your nose and a polarimeter to determine
(a) whether it is the (+) or (−) enantiomer that smells like spearmint?
(b) whether it is the (R) or (S) enantiomer that smells like spearmint?
(c) With the information given in the drawings of carvone above, what can you add to your answers to (a) or (b)?

5-6
Racemic Mixtures

Suppose we had a mixture of equal amounts of (+)-2-butanol and (−)-2-butanol. The (+) isomer would rotate polarized light clockwise with a specific rotation of +13.5°, and the (−) isomer would rotate the polarized light counterclockwise by exactly the same amount. We would observe a rotation of zero, just as though 2-butanol were achiral. A solution of equal amounts of two enantiomers, so that the mixture is optically inactive, is called a **racemic mixture**.

Sometimes a racemic mixture is called a **racemate**, a (±) **pair**, or a (*d,l*) **pair**. A racemic mixture is symbolized by placing (±) or (*d,l*) in front of the name of the compound. For example, racemic 2-butanol would be symbolized by "(±)-2-butanol" or "(*d,l*)-2-butanol."

(*S*)-(+)-2-butanol
+13.5° rotation

(*R*)-(−)-2-butanol
−13.5° rotation

A racemic mixture contains equal amounts of the two enantiomers.

You might think that a racemic mixture would be unusual, since it requires exactly equal amounts of the two enantiomers. This is not the case, however. Many reactions lead to racemic products, especially when an achiral molecule is converted to a chiral molecule.

A reaction that uses optically inactive reactants and catalysts cannot produce a product that is optically active. Any chiral product must be formed as a racemic mixture.

For example, hydrogen adds across the C=O double bond of a ketone to produce an alcohol.

Because the carbonyl group is flat, a simple ketone such as 2-butanone is achiral. Hydrogenation of 2-butanone gives 2-butanol, a chiral molecule (Figure 5-16). This reaction involves adding hydrogen atoms to the C=O carbon atom and oxygen atom. If the hydrogen atoms are added to one face of the double bond, the (*S*) enantiomer results. Addition of hydrogen to the other face forms the (*R*) enantiomer. It is equally probable for hydrogen to add to either face of the double bond, and equal amounts of the (*R*) and (*S*) enantiomers are formed.

Logically, it makes sense that optically inactive reagents and catalysts cannot form optically active products. If the starting materials and reagents are optically inactive, there is no reason for the dextrorotatory product to be favored over a levorotatory one or vice versa. The (+) product and the (−) product are favored equally, and they are formed in equal amounts: a racemic mixture.

Many drugs currently on the market are racemic mixtures. Ketamine, for example, is a potent anesthetic agent, but its use is limited because it is hallucinogenic (making it a drug of abuse widely known as "K"). The (*S*) isomer is responsible for the anesthetic effects, while the (*R*) isomer causes the hallucinogenic effects.

ketamine

◀ **Figure 5-16**
Hydrogenation of 2-butanone forms racemic 2-butanol. Hydrogen adds to either face of the double bond. Addition of H_2 to one face gives the (*R*) product, while addition to the other face gives the (*S*) product.

5-7
Enantiomeric Excess and Optical Purity

Sometimes we deal with mixtures that are neither optically pure (all one enantiomer) nor racemic (equal amounts of two enantiomers). In these cases, we specify the **optical purity** (**o.p.**) of the mixture. The optical purity of a mixture is defined as the ratio of its rotation to the rotation of a pure enantiomer. For example, if we have some [mostly (+)] 2-butanol with a specific rotation of +9.72°, we compare this rotation with the +13.5° rotation of the pure (+) enantiomer.

$$\text{o.p.} = \frac{\text{observed rotation}}{\text{rotation of pure enantiomer}} \times 100\% = \frac{9.72°}{13.5°} \times 100\% = 72.0\%$$

The **enantiomeric excess** (**e.e.**) is a similar method for expressing the relative amounts of enantiomers in a mixture. To compute the enantiomeric excess of a mixture, we calculate the *excess* of the predominant enantiomer as a percentage of the entire mixture. For a chemically pure compound, the calculation of enantiomeric excess generally gives the same result as the calculation of optical purity, and we often use the two terms interchangeably. Algebraically, we use the following formula:

$$\text{o.p.} = \text{e.e.} = \frac{|d - l|}{d + l} \times 100\% = \frac{(\text{excess of one over the other})}{(\text{entire mixture})} \times 100\%$$

The units cancel out in the calculation of either e.e. or o.p., so these formulas can be used whether the amounts of the enantiomers are expressed in concentrations, grams, or percentages. For the 2-butanol mixture described above, the optical purity of 72% (+) implies that $d - l = 72\%$, and we know that $d + l = 100\%$. Adding the equations gives $2d = 172\%$. We conclude that the mixture contains 86% of the d or (+) enantiomer and 14% of the l or (−) enantiomer.

SOLVED PROBLEM 5-5

Calculate the e.e. and the specific rotation of a mixture containing 6 g of (+)-2-butanol and 4 g of (−)-2-butanol.

SOLUTION

In this mixture, there is a 2 g excess of the (+) isomer and a total of 10 g, for an e.e. of 20%. We can envision this mixture as 80% racemic [4 g(+) and 4 g(−)] and 20% pure (+).

$$\text{o.p.} = \text{e.e.} = \frac{|6 - 4|}{6 + 4} = \frac{2}{10} = 20\%$$

The specific rotation of enantiomerically pure (+)-2-butanol is +13.5°. The rotation of this mixture is

$$\text{observed rotation} = (\text{rotation of pure enantiomer}) \times (\text{o.p.})$$

$$= (+13.5°) \times (20\%) = +2.7°$$

PROBLEM 5-12

When optically pure (*R*)-2-bromobutane is heated with water, 2-butanol is the product. The reaction forms twice as much (*S*)-2-butanol as (*R*)-2-butanol. Calculate the e.e. and the specific rotation expected for the product.

PROBLEM 5-13

A chemist finds that the addition of (+)-epinephrine to the catalytic reduction of 2-butanone (Fig. 5-16) gives a product that is slightly optically active, with a specific rotation of +0.45°. Calculate the percentages of (+)-2-butanol and (−)-2-butanol formed in this reaction.

Let's consider whether *cis*-1,2-dibromocyclohexane is chiral. If we did not know about chair conformations, we might draw a flat cyclohexane ring. With a flat ring, the molecule has an internal mirror plane of symmetry (σ), and it is achiral.

But we know the ring is puckered into a chair conformation with one bromine atom axial and one equatorial. A chair conformation of *cis*-1,2-dibromocyclohexane and its mirror image are shown below. These two mirror-image structures are non-superimposable. You may be able to see the difference more easily if you make models of these two conformations.

Does this mean that *cis*-1,2-dibromocyclohexane is chiral? No, it does not, because the chair–chair interconversion is rapid at room temperature. If we had a bottle of just the conformation on the left, the molecules would quickly undergo chair–chair interconversions. Since the two mirror-image conformations interconvert and have identical energies, any sample of *cis*-1,2-dibromocyclohexane must contain equal amounts of the two mirror images.

A molecule cannot be optically active if its chiral conformations are in equilibrium with their mirror images. We consider such a molecule to be achiral.

cis-1,2-Dibromocyclohexane appears to exist as a racemic mixture, but with a major difference: It is impossible to create an optically active sample of *cis*-1,2-dibromocyclohexane. The molecule is incapable of showing optical activity. We could have predicted the correct result by imagining that the cyclohexane ring is flat.

This finding leads to a general principle we can use with conformationally mobile systems:

To determine whether a conformationally mobile molecule can be optically active, consider its most symmetric conformation.

An alternative statement of this rule is that a molecule cannot be optically active if it is in equilibrium with a structure (or a conformation) that is achiral. Because **conformers** differ only by rotations about single bonds, they are generally in equilibrium at room temperature. We can consider cyclohexane rings as though they were flat (the most symmetric conformation), and we should consider straight-chain compounds in their most symmetric conformations (often an eclipsed conformation).

For example, the gauche conformations of butane are chiral, but they quickly interconvert. They are in equilibrium with the totally eclipsed conformation, which is symmetric, implying that butane must be achiral.

gauche
(chiral)

totally eclipsed
(achiral)

gauche
(chiral)

SOLVED PROBLEM 5-6

Draw each compound in its most symmetric conformation, and determine whether it is capable of showing optical activity.

(a) 1,2-dichloroethane

SOLUTION

The most symmetric conformation is the one with the two chlorine atoms eclipsed. This conformation has two internal mirror planes of symmetry, and it is achiral.

(Newman projection)

(b) *trans*-1,2-dibromocyclohexane

SOLUTION

Drawn in its most symmetric conformation, this molecule is still nonsuperimposable on its mirror image. It is a chiral compound, and no ring-flip can interconvert the two enantiomers.

PROBLEM-SOLVING HINT

Consider the most symmetric accessible conformation. You can also consider the most stable conformation and see if it can interconvert with its mirror image.

PROBLEM 5-14

(1) Make a model of each compound, draw it in its most symmetric conformation, and determine whether it is capable of showing optical activity.

(2) Star (*) each asymmetric carbon atom, and compare your result from part (1) with the prediction you would make based on the asymmetric carbons.

(a) 1-bromo-1-chloroethane
(b) 1-bromo-2-chloroethane
(c) 1-bromo-1,2-dichloroethane
(d) *cis*-1,3-dibromocyclohexane
(e) *trans*-1,3-dibromocyclohexane
(f) *trans*-1,4-dibromocyclohexane

Most chiral organic compounds have at least one asymmetric carbon atom. There are some compounds that are chiral because they have another asymmetric atom, such as phosphorus, sulfur, or nitrogen, serving as a chirality center. Some compounds are chiral even though they have no asymmetric atoms at all. In these types of compounds, there are characteristics of the molecules' shapes that lend chirality to the structure.

5-9
Chiral Compounds without Asymmetric Atoms

5-9A Conformational Enantiomerism

Some molecules are so bulky or so highly strained that they cannot easily convert from one chiral conformation to the mirror-image conformation. They cannot achieve the most symmetric conformation because it has too much steric hindrance or ring strain. Since these molecules are "locked" into a conformation, we must evaluate the individual locked-in conformation to determine whether the molecule is chiral.

staggered conformation
(chiral)

impossible,
too crowded

eclipsed conformation
(symmetric, achiral)

staggered conformation
(chiral)

◀ **Figure 5-17**
This biphenyl cannot pass through its symmetric conformation because there is too much crowding of the iodine and bromine atoms. The molecule is "locked" into one of the two chiral, enantiomeric, staggered conformations.

Figure 5-17 shows three conformations of a sterically crowded derivative of biphenyl. The center drawing shows the molecule in its most symmetric conformation. This conformation is planar, and it has a mirror plane of symmetry. If the molecule could achieve this conformation, or even pass through it for an instant, it would not be optically active. This planar conformation is very high in energy, however, because the iodine and bromine atoms are too large to be forced so close together. The molecule is *conformationally locked*. It can exist only in one of the two staggered conformations shown on the left and right. These conformations are nonsuperimposable mirror images, and they do not interconvert. They are enantiomers, and they can be separated and isolated. Each of them is optically active, and they have equal and opposite specific rotations.

Even a simple strained molecule can show conformational enantiomerism. *trans*-Cyclooctene is the smallest stable *trans*-cycloalkene, and it is strained. If *trans*-cyclooctene existed as a planar ring, even for an instant, it could not be chiral. Make a molecular model of *trans*-cyclooctene, however, and you will see that it cannot exist as a planar ring. In fact, its ring is folded into the three-dimensional structure pictured in Figure 5-18. The mirror image of this structure is different, and *trans*-cyclooctene is a chiral molecule. In fact, the enantiomers of *trans*-cyclooctene have been separated and characterized, and they are optically active.

▲ **Figure 5-18**
trans-Cyclooctene is strained, unable to achieve a symmetric planar conformation. It is locked into one of these two enantiomeric conformations.

5-9B Allenes

The *sp* hybrid carbon atom requires a linear arrangement of atoms (Section 2-4). There are two kinds of *sp* hybrid carbon atoms. The most common are those that participate in triple bonds, such as in alkynes or nitriles:

$$H-C\equiv C-H \qquad CH_3-C\equiv N:$$

sp hybridized

▲ Figure 5-19
The two ends of the allene molecule
are perpendicular.

A carbon atom that participates simultaneously in two double bonds is also *sp* hybridized. Compounds containing a C=C=C unit are called **allenes**. The parent compound, propadiene, has the common name *allene*.

sp hybridized

$$H_2C=C=CH_2$$
allene

In allene, the central carbon atom is *sp* hybridized and linear, while the two outer carbon atoms are sp^2 hybridized and trigonal. We might imagine that the whole molecule lies in a plane, but this is not correct. The central *sp* hybrid carbon atom must use different *p* orbitals to form the pi bonds with the two outer carbon atoms. The two unhybridized *p* orbitals on the *sp* hybrid carbon atom are perpendicular, so the two pi bonds must also be perpendicular. Figure 5-19 shows the bonding and three-dimensional structure of allene.

Allene itself is achiral. If you make a model and its mirror image, you will find it identical with the original molecule. If we add some substituents to allene, however, the molecule may be chiral.

Make a model of the following compound:

$$\overset{1}{CH_3}-\overset{2}{CH}=\overset{3}{C}=\overset{4}{CH}-\overset{5}{CH_3}$$
2,3-pentadiene

Carbon atom 3 is the *sp* hybrid allene carbon atom. Carbons 2 and 4 are both sp^2 and planar, but their planes are perpendicular to each other. None of the carbon atoms are attached to four different atoms, so there is no chiral carbon atom. Nevertheless, 2,3-pentadiene is chiral, as you should see from your models and from the following drawings of the enantiomers.

PROBLEM-SOLVING HINT

Dienes are compounds with two double bonds. In the name, each double bond is given the lower number of its two carbon atoms. *Allenes* are dienes with the two double bonds next to each other, joined at one carbon atom. An allene is chiral if each end has two distinct substituents.

mirror

enantiomers of 2,3-pentadiene

PROBLEM 5-15

Draw three-dimensional representations of the following compounds. Which have asymmetric carbon atoms? Which have no asymmetric carbons but are chiral anyway? Use your models for parts (a) through (d) and any others that seem unclear.
(a) 1,3-dichloropropadiene   **(b)** 1-chloro-1,2-butadiene
(c) 1-chloro-3-methyl-1,2-butadiene   **(d)** 1-chloro-1,3-butadiene

(e) bromocyclohexane   **(f)**

(g)   **(h)**

We have been using dashed lines and wedges to indicate perspective in drawing the stereochemistry of asymmetric carbon atoms. When we draw molecules with several asymmetric carbons, perspective drawings become time-consuming and cumbersome. In addition, the complicated drawings make it difficult to see the similarities and differences in groups of stereoisomers.

At the turn of the 20th century, Emil Fischer was studying the stereochemistry of sugars (Chapter 23), which contain as many as seven asymmetric carbon atoms. To draw these structures in perspective would have been difficult, and to pick out minor stereochemical differences in the drawings would have been nearly impossible. Fischer developed a symbolic way of drawing asymmetric carbon atoms, allowing them to be drawn rapidly. The **Fischer projection** also facilitates comparison of stereoisomers, holding them in their most symmetric conformation and emphasizing any differences in stereochemistry.

5-10
Fischer Projections

5-10A   Drawing Fischer Projections

The Fischer projection looks like a cross, with the asymmetric carbon (usually not drawn in) at the point where the lines cross. The horizontal lines are taken to be wedges—that is, bonds that project out toward the viewer. The vertical lines are taken to project away from the viewer, as dashed lines. Figure 5-20 shows the perspective implied by the Fischer projection. The center drawing illustrates why this projection is sometimes called the "bow-tie convention." Problem 5-16 should help you to visualize how the Fischer projection is used.

view from
this angle

(S)-lactic acid
perspective drawing

(S)-lactic acid
Fischer projection

◀ **Figure 5-20**
The Fischer projection uses a cross to represent an asymmetric carbon atom. The horizontal lines project toward the viewer, and the vertical lines project away from the viewer.

PROBLEM 5-16

For each set of examples, make a model of the first structure and tell the relationship of each of the other structures to the first structure. Examples of relationships: same compound, enantiomer, structural isomer.

(a)

COOH	COOH	H	CH₃
H—OH	HO—H	H₃C—COOH	HO—H
CH₃	CH₃	OH	COOH

(b)

CH₂CH₃	CH₃	CH₂CH₃	CH₃
H—Br	Br—H	Br—H	H—Br
CH₃	CH₂CH₃	CH₃	CH₂CH₃

(c) (*R*)-2-butanol

	CH₃	CH₃	CH₂CH₃
	H—OH	HO—H	H—OH
	CH₂CH₃	CH₂CH₃	CH₃

In working Problem 5-16, you may have noticed that Fischer projections that differ by a 180° rotation are the same. When we rotate a Fischer projection by 180°, the vertical (dashed line) bonds still end up vertical, and the horizontal (wedged) lines still end up horizontal. The "horizontal lines forward, vertical lines back" convention is maintained.

Rotation by 180° is allowed.

On the other hand, if we were to rotate a Fischer projection by 90°, we would change the configuration and confuse the viewer. The original projection has the vertical groups back (dashed lines) and the horizontal groups forward. When we rotate the projection by 90°, the vertical bonds become horizontal and the horizontal bonds become vertical. The viewer assumes that the horizontal bonds come forward and that the vertical bonds go back. The viewer sees a different molecule (actually, the enantiomer of the original molecule).

A 90° rotation is NOT allowed.

In comparing Fischer projections, we cannot rotate them by 90° and we cannot flip them over. Either of these operations gives an incorrect representation of the molecule. The Fischer projection must be kept in the plane of the paper, and it may be rotated only by 180°.

The final rule for drawing Fischer projections helps to ensure that we do not rotate the drawing by 90°. This rule is that the carbon chain is drawn along the vertical line of the Fischer projection, usually with the IUPAC numbering from top to bottom. In most cases, this numbering places the most highly oxidized carbon substituent at the top. For example, to represent (*R*)-1,2-propanediol with a Fischer projection, we should arrange the three carbon atoms along the vertical. C1 is placed at the top, and C3 at the bottom.

$$^1CH_2OH \quad = \quad ^1CH_2OH \quad = \quad ^1CH_2OH$$

(*R*)-1,2-propanediol

viewed from this angle

PROBLEM 5-17

Draw a Fischer projection for each compound. Remember that the cross represents an asymmetric carbon atom, and the carbon chain should be along the vertical, with the IUPAC numbering from top to bottom.
(a) (*S*)-1,2-propanediol
(b) (*R*)-2-bromo-1-butanol
(c) (*S*)-1,2-dibromobutane
(d) (*R*)-2-butanol
(e) (*R*)-glyceraldehyde, HO—$^3CH_2$—^2CH—^1CHO, with OH on C2

PROBLEM-SOLVING HINT

Interchanging any two groups on a Fischer projection (or on a perspective drawing) inverts the configuration of that chiral carbon from (*R*) to (*S*) or from (*S*) to (*R*).

5-10B Drawing Mirror Images of Fischer Projections

How does one draw the mirror image of a molecule drawn in Fischer projection? With our perspective drawings, the rule was to reverse left and right while keeping the other directions (up and down, front and back) in their same positions. This rule still applies to Fischer projections. Interchanging the groups on the horizontal part of the cross reverses left and right while leaving the other directions unchanged.

(*R*)-lactic acid   (*S*)-lactic acid

Testing for enantiomerism is particularly simple using Fischer projections. If the Fischer projections are properly drawn (carbon chain along the vertical), and if the mirror image cannot be made to look the same as the original structure with a 180° rotation in the plane of the paper, the two mirror images are enantiomers. In the following examples, any groups that fail to superimpose after a 180° rotation are circled in red.

Original   *Mirror image*   *180° rotation*

2-propanol

These mirror images are the same. 2-Propanol is achiral.

$$(R)\text{-1,2-propanediol}$$

CH₂OH		CH₂OH			CH₃
H——OH		HO——H	180° =		H——OH
CH₃		CH₃			CH₂OH

These mirror images are different. 1,2-Propanediol is chiral.

(2S,3S)-2,3-dibromobutane

	CH₃		CH₃			CH₃
	H——Br		Br——H	180° =		Br——H
	Br——H		H——Br			H——Br
	CH₃		CH₃			CH₃

These mirror images are different. This structure is chiral.

Mirror planes of symmetry are particularly easy to identify from the Fischer projection because this projection is normally the most symmetric conformation. In the first preceding example and in the following example, the symmetry planes are indicated in red; these molecules with symmetry planes cannot be chiral.

(2S,3R)-2,3-dibromobutane

	CH₃		CH₃			CH₃
σ - - -	H——Br		Br——H	- - - σ		H——Br
	H——Br		Br——H	180° =		H——Br
	CH₃		CH₃			CH₃

These mirror images are the same. This structure is achiral.

PROBLEM 5-18

For each Fischer projection
(1) Make a model.
(2) Draw the mirror image.
(3) Determine whether the mirror image is the same as, or different from, the original structure.
(4) Draw in any mirror planes of symmetry that are apparent from the Fischer projections.

(a)

CHO
H——OH
CH₂OH

(b)

CH₂OH
H——Br
CH₂OH

(c)

CH₂Br
Br——Br
CH₃

(d)

CHO
H——OH
H——OH
CH₂OH

(e)

CH₂OH
H——OH
H——OH
CH₂OH

(f)

CH₂OH
HO——H
H——OH
CH₂OH

5-10C   Assigning (R) and (S) Configurations from Fischer Projections

The Cahn–Ingold–Prelog convention (Section 5-3) can be applied to structures drawn using Fischer projections. Let's review the two rules for assigning (R) and (S): (1) Assign priorities to the groups bonded to the asymmetric carbon atom; (2) put the lowest-priority group (usually H) in back, and draw an arrow from group 1 to group 2 to group 3. Clockwise is (R), and counterclockwise is (S).

The (R) or (S) configuration can also be determined directly from the Fischer projection, without having to convert it to a perspective drawing. The lowest-priority atom is usually hydrogen. In the Fischer projection, the carbon chain is along the vertical line, so the hydrogen atom is on the horizontal line and projects out in front. Once we have assigned priorities, we can draw an arrow from group 1 to group 2 to group 3 and see which way it goes. If the molecule were turned around so that the hydrogen would be in back (as in the definition of (R) and (S)), the arrow would rotate in the other direction. By mentally turning the arrow around, we can assign the configuration.

As an example, consider the Fischer projection formula of one of the enantiomers of glyceraldehyde. First priority goes to the —OH group, followed by the —CHO group and the —CH$_2$OH group. The hydrogen atom receives the lowest priority. The arrow from group 1 to group 2 to group 3 appears counterclockwise in the Fischer projection. If the molecule is turned over so the hydrogen is in back, the arrow is clockwise, so this is the (R) enantiomer of glyceraldehyde.

counterclockwise   clockwise

Fischer projection
(R)-(+)-glyceraldehyde   hydrogen in front   perspective drawing
(R)-(+)-glyceraldehyde

PROBLEM 5-19

For each Fischer projection, label each asymmetric carbon atom as (R) or (S).
(a)–(f) the structures in Problem 5-18

(g)
$$\begin{array}{c} CH_2CH_3 \\ H \text{——} Br \\ CH_3 \end{array}$$

(h)
$$\begin{array}{c} COOH \\ H_2N \text{——} H \\ CH_3 \end{array}$$

(i)
$$\begin{array}{c} CH_2OH \\ Br \text{——} Cl \\ CH_3 \end{array}$$
(careful—no hydrogen)

PROBLEM-SOLVING $\mathbf{HINT}$

When naming (R) and (S) from Fischer projections with the hydrogen on a horizontal bond (toward you instead of away from you), just apply the normal rules backward.

SUMMARY Fischer Projections and Their Use

1. They are most useful for compounds with two or more asymmetric carbon atoms.
2. Asymmetric carbons are at the centers of crosses.
3. The vertical lines project away from the viewer, the horizontal lines toward the viewer.
4. The carbon chain is placed along the vertical, with the IUPAC numbering from top to bottom. In most cases, this places the most oxidized end (the carbon with the most bonds to O or halogen) at the top.
5. The entire projection can be rotated 180° (but not 90°) in the plane of the paper without changing its stereochemistry.
6. Interchanging any two groups on an asymmetric carbon (for example, those on the horizontal line) inverts its stereochemistry.

We have defined *stereoisomers* as isomers whose atoms are bonded together in the same order but differ in how the atoms are directed in space. We have also considered enantiomers (mirror-image isomers) in detail. All other stereoisomers are classified as **diastereomers**, which are defined as *stereoisomers that are not mirror images*. Most diastereomers are either geometric isomers or compounds containing two or more chirality centers (usually asymmetric carbons).

5-11
Diastereomers

5-11A   cis-trans Isomerism on Double Bonds

We have already seen one class of diastereomers, the **cis-trans isomers**, or **geometric isomers**. For example, there are two isomers of 2-butene:

cis-2-butene   *trans*-2-butene

These stereoisomers are not mirror images of each other, so they are not enantiomers. They are diastereomers.

5-11B   cis-trans Isomerism on Rings

cis-trans isomerism is also possible when there is a ring present. *cis*- and *trans*-1,2-dimethylcyclopentane are geometric isomers, and they are also diastereomers. The trans diastereomer has an enantiomer, but the cis diastereomer has an internal mirror plane of symmetry, so it is achiral.

enantiomers of *trans*-1,2-dimethylcyclopentane   *cis*-1,2-dimethylcyclopentane (achiral)

diastereomers

5-11C   Diastereomers of Molecules with Two or More Chirality Centers

Apart from geometric isomers, most other compounds that show diastereomerism have two or more chirality centers, usually asymmetric carbon atoms. For example, 2-bromo-3-chlorobutane has two asymmetric carbon atoms, and it exists in two diastereomeric forms (shown below). Make molecular models of these two stereoisomers.

diastereomers

These two structures are not the same; they are stereoisomers because they differ in the orientation of their atoms in space. They are not enantiomers, however, because they are not mirror images of each other: C2 has the (*S*) configuration in both structures, while C3 is (*R*) in the structure on the left and (*S*) in the structure on the right. The C3 carbon atoms are mirror images of each other, but the C2 carbon atoms are not. If these two compounds were mirror images of each other, both asymmetric carbons would have to be mirror images of each other.

Since these compounds are stereoisomers but not enantiomers, they must be diastereomers. In fact, both of these diastereomers are chiral and each has an enantiomer. Thus, there are a total of four stereoisomeric 2-bromo-3-chlorobutanes: two pairs of enantiomers. Either member of one pair of enantiomers is a diastereomer of either member of the other pair.

We have now seen all the types of isomers we need to study, and we can diagram their relationships and summarize their definitions.

SUMMARY   Types of Isomers

```
                          all isomers
                               |
             ┌─────────────────┴─────────────────┐
   constitutional isomers                    stereoisomers
   (structural isomers)                           |
                              ┌──────────────────┴──────────────┐
                         diastereomers                      enantiomers
                              |
                 ┌────────────┴────────────┐
          cis-trans isomers          other diastereomers
          (geometric isomers)      (two or more chirality centers)
```

Isomers are different compounds with the same molecular formula.
Constitutional isomers are isomers that differ in the order in which atoms are bonded together. Constitutional isomers are sometimes called **structural isomers** because they have different connections among their atoms.
Stereoisomers are isomers that differ only in the orientation of the atoms in space.
Enantiomers are mirror-image isomers.
Diastereomers are stereoisomers that are not mirror images of each other.
cis-trans isomers (geometric isomers) are diastereomers that differ in their cis-trans arrangement on a ring or double bond.

PROBLEM 5-20

For each pair, give the relationship between the two compounds. Making models will be helpful.
(a) (2R,3S)-2,3-dibromohexane and (2S,3R)-2,3-dibromohexane
(b) (2R,3S)-2,3-dibromohexane and (2R,3R)-2,3-dibromohexane
(c)

5-12
Stereochemistry of Molecules with Two or More Asymmetric Carbons

In the preceding section, we saw there are four stereoisomers (two pairs of enantiomers) of 2-bromo-3-chlorobutane. These four isomers are simply all the permutations of (R) and (S) configurations at the two asymmetric carbon atoms, C2 and C3:

A compound with n asymmetric carbon atoms often has 2^n stereoisomers. This formula is called the 2^n **rule**, where n is the number of chirality centers (usually asymmetric carbon atoms). The 2^n rule suggests we should look for a *maximum* of 2^n stereoisomers. We may not always find 2^n isomers, especially when two of the asymmetric carbon atoms have identical substituents.

2,3-Dibromobutane has fewer than 2^n stereoisomers. It has two asymmetric carbons (C2 and C3), so the 2^n rule predicts a maximum of four stereoisomers. The four permutations of (R) and (S) configurations at C2 and C3 are shown below. Make molecular models of these structures to compare them.

There are only three stereoisomers of 2,3-dibromobutane because two of the four structures are identical. The diastereomer on the right is achiral, having a mirror plane of symmetry. The asymmetric carbon atoms have identical substituents, and the one with (*R*) configuration reflects into the other having (*S*) configuration. It seems almost as though the molecule were a racemic mixture within itself.

Compounds that are achiral even though they have asymmetric carbon atoms are called **meso compounds**. The (2*R*,3*S*) isomer of 2,3-dibromobutane is a meso compound; most meso compounds have this kind of symmetric structure, with two similar halves of the molecule having opposite configurations. In speaking of the two diastereomers of 2,3-dibromobutane, the symmetric one is called the *meso diastereomer*, and the chiral one is called the (±) *diastereomer*, since one enantiomer is (+) and the other is (−).

5-13
Meso Compounds

MESO COMPOUND: An achiral compound that has chirality centers.

The term *meso* (Greek, "middle") was used to describe an achiral member of a set of diastereomers, some of which are chiral. The optically inactive isomer seemed to be in the "middle" between the dextrorotatory and levorotatory isomers. The definition given above ("an achiral compound with chirality centers") is more easily applied, especially when you remember that chirality centers are usually asymmetric carbon atoms.

We have already seen other meso compounds, although we have not yet called them that. For example, the cis isomer of 1,2-dichlorocyclopentane has two asymmetric carbon atoms, yet it is achiral. Thus it is a meso compound. *cis*-1,2,-Dibromocyclohexane is not symmetric in its chair conformation, but it consists of equal amounts of two enantiomeric chair conformations in a rapid equilibrium. We are justified in looking at the molecule in its symmetric flat conformation to show that it is achiral and meso. For acyclic compounds, the Fischer projection helps to show the symmetry of meso compounds.

cis-1,2-dichlorocyclopentane   *cis*-1,2-dibromocyclohexane   *meso*-2,3-dibromobutane   *meso* tartaric acid

SOLVED PROBLEM 5-7

Determine which of the following compounds are chiral. Star (*) any asymmetric carbon atoms, and draw in any mirror planes. Label any meso compounds. (Use your molecular models to follow along.)

SOLUTION

(a)

CH₃
H——*——OH
HO——*——H
CH₃

CH₃
HO——*——H
H——*——OH
CH₃

This compound does *not* have a plane of symmetry, and we might suspect that it is chiral. Drawing the mirror image shows that it is nonsuperimposable on the original structure. These are the enantiomers of a chiral compound.

(b)

CH₂OH
Br——*——Cl
Br——*——Cl   - - - σ
CH₂OH
meso

(c)

CH₃
H——*——OH
HO——*——H
HO——*——H   - - - - - - σ
H——*——OH
CH₃
meso

Both (b) and (c) have mirror planes of symmetry and are achiral. Because they have asymmetric carbon atoms yet are achiral, they are meso.

(d)

[hexagon ring with H H and *Cl *Br]   [hexagon ring with H H and *Br *Cl]

Drawing this compound in its most symmetric conformation (flat) shows that it does not have a mirror plane of symmetry. When we draw the mirror image, it is found to be an enantiomer.

SOLVED PROBLEM 5-8

One source defines a meso compound as "an achiral compound with stereocenters." Why is this a poor definition?

SOLUTION

A stereocenter is an atom at which the interchange of two groups gives a stereoisomer. Stereocenters include both chirality centers and double-bonded carbons giving rise to cis-trans isomers. For example, the isomers of 2-butene are achiral, and they contain stereocenters (circled), so they would meet this definition. They have no chiral diastereomers, however, so they are not correctly called meso.

PROBLEM 5-21

Which of the following compounds are chiral? Draw each compound in its most symmetric conformation, star (*) any asymmetric carbon atoms, and draw in any mirror planes. Label any meso compounds. You may use Fischer projections if you prefer.

(a) *meso*-2,3-dibromo-2,3-dichlorobutane   **(b)** (±)-2,3-dibromo-2,3-dichlorobutane

(c) butane

(d) (2R,3S) HOCH₂—CHBr—CHOH—CH₂OH
$$\overset{4}{} \quad \overset{3}{} \quad \overset{2}{} \quad \overset{1}{}$$

(e)

CHO
H——OH
HO——H
H——OH
H——OH
CH₂OH

(f)

[cyclohexane ring with H, CH₃, Br, H, Br, CH₃, H]

(g)

[cyclohexane ring with Br, H, H, H, Br, CH₃, CH₃, H]

PROBLEM 5-22

Draw all the distinct stereoisomers for each structure. Show the relationships (enantiomers, diastereomers, etc.) between the isomers. Label any meso isomers, and draw in any mirror planes of symmetry.
(a) CH_3—$CHCl$—$CHOH$—$COOH$
(b) tartaric acid, $HOOC$—$CHOH$—$CHOH$—$COOH$
(c) $HOOC$—$CHBr$—$CHOH$—$CHOH$—$COOH$

(d) HO—◁ with CH_3 groups

5-14
Absolute and Relative Configuration

Throughout our study of stereochemistry, we have drawn three-dimensional representations, and we have spoken of asymmetric carbons having the (R) or (S) configuration. These ways of describing the configuration of a chirality center are *absolute*; that is, they give the actual orientation of the atoms in space. We say that these methods specify the **absolute configuration** of the molecule. For example, given the name "(R)-2-butanol," any chemist can construct an accurate molecular model or draw a three-dimensional representation.

> ABSOLUTE CONFIGURATION: The detailed stereochemical picture of a molecule, including how the atoms are arranged in space. Alternatively, the (R) or (S) configuration at each chirality center.

Chemists have determined the absolute configurations of many chiral compounds since 1951, when X-ray crystallography was first used to find the orientation of atoms in space. Before 1951, there was no way to link the stereochemical drawings with the actual enantiomers and their observed rotations. No absolute configurations were known. It was possible, however, to correlate the configuration of one compound with another and to show that two compounds had the same or opposite configurations. When we convert one compound into another using a reaction that does not break bonds at the asymmetric carbon atom, we know that the product must have the same **relative configuration** as the reactant, even if we cannot determine the absolute configuration of either compound.

> RELATIVE CONFIGURATION: The experimentally determined relationship between the configurations of two molecules, even though we may not know the absolute configuration of either.

For example, optically active 2-methyl-1-butanol reacts with PBr_3 to give optically active 1-bromo-2-methylbutane. None of the bonds to the asymmetric carbon atom are broken in this reaction, so the product must have the same configuration at the asymmetric carbon as the starting material does.

$$CH_3CH_2\overset{*}{C}HCH_2OH + PBr_3 \longrightarrow CH_3CH_2\overset{*}{C}HCH_2Br$$
$$\underset{CH_3}{|} \qquad\qquad\qquad\qquad \underset{CH_3}{|}$$

(+)-2-methyl-1-butanol
$[\alpha]_D^{25} = +5.8°$

(−)-1-bromo-2-methylbutane
$[\alpha]_D^{25} = -4.0°$

We say that (+)-2-methyl-1-butanol and (−)-1-bromo-2-methylbutane have the same relative configuration, even though we don't have the foggiest idea whether either of these is (R) or (S) unless we relate them to a compound whose absolute configuration has been established by X-ray crystallography.

Before the advent of X-ray crystallography, several systems were used to compare the relative configurations of chiral compounds with those of standard compounds. Only one of these systems is still in common use today: the **D–L system**, also known as the *Fischer–Rosanoff convention*. The configurations of sugars and amino acids were related to the enantiomers of glyceraldehyde. Compounds with the same relative configuration as (+)-glyceraldehyde were assigned the D prefix, and those with the relative configuration of (−)-glyceraldehyde were given the L prefix.

CHO
HO——H
(S)
CH₂OH
L-(−)-glyceraldehyde

COOH
H₂N——H
R
an L-amino acid

COOH
H₂N——H
CH₂OH
L-(−)-serine

COOH
H₂N——H
CH₂CH₂COOH
L-(+)-glutamic acid

We now know the absolute configurations of the glyceraldehyde enantiomers: The (+) enantiomer has the (*R*) configuration, with the hydroxyl (OH) group on the right in the Fischer projection. The (−) enantiomer has the (*S*) configuration, with the hydroxyl group on the left. Most naturally occurring amino acids have the L configuration, with the amino (NH₂) group on the left in the Fischer projection.

Sugars have several asymmetric carbons, but they can all be degraded to glyceraldehyde by oxidizing them from the aldehyde end. (We discuss these reactions in Chapter 23.) Most naturally occurring sugars degrade to (+)-glyceraldehyde, so they are given the D prefix. This means that the bottom asymmetric carbon of the sugar has its hydroxyl (OH) group on the right in the Fischer projection.

CHO
H——OH
HO——H
H——OH
H——OH
CH₂OH
D-(+)-glucose

degrade →

CHO
H——OH
(*R*) CH₂OH
D-(+)-glyceraldehyde

← degrade

CHO
HO——H
H——OH
CH₂OH
D-(−)-threose

5-15
Physical Properties of Diastereomers

We have seen that enantiomers have identical physical properties except for the direction in which they rotate polarized light. Diastereomers, on the other hand, generally have different physical properties. For example, consider the diastereomers of 2-butene (shown below). The symmetry of *trans*-2-butene causes the dipole moments of the bonds to cancel. The dipole moments in *cis*-2-butene do not cancel but add together to create a molecular dipole moment. The dipole–dipole attractions of *cis*-2-butene give it a higher boiling point than *trans*-2-butene.

$\mu = 0$
trans-2-butene
bond dipoles cancel
bp = 0.9°C

$\mu = 0.33$ D
cis-2-butene
vector sum dipole↓
bp = 3.7°C

Diastereomers that are not geometric isomers also have different physical properties. The two diastereomers of 2,3-dibromosuccinic acid have melting points that differ by nearly 100°C!

(+) and (−)-2,3-dibromosuccinic acid
mp for either is 158°C

meso-2,3-dibromosuccinic acid
mp 256°C

Most of the common sugars are diastereomers of glucose. All these diastereomers have different physical properties. For example, glucose and galactose are diastereomeric sugars that differ only in the stereochemistry of one asymmetric carbon atom, C4.

D-(+)-glucose, mp 148°C   D-(+)-galactose, mp 167°C

Because diastereomers have different physical properties, we can separate them by ordinary means such as distillation, recrystallization, and chromatography. As we will see in the next section, the separation of enantiomers is a more difficult process.

PROBLEM 5-23

Which of the following pairs of compounds could be separated by recrystallization or distillation?

(a) *meso*-tartaric acid and (±)-tartaric acid (HOOC—CHOH—CHOH—COOH)

(b)

(c)

(d)*

5-16
Resolution
of Enantiomers

Pure enantiomers of optically active compounds are often obtained by isolation from biological sources. Most optically active molecules are found as only one enantiomer in living organisms. For example, pure (+)-tartaric acid can be isolated from the precipitate formed by yeast during the fermentation of wine. Pure (+)-glucose is obtained from many different sugar sources, such as grapes, sugar beets, sugarcane, and honey. Alanine is a common amino acid found in protein as the pure (+) enantiomer.

L-(+)-tartaric acid · D-(+)-glucose · L-(+)-alanine
(2R,3R)-tartaric acid

When a chiral compound is synthesized from achiral reagents, however, a racemic mixture of enantiomers is obtained. For example, we saw that the reduction of 2-butanone (achiral) to 2-butanol (chiral) gives a racemic mixture:

Wine sediments often contain salts of tartaric acid. This wine cork bears crystals of the potassium salt of L-(+)-tartaric acid.

If we need one pure enantiomer of 2-butanol, we must find a way of separating it from the other enantiomer. The separation of enantiomers is called **resolution**, and it is a different process from the usual physical separations. A chiral probe is necessary for the resolution of enantiomers; such a chiral compound or apparatus is called a **resolving agent**.

In 1848, Louis Pasteur noticed that a salt of racemic (±)-tartaric acid crystallizes into mirror-image crystals. Using a microscope and a pair of tweezers, he physically separated the enantiomeric crystals. He found that solutions made from the "left-handed" crystals rotate polarized light in one direction, and solutions made from the "right-handed" crystals rotate polarized light in the opposite direction. Pasteur had accomplished the first artificial resolution of enantiomers. Unfortunately, few racemic compounds crystallize as separate enantiomers, and other methods of separation are required.

An illustration of Louis Pasteur working in the laboratory. He is, no doubt, contemplating the implications of enantiomerism in tartaric acid crystals.

5-16A   Chemical Resolution of Enantiomers

The traditional method for resolving a racemic mixture into its enantiomers is to use an enantiomerically pure natural product that bonds with the compound to be resolved. When the enantiomers of the racemic compound bond to the pure resolving agent, a pair of diastereomers results. The diastereomers are separated, then the resolving agent is cleaved from the separated enantiomers.

Let's consider how we might resolve a racemic mixture of (R)- and (S)-2-butanol. We need a resolving agent that reacts with an alcohol and that is readily available in an enantiomerically pure state. A carboxylic acid combines with an alcohol to form an ester. Although we have not yet studied the chemistry of esters (Chapter 21), the following equation shows how an acid and an alcohol can combine with the loss of water:

▲ **Figure 5-21**

Formation of (*R*)- and (*S*)-2-butyl tartrate. The reaction of a pure enantiomer of one compound with a racemic mixture of another compound produces a mixture of diastereomers. Separation of the diastereomers, followed by hydrolysis, gives the resolved enantiomers.

For our resolving agent, we need an optically active chiral acid to react with 2-butanol. Any winery can provide large amounts of pure (+)-tartaric acid. Figure 5-21 shows that diastereomeric esters are formed when (*R*)- and (*S*)-2-butanol react with (+)-tartaric acid. We can represent the reaction schematically as follows:

The diastereomers of 2-butyl tartrate have different physical properties, and they can be separated by conventional distillation, recrystallization, or chromatography. Separation of the diastereomers leaves us with two flasks, each containing one of the diastereomeric esters. The resolving agent is then cleaved from the separated enantiomers of 2-butanol by the reverse of the reaction used to make the ester. Adding an acid catalyst and an excess of water to an ester drives the equilibrium toward the acid and the alcohol:

Hydrolysis of (*R*)-2-butyl tartrate gives (*R*)-2-butanol and (+)-tartaric acid, and hydrolysis of (*S*)-2-butyl tartrate gives (*S*)-2-butanol and (+)-tartaric acid. The recovered tartaric acid would probably be thrown away, since it is cheap and nontoxic. Many other chiral resolving agents are expensive, so they must be carefully recovered and recycled.

PROBLEM 5-24

To show that (R)-2-butyl-(R,R)-tartrate and (S)-2-butyl-(R,R)-tartrate are not enantiomers, draw and name the mirror images of these compounds.

5-16B   Chromatographic Resolution of Enantiomers

Chromatography is a powerful method for separating compounds. One type of chromatography involves passing a solution through a column containing particles whose surface tends to adsorb organic compounds. Compounds that are adsorbed strongly spend more time on the stationary particles; they come off the column later than less strongly adsorbed compounds, which spend more time in the mobile solvent phase.

In some cases, enantiomers may be resolved by passing the racemic mixture through a column containing particles whose surface is coated with chiral molecules (Figure 5-22). As the solution passes through the column, the enantiomers form weak complexes, usually through hydrogen bonding, with the chiral column packing. The solvent flows continually through the column, and the dissolved enantiomers gradually move along, retarded by the time they spend complexed with the column packing.

The special feature of this chromatography is the fact that the enantiomers form diastereomeric complexes with the chiral column packing. These diastereomeric complexes have different physical properties. They also have different binding energies and different equilibrium constants for complexation. One of the two enantiomers will spend more time complexed with the chiral column packing. The more strongly complexed enantiomer passes through the column more slowly and emerges from the column after the faster-moving (more weakly complexed) enantiomer.

> Enzymes can also be used to eliminate an undesired stereoisomer. The enzyme will process only one isomer in a racemic mixture and leave the other stereoisomer untouched.

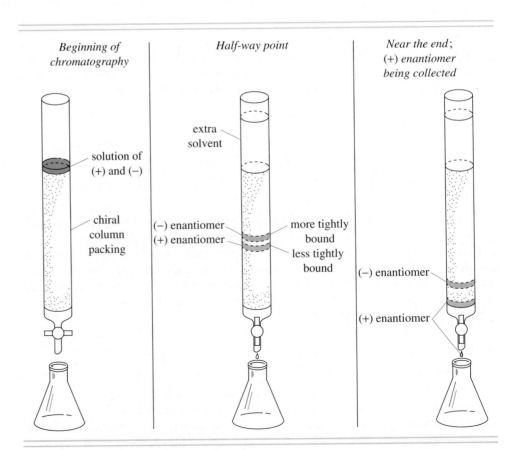

▶ **Figure 5-22**
Chromatographic resolution of enantiomers. The enantiomers of the racemic compound form diastereomeric complexes with the chiral material on the column packing. One of the enantiomers binds more tightly than the other, so it moves more slowly through the column.

absolute configuration The detailed stereochemical picture of a molecule, including how the atoms are arranged in space. Alternatively, the (R) or (S) configuration at each asymmetric carbon atom. (p. 201)

achiral Not chiral. (p. 168)

asymmetric carbon atom (**chiral carbon atom**) A carbon atom that is bonded to four different groups. (p. 170)

Cahn–Ingold–Prelog convention The accepted method for designating the absolute configuration of a chirality center (usually an asymmetric carbon) as either (R) or (S). (p. 175)

chiral Different from its mirror image. (p. 168)

chiral carbon atom (asymmetric carbon atom) A carbon atom that is bonded to four different groups. (p. 170)

chirality center The IUPAC term for an atom holding a set of ligands in a spatial arrangement that is not superimposable on its mirror image. Asymmetric carbon atoms are the most common chirality centers. (p. 170)

chiral probe A molecule or an object that is chiral and can use its own chirality to differentiate between mirror images. (p. 183)

cis On the same side of a ring or double bond. (p. 196)

cis-trans isomers (geometric isomers) Isomers that differ in their geometric arrangement on a ring or double bond. cis-trans isomers are a subclass of diastereomers. (p. 196)

configurations The two possible spatial arrangements around a chirality center or other stereocenter. (p. 201)

configurational isomers (see **stereoisomers**)

conformers (conformational isomers) Structures that differ only by rotations about single bonds. In most cases, conformers interconvert at room temperature; thus, they are not different compounds and not true isomers. (p. 187)

constitutional isomers (structural isomers) Isomers that differ in the order in which their atoms are bonded together. (p. 167)

D–L configurations (Fischer–Rosanoff convention) D has the same relative configuration as (+)-glyceraldehyde. L has the same relative configuration as (−)-glyceraldehyde. (p. 202)

dextrorotatory, (+), or (*d*) Rotating the plane of polarized light clockwise. (p. 181)

diastereomers Stereoisomers that are not mirror images. (p. 195)

enantiomeric excess (e.e.) The excess of one enantiomer in a mixture of enantiomers expressed as a percentage of the mixture. Similar to optical purity. (p. 186) Algebraically,

$$\text{e.e} = \frac{|R - S|}{R + S} \times 100\%$$

enantiomers A pair of nonsuperimposable mirror-image molecules: mirror-image isomers. (p. 169)

Fischer projection A method for drawing an asymmetric carbon atom as a cross. The carbon chain is kept along the vertical, with the IUPAC numbering from top to bottom. Vertical bonds project away from the viewer, and horizontal bonds project toward the viewer. (p. 191)

geometric isomers (see **cis-trans isomers**) (p. 196)

internal mirror plane (σ) A plane of symmetry through the middle of a molecule, dividing the molecule into two mirror-image halves. A molecule with an internal mirror plane of symmetry cannot be chiral. (p. 173)

isomers Different compounds with the same molecular formula. (p. 197)

levorotatory, (−), or (*l*) Rotating the plane of polarized light counterclockwise. (p. 181)

meso compound An achiral compound that contains chirality centers (usually asymmetric carbon atoms). (p. 199)

optical isomers (archaic; see **enantiomers**) Compounds with identical properties except for the direction in which they rotate polarized light. (p. 180)

optically active Capable of rotating the plane of polarized light. (p. 180)

optical purity (o.p.) The specific rotation of a mixture of two enantiomers, expressed as a percentage of the specific rotation of one of the pure enantiomers. Similar to enantiomeric excess. (p. 186) Algebraically,

$$\text{o.p.} = \frac{\text{observed rotation}}{\text{rotation of pure enantiomer}} \times 100\%$$

plane-polarized light Light composed of waves that vibrate in only one plane. (p. 179)

polarimeter An instrument that measures the rotation of plane-polarized light by an optically active compound. (p. 180)

racemic mixture [racemate, racemic modification, ($\pm$) pair, (*d,l*) pair] A mixture of equal quantities of enantiomers, such that the mixture is optically inactive. (p. 184)

relative configuration The experimentally determined relationship between the configurations of two molecules, even though the absolute configuration of either may not be known. (p. 201)

resolution The process of separating a racemic mixture into the pure enantiomers. Resolution requires a chiral resolving agent. (p. 204)

resolving agent A chiral compound (or chiral material on a chromatographic column) used for separating enantiomers. (p. 204)

2^n rule A molecule with *n* chiral carbon atoms might have as many as 2^n stereoisomers. (p. 198)

specific rotation A measure of a compound's ability to rotate the plane of polarized light, given by

$$[\alpha]_D^{25} = \frac{\alpha(\text{observed})}{c \cdot l}$$

where *c* is concentration in g/mL, and *l* is length of sample cell (path length) in decimeters. (p. 182)

stereocenter (stereogenic atom) An atom that gives rise to stereoisomers when its groups are interchanged. Asymmetric carbon atoms and double-bonded carbons in cis-trans alkenes are the most common stereocenters. (p. 170)

stereochemistry The study of the three-dimensional structure of molecules. (p. 167)

stereoisomers (configurational isomers) Isomers whose atoms are bonded together in the same order, but they differ in how the atoms are oriented in space. (p. 167)

structural isomers (see **constitutional isomers**) Isomers that differ in the order in which their atoms are bonded together. (p. 167)

superimposable Identical in all respects. The three-dimensional positions of all atoms coincide when the molecules are placed on top of each other. (p. 169)

trans On opposite sides of a ring or double bond. (p. 196)

Essential Problem-Solving Skills in Chapter 5

1. Classify molecules as chiral or achiral, and identify mirror planes of symmetry.
2. Identify asymmetric carbon atoms, and name them using the (*R*) and (*S*) nomenclature.
3. Calculate specific rotations from polarimetry data.
4. Draw all stereoisomers of a given structure.
5. Identify enantiomers, diastereomers, and meso compounds.
6. Draw correct Fischer projections of asymmetric carbon atoms.
7. Explain how the physical properties differ for different types of stereoisomers.
8. Suggest how to separate different types of stereoisomers.

Study Problems

5-25   Briefly define each term and give an example.

(a) (*R*) and (*S*)	**(b)** chiral and achiral	**(c)** chiral carbon atom
(d) cis and trans	**(e)** asymmetric carbon atom	**(f)** isomers
(g) constitutional isomers	**(h)** stereoisomers	**(i)** enantiomers
(j) Fischer projection	**(k)** diastereomers	**(l)** chirality center
(m) optical isomers	**(n)** meso	**(o)** stereocenter
(p) racemic mixture	**(q)** specific rotation	**(r)** dextrorotatory
(s) ($\pm$) and (*d,l*)	**(t)** absolute configuration	**(u)** relative configuration
(v) D and L configurations		

5-26   For each structure
(1) Star (*) any asymmetric carbon atoms.
(2) Label each asymmetric carbon as (*R*) or (*S*).
(3) Draw in any internal mirror planes of symmetry.

(4) Label the structure as chiral or achiral.
(5) Label any meso structures.

(a)   **(b)**   **(c)**   **(d)**

(e)   **(f)**   **(g)**   **(h)**

(i)   **(j)**   **(k)**   **(l)**

(m)   **(n)**   **(o)** (menthol)   **(p)**

5-27   For each of the compounds described by the following names,
(1) Draw a three-dimensional representation.
(2) Star (*) each chirality center.
(3) Does the structure have a plane of symmetry? If so, draw it.
(4) Does the structure have an enantiomer? If so, draw it.
(5) Does the structure have a diastereomer? If so, draw one.
(6) Label each structure you have drawn as chiral or achiral.

(a) (S)-2-chlorobutane
(c) (2R,3S)-2,3-dibromohexane
(e) meso-3,4-hexanediol, $CH_3CH_2CH(OH)CH(OH)CH_2CH_3$

(b) (R)-1,1,2-trimethylcyclohexane
(d) (1R,2R)-1,2-dibromocyclohexane
(f) (±)-3,4-hexanediol

5-28   Convert the following perspective formulas to Fischer projections.

(a)   **(b)**   **(c)**   **(d)**

5-29   Convert the following Fischer projections to perspective formulas.

(a) H₂N—┼—H
COOH
CH₃

(b) H—┼—OH
CHO
CH₂OH

(c) Br—┼—Cl
CH₂OH
CH₃

(d)
CH₂OH
H—┼—Br
H—┼—Cl
CH₃

5-30   Give the stereochemical relationships between each pair of isomers. Examples are same compound, structural isomers, enantiomers, diastereomers.

(a)   **(b)**   **(c)**

(d)   **(e)**   **(f)**

(g)   **(h)**

5-31   Draw the enantiomer, if any, for each structure.

(a)   **(b)** H——Br   **(c)**

(d)   **(e)**   **(f)**

(g)   **(h)**   **(i)**

5-32   Calculate the specific rotations of the following samples taken at 25°C using the sodium D line.
   (a)   1.00 g of sample is dissolved in 20.0 mL of ethanol. Then 5.00 mL of this solution is placed in a 20.0-cm polarimeter tube. The observed rotation is 1.25° counterclockwise.
   (b)   0.050 g of sample is dissolved in 2.0 mL of ethanol, and this solution is placed in a 2.0-cm polarimeter tube. The observed rotation is clockwise 0.043°.

5-33   (+)-Tartaric acid has a specific rotation of +12.0°. Calculate the specific rotation of a mixture of 68% (+)-tartaric acid and 32% (−)-tartaric acid.

5-34   The specific rotation of (S)-2-iodobutane is +15.90°.
   (a)   Draw the structure of (S)-2-iodobutane.
   (b)   Predict the specific rotation of (R)-2-iodobutane.
   (c)   Determine the percentage composition of a mixture of (R)- and (S)-2-iodobutane with a specific rotation of 7.95°.

*__5-35__   For each structure,
   (1) Draw all the stereoisomers.
   (2) Label each structure as chiral or achiral.
   (3) Give the relationships between the stereoisomers (enantiomers, diastereomers).

(a)   **(b)**   **(c)**

*__5-36__   Draw all the stereoisomers of 1,2,3-trimethylcyclopentane, and give the relationships between them.

*__5-37__   If you think you know your definitions, try this difficult problem.
   (a)   Draw all the stereoisomers of 2,3,4-tribromopentane. (Using Fischer projections may be helpful.) You should find two meso structures and one pair of enantiomers.
   (b)   Star (*) the asymmetric carbon atoms, and label each as (R) or (S).

    (c) In the meso structures, show how C3 is not asymmetric, nor is it a chirality center, yet it is stereogenic.

    (d) In the enantiomers, show how C3 is not stereogenic in this diastereomer.

*5-38   3,4-Dimethyl-1-pentene has the formula CH_2=CH—$CH(CH_3)$—$CH(CH_3)_2$. When pure (R)-3,4-dimethyl-1-pentene is treated with hydrogen over a platinum catalyst, the product is (S)-2,3-dimethylpentane.

    (a) Draw the equation for this reaction. Show the stereochemistry of the reactant and the product.

    (b) The reactant is named (R), but the product is named (S). Does this name change imply a change in the spatial arrangement of the groups around the chirality center? So why does the name switch from (R) to (S)?

    (c) How useful is the (R) or (S) designation for predicting the sign of an optical rotation? Can you predict the sign of the rotation of the reactant? Of the product? (*Hint* from Juliet Capulet: "What's in a name? That which we call a rose/By any other name would smell as sweet.")

*5-39   A graduate student was studying enzymatic reductions of cyclohexanones when she encountered some interesting chemistry. When she used an enzyme and NADPH to reduce the following ketone, she was surprised to find that the product was optically active. She carefully repurified the product so that no enzyme, NADPH, or other contaminants were present. Still, the product was optically active.

    (a) Does the product have any asymmetric carbon atoms or other stereocenters?

    (b) Is the product capable of showing optical activity? If it is, explain how.

    (c) If this reaction could be accomplished using H_2 and a nickel catalyst, would the product be optically active? Explain.

CHAPTER 6

Alkyl Halides: Nucleophilic Substitution and Elimination

6-1
Introduction

Our study of organic chemistry is organized into families of compounds according to their functional groups. In this chapter we consider the properties and reactions of alkyl halides. We use their reactions to introduce substitution and elimination, two of the most important types of reactions in organic chemistry. Stereochemistry (Chapter 5) will play a major role in our study of these reactions. Many other reactions show similarities to substitution and elimination, and the techniques introduced in this chapter will be used throughout our study of organic reactions.

There are three major classes of organohalogen compounds: the alkyl halides, the vinyl halides, and the aryl halides. An **alkyl halide** simply has a halogen atom bonded to one of the sp^3 hybrid carbon atoms of an alkyl group. A **vinyl halide** has a halogen atom bonded to one of the sp^2 hybrid carbon atoms of an alkene. An **aryl halide** has a halogen atom bonded to one of the sp^2 hybrid carbon atoms of an aromatic ring. The chemistry of vinyl halides and aryl halides is different from that of alkyl halides because their bonding and hybridization are different. We consider the reactions of vinyl halides and aryl halides in later chapters. The structures of some representative alkyl halides, vinyl halides, and aryl halides are shown below, with their most common names and uses.

Alkyl halides

$CHCl_3$	$CHClF_2$	$CCl_3 — CH_3$	$CF_3 — CHClBr$
chloroform	Freon-22®	1,1,1-trichloroethane	Halothane
solvent	refrigerant	cleaning fluid	nonflammable anesthetic

Vinyl halides

vinyl chloride
monomer for poly(vinyl chloride)

tetrafluoroethylene (TFE)
monomer for Teflon®

Aryl halides

para-dichlorobenzene
mothballs

thyroxine
thyroid hormone

chloromethane   EPM of chloromethane

◀ **Figure 6-1**
Chloromethane and its electrostatic
potential map (EPM). The polar
C—Cl bond is seen in the EPM as
an electron-rich (red) region around
chlorine and an electron-poor (blue)
region around carbon and the
hydrogen atoms.

The carbon–halogen bond in an alkyl halide is polar because halogen atoms are more electronegative than carbon atoms. Most reactions of alkyl halides result from breaking this polarized bond. The electrostatic potential map of chloromethane (Figure 6-1) shows higher electron density (red) around the chlorine atom and relatively low electron density (blue) around the carbon and hydrogen atoms. The carbon atom has a partial positive charge, making it somewhat electrophilic. A nucleophile can attack this electrophilic carbon, or the halogen atom can leave as a halide ion, taking the bonding pair of electrons with it. By serving as a leaving group, the halogen can be eliminated from the alkyl halide, or it can be replaced (substituted for) by a wide variety of functional groups. This versatility allows alkyl halides to serve as intermediates in the synthesis of many other functional groups.

PROBLEM 6-1

Classify each compound as an alkyl halide, a vinyl halide, or an aryl halide.

(a) $CH_3CHCFCH_3$ (b) $(CH_3)_3C$—Br   (c) ⬡—Br

(d) bromocyclohexane   (e) 1-bromocyclohexene   (f) Cl— a PCB (polychlorinated biphenyl)

There are two ways of naming alkyl halides. The systematic (IUPAC) nomenclature treats an alkyl halide as an alkane with a *halo-* substituent: Fluorine is *fluoro-*, chlorine is *chloro-*, bromine is *bromo-*, and iodine is *iodo-*. The result is a systematic **haloalkane** name, as in 1-chlorobutane or 2-bromopropane. Common or "trivial" names are constructed by naming the alkyl group and then the halide, as in "isopropyl bromide." This is the origin of the term *alkyl halide*. Common names are useful only for simple alkyl halides, such as the following:

6-2
Nomenclature of Alkyl Halides

	CH_3CH_2—F	$CH_2CH_2CH_2CH_3$	CH_3—CH—CH_3
IUPAC name:	fluoroethane	1-chlorobutane	2-bromopropane
common name:	ethyl fluoride	*n*-butyl chloride	isopropyl bromide

IUPAC name:	iodocyclohexane	*trans*-1-chloro-3-methylcyclopentane
common name:	cyclohexyl iodide	(none)

$$\underset{\text{3-(iodomethyl)pentane}}{\underset{\displaystyle \text{IUPAC name:}}{CH_3CH_2-\overset{\displaystyle \overset{CH_2-I}{|}}{CH}-CH_2CH_3}} \qquad \underset{\text{4-(2-fluoroethyl)heptane}}{CH_3CH_2CH_2-\overset{\displaystyle \overset{CH_2CH_2-F}{|}}{CH}-CH_2CH_2CH_3}$$

Some of the halomethanes have acquired common names that are not clearly related to their structures. A compound of formula CH_2X_2 (a methylene group with two halogens) is called a *methylene halide*; a compound of formula CHX_3 is called a *haloform*; and a compound of formula CX_4 is called a *carbon tetrahalide*.

	CH_2Cl_2	$CHCl_3$	CCl_4
IUPAC name:	dichloromethane	trichloromethane	tetrachloromethane
common name:	methylene chloride	chloroform	carbon tetrachloride

---

PROBLEM 6-2

Give the structures of the following compounds.

(a) methylene iodide

(b) carbon tetrabromide

(c) 3-bromo-2-methylpentane

(d) iodoform

(e) 2-bromo-3-ethyl-2-methylhexane

(f) isobutyl bromide

(g) *cis*-1-fluoro-3-(fluoromethyl)cyclohexane

(h) *t*-butyl chloride

Alkyl halides are classified according to the nature of the carbon atom bonded to the halogen. If the halogen-bearing carbon is bonded to one carbon atom, it is primary (1°) and the alkyl halide is a **primary halide**. If two carbon atoms are bonded to the halogen-bearing carbon, it is secondary (2°) and the compound is a **secondary halide**. A **tertiary halide** (3°) has three other carbon atoms bonded to the halogen-bearing carbon atom. If the halogen-bearing carbon atom is a methyl group (bonded to no other carbon atoms), the compound is a *methyl halide*.

$$\underset{\text{methyl halide}}{CH_3-X} \qquad \underset{\text{primary (1°) halide}}{R-CH_2-X} \qquad \underset{\text{secondary (2°) halide}}{R-\overset{\displaystyle\overset{R}{|}}{C}H-X} \qquad \underset{\text{tertiary (3°) halide}}{R-\overset{\displaystyle\overset{R}{|}}{\underset{\displaystyle\underset{R}{|}}{C}}-X}$$

Examples

	CH_3-Br	$\underset{1°}{CH_3CH_2CH_2-F}$	$\underset{2°}{CH_3-\overset{\displaystyle\overset{I}{|}}{C}H-CH_2CH_3}$	$\underset{3°}{(CH_3)_3C-Cl}$
IUPAC name:	bromomethane	1-fluoropropane	2-iodobutane	2-chloro-2-methylpropane
common name:	methyl bromide	*n*-propyl fluoride	*sec*-butyl iodide	*t*-butyl chloride

A **geminal dihalide** (Latin, *geminus* "twin") has the two halogen atoms bonded to the same carbon atom. A **vicinal dihalide** (Latin, *vicinus*, "neighboring") has the two halogens bonded to adjacent carbon atoms.

$$\underset{\text{a geminal dibromide}}{\overset{\displaystyle Br\diagdown\quad\diagup Br}{\underset{\diagup\quad\diagdown}{C}}} \qquad \underset{\text{a vicinal dichloride}}{-\overset{\displaystyle\overset{Cl}{|}}{C}-\overset{\displaystyle\overset{Cl}{|}}{C}-}$$

PROBLEM 6-3

For each of the following compounds

(1) Give the IUPAC name.

(2) Give the common name (if possible).

(3) Classify the compound as a methyl, primary, secondary, or tertiary halide.

(a) $(CH_3)_2CH—CH_2Cl$

(b) CH_2I_2

(c) $CH_3—CHCl_2$

(d) $BrCH_2—CCl_3$

(e) $CHCl_3$

(f) $(CH_3)_3C—Br$

(g) $CH_3—CH_2—CHBr—CH_3$

(h) $CH_3—CH—CH_2Cl$
$\qquad\qquad\quad |$
$\qquad\qquad\quad CH_2CH_3$

(i)

$$\begin{array}{c} Br \\ \text{\scriptsize\vdots} H \\ \square \\ \text{\textbf{—}} Cl \\ H \end{array}$$

(j) $CH_3—CH—CH_2CH_3$
$\qquad\qquad |$
$\quad Br—CH—CH_2—CH_3$

(k)

$$\begin{array}{c} CH_3 \\ CH_3 \\ H—\bigcirc \\ F \end{array}$$

(l)

$$\begin{array}{c} Cl \quad H \\ \bigcirc \\ H \quad Cl \end{array}$$

6-3A   Solvents

Alkyl halides are used primarily as industrial and household solvents. Carbon tetrachloride (CCl_4) was once used for dry cleaning, spot removing, and other domestic cleaning. Carbon tetrachloride is toxic and carcinogenic (causes cancer), however, so dry cleaners now use 1,1,1-trichloroethane and other solvents instead.

Methylene chloride (CH_2Cl_2) and chloroform ($CHCl_3$) are also good solvents for cleaning and degreasing work. Methylene chloride was once used to dissolve the caffeine from coffee beans to produce decaffeinated coffee. Concerns about the safety of coffee with residual traces of methylene chloride prompted coffee producers to use liquid carbon dioxide instead. Chloroform is more toxic and carcinogenic than methylene chloride; it has been replaced by methylene chloride and other solvents in most industrial degreasers and paint removers.

Even the safest halogenated solvents, such as methylene chloride and 1,1,1-trichloroethane, should be used carefully, however. They are all potentially toxic and carcinogenic, and they dissolve the fatty oils that protect skin, causing a form of dermatitis.

6-3B   Reagents

We will often encounter syntheses using alkyl halides as starting materials for making more complex molecules. The conversion of alkyl halides to organometallic reagents (compounds containing carbon–metal bonds) is a particularly important tool for organic synthesis. We discuss the formation of organometallic compounds in Section 10-8.

6-3C   Anesthetics

Chloroform ($CHCl_3$) was the first substance found to produce general anesthesia, opening new possibilities for careful surgery with a patient who is unconscious and relaxed. Chloroform is toxic and carcinogenic, however, and it was soon abandoned in favor of safer anesthetics. A less toxic anesthetic is a mixed alkyl halide, $CF_3CHClBr$, which goes

6-3
Common Uses of Alkyl Halides

Halogenated compounds persist in the environment because they resist to breakdown by soil bacteria. Many are chemically non-reactive and poorly soluble in water, hindering bacterial degradation. Nonetheless, there are strains of bacteria that can to use halogenated compounds as their sources of food.

Chloroform can also be converted to phosgene in the presence of oxygen at room temperature. Phosgene reacts with various biological molecules, leading to cell death. A small amount of alcohol is sometimes added to bottles of chloroform to react with the toxic compound and render it harmless.

$$\begin{array}{c} O \\ \| \\ C \\ Cl \quad\quad Cl \end{array}$$

acetone

Use of the Dubois chloroform inhaler to produce surgical anesthesia, around 1850.

by the trade name Halothane. Ethyl chloride is often used as a topical anesthetic for minor procedures. When sprayed on the skin, its evaporation (bp 12°C) cools the area and enhances the numbing effect.

6-3D   Freons: Refrigerants and Foaming Agents

The **freons** (also called *chlorofluorocarbons*, or CFCs) are fluorinated haloalkanes that were developed to replace ammonia as a refrigerant gas. Ammonia is toxic, and leaking refrigerators often killed people who were working or sleeping nearby. Freon-12®, CF_2Cl_2, was at one time the most widely used refrigerant. Low-boiling freons (such as Freon-11®, CCl_3F) were once used as *foaming agents* that are added to a plastic to vaporize and form a froth that hardens into a plastic foam. The release of freons into the atmosphere has raised concerns about their reactions with the earth's protective ozone layer. It appears that CFCs gradually diffuse up into the stratosphere, where the chlorine atoms catalyze the decomposition of ozone (O_3) into oxygen (O_2). The freon-catalyzed depletion of ozone has been blamed for the "hole" in the ozone layer that has been found to occur over the South Pole.

International treaties have limited the future production and use of the ozone-destroying freons. Freon-12® has been replaced in aerosol cans by low-boiling hydrocarbons or carbon dioxide. In refrigerators and automotive air conditioners, Freon-12® has been replaced by Freon-22®, $CHClF_2$. Freons with C—H bonds (such as Freon-22®), called HCFCs, are generally destroyed at lower altitudes before they reach the stratosphere. HCFC-123 ($CHCl_2CF_3$) is used as a substitute for Freon-11® in making plastic foams.

6-3E   Pesticides

Alkyl halides have contributed to human health through their use as insecticides. Since antiquity, people have died from famine and disease caused or carried by mosquitoes, fleas, lice, and other vermin. The "black death" of the Middle Ages wiped out nearly a third of the population of Europe through infection by the flea-borne bubonic plague. Whole regions of Africa and tropical America were uninhabited and unexplored because people could not survive insect-borne diseases such as malaria, yellow fever, and sleeping sickness.

Arsenic compounds, nicotine, and other crude insecticides were developed in the nineteenth century, but these compounds are just as toxic to birds, animals, and people as they are to insects. Their use is extremely hazardous; but a hazardous insecticide was still preferable to certain death by disease or starvation.

The war against insects changed dramatically in 1939 with the discovery of DDT (Figure 6-2). DDT is extremely toxic to insects, but its toxicity in mammals is quite low. About an ounce of DDT is required to kill a person, but that same amount of insecticide protects an acre of land against locusts or mosquitos. In 1970, the U.S. National Academy of Sciences reported "In little more than two decades DDT has prevented 500 million deaths due to malaria." Similar advances were made against the mosquitoes carrying yellow fever and the tsetse flies carrying sleeping sickness. Using DDT as a body dust protected people against louse-borne typhus, and dusting rodent burrows controlled the threat of plague. As with many inventions, DDT showed undesired side effects. It is a long-lasting insecticide, and its residues accumulate in the environment. The widespread use of DDT as an agricultural insecticide led to the development of substantial concentrations in wildlife, causing declines in several species. In 1972, DDT was banned by the U.S. Environmental Protection Agency for use as an agricultural insecticide. It is still used as a last resort, however, in countries where insect-borne diseases threaten human life.

Many other chlorinated insecticides have been developed. Some of them also accumulate in the environment, gradually producing toxic effects in wildlife. Others can be used with little adverse impact if they are applied properly. Because of their persistent toxic effects, chlorinated insecticides are rarely used in agriculture. They are generally used when a potent insecticide is needed to protect life or property. For example, lindane is

DDT

▲ **Figure 6-2**
DDT is DichloroDiphenyl-Trichloroethane, or 1,1,1-trichloro-2,2-bis-(*p*-chlorophenyl)ethane. DDT was the first chlorinated insecticide. Its use rendered large parts of the world safe from insect-borne disease and starvation, but it accumulated in the environment.

used in shampoos to kill lice, and chlordane is used to protect wooden buildings from termites. The structures of some chlorinated insecticides are shown below.

lindane	kepone	aldrin	chlordane

PROBLEM 6-4

Kepone and chlordane are synthesized from hexachlorocyclopentadiene and other five-membered-ring compounds. Show how these two pesticides are composed of two five-membered rings.

hexachlorocyclopentadiene

6-4
Structure of Alkyl Halides

In an alkyl halide, the halogen atom is bonded to an sp^3 hybrid carbon atom. The halogen is more electronegative than carbon, and the C—X bond is polarized with a partial positive charge on carbon and a partial negative charge on the halogen.

$\mu = 4.8 \times \delta \times d$
where δ is the charge and d is the bond length.

The electronegativities of the halogens *increase* in the order

	I	<	Br	<	Cl	<	F
electronegativity:	2.7		3.0		3.2		4.0

The carbon–halogen bond lengths *increase* as the halogen atoms become bigger (larger atomic radii) in the order

	C—F	<	C—Cl	<	C—Br	<	C—I
bond length:	1.38 Å		1.78 Å		1.94 Å		2.14 Å

These two effects oppose each other, with the larger halogens having longer bonds but weaker electronegativities. The overall result is that the bond dipole moments increase in the order

	C—I	<	C—Br	<	C—F	<	C—Cl
dipole moment, μ:	1.29 D		1.48 D		1.51 D		1.56 D

TABLE 6-1 Molecular Dipole Moments of Methyl Halides				
X	**CH₃X**	**CH₂X₂**	**CHX₃**	**CX₄**
F	1.82 D	1.97 D	1.65 D	0
Cl	1.94 D	1.60 D	1.03 D	0
Br	1.79 D	1.45 D	1.02 D	0
I	1.64 D	1.11 D	1.00 D	0

carbon tetrachloride

A *molecular* dipole moment is the vector sum of the individual bond dipole moments. Molecular dipole moments are not easy to predict because they depend on the bond angles and other factors that vary with the specific molecule. Table 6-1 lists the experimentally measured dipole moments of the halogenated methanes. Notice how the four symmetrically oriented polar bonds of the carbon tetrahalides cancel to give a molecular dipole moment of zero.

> **PROBLEM 6-5**
>
> Although the C—I bond is longer than a C—Cl bond, the C—Cl bond has a larger dipole moment. Explain.

6-5
Physical Properties of Alkyl Halides

6-5A   Boiling Points

Two types of intermolecular forces influence the boiling points of alkyl halides. The London force attraction is the strongest intermolecular force, and the dipole–dipole attraction (arising from the polar C—X bond) is an additional force. Because the London force is a *surface* attraction, molecules with larger surface areas have larger London attractions, resulting in higher boiling points.

Molecules with higher molecular weights generally have higher boiling points because they are heavier (and therefore slower moving), and they have greater surface area. The surface areas of the alkyl halides vary with the surface areas of halogens. We can get an idea of the relative surface areas of halogen atoms by considering their van der Waals radii. Figure 6-3 shows that an alkyl fluoride has nearly the same surface

Halogen	van der Waals Radius (10^{-8} cm)
F	1.35
Cl	1.8
Br	1.95
I	2.15
H (for comparison)	1.2

ethyl fluoride, bp −38°C

ethyl chloride, bp 12°C

ethyl bromide, bp 38°C

ethyl iodide, bp 72°C

▶ **Figure 6-3**
Space-filling drawings of the ethyl halides. The heavier halogens are larger, with much greater surface areas.

area as the corresponding alkane; thus its London attractive forces are similar. The alkyl fluoride has a larger dipole moment, however, so the total attractive forces are slightly greater in the alkyl fluoride, giving it a higher boiling point. For example, the boiling point of *n*-butane is 0°C, while that of *n*-butyl fluoride is 33°C.

The other halogens are considerably larger than fluorine, giving them more surface area and raising the boiling points of their alkyl halides. With a boiling point of 78°C, *n*-butyl chloride shows the influence of chlorine's much larger surface area. This trend continues with *n*-butyl bromide (bp 102°C) and *n*-butyl iodide (bp 131°C). Table 6-2 lists the boiling points and densities of some simple alkyl halides. Notice that compounds with branched, more spherical shapes have lower boiling points as a result of their smaller surface areas. For example, *n*-butyl bromide has a boiling point of 102°C, while the more spherical *t*-butyl bromide has a boiling point of only 73°C. This effect is similar to the one we saw with alkanes.

PROBLEM 6-6

For each pair of compounds, predict which compound has the higher boiling point. Check Table 6-2 to see if your prediction was right, then explain why that compound has the higher boiling point.
(a) isopropyl bromide and *n*-butyl bromide
(b) isopropyl chloride and *t*-butyl bromide
(c) *n*-butyl bromide and *n*-butyl chloride

TABLE 6-2 Properties of Some Simple Alkyl Halides

Compound	Molecular Weight	Boiling Point (°C)	Density (g/mL)
CH_3—F	34	−78	
CH_3—Cl	50.5	−24	0.92
CH_3—Br	95	4	1.68
CH_3—I	142	42	2.28
CH_2Cl_2	85	40	1.34
$CHCl_3$	119	61	1.50
CCl_4	154	77	1.60
CH_3CH_2—F	48	−38	0.72
CH_3CH_2—Cl	64.5	12	0.90
CH_3CH_2—Br	109	38	1.46
CH_3CH_2—I	156	72	1.94
$CH_3CH_2CH_2$—F	62	3	0.80
$CH_3CH_2CH_2$—Cl	78.5	47	0.89
$CH_3CH_2CH_2$—Br	123	71	1.35
$CH_3CH_2CH_2$—I	170	102	1.75
$(CH_3)_2CH$—Cl	78.5	36	0.86
$(CH_3)_2CH$—Br	123	59	1.31
$(CH_3)_2CH$—I	170	89	1.70
$CH_3CH_2CH_2CH_2$—F	76	33	0.78
$CH_3CH_2CH_2CH_2$—Cl	92.5	78	0.89
$CH_3CH_2CH_2CH_2$—Br	137	102	1.28
$CH_3CH_2CH_2CH_2$—I	184	131	1.62
$(CH_3)_3C$—Cl	92.5	52	0.84
$(CH_3)_3C$—Br	137	73	1.23
$(CH_3)_3C$—I	184	100	1.54

6-5B Densities

Table 6-2 also shows the densities of common alkyl halides. Like their boiling points, their densities follow a predictable trend. Alkyl fluorides and alkyl chlorides (those with just one chlorine atom) are less dense than water (1.00 g/mL). Alkyl chlorides with two or more chlorine atoms are denser than water, and all alkyl bromides and alkyl iodides are denser than water.

> ### PROBLEM 6-7
>
> When water is shaken with hexane, the two liquids separate into two phases. Show which compound is present in the top and which in the bottom phase. When water is shaken with chloroform, a similar two-phase system results. Again, show which compound is present in each phase. Explain the difference in the two experiments.

6-6
Preparation of Alkyl Halides

Many syntheses of alkyl halides exploit the chemistry of functional groups we have not yet covered. For now, we review free-radical halogenation and summarize the other, often more useful, syntheses of alkyl halides. The other syntheses are discussed in subsequent chapters.

6-6A Free-Radical Halogenation

Although we discussed its mechanism at length in Section 4-3, free-radical halogenation is rarely an effective method for the synthesis of alkyl halides. It usually produces mixtures of products because there are different kinds of hydrogen atoms that can be abstracted. Also, more than one halogen atom may react, giving multiple substitution. For example, the chlorination of propane can give a messy mixture of products.

$$CH_3-CH_2-CH_3 \ + \ Cl_2 \ \xrightarrow{h\nu} \ \begin{cases} CH_3-CH_2-CH_2Cl \ + \ CH_3-CHCl-CH_3 \\ + \ CH_3-CHCl-CH_2Cl \ + \ CH_3-CCl_2-CH_3 \\ + \ CH_3-CH_2-CHCl_2 \ + \ \text{others} \end{cases}$$

In industry, free-radical halogenation is sometimes useful because the reagents are cheap, the mixture of products can be separated by distillation, and each of the individual products is sold separately. In a laboratory, however, we need a good yield of one particular product. Free-radical halogenation rarely provides good selectivity and yield, so it is seldom used in the laboratory. Laboratory syntheses using free-radical halogenation are generally limited to specialized compounds that give a single major product, such as the following examples.

cyclohexane   chlorocyclohexane
                        (50%)

isobutane   t-butyl bromide
                    (90%)

All the hydrogen atoms in cyclohexane are equivalent, and free-radical chlorination gives a usable yield of chlorocyclohexane. Formation of dichlorides and trichlorides is possible, but these side reactions are controlled by using only a small amount of chlorine and an excess of cyclohexane. Free-radical bromination is highly selective

(Section 4-14), and it gives good yields of products that have one type of hydrogen atom that is more reactive than the others. Isobutane has only one tertiary hydrogen atom, and this atom is preferentially abstracted to give a tertiary free radical.

6-6B Allylic Halogenation

Under the right conditions, free-radical bromination of cyclohexene can give a good yield of 3-bromocyclohexene, where bromine has substituted for an **allylic** hydrogen on the carbon atom next to the double bond.

cyclohexene

3-bromocyclohexene
(80%)

This allylic bromination is selective because the allylic intermediate is resonance-stabilized. Abstraction of an allylic hydrogen atom gives a resonance-stabilized allylic radical. This radical reacts with Br_2, regenerating a bromine radical.

MECHANISM Allylic Bromination

cyclohexene

allylic radical

allylic bromide

Overall reaction

an allylic hydrogen

an allylic bromide

A large concentration of bromine must be avoided because bromine can add to the double bond (Chapter 8). *N*-Bromosuccinimide is often used as the bromine source in free-radical brominations because it combines with the HBr side product to regenerate a constant low concentration of bromine.

N-bromosuccinimide (NBS)

regenerates a low concentration of Br_2

Allylic halogenation is discussed in more detail in Chapter 15.

PROBLEM 6-8

(a) Propose a mechanism for the following reaction:

$$H_2C=CH-CH_3 + Br_2 \xrightarrow{h\nu} H_2C=CH-CH_2Br + HBr$$

(b) Use the bond-dissociation energies given in Table 4-2 (page 134) to calculate the value of ΔH for each step shown in your mechanism. (The BDE for $CH_2=CHCH_2-Br$ is about 67 kcal/mol.) Calculate the overall value of ΔH for the reaction. Are these values consistent with a rapid free-radical chain reaction?

PROBLEM 6-9

Show how free-radical halogenation might be used to synthesize each of the following compounds. In each case, explain why we expect to get a single major product.
(a) 1-chloro-2,2-dimethylpropane (neopentyl chloride)
(b) 2-bromo-2-methylbutane

(c)

Br
|
⟨benzene⟩—CH—CH$_2$CH$_2$CH$_3$

1-bromo-1-phenylbutane

(d)

CH$_3$
⟨cyclopentane⟩—Br

1-bromo-1-methylcyclopentane

PROBLEM 6-10

The light-catalyzed reaction of 2,3-dimethyl-2-butene with a small concentration of bromine gives two products:

$$\begin{array}{c} H_3C \\ \\ H_3C \end{array} C=C \begin{array}{c} CH_3 \\ \\ CH_3 \end{array} \xrightarrow{Br_2,\ h\nu} \begin{array}{c} H_3C \\ \\ H_3C \end{array} C=C \begin{array}{c} CH_2-Br \\ \\ CH_3 \end{array} + \quad Br-\overset{\overset{\displaystyle CH_3}{|}}{C}-\overset{\overset{\displaystyle CH_2}{}}{\underset{\underset{\displaystyle CH_3}{}}{C}}$$

2,3-dimethyl-2-butene

Give a mechanism for this reaction, showing how the two products arise as a consequence of the resonance-stabilized intermediate.

Following is a brief summary of the most important methods of making alkyl halides. Several of these methods are not discussed until later, but they are listed here so that you can use this summary for reference throughout the course.

SUMMARY **Methods for Preparing Alkyl Halides**

1. *From alkanes: free-radical halogenation (synthetically useful only in certain cases)* (Sections 4-13 and 6-6)

$$R-H \xrightarrow[\text{heat or light}]{X_2} R-X + H-X$$

Example

$$CH_3-\overset{\overset{\displaystyle CH_3}{|}}{\underset{\underset{\displaystyle H}{|}}{C}}-CH_3 \xrightarrow[\text{heat or light}]{Br_2} CH_3-\overset{\overset{\displaystyle CH_3}{|}}{\underset{\underset{\displaystyle Br}{|}}{C}}-CH_3$$

isobutane   *t*-butyl bromide

2. *From alkenes and alkynes*

$$\mathrm{C=C} \xrightarrow{\text{HX}} -\overset{|}{\underset{H}{C}}-\overset{|}{\underset{X}{C}}- \qquad \text{(Section 8-3)}$$

$$\mathrm{C=C} \xrightarrow{X_2} -\overset{|}{\underset{X}{C}}-\overset{|}{\underset{X}{C}}-$$

$$-\mathrm{C}\equiv\mathrm{C}- \xrightarrow{\text{2 HX}} -\overset{H}{\underset{H}{C}}-\overset{X}{\underset{X}{C}}- \qquad \text{(Section 9-9)}$$

$$-\mathrm{C}\equiv\mathrm{C}- \xrightarrow{\text{2 X}_2} -\overset{X}{\underset{X}{C}}-\overset{X}{\underset{X}{C}}-$$

$$\mathrm{C=C}\overset{|}{\underset{C-H}{|}} \xrightarrow[\text{light}]{\text{NBS:}} \mathrm{C=C}\overset{|}{\underset{C-Br}{|}} \qquad \text{(Sections 6-6, 15-7)}$$

(allylic bromination)

Examples

$$\mathrm{CH_3-CH=C(CH_3)_2} \xrightarrow{\text{HBr}} \mathrm{CH_3-CH_2-\overset{CH_3}{\underset{Br}{C}}-CH_3}$$

2-methyl-2-butene   2-bromo-2-methylbutane

$$\mathrm{CH_3-CH=CH-CH_3} \xrightarrow{\text{Cl}_2} \mathrm{CH_3-CHCl-CHCl-CH_3}$$

2-butene   2,3-dichlorobutane

$$\mathrm{H-C\equiv C-CH_2CH_2CH_3} \xrightarrow{\text{2 HBr}} \mathrm{CH_3-CBr_2-CH_2CH_2CH_3}$$

1-pentyne   2,2-dibromopentane

3. *From alcohols* (Sections 11-7, 11-8, 11-9)

$$\mathrm{R-OH} \xrightarrow{\text{HX, PX}_3,\ \text{or others}} \mathrm{R-X}$$

Example

$$\mathrm{CH_3CH_2CH_2CH_2OH} \xrightarrow{\text{HBr, H}_2\text{SO}_4} \mathrm{CH_3CH_2CH_2CH_2Br}$$

1-butanol   1-bromobutane

4. *From other halides* (Section 6-9)

$$\mathrm{R-X + I^-} \xrightarrow{\text{acetone}} \mathrm{R-I + X^-}$$

$$\mathrm{R-Cl + KF} \xrightarrow[\text{CH}_3\text{CN}]{\text{18-crown-6}} \mathrm{R-F}$$

Example

$$\mathrm{H_2C=CH-CH_2Cl + NaI} \xrightarrow{\text{acetone}} \mathrm{H_2C=CH-CH_2I}$$

allyl chloride   allyl iodide

6-7

Reactions of Alkyl Halides: Substitution and Elimination

Alkyl halides are easily converted to many other functional groups. The halogen atom can leave with its bonding pair of electrons to form a stable halide ion; we say that a halide is a good *leaving group*. When another atom replaces the halide ion, the reaction is a **substitution**. When the halide ion leaves with another atom or ion (often H^+), the reaction is an **elimination**. In many eliminations, a molecule of $H—X$ is lost from the alkyl halide to give an alkene. These eliminations are called **dehydrohalogenations** because a hydrogen halide has been removed from the alkyl halide. Substitution and elimination reactions often compete with each other.

In a **nucleophilic substitution**, a nucleophile ($Nuc:^-$) replaces a leaving group ($:\ddot{X}:^-$) from a carbon atom, using its lone pair of electrons to form a new bond to the carbon atom.

Nucleophilic substitution

In an **elimination**, both the halide ion and another substituent are lost. A new π bond is formed.

Elimination

In the elimination (a dehydrohalogenation), the reagent ($B:^-$) reacts as a base, abstracting a proton from the alkyl halide. Most nucleophiles are also basic and can engage in either substitution or elimination, depending on the alkyl halide and the reaction conditions.

In addition to alkyl halides, many other types of compounds undergo substitution and elimination reactions. Substitutions and eliminations are introduced in this chapter using the alkyl halides as examples. In later chapters, we encounter substitutions and eliminations of other types of compounds.

PROBLEM 6-11

Classify each reaction as a substitution, elimination, or neither.

PROBLEM 6-12

Give the structures of the substitution products expected when 1-bromohexane reacts with
(a) $Na^{+-}OCH_2CH_3$
(b) NaCN
(c) NaOH

A nucleophilic substitution has the general form

where Nuc:$^-$ is the nucleophile and $:\ddot{X}:^-$ is the leaving halide ion. An example is the reaction of iodomethane (CH$_3$I) with hydroxide ion. The product is methanol.

hydroxide   iodomethane   methanol   iodide

Hydroxide ion is a strong nucleophile (donor of an electron pair) because the oxygen atom has unshared pairs of electrons and a negative charge. Iodomethane is called the **substrate**, meaning the compound that is attacked by the reagent. The carbon atom of iodomethane is electrophilic because it is bonded to an electronegative iodine atom. Electron density is drawn away from carbon by the halogen atom, giving the carbon atom a partial positive charge. The negative charge of hydroxide ion is attracted to this partial positive charge.

nucleophile   electrophile

Hydroxide ion attacks the back side of the electrophilic carbon atom, donating a pair of electrons to form a new bond. (In general, nucleophiles are said to attack electrophiles, not the other way around.) Notice that arrows are used to show the movement of electron pairs, from the electron-rich nucleophile to the electron-poor carbon atom of the electrophile. Carbon can accommodate only eight electrons in its valence shell, so the carbon–iodine bond must begin to break as the carbon–oxygen bond begins to form. Iodide ion is the **leaving group**; it leaves with the pair of electrons that once bonded it to the carbon atom.

The following mechanism shows attack by the nucleophile (hydroxide), the transition state, and the departure of the leaving group (iodide).

KEY MECHANISM   The S$_N$2 Reaction

The S$_N$2 reaction takes place in a single (concerted) step:

nucleophile   electrophile   transition state   product   leaving
                                                                                        group

▶ **Figure 6-4**

The reaction-energy diagram for the reaction of methyl iodide with hydroxide shows only one energy maximum: the transition state. There are no intermediates.

PROBLEM-SOLVING HINT

A transition state is unstable and cannot be isolated. It exists for only an instant.

The anticancer drugs known as the nitrogen mustards are believed to alkylate DNA using two S_N2 reactions. First, the nitrogen nucleophile displaces the chloride on the primary alkyl chloride portion to generate a reactive intermediate that alkylates a nitrogen atom of DNA. The process is repeated, linking the two strands of the double helix DNA, and thereby preventing replication of the DNA.

The reaction of methyl iodide (iodomethane) with hydroxide ion is a **concerted reaction**, taking place in a single step with bonds breaking and forming at the same time. The middle structure is a transition state, a point of maximum energy, rather than an intermediate. In this transition state, the bond to the nucleophile (hydroxide) is partially formed, and the bond to the leaving group (iodide) is partially broken. Remember that a transition state is not a discrete molecule that can be isolated; it exists for only an instant.

The reaction-energy diagram for this substitution (Figure 6-4) shows only one transition state and no intermediates between the reactants and the products. The reactants are shown slightly higher in energy than the products because this reaction is known to be exothermic. The transition state is much higher in energy because it involves a five-coordinate carbon atom with two partial bonds.

The one-step mechanism shown in Figure 6-4 is supported by kinetic information. One can vary the concentrations of the reactants and observe the effects on the reaction rate (how much methanol is formed per second). The rate is found to double when the concentration of *either* reactant is doubled. The reaction is therefore first order in each of the reactants, and second order overall. The rate equation has the following form:

$$\text{rate} = k_r[\text{CH}_3\text{I}][^-\text{OH}]$$

This rate equation is consistent with the one-step mechanism shown in Figure 6-4. This mechanism requires a collision between a molecule of methyl iodide and a hydroxide ion. Both of these species are present in the transition state, and the collision frequency is proportional to both concentrations. The rate constant k_r depends on several factors, including the energy of the transition state and the temperature (Section 4-9).

This one-step nucleophilic substitution is an example of the **S_N2 mechanism**. The abbreviation S_N2 stands for *Substitution, Nucleophilic, bimolecular*. The term *bimolecular* means that the transition state of the rate-limiting step (the only step in this reaction) involves the collision of *two* molecules. Bimolecular reactions usually have rate equations that are second order overall.

PROBLEM 6-13

Under certain conditions, the reaction of 0.5 M methyl iodide with 1.0 M sodium hydroxide forms methanol at a rate of 0.05 mol/L per second. What would be the rate if 0.1 M methyl iodide and 2.0 M NaOH were used?

Many useful reactions take place by the S$_N$2 mechanism. The reaction of an alkyl halide, such as methyl iodide, with hydroxide ion gives an alcohol. Other nucleophiles convert alkyl halides to a wide variety of functional groups. The following table summarizes some of the types of compounds that can be formed by nucleophilic displacement of alkyl halides.

6-9
Generality of the S$_N$2 Reaction

SUMMARY S$_N$2 Reactions of Alkyl Halides

$$\text{Nuc}:^- \; + \; \text{R—X} \; \longrightarrow \; \text{Nuc—R} \; + \; \text{X}^-$$

Nucleophile		Product	Class of Product
R—X + $^-$:Ï:	$\longrightarrow$	R—Ï:	alkyl halide
R—X + $^-$:ÖH	$\longrightarrow$	R—ÖH	alcohol
R—X + $^-$:ÖR′	$\longrightarrow$	R—ÖR′	ether
R—X + $^-$:S̈H	$\longrightarrow$	R—S̈H	thiol (mercaptan)
R—X + $^-$:S̈R′	$\longrightarrow$	R—S̈R′	thioether (sulfide)
R—X + :NH$_3$	$\longrightarrow$	R—NH$_3^+$ X$^-$	amine
R—X + $^-$:N̈=N⁺=N̈:$^-$	$\longrightarrow$	R—N̈=N⁺=N̈:$^-$	azide
R—X + $^-$:C≡C—R′	$\longrightarrow$	R—C≡C—R′	alkyne
R—X + $^-$:C≡N:	$\longrightarrow$	R—C≡N:	nitrile
R—X + R′—CÖÖ:$^-$	$\longrightarrow$	R′—COO—R	ester
R—X + :P(Ph)$_3$	$\longrightarrow$	[R—PPh$_3$]$^+$ $^-$X	phosphonium salt

Examples

benzyl chloride (CHCl$_2$)   + $^-$:ÖH (hydroxide) $\longrightarrow$ benzyl alcohol (CH$_2$OH)

CH$_3$I   +   phenoxide ($\ddot{\text{O}}$:$^-$) $\longrightarrow$ methoxybenzene (methylphenyl ether) (OCH$_3$)

iodomethane (methyl iodide)

CH$_3$CH$_2$CH$_2$CH$_2$CH$_2$Br   + $^-$:S̈H $\longrightarrow$ CH$_3$CH$_2$CH$_2$CH$_2$CH$_2$SH
1-bromopentane ⟶ 1-pentanethiol

CH$_3$CH$_2$CH$_2$CH$_2$Cl   +   :NH$_3$ $\longrightarrow$ CH$_3$CH$_2$CH$_2$CH$_2$NH$_2$
1-chlorobutane (*n*-butyl chloride)   ammonia (excess)   1-butanamine (*n*-butylamine)

CH$_3$CH$_2$Br   +   Na$^+$ $^-$:C≡C—H $\longrightarrow$ CH$_3$CH$_2$—C≡C—H
bromoethane (ethyl bromide)   sodium acetylide   1-butyne (ethylacetylene)

CH$_3$CH$_2$CH$_2$I   + $^-$:C≡N $\longrightarrow$ CH$_3$CH$_2$CH$_2$—C≡N
1-iodopropane (*n*-propyl iodide)   cyanide   butanenitrile (butyronitrile)

Halogen Exchange Reactions   The S_N2 reaction provides a useful method for synthesizing alkyl iodides and fluorides, which are more difficult to make than alkyl chlorides and bromides. Halides can be converted to other halides by **halogen exchange reactions**, in which one halide displaces another.

Iodide is a good nucleophile, and many alkyl chlorides react with sodium iodide to give alkyl iodides. Alkyl fluorides are difficult to synthesize directly, and they are often made by treating alkyl chlorides or bromides with KF under conditions that use a crown ether and an aprotic solvent to enhance the normally weak nucleophilicity of the fluoride ion (see Section 6-10).

$$R{-}X \ + \ I^- \ \longrightarrow \ R{-}I \ + \ X^-$$

$$R{-}X \ + \ KF \ \xrightarrow[\text{CH}_3\text{CN}]{\text{18-crown-6}} \ R{-}F \ + \ KX$$

Examples

$$H_2C{=}CH{-}CH_2Cl \ + \ NaI \ \longrightarrow \ H_2C{=}CH{-}CH_2I \ + \ NaCl$$
allyl chloride   allyl iodide

$$CH_3CH_2Cl \ + \ KF \ \xrightarrow[\text{CH}_3\text{CN}]{\text{18-crown-6}} \ CH_3CH_2F \ + \ KCl$$
ethyl chloride   ethyl fluoride

PROBLEM 6-14

Predict the major products of the following substitutions.
(a) $CH_3CH_2Br \ + \ (CH_3)_3C{-}O^- {}^+K \longrightarrow$
ethyl bromide   potassium *t*-butoxide

(b) $H{-}C{\equiv}C{:}^- {}^+Na \ + \ CH_3CH_2CH_2CH_2{-}Cl \longrightarrow$
sodium acetylide   1-chlorobutane

(c) $(CH_3)_2CH{-}CH_2{-}Br \ + \ \text{excess } NH_3 \longrightarrow$
(d) $CH_3CH_2I \ + \ NaCN \longrightarrow$
(e) 1-chloropentane $+ \ NaI \longrightarrow$
(f) 1-chloropentane $+ \ KF \ \xrightarrow[\text{CH}_3\text{CN}]{\text{18-crown-6}}$

PROBLEM 6-15

Show how you would convert 1-chlorobutane into the following compounds.
(a) 1-butanol   (b) 1-fluorobutane   (c) 1-iodobutane
(d) $CH_3{-}(CH_2)_3{-}CN$ (e) $CH_3{-}(CH_2)_3{-}C{\equiv}CH$
(f) $CH_3CH_2{-}O{-}(CH_2)_3{-}CH_3$ (g) $CH_3{-}(CH_2)_3{-}NH_2$

6-10
Factors Affecting S_N2 Reactions: Strength of the Nucleophile

We will use the S_N2 reaction as an example of how we study the properties of the species that participate in the reaction. Both the nucleophile and the substrate (the alkyl halide) are important, as well as the type of solvent used. We begin by considering what makes a good nucleophile.

The nature of the nucleophile strongly affects the rate of the S_N2 reaction. A strong nucleophile is much more effective than a weak one in attacking an electrophilic carbon atom. For example, both methanol (CH_3OH) and methoxide ion (CH_3O^-) have easily shared pairs of nonbonding electrons, but methoxide ion reacts with electrophiles in the S$_N$2 reaction about 1 million times faster than methanol. It is generally true that a species with a negative charge is a stronger nucleophile than a similar, neutral species.

Methoxide ion has nonbonding electrons that are readily available for bonding. In the transition state, the negative charge is shared by the oxygen of methoxide ion and by the halide leaving group. Methanol, however, has no negative charge; the transition state has a partial negative charge on the halide but a partial positive charge on the methanol oxygen atom. We can generalize the case of methanol and the methoxide ion to say that

A base is always a stronger nucleophile than its conjugate acid.

conjugate base
(stronger nucleophile)

lower E_a

conjugate acid
(weaker nucleophile)

higher E_a

We might be tempted to say that methoxide is a much better nucleophile because it is much more basic. This would be a mistake because basicity and nucleophilicity are different properties. **Basicity** is defined by the *equilibrium constant* for abstracting a *proton*. **Nucleophilicity** is defined by the *rate* of attack on an electrophilic *carbon atom*. In both cases, the nucleophile (or base) forms a new bond. If the new bond is to a proton, it has reacted as a **base**; if the new bond is to carbon, it has reacted as a **nucleophile**. Predicting which way a species will react may be difficult; most (but not all) good nucleophiles are also strong bases, and vice versa.

Basicity

Nucleophilicity

Table 6-3 lists some common nucleophiles in decreasing order of their nucleophilicity in common solvents such as water and alcohols. The strength of nucleophiles shows three major trends:

TABLE 6-3 Some Common Nucleophiles.
Listed in Decreasing Order of Nucleophilicity in Hydroxylic Solvents Such as Water and Alcohols

strong nucleophiles	$(CH_3CH_2)_3P:$	moderate nucleophiles	$:\ddot{B}r:^-$
	$^-:\ddot{S}-H$		$:NH_3$
	$:\ddot{I}:^-$		$CH_3-\ddot{S}-CH_3$
	$(CH_3CH_2)_2\ddot{N}H$		$:\ddot{C}l:^-$
	$^-:C\equiv N$		
	$(CH_3CH_2)_3N:$		$CH_3\overset{\overset{O}{\|\|}}{C}-\ddot{O}:^-$
	$H-\ddot{O}:^-$	weak nucleophiles	$:\ddot{F}:^-$
	$CH_3-\ddot{O}:^-$		$H-\ddot{O}-H$
			$CH_3-\ddot{O}-H$

SUMMARY   **Trends in Nucleophilicity**

1. A species with a negative charge is a stronger nucleophile than a similar neutral species. In particular, a base is a stronger nucleophile than its conjugate acid.

$$^-:\ddot{O}H > H_2\ddot{O}: \qquad ^-:\ddot{S}H > H_2\ddot{S}: \qquad ^-:\ddot{N}H_2 > :NH_3$$

2. Nucleophilicity decreases from left to right in the periodic table, following the increase in electronegativity from left to right. The more electronegative elements have more tightly held nonbonding electrons that are less reactive toward forming new bonds.

$$^-:\ddot{O}H > :\ddot{F}:^- \qquad :NH_3 > H_2\ddot{O}: \qquad (CH_3CH_2)_3P: > (CH_3CH_2)_2\ddot{S}:$$

3. Nucleophilicity increases down the periodic table, following the increase in size and polarizability.

$$:\ddot{I}:^- > :\ddot{B}r:^- > :\ddot{C}l:^- > :\ddot{F}:^- \qquad ^-:\ddot{S}eH > ^-:\ddot{S}H > ^-:\ddot{O}H \qquad (CH_3CH_2)_3P: > (CH_3CH_2)_3N:$$

The third trend reflects an atom's ability to engage in partial bonding as it begins to attack an electrophilic carbon atom. Down a column in the periodic table, the atoms become larger, with more electrons at a greater distance from the nucleus. The electrons are more loosely held, and the atom is more **polarizable**: Its electrons can move more freely toward a positive charge, resulting in stronger bonding in the transition state. The increased mobility of its electrons enhances the atom's ability to begin to form a bond at a relatively long distance.

Figure 6-5 illustrates this polarizability effect by comparing the attack of iodide ion and fluoride ion on a methyl halide. The outer shell of the fluoride ion is the second shell. These electrons are tightly held, close to the nucleus. Fluoride is a "hard" (low-polarizability) nucleophile, and its nucleus must approach the carbon nucleus quite closely before the electrons can begin to overlap and form a bond. In the transition state, there is little bonding between fluorine and carbon. In contrast, the outer shell of the iodide ion is the fifth shell. These electrons are loosely held, making the iodide ion a "soft" (high-polarizability) nucleophile. The outer electrons begin to shift and overlap with the carbon atom from farther away. There is a great deal of bonding between iodine and carbon in the transition state.

"hard," small valence shell   transition state

"soft," large valence shell   transition state

◀ **Figure 6-5**

Comparison of fluoride ion and iodide ion as nucleophiles in the S_N2 reaction. Fluoride has tightly bound electrons that cannot begin to form a C—F bond until the atoms are close together. Iodide has more loosely bound outer electrons that begin bonding earlier in the reaction.

6-10A Steric Effects on Nucleophilicity

To serve as a nucleophile, an ion or molecule must get in close to a carbon atom to attack it. Bulky groups on the nucleophile hinder this close approach, and they slow the reaction rate. For example, the *t*-butoxide ion is a stronger *base* (for abstracting protons) than ethoxide ion, but *t*-butoxide ion has three methyl groups that hinder any close approach to a carbon atom. Therefore, ethoxide ion is a stronger nucleophile than *t*-butoxide ion. When bulky groups interfere with a reaction by virtue of their size, we call the effect **steric hindrance**.

t-butoxide (hindered)
stronger base, yet weaker nucleophile

ethoxide (unhindered)
weaker base, yet stronger nucleophile

Steric hindrance has little effect on *basicity* because basicity involves attack on an unhindered proton. When a *nucleophilic* attack at a carbon atom is involved, however, a bulky base cannot approach the carbon atom so easily. Most bases are also nucleophiles, capable of attacking either a proton or an electrophilic carbon atom. If we want a species to act as a base, we use a bulky reagent like *t*-butoxide ion. If we want it to react as a nucleophile, we use a less hindered reagent, like ethoxide.

PROBLEM 6-16

For each pair, predict the stronger nucleophile in the S_N2 reaction (using an alcohol as the solvent). Explain your prediction.

(a) $(CH_3CH_2)_3N$ or $(CH_3CH_2)_2NH$
(b) $(CH_3)_2O$ or $(CH_3)_2S$
(c) NH_3 or PH_3
(d) CH_3S^- or H_2S
(e) $(CH_3)_3N$ or $(CH_3)_2O$
(f) CH_3S^- or CH_3OH
(g) $(CH_3)_2CHO^-$ or $CH_3CH_2CH_2O^-$
(h) I^- or Cl^-

PROBLEM-SOLVING HINT

Steric hindrance (bulkiness) hinders nucleophilicity (S_N2) more than it does basicity.

6-10B   Solvent Effects on Nucleophilicity

Another factor in the nucleophilicity of these ions is their solvation, particularly in protic solvents. A **protic solvent** is one that has acidic protons, usually in the form of O—H or N—H groups. These groups form hydrogen bonds to negatively charged nucleophiles. Protic solvents, especially alcohols, are convenient solvents for nucleophilic substitutions because the reagents (alkyl halides, nucleophiles, etc.) tend to be quite soluble.

Small anions are solvated more strongly than large anions in a protic solvent because the solvent approaches a small anion more closely and forms stronger hydrogen bonds. When an anion reacts as a nucleophile, energy is required to "strip off" some of the solvent molecules, breaking some of the hydrogen bonds that stabilized the solvated anion. More energy is required to strip off solvent from a small, strongly solvated ion such as fluoride than from a large, diffuse, less strongly solvated ion like iodide.

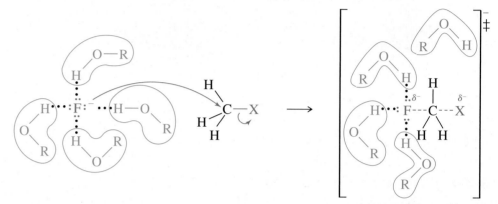

solvent partially stripped
off in the transition state

The enhanced solvation of smaller anions in protic solvents, requiring more energy to strip off their solvent molecules, reduces their nucleophilicity. This trend reinforces the trend in polarizability: The polarizability increases with increasing atomic number, and the solvation energy (in protic solvents) decreases with increasing atomic number. Therefore, nucleophilicity (in protic solvents) generally increases down a column in the periodic table, as long as we compare similar species with similar charges.

In contrast with protic solvents, aprotic solvents (solvents without O—H or N—H groups) enhance the nucleophilicity of anions. An anion is more reactive in an aprotic solvent because it is not so strongly solvated. There are no hydrogen bonds to be broken when solvent must make way for the nucleophile to approach an electrophilic carbon atom. The relatively weak solvating ability of aprotic solvents is also a disadvantage: Most polar, ionic reagents are not soluble in simple aprotic solvents such as alkanes.

Polar aprotic solvents have strong dipole moments to enhance solubility, yet they have no O—H or N—H groups to form hydrogen bonds with anions. Examples of useful polar aprotic solvents are acetonitrile, dimethylformamide, and acetone. We can add specific solvating reagents to enhance solubility without affecting the reactivity of the nucleophile. For example, the "crown ether" 18-crown-6 solvates potassium ions. Using the potassium salt of a nucleophile and solvating the potassium ions causes the nucleophilic anion to be dragged along into solution.

$CH_3—C{\equiv}N:$
acetonitrile

dimethylformamide
(DMF)

acetone

18-crown-6
solvates K^+ ions

The following example shows how fluoride ion, normally a poor nucleophile in hydroxylic (protic) solvents, can be a good nucleophile in an aprotic solvent. Although KF is not very soluble in acetonitrile, 18-crown-6 solvates the potassium ions, and the poorly solvated (and therefore nucleophilic) fluoride ion follows.

$$\text{(benzyl-CH}_2\text{Cl)} \xrightarrow[\text{CH}_3\text{CN}]{\text{KF, 18-crown-6}} \text{(benzyl-CH}_2\text{F)} + \text{Cl}^-$$

Just as the nucleophile is important in the S_N2 reaction, the structure of the alkyl halide is equally important. We will often refer to the alkyl halide as the **substrate**: literally, the compound that is being attacked by the reagent. In addition to alkyl halides, a variety of other types of compounds serve as substrates in S_N2 reactions. To be a good substrate for S_N2 attack by a nucleophile, a molecule must have an electrophilic carbon atom with a good **leaving group**, and that carbon atom must not be too **sterically hindered** for a nucleophile to attack.

6-11
Reactivity of the Substrate in S_N2 Reactions

6-11A   Leaving-Group Effects on the Substrate

A leaving group serves two purposes in the S_N2 reaction: It polarizes the $C-X$ bond—making the carbon atom electrophilic—and it leaves with the pair of electrons that once bonded it to the electrophilic carbon atom. To fill these roles, a good leaving group should be

1. Electron withdrawing, to polarize the carbon atom
2. Stable (not a strong base) once it has left
3. Polarizable, to stabilize the transition state

1.   The leaving group must be *electron withdrawing* to create a partial positive charge on the carbon atom, making the carbon electrophilic. An electron-withdrawing leaving group also stabilizes the negatively charged transition state. Halogen atoms are strongly electronegative, so alkyl halides are common substrates for S_N2 reactions. Oxygen, nitrogen, and sulfur also form strongly polarized bonds with carbon; given the right substituents, they can form the basis for excellent leaving groups.

Strongly polarized

$$C \rightleftharpoons X \ (X = halogen) \qquad C \rightleftharpoons O \quad C \rightleftharpoons N \quad C \rightleftharpoons S$$

2.   The leaving group must be *stable* once it has left with the pair of electrons that bonded it to carbon. A stable leaving group is needed for favorable energetics. The leaving group is leaving in the transition state; a reactive leaving group would raise the energy of the transition state, slowing the reaction. Also, the energy of the leaving group is reflected in the energy of the products. A reactive leaving group would raise the energy of the products, driving the equilibrium toward the reactants.

bond forming   bond breaking

$$\text{Nuc:}^- + \ \overset{\diagup}{\underset{\diagdown}{C}}-X \longrightarrow \left[\overset{\delta^-}{\text{Nuc}} \cdots \overset{|}{\underset{\diagup\diagdown}{C}} \cdots \overset{\delta^-}{X} \right]^{\ddagger} \longrightarrow \text{Nuc}-\overset{\diagup}{\underset{\diagdown}{C}} + \ :X^-$$

transition state

TABLE 6-4 Weak Bases That Are Common Leaving Groups

Ions:	:Cl:⁻ :Br:⁻ :I:⁻		⁻:O—S—R	⁻:O—S—OR	:O:⁻ ⁻:O—P—OR
	halides		sulfonate	sulfate	phosphate

Neutral molecules:	:O—H with H	:O—R with H	:N—R with R, R	:P—R with R, R
	water	alcohols	amines	phosphines

Good leaving groups should be *weak bases*; therefore, they are the conjugate bases of strong acids. The hydrohalic acids HCl, HBr, and HI are strong, and their conjugates (Cl⁻, Br⁻, and I⁻) are all weak bases. Other weak bases, such as sulfate ions, sulfonate ions, and phosphate ions, can also serve as good leaving groups. Table 6-4 lists examples of good leaving groups.

Hydroxide ion, alkoxide ions, and other strong bases are poor leaving groups for S_N2 reactions. For example, the —OH group of an alcohol is a poor leaving group because it would have to leave as hydroxide ion.

$$:Br:⁻ \quad CH_3—\ddot{O}H \quad \times\!\!\!\to \quad Br—CH_3 \;+\; ⁻:\ddot{O}H \quad \text{(strong base)}$$

Ions that are strong bases and poor leaving groups:

$$⁻:\ddot{O}H \qquad ⁻:\ddot{O}R \qquad ⁻:\ddot{N}H_2$$
hydroxide   alkoxide   amide

Table 6-4 also lists some neutral molecules that can be good leaving groups. A neutral molecule often serves as the leaving group from a *positively charged* electrophile. For example, if an alcohol is placed in an acidic solution, the hydroxyl group is protonated. Water then serves as the leaving group. Note that the need to protonate the alcohol (requiring acid) limits the choice of nucleophiles to those few that are weak bases, such as bromide and iodide. A strongly basic nucleophile would become protonated in acid.

$$CH_3—\ddot{O}H \;+\; H^+ \;\rightleftharpoons\; :Br:⁻ \quad CH_3—\overset{H}{\underset{}{\overset{|}{O}}}{}^+\!—H \;\longrightarrow\; :Br—CH_3 \;+\; :\overset{H}{\underset{}{O}}—H$$
protonated alcohol   water

3. Finally, a good leaving group should be *polarizable*, to maintain partial bonding with the carbon atom in the transition state. This bonding helps to stabilize the transition state and reduce the activation energy. The departure of a leaving group is much like the attack of a nucleophile, except that the bond is breaking rather than forming. Polarizable nucleophiles and polarizable leaving groups both stabilize the transition state by engaging in more bonding at a longer distance. Iodide ion, one of the most polarizable ions, is both a good nucleophile and a good leaving group. In contrast, fluoride ion is a small, "hard" ion. Fluoride is both a poor nucleophile (in protic solvents) and a poor leaving group in S_N2 reactions.

PROBLEM 6-17

When dimethyl ether ($CH_3 - O - CH_3$) is treated with concentrated HBr, the initial products are CH_3Br and CH_3OH. Propose a mechanism to account for this reaction.

PROBLEM-SOLVING HINT

Do not write S_N2 reactions that show hydroxide ions, alkoxide ions, or other strong bases serving as leaving groups.

6-11B   Steric Effects on the Substrate

Different alkyl halides undergo S_N2 reactions at vastly different rates. The structure of the substrate is the most important factor in its reactivity toward S_N2 displacement. The reaction goes rapidly with methyl halides and with most primary substrates. It is more sluggish with secondary halides. Tertiary halides fail to react at all by the S_N2 mechanism. Table 6-5 shows the effect of alkyl substitution on the rate of S_N2 displacements.

For simple alkyl halides, the relative rates for S_N2 displacement are

Relative rates for S_N2: $CH_3X > 1° > 2° > 3°$

The physical explanation for this order of reactivity is suggested by the information in Table 6-5. All the slow-reacting compounds have one property in common: The back side of the electrophilic carbon atom is crowded by the presence of bulky groups. Tertiary halides are more hindered than secondary halides, which are more hindered than primary halides. Even a bulky primary halide (like neopentyl bromide) undergoes S_N2 reaction at a rate similar to that of a tertiary halide. The relative rates show that it is the bulk of the alkyl groups, rather than an electronic effect, that hinders the reactivity of bulky alkyl halides in the S_N2 displacement.

This effect on the rate is another example of **steric hindrance**. When the nucleophile approaches the back side of the electrophilic carbon atom, it must come within bonding distance of the back lobe of the $C - X$ sp^3 orbital. If there are two alkyl groups bonded to the carbon atom, this is difficult. Three alkyl groups make it impossible. Just one alkyl group can produce a large amount of steric hindrance if it is unusually bulky, like the t-butyl group of neopentyl bromide.

Figure 6-6 shows the S_N2 reaction of hydroxide ion with ethyl bromide (1°), isopropyl bromide (2°), and t-butyl bromide (3°). The nucleophile can easily approach the electrophilic carbon atom of ethyl bromide. In isopropyl bromide, the approach is hindered, but still possible. In contrast, S_N2 approach to the tertiary carbon of t-butyl bromide is impossible because of the steric hindrance of the three methyl groups. Make models of ethyl bromide, isopropyl bromide, and t-butyl bromide, and compare the ease of bringing in an atom for a back-side attack.

PROBLEM 6-18

Rank the following compounds in decreasing order of their reactivity toward the S_N2 reaction with sodium ethoxide (Na^+ $^-OCH_2CH_3$) in ethanol.

methyl chloride	t-butyl iodide	neopentyl bromide
isopropyl bromide	methyl iodide	ethyl chloride

PROBLEM-SOLVING HINT

Do not write S_N2 reactions occurring on tertiary alkyl halides.

TABLE 6-5   Effect of Substituents on the Rates of S_N2 Reactions

Class of Halide	Example	Relative Rate
methyl	$CH_3 - Br$	>1000
primary(1°)	$CH_3CH_2 - Br$	50
secondary (2°)	$(CH_3)_2CH - Br$	1
tertiary (3°)	$(CH_3)_3C - Br$	<0.001
n-butyl (1°)	$CH_3CH_2CH_2CH_2 - Br$	20
isobutyl (1°)	$(CH_3)_2CHCH_2 - Br$	2
neopentyl (1°)	$(CH_3)_3CCH_2 - Br$	0.0005

Note: Two or three alkyl groups, or even a single bulky alkyl group, slow the reaction rate. The rates listed are compared to the secondary case (isopropyl bromide), assigned a relative rate of 1.

▲ Figure 6-6
S_N2 attack on a simple primary alkyl halide is unhindered. Attack on a secondary halide is hindered, and attack on a tertiary halide is impossible.

PROBLEM 6-19

For each pair of compounds, state which compound is the better S_N2 substrate.
(a) 2-methyl-1-iodopropane or *t*-butyl iodide
(b) cyclohexyl bromide or 1-bromo-1-methylcyclohexane
(c) 2-bromobutane or isopropyl bromide
(d) 2,2-dimethyl-1-chlorobutane or 2-chlorobutane
(e) 1-iodo-2,2-dimethylpropane or isopropyl iodide

6-12
Stereochemistry of the S$_N$2 Reaction

As we have seen, the S_N2 reaction requires attack by a nucleophile on the back side of an electrophilic carbon atom (Fig. 6-7). A carbon atom can have only four filled bonding orbitals (an octet), so the leaving group must leave as the nucleophile bonds to the carbon atom. The nucleophile's electrons insert into the back lobe of carbon's sp^3 hybrid orbital in its antibonding combination with the orbital of the leaving group (because the bonding MO is already filled). These electrons in the antibonding MO help to weaken the C—Br bond as bromine leaves. The transition state shows partial bonding to both the nucleophile and the leaving group.

Back-side attack literally turns the tetrahedron of the carbon atom inside out, like an umbrella caught by the wind (Fig. 6-7). In the product, the nucleophile assumes a stereochemical position opposite the position the leaving group originally occupied. We call this result an **inversion of configuration** at the carbon atom.

In the case of an asymmetric carbon atom, back-side attack gives the opposite configuration of the carbon atom. The S_N2 displacement is the most common example of a **Walden inversion**, a step (in a reaction sequence) where an asymmetric carbon atom undergoes inversion of configuration. In the 1890s, Paul Walden, of the University of Tübingen (Germany), was one of the first to study reactions giving inversion of configuration.

▲ **Figure 6-7**
The S$_N$2 reaction takes place through nucleophilic attack on the back lobe of carbon's sp^3 hybrid orbital. This back-side attack inverts the carbon atom's tetrahedron, like the wind inverts an umbrella.

MECHANISM   Inversion of Configuration in the S$_N$2 Reaction

In some cases, inversion of configuration is readily apparent. For example, when *cis*-1-bromo-3-methylcyclopentane undergoes S$_N$2 displacement by hydroxide ion, inversion of configuration gives *trans*-3-methylcyclopentanol.

The S$_N$2 displacement is a good example of a **stereospecific reaction**: one in which different stereoisomers react to give different stereoisomers of the product. To study the mechanism of a nucleophilic substitution, we often look at the product to see if the reaction is stereospecific, with inversion of configuration. If it is, the S$_N$2 mechanism is a good possibility, especially if the reaction kinetics are second order. In many cases (no asymmetric carbon or ring, for example), it is impossible to determine whether inversion has occurred. In these cases, we use kinetics and other evidence to help determine the reaction mechanism.

PROBLEM 6-20

Draw a perspective structure or a Fischer projection for the products of the following S_N2 reactions.

(a) *trans*-1-bromo-3-methylcyclopentane + KOH

(b) (*R*)-2-bromopentane + KCN

(c)

$$
\begin{array}{c}
CH_3 \\
Br\!-\!\!|\!\!-\!H \\
H\!-\!\!|\!\!-\!CH_3 \\
CH_2CH_3
\end{array}
\quad + \quad NaI \quad \xrightarrow{\text{acetone}}
$$

(d)

$$
\begin{array}{c}
Br \\
H\!-\!\!C\!\!-\!CH_3 \\
F
\end{array}
\quad + \quad NaSH
$$

(e)

$$
\begin{array}{c}
CH_3 \\
H\!-\!\!|\!\!-\!Br \\
CH_2CH_2CH_3
\end{array}
\quad + \quad NaOCH_3 \quad \xrightarrow{CH_3OH}
$$

(f)

$$
\begin{array}{c}
H \\
C\!\!-\!\!D \\
Cl
\end{array}
\quad + \quad NH_3
$$

PROBLEM 6-21

Under appropriate conditions, (*S*)-1-bromo-1-fluoroethane reacts with sodium methoxide to give pure (*S*)-1-fluoro-1-methoxyethane.

$$CH_3CHBrF + NaOCH_3 \longrightarrow CH_3CHFOCH_3 + NaBr$$
$$\quad\quad (S) \quad\quad\quad\quad\quad\quad\quad\quad (S)$$

(a) Why is bromide rather than fluoride replaced?
(b) Draw perspective structures (as shown on the previous page for 2-bromobutane) for the starting material, the transition state, and the product.
(c) Does the product show retention or inversion of configuration?
(d) Is this result consistent with reaction by the S_N2 mechanism?

PROBLEM-SOLVING HINT

(*R*) and (*S*) are just names. Don't rely on names to determine the stereochemistry of a reaction.

6-13
First-Order Nucleophilic Substitution: The S_N1 Reaction

When *t*-butyl bromide is placed in boiling methanol, methyl *t*-butyl ether can be isolated from the reaction mixture. Because this reaction takes place with the solvent acting as the nucleophile, it is called a **solvolysis** (*solvo* for "solvent," plus *lysis*, meaning "cleavage").

$$(CH_3)_3C\!-\!Br \;+\; CH_3\!-\!OH \xrightarrow{\text{boil}} (CH_3)_3C\!-\!O\!-\!CH_3 \;+\; HBr$$

$$\text{\textit{t}-butyl bromide} \quad\quad \text{methanol} \quad\quad\quad \text{methyl \textit{t}-butyl ether}$$

This solvolysis is a substitution because methoxide has replaced bromide on the *t*-butyl group. It does not go through the S_N2 mechanism, however. The S_N2 requires a strong nucleophile and a substrate that is not too hindered. Methanol is a weak nucleophile, and *t*-butyl bromide is a hindered tertiary halide: a poor S_N2 substrate.

An interesting characteristic of this reaction is that its rate does not depend on the concentration of methanol, the nucleophile. The rate depends only on the concentration of the substrate, *t*-butyl bromide.

$$\text{rate} = k_r[(CH_3)_3C\!-\!Br]$$

This rate equation is first order overall: first order in the concentration of the alkyl halide and zeroth order in the concentration of the nucleophile. Because the rate does not depend on the concentration of the nucleophile, we infer that the nucleophile is not present in the transition state of the rate-limiting step. The nucleophile must react *after* the slow step.

This type of substitution is called the **S_N1 reaction**, for *Substitution, Nucleophilic, unimolecular*. The term *unimolecular* means there is only one molecule involved in the transition state of the rate-limiting step. The mechanism of the S_N1 reaction of *t*-butyl bromide with methanol is shown here. Ionization of the alkyl halide (first step) is the rate-limiting step.

Step 1: Formation of carbocation (rate limiting)

$$(CH_3)_3C—\ddot{B}r: \quad \rightleftharpoons \quad (CH_3)_3C^+ \ + \ :\ddot{B}r:^- \quad (slow)$$

Step 2: Nucleophilic attack on the carbocation

$$(CH_3)_3C^+ \quad :\ddot{O}—CH_3 \quad \rightleftharpoons \quad (CH_3)_3C—\overset{..+}{\underset{|}{O}}—CH_3 \quad (fast)$$
$$\qquad\qquad\quad \underset{H}{|} \qquad\qquad\qquad\qquad\qquad\quad \underset{H}{}$$

Final Step: Loss of proton to solvent

$$(CH_3)_3C—\overset{..+}{\underset{|}{O}}—CH_3 \ + \ CH_3—\ddot{O}H \quad \rightleftharpoons \quad (CH_3)_3C—\ddot{O}—CH_3 \ + \ CH_3—\overset{..+}{\underset{|}{O}}—H \quad (fast)$$
$$\qquad\quad \underset{H}{} \qquad\qquad\qquad\qquad\qquad\qquad\qquad\qquad\qquad\qquad \underset{H}{}$$

The S$_N$1 mechanism is a multistep process. The first step is a slow ionization to form a carbocation. The second step is a fast attack on the carbocation by a nucleophile. The carbocation is a strong electrophile; it reacts very fast with both strong and weak nucleophiles. In the case of attack by an alcohol or water, loss of a proton gives the final uncharged product. The general mechanism for the S$_N$1 reaction is as follows:

KEY MECHANISM   The S$_N$1 Reaction

Step 1: Formation of a carbocation (rate limiting).

$$R—\ddot{X}: \quad \rightleftharpoons \quad R^+ \ + \ :\ddot{X}:^-$$

Step 2: Nucleophilic attack on the carbocation (fast).

$$R^+ \ + \ Nuc:^- \quad \longrightarrow \quad R—Nuc$$

The reaction-energy diagram of the S$_N$1 reaction (Figure 6-8) shows why the rate does not depend on the strength or concentration of the nucleophile. The ionization (first step) is highly endothermic, and its large activation energy determines the overall reaction rate. The nucleophilic attack (second step) is strongly exothermic, with a lower-energy transition state. In effect, a nucleophile reacts with the carbocation almost as soon as it forms.

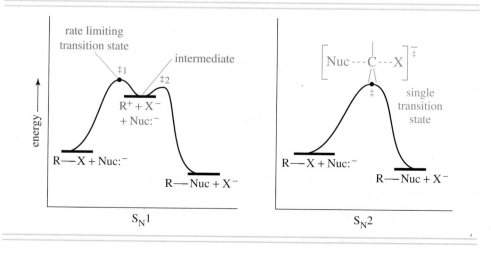

◀ **Figure 6-8**
Reaction-energy diagrams of the S$_N$1 and S$_N$2 reactions. The S$_N$1 is a two-step mechanism with two transition states (‡1 and ‡2) and a carbocation intermediate. The S$_N$2 has only one transition state and no intermediate.

The reaction-energy diagrams of the S_N1 mechanism and the S_N2 mechanism are compared in Figure 6-8. The S_N1 has a true intermediate, the carbocation. The intermediate appears as a relative minimum (a low point) in the reaction-energy diagram. Reagents and conditions that favor formation of the carbocation (the slow step) accelerate the S_N1 reaction; reagents and conditions that hinder its formation retard the reaction.

6-13A   Substituent Effects

The rate-limiting step of the S_N1 reaction is ionization to form a carbocation: a strongly endothermic process. The transition state resembles the carbocation (Hammond postulate, Sec. 4-14); consequently, rates of S_N1 reactions depend strongly on carbocation stability. In Section 4-16A, we saw that alkyl groups stabilize carbocations by donating electrons through sigma bonds (the *inductive effect*) and through overlap of filled orbitals with the empty *p* orbital of the carbocation *(hyperconjugation)*. Highly substituted carbocations are therefore more stable.

carbocation stability:
$3° > 2° > 1° > {}^+CH_3$

inductive effect

hyperconjugation

Reactivity toward S_N1 substitution mechanisms follows the stability of carbocations:

S_N1 reactivity: $3° > 2° > 1° > CH_3X$

This order is *opposite* that of the S_N2 reaction. Alkyl groups hinder the S_N2 by blocking attack of the strong nucleophile, but alkyl groups enhance the S_N1 by stabilizing the carbocation intermediate.

Resonance stabilization of the carbocation can also promote the S_N1 reaction. For example, allyl bromide is a primary halide, but it undergoes the S_N1 reaction about as fast as a secondary halide. The carbocation formed by ionization is resonance stabilized, with the positive charge spread equally over two carbon atoms.

allyl bromide

resonance-stabilized carbocation

PROBLEM 6-22

Choose the member of each pair that will react faster by the S_N1 mechanism.
(a) 1-bromopropane or 2-bromopropane
(b) 2-bromo-2-methylbutane or 2-bromo-3-methylbutane
(c) bromocyclohexane or 3-bromocyclohexene
(d) 1-bromo-2,2-dimethylpropane or 2-bromopropane

PROBLEM 6-23

3-Bromocyclohexene is a secondary halide, and benzyl bromide is a primary halide. Both halides undergo S_N1 substitution about as fast as most tertiary halides. Use resonance structures to explain this enhanced reactivity.

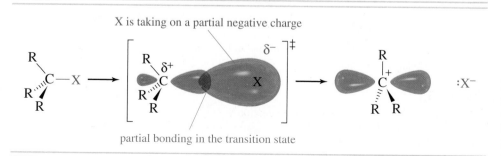

3-bromocyclohexene   benzyl bromide

6-13B Leaving-Group Effects

The leaving group is breaking its bond to carbon in the rate-limiting ionization step of the S_N1 mechanism. A highly polarizable leaving group helps stabilize the rate-limiting transition state through partial bonding as it leaves. The leaving group should be a weak base, very stable after it leaves with the pair of electrons that bonded it to carbon.

Figure 6-9 shows the transition state of the ionization step of the S_N1 reaction. Notice how the leaving group is taking on a negative charge while it stabilizes the new carbocation through partial bonding. The leaving group should be stable as it takes on this negative charge, and it should be polarizable to engage in effective partial bonding as it leaves. A good leaving group is just as necessary in the S_N1 reaction as it is in the S_N2, and similar leaving groups are effective for either reaction. Table 6-4 (page 234) lists some common leaving groups.

PROBLEM 6-24

Choose the member of each pair that will react faster by the S_N1 mechanism.

(a) 2-iodo-2-methylbutane or *t*-butyl chloride
(b) 2-bromo-2-methylbutane or ethyl iodide
(c) *n*-propyl bromide or 3-bromocyclohexene
(d) methyl iodide or cyclohexyl bromide

6-13C Solvent Effects

The S_N1 reaction goes much more readily in polar solvents that stabilize ions. The rate-limiting step forms two ions, and ionization is taking place in the transition state. Polar solvents solvate these ions by an interaction of the solvent's dipole moment with the charge of the ion. Protic solvents such as alcohols and water are even more effective solvents because anions form hydrogen bonds with the —OH hydrogen atom, and cations complex with the nonbonding electrons of the —OH oxygen atom.

X is taking on a partial negative charge

partial bonding in the transition state

◀ **Figure 6-9**
In the transition state of the S_N1 ionization, the leaving group is taking on a negative charge. The C—X bond is breaking, and a polarizable leaving group can still maintain substantial overlap.

$$R \overset{..}{\underset{..}{-X}} : \; \rightleftharpoons \; R^+ \quad : \overset{..}{\underset{..}{X}} : {}^-$$

ionization

solvated ions

TABLE 6-6 Dielectric Constants (ϵ) and Ionization Rates of *t*-Butyl Chloride in Common Solvents

Solvent	ϵ	Relative Rate
water	78	8000
methanol	33	1000
ethanol	24	200
acetone	21	1
diethyl ether	4.3	0.001
hexane	2.0	<0.0001

Ionization of an alkyl halide requires formation and separation of positive and negative charges, similar to what happens when sodium chloride dissolves in water. Therefore, S_N1 reactions require highly polar solvents that strongly solvate ions. One measure of a solvent's ability to solvate ions is its *dielectric constant* (ϵ), a measure of the solvent's polarity. Table 6-6 lists the dielectric constants of some common solvents and the relative ionization rates for *t*-butyl chloride in these solvents. Note that ionization occurs much faster in highly polar solvents.

6-14
Stereochemistry of the S_N1 Reaction

We saw (Section 6-12) that the S_N2 reaction is stereospecific: The nucleophile attacks from the back side of the electrophilic carbon atom, giving inversion of configuration. In contrast, the S_N1 reaction is not stereospecific. In the S_N1 mechanism, the carbocation intermediate is sp^2 hybridized and planar. A nucleophile can attack the carbocation from either face. Figure 6-10 shows the S_N1 solvolysis of a chiral compound, (*S*)-3-bromo-2,3-dimethylpentane, in ethanol. The carbocation is planar and achiral; attack from both faces gives both enantiomers of the product. Such a process, giving both enantiomers of the product (whether or not the two enantiomers are produced in equal amounts), is called **racemization**. The product is either racemic, or at least less optically pure than the starting material.

If a nucleophile attacks the carbocation in Figure 6-10 from the front side (the side the leaving group left), the product molecule shows **retention of configuration**. Attack from the back side gives a product molecule showing **inversion of configuration**. Racemization is simply a combination of retention and inversion. When racemization occurs,

---

MECHANISM Racemization in the S_N1 Reaction

▲ **Figure 6-10**
An asymmetric carbon atom undergoes racemization when it ionizes to a planar, achiral carbocation. A nucleophile can attack the carbocation from either face, giving either enantiomer of the product.

▲ Figure 6-11
In the S_N1 reaction of *cis*-1-bromo-3-deuteriocyclopentane with methanol, the carbocation can be attacked from either face. Because the leaving group (bromide) partially blocks the front side as it leaves, back-side attack (inversion of configuration) is slightly favored.

the product is rarely completely racemic, however; there is often more inversion than retention of configuration. As the leaving group leaves, it partially blocks the front side of the carbocation. The back side is unhindered, so attack is more likely there.

   Figure 6-11 shows a cyclic case where one of the faces of a cyclopentane ring has been "labeled" by a deuterium atom. Deuterium has the same size and shape as hydrogen, and it undergoes the same reactions. It distinguishes between the two faces of the ring: The bromine atom is cis to the deuterium in the reactant, so the nucleophile is cis to the deuterium in the retention product. The nucleophile is trans to the deuterium in the inversion product. The product mixture contains both cis and trans isomers, with the trans isomer slightly favored because the leaving group hinders approach of the nucleophilic solvent from the front side.

Carbocations frequently undergo structural changes, called **rearrangements**, to form more stable ions. A rearrangement may occur after a carbocation has formed, or it may occur as the leaving group is leaving. Rearrangements are not seen in S_N2 reactions, where no carbocation is formed and the one-step mechanism allows no opportunity for rearrangement.

   An example of a reaction with rearrangement is the S_N1 reaction of 2-bromo-3-methylbutane in boiling ethanol. The product is a mixture of 2-ethoxy-3-methylbutane (not rearranged) and 2-ethoxy-2-methylbutane (rearranged).

6-15
Rearrangements in S_N1 Reactions

$$CH_3-\overset{\overset{Br}{|}}{CH}-\overset{\overset{}{|}}{CH}-CH_3 \quad \xrightarrow[\text{heat}]{CH_3CH_2OH} \quad CH_3-\overset{\overset{OCH_2CH_3}{|}}{CH}-\overset{\overset{}{|}}{CH}-CH_3 \quad + \quad CH_3-CH_2-\overset{\overset{OCH_2CH_3}{|}}{\underset{\underset{CH_3}{|}}{C}}-CH_3 \quad + \quad HBr$$

2-bromo-3-methylbutane   (not rearranged)   2-ethoxy-3-methylbutane   (rearranged)   2-ethoxy-2-methylbutane

> **PROBLEM 6-25**
>
> Give the S_N1 mechanism for the formation of 2-ethoxy-3-methylbutane, the unrearranged product in this reaction.

The rearranged product, 2-ethoxy-2-methylbutane, results from a **hydride shift**: the movement of a hydrogen atom with its bonding pair of electrons. A hydride shift is represented by the symbol $\sim H$. In this case, the hydride shift converts the initially formed secondary carbocation to a more stable tertiary carbocation. Attack by the solvent gives the rearranged product.

MECHANISM Hydride Shift in an S_N1 Reaction

Step 1: Formation of the carbocation and rearrangement.

This rearrangement involves movement of a hydrogen atom with its bonding pair of electrons over to the empty *p* orbital of the carbocation. In three dimensions, the rearrangement looks like this:

Step 2: Solvent attack and loss of a proton

When neopentyl bromide is boiled with ethanol, it gives *only* a rearranged substitution product. This product results from a **methyl shift** (represented by the symbol $\sim CH_3$), the migration of a methyl group together with its pair of electrons. Without rearrangement, ionization of neopentyl bromide would give a very unstable primary carbocation.

The methyl shift occurs *while* bromide ion is leaving, so that only the more stable tertiary carbocation is formed.

PROBLEM-SOLVING HINT

Primary halides and methyl halides rarely ionize to carbocations in solution. If they do, they usually ionize with rearrangement.

MECHANISM   Methyl Shift in an S_N1 Reaction

In three dimensions,

Attack by ethanol on the tertiary carbocation gives the product. Because rearrangement is required for ionization, only rearranged products are observed.

In general, we should expect rearrangements in reactions involving carbocations whenever a hydride shift or an alkyl shift can form a more stable carbocation.

PROBLEM 6-26

Propose a mechanism involving a hydride shift or an alkyl shift for each solvolysis reaction. Explain how each rearrangement forms a more stable intermediate.

The rearrangements of carbocations also play a prominent role in the formation of the terpene natural products. Menthol, camphor, and α-pinene are common examples of terpene plant oils. They are constructed using a common building block, and undergo a series of rearrangements in the course of construction to generate the most stable carbocation.

menthol

camphor

α-pinene

PROBLEM-SOLVING HINT

Most rearrangements convert 2° (or incipient 1°) carbocations to 3° or resonance-stabilized carbocations.

6-16
Comparison of S$_N$1 and S$_N$2 Reactions

Let's compare what we know about the S$_N$1 and S$_N$2 reactions, then organize this material into a brief table.

Effect of the Nucleophile   The nucleophile takes part in the slow step (the only step) of the S$_N$2 reaction but not in the slow step of the S$_N$1. Therefore, a strong nucleophile promotes the S$_N$2 but not the S$_N$1. Weak nucleophiles fail to promote the S$_N$2 reaction; therefore, reactions with weak nucleophiles often go by the S$_N$1 mechanism if the substrate is secondary or tertiary.

> S$_N$1:   Nucleophile strength is unimportant.
> S$_N$2:   Strong nucleophiles are required.

Effect of the Substrate   The structure of the substrate (the alkyl halide) is an important factor in determining which of these substitution mechanisms might operate. Methyl halides and primary halides cannot easily ionize and undergo S$_N$1 substitution because methyl and primary carbocations are high in energy. They are relatively unhindered, however, so they make good S$_N$2 substrates.

Tertiary halides are too hindered to undergo S$_N$2 displacement, but they can ionize to form tertiary carbocations. Tertiary halides undergo substitution exclusively through the S$_N$1 mechanism. Secondary halides can undergo substitution by either mechanism, depending on the conditions.

> S$_N$1 substrates: $3° > 2°$ (1° and CH$_3$X are unlikely)
> S$_N$2 substrates: CH$_3$X $> 1° > 2°$ (3° is not suitable)

If silver nitrate (AgNO$_3$) is added to an alkyl halide in a good ionizing solvent, it removes the halide ion to give a carbocation. This technique can force some unlikely ionizations, often giving interesting rearrangements (see Problem 6-29.)

Effect of the Solvent   The slow step of the S$_N$1 reaction involves formation of two ions. Solvation of these ions is crucial to stabilizing them and lowering the activation energy for their formation. Very polar ionizing solvents such as water and alcohols are needed for the S$_N$1. The solvent may be heated to reflux (boiling) to provide the energy needed for ionization.

Less charge separation is generated in the transition state of the S$_N$2 reaction. Strong solvation may weaken the strength of the nucleophile because of the energy needed to strip off the solvent molecules. Thus, the S$_N$2 reaction often goes faster in less polar solvents if the nucleophile will dissolve. Polar aprotic solvents may enhance the strength of weak nucleophiles.

> S$_N$1:   Good ionizing solvent required.
> S$_N$2:   May go faster in a less polar solvent.

Kinetics   The rate of the S$_N$1 reaction is proportional to the concentration of the alkyl halide but not the concentration of the nucleophile. It follows a first-order rate equation.

The rate of the S$_N$2 reaction is proportional to the concentrations of both the alkyl halide [R—X] and the nucleophile [Nuc:$^-$]. It follows a second-order rate equation.

> S$_N$1 rate $= k_r$[R—X]
> S$_N$2 rate $= k_r$[R—X][Nuc:$^-$]

Stereochemistry   The S$_N$1 reaction involves a flat carbocation intermediate that can be attacked from either face. Therefore, the S$_N$1 usually gives a mixture of inversion and retention of configuration.

The S$_N$2 reaction takes place through a back-side attack, which inverts the stereochemistry of the carbon atom. Complete inversion of configuration is the result.

> S$_N$1 stereochemistry: Mixture of retention and inversion; racemization.
> S$_N$2 stereochemistry: Complete inversion.

Rearrangements   The S_N1 reaction involves a carbocation intermediate. This intermediate can rearrange, usually by a hydride shift or an alkyl shift, to give a more stable carbocation.

The S_N2 reaction takes place in one step with no intermediates. No rearrangement is possible in the S_N2 reaction.

S_N1:   Rearrangements are common.
S_N2:   Rearrangements are impossible.

SUMMARY   **Nucleophilic Substitutions**

	S_N1	S_N2
promoting factors		
nucleophile	weak nucleophiles are OK	strong nucleophile needed
substrate (RX)	3° > 2°	CH_3X > 1° > 2°
solvent	good ionizing solvent needed	wide variety of solvents
leaving group	good one required	good one required
other	$AgNO_3$ forces ionization	
characteristics		
kinetics	first order, $k_r[RX]$	second order, $k_r[RX][Nuc:^-]$
stereochemistry	mixture of inversion and retention	complete inversion
rearrangements	common	impossible

PROBLEM 6-27

For each reaction, give the expected substitution product, and predict whether the mechanism will be predominantly first order or second order.
(a) 2-chloro-2-methylbutane + CH_3COOH
(b) isobutyl bromide + sodium methoxide
(c) 1-iodo-1-methylcyclohexane + ethanol
(d) cyclohexyl bromide + methanol
(e) cyclohexyl bromide + sodium ethoxide

PROBLEM-SOLVING HINT

The strength of the nucleophile (or base) usually determines the order of the reaction. Also, S_N2 is unlikely on 3° halides, and S_N1 is unlikely on 1° halides.

PROBLEM 6-28

When (R)-2-bromobutane is heated with water, the S_N1 substitution proceeds twice as fast as the S_N2. Calculate the e.e. and the specific rotation expected for the product. The specific rotation of (R)-2 butanol is $-13.5°$.

PROBLEM 6-29

A reluctant first-order substrate can be forced to ionize by adding some silver nitrate (one of the few soluble silver salts) to the reaction. Silver ion reacts with the halogen to form a silver halide (a highly exothermic reaction), generating the cation of the alkyl group.

$$R-X + Ag^+ \longrightarrow R^+ + AgX \downarrow$$

Give mechanisms for the following silver-promoted rearrangements.

6-17 First-Order Elimination: The E1 Reaction

An **elimination** involves the loss of two atoms or groups from the substrate, usually with formation of a pi bond. Depending on the reagents and conditions involved, an elimination might be a first-order (E1) or second-order (E2) process. The following examples illustrate the types of eliminations we cover in this chapter.

E1:

E2:

In many cases, substitution and elimination reactions can take place at the same time and under the same conditions. We will consider these possibilities later, once we have discussed the E1 and E2 mechanisms.

6-17A   Mechanism and Kinetics of the E1 Reaction

The abbreviation **E1** stands for *Elimination, unimolecular*. The mechanism is called *unimolecular* because the rate-limiting transition state involves a single molecule rather than a collision between two molecules. The slow step of an E1 reaction is the same as in the S_N1 reaction: unimolecular ionization to form a carbocation. In a fast second step, a base abstracts a proton from the carbon atom adjacent to the C^+. The electrons that once formed the carbon–hydrogen bond now form a pi bond between two carbon atoms. The general mechanism for the E1 reaction is

KEY MECHANISM   The E1 Reaction

Step 1: Formation of the carbocation (rate limiting).

Step 2: A base abstracts a proton (fast).

6-17B   Competition with the S$_N$1 Reaction

The E1 reaction almost always occurs together with the S$_N$1. Whenever a carbocation is formed, it can undergo either substitution or elimination, and mixtures of products often result. The following reaction shows the competition between elimination and substitution in the reaction of *t*-butyl bromide with boiling ethanol.

CH$_3$—C(CH$_3$)(CH$_3$)—Br   +   CH$_3$—CH$_2$—OH $\xrightarrow{\text{heat}}$ H$_2$C=C(CH$_3$)(CH$_3$)   +   CH$_3$—C(CH$_3$)(CH$_3$)—O—CH$_2$—CH$_3$

t-butyl bromide   ethanol   2-methylpropene   ethyl *t*-butyl ether
                                                        (E1 product)   (S$_N$1 product)

The 2-methylpropene product results from **dehydrohalogenation**, an elimination of hydrogen and a halogen atom. Under these first-order conditions (the absence of a strong base), dehydrohalogenation takes place by the E1 mechanism: Ionization of the alkyl halide gives a carbocation intermediate, which loses a proton to give the alkene. Substitution results from nucleophilic attack on the carbocation. Ethanol serves as a base in the elimination and as a nucleophile in the substitution.

Step 1: Ionization to form a carbocation

Step 2: Basic attack by the solvent (E1 reaction)

or: Nucleophilic attack by the solvent (S$_N$1 reaction)

Under ideal conditions, one of these first-order reactions provides a good yield of one product or the other. Often, however, carbocation intermediates react in two or more ways to give mixtures of products. For this reason, S$_N$1 and E1 reactions of alkyl halides are not often used for organic synthesis. They have been studied in great detail to learn about the properties of carbocations, however.

6-17C   Orbitals and Energetics

In the second step of the E1 mechanism, the carbon atom next to the C$^+$ must rehybridize to sp^2 as the base attacks the proton and electrons flow into the new pi bond.

The potential-energy diagram for the E1 reaction (Figure 6-12) is similar to that for the S_N1 reaction. The ionization step is strongly endothermic, with a rate-limiting transition state. The second step is a fast exothermic deprotonation by a base. The base is not involved in the reaction until *after* the rate-limiting step, so the rate depends only on the concentration of the alkyl halide. Weak bases are common in E1 reactions.

Like other carbocation reactions, the E1 may be accompanied by rearrangement. Compare the following E1 reaction (with rearrangement) with the S_N1 reaction of the same substrate, shown on page 244.

MECHANISM   Rearrangement in an E1 Reaction

Formation of the carbocation and rearrangement.

secondary carbocation   tertiary carbocation

Removal of either adjacent proton.

PARTIALLY SOLVED PROBLEM 6-1

When the following compound is heated in methanol, several different products are formed. Propose mechanisms to account for the four products shown below.

SOLUTION

With no strong base and a good ionizing solvent, we would expect a first-order reaction. But this is a primary alkyl halide, so ionization is difficult unless it rearranges. It might rearrange as it forms, but we'll imagine the cation forming then rearranging.

◄ **Figure 6-12**
Reaction-energy diagram of the E1 reaction. The first step is a rate-limiting ionization. Compare this energy profile with that of the S_N1 reaction, Figure 6-8.

From these rearranged intermediates, either loss of a proton (E1) or attack by the solvent (S_N1) gives the observed products. Note that the actual reaction may give more than just these products, but the other products are not required for the problem.

PROBLEM 6-30

Finish Partially Solved Problem 6-1 by showing how the rearranged carbocations give the four products shown in the problem. Be careful when using curved arrows to show deprotonation and/or nucleophilic attack by the solvent. The curved arrows always show movement of electrons, not movement of protons or other species.

We can now summarize four ways that a carbocation can react to become more stable.

SUMMARY   Carbocation Reactions

A carbocation can
1. React with its own leaving group to return to the reactant.
2. React with a nucleophile to form a substitution product (S_N1).
3. Lose a proton to form an elimination product (an alkene) (E1).
4. Rearrange to a more stable carbocation, then react further.
The order of stability of carbocations is: resonance stabilized, 3° > 2° > 1°.

PROBLEM 6-31

Give the substitution and elimination products you would expect from the following reactions.
(a) 3-bromo-3-ethylpentane heated in methanol
(b) 1-iodo-1-methylcyclopentane heated in ethanol
(c) 3-bromo-2,2-dimethylbutane heated in ethanol
(d) iodocyclohexane + silver nitrate in water (see Problem 6-29)

6-18
Second-Order Elimination: The E2 Reaction

Elimination can also take place under second-order conditions with a strong base present. As an example, consider the reaction of *t*-butyl bromide with methoxide ion in methanol.

$$\text{Rate} = k_r[(CH_3)_3C\text{—Br}][^-OCH_3]$$

This is a second-order reaction because methoxide ion is a strong base as well as a strong nucleophile. It attacks the alkyl halide faster than the halide can ionize to give a first-order reaction. No substitution product (methyl *t*-butyl ether) is observed, however. The S_N2 mechanism is blocked because the tertiary alkyl halide is too hindered. The observed product is 2-methylpropene, resulting from elimination of HBr and formation of a double bond.

In the E2 reaction, methoxide reacts as a *base* rather than as a *nucleophile*. Most strong nucleophiles are also strong bases, and elimination commonly results when a strong base/nucleophile is used with a poor S_N2 substrate. Instead of attacking the back side of the hindered electrophilic carbon, methoxide abstracts a proton from one of the methyl groups. This reaction takes place in one step, with bromide leaving as the base is abstracting a proton.

The rate of this elimination is proportional to the concentrations of both the alkyl halide and the base, giving a second-order rate equation. This is a *bimolecular* process, with both the base and the alkyl halide participating in the transition state, so this mechanism is abbreviated **E2** for *Elimination, bimolecular.*

The E2 reaction requires abstraction of a proton on a carbon atom next to the carbon bearing the halogen. If there are two or more possibilities, mixtures of products

---

KEY MECHANISM   The E2 Reaction

The concerted E2 reaction takes place in a single step.

$$\text{E2 rate} = k_r[R\text{—X}][B^-]$$

may result. The following examples show how abstraction of different protons may lead to different products:

2-bromopentane → 1-pentene + 2-pentene (cis and trans)

$$+ \quad CH_3OH \quad + \quad NaBr$$

1-bromo-1-methylcyclohexane → methylenecyclohexane + 1-methylcyclohexene

$$+ \quad CH_3CH_2OH \quad + \quad NaBr$$

PROBLEM 6-32

Under the conditions given, one of the two preceding examples can also undergo substitution by the S_N2 mechanism. Show the product that would result from this S_N2 reaction.

PROBLEM 6-33

Predict the elimination products of the following reactions.
(a) *sec*-butyl bromide + $NaOCH_2CH_3$
(b) 3-bromo-3-ethylpentane + methanol
(c) 2-bromo-3-ethylpentane + $NaOCH_3$
(d) 1-bromo-2-methylcyclohexane + $NaOCH_2CH_3$

Many compounds can eliminate in more than one way, to give mixtures of products. In the examples shown in the previous section, both 2-bromopentane and 1-bromo-1-methylcyclohexane can eliminate in two ways. In most cases, we can predict which elimination product will predominate. For example, consider the E2 reaction of 2-bromobutane with potassium hydroxide:

6-19
Positional Orientation of Elimination: The Saytzeff Rule

$$CH_3—CH_2—\underset{\underset{Br}{|}}{CH}—CH_3 \xrightarrow{\;:\overset{..}{\underset{..}{O}}H\;} \underset{\underset{\text{monosubstituted}}{19\%}}{CH_3—CH_2—CH{=}CH_2} + \underset{\underset{\text{disubstituted}}{81\%}}{CH_3—CH{=}CH—CH_3}$$

The first product has a *monosubstituted* double bond, with one substituent on the doubly bonded carbons. It has the general formula $R—CH{=}CH_2$. The second product has a *disubstituted* double bond, with general formula $R—CH{=}CH—R$ (or $R_2C{=}CH_2$). In most E1 and E2 eliminations where there are two or more possible elimination products, *the product with the most highly substituted double bond will predominate*. This general principle is called the **Saytzeff rule***, and reactions that give the most highly substituted alkene are said to follow **Saytzeff orientation**.

*Saytzeff is transliterated from the Russian, and may also be spelled *Zaitsev*.

SAYTZEFF RULE: In elimination reactions, the most highly substituted alkene usually predominates.

$$R_2C{=}CR_2 \quad > \quad R_2C{=}CHR \quad > \quad RHC{=}CHR \quad \text{and} \quad R_2C{=}CH_2 \quad > \quad RHC{=}CH_2$$

tetrasubstituted   trisubstituted   disubstituted   monosubstituted

This order of preference is the same as the order of stability of alkenes. We consider the stability of alkenes in more detail in Section 7-7, but for now, it is enough just to know that more highly substituted alkenes are more stable. In Chapter 7, we will study some unusual reactions where the Saytzeff rule does not apply.

PROBLEM 6-34

In Section 6-18, two examples were shown to give mixtures of products: the elimination of 2-bromopentane with methoxide and the elimination of 1-bromo-1-methylcyclohexane with ethoxide. In each case, predict which product is the major product. Explain your answers, showing the degree of substitution of each double bond in the products.

PROBLEM 6-35

The reaction of 2-bromobutane with potassium hydroxide (above) can also give substitution. Show the substitution product, and give the mechanism for its formation.

SOLVED PROBLEM 6-2

When 3-iodo-2,2-dimethylbutane is treated with silver nitrate in ethanol, three elimination products are formed. Give their structures, and predict which ones are formed in larger amounts.

SOLUTION

Silver nitrate reacts with the alkyl iodide to give silver iodide and a cation.

This secondary carbocation can lose a proton to give an unrearranged alkene (**A**), or it can rearrange to a more stable tertiary cation.

Loss of a proton

Product (**A**) + $CH_3CH_2\overset{+}{O}H_2$

PROBLEM-SOLVING HINT

Whenever a carbocation is formed next to a more highly substituted carbon, consider whether a rearrangement might occur.

Rearrangement

2° carbocation   3° carbocation

The tertiary cation can lose a proton in either of two positions. One of the products (**B**) is a tetrasubstituted alkene, and the other (**C**) is disubstituted.

Formation of a tetrasubstituted alkene

(B)
(tetrasubstituted)

Formation of a disubstituted alkene

(C)
(disubstituted)

Product **B** predominates over product **C** because the double bond in **B** is more highly sub-stitued. Whether product **A** is a major product will depend on the specific reaction conditions and whether proton loss or rearrangement occurs faster.

Reactivity of the Substrate in the E2   The order of reactivity of alkyl halides to-ward E2 dehydrohalogenation is found to be

$$3° \; > \; 2° \; > \; 1°$$

This reactivity order reflects the greater stability of highly substituted double bonds. Elimination of a tertiary halide gives a more highly substituted alkene than elimi-nation of a secondary halide, which gives a more highly substituted alkene than a primary halide. The stabilities of the alkene products are reflected in the transition states, giving lower activation energies and higher rates for elimination of alkyl halides that lead to highly substituted alkenes.

PROBLEM 6-36

Each of the two carbocations in Solved Problem 6-2 can also react with ethanol to give a sub-stitution product. Give the structures of the two substitution products formed in this reaction.

PROBLEM 6-37

Give the structure of the elimination products for the following reactions, and label the major products. When a rearrangement occurs, show how a more stable intermediate is formed.
(a) 2-bromohexane + NaOH
(b) 1-(bromomethyl)-1-methylcyclopentane heated in methanol
(c) *cis*-1-bromo-2-methylcyclohexane + AgNO$_3$ in heated ethanol
(d) *cis*-1-bromo-2-methylcyclohexane + NaOEt
(e) neopentyl bromide + AgNO$_3$ heated in methanol
(f) neopentyl bromide + NaOCH$_3$

Like the S$_N$2 reaction, the E2 follows a **concerted mechanism**: Bond breaking and bond formation take place at the same time, and the partial formation of new bonds lowers the energy of the transition state. Concerted mechanisms require specific geo-metric arrangements so that the orbitals of the bonds being broken can overlap with those being formed and the electrons can flow smoothly from one bond to another. The geometric arrangement required by the S$_N$2 reaction is a back-side attack; with the E2 reaction, a coplanar arrangement of orbitals is needed.

**6-20
Stereochemistry of
the E2 Reaction**

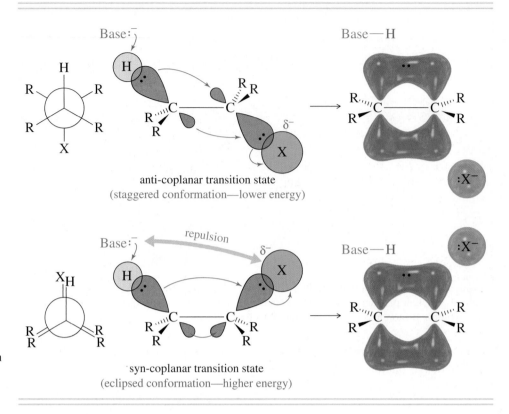

anti-coplanar transition state
(staggered conformation—lower energy)

syn-coplanar transition state
(eclipsed conformation—higher energy)

▶ **Figure 6-13**
Concerted transition states of the E2 reaction. The orbitals of the hydrogen atom and the halide must be aligned so they can begin to form a pi bond in the transition state.

Enzyme-catalyzed eliminations generally proceed by E2 mechanisms and produce only one stereoisomer. Two catalytic groups are involved: one abstracts the hydrogen, and the other assists in the departure of the leaving group. The groups are positioned in order to allow an anti-coplanar elimination.

E2 elimination requires partial formation of a new pi bond, with its parallel *p* orbitals, in the transition state. The electrons that once formed a C—H bond must begin to overlap with the orbital that the leaving group is vacating. Formation of this new pi bond implies that these two sp^3 orbitals must be parallel so that pi overlap is possible as the hydrogen and halogen leave and the orbitals rehybridize to the *p* orbitals of the new pi bond.

Figure 6-13 shows two conformations that provide the necessary coplanar alignment of the leaving group, the departing hydrogen, and the two carbon atoms. When the hydrogen and the halogen are **anti** to each other $(\theta = 180°)$, their orbitals are aligned. This is called the **anti-coplanar** conformation. When the hydrogen and the halogen eclipse each other $(\theta = 0°)$, their orbitals are once again aligned. This is called the **syn-coplanar** conformation. Make a model corresponding to Figure 6-13, and use it to follow along with this discussion.

Of these possible conformations, the anti-coplanar arrangement is most commonly seen in E2 reactions. The transition state for the anti-coplanar arrangement is a staggered conformation, with the base far away from the leaving group. In most cases, this transition state is lower in energy than that for the syn-coplanar elimination.

The transition state for syn-coplanar elimination is an eclipsed conformation. In addition to the higher energy resulting from eclipsing interactions, the transition state suffers from interference between the attacking base and the leaving group. To abstract the proton, the base must approach quite close to the leaving group. In most cases, the leaving group is bulky and negatively charged, and the repulsion between the base and the leaving group raises the energy of the syn-coplanar transition state.

Some molecules are rigidly held in eclipsed (or nearly eclipsed) conformations, with a hydrogen atom and a leaving group in a syn-coplanar arrangement. Such compounds are likely to undergo E2 elimination by a concerted syn-coplanar mechanism. Deuterium labeling (using D, the hydrogen isotope with mass number 2) is used in the following reaction to show which atom is abstracted by the base. Only the hydrogen atom is abstracted, because it is held in a syn-coplanar position with the bromine atom. Remember that syn-coplanar eliminations are unusual, however; anti-coplanar eliminations are more common.

The E2 is a **stereospecific** reaction, because different stereoisomers of the starting material react to give different stereoisomers of the product. This stereospecificity results from the anti-coplanar transition state that is usually involved in the E2. We consider more of the implications of the anti-coplanar transition state in Chapter 7. For now, Problem 6-38 will give you an opportunity to build models and see how the stereochemistry of an E2 elimination converts different stereoisomers into different stereoisomers of the product.

PROBLEM 6-38

When the first compound shown below is treated with sodium methoxide, the only elimination product is the trans isomer. The second diastereomer (blue) gives only the cis product. Use your models and careful drawings of the transition states to explain these results.

PROBLEM-SOLVING HINT

Models are helpful whenever complex stereochemistry is involved.

Let's summarize the major points to remember about the E1 and E2 reactions, concentrating on the factors that help us predict which of these mechanisms will operate under a given set of experimental conditions. Then we will organize these factors into a short table.

Effect of the Base   The nature of the base is the single most important factor in determining whether an elimination will go by the E1 or E2 mechanism. If a strong base is present, the rate of the bimolecular reaction will be greater than the rate of ionization, and the E2 reaction will predominate (perhaps accompanied by the S_N2).

If no strong base is present, with a good solvent a unimolecular ionization is likely, followed by loss of a proton to a weak base such as the solvent. Under these conditions, the E1 reaction usually predominates (always accompanied by the S_N1).

E1:   Base strength is unimportant. (usually weak)
E2:   Strong bases are required.

Effect of the Solvent   The slow step of the E1 reaction is the formation of two ions. Like the S_N1, the E1 reaction critically depends on polar ionizing solvents such as water and the alcohols.

In the E2 reaction, the transition state spreads out the negative charge of the base over the entire molecule. There is no more need for solvation in the E2 transition state than in the reactants. The E2 is therefore less sensitive to the solvent; in fact, some reagents are stronger bases in less polar solvents.

E1:   Good ionizing solvent required.
E2:   Solvent polarity is not so important.

6-21
Comparison of E1 and E2 Elimination Mechanisms

Effect of the Substrate   For both the E1 and the E2 reactions, the order of reactivity is

$$\text{E1, E2:}\quad 3° \;>\; 2° \;>\; 1° \qquad \text{(1° usually will not go E1)}$$

In the E1 reaction, the rate-limiting step is formation of a carbocation, and the reactivity order reflects the stability of carbocations. In the E2 reaction, the more highly substituted halides generally form more highly substituted, more stable alkenes.

Kinetics   The rate of the E1 reaction is proportional to the concentration of the alkyl halide [RX] but not to the concentration of the base. It follows a first-order rate equation.

The rate of the E2 reaction is proportional to the concentrations of both the alkyl halide [RX] and the base [B:$^-$]. It follows a second-order rate equation.

$$\text{E1 rate} = k_r[\text{RX}]$$
$$\text{E2 rate} = k_r[\text{RX}][\text{B:}^-]$$

Orientation of Elimination   In most E1 and E2 eliminations with two or more possible products, the product with the most highly substituted double bond (the most stable product) predominates. This principle is called the **Saytzeff rule**, and the most highly substituted product is called the **Saytzeff product**.

$$\text{E1, E2:}\quad \text{Usually Saytzeff orientation.}$$

Stereochemistry   The E1 reaction begins with an ionization to give a flat carbocation. No particular geometry is required.

The E2 reaction takes place through a concerted mechanism that requires a coplanar arrangement of the bonds to the atoms being eliminated. The transition state is usually anti-coplanar, although it may be syn-coplanar in rigid systems.

E1:   No particular geometry required for the slow step.
E2:   Coplanar arrangement (usually anti) required for the transition state.

Rearrangements   The E1 reaction involves a carbocation intermediate. This intermediate can rearrange, usually by the shift of a hydride or an alkyl group, to give a more stable carbocation.

The E2 reaction takes place in one step with no intermediates. No rearrangement is possible in the E2 reaction.

E1:   Rearrangements are common.
E2:   No rearrangements.

SUMMARY	Elimination Reactions

	E1	*E2*
promoting factors		
base	weak bases work	strong base required
solvent	good ionizing solvent	wide variety of solvents
substrate	3° > 2°	3° > 2° > 1°
leaving group	good one required	good one required
characteristics		
kinetics	first order, $k_r[\text{RX}]$	second order, $k_r[\text{RX}][\text{B:}^-]$
orientation	most highly substituted alkene	most highly substituted alkene
stereochemistry	no special geometry	coplanar transition state required
rearrangements	common	impossible

PROBLEM-SOLVING STRATEGY
Predicting Substitutions and Eliminations

Given a set of reagents and solvents, how can we predict what products will result and which mechanisms will be involved? Should you memorize all this theory about substitutions and eliminations? Students sometimes feel overwhelmed at this point.

Memorizing is not the best way to approach this material because it is too much and there are too many factors. Besides, the real world with its real reagents and solvents is not as clean as our equations on paper. Most nucleophiles are also basic, and most bases are also nucleophilic. Most solvents can solvate ions or react as nucleophiles, or both. We will review the most important factors that determine the reaction pathway, organized in a sequence that allows you to predict as much as can be predicted.

The first principle you must understand is that *we cannot always predict one unique product or one unique mechanism*. Often, the best we can do is to eliminate some of the possibilities and make some good predictions. Remembering this limitation, here are some general guidelines:

1. **The strength of the base or nucleophile determines the order of the reaction.**
   If a strong nucleophile (or base) is present, it will force second-order kinetics: either S_N2 or E2. A strong nucleophile attacks the electrophilic carbon atom or abstracts a proton faster than the molecule can ionize for first-order reactions.

   If no strong base or nucleophile is present, the fastest reaction will probably be a first-order reaction, either S_N1 or E1. Addition of silver salts to the reaction can force some kind of ionization.

   This is the most important of these guidelines. Consider the following examples:

$$(CH_3)_3C-Br \xrightarrow[CH_3OH]{NaOCH_3} E2 \quad (no \ S_N2 \ on \ 3° \ carbon)$$

2. **Primary halides usually undergo the S_N2 reaction, occasionally the E2 reaction.**
   Primary halides rarely undergo first-order reactions, since primary carbocations are rarely formed. With good nucleophiles, S_N2 substitution is usually observed. With a strong base, E2 elimination may also be observed.

   Sometimes silver salts or high temperatures are used to force a primary halide to ionize, usually with rearrangement to give a more stable carbocation. In such a case, the rearranged S_N1 and E1 products may be observed.

3. **Tertiary halides usually undergo the E2 reaction (strong base) or a mixture of S_N1 and E1 (weak base).**
   Tertiary halides cannot undergo the S_N2 reaction. A strong base forces second-order kinetics, resulting in elimination by the E2 mechanism. In the absence of a strong base, tertiary halides react by first-order processes, usually a mixture of S_N1 and E1. The specific reaction conditions determine the ratio of substitution to elimination.

$$(CH_3)_3C-Br \xrightarrow[CH_3OH]{NaOCH_3} E2 \quad (no \ S_N2 \ on \ 3° \ carbon)$$

$$(CH_3)_3C-Br \xrightarrow[heat]{CH_3OH} S_N1 \ and \ E1$$

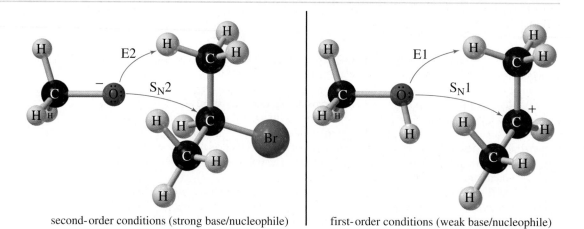

second-order conditions (strong base/nucleophile) | first-order conditions (weak base/nucleophile)

▲ **Figure 6-14**
Under second-order conditions (strong base/nucleophile), a secondary alkyl halide might undergo either substitution (S_N2) or elimination (E2). Under first-order conditions (weak base/nucleophile), S_N1 and E1 are possible.

4. **The reactions of secondary halides are the most difficult to predict.**
With a strong base, either the S_N2 or the E2 reaction is possible. With a weak base and a good ionizing solvent, either the S_N1 or the E1 reaction is possible. Mixtures of products are common. Figure 6-14 shows these possibilities with a secondary halide under second-order and first-order conditions.

$$\xrightarrow[\text{CH}_3\text{OH}]{\text{NaOCH}_3} \quad S_N2 \text{ and } E2$$

$$\xrightarrow[\text{heat}]{\text{CH}_3\text{OH}} \quad S_N1 \text{ and } E1$$

5. **Some bases favor substitution or elimination.**
To promote elimination, the base should readily abstract a proton, but not readily attack a carbon atom. A bulky strong base, such as *t*-butoxide [$^-\text{OC(CH}_3)_3$], enhances elimination. To promote substitution, we need a good nucleophile with limited basicity: a highly polarizable species that is the conjugate base of a strong acid. Bromide (Br^-) and iodide (I^-) are examples of good nucleophiles that are weak bases and favor substitution.

$$\xrightarrow{\text{NaI}} \quad \text{mostly } S_N2$$

$$\xrightarrow[\text{(CH}_3)_3\text{COH}]{\text{NaOC(CH}_3)_3} \quad \text{mostly } E2$$

PROBLEM 6-39
Give the structures of the products expected from the indicated mechanisms in the examples shown above.

SOLVED PROBLEM 6-3

Predict the mechanisms and products of the following reactions.

(a)

1-bromo-1-methylcyclohexane

(b) CH_3—$\overset{\overset{\textstyle Br}{|}}{CH}$—$CH_2CH_2CH_2CH_3$ $\xrightarrow[\text{CH}_3\text{OH}]{\text{NaOCH}_3}$

2-bromohexane

SOLUTION

(a) There is no strong base or nucleophile present, so this reaction must be first order, with an ionization of the alkyl halide as the slow step. Deprotonation of the carbocation gives either of two elimination products, and nucleophilic attack gives a substitution product.

(b) This reaction takes place with a strong base, so it is second order. This secondary halide can undergo both S_N2 substitution and E2 elimination. Both products will be formed, with the relative proportions of substitution and elimination depending on the reaction conditions.

$$CH_3—CH{=}CH—CH_2CH_2CH_3 \quad CH_2{=}CH—CH_2CH_2CH_2CH_3$$

major   minor

E2 products

$$CH_3—\overset{\overset{\textstyle OCH_3}{|}}{CH}—CH_2CH_2CH_2CH_3$$

S_N2 product

PROBLEM 6-40

Predict the products and mechanisms of the following reactions. When more than one product or mechanism is possible, explain which are most likely.
(a) ethyl bromide + sodium ethoxide
(b) *t*-butyl bromide + sodium ethoxide
(c) isopropyl bromide + sodium ethoxide
(d) isobutyl bromide + potassium hydroxide in ethanol/water
(e) isobutyl bromide + silver nitrate in ethanol/water
(f) 1-bromo-1-methylcyclopentane heated in methanol
(g) (bromomethyl)cyclopentane + silver nitrate in methanol

PROBLEM-SOLVING HINT

The strength of the base/nucleophile usually determines the order of the reaction.

SUMMARY   Reactions of Alkyl Halides

Some of these reactions have not yet been covered, but they are included here for completeness and for later reference. Notice the section numbers, indicating where each reaction is covered.

1. *Nucleophilic substitutions* (Section 6-9)
   a. *Alcohol formation*

   $$R—X + {}^-{:}\overset{..}{O}H \longrightarrow R—OH + {:}X^-$$

Example

$$CH_3CH_2\text{—}Br + NaOH \longrightarrow CH_3CH_2\text{—}OH + NaBr$$
ethyl bromide   ethyl alcohol

b. *Halide exchange*

$$R\text{—}X + :\ddot{I}:^- \longrightarrow R\text{—}I + :X^-$$

$$R\text{—}Cl + KF \xrightarrow[CH_3CN]{\text{18-crown-6}} R\text{—}F + KCl$$

Example

$$H_2C{=}CH\text{—}CH_2Cl + NaI \longrightarrow H_2C{=}CH\text{—}CH_2I + NaCl$$
allyl chloride   allyl iodide

c. *Williamson ether synthesis*

$$R\text{—}X + R'\ddot{O}:^- \longrightarrow R\text{—}\ddot{O}\text{—}R' + :X^-$$ ether synthesis

$$R\text{—}X + R'\ddot{S}:^- \longrightarrow R\text{—}\ddot{S}\text{—}R' + :X^-$$ thioether synthesis

Example

$$CH_3\text{—}I + CH_3CH_2\text{—}O^- Na^+ \longrightarrow CH_3\text{—}O\text{—}CH_2CH_3 + Na^+ I^-$$
methyl iodide   sodium ethoxide   methyl ethyl ether

d. *Amine synthesis*

$$R\text{—}X + \text{excess} :NH_3 \longrightarrow R\text{—}NH_3^+ X^- \xrightarrow{:NH_3} R\text{—}\ddot{N}H_2 + NH_4^+ :X^-$$
amine

Example

$$CH_3CH_2CH_2\text{—}Br + :NH_3 \longrightarrow CH_3CH_2CH_2\text{—}NH_2 + NH_4^+ {}^-Br$$
n-propyl bromide   *n*-propylamine

e. *Nitrile synthesis*

$$R\text{—}X + {}^-:C{\equiv}N: \longrightarrow R\text{—}C{\equiv}N: + :X^-$$
cyanide   nitrile

Example

$$(CH_3)_2CHCH_2CH_2\text{—}Cl + NaCN \longrightarrow (CH_3)_2CHCH_2CH_2\text{—}CN + NaCl$$
1-chloro-3-methylbutane   4-methylpentanenitrile

f. *Alkyne synthesis*

$$R\text{—}C{\equiv}C:^- + R'\text{—}X \longrightarrow R\text{—}C{\equiv}C\text{—}R' + :X^-$$
acetylide ion   alkyne

Example

$$CH_3\text{—}C{\equiv}C\text{—}H + NaNH_2 \longrightarrow CH_3\text{—}C{\equiv}C:^- Na^+ + NH_3$$
propyne   sodium amide   sodium propynide

$$CH_3\text{—}C{\equiv}C:^- Na^+ + CH_3CH_2\text{—}I \longrightarrow CH_3\text{—}C{\equiv}C\text{—}CH_2CH_3 + NaI$$
propynide ion   ethyl iodide   2-pentyne

2. *Eliminations*

  a. *Dehydrohalogenation* (Sections 6-18 and 7-9A)

$$-\overset{\displaystyle H}{\underset{\displaystyle |}{C}}-\overset{\displaystyle |}{\underset{\displaystyle X}{C}}- \quad\xrightarrow{\text{KOH}}\quad \overset{\diagup}{}C=C\overset{\diagdown}{} \quad + \quad :X^-$$

Example

2-bromohexane $\xrightarrow[\text{CH}_3\text{OH}]{\text{NaOCH}_3}$ 2-hexene + 1-hexene

  b. *Dehalogenation* (Section 7-9D)

$$-\overset{\displaystyle Br}{\underset{\displaystyle |}{C}}-\overset{\displaystyle |}{\underset{\displaystyle Br}{C}}- \quad\xrightarrow{\text{KI}}\quad \overset{\diagup}{}C=C\overset{\diagdown}{} \quad + \quad I-Br \quad + \quad KBr$$

Example

trans-1,2-dibromocyclohexane $\xrightarrow{\text{KI}}$ cyclohexene

3. *Formation of organometallic reagents* (Section 10-8)

  a. *Grignard reagents*

$$R-X \ + \ Mg \quad\xrightarrow{\text{CH}_3\text{CH}_2-\text{O}-\text{CH}_2\text{CH}_3}\quad R-Mg-X$$

(X = Cl, Br, or I)　　　　　　　　　　　　　　organomagnesium halide
　　　　　　　　　　　　　　　　　　　　　　　(Grignard reagent)

Example

bromocyclohexane + Mg $\xrightarrow{\text{ether}}$ cyclohexylmagnesium bromide

  b. *Organolithium reagents*

$$R-X \ + \ 2\,Li \quad\longrightarrow\quad R-Li \ + \ Li^+\,X^-$$

(X = Cl, Br, or I)　　　　　organolithium

Example

$$CH_3CH_2CH_2CH_2-Br \ + \ 2\,Li \quad\xrightarrow{\text{hexane}}\quad CH_3CH_2CH_2CH_2-Li \ + \ LiBr$$
n-butyl bromide　　　　　　　　　　　　　　　*n*-butyllithium

4. *Coupling of organocopper reagents* (Section 10-9)

$$2\,R-Li \ + \ CuI \quad\longrightarrow\quad R_2CuLi \ + \ LiI$$

$$R_2CuLi \ + \ R'-X \quad\longrightarrow\quad R-R' \ + \ R-Cu \ + \ LiX$$

Example

$$2 \text{ CH}_3\text{I} \xrightarrow{4 \text{ Li}} 2 \text{ CH}_3\text{Li} + 2 \text{ LiI} \xrightarrow{\text{CuI}} (\text{CH}_3)_2\text{CuLi}$$

5. *Reduction (Section 10-10)*

$$\text{R}{-}\text{X} \xrightarrow[\text{(2) H}_2\text{O}]{\text{(1) Mg or Li}} \text{R}{-}\text{H}$$

Example

$$\text{C}_9\text{H}_{19}{-}\text{CH}_2{-}\text{Br} \xrightarrow[\text{(2) H}_2\text{O}]{\text{(1) Mg, ether}} \text{C}_9\text{H}_{19}{-}\text{CH}_3$$
$$\text{\textit{n}-decyl bromide} \qquad\qquad\qquad \text{\textit{n}-decane}$$

Chapter 6 Glossary

acid A species that can donate a proton.

 acidity (acid strength): The thermodynamic reactivity of an acid, expressed quantitatively by the acid-dissociation constant K_a.

alkyl halide (haloalkane) A derivative of an alkane in which one (or more) of the hydrogen atoms has been replaced by a halogen. (p. 212)

allylic The saturated position adjacent to a carbon–carbon double bond. (p. 221)

allylic halogenation Substitution of a halogen for a hydrogen at the allylic position. (p. 221)

anti Adding to (or eliminating from) opposite faces of a molecule. (p. 256)

 anti-coplanar: Having a dihedral angle of 180°.

 syn-coplanar: Having a dihedral angle of 0°.

anti-coplanar   syn-coplanar

aprotic solvent A solvent that has no acidic protons; a solvent with no O—H or N—H groups. (p. 232)

aryl halide An aromatic compound (benzene derivative) in which a halogen is bonded to one of the carbon atoms of the aromatic ring. (p. 212)

base An electron-rich species that can abstract a proton. (p. 229)

 basicity (base strength): The thermodynamic reactivity of a base, expressed quantitatively by the base-dissociation constant K_b.

concerted reaction A reaction in which the breaking of bonds and the formation of new bonds occur at the same time (in one step). (pp. 226, 255)

dehydrohalogenation An elimination in which the two atoms lost are a hydrogen atom and a halogen atom. (pp. 224, 249)

electrophile (Lewis acid) A species that can accept an electron pair from a nucleophile, forming a bond. (p. 225)

electrophilicity (electrophile strength) The kinetic reactivity of an electrophile.

elimination A reaction that involves the loss of two atoms or groups from the substrate, usually resulting in the formation of a pi bond. (p. 224)

   E2 reaction (elimination, bimolecular): A concerted elimination involving a transition state where the base is abstracting a proton at the same time that the leaving group is leaving. The anti-coplanar transition state is generally preferred. (p. 252)

   E1 reaction (elimination, unimolecular): A multistep elimination where the leaving group is lost in a slow ionization step, then a proton is lost in a second step. Saytzeff orientation is generally preferred. (p. 248)

freons A generic name for a group of chlorofluorocarbons used as refrigerants, propellants, and solvents. Freon-12® is CF_2Cl_2, and Freon-22® is $CHClF_2$. (p. 216)

geminal dihalide A dihalide with both halogens on the same carbon atom. (p. 214)

$$CH_3-CH_2-CBr_2-CH_3$$
a geminal dibromide

halogen exchange reaction A substitution where one halogen atom replaces another; commonly used to form fluorides and iodides. (p. 228)

hydride shift (symbolized ~H) Movement of a hydrogen atom with a pair of electrons from one atom (usually carbon) to another. Hydride shifts are examples of rearrangements that convert carbocations into more stable carbocations. (p. 244)

hydroxylic solvent A solvent containing OH groups (the most common type of protic solvents). (p. 232)

inversion of configuration (see also **Walden inversion**) A process in which the groups around a chiral carbon atom are changed to the opposite spatial configuration, usually as a result of **back-side attack**. (p. 236)

The S_N2 reaction goes with *inversion of configuration*.

leaving group The atom or group of atoms that departs during a substitution or elimination. The leaving group can be charged or uncharged, but it leaves with the pair of electrons that originally bonded the group to the remainder of the molecule. (p. 225)

methyl shift (symbolized ~CH_3) Rearrangement of a methyl group with a pair of electrons from one atom (usually carbon) to another. A methyl shift (or any alkyl shift) in a carbocation generally results in a more stable carbocation. (p. 244)

nucleophile (Lewis base) An electron-rich species that can donate a pair of electrons to form a bond. (p. 229)

   nucleophilicity (nucleophile strength): The kinetic reactivity of a nucleophile; a measure of the rate of substitution in a reaction with a standard substrate.

nucleophilic substitution A reaction where a nucleophile replaces another group or atom (the leaving group) in a molecule. (p. 225)

organic synthesis The preparation of desired organic compounds from readily available starting materials.

polarizable Having electrons that are easily displaced toward a positive charge. Polarizable atoms can begin to form a bond at a relatively long distance. (p. 230)

primary halide, secondary halide, tertiary halide These terms specify the substitution of the halogen-bearing carbon atom (sometimes called the **head carbon**). If the head carbon is bonded to one other carbon, it is **primary**. If it is bonded to two carbons, it is **secondary**, and if bonded to three carbons, it is **tertiary**. (p. 214)

$$CH_3-CH_2-Br \qquad CH_3-\underset{\underset{CH_3}{|}}{CH}-Br \qquad CH_3-\underset{\underset{CH_3}{|}}{\overset{\overset{CH_3}{|}}{C}}-Br$$

a primary halide (1°)   a secondary halide (2°)   a tertiary halide (3°)

protic solvent A solvent containing acidic protons, usually $O-H$ or $N-H$ groups. (p. 232)

racemization The loss of optical activity that occurs when a reaction shows neither clean retention of configuration nor clean inversion of configuration. (p. 242)

reagent The compound that serves as the attacking species in a reaction.

rearrangement A reaction involving a change in the bonding sequence within a molecule. Rearrangements are common in reactions such as the S_N1 and E1 involving carbocation intermediates. (p. 243)

retention of configuration Formation of a product with the same configuration as the reactant. In a nucleophilic substitution, retention of configuration occurs when the nucleophile assumes the same stereochemical position in the product as the leaving group occupied in the reactant. (p. 242)

Saytzeff rule (Zaitsev rule) An elimination usually gives the most highly substituted alkene product. The Saytzeff rule does not always apply, but when it does, the reaction is said to give **Saytzeff orientation**. (p. 253)

solvolysis A nucleophilic substitution where the solvent serves as the attacking reagent. Solvolysis literally means "cleavage by the solvent." (p. 238)

$$(CH_3)_3C-Br \xrightarrow{\text{CH}_3\text{OH, heat}} (CH_3)_3C-OCH_3 + (CH_3)_2C=CH_2 + HBr$$

stereocenter An atom that gives rise to stereoisomers when its groups are interchanged. Asymmetric carbon atoms are the most common stereocenters.

stereospecific reaction A reaction in which different stereoisomers react to give different stereoisomers of the product. (p. 237)

steric hindrance (steric strain) Interference by bulky groups when they approach a position where their electron clouds begin to repel each other. (p. 235)

substitution (displacement) A reaction in which an attacking species (nucleophile, electrophile, or free radical) replaces another group. (p. 224)

> **S_N2 reaction** (substitution, nucleophilic, bimolecular): The concerted displacement of one nucleophile by another on an sp^3 hybrid carbon atom. (p. 226)
>
> **S_N1 reaction** (substitution, nucleophilic, unimolecular): A two-step interchange of nucleophiles, with bond breaking preceding bond formation. The first step is ionization to form a carbocation. The second step is the reaction of the carbocation with a nucleophile. (p. 238)

substrate The compound that is attacked by the reagent. (p. 233)

syn Adding to (or eliminating from) the same face of a molecule. (p. 256)

> **syn-coplanar** Having a dihedral angle of 0°. See **anti-coplanar** for a diagram.

vicinal dihalide A dihalide with the halogens on adjacent carbon atoms. (p. 214)

$$CH_3-CHBr-CHBr-CH_3 \qquad CH_3-CH=\underset{\underset{CH_3}{|}}{\overset{\overset{Br}{|}}{C}}-CH_3$$

a vicinal dibromide   a vinyl bromide

vinyl halide A derivative of an alkene in which one (or more) of the hydrogen atoms on the double-bonded carbon atoms has been replaced by a halogen. (p. 212)

Walden inversion (see also **inversion of configuration**): A step in a reaction sequence in which an asymmetric carbon atom undergoes inversion of configuration. (p. 236)

Essential Problem-Solving Skills in Chapter 6

1. Correctly name alkyl halides, and identify them as 1°, 2°, or 3°.
2. Predict the products of S_N1, S_N2, E1, and E2 reactions, including stereochemistry.
3. Draw the mechanisms and energy profiles of S_N1, S_N2, E1, and E2 reactions.
4. Predict and explain the rearrangement of cations in first-order reactions.
5. Predict which substitutions or eliminations will be faster, based on differences in substrate, base/nucleophile, leaving group, or solvent.
6. Predict whether a reaction will be first order or second order.
7. When possible, predict predominance of substitution or elimination.
8. Use the Saytzeff rule to predict major and minor elimination products.

PROBLEM-SOLVING $\mathbb{HINT}$

Don't try to memorize your way through this chapter. Try to understand what is happening in the different reactions. Some memorizing is necessary, but simply memorizing everything won't allow you to predict new reactions.

Study Problems

6-41   Define and give an example for each of the following terms.
(a) nucleophile	(b) electrophile	(c) leaving group
(d) substitution	(e) S_N2 reaction	(f) S_N1 reaction
(g) solvolysis	(h) elimination	(i) E2 reaction
(j) E1 reaction	(k) rearrangement	(l) base
(m) steric hindrance	(n) alkyl halide	(o) aryl halide
(p) vinyl halide	(q) allylic halide	(r) primary halide
(s) secondary halide	(t) tertiary halide	(u) anti elimination
(v) syn elimination	(w) stereospecific reaction	

6-42   Give the structures of the following compounds.
(a) isopropyl chloride	(b) isobutyl bromide	(c) 1,2-dibromo-3-methylpentane
(d) 2,2,2-trichloroethanol	(e) *trans*-1,4-diiodocyclohexane	(f) methylene chloride
(g) chloroform		

6-43   Give systematic (IUPAC) names for the following compounds.

6-44   Predict the compound in each pair that will undergo the S_N2 reaction faster.

6-45   Predict the compound in each pair that will undergo solvolysis (in aqueous ethanol) more rapidly.

(a) $(CH_3CH_2)_2CH—Cl$ or $(CH_3)_3C—Cl$

(b) ![structure] Cl or ![structure] Cl

(c) ![cyclohexane-Br] or ![cyclohexane-CH₂Br]

(d) ![cyclohexane-Cl] or ![cyclohexane-I]

(e) ![structure] or ![structure Br]

(f) ![cyclohexane Br] or ![cyclohexene Br]

6-46   Show how each compound might be synthesized by the S_N2 displacement of an alkyl halide.

(a) ![cyclohexyl]—CH_2OH

(b) ![cyclohexyl]—SCH_2CH_3

(c) ![tetrahydropyran]

(d) ![cyclohexyl]—CH_2NH_2

(e) $H_2C=CH—CH_2CN$ (f) $(CH_3)_3C—OCH_3$ (g) $H—C≡C—CH_2CH_2CH_3$

6-47   (a) Give two syntheses for $(CH_3)_2CH—O—CH_2CH_3$, and explain which synthesis is better.

(b) A student wanted to synthesize methyl *t*-butyl ether, $CH_3—O—C(CH_3)_3$. He attempted the synthesis by adding sodium methoxide (CH_3ONa) to *t*-butyl chloride, but he obtained none of the desired product. Show what product is formed in this reaction, and give a better synthesis for methyl *t*-butyl ether.

6-48   When ethyl bromide is added to potassium *t*-butoxide, the product is ethyl *t*-butyl ether.

$$CH_3CH_2—Br \;+\; (CH_3)_3C—O^- \;{}^+K \longrightarrow (CH_3)_3C—O—CH_2CH_3$$

ethyl bromide   potassium *t*-butoxide   ethyl *t*-butyl ether

(a) What happens to the reaction rate if the concentration of ethyl bromide is doubled?

(b) What happens to the rate if the concentration of potassium *t*-butoxide is tripled and the concentration of ethyl bromide is doubled?

(c) What happens to the rate if the temperature is raised?

6-49   When *t*-butyl bromide is heated with ethanol, one of the products is ethyl *t*-butyl ether.

(a) What happens to the reaction rate if the concentration of ethanol is doubled?

(b) What happens to the rate if the concentration of *t*-butyl bromide is tripled and the concentration of ethanol is doubled?

(c) What happens to the rate if the temperature is raised?

6-50   Chlorocyclohexane reacts with sodium cyanide (NaCN) in ethanol to give cyanocyclohexane. The rate of formation of cyanocyclohexane increases when a small amount of sodium iodide is added to the solution. Explain this acceleration in the rate.

6-51   Give the solvolysis products expected when each compound is heated in ethanol.

(a) ![structure Br]   (b) ![structure Br]   (c) ![cyclohexane Cl]   (d) ![structure Br]

6-52   Allylic halides have the structure

$$\underset{/}{\overset{\backslash}{}}C=C-\overset{|}{\underset{|}{C}}-X$$

(a) Show how the first-order ionization of an allylic halide leads to a resonance-stabilized cation.

(b) Draw the resonance structures of the allylic cations formed by ionization of the following allylic halides.

(c) Show the products expected from S_N1 solvolysis of these halides in ethanol.

(i) ![cyclohexene CH₂Br]   (ii) ![cyclohexene Br]   (iii) ![structure Br]

6-53   List the following carbocations in decreasing order of their stability.

6-54   Two of the carbocations in Problem 6-53 are prone to rearrangement. Show how they might rearrange to more stable carbocations.

6-55   Draw perspective structures or Fischer projections for the substitution products of the following reactions.

(a)

$$
\begin{array}{c}
CH_3 \\
| \\
H\!-\!\!\!-\!Br \\
| \\
CH_2CH_3
\end{array}
\quad + \; NaOH \; \xrightarrow{\;water\;}
$$

(b)

$$
\begin{array}{c}
CH_3 \\
| \\
Br\!-\!\!\!-\!H \\
| \\
H\!-\!\!\!-\!CH_3 \\
| \\
CH_2CH_3
\end{array}
\quad + \; NaI \; \xrightarrow{\;acetone\;}
$$

(c)

$$
\begin{array}{c}
CH_2CH_3 \\
| \\
Br\!-\!\!\!-\!CH_3 \\
| \\
CH(CH_3)_2
\end{array}
\quad \xrightarrow{\;EtOH,\; heat\;}
$$

6-56   Predict the products of the following S_N2 reactions.

(a) $CH_3CH_2ONa + CH_3CH_2Cl \xrightarrow{CH_3CH_2OH}$

(b) [phenyl]$-CH_2CH_2Br \; + \; NaCN \; \xrightarrow{acetone}$

(c) [cyclohexyl]$-Cl \; + \; NaSCH_3 \; \longrightarrow$

(d) $CH_3(CH_2)_8CH_2Cl + NaI \xrightarrow{acetone}$

(e) [pyridine]$N{:} \; + \; CH_3I \; \longrightarrow$

(f) $(CH_3)_3C\!-\!CH_2CH_2Br + excess\; NH_3 \longrightarrow$

(g) [ring with Cl and OH] $+ \; NaOH \; \longrightarrow$

(h) $Br\!\blacktriangleright$[cyclohexane]$\cdots CH_3 \; \xrightarrow[CH_3OH]{NaOH}$

***6-57**   When $(\pm)$-2,3-dibromobutane reacts with potassium hydroxide, some of the products are (2S,3R)-3-bromo-2-butanol and its enantiomer and *trans*-2-bromo-2-butene. Give mechanisms to account for these products.

$$
\begin{array}{c}
CH_3 \\
| \\
H\!-\!\!\!-\!OH \\
| \\
H\!-\!\!\!-\!Br \\
| \\
CH_3 \\
(2S,3R)
\end{array}
\qquad
\begin{array}{c}
CH_3 \\
| \\
HO\!-\!\!\!-\!H \\
| \\
Br\!-\!\!\!-\!H \\
| \\
CH_3 \\
(2R,3S)
\end{array}
\qquad
\begin{array}{c}
H_3C \qquad\quad Br \\
C\!=\!C \\
H \qquad\quad CH_3
\end{array}
$$

3-bromo-2-butanol   *trans*-2-bromo-2-butene

6-58   A solution of pure (S)-2-iodobutane ($[\alpha] = +15.90°$) in acetone is allowed to react with radioactive iodide, $^{131}I^-$, until 1.0% of the iodobutane contains radioactive iodine. The specific rotation of this recovered iodobutane is found to be +15.58°.
(a) Determine the percentages of (R)- and (S)-2-iodobutane in the product mixture.
(b) What does this result suggest about the mechanism of the reaction of 2-iodobutane with iodide ion?

6-59   (a) Optically active 2-bromobutane undergoes racemization on treatment with a solution of KBr. Give a mechanism for this racemization.
(b) In contrast, optically active 2-butanol does not racemize on treatment with a solution of KOH. Explain why a reaction like that in part (a) does not occur.
(c) Optically active 2-butanol racemizes in dilute acid. Propose a mechanism for this racemization.

6-60   Predict the products of E1 elimination of the following compounds. Label the major products.

(a) $(CH_3)_3C\!-\!\underset{\underset{Br}{|}}{CH}\!-\!CH_3$

(b) [cyclohexane with CH_3, Br, CH_3 substituents]

(c) [cyclohexane with Br, CH_3 substituents]

6-61   When 1-bromomethylcyclohexene undergoes solvolysis in ethanol, three major products are formed. Give mechanisms to account for these three products.

[cyclohexene]$-CH_2Br$ $\xrightarrow[heat]{ethanol}$ [methylenecyclohexene] $+$ [methylenecyclohexane with OC_2H_5] $+$ [cyclohexene]$-CH_2OC_2H_5$

1-bromomethylcyclohexene

*6-62   Protonation converts the hydroxyl group of an alcohol to a good leaving group. Suggest a mechanism for each reaction.

(a) [cyclohexanol] $\xrightarrow[\text{(E1)}]{\text{H}_2\text{SO}_4,\text{ heat}}$ [cyclohexene] + H_2O (b) [cyclohexanol] $\xrightarrow[\text{(S}_\text{N}2\text{ or S}_\text{N}1\text{)}]{\text{HBr, heat}}$ [bromocyclohexane] + H_2O

6-63   Give a mechanism to explain the two products formed in the following reaction.

[structure: 3-methyl-1-butene] $\xrightarrow{\text{NBS, } hv}$ [structure: not rearranged] + [structure: rearranged]

3-methyl-1-butene   not rearranged   rearranged

6-64   Predict the major product of the following reaction, and give a mechanism to support your prediction.

[structure]—CH_2CH_3 $\xrightarrow{\text{NBS, } hv}$

ethylbenzene

6-65   Because the $\text{S}_\text{N}1$ reaction goes through a flat carbocation, we might expect an optically active starting material to give a completely racemized product. In most cases, however, $\text{S}_\text{N}1$ reactions actually give more of the *inversion* product. In general, as the stability of the carbocation increases, the excess inversion product decreases. Extremely stable carbocations give completely racemic products. Explain these observations.

6-66   When 1-bromo-2-methylcyclohexane undergoes solvolysis in methanol, four major products are formed. Give mechanisms to account for these products.

[structure] $\xrightarrow[\text{heat}]{\text{CH}_3\text{OH}}$ [structure with OCH$_3$, CH$_3$] + [structure with OCH$_3$, CH$_3$] + [structure with CH$_3$] + [structure with CH$_2$]

*6-67   Deuterium (D) is the isotope of hydrogen of mass number 2, with a proton and a neutron in its nucleus. The chemistry of deuterium is nearly identical to the chemistry of hydrogen, except that the C—D bond is slightly (1.2 kcal/mol or 5.0 kJ/mol) stronger than the C—H bond. Reaction rates tend to be slower if a C—D bond (as opposed to a C—H bond) is broken in a rate-limiting step. This effect on the rate is called a *kinetic isotope effect*.
(a)   Propose a mechanism to explain each product in the following reaction.

$$\text{CH}_3\overset{\overset{\displaystyle\text{Br}}{|}}{-}\text{CH}-\text{CH}_3 \xrightarrow{\text{KOH, alcohol}} \text{CH}_2{=}\text{CH}-\text{CH}_3 \ + \ \text{CH}_3\overset{\overset{\displaystyle\text{OH}}{|}}{-}\text{CH}-\text{CH}_3$$

                                elimination product   substitution product

(b)   When the following deuterated compound reacts under the same conditions, the rate of formation of the substitution product is unchanged, while the rate of formation of the elimination product is slowed by a factor of 7.

$$\text{CD}_3\overset{\overset{\displaystyle\text{Br}}{|}}{-}\text{CH}-\text{CD}_3 \xrightarrow{\text{KOH, alcohol}} \text{CD}_2{=}\text{CH}-\text{CD}_3 \ + \ \text{CD}_3\overset{\overset{\displaystyle\text{OH}}{|}}{-}\text{CH}-\text{CD}_3$$

                                seven times slower   rate unchanged

Explain why the elimination rate is slowed but the substitution rate is unchanged.
(c)   A similar reaction takes place on heating the alkyl halide in an acetone/water mixture.

$$\text{CH}_3\overset{\overset{\displaystyle\text{Br}}{|}}{-}\text{CH}-\text{CH}_3 \xrightarrow[\text{heat}]{\text{water/acetone}} \text{CH}_2{=}\text{CH}-\text{CH}_3 \ + \ \text{CH}_3\overset{\overset{\displaystyle\text{OH}}{|}}{-}\text{CH}-\text{CH}_3$$

                                elimination product   substitution product

Give a mechanism for the formation of each product under these conditions, and predict how the rate of formation of each product will change when the deuterated halide reacts. Explain your prediction.

$$\text{CD}_3\overset{\overset{\displaystyle\text{Br}}{|}}{-}\text{CH}-\text{CD}_3 \xrightarrow[\text{heat}]{\text{water/acetone}} \text{CD}_2{=}\text{CH}-\text{CD}_3 \ + \ \text{CD}_3\overset{\overset{\displaystyle\text{OH}}{|}}{-}\text{CH}-\text{CD}_3$$

                                rate changed?   rate changed?

*6-68   When the following compound is treated with sodium methoxide in methanol, two elimination products are possible. Explain why the deuterated product predominates by about a 7 : 1 ratio (refer to Problem 6-67).

87%   13%

*6-69   The reaction of an amine with an alkyl halide gives an ammonium salt.

$$R_3N: + R'{-}X \longrightarrow R_3\overset{+}{N}{-}R'\ X^-$$

amine   alkyl halide   ammonium salt

The rate of this S_N2 reaction is sensitive to the polarity of the solvent. Draw an energy diagram for this reaction in a nonpolar solvent and another in a polar solvent. Consider the nature of the transition state, and explain why this reaction should be sensitive to the polarity of the solvent. Predict whether it will be faster or slower in a more polar solvent.

*6-70   The following reaction takes place under second-order conditions (strong nucleophile), yet the structure of the product shows rearrangement. Also, the rate of this reaction is several thousand times faster than the rate of substitution of hydroxide ion on 2-chlorobutane under similar conditions. Propose a mechanism to explain the enhanced rate and rearrangement observed in this unusual reaction. ("Et" is the abbreviation for ethyl.)

*6-71   **(a)** Design an alkyl halide that will give *only* 2,4-diphenyl-2-pentene upon treatment with potassium *t*-butoxide (a bulky base that promotes E2 elimination).

   (b) What stereochemistry is required in your alkyl halide so that *only* the following stereoisomer of the product is formed?

*6-72   When 2-bromo-3-phenylbutane is treated with sodium methoxide, two alkenes result (by E2 elimination). The Saytzeff product predominates.

   (a) Draw the reaction, showing the major and minor products.

   (b) When one pure stereoisomer of 2-bromo-3-phenylbutane reacts, one pure stereoisomer of the major product results. For example, when (2R,3R)-2-bromo-3-phenylbutane reacts, the product is the stereoisomer with the methyl groups cis. Use your models to draw a Newman projection of the transition state to show why this stereospecificity is observed.

   (c) Use a Newman projection of the transition state to predict the major product of elimination of (2S,3R)-2-bromo-3-phenylbutane.

   (d) Predict the major product from elimination of (2S,3S)-2-bromo-3-phenylbutane. This prediction can be made without drawing any structures, by considering the results in part (b).

*6-73   Solvolysis of bromomethylcyclopentane in methanol gives a complex product mixture of the following five compounds. Propose mechanisms to account for these products.

CHAPTER 7

Structure and Synthesis of Alkenes

7-1
Introduction

Alkenes are hydrocarbons with carbon–carbon double bonds. Alkenes are sometimes called **olefins**, a term derived from *olefiant gas*, meaning "oil-forming gas." This term originated with early experimentalists who noticed the oily appearance of alkene derivatives. Alkenes are among the most important industrial compounds (see Section 7-6), and many alkenes are also found in plants and animals. *Ethylene* is the largest-volume industrial organic compound, used to make polyethylene and a variety of other industrial and consumer chemicals. *Pinene* is a major component of *turpentine*, the paint solvent distilled from extracts of evergreen trees. *Muscalure* (*cis*-9-tricosene) is the sex attractant of the common housefly.

ethylene (ethene)   α-pinene   *cis*-9-tricosene, "muscalure"

The bond energy of a carbon–carbon double bond is about 146 kcal/mol (611 kJ/mol), compared with the single-bond energy of about 83 kcal/mol (347 kJ/mol). From these energies, we can calculate the approximate energy of a pi bond:

double-bond dissociation energy	146 kcal/mol	(611 kJ/mol)
subtract sigma bond dissociation energy	(−) 83 kcal/mol	(−)(347 kJ/mol)
pi bond dissociation energy	63 kcal/mol	(264 kJ/mol)

This value of 63 kcal/mol is much less than the sigma bond energy of 83 kcal/mol, showing that pi bonds should be more reactive than sigma bonds.

Because a carbon–carbon double bond is relatively reactive, it is considered to be a *functional group*, and its reactions are characteristic of alkenes. In previous chapters, we saw alkene synthesis by elimination reactions and we encountered a few reactions of alkenes. In this chapter, we study alkenes in more detail, concentrating on their properties and the ways they are synthesized.

7-2
The Orbital Description of the Alkene Double Bond

In a Lewis structure, the double bond of an alkene is represented by two pairs of electrons between the carbon atoms. The Pauli exclusion principle tells us that two pairs of electrons can go into one region of space between the carbon nuclei only if each pair has its own molecular orbital. Using ethylene as an example, let's consider how the electrons are distributed in the double bond.

7-2A   The Sigma Bond Framework

In Chapter 2, we saw how we can visualize the sigma bonds of organic molecules using hybrid atomic orbitals. In ethylene, each carbon atom is bonded to three other atoms (one carbon and two hydrogens), and there are no nonbonding electrons. Three hybrid orbitals are needed, implying sp^2 hybridization. We have seen (Section 2–4) that sp^2 hybridization corresponds to bond angles of about 120°, giving optimum separation of three atoms bonded to the carbon atom.

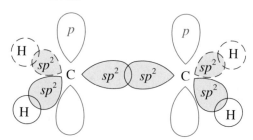

sigma bonding orbitals of ethylene

Each of the carbon–hydrogen bonds is formed by overlap of an sp^2 hybrid orbital on carbon with the $1s$ orbital of a hydrogen atom. The C—H bond length in ethylene (1.08 Å) is slightly shorter than the C—H bond in ethane (1.09 Å) because the sp^2 orbital in ethylene has more s character (one-third s) than an sp^3 orbital (one-fourth s). The s orbital is closer to the nucleus than the p orbital, contributing to shorter bonds.

ethylene   ethane

The remaining sp^2 orbitals overlap in the region between the carbon nuclei, providing a bonding orbital. The pair of electrons in this bonding orbital forms one bond between the double-bonded carbon atoms. This bond is a sigma bond because its electron density is centered along the line joining the nuclei. The C=C bond in ethylene (1.33 Å) is much shorter than the C—C bond (1.54 Å) in ethane, partly because the sigma bond of ethylene is formed from sp^2 orbitals (with more s character) and partly because there are two bonds drawing the atoms together.

7-2B   The Pi Bond

Two more electrons must go into the carbon–carbon bonding region. Each carbon atom still has an unhybridized p orbital, and these overlap to form a pi-bonding molecular orbital. The two electrons in this orbital form the second bond between the double-bonded carbon atoms. For pi overlap to occur, these p orbitals must be parallel, which requires that the two carbon atoms be oriented with all their C—H bonds in a single plane (Figure 7-1).

electrostatic potential map

◀ **Figure 7-1**
The pi bond in ethylene is formed by overlap of the unhybridized p orbitals on the sp^2 hybrid carbon atoms. This overlap requires the two ends of the molecule to be coplanar.

▶ Figure 7-2
The two isomers of 2-butene cannot interconvert by rotation about the carbon–carbon double bond without breaking the pi bond.

Half of the pi-bonding orbital is above the C—C sigma bond, and the other half is below the sigma bond. The pi-bonding electrons give rise to regions of high electron density (red) in the electrostatic potential map of ethylene shown in Figure 7-1.

Figure 7-2 shows that the two ends of the ethylene molecule cannot be twisted with respect to each other without disrupting the pi bond. Unlike single bonds, a carbon–carbon double bond does not permit rotation. This is the origin of cis-trans isomerism. If two groups are on the same side of a double bond (cis), they cannot rotate to opposite sides (trans). Figure 7-2 shows that there are two distinct isomers of 2-butene: *cis*-2-butene and *trans*-2-butene.

7-3
Elements
of Unsaturation

7-3A   Elements of Unsaturation in Hydrocarbons

Alkenes are said to be **unsaturated** because they are capable of adding hydrogen in the presence of a catalyst. The product, an alkane, is called **saturated** because it cannot react with any more hydrogen. The presence of a pi bond of an alkene (or an alkyne) or the ring of a cyclic compound decreases the number of hydrogen atoms in a molecular formula. These structural features are called **elements of unsaturation**.* Each element of unsaturation corresponds to two fewer hydrogen atoms than in the "saturated" formula.

$$CH_3—CH_2—CH_3 \qquad CH_3—CH=CH_2 \qquad \overset{\displaystyle CH_2}{\overset{\displaystyle /\backslash}{CH_2—CH_2}}$$

propane, C_3H_8 propene, C_3H_6 cyclopropane, C_3H_6
saturated   one element of unsaturation   one element of unsaturation

Consider, for example, the formula C_4H_8. A saturated alkane would have a $C_nH_{(2n+2)}$ formula, or C_4H_{10}. The formula C_4H_8 is missing two hydrogen atoms, so it has one element of unsaturation, either a pi bond or a ring. There are five constitutional isomers of formula C_4H_8:

$$CH_2=CH—CH_2CH_3 \qquad CH_3—CH=CH—CH_3$$
1-butene   2-butene

$$CH_2=\overset{\displaystyle CH_3}{\underset{}{\overset{|}{C}}}—CH_3 \qquad \overset{\displaystyle CH_2—CH_2}{\underset{\displaystyle CH_2—CH_2}{|\qquad\quad|}} \qquad \overset{\displaystyle CH_2}{\overset{\displaystyle /\backslash}{CH_2—CH—CH_3}}$$

isobutylene   cyclobutane   methylcyclopropane

When you need a structure for a particular molecular formula, it helps to find the number of elements of unsaturation. Calculate the number of hydrogen atoms from

Degree of unsaturation and *index of hydrogen deficiency* are equivalent terms.

the saturated formula, $C_nH_{(2n+2)}$, and see how many are missing. The number of elements of unsaturation is simply half the number of missing hydrogens. This simple calculation allows you to consider possible structures quickly, without always having to check for the correct molecular formula.

PROBLEM-SOLVING HINT

If you prefer to use a formula,

elements of unsaturation

$$= \frac{1}{2}(2C + 2 - H)$$

C = number of carbons

H = number of hydrogens

PROBLEM 7-1

(a) Calculate the number of elements of unsaturation implied by the molecular formula C_6H_{12}.
(b) Give five examples of structures with this formula (C_6H_{12}). At least one should contain a ring, and at least one a double bond.

PROBLEM 7-2

Determine the number of elements of unsaturation in the molecular formula C_4H_6. Give all nine possible structures having this formula. Remember that

a double bond = one element of unsaturation

a ring = one element of unsaturation

a triple bond = two elements of unsaturation

7-3B   Elements of Unsaturation with Heteroatoms

Heteroatoms (*hetero,* "different") are any atoms other than carbon and hydrogen. The rule for calculating elements of unsaturation in hydrocarbons can be extended to include heteroatoms. Let's consider how the addition of a heteroatom affects the number of hydrogen atoms in the formula.

Halogens   Halogens simply substitute for hydrogen atoms in the molecular formula. The formula C_2H_6 is saturated, so the formula $C_2H_4F_2$ is also saturated. C_4H_8 has one element of unsaturation, and $C_4H_5Br_3$ also has one element of unsaturation. In calculating the number of elements of unsaturation, simply *count halogens as hydrogen atoms.*

$$CH_3-CHF_2 \qquad CH_3-CH=CH-CBr_3 \qquad \begin{matrix} CH_2-CBr_2 \\ | \quad\quad | \\ CH_2-CHBr \end{matrix}$$

$C_2H_4F_2$ $C_4H_5Br_3$ $C_4H_5Br_3$
saturated   one element of unsaturation   one element of unsaturation

Oxygen   An oxygen atom can be added to the chain without changing the number of hydrogen atoms or carbon atoms. In calculating the number of elements of unsaturation, *ignore the oxygen atoms.*

$$CH_3-CH_3 \qquad CH_3-O-CH_3 \qquad \overset{\displaystyle O}{\overset{\|}{CH_3-C-H}} \quad \text{or} \quad \overset{O}{CH_2-CH_2}$$

C_2H_6, saturated C_2H_6O, saturated C_2H_4O, one element of unsaturation

Nitrogen   A nitrogen atom can take the place of a carbon atom in the chain, but nitrogen is trivalent, having only one additional hydrogen atom, compared with two hydrogens for each additional carbon atom. In computing the elements of unsaturation, *count nitrogen as half a carbon atom.*

$$H-\overset{\overset{\displaystyle H}{|}}{\underset{\underset{\displaystyle H}{|}}{C}}-\overset{\overset{\displaystyle H}{|}}{\underset{\underset{\displaystyle H}{|}}{C}}-\overset{\overset{\displaystyle H}{|}}{\underset{\underset{\displaystyle H}{|}}{C}}-\overset{\overset{\displaystyle \cdot\cdot}{}}{\underset{\underset{\displaystyle H}{|}}{N}}-\overset{\overset{\displaystyle H}{|}}{\underset{\underset{\displaystyle H}{|}}{C}}-H$$

carbon + 2 H   nitrogen + 1 H

The formula C_4H_9N is like a formula with $4\frac{1}{2}$ carbon atoms, with saturated formula $C_{4.5}H_{9+2}$. The formula C_4H_9N has one element of unsaturation, because it is two hydrogen atoms short of the saturated formula.

examples of formula C_4H_9N, one element of unsaturation

PROBLEM SOLVING $\mathbf{HINT}$

In figuring elements of unsaturation:
Count halogens as hydrogens.
Ignore oxygen.
Count nitrogen as half a carbon.

SOLVED PROBLEM 7-1

Draw at least four compounds of formula C_4H_6NOCl.

SOLUTION

Counting the nitrogen as $\frac{1}{2}$ carbon, ignoring the oxygen, and counting chlorine as a hydrogen shows the formula is equivalent to $C_{4.5}H_7$. The saturated formula for 4.5 carbon atoms is $C_{4.5}H_{11}$, so C_4H_6NOCl has two elements of unsaturation. These could be two double bonds, two rings, one triple bond, or a ring and a double bond. There are many possibilities, four of which are listed below.

two double bonds	two rings	one triple bond	one ring, one double bond

PROBLEM 7-3

Draw five more compounds of formula C_4H_6NOCl.

PROBLEM 7-4

For each of the following molecular formulas, determine the number of elements of unsaturation, and give three examples.
(a) $C_3H_4Cl_2$ (b) C_4H_8O (c) $C_4H_4O_2$ (d) $C_5H_5NO_2$ (e) C_6H_3NClBr

7-4
Nomenclature of Alkenes

Simple alkenes are named much like alkanes, using the root name of the longest chain containing the double bond. The ending is changed from -*ane* to -*ene*. For example, "ethane" becomes "ethene," "propane" becomes "propene," and "cyclohexane" becomes "cyclohexene."

	$CH_2{=}CH_2$	$CH_2{=}CH{-}CH_3$	
IUPAC names:	ethene	propene	cyclohexene
Common names:	ethylene	propylene	

When the chain contains more than three carbon atoms, a number is used to give the location of the double bond. The chain is numbered starting from the end closest to the double bond, and the double bond is given the *lower* number of its two double-bonded carbon atoms. Cycloalkenes are assumed to have the double bond in the number 1 position.

$$\overset{1}{C}H_2=\overset{2}{C}H-\overset{3}{C}H_2-\overset{4}{C}H_3 \qquad \overset{1}{C}H_2=\overset{2}{C}H-\overset{3}{C}H_2-\overset{4}{C}H_2-\overset{5}{C}H_3$$

IUPAC names: 1-butene   1-pentene
new IUPAC names: but-1-ene   pent-1-ene

cyclohexene

$$\overset{1}{C}H_3-\overset{2}{C}H=\overset{3}{C}H-\overset{4}{C}H_3 \qquad \overset{1}{C}H_3-\overset{2}{C}H=\overset{3}{C}H-\overset{4}{C}H_2-\overset{5}{C}H_3$$

IUPAC names: 2-butene   2-pentene
new IUPAC names: but-2-ene   pent-2-ene

In 1993, the IUPAC recommended a logical change in the positions of the numbers used in names. Instead of placing the numbers before the root name (1-butene), they recommended placing them immediately before the part of the name they locate (but-1-ene). You should be prepared to recognize names using either placement of the numbers, because both are widely used. In this section, names using the new number placement are printed in green. Throughout this book, we will use the new number placement whenever it helps to clarify a name.

A compound with two double bonds is a **diene**. A **triene** has three double bonds, and a **tetraene** has four. Numbers are used to specify the locations of the double bonds.

$$\overset{1}{C}H_2=\overset{2}{C}H-\overset{3}{C}H=\overset{4}{C}H_2 \qquad \overset{7}{C}H_3-\overset{6}{C}H=\overset{5}{C}H-\overset{4}{C}H=\overset{3}{C}H-\overset{2}{C}H=\overset{1}{C}H_2$$

IUPAC names: 1,3-butadiene   1,3,5-heptatriene   1,3,5,7-cyclooctatetraene
new IUPAC names: buta-1,3-diene   hepta-1,3,5-triene   cycloocta-1,3,5,7-tetraene

Each alkyl group attached to the main chain is listed with a number to give its location. Note that the double bond is still given preference in numbering, however.

$$\overset{4}{C}H_3-\overset{3}{C}H=\overset{2}{C}-\overset{1}{C}H_3 \qquad \overset{4}{C}H_3-\overset{3}{C}H-\overset{2}{C}H=\overset{1}{C}H_2 \qquad \overset{1}{C}H_3-\overset{2}{C}H=\overset{3}{C}-\overset{4}{C}H_2-\overset{5}{C}H_2-\overset{6}{C}H-\overset{7}{C}H_3$$

CH_3	CH_3

2-methyl-2-butene   3-methyl-1-butene   3,6-dimethyl-2-heptene
2-methylbut-2-ene   3-methylbut-1-ene   3,6-dimethylhept-2-ene

1-methylcyclopentene   2-ethyl-1,3-cyclohexadiene   7-bromo-1,3,5-cycloheptatriene   3-propyl-1-heptene
                          2-ethylcyclohexa-1,3-diene   7-bromocyclohepta-1,3,5-triene   3-propylhept-1-ene

Alkenes as Substituents   Alkenes named as substituents are called *alkenyl groups*. They can be named systematically (ethenyl, propenyl, etc.), or by common names. Common substituents are the vinyl, allyl, methylene, and phenyl groups. The phenyl group (Ph) is different from the others because it is aromatic (see Chapter 16) and does not undergo the typical reactions of alkenes.

$$\{=CH_2$$

methylene group
(methylidene group)

$$\{-CH=CH_2$$

vinyl group
(ethenyl group)

3-methylenecyclohexene

$$CH=CH_2$$
$$CH_2=CHCHCH_2CH=CH_2$$

3-vinyl-1,5-hexadiene
3-vinylhexa-1,5-diene

> The polyene antifungals are a group of drugs with a nonpolar region consisting of 4–7 sets of alternating single and double bonds. They insert themselves in the cell membranes of fungi, causing disruption and leakiness that results in fungal cell death.

$$\{-CH_2-CH=CH_2$$

allyl group
(2-propenyl group)

phenyl group
(Ph)

$$CH_2=CH-CH_2-Cl$$

allyl chloride
(3-chloropropene)

2-phenyl-1,3-cyclopentadiene
2-phenylcyclopenta-1,3-diene

Common Names   Most alkenes are conveniently named by the IUPAC system, but common names are sometimes used for the simplest compounds.

	$CH_2=CH_2$	$CH_2=CH-CH_3$
IUPAC names:	ethene	propene
common name:	ethylene	propylene

$$CH_2=\overset{\overset{\displaystyle CH_3}{|}}{C}-CH_3$$

2-methylpropene
isobutylene

$$CH_2=\overset{\overset{\displaystyle CH_3}{|}}{C}-CH=CH_2$$

2-methylbuta-1,3-diene
isoprene

ethenylbenzene
styrene

7-5
Nomenclature of Cis-trans Isomers

7-5A   Cis-trans Nomenclature

In Chapters 2 and 5, we saw how the rigidity and lack of rotation of carbon–carbon double bonds give rise to **cis-trans isomerism**, also called **geometric isomerism**. If two similar groups bonded to the carbons of the double bond are on the same side of the bond, the alkene is the **cis** isomer. If the similar groups are on opposite sides of the bond, the alkene is **trans**. Not all alkenes are capable of showing cis-trans isomerism. If either carbon of the double bond holds two identical groups, the molecule cannot have cis and trans forms. Following are some cis and trans alkenes and some alkenes that cannot show cis-trans isomerism.

cis-2-pentene
cis-pent-2-ene

trans-2-pentene
trans-pent-2-ene

2-methyl-2-pentene
2-methylpent-2-ene

1-pentene
pent-1-ene

(neither cis nor trans)

Trans cycloalkenes are unstable unless the ring is large enough (at least eight carbon atoms) to accommodate the trans double bond (Section 7-7D). Therefore, all cycloalkenes are assumed to be cis unless they are specifically named trans. The cis name is rarely used with cycloalkenes, except to distinguish a large cycloalkene from its trans isomer.

cyclohexene   cyclooctene   *trans*-cyclodecene   *cis*-cyclodecene

cis-cyclooctene

7-5B *E-Z* Nomenclature

The cis-trans nomenclature for geometric isomers sometimes fails to give an unambiguous name. For example, the isomers of 1-bromo-1-chloropropene are not clearly cis or trans because it is not obvious which substituents are referred to as being cis or trans.

geometric isomers of 1-bromo-1-chloropropene

To deal with this problem, we use the **_E-Z_ system** of nomenclature for cis-trans isomers, which is patterned after the Cahn–Ingold–Prelog convention for asymmetric carbon atoms (Section 5-3). It assigns a unique configuration of either *E* or *Z* to any double bond capable of geometric isomerism.

To name an alkene by the *E-Z* system, mentally separate the double bond into its two ends. Remember how you used the Cahn–Ingold–Prelog rules (page 175) to assign priorities to groups on an asymmetric carbon atom so you could name it (*R*) or (*S*). Consider each end of the double bond separately, and use those same rules to assign first and second priorities to the two substituent groups on that end. Do the same for the other end of the double bond. If the two first priority atoms are *together (cis)* on the same side of the double bond, you have the **Z** isomer, from the German word *zusammen*, "together." If the two first priority atoms are on *opposite (trans)* sides of the double bond, you have the **E** isomer, from the German word *entgegen*, "opposite."

Zusammen   Entgegen

For example,

(Z)-1-bromo-1-chloropropene

The other isomer is named similarly:

(E)-1-bromo-1-chloropropene

If the alkene has more than one double bond, the stereochemistry about each double bond should be specified. The following compound is properly named 3-bromo-(3Z,5E)-octadiene:

3-bromo-(3Z,5E)-octadiene

3-bromo-(3Z,5E)-octa-3,5-diene

The use of *E-Z* names (rather than cis and trans) is always an option, but it is required whenever a double bond is not clearly cis or trans. Most trisubstituted and tetrasubstituted double bonds are more clearly named *E* or *Z* rather than cis or trans.

SUMMARY **Rules for Naming Alkenes**

The following rules summarize the IUPAC system for naming alkenes.
1. Select the longest chain or largest ring that contains the *largest possible number of double bonds*, and name it with the *-ene* suffix. If there are two double bonds, the suffix is *-diene*; for three, *-triene*; for four, *-tetraene*; and so on.
2. Number the chain from the end closest to the double bonds. Number a ring so that the double bond is between carbons 1 and 2. Place the numbers giving the locations of the double bonds in front of the root name (old system) or in front of the suffix *-ene, -diene*, etc. (new system).
3. Name substituent groups as in alkanes, indicating their locations by the number of the main-chain carbon to which they are attached. The ethenyl group and the propenyl group are usually called the *vinyl* group and the *allyl* group, respectively.
4. For compounds that show geometric isomerism, add the appropriate prefix: *cis-* or *trans-*, or *E-* or *Z-*. Cycloalkenes are assumed to be cis unless named otherwise.

PROBLEM-SOLVING HINT

To see whether a compound can have cis and trans isomers, draw the structure, then draw it again with the groups on one end of the double bond reversed. See if you can describe a difference between the two.

PROBLEM 7-5

Give the systematic (IUPAC) names of the following alkenes.

(a) $CH_2=CH-CH_2-CH(CH_3)_2$ (b) $CH_3(CH_2)_3-\overset{\displaystyle ||}{\underset{\displaystyle CH_2}{C}}-CH_2CH_3$

(c) $CH_2=CH-CH_2-CH=CH_2$ (d) $CH_2=C=CH-CH=CH_2$

PROBLEM 7-6

(1) Determine which of the following compounds show cis-trans isomerism.
(2) Draw and name the cis and trans isomers of those that do.

(a) 3-hexene   (b) 1,3-butadiene   (c) 2,4-hexadiene
(d) 3-methylpent-2-ene   (e) 2,3-dimethylpent-2-ene   (f) cyclopentene

PROBLEM 7-7

Each of the following names is incorrect. Draw the structure represented by the incorrect name
(or a consistent structure if the name is ambiguous), and give your drawing the correct name.

(a) *cis*-2,3-dimethyl-2-pentene   (b) 3-vinylhex-4-ene
(c) 2-methylcyclopentene   (d) 6-chlorocyclohexadiene
(e) 3,4-dimethylcyclohexene   (f) *cis*-2,5-dibromo-3-ethylpent-2-ene

PROBLEM 7-8

Some of the following examples can show geometric isomerism, and some cannot. For the
ones that can, draw all the geometric isomers, and assign complete names using the *E-Z* system.

(a) 3-bromo-2-chloro-2-pentene   (b) 3-ethylhexa-2,4-diene
(c) 3-bromo-2-methyl-2-butene   (d) 1,3-pentadiene
(e) 4-*t*-butyl-5-methyl-4-octene   (f) 3,7-dichloroocta-2,5-diene

(g)
cyclohexene

(h)
cyclodecene

(i)
1,5-cyclodecadiene

Because the carbon–carbon double bond is readily converted to other functional groups,
alkenes are important intermediates in the synthesis of polymers, drugs, pesticides, and
other valuable chemicals. Ethylene is the organic compound produced in the largest
volume, at around 158 billion pounds per year worldwide.

Most of this ethylene is polymerized to form 85 billion pounds of polyethylene
per year. The remainder is used to synthesize a wide variety of organic chemicals including ethanol, acetic acid, ethylene glycol, and vinyl chloride (Figure 7-3).

7-6 Commercial Importance of Alkenes

◀ **Figure 7-3**
Ethylene and propylene are the
largest-volume industrial organic
chemicals. They can be used to
synthesize a wide variety of useful
compounds. *(continues on next page)*

▶ Figure 7-3 (continued)

Ethylene also serves as a plant hormone, accelerating the ripening of fruit. For example, tomatoes are harvested and shipped while green, then treated with ethylene to make them ripen and turn red just before they are placed on display. Propylene is produced at the rate of about 86 billion pounds per year worldwide, with much of that going to make about 40 billion pounds of polypropylene. The rest is used to make propylene glycol, acetone, isopropyl alcohol, and a variety of other important organic chemicals.

The largest use of alkenes is for the production of polymers, which are used in consumer products from shoes to plastic bags to car bumpers. A **polymer** (Greek, *poly*, "many," and *meros*, "parts") is a large molecule made up of many **monomer** (Greek, *mono*, "one") molecules. Many common polymers are made simply by polymerizing alkenes. An alkene monomer can **polymerize** by a chain reaction where additional alkene molecules add to the end of the growing polymer chain. Because these polymers result from addition of many individual alkene units, they are called **addition polymers. Polyolefins** are polymers made from monofunctional (single functional group) alkenes such as ethylene and propylene. Figure 7-4 shows some addition polymers made from simple alkenes and haloalkenes. We discuss polymerization reactions in Chapters 8 and 26.

This plant in Tokuyama, Japan, passes ethane rapidly over a hot catalyst. The products are ethylene and hydrogen.

▶ Figure 7-4
Alkenes polymerize to form addition polymers. Many common polymers are produced this way.

In making alkenes, we often find that the major product is the most stable alkene. Many reactions also provide opportunities for double bonds to rearrange to more stable isomers. Therefore, we need to know how the stability of an alkene depends on its structure. Stabilities can be compared by converting different compounds to a common product and comparing the amounts of heat given off. One possibility would be to measure heats of combustion from converting alkenes to CO_2 and H_2O. Heats of combustion are large numbers (several hundred kilocalories per mole), however, and measuring small differences in these large numbers is difficult. Instead, alkene energies are often compared by measuring the **heat of hydrogenation**: the heat given off ($\Delta H°$) during catalytic hydrogenation. Heats of hydrogenation can be measured about as easily as heats of combustion, yet they are smaller numbers and provide more accurate energy differences.

7-7
Stability of Alkenes

7-7A   Heats of Hydrogenation

When an alkene is treated with hydrogen in the presence of a platinum catalyst, hydrogen adds to the double bond, reducing the alkene to an alkane. Hydrogenation is mildly exothermic, evolving about 20 to 30 kcal (80 to 120 kJ) of heat per mole of hydrogen consumed. Consider the hydrogenation of 1-butene and *trans*-2-butene:

$$H_2C{=}CH{-}CH_2{-}CH_3 \ + \ H_2 \ \xrightarrow{Pt} \ CH_3{-}CH_2{-}CH_2{-}CH_3 \qquad \Delta H° = -30.3 \ \text{kcal/mol}$$

1-butene   butane $(-127 \ \text{kJ/mol})$
(monosubstituted)

trans-2-butene   butane $\Delta H° = -27.6 \ \text{kcal/mol}$
(disubstituted) $(-115 \ \text{kJ/mol})$

Figure 7-5 shows these heats of hydrogenation on a reaction-energy diagram. The difference in the stabilities of 1-butene and *trans*-2-butene is the difference in their heats of hydrogenation. *trans*-2-Butene is more stable by

$$30.3 \ \text{kcal/mol} \ - \ 27.6 \ \text{kcal/mol} = 2.7 \ \text{kcal/mol} \ (11 \ \text{kJ/mol})$$

7-7B   Substitution Effects

A 2.7 kcal/mol (11 kJ/mol) stability difference is typical for a monosubstituted alkene (1-butene) and a trans-disubstituted alkene (*trans*-2-butene). In the following equations, we compare the monosubstituted double bond of 3-methyl-1-butene with the trisubstituted

◀ **Figure 7-5**
Trans-2-butene is more stable than 1-butene by 2.7 kcal/mol (11 kJ/mol).

double bond of 2-methyl-2-butene. The trisubstituted alkene is more stable by 3.4 kcal/mol (14 kJ/mol).

$$CH_2\!\!=\!\!CH\!-\!\underset{\underset{CH_3}{|}}{CH}\!-\!CH_3 \quad \xrightarrow{\text{H}_2,\ \text{Pt}} \quad CH_3\!-\!CH_2\!-\!\underset{\underset{CH_3}{|}}{CH}\!-\!CH_3 \qquad \Delta H° = -30.3\ \text{kcal}\ (-127\ \text{kJ})$$

3-methyl-1-butene (monosubstituted) → 2-methylbutane

$$CH_3\!-\!CH\!\!=\!\!\underset{\underset{CH_3}{|}}{C}\!-\!CH_3 \quad \xrightarrow{\text{H}_2,\ \text{Pt}} \quad CH_3\!-\!CH_2\!-\!\underset{\underset{CH_3}{|}}{CH}\!-\!CH_3 \qquad \Delta H° = -26.9\ \text{kcal}\ (-113\ \text{kJ})$$

2-methyl-2-butene (trisubstituted) → 2-methylbutane

To be completely correct, we should compare heats of hydrogenation only for compounds that give the same alkane, as 3-methyl-1-butene and 2-methyl-2-butene do. Yet most alkenes with similar substitution patterns give similar heats of hydrogenation. For example, 3,3-dimethyl-1-butene (below) hydrogenates to give a different alkane than does 3-methyl-1-butene or 1-butene (above); yet these three monosubstituted alkenes have similar heats of hydrogenation because the alkanes formed have similar energies. In effect, the heat of hydrogenation is a measure of the energy content of the π bond.

$$CH_2\!\!=\!\!CH\!-\!\underset{\underset{CH_3}{|}}{\overset{\overset{CH_3}{|}}{C}}\!-\!CH_3 \quad \xrightarrow{\text{H}_2,\ \text{Pt}} \quad CH_3\!-\!CH_2\!-\!\underset{\underset{CH_3}{|}}{\overset{\overset{CH_3}{|}}{C}}\!-\!CH_3 \qquad \Delta H° = -30.3\ \text{kcal}\ (-127\ \text{kJ})$$

3,3-dimethyl-1-butene (monosubstituted) → 2,2-dimethylbutane

In practice, we can use heats of hydrogenation to compare the stabilities of different alkenes as long as they hydrogenate to give alkanes of similar energies. Most open-chain alkanes and unstrained cycloalkanes have similar energies, and we can use this approximation. Table 7-1 shows the heats of hydrogenation of a variety of alkenes with different substitution. The compounds are ranked in decreasing order of their heats of hydrogenation, that is, from the least stable double bonds to the most stable. Note that the values are similar for alkenes with similar substitution patterns.

The most stable double bonds are those with the most alkyl groups attached. For example, hydrogenation of ethylene (no alkyl groups attached) evolves 32.8 kcal/mol, while propene and 1-pentene (one alkyl group for each) give off 30.1 kcal/mol. Double bonds with two alkyl groups hydrogenate to produce about 28 kcal/mol. Three or four alkyl substituents further stabilize the double bond, as with 2-methyl-2-butene (trisubstituted, 26.9 kcal) and 2,3-dimethyl-2-butene (tetrasubstituted, 26.6 kcal).

The values in Table 7-1 confirm the **Saytzeff rule** (**Zaitsev rule**):

More highly substituted double bonds are usually more stable.

In other words, the alkyl groups attached to the double-bonded carbons stabilize the alkene.

Two factors are probably responsible for the stabilizing effect of alkyl groups on a double bond. Alkyl groups are electron-donating, and they contribute electron density to the pi bond. In addition, bulky substituents like alkyl groups are best situated as far apart as possible. In an alkane, they are separated by the tetrahedral bond angle, about 109.5°. A double bond increases this separation to about 120°. In general, alkyl groups are separated best by the most highly substituted double bond. This steric effect is illustrated in Figure 7-6 for two **double-bond isomers** (isomers that differ only in the position of the double bond). The isomer with the monosubstituted double bond separates the alkyl groups by only 109.5°, while the trisubstituted double bond separates them by about 120°.

less highly substituted

109.5° separation

closer groups

more highly substituted

120° separation

wider separation

▲ Figure 7-6
The isomer with the more highly substituted double bond has a larger angular separation between the bulky alkyl groups.

TABLE 7-1 Molar Heats of Hydrogenation of Alkenes

Name	Structure	Molar Heat of Hydrogenation ($-\Delta H°$)		General Structure
		kcal	kJ	
ethene (ethylene)	$H_2C{=}CH_2$	32.8	137}	unsubstituted
propene (propylene)	$CH_3{-}CH{=}CH_2$	30.1	126	
1-butene	$CH_3{-}CH_2{-}CH{=}CH_2$	30.3	127	
1-pentene	$CH_3{-}CH_2{-}CH_2{-}CH{=}CH_2$	30.1	126	monosubstituted
1-hexene	$CH_3{-}(CH_2)_3{-}CH{=}CH_2$	30.1	126	$R{-}CH{=}CH_2$
3-methyl-1-butene	$(CH_3)_2CH{-}CH{=}CH_2$	30.3	127	
3,3-dimethyl-1-butene	$(CH_3)_3C{-}CH{=}CH_2$	30.3	127	
cis-2-butene	(structure)	28.6	120	disubstituted (cis)
cis-2-pentene	(structure)	28.6	120	(structure)
2-methylpropene (isobutylene)	$(CH_3)_2C{=}CH_2$	28.0	117	disubstituted (geminal)
2-methyl-1-butene	(structure)	28.5	119	
2,3-dimethyl-1-butene	(structure)	28.0	117	(structure)
trans-2-butene	(structure)	27.6	116	disubstituted (trans)
trans-2-pentene	(structure)	27.6	116	(structure)
2-methyl-2-butene	(structure)	26.9	113}	trisubstituted $R_2C{=}CHR$
2,3-dimethyl-2-butene	$(CH_3)_2C{=}C(CH_3)_2$	26.6	111}	tetrasubstituted $R_2C{=}CR_2$

Note: A lower heat of hydrogenation corresponds to a lower energy and greater stability in the alkene.

PROBLEM 7-9

Use the data in Table 7-1 to determine the energy difference between 2,3-dimethyl-1-butene and 2,3-dimethyl-2-butene. Which of these double-bond isomers is more stable?

PROBLEM-SOLVING HINT

Heats of hydrogenation are always exothermic. A larger amount of heat given off implies a less stable alkene.

7-7C Energy Differences in cis-trans Isomers

The heats of hydrogenation in Table 7-1 show that trans isomers are generally more stable than the corresponding cis isomers. This trend seems reasonable because the alkyl substituents are separated farther in trans isomers than they are in cis isomers. The greater stability of the trans isomer is evident in the 2-pentenes, which show a 1.0 kcal (4 kJ) difference between the cis and trans isomers.

H$_3$C   CH$_2$CH$_3$
   \\   /
    C=C   + H$_2$ $\xrightarrow{\text{Pt}}$ CH$_3$—CH$_2$—CH$_2$—CH$_2$CH$_3$ $\Delta H° = -28.6$ kcal/mol
   /   \\ $(-120$ kJ/mol)
  H   H

cis

H$_3$C   H
   \\   /
    C=C   + H$_2$ $\xrightarrow{\text{Pt}}$ CH$_3$—CH$_2$—CH$_2$—CH$_2$CH$_3$ $\Delta H° = -27.6$ kcal/mol
   /   \\ $(-116$ kJ/mol)
  H   CH$_2$CH$_3$

trans

A 1 kcal/mol difference between cis and trans isomers is typical for disubstituted alkenes. Figure 7-7 summarizes the relative stabilities of alkenes, comparing them with ethylene, the least stable of the simple alkenes.

> **PROBLEM 7-10**
>
> Tell which member of each pair is more stable, and by about how many kcal/mol or kJ/mol.
> (a) *cis, cis* 2,4-hexadiene or *trans, trans*-2,4-hexadiene
> (b) 2-methyl-1-butene or 3-methyl-1-butene
> (c) 2-methyl-1-butene or 2-methyl-2-butene
> (d) 2,3-dimethyl-1-butene or 2,3-dimethyl-2-butene

7-7D   Stability of Cycloalkenes

Most cycloalkenes react like acyclic (noncyclic) alkenes. The presence of a ring makes a major difference only if there is ring strain, either because of a small ring or because of a trans double bond. Rings that are five-membered or larger can easily accommodate double bonds, and these cycloalkenes react much like straight-chain alkenes. Three- and four-membered rings show evidence of ring strain, however.

▲ **Figure 7-7**
Relative energies of typical π bonds compared with ethylene. (The numbers are approximate.)

Cyclobutene   Cyclobutene has a heat of hydrogenation of -30.7 kcal/mol (-128 kJ/mol), compared with -26.6 kcal/mol (111 kJ/mol) for cyclopentene.

$\Delta H° = -30.7$ kcal/mol
(-128 kJ/mol)

$\Delta H° = -26.6$ kcal/mol
(-111 kJ/mol)

The double bond in cyclobutene has about 4 kcal/mol of *extra* ring strain (in addition to the ring strain in cyclobutane) by virtue of the small ring. The 90° bond angles in cyclobutene compress the angles of the sp^2 hybrid carbons (normally 120°) more than they compress the sp^3 hybrid angles (normally 109.5°) in cyclobutane. The extra ring strain in cyclobutene makes its double bond more reactive than a typical double bond.

Cyclopropene   Cyclopropene has bond angles of about 60°, compressing the bond angles of the carbon–carbon double bond to half their usual value of 120°. The double bond in cyclopropene is highly strained.

Many chemists once believed that a cyclopropene could never be made because it would snap open immediately from the large ring strain. Cyclopropene was eventually synthesized, however, and it can be stored in the cold. Cyclopropenes were still considered to be strange, highly unusual compounds. Natural-product chemists were surprised when they found that the kernel oil of *Sterculia foelida*, a tropical tree, contains *sterculic acid*, a carboxylic acid with a cyclopropene ring.

Sterculic acid is a potent inhibitor of several desaturases, which are the enzymes responsible for the formation of double bonds in long-chain acids used as fuels, membrane components, and other critical biological molecules. Consequently, vegetable oils containing sterculic acid must be hydrogenated or processed at high temperatures to reduce or destroy the cyclopropene ring.

sterculic acid

Trans Cycloalkenes   Another difference between cyclic and acyclic alkenes is the relationship between cis and trans isomers. In acyclic alkenes, the trans isomers are usually more stable; but the trans isomers of small cycloalkenes are rare, and those with fewer than eight carbon atoms are unstable at room temperature. The problem with making a trans cycloalkene lies in the geometry of the trans double bond. The two alkyl groups on a trans double bond are so far apart that several carbon atoms are needed to complete the ring.

Try to make a model of trans-cyclohexene, being careful that the large amount of ring strain does not break your models. *Trans*-cyclohexene is too strained to be isolated, while *trans*-cycloheptene can be isolated at low temperatures. *Trans*-cyclooctene is stable at room temperature, although its cis isomer is still more stable.

ring connects behind the double bond

trans cyclic system	*trans*-cycloheptene	*trans*-cyclooctene	*cis*-cyclooctene
	marginally stable	stable	more stable

Once a cycloalkene contains at least ten or more carbon atoms, it can easily accommodate a trans double bond. For cyclodecene and larger cycloalkenes, the trans isomer is nearly as stable as the cis isomer.

cis-cyclodecene   *trans*-cyclodecene

7-7E   Bredt's Rule

We have seen that a trans cycloalkene is not stable unless there are at least eight carbon atoms in the ring. An interesting extension of this principle is called **Bredt's rule**.

> BREDT'S RULE: A bridged bicyclic compound cannot have a double bond at a bridgehead position unless one of the rings contains at least eight carbon atoms.

Let's review exactly what Bredt's rule means. A **bicyclic** compound is one that contains two rings. The **bridgehead carbon atoms** are part of both rings, with three links connecting them. A **bridged bicyclic** compound has at least one carbon atom in each of the three links between the bridgehead carbons. In the following examples, the bridgehead carbon atoms are circled in red.

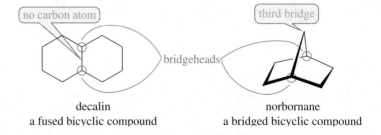

decalin
a fused bicyclic compound

norbornane
a bridged bicyclic compound

If there is a double bond at the bridgehead carbon of a bridged bicyclic system, one of the two rings contains a cis double bond and the other must contain a trans double bond. For example, the following structures show that norbornane contains a five-membered ring and a six-membered ring. If there is a double bond at the bridgehead carbon atom, the five-membered ring contains a cis double bond, and the six-membered ring contains a trans double bond. This unstable arrangement is called a "Bredt's rule violation." If the larger ring contains at least eight carbon atoms, then it can contain a trans double bond, and the bridgehead double bond is stable.

Bredt's rule violation

Stable: trans in an eight-membered ring

SOLVED PROBLEM 7-2

Which of the following alkenes are stable?

SOLUTION

Compound (a) is stable. Although the double bond is at a bridgehead, it is not a bridged bicyclic system. The trans double bond is in a ten-membered ring. Compound (b) is a Bredt's rule violation and is not stable. The largest ring contains six carbon atoms, and the trans double bond cannot be stable in this bridgehead position.

Compound (c) (norbornene) is stable. The (cis) double bond is not at a bridgehead carbon.

Compound (d) is stable. Although the double bond is at the bridgehead of a bridged bicyclic system, there is an eight-membered ring to accommodate the trans double bond.

PROBLEM 7-11

Explain why each of the following alkenes is stable or unstable.
(a) 1,2-dimethylcyclobutene **(b)** *trans*-1,2-dimethylcyclobutene
(c) *trans*-3,4-dimethylcyclobutene **(d)** *trans*-1,2-dimethylcyclodecene

---

7-8A Boiling Points and Densities

Most physical properties of alkenes are similar to those of the corresponding alkanes. For example, the boiling points of 1-butene, *cis*-2-butene, *trans*-2-butene, and *n*-butane are all close to 0°C. Also like the alkanes, alkenes have densities around 0.6 or 0.7 g/cm^3. The boiling points and densities of some representative alkenes are listed in Table 7-2. The table shows that boiling points of alkenes increase smoothly with molecular weight. As with alkanes, increased branching leads to greater volatility and lower boiling points. For example, 2-methylpropene (isobutylene) has a boiling point of −7°C, which is lower than the boiling point of any of the unbranched butenes.

7-8
Physical Properties of Alkenes

TABLE 7-2 Physical Properties of Some Representative Alkenes

Name	Structure	Carbons	Boiling Point (°C)	Density (g/cm³)
ethene (ethylene)	$CH_2{=}CH_2$	2	−104	
propene (propylene)	$CH_3CH{=}CH_2$	3	−47	0.52
2-methylpropene (isobutylene)	$(CH_3)_2C{=}CH_2$	4	−7	0.59
1-butene	$CH_3CH_2CH{=}CH_2$	4	−6	0.59
trans-2-butene	$\begin{array}{c}H_3C \qquad H\\ \diagdown \quad / \\ C{=}C \\ / \quad \diagdown \\ H \qquad CH_3\end{array}$	4	1	0.60
cis-2-butene	$\begin{array}{c}H_3C \qquad CH_3\\ \diagdown \quad / \\ C{=}C \\ / \quad \diagdown \\ H \qquad H\end{array}$	4	4	0.62
3-methyl-1-butene	$(CH_3)_2CH{-}CH{=}CH_2$	5	25	0.65
1-pentene	$CH_3CH_2CH_2{-}CH{=}CH_2$	5	30	0.64
trans-2-pentene	$\begin{array}{c}H_3C \qquad H\\ \diagdown \quad / \\ C{=}C \\ / \quad \diagdown \\ H \qquad CH_2CH_3\end{array}$	5	36	0.65
cis-2-pentene	$\begin{array}{c}H_3C \qquad CH_2CH_3\\ \diagdown \quad / \\ C{=}C \\ / \quad \diagdown \\ H \qquad H\end{array}$	5	37	0.66
2-methyl-2-butene	$(CH_3)_2C{=}CH{-}CH_3$	5	39	0.66
1-hexene	$CH_3(CH_2)_3{-}CH{=}CH_2$	6	64	0.68
2,3-dimethyl-2-butene	$(CH_3)_2C{=}C(CH_3)_2$	6	73	0.71
1-heptene	$CH_3(CH_2)_4{-}CH{=}CH_2$	7	93	0.70
1-octene	$CH_3(CH_2)_5{-}CH{=}CH_2$	8	122	0.72
1-nonene	$CH_3(CH_2)_6{-}CH{=}CH_2$	9	146	0.73
1-decene	$CH_3(CH_2)_7{-}CH{=}CH_2$	10	171	0.74

7-8B   Polarity

Like alkanes, alkenes are relatively nonpolar. They are insoluble in water but soluble in non-polar solvents such as hexane, gasoline, halogenated solvents, and ethers. Alkenes tend to be slightly more polar than alkanes, however, for two reasons: The more weakly held electrons in the pi bond are more polarizable (contributing to instantaneous dipole moments), and the vinylic bonds tend to be slightly polar (contributing to a permanent dipole moment).

Alkyl groups are slightly electron donating toward a double bond, helping to stabilize it. This donation slightly polarizes the vinylic bond, with a small partial positive charge on the alkyl group and a small negative charge on the double-bond carbon atom. For example, propene has a small dipole moment of 0.35 D.

propene, $\mu = 0.35$ D

vector sum = ↕
$\mu = 0.33$ D
cis-2-butene, bp 4°C

vector sum = 0
$\mu = 0$
trans-2-butene, bp 1°C

In a cis-disubstituted alkene, the vector sum of the two dipole moments is directed perpendicular to the double bond. In a trans-disubstituted alkene, the two dipole moments tend to cancel out. If an alkene is symmetrically trans-disubstituted, the dipole moment is zero. For example, cis-2-butene has a nonzero dipole moment, while trans-2-butene has no measurable dipole moment.

Compounds with permanent dipole moments engage in dipole–dipole attractions, while those without permanent dipole moments engage only in van der Waals attractions. *Cis-* and *trans*-2-butene have similar van der Waals attractions, but only the cis isomer has dipole–dipole attractions. Because of its increased intermolecular attractions, *cis*-2-butene must be heated to a slightly higher temperature (4°C versus 1°C) before it begins to boil.

The effect of bond polarity is even more apparent in the 1,2-dichloroethenes, with their strongly polar carbon–chlorine bonds. The cis isomer has a large dipole moment (2.4 D), giving it a boiling point 12°C higher than the trans isomer, with zero dipole moment.

cis
vector sum = $\updownarrow$
μ = 2.4 D
bp = 60°C

trans
vector sum = 0
μ = 0
bp = 48°C

PROBLEM 7-12

For each pair of compounds, predict the one with a higher boiling point. Which compounds have zero dipole moments?
(a) *cis*-1,2-dichloroethene or *cis*-1,2-dibromoethene
(b) *cis*- or *trans*-2,3-dichloro-2-butene
(c) cyclohexene or 1,2-dichlorocyclohexene

Dehydrohalogenation is the elimination of a hydrogen and a halogen from an alkyl halide to form an alkene. In Sections 6-17 through 6-21 we saw how dehydrohalogenation can take place by the E1 and E2 mechanisms. The second-order elimination (E2) is usually better for synthetic purposes because the E1 has more competing reactions.

7-9
Alkene Synthesis by Elimination of Alkyl Halides

7-9A Dehydrohalogenation by the E2 Mechanism

Second-order elimination is a reliable synthetic reaction, especially if the alkyl halide is a poor S_N2 substrate. E2 dehydrohalogenation takes place in one step, in which a strong base abstracts a proton from one carbon atom as the leaving group leaves the adjacent carbon.

MECHANISM Dehydrohalogenation by the E2 Mechanism

E2 Elimination takes place by a concerted one-step reaction.

The E2 dehydrohalogenation gives excellent yields with bulky secondary and tertiary alkyl halides that are poor S_N2 substrates. A strong base forces second-order elimination (E2) by abstracting a proton. The molecule's bulkiness hinders second-order substitution (S$_N$2), and a relatively pure elimination product results. Tertiary halides are the best E2 substrates because they are prone to elimination and cannot undergo S$_N$2 substitution.

$$CH_3-\underset{\underset{\displaystyle CH_3}{|}}{\overset{\overset{\displaystyle CH_3}{|}}{C}}-Br \ + \ OH^- \ \longrightarrow \ H_2C=C\underset{\displaystyle CH_3}{\overset{\displaystyle CH_3}{<}} \ + \ H_2O \ + \ Br^-$$

$(>90\%)$

Use of a Bulky Base   If the substrate is prone to substitution, a bulky base can minimize the amount of substitution. Large alkyl groups on a bulky base hinder its approach to attack a carbon atom (substitution), yet it can easily abstract a proton (elimination). Some of the bulky strong bases commonly used for elimination are *t*-butoxide ion, diisopropylamine, triethylamine, and 2,6-dimethylpyridine.

$$CH_3-\underset{\underset{\displaystyle CH_3}{|}}{\overset{\overset{\displaystyle CH_3}{|}}{C}}-O^- \qquad (CH_3)_2CH-\underset{\underset{\displaystyle H}{|}}{\overset{\overset{\displaystyle (CH_3)_2CH}{|}}{N}}\colon \qquad (CH_3CH_2)_3N\colon$$

t-butoxide   diisopropylamine   triethylamine   2,6-dimethylpyridine

The dehydrohalogenation of bromocyclohexane illustrates the use of a bulky base for elimination. Bromocyclohexane, a secondary alkyl halide, can undergo both substitution and elimination. Elimination (E2) is favored over substitution (S$_N$2) by using a bulky base such as diisopropylamine. Diisopropylamine is too bulky to be a good nucleophile, but it acts as a strong base to abstract a proton.

bromocyclohexane $\xrightarrow{\text{(i-Pr)}_2\ddot{\text{N}}\text{H, heat}}$ cyclohexene $+ \ [(CH_3)_2CH]_2\overset{+}{N}H_2 \ Br^-$
(93%)

Formation of the Hofmann Product   Bulky bases can also accomplish dehydrohalogenations that do not follow the Saytzeff rule. Steric hindrance often prevents a bulky base from abstracting the proton that leads to the most highly substituted alkene. In these cases, it abstracts a less hindered proton, often the one that leads to formation of the least highly substituted product, called the **Hofmann product**. The following reaction gives mostly the **Saytzeff product** with (relatively unhindered) ethoxide ion, but mostly the Hofmann product with the bulky *t*-butoxide ion.

Saytzeff product   *Hofmann product*

$$CH_3-\underset{\underset{\displaystyle H}{|}}{\overset{\overset{\displaystyle H}{|}}{C}}-\underset{\underset{\displaystyle Br}{|}}{\overset{\overset{\displaystyle CH_3}{|}}{C}}-\underset{\underset{\displaystyle H}{|}}{CH_2} \qquad \xrightarrow[\text{CH}_3\text{CH}_2\text{OH}]{^-\text{OCH}_2\text{CH}_3}$$

71%   29%

$$CH_3-\underset{\underset{\displaystyle H}{|}}{\overset{\overset{\displaystyle H}{|}}{C}}-\underset{\underset{\displaystyle Br}{|}}{\overset{\overset{\displaystyle CH_3}{|}}{C}}-\underset{\underset{\displaystyle H}{|}}{CH_2} \qquad \xrightarrow[\text{(CH}_3)_3\text{COH}]{^-\text{OC(CH}_3)_3}$$

less hindered

28%   72%

PROBLEM 7-13

For each reaction, decide whether substitution or elimination (or both) is possible, and predict the products you expect. Label the major products.
(a) 1-bromo-1-methylcyclohexane + NaOH in acetone
(b) 1-bromo-1-methylcyclohexane + triethylamine, $(CH_3CH_2)_3N$:
(c) chlorocyclohexane + $NaOCH_3$ in CH_3OH
(d) chlorocyclohexane + $NaOC(CH_3)_3$ in $(CH_3)_3COH$

7-9B Stereospecific E2 Reactions

Like the S_N2 reaction, the E2 is **stereospecific**: Different stereoisomers of the reactant give different stereoisomers of the product. The E2 is stereospecific because it normally goes through an *anti* and coplanar transition state. The products are alkenes, and different diastereomers of starting materials commonly give different diastereomers of alkenes. In Problem 6-38, you showed why the E2 elimination of one diastereomer of 1-bromo-1,2-diphenylpropane gives only the trans isomer of the alkene product.

If we look at this reaction from the left end of the molecule, the *anti* and coplanar arrangement of the H and Br is apparent.

MECHANISM   Stereochemistry of the E2 Reaction

Most E2 reactions go through an anti and coplanar transition state.

The following reaction shows how the anti-coplanar elimination of the other diastereomer gives only the cis isomer of the product. In effect, the two different diastereomers of the reactant give two different diastereomers of the product: a stereospecific result.

Viewed from the left end of the molecule,

PROBLEM-SOLVING HINT

Don't try to memorize your way through these reactions. Look at each one, and consider what it might do. Use your models for the ones that involve stereochemistry.

PROBLEM 7-14

Show that the other enantiomer (*S,S*) of this second diastereomer of 1-bromo-1,2-diphenyl-propane also undergoes E2 elimination to give the cis diastereomer of the product. (We do not expect these achiral reagents to distinguish between enantiomers.)

PROBLEM 7-15

Make models of the following compounds, and predict the products formed when they react with the strong bases shown.

(a)

$+$ KOH $\longrightarrow$ (substitution and elimination)

(b) *meso*-1,2-dibromo-1,2-diphenylethane $+$ $(CH_3CH_2)_3N:$
(c) (*d,l*)-1,2-dibromo-1,2-diphenylethane $+$ $(CH_3CH_2)_3N:$

(d)

$+$ NaOH in acetone

(e)

$+$ $(CH_3)_3C-O^-$ in $(CH_3)_3C-OH$

7-9C   E2 Reactions in Cyclohexane Systems

Nearly all cyclohexanes are most stable in chair conformations. In the chair, all the carbon–carbon bonds are staggered, and any two adjacent carbon atoms have axial bonds in an anti-coplanar conformation, ideally oriented for the E2 reaction. (As drawn below, the axial bonds are vertical.) On any two adjacent carbon atoms, one has its axial bond pointing up and the other has its axial bond pointing down. These two bonds are trans to each other, and we refer to their geometry as **trans-diaxial**.

perspective view

Newman projection

▲ **Figure 7-8**
E2 elimination of bromocyclohexane requires that the proton and the leaving group be trans
and both be axial.

An E2 elimination can take place on this chair conformation only if the proton and the
leaving group can get into a trans-diaxial arrangement. Figure 7-8 shows the E2 dehydrohalogenation of bromocyclohexane. The molecule must flip into the chair conformation with the bromine atom axial before elimination can occur.

(You should make models of the structures in the following examples and problems so you can follow along more easily.)

SOLVED PROBLEM 7-3

Explain why the following deuterated 1-bromo-2-methylcyclohexane undergoes dehydrohalogenation by the E2 mechanism, to give only the indicated product. Two other alkenes are
not observed.

SOLUTION

In an E2 elimination, the hydrogen atom and the leaving group must have a trans-diaxial
relationship. In this compound, only one hydrogen atom—the deuterium—is trans to the
bromine atom. When the bromine atom is axial, the adjacent deuterium is also axial, providing
a trans-diaxial arrangement.

PROBLEM 7-16

Predict the elimination products of the following reactions, and label the major products.
(a) *cis*-1-bromo-2-methylcyclohexane + $NaOCH_3$ in CH_3OH
(b) *trans*-1-bromo-2-methylcyclohexane + $NaOCH_3$ in CH_3OH

PROBLEM 7-17

When the following stereoisomer of 2-bromo-1,3-dimethylcyclohexane is treated with sodium methoxide, no E2 reaction is observed. Explain why this compound cannot undergo the E2 reaction in the chair conformation.

Problem 7-17

fast elimination   slow elimination
Problem 7-18

PROBLEM 7-18

Two stereoisomers of a bromodecalin are shown above. Although the difference between these stereoisomers may seem trivial, one isomer undergoes elimination with KOH much faster than the other.
(a) Predict the products of these eliminations.
(b) Explain the large difference in the ease of elimination.

PROBLEM 7-19

PROBLEM-SOLVING HINT
Look for a hydrogen trans to the leaving group; then see if the hydrogen and the leaving group can become diaxial.

Give the expected product(s) of E2 elimination for each reaction. (*Hint*: Use models!)

(a) $\xrightarrow{NaOCH_3}$ one product

(b) $\xrightarrow{NaOCH_3}$ two products

7-9D   Dehalogenation of Vicinal Dibromides

Vicinal dibromides (two bromines on adjacent carbon atoms) are converted to alkenes by reduction with either iodide ion or zinc in acetic acid. This **dehalogenation** is rarely an important synthetic reaction, because the most likely origin of a vicinal dibromide is from bromination of an alkene (Section 8-10). We discuss this reaction with dehydrohalogenation because the mechanisms are similar.

$$-\overset{\displaystyle Br}{\underset{\displaystyle Br}{C}}-\overset{\displaystyle}{\underset{\displaystyle}{C}}- \ + \ Zn \ \xrightarrow{CH_3COOH} \ \diagup C{=}C\diagup \ + \ ZnBr_2$$

Dehalogenation is formally a reduction because a molecule of Br_2 (an oxidizing agent) is removed. The reaction with iodide takes place by the E2 mechanism, with the same geometric constraints as the E2 dehydrohalogenation. Elimination usually takes place through an anti-coplanar arrangement, as shown in the following example.

MECHANISM   E2 Dibromination of a Vicinal Dibromide

E2 Debromination takes place by a concerted, stereospecific mechanism.

meso-1,2-dibromo-1,2-diphenylethane

trans-stilbene (89%)

Use your models to show that only the trans isomer of stilbene is formed in this example by elimination through the anti-coplanar transition state.

Zinc serves as a reducing agent in zinc/acetic acid dehalogenation. The reaction is heterogeneous (part solid and part liquid), with the actual reduction taking place at the surface of the metallic zinc. Zinc is oxidized from the 0 oxidation state to the +2 oxidation state, forming $ZnBr_2$.

$$CH_3-\underset{\underset{\displaystyle Br}{|}}{CH}-\underset{\underset{\displaystyle Br}{|}}{CH}-CH_3 \ + \ Zn \ \xrightarrow{CH_3COOH}$$

$$CH_3-CH{=}CH-CH_3 \ + \ ZnBr_2$$

A derivative of trans-stilbene known as diethylstilbesterol or DES was once taken by women during pregnancy to prevent miscarriages. The use of DES was discontinued because studies showed that DES increases the risk of cervical cancer in the children women who take it.

diethylstilbesterol (DES)

PROBLEM 7-20

The example above shows *meso*-1,2-dibromo-1,2-diphenylethane reacting with iodide ion to give *trans*-stilbene. Show how the other diastereomer of the starting material gives a different stereoisomer of the product.

SOLVED PROBLEM 7-4

Show that the dehalogenation of 2,3-dibromobutane by iodide ion is stereospecific, by showing that the two diastereomers of the starting material give different diastereomers of the product.

SOLUTION

Rotating *meso*-2,3-dibromobutane into a conformation where the bromine atoms are anti and coplanar, we find that the product will be *trans*-2-butene. A similar conformation of either enantiomer of the ($\pm$) diastereomer shows that the product will be *cis*-2-butene. (*Hint*: Using your models will be helpful.)

meso-2,3-dibromobutane → *trans*-2-butene

(*R,R*)-2,3-dibromobutane → *cis*-2-butene

PROBLEM 7-21

Solved Problem 7-4 showed that the debromination of (*R,R*)-2,3-dibromobutane gives *cis*-2-butene. Draw the same reaction using the other enantiomer (*S,S*) and show that it gives the same diastereomer of the product.

PROBLEM-SOLVING HINT

Make a model of each compound, and place it in the conformation where the groups to be eliminated are *anti* and coplanar. The positions of the other groups will be near their positions in the alkene product.

PROBLEM 7-22

Predict the elimination products formed by debromination of the following compounds with iodide ion in acetone. Include stereochemistry, and give a correct name for each product.
(a) *trans*-1,2-dibromocyclohexane
(b) (3*R*,4*R*)-3,4-dibromoheptane

(c) (d) (e)

PROBLEM 7-23

The following compounds show different rates of debromination. One reacts quite fast, and the other seems not to react at all. Explain this surprising difference in rates.

$(CH_3)_3C$ → KI, acetone → $(CH_3)_3C$

$(CH_3)_3C$ → KI, acetone → no reaction

7-9E   Dehydrohalogenation by the E1 Mechanism

First-order dehydrohalogenation usually takes place in a good ionizing solvent (such as an alcohol or water), without a strong nucleophile or base to force second-order kinetics. The substrate is usually a secondary or tertiary alkyl halide. First-order elimination requires ionization to form a carbocation, which loses a proton to a weak base (usually the solvent). E1 dehydrohalogenation is generally accompanied by S_N1 substitution, because the nucleophilic solvent can also attack the carbocation directly to form the substitution product.

Elimination by the E1 mechanism

Accompanied by S_N1 substitution

Like all reactions involving carbocation intermediates, E1 dehydrohalogenations are prone to rearrangement, as shown in Problem 7-24.

PROBLEM 7-24

Propose mechanisms for the following reactions.

7-10
Alkene Synthesis by Dehydration of Alcohols

Dehydration of alcohols is a common method for making alkenes. The word *dehydration* literally means "removal of water."

$$\begin{array}{c} | \quad | \\ -C-C- \\ | \quad | \\ H \quad OH \end{array} \quad \xrightleftharpoons[]{\text{acidic catalyst, heat}} \quad \diagdown C=C \diagup + \; H_2O$$

Dehydration is reversible, and in most cases the equilibrium constant is not large. In fact, the reverse reaction (hydration) is a method for converting alkenes to alcohols (see Section 8-4). Dehydration can be forced to completion by removing the products from the reaction mixture as they form. The alkene boils at a lower temperature than the alcohol because the alcohol is hydrogen bonded. A carefully controlled distillation removes the alkene while leaving the alcohol in the reaction mixture.

Concentrated sulfuric acid and/or concentrated phosphoric acid are often used as reagents for dehydration because these acids act both as acidic catalysts and as dehydrating agents. Hydration of these acids is highly exothermic. The overall reaction (using sulfuric acid) is

$$\begin{array}{c} | \quad | \\ -C-C- \\ | \quad | \\ H \quad OH \end{array} + \; H_2SO_4 \; \rightleftharpoons \; \diagdown C=C \diagup + \; H_3O^+ + \; HSO_4^-$$

The mechanism of dehydration resembles the E1 mechanism introduced in Chapter 6. The hydroxyl group of the alcohol is a poor leaving group (^-OH), but protonation by the acidic catalyst converts it to a good leaving group (H_2O). In the second step, loss of water from the protonated alcohol gives a carbocation. The carbocation is a very strong acid: Any weak base such as H_2O or HSO_4^- can abstract the proton in the final step to give the alkene.

Hydration and dehydration reactions are common in many biological pathways. The enzyme fumarase catalyzes the reversible addition of water to the double bond of fumarate to form malate. In contrast to the harsh conditions used in the chemical reaction, the enzymatic reaction takes place at neutral pH and at 37°C.

$$\begin{array}{c} H \diagdown \quad \diagup COO^- \\ C \\ \| \\ C \\ ^-OOC \diagup \quad \diagdown H \end{array} \quad \xrightleftharpoons[\text{fumarase}]{H_2O}$$

fumarate

$$\begin{array}{c} CH_2COO^- \\ | \\ ^-OOC-C\text{''''}OH \\ \diagup \\ H \end{array}$$

(*s*)-malate

KEY MECHANISM   Acid-Catalyzed Dehydration of an Alcohol

Step 1: Protonation of the hydroxyl group (fast equilibrium).

Step 2: Ionization (slow; rate limiting).

Step 3: Proton abstraction (fast).

Like other E1 reactions, alcohol dehydration follows an order of reactivity that reflects carbocation stability: 3° alcohols react faster than 2° alcohols, and 1° alcohols are the least reactive. Rearrangements of the carbocation intermediates are common in alcohol dehydrations.

SOLVED PROBLEM 7-5

Propose a mechanism for the sulfuric acid–catalyzed dehydration of *t*-butyl alcohol.

SOLUTION

The first step is protonation of the hydroxyl group, which converts it to a good leaving group.

The second step is ionization of the protonated alcohol to give a carbocation.

Abstraction of a proton completes the mechanism.

PROBLEM-SOLVING $\mathbb{HINT}$

Alcohol dehydrations usually go through E1 elimination of the protonated alcohol. Reactivity is:

$$3° > 2° \gg 1°$$

Rearrangements are common.

PROBLEM 7-25

Propose mechanisms for the following reactions:

(a)

cyclopentanol   cyclopentene

(b)

2-methylcyclohexanol   1-methylcyclohexene   3-methylcyclohexene   methylenecyclohexane

---

7-11A Catalytic Cracking of Alkanes

The least expensive way to make alkenes on a large scale is by the **catalytic cracking** of petroleum: heating a mixture of alkanes in the presence of a catalyst (usually Al-silicates). Alkenes are formed by bond cleavage to give an alkene and a shortened alkane.

7-11
Alkene Synthesis by High-Temperature Industrial Methods

long-chain alkane   shorter alkane   alkene

Cracking is used primarily to make small alkenes, up to about six carbon atoms. Its value depends on having a market for all the different alkenes and alkanes produced. The average molecular weight and the relative amounts of alkanes and alkenes can be controlled by varying the temperature, catalyst, and concentration of hydrogen in the cracking process. A careful distillation on a huge column separates the mixture into its pure components, ready to be packaged and sold.

Because the products are always mixtures, catalytic cracking is not suitable for laboratory synthesis of alkenes. Better methods are available for synthesizing relatively pure alkenes from a variety of other functional groups. Several of these methods are discussed in Sections 7-9, 7-10, and later sections listed in the Summary.

7-11B   Dehydrogenation of Alkanes

Dehydrogenation is the removal of H_2 from a molecule, just the reverse of hydrogenation. Dehydrogenation of an alkane gives an alkene. This reaction has an unfavorable enthalpy change but a favorable entropy change.

$$\Delta H^\circ \;=\; +20 \text{ to } +30 \text{ kcal } (+80 \text{ to } +120 \text{ kJ}) \qquad \Delta S^\circ \;=\; +30 \text{ eu}$$

The hydrogenation of alkenes (Section 7-7) is exothermic, with values of ΔH° around -20 to -30 kcal (-80 to -120 kJ). Therefore, dehydrogenation is endothermic and has an unfavorable (positive) value of ΔH°. The entropy change for dehydrogenation is strongly favorable ($\Delta S^\circ \;=\; +30$ eu), however, because one alkane molecule is converted into two molecules (the alkene and hydrogen), and two molecules are more disordered than one.

The equilibrium constant for the hydrogenation–dehydrogenation equilibrium depends on the change in free energy, $\Delta G \;=\; \Delta H \;-\; T\Delta S$. At room temperature, the enthalpy term predominates and hydrogenation is favored. When the temperature is raised, however, the ($-T\Delta S$) entropy term becomes larger and eventually dominates the expression. At a sufficiently high temperature, dehydrogenation is favored.

PROBLEM 7-26

The dehydrogenation of butane to *trans*-2-butene has $\Delta H^\circ \;=\; +27.6$ kcal/mol ($+116$ kJ/mol) and $\Delta S^\circ \;=\; +28.0$ eu (1 eu = 1 cal/kelvin·mol).
(a) Compute the value of ΔG° for dehydrogenation at room temperature (25°C or 298 K). Is dehydrogenation favored or disfavored?
(b) Compute the value of ΔG for dehydrogenation at 1000°C, assuming ΔS and ΔH are constant. Is dehydrogenation favored or disfavored?

In many ways, dehydrogenation is similar to catalytic cracking. In both cases, a catalyst is used to lower the activation energy, and both reactions use high temperatures to increase a favorable entropy term ($-T\Delta S$) and overcome an unfavorable enthalpy term (ΔH). Unfortunately, dehydrogenation and catalytic cracking also share a tendency to produce mixtures of products, and neither reaction is well suited for laboratory synthesis of alkenes.

PROBLEM-SOLVING STRATEGY
Proposing Reaction Mechanisms

At this point, we have seen examples of three major classes of reaction mechanisms:

Those involving strong bases and strong nucleophiles

Those involving strong acids and strong electrophiles

Those involving free radicals

Many students have difficulty proposing mechanisms. We can use some general principles to approach this process, however, by breaking it down into a series of logical steps. Using a systematic approach, we can usually come up with a mechanism that is at least possible and that explains the products, without requiring any unusual steps. Appendix 4 contains more complete methods for approaching mechanism problems.

First, Classify the Reaction

Before you begin to propose a mechanism, you must determine what kind of reaction you are dealing with. Examine what you know about the reactants and the reaction conditions:

A free-radical initiator such as chlorine, bromine, or a peroxide (with heat or light) suggests that a free-radical chain reaction is most likely. Free-radical reactions were discussed in detail in Chapter 4.

Strong acids or strong electrophiles (or a reactant that can dissociate to give a strong electrophile) suggest mechanisms such as the S_N1, E1, alcohol dehydration, etc., that involve carbocations and other strongly acidic intermediates.

Strong bases or strong nucleophiles suggest mechanisms such as the S_N2 or E2, involving attack by the strong base or nucleophile on a substrate.

General Principles for Drawing Mechanisms

Once you have decided which type of mechanism is most likely (acidic, basic, or free-radical), some general principles can guide you in proposing the mechanism. Some of these principles for free-radical reactions were discussed in Chapter 4. Now we consider reactions that involve either strong nucleophiles or strong electrophiles as intermediates. In later chapters, we will apply these principles to more complex mechanisms.

Whenever you start to work out a mechanism, **draw all the bonds** and all the substituents of each carbon atom affected throughout the mechanism. Three-bonded carbon atoms are likely to be the reactive intermediates. If you attempt to draw condensed formulas or line–angle formulas, you will likely misplace a hydrogen atom and show the wrong carbon atom as a radical, cation, or anion.

Show only one step at a time; never combine steps, unless two or more bonds really do change position in one step (as in the E2 reaction, for example). Protonation of an alcohol and loss of water to give a carbocation, for example, must be shown as two steps. You must not simply circle the hydroxyl and the proton to show water falling off.

Use curved arrows to show the movement of electrons in each step of the reaction. This movement is always *from* the nucleophile (electron donor) to the electrophile (electron acceptor). For example, protonation of an alcohol must show the arrow going from the electrons of the hydroxyl oxygen to the proton—never from the proton to the hydroxyl group. *Don't* use curved arrows to try to "point out" where the proton (or other reagent) goes.

Reactions Involving Strong Nucleophiles

When a strong base or nucleophile is present, we expect to see intermediates that are also strong bases and strong nucleophiles; anionic intermediates are common. Acids and electrophiles in such a reaction are generally weak. Avoid drawing carbocations, H_3O^+, and other strong acids. They are unlikely to coexist with strong bases and strong nucleophiles.

Functional groups are often converted to alkoxides, carbanions, or other strong nucleophiles by deprotonation or reaction with a strong nucleophile. Then the carbanion or other strong nucleophile reacts with a weak electrophile such as a carbonyl group or an alkyl halide.

Consider, for example, the mechanism for the dehydrohalogenation of 3-bromopentane.

$$CH_3-CH_2-CH-CH_2-CH_3 \xrightarrow{CH_3CH_2O^-} CH_3-CH=CH-CH_2-CH_3$$
$$\underset{Br}{|}$$

Someone who has not read Chapter 6 or these guidelines for classifying mechanisms might propose an ionization, followed by loss of a proton:

Incorrect mechanism

This mechanism would violate several general principles of proposing mechanisms. First, in the presence of ethoxide ion (a strong base), both the carbocation and the H^+ ion are unlikely. Second, the mechanism fails to explain why the strong base is required; the rate of ionization would be unaffected by the presence of ethoxide ion. Also, H^+ doesn't just fall off (even in an acidic reaction); it must be removed by a base.

The presence of ethoxide ion (a strong base and a strong nucleophile) in the reaction suggests that the mechanism involves only strong bases and nucleophiles and not any strongly acidic intermediates. As shown in Section 7-9A, the reaction occurs by the E2 mechanism, an example of a reaction involving a strong nucleophile. In this concerted reaction, ethoxide ion removes a proton as the electron pair left behind forms a pi bond and expels bromide ion.

Correct mechanism

Reactions Involving Strong Electrophiles

When a strong acid or electrophile is present, expect to see intermediates that are also strong acids and strong electrophiles; cationic intermediates are common. Bases and nucleophiles in such a reaction are generally weak. Avoid drawing carbanions, alkoxide ions, and other strong bases. They are unlikely to coexist with strong acids and strong electrophiles.

Functional groups are often converted to carbocations or other strong electrophiles by protonation or by reaction with a strong electrophile; then the carbocation or other strong electrophile reacts with a weak nucleophile such as an alkene or the solvent.

For example, consider the dehydration of 2,2-dimethyl-1-propanol:

The presence of sulfuric acid indicates that the reaction is acidic and should involve strong electrophiles. The carbon skeleton of the product is different from the reactant. Under these acidic conditions, formation and rearrangement of a carbocation would be likely. The hydroxyl group is a poor leaving group; it certainly cannot ionize to give a carbocation and $^-$OH (and we do not expect to see a strong base like ¯OH in this acidic reaction). The hydroxyl group is weakly basic, however, and it can become protonated in the presence of a strong acid.

Step 1: Protonation of the hydroxyl group

The protonated hydroxyl group —$\overset{+}{O}H_2$ is a good leaving group. A simple ionization to a carbocation would form a primary carbocation. Primary carbocations are very unstable however. Thus, a methyl shift occurs as water leaves, so that a primary carbocation is never formed. A tertiary carbocation results. (You can visualize this as two steps if you prefer.)

Step 2: Ionization with rearrangement

The final step is loss of a proton to a weak base, such as HSO_4^- or H_2O (but *not* ¯OH, which is incompatible with the acidic solution). Either of two types of protons, labeled 1 and 2 below, could be lost to give alkenes. Loss of proton 2 gives the required product.

Step 3: Abstraction of a proton to form the required product

Because abstraction of proton 2 gives the more highly substituted (therefore more stable) product, the Saytzeff rule predicts it will be the major product. Note that in other problems, however, you may be asked to propose mechanisms to explain unusual compounds that are only minor products.

PROBLEM 7-27

For practice in recognizing mechanisms, classify each reaction according to the type of mechanism you expect:
(1) Free-radical chain reaction
(2) Reaction involving strong bases and strong nucleophiles
(3) Reaction involving strong acids and strong electrophiles

(b) $\xrightarrow[\text{H}_2\text{O}]{\text{H}^+}$

(c) styrene $\xrightarrow[\text{heat}]{}$ polystyrene

(d) ethylene $\xrightarrow{\text{BF}_3}$ polyethylene

PROBLEM-SOLVING HINT

Alcohol dehydrations usually go through E1 elimination of the protonated alcohol, with a carbocation intermediate.

PROBLEM 7-28

Propose mechanisms for the following reactions:

(a) $CH_3—CH_2—CH_2—CH_2—OH \xrightarrow{H_2SO_4,\ 140°C} CH_3—CH=CH—CH_3$

(*Hint:* Hydride shift)

(b) $\xrightarrow{\text{NaOCH}_3}$

PROBLEM 7-29

Propose mechanisms for the following reactions:

(a) $\xrightarrow[\text{heat}]{\text{H}_3\text{PO}_4}$

(b) $\xrightarrow[\text{heat}]{\text{H}_2\text{SO}_4}$

(c) $\xrightarrow[\text{heat}]{\text{H}_2\text{SO}_4}$

SUMMARY   Methods for Synthesis of Alkenes

1. *Dehydrohalogenation of alkyl halides* (Section 7-9)

Example

chlorocyclooctane   cyclooctene

2. *Dehalogenation of vicinal dibromides* (Section 7-9D)

Example

3. *Dehydration of alcohols* (Section 7-10)

Example

cyclohexanol   cyclohexene

4. *Dehydrogenation of alkanes* (Section 7-11B)

(Useful only for small alkenes; commonly gives mixtures.)

Example

$$CH_3CH_2CH_2CH_3 \xrightarrow{\text{Pt, 500°C}} \begin{cases} \text{1-butene} + cis\text{- and } trans\text{-2-butene} + \\ \text{1,3-butadiene} + H_2 \end{cases}$$

5. *Hofmann and Cope eliminations* (Sections 19-15 and 19-16)

(Usually gives the least highly substituted alkene.)

Example

$$CH_3-CH_2-\underset{\underset{I^-}{\overset{|}{\underset{^+N(CH_3)_3}{}}}}{CH}-CH_3 \xrightarrow{\text{Ag}_2\text{O, heat}} CH_3-CH_2-CH=CH_2 + :N(CH_3)_3$$

6. *Reduction of alkynes* (Section 9-9)

cis-alkene

$$R-C\equiv C-R' \xrightarrow{\text{Na, NH}_3} \underset{H}{\overset{R}{\diagdown}} C=C \underset{R'}{\overset{H}{\diagup}} \quad \text{trans-alkene}$$

Examples

$$CH_3CH_2-C\equiv C-CH_2CH_3 \xrightarrow[\text{quinoline}]{\text{H}_2, \text{Pd/BaSO}_4} \underset{H}{\overset{CH_3CH_2}{\diagdown}} C=C \underset{H}{\overset{CH_2CH_3}{\diagup}}$$

$$CH_3CH_2-C\equiv C-CH_2CH_3 \xrightarrow{\text{Na, NH}_3} \underset{H}{\overset{CH_3CH_2}{\diagdown}} C=C \underset{CH_2CH_3}{\overset{H}{\diagup}}$$

7. *Wittig reaction* (Section 18-13)

$$\underset{R}{\overset{R'}{\diagdown}} C=O \ + \ Ph_3P=CHR'' \ \longrightarrow \ \underset{R}{\overset{R'}{\diagdown}} C=CHR'' \ + \ Ph_3P=O$$

Example

$$\bigⓅentagon C=O \ + \ Ph_3P=CHCH_3 \ \longrightarrow \ \bigⓅentagon C=C\underset{CH_3}{\overset{H}{\diagup}} \ + \ Ph_3P=O$$

cyclopentanone

Chapter 7 Glossary

alkene (olefin) A hydrocarbon with one or more carbon–carbon double bonds. (p. 272)

diene: A compound with two carbon–carbon double bonds. (p. 277)

triene: A compound with three carbon–carbon double bonds. (277)

tetraene: A compound with four carbon–carbon double bonds. (277)

allyl group A vinyl group plus a methylene group: $CH_2{=}CH{-}CH_2{-}$ (p. 278)

Bredt's rule A stable bridged bicyclic compound cannot have a double bond at a bridgehead position unless one of the rings contains at least eight carbon atoms. (p. 288)

bicyclic: Containing two rings.

bridged bicyclic: Having at least one carbon atom in each of the three links connecting the bridgehead carbons.

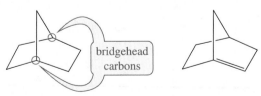

a bridged bicyclic compound   a Bredt's rule violation

bridgehead carbons: Those carbon atoms that are part of both rings, with three bridges of bonds connecting them.

catalytic cracking The heating of petroleum products in the presence of a catalyst (usually an aluminosilicate mineral), causing bond cleavage to form alkenes and alkanes of lower molecular weight. (p. 301)

cis-trans isomers (geometric isomers) Isomers that differ in their cis-trans arrangement on a ring or double bond. Cis-trans isomers are a subclass of diastereomers. (p. 278)

    cis: Having similar groups on the same side of a double bond or a ring.

    trans: Having similar groups on opposite sides of a double bond or a ring.

    Z: Having the higher-priority groups on the same side of a double bond.

    E: Having the higher-priority groups on opposite sides of a double bond.

dehalogenation The elimination of a halogen (X_2) from a compound. Dehalogenation is formally a reduction. (p. 296)

$$-\overset{|}{\underset{|}{C}}\overset{X}{}-\overset{X}{\underset{|}{C}}-\quad\xrightarrow[\text{(or NaI in acetone)}]{\text{Zn, }CH_3COOH}\quad \overset{\diagdown}{\diagup}C=C\overset{\diagup}{\diagdown}\ +\ ZnX_2$$

dehydration The elimination of water from a compound; usually acid-catalyzed. (p. 300)

$$-\overset{|}{\underset{|}{C}}\overset{H}{}-\overset{OH}{\underset{|}{C}}-\quad\xrightleftharpoons{H^+}\quad \overset{\diagdown}{\diagup}C=C\overset{\diagup}{\diagdown}\ +\ H_2O$$

dehydrogenation The elimination of hydrogen (H_2) from a compound; usually done in the presence of a catalyst. (p. 302)

$$-\overset{|}{\underset{|}{C}}\overset{H}{}-\overset{H}{\underset{|}{C}}-\quad\xrightleftharpoons{\text{Pt, high temperature}}\quad \overset{\diagdown}{\diagup}C=C\overset{\diagup}{\diagdown}\ +\ H_2$$

dehydrohalogenation The elimination of a hydrogen halide (HX) from a compound; usually base-promoted. (p. 291)

$$-\overset{|}{\underset{|}{C}}\overset{H}{}-\overset{X}{\underset{|}{C}}-\quad\xrightarrow{\text{KOH}}\quad \overset{\diagdown}{\diagup}C=C\overset{\diagup}{\diagdown}\ +\ H_2O\ +\ K^+X^-$$

double-bond isomers Constitutional isomers that differ only in the position of a double bond. Double-bond isomers hydrogenate to give the same alkane. (p. 284)

element of unsaturation A structural feature that causes a reduction of two hydrogen atoms in the molecular formula. A double bond or a ring is one element of unsaturation; a triple bond is two elements of unsaturation. (p. 274)

geminal dihalide A compound with two halogen atoms on the same carbon atom.

geometric isomers See **cis-trans isomers**. (p. 278)

heteroatom Any atom other than carbon or hydrogen. (p. 275)

Hofmann product The least highly substituted alkene product. (p. 292)

hydrogenation Addition of hydrogen to a molecule. The most common hydrogenation is the addition of H_2 across a double bond in the presence of a catalyst (*catalytic hydrogenation*). The value of $(-\Delta H°)$ for this reaction is called the **heat of hydrogenation**. (p. 283)

$$\overset{\diagdown}{\diagup}C=C\overset{\diagup}{\diagdown}\ +\ H_2\quad\xrightarrow{\text{Pt}}\quad -\overset{|}{\underset{|}{C}}\overset{H}{}-\overset{H}{\underset{|}{C}}-\qquad -\Delta H°\ =\ \text{heat of hydrogenation}$$

olefin An alkene (p. 272)

polymer A substance of high molecular weight made by linking many small molecules, called **monomers**. (p. 282)

    addition polymer: A polymer formed by simple addition of monomer units.

    polyolefin: A type of addition polymer with an olefin serving as the monomer.

saturated Having only single bonds; incapable of undergoing addition reactions. (p. 274)

Saytzeff rule (Zaitsev rule) An elimination usually gives the most stable alkene product, commonly the most highly substituted alkene. The Saytzeff rule does not always apply, especially with a bulky base or a bulky leaving group. (pp. 284)

    Saytzeff elimination: An elimination that gives the Saytzeff product.

    Saytzeff product: The most highly substituted alkene product.

stereospecific reaction A reaction in which different stereoisomers react to give different stereoisomers of the product. (p. 293).

trans-diaxial An anti and coplanar arrangement allowing E2 elimination of two adjacent substituents on a cyclohexane ring. The substituents must be trans to each other, and both must be in axial positions on the ring. (p. 294)

unsaturated Having multiple bonds that can undergo addition reactions. (p. 274)

vicinal dihalide A compound with two halogens on adjacent carbon atoms. (p. 296)

vinyl group An ethenyl group, $CH_2\!=\!CH\!-$. (p. 278)

Zaitsev Alternate spelling of **Saytzeff**.

Essential Problem-Solving Skills in Chapter 7

1. Draw and name all alkenes with a given molecular formula.
2. Use the *E-Z* and cis-trans systems to name geometric isomers.
3. Use heats of hydrogenation to compare stabilities of alkenes.
4. Predict relative stabilities of alkenes and cycloalkenes, based on structure and stereochemistry.
5. Predict the products of dehydrohalogenation of alkyl halides, dehalogenation of dibromides, and dehydration of alcohols, including major and minor products.
6. Propose logical mechanisms for dehydrohalogenation, dehalogenation, and dehydration reactions.
7. Predict and explain the stereochemistry of E2 eliminations to form alkenes. Predict E2 reactions on cyclohexane systems.
8. Propose effective single-step and multistep syntheses of alkenes.

Study Problems

7-30 Define each term and give an example.
  (a) double-bond isomers
  (b) Saytzeff elimination
  (c) element of unsaturation
  (d) Hofmann product
  (e) Bredt's rule violation
  (f) hydrogenation
  (g) dehydrogenation
  (h) dehydrohalogenation
  (i) dehydration
  (j) dehalogenation
  (k) geminal dihalide
  (l) vicinal dihalide
  (m) heteroatom
  (n) polymer

7-31 Draw a structure for each compound.
  (a) 3-methyl-1-pentene
  (b) 3,4-dibromo-1-butene
  (c) 1,3-cyclohexadiene
  (d) (Z)-3-methyl-2-octene
  (e) vinylcyclopropane
  (f) (Z)-2-bromo-2-pentene
  (g) (3Z,6E)-1,3,6-octatriene

7-32 Give a correct name for each compound.

(a) $CH_3\!-\!CH_2\!-\!\underset{\underset{\displaystyle CH_2}{\|}}{C}\!-\!CH_2\!-\!CH_2\!-\!CH_3$

(b) $(CH_3CH_2)_2C\!=\!CHCH_3$

(c)

(d)

(e)

7-33 Label each structure as *Z*, *E*, or neither.

(a)   (b)   (c) (structure) (d) (structure)

7-34 (a) Draw and name all five isomers of formula C_3H_5F.
  (b) Cholesterol, $C_{27}H_{46}O$, has only one pi bond. What else can you say about its structure?

7-35   Draw and name all stereoisomers of 3-methyl-2,4-hexadiene
    (a) using the cis-trans nomenclature   (b) using the *E-Z* nomenclature

7-36   Determine which compounds show cis-trans isomerism. Draw and label the isomers, using both the cis-trans and *E-Z* nomenclatures where applicable.
    (a) 1-pentene   (b) 2-pentene   (c) 3-hexene
    (d) 1,1-dibromopropane   (e) 1,2-dibromopropane   (f) 2,4-hexadiene

7-37   For each alkene, indicate the direction of the dipole moment. For each pair, determine which compound has the larger dipole moment.
    (a) *cis*-1,2-difluoroethene   or   *trans*-1,2-difluoroethene
    (b) *cis*-1,2-dibromoethene   or   *trans*-2,3-dibromo-2-butene
    (c) *cis*- or *trans*-1,2-dibromo-1,2-dichloroethene
    (d) *cis*-1,2-dibromo-1,2-dichloroethene   or   *cis*-1,2-dichloroethene

7-38   Predict the products of the following reactions. When more than one product is expected, predict which will be the major product.

7-39   Write a balanced equation for each reaction.

7-40   Show how you would prepare cyclopentene from each compound.
    (a) *trans*-1,2-dibromocyclopentane   (b) cyclopentanol
    (c) cyclopentyl bromide   (d) cyclopentane (not by dehydrogenation)

7-41   Predict the products formed by sodium hydroxide–promoted dehydrohalogenation of the following compounds. In each case, predict which will be the major product.
    (a) 1-bromobutane   (b) 2-chlorobutane   (c) 3-bromopentane
    (d) 1-bromo-1-methylcyclohexane   (e) 1-bromo-2-methylcyclohexane

7-42   What halides would undergo dehydrohalogenation to give the following pure alkenes?
    (a) 1-butene   (b) isobutylene   (c) 2-pentene
    (d) methylenecyclohexane   (e) 4-methylcyclohexene

7-43   In the dehydrohalogenation of alkyl halides, a strong base such as *t*-butoxide usually gives the best results via the E2 mechanism.
    (a) Explain why a strong base such as *t*-butoxide cannot dehydrate an alcohol through the E2 mechanism.
    (b) Explain why strong acid, used in the dehydration of an alcohol, is not effective in the dehydrohalogenation of an alkyl halide.

7-44   Predict the major products of dehydration of the following alcohols.
    (a) 2-pentanol   (b) 1-methyleyclopentanol   (c) 2-methylcyclohexanol   (d) 2,2-dimethyl-1-propanol

7-45   Dehydration of 2-methylcyclopentanol gives a mixture of three alkenes. Propose mechanisms to account for these three products.

7-46   Predict the dehydrohalogenation product(s) that result when the following alkyl halides are heated in alcoholic KOH. When more than one product is formed, predict the major and minor products.

(a) $(CH_3)_2CH-C(CH_3)_2$
            |
            Br

(b) $(CH_3)_2CH-CH-CH_3$
                     |
                     Br

(c) $(CH_3)_2C-CH_2-CH_3$
                   |
                   Br

(d) [cyclohexane ring with CH₃ and Cl substituents]

(e) [cyclohexane ring with Cl and CH₃ substituents]

7-47   E1 eliminations of alkyl halides are rarely useful for synthetic purposes, because they give mixtures of substitution and elimination products. Explain why the sulfuric acid–catalyzed dehydration of cyclohexanol gives a good yield of cyclohexene even though the reaction goes by an E1 mechanism. (*Hint*: What are the nucleophiles in the reaction mixture? What products are formed if these nucleophiles attack the carbocation? What further reactions can these substitution products undergo?)

7-48   The following reaction is called the **pinacol rearrangement**. The reaction begins with an acid-promoted ionization to give a carbocation. This carbocation undergoes a methyl shift to give a more stable, resonance-stabilized cation. Loss of a proton gives the observed product. Propose a mechanism for the pinacol rearrangement.

$$\begin{array}{ccc} H_3C & CH_3 \\ | & | \\ CH_3-C-C-CH_3 & \xrightarrow{H_2SO_4,\ heat} & CH_3-C-C-CH_3 \\ | & | \\ HO & OH \end{array}$$

pinacol   pinacolone

7-49   Propose a mechanism to explain the formation of two products in the following reaction:

[methylenecyclohexane] $\xrightarrow{NBS,\ hv}$ [bromomethylenecyclohexane with Br] + [cyclohexene-CH₂Br]

7-50   A chemist allows some pure (2S,3R)-3-bromo-2,3-diphenylpentane to react with a solution of sodium ethoxide (NaOCH₂CH₃) in ethanol. The products are two alkenes: **A** (cis-trans mixture) and **B**, a single pure isomer. Under the same conditions, the reaction of (2S,3S)-3-bromo-2,3-diphenylpentane gives two alkenes, **A** (cis-trans mixture) and **C**. Upon catalytic hydrogenation, all three of these alkenes (**A**, **B**, and **C**) give 2,3-diphenylpentane. Determine the structures of **A**, **B**, and **C**, give equations for their formation, and explain the stereospecificity of these reactions.

7-51   The energy difference between *cis*- and *trans*-2-butene is about 1 kcal/mol; however, the trans isomer of 4,4-dimethyl-2-pentene is 3.8 kcal/mol more stable than the cis isomer. Explain this large difference.

7-52   A double bond in a six-membered ring is usually more stable in an endocyclic position than in an exocyclic position. Hydrogenation data on two pairs of compounds are given below. One pair suggests that the energy difference between endocyclic and exocyclic double bonds is about 2.1 kcal. The other pair suggests an energy difference of about 1.2 kcal. Which number do you trust as being more representative of the actual energy difference? Explain your answer.

[six ring structures with double bonds]

endocyclic   exocyclic   25.7   27.8   25.1   26.3

heats of hydrogenation (kcal/mol)

7-53   Predict the products of the following eliminations of vicinal dibromides with potassium iodide. Remember to consider the geometric constraints of the E2 reaction.

(a) $CH_3-CH_2-CH-CH_2-Br$
                        |
                        Br

(b) [cyclohexane with two Br, trans]

(c) [cyclohexane with two Br]

(d) [decalin ring system with H, Br, Br, H]

(e) [decalin ring system with H, Br, H, H, Br]

7-54   One of the following dichloronorbornanes undergoes elimination much faster than the other. Determine which one reacts faster, and explain the large difference in rates.

cis   trans

*7-55   A graduate student wanted to make methylenecyclobutane, and he tried the following reaction:

$$\triangleright\!\!-CH_2OH \quad \xrightarrow[\text{heat}]{H_2SO_4} \quad \square\!\!= \quad + \quad \text{other products}$$

methylenecyclobutane
(minor)

Propose structures for the other products, and give mechanisms to account for their formation.

*7-56   Give a mechanism to explain the formation of the following product. In your mechanism, explain the cause of the rearrangement, and explain the failure to form the Saytzeff product.

Reactions of Alkenes

All alkenes have a common feature: a carbon–carbon double bond. The reactions of alkenes arise from the reactivity of the carbon–carbon double bond. Once again, the concept of the functional group helps to organize and simplify the study of chemical reactions. By studying the characteristic reactions of the double bond, we can predict the reactions of alkenes we have never seen before.

8-1
Reactivity of the Carbon–Carbon Double Bond

Because single bonds (sigma bonds) are more stable than pi bonds, the most common reactions of double bonds are to transform the pi bond into a sigma bond. For example, catalytic hydrogenation converts the $C=C$ pi bond and the $H-H$ sigma bond into two $C-H$ sigma bonds (Section 7-7). The reaction is exothermic ($\Delta H° =$ about -20 to -30 kcal/mol or about -80 to -120 kJ/mol), showing that the product is more stable than the reactants.

$$\begin{array}{ccc}\diagdown \!\!\!\!\diagup \\ C=C \\ \diagup \!\!\!\!\diagdown \end{array} \quad + \quad H-H \quad \xrightarrow{\text{catalyst}} \quad \begin{array}{c} | \quad | \\ -C-C- \\ | \quad | \\ H \quad H \end{array} \quad + \quad \text{energy}$$

Hydrogenation of an alkene is an example of an **addition**, one of the three major reaction types we have studied: addition, elimination, and substitution. In an addition, two molecules combine to form one product molecule. When an alkene undergoes addition, two groups add to the carbon atoms of the double bond, and the carbons become saturated. In many ways, addition is the reverse of **elimination**, in which one molecule is split into two fragment molecules. In a **substitution**, one fragment replaces another fragment in a molecule.

Addition

$$\begin{array}{ccc}\diagdown \!\!\!\!\diagup \\ C=C \\ \diagup \!\!\!\!\diagdown \end{array} \quad + \quad X-Y \quad \longrightarrow \quad \begin{array}{c} | \quad | \\ -C-C- \\ | \quad | \\ X \quad Y \end{array}$$

Elimination

$$\begin{array}{c} | \quad | \\ -C-C- \\ | \quad | \\ X \quad Y \end{array} \quad \longrightarrow \quad \begin{array}{ccc}\diagdown \!\!\!\!\diagup \\ C=C \\ \diagup \!\!\!\!\diagdown \end{array} \quad + \quad X-Y$$

Substitution

$$\begin{array}{c} | \\ -C-X \\ | \end{array} \quad + \quad Y^- \quad \longrightarrow \quad \begin{array}{c} | \\ -C-Y \\ | \end{array} \quad + \quad X^-$$

Addition is the most common reaction of alkenes, and we will consider additions in detail. A wide variety of functional groups can be formed by adding suitable reagents to the double bonds of alkenes.

In principle, many different reagents could add to a double bond to form more stable products; that is, the reactions are energetically favorable. Not all of these reactions have convenient rates, however. For example, the reaction of ethylene with hydrogen (to give ethane) is strongly exothermic, but the rate is very slow. A mixture of ethylene and hydrogen can remain for years without appreciable reaction. Adding a catalyst such as platinum, palladium, or nickel allows the reaction to take place at a rapid rate.

Some reagents react with carbon–carbon double bonds without the aid of a catalyst. To understand what types of reagents react with double bonds, consider the structure of the pi bond. Although the electrons in the sigma bond framework are tightly held, the pi bond is delocalized above and below the sigma bond (Figure 8-1). The pi–bonding electrons are spread farther from the carbon nuclei, and they are more loosely held. A strong electrophile has an affinity for these loosely held electrons; it can pull them away to form a new bond (Figure 8-2), leaving one of the carbon atoms with only three bonds and a positive charge: a carbocation. In effect, the double bond has reacted as a nucleophile, donating a pair of electrons to the electrophile.

In most additions, a nucleophile attacks the carbocation (as in the second step of the S_N1 reaction), forming a stable addition product. In the product, both the electrophile and the nucleophile are bonded to the carbon atoms that were connected by the double bond. The following schematic reaction uses E^+ as the electrophile and $Nuc:^-$ as the nucleophile.

This type of reaction requires a strong electrophile to attract the electrons of the pi bond and generate a carbocation in the rate-limiting step. Most alkene reactions fall into this large class of **electrophilic additions** to alkenes.

8-2
Electrophilic Addition to Alkenes

▲ **Figure 8-1**
The electrons in the pi bond are spread farther from the carbon nuclei and are more loosely held than the sigma electrons.

KEY MECHANISM   Electrophilic Addition to Alkenes

Step 1: Attack of the pi bond on the electrophile forms a carbocation.

$$\overset{}{C}=\overset{}{C} \quad + \quad E^+ \quad \longrightarrow \quad -\overset{}{\underset{E}{C}}-\overset{+}{C}$$

+ on the more substituted carbon

Step 2: Attack by a nucleophile gives the product of addition.

$$-\overset{}{\underset{E}{C}}-\overset{+}{C} \quad + \quad Nuc:^- \quad \longrightarrow \quad -\overset{}{\underset{E}{C}}-\overset{}{\underset{Nuc}{C}}-$$

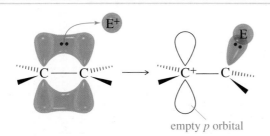

empty *p* orbital

◀ **Figure 8-2**
A strong electrophile attracts the electrons out of the pi bond to form a new sigma bond, generating a carbocation. The (red) curved arrow shows the movement of electrons, from the electron-rich pi bond to the electron-poor electrophile.

To illustrate an electrophilic addition, consider what happens when gaseous HBr adds to 2-butene. The proton in HBr is electrophilic; it reacts with the alkene to form a carbocation. Bromide ion reacts rapidly with the carbocation to give a stable product in which the elements of HBr have added to the ends of the double bond.

MECHANISM   Ionic Addition of HBr to an Alkene

Step 1: Protonation of the double bond forms a carbocation.

$$CH_3-\overset{\overset{\displaystyle H}{|}}{C}=\overset{\overset{\displaystyle H}{|}}{C}-CH_3 \quad\rightleftharpoons\quad CH_3-\overset{\overset{\displaystyle H}{|}}{\underset{\underset{\displaystyle H}{|}}{C}}-\overset{\overset{\displaystyle H}{|}}{\underset{+}{C}}-CH_3 \quad+\quad :\ddot{Br}:^-$$

$$H-\ddot{Br}:$$

Step 2: Bromide ion attacks the carbocation.

$$CH_3-\overset{\overset{\displaystyle H}{|}}{\underset{\underset{\displaystyle H}{|}}{C}}-\overset{\overset{\displaystyle H}{|}}{\underset{+}{C}}-CH_3 \quad+\quad :\ddot{Br}:^- \quad\rightleftharpoons\quad CH_3-\overset{\overset{\displaystyle H}{|}}{\underset{\underset{\displaystyle H}{|}}{C}}-\overset{\overset{\displaystyle H}{|}}{\underset{\underset{\displaystyle :\ddot{Br}:}{|}}{C}}-CH_3$$

We will consider several types of additions to alkenes, using a wide variety of reagents: water, borane, hydrogen, carbenes, halogens, oxidizing agents, and even other alkenes. Most, but not all, of these will be electrophilic additions. Table 8-1 summarizes the classes of additions we will cover. Note that the table shows what elements have added across the double bond in the final product, but it says nothing about reagents or mechanisms. As we study these reactions, you should note the *regiochemistry* of each reaction, also called the *orientation of addition*, meaning which part of the reagent adds to which end of the double bond. Also note the *stereochemistry* if the reaction is stereospecific.

TABLE 8-1 Types of Additions to Alkenes

$\diagdown \atop \diagup$ C=C $\diagup \atop \diagdown$	Type of Addition [Elements Added]ᵃ →	Product
	hydration [H₂O] →	$-\overset{H}{\underset{\mid}{\overset{\mid}{C}}}-\overset{OH}{\underset{\mid}{\overset{\mid}{C}}}-$
	hydrogenation [H₂], a reduction →	$-\overset{H}{\underset{\mid}{\overset{\mid}{C}}}-\overset{H}{\underset{\mid}{\overset{\mid}{C}}}-$
	hydroxylation [HOOH], an oxidation →	$-\overset{OH}{\underset{\mid}{\overset{\mid}{C}}}-\overset{OH}{\underset{\mid}{\overset{\mid}{C}}}-$
	oxidative cleavage [O₂], an oxidation →	$\diagup \atop \diagdown$ C=O O=C $\diagdown \atop \diagup$
	epoxidation [O], an oxidation →	$-\overset{O}{\overset{\diagup\diagdown}{C-C}}-$
	halogenation [X₂], an oxidation →	$-\overset{X}{\underset{\mid}{\overset{\mid}{C}}}-\overset{X}{\underset{\mid}{\overset{\mid}{C}}}-$
	halohydrin formation [HOX], an oxidation →	$-\overset{X}{\underset{\mid}{\overset{\mid}{C}}}-\overset{OH}{\underset{\mid}{\overset{\mid}{C}}}-$
	HX addition [HX] →	$-\overset{H}{\underset{\mid}{\overset{\mid}{C}}}-\overset{X}{\underset{\mid}{\overset{\mid}{C}}}-$
	cyclopropanation [CH₂] →	$-\overset{\overset{\displaystyle H\diagup\overset{C}{\diagdown}H}{}}{C-C}-$

ᵃThese are not the reagents used but simply the groups that appear in the product.

8-3A Orientation of Addition: Markovnikov's Rule

The simple mechanism shown for addition of HBr to 2-butene applies to a large number of electrophilic additions. We can use this mechanism to predict the outcome of some fairly complicated reactions. For example, the addition of HBr to 2-methyl-2-butene could lead to either of two products, yet only one is observed.

The first step is protonation of the double bond. If the proton adds to the secondary carbon, the product will be different from the one formed if the proton adds to the tertiary carbon.

PROBLEM-SOLVING HINT

Stability of carbocations:

$$3° > 2° > 1° > {}^+CH_3$$

When the proton adds to the secondary carbon, a tertiary carbocation results. When the proton adds to the tertiary carbon atom, a secondary carbocation results. The tertiary carbocation is more stable (see Section 4-16A), so the first reaction is favored.

The second half of the mechanism shows the final product of the reaction of 2-methyl-2-butene with HBr:

Note that protonation of one carbon atom of a double bond gives a carbocation on the carbon atom that was *not* protonated. Therefore, the proton adds to the end of the double bond that is *less* substituted to give the *more substituted carbocation* (the more stable carbocation).

Figure 8-3 shows examples of additions where the proton has added to the less highly substituted carbon atom of the double bond. The addition of HBr is said to be **regioselective** because in each case, one of the two possible orientations of addition is formed preferentially over the other.

Markovnikov's Rule   A Russian chemist, Vladimir Markovnikov, first showed the orientation of addition of HBr to alkenes in 1869. Markovnikov stated:

> MARKOVNIKOV'S RULE: The addition of a proton acid to the double bond of an alkene results in a product with the acid proton bonded to the carbon atom that already holds the greater number of hydrogen atoms.

▶ **Figure 8-3**
An electrophile adds to the less substituted end of the double bond to give the more highly substituted (and therefore more stable) carbocation.

This is the original statement of **Markovnikov's rule**. Reactions that follow this rule are said to follow **Markovnikov orientation** and give the **Markovnikov product**. We are often interested in adding electrophiles other than proton acids to the double bonds of alkenes. Markovnikov's rule can be extended to include a wide variety of other additions, based on the addition of the electrophile in such a way as to produce the most stable carbocation.

 MARKOVNIKOV'S RULE (extended):   In an electrophilic addition to an alkene, the electrophile adds in such a way as to generate the most stable intermediate.

 Like HBr, both HCl and HI add to the double bonds of alkenes, and they also follow Markovnikov's rule; for example,

---

PROBLEM 8-1

Predict the major products of the following reactions.

(a) CH_3—CH=CH_2 + HBr   (b) 2-methylpropene + HCl

(c) 1-methylcyclohexene + HI   (d) 4-methylcyclohexene + HBr

PROBLEM 8-2

When 1,3-butadiene reacts with 1 mol of HBr, both 3-bromo-1-butene and 1-bromo-2-butene are formed. Propose a mechanism to account for this mixture of products.

8-3B   Free-Radical Addition of HBr: Anti-Markovnikov Addition

In 1933, M. S. Kharasch and F. W. Mayo showed that **anti-Markovnikov products** result from addition of HBr (but not HCl or HI) in the presence of peroxides. Peroxides give rise to free radicals that act as catalysts to accelerate the addition, causing it to occur by a different mechanism. The oxygen–oxygen bond in peroxides is rather weak. It can break to give two radicals.

$$R—\ddot{O}—\ddot{O}—R \xrightarrow{\text{heat}} R—\ddot{O}\cdot \ + \ \cdot\ddot{O}—R \qquad \Delta H° \ = \ +36 \text{ kcal } (+150 \text{ kJ})$$

Alkoxy radicals (R—O·) catalyze the anti-Markovnikov addition of HBr. The mechanism of this free-radical chain reaction is shown below.

MECHANISM   Free-Radical Addition of HBr to an Alkene

Initiation

$$R—O—O—R \xrightarrow{\text{heat}} R—O\cdot + \cdot O—R$$

$$R—O\cdot + H—Br \longrightarrow R—O—H + Br\cdot$$

Propagation

radical on the more substituted carbon

$$\text{C}=\text{C} + Br\cdot \longrightarrow \overset{\displaystyle |}{\underset{\displaystyle Br}{-C}} - C\colon$$

$$\overset{\displaystyle |}{\underset{\displaystyle Br}{-C}} - C\colon \ + \ H—Br \longrightarrow \underset{\displaystyle Br \ H}{-C—C-} \ + \ Br\cdot$$

Let's consider the individual steps. In the initiation step, free radicals generated from the peroxide react with HBr to form bromine radicals.

$$R—\ddot{O}\cdot + H—\ddot{Br}\colon \longrightarrow R—\ddot{O}—H + \colon\ddot{Br}\cdot \qquad \Delta H° = -15 \text{ kcal } (-63 \text{ kJ})$$

The bromine radical lacks an octet of electrons in its valence shell, making it electron-deficient and electrophilic. It adds to a double bond, forming a new free radical with the odd electron on a carbon atom.

$$\colon\ddot{Br}\cdot + \text{C}=\text{C} \longrightarrow \underset{\displaystyle \colon\ddot{Br}\colon}{-C—C\colon} \qquad \Delta H° = -3 \text{ kcal } (-12 \text{ kJ})$$

This free radical reacts with an HBr molecule to form a C—H bond and generate another bromine radical.

$$\text{—C—C·} + \text{H—Br:} \longrightarrow \text{—C—C—} + \text{:Br·} \qquad \Delta H^\circ = -6 \text{ kcal } (-25 \text{ kJ})$$

$$\underset{\text{Br}}{} \qquad\qquad\qquad \underset{\text{Br H}}{}$$

The regenerated bromine radical reacts with another molecule of the alkene, continuing the chain reaction. Note that each propagation step starts with one free radical and ends with another free radical. The number of free radicals is constant, until free radicals come together and terminate the chain reaction.

Radical Addition of HBr to Unsymmetrical Alkenes Now we must explain the anti-Markovnikov orientation found in the products of the peroxide-catalyzed reaction. When the alkene is unsymmetrical, adding the bromine radical to the secondary end of the double bond forms a tertiary radical.

$$\underset{\text{CH}_3}{\text{CH}_3-\overset{|}{\text{C}}=\text{CH}-\text{CH}_3} + \text{Br·} \longrightarrow \underset{\text{Br}}{\underset{|}{\text{CH}_3-\overset{\overset{\text{CH}_3}{|}}{\text{C}}-\text{CH}-\text{CH}_3}} \quad \text{but not} \quad \underset{\text{Br}}{\underset{|}{\text{CH}_3-\overset{\overset{\text{CH}_3}{|}}{\text{C}}-\text{CH}-\text{CH}_3}}$$

tertiary radical (more stable)   secondary radical (less stable)

As we saw in the protonation of an alkene, the electrophile (in this case, Br·) adds to the less substituted end of the double bond, and the radical electron appears on the more substituted carbon to give the more stable free radical. This intermediate reacts with HBr to give the anti-Markovnikov product, in which H has added to the more substituted end of the double bond: the end that started with *fewer* hydrogens.

$$\underset{\text{Br}}{\underset{|}{\text{CH}_3-\overset{\overset{\text{CH}_3}{|}}{\text{C}}-\text{CH}-\text{CH}_3}} + \text{H—Br} \longrightarrow \underset{\text{H Br}}{\underset{|\ \ |}{\text{CH}_3-\overset{\overset{\text{CH}_3}{|}}{\text{C}}-\text{CH}-\text{CH}_3}} + \text{Br·}$$

anti-Markovnikov product

PROBLEM-SOLVING HINT

Stability of radicals

$$3° > 2° > 1° > ·\text{CH}_3$$

Note that *both* mechanisms for the addition of HBr to an alkene (with and without peroxides) follow our extended statement of Markovnikov's rule: In both cases, the electrophile adds to the less substituted end of the double bond to give the more stable carbocation or free radical. In the ionic reaction, the electrophile is H⁺. In the peroxide-catalyzed free-radical reaction, the electrophile is Br·.

Many students wonder why the reaction with Markovnikov orientation does not take place in the presence of peroxides, together with the free-radical chain reaction. It actually does take place, but the peroxide-catalyzed reaction is faster. If just a tiny bit of peroxide is present, a mixture of Markovnikov and anti-Markovnikov products results. If an appreciable amount of peroxide is present, the radical chain reaction is so much faster than the uncatalyzed ionic reaction that only the anti-Markovnikov product is observed.

The reversal of orientation in the presence of peroxides is called the **peroxide effect**. It occurs only with the addition of HBr to alkenes. The reaction of an alkyl radical with HCl is strongly endothermic, so the free-radical chain reaction is not effective for the addition of HCl.

$$\underset{|}{\text{Cl—C—C·}} + \text{H—Cl} \longrightarrow \underset{|\ \ |}{\text{Cl—C—C—H}} + \text{Cl·} \qquad \Delta H^\circ = +10 \text{ kcal } (+42 \text{ kJ})$$

Similarly, the reaction of an iodine atom with an alkene is strongly endothermic, and the free-radical addition of HI is not observed. Only HBr has just the right reactivity for each step of the free-radical chain reaction to take place.

$$I\cdot \ + \ \underset{}{\overset{}{C}}{=}\underset{}{\overset{}{C}} \quad \longrightarrow \quad I{-}\underset{|}{\overset{|}{C}}{-}\overset{}{C}\cdot \qquad \Delta H^\circ = +13 \text{ kcal } (+54 \text{ kJ})$$

PROBLEM 8-3

Predict the major products of the following reactions, and propose mechanisms to support your predictions.

(a) 2-methylpropene + HBr + $CH_3{-}\overset{\overset{\displaystyle O}{\|}}{C}{-}O{-}O{-}\overset{\overset{\displaystyle O}{\|}}{C}{-}CH_3$

(b) 1-methylcyclopentene + HBr + $CH_3CH_2{-}O{-}O{-}CH_2CH_3$

(c) 1-phenylpropene + HBr + di-*t*-butyl peroxide $\left(\text{phenyl} \ = \ Ph \ = \ \text{[ring]}{-}\right)$

PROBLEM-SOLVING HINT
Remember to write out complete structures, including all bonds and charges, when writing a mechanism or determining the course of a reaction.

SOLVED PROBLEM 8-1

Show how you would accomplish the following synthetic conversions.
(a) Convert 1-methylcyclohexene to 1-bromo-1-methylcyclohexane.

SOLUTION

This synthesis requires the addition of HBr to an alkene with Markovnikov orientation. Ionic addition of HBr gives the correct product.

1-methylcyclohexene   +   HBr   ⟶   1-bromo-1-methylcyclohexane

(b) Convert 1-methylcyclohexanol to 1-bromo-2-methylcyclohexane.

SOLUTION

This synthesis requires the conversion of an alcohol to an alkyl bromide with the bromine atom at the neighboring carbon atom. This is the anti-Markovnikov product, which could be formed by the radical-catalyzed addition of HBr to 1-methylcyclohexene.

1-methylcyclohexene   +   HBr $\xrightarrow[\text{heat}]{R{-}O{-}O{-}R}$ 1-bromo-2-methylcyclohexane

1-Methylcyclohexene is easily synthesized by the dehydration of 1-methylcyclohexanol. The most highly substitued alkene is the desired product.

1-methylcyclohexanol $\xrightarrow[\text{heat}]{H_2SO_4}$ 1-methylcyclohexene   + H_2O

The two-step synthesis is summarized as follows:

1-methylcyclohexanol   1-methylcyclohexene   1-bromo-2-methylcyclohexane

PROBLEM 8-4

Show how you would accomplish the following synthetic conversions.
(a) 1-butene ⟶ 1-bromobutane
(b) 1-butene ⟶ 2-bromobutane
(c) 2-methylcyclohexanol → 1-bromo-1-methylcyclohexane
(d) 2-methyl-2-butanol ⟶ 2-bromo-3-methylbutane

8-4
Addition of Water: Hydration of Alkenes

An alkene may react with water in the presence of a strongly acidic catalyst to form an alcohol. Formally, this reaction is a **hydration** (the addition of water), with a hydrogen atom adding to one carbon and a hydroxyl group adding to the other. Hydration of an alkene is the reverse of the dehydration of alcohols we studied in Section 7-10.

Hydration of an alkene

Dehydration of an alcohol

For dehydrating alcohols, a concentrated dehydrating acid (such as H_2SO_4 or H_3PO_4) is used to drive the equilibrium to favor the alkene. Hydration of an alkene, on the other hand, is accomplished by adding excess water to drive the equilibrium toward the alcohol.

8-4A Mechanism of Hydration

The *principle of microscopic reversibility* states that a forward reaction and a reverse reaction taking place under the same conditions (as in an equilibrium) must follow the same reaction pathway in microscopic detail. The hydration and dehydration reactions are the two complementary reactions in an equilibrium; therefore, they must follow the same reaction pathway. It makes sense that the lowest-energy transition states and intermediates for the reverse reaction are the same as those for the forward reaction, except in reverse order.

According to the principle of microscopic reversibility, we can write the hydration mechanism by reversing the order of the steps of the dehydration (Section 7-10). Protonation of the double bond forms a carbocation. Nucleophilic attack by water, followed by loss of a proton, gives the alcohol.

MECHANISM   Acid-Catalyzed Hydration of an Alkene

Step 1: Protonation of the double bond forms a carbocation.

$$\diagdown C = C \diagdown \quad + \quad H - \overset{H}{\underset{\cdot\cdot}{\overset{|}{O^+}}} - H \quad \rightleftharpoons \quad -\overset{|}{\underset{|}{C}} - C^+ \diagdown \quad + \quad H_2\overset{\cdot\cdot}{O}:$$

Step 2: Nucleophilic attack by water.

$$\overset{H}{\underset{|}{-\overset{|}{C}}} - C^+ \diagdown \quad + \quad H_2\overset{\cdot\cdot}{O}: \quad \rightleftharpoons \quad \overset{H \quad :\overset{H}{\underset{|}{O^+}} - H}{-\overset{|}{C} - \overset{|}{C} -}$$

Step 3: Deprotonation to the alcohol.

$$\overset{H}{\underset{-\overset{|}{C} - \overset{|}{C} -}{\overset{|}{H} :\overset{|}{O^+} - H}} \quad + \quad H_2\overset{\cdot\cdot}{O}: \quad \rightleftharpoons \quad \overset{H \quad :\overset{\cdot\cdot}{O}H}{-\overset{|}{C} - \overset{|}{C} -} \quad + \quad H_3\overset{\cdot\cdot}{O}^+$$

8-4B   Orientation of Hydration

Step 1 of the hydration mechanism is similar to the first step in the addition of HBr. The proton adds to the *less* highly substituted end of the double bond to form the *more* highly substituted carbocation. Water attacks the carbocation to give (after loss of a proton) the alcohol with the —OH group on the more highly substituted carbon. Like the addition of hydrogen halides, hydration is *regioselective*: It follows Markovnikov's rule, giving a product in which the new hydrogen has added to the less highly substituted end of the double bond. Consider the hydration of 2-methyl-2-butene:

$$\underset{CH_3}{\overset{CH_3}{\underset{|}{CH_3 - C = CH - CH_3}}} + H - \overset{H}{\underset{\cdot\cdot}{\overset{|}{O^+}}} - H \rightleftharpoons \underset{\overset{|}{H}}{\overset{CH_3}{\underset{|}{CH_3 - \overset{+}{C} - CH - CH_3}}} \text{ but not } \underset{\overset{|}{H}}{\overset{CH_3}{\underset{|}{CH_3 - C - \overset{+}{CH} - CH_3}}}$$

$$\text{3°, more stable} \qquad\qquad \text{2°, less stable}$$

The proton adds to the less highly substituted end of the double bond, so the positive charge appears at the more highly substituted end. Water attacks the carbocation to give the protonated alcohol.

$$\underset{H_2\overset{\cdot\cdot}{O}:}{\overset{CH_3}{\underset{+}{\overset{|}{CH_3 - C - CH - CH_3}}}} \rightleftharpoons \underset{H_2\overset{\cdot\cdot}{O}:}{\overset{CH_3}{\underset{\overset{+}{O}:}{\overset{|}{CH_3 - C - CH - CH_3}}}} \rightleftharpoons \underset{H_3\overset{\cdot\cdot}{O}^+}{\overset{CH_3}{\underset{OH \quad H}{\overset{|}{CH_3 - C - CH - CH_3}}}}$$

The reaction has obeyed Markovnikov's rule. The proton has added to the end of the double bond that already had more hydrogens (that is, the less highly substituted end), and the —OH group has added to the more highly substituted end.

Like other reactions that involve carbocation intermediates, hydration may take place with rearrangement. For example, when 3,3-dimethyl-1-butene undergoes acid-catalyzed hydration, the major product results from rearrangement of the carbocation intermediate.

$$
\begin{array}{ccc}
& \underset{\displaystyle \overset{|}{\underset{|}{CH_3}}}{CH_3-\overset{CH_3}{\underset{}{C}}-CH=CH_2} & \xrightarrow{\ 50\%\ H_2SO_4\ } & CH_3-\overset{CH_3}{\underset{OH}{C}}-\overset{}{\underset{CH_3}{CH}}-CH_3 \\
\end{array}
$$

3,3-dimethyl-1-butene   2,3-dimethyl-2-butanol
(major product)

PROBLEM 8-5

Propose a mechanism to show how 3,3-dimethyl-1-butene reacts with dilute aqueous H_2SO_4 to give 2,3-dimethyl-2-butanol and a small amount of 2,3-dimethyl-2-butene.

PROBLEM 8-6

Predict the products of the following hydration reactions.
(a) 1-methylcyclopentene + dilute acid
(b) 2-phenylpropene + dilute acid
(c) 1-phenylcyclohexene + dilute acid

PROBLEM-SOLVING $\mathbf{HINT}$

When predicting products for electrophilic additions, first draw the structure of the carbocation (or other intermediate) that results from electrophilic attack.

8-5
Hydration by Oxymercuration–Demercuration

Many alkenes do not easily undergo hydration in aqueous acid. Some alkenes are nearly insoluble in aqueous acid, and others undergo side reactions such as polymerization or charring under these strongly acidic conditions. In some cases, the overall equilibrium favors the alkene rather than the alcohol. No amount of catalysis can cause a reaction to occur if the energetics are unfavorable.

Oxymercuration–demercuration is another method for converting alkenes to alcohols with Markovnikov orientation. Oxymercuration–demercuration works with many alkenes that do not easily undergo direct hydration, and it takes place under milder conditions. No free carbocation is formed, so there is no opportunity for rearrangements.

Oxymercuration–Demercuration

$$
\begin{array}{ccccccc}
\overset{\diagdown}{\underset{\diagup}{C}}=\overset{\diagup}{\underset{\diagdown}{C}} & + & Hg(OAc)_2 & \xrightarrow{\ H_2O\ } & \underset{\underset{HO}{|}\ \ \underset{HgOAc}{|}}{-C-C-} & \xrightarrow{\ NaBH_4\ } & \underset{\underset{HO}{|}\ \ \underset{H}{|}}{-C-C-}
\end{array}
$$

(Markovnikov orientation)

The reagent for mercuration is mercuric acetate, $Hg(OCOCH_3)_2$, abbreviated $Hg(OAc)_2$. There are several theories as to how this reagent acts as an electrophile, but the simplest one is that mercuric acetate dissociates slightly to form a positively charged mercury species, $^+Hg(OAc)$.

$$
\underset{Hg(OAc)_2}{CH_3-\overset{O}{\overset{||}{C}}-O-Hg-O-\overset{O}{\overset{||}{C}}-CH_3} \ \rightleftharpoons \ \underset{^+Hg(OAc)}{CH_3-\overset{O}{\overset{||}{C}}-O-Hg^+} \ + \ \underset{^-OAc}{CH_3-\overset{O}{\overset{||}{C}}-O^-}
$$

Oxymercuration involves an electrophilic attack on the double bond by the positively charged mercury species. The product is a *mercurinium ion*, an organometallic cation containing a three-membered ring. In the second step, water from the solvent attacks the mercurinium ion to give (after deprotonation) an organomercurial alcohol.

MECHANISM   Oxymercuration of an Alkene

Step 1: Electrophilic attack forms a mercurinium ion.

Step 2: Water opens the ring to give an organomercurial alcohol.

organomercurial alcohol

The second step is **demercuration**, to remove the mercury. Sodium borohydride ($NaBH_4$, a reducing agent) replaces the mercuric acetate fragment with hydrogen.

organomercurial alcohol   alcohol

Oxymercuration–demercuration of an unsymmetrical alkene generally gives Markovnikov orientation of addition, as shown by the oxymercuration of 2-methyl-2-butene. In this unsymmetrical case, the mercurinium ion has a considerable amount of positive charge on the more highly substituted carbon atom. Attack by water occurs on this more electrophilic carbon, giving Markovnikov orientation. The electrophile, $^{+}Hg(OAc)$, remains bonded to the less highly substituted end of the double bond. Reduction of the organomercurial alcohol gives the Markovnikov alcohol: 2-methyl-2-butanol.

Although mercury and its compounds were used for centuries as ingredients in antibacterial drugs, skin creams, and antiseptics, it is toxic. In the body, it combines with the thiol groups of critical enzymes, inactivating them. Brain and kidney damage result and exposure can be fatal.

2-methyl-2-butene   mercurinium ion   Markovnikov product

Markovnikov product   2-methyl-2-butanol
                                       (90% overall)

The following reaction shows the oxymercuration–demercuration of cyclopentene to cyclopentanol. Attack by water on the mercurinium ion comes from the opposite side of the ring, resulting in addition of the hydroxyl group and the mercury atom to opposite sides of the ring. An addition to opposite faces of a double bond is called an **anti addition**. Conversely, addition of two groups to the same face of a double bond is called a **syn addition**.

Of the methods we have seen for Markovnikov hydration of alkenes, oxymercuration–demercuration is most commonly used in the laboratory. It gives better yields than direct acid–catalyzed hydration, it avoids the possibility of rearrangements, and it does not involve harsh conditions. There are also disadvantages, however. Organomercurial compounds are highly toxic. They must be used with great care, then disposed of properly.

8-6
Alkoxymercuration–Demercuration

When mercuration takes place in an alcohol solvent, the alcohol serves as a nucleophile to attack the mercurinium ion. The resulting product contains an alkoxyl ($-O-R$) group. In effect, **alkoxymercuration**–demercuration converts alkenes to ethers by adding an alcohol across the double bond of the alkene.

As we have seen, an alkene reacts to form a mercurinium ion that is attacked by the nucleophilic solvent. Attack by an alcohol solvent gives an organomercurial ether that can be reduced to the ether.

organomercurial ether

organomercurial ether $\xrightarrow{NaBH_4}$ an ether

The solvent attacks the mercurinium ion at the *more* highly substituted end of the double bond (where there is more δ^+ charge), giving Markovnikov orientation of addition. The Hg(OAc) group appears at the *less* highly substituted end of the double bond. Reduction gives the Markovnikov product, with hydrogen at the less highly substituted end of the double bond.

SOLVED PROBLEM 8-2

Show the intermediates and products that result from alkoxymercuration-demercuration of 1-methylcyclopentene, using methanol as the solvent.

SOLUTION

Mercuric acetate adds to 1-methylcyclopentene to give the cyclic mercurinium ion. This ion has a considerable amount of positive charge on the more highly substituted tertiary carbon atom. Methanol attacks this carbon.

1-methylcyclopentene   mercurinium ion   trans intermediate (product of anti addition)

Reduction of the intermediate gives the Markovnikov product, 1-methoxy-1-methylcyclopentane.

intermediate $\xrightarrow{NaBH_4}$ 1-methoxy-1-methylcyclopentane

Mercury and its compounds were used for centuries as ingredients in antibacterial drugs, skin creams, and antiseptics. Mercury compounds are quite toxic, however. In the body, mercury combines with the thiol groups of critical enzymes, inactivating them. Mercury poisoning causes brain and kidney damage, often leading to death.

PROBLEM 8-7

(a) Propose a mechanism for the following reaction.

(b) Give the structure of the product that results when this intermediate is reduced by sodium borohydride.

PROBLEM 8-8

Predict the major products of the following reactions.
(a) 1-methylcyclohexene + aqueous $Hg(OAc)_2$
(b) the product from part (a), treated with $NaBH_4$
(c) 4-chlorocycloheptene + $Hg(OAc)_2$ in CH_3OH
(d) the product from part (c), treated with $NaBH_4$

PROBLEM 8-9

Show how you would accomplish the following synthetic conversions.
(a) 2-methyl-1-pentene $\rightarrow$ 2-methyl-2-pentanol
(b) 1-butene $\rightarrow$ 2-methoxybutane
(c) 1-iodo-2-methylcyclopentane $\rightarrow$ 1-methylcyclopentanol

8-7
Hydroboration
of Alkenes

We have seen two methods for hydrating an alkene with Markovnikov orientation. What if we need to convert an alkene to the anti-Markovnikov alcohol? For example, the following transformation cannot be accomplished using the hydration procedures covered thus far.

2-methyl-2-butene   3-methyl-2-butanol

Such an anti-Markovnikov hydration was impossible until H. C. Brown, of Purdue University, discovered that diborane (B_2H_6) adds to alkenes with anti-Markovnikov orientation to form alkylboranes, which can be oxidized to give anti-Markovnikov alcohols. This discovery led to the development of a large field of borane chemistry, for which Brown received the Nobel Prize in chemistry in 1979.

2-methyl-2-butene   an alkylborane   3-methyl-2-butanol (>90%)

Diborane (B_2H_6) is a dimer composed of two molecules of borane (BH_3). The bonding in diborane is unconventional, using three-centered (banana-shaped) bonds with protons in the middle of them. Diborane is in equilibrium with a small amount of borane (BH_3), a strong Lewis acid with only six valence electrons.

diborane   borane

Diborane is an inconvenient reagent. It is a toxic, flammable, and explosive gas. It is more easily used as a complex with tetrahydrofuran (THF), a cyclic ether. This complex reacts like diborane, yet the solution is easily measured and transferred.

$$2 \quad \begin{matrix} CH_2-CH_2 \\ \quad \quad \quad O: \\ CH_2-CH_2 \end{matrix} \quad + \quad B_2H_6 \quad \longrightarrow \quad 2 \begin{matrix} CH_2-CH_2 \quad \quad H \\ \quad \quad \quad :O^+\!\!-\!\!B\!\!-\!\!H \\ CH_2-CH_2 \quad \quad H \end{matrix} \quad = \quad \begin{matrix} \\ O^+\!\!-\!\bar{B}H_3 \end{matrix} \quad$$

| tetrahydrofuran | diborane | borane–THF complex | $= \quad BH_3 \cdot THF$ |
| (THF) | | | |

The $BH_3 \cdot THF$ reagent is the form of borane commonly used in organic reactions. BH_3 adds to the double bond of an alkene to give an alkylborane. Basic hydrogen peroxide oxidizes the alkylborane to an alcohol. In effect, hydroboration–oxidation converts alkenes to alcohols by adding water across the double bond, with anti-Markovnikov orientation.

Hydroboration–oxidation:

$$☞ \quad \begin{matrix} \backslash \quad \quad / \\ C=C \\ / \quad \quad \backslash \end{matrix} \quad + \quad BH_3 \cdot THF \quad \longrightarrow \quad \begin{matrix} | \quad \quad | \\ -C-C- \\ | \quad \quad | \\ H \quad B-H \\ \quad \quad | \\ \quad \quad H \end{matrix} \quad \xrightarrow{H_2O_2, \ ^-OH} \quad \begin{matrix} | \quad \quad | \\ -C-C- \\ | \quad \quad | \\ H \quad OH \end{matrix}$$

anti-Markovnikov orientation
(syn stereochemistry)

8-7A Mechanism of Hydroboration

Borane is an electron-deficient compound. It has only six valence electrons, so the boron atom lacks an octet. Acquiring an octet is the driving force for the unusual bonding structures ("banana" bonds, for example) found in boron compounds. As an electron-deficient compound, BH_3 is a strong electrophile, capable of adding to a double bond (Figure 8-4). This **hydroboration** of the double bond is thought to occur in one step, with the boron atom adding to the less highly substituted end of the double bond.

In the transition state, the electrophilic boron atom withdraws electrons from the pi bond, and the carbon at the other end of the double bond acquires a partial positive charge. This partial charge is more stable on the more highly substituted carbon atom. The product shows boron bonded to the less highly substituted end of the double bond and hydrogen bonded to the more highly substituted end. Also, steric hindrance favors boron adding to the less hindered, less substituted end of the double bond.

MECHANISM   Hydroboration of an Alkene

more stable transition state

less stable transition state

◀ Figure 8-4
Borane adds to the double bond in a single step. Boron adds to the less hindered, less substituted carbon, and hydrogen adds to the more substituted carbon.

The second step is the oxidation of the boron atom, removing it from carbon and replacing it with a hydroxyl (—OH) group. Aqueous sodium hydroxide and hydrogen peroxide (HOOH or H_2O_2) are used for the oxidation.

$$CH_3\underset{\underset{H}{|}}{\overset{\overset{CH_3}{|}}{C}}\!-\!\underset{\underset{BH_2}{|}}{\overset{\overset{H}{|}}{C}}\!-\!CH_3 \xrightarrow[\;H_2O\;]{H_2O_2,\ NaOH} CH_3\!-\!\underset{\underset{H}{|}}{\overset{\overset{CH_3}{|}}{C}}\!-\!\underset{\underset{OH}{|}}{C}H\!-\!CH_3$$

This hydration of an alkene by hydroboration–oxidation is another example of a reaction that does not follow the original statement of Markovnikov's rule (the product is anti-Markovnikov), but still follows our understanding of the reasoning behind Markovnikov's rule. The electrophilic boron atom adds to the *less* highly substituted end of the double bond, placing the positive charge (and the hydrogen atom) at the more highly substituted end.

SOLVED PROBLEM 8-3

Show how you would convert 1-methylcyclopentanol to 2-methylcyclopentanol.

SOLUTION

Working backward, use hydroboration–oxidation to form 2-methyl–cyclopentanol from 1-methylcyclopentene. The use of (1) and (2) above and below the reaction arrow indicates individual steps in a two-step sequence.

1-methylcyclopentene   (1) $BH_3 \cdot THF$ (2) H_2O_2, ^-OH *trans*-2-methylcyclopentanol

(The 2-methylcyclopentanol that results from this synthesis is the pure trans isomer. This stereochemical result is discussed in Section 8-7C.

1-Methylcyclopentene is the most highly substituted alkene that results from dehydration of 1-methylcyclopentanol.

1-methylcyclopentanol H_2SO_4 1-methylcyclopentene   + H_2O

PROBLEM 8-10

Predict the major products of the following reactions.
(a) propene + $BH_3 \cdot THF$
(b) the product from part (a) + H_2O_2/OH^-
(c) 2-methyl-2-pentene + $BH_3 \cdot THF$
(d) the product from part (c) + H_2O_2/OH^-
(e) 1-methylcyclohexene + $BH_3 \cdot THF$
(f) the product from part (e) + H_2O_2/OH^-

PROBLEM 8-11

Show how you would accomplish the following synthetic conversions.
(a) 1-butene → 1-butanol   (b) 1-butene → 2-butanol
(c) 2-bromo-2,4-dimethylpentane → 2,4-dimethyl-3-pentanol

8-7B Stoichiometry of Hydroboration

For simplicity, we have neglected the fact that 3 moles of an alkene react with each mole of BH_3. Each $B\text{—}H$ bond in BH_3 can add across the double bond of an alkene. The first addition forms an alkylborane, the second a dialkylborane, and the third a trialkylborane.

alkylborane   dialkylborane   trialkylborane

Summary

Trialkylboranes react exactly as we have discussed, and they oxidize to give anti-Markovnikov alcohols. Trialkylboranes are quite bulky, further reinforcing the preference for boron to add to the less hindered carbon atom of the double bond. Boranes are often drawn as the 1:1 monoalkylboranes to simplify their structure and emphasize the organic part of the molecule.

8-7C Stereochemistry of Hydroboration

The simultaneous addition of boron and hydrogen to the double bond, as shown in Figure 8-4, leads to a *syn addition*: Boron and hydrogen add across the double bond on the *same side* of the molecule. (If they added to opposite sides of the molecule, the process would be an *anti addition*.)

The stereochemistry of the hydroboration–oxidation of 1-methylcyclopentene is shown below. Boron and hydrogen add to the same face of the double bond (syn) to form a trialkylborane. Oxidation of the trialkylborane replaces boron with a hydroxyl group in the same stereochemical position. The product is *trans*-2-methylcyclopentanol. A racemic mixture is expected because a chiral product is formed from achiral reagents.

transition state

trans-2-methylcyclopentanol
(85% overall)
(racemic mixture of enantiomers)

The second step (oxidation of the borane to the alcohol) takes place with retention of configuration. Hydroperoxide ion adds to the borane, causing the alkyl group to migrate from boron to oxygen. The alkyl group migrates with retention of configuration because it moves with its electron pair and does not alter the tetrahedral structure of the migrating carbon atom. Hydrolysis of the borate ester gives the alcohol.

MECHANISM Oxidation of a Trialkylborane

Formation of hydroperoxide ion

$$H\!-\!\ddot{O}\!-\!\ddot{O}\!-\!H + {}^{-}\!\ddot{O}H \rightleftharpoons H\!-\!\ddot{O}\!-\!\ddot{O}{}^{-} + H_2\ddot{O}$$

Addition of hydroperoxide and migration of the alkyl group

R—B with two R groups, :Ö—Ö—H → R—B—Ö—Ö—H → R—B—Ö: + ⁻ÖH

hydroperoxide   R migrates   borate ester

Twice more to oxidize the other two alkyl groups

R—B—O with R below $\xrightarrow{\text{⁻OOH}}$ $\xrightarrow{\text{⁻OOH}}$ trialkyl borate ester (O—B with O—R, R, O—R groups)

trialkyl borate ester

Hydrolysis of the borate ester

:Ö—B—Ö: $\rightleftharpoons$ O—B⁻—Ö: $\rightleftharpoons$ O—B :Ö⁻ $\xrightarrow{H_2O}$ O—B

+ R—OH

+ ⁻OH

(The other two OR groups hydrolyze similarly)

Hydroboration of alkenes is another example of a **stereospecific reaction**, in which different stereoisomers of the starting compound react to give different stereoisomers of the product. Problem 8-14 considers the different products formed by the hydroboration–oxidation of two acyclic diastereomers.

SOLVED PROBLEM 8-4

A norbornene molecule labeled with deuterium is subjected to hydroboration–oxidation. Give the structures of the intermediates and products.

exo (outside) face

$\xrightarrow{\text{BH}_3 \cdot \text{THF}}$

exo BH₂

$\xrightarrow{\text{H}_2\text{O}_2, \ {}^{-}\text{OH}}$

endo (inside) face

deuterium-labeled norbornene   alkylborane   alcohol
(racemic mixture)

SOLUTION

The syn addition of BH_3 across the double bond of norbornene takes place mostly from the more accessible outside *(exo)* face of the double bond. Oxidation gives a product with both the hydrogen atom and the hydroxyl group in exo positions. (The less accessible inner face of the double bond is called the *endo* face.)

PROBLEM 8-12

In the hydroboration of 1-methylcyclopentene shown in Solved Problem 8-3, the reagents are achiral, and the products are chiral. The product is a racemic mixture of *trans*-2-methylcyclopentanol, but only one enantiomer is shown. Show how the other enantiomer is formed.

PROBLEM 8-13

Predict the major products of the following reactions. Include stereochemistry where applicable.
(a) 1-methylcycloheptene $+$ $BH_3 \cdot THF$, then H_2O_2, OH^-
(b) *trans*-4,4-dimethyl-2-pentene $+$ $BH_3 \cdot THF$, then H_2O_2, OH^-

(c) $+$ $BH_3 \cdot THF$, then H_2O_2, OH^-

PROBLEM 8-14

(a) When (*Z*)-3-methyl-3-hexene undergoes hydroboration–oxidation, two isomeric products are formed. Give their structures, and label each asymmetric carbon atom as (*R*) or (*S*). What is the relationship between these isomers?
(b) Repeat part (a) for (*E*)-3-methyl-3-hexene. What is the relationship between the products formed from (*Z*)-3-methyl-3-hexene and those formed from (*E*)-3-methyl-3-hexene?

PROBLEM 8-15

Show how you would accomplish the following transformations.

(a)

(b)

(c) 1-methylcycloheptanol $\rightarrow$ 2-methylcycloheptanol

PROBLEM 8-16

When HBr adds across the double bond of 1,2-dimethylcyclopentene, the product is a mixture of the cis and trans isomers. Show why this addition is not stereospecific.

Although we have mentioned **catalytic hydrogenation** before (Sections 7-7 and 8-1), we now consider the mechanism and stereochemistry in more detail. Hydrogenation of an alkene is formally a reduction, with H_2 adding across the double bond to give an alkane. The process usually requires a catalyst containing Pt, Pd, or Ni.

8-8
Catalytic Hydrogenation of Alkenes

Example

$$CH_3-CH{=}CH-CH_3 \;+\; H_2 \quad\xrightarrow{\;Pt\;}\quad CH_3-CH_2-CH_2-CH_3$$

The Parr hydrogenation apparatus shakes the reaction vessel (containing the alkene and the solid catalyst), while a pressurized cylinder supplies hydrogen.

For most alkenes, hydrogenation takes place at room temperature, using hydrogen gas at atmospheric pressure. The alkene is usually dissolved in an alcohol, an alkane, or acetic acid. A small amount of platinum, palladium, or nickel catalyst is added, and the container is shaken or stirred while the reaction proceeds. Hydrogenation actually takes place at the surface of the metal, where the liquid solution of the alkene comes into contact with hydrogen and the catalyst.

Hydrogen gas is adsorbed onto the surface of these metal catalysts, and the catalyst weakens the H—H bond. In fact, if H_2 and D_2 are mixed in the presence of a platinum catalyst, the two isotopes quickly scramble to produce a random mixture of HD, H_2, and D_2. (No scrambling occurs in the absence of the catalyst.) Hydrogenation is an example of **heterogeneous catalysis**, because the (solid) catalyst is in a different phase from the reactant solution. In contrast, **homogeneous catalysis** involves reactants and catalyst in the same phase, as in the acid-catalyzed dehydration of an alcohol.

Because the two hydrogen atoms add from a solid surface, they add with **syn** stereochemistry. For example, when 1,2-dideuteriocyclohexene is treated with hydrogen gas over a catalyst, the product is the cis isomer resulting from syn addition (Figure 8-5)

One face of the alkene pi bond binds to the catalyst, which has hydrogen adsorbed on its surface. Hydrogen inserts into the pi bond, and the product is freed from the catalyst. Both hydrogen atoms add to the face of the double bond that is complexed with the catalyst.

Soluble homogeneous catalysts, such as *Wilkinson's catalyst*, also catalyze the hydrogenation of carbon–carbon double bonds.

▲ **Figure 8-5**
A solid heterogeneous catalyst adds two hydrogen atoms to the same face of the pi bond (syn stereochemistry).

◀ **Figure 8-6**
Rhodium and ruthenium phosphines are used as homogeneous hydrogenation catalysts. Chiral ligands can be attached to accomplish asymmetric induction, the creation of a new asymmetric carbon as mostly one enantiomer.

Wilkinson's catalyst is not chiral, but its triphenylphosphine (PPh_3) ligands can be replaced by chiral ligands to give chiral catalysts that are capable of converting optically inactive starting materials to optically active products. Such a process is called **asymmetric induction** or **enantioselective synthesis**. For example, Figure 8-6 shows a chiral ruthenium complex catalyzing an enantioselective hydrogenation of a carbon–carbon double bond to give a large excess of one enantiomer. Because the catalyst is chiral, the transition states leading to the two enantiomers of product are diastereomeric. They have different energies, and the transition state leading to the (R) enantiomer is favored. Ryoji Noyori and William Knowles shared the 2001 Nobel Prize in Chemistry for their work on chirally catalyzed hydrogenation reactions.

Enantioselective synthesis is particularly important in the pharmaceutical industry, because only one enantiomer of a chiral drug is likely to have the desired effect. For example, levodopa [(−)-dopa or l-dopa] is used in patients with Parkinson's disease to counteract a deficiency of dopamine, one of the neurotransmitters in the brain. Dopamine itself is useless as a drug because it cannot cross the "blood–brain barrier"; that is, it cannot get into the cerebrospinal fluid from the bloodstream. (−)-Dopa on the other hand, is an amino acid related to tyrosine. It crosses the blood–brain barrier into the cerebrospinal fluid, where it undergoes enzymatic conversion to dopamine. Only the (−) enantiomer of dopa can be transformed into dopamine; the other enantiomer, (+)-dopa, is toxic to the patient.

The correct enantiomer can be synthesized from an achiral starting material by catalytic hydrogenation using a complex of rhodium with a chiral ligand called DIOP. Such an enantioselective synthesis is more efficient than making a racemic mixture, resolving it into enantiomers, and discarding the unwanted enantiomer.

The enzymatic reduction of a double bond is a key step in the formation of a fatty acid that is ultimately incorporated into the cell wall of the bacterium that causes tuberculosis. The antituberculosis drug isoniazid blocks this enzyme, preventing reduction of the double bond. Without an intact cell wall, the bacteria die.

isoniazid

PROBLEM 8-17

Give the expected major product for each reaction, including stereochemistry where applicable.

(a) 1-butene + H_2/Pt

(b) *cis*-2-butene + H_2/Ni

(c) + H_2/Pt

(d) + excess H_2/Pt

PROBLEM 8-18

One of the principal components of lemon grass oil is *limonene*, $C_{10}H_{16}$. When limonene is treated with excess hydrogen and a platinum catalyst, the product is an alkane of formula $C_{10}H_{20}$. What can you conclude about the structure of limonene?

PROBLEM 8-19

The chiral BINAP ligand shown in Figure 8-6 contains no asymmetric carbon atoms. Explain how this ligand is chiral.

8-9
Addition of Carbenes to Alkenes

Methylene ($:CH_2$) is the simplest of the **carbenes**: uncharged, reactive intermediates that have a carbon atom with two bonds and two nonbonding electrons. Like borane (BH_3), methylene is a potent electrophile because it has an unfilled octet. It adds to the electron-rich pi bond of an alkene to form a cyclopropane.

methylene

Heating or photolysis of diazomethane (CH_2N_2) gives nitrogen gas and methylene:

diazomethane

methylene

There are two difficulties with using diazomethane to cyclopropanate double bonds. First, it is extremely toxic and explosive. A safer reagent would be more convenient for routine use. Second, methylene generated from diazomethane is so reactive that it inserts into C—H bonds as well as C=C bonds. In the reaction of propene with diazomethane-generated methylene, for example, several side products are formed.

propene

PROBLEM 8-20

Show how the insertion of methylene into a bond of cyclohexene can produce the following.

(a) 1-methylcyclohexene   **(b)** 3-methylcyclohexene   **(c)** norcarane,

8-9A   The Simmons–Smith Reaction

$$\text{☞} \quad \diagdown C {=} C \diagup \quad + \quad ICH_2ZnI \quad \longrightarrow \quad -\underset{\diagdown \;\; \diagup}{\overset{|\qquad\;\; |}{C{-}\underset{CH_2}{}{-}C}}- \quad + \quad ZnI_2$$

Two DuPont chemists discovered a reagent that converts alkenes to cyclopropanes in better yields than diazomethane, with fewer side reactions. The **Simmons–Smith reaction**, named in their honor, is one of the best ways of making cyclopropanes.

    The Simmons–Smith reagent is made by adding methylene iodide to the "zinc–copper couple," zinc dust that has been activated with an impurity of copper. The reagent probably resembles iodomethyl zinc iodide, ICH_2ZnI. This kind of reagent is called a *carbenoid* because it reacts much like a carbene, but it does not actually contain a divalent carbon atom.

$$CH_2I_2 \quad + \quad Zn(Cu) \quad \longrightarrow \quad ICH_2ZnI$$
Simmons–Smith reagent
(a carbenoid)

$$\text{(cyclohexene)} \xrightarrow[\text{Zn, CuCl}]{CH_2I_2} \text{(bicyclic product)}$$

(59%)

8-9B   Formation of Carbenes by Alpha Elimination

Carbenes are also formed by reactions of halogenated compounds with bases. If a carbon atom has bonds to at least one hydrogen and to enough halogen atoms to make the hydrogen slightly acidic, it may be possible to form a carbene. For example, bromoform ($CHBr_3$) reacts with a 50% aqueous solution of potassium hydroxide to form dibromocarbene.

$$\underset{\text{bromoform}}{CHBr_3} \quad + \quad K^+ \; {}^-OH \quad \rightleftharpoons \quad {}^-{:}CBr_3 \quad K^+ \quad + \quad H_2O$$

$$Br{-}\overset{\displaystyle ..}{\underset{\displaystyle |}{\underset{Br}{C}}}{}^-{-}Br \quad \rightleftharpoons \quad \underset{\text{dibromocarbene}}{:CBr_2} \quad + \quad :Br^-$$

This dehydrohalogenation is called an **alpha elimination** (α elimination) because the hydrogen and the halogen are lost from the same carbon atom. The more common dehydrohalogenations (to form alkenes) are called **beta eliminations** because the hydrogen and the halogen are lost from adjacent carbon atoms. Dibromocarbene formed from $CHBr_3$ can add to a double bond to form a dibromocyclopropane.

$$\text{(cyclopentene)} \xrightarrow[\text{KOH/H}_2\text{O}]{CHBr_3} \text{(dibromobicyclic product)} \underset{Br}{\overset{Br}{}}$$

The products of these cyclopropanations retain any cis or trans stereochemistry of the reactants.

PROBLEM 8-21

Predict the major products of the following reactions.
(a) cyclohexene + $CHCl_3$, 50% $NaOH/H_2O$

(b)

$+ CH_2I_2, Zn(Cu)$

(c)

$+$ 50% $NaOH/H_2O$

PROBLEM 8-22

Show how you would accomplish each of the following synthetic conversions.
(a) *trans*-2-butene $\longrightarrow$ *trans*-1,2-dimethylcyclopropane

(b) cyclopentene $\longrightarrow$

(c) cyclohexanol $\longrightarrow$

8-10
Addition
of Halogens
to Alkenes

Halogens add to alkenes to form vicinal dihalides.

$(X_2 = Cl_2, Br_2, \text{ sometimes } I_2)$ usually anti addition

8-10A   Mechanism of Halogen Addition

A halogen molecule (Br_2, Cl_2, or I_2) is electrophilic; a nucleophile can react, displacing a halide ion:

The nucleophile attacks the electrophilic nucleus of one bromine atom, and the other bromine serves as the leaving group, departing as bromide ion. Many reactions fit this general pattern; for example:

$$HO:^- \quad + \quad :Br{-}Br: \quad \longrightarrow \quad HO{-}Br: \quad + \quad :Br:^-$$

$$H_3N: \quad + \quad :Cl{-}Cl: \quad \longrightarrow \quad \overset{+}{H_3N}{-}Cl: \quad + \quad :Cl:^-$$

In the last reaction, the pi electrons of an alkene attack the bromine molecule, expelling bromide ion. A **bromonium ion** results, containing a three-membered ring with a positive charge on the bromine atom (similar in structure to the mercurinium ion discussed in Section 8-5B). The following reaction shows the formation and opening of a general **halonium ion** together with the structures of a **chloronium ion**, a **bromonium ion**, and an **iodonium ion**.

MECHANISM Addition of Halogens to Alkenes

Step 1: Formation of a halonium ion.

Step 2: Opening of the halonium ion.

X⁻ attacks from the back side

Examples

chloronium ion   bromonium ion   iodonium ion

Unlike a normal carbocation, all the atoms in a halonium ion have filled octets. The three-membered ring has considerable ring strain, however, which combines with a positive charge on an electronegative halogen atom to make the halonium ion strongly electrophilic. Attack by a nucleophile, such as a halide ion, opens the halonium ion to give a stable product.

Chlorine and bromine commonly add to alkenes by the halonium ion mechanism. Iodination is used less frequently because diiodide products decompose easily. Any solvents used must be inert to the halogens; methylene chloride (CH_2Cl_2), chloroform ($CHCl_3$), and carbon tetrachloride (CCl_4) are the most frequent choices.

8-10B Stereochemistry of Halogen Addition

The addition of bromine to cyclopentene is a stereospecific **anti addition**.

cyclopentene   trans-1,2-dibromocyclopentane   cis-1,2-dibromocyclopentane
                                        (92%)   (not formed)

Anti stereochemistry results from the bromonium ion mechanism. When a nucleophile attacks a halonium ion, it must do so from the back side, in a manner similar to the S_N2 displacement. This back-side attack assures anti stereochemistry of addition.

Halogen addition is another example of a stereospecific reaction, in which different stereoisomers of the starting material give different stereoisomers of the product. Figure 8-7 shows additional examples of the anti addition of halogens to alkenes.

cyclohexene   racemic trans-1,2-dichlorocyclohexane

cis-2-butene   (+ enantiomer)   (±)-2,3-dibromobutane

trans-2-butene   meso-2,3-dibromobutane

▲ **Figure 8-7**
The stereospecific addition of halogens to alkenes gives anti addition to the double bond.

The addition of bromine has been used as a simple chemical test for the presence of olefinic double bonds. A solution of bromine in carbon tetrachloride is a clear, deep red color. When this red solution is added to an alkene, the red bromine color disappears (we say it is "decolorized"), and the solution becomes clear and colorless. (Although there are other functional groups that decolorize bromine, few do it as quickly as alkenes.)

PROBLEM 8-23

Give mechanisms to account for the stereochemistry of the products observed from the addition of bromine to *cis*- and *trans*-2-butene (Figure 8-7). Why are two products formed from the cis isomer but only one from the trans? (Making models will be helpful.)

PROBLEM 8-24

Propose mechanisms and predict the major products of the following reactions. Include stereo-chemistry where appropriate.

(a) cycloheptene + Br_2 in CH_2Cl_2 (b) (*E*)-3-decene + Br_2 in CCl_4

(c) [structure] + Cl_2 in $CHCl_3$ (d) [structure] + 2 Cl_2 in CCl_4

When a solution of bromine (red-brown) is added to cyclohexene, the bromine color quickly disappears because bromine adds across the double bond. When bromine is added to cyclohexane (at right), the color persists.

PROBLEM-SOLVING $\mathbb{HINT}$

Models may be helpful whenever stereochemistry is involved. Write complete structures, including all bonds and charges, when writing mechanisms.

8-11
Formation of Halohydrins

A **halohydrin** is an alcohol with a halogen on the adjacent carbon atom. In the presence of water, halogens add to alkenes to form halohydrins. The halogen adds to the alkene to give a halonium ion, which is strongly electrophilic. Water acts as a nucleophile to open the halonium ion and form the halohydrin.

When halogenation takes place with no solvent or with an inert solvent such as carbon tetrachloride (CCl_4) or chloroform, ($CHCl_3$), only the halide ion is available as

MECHANISM Formation of Halohydrins

Step 1: Formation of a halonium ion.

$$\text{C}=\text{C} + \text{X}-\text{X} \longrightarrow \begin{array}{c} \text{X}^+ \\ \triangle \\ -\text{C}-\text{C}- \end{array} \quad \text{X}^-$$

(X = Cl, Br, or I)   halonium ion

Step 2: Water opens the halonium ion; deprotonation gives the halohydrin.

$$\begin{array}{c} \text{X}^+ \\ \triangle \\ -\text{C}-\text{C}- \\ \\ H_2\ddot{O}: \end{array} \longrightarrow \begin{array}{c} \text{X} \\ | \ \ | \\ -\text{C}-\text{C}- \\ | \ \ | \\ :\text{O}^+ \\ H_2\ddot{O}: \ \ H \quad H \end{array} \longrightarrow \begin{array}{c} \text{X} \\ | \ \ | \\ -\text{C}-\text{C}- + \ H_3O^+X^- \\ | \ \ | \\ :\text{OH} \end{array}$$

back-side attack

halohydrin
Markovnikov orientation
anti stereochemistry

a nucleophile to attack the halonium ion. A dihalide results. But when an alkene reacts with a halogen in the presence of a nucleophilic solvent such as water, a solvent molecule is the most likely nucleophile to attack the halonium ion. When a water molecule attacks the halonium ion, the final product is a halohydrin, with a halogen on one carbon atom and a hydroxyl group on the adjacent carbon. The product may be a *chlorohydrin*, a *bromohydrin*, or an *iodohydrin*, depending on the halogen.

Stereochemistry of Halohydrin Formation   Because the mechanism involves a halonium ion, the stereochemistry of addition is anti, as in halogenation. For example, the addition of bromine water to cyclopentene gives *trans*-2-bromocyclopentanol, the product of anti addition across the double bond.

cyclopentene

trans-2-bromocyclopentanol
(cyclopentene bromohydrin)

+   enantiomer

PROBLEM 8-25

Propose a mechanism for the addition of bromine water to cyclopentene, being careful to show why the trans product results and how both enantiomers are formed.

Orientation of Halohydrin Formation   Even though a halonium ion is involved, rather than a carbocation, the extended version of Markovnikov's rule applies to halohydrin formation. When propene reacts with chlorine water, the major product has the electrophile (the chlorine atom) bonded to the less highly substituted carbon of the double bond. The nucleophile (the hydroxyl group) is bonded to the more highly substituted carbon.

$$H_2C{=}CH{-}CH_3 \ + \ Cl_2 \ + \ H_2O \ \longrightarrow \ \underset{\underset{Cl \ \ \ OH}{|\ \ \ \ \ |}}{H_2C{-}CH{-}CH_3} \ + \ HCl$$

The Markovnikov orientation observed in halohydrin formation is explained by the structure of the halonium ion intermediate. The two carbon atoms bonded to the halogen have partial positive charges, with a larger charge (and a weaker bond to the halogen) on the more highly substituted carbon atom (Figure 8-8). The nucleophile (water) attacks this more highly substituted, more electrophilic carbon atom. The result is both anti stereochemistry and Markovnikov orientation. This halonium ion mechanism can be used to explain and predict a wide variety of reactions in both nucleophilic and nonnucleophilic solvents. The halonium ion mechanism is similar to the mercurinium ion mechanism for oxymercuration of an alkene with Markovnikov orientation (Section 8-5).

▶ **Figure 8-8**
The more highly substituted carbon of the chloronium ion bears more positive charge than the less substituted carbon. Attack by water occurs on the more highly substituted carbon to give the Markovnikov product.

SOLVED PROBLEM 8-5

Propose a mechanism for the reaction of 1-methylcyclopentene with bromine water.

SOLUTION

1-Methylcyclopentene reacts with bromine to give a bromonium ion. Attack by water could occur at either the secondary carbon or the tertiary carbon of the bromonium ion. Attack actually occurs at the more highly substituted carbon, which bears more positive charge. The product is formed as a racemic mixture.

SOLVED PROBLEM 8-6

When cyclohexene is treated with bromine in saturated aqueous sodium chloride, a mixture of *trans*-2-bromocyclohexanol and *trans*-1-bromo-2-chlorocyclohexane results. Propose a mechanism to account for these two products.

SOLUTION

Cyclohexene reacts with bromine to give a bromonium ion, which will react with any available nucleophile. The most abundant nucleophiles in saturated aqueous sodium chloride solution are water and chloride ions. Attack by water gives the bromohydrin, and attack by chloride gives the dihalide. Either of these attacks gives anti stereochemistry.

PROBLEM 8-26

The solutions to Solved Problem 8-5 and Solved Problem 8-6 showed only how one enantiomer of the product is formed. For each product, show how an equally probable reaction forms the other enantiomer.

PROBLEM 8-27

Predict the major product(s) for each reaction. Include stereochemistry where appropriate.
(a) 1-methylcyclopentene + Cl_2/H_2O (b) 2-methyl-2-butene + Br_2/H_2O
(c) *cis*-2-butene + Cl_2/H_2O (d) *trans*-2-butene + Cl_2/H_2O
(e) 1-methylcyclopentene + Br_2 in saturated aqueous NaCl

PROBLEM 8-28

Show how you would accomplish the following synthetic conversions.
(a) 3-methyl-2-pentene → 2-chloro-3-methyl-3-pentanol
(b) chlorocyclohexane → *trans*-2-chlorocyclohexanol
(c) 1-methylcyclopentanol → 2-chloro-1-methylcyclopentanol

PROBLEM-SOLVING HINT

The opening of a halonium ion is driven by its electrophilic nature. The weak nucleophile attacks the carbon bearing more positive charge.

8-12
Epoxidation of Alkenes

Some of the most important reactions of alkenes involve oxidation. When we speak of oxidation, we usually mean reactions that form carbon–oxygen bonds. (Halogens are oxidizing agents, and the addition of a halogen molecule across a double bond is formally an oxidation as well.) Oxidations are particularly important because many common functional groups contain oxygen, and alkene oxidations are some of the best methods for introducing oxygen into organic molecules. We will consider methods for epoxidation, hydroxylation, and oxidative cleavage of alkene double bonds.

An **epoxide** is a three-membered cyclic ether, also called an **oxirane**. Epoxides are valuable synthetic intermediates used for converting alkenes to a variety of other functional groups. An alkene is converted to an epoxide by a **peroxyacid**, a carboxylic acid that has an extra oxygen atom in a —O—O—(peroxy) linkage.

The epoxidation of an alkene is clearly an oxidation, since an oxygen atom is added. Peroxyacids are highly selective oxidizing agents. Some simple peroxyacids (sometimes called *peracids*) and their corresponding carboxylic acids are shown below.

Epoxides are often found as components of natural products that plants use as defense mechanisms against insects. Typically, the epoxide reacts with critical cellular enzymes or DNA, preventing them from carrying out their normal functions. As a result, the insect is killed.

A peroxyacid epoxidizes an alkene by a concerted electrophilic reaction where several bonds are broken and several are formed at the same time. Starting with the alkene and the peroxyacid, a one-step reaction gives the epoxide and the acid directly, without any intermediates.

MECHANISM   Epoxidation of Alkenes

Peroxyacids epoxidize alkenes in a one-step (concerted) process.

Because the epoxidation takes place in one step, there is no opportunity for the alkene molecule to rotate and change its cis or trans geometry. The epoxide retains whatever stereochemistry is present in the alkene.

The following examples use *m*-chloroperoxybenzoic acid (MCPBA), a common epoxidizing reagent, to convert alkenes to epoxides having the same cis or trans stereochemistry. MCPBA is used for its desirable solubility properties: The peroxyacid dissolves, then the spent acid precipitates out of solution.

PROBLEM 8-29

Predict the products, including stereochemistry where appropriate, for the *m*-chloroperoxybenzoic acid epoxidations of the following alkenes.
(a) *cis*-2-hexene   (b) *trans*-2-hexene
(c) *cis*-cyclodecene   (d) *trans*-cyclodecene

Most epoxides are easily isolated as stable products if the solution is not too acidic. Any moderately strong acid protonates the epoxide, however. Water attacks the protonated epoxide, opening the ring and forming a 1,2-diol, commonly called a **glycol**.

8-13
Acid-Catalyzed Opening of Epoxides

MECHANISM   Acid-Catalyzed Opening of Epoxides

Water attacks the back side of the protonated epoxide.

Because glycol formation involves a back-side attack on a protonated epoxide, the result is anti orientation of the hydroxyl groups on the double bond. For example, when 1,2-epoxycyclopentane ("cyclopentene oxide") is treated with dilute mineral acid, the product is pure *trans*-1,2-cyclopentanediol.

cyclopentene oxide   *trans*-1,2-cyclopentanediol
                                                                                      (racemic)

PROBLEM 8-30

(a) Propose a detailed mechanism for the conversion of *cis*-3-hexene to the epoxide (3,4-epoxyhexane) and the ring-opening reaction to give the glycol, 3,4-hexanediol. In your mechanism, pay particular attention to the stereochemistry of the intermediates and products.

(b) Repeat part (a) for *trans*-3-hexene. Compare the products obtained from *cis*- and *trans*-3-hexene. Is this reaction sequence stereospecific?

Epoxidation reagents can be chosen to favor either the epoxide or the glycol. Peroxyacetic acid is used in strongly acidic water solutions. The acidic solution protonates the epoxide and converts it to the glycol. Peroxybenzoic acids are weak acids that can be used in non-nucleophilic solvents such as carbon tetrachloride. *m*-Chloroperoxybenzoic acid in CCl$_4$ generally gives good yields of epoxides. Figure 8-9 compares the uses of these reagents.

▶ **Figure 8-9**
Peroxyacetic acid is used in strongly acidic aqueous solutions. Alkenes are epoxidized, then opened to glycols in one step. Weakly acidic peroxyacids, such as peroxybenzoic acid, can be used in nonaqueous solutions to give good yields of epoxides.

PROBLEM 8-31

Magnesium monoperoxyphthalate (MMPP) epoxidizes alkenes much like MCPBA. MMPP is more stable, however, and it may be safer to use for large-scale and industrial reactions. Propose a mechanism for the reaction of *trans*-2-methyl-3-heptene with MMPP, and predict the structure of the product(s).

Magnesium monoperoxyphthalate, MMPP

The body oxidizes the alkene components of drugs and other substances to epoxides, which are then hydrolyzed to diols by an epoxide hydrolase enzyme. The more reactive epoxides are rapidly converted to water-soluble diols and eliminated in the urine. Epoxide hydrolase enzymes are sometimes used in organic synthesis to produce chiral diols.

PROBLEM 8-32

Predict the major products of the following reactions.
(a) *cis*-2-hexene + MCPBA in chloroform
(b) *trans*-3-hexene + peroxyacetic acid (CH_3CO_3H) in water
(c) 1-methylcyclohexene + MMPP in ethanol
(d) *trans*-cyclodecene + peroxyacetic acid in water
(e) *cis*-cyclodecene + MCPBA in CH_2Cl_2, then dilute aqueous acid

PROBLEM 8-33

When 1,2-epoxycyclohexane (cyclohexene oxide) is treated with anhydrous HCl in methanol, the principal product is *trans*-2-methoxycyclohexanol. Give a detailed mechanism to account for the formation of this product.

Converting an alkene to a glycol requires adding a hydroxyl group to each end of the double bond. This addition is called **hydroxylation** of the double bond. We have seen that epoxidation of an alkene, followed by acidic hydrolysis, gives *anti* hydroxylation of the double bond. Reagents are also available for the hydroxylation of alkenes with *syn* stereochemistry. The two most common reagents for this purpose are osmium tetroxide and potassium permanganate.

8-14
Syn Hydroxylation of Alkenes

8-14A Osmium Tetroxide Hydroxylation

Osmium tetroxide (OsO_4, sometimes called *osmic acid*) reacts with alkenes in a concerted step to form a cyclic osmate ester. Hydrogen peroxide hydrolyzes the osmate ester and reoxidizes osmium to osmium tetroxide. The regenerated osmium tetroxide catalyst continues to hydroxylate more molecules of the alkene.

Because the two carbon–oxygen bonds are formed simultaneously with the cyclic osmate ester, the oxygen atoms add to the same face of the double bond; that is, they add with syn stereochemistry. The following reactions show the use of OsO_4 and H_2O_2 for the syn hydroxylation of alkenes.

concerted formation of osmate ester

cis-glycol
(65%)

cis-3-hexene

meso-3,4-hexanediol

8-14B   Permanganate Hydroxylation

Osmium tetroxide is expensive, highly toxic, and volatile. A cold, dilute solution of potassium permanganate also hydroxylates alkenes with syn stereochemistry, with slightly reduced yields in most cases. Like osmium tetroxide, permanganate adds to the alkene double bond to form a cyclic ester: a manganate ester in this case. The basic solution hydrolyzes the manganate ester, liberating the glycol and producing a brown precipitate of manganese dioxide, MnO_2.

concerted formation of manganate ester

cis-glycol
(49%)

In addition to its synthetic value, the permanganate oxidation of alkenes provides a simple chemical test for the presence of an alkene. When an alkene is added to a clear, deep purple aqueous solution of potassium permanganate, the solution loses its purple color and becomes the murky, opaque brown color of MnO_2. (Although there are other functional groups that decolorize permanganate, few do it as quickly as alkenes.)

8-14C   Choosing a Reagent

To hydroxylate an alkene with syn stereochemistry, which is the better reagent: osmium tetroxide or potassium permanganate? Osmium tetroxide gives better yields, but permanganate is cheaper and safer to use. The answer depends on the circumstances.

If the starting material is only 2 mg of a compound 15 steps along in a difficult synthesis, we use osmium tetroxide. The better yield is crucial because the starting material is precious, and little osmic acid is needed. If the hydroxylation is the first step in a synthesis and involves 5 kg of the starting material, we use potassium permanganate. The cost of buying enough osmium tetroxide would be prohibitive, and dealing with such a large amount of a volatile, toxic reagent would be inconvenient. On such a large scale, we can accept the lower yield of the permanganate oxidation.

PROBLEM 8-34

Predict the major products of the following reactions, including stereochemistry.
(a) cyclohexene + OsO_4/H_2O_2
(b) cyclohexene + peroxyacetic acid in water
(c) *cis*-2-pentene + OsO_4/H_2O_2
(d) *cis*-2-pentene + peroxyacetic acid in water
(e) *trans*-2-pentene + OsO_4/H_2O_2
(f) *trans*-2-pentene + peroxyacetic acid in water

PROBLEM 8-35

Show how you would accomplish the following conversions.
(a) *cis*-3-hexene to *meso*-3,4-hexanediol   (b) *cis*-3-hexene to (*d,l*)-3,4-hexanediol
(c) *trans*-3-hexene to *meso*-3,4-hexanediol   (d) *trans*-3-hexene to (*d,l*)-3,4-hexanediol

8-15A   Cleavage by Permanganate

8-15 Oxidative Cleavage of Alkenes

In a potassium permanganate hydroxylation, if the solution is warm or acidic or too concentrated, **oxidative cleavage** of the glycol may occur. Mixtures of ketones and carboxylic acids are formed, depending on whether there are any oxidizable aldehydes in the initial fragments. A terminal $=CH_2$ group is oxidized to CO_2 and water. Figure 8-10 shows the oxidative cleavage of a double bond by warm or concentrated permanganate.

▲ **Figure 8-10**
Warm, concentrated $KMnO_4$ oxidizes alkenes to glycols, then cleaves the glycols. The products are initially ketones and aldehydes, but aldehydes are oxidized to carboxylic acids under these conditions.

8-15B Ozonolysis

Like permanganate, ozone cleaves double bonds to give ketones and aldehydes. However, ozonolysis is milder, and both ketones and aldehydes can be recovered without further oxidation.

Ozone (O_3) is a high-energy form of oxygen produced when ultraviolet light or an electrical discharge passes through oxygen gas. Ultraviolet light from the sun converts oxygen to ozone in the upper atmosphere, where the "ozone layer" shields the earth from some of the high-energy ultraviolet radiation it would otherwise receive.

$$\tfrac{3}{2}O_2 + 34 \text{ kcal} \, (142 \text{ kJ}) \longrightarrow O_3$$

Ozone has 34 kcal/mol (142 kJ/mol) of excess energy over oxygen, and it is much more reactive. A Lewis structure of ozone shows that the central oxygen atom bears a positive charge, and each of the outer oxygen atoms bears half a negative charge.

Ozone reacts with an alkene to form a cyclic compound called a *primary ozonide* or *molozonide* (because 1 mole of ozone has been added). The molozonide has two peroxy (—O—O—) linkages, so it is quite unstable. It rearranges rapidly, even at low temperatures, to form an ozonide.

Ozone is a powerful lung irritant, causing a cough, sore throat, and tiredness. It can also increase a person's sensitivity to allergens. The mechanism may involve peroxidation of the double bonds of the fatty acids that make up the surfactants and the membranes of the cells lining the bronchial airways and lungs.

molozonide
(primary ozonide)

ozonide

Ozonides are not very stable, and they are rarely isolated. In most cases they are immediately reduced by a mild reducing agent such as zinc or (more recently) dimethyl sulfide. The products of this reduction are ketones and aldehydes.

The following reactions show the products obtained from ozonolysis of some representative alkenes. Note the use of (1) and (2) to denote the steps in a two-step sequence used with a single reaction arrow.

$$\underset{\text{3-nonene}}{\diagup\!\!\diagdown\!\!\diagup\!\!\diagdown\!\!\diagup} \xrightarrow[\text{(2) }(CH_3)_2S]{\text{(1) }O_3} \underset{(65\%)}{CH_3CH_2CHO \ + \ CH_3(CH_2)_4CHO}$$

$$\xrightarrow[\text{(2) }(CH_3)_2S]{\text{(1) }O_3}$$

$$\xrightarrow[\text{(2) }(CH_3)_2S]{\text{(1) }O_3} \qquad + \ H_2C{=}O$$

One of the most common uses of ozonolysis has been for determining the positions of double bonds in alkenes. For example, if we were uncertain of the position of the methyl group in a methylcyclopentene, the products of ozonolysis–reduction would confirm the structure of the original alkene.

$$\underset{\text{1-methylcyclopentene}}{} \xrightarrow[\text{(2) }(CH_3)_2S]{\text{(1) }O_3}$$

$$\underset{\text{3-methylcyclopentene}}{} \xrightarrow[\text{(2) }(CH_3)_2S]{\text{(1) }O_3}$$

SOLVED PROBLEM 8-7

Ozonolysis–reduction of an unknown alkene gives an equimolar mixture of cyclohexanecarbaldehyde and 2-butanone. Determine the structure of the original alkene.

cyclohexanecarbaldehyde

$$CH_3{-}\overset{\displaystyle O}{\overset{\|}{C}}{-}CH_2{-}CH_3$$

2-butanone

SOLUTION

We can reconstruct the alkene by removing the two oxygen atoms of the carbonyl groups (C=O) and connecting the remaining carbon atoms with a double bond. One uncertainty remains, however: The original alkene might be either of two possible geometric isomers.

came from

remove oxygen atoms and reconnect the double bond

PROBLEM 8-36

Give structures of the alkenes that would give the following products upon ozonolysis–reduction.

(a) $CH_3-\overset{O}{\overset{\|}{C}}-CH_2-CH_2-CH_2-\overset{O}{\overset{\|}{C}}-CH_2-CH_3$

(b)   [cyclohexanone structure with $=O$]   and $CH_3-CH_2-CH_2-\overset{O}{\overset{\|}{C}}-H$

cyclohexanone

(c) $CH_3-CH_2-\overset{O}{\overset{\|}{C}}-CH_2-CH_2-CH_2-CH_3$ and $CH_3-CH_2-\overset{O}{\overset{\|}{C}}-H$

8-15C   Comparison of Permanganate Cleavage and Ozonolysis

Both permanganate and ozonolysis break the carbon–carbon double bond and replace it with carbonyl (C=O) groups. In the permanganate cleavage, any aldehyde products are further oxidized to carboxylic acids. In the ozonolysis–reduction procedure, the aldehyde products are generated in the dimethyl sulfide reduction step (and not in the presence of ozone), and they are not oxidized.

(not isolated)

PROBLEM 8-37

Predict the major products of the following reactions.
(a) (E)-3-methyl-3-octene + ozone, then $(CH_3)_2S$
(b) (Z)-3-methyl-3-octene + warm, concentrated $KMnO_4$

(c) [cyclohexylidenecyclopentane structure]   + O_3, then $(CH_3)_2S$

(d) 1-ethylcycloheptene + ozone, then $(CH_3)_2S$
(e) 1-ethylcycloheptene + warm, concentrated $KMnO_4$
(f) 1-ethylcycloheptene + cold, dilute $KMnO_4$

8-16
Polymerization
of Alkenes

A **polymer** is a large molecule composed of many smaller repeating units (the **monomers**) bonded together. Alkenes serve as monomers for some of the most common polymers, such as polyethylene, polypropylene, polystyrene, poly(vinyl chloride), and many others. Alkenes generally undergo **addition polymerization**, the rapid addition of one molecule at a time to a growing polymer chain. There is generally a reactive intermediate (cation, anion, or radical) at the growing end of the chain; for that reason, addition polymers are also called **chain-growth polymers**.

Many alkenes undergo addition polymerization under the right conditions. The chain-growth mechanism involves addition of the reactive end of the growing chain across the double bond of the alkene monomer. Depending on the structure of the monomer, the reactive intermediates may be carbocations, free radicals, or carbanions.

8-16A   Cationic Polymerization

Alkenes that easily form carbocations are good candidates for **cationic polymerization**, which is just another example of electrophilic addition to an alkene. Consider what happens when pure isobutylene is treated with a trace of concentrated sulfuric acid. Protonation of the alkene forms a carbocation. If a large concentration of isobutylene is available, another molecule of the alkene may act as the nucleophile and attack the carbocation to form the *dimer* (two monomers joined together) and give another carbocation. If the conditions are right, the growing cationic end of the chain will keep adding across more molecules of the monomer. The polymer of isobutylene is *polyisobutylene*, one of the constituents of *butyl rubber* used in inner tubes and other synthetic rubber products.

Protonation   *Attack by the second molecule of isobutylene*

Attack by a third molecule to give a trimer

Loss of a proton is the most common side reaction that terminates chain growth:

Boron trifluoride (BF_3) is an excellent catalyst for cationic polymerization because it leaves no counterion that might attack the carbocation at the end of the chain. Boron trifluoride is electron-deficient and a strong Lewis acid. It adds to the less highly substituted end of an alkene double bond to give the more stable carbocation. Each additional monomer molecule adds with the same orientation, always giving the more stable carbocation. The following reaction shows the polymerization of styrene (vinylbenzene) using BF_3 as the catalyst.

First chain-lengthening step

The polymerization continues

(P)— = growing polymer chain

The most likely ending of this BF_3-catalyzed polymerization is the loss of a proton from the carbocation at the end of the chain. This side reaction protonates another molecule of styrene, however, initiating a new polymer chain.

Termination of a polymer chain

polystyrene   starts another chain

The product of this polymerization is polystyrene: a clear, brittle plastic that is often used for inexpensive lenses, transparent containers, and styrofoam insulation. Polystyrene is also the major component of the resin beads that are used to make synthetic proteins (See Section 24-11).

PROBLEM 8-38

Propose a mechanism for the following reaction.

$$2 \ (CH_3)_2C{=}CH{-}CH_3 \ + \ cat. \ H^+ \longrightarrow 2,3,4,4\text{-tetramethyl-2-hexene}$$

PROBLEM 8-39

Show the first three steps (as far as the tetramer) in the BF_3-catalyzed polymerization of propylene to form polypropylene.

PROBLEM 8-40

When cyclohexanol is dehydrated to cyclohexene, a gummy green substance forms on the bottom of the flask. Suggest what this residue might be, and propose a mechanism for its formation (as far as the dimer).

8-16B   Free-Radical Polymerization

Many alkenes undergo **free-radical polymerization** when they are heated with radical initiators. For example, styrene polymerizes to polystyrene when it is heated to 100°C with a peroxide initiator. A radical adds to styrene to give a resonance-stabilized radical, which then attacks another molecule of styrene to give an elongated radical.

Initiation step $ROOR \xrightarrow{\text{heat}} 2 \ RO\cdot$

Propagation step

Each propagation step adds another molecule of styrene to the radical end of the growing chain. This addition always takes place with the orientation that gives another resonance-stabilized benzylic (next to a benzene ring) radical.

Propagation step

Chain growth may continue with addition of several hundred or several thousand styrene units. Eventually, the chain reaction stops, either by the coupling of two chains or by reaction with an impurity (such as oxygen) or simply by running out of monomer.

PROBLEM 8-41

Show the intermediate that would result if the growing chain added to the other end of the styrene double bond. Explain why the final polymer has phenyl groups substituted on alternate carbon atoms rather than randomly distributed.

Ethylene is also polymerized by free-radical chain-growth polymerization. With ethylene, the free-radical intermediates are less stable, so stronger reaction conditions are required. Ethylene is commonly polymerized by free-radical initiators at pressures around 3000 atm and temperatures of about 200°C. The product, called *low-density polyethylene*, is the material commonly used in polyethylene bags.

PROBLEM 8-42

Propose a mechanism for reaction of the first three ethylene units in the polymerization of ethylene in the presence of a peroxide.

$$n \ H_2C{=}CH_2 \xrightarrow[\text{high pressure}]{\text{ROOR}} \left[\begin{array}{cc} H & H \\ | & | \\ C & - C \\ | & | \\ H & H \end{array}\right]_n$$

ethylene   polyethylene

8-16C Anionic Polymerization

Like cationic polymerization, **anionic polymerization** depends on the presence of a stabilizing group. To stabilize anions, the double bond should have a strong electron-withdrawing group such as a carbonyl group, a cyano group, or a nitro group. Methyl α-cyanoacrylate contains two powerful electron-withdrawing groups, and it undergoes nucleophilic additions very easily. If this liquid monomer is spread in a thin film between two surfaces, traces of basic impurities (metal oxides, etc.) can catalyze its rapid polymerization. The solidified polymer joins the two surfaces. The chemists who first made this monomer noticed how easily it polymerizes and realized that it could serve as a fast-setting glue. Methyl α-cyanoacrylate is sold commercially as Super Glue.

Initiation step

trace of base   Super Glue   highly stabilized anion

Propagation step

growing chain   monomer   elongated chain   polymer

PROBLEM 8-43

Draw a mechanism for a base-catalyzed polymerization of methyl α-methacrylate to give the Plexiglas® polymer.

methyl α-methacrylate

PROBLEM-SOLVING STRATEGY
Organic Synthesis

Alkyl halides and alkenes are readily made from other compounds, and they are easily converted to other functional groups. This flexibility makes them useful as reagents and intermediates for organic synthesis. Alkenes are particularly important for industrial syntheses because they are inexpensive and available in large quantities from cracking and dehydrogenation of petroleum fractions.

Organic synthesis is the preparation of desired compounds from readily available materials. Synthesis is one of the major areas of organic chemistry, and nearly every chapter of this book involves organic synthesis in some way. A synthesis may be a simple one-step reaction, or it may involve many steps and incorporate a subtle strategy for assembling the correct carbon skeleton with all the functional groups in the right positions.

Many of the problems in this book are synthesis problems. In some synthesis problems, you are asked to show how to convert a given starting material to the desired product. There are obvious one-step answers to some of these problems, while others may require several steps and there may be many correct answers. In solving multistep synthetic problems, it is often helpful to analyze the problem backward: Begin with the desired product (called the *target compound*) and see how it might be mentally changed or broken down to give the starting materials. This backward approach to synthesis is called a **retrosynthetic analysis**.

Some problems allow you to begin with any compounds that meet a certain restriction; for example, you might be allowed to use any alcohols containing no more than four carbon atoms. A retrosynthetic analysis can be used to break down the target compound into fragments no larger than four carbon atoms; then those fragments could be formed from the appropriate alcohols by functional group chemistry.

The following suggestions should help you solve synthesis problems:

1. Do not guess a starting material and try every possible reaction to convert it to the target compound. Rather, begin with the target compound and use a retrosynthetic analysis to simplify it.

2. Use simple equations, with reagents written above and below the arrows, to show the reactions. The equations do not have to be balanced, but they should include all the reagents and conditions that are important to the success of the reaction.

$$\text{A} \xrightarrow{\text{Br}_2, \text{ light}} \text{B} \xrightarrow[\text{heat}]{\text{NaOH, alcohol}} \text{C} \xrightarrow{\text{H}^+, \text{H}_2\text{O}} \text{D}$$

3. Focus your attention on the functional groups, since that is generally where reactions occur. Do not use any reagents that react with a functional group you don't intend to modify.

In solving multistep synthesis problems, you will rarely be able to "see" the solution immediately. These problems are best approached systematically, working backward and considering alternative routes. To illustrate a systematic approach that can guide you in solving synthesis problems, we will work through the synthesis of a complex ether starting from alkenes. The problem-solving method described here will be extended in future chapters to multistep syntheses based on the reactions of additional functional groups.

A systematic retrosynthetic analysis begins with an examination of the structure of the product. We will consider the synthesis of the following compound from alkenes containing up to five carbon atoms.

1. **Review the functional groups and carbon skeleton of the target compound.**
   The target compound is an ether. One alkyl group is a five-carbon cyclopentane ring with two oxygen atoms situated trans. The other group has three carbons containing a reactive epoxide ring.

2. **Review the functional groups and carbon skeletons of the starting materials (if specified), and see how their skeletons might fit together in the target compound.**
   The synthesis is to begin with alkenes containing up to five carbon atoms, so all the functional groups in the product must be derived from alkenes. Most likely, we will start with cyclopentene to give the five-carbon ring and propene to give the three-carbon chain.

3. **Compare methods for synthesizing the functional groups in the target compound, and select the reactions that are most likely to give the correct product.**
   This step may require writing several possible reactions and evaluating them.
   Ethers can be synthesized by nucleophilic reactions between alkyl halides and alkoxides (Section 6-9). The target compound might be formed by S_N2 attack of an alkoxide ion on an alkyl halide in either of two ways:

   The first reaction is better because the S_N2 attack is on a primary alkyl halide, while the second is on a secondary halide. Also, in the second reaction the alkoxide might simply deprotonate the alcohol on the left and cause the reaction to fail.

4. **In general, reactive functional groups are best put into place toward the end of a synthesis.**
   The target compound contains a reactive epoxide ring. Epoxides react with acids and bases, and the epoxide might not survive the crucial ether-forming reaction shown above. Perhaps the epoxide is best added after formation of the ether. That gives us the following final two steps in the synthesis:

5. **Working backward through as many steps as necessary, compare methods for synthesizing the reactants needed for the final step.**
   This process may require writing several possible reaction sequences and evaluating them, keeping in mind the specified starting materials.
   Two reactants are needed to form the ether: an allylic halide and an alkoxide ion. Alkoxide ions are commonly formed by the reaction of an alcohol with sodium metal:

$$\text{R—O—H} + \text{Na} \longrightarrow \text{Na}^+ \ ^-\text{O—R} + \tfrac{1}{2}\text{H}_2 \uparrow$$

   The alkoxide needed in the final step is formed by adding sodium to a trans diol. Trans diols are formed by epoxidation and hydrolysis of alkenes (Section 8-13).

The other piece we need is an allylic bromide. Allylic bromides are formed by allylic bromination of alkenes (Section 6-6B).

$$\text{CH}_3 \xrightarrow[hv]{\text{NBS (trace Br}_2)} \text{Br}$$

6. **Summarize the complete synthesis in the forward direction, including all steps and all reagents, and check it for errors and omissions.**
   This summary is left to you (Problem 8-44), as a review of both the chemistry involved in the synthesis and the method used to develop multistep syntheses.

PROBLEM 8-44

Summarize the synthesis outlined above in the problem-solving exercise. This summary should be in the synthetic (forward) direction, showing each step and all reagents.

Problem 8-45 requires devising several multistep syntheses. As practice in working such problems, we suggest that you proceed in order through the five steps outlined above.

PROBLEM 8-45

Show how you would synthesize each compound, starting with alkenes or cycloalkenes that contain no more than six carbon atoms. You may use any additional reagents you need.

(a) ![structure a]   (b) ![structure b]   (c) ![structure c]

| SUMMARY | **Reactions of Alkenes** |

1. *Electrophilic Additions*

a. *Addition of hydrogen halides* (Section 8-3)

$$\underset{\diagup}{\overset{\diagdown}{\text{C}}}=\underset{\diagdown}{\overset{\diagup}{\text{C}}} \;+\; \text{H}-\text{X} \longrightarrow -\underset{\text{H}}{\overset{|}{\text{C}}}-\underset{\text{X}}{\overset{|}{\text{C}}}-$$

(HX = HCl, HBr, or HI)   Markovnikov orientation
(anti-Markovnikov with HBr and peroxides)

Example

$$\underset{\text{2-methylpropene}}{\overset{\overset{\text{CH}_3}{|}}{\text{CH}_3-\text{C}=\text{CH}_2}} + \text{HBr}$$

no peroxides →

$$\underset{t\text{-butyl bromide}}{\underset{\text{(Markovnikov orientation)}}{\overset{\overset{\text{CH}_3}{|}}{\underset{\overset{|}{\text{Br}}}{\text{CH}_3-\text{C}-\text{CH}_3}}}}$$

peroxides →

$$\underset{\text{isobutyl bromide}}{\underset{\text{(anti-Markovnikov orientation)}}{\overset{\overset{\text{CH}_3}{|}}{\text{CH}_3-\text{CH}-\text{CH}_2\text{Br}}}}$$

b. *Acid-catalyzed hydration* (Section 8-4)

$$\ce{C=C} + H_2O \xrightarrow{H^+} \underset{\substack{| \quad | \\ H \quad OH}}{-\overset{|}{C}-\overset{|}{C}-}$$

(Markovnikov orientation)

Example

$$CH_3-CH=CH_2 + H_2O \xrightarrow{H_2SO_4} CH_3-\overset{\overset{\displaystyle OH}{|}}{C}H-CH_3$$

propene   2-propanol

c. *Oxymercuration–demercuration* (Section 8-5)

$$\ce{C=C} + Hg(OAc)_2 \xrightarrow{H_2O} \underset{\substack{| \quad | \\ HO \quad HgOAc}}{-\overset{|}{C}-\overset{|}{C}-} \xrightarrow{NaBH_4} \underset{\substack{| \quad | \\ HO \quad H}}{-\overset{|}{C}-\overset{|}{C}-}$$

(Markovnikov orientation)

Example

$$H_2C=CHCH_2CH_3 \xrightarrow[H_2O]{Hg(OAc)_2} \underset{\substack{| \quad | \\ CH_2 \quad CHCH_2CH_3}}{\overset{\overset{\displaystyle AcOHg \quad OH}{|\qquad |}}{}} \xrightarrow{NaBH_4} CH_3-\overset{\overset{\displaystyle OH}{|}}{C}HCH_2CH_3$$

1-butene   2-butanol

d. *Alkoxymercuration–demercuration* (Section 8-6)

$$\ce{C=C} + Hg(OAc)_2 \xrightarrow{ROH} \underset{\substack{| \quad | \\ RO \quad HgOAc}}{-\overset{|}{C}-\overset{|}{C}-} \xrightarrow{NaBH_4} \underset{\substack{| \quad | \\ RO \quad H}}{-\overset{|}{C}-\overset{|}{C}-}$$

(Markovnikov orientation)

Example

$$H_2C=CH-CH_2-CH_3 \xrightarrow[\text{(2) } NaBH_4]{\text{(1) } Hg(OAc)_2,\ CH_3OH} CH_3-\underset{\underset{\displaystyle OCH_3}{|}}{C}H-CH_2-CH_3$$

1-butene   2-methoxybutane

e. *Hydroboration–oxidation* (Section 8-7)

$$\ce{C=C} + BH_3 \cdot THF \longrightarrow \underset{\substack{| \quad | \\ H \quad B-H \\ \quad\ | \\ \quad\ H}}{-\overset{|}{C}-\overset{|}{C}-} \xrightarrow{H_2O_2,\ ^-OH} \underset{\substack{| \quad | \\ H \quad OH}}{-\overset{|}{C}-\overset{|}{C}-}$$

anti-Markovnikov orientation
(syn stereochemistry)

Example

$$\xrightarrow[\text{(2) } H_2O_2,\ ^-OH]{\text{(1) } BH_3 \cdot THF}$$

f. *Polymerization* (Section 8-16)

$$R^+ + \ce{C=C} \longrightarrow R-\underset{\substack{|}}{\overset{|}{C}}-\overset{|}{C}+ \xrightarrow{\ce{C=C}} R-\overset{|}{C}-\overset{|}{C}-\overset{|}{C}-\overset{|}{C}+ \longrightarrow$$

Example

$$n \quad CH_3-CH=CH_2 \xrightarrow{BF_3} \begin{bmatrix} \overset{H}{\underset{H}{C}} - \overset{H}{\underset{CH_3}{C}} \end{bmatrix}_n$$

propylene polypropylene

2. *Reduction: Catalytic Hydrogenation* (Section 8-8)

$$\text{C=C} + H_2 \xrightarrow{\text{Pt, Pd, or Ni}} \text{—C—C—}$$
(syn addition)

3. *Addition of Carbenes: Cyclopropanation* (Section 8-9)

$$\text{C=C} + :C\overset{X}{\underset{Y}{\diagdown}} \longrightarrow$$

(X,Y = H, Cl, Br, I, or —COOEt)

Example

cyclohexene + CHBr₃ $\xrightarrow{\text{NaOH/H}_2\text{O}}$ (Br, Br)

4. *Oxidative Additions*

a. *Addition of halogens* (Section 8-10)

$$\text{C=C} + X_2 \longrightarrow \text{—C—C—}$$
(X_2 = Cl₂, Br₂, sometimes I₂)
(anti addition)

Example

cyclohexene + Br₂ ⟶ *trans*-1,2-dibromocyclohexane

b. *Halohydrin formation* (Section 8-11)

+ Br₂ $\xrightarrow[\substack{\text{anti addition}\\\text{(Markovnikov orientation)}}]{\text{H}_2\text{O}}$ (CH₃, OH, Br)

c. *Epoxidation* (Section 8-12)

$$\underset{\text{alkene}}{\ce{>C=C<}} + \underset{\text{peroxyacid}}{R-\overset{\displaystyle O}{\underset{\|}{C}}-O-O-H} \longrightarrow \underset{\text{syn addition}}{-\overset{|}{\underset{O}{C}}-\overset{|}{\underset{\triangle}{C}}-} + R-\overset{\displaystyle O}{\underset{\|}{C}}-O-H$$

Example

cyclohexene   +   [3-chlorobenzene—C—OOH with ‖O]   →   epoxycyclohexane (cyclohexene oxide)   +   [3-chlorobenzene—C—OH with ‖O]

d. *Anti hydroxylation* (Section 8-13)

$$\ce{>C=C<} + R-\overset{\displaystyle O}{\underset{\|}{C}}-O-O-H \longrightarrow -\overset{|}{\underset{O}{C}}-\overset{|}{\underset{\triangle}{C}}- \xrightarrow{H^+, H_2O} \underset{\text{(anti addition)}}{-\overset{|}{\underset{OH}{C}}-\overset{OH}{\underset{|}{C}}-}$$

Example

cyclohexene $\xrightarrow{\text{H—C—OOH, } H_3O^+}$ *trans*-cyclohexane-1,2-diol

e. *Syn hydroxylation* (Section 8-14)

$$\ce{>C=C<} + KMnO_4 + \,^-OH, H_2O \longrightarrow \underset{\text{(syn addition)}}{-\overset{|}{\underset{OH}{C}}-\overset{|}{\underset{OH}{C}}-}$$
$$(or\ OsO_4, H_2O_2)$$

Example

cyclohexene $\xrightarrow{OsO_4, H_2O_2}$ *cis*-cyclohexane-1,2-diol

5. *Oxidative Cleavage of Alkenes* (Section 8-15)

a. *Ozonolysis*

$$\underset{R}{\overset{R}{>}}C=C\underset{H}{\overset{R'}{<}} + O_3 \longrightarrow \underset{\text{ozonide}}{\left[R,R-C\overset{O}{\underset{O-O}{\big\rangle}}C\overset{R'}{\underset{H}{\big\langle}}\right]} \xrightarrow{(CH_3)_2S} \underset{\text{ketones and aldehydes}}{\overset{R}{\underset{R}{>}}C=O + O=C\overset{R'}{\underset{H}{<}}}$$

Example

$$\underset{\text{2-methyl-2-butene}}{CH_3-\overset{\displaystyle H}{C}=\overset{\displaystyle CH_3}{C}-CH_3} \xrightarrow[\text{(2) } (CH_3)_2S]{\text{(1) } O_3} \underset{\text{acetaldehyde}}{CH_3-\overset{\displaystyle H}{C}=O} + \underset{\text{acetone}}{O=\overset{\displaystyle CH_3}{C}-CH_3}$$

b. *Potassium permanganate*

$$\underset{R}{\overset{R}{\diagdown}}C=C\underset{H}{\overset{R'}{\diagup}} + KMnO_4 \xrightarrow{\text{warm}} \underset{R}{\overset{R}{\diagdown}}C=O + O=C\underset{OH}{\overset{R'}{\diagup}}$$

ketones and acids
(aldehydes are oxidized)

Example

$$CH_3-\overset{\overset{\displaystyle H}{|}}{C}=\overset{\overset{\displaystyle CH_3}{|}}{C}-CH_3 + KMnO_4 \xrightarrow{\text{warm}} CH_3-\overset{\overset{\displaystyle OH}{|}}{C}=O + O=\overset{\overset{\displaystyle CH_3}{|}}{C}-CH_3$$

2-methyl-2-butene   acetic acid   acetone

Chapter 8 Glossary

addition A reaction involving an increase in the number of groups attached to the alkene and a decrease in the number of elements of unsaturation. (p. 314)

  anti addition: An addition in which two groups add to opposite faces of the double bond (as in addition of Br_2). (p. 326)

  electrophilic addition: An addition in which the electrophile (electron-pair acceptor) bonds to one of the double-bonded carbons first, followed by the nucleophile. (p. 315)

  syn addition: An addition in which two groups add to the same face of the double bond (as in osmium tetroxide hydroxylation). (p. 326)

addition polymer (chain-growth polymer) A polymer that results from rapid addition of one molecule at a time to a growing polymer chain, usually with a reactive intermediate (cation, radical, or anion) at the growing end of the chain. (p. 352)

alkoxymercuration The addition of mercuric acetate to an alkene in an alcohol solution, forming an alkoxymercurial intermediate. Demercuration gives an ether. (p. 326)

$$\overset{\diagdown}{\underset{\diagup}{}}C=C\overset{\diagup}{\underset{\diagdown}{}} + Hg(OAc)_2 \xrightarrow{R-OH} \underset{HgOAc}{-\overset{\overset{\displaystyle R-O}{|}}{C}-\overset{|}{C}-} \xrightarrow{NaBH_4} \underset{H}{-\overset{\overset{\displaystyle R-O}{|}}{C}-\overset{|}{C}-}$$

alpha elimination (α elimination) The elimination of two atoms or groups from the same carbon atom. Alpha eliminations are frequently used to form carbenes. (p. 337)

$$CHBr_3 + KOH \longrightarrow :CBr_2 + H_2O + KBr$$

anionic polymerization The process of forming an addition polymer by chain-growth polymerization involving an anion at the end of the growing chain. (p. 356)

asymmetric induction (enantioselective synthesis) The formation of an optically active product from an optically inactive starting material. Such a process requires the use of an optically active reagent or catalyst. (p. 335)

beta elimination (β elimination) The elimination of two atoms or groups from adjacent carbon atoms. This is the most common type of elimination. (p. 337)

$$-\overset{\overset{\displaystyle H}{|}}{C}-\overset{\overset{\displaystyle Br}{|}}{C}- + KOH \longrightarrow \overset{\diagdown}{\underset{\diagup}{}}C=C\overset{\diagup}{\underset{\diagdown}{}} + H_2O + KBr$$

carbene A reactive intermediate with a neutral carbon atom having only two bonds and two nonbonding electrons. Methylene ($:CH_2$) is the simplest carbene. (p. 336)

cationic polymerization The process of forming an addition polymer by chain-growth polymerization involving a cation at the end of the growing chain. (p. 353)

chain-growth polymer See **addition polymer**. (p. 353)

demercuration The removal of a mercury species from a molecule. Demercuration of the products of oxymercuration and alkoxymercuration is usually accomplished using sodium borohydride. (p. 325)

epoxide (oxirane) A three-membered cyclic ether. (p. 344)

  epoxidation: Formation of an epoxide, usually from an alkene. A peroxyacid is generally used for alkene epoxidations.

free-radical polymerization The process of forming an addition polymer by chain-growth polymerization involving a free radical at the end of the growing chain. (p. 355)

glycol A 1,2-diol. (p. 345)

halogenation The addition of a halogen (X_2) to a molecule, or the free-radical substitution of an X for an H (p. 338)

halohydrin A beta-haloalcohol, with a halogen and a hydroxyl group on adjacent carbon atoms. (p. 341)

halonium ion A reactive, cationic intermediate with a three-membered ring containing a halogen atom. Usually a **chloronium ion**, a **bromonium ion**, or an **iodonium ion**. (p. 339)

heterogeneous catalysis Use of a catalyst that is in a separate phase from the reactants. For example, a platinum hydrogenation catalyst is a solid, a separate phase from the liquid alkene. (p. 334)

homogeneous catalysis Use of a catalyst that is in the same phase as the reactants. For example, the acid catalyst in hydration is in the liquid phase with the alkene. (p. 334)

hydration The addition of water to a molecule. Hydration of an alkene forms an alcohol. (p. 322)

$$\text{C=C} \;+\; H_2O \;\xrightarrow{H^+}\; \overset{H\;\;\;OH}{-C-C-}$$

hydroboration The addition of borane (BH_3) or one of its derivatives ($BH_3 \cdot THF$, for example) to a molecule. (p. 329)

hydrogenation The addition of hydrogen to a molecule. The most common hydrogenation is the addition of H_2 across a double bond in the presence of a catalyst (**catalytic hydrogenation** or **catalytic reduction**). (p. 333)

hydroxylation The addition of two hydroxyl groups, one at each carbon of the double bond. Formally an oxidation. (p. 347)

$$\text{C=C} \;+\; H_2O_2 \;\xrightarrow{OsO_4}\; \overset{HO\;\;\;OH}{-C-C-}$$

Markovnikov's rule (*original statement*) When a proton acid adds to the double bond of an alkene, the proton bonds to the carbon atom that already has more hydrogen atoms.

(*extended statement*) In an electrophilic addition to an alkene, the electrophile adds in such a way as to generate the most stable intermediate. (p. 317)

> **Markovnikov orientation:** An orientation of addition that obeys the original statement of Markovnikov's rule; one that gives the **Markovnikov product**. (p. 318)

> **anti-Markovnikov orientation:** An orientation of addition that is the opposite of that predicted by the original statement of Markovnikov's rule; one that gives the **anti-Markovnikov product**. (p. 319)

monomer One of the small molecules that bond together to form a polymer. (p. 352)

organic synthesis The preparation of desired organic compounds from readily available materials. (p. 357)

oxidative cleavage The cleavage of a carbon–carbon bond through oxidation. Carbon–carbon double bonds are commonly cleaved by ozonolysis/reduction or by warm, concentrated permanganate. (p. 349)

oxymercuration The addition of aqueous mercuric acetate to an alkene. (p. 324)

$$\text{C=C} \;+\; Hg(OAc)_2 \;\xrightarrow{H_2O}\; \overset{HO}{\underset{HgOAc}{-C-C-}} \;+\; HOAc$$

ozonolysis The use of ozone, usually followed by reduction, to cleave a double bond. (p. 350)

peroxide effect The reversal of orientation of HBr addition to alkenes in the presence of peroxides. A free-radical mechanism is responsible for the peroxide effect. (p. 320)

peroxyacid (peracid) A carboxylic acid with an extra oxygen atom and a peroxy ($-O-O-$) linkage. The general formula is RCO_3H. (p. 344)

polymer A high molecular weight compound composed of many molecules of a smaller, simpler compound called the **monomer**. (p. 352)

> **polymerization:** The reaction of monomer molecules to form a polymer.

regioselective reaction A reaction in which one direction of bond making or bond breaking occurs preferentially over all other directions. For example, the addition of HCl is regioselective, predicted by Markovnikov's rule. Hydroboration–oxidation is regioselective because it consistently gives anti-Markovnikov orientation. (p. 317)

retrosynthetic analysis A method of working backward to solve multistep synthetic problems. (p. 357)

Simmons–Smith reaction A cyclopropanation of an alkene using the carbenoid reagent generated from diiodomethane and the zinc–copper couple. (p. 337)

sterospecific reaction A reaction that converts different stereoisomers of the starting material into different stereoisomers of the product. (p. 332)

Essential Problem-Solving Skills in Chapter 8

1. Predict the products of additions, oxidations, reductions, and cleavages of alkenes, including
   (a) Orientation of reaction (regiochemistry)
   (b) Stereochemistry.

2. Propose logical mechanisms to explain the observed products of alkene reactions, including regiochemistry and stereochemistry.

3. Use retrosynthetic analysis to solve multistep synthesis problems with alkenes as reagents, intermediates, or products.

4. When more than one method is usable for a chemical transformation, choose the better method and explain its advantages.

5. Use clues provided by products of reactions such as ozonolysis to determine the structure of an unknown alkene.

In studying these reaction-intensive chapters, students ask whether they should "memorize" all the reactions. Doing organic chemistry is like speaking a foreign language, and the reactions are our vocabulary. Without knowing the words, how can you construct sentences? Making flash cards often helps.

In organic chemistry, the mechanisms, regiochemistry, and stereochemistry are our grammar. You must develop *facility* with the reactions, as you develop facility with the words and grammar you use in speaking. Problems and multistep syntheses are the sentences of organic chemistry. You must *practice* combining all aspects of your vocabulary in solving these problems.

Students who fail exams often do so because they have memorized the vocabulary, but they have not practiced doing problems. Others fail because they think they can do problems, but they lack the vocabulary. If you understand the reactions and can do the end-of-chapter problems without looking back, you should do well on your exams.

Study Problems

8-46   Define each term, and give an example.

(a) dimerization	**(b)** polymerization	**(c)** electrophilic addition
(d) stereospecific addition	**(e)** syn addition	**(f)** anti addition
(g) Markovnikov addition	**(h)** anti-Markovnikov addition	**(i)** peroxide effect
(j) hydrogenation	**(k)** hydration	**(l)** homogeneous catalysis
(m) heterogeneous catalysis	**(n)** halogenation	**(o)** halohydrin
(p) hydroxylation	**(q)** epoxidation	**(r)** oxidative cleavage
(s) hydroboration	**(t)** oxymercuration–demercuration	**(u)** alkoxymercuration–demercuration
(v) carbene addition	**(w)** alpha elimination	**(x)** beta elimination
(y) addition polymer	**(z)** monomer	**(aa)** cationic polymerization

8-47   Predict the major products of the following reactions, and give the structures of any intermediates. Include stereochemistry where appropriate.

(a) $\xrightarrow{\text{HCl}}$ **(b)** $\xrightarrow[\text{CCl}_4]{\text{Br}_2}$ **(c)** $\xrightarrow[\text{(2) H}_2\text{O}_2,\ ^-\text{OH}]{\text{(1) BH}_3 \cdot \text{THF}}$

(d) $\xrightarrow[\text{(2) (CH}_3)_2\text{S}]{\text{(1) O}_3}$

(e) $\xrightarrow[\text{ROOR}]{\text{HBr}}$

(f) $\xrightarrow[\text{ROOR}]{\text{HCl}}$

(g) $\xrightarrow{\text{PhCO}_3\text{H}}$

(h) $\xrightarrow[\text{H}_2\text{O}_2]{\text{OsO}_4}$

(i) $\xrightarrow[\text{(cold, dil)}]{\text{KMnO}_4,\ ^-\text{OH}}$

(j) $\xrightarrow[\text{H}^+,\ \text{H}_2\text{O}]{\text{CH}_3\text{CO}_3\text{H}}$

(k) $\xrightarrow[\text{(warm, concd.)}]{\text{KMnO}_4,\ ^-\text{OH}}$

(l) $\xrightarrow[\text{(2) (CH}_3)_2\text{S}]{\text{(1) O}_3}$

(m) $\xrightarrow[\text{Pt}]{\text{H}_2}$

(n) $\xrightarrow{\text{H}^+,\ \text{H}_2\text{O}}$

(o) $\xrightarrow[\text{(2) NaBH}_4]{\text{(1) Hg(OAc)}_2,\ \text{H}_2\text{O}}$

(p) $\xrightarrow[\text{H}_2\text{O}]{\text{Cl}_2}$

8-48   Propose mechanisms consistent with the following reactions.

(a) $\xrightarrow[\text{ROOR}]{\text{HBr}}$

(b) $\xrightarrow[\text{H}_2\text{O}]{\text{H}_2\text{SO}_4}$

(c) $\xrightarrow{\text{HBr}}$ +

(d) $\xrightarrow[\text{NaOH}]{\text{CHBr}_3}$

(e) $\xrightarrow[\text{CH}_3\text{OH}]{\text{HCl}}$ +

(f) $\xrightarrow[\text{LiCl in CH}_3\text{OH}]{\text{Br}_2}$ + +

(g) 2 $\xrightarrow{\text{H}^+}$

8-49   Show how you would synthesize each compound using methylenecyclohexane as your starting material.

methylenecyclohexane

(a) OH   **(b)** Br   **(c)** OH

(d) O   **(e)** OCH₃   **(f)** OH OH

(g)   **(h)** OH Cl   **(i)** Br Br

8-50   Limonene is one of the compounds that give lemons their tangy odor. Show the structures of the products expected when limonene reacts with an excess of each of these reagents.

limonene

   (a) borane in tetrahydrofuran, followed by basic hydrogen peroxide
   (b) *m*-chloroperoxybenzoic acid
   (c) ozone, then dimethyl sulfide

(d) a mixture of osmic acid and hydrogen peroxide
(e) hot, concentrated potassium permanganate
(f) peroxyacetic acid in water
(g) hydrogen and a platinum catalyst
(h) hydrogen bromide gas
(i) hydrogen bromide gas in a solution containing dimethyl peroxide
(j) bromine water
(k) chlorine gas
(l) mercuric acetate in methanol, followed by sodium borohydride
(m) methylene iodide pretreated with the zinc–copper couple

8-51 The structures of three monomers are shown below. In each case, show the structure of the polymer that would result from polymerization of the monomer. Vinyl chloride is polymerized to "vinyl" plastics and PVC pipe. Tetrafluoroethylene polymerizes to Teflon®, used as non-stick coatings and PTFE valves and gaskets. Acrylonitrile is polymerized to Orlon®, used in sweaters and carpets.

*8-52 When styrene (vinylbenzene) is commercially polymerized, about 1–3% of 1,4-divinylbenzene is often added to the styrene. The incorporation of some divinylbenzene gives a polymer with more strength and better resistance to organic solvents. Explain how a very small amount of divinylbenzene has a marked effect on the properties of the polymer.

8-53 The cationic polymerization of isobutylene (2-methylpropene) is shown in Section 8-16A. Isobutylene is often polymerized under free-radical conditions. Propose a mechanism for the free-radical polymerization of isobutylene.

8-54 Poly(ethyl acrylate) has the formula

$$\left(CH_2 - CH \right)_n \quad \overset{\displaystyle O}{\underset{\displaystyle |}{\overset{\displaystyle \|}{C}} - OCH_2CH_3}$$

Give the structure of the ethyl acrylate monomer.

8-55 Draw the structures of the following compounds, and determine which member of each pair is more reactive toward the addition of HBr.
(a) propene or 2-methylpropene   (b) cyclohexene or 1-methylcyclohexene
(c) 1-butene or 1,3-butadiene

8-56 Cyclohexene is dissolved in a solution of lithium chloride in chloroform. To this solution is added one equivalent of bromine. The material isolated from this reaction contains primarily a mixture of *trans*-1,2-dibromocyclohexane and *trans*-1-bromo-2-chlorocyclohexane. Propose a mechanism to show how these compounds are formed.

8-57 Draw a reaction-energy diagram for the propagation steps of the free-radical addition of HBr to isobutylene. Draw curves representing the reactions leading to both the Markovnikov and the anti-Markovnikov products. Compare the values of $\Delta G°$ and E_a for the rate-limiting steps, and explain why only one of these products is observed.

8-58 Give the products expected when the following compounds are ozonized and reduced.

8-59 Show how you would make the following compounds from a suitable cyclic alkene.

8-60 Unknown **X**, C_5H_9Br, does not react with bromine or with dilute KMnO$_4$. Upon treatment with potassium *t*-butoxide, **X** gives only one product, **Y**, C_5H_8. Unlike **X**, **Y** decolorizes bromine and changes KMnO$_4$ from purple to brown. Ozonolysis–reduction of **Y** gives **Z**, $C_5H_8O_2$. Propose consistent structures for **X**, **Y**, and **Z**.

8-61   One of the constituents of turpentine is α-pinene, formula $C_{10}H_{16}$. The following scheme (called a "road map") gives some reactions of α-pinene. Determine the structure of α-pinene and of the reaction products **A** through **E**.

8-62   The sex attractant of the housefly has the formula $C_{23}H_{46}$. When treated with warm potassium permanganate, this pheromone gives two products: $CH_3(CH_2)_{12}COOH$ and $CH_3(CH_2)_7COOH$. Suggest a structure for this sex attractant. Explain which part of the structure is uncertain.

8-63   In contact with a platinum catalyst, an unknown alkene reacts with 3 equivalents of hydrogen gas to give 1-isopropyl-4-methylcyclohexane. When the unknown alkene is ozonized and reduced, the products are the following:

Deduce the structure of the unknown alkene.

*8-64   Propose a mechanism for the following reaction.

8-65   The two butenedioic acids are called *fumaric acid* (trans) and *maleic acid* (cis). 2,3-Dihydroxybutanedioic acid is called *tartaric acid*.

Show how you would convert
(a) fumaric acid to ($\pm$)-tartaric acid.   (b) fumaric acid to *meso*-tartaric acid.
(c) maleic acid to ($\pm$)-tartaric acid.   (d) maleic acid to *meso*-tartaric acid.

8-66   The compound BD_3 is a deuterated form of borane. Predict the product formed when 1-methylcyclohexene reacts with $BD_3 \cdot THF$, followed by basic hydrogen peroxide.

8-67   A routine addition of HBr across the double bond of a vinylcyclopentane gave an unexpected rearranged product. Propose a mechanism for the formation of this product, and explain why the rearrangement occurs.

8-68   An unknown compound decolorizes bromine in carbon tetrachloride, and it undergoes catalytic reduction to give decalin. When treated with warm, concentrated potassium permanganate, this compound gives *cis*-cyclohexane-1,2-dicarboxylic acid and oxalic acid. Propose a structure for the unknown compound.

decalin

unknown compound

$KMnO_4$ (warm, concd.)

cis-cyclohexane-1,2-dicarboxylic acid

+ oxalic acid (→ further oxidation)

*8-69   Many enzymes catalyze reactions that are similar to reactions we might use for organic synthesis. Enzymes tend to be stereospecific in their reactions, and asymmetric induction is common. The following reaction, part of the tricarboxylic acid cycle of cell respiration, resembles a reaction we might use in the laboratory; however, the enzyme-catalyzed reaction gives only the (S) enantiomer of the product, malic acid.

fumaric acid

fumarase H_2O, pH 7.4

(S)-malic acid

product in D_2O

(a) What type of reaction does fumarase catalyze?

(b) Is fumaric acid chiral? Is malic acid chiral? In the enzyme-catalyzed reaction, is the product (malic acid) optically active?

(c) If we could run the above reaction in the laboratory using sulfuric acid as the catalyst, would the product (malic acid) be optically active?

(d) Do you expect the fumarase enzyme to be a chiral molecule?

(e) When the enzyme-catalyzed reaction takes place in D_2O, the *only* product is the stereoisomer pictured above. No enantiomer or diastereomer of this compound is formed. Is the enzyme-catalyzed reaction a syn or anti addition?

(f) Assume we found conditions to convert fumaric acid to deuterated malic acid using hydroboration with $BD_3 \cdot THF$, followed by oxidation with D_2O_2 and NaOD. Use Fischer projections to show the stereoisomer(s) of deuterated malic acid you would expect to be formed.

*8-70   (a) The following cyclization has been observed in the oxymercuration–demercuration of this unsaturated alcohol. Propose a mechanism for this reaction.

(1) $Hg(OAc)_2$
(2) $NaBH_4$

(b) Predict the product of formula $C_7H_{13}BrO$ from the reaction of this same unsaturated alcohol with bromine. Propose a mechanism to support your prediction.

*8-71   An inexperienced graduate student treated 5-decene with borane in THF, placed the flask in a refrigerator, and left for a party. When he returned from the party, he discovered that the refrigerator was broken, and it had gotten quite warm inside. Although all the THF had evaporated from the flask, he treated the residue with basic hydrogen peroxide. To his surprise, he recovered a fair yield of 1-decanol. Use a mechanism to show how this reaction might have occurred. (*Hint*: The addition of BH_3 is reversible.)

*8-72   We have seen many examples where halogens add to alkenes with anti stereochemistry via the halonium ion mechanism. However, when 1-phenylcyclohexene reacts with chlorine in carbon tetrachloride, a mixture of the cis and trans isomers of the product is recovered. Propose a mechanism, and explain this lack of stereospecificity.

1-phenylcyclohexene

Cl_2
CCl_4

cis- and trans-
1,2-dichloro-1-phenylcyclohexane

Alkynes

9-1
Introduction

Alkynes are hydrocarbons that contain carbon–carbon triple bonds. Alkynes are also called **acetylenes** because they are derivatives of acetylene, the simplest alkyne.

$$H-C \equiv C-H \qquad CH_3CH_2-C \equiv C-H \qquad CH_3-C \equiv C-CH_3$$

acetylene	ethylacetylene	dimethylacetylene
ethyne	1-butyne	2-butyne

The chemistry of the carbon–carbon triple bond is similar to that of the double bond. In this chapter, we see that alkynes undergo most of the reactions of alkenes, especially the additions and the oxidations. We also consider reactions that are specific to alkynes: some that depend on the unique characteristics of the $C \equiv C$ triple bond, and others that depend on the unusual acidity of the acetylenic $C-H$ bond.

A triple bond gives an alkyne four fewer hydrogens than the corresponding alkane. Its molecular formula is like that of a molecule with two double bonds: C_nH_{2n-2}. Therefore, the triple bond contributes two elements of unsaturation (eu) (Section 7-3).

ethane, C_2H_6	ethene, C_2H_4	ethyne, C_2H_2
0 eu, C_nH_{2n+2}	1 eu, C_nH_{2n}	2 eu, C_nH_{2n-2}

Alkynes are not as common in nature as alkenes, but some plants do use alkynes to protect themselves against disease or predators. Cicutoxin is a toxic compound found in water hemlock, and capillin protects a plant against fungal diseases. The alkyne functional group is not common in drugs, but parsalmide is used as an analgesic, and ethynyl estradiol (a synthetic female hormone) is a common ingredient in birth control pills. Dynemicin A is an antibacterial compound that is being tested as an antitumor agent.

$$HOCH_2CH_2CH_2-C \equiv C-C \equiv C-CH=CH-CH=CH-CH=CH-CHCH_2CH_2CH_3$$
$$|$$
$$OH$$

Cicutoxin

$$CH_3-C \equiv C-C \equiv C-\overset{\displaystyle O}{\overset{\displaystyle \|}{C}}-$$

capillin

parsalmide

ethynyl estradiol

Dynemicin A

PROBLEM 9-1

Draw structural formulas of at least two alkynes of each molecular formula.

(a) C_6H_{10} (b) C_8H_{12} (c) C_7H_{10}

9-2 Nomenclature of Alkynes

IUPAC Names The IUPAC nomenclature for alkynes is similar to that for alkenes. We find the longest continuous chain of carbon atoms that includes the triple bond and change the *-ane* ending of the parent alkane to *-yne*. The chain is numbered from the end closest to the triple bond, and the position of the triple bond is designated by its lower-numbered carbon atom. Substituents are given numbers to indicate their locations.

$$H-C\equiv C-H \quad CH_3-C\equiv C-H \quad CH_3-C\equiv C-CH_3 \quad CH_3-\underset{\underset{CH_3}{|}}{CH}-C\equiv C-CH_2-\underset{\underset{Br}{|}}{CH}-CH_3$$

IUPAC name:	ethyne	propyne	2-butyne	6-bromo-2-methyl-3-heptyne
	(acetylene)		but-2-yne	6-bromo-2-methylhept-3-yne

When additional functional groups are present, the suffixes are combined to produce the compound names of the *alkenynes* (a double bond and a triple bond), *alkynols* (a triple bond and an alcohol), and so on. The new IUPAC system (placing the number right before the group) helps with these names.

$$H_2C=\underset{\underset{CH_3}{|}}{C}-C\equiv C-CH_3 \quad CH_3-\underset{\underset{OH}{|}}{CH}-C\equiv C-H \quad CH_3-C\equiv C-\underset{\underset{OCH_3}{|}}{CH}-CH_2CH_3$$

IUPAC name:	2-methyl-1-penten-3-yne	3-butyn-2-ol	4-methoxy-2-hexyne
new IUPAC name:	2-methylpent-1-en-3-yne	but-3-yn-2-ol	4-methoxyhex-2-yne

Common Names The common names of alkynes describe them as derivatives of acetylene. Most alkynes can be named as a molecule of acetylene with one or two alkyl substituents. This nomenclature is like the common nomenclature for ethers, where we name the two alkyl groups bonded to oxygen.

$$H-C\equiv C-H$$
acetylene

$$R-C\equiv C-H$$
an alkylacetylene

$$R-C\equiv C-R'$$
a dialkylacetylene

$$CH_3-C\equiv C-H$$
methylacetylene

$$Ph-C\equiv C-H$$
phenylacetylene

$$CH_3-C\equiv C-CH_2CH_3$$
ethylmethylacetylene

$$(CH_3)_2CH-C\equiv C-CH(CH_3)_2$$
diisopropylacetylene

$$Ph-C\equiv C-Ph$$
diphenylacetylene

$$H-C\equiv C-CH_2OH$$
hydroxymethylacetylene
(propargyl alcohol)

Many of an alkyne's chemical properties depend on whether there is an acetylenic hydrogen ($H-C\equiv C$); that is, whether the triple bond comes at the end of a carbon chain. Such an alkyne is called a **terminal alkyne** or a **terminal acetylene**. If the triple bond is located somewhere other than the end of the carbon chain, the alkyne is called an **internal alkyne** or an **internal acetylene**.

acetylenic hydrogen

(no acetylenic hydrogen)

$$H-C\equiv C-CH_2CH_3 \qquad\qquad CH_3-C\equiv C-CH_3$$

1-butyne, a *terminal* alkyne   2-butyne, an *internal* alkyne

PROBLEM 9-2

For each molecular formula, draw all the isomeric alkynes, and give their IUPAC names. Circle the acetylenic hydrogen of each terminal alkyne.
(a) C_4H_6 (two isomers)   (b) C_5H_8 (three isomers)

9-3
Physical Properties of Alkynes

The physical properties of alkynes (Table 9-1) are similar to those of alkanes and alkenes of similar molecular weights. Alkynes are relatively nonpolar and nearly insoluble in water. They are quite soluble in most organic solvents, including acetone, ether, methylene chloride, chloroform, and alcohols. Many alkynes have characteristic, mildly offensive odors.

Acetylene, propyne, and the butynes are gases at room temperature, just like the corresponding alkanes and alkenes. In fact, the boiling points of alkynes are nearly the same as those of alkanes and alkenes with similar carbon skeletons.

TABLE 9-1   Physical Properties of Selected Alkynes

Name	Structure	mp (°C)	bp (°C)	Density (g/cm^3)
ethyne (acetylene)	$H-C\equiv C-H$	−82	−84	0.62
propyne	$H-C\equiv C-CH_3$	−101	−23	0.67
1-butyne	$H-C\equiv C-CH_2CH_3$	−126	8	0.67
2-butyne	$CH_3-C\equiv C-CH_3$	−32	27	0.69
1-pentyne	$H-C\equiv C-CH_2CH_2CH_3$	−90	40	0.70
2-pentyne	$CH_3-C\equiv C-CH_2CH_3$	−101	55	0.71
3-methyl-1-butyne	$CH_3-CH(CH_3)-C\equiv C-H$		28	0.67
1-hexyne	$H-C\equiv C-(CH_2)_3-CH_3$	−132	71	0.72
2-hexyne	$CH_3-C\equiv C-CH_2CH_2CH_3$	−90	84	0.73
3-hexyne	$CH_3CH_2-C\equiv C-CH_2CH_3$	−101	82	0.73
3,3-dimethyl-1-butyne	$(CH_3)_3C-C\equiv C-H$	−81	38	0.67
1-heptyne	$H-C\equiv C-(CH_2)_4CH_3$	−81	100	0.73
1-octyne	$H-C\equiv C-(CH_2)_5CH_3$	−79	125	0.75
1-nonyne	$H-C\equiv C-(CH_2)_6CH_3$	−50	151	0.76
1-decyne	$H-C\equiv C-(CH_2)_7CH_3$	−36	174	0.77

9-4
Commercial Importance of Alkynes

9-4A   Uses of Acetylene and Methylacetylene

Acetylene is by far the most important commercial alkyne. Acetylene is an important industrial feedstock, but its largest use is as the fuel for the oxyacetylene welding torch. Acetylene is a colorless, foul-smelling gas that burns in air with a yellow, sooty flame. When the flame is supplied with pure oxygen, however, the color turns to light blue, and flame temperature increases dramatically. A comparison of the heat of combustion for acetylene with those of ethene and ethane shows why this gas makes an excellent fuel for a high-temperature flame.

$$CH_3CH_3 + \tfrac{7}{2}O_2 \longrightarrow 2\,CO_2 + 3\,H_2O \qquad \Delta H° = -373 \text{ kcal } (-1561 \text{ kJ})$$

$$-373 \text{ kcal divided by 5 moles of products } = -75 \text{ kcal/mol of products}$$
$$(-312 \text{ kJ/mol})$$

$$H_2C{=}CH_2 + 3\,O_2 \longrightarrow 2\,CO_2 + 2\,H_2O \qquad \Delta H° = -337 \text{ kcal } (-1410 \text{ kJ})$$

$$-337 \text{ kcal divided by 4 moles of products } = -84 \text{ kcal/mol of products}$$
$$(-352 \text{ kJ/mol})$$

$$HC{\equiv}CH + \tfrac{5}{2}O_2 \longrightarrow 2\,CO_2 + 1\,H_2O \qquad \Delta H° = -317 \text{ kcal } (-1326 \text{ kJ})$$

$$-317 \text{ kcal divided by 3 moles of products } = -106 \text{ kcal/mol of products}$$
$$(-442 \text{ kJ/mol})$$

If we were simply heating a house by burning one of these fuels, we might choose ethane as our fuel because it produces the most heat per mole of gas consumed. In the welding torch, we want the highest possible *temperature* of the gaseous products. The heat of reaction must raise the temperature of the products to the flame temperature. Roughly speaking, the increase in temperature of the products is proportional to the heat given off *per mole of products* formed. This rise in temperature is largest with acetylene, which gives off the most heat per mole of products. The oxyacetylene flame reaches temperatures as high as 2800°C.

When acetylene was first used for welding, it was considered a dangerous, explosive gas. Acetylene is thermodynamically unstable. When the compressed gas is subjected to thermal or mechanical shock, it decomposes to its elements, releasing 56 kcal (234 kJ) of energy per mole. This initial decomposition often splits the container, allowing the products (hydrogen and finely divided carbon) to burn in the air.

An oxygen-acetylene flame is hot enough to melt steel for welding. A cutting torch uses an extra jet of oxygen to burn away hot steel.

$$H{-}C{\equiv}C{-}H \longrightarrow 2\,C + H_2 \qquad \Delta H° = -56 \text{ kcal/mol } (-234 \text{ kJ/mol})$$

$$2\,C + H_2 \xrightarrow{\tfrac{5}{2}O_2} 2\,CO_2 + H_2O \qquad \Delta H° = -261 \text{ kcal/mol } (-1090 \text{ kJ/mol})$$

Acetylene is safely stored and handled in cylinders that are filled with crushed firebrick wet with acetone. Acetylene dissolves freely in acetone, and the dissolved gas is not so prone to decomposition. Firebrick helps to control the decomposition by minimizing the free volume of the cylinder, cooling and controlling any decomposition before it gets out of control.

Methylacetylene also is used in welding torches. Methylacetylene does not decompose as easily as acetylene, and it burns better in air (as opposed to pure oxygen). Methylacetylene is well suited for household soldering and brazing that requires higher temperatures than propane torches can reach. The industrial synthesis of methylacetylene gives a mixture with its isomer, propadiene (allene). This mixture is sold commercially under the name MAPP® gas (*M*ethyl*A*cetylene-*P*ro*P*adiene).

$$CH_3{-}C{\equiv}C{-}H \qquad H_2C{=}C{=}CH_2$$
$$\text{methylacetylene} \qquad \text{propadiene (allene)}$$

9-4B Manufacture of Acetylene

Acetylene, one of the cheapest organic chemicals, is made from coal or from natural gas. The synthesis from coal involves heating lime and coke (roasted coal) in an electric furnace to produce calcium carbide. Addition of water to calcium carbide produces acetylene and hydrated lime.

$$3\ C\ +\ CaO\ \xrightarrow{\text{electric furnace, 2500°C}}\ CaC_2\ +\ CO$$

coke   lime   calcium carbide

$$CaC_2\ +\ 2\ H_2O\ \longrightarrow\ H-C\equiv C-H\ +\ Ca(OH)_2$$

acetylene   hydrated lime

This second reaction once served as a light source in coal mines until battery-powered lights became available. A miner's lamp works by allowing water to drip slowly onto some calcium carbide. Acetylene is generated, feeding a small flame where the gas burns in air with a yellow flickering light. Unfortunately, this flame ignites the methane gas commonly found in coal seams, causing explosions. Battery-powered miner's lamps provide better light and reduce the danger of methane explosions.

The synthesis of acetylene from natural gas is a simple process. Natural gas consists mostly of methane, which forms acetylene when it is heated for a very short period of time.

$$2\ CH_4\ \xrightarrow[0.01\ \text{sec}]{1500°C}\ H-C\equiv C-H\ +\ 3\ H_2$$

Although this reaction is endothermic, there are twice as many moles of products as reactants. The increase in the number of moles results in an increase in entropy, and the $(-T\Delta S)$ term in the free energy $(\Delta G = \Delta H - T\Delta S)$ predominates at this high temperature.

> **PROBLEM 9-3**
> What reaction would acetylene likely undergo if it were kept at 1500°C for too long?

9-5
Electronic Structure of Alkynes

In Section 2-4 we studied the electronic structure of a triple bond. Let's review this structure, using acetylene as the example. The Lewis structure of acetylene shows three pairs of electrons in the region between the carbon nuclei:

$$H:C:::C:H$$

Each carbon atom is bonded to two other atoms, and there are no nonbonding valence electrons. Each carbon atom needs two hybrid orbitals to form the sigma bond framework. Hybridization of the *s* orbital with one *p* orbital gives two hybrid orbitals, directed 180° apart, for each carbon atom. Overlap of these *sp* hybrid orbitals with each other and with the hydrogen *s* orbitals gives the sigma bond framework. Experimental results have confirmed this linear (180°) structure.

Two pi bonds result from overlap of the two remaining unhybridized *p* orbitals on each carbon atom. These orbitals overlap at right angles to each other, forming one pi bond with electron density above and below the C—C sigma bond, and the other with electron density in front and in back of the sigma bond. The shape of these pi bonds is such that they blend to form a cylinder of electron density encircling the sigma bond between the two carbon atoms.

overlap of *p* orbitals   cylinder of electron density

The carbon–carbon bond length in acetylene is 1.20 Å, and each carbon–hydrogen bond is 1.06 Å. Both bonds are shorter than the corresponding bonds in ethane and in ethene.

ethane   ethene   ethyne

The triple bond is relatively short because of the attractive overlap of three bonding pairs of electrons and the high *s* **character** of the *sp* hybrid orbitals. The *sp* hybrid orbitals are about one-half *s* character (as opposed to one-third *s* character of sp^2 hybrids and one-fourth of sp^3 hybrids), using more of the closer, tightly held *s* orbitals. The *sp* hybrid orbitals also account for the slightly shorter C—H bonds in acetylene compared with ethylene.

Terminal alkynes are much more acidic than other hydrocarbons. Removal of an acetylenic proton forms an acetylide ion, which plays a central role in alkyne chemistry. The acidity of an acetylenic hydrogen stems from the nature of the *sp* hybrid ≡C—H bond.

Table 9-2 shows how the acidity of a C—H bond varies with its hybridization, increasing with the increasing *s* character of the orbitals: $sp^3 < sp^2 < sp$. (Remember

9-6
Acidity of Alkynes

TABLE 9-2

Compound	Conjugate Base	Hybridization	*s* Character	pK_a	
		sp^3	25%	50	weakest acid
		sp^2	33%	44	
$:NH_3$	$:\ddot{N}H_2^-$	(ammonia)		35	
H—C≡C—H	H—C≡C: $^-$	*sp*	50%	25	
R—OH	R—$\ddot{O}$: $^-$	(alcohols)		16–18	stronger acid

that a *smaller* value of pK_a corresponds to a stronger acid.) The acetylenic proton is about 10^{19} times as acidic as a vinyl proton. When an acetylenic proton is abstracted, the resulting carbanion has the lone pair of electrons in the *sp* hybrid orbital. Electrons in this orbital are close to the nucleus, and there is less charge separation than in carbanions with the lone pair in sp^2 or sp^3 hybrid orbitals. Ammonia and alcohols are included for comparison; note that acetylene can be deprotonated by the amide ($^-NH_2$) ion, but not by an alkoxide ion.

9-6A   Formation of Acetylide Ions

Unlike alkanes and alkenes, terminal acetylenes can be deprotonated by a very strong base, such as sodium amide, to form carbanions called **acetylide ions** (or **alkynide ions**). Hydroxide ion and alkoxide ions are not strong enough bases to deprotonate alkynes. Internal alkynes do not have acetylenic protons, so they do not react.

$$CH_3CH_2-C\equiv C-H \;+\; NaNH_2 \longrightarrow CH_3CH_2-C\equiv C:^- \;^+Na \;+\; NH_3$$

1-butyne, a terminal alkyne   sodium amide   sodium butynide

(acidic proton)

$$\bigcirc-C\equiv C-H \;+\; NaNH_2 \longrightarrow \bigcirc-C\equiv C:^- \;^+Na \;+\; NH_3$$

cyclohexylacetylene   sodium amide   sodium cyclohexylacetylide

$$CH_3-C\equiv C-CH_3 \xrightarrow{NaNH_2} \text{no reaction}$$

(no acetylenic proton)
2-butyne, an internal alkyne

Sodium amide ($Na^+ \;^-:\ddot{N}H_2$) is frequently used as the base in forming acetylide salts. The amide ion ($^-NH_2$) is the conjugate base of ammonia, a compound that is itself a base. Ammonia is also a very weak acid, however, with $K_a = 10^{-35}$ (p$K_a = 35$). One of its hydrogens can be reduced by sodium to give the sodium salt of the amide ion, a very strong conjugate base.

$$H-\underset{\substack{|\\H}}{\overset{\substack{H\\|}}{N}}-H \;+\; Na \xrightarrow{Fe^{3+} \text{ catalyst}} Na^+ \;^-:\underset{\substack{|\\H}}{\overset{\substack{H\\|}}{N}}-H \;+\; \tfrac{1}{2}H_2\uparrow$$

ammonia   sodium amide
                                    ("sodamide")

☞ $$R-C\equiv C-H \;+\; Na^+ \;^-:\ddot{N}H_2 \longrightarrow R-C\equiv C:^- \;^+Na \;+\; :NH_3$$

a sodium acetylide

Acetylide ions are strong nucleophiles. In fact, one of the best methods for synthesizing substituted alkynes is a nucleophilic attack by an acetylide ion on an unhindered alkyl halide. We consider this displacement reaction in detail in Section 9-7A.

$$CH_3CH_2-C\equiv C:^- \;^+Na \;+\; H_3C-I \longrightarrow CH_3CH_2-C\equiv C-CH_3 \;+\; NaI$$

PROBLEM 9-4

The boiling points of 1-hexene (64°C) and 1-hexyne (71°C) are sufficiently close that it is difficult to achieve a clean separation by distillation. Show how you might use the acidity of 1-hexyne to remove the last trace of it from a sample of 1-hexene.

PROBLEM 9-5

Predict the products of the following acid–base reactions, or indicate if no significant reaction would take place.

(a) $H-C\equiv C-H + NaNH_2$ **(b)** $H-C\equiv C-H + CH_3Li$
(c) $H-C\equiv C-H + NaOCH_3$ **(d)** $H-C\equiv C-H + NaOH$
(e) $H-C\equiv C:^- {}^+Na + CH_3OH$ **(f)** $H-C\equiv C:^- {}^+Na + H_2O$
(g) $H-C\equiv C:^- {}^+Na + H_2C=CH_2$ **(h)** $H_2C=CH_2 + NaNH_2$
(i) $CH_3OH + NaNH_2$

9-6B Heavy-Metal Acetylides

Silver(I) and copper(I) salts react with terminal alkynes to form silver and copper acetylides. Silver and copper acetylides are bonded more covalently than other acetylides, however, and they are much less basic and less nucleophilic. Silver and copper acetylides are not very soluble; they form characteristic precipitates:

$$R-C\equiv C-H \ + \ Ag^+ \ \longrightarrow \ R-C\equiv C-Ag \downarrow \ + \ H^+$$
(light colored precipitate)

$$R-C\equiv C-H \ + \ Cu^+ \ \longrightarrow \ R-C\equiv C-Cu \downarrow \ + \ H^+$$
(brick red precipitate)

This reaction provides a simple chemical test for terminal alkynes. Internal acetylenes are unreactive toward Ag^+ and Cu^+ because they have no acetylenic protons. Adding a silver reagent or a copper(I) reagent to a solution of an alkyne shows whether the alkyne is terminal or internal. The terminal alkyne forms a precipitate, but the internal alkyne does not.

$$CH_3-CH_2-C\equiv C-H \ + \ Ag^+ \ (or \ Cu^+) \ \longrightarrow \ CH_3-CH_2-C\equiv C-Ag \downarrow \ (or \ Cu) \ + \ H^+$$
precipitate

$$CH_3-C\equiv C-CH_3 \ + \ Ag^+ \ (or \ Cu^+) \ \longrightarrow \ no \ reaction$$

This qualitative test commonly uses $AgNO_3$ or $CuNO_3$, often in an alcoholic solution, or ammonia complexes of Ag(I) and Cu(I) ions made by adding a small amount of aqueous ammonia to the solution of silver nitrate or cuprous nitrate.

$$Ag^+ \ {}^-NO_3 + 2\ NH_3 \longrightarrow Ag(NH_3)_2{}^+ \ {}^-NO_3$$

$$Cu^+ \ {}^-NO_3 + 2\ NH_3 \longrightarrow Cu(NH_3)_2{}^+ \ {}^-NO_3$$

In addition to the qualitative test, the formation of heavy-metal acetylides can be used for purifying alkynes—for example, when a mixture of terminal and internal alkyne isomers is formed in a double dehydrohalogenation (as in Problem 9-35). Addition of Ag(I) or Cu(I) causes the terminal alkyne to precipitate as its acetylide, which can be filtered off from the internal alkyne.

$$R-CH_2-C\equiv C-H \ + \ Cu^+ \ \longrightarrow \ R-CH_2-C\equiv C-Cu \downarrow \ (precipitate, \ filtered \ out)$$

$$R-C\equiv C-CH_3 \ + \ Cu^+ \ \longrightarrow \ no \ reaction, \ remains \ in \ solution$$

Addition of dilute acid regenerates the terminal alkyne from its acetylide.

$$R-CH_2-C\equiv C-Cu \ + \ HCl \ \longrightarrow \ R-CH_2-C\equiv C-H \ + \ CuCl$$

Because they are not strong nucleophiles, silver and copper acetylides are not commonly used in the alkylation and carbonyl addition reactions discussed in Section 9-7. Heavy-metal acetylides tend to explode when dry, so they are always acidified while still wet. Because copper acetylides are explosive, copper tubing is never used with acetylene for fear that a coating of copper acetylide might build up and explode.

9-7
Synthesis of Alkynes from Acetylides

Two different approaches are commonly used for the synthesis of alkynes. In the first, an appropriate electrophile undergoes nucleophilic attack by an acetylide ion. The electrophile may be an unhindered primary alkyl halide (undergoes S_N2), or it may be a carbonyl compound (undergoes addition to give an alcohol). Either reaction joins two fragments and gives a product with a lengthened carbon skeleton. This approach is used in many laboratory syntheses of alkynes.

The second approach forms the triple bond by a double dehydrohalogenation of a dihalide. This reaction does not enlarge the carbon skeleton. Isomerization of the triple bond may occur (see Section 9-8), so dehydrohalogenation is useful only when the desired product has the triple bond in a thermodynamically favored position.

9-7A   Alkylation of Acetylide Ions

An acetylide ion is a strong base and a powerful nucleophile. It can displace a halide ion from a suitable substrate, giving a substituted acetylene.

$$R—C\equiv C\!:^- \ + \ R'—X \ \xrightarrow{\ S_N2\ } \ R—C\equiv C—R' \ + \ X^-$$

(R'—X must be a primary alkyl halide)

If this S_N2 reaction is to produce a good yield, the alkyl halide must be an excellent S_N2 substrate: It must be primary, with no bulky substituents or branches close to the reaction center. In the following examples, acetylide ions displace primary halides to form elongated alkynes.

$$H—C\equiv C\!:^- \ ^+Na \ + \ CH_3CH_2CH_2CH_2—Br \ \longrightarrow \ H—C\equiv C—CH_2CH_2CH_2CH_3 \ + \ NaBr$$

sodium acetylide   1-bromobutane   1-hexyne (butylacetylene) (75%)

ethynylcyclohexane
(cyclohexylacetylene)

$$\xrightarrow[\text{(2) ethyl bromide}]{\text{(1) NaNH}_2}$$

1-cyclohexyl-1-butyne
(ethylcyclohexylacetylene)
(70%)

If the back-side approach is hindered, the acetylide ion may abstract a proton, giving elimination by the E2 mechanism.

$$CH_3CH_2—C\equiv C\!:^- \ + \ \underset{\text{isopropyl bromide}}{H_3C—\overset{\overset{\displaystyle Br}{|}}{C}H—CH_3} \ \xrightarrow{\ E2\ } \ \underset{\text{butyne}}{CH_3CH_2—C\equiv C—H} \ + \ \underset{\text{propene}}{H_2C=CH—CH_3} \ + \ Br^-$$

butynide ion

SOLVED PROBLEM 9-1

Show how to synthesize 3-decyne from acetylene and any necessary alkyl halides.

SOLUTION

Another name for 3-decyne is ethyl *n*-hexylacetylene. It can be made by adding an ethyl group and a hexyl group to acetylene. This can be done in either order; we begin by adding the hexyl group.

$$H-C\equiv C-H \xrightarrow[\text{(2) CH}_3\text{(CH}_2\text{)}_5\text{Br}]{\text{(1) NaNH}_2} CH_3(CH_2)_5-C\equiv C-H$$

acetylene   1-octyne

$$CH_3(CH_2)_5-C\equiv C-H \xrightarrow[\text{(2) CH}_3\text{CH}_2\text{Br}]{\text{(1) NaNH}_2} CH_3(CH_2)_5-C\equiv C-CH_2CH_3$$

1-octyne   3-decyne

PROBLEM 9-6

Show the reagents and intermediates involved in the other order of synthesis of 3-decyne, by adding the ethyl group first and the hexyl group last.

PROBLEM 9-7

Show how you might synthesize the following compounds, using acetylene and any suitable alkyl halides as your starting materials. If the compound given cannot be synthesized by this method, explain why.
(a) 1-hexyne   (b) 2-hexyne
(c) 3-hexyne   (d) 4-methyl-2-hexyne
(e) 5-methyl-2-hexyne   (f) cyclodecyne

PROBLEM-SOLVING HINT
Alkylation of acetylide ions is an excellent way of lengthening a carbon chain. The triple bond can later be reduced (to an alkane or an alkene) if needed.

9-7B Addition of Acetylide Ions to Carbonyl Groups

Like other carbanions, acetylide ions are strong nucleophiles and strong bases. In addition to displacing halide ions in S$_N$2 reactions, they can add to carbonyl (C=O) groups, Figure 9-1 shows the structure of the carbonyl group. Because oxygen is more electronegative than carbon, the C=O double bond is polarized. The oxygen atom has a partial negative charge balanced by an equal amount of positive charge on the carbon atom.

The positively charged carbon is electrophilic; attack by a nucleophile places a negative charge on the electronegative oxygen atom.

$$Nuc:^- \quad \overset{\delta+}{\underset{}{C}}=\overset{\delta-}{\ddot{O}}: \quad \longrightarrow \quad Nuc-\overset{|}{\underset{|}{C}}-\ddot{O}:^-$$

half of π bond

$$R\overset{\delta+}{\underset{R}{C}} \xrightarrow{\sigma \text{ bond}} \overset{\delta-}{O} \qquad R\overset{\delta+}{\underset{R}{C}}=\overset{\delta-}{\ddot{O}}:$$

half of π bond

$$\mu \cong 2.4\ D$$

◀ **Figure 9-1**
The C=O double bond of a carbonyl group resembles the C=C double bond of an alkene; however, the carbonyl double bond is strongly polarized. The oxygen atom bears a partial negative charge, and the carbon atom bears a partial positive charge.

The product of this nucleophilic attack is an alkoxide ion, a strong base. (An **alkoxide ion** is the conjugate base of an alcohol, a weak acid.) Addition of water or a dilute acid protonates the alkoxide to give the alcohol.

$$Nuc-\overset{|}{\underset{|}{C}}-\overset{\cdot\cdot}{\underset{\cdot\cdot}{O}}:^- \quad H-\overset{\cdot\cdot}{\underset{\cdot\cdot}{O}}-H \longrightarrow Nuc-\overset{|}{\underset{|}{C}}-\overset{\cdot\cdot}{O}H \ + \ ^-:\overset{\cdot\cdot}{O}H$$

alkoxide   (or H₃O⁺)

An acetylide ion can serve as the nucleophile in this addition to a carbonyl group. The acetylide ion adds to the carbonyl group to form an alkoxide ion. Addition of dilute acid (in a separate step) protonates the alkoxide to give the alcohol.

$$R'-C\equiv C:^- \ + \ \underset{R}{\overset{R}{>}}C=\overset{\cdot\cdot}{\underset{\cdot\cdot}{O}}: \longrightarrow R'-C\equiv C-\overset{R}{\underset{R}{\overset{|}{\underset{|}{C}}}}-\overset{\cdot\cdot}{\underset{\cdot\cdot}{O}}:^-$$

acetylide   aldehyde
or ketone

$$R'-C\equiv C-\overset{R}{\underset{R}{\overset{|}{\underset{|}{C}}}}-\overset{\cdot\cdot}{O}:^- \ \xrightarrow{H_3O^+} \ R'-C\equiv C-\overset{R}{\underset{R}{\overset{|}{\underset{|}{C}}}}-OH$$

an acetylenic alcohol

An acetylide adds to formaldehyde ($H_2C=O$) to give (after the protonation step) a primary alcohol with one more carbon atom than there was in the acetylide.

The addition of an acetylide ion to a carbonyl group is used in the synthesis of ethchlorvynol, a drug used to cause drowsiness and induce sleep. Ethchlorvynol is relatively non-polar, enhancing its distribution into the fatty tissue of the central nervous system.

$$H_2CH_3C-\overset{OH}{\underset{\underset{H}{\overset{|}{C}}\diagdown}{\overset{|}{C}}}-C\equiv CH$$
$$\qquad\qquad\quad CHCl$$
ethchlorvynol

$$R'-C\equiv C:^- \ + \ \underset{H}{\overset{H}{>}}C=\overset{\cdot\cdot}{\underset{\cdot\cdot}{O}}: \longrightarrow R'-C\equiv C-\overset{H}{\underset{H}{\overset{|}{\underset{|}{C}}}}-\overset{\cdot\cdot}{\underset{\cdot\cdot}{O}}:^-$$

formaldehyde

$$R'-C\equiv C-\overset{H}{\underset{H}{\overset{|}{\underset{|}{C}}}}-\overset{\cdot\cdot}{O}:^- \ \xrightarrow{H_3O^+} \ R'-C\equiv C-\overset{H}{\underset{H}{\overset{|}{\underset{|}{C}}}}-OH$$

a 1° alcohol

Example

$$CH_3-C\equiv C-H \ \xrightarrow[\text{(3) } H_3O^+]{\overset{\text{(1) } NaNH_2}{\text{(2) } H_2C=O}} \ CH_3-C\equiv C-CH_2-OH$$

propyne   2-butyn-1-ol (1°)

An acetylide adds to an aldehyde to give, after protonation, a secondary alcohol. The two groups of the secondary alcohol are the acetylide and the alkyl group that was bonded to the carbonyl group of the aldehyde.

$$R'-C\equiv C:^- \; + \; \underset{H}{\overset{R}{C}}=\overset{..}{\underset{..}{O}}: \longrightarrow R'-C\equiv C-\underset{H}{\overset{R}{\underset{|}{\overset{|}{C}}}}-\overset{..}{\underset{..}{O}}:^-$$

an aldehyde

$$R'-C\equiv C-\underset{H}{\overset{R}{\underset{|}{\overset{|}{C}}}}-\overset{..}{\underset{..}{O}}:^- \; \xrightarrow{H_3O^+} \; R'-C\equiv C-\underset{H}{\overset{R}{\underset{|}{\overset{|}{C}}}}-OH$$

a 2° alcohol

Example

$$CH_3-\underset{\overset{|}{CH_3}}{CH}-C\equiv C-H \quad \xrightarrow[\text{(3) } H_3O^+]{\substack{\text{(1) NaNH}_2 \\ \text{(2) PhCHO}}} \quad CH_3-\underset{\overset{|}{CH_3}}{CH}-C\equiv C-\underset{\overset{|}{Ph}}{CH}-OH$$

3-methyl-1-butyne   4-methyl-1-phenyl-2-pentyn-1-ol (2°)

A ketone has two alkyl groups bonded to its carbonyl carbon atom. Addition of an acetylide, followed by protonation, gives a tertiary alcohol. The three alkyl groups bonded to the carbinol carbon atom (the carbon bearing the —OH group) are the acetylide and the two alkyl groups originally bonded to the carbonyl group in the ketone.

$$R'-C\equiv C:^- \; + \; \underset{R}{\overset{R}{C}}=\overset{..}{\underset{..}{O}}: \longrightarrow R'-C\equiv C-\underset{R}{\overset{R}{\underset{|}{\overset{|}{C}}}}-\overset{..}{\underset{..}{O}}:^-$$

a ketone

$$R'-C\equiv C-\underset{R}{\overset{R}{\underset{|}{\overset{|}{C}}}}-\overset{..}{\underset{..}{O}}:^- \; \xrightarrow{H_3O^+} \; R'-C\equiv C-\underset{R}{\overset{R}{\underset{|}{\overset{|}{C}}}}-OH$$

a 3° alcohol

Example

cyclohexanone

$$\xrightarrow[\text{(2) } H_3O^+]{\text{(1) Na}^{+\,-}:C\equiv C-H}$$

1-ethynylcyclohexanol (3°)

SOLVED PROBLEM 9-2

Show how you would synthesize the following compound, beginning with acetylene and any necessary additional reagents.

$$\underset{\text{(cyclopentyl)}}{}-\underset{\overset{|}{OH}}{CH}-C\equiv C-CH_2CH_3$$

SOLUTION

We need to add two groups to acetylene: an ethyl group and a six-carbon aldehyde (to form the secondary alcohol). If we formed the alcohol group first, the weakly acidic —OH group would interfere with the alkylation by the ethyl group. Therefore, we should add the less reactive ethyl group first, and add the alcohol group later in the synthesis.

$$H—C≡C—H \xrightarrow[\text{(2) } CH_3CH_2Br]{\text{(1) } NaNH_2} H—C≡C—CH_2CH_3$$

The ethyl group is not acidic, and it does not interfere with the addition of the second group:

$$H—C≡C—CH_2CH_3 \xrightarrow{NaNH_2} Na^+ \;^-:C≡C—CH_2CH_3 \xrightarrow[\text{(2) } H_3O^+]{\text{(1)}}$$

PROBLEM-SOLVING **HINT**

If a synthesis requires both alkylation of an acetylide and addition to a carbonyl, add the less reactive group first: alkylate, then add to a carbonyl. In general, you should add reactive functional groups late in a synthesis.

PROBLEM 9-8

Show how you would synthesize each compound, beginning with acetylene and any necessary additional reagents.

(a) 2-propyn-1-ol (propargyl alcohol)

$$H—C≡C—CH_2OH$$

(b) 2-phenyl-3-butyn-2-ol

$$CH_3—\overset{\overset{\displaystyle OH}{|}}{\underset{\underset{\displaystyle Ph}{|}}{C}}—C≡C—H$$

(c) 2-heptyn-4-ol

$$CH_3—C≡C—\overset{\overset{\displaystyle OH}{|}}{C}H—CH_2CH_2CH_3$$

(d) 3-methyl-4-hexyn-3-ol

$$CH_3CH_2—\overset{\overset{\displaystyle OH}{|}}{\underset{\underset{\displaystyle CH_3}{|}}{C}}—C≡C—CH_3$$

9-8
Synthesis of Alkynes by Elimination Reactions

In some cases, we can generate a carbon–carbon triple bond by eliminating two molecules of HX from a dihalide. Dehydrohalogenation of a *geminal* or *vicinal* dihalide gives a vinyl halide. Under strongly basic conditions, a second dehydrohalogenation may occur to form an alkyne.

Conditions for Elimination We have already seen (Section 7-9A) many examples of dehydrohalogenation of alkyl halides. The second step is new, however, because it involves dehydrohalogenation of a vinyl halide to give an alkyne. This second dehydrohalogenation occurs only under extremely basic conditions—for example, fused (molten) KOH or alcoholic KOH in a sealed tube, usually heated to temperatures close to 200°C. Sodium amide is also used for the double dehydrohalogenation. Since the amide ion ($^-:\ddot{N}H_2$) is a much stronger base than hydroxide, the amide reaction takes place at a lower temperature. The following reactions are carefully chosen to form products that do not rearrange (see below).

$$CH_3-CH_2-\overset{\overset{\displaystyle Br}{|}}{CH}-\overset{\overset{\displaystyle Br}{|}}{CH}-CH_3 \xrightarrow[200°C]{KOH \text{ (fused)}} CH_3-CH_2-C\equiv C-CH_3$$
2,3-dibromopentane   2-pentyne (45%)

$$CH_3-CH_2-CH_2-CH_2-CHCl_2 \xrightarrow[\text{(2) } H_2O]{\text{(1) NaNH}_2, \, 150°C} CH_3-CH_2-CH_2-C\equiv C-H$$
1,1-dichloropentane   1-pentyne (55%)

Base-Catalyzed Rearrangements   Unfortunately, the double dehydrohalogenation is limited by the severe conditions. Any functional groups that are sensitive to strong bases cannot survive; also, the alkyne products may rearrange under these extremely basic conditions. Figure 9-2 shows how the loss of protons at one carbon atom and their replacement elsewhere leads to isomerization of the triple bond. This ability to isomerize implies that all the possible triple-bond isomers will equilibrate, and the most stable isomer will predominate. The most stable alkyne isomer is usually the internal alkyne or a mixture of internal alkynes.

Any of several isomers of dibromopentane give mostly 2-pentyne on dehydrohalogenation with fused KOH at 200°C. In each case the alkyne initially formed rearranges to the most stable isomer, 2-pentyne.

$$CH_3-\overset{\overset{\displaystyle |}{Br}}{CH}-\overset{\overset{\displaystyle |}{Br}}{CH}-CH_2CH_3 \qquad Br-\overset{\overset{\displaystyle |}{Br}}{CH}-CH_2CH_2CH_2CH_3$$
2,3-dibromopentane   1,1-dibromopentane

$$CH_3-\overset{\overset{\displaystyle Br}{|}}{\underset{\underset{\displaystyle Br}{|}}{C}}-CH_2CH_2CH_3 \qquad \overset{\overset{\displaystyle |}{Br}}{CH_2}-\overset{\overset{\displaystyle |}{Br}}{CH}-CH_2CH_2CH_3$$
2,2-dibromopentane   1,2-dibromopentane

$$\xrightarrow{KOH, \, 200°C} CH_3-C\equiv C-CH_2CH_3$$
2-pentyne

$$R-\overset{\overset{\displaystyle H}{|}}{\underset{\underset{\displaystyle H}{|}}{C}}-C\equiv C-R \; + \; B:^- \; \rightleftharpoons \; \left[\overset{\displaystyle R}{\underset{\displaystyle H}{{}^-:C}}-C\equiv C-R \; \longleftrightarrow \; \overset{\displaystyle R}{\underset{\displaystyle H}{}}C=C=\overset{..}{C}-R \right] \; + \; H-B$$
an acetylene   resonance-stabilized carbanion

$$\overset{R}{\underset{H}{}}C=C=C\overset{H}{\underset{R}{}} \; + \; B:^-$$
an allene

$$R-C\equiv C-\overset{\overset{\displaystyle H}{|}}{\underset{\underset{\displaystyle H}{|}}{C}}-R \; + \; B:^- \; \rightleftharpoons \; \left[R-\overset{..}{C}=C=C\overset{H}{\underset{R}{}} \; \longleftrightarrow \; R-C\equiv C-C\overset{H}{\underset{R}{}}:^- \right] \; + \; H-B$$
an isomerized acetylene   resonance-stabilized carbanion

▲ **Figure 9-2**
Under extremely basic conditions, an acetylenic triple bond can migrate along the carbon chain by repeated deprotonation and reprotonation.

PROBLEM 9-9

Give a mechanism to show how 1,1-dibromopentane reacts with fused KOH at 200°C to give 2-pentyne.

$$\underset{\substack{\text{Br} \\ | \\ \text{Br}-\text{CH}-\text{CH}_2\text{CH}_2\text{CH}_2\text{CH}_3}}{} \xrightarrow{\text{KOH, 200°C}} \text{CH}_3-\text{C}\equiv\text{C}-\text{CH}_2\text{CH}_3$$

PROBLEM 9-10*

Using heats of hydrogenation, we can show that most internal alkynes are about 4 kcal/mol (17 kJ/mol) more stable than their corresponding terminal alkynes. Calculate the ratio of terminal alkyne to internal alkyne present at equilibrium at 200°C.

Isomerization also results when sodium amide is used as the base in the double dehydrohalogenation. All possible triple-bond isomers are formed, but sodium amide is such a strong base that it deprotonates the terminal acetylene, removing it from the equilibrium. The acetylide ion becomes the favored product. When water is added to quench the reaction, the acetylide ion is protonated to give the terminal alkyne.

$$\underset{\substack{\text{one component of} \\ \text{the mixture}}}{\text{R}-\text{C}\equiv\text{C}-\text{H}} + \overset{..}{^-\text{:NH}_2} \quad \rightleftarrows \quad \underset{\substack{\text{acetylide ion} \\ \text{(major component)}}}{\text{R}-\text{C}\equiv\text{C:}^-} + \text{:NH}_3 \xrightarrow{\text{H}_2\text{O}} \underset{\text{major product}}{\text{R}-\text{C}\equiv\text{C}-\text{H}}$$

$$\underset{\text{2-octyne}}{\text{CH}_3(\text{CH}_2)_4-\text{C}\equiv\text{C}-\text{CH}_3} \xrightarrow[\text{(2) H}_2\text{O}]{\text{(1) NaNH}_2, 150°\text{C}} \underset{\substack{\text{1-octyne} \\ (80\%)}}{\text{CH}_3(\text{CH}_2)_4-\text{CH}_2-\text{C}\equiv\text{C}-\text{H}}$$

PROBLEM-SOLVING $\mathbf{HINT}$

PROBLEM 9-11

(a) Propose a mechanism to show how 2-pentyne reacts with sodium amide to give 1-pentyne.
(b) Explain how this reaction converts a more stable isomer (2-pentyne) to a less stable isomer (1-pentyne).
(c) How would you accomplish the opposite reaction, converting 1-pentyne into 2-pentyne?

PROBLEM 9-12

Show which of the following compounds could be synthesized in good yield by a double dehydrohalogenation from a dihalide. In each case

(1) Show which base you would use (KOH or NaNH$_2$).
(2) Show how your starting material might be synthesized from an alkene.

(a) 2-butyne   **(b)** 1-octyne   **(c)** 2-octyne   **(d)** cyclodecyne

---

SUMMARY   **Syntheses of Alkynes**

1. *Alkylation of acetylide ions* (Section 9-7A)

$$\text{R}-\text{C}\equiv\text{C:}^- + \text{R}'-\text{X} \xrightarrow{\text{S}_\text{N}2} \text{R}-\text{C}\equiv\text{C}-\text{R}' + \text{X}^-$$

(R'—X must be an unhindered primary halide or tosylate)

Example

$$\underset{\text{sodium propynide}}{\text{H}_2\text{C}-\text{C}\equiv\text{C:}^- {}^+\text{Na}} + \underset{\text{1-bromopropane}}{\text{CH}_3\text{CH}_2\text{CH}_2-\text{Br}} \longrightarrow \underset{\text{2-hexyne}}{\text{H}_3\text{C}-\text{C}\equiv\text{C}-\text{CH}_2\text{CH}_2\text{CH}_3}$$

2. *Additions to carbonyl groups* (Section 9-7B)

$$R—C≡C:^- \quad + \quad \underset{R'}{\overset{R'}{C}}=\ddot{O}: \quad \longrightarrow \quad R—C≡C—\underset{R'}{\overset{R'}{C}}—\ddot{O}:^- \quad \xrightarrow[\text{(or } H_3O^+)]{H_2O} \quad R—C≡C—\underset{R'}{\overset{R'}{C}}—OH$$

Example

$$\underset{\text{sodium acetylide}}{H—C≡C:^-} \quad + \quad \underset{\text{propanal}}{CH_3CH_2—\overset{\overset{O}{\|}}{C}—H} \quad \longrightarrow \quad \xrightarrow{H_2O} \quad \underset{\text{1-pentyn-3-ol}}{H—C≡C—\overset{OH}{\overset{|}{C}H}—CH_2CH_3}$$

3. *Double dehydrohalogenation of alkyl dihalides* (Section 9-8)

$$R—\underset{H}{\overset{X}{C}}—\underset{H}{\overset{X}{C}}—R' \quad or \quad R—\underset{H}{\overset{H}{C}}—\underset{X}{\overset{X}{C}}—R' \quad \xrightarrow[\text{or } NaNH_2]{\text{fused KOH}} \quad R—C≡C—R'$$

(KOH forms internal alkynes; $NaNH_2$ forms terminal alkynes.)

Examples

$$\underset{\text{2,2-dichloropentane}}{CH_3CH_2—CH_2—CCl_2—CH_3} \quad \xrightarrow{\text{KOH, 200°C (fused)}} \quad \underset{\text{2-pentyne}}{CH_3CH_2—C≡C—CH_3}$$

$$\underset{\text{2,2-dichloropentane}}{CH_3CH_2—CH_2—CCl_2—CH_3} \quad \xrightarrow{NaNH_2, \text{150°C}} \quad \underset{\text{1-pentyne}}{CH_3CH_2—CH_2—C≡C—H}$$

9-9 Addition Reactions of Alkynes

We have already discussed some of the most important reactions of alkynes. The nucleophilic attack of acetylide ions on electrophiles, for example, is one of the best methods for making more complicated alkynes (Section 9–7). Now we consider reactions that involve transformations of the carbon–carbon triple bond itself.

Many of the reactions of alkynes are similar to the corresponding reactions of alkenes because both involve pi bonds between two carbon atoms. Like the pi bond of an alkene, the pi bonds of an alkyne are electron-rich, and they readily undergo addition reactions. Table 9-3 shows how the energy differences between the kinds of carbon–carbon bonds can be used to estimate how much energy it takes to break a particular bond. The bond energy of the alkyne triple bond is only about 54 kcal (226 kJ) more than the bond energy of an alkene double bond. This is the energy needed to break one of the pi bonds of an alkyne.

TABLE 9-3 Approximate Bond Energies of Carbon–Carbon Bonds

Bond	Total Energy	Class of Bond	Approximate Energy
C—C	83 kcal (347 kJ)	alkane sigma bond	83 kcal (347 kJ)
C=C	146 kcal (611 kJ)	alkene pi bond	63 kcal (264 kJ)
C≡C	200 kcal (837 kJ)	second alkyne pi bond	54 kcal (226 kJ)

Reagents add across the triple bonds of alkynes just as they add across the double bonds of alkenes. In effect, this reaction converts a pi bond and a sigma bond into two sigma bonds. Since sigma bonds are generally stronger than pi bonds, the reaction is usually exothermic. Alkynes have two pi bonds, so up to two molecules can add across the triple bond, depending on the reagents and the conditions.

$$R-C\equiv C-R' \quad + \quad A-B \quad \longrightarrow \quad \underset{R}{\overset{A}{\underset{\displaystyle C=C}{}}}\overset{B}{\underset{R'}{}} \quad \xrightarrow{A-B} \quad R-\overset{A}{\underset{A}{C}}-\overset{B}{\underset{B}{C}}-R'$$

We must consider the possibility of a double addition whenever a reagent adds across the triple bond of an alkyne. Some conditions may allow the reaction to stop after a single addition, while other conditions give double addition.

9-9A Catalytic Hydrogenation to Alkanes

In the presence of a suitable catalyst, hydrogen adds to an alkyne, reducing it to an alkane. For example, when either of the butyne isomers reacts with hydrogen and a platinum catalyst, the product is *n*-butane. Platinum, palladium, and nickel catalysts are commonly used in this reduction.

$$R-C\equiv C-R' \quad + \quad 2\,H_2 \quad \xrightarrow{Pt,\ Pd,\ or\ Ni} \quad R-\overset{H}{\underset{H}{C}}-\overset{H}{\underset{H}{C}}-R'$$

Examples

$$H-C\equiv C-CH_2CH_3 \quad + \quad 2\,H_2 \quad \xrightarrow{Pt} \quad H-CH_2-CH_2-CH_2CH_3$$
1-butyne   butane
                                                                                  (100%)

$$CH_3-C\equiv C-CH_3 \quad + \quad 2\,H_2 \quad \xrightarrow{Pt} \quad CH_3-CH_2-CH_2-CH_3$$
2-butyne   butane
                                                                                  (100%)

Catalytic hydrogenation takes place in two steps, with an alkene intermediate. With efficient catalysts such as platinum, palladium, or nickel, it is usually impossible to stop the reaction at the alkene stage.

$$R-C\equiv C-R' \quad \xrightarrow{H_2,\ Pt} \quad \left[\underset{H}{\overset{R}{\underset{\displaystyle C=C}{}}}\overset{R'}{\underset{H}{}}\right] \quad \xrightarrow{H_2,\ Pt} \quad R-\overset{H}{\underset{H}{C}}-\overset{H}{\underset{H}{C}}-R'$$

9-9B Catalytic Hydrogenation to cis Alkenes

Hydrogenation of an alkyne can be stopped at the alkene stage by using a "poisoned" (partially deactivated) catalyst made by treating a good catalyst with a compound that makes the catalyst less effective. **Lindlar's catalyst** is a poisoned palladium catalyst, composed of powdered barium sulfate coated with palladium, poisoned with quinoline. Nickel boride (Ni_2B) is a newer alternative to Lindlar's catalyst that is more easily made and often gives better yields.

$$R-C\equiv C-R' \quad \xrightarrow[\substack{\text{quinoline}\\ \text{(Lindlar's catalyst)}}]{\substack{H_2,\ Pd/BaSO_4\\ CH_3OH}} \quad \underset{H}{\overset{R}{\underset{\displaystyle C=C}{}}}\overset{R'}{\underset{H}{}}$$
alkyne   cis alkene

R—C≡C—R

The catalytic hydrogenation of alkynes is similar to the hydrogenation of alkenes, and both proceed with syn stereochemistry. In catalytic hydrogenation, the face of a pi bond contacts the solid catalyst, and the catalyst weakens the pi bond, allowing two hydrogen atoms to add (Figure 9-3). This simultaneous (or nearly simultaneous) addition of two hydrogen atoms on the same face of the alkyne ensures syn stereochemistry.

In an internal alkyne, syn addition gives a cis product. For example, when 2-hexyne is hydrogenated using the Lindlar catalyst, the product is *cis*-2-hexene.

$$CH_3—C≡C—CH_2CH_2CH_3 \ + \ H_2 \xrightarrow[\text{(or Ni}_2\text{B)}]{\text{Lindlar's catalyst}} \ \underset{\text{\textit{cis}-2-hexene}}{\overset{H_3C\diagdown\,\diagup CH_2CH_2CH_3}{\underset{H\diagup\,\diagdown H}{C=C}}}$$

2-hexyne

$$H—C≡C—CH_2CH_2CH_2CH_3 \ + \ H_2 \xrightarrow[\text{(or Ni}_2\text{B)}]{\text{Lindlar's catalyst}} \ \underset{\text{1-hexene}}{\overset{H\diagdown\,\diagup (CH_2)_3CH_3}{\underset{H\diagup\,\diagdown H}{C=C}}}$$

1-hexyne

9-9C   Metal-Ammonia Reduction to trans Alkenes

To form a trans alkene, two hydrogens must be added to the alkyne with anti stereochemistry. Sodium metal in liquid ammonia reduces alkynes with anti stereochemistry, so this reduction is used to convert alkynes to trans alkenes.

$$☞ \quad R—C≡C—R \ + \ Na/NH_3 \ \longrightarrow \ \underset{\text{trans alkene}}{\overset{R\diagdown\,\diagup H}{\underset{H\diagup\,\diagdown R}{C=C}}}$$

alkyne

Example

$$CH_3—C≡C—(CH_2)_4CH_3 \ \xrightarrow{Na/NH_3} \ \underset{\substack{\text{\textit{trans}-2-octene}\\(80\%)}}{\overset{H_3C\diagdown\,\diagup H}{\underset{H\diagup\,\diagdown (CH_2)_4CH_3}{C=C}}}$$

2-octyne

Ammonia (bp −33°C) is a gas at room temperature, but it is kept liquid by using dry ice to cool the reaction vessel. As sodium dissolves in liquid ammonia, it gives up electrons, which produce a deep blue color. It is these solvated electrons that actually reduce the alkyne.

$$NH_3 \ + \ Na \ \longrightarrow \ NH_3{\cdot}e^- \ (\text{deep blue solution}) \ + \ Na^+$$

solvated electron

The metal-ammonia reduction proceeds by addition of an electron to the alkyne to form a radical anion, followed by protonation to give a neutral radical. Protons are provided by the ammonia solvent or by an alcohol added as a cosolvent. Addition of another electron, followed by another proton, gives the product.

MECHANISM   Metal-Ammonia Reduction of an Alkyne

Step 1: An electron adds to the alkyne, forming a radical anion.

alkyne   radical anion

Step 2: The radical anion is protonated to give a radical.

vinyl radical

Step 3: An electron adds to the radical, forming an anion.

most stable *trans*   vinyl anion

Step 4: Protonation of the anion gives an alkene.

trans alkene

The anti stereochemistry of the sodium–ammonia reduction appears to result from the greater stability of the vinyl radical in the trans configuration, where the alkyl groups are farther apart. An electron is added to the trans radical to give a trans vinyl anion, which is quickly protonated to the trans alkene.

PROBLEM 9-13

Show how you would convert
(a) 3-octyne to *cis*-3-octene
(b) 2-pentyne to *trans*-2-pentene
(c) *cis*-cyclodecene to *trans*-cyclodecene
(d) *trans*-2-pentene to *cis*-2-pentene

9-9D   Addition of Halogens

Bromine and chlorine add to alkynes just as they add to alkenes. If 1 mole of halogen adds to an alkyne, the product is a dihaloalkene. The stereochemistry of addition may be either syn or anti, and the products are often mixtures of cis and trans isomers.

Example

If 2 moles of halogen add to an alkyne, a tetrahalide results. Sometimes it is difficult to keep the reaction from proceeding all the way to the tetrahalide even when we want it to stop at the dihalide.

$$R-C\equiv C-R' + 2\,X_2 \xrightarrow{\hspace{1cm}} R-\overset{\displaystyle X}{\underset{\displaystyle X}{\overset{|}{\underset{|}{C}}}}-\overset{\displaystyle X}{\underset{\displaystyle X}{\overset{|}{\underset{|}{C}}}}-R'$$
$$(X_2 = Cl_2 \text{ or } Br_2)$$

Example

$$CH_3(CH_2)_3-C\equiv C-H + 2\,Cl_2 \xrightarrow{\hspace{1cm}} CH_3(CH_2)_3-\overset{\displaystyle Cl}{\underset{\displaystyle Cl}{\overset{|}{\underset{|}{C}}}}-\overset{\displaystyle Cl}{\underset{\displaystyle Cl}{\overset{|}{\underset{|}{C}}}}-H$$
$$(100\%)$$

PROBLEM 9-14

In the addition of just 1 mole of bromine to 1-hexyne, should the 1-hexyne be added to a bromine solution or should the bromine be added to the 1-hexyne? Explain your answer.

9-9E   Addition of Hydrogen Halides

Hydrogen halides add across the triple bond of an alkyne in much the same way they add across the alkene double bond. The initial product is a vinyl halide. When a hydrogen halide adds to a terminal alkyne, the product has the orientation predicted by Markovnikov's rule. A second molecule of HX can add, usually with the same orientation as the first.

$$R-C\equiv C-H + \boxed{H}-X \xrightarrow{\hspace{1cm}} \overset{R}{\underset{X}{>}}C=C\overset{H}{\underset{\boxed{H}}{<}}$$
$$(HX = HCl, HBr, \text{ or } HI)$$

$$\overset{R}{\underset{X}{>}}C=C\overset{H}{\underset{\boxed{H}}{<}} \xrightarrow{\boxed{H}-X} R-\overset{\displaystyle X}{\underset{\displaystyle X}{\overset{|}{\underset{|}{C}}}}-\overset{\displaystyle \boxed{H}}{\underset{\displaystyle \boxed{H}}{\overset{|}{\underset{|}{C}}}}-H$$

For example, the reaction of 1-pentyne with HBr gives the Markovnikov product. In an internal alkyne such as 2-pentyne, however, the acetylenic carbon atoms are equally substituted, and a mixture of products results.

$$H-C\equiv C-CH_2CH_2CH_3 + HBr \xrightarrow{\hspace{1cm}} \overset{H}{\underset{H}{>}}C=C\overset{CH_2CH_2CH_3}{\underset{Br}{<}}$$

1-pentyne   2-bromo-1-pentene
                             (Markovnikov product)

$$CH_3-C\equiv C-CH_2CH_3 + HBr \xrightarrow{\hspace{1cm}}$$
2-pentyne

$$CH_3-\overset{\displaystyle Br}{\overset{|}{C}}=\overset{\displaystyle H}{\overset{|}{C}}-CH_2CH_3 + CH_3-\overset{\displaystyle H}{\overset{|}{C}}=\overset{\displaystyle Br}{\overset{|}{C}}-CH_2CH_3$$

2-bromo-2-pentene   3-bromo-2-pentene
(*E* and *Z* isomers)   (*E* and *Z* isomers)

The mechanism is similar to the mechanism of hydrogen halide addition to alkenes. The **vinyl cation** formed in the first step is more stable with the positive charge on the more highly substituted carbon atom. Attack by halide ion completes the reaction.

When 2 moles of a hydrogen halide add to an alkyne, the second mole usually adds with the same orientation as the first. This consistent orientation leads to a geminal dihalide. For example, a double Markovnikov addition of HBr to 1-pentyne gives 2,2-dibromopentane.

PROBLEM 9-15

Propose a mechanism for the entire reaction of 1-pentyne with 2 moles of HBr. Show why Markovnikov's rule should be observed in both the first and second additions of HBr.

PROBLEM 9-16

The reaction of 2-octyne with 2 equivalents of HCl gives a mixture of two products.
(a) Give the structures of the two products.
(b) Show why the second equivalent of HCl adds with the same orientation as the first in each case.

In Section 8-3B, we saw the effect of peroxides on the addition of HBr to alkenes. Peroxides catalyze a free-radical chain reaction that adds HBr across the double bond of an alkene in the anti-Markovnikov sense. A similar reaction occurs with alkynes, with HBr adding with anti-Markovnikov orientation.

1-bromo-1-pentene
(mixture of *E* and *Z* isomers)

PROBLEM 9-17

Propose a mechanism for the reaction of 1-pentyne with HBr in the presence of peroxides. Show why anti-Markovnikov orientation results.

PROBLEM 9-18

Show how 1-hexyne might be converted to
(a) 1,2-dichloro-1-hexene   (b) 1-bromo-1-hexene
(c) 2-bromo-1-hexene   (d) 1,1,2,2-tetrabromohexane
(e) 2-bromohexane   (f) 2,2-dibromohexane

9-9F Hydration of Alkynes to Ketones and Aldehydes

Mercuric Ion–Catalyzed Hydration Alkynes undergo acid-catalyzed addition of water across the triple bond in the presence of mercuric ion as a catalyst. A mixture of mercuric sulfate in aqueous sulfuric acid is commonly used as the reagent. The hydration of alkynes is similar to the hydration of alkenes, and it also goes with Markovnikov orientation. The products are not the alcohols we might expect, however.

a vinyl alcohol (enol)   ketone

Electrophilic addition of mercuric ion gives a vinyl cation, which reacts with water and loses a proton to give an organomercurial alcohol. Under the acidic reaction conditions, mercury is replaced by hydrogen to give a vinyl alcohol, called an **enol**.

A number of biological reactions involve the formation of an enol. Researchers are focusing on ways to use these reactions for therapeutic purposes. Several investigators have synthesized stable enols by placing bulky substituents around the double bond.

methyl ketone   vinyl alcohol   organomercurial alcohol
(keto)   (enol)

Enols tend to be unstable, and isomerize to the ketone form. As shown below, this isomerization involves the shift of a proton and a double bond. The (boxed) hydroxyl proton is lost, and a proton is regained at the methyl position, while the pi bond shifts from the C=C position to the C=O position. This type of rapid equilibrium is called a **tautomerism**. The one shown is the **keto–enol tautomerism**, which is covered in more detail in Chapter 22. The keto form usually predominates.

keto   enol

keto–enol tautomerism

In acidic solution, the keto–enol tautomerism takes place by addition of a proton to the adjacent carbon atom, followed by loss of the hydroxyl proton from oxygen.

MECHANISM Acid-Catalyzed Keto–Enol Tautomerism

Step 1: Addition of a proton at the methylene group.

enol form   resonance-stabilized intermediate

Step 2: Loss of the hydroxyl proton.

resonance-stabilized intermediate   keto form

For example, the mercuric-catalyzed hydration of 1-butyne gives 1-buten-2-ol as an intermediate. In the acidic solution, the intermediate quickly equilibrates to its more stable keto tautomer, 2-butanone.

1-butyne   1-buten-2-ol   2-butanone

PROBLEM-SOLVING $\mathbb{HINT}$

To move a proton (as in a tautomerism) under acidic conditions, try adding a proton in the new position, then removing it from the old position.

PROBLEM 9-19

When 2-pentyne reacts with mercuric sulfate in dilute sulfuric acid, the product is a mixture of two ketones. Give the structures of these products, and use mechanisms to show how they are formed.

Hydroboration–Oxidation   In Section 8-7 we saw that hydroboration–oxidation adds water across the double bonds of alkenes with anti-Markovnikov orientation. A similar reaction takes place with alkynes, except that a hindered dialkylborane must be used to prevent addition of two molecules of borane across the triple bond. Di(secondary isoamyl)borane, called "disiamylborane," adds to the triple bond only once to give a vinylborane. (**Amyl** is an older common name for pentyl.) In a terminal alkyne, the boron atom bonds to the terminal carbon atom.

terminal alkyne   disiamylborane   a vinylborane   "*sec*-isoamyl" or "**siamyl**"

Oxidation of the vinylborane (using basic hydrogen peroxide) gives a vinyl alcohol (enol), resulting from anti-Markovnikov addition of water across the triple bond. This enol quickly tautomerizes to its more stable carbonyl (keto) form. In the case of a terminal alkyne, the keto product is an aldehyde. This sequence is an excellent method for converting terminal alkynes to aldehydes.

vinylborane   unstable enol form   aldehyde

SOLVED PROBLEM 9-3

Under basic conditions, the keto–enol tautomerism operates by a different mechanism. Propose a base-catalyzed mechanism for the tautomerism of the enol formed in the hydroboration–oxidation to its keto form, the aldehyde.

SOLUTION

In acid, the enol was first protonated, and then it lost a proton. Under basic conditions, the enol first loses its hydroxyl proton, then regains a proton on the adjacent carbon atom.

enol form   stabilized "enolate" ion   keto form

Hydroboration of 1-hexyne, for example, gives the vinylborane with boron on the less highly substituted carbon. Oxidation of this intermediate gives an enol that quickly tautomerizes to hexanal.

1-hexyne   a vinylborane

vinylborane   enol

enol   hexanal
(65%)

PROBLEM 9-20

The hydroboration–oxidation of internal alkynes produces ketones.
(a) When hydroboration–oxidation is applied to 2-butyne, a single pure product is obtained. Determine the structure of this product, and show the intermediates in its formation.
(b) When hydroboration–oxidation is applied to 2-pentyne, two products are obtained. Show why a mixture of products should be expected with any unsymmetrical internal alkyne.

PROBLEM 9-21

For each compound, give the product(s) expected from (1) $HgSO_4/H_2SO_4$-catalyzed hydration and (2) hydroboration–oxidation.
(a) 1-hexyne   **(b)** 2-hexyne   **(c)** 3-hexyne   **(d)** cyclodecyne

PROBLEM 9-22

Disiamylborane adds only once to alkynes by virtue of its two bulky secondary isoamyl groups. Disiamylborane is prepared by the reaction of $BH_3 \cdot THF$ with an alkene.
(a) Draw the structural formulas of the reagents and the products in the preparation of disiamylborane.
(b) Explain why the reaction in part (a) goes only as far as the dialkylborane. Why is Sia_3B not formed?

9-10 Oxidation of Alkynes

9-10A Permanganate Oxidations

Under mild conditions, potassium permanganate oxidizes alkenes to diols (Section 8-14B). A similar reaction occurs with alkynes. If an alkyne is treated with aqueous potassium permanganate under nearly neutral conditions, an α-diketone results. This is conceptually the same as hydroxylating each of the two pi bonds of the alkyne, then losing two molecules of water to give the diketone.

$$R-C{\equiv}C-R' \xrightarrow[\text{H}_2\text{O, neutral}]{\text{KMnO}_4} \left[\begin{array}{c} \text{OH OH} \\ | \quad | \\ R-\overset{|}{\underset{|}{C}}-\overset{|}{\underset{|}{C}}-R' \\ | \quad | \\ \text{OH OH} \end{array} \right] \xrightarrow{(-2\,\text{H}_2\text{O})} \underset{\text{O} \quad \text{O}}{R-\overset{\|}{C}-\overset{\|}{C}-R'}$$

For example, when 2-pentyne is treated with a cold, dilute solution of neutral permanganate, the product is 2,3-pentanedione.

$$\underset{\text{2-pentyne}}{CH_3-C{\equiv}C-CH_2CH_3} \xrightarrow[\text{H}_2\text{O, neutral}]{\text{KMnO}_4} \underset{\substack{\text{2,3-pentanedione} \\ (90\%)}}{CH_3-\overset{\text{O}}{\overset{\|}{C}}-\overset{\text{O}}{\overset{\|}{C}}-CH_2CH_3}$$

If the reaction mixture becomes warm or too basic, the diketone undergoes oxidative cleavage. The products are the salts of carboxylic acids, which can be converted to the free acids by adding dilute acid.

$$R-C{\equiv}C-R' \xrightarrow[\text{H}_2\text{O, heat}]{\text{KMnO}_4,\ \text{KOH}} \underset{\text{O}}{R-\overset{\|}{C}-O^-} + \underset{\text{O}}{{}^-O-\overset{\|}{C}-R'}$$

$$\xrightarrow[\text{H}_2\text{O}]{\text{HCl}} \underset{\text{O}}{R-\overset{\|}{C}-OH} + \underset{\text{O}}{HO-\overset{\|}{C}-R'}$$

For example, warm, basic permanganate cleaves the triple bond of 2-pentyne to give acetate and propionate ions. Acidification reprotonates these anions to acetic acid and propionic acid.

$$CH_3-C\equiv C-CH_2CH_3 \xrightarrow[\text{H}_2\text{O, heat}]{\text{KMnO}_4,\ \text{KOH}} \underset{\text{acetate}}{CH_3-\overset{\displaystyle O}{\overset{\|}{C}}-O^-} + \underset{\text{propionate}}{^-O-\overset{\displaystyle O}{\overset{\|}{C}}-CH_2CH_3}$$

2-pentyne

$$\xrightarrow{\text{H}^+} \underset{\text{acetic acid}}{CH_3-\overset{\displaystyle O}{\overset{\|}{C}}-OH} + \underset{\text{propionic acid}}{HO-\overset{\displaystyle O}{\overset{\|}{C}}-CH_2CH_3}$$

Terminal alkynes are cleaved similarly to give a carboxylic acid and CO_2.

$$\underset{\text{1-hexyne}}{CH_3(CH_2)_3-C\equiv C-H} \xrightarrow[\text{(2) H}^+]{\text{(1) KMnO}_4,\ \text{KOH, H}_2\text{O}} \underset{\text{pentanoic acid}}{CH_3(CH_2)_3-\overset{\displaystyle O}{\overset{\|}{C}}-OH} + CO_2\uparrow$$

9-10B   Ozonolysis

Ozonolysis of an alkyne, followed by hydrolysis, gives products similar to those obtained from oxidative cleavage by permanganate. Either cleavage can be used to determine the position of the triple bond in an unknown alkyne (see Problem 9-24).

$$R-C\equiv C-R' \xrightarrow[\text{(2) H}_2\text{O}]{\text{(1) O}_3} R-COOH + R'-COOH$$

Example

$$CH_3-C\equiv C-CH_2CH_3 \xrightarrow[\text{(2) H}_2\text{O}]{\text{(1) O}_3} CH_3-COOH + CH_3CH_2-COOH$$

PROBLEM 9-23

Predict the product(s) you would expect from treatment of each compound with (1) dilute, neutral $KMnO_4$ and (2) warm basic $KMnO_4$, then dilute acid.
(a) 1-hexyne   (b) 2-hexyne   (c) 3-hexyne
(d) 2-methyl-3-hexyne   (e) cyclodecyne

PROBLEM 9-24

Oxidative cleavages can help to determine the positions of the triple bonds in alkynes.
(a) An unknown alkyne undergoes oxidative cleavage to give adipic acid and two equivalents of acetic acid. Propose a structure for the alkyne.

$$\text{unknown alkyne} \xrightarrow[\text{(2) H}_2\text{O}]{\text{(1) O}_3} \underset{\text{adipic acid}}{HOOC-(CH_2)_4-COOH} + 2\ CH_3COOH$$

(b) An unknown alkyne undergoes oxidative cleavage to give the following triacid plus one equivalent of propanoic acid. Propose a structure for the alkyne.

$$\text{unknown alkyne} \xrightarrow[\text{(2) H}_2\text{O}]{\text{(1) O}_3} \underset{\text{a triacid}}{HOOC-(CH_2)_7-\overset{\displaystyle COOH}{\overset{|}{CH}}-COOH} + \underset{\text{propionic acid}}{CH_3CH_2COOH}$$

PROBLEM-SOLVING STRATEGY
Multistep Synthesis

Multistep synthesis problems are useful for exercising your knowledge of organic reactions, and in Chapter 8 we illustrated a systematic approach to synthesis. Now we apply this approach to a fairly difficult problem emphasizing alkyne chemistry. The compound to be synthesized is *cis*-2-methylhex-4-en-3-ol. (The "3-ol" means there is an alcohol —OH group on C3.)

cis-2-methylhex-4-en-3-ol

The starting materials are acetylene and compounds containing no more than four carbon atoms. In this problem, it is necessary to consider not only how to assemble the carbon skeleton and how to introduce the functional groups, but also when it is best to put in the functional groups. We begin with an examination of the target compound, and then we examine possible intermediates and synthetic routes.

1. **Review the functional groups and carbon skeleton of the target compound.**
   The target compound contains seven carbon atoms and two functional groups: a *cis* carbon–carbon double bond and an alcohol. The best method for generating a cis double bond is the catalytic hydrogenation of a triple bond (Section 9-9B).

Using this hydrogenation as the final step reduces the problem to a synthesis of this acetylenic alcohol. We know how to form carbon–carbon bonds next to triple bonds, and we have seen the formation of acetylenic alcohols (Section 9-7B).

2. **Review the functional groups and carbon skeletons of the starting materials, and see how their skeletons might fit together in the target compound.**
   Acetylene is listed as one of the starting materials, and we have good methods (Section 9-7) for making carbon–carbon bonds next to triple bonds, by using acetylide ions as nucleophiles. We can break the target structure into three pieces, each containing no more than four carbon atoms.

3. **Compare methods for assembling the carbon skeleton of the target compound. Which ones provide a key intermediate with the correct carbon skeleton and functional groups correctly positioned for conversion to the functionality in the target molecule?**
   Acetylenic alcohols result when acetylides add to ketones and aldehydes (Section 9-7B). Reaction of the acetylide ion with 2-methylpropanal gives one of the groups needed on the triple bond.

A methyl group is needed on the other end of the double bond of the target compound. Methylation requires formation of an acetylide, however (Section 9-7A):

$$CH_3I + {}^-:C \equiv C-R \longrightarrow H_3C-C \equiv C-R + I^-$$

The hydroxyl group in the acetylenic alcohol is much more acidic than the acetylenic proton, however. Any attempt to form the acetylide would fail.

$$H-C \equiv C-\overset{\overset{\displaystyle OH}{|}}{CH}-CH(CH_3)_2 + NaNH_2 \longrightarrow H-C \equiv C-\overset{\overset{\displaystyle O^-}{|}}{CH}-CH(CH_3)_2 + NH_3$$

This problem can be overcome by adding the methyl group first and then the alcohol portion. *In general, we try to add less reactive groups earlier in a synthesis, and more reactive groups later.* In this case, we add the functionalized group after adding the alkyl group because the alkyl group is less likely to be affected by subsequent reactions.

$$H-C \equiv C-H \xrightarrow[\text{(2) } CH_3I]{\text{(1) } NaNH_2} H_3C-C \equiv C-H \xrightarrow{NaNH_2} H_3C-C \equiv C:^- \; Na^+$$

$$H_3C-C \equiv C:^- \; + \; H-\overset{\overset{\displaystyle O}{||}}{C}-CH(CH_3)_2 \longrightarrow \xrightarrow{H_3O^+} H_3C-C \equiv C-\overset{\overset{\displaystyle OH}{|}}{CH}-CH(CH_3)_2$$

4. **Working backward through as many steps as necessary, compare methods for synthesizing the reactants needed for assembly of the key intermediate with the correct carbon skeleton and functionality.**
   These compounds are all allowed as starting materials. Later, when we have covered more synthetic reactions, we will encounter problems that require us to evaluate how to make the compounds needed to assemble the key intermediates.

5. **Summarize the complete synthesis in the forward direction, including all steps and all reagents, and check it for errors and omissions.**
   This final step is left to you as an exercise. Try to do it without looking at this solution, reviewing each thought process as you summarize the synthesis.
   Now practice using a systematic approach with the syntheses in Problem 9-25.

PROBLEM 9-25

Develop syntheses for the following compounds, using acetylene and compounds containing no more than four carbon atoms as your organic starting materials.
(a) 3-methylnon-4-yn-3-ol ("3-ol" means there is an OH group on C3.)
(b) *cis*-1-ethyl-2-methylcyclopropane
(c)

$$CH_3CH_2 \overset{O}{\underset{H}{\diagup \diagdown}} H \atop CH_2CH_2CH_3$$

SUMMARY **Reactions of Alkynes**

Acetylide Chemistry
1. *Formation of acetylide anions (alkynides)*
   a. *Sodium, lithium, and magnesium acetylides* (Section 9-6A)

$$R-C \equiv C-H + NaNH_2 \longrightarrow R-C \equiv C:^- \; {}^+Na + NH_3$$

$$R-C \equiv C-H + R'Li \longrightarrow R-C \equiv CLi + R'H$$

$$R-C \equiv C-H + R'MgX \longrightarrow R-C \equiv CMgX + R'$$

Example

$$CH_3-C\equiv C-H + NaNH_2 \longrightarrow CH_3-C\equiv C:^- {}^+Na + NH_3$$

propyne	sodium amide	sodium propynide
		(propynyl sodium)

b. **Heavy-metal acetylides** (Section 9-6B)

$$R-C\equiv C-H + Ag^+ \longrightarrow R-C\equiv C-Ag\downarrow$$

$$R-C\equiv C-H + Cu^+ \longrightarrow R-C\equiv C-Cu\downarrow$$

(These reactions are used to test for the presence of a terminal alkyne.)

2. *Alkylation of acetylide ions* (Section 9-7A)

$$R-C\equiv C:^- + R'-X \longrightarrow R-C\equiv C-R'$$

(R′—X must be an unhindered primary halide or tosylate.)

Example

$$CH_3CH_2-C\equiv C:^- {}^+Na + CH_3CH_2CH_2-Br \longrightarrow CH_3CH_2-C\equiv C-CH_2CH_2CH_3$$

sodium butynide	1-bromopropane	3-heptyne

3. *Reactions with carbonyl groups* (Section 9-7B)

Example

3-methyl-4-hexyn-3-ol

II. *ADDITIONS TO THE TRIPLE BOND*

1. *Reduction to alkanes* (Section 9-9A)

Example

$$CH_3CH_2-C\equiv C-CH_2-OH + 2 H_2 \xrightarrow{Pt} CH_3CH_2-CH_2-CH_2-CH_2-OH$$

2-pentyn-1-ol	1-pentanol

2. *Reduction to alkenes* (Sections 9-9B and 9-9C)

cis

trans

Examples

$$CH_3CH_2-C\equiv C-CH_2CH_3 \quad \xrightarrow[\text{quinoline}]{H_2,\ Pd/BaSO_4} \quad$$

3-hexyne

$$\underset{\text{H}}{\overset{CH_3CH_2}{>}}C=C\underset{\text{H}}{\overset{CH_2CH_3}{<}}$$

cis-3-hexene

$$CH_3CH_2-C\equiv C-CH_2CH_3 \quad \xrightarrow{Na,\ NH_3} \quad$$

3-hexyne

$$\underset{\text{H}}{\overset{CH_3CH_2}{>}}C=C\underset{CH_2CH_3}{\overset{H}{<}}$$

trans-3-hexene

3. *Addition of halogens (X$_2$ = Cl$_2$, Br$_2$) (Section 9-9D)*

$$R-C\equiv C-R' \xrightarrow{X_2} R-CX=CX-R' \xrightarrow{X_2} R-\overset{\overset{\displaystyle X}{|}}{\underset{\underset{\displaystyle X}{|}}{C}}-\overset{\overset{\displaystyle X}{|}}{\underset{\underset{\displaystyle X}{|}}{C}}-R'$$

Example

$$CH_3C\equiv CCH_2CH_3 \xrightarrow{Br_2} CH_3CBr=CBrCH_2CH_3 \xrightarrow{Br_2} CH_3-\overset{\overset{\displaystyle Br}{|}}{\underset{\underset{\displaystyle Br}{|}}{C}}-\overset{\overset{\displaystyle Br}{|}}{\underset{\underset{\displaystyle Br}{|}}{C}}-CH_2CH_3$$

2-pentyne   *cis-* and *trans-* 2,3-dibromo-2-pentene   2,2,3,3-tetrabromopentane

4. *Addition of hydrogen halides (where HX = HCl, HBr, or HI) (Section 9-9E)*

$$R-C\equiv C-R' \xrightarrow{H-X} \underset{\text{(Markovnikov orientation)}}{R-CH=CX-R'} \xrightarrow{H-X} R-\overset{\overset{\displaystyle H}{|}}{\underset{\underset{\displaystyle H}{|}}{C}}-\overset{\overset{\displaystyle X}{|}}{\underset{\underset{\displaystyle X}{|}}{C}}-R'$$

Example

$$CH_3CH_2-C\equiv C-H \xrightarrow{HCl} \underset{\text{Cl}}{\overset{CH_3CH_2}{>}}C=C\underset{\text{H}}{\overset{H}{<}} \xrightarrow{HCl} CH_3CH_2-\overset{\overset{\displaystyle Cl}{|}}{\underset{\underset{\displaystyle Cl}{|}}{C}}-CH_3$$

1-butyne   2-chloro-1-butene   2,2-dichlorobutane

5. *Addition of water (Section 9-9F)*
    a. *Catalyzed by HgSO$_4$/H$_2$SO$_4$*

$$R-C\equiv C-H + H_2O \xrightarrow{HgSO_4,\ H_2SO_4} \left[\underset{HO}{\overset{R}{>}}C=C\underset{\text{H}}{\overset{H}{<}}\right] \longrightarrow R-\overset{}{\underset{\underset{\displaystyle O}{\parallel}}{C}}-\overset{\overset{\displaystyle H}{|}}{\underset{\underset{\displaystyle H}{|}}{C}}-H$$

(Markovnikov orientation)   vinyl alcohol (unstable)   ketone (stable)

Example

$$CH_3-C\equiv C-H + H_2O \xrightarrow{HgSO_4,\ H_2SO_4} CH_3-\overset{\overset{\displaystyle O}{\parallel}}{C}-CH_3$$

propyne   2-propanone (acetone)

    b. *Hydroboration–oxidation*

$$R-C\equiv C-R' \xrightarrow[\text{(2) } H_2O_2,\ NaOH]{\text{(1) } Sia_2BH\cdot THF} \left[\underset{H}{\overset{R}{>}}C=C\underset{OH}{\overset{H}{<}}\right] \longrightarrow R-\overset{\overset{\displaystyle H}{|}}{\underset{\underset{\displaystyle H}{|}}{C}}-\overset{}{\underset{\underset{\displaystyle O}{\parallel}}{C}}-H$$

(anti-Markovnikov orientation)   vinyl alcohol (unstable)   ketone or aldehyde (stable)

Example

$$CH_3-C\equiv C-H \xrightarrow[\text{(2) } H_2O_2, \text{ NaOH}]{\text{(1) } Sia_2BH \cdot THF} CH_3-CH_2-\overset{\overset{\displaystyle O}{\|}}{C}-H$$

propyne   propanal

III. *OXIDATION OF ALKYNES (SECTION 9-10)*

1. *Oxidation to α-diketones (Section 9-10A)*

$$R-C\equiv C-R' \xrightarrow[\text{H}_2\text{O, neutral}]{\text{KMnO}_4} R-\overset{\overset{\displaystyle O}{\|}}{C}-\overset{\overset{\displaystyle O}{\|}}{C}-R'$$

Example

$$CH_3-C\equiv C-CH_2CH_3 \xrightarrow[\text{H}_2\text{O, neutral}]{\text{KMnO}_4} CH_3-\overset{\overset{\displaystyle O}{\|}}{C}-\overset{\overset{\displaystyle O}{\|}}{C}-CH_2CH_3$$

2-pentyne   pentane-2,3-dione

2. *Oxidative cleavage (Section 9-10B)*

$$R-C\equiv C-R' \xrightarrow[\text{(or O}_3\text{, then H}_2\text{O)}]{\substack{\text{(1) KMnO}_4, \text{ }^-\text{OH} \\ \text{(2) H}^+}} R-\overset{\overset{\displaystyle O}{\|}}{C}-OH \quad + \quad HO-\overset{\overset{\displaystyle O}{\|}}{C}-R'$$

Examples

$$CH_3-C\equiv C-CH_2CH_3 \xrightarrow[\text{(2) H}^+]{\text{(1) KMnO}_4, \text{ NaOH}} CH_3-\overset{\overset{\displaystyle O}{\|}}{C}-OH \quad + \quad HO-\overset{\overset{\displaystyle O}{\|}}{C}-CH_2CH_3$$

$$CH_3CH_2CH_2-C\equiv C-H \xrightarrow[\text{(2) H}^+]{\text{(1) KMnO}_4, \text{ NaOH}} CH_3CH_2CH_2-\overset{\overset{\displaystyle O}{\|}}{C}-OH \quad + \quad CO_2\uparrow$$

Chapter 9 Glossary

acetylene The simplest alkyne, $H-C\equiv C-H$. Also used as a synonym for *alkyne*, a generic term for a compound containing a $C\equiv C$ triple bond. (p. 370)

acetylide ion (alkynide ion) The anionic salt of a terminal alkyne. Metal acetylides are organometallic compounds with a metal atom in place of the acetylenic hydrogen of a terminal alkyne. The metal–carbon bond may be covalent or ionic or partially covalent and partially ionic. (p. 377)

$$R-C\equiv C:^-Na^+ \qquad R-C\equiv C-Ag$$
a sodium acetylide   a silver acetylide

alkoxide ion $R-O^-$, the conjugate base of an alcohol. (p. 376)

$$R-\ddot{O}:^- + H_2O \rightleftharpoons R-\ddot{O}-H + {}^-OH$$
alkoxide   alcohol

alkyne Any compound containing a carbon–carbon triple bond. (pp. 370, 372)
A **terminal alkyne** has a triple bond at the end of a chain, with an **acetylenic hydrogen**.
An **internal alkyne** has the triple bond somewhere other than at the end of the chain.

acetylenic hydrogen   (no acetylenic hydrogen)

$$\boxed{H}-C\equiv C-CH_2CH_3 \qquad\qquad CH_3-C\equiv C-CH_3$$
1-butyne, a terminal alkyne   2-butyne, an internal alkyne

amyl An older common name for pentyl. (p. 392)

enol An alcohol with the hydroxyl group bonded to a carbon atom of a carbon–carbon double bond. Most enols are unstable, spontaneously isomerizing to their carbonyl tautomers, called the **keto** form of the compound. See **tautomers**. (p. 391)

Lindlar's catalyst A heterogeneous catalyst for the hydrogenation of alkynes to cis alkenes. In its most common form, it consists of a thin coating of palladium on barium sulfate, with quinoline added to decrease the catalytic activity. (p. 386)

s character The fraction of a hybrid orbital that corresponds to an *s* orbital; about one-half for *sp* hybrids, one-third for sp^2 hybrids, and one-fourth for sp^3 hybrids. (p. 375)

siamyl group A contraction for secondary isoamyl, abbreviated "Sia." This is the 1,2-dimethyl-propyl group. Disiamylborane is used for hydroboration of terminal alkynes because this bulky borane adds only once to the triple bond. (p. 392)

"*sec*-isoamyl" or "siamyl"	alkyne	disiamylborane	a vinylborane

tautomers Isomers that can quickly interconvert by the movement of a proton (and a double bond) from one site to another. An equilibrium between tautomers is called a **tautomerism**. (p. 391)

enol form	keto form

The keto–enol tautomerism is the equilibrium between these two tautomers.

vinyl cation A cation with a positive charge on one of the carbon atoms of a C=C double bond. The cationic carbon atom is usually *sp* hybridized. Vinyl cations are often generated by the addition of an electrophile to a carbon–carbon triple bond. (p. 390)

a vinyl cation

Essential Problem-Solving Skills in Chapter 9

1. Name alkynes, and draw the structures from their names.
2. Explain why alkynes are more acidic than alkanes and alkenes. Show how to generate nucleophilic acetylide ions and heavy-metal acetylides.
3. Propose effective single-step and multistep syntheses of alkynes.
4. Predict the products of additions, oxidations, reductions, and cleavages of alkynes, including orientation of reaction (regiochemistry) and stereochemistry.
5. Use alkynes as starting materials and intermediates in one-step and multistep syntheses.
6. Show how the reduction of an alkyne leads to an alkene or alkene derivative with the desired stereochemistry.

Study Problems

9-26   Briefly define each term, and give an example.

(a) alkyne   (b) acetylide ion   (c) enol

(d) tautomerism   (e) Lindlar's catalyst   (f) disiamylborane

(g) vinyl cation   (h) oxidative cleavage of an alkyne   (i) hydration of an alkyne

(j) hydroboration of an alkyne

9-27   Write structural formulas for the following compounds.
   (a) 3-nonyne
   (b) methyl-*n*-pentylacetylene
   (c) ethynylbenzene
   (d) cyclohexylacetylene
   (e) 5-methyl-3-octyne
   (f) *trans*-3,5-dibromocyclodecyne
   (g) 3-octyn-2-ol
   (h) *cis*-6-ethyl-2-octen-4-yne
   (i) 1,4-heptadiyne
   (j) vinylacetylene
   (k) (*S*)-3-methyl-1-penten-4-yne

9-28   Give common names for the following compounds.
   (a) $CH_3-C\equiv C-CH_2CH_3$
   (b) $Ph-C\equiv C-H$
   (c) 3-methyl-4-octyne
   (d) $(CH_3)_3C-C\equiv C-CH(CH_3)CH_2CH_3$

9-29   Give IUPAC names for the following compounds.

   (a) $CH_3-C\equiv C-\overset{\overset{\textstyle Ph}{|}}{C}H-CH_3$

   (b) $CH_3-CBr_2-C\equiv C-CH_3$
   (c) $(CH_3)_2CH-C\equiv C-CH_2C(CH_3)_3$

   (d) $\underset{H}{\overset{H_3C}{>}}C=C\underset{C\equiv C-CH_2CH_3}{\overset{CH_3}{<}}$

   (e) $CH_3-C\equiv C-\overset{\overset{\textstyle CH_3}{|}}{\underset{\underset{\textstyle CH_2CH_3}{|}}{C}}-OH$

   (f) cycloheptyl$-C\equiv C-CH_3$

9-30   **(a)** Draw and name the seven alkynes of formula C_6H_{10}.
           (b) Which compounds in part (a) will form precipitates when treated with a solution of cuprous ions?

9-31   When we synthesize an internal alkyne, it is often contaminated with small amounts of a terminal isomer. The boiling points are usually too close for a clean separation by distillation. Give equations to show how you might remove small amounts of 1-decyne from a sample of 2-decyne.

9-32   *Muscalure*, the sex attractant of the common housefly, is *cis*-9-tricosene. Most syntheses of alkenes give the more stable trans isomer as the major product. Devise a synthesis of muscalure from acetylene and other compounds of your choice. Your synthesis must give specifically the cis isomer of muscalure.

$$\underset{H}{\overset{CH_3(CH_2)_7}{>}}C=C\underset{H}{\overset{(CH_2)_{12}CH_3}{<}}$$

cis-9-tricosene, "muscalure"

9-33   Predict the products of reaction of 1-pentyne with the following reagents.
   (a) 1 equivalent of HCl
   (b) 2 equivalents of HCl
   (c) excess H_2, Ni
   (d) H_2, Pd/BaSO$_4$, quinoline
   (e) 1 equivalent of Br$_2$
   (f) 2 equivalents of Br$_2$
   (g) cold, dilute KMnO$_4$
   (h) warm, concd KMnO$_4$, NaOH
   (i) Na, liquid ammonia
   (j) NaNH$_2$
   (k) Ag(NH$_3$)$_2^+$
   (l) H_2SO_4/HgSO$_4$, H_2O
   (m) Sia$_2$BH, then H_2O_2, $^-$OH

9-34   Show how you would accomplish the following synthetic transformations. Show all intermediates.
   (a) 2,2-dibromobutane $\longrightarrow$ 1-butyne
   (b) 2,2-dibromobutane $\longrightarrow$ 2-butyne
   (c) 1-butyne $\longrightarrow$ 3-octyne
   (d) *trans*-2-hexene $\longrightarrow$ 2-hexyne
   (e) *cis*-2-hexene $\longrightarrow$ 1-hexyne
   (f) cyclodecyne $\longrightarrow$ *cis*-cyclodecene
   (g) cyclodecyne $\longrightarrow$ *trans*-cyclodecene
   (h) 1-hexyne $\longrightarrow$ 2-hexanone, $CH_3COCH_2CH_2CH_2CH_3$
   (i) 1-hexyne $\longrightarrow$ hexanal, $CH_3(CH_2)_4CHO$
   (j) *trans*-2-hexene $\longrightarrow$ *cis*-2-hexene

9-35   Potassium hydroxide is mixed with 2,3-dibromohexane, and the mixture is heated at 200°C in a sealed tube for 1 hour. The product mixture (**A**) is mixed with a copper(I)–ammonia complex, and a precipitate forms. The precipitate (**B**) and the liquid phase (**C**) are separated. The precipitate is acidified, and the product (**D**) is distilled (bp 71°C). Product (**D**) is treated with sodium amide, followed by acetone, and then by dilute acid to give alcohol (**F**).

Note the liquid phase (**C**) is distilled, and the products are collected over a boiling range of 80°C to 85°C. This distillate is treated with sodium amide at 150°C for 1 hour, and the product mixture is distilled to give a pure alkyne (**E**) of boiling point 71°C. Give the structures of the alkynes present in products (**A**) through (**E**), and give the structure of alcohol (**F**).

9-36   Predict the products formed when $CH_3CH_2\!-\!C\!\equiv\!C:^-\ ^+Na$ reacts with the following compounds.
(**a**) ethyl bromide
(**b**) *t*-butyl bromide
(**c**) formaldehyde
(**d**) cyclohexanone
(**e**) $CH_3CH_2CH_2CHO$
(**f**) cyclohexanol
(**g**) 2-butanone, $CH_3CH_2COCH_3$

9-37   Show how you would synthesize the following compounds, starting with acetylene and any compounds containing no more than four carbon atoms.
(**a**) 1-hexyne
(**b**) 2-hexyne
(**c**) *cis*-2-hexene
(**d**) *trans*-2-hexene
(**e**) hexane
(**f**) 2,2-dibromohexane
(**g**) pentanal, $CH_3CH_2CH_2CH_2CHO$
(**h**) 2-pentanone, $CH_3\!-\!CO\!-\!CH_2CH_2CH_3$
(**i**) (±)-3,4-dibromohexane
(**j**) *meso*-2,3-butanediol

9-38   When treated with hydrogen and a platinum catalyst, an unknown compound (**X**) absorbs 5 equivalents of hydrogen to give *n*-butylcyclohexane. When **X** is treated with silver nitrate in ethanol, a white precipitate forms. This precipitate is found to be soluble in dilute acid. Treatment of **X** with an excess of ozone, followed by dimethyl sulfide and water, gives the following products:

$$H\!-\!\overset{O}{\overset{\|}{C}}\!-\!CH_2\!-\!CH_2\!-\!\overset{O}{\overset{\|}{C}}\!-\!\overset{O}{\overset{\|}{C}}\!-\!H \qquad H\!-\!\overset{O}{\overset{\|}{C}}\!-\!\overset{O}{\overset{\|}{C}}\!-\!H \qquad H\!-\!\overset{O}{\overset{\|}{C}}\!-\!\overset{O}{\overset{\|}{C}}\!-\!OH \qquad H\!-\!\overset{O}{\overset{\|}{C}}\!-\!OH$$

Propose a structure for the unknown compound (**X**). Is there any uncertainty in your structure?

9-39   When compound (**Z**) is treated with a silver–ammonia complex, a white precipitate forms. Treatment of compound (**Z**) with ozone, followed by dimethyl sulfide and washing with water, gives formic acid, 3-oxobutanic acid, and hexanal.

$$(\mathbf{Z}) \xrightarrow[\text{(2) }(CH_3)_2S,\ H_2O]{\text{(1) }O_3} \underset{\text{formic acid}}{H\!-\!\overset{O}{\overset{\|}{C}}\!-\!OH} + \underset{\text{3-oxobutanoic acid}}{CH_3\!-\!\overset{O}{\overset{\|}{C}}\!-\!CH_2\!-\!\overset{O}{\overset{\|}{C}}\!-\!OH} + \underset{\text{hexanal}}{CH_3(CH_2)_4\!-\!\overset{O}{\overset{\|}{C}}\!-\!H}$$

Propose a structure for compound (**Z**). What uncertainty is there in the structure you have proposed?

9-40   Propose structures for intermediates and products (**A**) through (**L**).

1-bromobutane $\xrightarrow{\text{Mg, ether}}$ (**D**) $\xrightarrow[\text{(2) H}_3\text{O}^+]{\text{(1) CH}_3(\text{CH}_2)_3\text{CHO}}$ (**E**)

$\downarrow$ KOH, H$_2$O

(**A**)

$\downarrow$ Na

(**B**)

$\downarrow$ CH$_3$CH$_2$Br

(**C**)

(E) $\xrightarrow{\text{H}_2\text{SO}_4,\text{ heat}}$ (**F**)

(F) $\xrightarrow[\text{(2) (CH}_3)_2\text{S}]{\text{(1) O}_3}$ CH$_3$CH$_2$CH$_2$CHO + CH$_3$(CH$_2$)$_3$CHO

(F) $\xrightarrow{\text{Br}_2}$ (**G**)

(G) $\xrightarrow{\text{fused KOH}}$ (**H**) mixture

(H) $\xrightarrow[\text{(2) H}_2\text{O}]{\text{(1) NaNH}_2, 150°\text{C}}$ 1-nonyne

1-nonyne $\xrightarrow[\text{(2) H}_2\text{O}_2, \text{NaOH}]{\text{(1) Sia}_2\text{BH}}$ (**I**)

1-nonyne $\xrightarrow{\text{NaNH}_2}$ (**J**)

(J) $\xrightarrow[\text{(2) H}_3\text{O}^+]{\text{(1) CH}_3\text{—C(=O)—CH}_2\text{CH}_3}$ (**K**)

1-nonyne $\xrightarrow{\text{2 HBr}}$ (**L**)

*9-41   The following functional group interchange is a useful synthesis of aldehydes.

$$\text{R—C}\equiv\text{C—H} \longrightarrow \text{R—CH}_2\text{—C(=O)—H}$$
terminal alkyne   aldehyde

(a) What reagents were used in this chapter for this transformation? Give an example to illustrate this method.
(b) This functional group interchange can also be accomplished using the following sequence.

$$\text{R—C}\equiv\text{C—H} \xrightarrow[\text{CH}_3\text{CH}_2\text{OH}]{\text{NaOCH}_2\text{CH}_3} \quad \underset{R}{\overset{H}{>}}\text{C=C}\underset{H}{\overset{OCH_2CH_3}{<}} \xrightarrow{\text{H}_3\text{O}^+} \text{R—CH}_2\text{—C(=O)—H}$$

Propose mechanisms for these steps.
(c) Explain why a nucleophilic reagent such as ethoxide adds to an alkyne more easily than it adds to an alkene.

*9-42   Using any necessary inorganic reagents, show how you would convert acetylene and isobutyl bromide to
(a) *meso*-2,7-dimethyl-4,5-octanediol, (CH$_3$)$_2$CHCH$_2$CH(OH)CH(OH)CH$_2$CH(CH$_3$)$_2$
(b) (±)-2,7-dimethyl-4,5-octanediol

CHAPTER 10

Structure and Synthesis of Alcohols

Introduction

Alcohols are organic compounds containing hydroxyl (—OH) groups. They are some of the most common and useful compounds in nature, in industry, and around the house. The word *alcohol* is one of the oldest chemical terms, derived from the early Arabic *al-kuhl*. Originally it meant "the powder," and later "the essence." Ethyl alcohol, distilled from wine, was considered to be "the essence" of wine. Ethyl alcohol (grain alcohol) is found in alcoholic beverages, cosmetics, and drug preparations. Methyl alcohol (wood alcohol) is used as a fuel and solvent. Isopropyl alcohol (rubbing alcohol) is used as a skin cleanser for injections and minor cuts.

$$CH_3-CH_2-OH \qquad CH_3-OH \qquad CH_3-\overset{\overset{\displaystyle OH}{|}}{CH}-CH_3$$

ethyl alcohol   methyl alcohol   isopropyl alcohol

Alcohols are synthesized by a wide variety of methods, and the hydroxyl group may be converted to most other functional groups. For these reasons, alcohols are versatile synthetic intermediates. In this chapter, we discuss the physical properties of alcohols and summarize the methods used to synthesize them. In Chapter 11 (Reactions of Alcohols), we continue our study of the central role that alcohols play in organic chemistry as reagents, solvents, and synthetic intermediates.

10-2 Structure and Classification of Alcohols

The structure of an alcohol resembles the structure of water, with an alkyl group replacing one of the hydrogen atoms of water. Figure 10-1 compares the structures of water and methanol. Both have sp^3 hybridized oxygen atoms, but the C—O—H bond angle in methanol (108.9°) is considerably larger than the H—O—H bond angle in water (104.5°) because the methyl group is much larger than a hydrogen atom. The bulky

◀ **Figure 10-1**
Comparison of the structures of water and methyl alcohol.

405

methyl group counteracts the bond angle compression caused by oxygen's nonbonding pairs of electrons. The O—H bond lengths are about the same in water and methanol (0.96 Å), but the C—O bond is considerably longer (1.4 Å), reflecting the larger covalent radius of carbon compared to hydrogen.

One way of organizing the alcohol family is to classify each alcohol according to the type of **carbinol carbon atom**: the one bonded to the —OH group. If this carbon atom is primary (bonded to one other carbon atom), the compound is a **primary alcohol**. A **secondary alcohol** has the —OH group attached to a secondary carbon atom, and a **tertiary alcohol** has it bonded to a tertiary carbon. When we studied alkyl halides (Chapter 6), we saw that primary, secondary, and tertiary halides react differently. The same is true for alcohols. We need to learn how these classes of alcohols are similar and under what conditions they react differently. Figure 10-2 shows examples of primary, secondary, and tertiary alcohols.

Compounds with a hydroxyl group bonded directly to an aromatic (benzene) ring are called **phenols**. Phenols have many properties similar to those of alcohols, while other properties derive from their aromatic character. In this chapter, we consider the properties of phenols that are similar to those of alcohols and note some of the differences. In Chapter 16, we consider the aromatic nature of phenols and the reactions that result from that aromaticity.

▲ **Figure 10-2**
Alcohols are classified according to the type of carbon atom (primary, secondary, or tertiary) bonded to the hydroxyl group. Phenols have a hydroxyl group bonded to a carbon atom in a benzene ring.

10-3A   IUPAC Names ("Alkanol" Names)

The IUPAC system provides unique names for alcohols, based on rules that are similar to those for other classes of compounds. In general, the name carries the *-ol* suffix, together with a number to give the location of the hydroxyl group. The formal rules are summarized in the following three steps:

1. Name the longest carbon chain that contains the carbon atom bearing the —OH group. Drop the final *-e* from the alkane name and add the suffix *-ol* to give the root name.

2. Number the longest carbon chain starting at the end nearest the hydroxyl group, and use the appropriate number to indicate the position of the —OH group. (The hydroxyl group takes precedence over double and triple bonds.)

3. Name all the substituents and give their numbers, as you would for an alkane or an alkene.

In the following example, the longest carbon chain has four carbons, so the root name is *butanol*. The —OH group is on the second carbon atom, so this is a 2-butanol. The complete IUPAC name is 1-bromo-3,3-dimethyl-2-butanol. The new IUPAC positioning of numbers would place the 2 next to the group it locates (-ol), giving the name 1-bromo-3,3-dimethylbutan-2-ol.

$$^4CH_3 - {^3C} - {^2CH} - {^1CH_2} - Br$$

with CH_3 and OH on the 2-carbon and 3-carbon positions, and CH_3 below C3.

Cyclic alcohols are named using the prefix *cyclo-*; the hydroxyl group is assumed to be on C1.

IUPAC name:	*trans*-2-chlorocyclohexanol 1-ethylcyclopropanol
new IUPAC name:	*trans*-2-bromocyclohexan-1-ol 1-ethylcyclopropan-1-ol

SOLVED PROBLEM 10-1

Give the systematic (IUPAC) name for the following alcohol.

$$CH_3 - CH_2 - CH - CH - CH - CH_3$$

with CH_2I and $CH_2{-}OH$ on the respective carbons and CH_3 below.

SOLUTION

The longest chain contains six carbon atoms, but it does not contain the carbon bonded to the hydroxyl group. The longest chain containing the carbon bonded to the —OH group is the one indicated by the green box, containing five carbon atoms. This chain is numbered from right to left so as to give the hydroxyl-bearing carbon atom the lowest possible number.

$$CH_2I \quad {}^1CH_2{-}OH$$
$${}^5CH_3{-}{}^4CH_2{-}{}^3CH{-}{}^2CH{-}CH{-}CH_3$$
$$CH_3$$

The correct name for this compound is 3-(iodomethyl)-2-isopropylpentan-1-ol.

**Main Groups
(decreasing priority)**

acids
esters
aldehydes
ketones
alcohols
amines
alkenes
alkynes
alkanes
ethers
halides

In naming alcohols containing double and triple bonds, use the *-ol* suffix after the alkene or alkyne name. The alcohol functional group takes precedence over double and triple bonds, so the chain is numbered in order to give the lowest possible number to the carbon atom bonded to the hydroxyl group. The position of the —OH group is given by putting its number before the *-ol* suffix. Numbers for the multiple bonds were once given early in the name, but the 1997 revision of the IUPAC rules puts them next to the *-en* or *-yn* suffix they describe. Both the new and old placements of the numbers are shown below.

IUPAC name:	*trans*-2-penten-1-ol	(Z)-4-chloro-3-buten-2-ol	2-cyclohexen-1-ol
new IUPAC name:	*trans*-pent-2-en-1-ol	(Z)-4-chlorobut-3-en-2-ol	cyclohex-2-en-1-ol

At left is a partial table showing the order of precedence of functional groups for assigning IUPAC names. A more complete table, titled "Summary of Functional Group Nomenclature," appears inside the back cover. In general, the highest priority functional group is considered the "main" group, and the others are treated as substituents.

The —OH functional group is named as a **hydroxy** substituent when it appears on a structure with a higher priority functional group, or when the structure is too difficult to name as a simple alcohol.

2-hydroxymethylcyclohexanone   *trans*-3-(2-hydroxyethyl)cyclopentanol   3-hydroxybutanoic acid

PROBLEM 10-1

Give the IUPAC names of the following alcohols.

(a)

(b)

(c)

(d)

(e)

(f)

10-3B Common Names of Alcohols

The common name of an alcohol is derived from the common name of the alkyl group and the word alcohol. This system pictures an alcohol as a molecule of water with an alkyl group replacing one of the hydrogen atoms. If the structure is complex, the common nomenclature becomes awkward, and the IUPAC nomenclature should be used.

	CH_3—OH	$CH_3CH_2CH_2$—OH	CH_3—$\overset{\displaystyle OH}{\underset{\displaystyle |}{CH}}$—$CH_3$	H_2C=CH—CH_2—OH
common name:	methyl alcohol	*n*-propyl alcohol	isopropyl alcohol	allyl alcohol
IUPAC name:	methanol	1-propanol	2-propanol	2-propen-1-ol

	$CH_3CH_2CH_2CH_2$—OH	CH_3—$\overset{\displaystyle OH}{\underset{\displaystyle |}{CH}}$—$CH_2CH_3$	CH_3—$\overset{\displaystyle CH_3}{\underset{\displaystyle CH_3}{\overset{\displaystyle |}{\underset{\displaystyle |}{C}}}}$—OH	CH_3—$\overset{\displaystyle CH_3}{\underset{\displaystyle |}{CH}}$—$CH_2$—OH
common name:	*n*-butyl alcohol	*sec*-butyl alcohol	*t*-butyl alcohol	isobutyl alcohol
IUPAC name:	1-butanol	2-butanol	2-methyl-2-propanol	2-methyl-1-propanol

PROBLEM 10-2

Give both the IUPAC name and the common name for each alcohol.

(a) $\overset{\displaystyle OH}{\triangle}$ (b) $(CH_3)_3COH$ (c) ▢⁀$\overset{}{OH}$ (d) $(CH_3)_2CHCH_2CH_2OH$

PROBLEM 10-3

For each molecular formula, draw all the possible alcohols with that formula. Give the IUPAC name for each alcohol.

(a) C_3H_8O (b) $C_4H_{10}O$ (c) C_4H_8O

10-3C Names of Diols

Alcohols with two —OH groups are called **diols** or **glycols**. They are named like other alcohols except that the suffix *diol* is used and two numbers are needed to tell where the two hydroxyl groups are located. This is the preferred, systematic (IUPAC) method for naming diols.

IUPAC name:	CH_3—$\overset{\displaystyle OH}{\underset{\displaystyle |}{CH}}$—$CH_2OH$		
	propane-1,2-diol	1-cyclohexylbutane-1,3-diol	*trans*-cyclopentane-1,2-diol

The term *glycol* generally means a 1,2-diol, or **vicinal diol**, with its two hydroxyl groups on adjacent carbon atoms. Glycols are usually synthesized by the hydroxylation of alkenes, using peroxyacids, osmium tetroxide, or potassium permanganate (Section 8–14).

$$\underset{\text{alkene}}{\overset{\diagdown}{\diagup}C=C\overset{\diagup}{\diagdown}} \xrightarrow[\substack{\text{or KMnO}_4,\ ^-\text{OH} \\ \text{or OsO}_4,\ \text{H}_2\text{O}_2}]{\text{RCO}_3\text{H},\ \text{H}^+} \underset{\substack{| \quad | \\ \text{HO} \quad \text{OH} \\ \text{vicinal diol (glycol)}}}{-\overset{|}{C}-\overset{|}{C}-}$$

This synthesis of glycols is reflected in their common names. The glycol is named for the alkene from which it is synthesized:

CH₂—CH₂ | | OH OH	CH₂—CH—CH₃ | | OH OH	(cyclohexane with two OH)
IUPAC name: ethane-1,2-diol	propane-1,2-diol	*cis*-cyclohexane-1,2-diol
common name: ethylene glycol	propylene glycol	*cis*-cyclohexene glycol

The common names of glycols can be awkward and confusing because the *ene* portion of the name implies the presence of an alkene double bond, but the glycol does not contain a double bond. We will generally use the IUPAC "diol" nomenclature for diols; however, you should be aware that the names "ethylene glycol" (automotive antifreeze) and "propylene glycol" (used in medicines and foods) are universally accepted for these common diols.

PROBLEM 10-4

Give a systematic (IUPAC) name for each diol.
(a) CH₃CH(OH)(CH₂)₄CH(OH)C(CH₃)₃   (b) HO—(CH₂)₈—OH

(c) (cyclohexene diol structure)

(d) (heptane diol with cyclopentyl structure)

10-3D   Names of Phenols

Phenol was given the common name "carbolic acid" because it was originally isolated from the acidic fraction of coal tar. Phenol is more acidic than water, and it is highly corrosive to skin.

Because the phenol structure involves a benzene ring, the terms *ortho* (1,2-disubstituted), *meta* (1,3-disubstituted), and *para* (1,4-disubstituted) are often used in the common names. The following examples illustrate the systematic names and the common names of some simple phenols.

IUPAC name: 2-bromophenol	3-nitrophenol	4-ethylphenol
common name: *ortho*-bromophenol	*meta*-nitrophenol	*para*-ethylphenol

Soil bacteria convert many aromatic compounds to catechol. Then they cleave the ring and degrade the products further to water and carbon dioxide. The soil around highways is often enriched in these bacteria feeding on the runoff of hydrocarbons that vehicles deposit on the road.

The methylphenols are called *cresols*, while the benzenediols have names based on their historical uses and sources rather than their structures. We will generally use the systematic names of phenolic compounds.

IUPAC name: 2-methylphenol	benzene-1,2-diol	benzene-1,3-diol	benzene-1,4-diol
common name: *ortho*-cresol	catechol	resorcinol	hydroquinone

Most of the common alcohols, up to about 11 or 12 carbon atoms, are liquids at room temperature. Methanol and ethanol are free-flowing volatile liquids with characteristic fruity odors. The higher alcohols (the butanols through the decanols) are somewhat viscous, and some of the highly branched isomers are solids at room temperature. These higher alcohols have heavier but still fruity odors. 1-Propanol and 2-propanol fall in the middle, with a barely noticeable viscosity and a characteristic odor often associated with a physician's office. Table 10-1 lists the physical properties of some common alcohols.

10-4
Physical Properties of Alcohols

10-4A Boiling Points of Alcohols

Because we often deal with liquid alcohols, we forget how surprising it *should* be that the lower molecular weight alcohols are liquids. For example, ethyl alcohol and propane have similar molecular weights, yet their boiling points differ by about 120°C. Dimethyl ether has an intermediate boiling point.

$\updownarrow \mu = 1.69$ D

O

H CH_2CH_3

ethanol, MW 46
bp 78°C

$\updownarrow \mu = 1.30$ D

O

H_3C CH_3

dimethyl ether, MW 46
bp −25°C

$\mu = 0.08$ D

H   H

C

H_3C CH_3

propane, MW 44
bp −42°C

Such a large difference in boiling points suggests that ethanol molecules are attracted to each other much more strongly than propane molecules. Two important intermolecular forces are responsible: hydrogen bonding and dipole–dipole attractions (Section 2-10).

Hydrogen bonding is the major intermolecular attraction responsible for ethanol's high boiling point. The hydroxyl hydrogen of ethanol is strongly polarized by its bond to oxygen, and it forms a hydrogen bond with a pair of nonbonding electrons from the oxygen atom of another alcohol molecule (Section 2-10C). Ethers have two alkyl groups

TABLE 10-1 Physical Properties of Selected Alcohols

IUPAC Name	Common Name	Formula	mp(°C)	bp(°C)	Density
methanol	methyl alcohol	CH_3OH	−97	65	0.79
ethanol	ethyl alcohol	CH_3CH_2OH	−114	78	0.79
1-propanol	n-propyl alcohol	$CH_3CH_2CH_2OH$	−126	97	0.80
2-propanol	isopropyl alcohol	$(CH_3)_2CHOH$	−89	82	0.79
1-butanol	n-butyl alcohol	$CH_3(CH_2)_3OH$	−90	118	0.81
2-butanol	sec-butyl alcohol	$CH_3CH(OH)CH_2CH_3$	−114	100	0.81
2-methyl-1-propanol	isobutyl alcohol	$(CH_3)_2CHCH_2OH$	−108	108	0.80
2-methyl-2-propanol	t-butyl alcohol	$(CH_3)_3COH$	25	83	0.79
1-pentanol	n-pentyl alcohol	$CH_3(CH_2)_4OH$	−79	138	0.82
3-methyl-1-butanol	isopentyl alcohol	$(CH_3)_2CHCH_2CH_2OH$	−117	132	0.81
2,2-dimethyl-1-propanol	neopentyl alcohol	$(CH_3)_3CCH_2OH$	52	113	0.81
cyclopentanol	cyclopentyl alcohol	cyclo-C_5H_9OH	−19	141	0.95
1-hexanol	n-hexanol	$CH_3(CH_2)_5OH$	−52	156	0.82
cyclohexanol	cyclohexyl alcohol	cyclo-$C_6H_{11}OH$	25	162	0.96
1-heptanol	n-heptyl alcohol	$CH_3(CH_2)_6OH$	−34	176	0.82
1-octanol	n-octyl alcohol	$CH_3(CH_2)_7OH$	−16	194	0.83
1-nonanol	n-nonyl alcohol	$CH_3(CH_2)_8OH$	−6	214	0.83
1-decanol	n-decyl alcohol	$CH_3(CH_2)_9OH$	6	233	0.83
2-propen-1-ol	allyl alcohol	$H_2C{=}CH{-}CH_2OH$	−129	97	0.86
phenylmethanol	benzyl alcohol	$Ph{-}CH_2OH$	−15	205	1.05
diphenylmethanol	diphenylcarbinol	Ph_2CHOH	69	298	
triphenylmethanol	triphenylcarbinol	Ph_3COH	162	380	1.20
1,2-ethanediol	ethylene glycol	$HOCH_2CH_2OH$	−13	198	1.12
1,2-propanediol	propylene glycol	$CH_3CH(OH)CH_2OH$	−59	188	1.04
1,2,3-propanetriol	glycerol	$HOCH_2CH(OH)CH_2OH$	18	290	1.26

bonded to their oxygen atoms, so they have no O—H hydrogen atoms to form hydrogen bonds. Hydrogen bonds have a strength of about 5 kcal (21 kJ) per mole: weaker than typical covalent bonds of 70 to 110 kcal, but much stronger than dipole–dipole attractions.

Dipole–dipole attractions also contribute to the relatively high boiling points of alcohols and ethers. The polarized C—O and H—O bonds and the nonbonding electrons add to produce a dipole moment of 1.69 D in ethanol, compared with a dipole moment of only 0.08 D in propane. In liquid ethanol, the positive and negative ends of these dipoles align to produce attractive interactions.

We can compare the effects of hydrogen bonding and dipole–dipole attractions by comparing ethanol with dimethyl ether. Like ethanol, dimethyl ether has a large dipole moment (1.30 D), but dimethyl ether cannot engage in hydrogen bonding because it has no —O—H hydrogens.

The boiling point of dimethyl ether is −25°C, about 17° higher than that of propane, but still 103° lower than that of ethanol. Hydrogen bonds are clearly much stronger intermolecular attractions than dipole-dipole attractions.

10-4B Solubility Properties of Alcohols

Water and alcohols have similar properties because they all contain hydroxyl groups that can form hydrogen bonds. Alcohols form hydrogen bonds with water, and several of the lower molecular weight alcohols are **miscible** (soluble in any proportions) with water. Similarly, alcohols are much better solvents than hydrocarbons for polar substances. Significant amounts of ionic compounds such as sodium chloride can dissolve in some of the lower alcohols. We call the hydroxyl group **hydrophilic**, meaning "water loving," because of its affinity for water and other polar substances.

The alcohol's alkyl group is called **hydrophobic** ("water hating") because it acts like an alkane: It disrupts the network of hydrogen bonds and dipole–dipole attractions of a polar solvent such as water. The alkyl group makes the alcohol less hydrophilic, yet it lends solubility in nonpolar organic solvents. As a result, many alcohols are miscible with a wide range of nonpolar organic solvents.

Table 10-2 lists the solubility of some simple alcohols in water. The water solubility decreases as the alkyl group becomes larger. Alcohols with one-, two-, or three-carbon alkyl groups are miscible with water. A four-carbon alkyl group is large enough that some isomers are not miscible, yet *t*-butyl alcohol, with a compact spherical shape, is miscible. Phenol is unusually soluble for a six-carbon alcohol because of its compact shape and the particularly strong hydrogen bonds formed between phenolic —OH groups and water molecules.

TABLE 10-2 Solubility of Alcohols in Water (at 25°C)

Alcohol	Solubility in Water
methyl	miscible
ethyl	miscible
n-propyl	miscible
t-butyl	miscible
isobutyl	10.0%
n-butyl	9.1%
n-pentyl	2.7%
cyclohexyl	3.6%
n-hexyl	0.6%
phenol	9.3%
hexane-1,6-diol	miscible

PROBLEM 10-5

Predict which member of each pair will be more soluble in water. Explain the reasons for your answers.

(a) 1-hexanol or cyclohexanol
(b) 1-heptanol or 4-methylphenol
(c) 3-ethyl-3-hexanol or 2-octanol
(d) 2-hexanol or cyclooctane-1,4-diol

(e)

PROBLEM 10-6

Dimethylamine, $(CH_3)_2NH$, has a molecular weight of 45 and a boiling point of 7.4°C. Trimethylamine, $(CH_3)_3N$, has a higher molecular weight (59) but a *lower* boiling point (3.5°C). Explain this apparent discrepancy.

10-5A Methanol

10-5
Commercially Important Alcohols

Methanol (methyl alcohol) was originally produced by the destructive distillation of wood chips in the absence of air. This source led to the name **wood alcohol**. During Prohibition (1919–1933), when the manufacture of alcoholic beverages was prohibited, anything called "alcohol" was often used for mixing drinks. Since methanol is more toxic than ethanol, this practice resulted in many cases of blindness and death.

Today, most methanol is synthesized by a catalytic reaction of carbon monoxide with hydrogen. This reaction uses high temperatures and pressures and requires large, complicated industrial reactors.

$$CO + 2\,H_2 \xrightarrow[\text{CuO–ZnO/Al}_2\text{O}_3]{\text{300-400 °C, 200-300 atm H}_2} CH_3OH$$

synthesis gas

Synthesis gas, containing the hydrogen and carbon monoxide needed to make methanol, can be generated by the partial burning of coal in the presence of water. Careful regulation of the amount of water added allows production of synthesis gas with the correct ratio of carbon monoxide to hydrogen.

$$3\,C + 4\,H_2O \xrightarrow{\text{high temperature}} CO_2 + 2\,CO + 4\,H_2$$

synthesis gas

Methanol is one of the most common industrial solvents. It is cheap, relatively less toxic (compared with halogenated solvents), and it dissolves a wide variety of polar and nonpolar substances. Methanol is also a starting material for a wide variety of methyl ethers, methyl esters, and other compounds used in plastics, medicines, fuels, and solvents.

Methanol is a good fuel for internal combustion engines. Since 1965, all the cars at the Indianapolis 500 have used methanol-fueled engines. The switch from gasoline to methanol was driven by a bad fire after a crash in 1964. Methanol is less flammable than gasoline, and water is effective against methanol fires (water mixes with and dilutes methanol). As with any alternative fuel, there are advantages and disadvantages to the use of methanol. Its high octane rating, low pollutant emissions, and lower flammability must be weighed against its lower energy content (smaller ΔH of combustion per gram), requiring 1.7 g of methanol to produce the same energy as 1 g of gasoline.

Because of its excellent solvent properties, methanol is hard on rings, seals, and plastic fuel-system parts. Its tendency to burn with little or no visible flame can allow dangerous methanol fires to go undetected.

10-5B   Ethanol

The prehistoric discovery of ethanol probably occurred when rotten fruit was consumed and found to have an intoxicating effect. This discovery presumably led to the intentional fermentation of fruit juices. The primitive wine that resulted could be stored (in a sealed container) without danger of decomposition, and it also served as a safe, unpolluted source of water to drink.

Ethanol can be produced by the fermentation of sugars and starches from many different sources. Grains such as corn, wheat, rye, and barley are common sources, resulting in the name **grain alcohol** for ethanol. Cooking the grain, followed by addition of sprouted barley, called *malt*, converts some of the starches to simpler sugars. Brewer's yeast is then added, and the solution is incubated while the yeast cells convert simple sugars such as glucose to ethanol and carbon dioxide.

$$\underset{\text{glucose}}{C_6H_{12}O_6} \xrightarrow{\text{yeast enzymes}} \underset{\text{ethanol}}{2\,C_2H_5OH} + 2\,CO_2$$

The alcoholic solution that results from fermentation contains only 12–15% alcohol, because yeast cells cannot survive higher concentrations. Distillation increases the alcohol concentration to about 40–50% (80 to 100 "proof") for "hard" liquors. Distillation of ethanol–water solutions cannot increase the ethanol concentration above 95% because the solution of 95% ethanol and 5% water boils at a lower temperature (78.15°C) than either pure water (100°C) or pure ethanol (78.3°C). Such a mixture of liquids that boils at a lower temperature than either of its components is called a minimum-boiling **azeotrope**.

The 95% alcohol produced by distillation is well suited for use as a solvent and a reagent when traces of water do not affect the reaction. When **absolute alcohol** (100% ethanol) is required, the 95% azeotrope is passed through a dehydrating agent such as anhydrous calcium oxide (CaO), which removes the final 5% of water.

Since World War II, most industrial ethanol has been synthesized directly by the catalyzed high-temperature, high-pressure, gas-phase reaction of water with ethylene. This process uses catalysts such as P_2O_5, tungsten oxide, or various specially treated clays.

$$H_2C{=}CH_2 + H_2O \xrightarrow[\text{catalyst}]{\text{100-300 atm, 300°C}} CH_3{-}CH_2{-}OH$$

Like methanol, ethanol is an excellent solvent of low toxicity that is cheap to produce. Unfortunately, the liquor tax makes ethanol relatively expensive. Use of untaxed ethanol is possible, but it requires extensive recordkeeping and purchase of a special license. **Denatured alcohol** is ethanol that contains impurities that make it undrinkable. Denatured ethanol is untaxed, but the impurities (methanol, methyl isobutyl ketone, aviation gasoline, etc.) also make it unsuitable for many laboratory uses.

Like methanol, ethanol is a good motor fuel, with similar advantages and disadvantages. A car's carburetor must be adjusted (for a richer mixture) and fitted with alcohol-resistant seals if it is to run on pure ethanol. Solutions of about 10% ethanol in gasoline ("gasohol") work well without any adjustments, however.

Many people imagine ethanol to be nontoxic, and methanol to be horribly toxic. Actually, methanol is about twice as toxic as ethanol: Typical fatal doses for adults are about 100 mL of methanol or about 200 mL of ethanol, although smaller doses of methanol may damage the optic nerve. Many people die each year from underestimating ethanol's toxicity. In the lab, we would never ingest even a tiny fraction of these amounts, so as chemists, we consider these solvents to be relatively nontoxic compared with hazardous solvents such as benzene and chloroform.

Experience at Indianapolis has proved methanol (derived from coal) to be an excellent fuel for automotive engines.

Both ethanol and 2-propanol are effective topical antiseptics. Ethanol is also an ingredient in many mouthwashes. These alcohols kill the microorganisms on the wound surface and in the mouth, but their low toxicity does not kill the cells of the skin or mouth tissues.

FRIENDLY REMINDER

Everything is toxic in large enough amounts.

10-5C 2-Propanol

2-Propanol (isopropyl alcohol) is made by the catalytic hydration of propylene. Isopropyl alcohol is commonly used as **rubbing alcohol** (rather than ethanol) because it has less of a drying effect on the skin, and it is not regulated and taxed by the government. 2-Propanol is about as toxic as methanol when taken orally, but it is safer for use on the skin because it does not pass through skin as easily as methanol.

$$CH_3-CH{=}CH_2 \; + \; H_2O \quad \xrightarrow[\text{catalyst}]{100-300\text{ atm, }300°C} \quad CH_3-\underset{\underset{\displaystyle OH}{|}}{CH}-CH_3$$

propylene

2-propanol

10-6
Acidity of Alcohols and Phenols

Like the hydroxyl proton of water, the hydroxyl proton of an alcohol is weakly acidic. A strong base can remove the hydroxyl proton to give an **alkoxide ion**.

☞ $\quad R-\overset{..}{\underset{..}{O}}-H \; + \; B{:}^- \; \rightleftharpoons \; R-\overset{..}{\underset{..}{O}}{:}^- \; + \; B-H$

alcohol $\qquad\qquad\qquad\qquad$ alkoxide ion

Example

$\quad CH_3CH_2-\overset{..}{\underset{..}{O}}-H \; + \; B{:}^- \; \rightleftharpoons \; CH_3CH_2-\overset{..}{\underset{..}{O}}{:}^- \; + \; B-H$

ethanol $\qquad\qquad\qquad\qquad$ ethoxide ion

The acidities of alcohols vary widely, from alcohols that are about as acidic as water to some that are much less acidic. The acid-dissociation constant, K_a, of an alcohol is defined by the equilibrium

$$R-O-H + H_2O \;\overset{K_a}{\rightleftharpoons}\; R-O^- + H_3O^+$$

$$K_a = \frac{[H_3O^+][RO^-]}{[ROH]} \qquad pK_a = -\log(K_a)$$

Table 10-3 compares the acid-dissociation constants of some alcohols with those of water and other acids.

TABLE 10-3 Acid-Dissociation Constants of Representative Alcohols

Alcohol	Structure	K_a	pK_a
methanol	CH_3-OH	3.2×10^{-16}	15.5
ethanol	CH_3CH_2-OH	1.3×10^{-16}	15.9
2-chloroethanol	$Cl-CH_2CH_2-OH$	5.0×10^{-15}	14.3
2,2,2-trichloroethanol	Cl_3C-CH_2-OH	6.3×10^{-13}	12.2
isopropyl alcohol	$(CH_3)_2CH-OH$	3.2×10^{-17}	16.5
t-butyl alcohol	$(CH_3)_3C-OH$	1.0×10^{-18}	18.0
cyclohexanol	$C_6H_{11}-OH$	1.0×10^{-18}	18.0
phenol	C_6H_5-OH	1.0×10^{-10}	10.0
Comparison with Other Acids			
water	H_2O	1.8×10^{-16}	15.7
acetic acid	CH_3COOH	1.6×10^{-5}	4.8
hydrochloric acid	HCl	$1.6 \times 10^{+2}$	-2.2

10-6A   Effects on Acidity

The acid-dissociation constants for alcohols vary according to their structure, from about 10^{-16} for methanol down to about 10^{-18} for most tertiary alcohols. The acidity decreases as the substitution on the alkyl group increases, because a more highly substituted alkyl group inhibits solvation of the alkoxide ion and drives the dissociation equilibrium to the left.

Table 10-3 shows that substitution by electron-withdrawing halogen atoms enhances the acidity of alcohols. For example, 2-chloroethanol is more acidic than ethanol because the electron-withdrawing chlorine atom helps to stabilize the 2-chloroethoxide ion.

$$CH_3-CH_2-OH \; + \; H_2O \; \rightleftharpoons \; CH_3-CH_2-O^- \; + \; H_3O^+ \qquad K_a = 1.3 \times 10^{-16}$$

ethanol   ethoxide ion
(less stable)

$$Cl-CH_2-CH_2-OH \; + \; H_2O \; \rightleftharpoons \; Cl-CH_2-CH_2-O^- \; + \; H_3O^+ \qquad K_a = 5.0 \times 10^{-15}$$

2-chloroethanol   2-chloroethoxide ion
(stabilized by Cl)

PROBLEM 10-7

Predict which member of each pair will be more acidic. Explain your answers.
(a) methanol or *t*-butyl alcohol   **(b)** 1-chloroethanol or 2-chloroethanol
(c) 2-chloroethanol or 2,2-dichloroethanol

PROBLEM 10-8

Without looking them up, rank the following compounds in decreasing order of acidity. These examples represent large classes of compounds that differ widely in acidity.

water, ethanol, 2-chloroethanol, *t*-butyl alcohol, ammonia, sulfuric acid, hexane

---

10-6B   Formation of Sodium and Potassium Alkoxides

In Chapter 11, we will see many useful reactions of alkoxide ions. When an alkoxide ion is needed in a synthesis, it is often formed by the reaction of sodium or potassium metal with the alcohol. This is an oxidation-reduction, with the metal being oxidized and the hydrogen ion reduced to form hydrogen gas, which bubbles out of the solution, leaving the sodium or potassium salt of the alkoxide ion.

$$R-O-H \; + \; Na \; \longrightarrow \; R-O^- \, {}^+Na \; + \; \tfrac{1}{2}H_2 \uparrow$$

Example

$$CH_3CH_2OH \; + \; Na \; \longrightarrow \; CH_3CH_2O^- \, {}^+Na \; + \; \tfrac{1}{2}H_2 \uparrow$$

ethanol   sodium metal   sodium ethoxide   hydrogen gas

The more acidic alcohols, like methanol and ethanol, react rapidly with sodium to form sodium methoxide and sodium ethoxide. Secondary alcohols, such as 2-propanol, react more slowly. Tertiary alcohols, such as *t*-butyl alcohol, react very slowly with sodium. Potassium is often used with secondary and tertiary alcohols because it is more reactive than sodium, and the reaction can be completed in a convenient amount of time.

$$(CH_3)_3C-OH \; + \; K \; \longrightarrow \; (CH_3)_3C-O^- \, {}^+K \; + \; \tfrac{1}{2}H_2 \uparrow$$

t-butyl alcohol   potassium   potassium *t*-butoxide
metal

Some alcohols react slowly with both sodium and potassium. In these cases, a useful alternative is sodium hydride, often in tetrahydrofuran (THF) solution. Sodium hydride reacts quickly to form the alkoxide, even with difficult compounds.

$$R—\ddot{O}—H \; + \quad NaH \quad \xrightarrow{\text{THF}} \quad R—\ddot{O}{:}^- \; {}^+Na \; + \quad H_2{\uparrow}$$

alcohol   sodium hydride   sodium alkoxide   hydrogen

10-6C   Acidity of Phenols

We might expect that phenol would have about the same acidity as cyclohexanol, since their structures are similar. This prediction is wrong: Phenol is nearly 100 million (10^8) times more acidic than cyclohexanol.

$$H_2O \; + \quad \text{(cyclohexanol, OH)} \quad \underset{}{\overset{K_a = 10^{-18}}{\rightleftharpoons}} \quad \text{(alkoxide ion, } O^-\text{)} \; + \; H_3O^+$$

cyclohexanol   alkoxide ion

$$H_2O \; + \quad \text{(phenol, OH)} \quad \underset{}{\overset{K_a = 10^{-10}}{\rightleftharpoons}} \quad \text{(phenoxide ion, } O^-\text{)} \; + \; H_3O^+$$

phenol   phenoxide ion
(phenolate ion)

The famous British surgeon Joseph Lister (1827–1912) used dilute solutions of phenol to kill germs in wounds. He found that antiseptic surgery using phenol reduced the mortality rate from amputations by nearly 50 percent. Within the next decade, Lister's phenol-based antiseptic techniques revolutionized surgery.

Cyclohexanol is a typical secondary alcohol, with a typical acid-dissociation constant. There must be something special about phenol that makes it unusually acidic. The phenoxide ion is more stable than a typical alkoxide ion because the negative charge is not confined to the oxygen atom but is delocalized over the oxygen and three carbon atoms of the ring.

A large part of the negative charge in the resonance hybrid still resides on the oxygen atom, since it is the most electronegative of the four atoms sharing the charge. But the ability to spread the negative charge over four atoms rather than concentrating it on just one atom produces a more stable ion. The reaction of phenol with sodium hydroxide is exothermic, and the following equilibrium lies to the right.

$$\text{(phenol, } \ddot{O}—H) \; + \; Na^+ \; {}^-{:}\ddot{O}H \quad \rightleftharpoons \quad \text{(sodium phenoxide, } \ddot{O}{:}^- \; {}^+Na) \; + \; H_2O$$

phenol   sodium phenoxide

EPM of phenoxide ion

Phenoxide anions are prepared simply by adding the phenol to an aqueous solution of sodium hydroxide or potassium hydroxide. There is no need to use sodium or potassium metal. Phenol was once called *carbolic acid* because of its ability to neutralize common bases.

PROBLEM 10-9

A nitro group ($—NO_2$) effectively stabilizes a negative charge on an adjacent carbon atom through resonance:

Two of the following nitrophenols are much more acidic than phenol itself. The third compound is only slightly more acidic than phenol. Use resonance structures of the appropriate phenoxide ions to show why two of these anions should be unusually stable.

2-nitrophenol   3-nitrophenol   4-nitrophenol

PROBLEM 10-10

The following compounds are only slightly soluble in water, but one of them is very soluble in a dilute aqueous solution of sodium hydroxide. The other is still only slightly soluble.

(a) Explain the difference in solubility of these compounds in dilute sodium hydroxide.
(b) Show how this difference might be exploited to separate a mixture of these two compounds using a separatory funnel.

> Phenol is commonly used to separate cellular DNA from proteins, one step in the process that leads to the cloning of a gene. Phenol is very toxic to cells, however, and it is readily absorbed through the skin, so it must be handled with care.

10-7
Synthesis of Alcohols: Introduction and Review

One of the reasons alcohols are important synthetic intermediates is that they can be synthesized directly from a wide variety of other functional groups. In Chapters 6 and 8, we examined the conversion of alkyl halides to alcohols by substitution and the conversion of alkenes to alcohols by hydration, hydroboration, and hydroxylation. These reactions are summarized below, with references for review if needed.

Following this review, we will consider the largest and most versatile group of alcohol syntheses: nucleophilic additions to carbonyl compounds.

SUMMARY   **Previous Alcohol Syntheses**

Nucleophilic Substitution on an Alkyl Halide (See Chapter 6.)
Usually via the S_N2 mechanism; competes with elimination.

transition state

Example

H   Br
 \ |
  C
 / \
CH₃   CH₂CH₃

(S)-2-bromobutane

$\xrightarrow{\text{KOH}}$

HO   H
 \ |
  C
 / \
CH₃   CH₂CH₃

(R)-2-butanol, 100% inverted configuration
(plus elimination products)

Synthesis of Alcohols from Alkenes *(see Chapter 8.)*

1. Acid-catalyzed hydration (Section 8–4)

$$C=C + H_2O \xrightarrow{H^+}$$

—C—C—
  | |
  H   OH

Markovnikov orientation

2. Oxymercuration–demercuration (Section 8–5)

$$C=C + Hg(OAc)_2 \xrightarrow{H_2O}$$

—C—C—
  | |
(AcO)Hg   OH

$\xrightarrow{\text{NaBH}_4}$

—C—C—
  | |
  H   OH

Markovnikov orientation

Example

H₃C   CH₃
   \   /
    C=C
   /   \
  H   CH₃

$\xrightarrow[\text{H}_2\text{O}]{\text{Hg(OAc)}_2}$

H₃C   OH
  | |
H—C———C—CH₃
  | |
(AcO)Hg   CH₃

$\xrightarrow{\text{NaBH}_4}$

H₃C   OH
  | |
H—C———C—CH₃
  | |
  H   CH₃

(90% overall)

3. Hydroboration–oxidation (Section 8–7)

$$C=C \xrightarrow[\text{(2) H}_2\text{O}_2,\ \text{NaOH}]{\text{(1) BH}_3 \cdot \text{THF}}$$

—C—C—
  | |
  H   OH

syn addition, anti-Markovnikov orientation

Example

1-methylcyclopentene

$\xrightarrow{\text{BH}_3 \cdot \text{THF}}$

CH₃
  ⋯H

  ⋯BH₂
H

$\xrightarrow{\text{H}_2\text{O}_2/\text{NaOH}}$

CH₃
  ◂H

  ⋯OH
H

trans-2-methylcyclopentanol
(85%)

4. Hydroxylation: synthesis of 1,2-diols from alkenes (Sections 8–13 and 8–14)

$$C=C \xrightarrow[\substack{\text{or KMnO}_4,\ ^-\text{OH} \\ \text{(cold, dilute)}}]{\text{OsO}_4,\ \text{H}_2\text{O}_2}$$

—C—C—
  | |
 HO   OH

syn hydroxylation

Example

cis-3-hexene $\xrightarrow{\text{KMnO}_4}$ → $\xrightarrow[\text{H}_2\text{O}]{^-\text{OH}}$ → *meso*-3,4-hexanediol (60%)

$\xrightarrow{\text{R}-\overset{\text{O}}{\overset{\|}{\text{C}}}-\text{OOH, H}_3\text{O}^+}$

anti hydroxylation

Example

cis-3-hexene $\xrightarrow{\text{HCO}_3\text{H}}$ → $\xrightarrow{\text{H}_3\text{O}^+}$ → (±)-3,4-hexanediol (70%)

10-8
Organometallic Reagents for Alcohol Synthesis

Organometallic compounds contain covalent bonds between carbon atoms and metal atoms. Organometallic reagents are useful because they have nucleophilic carbon atoms, in contrast to the electrophilic carbon atoms of alkyl halides. Most metals (M) are more electropositive than carbon, and the C—M bond is polarized with a partial positive charge on the metal and a partial negative charge on carbon. The following partial periodic table shows the electronegativities of some metals used in making organometallic compounds.

Electronegativities					C—M bond
Li 1.0			C 2.5		$\underset{\delta^-}{\text{C}}\overset{\leftrightarrow}{-}\underset{\delta^+}{\text{Li}}$
Na 0.9	Mg 1.3	Al 1.6			$\underset{\delta^-}{\text{C}}\overset{\leftrightarrow}{-}\underset{\delta^+}{\text{Mg}}$
K 0.8					

We have already encountered one type of organometallic compound with a negative charge on carbon: sodium acetylides, covered in Section 9-7. Terminal alkynes are weakly acidic, and they are converted to sodium acetylides by treatment with an unusually strong base, sodium amide. These sodium acetylides are useful nucleophiles, reacting with alkyl halides and carbonyl compounds to form new carbon–carbon bonds.

$$\underset{\text{terminal alkyne}}{\text{R}-\text{C}\equiv\text{C}-\text{H}} \quad + \quad \underset{\text{sodium amide}}{\text{NaNH}_2} \quad \longrightarrow \quad \underset{\text{a sodium acetylide}}{\text{R}-\text{C}\equiv\text{C}:^- \text{ Na}^+} \quad + \quad \underset{\text{ammonia}}{\text{NH}_3}$$

$$\underset{\text{acetylide}}{\text{R}-\text{C}\equiv\text{C}:^-} \quad + \quad \underset{\text{alkyl halide}}{\text{R}'-\text{CH}_2-\text{X}} \quad \longrightarrow \quad \underset{\text{substituted alkyne}}{\text{R}-\text{C}\equiv\text{C}-\text{CH}_2-\text{R}'} \quad + \quad \text{X}^-$$

$$R-C\equiv C\colon^{-} \quad + \quad \underset{R'}{\overset{R'}{>}}C=\ddot{O}\colon \longrightarrow R-C\equiv C-\underset{\underset{R'}{|}}{\overset{\overset{R'}{|}}{C}}-\ddot{O}\colon^{-} \xrightarrow{H_3O^+} R-C\equiv C-\underset{\underset{R'}{|}}{\overset{\overset{R'}{|}}{C}}-OH$$

acetylide   ketone or aldehyde   alkoxide   acetylenic alcohol

Most alkyl and alkenyl groups are not acidic enough to be deprotonated by sodium amide, but they can be made into Grignard reagents and organolithium reagents. These reagents are extremely versatile, providing some of our best ways of forming carbon–carbon bonds.

10-8A   Grignard Reagents

Organometallic compounds of lithium and magnesium are most frequently used for the synthesis of alcohols. The organomagnesium halides, of empirical formula R—Mg—X, are called **Grignard reagents** in honor of the French chemist Victor Grignard, who discovered their utility around 1905 and received the Nobel Prize in chemistry in 1912. Grignard reagents result from the reaction of an alkyl halide with magnesium metal. This reaction is always carried out in an ether solvent, which is needed to solvate and stabilize the Grignard reagent as it forms. Although we write the Grignard reagent as R—Mg—X, the actual species in solution usually contains two, three, or four of these units associated together with several molecules of the ether solvent. Diethyl ether, $CH_3CH_2-O-CH_2CH_3$, is the most common solvent for these reactions, although other ethers are also used.

$$\text{R—X} \quad + \quad \text{Mg} \quad \xrightarrow{CH_3CH_2OCH_2CH_3} \quad \overset{\delta^-}{R}-\overset{\delta^+}{Mg}-X \qquad \text{reacts like } R\colon^{-} \overset{+}{Mg}X$$

(X = Cl, Br, or I)   organomagnesium halide
                                                              (Grignard reagent)

Grignard reagents may be made from primary, secondary, and tertiary alkyl halides, as well as from vinyl and aryl halides. Alkyl iodides are the most reactive halides, followed by bromides and chlorides. Alkyl fluorides generally do not react.

reactivity   R—I > R—Br > R—Cl ≫ R—F

The following reactions show the formation of some typical Grignard reagents.

$$CH_3-I \quad + \quad Mg \quad \xrightarrow{ether} \quad CH_3-Mg-I$$
iodomethane   methylmagnesium iodide

bromocyclohexane + Mg $\xrightarrow{ether}$ cyclohexylmagnesium bromide

$$H_2C=CH-CH_2-Br \quad + \quad Mg \quad \xrightarrow{ether} \quad H_2C=CH-CH_2-MgBr$$
allyl bromide   allylmagnesium bromide

10-8B   Organolithium Reagents

Like magnesium, lithium reacts with alkyl halides, vinyl halides, and aryl halides to form organometallic compounds. Ether is not necessary for this reaction: **organolithium reagents** are made and used in a wide variety of solvents.

$$R-X + 2\,Li \longrightarrow Li^+\,{}^-X + R-Li \quad \text{reacts like } R\colon^{-} Li^+$$

(X = Cl, Br, or I)   organolithium

Examples

$$CH_3CH_2CH_2CH_2\text{—Br} + 2\ Li \xrightarrow{\text{hexane}} CH_3CH_2CH_2CH_2\text{—Li} + LiBr$$

n-butyl bromide   *n*-butyllithium

$$H_2C\!\!=\!\!CH\text{—Cl} + 2\ Li \xrightarrow{\text{pentane}} H_2C\!\!=\!\!CH\text{—Li} + LiCl$$

vinyl chloride   vinyllithium

Br + 2 Li $\xrightarrow{\text{ether}}$ Li + LiBr

bromobenzene   phenyllithium

The electrostatic potential map (EPM) of methyllithium is shown at left. The blue (electron-poor) color of the metal results from its partial positive charge, and the red (electron-rich) color of the methyl group shows its partial negative charge.

EPM of CH_3Li

PROBLEM 10-11

Which of the following compounds are suitable solvents for Grignard reactions?
(a) *n*-hexane   (b) $CH_3\text{—O—}CH_3$ (c) $CHCl_3$
(d) cyclohexane   (e) benzene   (f) $CH_3OCH_2CH_2OCH_3$

(g)   (h)

THF
(tetrahydrofuran)   1,4-dioxane

PROBLEM 10-12

Predict the products of the following reactions.

(a) $CH_3CH_2Br + Mg \xrightarrow{\text{ether}}$

(b) isobutyl iodide $+\ Li \xrightarrow{\text{hexane}}$

(c) 1-bromo-4-fluorocyclohexane $+\ Mg \xrightarrow{\text{THF}}$

(d) $CH_2\!\!=\!\!CCl\text{—}CH_2CH_3 + Li \xrightarrow{\text{ether}}$

10-9
Addition of Organometallic Reagents to Carbonyl Compounds

Because they resemble carbanions, Grignard and organolithium reagents are strong nucleophiles and strong bases. Their most useful nucleophilic reactions are additions to carbonyl (C=O) groups, much like we saw with acetylide ions (Section 9-7B). The carbonyl group is polarized, with a partial positive charge on carbon and a partial negative charge on oxygen. The positively charged carbon is electrophilic; attack by a nucleophile places a negative charge on the electronegative oxygen atom.

$R\!:^{-}$ ⤻ $\overset{\delta^+}{\underset{}{C}}\!\!=\!\!\overset{\delta^-}{\underset{}{O}}\!:\ \longrightarrow\ R\text{—}\overset{|}{\underset{|}{C}}\text{—}\ddot{O}\!:^{-}$

The product of this nucleophilic attack is an alkoxide ion, a strong base. Addition of water or a dilute acid protonates the alkoxide to give the alcohol.

$R\text{—}\overset{|}{\underset{|}{C}}\text{—}\ddot{O}\!:^{-}\ \ \overset{\curvearrowleft}{H}\text{—}\ddot{O}\text{—}H\ \longrightarrow\ R\text{—}\overset{|}{\underset{|}{C}}\text{—}\ddot{O}H\ +\ ^{-}\!:\ddot{O}H$

alkoxide   (or H_3O^+)

Either a Grignard reagent or an organolithium reagent can serve as the nucleophile in this addition to a carbonyl group. The following discussions refer to Grignard reagents, but they also apply to organolithium reagents. The Grignard reagent adds to the carbonyl group to form an alkoxide ion. Addition of dilute acid (in a separate step) protonates the alkoxide to give the alcohol.

We are interested primarily in the reactions of Grignard reagents with ketones and aldehydes. **Ketones** are compounds with two alkyl groups bonded to a carbonyl group. **Aldehydes** have one alkyl group and one hydrogen atom bonded to the carbonyl group. **Formaldehyde** has two hydrogen atoms bonded to the carbonyl group. The electrostatic potential map (EPM) of formaldehyde shows the polarization of the carbonyl group, with an electron-rich (red) region around oxygen and an electron-poor (blue) region near carbon.

a ketone   an aldehyde   formaldehyde

EPM of formaldehyde

KEY MECHANISM   Grignard Reactions

First reaction: The Grignard reagent attacks a carbonyl compound to form an alkoxide salt.

magnesium alkoxide salt

Second reaction: After the first reaction is complete, water or dilute acid is added to protonate the alkoxide.

magnesium alkoxide salt   alcohol

10-9A   Addition to Formaldehyde: Formation of Primary Alcohols

Grignard reagent   formaldehyde   primary alcohol

Addition of a Grignard reagent to formaldehyde, followed by protonation, gives a primary alcohol with one more carbon atom than in the Grignard reagent.

$$CH_3CH_2CH_2CH_2{-}MgBr \ + \ \overset{H}{\underset{H}{}}C{=}O \quad \xrightarrow[\text{(2) } H_3O^+]{\text{(1) ether solvent}} \quad CH_3CH_2CH_2CH_2{-}\overset{H}{\underset{H}{\overset{|}{\underset{|}{C}}}}{-}OH$$

butylmagnesium bromide   formaldehyde   1-pentanol (92%)

PROBLEM-SOLVING HINT

Note the use of $\xrightarrow[(2)]{(1)}$ to show separate reactions with one reaction arrow.

PROBLEM 10-13

Show how you would synthesize the following alcohols by adding an appropriate Grignard reagent to formaldehyde.

(a) [cyclohexyl-CH$_2$OH structure]

(b) [3-methylbutan-1-ol structure with OH]

(c) [cyclopentenyl-CH$_2$OH structure]

10-9B   Addition to Aldehydes: Formation of Secondary Alcohols

☞ $\boxed{R{-}MgX}$ + $\underset{H}{\overset{R'}{C}}{=}O$ $\xrightarrow{\text{ether}}$ $\boxed{R}{-}\underset{H}{\overset{R'}{C}}{-}O^-\ ^+MgX$

Grignard reagent   aldehyde

$\boxed{R}{-}\underset{H}{\overset{R'}{C}}{-}O^-\ ^+MgX$ $\xrightarrow{H_3O^+}$ $\boxed{R}{-}\underset{H}{\overset{R'}{C}}{-}OH$

secondary alcohol

Grignard reagents add to aldehydes to give, after protonation, secondary alcohols. The two alkyl groups of the secondary alcohol are the alkyl group from the Grignard reagent and the alkyl group that was bonded to the carbonyl group of the aldehyde.

$CH_3CH_2{-}MgBr$ + $\underset{H}{\overset{H_3C}{C}}{=}O$ $\xrightarrow{\text{ether}}$ $CH_3{-}CH_2{-}\underset{H}{\overset{CH_3}{C}}{-}O^-\ ^+MgBr$

acetaldehyde

$CH_3{-}CH_2{-}\underset{H}{\overset{CH_3}{C}}{-}O^-\ ^+MgBr$ $\xrightarrow{H_3O^+}$ $CH_3CH_2{-}\underset{H}{\overset{CH_3}{C}}{-}OH$

2-butanol
(85%)

PROBLEM-SOLVING HINT

A secondary alcohol has two groups on the carbinol carbon atom. Consider two possible reactions, with either group as the Grignard reagent.

PROBLEM 10-14

Show two ways you could synthesize each of the following alcohols by adding an appropriate Grignard reagent to an aldehyde.

(a) [4-methylpentan-2-ol structure with OH]

(b) [1-phenylpropan-1-ol structure with OH]

(c) [dicyclohexyl carbinol structure: H, C, OH with two cyclohexyl groups]

10-9C   Addition to Ketones: Formation of Tertiary Alcohols

☞ $\boxed{R}{-}MgX$ + $\overset{R'}{\underset{R''}{>}}C{=}O$ $\xrightarrow{\text{ether}}$ $\boxed{R}{-}\overset{R'}{\underset{R''}{\overset{|}{\underset{|}{C}}}}{-}O^-\ {}^+MgX$ $\xrightarrow{H_3O^+}$ $\boxed{R}{-}\overset{R'}{\underset{R''}{\overset{|}{\underset{|}{C}}}}{-}OH$

Grignard reagent   ketone   tertiary alcohol

A ketone has two alkyl groups bonded to its carbonyl carbon atom. Addition of a Grignard reagent, followed by protonation, gives a tertiary alcohol, with three alkyl groups bonded to the carbinol carbon atom. Two of the alkyl groups are the two originally bonded to the ketone carbonyl group. The third alkyl group comes from the Grignard reagent.

$CH_3CH_2{-}MgBr$ + $\overset{CH_3CH_2CH_2}{\underset{H_3C}{>}}C{=}O$ $\xrightarrow[\text{(2) } H_3O^+]{\text{(1) ether solvent}}$ $CH_3CH_2{-}\overset{CH_3CH_2CH_2}{\underset{CH_3}{\overset{|}{\underset{|}{C}}}}{-}OH$

                                    2-pentanone   3-methyl-3-hexanol
                                                                                                        (90%)

SOLVED PROBLEM 10-2

Show how you would synthesize the following alcohol from compounds containing no more than five carbon atoms.

SOLUTION

This is a tertiary alcohol; any one of the three alkyl groups might be added in the form of a Grignard reagent. We can propose three combinations of Grignard reagents with ketones:

Any of these three syntheses would probably work, but only the third begins with fragments containing no more than five carbon atoms. The other two syntheses would require further steps to generate the ketones from compounds containing no more than five carbon atoms.

PROBLEM-SOLVING HINT

A tertiary alcohol has three groups on the carbinol carbon atom. Consider three possible reactions (as in Solved Problem 10-2), with each of these groups as the Grignard reagent.

PROBLEM 10-15

Show how you would synthesize each alcohol by adding an appropriate Grignard reagent to a ketone.

(a) 3-phenylhexan-3-ol (3 ways)   (b) Ph_3COH
(c) 1-methylcyclohexanol   (d) 1,1-dicyclohexyl-1-butanol

10-9D Addition to Acid Chlorides and Esters

Acid chlorides and **esters** are derivatives of carboxylic acids. In such **acid derivatives**, the —OH group of a carboxylic acid is replaced by other electron-withdrawing groups. In acid chlorides, the hydroxyl group of the acid is replaced by a chlorine atom. In esters, the hydroxyl group is replaced by an alkoxyl (—O—R) group.

$$R-\overset{\overset{\textstyle O}{\|}}{C}-OH \qquad R-\overset{\overset{\textstyle O}{\|}}{C}-Cl \qquad R-\overset{\overset{\textstyle O}{\|}}{C}-O-R'$$

carboxylic acid   acid chloride   ester

Acid chlorides and esters react with two equivalents of Grignard reagents to give (after protonation) tertiary alcohols.

☞ $2 \boxed{R}-MgX + R'-\overset{\overset{\textstyle O}{\|}}{C}-Cl \xrightarrow[\text{(2) } H_3O^+]{\text{(1) ether solvent}} R'-\overset{\boxed{R}}{\underset{\boxed{R}}{C}}-OH$

acid chloride   tertiary alcohol

☞ $2 \boxed{R}-MgX + R'-\overset{\overset{\textstyle O}{\|}}{C}-OR'' \xrightarrow[\text{(2) } H_3O^+]{\text{(1) ether solvent}} R'-\overset{\boxed{R}}{\underset{\boxed{R}}{C}}-OH$

ester   tertiary alcohol

Addition of the first equivalent of the Grignard reagent produces an unstable intermediate that expels a chloride ion (in the acid chloride) or an alkoxide ion (in the ester), giving a ketone. The alkoxide ion is a suitable leaving group in this reaction because its leaving stabilizes a negatively charged intermediate in a strongly exothermic step.

Attack on an acid chloride

$\boxed{R}-MgX \quad \overset{R'}{\underset{Cl}{C}}=\ddot{O} \longrightarrow R-\overset{R'}{\underset{:\ddot{C}l:}{\boxed{C}}}-\ddot{O}:^- \longrightarrow \overset{R'}{\underset{\boxed{R}}{C}}=\ddot{O} \quad :\ddot{C}l:^-$

acid chloride   intermediate   ketone

Attack on an ester

$\boxed{R}-MgX \quad \overset{R'}{\underset{R''-\ddot{O}:}{C}}=\ddot{O}: \longrightarrow R-\overset{R'}{\underset{:\ddot{O}-R''}{\boxed{C}}}-\ddot{O}:^- \longrightarrow \overset{R'}{\underset{\boxed{R}}{C}}=O \quad ^-:\ddot{O}-R''$

ester   intermediate   ketone

The ketone reacts with a second equivalent of the Grignard reagent, forming the magnesium salt of a tertiary alkoxide. Protonation gives a tertiary alcohol with one of its alkyl groups derived from the acid chloride or ester, and the other two derived from the Grignard reagent.

Grignard   ketone   alkoxide   tertiary alcohol
(second equivalent)   intermediate

Consider an example using an ester. When an excess of ethylmagnesium bromide is added to methyl benzoate, the first equivalent adds and methoxide is expelled, giving propiophenone. Addition of a second equivalent, followed by protonation, gives a tertiary alcohol: 3-phenyl-3-pentanol.

first equivalent   methyl benzoate

propiophenone

second equivalent

propiophenone

3-phenyl-3-pentanol (82%)

PROBLEM-SOLVING HINT

When making a tertiary alcohol with two identical alkyl groups, consider using an acid chloride or ester.

PROBLEM 10-16

Propose a mechanism for the reaction of acetyl chloride with phenylmagnesium bromide to give 1,1-diphenylethanol.

acetyl chloride   phenylmagnesium bromide   1,1-diphenylethanol

PROBLEM 10-17

Show how you would add Grignard reagents to acid chlorides and esters to synthesize the following alcohols.

(a) $Ph_3C—OH$ (b) 3-ethyl-2-methyl-3-pentanol (c) dicyclohexylphenylmethanol

PROBLEM 10-18

A formate ester, such as ethyl formate, reacts with an excess of a Grignard reagent to give (after protonation) secondary alcohols with two identical alkyl groups.

(a) Propose a mechanism to show how the reaction of ethyl formate with an excess of allylmagnesium bromide gives, after protonation, 1,6-heptadien-4-ol.

(b) Show how you would use reactions of Grignard reagents with ethyl formate to synthesize the following secondary alcohols.

(i) 3-pentanol (ii) diphenylmethanol (iii) *trans, trans*-2,7-nonadien-5-ol

PROBLEM-SOLVING HINT

When making a secondary alcohol with identical alkyl groups, consider using a formate ester.

10-9E Addition to Ethylene Oxide

Grignard reagents usually do not react with ethers, but **epoxides** are unusually reactive ethers because of their ring strain. Ethylene oxide reacts with Grignard reagents to give, after protonation, primary alcohols with *two* additional carbon atoms. Notice that the nucleophilic attack by the Grignard reagent opens the ring and relieves the ring strain.

Example

PROBLEM 10-19

Show how you would synthesize the following alcohols by adding Grignard reagents to ethylene oxide.

(a) 2-phenylethanol (b) 4-methyl-1-pentanol (c)

PROBLEM 10-20

In Section 9-7B, we saw how acetylide ions add to carbonyl groups in much the same way that Grignard and organolithium reagents add. Acetylide ions also add to ethylene oxide much like Grignard and organolithium reagents. Predict the products obtained by adding the following acetylide ions to ethylene oxide, followed by a dilute acid workup.

(a) $HC\equiv C:^-$ (b) $CH_3CH_2-C\equiv C:^-$

SUMMARY   **Grignard Reactions**

$$R-MgX + \begin{matrix} H \\ \backslash \\ C=O \\ / \\ H \end{matrix} \quad \xrightarrow[\text{(2) } H_3O^+]{\text{(1) ether solvent}} \quad \begin{matrix} H \\ | \\ R-C-OH \\ | \\ H \end{matrix}$$

formaldehyde   1° alcohol

$$R-MgX + \begin{matrix} R' \\ \backslash \\ C=O \\ / \\ H \end{matrix} \quad \xrightarrow[\text{(2) } H_3O^+]{\text{(1) ether solvent}} \quad \begin{matrix} R' \\ | \\ R-C-OH \\ | \\ H \end{matrix}$$

aldehyde   2° alcohol

$$R-MgX + \begin{matrix} R' \\ \backslash \\ C=O \\ / \\ R'' \end{matrix} \quad \xrightarrow[\text{(2) } H_3O^+]{\text{(1) ether solvent}} \quad \begin{matrix} R' \\ | \\ R-C-OH \\ | \\ R'' \end{matrix}$$

ketone   3° alcohol

$$2\ R-MgX + \begin{matrix} O \\ \| \\ R'-C-OR'' \end{matrix} \quad \xrightarrow[\text{(2) } H_3O^+]{\text{(1) ether solvent}} \quad \begin{matrix} R \\ | \\ R'-C-OH \\ | \\ R \end{matrix}$$

ester
or acid chloride

3° alcohol
two groups added

$$R-MgX + \begin{matrix} O \\ / \backslash \\ CH_2-CH_2 \end{matrix} \quad \xrightarrow[\text{(2) } H_3O^+]{\text{(1) ether solvent}} \quad R-CH_2CH_2-OH$$

ethylene oxide   1° alcohol
two carbons added

PROBLEM 10-21

Recall from Chapter 9 how acetylide ions are alkylated by displacing unhindered alkyl halides.

$$H-C\equiv C:^- \quad R-CH_2-Br \quad \longrightarrow \quad H-C\equiv C-CH_2-R + Br^-$$

Like acetylide ions, Grignard and organolithium reagents are strong bases and strong nucleophiles. Luckily, however, they do not displace halides as easily as acetylide ions do. If they did displace alkyl halides, it would be impossible to form the reagents from alkyl halides because whenever a molecule of reagent formed, it would react with a molecule of starting material. All that would be formed is a coupling product. In fact, coupling is a side reaction that hurts the yield of many Grignard reactions.

$$R-Br + Mg \quad \xrightarrow{\text{ether}} \quad R-Mg-Br \quad \xrightarrow{R-Br} \quad R-R + MgBr_2$$

If we *want* to couple two groups together, we can do it by using an organocopper reagent, a **lithium dialkylcuprate**, to couple with an alkyl halide.

☞ $R_2CuLi + R'—X \longrightarrow R—R' + R—Cu + LiX$

a lithium dialkylcuprate

The lithium dialkylcuprate (also called a *Gilman reagent*) is formed by the reaction of two equivalents of the corresponding organolithium reagent (Section 10-8B) with cuprous iodide:

$$2\,R—Li + CuI \longrightarrow R_2CuLi + LiI$$

The coupling takes place as if a carbanion ($R:^-$) were present and the carbanion attacked the alkyl halide to displace the halide ion. This is not necessarily the actual mechanism, however.

$$R—Cu^{-\,+}Li \quad \overset{|}{\underset{|}{-C}}—X \longrightarrow R—\overset{|}{\underset{|}{C}}— \quad ^-:X$$
$$\underset{R}{|}$$

(hypothetical mechanism)

Example

$$CH_3CH_2\overset{|}{\underset{CH_3}{CH}}—Cl \xrightarrow[\text{(2) CuI}]{\text{(1) Li}} \left(CH_3CH_2\overset{|}{\underset{CH_3}{CH}}—\right)_2 CuLi$$

2-chlorobutane   lithium dialkylcuprate

$$\left(CH_3CH_2\overset{|}{\underset{CH_3}{CH}}—\right)_2 CuLi \xrightarrow{CH_3CH_2CH_2CH_2—Br} CH_3CH_2\overset{|}{\underset{CH_3}{CH}}—CH_2CH_2CH_2CH_3$$

lithium dialkylcuprate

3-methylheptane
(70%)

Show how you would synthesize the following compounds from alkyl halides, vinyl halides, and aryl halides containing no more than six carbon atoms.
(a) *n*-octane   (b) 3-methylheptane
(c) *n*-butylcyclohexane   (d) *trans*-3-octene

10-10
Side Reactions of Organometallic Reagents: Reduction of Alkyl Halides

Grignard and organolithium reagents are strong nucleophiles and strong bases. Besides their additions to carbonyl compounds, they react with other acidic or electrophilic compounds. In some cases these are useful reactions, but they are often seen as annoying side reactions where a small impurity of water or an alcohol destroys the reagent.

10-10A   Reactions with Acidic Compounds

Grignard and organolithium reagents react vigorously and irreversibly with water. Therefore, all reagents and solvents used in these reactions must be dry.

$$\overset{\delta^-}{R}—\overset{\delta^+}{MgX} + H—O—H \longrightarrow R—H + XMgOH$$

For example, consider the reaction of ethyllithium with water:

$$\overset{\delta^-}{CH_3}—\overset{\delta^+}{CH_2}—Li + H—O—H \longrightarrow CH_3—CH_2—H + Li^+\ {}^-OH$$

ethyllithium   ethane

The products are strongly favored in this reaction. Ethane is a *very* weak acid (K_a of about 10^{-50}), so the reverse reaction (abstraction of a proton from ethane by lithium hydroxide) is unlikely. When ethyllithium is added to water, ethane instantly bubbles to the surface.

Why would we ever want to add an organometallic reagent to water? This is a method for reducing an alkyl halide to an alkane:

$$R\!-\!X + Mg \xrightarrow[\text{ether}]{} R\!-\!MgX \xrightarrow{H_2O} R\!-\!H + XMgOH$$

$$R\!-\!X + 2\,Li \longrightarrow R\!-\!Li + LiX \xrightarrow{H_2O} R\!-\!H + LiOH$$

The overall reaction is a *reduction* because it replaces the electronegative halogen atom with a hydrogen atom. In particular, this reaction provides a way to "label" a compound with deuterium (^2H, a heavy isotope of hydrogen) at any position where a halogen is present.

In addition to O—H groups, the protons of N—H and S—H groups and the hydrogen atom of a terminal alkyne, $-C\!\equiv\!C\!-\!H$, are sufficiently acidic to protonate Grignard and organolithium reagents. Unless we want to protonate the reagent, compounds with these groups are considered incompatible with Grignard and organolithium reagents.

PROBLEM 10-22

Predict the products of the following reactions.
(a) methylmagnesium iodide + D_2O $\longrightarrow$
(b) *n*-butyllithium + CH_3CH_2OH $\longrightarrow$
(c) isobutylmagnesium bromide + 1-butyne $\longrightarrow$

10-10B   Reactions with Electrophilic Multiple Bonds

Grignard reagents are useful because they add to the electrophilic double bonds of carbonyl groups. However, we must make sure that the *only* electrophilic double bond in the solution is the one we want the reagent to attack. There must not be any electrophilic double (or triple) bonds in the solvent or in the Grignard reagent itself, or they will be attacked as well. Any multiple bond involving a strongly electronegative element is likely to be attacked, including $C\!=\!O$, $S\!=\!O$, $C\!=\!N$, $N\!=\!O$, and $C\!\equiv\!N$ bonds.

In later chapters, we will encounter methods for *protecting* susceptible groups to prevent the reagent from attacking them. For now, simply remember that the following groups react with Grignard and organolithium reagents; avoid compounds containing these groups except for the one carbonyl group that gives the desired reaction.

Protonate the Grignard or organolithium: O—H, N—H, S—H, $-C\!\equiv\!C\!-\!H$

Attacked by the Grignard or organolithium: $C\!=\!O$, $C\!=\!N$, $C\!\equiv\!N$, $S\!=\!O$, $N\!=\!O$

PROBLEM-SOLVING $\mathbb{HINT}$

Grignard reagents are incompatible with water or acid. Dilute acid is used *in a separate step* to hydrolyze the magnesium alkoxide.

correct: $\xrightarrow{\text{(1)RMgX}}_{\text{(2)H}_3\text{O}^+}$

correct: $\xrightarrow[\text{ether}]{\text{RMgX}} \xrightarrow{\text{H}_3\text{O}^+}$

incorrect: $\xrightarrow[\text{H}_3\text{O}^+]{\text{RMgX}}$

(The incorrect example means a Grignard in aqueous acid.)

PROBLEM 10-23

Point out the flaws in the following incorrect Grignard syntheses.

(a)

$$\text{(1) Mg, ether}$$
$$\text{(2) Ph—CHO}$$
$$\text{(3) H}_3\text{O}^+$$

(b)

$$\text{(1) Mg, ether}$$
$$\text{(2) H}_2\text{C}\text{=}\text{O}$$
$$\text{(3) H}_3\text{O}^+$$

(c)

$$\text{CH}_3\text{CCH}_3, \text{H}_3\text{O}^+$$

(d)

$$\text{(1) CH}_3\text{CH}_2\text{MgBr}$$
$$\text{(2) H}_3\text{O}^+$$

10-11
Reduction of the Carbonyl Group: Synthesis of 1° and 2° Alcohols

Grignard reagents convert carbonyl compounds to alcohols by adding alkyl groups. **Hydride reagents** add a hydride ion ($\text{H}{:}^-$), reducing the carbonyl group to an alkoxide ion with no additional carbon atoms. Subsequent protonation gives the alcohol. Converting a ketone or an aldehyde to an alcohol involves adding two hydrogen atoms across the $\text{C}\text{=}\text{O}$ bond: a reduction.

MECHANISM   Hydride Reduction of a Carbonyl Group

Nucleophilic attack by hydride ion forms an alkoxide ion.

hydride ion   alkoxide ion

The two most useful hydride reagents, sodium borohydride (NaBH_4) and lithium aluminum hydride (LiAlH_4), reduce carbonyl groups in excellent yields. These reagents are called *complex hydrides* because they do not have a simple hydride structure such as $\text{Na}^{+}\text{ }^-\text{H}$ or $\text{Li}^{+}\text{ }^-\text{H}$. Instead, their hydrogen atoms, bearing partial negative charges, are covalently bonded to boron and aluminum atoms. This arrangement makes the hydride a better nucleophile while reducing its basicity.

$$\text{Na}^+ \quad \text{H}\text{—}\overset{\displaystyle H}{\underset{\displaystyle H}{\text{B}}}\text{—H} \qquad\qquad \text{Li}^+ \quad \text{H}\text{—}\overset{\displaystyle H}{\underset{\displaystyle H}{\text{Al}}}\text{—H}$$

sodium borohydride   lithium aluminum hydride

Of these reducing agents, lithium aluminum hydride (LAH) is much stronger and more difficult to work with. LAH reacts explosively with water and alcohols, liberating hydrogen gas and sometimes starting fires. Sodium borohydride reacts slowly with alcohols and with water as long as the pH is high (basic). Sodium borohydride is a convenient and highly selective reducing agent.

10-11A   Uses of Sodium Borohydride

Sodium borohydride ($NaBH_4$) reduces aldehydes to primary alcohols, and ketones to secondary alcohols. The reactions take place in a wide variety of solvents, including alcohols, ethers, and water. The yields are generally excellent.

cyclohexane carbaldehyde

cyclohexyl methanol
(95%)

2-butanone

2-butanol (100%)

Sodium borohydride is selective; it usually does not react with carbonyl groups that are less reactive than ketones and aldehydes. For example, carboxylic acids and esters are unreactive toward borohydride reduction. Thus, sodium borohydride can reduce a ketone or an aldehyde in the presence of an acid or an ester.

10-11B   Uses of Lithium Aluminum Hydride

Lithium aluminum hydride ($LiAlH_4$, abbreviated LAH) is a much stronger reagent than sodium borohydride. It easily reduces ketones and aldehydes and also the less reactive carbonyl groups: in acids, esters, and other acid derivatives (see Chapter 21). LAH reduces ketones to secondary alcohols, and it reduces aldehydes, acids, and esters to primary alcohols. The lithium salt of the alkoxide ion is initially formed, then the (cautious!) addition of dilute acid protonates the alkoxide. For example, LAH reduces both functional groups of the keto ester in the previous example.

In summary, sodium borohydride is the best reagent for reduction of a simple ketone or aldehyde. Using $NaBH_4$, we can reduce a ketone or an aldehyde in the presence of an acid or an ester, but we do not have a method (so far) for reducing an acid or an ester in the presence of a ketone or an aldehyde. The sluggish acid or ester requires the use of $LiAlH_4$, and this reagent also reduces the ketone or aldehyde.

PROBLEM-SOLVING HINT

LAH and water are incompatible. Water is added in a *separate* hydrolysis step. An explosion and fire would result from the process indicated by $\xrightarrow{LiAlH_4}{H_3O^+}$.

ease of reduction

SUMMARY **SUMMARY** Reactions of LiAlH$_4$ and NaBH$_4$

		$NaBH_4$	$LiAlH_4$
aldehyde	$R-\overset{O}{\underset{\|\|}{C}}-H$	$R-CH_2-OH$	$R-CH_2-OH$
ketone	$R-\overset{O}{\underset{\|\|}{C}}-R'$	$\overset{OH}{\underset{\|}{R-CH-R'}}$	$\overset{OH}{\underset{\|}{R-CH-R'}}$
alkene	$\underset{}{\overset{}{C=C}}$	no reaction	no reaction
acid anion	$R-\overset{O}{\underset{\|\|}{C}}-O^-$ anion in base	no reaction	$R-CH_2-OH$
ester	$R-\overset{O}{\underset{\|\|}{C}}-OR'$	no reaction	$R-CH_2-OH$

Note: The products shown are the final products, after hydrolysis of the alkoxide.

PROBLEM-SOLVING HINT

When making a primary or secondary alcohol, you can consider adding an alkyl group last (as a Grignard reagent) or adding a hydrogen last (by reducing a ketone or aldehyde).

PROBLEM 10-24

Predict the products you would expect from the reaction of NaBH$_4$ with the following compounds.

(a) $CH_3-(CH_2)_8-CHO$ (b) $CH_3CH_2-\overset{O}{\underset{\|\|}{C}}-OCH_3$ (c) $Ph-COOH$

(d) [cyclohexanone structure]

(e) [cyclohexane ring with CHO and COOCH$_3$ substituents and a ketone]

(f) [lactone ring with aldehyde substituent]

PROBLEM 10-25

Repeat Problem 10-24 using LiAlH$_4$ (followed by hydrolysis) as the reagent.

PROBLEM 10-26

Show how you would synthesize the following alcohols by reducing appropriate carbonyl compounds.

(a) 1-heptanol   (b) 2-heptanol   (c) 2-methyl-3-hexanol   (d) [bicyclic lactone structure with OH]

10-11C   Catalytic Hydrogenation of Ketones and Aldehydes

Reducing a ketone or an aldehyde to an alcohol involves adding two hydrogen atoms across the C=O bond. This addition can be accomplished by catalytic hydrogenation, commonly using Raney nickel as the catalyst.

$$-\overset{O}{\underset{\|\|}{C}}- + H_2 \xrightarrow{\text{Raney Ni}} \overset{OH}{\underset{\|}{-CH-}}$$

Raney nickel is a finely divided hydrogen-bearing form of nickel made by treating a nickel–aluminum alloy with a strong sodium hydroxide solution. The aluminum in the alloy reacts to form hydrogen, leaving behind a finely divided nickel powder saturated with hydrogen. Raney nickel is an effective catalyst for the hydrogenation of ketones and aldehydes to alcohols. Carbon–carbon double bonds are also reduced under these conditions, however, so any alkene double bonds in the starting material will also be reduced. In most cases, sodium borohydride is more convenient for reducing simple ketones and aldehydes.

2,2-dimethyl-4-pentenal

NaBH₄
(for comparison)

2,2-dimethyl-1-pentanol (94%)

2,2-dimethylpent-4-en-1-ol

SUMMARY   Alcohol Syntheses

I. FROM ALKENES

1. *Hydration* (Sections 8-4 through 8-7)
   a. *Acid catalyzed: forms Markovnikov alcohols;*
   b. *Oxymercuration–demercuration: forms Markovnikov alcohols*

   c. *Hydroboration–oxidation: forms anti-Markovnikov alcohols*

2. *Hydroxylation: forms vicinal diols (glycols)* (Sections 8-13 and 8-14)
   a. *Syn hydroxylation, using KMnO₄/NaOH or using OsO₄/H₂O₂*

cyclopentene

cis-cyclopentane-1,2-diol

   b. *Anti hydroxylation, using peracids*

cyclopentene

trans-cyclopentane-1,2-diol

II. *FROM ALKYL HALIDES: NUCLEOPHILIC SUBSTITUTION* (SECTIONS 6-9 AND 6-13)

1. *Second-order substitution: primary (and some secondary) halides*

$$(CH_3)_2CHCH_2CH_2\text{—}Br \xrightarrow[H_2O]{KOH} (CH_3)_2CHCH_2CH_2\text{—}OH$$

2. *First-order substitution: tertiary (and some secondary) halides*

t-butyl chloride   *t*-butyl alcohol   isobutylene

III. *FROM CARBONYL COMPOUNDS: NUCLEOPHILIC ADDITION TO THE CARBONYL GROUP*

1. *Addition of a Grignard or organolithium reagent* (Section 10-9)

a. *Addition to formaldehyde gives a primary alcohol*

$$CH_3CH_2MgBr \quad + \quad H_2C{=}O \xrightarrow[\text{(2) } H_3O^+]{\text{(1) ether solvent}} CH_3CH_2\text{—}CH_2\text{—}OH$$

ethylmagnesium bromide   1-propanol

b. *Addition to an aldehyde gives a secondary alcohol*

phenylmagnesium bromide   acetaldehyde   1-phenylethanol

c. *Addition to a ketone gives a tertiary alcohol*

cyclohexanone   1-ethylcyclohexanol

d. *Addition to an acid halide or an ester gives a tertiary alcohol*

acetyl chloride

or

methyl acetate

1,1-dicyclohexylethanol

e. *Addition to ethylene oxide gives a primary alcohol (with two carbon atoms added)*

cyclohexylmagnesium bromide → 2-cyclohexylethanol

2. *Reduction of carbonyl compounds* (Section 10-11)
   a. *Catalytic hydrogenation of aldehydes and ketones*

$$-\overset{\overset{\displaystyle O}{\|}}{C}- \ + \ H_2 \ \xrightarrow{\text{Raney Ni}} \ -\overset{\overset{\displaystyle OH}{|}}{C}H-$$

This method is usually not as selective or as effective as the use of hydride reagents.

   b. *Use of hydride reagents*
      (1) Reduction of an aldehyde gives a primary alcohol

benzaldehyde → benzyl alcohol

      (2) Reduction of a ketone gives a secondary alcohol

cyclohexanone → cyclohexanol

      (3) Reduction of an acid or ester gives a primary alcohol

$$CH_3-(CH_2)_8-\overset{\overset{\displaystyle O}{\|}}{C}-OH$$
decanoic acid

$$CH_3-(CH_2)_8-\overset{\overset{\displaystyle O}{\|}}{C}-OCH_3$$
methyl decanoate

$$\xrightarrow[\text{(2) } H_3O^+]{\text{(1) } LiAlH_4}$$

$$CH_3-(CH_2)_8-CH_2-OH$$
1-decanol

IV. *SYNTHESIS OF PHENOLS* (CHAPTER 17)

Thiols are sulfur analogues of alcohols, with an —SH group in place of the alcohol —OH group. Oxygen and sulfur are in the same column of the periodic table (group VIA), with oxygen in the second row and sulfur in the third. IUPAC names for thiols are derived from the alkane names, using the suffix -*thiol*. Thiols are also called **mercaptans** ("captures mercury") because they form stable heavy-metal derivatives. Common names are formed like those of alcohols, using the name of the alkyl group with the word *mercaptan*. The —SH group itself is called a *mercapto* group.

10-12
Thiols (Mercaptans)

CH_3—SH	$CH_3CH_2CH_2CH_2$—SH	CH_3CH=$CHCH_2$—SH	HS—CH_2CH_2—OH
IUPAC name:	methanethiol	1-butanethiol	2-butene-1-thiol	2-mercaptoethanol
common name:	methyl mercaptan	*n*-butyl mercaptan		

Thiols' ability to complex heavy metals has proved useful for making antidotes to heavy-metal poisoning. For example, in World War II the Allies were concerned that the Germans would use lewisite, a volatile arsenic compound, as a chemical warfare agent. Thiols complex strongly with arsenic, and British scientists developed dimercaprol (2,3-dimercapto-1-propanol) as an effective antidote. The Allies came to refer to this compound as "British anti-lewisite" (BAL), a name that is still used. Dimercaprol is useful against a variety of heavy metals, including arsenic, mercury, and gold.

$$\underset{\substack{\text{lewisite}}}{\underset{H}{\overset{Cl}{\diagdown}}C=C\underset{AsCl_2}{\overset{H}{\diagup}}} \qquad \underset{\substack{\text{dimercaprol}\\\text{British anti-lewisite (BAL)}}}{\overset{CH_2-CH-CH_2}{\underset{SH\quad SH\quad OH}{|\quad\ |\quad\ |}}}$$

The odor of thiols is their strongest characteristic. **Skunk** scent is composed mainly of 3-methyl-1-butanethiol and 2-butene-1-thiol, with small amounts of other thiols, Ethanethiol is added to natural gas (odorless methane) to give it the characteristic "gassy" odor for detecting leaks.

Although oxygen is more electronegative than sulfur, thiols are more acidic than alcohols. Their enhanced acidity results from two effects: First, S—H bonds are generally weaker than O—H bonds, making S—H bonds easier to break. Second, the **thiolate ion** ($R-S^-$) has its negative charge on sulfur, which allows the charge to be delocalized over a larger region than the negative charge of an alkoxide ion, borne on a smaller oxygen atom. Thiolate ions are easily formed by treating the thiol with aqueous sodium hydroxide.

Offensive use of thiols. Skunks spray thiols to protect themselves from people, dogs, and other animals.

$$\underset{\substack{\text{ethanethiol}\\ pK_a = 10.5}}{CH_3-CH_2-SH} \ + \ ^-OH \ \rightleftharpoons \ \underset{\substack{\text{ethanethiolate}}}{CH_3-CH_2-S^-} \ + \ \underset{pK_a = 15.7}{H_2O}$$

$$\underset{\substack{\text{benzenethiol (thiophenol)}\\ pK_a = 7.8}}{\bigcirc\!\!-SH} \ + \ ^-OH \ \rightleftharpoons \ \underset{\substack{\text{benzenethiolate}}}{\bigcirc\!\!-S^-} \ + \ \underset{pK_a = 15.7}{H_2O}$$

For comparison

$$\underset{\substack{\text{ethanol}\\ pK_a = 15.9}}{CH_3-CH_2-OH} \ + \ ^-OH \ \rightleftharpoons \ \underset{\substack{\text{ethoxide}}}{CH_3-CH_2-O^-} \ + \ \underset{pK_a = 15.7}{H_2O}$$

Garlic has served throughout history as a remedy for numerous diseases. Movies have often depicted its power to repel werewolves and vampires. The characteristic garlic odor derives from the many thiols it contains. Scientists have isolated some of these thiols to evaluate their biological activities.

PROBLEM 10-27

Arrange the following compounds in order of decreasing acidity.

$$CH_3COOH \quad CH_3OH \quad CH_3CH_3 \quad CH_3SO_3H \quad CH_3NH_2 \quad CH_3SH \quad CH_3C\equiv CH$$

Thiols can be prepared by S_N2 reactions of sodium hydrosulfide with unhindered alkyl halides. The thiol product is still nucleophilic, so a large excess of hydrosulfide is used to prevent the product from undergoing a second alkylation to give a sulfide (R—S—R).

$$\underset{\substack{\text{sodium hydrosulfide}}}{Na^+ \ \ H-\overset{..}{\underset{..}{S}}:^-} \ + \ \underset{\substack{\text{alkyl halide}}}{R-X} \ \longrightarrow \ \underset{\substack{\text{thiol}}}{R-SH} \ + \ Na^+ \ ^-X$$

Unlike alcohols, thiols are easily oxidized to give a dimer called a **disulfide**. The reverse reaction, reduction of the disulfide to the thiol, takes place under reducing conditions. Formation and cleavage of disulfide linkages is an important aspect of protein chemistry (Chapter 24), where disulfide "bridges" between cysteine amino acid residues hold the protein chain in its active conformation.

$$\text{R—SH + HS—R} \underset{\text{Zn, HCl}}{\overset{\text{Br}_2}{\rightleftharpoons}} \text{R—S—S—R + 2 HBr}$$

two molecules of thiol   disulfide

Example

two cysteine residues   cystine disulfide bridge

More examples of disulfide bridges appear in Section 24-8C.

Just as mild oxidation converts thiols to disulfides, vigorous oxidation converts them to **sulfonic acids**. $KMnO_4$ or nitric acid (HNO_3), or even bleach, ($NaOCl$), can be used as the oxidant for this reaction. Any Lewis structure of a sulfonic acid requires either separation of formal charges or more than eight electrons around sulfur. Sulfur can have an expanded octet, as it does in SF_4 (10 electrons) and SF_6 (12 electrons). The three resonance forms shown below are most commonly used. Organic chemists tend to use the form with an expanded octet, and inorganic chemists tend to use the forms with charge separation.

$$\text{R—}\overset{\cdot\cdot}{\underset{\cdot\cdot}{S}}\text{—H} \xrightarrow[\text{(vigorous oxidation)}]{KMnO_4 \text{ or } HNO_3}$$

thiol

sulfonic acid

Example

benzenethiol   benzenesulfonic acid

A protein can be made that withstands higher temperatures by the introduction of additional disulfide bonds. For example, laundry detergents frequently contain enzymes to remove protein and blood stains. By replacing strategically located amino acid residues with cysteines, a modified enzyme results that stays in its active conformation in hot water.

Glutathione, a tripeptide containing a thiol group, serves as a mild reducing agent to detoxify peroxides and maintain the cysteine residues of hemoglobin in the reduced state. Glutathione can also detoxify alkylating agents. For example, the thiol of glutathione reacts with methyl iodide by an S_N2 reaction, rendering the methyl iodide harmless and preventing its reaction with other molecules in the body.

PROBLEM 10-28

Give IUPAC names for the following compounds.

(a) $CH_3CHCH_2CHCH_3$ (with CH_3 and SH substituents)

(b) [structure: CH_3CH_2 and CH_3 on one carbon, CH_2SH and CH_3 on the other, joined by C=C]

(c) [cyclohexene ring with SH substituent]

PROBLEM 10-29

Authentic skunk spray has become valuable for use in scent-masking products. Show how you would synthesize the two major components of skunk spray from any of the readily available butenes or from 1,3-butadiene.

Chapter 10 Glossary

acid derivatives Compounds that are related to carboxylic acids but have other electron-withdrawing groups in place of the —OH group of the acid. Three examples are **acid chlorides, esters**, and **amides**. (p. 426)

[structures]

R—C(=O)—OH   R—C(=O)—Cl   R—C(=O)—O—R′   R—C(=O)—NH₂
carboxylic acid   acid chloride   ester   amide

alcohol A compound in which a hydrogen atom of a hydrocarbon has been replaced by a hydroxyl group. —OH. (p. 405)
Alcohols are classified as **primary**, **secondary**, or **tertiary** depending on whether the hydroxyl group is bonded to a primary, secondary, or tertiary carbon atom. (p. 406)

[structures]

R—C(OH)(H)—H   R—C(OH)(H)—R   R—C(OH)(R)—R
primary alcohol   secondary alcohol   tertiary alcohol

aldehyde A carbonyl compound with one alkyl group and one hydrogen on the carbonyl group. (p. 423) **Formaldehyde** has two hydrogens on the carbonyl group.
alkoxide ion The anion ($R—\ddot{O}:$) formed by deprotonation of an alcohol. (p. 415)
azeotrope A mixture of two or more liquids that distills at a constant temperature and gives a distillate of definite composition. For example, a mixture of 95% ethanol and 5% water has a lower boiling point than pure ethanol or pure water. (p. 414)
carbinol carbon atom In an alcohol, the carbon atom bonded to the hydroxyl group. (p. 406)
denatured alcohol A form of ethanol containing toxic impurities, making it unfit for drinking. (p. 414)
diol A compound with two alcohol —OH groups. (p. 409)
disulfide The oxidized dimer of a thiol, R—S—S—R. (p. 439)
epoxides (oxiranes) Compounds containing oxygen in a three-membered ring. (p. 428)
glycol Synonymous with **diol**. The term "glycol" is most commonly applied to the 1,2-diols. also called **vicinal diols**. (p. 409)
grain alcohol Ethanol, ethyl alcohol. **Absolute alcohol** is 100% ethanol. (p. 414)
Grignard reagent An organomagnesium halide, written in the form R—Mg—X. The actual reagent is more complicated in structure, usually a dimer or trimer complexed with several molecules of ether. (p. 421)
hydride reagent A compound of hydrogen with a less electronegative element, so the hydrogen can be donated with its pair of electrons to an organic compound. Hydride transfer reduces the organic compound. Hydride reagents include simple hydrides such as NaH and LiH as well as complex hydrides such as $NaBH_4$ and $LiAlH_4$. (p. 432)

[reaction scheme]

$M^+ \quad H:^- \quad + \quad$ C=$\ddot{O}:$ $\longrightarrow$ H—C—$\ddot{O}:^- {}^+M$

hydride reagent   reduced

hydrophilic ("water loving") Attracted to water, water-soluble. (p. 412)
hydrophobic ("water hating") Repelled by water, water-insoluble. (p. 412)
hydroxy group (hydroxyl group) The —OH group, as in an alcohol. (p. 408)
ketone A carbonyl compound with two alkyl groups bonded to the carbonyl group. (p. 423)
lithium dialkylcuprate (*Gilman reagent*) An organometallic reagent used to couple with an alkyl halide. (p. 430)

$$R_2CuLi + R'{-}X \longrightarrow R{-}R' + R{-}Cu + LiX$$

  a lithium dialkylcuprate

mercaptan (thiol) The sulfur analogue of an alcohol, R—SH. (p. 437)
miscible Mutually soluble in any proportions. (p. 412)
organolithium reagent An organometallic reagent of the form R—Li. (p. 421)
organometallic compounds (organometallic reagents) Compounds containing metal atoms directly bonded to carbon. (p. 420)
phenol A compound with a hydroxyl group bonded directly to an aromatic ring. (p. 406)
Raney nickel A finely divided nickel/aluminum alloy that has been treated with NaOH to dissolve out most of the aluminum. (p. 435)
rubbing alcohol 2-Propanol, isopropyl alcohol. (p. 415)
skunk (*noun*) A digitigrade omnivorous quadruped that effectively synthesizes thiols; (*verb*) to prevent from scoring in a game or contest. (p. 438)
sulfonic acid A strongly acidic compound of formula R—SO_3H, formed by vigorous oxidation of a thiol. (p. 439)
thiol (mercaptan) The sulfur analogue of an alcohol, R—S—H. (p. 437)
thiolate ion (mercaptide) The anion (R—$\ddot{\text{S}}\text{:}^-$) formed by deprotonation of a thiol. (p. 438)
wood alcohol Methanol, methyl alcohol. (p. 413)

Essential Problem-Solving Skills in Chapter 10

1. Draw and name alcohols, phenols, diols, and thiols.
2. Predict relative boiling points, acidities, and solubilities of alcohols.
3. Show how to convert alkenes, alkyl halides, and carbonyl compounds to alcohols.
4. Predict the alcohol products of hydration, hydroboration, and hydroxylation of alkenes.
5. Use Grignard and organolithium reagents effectively for the synthesis of primary, secondary, and tertiary alcohols with the required carbon skeletons.
6. Propose syntheses and oxidation products of simple thiols.

Study Problems

10-30 Briefly define each term, and give an example.
  (a) primary alcohol **(b)** secondary alcohol **(c)** tertiary alcohol
  (d) phenol **(e)** diol **(f)** glycol
  (g) alkoxide ion **(h)** epoxide **(i)** Grignard reagent
  (j) organolithium reagent **(k)** ketone **(l)** aldehyde
  (m) carboxylic acid **(n)** acid chloride **(o)** ester
  (p) hydride reagents **(q)** thiol **(r)** disulfide

10-31 Give a systematic (IUPAC) name for each alcohol. Classify each as primary, secondary, or tertiary.

10-32 Give systematic (IUPAC) names for the following diols and phenols.

(a) (b) (c) (d)

10-33 Draw the structures of the following compounds.
  (a) triphenylmethanol
  (d) 3-cyclohexyl-3-pentanol
  (g) 4-iodophenol
  (j) dimethyl disulfide

  (b) 3-(bromomethyl)-4-octanol
  (e) *meso*-2,4-pentanediol
  (h) (2R, 3R)-2,3-hexanediol

  (c) 3-cyclopenten-1-ol
  (f) cyclopentene glycol
  (i) 3-cyclopentenethiol

10-34 Predict which member of each pair has the higher boiling point, and explain the reasons for your predictions.
  (a) 1-hexanol or 3,3-dimethyl-1-butanol
  (c) 2-hexanol or 1,5-hexanediol

  (b) 2-hexanone or 2-hexanol
  (d) 2-pentanol or 2-hexanol

10-35 Predict which member of each pair is more acidic, and explain the reasons for your predictions.
  (a) cyclopentanol or 3-chlorophenol
  (c) cyclohexanol or cyclohexanecarboxylic acid

  (b) cyclohexanol or 2-chlorocyclohexanol
  (d) 2,2-dimethyl-1-butanol or 1-butanol

10-36 Predict which member of each group is most soluble in water, and explain the reasons for your predictions.
  (a) 1-butanol, 2-methyl-1-propanol, or 2-methyl-2-propanol
  (b) chlorocyclohexane, cyclohexanol, or 1,2-cyclohexanediol
  (c) chlorocyclohexane, cyclohexanol, or 4-methylcyclohexanol

10-37 Show how you would synthesize the following alcohols from appropriate alkenes.

(a) (b) (c)

(d) (e) (f)

10-38 Draw the organic products you would expect to isolate from the following reactions (after hydrolysis).

(a) + $H_2C=O$ (b) + (c) $CH_3-CH + Ph-CHO$

(d) CH_3MgI + (e) 2 $MgCl + Ph-C-Cl$ (f) $Ph-MgBr$ +

(g) 2 $Ph-MgBr + CH_3CH_2-C-OCH_3$ (h) $CH_3-CH-CHO + NaBH_4$

(i) + $NaBH_4$ (j) + $LiAlH_4$

(k) $\xrightarrow{\text{(1) Hg(OAc)}_2, \text{H}_2\text{O}}_{\text{(2) NaBH}_4}$ (l) $\xrightarrow{\text{(1) BH}_3 \cdot \text{THF}}_{\text{(2) H}_2\text{O}_2, \text{ }^-\text{OH}}$

(m)

$$\underset{H}{\overset{H_3C}{}}C=C\underset{CH_2CH_2CH_3}{\overset{H}{}} \quad \xrightarrow[\text{$^-$OH}]{\substack{\text{cold, dilute} \\ \text{KMnO}_4}}$$

(n)

$$\underset{H}{\overset{H_3C}{}}C=C\underset{CH_2CH_2CH_3}{\overset{H}{}} \quad \xrightarrow[\text{H}_3\text{O}^+]{\text{HCO}_3\text{H}}$$

(o) $(CH_2\!=\!CH)_2\,CuLi \;+\; CH_3CH_2CH\!=\!CHCH_2Br$

10-39 Show how you would use Grignard syntheses to prepare the following alcohols from the indicated starting materials and any other necessary reagents.

    (a) 3-octanol from hexanal, $CH_3(CH_2)_4CHO$ **(b)** 1-octanol from 1-bromoheptane

    (c) 1-cyclohexylethanol from acetaldehyde, CH_3CHO **(d)** 2-cyclohexylethanol from bromocyclohexane

    (e) benzyl alcohol (Ph—CH_2—OH) from bromobenzene (Ph—Br)

(f)

    (g) cyclopentylphenylmethanol from benzaldehyde (Ph—CHO)

10-40 Show how you would accomplish the following transformations. You may use any additional reagents you need.

(a)

(b) Ph—CH_2CH_2Cl $\longrightarrow$ Ph—$CH_2CH_2CH_2OH$

(c)

(d)

(e)

$$CH_3\!-\!\overset{\overset{\displaystyle O}{\|}}{C}\!-\!CH_2CH_2\!-\!\overset{\overset{\displaystyle O}{\|}}{C}\!-\!OCH_2CH_3 \quad\longrightarrow\quad CH_3\!-\!\overset{\overset{\displaystyle OH}{|}}{CH}\!-\!CH_2CH_2\!-\!\overset{\overset{\displaystyle O}{\|}}{C}\!-\!OCH_2CH_3$$

(f)

$$CH_3\!-\!\overset{\overset{\displaystyle O}{\|}}{C}\!-\!CH_2CH_2\!-\!\overset{\overset{\displaystyle O}{\|}}{C}\!-\!OCH_2CH_3 \quad\longrightarrow\quad CH_3\!-\!\overset{\overset{\displaystyle OH}{|}}{CH}\!-\!CH_2CH_2\!-\!\overset{\overset{\displaystyle OH}{|}}{CH_2}$$

10-41 Show how you would synthesize

    (a) 2-phenylethanol by the addition of formaldehyde to a suitable Grignard reagent

    (b) 2-phenylethanol from a suitable alkene

    (c) cyclohexylmethanol from an alkyl halide using an S_N2 reaction

    (d) 3-cyclohexyl-1-propanol by the addition of ethylene oxide to a suitable Grignard reagent

    (e) *cis*-2-penten-1-thiol from a suitable alkenyl halide

    (f) 2,5-dimethylhexane from a four-carbon alkyl halide

10-42 Complete the following acid–base reactions. In each case, indicate whether the equilibrium favors the reactants or the products, and explain your reasoning.

(a) CH_3CH_2—O^- +

(b) KOH +

(c) KOH + CH_3CH_2OH $\rightleftharpoons$

(d) $(CH_3)_3C$—O^- + CH_3CH_2OH $\rightleftharpoons$

(e) $(CH_3)_3C$—O^- + H_2O $\rightleftharpoons$

(f)

$+\;CH_3O^-\;\rightleftharpoons$

10-43 Suggest carbonyl compounds and reducing agents that might be used to form the following alcohols.

    (a) *n*-octanol   **(b)** 1-cyclohexyl-1-propanol   **(c)** 1-phenyl-1-butanol

(d)   **(e)**   **(f)**

10-44   Show how you would synthesize the following compound from any starting materials containing no more than six carbon atoms.

*10-45   Geminal diols, or 1,1-diols, are usually unstable, spontaneously losing water to give carbonyl compounds. Therefore, geminal diols are regarded as hydrated forms of ketones and aldehydes. Propose a mechanism for the acid-catalyzed loss of water from propane-2,2-diol to give acetone.

*10-46   Vinyl alcohols are generally unstable, quickly isomerizing to carbonyl compounds. Propose mechanisms for the following isomerizations.

*10-47   Compound **A** ($C_7H_{11}Br$) is treated with magnesium in ether to give **B** ($C_7H_{11}MgBr$), which reacts violently with D_2O to give 1-methylcyclohexene with a deuterium atom on the methyl group (**C**). Reaction of **B** with acetone (CH_3COCH_3) followed by hydrolysis gives **D** ($C_{10}H_{18}O$). Heating **D** with concentrated H_2SO_4 gives **E** ($C_{10}H_{16}$), which decolorizes two equivalents of Br_2 to give **F** ($C_{10}H_{16}Br_4$). **E** undergoes hydrogenation with excess H_2 and a Pt catalyst to give isobutylcyclohexane. Determine the structures of compounds **A** through **F**, and show your reasoning throughout.

*10-48   Grignard reagents react slowly with oxetane to produce primary alcohols. Propose a mechanism for this reaction, and suggest why oxetane reacts with Grignard reagents even though most ethers do not.

Grignard reagent   oxetane   salt of 1° alcohol

*10-49   Determine the structures of compounds **A** through **G**, including stereochemistry where appropriate.

CHAPTER 11

Reactions of Alcohols

$$CH_3CH_2CH_2$$

Alcohols are important organic compounds because the hydroxyl group is easily converted to almost any other functional group. In Chapter 10, we studied reactions that form alcohols. In this chapter, we seek to understand how alcohols react and which reagents are best for converting them to other kinds of compounds. Table 11-1 summarizes the types of reactions alcohols undergo and the products that result.

TABLE 11-1 Types of Reactions of Alcohols

$$R—OH \xrightarrow{\text{type of reaction}} \text{Product}$$

$R—OH \xrightarrow{\text{dehydration}}$ alkenes	$R—OH \xrightarrow{\text{esterification}}$	$R—O—\overset{\overset{O}{\|\|}}{C}—R'$ esters
$R—OH \xrightarrow{\text{oxidation}}$ ketones, aldehydes, acids	$R—OH \xrightarrow{\text{tosylation}}$	$R—OTs$ tosylate esters (good leaving group)
$R—OH \xrightarrow{\text{substitution}}$ $R—X$ halides		
$R—OH \xrightarrow{\text{reduction}}$ $R—H$ alkanes	$R—OH \xrightarrow[\text{(2) R'X}]{\text{(1) form alkoxide}}$	$R—O—R'$ ethers

Oxidation of alcohols leads to ketones, aldehydes, and carboxylic acids. These functional groups, in turn, undergo a wide variety of additional reactions. For these reasons, alcohol oxidations are some of the most common organic reactions.

In inorganic chemistry, we think of oxidation as a loss of electrons and reduction as a gain of electrons. This picture works well for inorganic ions, as when Cr^{6+} is reduced to Cr^{3+}. Most organic compounds are uncharged, however, and gain or loss of electrons is not obvious. Organic chemists tend to think of oxidation as the result of adding an oxidizing agent (O_2, Br_2, etc.), and reduction as the result of adding a reducing agent (H_2, $NaBH_4$, etc.). Most organic chemists habitually use the following simple rules, based on the change in the formula of the substance:

OXIDATION: addition of O or O_2; loss of H_2; addition of X_2 (halogens)

REDUCTION: addition of H_2 (or H^-); loss of O or O_2; loss of X_2

NEITHER: addition or loss of H^+, H_2O, HX, etc. is neither an oxidation nor a reduction.

11-1
Oxidation States of Alcohols and Related Functional Groups

▶ **Figure 11-1**
An alcohol is more oxidized than an alkane, yet less oxidized than carbonyl compounds such as ketones, aldehydes, and acids. Oxidation of a primary alcohol leads to an aldehyde, and further oxidation leads to an acid. Secondary alcohols are oxidized to ketones. Tertiary alcohols cannot be oxidized without breaking carbon–carbon bonds.

We can tell that an oxidation or a reduction of an alcohol has taken place by counting the number of C—O bonds to the carbon atom. For example, in a primary alcohol, the carbinol (C—OH) carbon atom has one bond to oxygen; in an aldehyde, the carbonyl carbon has two (more oxidized); and in an acid, it has three. Oxidation of an alcohol usually converts C—H bonds to C—O bonds. If we convert an alcohol to an alkane, the carbinol carbon loses its bond to oxygen and gains another bond to hydrogen.

Figure 11-1 compares the oxidation states of primary, secondary, and tertiary alcohols with those obtained by oxidation or reduction. The symbol [O] indicates an unspecified oxidizing agent. Notice that oxidation of a primary or secondary alcohol forms a carbonyl (C=O) group by the removal of two hydrogen atoms: one from the carbinol carbon and one from the hydroxyl group. A tertiary alcohol cannot easily oxidize because there is no hydrogen atom available on the carbinol carbon.

PROBLEM 11-1

Classify each reaction as an oxidation, a reduction, or neither.

(a) $CH_3—CH_2OH \xrightarrow{CrO_3 \cdot pyridine} CH_3—\overset{\displaystyle O}{\overset{\|}{C}}—H \xrightarrow{H_2CrO_4} CH_3—\overset{\displaystyle O}{\overset{\|}{C}}—OH$

(b) $CH_4 \longrightarrow CH_3OH \longrightarrow H—\overset{\displaystyle O}{\overset{\|}{C}}—OH \longrightarrow H—\overset{\displaystyle O}{\overset{\|}{C}}—H \longrightarrow HO—\overset{\displaystyle O}{\overset{\|}{C}}—OH$

(c)

$$H_3C \quad CH_3$$
$$CH_3-\underset{\underset{HO}{|}}{C}-\underset{\underset{OH}{|}}{C}-CH_3 \xrightarrow{H^+} CH_3-\underset{\underset{O}{\|}}{C}-\underset{\underset{CH_3}{|}}{C}-CH_3 + H_2O$$

(d) $CH_3-CH_2-OH \xrightarrow{LiAlH_4/TiCl_4} CH_3-CH_3$

(e) $\xrightarrow{H^+,\ CH_3OH}$ $+ H_2O$

(f) $\xrightarrow{Br_2}$

(g) $\xrightarrow{HBr}$

(h) $\xrightarrow{H_2SO_4}$ $+ H_2O$

(i) $\xrightarrow{OsO_4 \atop H_2O_2}$

(j) $\xrightarrow{RCO_3H}$ $\xrightarrow{H_3O^+}$

(k) $\xrightarrow{BH_3 \cdot THF}$ $\xrightarrow[H_2O]{H_2O_2,\ ^-OH}$

11-2
Oxidation of Alcohols

Primary and secondary alcohols are easily oxidized by a variety of reagents, including chromium reagents, permanganate, nitric acid, and even household bleach (NaOCl, sodium hypochlorite). The choice of reagent depends on the amount and value of the alcohol. We use cheap oxidants for large-scale oxidations of simple, inexpensive alcohols. We use the most effective and selective reagents, regardless of cost, for delicate and valuable alcohols. In this chapter, we study only the oxidants that have the widest range of uses and the best selectivity. An understanding of the most common oxidants can later be extended to include additional reagents.

11-2A   Oxidation of Secondary Alcohols

Secondary alcohols are easily oxidized to give excellent yields of ketones. The **chromic acid reagent** is often best for laboratory oxidations of secondary alcohols.

$$\underset{\text{secondary alcohol}}{R-\underset{\underset{OH}{|}}{C}H-R'} \xrightarrow{Na_2Cr_2O_7/H_2SO_4} \underset{\text{ketone}}{R-\underset{\underset{O}{\|}}{C}-R'}$$

Example

$\xrightarrow[H_2SO_4]{Na_2Cr_2O_7}$

cyclohexanol   →   cyclohexanone
(90%)

The chromic acid reagent is prepared by dissolving sodium dichromate, $(Na_2Cr_2O_7)$ in a mixture of sulfuric acid and water. The active species in the mixture is probably chromic acid, H_2CrO_4, or the acid chromate ion, $HCrO_4^-$. Adding chromium trioxide (CrO_3) to dilute sulfuric acid achieves the same result.

$$Na_2Cr_2O_7 + H_2O + 2\,H_2SO_4 \longrightarrow 2\ \underset{\text{chromic acid }(H_2CrO_4)}{HO-\overset{\overset{O}{\|}}{\underset{\underset{O}{\|}}{Cr}}-OH} + 2\,Na^+ + 2\,HSO_4^-$$

sodium dichromate

$$\underset{\text{chromium trioxide}}{CrO_3} + H_2O \longrightarrow \underset{\text{chromic acid}}{HO-\overset{\overset{O}{\|}}{\underset{\underset{O}{\|}}{Cr}}-OH} \rightleftharpoons H^+ + \underset{\text{acid chromate ion}}{{}^-O-\overset{\overset{O}{\|}}{\underset{\underset{O}{\|}}{Cr}}-OH}$$

The mechanism of chromic acid oxidation probably involves the formation of a chromate ester. Elimination of the chromate ester gives the ketone. In the elimination, the carbinol carbon retains its oxygen atom but loses its hydrogen and gains the second bond to oxygen.

Formation of the chromate ester

$$\underset{\text{alcohol}}{R-\overset{\overset{R'}{|}}{\underset{\underset{H}{|}}{C}}-O-H} + \underset{\text{chromic acid}}{H-O-\overset{\overset{O}{\|}}{\underset{\underset{O}{\|}}{Cr}}-OH} \longrightarrow \underset{\text{chromate ester}}{R-\overset{\overset{R'}{|}}{\underset{\underset{H}{|}}{C}}-O-\overset{\overset{O}{\|}}{\underset{\underset{O}{\|}}{Cr}}-OH} + H_2O$$

Elimination of the chromate ester and oxidation of the carbinol carbon

$$\underset{\underset{Cr\,(VI)}{}}{\underset{H_2\ddot{O}:\ H}{R-\overset{\overset{R'}{|}}{\underset{\underset{}{|}}{C}}-\ddot{O}-\overset{\overset{\ddot{O}}{\|}}{\underset{\underset{.\ddot{O}.}{}}{Cr}}-OH}} \longrightarrow \underset{H_3O^+}{R-\overset{\overset{R'}{|}}{C}=\ddot{O}:} + \underset{\underset{Cr\,(IV)}{}}{\overset{\ddot{O}.}{\underset{{}^-\ddot{O}.}{Cr}}-OH}$$

The chromium(IV) species formed above reacts further to give the stable reduced form, chromium(III). Both sodium dichromate and chromic acid are orange, while chromic ion (Cr^{3+}) is a deep blue. One can follow the progress of a chromic acid oxidation by observing the color change from orange through various shades of green to a greenish blue. In fact, the color change observed with chromic acid can be used as a test for the presence of an oxidizable alcohol.

11-2B   Oxidation of Primary Alcohols

Oxidation of a primary alcohol initially forms an aldehyde. Unlike a ketone, however, an aldehyde is easily oxidized further to give a carboxylic acid.

$$\underset{\text{primary alcohol}}{R-\overset{\overset{OH}{|}}{\underset{\underset{}{}}{CH}}-H} \xrightarrow{[O]} \underset{\text{aldehyde}}{R-\overset{\overset{O}{\|}}{C}-H} \xrightarrow{[O]} \underset{\text{acid}}{R-\overset{\overset{O}{\|}}{C}-OH}$$

Obtaining the aldehyde is often difficult, since most oxidizing agents strong enough to oxidize primary alcohols also oxidize aldehydes. Chromic acid generally oxidizes a primary alcohol all the way to the carboxylic acid.

cyclohexyl methanol → cyclohexanecarboxylic acid (92%)

A better reagent for the limited oxidation of primary alcohols to aldehydes is **pyridinium chlorochromate (PCC)**, a complex of chromium trioxide with pyridine and HCl. PCC oxidizes most primary alcohols to aldehydes in excellent yields. Unlike most other oxidants, PCC is soluble in nonpolar solvents such as dichloromethane (CH_2Cl_2), which is an excellent solvent for most organic compounds. PCC can also serve as a mild reagent for oxidizing secondary alcohols to ketones.

Pyridinium chlorochromate (PCC):

pyridine · chromium trioxide · HCl
or $pyH^+ CrO_3Cl^-$

primary alcohol → aldehyde

Example

1-heptanol → heptanal (78%)

11-2C   Resistance of Tertiary Alcohols to Oxidation

Oxidation of tertiary alcohols is not an important reaction in organic chemistry. Tertiary alcohols have no hydrogen atoms on the carbinol carbon atom, so oxidation must take place by breaking carbon–carbon bonds. These oxidations require severe conditions and result in mixtures of products.

The **chromic acid test** for primary and secondary alcohols exploits the resistance of tertiary alcohols to oxidation. When a primary or secondary alcohol is added to the chromic acid reagent, the orange color changes to green or blue. When a nonoxidizable substance (such as a tertiary alcohol, a ketone, or an alkane) is added to the reagent, no immediate color change occurs.

Summary of Alcohol Oxidations		
To Oxidize	**Product**	**Reagent**
2° alcohol	ketone	chromic acid (or PCC)
1° alcohol	aldehyde	PCC
1° alcohol	acid	chromic acid

PROBLEM 11-2

Predict the products of the reactions of the following compounds with chromic acid and also with PCC.
(a) cyclohexanol
(b) 1-methylcyclohexanol
(c) 2-methylcyclohexanol
(d) cyclohexanone
(e) cyclohexane
(f) acetic acid, CH_3COOH
(g) ethanol
(h) acetaldehyde, CH_3CHO

11-3
Additional Methods for Oxidizing Alcohols

Many other reagents and procedures have been developed for oxidizing alcohols. Some are simply modifications of the procedures we have seen. For example, the **Collins reagent** is a complex of chromium trioxide and pyridine, the original version of PCC. The **Jones reagent** is a milder form of chromic acid: a solution of diluted chromic acid in acetone.

Two other strong oxidants are potassium permanganate and nitric acid. Both of these reagents are less expensive than the chromium reagents, and both of them give by-products that are less environmentally hazardous than spent chromium reagents. Both permanganate and nitric acid oxidize secondary alcohols to ketones and primary alcohols to carboxylic acids. If these strong oxidants are not carefully controlled, they will cleave carbon–carbon bonds.

1-phenylethanol → acetophenone (72%) + MnO_2 (KMnO$_4$ / H$_2$O)

$CH_3(CH_2)_4$—CH_2OH → $CH_3(CH_2)_4$—$\overset{O}{\overset{\|}{C}}$—OH   (71% HNO$_3$, 10–20°)
1-hexanol   hexanoic acid (80%)

Perhaps the least expensive method for oxidizing alcohols is dehydrogenation: literally the removal of two hydrogen atoms. This industrial reaction takes place at high temperature using a copper or copper oxide catalyst. The hydrogen by-product may be sold or used for reductions elsewhere in the plant. The primary limitation of dehydrogenation is the inability of many organic compounds to survive a reaction at 300°C. *Dehydrogenation is not well suited for laboratory syntheses.*

$$R-\underset{H}{\overset{O-H}{\underset{|}{\overset{|}{C}}}}-R' \xrightarrow{\text{heat, CuO}} R-\overset{O}{\overset{\|}{C}}-R' + H_2$$

Example

cyclohexanol $\xrightarrow[300°C]{\text{CuO}}$ cyclohexanone (90%) + H$_2$

The **Swern oxidation** uses dimethyl sulfoxide (DMSO) as the oxidizing agent to convert alcohols to ketones and aldehydes. DMSO and oxalyl chloride are added to the alcohol at low temperature, followed by a hindered base such as triethylamine. Secondary alcohols are oxidized to ketones, and primary alcohols are oxidized only as far as the aldehyde. The by-products of this reaction are all volatile, and are easily separated from the organic products.

A small fuel cell in this portable breath tester catalyzes the oxidation of ethanol by oxygen in the air. The oxidation generates an electric current that is proportional to the concentration of ethanol in the sample.

$$-\underset{|}{\overset{OH}{\underset{|}{C}}}-H + H_3C-\overset{O}{\overset{\|}{S}}-CH_3 + Cl-\overset{O}{\overset{\|}{C}}-\overset{O}{\overset{\|}{C}}-Cl \xrightarrow[CH_2Cl_2]{(CH_3CH_2)_3N:} -\overset{O}{\overset{\|}{C}}\diagdown + H_3C-S-CH_3 \begin{array}{l} + CO_2 \\ + CO \\ + 2\ HCl \end{array}$$

alcohol   DMSO   (COCl)$_2$ ketone   dimethyl sulfide
          dimethyl sulfoxide   oxalyl chloride   or aldehyde

Examples

OH → DMSO, (COCl)$_2$ / Et$_3$N, CH$_2$Cl$_2$, −60°C → O

cyclopentanol   cyclopentanone (90%)

OH
|
CH$_3$(CH$_2$)$_8$—C—H → DMSO, (COCl)$_2$ / Et$_3$N, CH$_2$Cl$_2$, −60°C → CH$_3$(CH$_2$)$_8$—C
| ‖ O
H   \ H

1-decanol   decanal (85%)

PROBLEM 11-3

The Swern oxidation provides a useful alternative to PCC that avoids using chromium reagents as oxidants.
(a) Determine which species are oxidized and which are reduced in the Swern oxidation.
(b) We have seen dimethyl sulfide before (Section 8-15B), added to an ozonide after ozonolysis. What was its function there?

PROBLEM 11-4

What is it about dehydrogenation that enables it to take place at 300°C but not at 25°C?
(a) Would you expect the kinetics, thermodynamics, or both to be unfavorable at 25°C? (*Hint*: Is the reverse reaction favorable at 25°C?)
(b) Which of these factors (kinetics or thermodynamics) improves as the temperature is raised?
(c) Explain the changes in the kinetics and thermodynamics of this reaction as the temperature increases.

PROBLEM 11-5

Give the structure of the principal product(s) when each of the following alcohols reacts with (1) Na$_2$Cr$_2$O$_7$/H$_2$SO$_4$; (2) PCC; (3) KMnO$_4$, ¯OH.
(a) 1-octanol   (b) 3-octanol
(c) 2-cyclohexen-1-ol   (d) 1-methylcyclohexanol

> **PROBLEM-SOLVING HINT**
> The summary table on page 449 is worth reviewing. Remember that permanganate oxidizes alkenes as well as alcohols.

SOLVED PROBLEM 11-1

Suggest the most appropriate method for each of the following laboratory syntheses.
(a) cyclopentanol ⟶ cyclopentanone
  Many reagents are available to oxidize a simple secondary alcohol to a ketone. For a laboratory synthesis, however, dehydrogenation is not practical, and cost is not as large a factor as it would be in industry. Most labs would have chromium trioxide or sodium dichromate available, and the chromic acid oxidation would be simple. PCC and the Swern oxidation would also work, although these reagents are more complicated to prepare and use.

OH → Na$_2$Cr$_2$O$_7$ / H$_2$SO$_4$ → O

cyclopentanol   (or PCC,   cyclopentanone
                  or Swern)

(b) 2-octen-l-ol $\longrightarrow$ 2-octenal (structure below)
This synthesis requires more finesse. The aldehyde is easily over-oxidized to a carboxylic acid, and the double bond reacts with oxidants such as $KMnO_4$. Our choices are limited to PCC or the Swern oxidation.

$$\text{CH}_2\text{OH} \quad \xrightarrow[\text{(or Swern)}]{\text{PCC, CH}_2\text{Cl}_2} \quad \underset{\text{O}}{\overset{\text{H}}{\text{C}}}$$

2-octen-1-ol   2-octenal

PROBLEM 11-6

Suggest the most appropriate method for each of the following *laboratory* syntheses.
(a) 1-butanol $\rightarrow$ butanal, $CH_3CH_2CH_2CHO$
(b) 1-butanol $\rightarrow$ butanoic acid, $CH_3CH_2CH_2COOH$
(c) 2-butanol $\rightarrow$ 2-butanone, $CH_3COCH_2CH_3$
(d) 2-buten-1-ol $\rightarrow$ 2-butenal, $CH_3CH{=}CH{-}CHO$
(e) 2-buten-1-ol $\rightarrow$ 2-butenoic acid, $CH_3CH{=}CH{-}COOH$
(f) 1-methylcyclohexanol $\rightarrow$ 2-methylcyclohexanone (several steps)

11-4
Biological Oxidation of Alcohols

Although it is the least toxic alcohol, ethanol is still a poisonous substance. When someone is suffering from a mild case of ethanol poisoning, we say that he or she is in*toxic*ated. Animals often consume food that has fermented and contains alcohol. Their bodies must detoxify any alcohol in the food to keep it from building up in the blood and poisoning the brain. To detoxify ethanol, the liver produces an enzyme called **alcohol dehydrogenase (ADH)**.

Alcohol dehydrogenase catalyzes an oxidation: the removal of two hydrogen atoms from the alcohol molecule. The oxidizing agent is called **nicotinamide adenine dinucleotide (NAD)**. NAD exists in two forms: the oxidized form, called NAD^+, and the reduced form, called NADH. The following equation shows that ethanol is oxidized to acetaldehyde, and NAD^+ is reduced to NADH. A subsequent oxidation, catalyzed by aldehyde dehydrogenase (ALDH), converts acetaldehyde to acetic acid, a normal metabolite.

NAD^+ is derived from vitamin B_3 known as niacin. As a result, there is a misconception that taking large supplements of niacin before a night of drinking ethanol will lessen the severity of the "hang-over" the next morning. Metabolism is more complex than this simple idea suggests.

nicotinic acid
(niacin)

ethanol
reduced form

NAD^+
oxidized form

ADH = alcohol dehydrogenase

acetaldehyde
oxidized form

NADH
reduced form

$$\text{CH}_3\text{—}\overset{\displaystyle O}{\overset{\|}{\text{C}}}\text{—H} \ + \ \text{H}_2\text{O} \ + \ \text{(pyridine ring with C(=O)NH}_2 \text{, N}^+ \text{, sugar)} \xrightarrow{\text{ALDH}}$$

ALDH = aldehyde dehydrogenase

NAD$^+$

$$\underset{\text{acetic acid}}{\text{CH}_3\text{—}\overset{\displaystyle O}{\overset{\|}{\text{C}}}\text{—OH}} \ + \ \text{(dihydropyridine ring with C(=O)NH}_2 \text{, N, sugar)} \ + \ \text{H}^+$$

NADH

These oxidations take place with most small primary alcohols. Unfortunately, the oxidation products of other alcohols may be more toxic than acetic acid. Methanol is oxidized first to formaldehyde and then to formic acid. Both of these compounds are more toxic than methanol itself.

$$\underset{\text{methanol}}{\text{H}\text{—}\overset{\displaystyle \text{H}}{\underset{\displaystyle \text{H}}{\text{C}}}\text{—O}\text{—H}} \xrightarrow[-2\,\text{H}]{[\text{O}]} \underset{\text{formaldehyde}}{\text{H}\text{—}\overset{\displaystyle }{\underset{\displaystyle \text{H}}{\text{C}}}\text{=O}} \xrightarrow{[\text{O}]} \underset{\text{formic acid}}{\text{H}\text{—}\overset{\displaystyle }{\underset{\displaystyle \text{O}\text{—H}}{\text{C}}}\text{=O}}$$

Ethylene glycol is a toxic diol. Its oxidation product is oxalic acid, the toxic compound found in the leaves of rhubarb and many other plants.

$$\underset{\text{ethylene glycol}}{\text{H}\text{—}\overset{\displaystyle \text{H}}{\underset{\displaystyle \text{HO}}{\text{C}}}\text{—}\overset{\displaystyle \text{H}}{\underset{\displaystyle \text{OH}}{\text{C}}}\text{—H}} \xrightarrow[-4\,\text{H}]{2[\text{O}]} \text{H}\text{—}\overset{\displaystyle }{\underset{\displaystyle \text{O}}{\text{C}}}\text{—}\overset{\displaystyle }{\underset{\displaystyle \text{O}}{\text{C}}}\text{—H} \ + \ 2\,\text{H}_2\text{O} \xrightarrow{2[\text{O}]} \underset{\text{oxalic acid}}{\text{HO}\text{—}\overset{\displaystyle }{\underset{\displaystyle \text{O}}{\text{C}}}\text{—}\overset{\displaystyle }{\underset{\displaystyle \text{O}}{\text{C}}}\text{—OH}}$$

Many poisonings by methanol and ethylene glycol occur each year. Alcoholics occasionally drink ethanol that has been denatured by the addition of methanol. Methanol is oxidized to formic acid, which may cause blindness and death. Dogs are often poisoned by sweet-tasting ethylene glycol when antifreeze is left in an open container. Once the glycol is metabolized to oxalic acid, the dog's kidneys fail, causing death.

The treatment for methanol or ethylene glycol poisoning is the same. The patient is given intravenous infusions of diluted ethanol. The ADH enzyme is swamped by all the ethanol, allowing time for the kidneys to excrete most of the methanol (or ethylene glycol) before it can be oxidized to formic acid (or oxalic acid). This is an example of *competitive inhibition* of an enzyme. The enzyme catalyzes oxidation of both ethanol and methanol, but a large quantity of ethanol ties up the enzyme, allowing time for excretion of most of the methanol before it is oxidized.

PROBLEM 11-7

A chronic alcoholic requires a much larger dose of ethanol as an antidote to methanol poisoning than does a nonalcoholic patient. Suggest a reason for the larger dose of the competitive inhibitor for an alcoholic.

PROBLEM 11-8

Unlike ethylene glycol, propylene glycol (propane-1,2-diol) is nontoxic because it oxidizes to a common metabolic intermediate. Give the structures of the biological oxidation products of propylene glycol.

11-5
Alcohols as Nucleophiles and Electrophiles; Formation of Tosylates

One reason alcohols are such versatile chemical intermediates is that they react as both nucleophiles and electrophiles. The following scheme shows an alcohol reacting as a weak nucleophile, bonding to a strong electrophile (in this case, a carbocation).

An alcohol is easily converted to a strong nucleophile by forming its alkoxide ion. The alkoxide ion can attack a weaker electrophile, such as an alkyl halide.

In both cases, the O—H bond is broken when an alcohol reacts as a weak nucleophile or in making the alkoxide to react as a strong nucleophile. In contrast, when an alcohol reacts as an electrophile, it is the C—O bond that is broken.

This bond is broken when alcohols react as nucleophiles.

This bond is broken when alcohols react as electrophiles.

An alcohol is a weak electrophile because the hydroxyl group is a poor leaving group. The hydroxyl group becomes a good leaving group (H_2O) when it is protonated. For example, HBr reacts with a primary alcohol by an S_N2 attack of bromide on the protonated alcohol. Note that the C—O bond is broken in this reaction.

The disadvantage of using a protonated alcohol is that a strongly acidic solution is required to protonate the alcohol. Although halide ions are stable in acid, few other good nucleophiles are stable in strongly acidic solutions. Most strong nucleophiles are also basic and will abstract a proton in acid. Once protonated, the reagent is no longer nucleophilic. For example, an acetylide ion would instantly become protonated if it were added to a protonated alcohol.

$$R-\overset{+}{\underset{\cdot\cdot}{O}}-H + \ :C\equiv C-H \longrightarrow R-\overset{\cdot\cdot}{O}-H + H-C\equiv C-H$$

no S_N2

How can we convert an alcohol to an electrophile that is compatible with basic nucleophiles? We can convert it to an alkyl halide, or we can simply make its tosylate ester. A **tosylate ester** (symbolized ROTs) is the product of condensation of an alcohol with *p*-toluenesulfonic acid (symbolized TsOH).

Ts = $-\overset{O}{\underset{O}{\overset{\|}{\underset{\|}{S}}}}-\langle\text{ring}\rangle-CH_3$
tosyl group

ROTs = $R-O-\overset{O}{\underset{O}{\overset{\|}{\underset{\|}{S}}}}-\langle\text{ring}\rangle-CH_3$
tosylate ester

TsOH = $H-O-\overset{O}{\underset{O}{\overset{\|}{\underset{\|}{S}}}}-\langle\text{ring}\rangle-CH_3$
tosic acid

TsCl = $Cl-\overset{O}{\underset{O}{\overset{\|}{\underset{\|}{S}}}}-\langle\text{ring}\rangle-CH_3$
tosyl chloride

$^-$OTs = $^-O-\overset{O}{\underset{O}{\overset{\|}{\underset{\|}{S}}}}-\langle\text{ring}\rangle-CH_3$
tosylate ion

$$R-O\overset{\xi}{\underset{\xi}{}}H + HO\overset{\xi}{\underset{\xi}{}}\overset{O}{\underset{O}{\overset{\|}{\underset{\|}{S}}}}-\langle\text{ring}\rangle-CH_3 \rightleftharpoons R-O-\overset{O}{\underset{O}{\overset{\|}{\underset{\|}{S}}}}-\langle\text{ring}\rangle-CH_3 + H_2O$$

alcohol   TsOH   alkyl tosylate, ROTs
p-toluenesulfonic acid   a *p*-toluenesulfonate ester

The tosylate group is an excellent leaving group, and alkyl tosylates undergo substitution and elimination much like alkyl halides. In many cases, a tosylate is more reactive than the equivalent alkyl halide.

☞ $-\overset{\overset{OH}{|}}{\underset{|}{C}}-\overset{|}{\underset{|}{C}}- \xrightarrow[\text{pyridine}]{\text{TsCl}} -\overset{\overset{OTs}{|}}{\underset{|}{C}}-\overset{|}{\underset{|}{C}}- \xrightarrow[\text{(substitution)}]{\text{Nuc:}^-} -\overset{|}{\underset{|}{C}}-\overset{|}{\underset{Nuc}{C}}- + \ ^-OTs$

or elimination:

☞ $-\overset{|}{\underset{\underset{H}{|}}{C}}-\overset{\overset{OTs}{|}}{\underset{|}{C}}- \xrightarrow{\text{(elimination)}} \ \overset{}{\underset{}{C}}=\overset{}{\underset{}{C} } + \ B-H + \ ^-OTs$

B:

Tosylates are made from alcohols using tosyl chloride (TsCl) in pyridine, as shown below. This reaction gives much higher yields than the reaction with TsOH itself. The mechanism of tosylate formation shows that the C—O bond of the alcohol remains intact throughout the reaction, and the alcohol retains its stereochemical configuration. Pyridine serves as an organic base to remove the HCl formed in the reaction, preventing it from protonating the alcohol and causing side reactions.

$$R-\overset{\cdot\cdot}{\underset{\underset{H}{|}}{O}}: + \ O=S=O \longrightarrow R-\overset{+}{\underset{\underset{H}{|}}{\overset{\cdot\cdot}{O}}}-\overset{O}{\underset{}{\overset{\|}{S}}}=O \ \ \overset{Cl^-}{\longrightarrow} R-\overset{\cdot\cdot}{\underset{}{O}}-\overset{O}{\underset{}{\overset{\|}{S}}}=O \ + \ \langle\text{pyridine}\rangle$$

p-toluenesulfonyl chloride   pyridine   ROTs, a tosylate ester
TsCl, "tosyl chloride"

The following reaction shows the S_N2 displacement of tosylate ion ($^-$OTs) from (*S*)-2-butyl tosylate with inversion of configuration. The tosylate ion is a particularly stable anion, with its negative charge delocalized over three oxygen atoms.

(*S*)-2-butyl tosylate   (*R*)-2-butyl iodide   tosylate ion

$^-$OTs
tosylate ion

resonance-stabilized anion

Like halides, the tosylate leaving group is displaced by a wide variety of nucleophiles. The S_N2 mechanism (strong nucleophile) is more commonly used than the S_N1 in synthetic preparations. The following reactions show the generality of S_N2 displacements of tosylates. In each case, R must be a primary or unhindered secondary alkyl group if substitution is to predominate over elimination.

SUMMARY S_N2 Reactions of Tosylate Esters

$$R\text{—OTs} + {}^-\text{OH} \longrightarrow R\text{—OH} + {}^-\text{OTs}$$
hydroxide   alcohol

$$R\text{—OTs} + {}^-\text{C}\equiv\text{N} \longrightarrow R\text{—C}\equiv\text{N} + {}^-\text{OTs}$$
cyanide   nitrile

$$R\text{—OTs} + \text{Br}^- \longrightarrow R\text{—Br} + {}^-\text{OTs}$$
halide   alkyl halide

$$R\text{—OTs} + R'\text{—O}^- \longrightarrow R\text{—O—}R' + {}^-\text{OTs}$$
alkoxide   ether

$$R\text{—OTs} + :\text{NH}_3 \longrightarrow R\text{—NH}_3{}^+{}^-\text{OTs}$$
ammonia   amine salt

$$R\text{—OTs} + \text{LiAlH}_4 \longrightarrow R\text{—H} + {}^-\text{OTs}$$
LAH   alkane

PROBLEM-SOLVING HINT

Tosylate esters are particularly useful: They are great leaving groups, often better than halides. Grignard reactions build alcohols, which are easily converted to tosylates for substitution or elimination.

PROBLEM 11-9

Predict the major products of the following reactions.
(a) ethyl tosylate + potassium *t*-butoxide
(b) isobutyl tosylate + NaI
(c) (*R*)-2-hexyl tosylate + NaCN
(d) the tosylate of cyclohexylmethanol + excess NH_3
(e) *n*-butyl tosylate + sodium acetylide, $H\text{—C}\equiv\text{C}:^-{}^+\text{Na}$

PROBLEM 11-10

Show how you would convert 1-propanol (and whatever reagents are needed) to the following compounds using tosylate intermediates.
(a) 1-bromopropane   (b) *n*-propylamine, $CH_3CH_2CH_2NH_2$
(c) $CH_3CH_2CH_2OCH_2CH_3$ (d) $CH_3CH_2CH_2CN$
    ethyl propyl ether   butyronitrile

The reduction of alcohols to alkanes is not a common reaction because it removes a functional group, leaving fewer options for further reactions.

$$R\text{—}OH \xrightarrow{\text{reduction}} R\text{—}H \qquad \text{(rare)}$$

We can reduce an alcohol in two steps, by dehydrating it to an alkene, then hydrogenating the alkene.

cyclopentanol → cyclopentene → cyclopentane

$$\xrightarrow[\text{heat}]{H_2SO_4}$$

$$\xrightarrow[Pt]{H_2}$$

Another method for reducing an alcohol involves converting the alcohol to the tosylate ester, then using a hydride reducing agent to displace the tosylate leaving group. This reaction works with most primary and secondary alcohols.

cyclohexanol   +   tosyl chloride, TsCl $\xrightarrow{\text{pyridine}}$ cyclohexyl tosylate $\xrightarrow{LiAlH_4}$ cyclohexane (75%)

11-6
Reduction of Alcohols

PROBLEM 11-11

Predict the products of the following reactions.
(a) cyclohexylmethanol + TsCl/pyridine
(b) product of (a) + LiAlH$_4$
(c) 1-methylcyclohexanol + H$_2$SO$_4$, heat
(d) product of (c) + H$_2$, Pt

11-7
Reactions of Alcohols with Hydrohalic Acids

Tosylation of an alcohol, followed by displacement of the tosylate by a halide ion, converts an alcohol to an alkyl halide. This is not the most common method for converting alcohols to alkyl halides, however, because simple, one-step reactions are available. A common method is to treat the alcohol with a hydrohalic acid, usually HCl or HBr.

In acidic solution, an alcohol is in equilibrium with its protonated form. Protonation converts the hydroxyl group from a poor leaving group ($^-$OH) to a good leaving group (H$_2$O). Once the alcohol is protonated, all the usual substitution and elimination reactions are feasible, depending on the structure (1°, 2°, 3°) of the alcohol.

$$R\text{—}\overset{..}{\underset{..}{O}}\text{—}H \quad + \quad H^+ \quad \rightleftharpoons \quad R\text{—}\overset{\overset{H}{|}}{\underset{..}{O}}{}^+\text{—}H \quad \xrightarrow[S_N1 \text{ or } S_N2]{X^-} \quad R\text{—}X$$

poor leaving group   good leaving group

Most good nucleophiles are basic, becoming protonated and losing their nucleophilicity in acidic solutions. Halide ions are exceptions, however. Halides are anions of strong acids, so they are weak bases. Solutions of HBr and HCl contain nucleophilic Br$^-$ and Cl$^-$ ions. These acids are commonly used to convert alcohols to the corresponding alkyl halides.

Reactions with Hydrobromic Acid

☞ $R—OH + HBr/H_2O \longrightarrow R—Br$

Concentrated hydrobromic acid rapidly converts *t*-butyl alcohol to *t*-butyl bromide. The strong acid protonates the hydroxyl group, converting it to a good leaving group. The hindered tertiary carbon atom cannot undergo S_N2 displacement, but it can ionize to a tertiary carbocation. Attack by bromide gives the alkyl bromide. The mechanism is similar to other S_N1 mechanisms we have studied, except that water serves as the leaving group from the protonated alcohol.

MECHANISM   Reaction of a Tertiary Alcohol with HBr (S_N1)

Protonation and loss of the hydroxyl group   *Attack by bromide*

t-butyl alcohol   protonated   tertiary carbocation   *t*-butyl bromide

---

Many other alcohols react with HBr, with the reaction mechanism depending on the structure of the alcohol. For example, 1-butanol reacts with sodium bromide in concentrated sulfuric acid to give 1-bromobutane by an S_N2 displacement. The sodium bromide/sulfuric acid reagent generates HBr in the solution.

$$CH_3(CH_2)_2—CH_2OH \xrightarrow{\text{NaBr, }H_2SO_4} CH_3(CH_2)_2—CH_2Br$$

1-butanol   1-bromobutane (90%)

Protonation converts the hydroxyl group to a good leaving group, but ionization to a primary carbocation is unfavorable. The protonated primary alcohol is well suited for the S_N2 displacement, however. Back-side attack by bromide ion gives 1-bromobutane.

MECHANISM   Reaction of a Primary Alcohol with HBr (S_N2)

Protonation of the alcohol   *Displacement by bromide*

Secondary alcohols also react with HBr to form alkyl bromides, usually by the S_N1 mechanism. For example, cyclohexanol is converted to bromocyclohexane using HBr as the reagent.

cyclohexanol   bromocyclohexane
                                                        (80%)

PROBLEM-SOLVING **HINT**

Memorizing all these mechanisms is not the best way to study this material. Depending on the substrate, these reactions can go by more than one mechanism. Gain experience working problems, then consider each individual case to propose a likely mechanism.

PROBLEM 11-12

Propose a mechanism for the reaction of
(a) cyclohexanol with HBr to form bromocyclohexane
(b) 2-cyclohexylethanol with HBr to form 1-bromo-2-cyclohexylethane

Reactions with Hydrochloric Acid

$$\text{☞} \quad R-OH + HCl/H_2O \xrightarrow{ZnCl_2} R-Cl$$

Hydrochloric acid (HCl) reacts with alcohols in much the same way that hydrobromic acid does. For example, concentrated aqueous HCl reacts with *t*-butyl alcohol to give *t*-butyl chloride.

$$(CH_3)_3C-OH + HCl/H_2O \longrightarrow (CH_3)_3C-Cl + H_2O$$

t-butyl alcohol · *t*-butyl chloride
(98%)

PROBLEM 11-13

The reaction of *t*-butyl alcohol with concentrated HCl goes by the S$_N$1 mechanism. Write a mechanism for this reaction.

Chloride ion is a weaker nucleophile than bromide ion because it is smaller and less polarizable. An additional Lewis acid, such as zinc chloride (ZnCl$_2$), is sometimes necessary to promote the reaction of HCl with primary and secondary alcohols. Zinc chloride coordinates with the oxygen of the alcohol in the same way a proton does—except that zinc chloride coordinates more strongly.

The reagent composed of HCl and ZnCl$_2$ is called the **Lucas reagent**. Secondary and tertiary alcohols react with the Lucas reagent by the S$_N$1 mechanism.

S$_N$1 reaction with the Lucas reagent (fast)

When a primary alcohol reacts with the Lucas reagent, ionization is not possible—the primary carbocation is too unstable. Primary substrates react by the S$_N$2 mechanisms, which is slower than the S$_N$1 reaction of secondary and tertiary substrates. For example, when 1-butanol reacts with the Lucas reagent, the chloride ion attacks the complex from the back, displacing the leaving group.

S$_N$2 reaction with the Lucas reagent (slow)

TABLE 11-2 Reactions of Alcohols with the Lucas Reagent

Alcohol Type	Time to React (min)
primary	> 6
secondary	1–5
tertiary	< 1

The Lucas Test   The Lucas reagent reacts with primary, secondary, and tertiary alcohols at fairly predictable rates, and these rates can be used to distinguish among the three types of alcohols. When the reagent is first added to the alcohol, the mixture forms a single homogeneous phase: The concentrated HCl solution is very polar, and the polar alcohol–zinc chloride complex dissolves. Once the alcohol has reacted to form the alkyl halide, the relatively nonpolar halide separates into a second phase.

The **Lucas test** involves adding the Lucas reagent to an unknown alcohol and watching for the second phase to separate (see Table 11-2). Tertiary alcohols react almost instantaneously because they form relatively stable tertiary carbocations. Secondary alcohols react in about 1 to 5 minutes because their secondary carbocations are less stable than tertiary ones. Primary alcohols react very slowly. Since the activated primary alcohol cannot form a carbocation, it simply remains in solution until it is attacked by the chloride ion. With a primary alcohol, the reaction may take from 10 minutes to several days.

PROBLEM 11-14

Show how you would use a simple chemical test to distinguish between the following pairs of compounds. Tell what you would observe with each compound.
(a) isopropyl alcohol and *t*-butyl alcohol
(b) isopropyl alcohol and 2-butanone, $CH_3COCH_2CH_3$
(c) 1-hexanol and cyclohexanol
(d) allyl alcohol and 1-propanol
(e) 2-butanone and *t*-butyl alcohol

Limitations on the Use of Hydrohalic Acids with Alcohols   The reactions of alcohols with hydrohalic acids do not always give good yields of the expected alkyl halides. Four principal limitations restrict the generality of this technique.

1. *Limited ability to make alkyl iodides.* Most alcohols do not react with HI to give acceptable yields of alkyl iodides. Alkyl iodides are valuable intermediates, however, because iodides are the most reactive of the alkyl halides. We will discuss a better technique for making alkyl iodides in the next section.

2. *Poor yields of alkyl chlorides from primary and secondary alcohols.* Primary and secondary alcohols react with HCl much more slowly than tertiary alcohols, even with zinc chloride added. Under these conditions, side reactions prevent good yields of the alkyl halides.

3. *Eliminations.* Heating an alcohol in a concentrated acid such as HCl or HBr often leads to elimination. Once the hydroxyl group of the alcohol has been protonated and converted to a good leaving group, it becomes a candidate for both substitution and elimination.

4. *Rearrangements.* Carbocation intermediates are always prone to rearrangements. We have seen (Section 6-15) that hydrogen atoms and alkyl groups can migrate from one carbon atom to another to form a more stable carbocation. This rearrangement may occur as the leaving group leaves, or it may occur once the cation has formed.

SOLVED PROBLEM 11-2

When 3-methyl-2-butanol is treated with concentrated HBr, the major product is 2-bromo-2-methylbutane. Propose a mechanism for the formation of this product.

$$CH_3-\underset{\underset{CH_3}{|}}{\overset{\overset{H}{|}}{C}}-\overset{\overset{OH}{|}}{C}H-CH_3 \quad \xrightarrow{HBr} \quad CH_3-\underset{\underset{CH_3}{|}}{\overset{\overset{Br}{|}}{C}}-CH_2-CH_3$$

3-methyl-2-butanol   2-bromo-2-methylbutane

SOLUTION

The alcohol is protonated by the strong acid. This protonated secondary alcohol loses water to form a secondary carbocation.

protonated alcohol   secondary carbocation

A hydride shift transforms the secondary carbocation into a more stable tertiary cation. Attack by bromide leads to the observed product.

secondary carbocation   tertiary carbocation   observed product

Although rearrangements are usually seen as annoying side reactions, a clever chemist can use a rearrangement to accomplish a synthetic goal. Problem 11-15 shows how an alcohol substitution with rearrangement might be used in a synthesis.

PROBLEM 11-15

Neopentyl alcohol, $(CH_3)_3CCH_2OH$, reacts with concentrated HBr to give 2-bromo-2-methylbutane, a rearranged product. Propose a mechanism for the formation of this product.

PROBLEM 11-16

When cyclohexylmethanol reacts with the Lucas reagent, one of the minor products is chlorocycloheptane. Propose a mechanism to explain the formation of this product.

PROBLEM 11-17

When *cis*-2-methylcyclohexanol reacts with the Lucas reagent, the major product is 1-chloro-1-methylcyclohexane. Propose a mechanism to explain the formation of this product.

Several phosphorus halides are useful for converting alcohols to alkyl halides. Phosphorus tribromide, phosphorus trichloride, and phosphorus pentachloride work well and are commercially available.

$$3\,R\!-\!OH + PCl_3 \longrightarrow 3\,R\!-\!Cl + P(OH)_3$$

$$3\,R\!-\!OH + PBr_3 \longrightarrow 3\,R\!-\!Br + P(OH)_3$$

$$R\!-\!OH + PCl_5 \longrightarrow R\!-\!Cl + POCl_3 + HCl$$

Phosphorus triiodide is not sufficiently stable to be stored, but it can be generated in situ (in the reaction mixture) by the reaction of phosphorus with iodine.

$$6\,R\!-\!OH + 2\,P + 3\,I_2 \longrightarrow 6\,R\!-\!I + 2\,P(OH)_3$$

Phosphorus halides produce good yields of most primary and secondary alkyl halides, but none works well with tertiary alcohols. The two phosphorus halides used

11-8
Reactions of Alcohols with Phosphorus Halides

most often are PBr_3 and the phosphorus/iodine combination. Phosphorus tribromide is often the best reagent for converting a primary or secondary alcohol to the alkyl bromide, especially if the alcohol might rearrange in strong acid. A phosphorus and iodine combination is one of the best reagents for converting a primary or secondary alcohol to the alkyl iodide. For the synthesis of alkyl chlorides, thionyl chloride (discussed below) generally gives better yields than PCl_3 or PCl_5, especially with tertiary alcohols.

The following examples show the conversion of primary and secondary alcohols to bromides and iodides by treatment with PBr_3 and P/I_2.

neopentyl alcohol   →   neopentyl bromide (60%)

$$CH_3(CH_2)_{14}\text{—}CH_2OH \quad + \quad P/I_2 \quad \longrightarrow \quad CH_3(CH_2)_{14}\text{—}CH_2I \quad (85\%)$$

PROBLEM 11-18

Write balanced equations for the two preceding reactions.

Mechanism of the Reaction with Phosphorus Trihalides   The mechanism of the reaction of alcohols with phosphorus trihalides explains why rearrangements are uncommon and why phosphorus halides work poorly with tertiary alcohols. The mechanism is shown here using PBr_3 as the reagent; PCl_3 and PI_3 (generated from phosphorus and iodine) react in a similar manner.

MECHANISM   Reaction of Alcohols with PBr_3

Step 1: Displacement of bromide ion, forming an excellent leaving group.

Step 2: S_N2 attack on the alkyl group.

leaving group

Rearrangements are uncommon because no carbocation is involved, so there is no opportunity for rearrangement. This mechanism also explains the poor yields with tertiary alcohols. The final step is an S_N2 displacement where bromide attacks the back side of the alkyl group. This attack is hindered if the alkyl group is tertiary. In the case of a tertiary alcohol, an ionization to a carbocation is needed. This ionization is slow, and it invites side reactions.

Thionyl chloride $(SOCl_2)$ is often the best reagent for converting an alcohol to an alkyl chloride. The by-products (gaseous SO_2 and HCl) leave the reaction mixture and ensure there can be no reverse reaction.

☞ $R-OH$ + $Cl-\overset{\overset{\displaystyle O}{\|}}{S}-Cl$ $\xrightarrow{\text{heat}}$ $R-Cl$ + SO_2 + HCl

Under the proper conditions, thionyl chloride reacts by the interesting mechanism summarized below. In the first step, the nonbonding electrons of the hydroxyl oxygen atom attack the electrophilic sulfur atom of thionyl chloride. A chloride ion is expelled, and a proton is lost to give a chlorosulfite ester. In the next step, the chlorosulfite ester ionizes (when R = 2° or 3°), and the sulfur atom quickly delivers chloride to the carbocation. When R is primary, chloride probably bonds to carbon at the same time that the C—O bond is breaking.

thionyl chloride   chlorosulfite ester

chlorosulfite ester   ion pair

This mechanism resembles the S_N1, except that the nucleophile is delivered to the carbocation by the leaving group, giving retention of configuration as shown in the following example. (Under different conditions, retention of configuration may not be observed.)

(R)-2-octanol   dioxane (solvent)   (R)-2-chlorooctane (84%)

Summary of the Best Reagents for Converting Alcohols to Alkyl Halides

Class of Alcohol	Chloride	Bromide	Iodide
primary	$SOCl_2$	PBr_3 or HBr*	P/I_2
secondary	$SOCl_2$	PBr_3	P/I_2*
tertiary	HCl	HBr	HI*

*Works only in selected cases.

PROBLEM 11-19

Suggest how you would convert *trans*-4-methylcyclohexanol to
(a) *trans*-1-chloro-4-methylcyclohexane
(b) *cis*-1-chloro-4-methylcyclohexane

PROBLEM 11-20

Two products are observed in the reaction.

(a) Suggest a mechanism to explain how these two products are formed.
(b) Your mechanism for part (a) should be different from the usual mechanism of the reaction of $SOCl_2$ with alcohols. Explain why the reaction follows a different mechanism in this case.

PROBLEM 11-21

Give the structures of the products you would expect when each alcohol reacts with (1) HCl, $ZnCl_2$; (2) HBr; (3) PBr_3; (4) P/I_2; (5) $SOCl_2$.
(a) 1-butanol   (b) 2-butanol   (c) 2-methyl-2-butanol
(d) 2,2-dimethyl-1-butanol   (e) *cis*-3-methylcyclopentanol

11-10
Dehydration Reactions of Alcohols

11-10A   Formation of Alkenes

We studied the mechanism for dehydration of alcohols to alkenes in Section 7-10 together with other syntheses of alkenes. Dehydration requires an acidic catalyst to protonate the hydroxyl group of the alcohol and convert it to a good leaving group. Loss of water, followed by loss of a proton, gives the alkene. An equilibrium is established between reactants and products.

MECHANISM   (Review): Acid-catalyzed Dehydration of an Alcohol

To drive this equilibrium to the right, we remove one or both of the products as they form; either by distilling the products out of the reaction mixture or by adding a dehydrating agent to remove water. In practice, we often use a combination of distillation and a dehydrating agent. The alcohol is mixed with a dehydrating acid, and the mixture is heated to boiling. The alkene boils at a lower temperature than the alcohol (because the alcohol is hydrogen-bonded), and the alkene distills out of the mixture. For example.

cyclohexanol, bp 161°C   cyclohexene, bp 83°C (80%)
                                          (distilled from the mixture)

◀ **Figure 11-2**
Reaction-energy diagram for
dehydration of an alcohol.

Alcohol dehydrations generally take place through the E1 mechanism. Protonation of the hydroxyl group converts it to a good leaving group. Water leaves, forming a carbocation. Loss of a proton gives the alkene.

Figure 11-2 shows the reaction-energy diagram for the E1 dehydration of an alcohol. The first step is a mildly exothermic protonation, followed by an endothermic, rate limiting ionization. A fast, strongly exothermic deprotonation gives the alkene. Because the rate-limiting step is formation of a carbocation, the case of dehydration follows the same order as the ease of formation of carbocations: $3° > 2° > 1°$. As in other carbocation reactions, rearrangements are common.

With primary alcohols, rearrangement and isomerization of the products are so common that acid-catalyzed dehydration is rarely a good method for converting them to alkenes. The following mechanism shows how 1-butanol undergoes dehydration with rearrangement to give a mixture of 1-butene and 2-butene. The more highly substituted product, 2-butene, is the major product, in accordance with the Saytzeff rule (Section 6-19).

Ionization of the protonated alcohol, with rearrangement

Loss of either proton to give two products

secondary carbocation

loss of H_a^+
2-butene (major, 70%)
a disubstituted alkene

loss of H_b^+
1-butene (minor, 30%)
a monosubstituted alkene

We can summarize the utility of dehydration and give guidelines for predicting the products:

1. Dehydration usually goes by the E1 mechanism. Rearrangements may occur to form more stable carbocations.

2. Dehydration works best with tertiary alcohols and almost as well with secondary alcohols. Rearrangements and poor yields are common with primary alcohols.

3. (Saytzeff rule) If two or more alkenes might be formed by deprotonation of the carbocation, the most highly substituted alkene usually predominates.

PROBLEM-SOLVING HINT

Most alcohol dehydrations go by E1 mechanisms involving protonation of the OH group, followed by loss of water.

Solved Problem 11-3 shows how these rules are used to predict the products of dehydrations. The carbocations are drawn to show how rearrangements occur and how more than one product may result.

SOLVED PROBLEM 11-3

Predict the products of sulfuric acid-catalyzed dehydration of the following alcohols.
(a) 1-methylcyclohexanol
(b) neopentyl alcohol

SOLUTION

(a) 1-Methylcyclohexanol reacts to form a tertiary carbocation. Abstraction of a proton may occur on any one of three carbon atoms. The two secondary atoms are equivalent, and abstraction of a proton from one of these carbons leads to the trisubstituted double bond of the major product. Abstraction of a methyl proton leads to the disubstituted double bond of the minor product.

1-methylcyclohexanol   protonated   cation   major product (trisubstituted)   minor product (disubstituted)

(b) Neopentyl alcohol cannot simply ionize to form a primary cation. Rearrangement occurs as the leaving group leaves, giving a tertiary carbocation. Loss of a proton from the adjacent secondary carbon gives the trisubstituted double bond of the major product. Loss of a proton from the methyl group gives the monosubstituted double bond of the minor product.

neopentyl alcohol (2,2-dimethyl-1-propanol)   ionization with rearrangement   3° cation

PROBLEM 11-22

Predict the products of the sulfuric acid-catalyzed dehydration of the following alcohols. When more than one product is expected, label the major and minor products.

(a) 2-methyl-2-butanol   (b) 1-pentanol   (c) 2-pentanol
(d) 1-isopropylcyclohexanol   (e) 2-methylcyclohexanol

PROBLEM-SOLVING HINT

Draw the carbocation, look for possible rearrangements, then consider all the ways that the original carbocation and any rearranged carbocation might lose protons to give alkenes. Saytzeff's rule usually predicts the major product.

PROBLEM 11-23

Some alcohols undergo rearrangement or other unwanted side reactions when they dehydrate in acid. Alcohols may be dehydrated under mildly *basic* conditions using phosphorus oxychloride (POCl$_3$) in pyridine. The alcohol reacts with phosphorus oxychloride much like it reacts with tosyl chloride (Section 11-5), displacing a chloride ion from phosphorus to give an alkyl dichlorophosphate ester. The dichlorophosphate group is an outstanding leaving group. Pyridine reacts as a base with the dichlorophosphate ester to give an E2 elimination. Propose a mechanism for the dehydration of cyclohexanol by POCl$_3$ in pyridine.

phosphorus oxychloride   pyridine

11-10B   Bimolecular Dehydration to Form Ethers (Industrial)

In some cases, a protonated primary alcohol may be attacked by another molecule of the alcohol and undergo an S$_N$2 displacement. The net reaction is a bimolecular dehydration to form an ether. For example, the attack by ethanol on a protonated molecule of ethanol gives diethyl ether.

nucleophilic   electrophilic   protonated ether   water   diethyl ether

Bimolecular dehydration can be used to synthesize symmetrical dialkyl ethers from simple, unhindered primary alcohols. This method is used for the industrial synthesis of diethyl ether (CH$_3$CH$_2$—O—CH$_2$CH$_3$) and dimethyl ether (CH$_3$—O—CH$_3$). Under the acidic dehydration conditions, two reactions compete: elimination (to give an alkene) competes with substitution (to give an ether).

Substitution to give the ether, a bimolecular dehydration

$$2 \text{ CH}_3\text{CH}_2\text{OH} \xrightarrow{\text{H}_2\text{SO}_4,\ 140°\text{C}} \text{CH}_3\text{CH}_2-\text{O}-\text{CH}_2\text{CH}_3 + \text{H}_2\text{O}$$

ethanol   diethyl ether

Elimination to give the alkene, a unimolecular dehydration

$$CH_3CH_2OH \xrightarrow{\text{H}_2\text{SO}_4,\ 180°\text{C}} CH_2{=}CH_2 + H_2O$$

ethanol   ethylene

PROBLEM 11-24

Contrast the mechanisms of the two preceding dehydrations of ethanol.

How can we control these two competing dehydrations? The ether synthesis (substitution) shows two molecules of alcohol giving two product molecules: one of diethyl ether and one of water. The elimination shows one molecule of alcohol giving two molecules: one of ethylene and one of water. The elimination results in an increase in the number of molecules and therefore an *increase* in the randomness (entropy) of the system. The elimination has a more positive change in entropy (ΔS) than the substitution, and the $-T\Delta S$ term in the Gibbs free energy becomes more favorable for the elimination as the temperature increases. Substitution (to give the ether) is favored around 140°C and below, and elimination is favored around 180°C and above. Diethyl ether is produced industrially by heating ethanol with an acidic catalyst at around 140°C.

PROBLEM 11-25

Explain why the acid-catalyzed dehydration is a poor method for the synthesis of an unsymmetrical ether such as ethyl methyl ether, $CH_3CH_2{-}O{-}CH_3$.

PROBLEM 11-26

Propose a detailed mechanism for the following reaction.

$$2\ CH_3OH \xrightarrow{\text{H}_2\text{SO}_4,\ \text{heat}} CH_3{-}O{-}CH_3 + H_2O$$

PROBLEM-SOLVING STRATEGY
Proposing Reaction Mechanisms

In view of the large number of reactions we've covered, proposing mechanisms for reactions you have never seen may seem nearly impossible. As you gain experience in working mechanism problems, you will start to see similarities to known reactions. Let's consider how an organic chemist systematically approaches a mechanism problem. (A more complete version of this method appears in Appendix 4.) Although this stepwise approach cannot solve all mechanism problems, it should provide a starting point to begin building your experience and confidence.

Determining the Type of Mechanism

First, determine what kinds of conditions and catalysts are involved. In general, reactions may be classified as involving (a) strong electrophiles (including acid-catalyzed reactions), (b) strong nucleophiles (including base-catalyzed reactions), or (c) free radicals. These three types of mechanisms are quite distinct, and you should first try to determine which type is involved.

(a) In the presence of a strong acid or a reactant that can dissociate to give a strong electrophile, the mechanism probably involves strong electrophiles as intermediates. Acid-catalyzed reactions and reactions involving carbocations (such as the S_N1, the E1, and most alcohol dehydrations) fall in this category.

(b) In the presence of a strong base or a strong nucleophile, the mechanism probably involves strong nucleophiles as intermediates. Base-catalyzed reactions and those depending on base strength (such as the S_N2 and the E2) generally fall in this category.

(c) Free-radical reactions usually require a free-radical initiator such as chlorine, bromine, NBS, or a peroxide. In most free-radical reactions, there is no need for a strong acid or base.

Once you have determined which type of mechanism you will write, there are general methods for approaching the problem. At this point, we consider mostly the electrophilic reactions covered in recent chapters. Suggestions for drawing the mechanisms of reactions involving strong nucleophiles and free-radical reactions are collected in Appendix 4.

Reactions Involving Strong Electrophiles

When a strong acid or electrophile is present, expect to see intermediates that are strong acids and strong electrophiles; cationic intermediates are common. Bases and nucleophiles in such a reaction are generally weak, however. Avoid drawing carbanions, alkoxide ions, and other strong bases. They are unlikely to coexist with strong acids and strong electrophiles.

Functional groups are often converted to carbocations or other strong electrophiles by protonation or reaction with a strong electrophile; then the carbocation or other strong electrophile reacts with a weak nucleophile such as an alkene or the solvent.

1. **Consider the carbon skeletons of the reactants and products, and decide which carbon atoms in the products are most likely derived from which carbon atoms in the reactants.**

2. **Consider whether any of the reactants is a strong enough electrophile to react without being activated. If not, consider how one of the reactants might be converted to a strong electrophile by protonation of a Lewis basic site (or complexation with a Lewis acid).**
   Protonation of an alcohol, for example, converts it to a strong electrophile, which can undergo attack or lose water to give a carbocation, an even stronger electrophile. Protonation of an alkene converts it to a carbocation.

3. **Consider how a nucleophilic site on another reactant (or, in a cyclization in another part of the same molecule) can attack the strong electrophile to form a bond needed in the product. Draw the product of this bond formation.**
   If the intermediate is a carbocation, consider whether it is likely to rearrange to form a bond in the product. If there isn't any possible nucleophilic attack that leads in the direction of the product, consider other ways of converting one of the reactants to a strong electrophile.

4. **Consider how the product of a nucleophilic attack might be converted to the final product (if it has the right carbon skeleton) or reactivated to form another bond needed in the product.**
   To move a proton from one atom to another (as in an isomerization), try adding a proton to the new position, then removing it from the old position.

5. **Draw out all steps of the mechanism using curved arrows to show the movement of electrons.**
   Be careful to show only one step at a time.

Common Mistakes to Avoid in Drawing Mechanisms

1. Do not use condensed or line-angle formulas for reaction sites. Draw all the bonds and all the substituents of each carbon atom affected throughout the mechanism. In reactions involving strong electrophiles and acidic conditions, three-bonded carbon atoms are likely to be carbocations. If you draw condensed formulas or line-angle formulas, you will likely misplace a hydrogen atom and show a reactive species on the wrong carbon.

2. Do not show more than one step occurring at once. Do not show two or three bonds changing position in one step unless the changes really are concerted (take place simultaneously). For example, protonation of an alcohol and loss of water to give a carbocation are two steps. You must not show the hydroxyl group "jumping" off the alcohol to join up with an anxiously waiting proton.

3. Remember that curved arrows show *movement of electrons*, always from the nucleophile (electron donor) to the electrophile (electron acceptor). For example, protonation of a double bond must show the arrow going from the electrons of the double bond to the proton— never from the proton to the double bond. Resist the urge to use an arrow to "point out" where the proton (or other reagent) goes.

SAMPLE PROBLEM

To illustrate the stepwise method for reactions involving strong electrophiles, we will develop a mechanism to account for the following cyclization:

The cyclized product is a minor product in this reaction. Note that a mechanism problem is different from a synthesis problem: In a mechanism problem, we are limited to the reagents given and are asked to explain how these reactants form these products under the conditions shown. Also, a mechanism problem may deal with how an unusual or unexpected minor product is formed.

In the presence of sulfuric acid, this is clearly an acid-catalyzed mechanism. We expect strong electrophiles, cationic intermediates (possibly carbocations), and strong acids. Carbanions, alkoxide ions, and other strong bases and strong nucleophiles are unlikely.

1. **Consider the carbon skeletons of the reactants and products, and decide which carbon atoms in the products are most likely derived from which carbon atoms in the reactants.**

   Drawing the starting material and the product with all the substituents of the affected carbon atoms, we see the major changes shown here. A vinyl hydrogen must be lost, a $=C-C$ bond must be formed, a methyl group must move over one carbon atom, and the hydroxyl group must be lost.

2. **Consider whether any of the reactants is a strong enough electrophile to react without being activated. If not, consider how one of the reactants might be converted to a strong electrophile by protonation of a Lewis basic site (or complexation with a Lewis acid).**

   The starting material is not a strong electrophile, so it must be activated. Sulfuric acid could generate a strong electrophile either by protonating the double bond or by protonating the hydroxyl group. Protonating the double bond would form the tertiary carbocation, activating the wrong end of the double bond. Also, there is no good nucleophilic site on the side chain to attack this carbocation to form the correct ring. Protonating the double bond is a dead end.

The other basic site is the hydroxyl group. An alcohol can protonate on the hydroxyl group and lose water to form a carbocation.

3. **Consider how a nucleophilic site on another reactant (or, in a cyclization, in another part of the same molecule) can attack the strong electrophile to form a bond needed in the product. Draw the product of this bond formation.**

The carbocation can be attacked by the electrons in the double bond to form a ring; but the positive charge is on the wrong carbon atom to give a six-membered ring. A favorable rearrangement of the secondary carbocation to a tertiary one shifts the positive charge to the correct carbon atom and accomplishes the methyl shift we identified in step 1. Attack by the (weakly) nucleophilic electrons in the double bond gives the correct six-membered ring.

4. **Consider how the product of nucleophilic attack might be converted to the final product (if it has the right carbon skeleton) or reactivated to form another bond needed in the product.**

Loss of a proton (to HSO_4^- or H_2O, but *not* to ^-OH, which is not compatible!) gives the observed product.

5. **Draw out all steps of the mechanism using curved arrows to show the movement of electrons.**

Combining the equations written immediately above gives the complete mechanism for this reaction.

The following problems require proposing mechanisms for reactions involving strong electrophiles. Work each one by completing the five steps described above.

PROBLEM 11-27

Propose a mechanism for each reaction.

11-11
Unique Reactions of Diols

11-11A   The Pinacol Rearrangement

Using our knowledge of alcohol reactions, we can explain results that seem strange at first glance. The following reaction is an example of the **pinacol rearrangement**.

pinacol
(2,3-dimethyl-2,3-butanediol)

pinacolone
(3,3-dimethyl-2-butanone)

The pinacol rearrangement is formally a dehydration. The reaction is acid-catalyzed, and the first step is protonation of one of the hydroxyl oxygens. Loss of water gives a tertiary carbocation, as expected for any tertiary alcohol.

Migration of a methyl group places the positive charge on the carbon atom bearing the second —OH group, where oxygen's nonbonding electrons help to stabilize the charge through resonance. This extra stability is the driving force for the rearrangement. Deprotonation of the resonance-stabilized cation gives the product, pinacolone.

MECHANISM   The Pinacol Rearrangement

Step 1: Protonation of a hydroxyl group.   *Step 2:* Loss of water gives a carbocation.

Step 3: Methyl migration forms a resonance-stabilized carbocation.

resonance-stabilized carbocation

Step 4: Deprotonation gives the product.

resonance-stabilized carbocation

pinacolone

Pinacol-like rearrangements are common in acid-catalyzed reactions of diols. One of the hydroxyl groups protonates and leaves as water, forming a carbocation. Rearrangement gives a resonance-stabilized cation with the remaining hydroxyl group helping to stabilize the positive charge. Problem 11-28 shows some additional examples of pinacol rearrangements.

PROBLEM 11-28

Propose a mechanism for each reaction.

(a)

(b)

PROBLEM 11-29

The following reaction involves a starting material with a double bond and a hydroxyl group, yet its mechanism resembles a pinacol rearrangement. Propose a mechanism, and point out the part of your mechanism that resembles a pinacol rearrangement.

PROBLEM-SOLVING HINT

By analogy with the pinacol rearrangement, watch for carbocation rearrangements that place the + charge on a carbinol carbon atom.

11-11B   Periodic Acid Cleavage of Glycols

1,2-Diols (glycols), such as those formed by hydroxylation of alkenes, are cleaved by periodic acid (HIO$_4$). The products are the same ketones and aldehydes that would be formed by ozonolysis–reduction of the alkene. Hydroxylation followed by periodic acid cleavage serves as a useful alternative to ozonolysis, and the periodate cleavage by itself is useful for determining the structures of sugars (Chapter 23).

Periodic acid cleavage of a glycol probably involves a cyclic periodate intermediate like that shown below.

PROBLEM 11-30

Predict the products formed by periodic acid cleavage of the following diols.

(a) $CH_3CH(OH)CH(OH)CH_3$ (b)

PROBLEM-SOLVING HINT

Periodic acid cleaves a 1,2-diol to give the same products as ozonolysis–reduction (O$_3$ followed by Me$_2$S) of the alkene.

OH
|
(c) Ph—C—CH(OH)CH$_2$CH$_3$
|
CH$_3$

(d)

OH
H
OH
H

11-12
Esterification
of Alcohols

Although we have used *tosylate esters* on several occasions, to an organic chemist the term **ester** normally means an ester of a carboxylic acid. Replacing the —OH group of a carboxylic acid with the —OR group of an alcohol gives a carboxylic ester. The following reaction, called the **Fischer esterification**, shows the relationship between the alcohol and the acid on the left, and the ester and water on the right.

$$R—O\boxed{—H} \quad + \quad \boxed{H—O}—\overset{\overset{\displaystyle O}{\|}}{C}—R' \quad \underset{}{\overset{H^+}{\rightleftharpoons}} \quad R—O—\overset{\overset{\displaystyle O}{\|}}{C}—R' \quad + \quad \boxed{H—O—H}$$

alcohol acid ester

For example, if we mix isopropyl alcohol with acetic acid and add a drop of sulfuric acid as a catalyst, the following equilibrium results.

CH$_3$
|
H—C—O—H   +   H—O—C—CH$_3$ ⇌ (H$_2$SO$_4$)   H—C—O—C—CH$_3$ +   H—O—H
| (O) | (O)
CH$_3$ CH$_3$

isopropyl alcohol acetic acid isopropyl acetate water

> The alcohol groups of unpleasant-tasting drugs are often converted to esters in order to mask the taste. In most cases, the ester has a less unpleasant taste than free alcohol.

Because the Fischer esterification is an equilibrium (often with an unfavorable equilibrium constant), clever techniques are often required to achieve good yields of esters. For example, we can use a large excess of the alcohol or the acid. Adding a dehydrating agent removes water (one of the products), driving the reaction to the right. There is a more powerful way to form an ester, however, without having to deal with an unfavorable equilibrium. An alcohol reacts with an acid chloride in an exothermic reaction to give an ester.

$$R—O\boxed{—H} \quad + \quad \boxed{Cl}—\overset{\overset{\displaystyle O}{\|}}{C}—R' \quad \longrightarrow \quad R—O—\overset{\overset{\displaystyle O}{\|}}{C}—R' \quad + \quad \boxed{HCl}$$

alcohol acid chloride ester

The mechanisms of these reactions are covered with similar mechanisms in Chapter 21.

PROBLEM 11-31

Show the alcohol and the acid chloride you would use to make the following esters.

(a)
O
‖
CH$_3$CH$_2$CH$_2$C—OCH$_2$CH$_2$CH$_3$
n-propyl butyrate

(b)
O
‖
CH$_3$(CH$_2$)$_3$—O—C—CH$_2$CH$_3$
n-butyl propionate

(c)
O
‖
H$_3$C—⟨benzene ring⟩—O—C—CH(CH$_3$)$_2$
p-tolyl isobutyrate

(d)
O
‖
▷—O—C—⟨benzene ring⟩
cyclopropyl benzoate

In addition to forming esters with carboxylic acids, alcohols form **inorganic esters** with inorganic acids such as nitric acid, sulfuric acid, and phosphoric acid. In each type of ester, the alkoxy (—OR) group of the alcohol replaces a hydroxyl group of the acid, with loss of water. We have already studied tosylate esters, composed of *para*-toluenesulfonic acid and alcohols (but made using tosyl chloride, Section 11-5). Tosylate esters are analogous to sulfate esters (Section 11-13A), which are composed of sulfuric acid and alcohols.

11-13
Esters of Inorganic Acids

alcohol   *para*-toluenesulfonic acid   *para*-toluenesulfonate ester
                        (TsOH)   (ROTs)

Made using tosyl chloride

alcohol   *para*-toluenesulfonyl chloride   tosylate ester
                        (TsCl)   (ROTs)

11-13A   Sulfate Esters

A **sulfate ester** is like a sulfonate ester, except there is no alkyl group directly bonded to the sulfur atom. In an alkyl sulfate ester, alkoxy groups are bonded to sulfur through oxygen atoms. Using methanol as the alcohol,

sulfuric acid   methyl sulfate   dimethyl sulfate

Sulfate ions are excellent leaving groups. Like sulfonate esters, sulfate esters are good electrophiles. Nucleophiles react with sulfate esters to give alkylated products. For example, the reaction of dimethyl sulfate with ammonia gives a sulfate salt of methylamine, $CH_3NH_3^+ \ CH_3OSO_3^-$.

The body converts the hydroxyl groups of some drugs to their sulfate derivatives in order to produce water-soluble compounds that are readily excreted. The reaction is not as common as it might be because of the body's limited availability of inorganic sulfate.

ammonia   dimethyl sulfate   methylammonium ion   methylsulfate ion

PROBLEM 11-32

Use resonance structures to show that the negative charge in the methylsulfate anion is shared equally by three oxygen atoms.

11-13B   Nitrate Esters

Nitrate esters are formed from alcohols and nitric acid.

Illustration of Alfred Nobel operating the apparatus used to make nitroglycerine. The temperature must be monitored and controlled carefully during this process; therefore, the operator's stool has only one leg to ensure that he stays awake.

Nitroglycerin is also used to relieve angina, a condition where the heart is not receiving enough oxygen. Angina is characterized by severe pain in the chest, often triggered by stress or exercise.

The best known nitrate ester is "nitroglycerine," whose systematic name is *glyceryl trinitrate*. Glyceryl trinitrate results from the reaction of glycerol (1,2,3-propanetriol) with three molecules of nitric acid.

Nitroglycerine was first made in 1847 and was found to be a much more powerful explosive than black powder, which is a physical mixture of potassium nitrate, sulfur, and charcoal. In black powder, potassium nitrate is the oxidizer, and sulfur and charcoal provide the fuel to be oxidized. The rate of a black powder explosion is limited by how fast oxygen from the grains of heated potassium nitrate can diffuse to the grains of sulfur and charcoal. A black powder explosion does its work by the rapid increase in pressure resulting from the reaction. The explosion must be confined, as in a cannon or a firecracker, to be effective.

In nitroglycerine, the nitro groups are the oxidizer and the CH and CH_2 groups are the fuel to be oxidized. This intimate association of fuel and oxidizer allows the explosion to proceed at a much faster rate, forming a shock wave that propagates through the explosive and initiates the reaction. The explosive shock wave can shatter rock or other substances without the need for confinement. Because of its unprecedented explosive power, nitroglycerine was called a *high explosive*. Many other high explosives have been developed, including picric acid, TNT (trinitrotoluene), PETN (pentaerythritol tetranitrate); and RDX (research department explosive). Nitroglycerine and PETN are nitrate esters.

picric acid   TNT   PETN   RDX

Pure nitroglycerine is hazardous to make, use, and transport. Alfred Nobel's family were experts at making and using nitroglycerine, yet his brother and several workers were killed by an explosion. In 1866, Nobel found that nitroglycerine soaks into diatomaceous earth to give a pasty mixture that can be molded into sticks that do not detonate so easily. He called the sticks *dynamite*, and founded the firm Dynamit Nobel, which is still one of the world's leading ammunition and explosives manufacturers. The Nobel prizes are funded from an endowment that originated with Nobel's profits from the dynamite business.

◄ **Figure 11-3**
Phosphate ester groups bond the individual nucleotides together in DNA. The "base" on each of the nucleotides corresponds to one of the four heteroxyclic bases of DNA (see Section 23-20).

11-13C   Phosphate Esters

Alkyl phosphates are composed of 1 mole of phosphoric acid combined with 1, 2, or 3 moles of an alcohol. For example, methanol forms three **phosphate esters**.

By controlling the formation of phosphate esters on key proteins, the body is able to regulate many cellular processes. Any disruption of these processes can result in numerous health problems including cancer, diabetes, and obesity.

Phosphate esters play a central role in biochemistry. Figure 11-3 shows how phosphate ester linkages compose the backbone of the nucleic acids RNA (ribonucleic acid) and DNA (deoxyribonucleic acid). These nucleic acids, which carry the genetic information in the cell, are discussed in Chapter 23.

In Section 10-6B, we learned to remove the hydroxyl proton from an alcohol by reduction with an "active" metal such as sodium or potassium. This reaction generates a sodium or potassium salt of an **alkoxide ion** and hydrogen gas.

11-14
Reactions of Alkoxides

$$R-\ddot{O}-H + Na \longrightarrow R-\ddot{O}:^- \ ^+Na + \tfrac{1}{2} H_2 \uparrow$$

$$R-\ddot{O}-H + K \longrightarrow R-\ddot{O}:^- \ ^+K + \tfrac{1}{2} H_2 \uparrow$$

The reactivity of alcohols toward sodium and potassium decreases in the order: methyl $> 1° > 2° > 3°$. Sodium reacts quickly with primary alcohols and some secondary alcohols. Potassium is more reactive than sodium and is commonly used with tertiary alcohols and some secondary alcohols.

Sodium metal reacts vigorously with simple primary alcohols such as ethanol.

Some alcohols react sluggishly with both sodium and potassium. In these cases, a useful alternative is sodium hydride, usually in tetrahydrofuran solution. Sodium hydride reacts quickly to form the alkoxide, even with difficult compounds.

$$\underset{\text{alcohol}}{R-\overset{..}{\underset{..}{O}}-H} \; + \; \underset{\text{sodium hydride}}{NaH} \; \xrightarrow{\text{THF}} \; \underset{\text{sodium alkoxide}}{R-\overset{..}{\underset{..}{O}}:^- \; Na^+} \; + \; \underset{\text{hydrogen}}{H_2 \uparrow}$$

The alkoxide ion is a strong nucleophile as well as a powerful base. Unlike the alcohol itself, the alkoxide ion reacts with primary alkyl halides and tosylates to form ethers. This general reaction, called the **Williamson ether synthesis**, is an S_N2 displacement. The alkyl halide (or tosylate) must be primary so that a back-side attack is not hindered. When the alkyl halide is not primary, elimination usually results.

KEY MECHANISM   The Williamson Ether Synthesis

$$\underset{\text{alkoxide ion}}{R-\overset{..}{\underset{..}{O}}:^- \; {}^+Na} \quad \underset{\text{primary alkyl halide}}{R'-CH_2 \overset{\frown}{} X} \quad \longrightarrow \quad \underset{\text{ether}}{R-O-CH_2-R'} \; + \; NaX$$

Example

$$\underset{\text{sodium ethoxide}}{CH_3CH_2-O^- \; {}^+Na} \; + \; \underset{\text{methyl iodide}}{CH_3I} \; \longrightarrow \; \underset{\text{ethyl methyl ether}}{CH_3CH_2-O-CH_3} \; + \; NaI$$

PROBLEM 11-33

What is wrong with the following proposed synthesis of ethyl methyl ether? First, ethanol is treated with acid to protonate the hydroxyl group, and then methoxide is added to displace water.

$$CH_3CH_2-OH \; + \; H^+ \; \longrightarrow \; CH_3CH_2-\overset{+}{O}H_2 \; \xrightarrow[]{CH_3O^-} \!\!\!\!\!\! \diagup \!\!\!\!\!\! \longrightarrow \; CH_3CH_2-O-CH_3$$

(incorrect synthesis of ethyl methyl ether)

PROBLEM 11-34

(a) Show how ethanol and cyclohexanol may be used to synthesize cyclohexyl ethyl ether (tosylation followed by the Williamson ether synthesis).
(b) Why can't we synthesize this product simply by mixing the two alcohols, adding some sulfuric acid, and heating?

PROBLEM 11-35

A student wanted to make (R)-2-ethoxybutane, using the Williamson ether synthesis. He remembered that the Williamson synthesis involves an S_N2 displacement, which takes place with inversion of configuration. He ordered a bottle of (S)-2-butanol for his chiral starting material. He also remembered that the S_N2 goes best on primary halides and tosylates, so he made ethyl tosylate and sodium (S)-2-butoxide. After warming these reagents together, he obtained an excellent yield of 2-ethoxybutane.

(a) What enantiomer of 2-ethoxybutane did he obtain? Explain how this enantiomer results from the S_N2 reaction of ethyl tosylate with sodium (S)-2-butoxide.
(b) What would have been the best synthesis of (R)-2-ethoxybutane?
(c) How can this student convert the rest of his bottle of (S)-2-butanol to (R)-2-ethoxybutane?

PROBLEM 11-36

The anions of phenols (phenoxide ions) may be used in the Williamson ether synthesis, especially with very reactive alkylating reagents such as dimethyl sulfate. Using phenol, dimethyl sulfate, and other necessary reagents, show how you would synthesize methyl phenyl ether.

PROBLEM-SOLVING HINT

In using the Williamson ether synthesis to make R—O—R', choose the less hindered alkyl group to serve as the alkyl halide (R'—X) because it will make a better S_N2 substrate. Choose the more hindered alkyl group to form the alkoxide (R—O⁻) because it is less sensitive to steric hindrance in the reaction.

$$R-O^- + R'-X \longrightarrow R-O-R'$$

PROBLEM-SOLVING STRATEGY
Multistep Synthesis

Chemists use organic syntheses both to make larger amounts of useful natural compounds and to invent totally new compounds in search of improved properties and biological effects. Synthesis also serves as one of the best methods for developing a firm command of organic chemistry. Planning a practical multistep synthesis requires a working knowledge of the applications and the limitations of a variety of organic reactions. We will often use synthesis problems for reviewing and reinforcing the reactions we have covered.

We use a systematic approach to solving multistep synthesis problems, working backward, in the "retrosynthetic" direction. We begin by studying the target molecule and considering what final reactions might be used to create it from simpler intermediate compounds. Most syntheses require comparison of two or more pathways and the intermediates involved. Eventually, this retrosynthetic analysis should lead back to starting materials that are readily available or meet the requirements defined in the problem.

We can now extend our systematic analysis to problems involving alcohols and the Grignard reaction. As examples, we consider the syntheses of an acyclic diol and a disubstituted cyclohexane, concentrating on the crucial steps that assemble the carbon skeletons and generate the final functional groups.

SAMPLE PROBLEM

Our first problem is to synthesize 3-ethyl-2,3-pentanediol from compounds containing no more than three carbon atoms.

3-ethyl-2,3-pentanediol

1. **Review the functional groups and carbon skeleton of the target compound.**
   The compound is a vicinal diol (glycol) containing seven carbon atoms. Glycols are commonly made by hydroxylation of alkenes, and this glycol would be made by hydroxylation of 3-ethyl-2-pentene, which effectively becomes the target compound.

2. **Review the functional groups and carbon skeletons of the starting materials (if specified), and see how their skeletons might fit together into the target compound.**
   The limitation is that the starting materials must contain no more than three carbon atoms. To form a seven-carbon product requires at least three fragments, probably a three-carbon fragment and two two-carbon fragments. A functional group that can be converted to an alkene will be needed on either C2 or C3 of the chain, since 3-ethyl-2-pentene has a double bond between C2 and C3.

3. **Compare methods for assembling the carbon skeleton of the target compound. Which ones provide a key intermediate with the correct carbon skeleton and functional groups correctly positioned for conversion to the functionality in the target molecule?**
   At this point, the Grignard reaction is our most powerful method for assembling a carbon skeleton, and Grignards can be used to make primary, secondary, and tertiary alcohols. The secondary alcohol 3-ethyl-2-pentanol has its functional group on C2, while the tertiary alcohol 3-ethyl-3-pentanol has it on C3. Either of these alcohols can be synthesized by an appropriate Grignard reaction, but 3-ethyl-2-pentanol may dehydrate to give a mixture of products. Because of its symmetry, 3-ethyl-3-pentanol dehydrates to give only the desired alkene, 3-ethyl-2-pentene. It also dehydrates more easily because it is a tertiary alcohol.

$$\underset{\underset{\displaystyle \text{3-ethyl-2-pentanol}}{}}{\underset{\displaystyle \overset{\displaystyle |}{OH}}{CH_3-CH-CH-CH_2-CH_3}}\overset{\displaystyle CH_2-CH_3}{\underset{|}{}} \xrightarrow{H_2SO_4} \left\{ \underset{\underset{\displaystyle \text{3-ethyl-2-pentene}}{\text{(major)}}}{CH_3-CH=C-CH_2-CH_3}\overset{\displaystyle CH_2-CH_3}{\underset{|}{}} + \underset{\underset{\displaystyle \text{3-ethyl-1-pentene}}{\text{(minor)}}}{CH_2=CH-CH-CH_2-CH_3}\overset{\displaystyle CH_2-CH_3}{\underset{|}{}} \right.$$

Preferred synthesis:

$$\underset{\underset{\displaystyle \text{3-ethyl-3-pentanol}}{}}{\underset{\displaystyle \overset{\displaystyle |}{OH}}{CH_3-CH_2-C-CH_2-CH_3}}\overset{\displaystyle CH_2-CH_3}{\underset{|}{}} \xrightarrow{H_2SO_4} \underset{\underset{\displaystyle \text{3-ethyl-2-pentene}}{\text{(only product)}}}{CH_3-CH=C-CH_2-CH_3}\overset{\displaystyle CH_2-CH_3}{\underset{|}{}}$$

4. **Working backward through as many steps as necessary, compare methods for synthesizing the reactants needed for assembly of the key intermediate. (This process may require writing several possible reaction sequences and evaluating them, keeping in mind the specified starting materials.)**

The key intermediate, 3-ethyl-3-pentanol, is simply methanol substituted by three ethyl groups. The last step in its synthesis must add an ethyl group. Addition of ethyl magnesium bromide to 3-pentanone gives 3-ethyl-3-pentanol.

$$\underset{\text{3-pentanone}}{\underset{\displaystyle \overset{\displaystyle ||}{O}}{CH_3-CH_2-C-CH_2-CH_3}} \xrightarrow[\text{(2) }H_3O^+]{\text{(1) }CH_3CH_2-MgBr} \underset{\text{3-ethyl-3-pentanol}}{\underset{\displaystyle \overset{\displaystyle |}{OH}}{CH_3-CH_2-C-CH_2-CH_3}}\overset{\displaystyle CH_2-CH_3}{\underset{|}{}}$$

The synthesis of 3-pentanone from a three-carbon fragment and a two-carbon fragment requires several steps (see Problem 11-37). Perhaps there is a better alternative, considering that the key intermediate has three ethyl groups on a carbinol carbon atom. Two similar alkyl groups can be added in one Grignard reaction with an acid chloride or an ester (Section 10-9D). Addition of 2 moles of ethyl magnesium bromide to a three-carbon acid chloride gives 3-ethyl-3-pentanol.

$$\underset{\text{propionyl chloride}}{\underset{\displaystyle \overset{\displaystyle ||}{O}}{CH_3-CH_2-C-Cl}} \xrightarrow[\text{(2) }H_3O^+]{\text{(1) 2 }CH_3CH_2-MgBr} \underset{\text{3-ethyl-3-pentanol}}{\underset{\displaystyle \overset{\displaystyle |}{OH}}{CH_3-CH_2-C-CH_2-CH_3}}\overset{\displaystyle CH_2-CH_3}{\underset{|}{}}$$

5. **Summarize the complete synthesis in the forward direction, including all steps and all reagents, and check it for errors and omissions.**

$$\underset{\text{propionyl chloride}}{\underset{\displaystyle \overset{\displaystyle ||}{O}}{CH_3-CH_2-C-Cl}} \xrightarrow[\text{(2) }H_3O^+]{\text{(1) 2 }CH_3CH_2-MgBr} \underset{\underset{\displaystyle \text{3-ethyl-3-pentanol}}{}}{\underset{\displaystyle \overset{\displaystyle |}{OH}}{CH_3-CH_2-C-CH_2-CH_3}}\overset{\displaystyle CH_2-CH_3}{\underset{|}{}} \xrightarrow{H_2SO_4}$$

$$\underset{\text{3-ethyl-2-pentene}}{CH_3-CH=C-CH_2-CH_3}\overset{\displaystyle CH_2-CH_3}{\underset{|}{}} \xrightarrow[\text{(cold, dilute)}]{KMnO_4} \underset{\underset{\displaystyle \text{3-ethyl-2,3-pentanediol}}{}}{\underset{\displaystyle \overset{\displaystyle |\quad|}{OH\;\,OH}}{CH_3-CH-C-CH_2-CH_3}}\overset{\displaystyle CH_2-CH_3}{\underset{|}{}}$$

PROBLEM 11-37

To practice working through the early parts of a multistep synthesis, devise syntheses of
(a) 3-ethyl-2-pentanol from compounds containing no more than three carbon atoms.
(b) 3-pentanone from alcohols containing no more than three carbon atoms.

SAMPLE PROBLEM

As another example of the systematic approach to multistep synthesis, let's consider the synthesis of 1-bromo-2-methylcyclohexane from cyclohexanol.

1. **Review the functional groups and carbon skeleton of the target compound.**
   The skeleton has seven carbon atoms: a cyclohexyl ring with a methyl group. It is an alkyl bromide, with the bromine atom on a ring carbon one atom removed from the methyl group.

2. **Review the functional groups and carbon skeletons of the starting materials (if specified), and see how their skeletons might fit together into the target compound.**
   The starting compound has only six carbon atoms; clearly, the methyl group must be added, presumably at the functional group. There are no restrictions on the methylating reagent, but it must provide a product with a functional group that can be converted to an adjacent halide.

3. **Compare methods for assembling the carbon skeleton of the target compound to determine which methods provide a key intermediate with the correct carbon skeleton and functional groups at the correct positions for being converted to the functionality in the target molecule.**
   Once again, the best choice is a Grignard reaction, but there are two possible reactions that give the methylcyclohexane skeleton. A cyclohexyl Grignard reagent can add to formaldehyde, or a methyl Grignard reagent can add to cyclohexanone. (There are other possibilities, but none that are more direct.)

Neither product has its alcohol functional group on the carbon atom that is functionalized in the target compound. Alcohol C needs its functional group moved two carbon atoms, while alcohol D needs it moved only one carbon atom. Converting alcohol D to an alkene functionalizes the correct carbon atom. Anti-Markovnikov addition of HBr converts the alkene to an alkyl halide with the bromine atom on the correct carbon atom.

4. **Working backward through as many steps as necessary, compare methods for synthesizing the reactants needed for assembly of the key intermediate.**
All that remains is to make cyclohexanone by oxidation of cyclohexanol.

5. **Summarize the complete synthesis in the forward direction, including all steps and all reagents, and check it for errors and omissions.**

Problem 11-38 provides practice in multistep syntheses and using alcohols as intermediates.

PROBLEM 11-38

Develop syntheses for the following compounds. As starting materials, you may use cyclopentanol, alcohols containing no more than four carbon atoms, and any common reagents and solvents.

(a) *trans*-cyclopentane-1,2-diol   (b) 1-chloro-1-ethylcyclopentane

(c)   (d)   (e)

SUMMARY **Reactions of Alcohols**

1. *Oxidation–reduction reactions*
   a. *Oxidation of secondary alcohols to ketones* (Section 11-2A)

Example

   b. *Oxidation of primary alcohols to carboxylic acids* (Section 11-2B)

Example

$$CH_3(CH_2)_4-CH_2-OH \xrightarrow{Na_2Cr_2O_7,\ H_2SO_4} CH_3(CH_2)_4-\overset{\overset{\displaystyle O}{\|}}{C}-OH$$

1-hexanol   hexanoic acid

c. *Oxidation of primary alcohols to aldehydes* (Section 11-2B)

$$R-CH_2-OH \xrightarrow{PCC} R-\overset{\overset{\displaystyle O}{\|}}{C}-H$$

Example

$$CH_3(CH_2)_4-CH_2-OH \xrightarrow{PCC} CH_3(CH_2)_4-\overset{\overset{\displaystyle O}{\|}}{C}-H$$

1-hexanol   hexanal

d. *Reduction of alcohols to alkanes* (Section 11-6)

$$R-OH \xrightarrow[\text{(2) LiAlH}_4]{\text{(1) TsCl/pyridine}} R-H$$

Example

cyclohexanol   cyclohexane

2. *Cleavage of the alcohol hydroxyl group* $-C\overset{\xi}{}O-H$

  a. *Conversion of alcohols to alkyl halides* (Sections 11-7 through 11-9)

$$R-OH \xrightarrow{\text{HCl or SOCl}_2/\text{pyridine}} R-Cl$$

$$R-OH \xrightarrow{\text{HBr or PBr}_3} R-Br$$

$$R-OH \xrightarrow{\text{HI or P/I}_2} R-I$$

Examples

$$(CH_3)_3C-OH \xrightarrow{HCl} (CH_3)_3C-Cl$$

t-butyl alcohol   *t*-butyl chloride

$$(CH_3)_2CH-CH_2OH \xrightarrow{PBr_3} (CH_3)_2CH-CH_2Br$$

isobutyl alcohol   isobutyl bromide

$$CH_3(CH_2)_4-CH_2OH \xrightarrow{P/I_2} CH_3(CH_2)_4-CH_2I$$

1-hexanol   1-iodohexane

b. *Dehydration of alcohols to form alkenes* (Section 11-10A)

$$\underset{\big|}{\overset{\big|}{-}}\overset{H}{\underset{\displaystyle C}{|}}-\overset{OH}{\underset{\displaystyle C}{|}}\underset{\big|}{\overset{\big|}{-}} \underset{\text{H}_2\text{SO}_4 \text{ or H}_3\text{PO}_4}{\rightleftharpoons} \quad C=C \quad + \quad H_2O$$

Example

cyclohexanol   cyclohexene

c. *Industrial dehydration of alcohols to form ethers* (Section 11-10B)

$$2\,R\!-\!OH \rightleftharpoons^{H^+} R\!-\!O\!-\!R + H_2O$$

Example

$$2\,CH_3CH_2OH \xrightarrow[140°C]{H_2SO_4} CH_3CH_2\!-\!O\!-\!CH_2CH_3 + H_2O$$

ethanol   diethyl ether

3. *Cleavage of the hydroxyl proton* $-\overset{|}{\underset{|}{C}}\!-\!O\!\!\frac{?}{?}\!H$

a. *Tosylation* (Section 11-5)

alcohol   tosyl chloride (TsCl)   alkyl tosylate

Example

$$(CH_3)_2CH\!-\!OH \xrightarrow{TsCl/pyridine} (CH_3)_2CH\!-\!OTs$$

isopropyl alcohol   isopropyl tosylate

b. *Acylation to form esters* (Section 11-12)

Example

cyclohexanol   cyclohexyl acetate

c. *Deprotonation to form an alkoxide* (Section 11-14)

$$R\!-\!OH + Na\ (or\ K) \longrightarrow R\!-\!O^-\ ^+Na + \tfrac{1}{2}H_2\!\uparrow$$

$$R\!-\!OH + NaH \longrightarrow R\!-\!O^-\ ^+Na + \tfrac{1}{2}H_2\!\uparrow$$

Example

$$CH_3\!-\!CH_2\!-\!OH + Na \longrightarrow Na^{+-}O\!-\!CH_2\!-\!CH_3$$

ethanol   sodium ethoxide

d. *Williamson ether synthesis* (Sections 11-14 and 14-5)

$$R\!-\!O^- + R'\!-\!X \longrightarrow R\!-\!O\!-\!R' + X^-$$

(R′ must be primary)

Example

$$Na^{+-}O\!-\!CH_2CH_3 + CH_3I \longrightarrow CH_3CH_2\!-\!O\!-\!CH_3 + NaI$$

sodium ethoxide   methyl iodide   ethyl methyl ether

alcohol dehydrogenase (ADH) An enzyme used by living cells to oxidize ethyl alcohol to acetaldehyde. (p. 452)

alkoxide ion The anion formed by deprotonating an alcohol. (p. 477)

$$R-\ddot{O}-H \; + \; Na \; \longrightarrow \; R-\ddot{O}\colon^- \; {}^+Na \; + \; \tfrac{1}{2} H_2 \uparrow$$

chromic acid reagent The solution formed by adding sodium or potassium dichromate (and a small amount of water) to concentrated sulfuric acid. (p. 447)

> **chromic acid test:** When a primary or secondary alcohol is warmed with the chromic acid reagent, the orange color changes to green or blue. A nonoxidizable compound (such as a tertiary alcohol, a ketone, or an alkane) produces no color change. (p. 449)

Collins reagent ($CrO_3 \cdot 2$ pyridine) A complex of chromium trioxide with pyridine, used to oxidize primary alcohols selectively to aldehydes. (p. 450)

ester An acid derivative formed by the reaction of an acid with an alcohol with loss of water. The most common esters are carboxylic esters, composed of carboxylic acids and alcohols. (p. 474)

> **Fischer esterification:** The acid-catalyzed reaction of an alcohol with a carboxylic acid to give an ester. (p. 474)

$$\underset{\text{acid}}{R-\overset{\displaystyle O}{\overset{\|}{C}}-\boxed{OH}} \; + \; \underset{\text{alcohol}}{\boxed{H}-O-R'} \; \underset{\longleftarrow}{\overset{H^+}{\longrightarrow}} \; \underset{\text{ester}}{R-\overset{\displaystyle O}{\overset{\|}{C}}-O-R'} \; + \; \boxed{H_2O}$$

ether A compound containing an oxygen atom bonded to two alkyl or aryl groups. (p. 478)

inorganic esters Compounds derived from alcohols and inorganic acids with loss of water. (p. 475) Examples are:

$$\underset{\text{sulfonate esters}}{R-O-\overset{\displaystyle O}{\underset{\displaystyle O}{\overset{\|}{\underset{\|}{S}}}}-R} \qquad \underset{\text{sulfate esters}}{R-O-\overset{\displaystyle O}{\underset{\displaystyle O}{\overset{\|}{\underset{\|}{S}}}}-O-R} \qquad \underset{\text{nitrate esters}}{R-O-\overset{+}{N}\!\!\begin{array}{c}{\nearrow O}\\[-2pt]{\searrow O^-}\end{array}} \qquad \underset{\text{phosphate esters}}{R-O-\overset{\displaystyle O}{\underset{\displaystyle O-R}{\overset{\|}{\underset{\|}{P}}}}-O-R}$$

Jones reagent A solution of dilute chromic acid dissolved in acetone, used for alcohol oxidations. (p. 450).

Lucas test A test used to determine whether an alcohol is primary, secondary, or tertiary. The test measures the rate of reaction with the **Lucas reagent**, $ZnCl_2$ in concentrated HCl. Tertiary alcohols react fast, secondary alcohols react more slowly, and primary alcohols react very slowly. (p. 459).

nicotinamide adenine dinucleotide (NAD) A biological oxidizing/reducing reagent that operates in conjunction with enzymes such as alcohol dehydrogenase. (p. 452).

oxidation Loss of H_2; addition of O or O_2; addition of X_2 (halogens). Alternatively, an increase in the number of bonds to oxygen or halogens or a decrease in the number of bonds to hydrogen. (p. 445).

pinacol rearrangement Dehydration of a glycol in which one of the groups migrates to give a ketone. (p. 472).

pyridinium chlorochromate (PCC) A complex of chromium trioxide with pyridine and HCl. PCC oxidizes primary alcohols to aldehydes. (p. 449).

reduction Addition of H_2 (or H^-); loss of O or O_2; loss of X_2 (halogens). Alternatively, a reduction in the number of bonds to oxygen or halogens, or an increase in the number of bonds to hydrogen. (p. 445).

Swern oxidation A mild oxidation, using DMSO and oxalyl chloride, that can oxidize primary alcohols to aldehydes and secondary alcohols to ketones. (p. 450)

tosylate ester An ester of an alcohol with *para*-toluenesulfonic acid. Like halide ions, the tosylate anion is an excellent leaving group. (p. 455)

Williamson ether synthesis The S_N2 reaction between an alkoxide ion and a primary alkyl halide or tosylate. The product is an ether. (p. 478)

$$R-\ddot{O}\colon \curvearrowright R'\!\!-\!\!X \; \longrightarrow \; R-\ddot{O}-R' \; + \; X^-$$

Essential Problem-Solving Skills in Chapter 11

1. Identify whether oxidation or reduction is needed to interconvert alkanes, alcohols, aldehydes, ketones, and acids, and identify reagents that will accomplish the conversion.

2. Predict the products of the reactions of alcohols with
   (a) Oxidizing and reducing agents
   (b) Carboxylic acids and acid chlorides
   (c) Dehydrating reagents, especially H_2SO_4 and H_3PO_4
   (d) Inorganic acids
   (e) Sodium, potassium, and sodium hydride

3. Predict the products of reactions of alkoxide ions.

4. Propose chemical tests to distinguish alcohols from the other types of compounds we have studied.

5. Use your knowledge of alcohol and diol reactions to propose mechanisms and products of similar reactions you have never seen before.

6. Show how to convert an alcohol to a related compound with a different functional group.

7. Predict the products of pinacol rearrangement and periodate cleavage of glycols.

8. Use retrosynthetic analysis to propose effective single-step and multistep syntheses of compounds using alcohols as intermediates (especially those using Grignard and organolithium reagents to assemble the carbon skeletons).

Study Problems

11-39 Briefly define each term, and give an example.
   (a) PCC oxidation
   (b) carboxylic ester
   (c) tosylate ester
   (d) ether
   (e) Williamson ether synthesis
   (f) alkyl phosphate ester
   (g) Lucas test
   (h) pinacol rearrangement
   (i) chromic acid oxidation
   (j) alkoxide ion
   (k) nitrate ester

11-40 In each case, show how you would synthesize the chloride, bromide, and iodide from the corresponding alcohol.
   (a) 1-halobutane (halo = chloro, bromo, iodo)
   (b) halocyclopentane
   (c) 1-halo-1-methylcyclohexane
   (d) 1-halo-2-methylcyclohexane

11-41 Predict the major products of the following reactions, including stereochemistry where appropriate.
   (a) (R)-2-butanol + TsCl in pyridine
   (b) (S)-2-butyl tosylate + NaBr
   (c) cyclooctanol + CrO_3/H_2SO_4
   (d) cyclopentylmethanol + CrO_3·pyridine·HCl
   (e) cyclopentylmethanol + $Na_2Cr_2O_7/H_2SO_4$
   (f) cyclopentanol + $HCl/ZnCl_2$
   (g) cycloheptanol + $LiAlH_4/TiCl_4$
   (h) cyclooctylmethanol + CH_3CH_2MgBr
   (i) potassium t-butoxide + methyl iodide
   (j) sodium methoxide + t-butyl iodide
   (k) n-butanol + HBr
   (l) cyclopentanol + H_2SO_4/heat
   (m) product from (l) + OsO_4/H_2O_2, then HIO_4
   (n) sodium ethoxide + 1-bromobutane
   (o) sodium ethoxide + 2-methyl-2-bromobutane

11-42 Show how you would accomplish the following synthetic conversions.

11-43   Predict the major products of dehydration catalyzed by sulfuric acid.

(a) 1-hexanol   (b) 2-hexanol   (c) 3-pentanol
(d) 1-methylcyclopentanol   (e) cyclopentylmethanol   (f) 2-methylcyclopentanol

11-44   Predict the esterification products of the following acid/alcohol pairs.

(a) $CH_3CH_2CH_2COOH + CH_3OH$

(b) [phenol with OH] + $CH_3 - \overset{\overset{O}{\|}}{C} - OH$

(c) [benzene with COOH] + CH_3CH_2OH

(d) $2\ CH_3CH_2OH + H_3PO_4$

(e) $CH_3OH + HNO_3$

11-45   Show how you would make the methanesulfonate ester of cyclohexanol, beginning with cyclohexanol and an appropriate acid chloride.

cyclohexyl methanesulfonate: [structure of cyclohexyl—O—S(=O)(=O)—CH₃]

11-46   Show how you would convert (S)-2-hexanol to
(a) (S)-2-chlorohexane   (b) (R)-2-bromohexane

11-47   When 1-cyclohexylethanol is treated with concentrated aqueous HBr, the major product is 1-bromo-1-ethylcyclohexane.

[structure: cyclohexyl-CH(OH)CH₃] $\xrightarrow[H_2O]{HBr}$ [structure: 1-bromo-1-ethylcyclohexane]

(a) Give a mechanism for this reaction.
(b) How would you convert 1-cyclohexylethanol to (1-bromoethyl)cyclohexane in good yield?

[structure: cyclohexyl-CH(OH)CH₃] $\xrightarrow{?}$ [structure: cyclohexyl-CH(Br)CH₃]

11-48   Show how you would make each compound, beginning with an alcohol of your choice.

(a) [cycloheptane with CHO]   (b) [cyclohexane with CH₂Br]   (c) [cyclohexane with OCH₃]   (d) [cyclohexane with Cl and CH₃ and H substituents]   (e) [structure with CH₃ and C=O groups]

(f) [cyclopentane with C(=O)—OH]   (g) [cyclohexanone with tert-butyl]   (h) [cyclohexene with tert-butyl]   (i) [cyclopentane with CH₃ and OTs and H substituents]   (j) [cyclopentane with I and H]

11-49   Predict the major products (including stereochemistry) when cis-3-methylcyclohexanol reacts with the following reagents.
(a) PBr₃   (b) SOCl₂   (c) Lucas reagent   (d) concentrated HBr   (e) TsCl/pyridine, then NaBr

11-50   Show how you would use simple chemical tests to distinguish between the following pairs of compounds. In each case, describe what you would do and what you would observe.
(a) 1-butanol and 2-butanol   (b) 2-butanol and 2-methyl-2-butanol
(c) cyclohexanol and cyclohexene   (d) cyclohexanol and cyclohexanone
(e) cyclohexanone and 1-methylcyclohexanol

11-51 Write the important resonance structures of the following anions.

(a) (b) (c)

11-52 Compound **A** is an optically active alcohol. Treatment with chromic acid converts **A** into a ketone, **B**. In a separate reaction, **A** is treated with PBr₃, converting **A** into compound **C**. Compound **C** is purified, and then it is allowed to react with magnesium in ether. Compound **B** is added to the resulting solution of the Grignard reagent. After hydrolysis, this solution is found to contain 3,4-dimethyl-3-hexanol. Propose structures for compounds **A**, **B**, and **C**.

11-53 Give the structures of the intermediates and products **V** through **Z**.

11-54 Under acid catalysis, tetrahydrofurfuryl alcohol reacts to give surprisingly good yields of dihydropyran. Propose a mechanism to explain this useful synthesis.

tetrahydrofurfuryl alcohol   dihydropyran

*11-55 The following compound is first treated with OsO₄/H₂O₂ and then with concentrated H₂SO₄. Propose structures for both the intermediate product and the final product, which has undergone a pinacol rearrangement.

$C_{10}H_{18}O_2$
intermediate

$C_{10}H_{16}O$
product

11-56 Show how you would synthesize the following compounds. As starting materials, you may use any alcohols containing four or fewer carbon atoms, cyclohexanol, and any necessary solvents and inorganic reagents.

(a) (b) (c) (d)

(e) (f) (g) *(h)

*11-57 Many alcohols undergo dehydration at 0°C when treated with phosphorus oxychloride (POCl₃) in the basic solvent pyridine. (Phosphorus oxychloride is the acid chloride of phosphoric acid, with chlorine atoms in place of the hydroxyl groups of phosphoric acid.)

(a) Propose a mechanism for the dehydration of cyclopentanol using POCl₃ and pyridine. The first half of the mechanism, formation of a dichlorophosphate ester, is similar to the first half of the mechanism of reaction of an alcohol with thionyl chloride. Like a tosylate, the dichlorophosphate group is a good leaving group. The second half of the mechanism might be either first order or second order; draw both alternatives for now.

(b) When *trans*-2-methylcyclopentanol undergoes dehydration using POCl₃ in pyridine, the major product is 3-methylcyclopentene, and not the Saytzeff product. What is the stereochemistry of the dehydration? What does this stereochemistry imply about the correct mechanism in part (a)? Explain your reasoning.

*11-58   Two unknowns, **X** and **Y**, both having the molecular formula C_4H_8O, give the following results with four chemical tests. Propose structures for **X** and **Y** consistent with this information.

	Bromine	Na Metal	Chromic Acid	Lucas Reagent
Compound **X**	decolorizes	bubbles	orange to green	no reaction
Compound **Y**	no reaction	no reaction	no reaction	no reaction

*11-59   Alcohols combine with ketones and aldehydes to form interesting derivatives, which we will discuss in Chapter 18. The following reactions show the hydrolysis of two such derivatives. Propose mechanisms for these reactions.

(a)

(b)

CHAPTER 12

Infrared Spectroscopy and Mass Spectrometry

12-1
Introduction

One of the most important tasks of organic chemistry is the determination of organic structures. When an interesting compound is isolated from a natural source, its structure must be completely determined before a synthesis is begun. Whenever we run a reaction, we must determine whether the product has the desired structure. The structure of an unwanted product must be known so the reaction conditions can be altered to favor the desired product.

In many cases, a compound can be identified by chemical means. We find the molecular formula by analyzing the elemental composition and determining the molecular weight. If the compound has been characterized before, we can compare its physical properties (melting point, boiling point, etc.) with the published values. Chemical tests can suggest the functional groups and narrow the range of possible structures before the physical properties are used to make an identification.

These procedures are not sufficient, however, for complex compounds that have never been synthesized and characterized. They are also impractical with compounds that are difficult to obtain, because a relatively large sample is required to complete the elemental analysis and all the functional group tests. We need analytical techniques that work with tiny samples and that do not destroy the sample.

Spectroscopic techniques often meet these requirements. **Absorption spectroscopy** is the measurement of the amount of light absorbed by a compound as a function of the wavelength of light. In general, a sample is irradiated by a light source, and the amount of light transmitted at various wavelengths is measured by a detector and plotted on a graph. Unlike chemical tests, most spectroscopic techniques are *nondestructive*; that is, the sample is not destroyed. Many different kinds of spectra can be measured with little or no loss of sample.

In this book, we cover four spectroscopic or related techniques that serve as powerful tools for structure determination in organic chemistry:

Infrared (IR) spectroscopy, covered in this chapter, observes the vibrations of bonds and provides evidence of the functional groups present.

Mass spectrometry (MS), also covered in this chapter, bombards molecules with electrons and breaks them apart. Analysis of the masses of the fragments gives the molecular weight, possibly the molecular formula, and clues to the structure and functional groups. Less than a milligram of sample is destroyed in this analysis.

Nuclear magnetic resonance (NMR) spectroscopy, covered in Chapter 13, observes the chemical environments of the hydrogen atoms (or the carbon atoms) and provides evidence for the structure of the alkyl groups and clues to the functional groups.

Ultraviolet (UV) spectroscopy, covered in Chapter 15, observes electronic transitions and provides information on the electronic bonding in the sample.

These spectroscopic techniques are complementary, and they are most powerful when used together. In many cases, an unknown compound cannot be completely identified from one spectrum without additional information, yet the structure can be determined with confidence using two or more different types of spectra. In Chapter 13, we consider how information from different types of spectroscopy is combined to provide a reliable structure.

12-2
The Electromagnetic Spectrum

Visible light, infrared light, ultraviolet light, microwaves, and radio waves are examples of electromagnetic radiation. They all travel at the speed of light, about 3×10^{10} cm/sec, but they differ in frequency and wavelength. The **frequency** of a wave is the number of complete wave cycles that pass a fixed point in a second. Frequency, represented by the Greek letter ν (nu), is usually given in hertz (Hz), meaning cycles per second. The **wavelength**, represented by the Greek letter λ (lambda), is the distance between any two peaks (or any two troughs) of the wave.

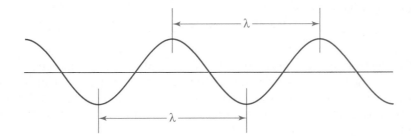

The wavelength and frequency, which are inversely proportional, are related by the equation

$$\nu\lambda = c \quad \text{or} \quad \lambda = \frac{c}{\nu}$$

where

$$c = \text{speed of light } (3 \times 10^{10} \text{ cm/sec})$$

$$\nu = \text{frequency in hertz}$$

$$\lambda = \text{wavelength in centimeters}$$

Electromagnetic waves travel as **photons**, which are massless packets of energy. The energy of a photon is proportional to its frequency and inversely proportional to its wavelength. A photon of frequency ν (or wavelength λ) has an energy given by

$$\varepsilon = h\nu = \frac{hc}{\lambda}$$

where h is Planck's constant, 1.58×10^{-37} kcal/sec or 6.62×10^{-37} kJ/sec.

Under certain conditions, a molecule struck by a photon may absorb the photon's energy. In this case, the molecule's energy is increased by an amount equal to the photon's energy $h\nu$. For this reason, we often represent the irradiation of a reaction mixture by the symbol $h\nu$.

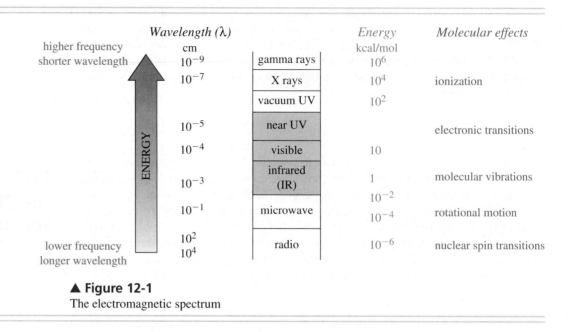

▲ **Figure 12-1**
The electromagnetic spectrum

The **electromagnetic spectrum** is the range of all possible frequencies, from zero to infinity. In practice, the spectrum ranges from the very low radio frequencies used to communicate with submarines to the very high frequencies of gamma rays. Figure 12-1 shows the wavelength and energy relationships of the various parts of the electromagnetic spectrum.

The electromagnetic spectrum is continuous, and the exact positions of the dividing lines between the different regions are somewhat arbitrary. Toward the top of the spectrum in Figure 12-1 are the higher frequencies, shorter wavelengths, and higher energies. Toward the bottom are the lower frequencies, longer wavelengths, and lower energies. X rays (very high energy) are so energetic that they excite electrons past all the energy levels, causing ionization. Energies in the ultraviolet–visible range excite electrons to higher energy levels within molecules. Infrared energies excite molecular vibrations, and microwave energies excite rotations. Radio-wave frequencies (very low energy) excite the nuclear spin transitions observed in NMR spectroscopy.

12-3
The Infrared Region

The infrared (from the Latin, *infra*, meaning "below" red) region of the spectrum corresponds to frequencies from just below the visible frequencies to just above the highest microwave and radar frequencies: wavelengths of about 8×10^{-5} cm to 1×10^{-2} cm. Common infrared spectrometers operate in the middle of this region, at wavelengths between 2.5×10^{-4} cm and 25×10^{-4} cm, corresponding to energies of about 1.1 to 11 kcal/mol (4.6 to 46 kJ/mol). Infrared photons do not have enough energy to cause electronic transitions, but they can cause groups of atoms to vibrate with respect to the bonds that connect them. Like electronic transitions, these vibrational transitions correspond to distinct energies, and molecules absorb infrared radiation only at certain wavelengths and frequencies.

The position of an infrared band is specified by its wavelength (λ), measured in *microns* (μm). A micron (or *micrometer*) corresponds to one millionth (10^{-6}) of a meter, or 10^{-4} cm. A more common unit, however, is the **wavenumber** ($\bar{\nu}$), which corresponds to the number of cycles (wavelengths) of the wave in a centimeter. The wavenumber is the reciprocal of the wavelength (in centimeters). Since 1 cm $= 10,000\ \mu$m, the wavenumber can be calculated by dividing 10,000 by the wavelength in microns. The units of the wavenumber are cm^{-1} (*reciprocal centimeters*).

$$\bar{\nu}\ (cm^{-1}) = \frac{1}{\lambda\ (cm)} = \frac{10,000\ \mu m/cm}{\lambda\ (\mu m)} \quad or \quad \lambda\ (\mu m) = \frac{10,000\ \mu m/cm}{\bar{\nu}\ (cm^{-1})}$$

For example, an absorption at a wavelength of 4 μm corresponds to a wavenumber of 2500 cm^{-1}.

$$\bar{\nu} = \frac{10,000\ \mu m/cm}{4\ \mu m} = 2500\ cm^{-1} \quad or \quad \lambda = \frac{10,000\ \mu m/cm}{2500\ cm^{-1}} = 4\ \mu m$$

Wavenumbers (in cm^{-1}) have become the most common method for specifying IR absorptions, and we will use wavenumbers throughout this book. The wavenumber is proportional to the frequency (ν) of the wave, so it is also proportional to the energy of a photon of this frequency ($E = h\nu$). Some reference works still use micron units, however, so you should know how to convert these units.

PROBLEM 12-1

Complete the following conversion table.

$\bar{\nu}\ (cm^{-1})$	4000				1700	1640	1600	400
$\lambda\ (\mu m)$	2.50	3.03	3.33	4.55				25.0

Before discussing characteristic infrared absorptions, we should understand some theory about the vibrational energies of molecules. The following drawing shows how a covalent bond between two atoms acts like a spring. If the bond is stretched, a restoring force pulls the two atoms together toward their equilibrium bond length. If the bond is compressed, the restoring force pushes the two atoms apart. If the bond is stretched or compressed and then released, the atoms vibrate.

12-4
Molecular
Vibrations

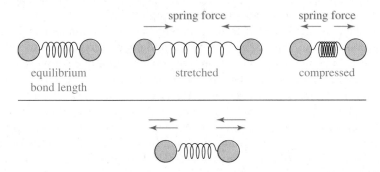

The frequency of the stretching vibration depends on the masses of the atoms and the stiffness of the bond. Heavier atoms vibrate more slowly than lighter ones; for example, a C—D bond has a lower characteristic frequency than a C—H bond. In a group of bonds with similar bond energies, the *frequency decreases with increasing atomic weight*.

Stronger bonds are generally stiffer, requiring more force to stretch or compress them. Thus, stronger bonds usually vibrate faster than weaker bonds (assuming the atoms have similar masses). For example, O—H bonds are stronger than C—H bonds, and O—H bonds vibrate at higher frequencies. Triple bonds are stronger than double bonds, so triple bonds vibrate at higher frequencies than double bonds. Similarly, double bonds vibrate at higher frequencies than single bonds. In a group of bonds having atoms of similar masses, the *frequency increases with bond energy*.

TABLE 12-1 Bond Stretching Frequencies

Bond		Bond Energy [kcal (kJ)]	Stretching Frequency (cm^{-1})
Frequency dependence on atomic masses			
C—H	heavier atoms	100 (420)	3000 $\bar{\nu}$ decreases
C—D		100 (420)	2100
C—C		83 (350)	1200
Frequency dependence on bond energies			
C—C	stronger bond	83 (350)	1200 $\bar{\nu}$ increases
C=C		146 (611)	1660
C≡C		200 (840)	2200
C—N		73 (305)	1200
C=N		147 (615)	1650
C≡N		213 (891)	2200
C—O		86 (360)	1100
C=O		178 (745)	1700

Note: In a group of bonds with similar bond energies, *the frequency decreases with increasing atomic weight.* In a group of bonds between similar atoms, the *frequency increases with bond energy.* Bond energies and frequencies are approximate.

Table 12-1 lists some common types of bonds, together with their stretching frequencies to show how frequency varies with the masses of the atoms and the strength of the bonds.

Even with simple compounds, infrared spectra contain many different absorptions, not just one absorption for each bond. Many of these absorptions result from stretching vibrations of the molecule as a whole, or from bending vibrations. In a bending vibration, the bond lengths stay constant, but the bond angles vibrate about their equilibrium values.

Consider the fundamental vibrational modes of a water molecule (Fig. 12-2). The two O—H bonds can stretch in phase with each other (symmetric stretching), or they can stretch out of phase (antisymmetric stretching). The H—O—H bond angle can also change in a bending vibration, making a scissoring motion.

A nonlinear molecule with *n* atoms generally has $3n - 6$ fundamental vibrational modes. Methane, then, has $3(5) - 6 = 9$ fundamental modes, and ethane has $3(8) - 6 = 18$ fundamental modes. Combinations and multiples of these simple fundamental modes of vibration are also observed. As you can see, the number of absorptions in an infrared spectrum can be quite large, even for simple molecules.

It is unlikely that two different compounds (except enantiomers) will have the same frequencies for all their various complex vibrations. For this reason, the infrared spectrum provides a "fingerprint" of a molecule. In fact, the region of the IR spectrum containing most of these complex vibrations (600 to 1400 cm^{-1}) is commonly called the **fingerprint region** of the spectrum.

▶ **Figure 12-2**
A nonlinear molecule with *n* atoms has $3n - 6$ fundamental vibrational modes. Water has $3(3) - 6 = 3$ modes. Two of these are stretching modes, and one is a bending mode.

symmetric stretching   antisymmetric stretching   bending (scissoring)

The simple stretching vibrations (in the 1600 to 3500 cm^{-1} region) are the most characteristic and predictable; therefore, our study of infrared spectroscopy will concentrate on them. Although we will largely ignore bending vibrations at this stage, you should remember that their absorptions generally appear in the 600 to 1400 cm^{-1} region of the spectrum. Experienced spectroscopists can tell a great deal about the structure of a molecule from these "wagging," "scissoring," "rocking," and "twisting" vibrations in the fingerprint region. The reference table of IR frequencies (Appendix 2) lists both stretching and bending characteristic frequencies.

12-5 IR-Active and IR-Inactive Vibrations

Not all molecular vibrations absorb infrared radiation. To understand which ones do and which do not, we need to consider how an electromagnetic field interacts with a molecular bond. The key to this interaction lies with the dipole moment of the bond. A bond with a dipole moment can be visualized as a positive charge and a negative charge separated by a spring. If this bond is placed in an electric field (Fig. 12-3), it is either stretched or compressed, depending on the direction of the field.

One of the components of an electromagnetic wave is a rapidly reversing electric field ($\mathbf{E}$). This field alternately stretches and compresses a polar bond, as shown in Figure 12-3. When the electric field is in the same direction as the dipole moment, the bond is compressed and its dipole moment decreases. When the field is opposite the dipole moment, the bond stretches and its dipole moment increases. If this alternate stretching and compressing of the bond occurs at the frequency of the molecule's natural rate of vibration, energy may be absorbed. Vibrations of bonds with dipole moments generally result in IR absorptions and are said to be **IR active**.

If a bond is symmetrical and has zero dipole moment, the electric field does not interact with the bond. For example, the triple bond of acetylene (H—C≡C—H) has zero dipole moment, and the dipole moment remains zero if the bond is stretched or compressed. Because the vibration produces no change in the dipole moment, there is no absorption of energy. This vibration is said to be **IR inactive**, and it produces no absorption in the IR spectrum. The key to an IR-active vibration is that *the vibration must change the dipole moment of the molecule.*

In general, if a bond has a dipole moment, its stretching frequency causes an absorption in the IR spectrum. If a bond is symmetrically substituted and has zero dipole moment, its stretching vibration is weak or absent in the spectrum. Bonds with zero dipole moments sometimes produce absorptions (usually weak) because molecular collisions, rotations, and vibrations make them unsymmetrical part of the time.

◄ Figure 12-3
A bond with a dipole moment (as in HF, for example) is either stretched or compressed by an electric field, depending on the direction of the field. Notice that the force on the positive charge is in the direction of the electric field ($\mathbf{E}$), and the force on the negative charge is in the opposite direction.

PROBLEM 12-2

Which of the bonds shown in red are expected to have IR-active stretching frequencies?

$$H—C≡C—H \qquad\qquad H—C≡C—H$$

$$H—C≡C—CH_3 \qquad H_3C—C≡C—CH_3 \qquad H_3C—C≡C—CH_3$$

$$H_3C—CH_3 \qquad\qquad H_3C—\overset{\displaystyle H}{\underset{\displaystyle H}{\overset{|}{\underset{|}{C}}}}—H$$

$$\begin{array}{cc} H_3C & CH_3 \\ \diagdown C{=}C\diagup \\ H & H \end{array} \qquad \begin{array}{cc} H_3C & CH_3 \\ \diagdown C{=}C\diagup \\ H & H \end{array} \qquad \begin{array}{cc} H & CH_3 \\ \diagdown C{=}C\diagup \\ H & H \end{array}$$

12-6
Measurement of the IR Spectrum

Infrared spectra can be measured using liquid, solid, or gaseous samples. A drop of a liquid can be placed as a thin film between two salt plates made of NaCl or KBr, which are transparent to infrared light at most important frequencies. A solid can be ground with KBr and pressed into a disk that is placed in the light beam. Alternatively, a solid sample can be ground into a pasty *mull* with paraffin oil. The mull is placed between two salt plates like a liquid. Solids can also be dissolved in common solvents such as CH_2Cl_2, CCl_4, or CS_2 that do not have absorptions in the areas of interest. Gases are placed in a longer cell with polished salt windows. These *gas cells* often contain mirrors that reflect the beam through the cell several times for stronger absorption.

An **infrared spectrometer** measures the frequencies of infrared light absorbed by a compound. In a simple infrared spectrometer (Fig. 12-4), two beams of light are used. The *sample beam* passes through the sample cell, while the *reference beam* passes through a reference cell that contains only the solvent. A rotating mirror alternately allows light from each of the two beams to enter the monochromator.

The monochromator uses prisms or diffraction gratings to allow only one frequency of light to enter the detector at a time. It scans the range of infrared frequencies as a pen moves along the corresponding frequencies on the x axis of the chart paper. Higher frequencies (shorter wavelengths) appear toward the left of the chart paper. The detector signal is proportional to the *difference* in the intensity of light in the sample and reference beams, with the reference beam compensating for any absorption by air or by

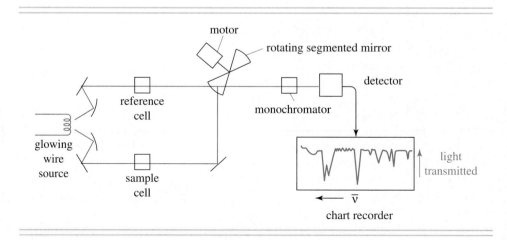

▶ **Figure 12-4**
Block diagram of an infrared spectrometer.

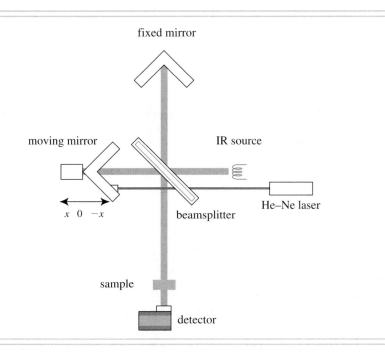

fixed mirror

moving mirror

IR source

He–Ne laser

x 0 −x

beamsplitter

sample

detector

◀ **Figure 12-5**
Block diagram of an interferometer
in an FT–IR spectrometer.

the solvent. The detector signal controls movement of the pen along the y axis, with 100% transmittance (no absorption) at the top of the paper, and 0% transmittance (absorption of all the light) at the bottom.

The spectrometer shown in Figure 12-4 is called a *dispersive* instrument because it disperses light into all the different frequencies and measures them individually. Dispersive instruments require expensive prisms and diffraction gratings, and they must be manually aligned and calibrated on a regular basis. Since only one frequency is observed at a time, dispersive instruments require strong IR sources, and they require 2 to 10 minutes to scan through a complete spectrum. Dispersive infrared spectrometers are being replaced by Fourier transform infrared (**FT–IR**) spectrometers.

A **Fourier transform infrared spectrometer (FT–IR)** uses an **interferometer**, like that shown in Figure 12-5, to measure a spectrum. The infrared light goes from the glowing source to a beamsplitter, usually made of polished KBr, placed at a 45° angle. Part of the beam passes through the beamsplitter, and part is reflected at a right angle. The reflected beam strikes a stationary mirror, while the transmitted beam strikes a mirror that moves at a constant speed. The beams return from the mirrors to recombine at the beamsplitter. The beam from the moving mirror has traveled a different distance than the beam from the fixed mirror, and the two beams combine to create an interference pattern called an **interferogram**. This interferogram, which simultaneously contains all frequencies, passes through the sample compartment to reach the detector.

The interferogram shown in the upper half of Figure 12-6 contains all the information contained in the spectrum shown in the lower half. The interferogram is said to be in the *time domain*, corresponding to the energy seen by the detector as the mirror moves through the signal. A standard computer algorithm called a Fourier transform converts the time domain to the *frequency domain* spectrum that allows us to see the strength of absorption as a function of the frequency (or wavelength). Figure 12-6 shows both the interferogram and the IR spectrum of *n*-octane.

The FT–IR spectrometer has several major advantages over the dispersive instrument: Its sensitivity is better because it measures all frequencies simultaneously rather than scanning through the individual frequencies. Less energy is needed from the source, and less time (typically 1 to 2 seconds) is needed for a scan. Several scans can be completed in a few seconds and averaged to improve the signal. Resolution and accuracy

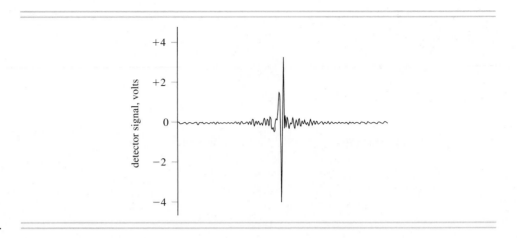

▶ **Figure 12-6(a)**
Interferogram generated by *n*-octane.

▲ **Figure 12-6(b)**
Infrared spectrum of *n*-octane. Notice that the frequencies shown in a routine IR spectrum range from about 600 cm^{-1} to about 4000 cm^{-1}.

PROBLEM-SOLVING HINT

Use *spectrum* (singular) and *spectra* (plural) correctly: "This spectrum is... " "These spectra are.... "

---

are also improved because a laser beam is used alongside the IR beam to control the speed of the moving mirror and to time the collection of data points. The laser beam is a precise frequency reference that keeps the spectrometer accurately calibrated.

In the **infrared spectrum** of *n*-octane (Fig. 12-6(b)) there are four major absorption bands. The broad band between 2800 and 3000 cm^{-1} results from C—H stretching vibrations, and the band at 1467 cm^{-1} results from a scissoring vibration of the CH_2 groups. The absorptions at 1378 and 722 cm^{-1} result from the rocking of CH_3 and CH_2 groups, respectively. Since most organic compounds contain at least some saturated C—H bonds and some CH_2 and CH_3 groups, all these bands are common. In fact, without an authentic spectrum for comparison, *we could not look at this spectrum and conclude that the compound is octane.* We could be fairly certain that it is an alkane, however, because no functional groups are present.

Another characteristic in the octane spectrum is the absence of any identifiable C—C stretching absorptions. (Table 12-1 shows that C—C stretching absorptions occur around 1200 cm^{-1}.) Although there are seven C—C bonds in octane, their dipole moments are small, and their absorptions are weak and indistinguishable. This result is common for alkanes with no functional groups to polarize the C—C bonds.

Hydrocarbons contain only carbon–carbon bonds and carbon–hydrogen bonds. An infrared spectrum does not provide enough information to identify a structure conclusively (unless an authentic spectrum is available to compare "fingerprints"), but the absorptions of the carbon–carbon and carbon–hydrogen bonds can indicate the presence of double and triple bonds.

12-7 Infrared Spectroscopy of Hydrocarbons

12-7A Carbon–Carbon Bond Stretching

Stronger bonds generally absorb at higher frequencies because of their greater stiffness. Carbon–carbon single bonds absorb around 1200 cm^{-1}, C=C double bonds absorb around 1660 cm^{-1}, and C≡C triple bonds absorb around 2200 cm^{-1}.

Carbon–carbon bond stretching frequencies

$$C—C \quad 1200 \text{ cm}^{-1}$$
$$C=C \quad 1660 \text{ cm}^{-1}$$
$$C≡C \quad 2200 \text{ cm}^{-1}$$

As discussed for the octane spectrum, C—C single bond absorptions (and most other absorptions in the fingerprint region) are not very reliable. We use the fingerprint region primarily to confirm the identity of an unknown compound by comparison with an authentic spectrum.

The absorptions of C=C double bonds, however, are useful for structure determination. Most unsymmetrically substituted double bonds produce observable stretching absorptions in the region of 1600 to 1680 cm^{-1}. The specific frequency of the double-bond stretching vibration depends on whether there is another double bond nearby. When two double bonds are one bond apart (as in 1,3-cyclohexadiene, below) they are said to be **conjugated**. As we will see in Chapter 15, conjugated double bonds are slightly more stable than isolated double bonds because there is a small amount of pi bonding between them. This overlap between the pi bonds leaves a little less electron density in the double bonds themselves; as a result, they are a little less stiff and vibrate a little more slowly than an isolated double bond. Isolated double bonds absorb around 1640 to 1680 cm^{-1}, and conjugated double bonds absorb around 1620 to 1640 cm^{-1}.

some pi overlap
less pi overlap than an isolated double bond

1645 cm^{-1} 1620 cm^{-1}
cyclohexene (isolated)   1,3-cyclohexadiene (conjugated)

The effect of conjugation is even more pronounced in aromatic compounds. Aromatic C=C bonds are more like $1\frac{1}{2}$ bonds than true double bonds, and their reduced pi bonding results in less stiff bonds with lower stretching frequencies, around 1600 cm^{-1}.

$$\bar{v} = 1600 \text{ cm}^{-1}$$

bond order $= 1\frac{1}{2}$

Characteristic C=C stretching frequencies

$$\text{isolated C=C} \quad 1640\text{–}1680 \text{ cm}^{-1}$$
$$\text{conjugated C=C} \quad 1620\text{–}1640 \text{ cm}^{-1}$$
$$\text{aromatic C=C} \quad \text{approx. } 1600 \text{ cm}^{-1}$$

Carbon–carbon triple bonds in alkynes are stronger (and stiffer) than carbon–carbon single or double bonds, and they absorb infrared light at higher frequencies. Most alkyne $C\equiv C$ triple bonds have stretching frequencies between 2100 and 2200 cm^{-1}. Terminal alkynes usually give sharp $C\equiv C$ stretching signals of moderate intensity. The $C\equiv C$ stretching absorption of an internal alkyne may be weak or absent, however, due to the symmetry of the disubstituted triple bond with a very small or zero dipole moment.

$$\overset{\overset{\mu}{\longrightarrow}}{R-C\equiv C-H}$$
terminal alkyne

$C\equiv C$ stretch observed around 2100 to 2200 cm^{-1}

$$\overset{\boxed{\mu \text{ is small or zero}}}{R-C\equiv C-R'}$$
internal alkyne

$C\equiv C$ stretch may be weak or absent

12-7B   Carbon–Hydrogen Bond Stretching

Alkanes, alkenes, and alkynes also have characteristic C—H stretching frequencies. Carbon–hydrogen bonds involving sp^3 hybrid carbon atoms generally absorb at frequencies just *below* (to the right of) 3000 cm^{-1}, while those involving sp^2 hybrid carbons absorb just *above* (to the left of) 3000 cm^{-1}. We explain this difference by the amount of s character in the carbon orbital used to form the bond. The s orbital is closer to the nucleus than the p orbitals, and stronger, stiffer bonds result from orbitals with more s character.

An sp^3 orbital is one-fourth s character, while an sp^2 orbital is one-third s character. We expect the bond using the sp^2 orbital to be slightly stronger, with a higher vibration frequency. The C—H bond of a terminal alkyne is formed using an sp hybrid orbital, with about one-half s character. This bond is stiffer than a C—H bond using an sp^3 or sp^2 hybrid carbon, and it absorbs at a higher frequency: about 3300 cm^{-1}.

C—H bond stretching frequencies: $sp > sp^2 > sp^3$

$$-\overset{|}{\underset{|}{C}}-\overset{|}{\underset{|}{C}}-H$$ sp^3 hybridized, one-fourth s character   2800–3000 cm^{-1}

$$\overset{\diagdown}{\diagup}C=C\overset{\diagup}{\underset{\diagdown}{}}_H$$ sp^2 hybridized, one-third s character   3000–3100 cm^{-1}

$$-C\equiv C-H$$ sp hybridized, one-half s character   3300 cm^{-1}

12-7C   Interpretation of the IR Spectra of Hydrocarbons

Figure 12-7 compares the IR spectra of *n*-hexane, 1-hexene, and *cis*-2-octene. The hexane spectrum is similar to that of *n*-octane (Fig. 12-6). The C—H stretching frequencies form a band between 2800 and 3000 cm^{-1}, and the bands in the fingerprint region are due to the bending vibrations discussed for Figure 12-6. This spectrum simply indicates the *absence* of any IR-active functional groups.

The spectrum of 1-hexene shows additional absorptions characteristic of a double bond. The C—H stretch at 3080 cm^{-1} corresponds to the alkene =C—H bonds involving sp^2 hybrid carbons. The absorption at 1642 cm^{-1} results from stretching of the C=C double bond.

PROBLEM-SOLVING $\mathbf{HINT}$

The unsaturated =C—H stretch, to the left of 3000 cm^{-1}, should alert you to look for a weak C=C stretch.

▲ Figure 12-7

Comparison of the IR spectra of (a) *n*-hexane, (b) 1-hexene, and (c) *cis*-2-octene. The most characteristic absorptions in the 1-hexene spectrum are the C=C stretch at 1642 cm^{-1} and the unsaturated =C—H stretch at 3080 cm^{-1}. The nearly symmetrically substituted double bond in *cis*-2-octene gives a weak C=C absorption at 1660 cm^{-1}. The unsaturated =C—H stretch at 3023 cm^{-1} is still apparent, however.

▲ **Figure 12-8**

(a) The IR spectrum of 1-octyne shows characteristic absorptions at 3313 cm^{-1} (alkynyl ≡C—H stretch) and at 2119 cm^{-1} (C≡C stretch). (b) Neither of these absorptions is seen in the spectrum of 4-octyne.

The spectrum of *cis*-2-octene (Figure 12-7c) resembles the spectrum of 1-hexene, except that the C=C stretching absorption at 1660 cm^{-1} is very weak in *cis*-2-octene because the disubstituted double bond has a very small dipole moment. Even without the weak C=C stretching absorption, the unsaturated =C—H stretching absorption just above 3000 cm^{-1} suggests the presence of an alkene double bond.

Figure 12-8 compares the IR spectra of 1-octyne and 4-octyne. In addition to the alkane absorptions, the 1-octyne spectrum shows sharp peaks at 3313 and 2119 cm^{-1}. The absorption at 3313 cm^{-1} results from stretching of the stiff ≡C—H bond formed by the *sp* hybrid alkyne carbon. The 2119 cm^{-1} absorption results from stretching of the C≡C triple bond.

The spectrum of 4-octyne is not very helpful. Since there is no acetylenic hydrogen, there is no ≡C—H stretching absorption around 3300 cm^{-1}. There is no visible C≡C stretching absorption around 2100 to 2200 cm^{-1} either, because the disubstituted triple bond has a very small dipole moment. This spectrum fails to alert us to the presence of a triple bond.

PROBLEM 12-3

For each hydrocarbon spectrum, determine whether the compound is an alkane, an alkene, an alkyne, or an aromatic hydrocarbon. More than one unsaturated group may be present.

12-8
Characteristic Absorptions of Alcohols and Amines

The O—H bonds of alcohols and the N—H bonds of amines are strong and stiff. The vibration frequencies of O—H and N—H bonds therefore occur at higher frequencies than those of most C—H bonds (except for alkynyl ≡C—H bonds).

$$R—O—H \qquad \underset{H}{\overset{|}{R—N—H}} \qquad \underset{H}{\overset{|}{R—N—R'}} \qquad \underset{R''}{\overset{|}{R—N—R'}}$$

alcohol   primary   secondary   tertiary

amines

O—H and N—H stretching frequencies

alcohol O—H	3300 cm^{-1}, broad
acid O—H	3300 cm^{-1}, broad
amine N—H	3300 cm^{-1}, broad with spikes

Alcohol O—H bonds absorb over a wide range of frequencies, centered around 3300 cm^{-1}. Alcohol molecules are involved in hydrogen bonding, with different molecules having different instantaneous arrangements. The O—H stretching frequencies reflect this diversity of hydrogen-bonding arrangements, resulting in very broad absorptions. Notice the broad O—H absorption centered around 3300 cm^{-1} in the infrared spectrum of 1-butanol (Fig. 12-9).

Figure 12-9 also shows a strong C—O stretching absorption centered near 1050 cm^{-1}. Compounds with C—O bonds (alcohols and ethers, for example) generally show strong absorptions in the range of 1000 to 1200 cm^{-1}; however, there are other functional groups that also absorb in this region. Therefore, a strong peak between 1000 and 1200 cm^{-1} does not necessarily imply a C—O bond, but the *absence* of an absorption in this region suggests the absence of a C—O bond. For simple ethers, this unreliable C—O absorption is usually the only clue that the compound might be an ether.

▲ **Figure 12-9**
The IR spectrum of 1-butanol shows a broad, intense O—H stretching absorption centered around 3300 cm^{-1}. The broad shape is due to the diverse nature of the hydrogen-bonding interactions of alcohol molecules.

▲ Figure 12-10
The IR spectrum of dipropylamine shows a broad N—H stretching absorption centered around 3300 cm^{-1}. Notice the spike in this broad absorption.

Amine N—H bonds also have stretching frequencies in the 3300 cm^{-1} region, or even slightly higher. Like alcohols, amines participate in hydrogen bonding that can broaden the N—H absorptions. With amines, however, the absorption is somewhat weaker, and there may be one or more sharp spikes superimposed on the broad N—H stretching absorption: often one N—H spike for a secondary amine (R$_2$NH) and two N—H spikes for a primary amine (RNH$_2$). These sharp spikes, combined with the presence of nitrogen in the molecular formula, help to distinguish amines from alcohols. Tertiary amines (R$_3$N) have no N—H bonds, and they do not give rise to N—H stretching absorptions in the IR spectrum. Figure 12-10 shows the spectrum of dipropylamine, a secondary amine.

Because it has a large dipole moment, the C=O double bond produces intense infrared stretching absorptions. Carbonyl groups absorb at frequencies around 1700 cm^{-1}, but the exact frequency depends on the specific functional group and the rest of the molecule. For these reasons, infrared spectroscopy is often the best method for detecting and identifying the type of carbonyl group in an unknown compound. To simplify our discussion of carbonyl absorptions, we first consider a "normal" stretching frequency for simple ketones, aldehydes, and carboxylic acids, then we examine the types of carbonyl groups that deviate from this frequency.

12-9
Characteristic Absorptions of Carbonyl Compounds

12-9A Simple Ketones, Aldehydes, and Acids

The C=O stretching vibrations of simple ketones, aldehydes, and carboxylic acids occur at frequencies around 1710 cm^{-1}. These frequencies are higher than those for C=C double bonds because the C=O double bond is stronger and stiffer.

$$\underset{\text{ketone}}{\overset{\displaystyle O}{\underset{\|}{R-C-R'}}} \text{1710 cm}^{-1} \qquad \underset{\substack{\text{aldehyde}\\ \text{2700, 2800 cm}^{-1}}}{\overset{\displaystyle O}{\underset{\|}{R-C-H}}} \text{1710 cm}^{-1} \qquad \underset{\substack{\text{acid}\\ \text{broad, 2500-3500 cm}^{-1}}}{\overset{\displaystyle O}{\underset{\|}{R-C-O-H}}} \text{1710 cm}^{-1}$$

▲ **Figure 12-11**

Infrared spectra of (a) 2-heptanone and (b) butyraldehyde. Both the ketone and the aldehyde show intense carbonyl absorptions near 1710 cm^{-1}. In the aldehyde spectrum, there are two peaks (2720 and 2820 cm^{-1}) characteristic of the aldehyde C—H stretch.

PROBLEM-SOLVING HINT

Real spectra are rarely perfect. Samples often contain traces of water, giving weak absorptions in the O—H region.

Many compounds oxidize in air. For example, alcohols often give weak C=O absorptions from oxidized impurities.

In addition to the strong C=O stretching absorption, an aldehyde shows a characteristic set of two low-frequency C—H stretching frequencies around 2700 and 2800 cm^{-1}. Neither a ketone nor an acid produces these absorptions. Figure 12-11 compares the IR spectra of a ketone and an aldehyde. Notice the characteristic carbonyl stretching absorptions in both spectra, as well as the aldehyde C—H absorptions at 2720 and 2820 cm^{-1} in the butyraldehyde spectrum.

A carboxylic acid produces a characteristic broad O—H absorption in addition to the intense carbonyl stretching absorption (Fig. 12-12). Because of the unusually strong hydrogen bonding in carboxylic acids, the broad O—H stretching frequency is shifted to about 3000 cm^{-1}, centered on top of the usual C—H absorption. This broad O—H absorption gives a characteristic overinflated shape to the peaks in the C—H stretching region. Participation of the acid carbonyl group in hydrogen bonding frequently results in broadening of the strong carbonyl absorption as well.

▲ **Figure 12-12**
Infrared spectrum of hexanoic acid. Carboxylic acids show a broad O—H absorption from about 2500 to 3500 cm^{-1}. This broad absorption gives the entire C—H stretching region a broad appearance, punctuated by sharper C—H stretching absorptions.

SOLVED PROBLEM 12-1

Determine the functional group(s) in the compound whose IR spectrum appears below.

SOLUTION

First, look at the spectrum and see what peaks (outside the fingerprint region) don't look like alkane peaks; a weak peak around 3400 cm^{-1}, a strong peak about 1720 cm^{-1}, and an unusual C—H stretching region. The C—H region has two additional peaks around 2720 and 2820 cm^{-1}. The strong peak at 1725 cm^{-1} must be a C=O, and the peaks at 2720 and 2820 cm^{-1} suggest an aldehyde. The weak peak around 3400 cm^{-1} might be mistaken for an alcohol O—H. From experience, we know alcohols give much stronger O—H absorptions. This small peak might be from an impurity of water or from a small amount of the hydrate of the aldehyde (see Chapter 18). Many IR spectra show small, unexplained absorptions in the O—H region.

PROBLEM 12-4

Spectra are given for three compounds. Each compound has one or more of the following functional groups: alcohol, amine, ketone, aldehyde, and carboxylic acid. Determine the functional group(s) in each compound.

12-9B   Resonance Lowering of Carbonyl Frequencies

In Section 12-7A we saw that conjugation of a $C{=}C$ double bond lowers its stretching frequency. This is also true of conjugated carbonyl groups, as shown below. Delocalization of the pi electrons reduces the electron density of the carbonyl double bond, weakening it and lowering the stretching frequency from about 1710 cm^{-1} to about 1685 cm^{-1} for conjugated ketones, aldehydes, and acids.

2-cyclohexenone   2-butenal   benzoic acid

The $C{=}C$ absorption of a conjugated carbonyl compound may not be apparent in the IR spectrum because it is so much weaker than the $C{=}O$ absorption. The presence of the $C{=}C$ double bond is often inferred from its effect on the $C{=}O$ frequency and the presence of unsaturated $={C}{-}H$ absorptions above 3000 cm^{-1}.

The carbonyl groups of amides absorb at particularly low IR frequencies: about 1640 to 1680 cm^{-1}. The dipolar resonance structure (shown below) places part of the pi bond between carbon and nitrogen, leaving less than a full $C{=}O$ double bond.

> **The IR absorption frequency of the amide N—H group is sensitive to the strength of hydrogen bonding. Therefore, IR spectroscopy provides structural information about peptide and protein conformations, which are stabilized by the hydrogen bonding of amide groups.**

The very low frequency of the amide carbonyl might be mistaken for an alkene $C{=}C$ stretch. For example, consider the spectra of butyramide ($C{=}O$ about 1640 cm^{-1}) and 1-methylcyclopentene ($C{=}C$ at 1658 cm^{-1}) in Figure 12-13. Three striking differences are evident in these spectra: (1) The amide carbonyl absorption is much stronger than the absorption of the alkene double bond; (2) there are prominent N—H stretching absorptions in the amide spectrum; and (3) there is an unsaturated C—H stretching (just to the left of 3000 cm^{-1}) in the alkene spectrum. These examples show that we can distinguish between $C{=}O$ and $C{=}C$ absorptions, even when they appear in the same region of the spectrum.

Like primary amines, most primary amides show two spikes in the N—H stretching region (about 3300 cm^{-1}), as in the butyramide spectrum (Fig. 12-13). Secondary amides (like secondary amines) generally show one N—H spike.

primary amide

R—NH$_2$
primary amine

secondary amide

R—NH—R'
secondary amine

▲ Figure 12-13

The carbonyl group of butyramide (a) and the C═C double bond of 1-methylcyclopentene (b) absorb in the same region, but three clues distinguish the alkene from the amide: (1) The C═O absorption is much stronger than the C═C; (2) there are N─H absorptions (near 3300 cm^{-1}) in the amide; and (3) there is an unsaturated ═C─H absorption in the alkene.

12-9C Carbonyl Absorptions Above 1710 cm^{-1}

Some carbonyl groups absorb at frequencies *higher* than 1710 cm^{-1}. For example, simple carboxylic esters absorb around 1735 cm^{-1}. These higher-frequency absorptions are also seen in strained cyclic ketones (in a five-membered ring or smaller). In a small ring, the angle strain on the carbonyl group forces more electron density into the C═O double bond, resulting in a stronger, stiffer bond.

IR spectroscopy can also be used to monitor the progress of biological reactions. For example, the hydrolysis of complex lipids (esters of glycerol) causes a characteristic decrease in intensity of the ester carbonyl absorption at 1735 cm^{-1}, with a corresponding appearance of a carboxylic acid absorption near 1710 cm^{-1}.

Infrared absorptions of carbon–nitrogen bonds are similar to those of carbon–carbon bonds, except that carbon–nitrogen bonds are more polar and give stronger absorptions. Carbon–nitrogen single bonds absorb around 1200 cm^{-1}, in a region close to many C—C and C—O absorptions. Therefore, the C—N single bond stretch is rarely useful for structure determination.

Carbon–nitrogen double bonds resemble C=C double bonds, absorbing around 1660 cm^{-1}; however, the C=N bond gives rise to stronger absorptions because of its greater dipole moment. A C=N stretch often resembles a carbonyl absorption in intensity, except at a lower frequency than most carbonyl absorptions.

The most readily recognized carbon–nitrogen bond is the triple bond of a nitrile (Fig. 12-14). The stretching frequency of the nitrile C≡N bond is close to that of an acetylenic C≡C triple bond, about 2200 cm^{-1}; however, nitriles generally absorb *above* 2200 cm^{-1} (2200 to 2300 cm^{-1}), while alkynes absorb *below* 2200 cm^{-1}. Also, nitrile triple bonds are more polar than C≡C triple bonds, so nitriles usually produces stronger absorptions than alkynes.

12-10
Characteristic Absorptions of C—N Bonds

C–N bond stretching frequencies

$$
\left.
\begin{array}{lll}
\text{C—N} & 1200\text{ cm}^{-1} \\
\text{C=N} & 1600\text{ cm}^{-1} \\
\text{C≡N} & >2200\text{ cm}^{-1}
\end{array}
\right\} \text{ usually strong}
$$

for comparison: C≡C   <2200 cm^{-1} (usually moderate or weak)

▲ **Figure 12-14**
Nitrile triple bond stretching absorptions are at slightly higher frequencies (and usually more intense) than those of alkyne triple bonds. Compare this spectrum of butyronitrile with that of 1-octyne in Figure 12-8.

PROBLEM 12-5

The infrared spectra for three compounds are provided. Each compound has one or more of the following functional groups: conjugated ketone, ester, amide, nitrile, and alkyne. Determine the functional group(s) in each compound.

It may seem there are too many numbers to memorize in learning about infrared spectroscopy. There are hundreds of characteristic absorptions for different kinds of compounds, and a detailed reference table of characteristic frequencies is given in Appendix 2. Please glance at Appendix 2, and note that Appendix 2A is organized visually, while Appendix 2B is organized by functional groups. For everyday use, it is necessary to memorize only a few stretching frequencies, such as those shown in Table 12-2. In using this table, remember that the numbers are approximate and they do not give ranges to cover all the unusual cases. Also, remember how frequencies change as a result of conjugation, ring strain, and other factors.

12-11
Simplified Summary of IR Stretching Frequencies

TABLE 12-2 Summary of IR Stretching Frequencies

Frequency (cm⁻¹)	Functional Group		Comments
3300	alcohol	O—H	always broad
	amine, amide	N—H	may be broad, sharp, or broad with spikes
	alkyne	≡C—H	always sharp, usually strong
3000	alkane	$-\overset{\|}{\underset{\|}{C}}-H$	just below 3000 cm⁻¹
	alkene	$=C\overset{H}{\diagdown}$	just above 3000 cm⁻¹
	acid	O—H	very broad
2200	alkyne	—C≡C—	just below 2200 cm⁻¹
	nitrile	—C≡N	just above 2200 cm⁻¹
1710 (very strong)	carbonyl	$\diagup C=O$	ketones, aldehydes, acids esters higher, about 1735 cm⁻¹ conjugation lowers frequency amides lower, about 1650 cm⁻¹
1660	alkene	$\diagup C=C\diagdown$	conjugation lowers frequency aromatic C=C about 1600 cm⁻¹
	imine	$\diagup C=N\diagup$	stronger than C=C
	amide	$\diagup C=O$	stronger than C=C (see above)

Ethers, esters, and alcohols also show C—O stretching between 1000 and 1200 cm⁻¹.

PROBLEM-SOLVING $HINT$

Table 12-2 provides the *numbers* but not the understanding and practice needed to work most IR problems. Learn to use the material in this table, then practice doing problems until you feel confident.

Strengths and Limitations of Infrared Spectroscopy   The most useful aspect of infrared spectroscopy is its ability to identify functional groups, but IR does not provide much information about the carbon skeleton or the alkyl groups in the compound. These aspects of the structure are more easily determined by NMR, as we will see in Chapter 13. Even an expert spectroscopist can rarely determine a structure based only on the IR spectrum.

Ambiguities often arise in the interpretation of IR spectra. For example, a strong absorption at 1680 cm^{-1} might arise from an amide, an isolated double bond, a conjugated ketone, a conjugated aldehyde, or a conjugated carboxylic acid. Familiarity with other regions of the spectrum usually enables us to determine which of these functional groups is present. In some cases, we cannot be entirely certain of the functional group without additional information, usually provided by other types of spectroscopy.

Infrared spectroscopy *can* provide conclusive proof that two compounds are either the same or different. The peaks in the fingerprint region depend on complex vibrations involving the entire molecule, and it is impossible for any two compounds (except enantiomers) to have precisely the same infrared spectrum.

To summarize, an infrared spectrum is valuable in three ways.

1. It indicates the functional groups in the compound.

2. It shows the *absence* of other functional groups that would give strong absorptions if they were present.

3. It can confirm the identity of a compound by comparison with a known sample.

SOLVED PROBLEM 12-2

You have an unknown with an absorption at 1680 cm^{-1}; it might be an amide, an isolated double bond, a conjugated ketone, a conjugated aldehyde, or a conjugated carboxylic acid. Describe what spectral characteristics you would look for to help you determine which of these possible functional groups might be causing the 1680 peak.

SOLUTION

Amide: (1680 peak is strong) Look for N—H absorptions (with spikes) around 3300 cm^{-1}.

Isolated double bond: (1680 peak is weak or moderate) Look for =C—H absorptions just above 3000 cm^{-1}.

Conjugated ketone: (1680 peak is strong) There must be a double bond nearby, conjugated with the C=O, to lower the C=O frequency to 1680 cm^{-1}. Look for the C=C of the nearby double bond (moderate, 1620 to 1640 cm^{-1}) and its =C—H above 3000 cm^{-1}.

Conjugated aldehyde: (1680 peak is strong) Look for the aldehyde C—H stretch about 2700 and 2800 cm^{-1}. Also look for the C=C and C—H of the nearby double bond (1620 to 1640 cm^{-1} and just above 3000 cm^{-1}).

Conjugated carboxylic acid: (1680 peak is strong) Look for the characteristic acid O—H stretch centered *on top of* the C—H stretch around 3000 cm^{-1}. Also look for the C=C and =C—H of the nearby double bond (1620 to 1640 cm^{-1} and just above 3000 cm^{-1}).

**12-12
Reading
and Interpreting
IR Spectra
(Solved Problems)**

Many students are unsure how much information they should be able to obtain from an infrared spectrum. In Chapter 13, we will use IR together with NMR and other information to determine the entire structure. For the present, concentrate on getting as much information as you can from the IR spectrum by itself. Several solved spectra are included in this section to show what information can be inferred. An experienced spectroscopist could obtain more information from these spectra, but we will concentrate on the major, most reliable, features.

Study this section by looking at each spectrum and writing down the important frequencies and your proposed functional groups. Then look at the solution and compare it with your solution. The actual structures of these compounds are shown at the end of this section. They are not given with the solutions because *you cannot determine these structures using only the infrared spectra*, so a complete structure is not a part of a realistic solution.

Compound 1 This spectrum is most useful for what it does *not* show. There is a carbonyl absorption at 1714 cm^{-1} and little else. There is no aldehyde C—H, no hydroxyl O—H, and no N—H. The carbonyl absorption could indicate an aldehyde, ketone, or acid, except that the lack of aldehyde C—H stretch eliminates an aldehyde, and the lack of O—H stretch eliminates an acid. There is no visible C=C stretch and no unsaturated C—H absorption above 3000 cm^{-1}, so the compound appears to be otherwise saturated. The compound is probably a simple ketone.

Compound 2 The absorption at 1650 cm^{-1} is so intense that it probably indicates a carbonyl group. A carbonyl group at this low frequency suggests an amide. The doublet (a pair of peaks) of N—H absorption around 3300 cm^{-1}, also suggests a primary amide, R—CONH$_2$. Since there is no C—H absorption above 3000 cm^{-1}, this is probably a saturated amide.

Compound 3   The sharp peak at 2246 cm^{-1} results from a nitrile C≡N stretch. (An alkyne C≡C absorption would be weaker and below 2200 cm^{-1}.) The absence of C═C stretch or C—H stretch above 3000 cm^{-1} suggests that the nitrile is otherwise saturated.

Compound 4   The carbonyl absorption at 1685 cm^{-1} is about right for a conjugated ketone, aldehyde, or acid. (An amide would be lower in frequency, and a C═C double bond would not be so strong.) The absence of any N—H stretch, O—H stretch, or aldehyde C—H stretch leaves a conjugated ketone as the best possibility. The C═C stretch at 1599 cm^{-1} indicates an aromatic ring. We presume that the aromatic ring is conjugated with the carbonyl group of the ketone.

Compound 5 The broad O—H stretch that spans most of the C—H stretching region suggests a carboxylic acid. (This acid is a solid, and its OH absorption is weaker than that of the liquid shown in Fig. 12-12.) The C=O stretch is low for an acid (1688 cm^{-1}), implying a conjugated acid. The aromatic C=C absorption at 1600 cm^{-1} suggests that the acid may be conjugated with an aromatic ring.

Compound 6 The carbonyl absorption at 1727 cm^{-1} suggests a ketone, aldehyde, or acid. The C—H stretching at 2710 and 2805 cm^{-1} confirms an aldehyde. Because all the C—H stretch is below 3000 cm^{-1} and there is no visible C=C stretch, the aldehyde is probably saturated.

Compound 7 The carbonyl absorption at 1739 cm^{-1} suggests an ester. The weak peak at 1600 cm^{-1} indicates an aromatic ring, but it is not conjugated with the ester because the ester absorption is close to its usual (unconjugated) position. The presence of both saturated (below 3000 cm^{-1}) and unsaturated (above 3000 cm^{-1}) C—H stretching in the 3000 cm^{-1} region confirms the presence of both alkyl and unsaturated portions of the molecule.

Structures of the compounds

(These structures cannot be determined from their IR spectra alone.)

compound 1   compound 2   compound 3   compound 4

compound 5   compound 6   compound 7

PROBLEM 12-6

For each spectrum, interpret all the significant stretching frequencies above 1580 cm^{-1}.

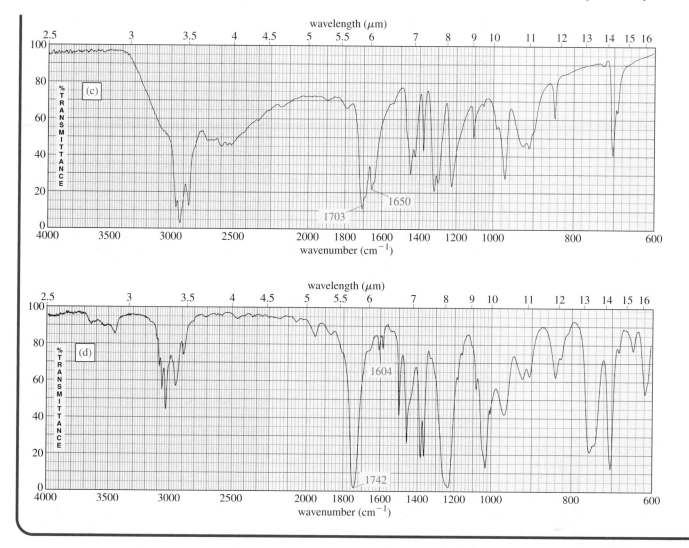

Infrared spectroscopy gives information about the functional groups in a molecule, but it tells little about the size of the molecule or what heteroatoms are present. To determine a structure, we need a molecular weight and a molecular formula. Molecular formulas were once obtained by careful analysis of the elemental composition, and a molecular weight was determined by freezing-point depression or some other technique. These are long and tedious processes, and they require a large amount of pure material. Many important compounds are available only in small quantities, and they may be impure.

Mass spectrometry (MS) provides the molecular weight and valuable information about the molecular formula, using a very small amount of sample. High-resolution mass spectrometry (HRMS) can provide an accurate molecular formula. The mass spectrum also provides structural information that can confirm a structure derived from NMR and IR spectroscopy.

Mass *spectrometry* is fundamentally different from *spectroscopy*. Spectroscopy involves the absorption (or emission) of light over a range of wavelengths. Mass spectrometry does not use light at all. In the mass spectrometer, a sample is struck by high-energy electrons, breaking the molecules apart. The masses of the fragments are measured, and this information is used to reconstruct the molecule. The process is similar to analyzing a vase by shooting it with a rifle, then weighing all the pieces.

12-13
Introduction to
Mass Spectrometry

12-13A   The Mass Spectrometer

A **mass spectrometer** ionizes molecules in a high vacuum, sorts the ion according to their masses, and records the abundance of ions of each mass. A **mass spectrum** is the graph plotted by the mass spectrometer, with the masses plotted as the x axis and the relative number of ions of each mass on the y axis. Several methods are used to ionize samples and then to separate ions according to their masses. We will discuss only the most common techniques, *electron impact ionization* for forming the ions, and *magnetic deflection* for separating the ions.

Electron Impact Ionization   In the **ion source**, the sample is bombarded by a beam of electrons. When an electron strikes a neutral molecule, it may ionize that molecule by knocking out an additional electron.

$$e^- + M \longrightarrow [M]^{\ddagger} + 2e^-$$

When a molecule loses one electron, it then has a positive charge and one unpaired electron. The ion is therefore a **radical cation**. The electron impact ionization of methane is shown below.

Most carbocations have a three-bonded carbon atom with six paired electrons in its valence shell. The radical cation shown above is not a normal carbocation. The carbon atom has seven electrons around it, and they bond it to four other atoms. This unusual cation is represented by the formula $[CH_4]^{\ddagger}$, with the $+$ indicating the positive charge and the $\cdot$ indicating the unpaired electron.

In addition to ionizing a molecule, the impact of an energetic electron may break it apart. This **fragmentation** process gives a characteristic mixture of ions. The radical cation corresponding to the mass of the original molecule is called the **molecular ion**, abbreviated $\mathbf{M^+}$. The ions of smaller molecular weights are called *fragments*. Bombardment of ethane molecules by energetic electrons, for example, produces the molecular ion and several fragments. Both charged and uncharged fragments are formed, but *only the positively charged fragments are detected by the mass spectrometer*. We will often use green type for the "invisible" uncharged fragments.

We discuss the common modes of fragmentation in Section 12-15.

Separation of Ions of Different Masses   Once ionization and fragmentation have formed a mixture of ions, these ions are separated and detected. The most common type of mass spectrometer, shown in Figure 12-15, separates ions by *magnetic deflection*.

After ionization, the positively charged ions are attracted to a negatively charged accelerator plate, which has a narrow slit to allow some of the ions to pass through. The ion beam enters an evacuated flight tube, with a curved portion positioned between the poles

Mass spectrometry can be used to determine the precise mass of a protein. Because of their large size and low volatility, proteins require specialized mass spectral techniques, such as electrospray ionization: spraying a charged, heated stream of droplets into the evacuated source chamber. The solvent evaporates to leave ions of the compound to be analyzed.

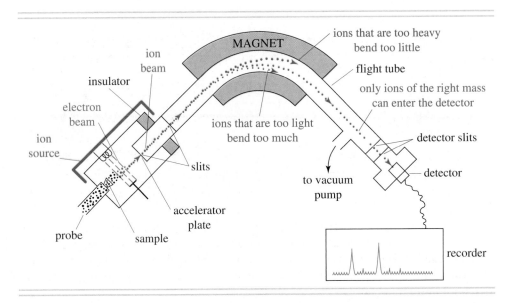

◀ **Figure 12-15**
Diagram of a mass spectrometer. A beam of electrons causes molecules to ionize and fragment. The mixture of ions is accelerated and passes through a magnetic field, where the paths of lighter ions are bent more than those of heavier ions. By varying the magnetic field, the spectrometer plots the abundance of ions of each mass.

of a large magnet. When a charged particle passes through a magnetic field, a transverse force bends its path. The path of a heavier ion bends less than the path of a lighter ion.

The exact radius of curvature of an ion's path depends on its mass-to-charge ratio, symbolized by m/z (or by m/e in the older literature). In this expression, m is the mass of the ion (in amu) and z is its charge in units of the electronic charge. The vast majority of ions have a charge of $+1$, so we consider their path to be curved by an amount that depends only on their mass.

At the end of the flight tube is another slit, followed by an ion detector connected to an amplifier. At any given magnetic field, only ions of one particular mass are bent exactly the right amount to pass through the slit and enter the detector. The detector signal is proportional to the number of ions striking it. By varying the magnetic field, the spectrometer scans all the possible ion masses and produces a graph of the number of ions of each mass.

A modern double-focusing mass spectrometer. This one is combined with a gas chromatograph to use as a GC-MS. The gas chromatograph separates a mixture into its components and injects the purified components into the ion source of the mass spectrometer.

12-13B   The Mass Spectrum

The mass spectrometer usually plots the spectrum as a graph on a computer screen. This information is tabulated, and the spectrum is printed as a bar graph or as a table of relative abundances (Fig. 12-16). In the printed mass spectrum, all the masses are rounded to the nearest whole-number mass unit. The peaks are assigned abundances as percentages of the strongest peak, called the **base peak**. Notice that *the base peak does not necessarily correspond to the mass of the molecular ion.* It is simply the strongest peak, making it easy for other peaks to be expressed as percentages.

m/z	Abundance (% of base peak)
39	62
41	100 (base peak)
42	24
43	90
56	23
57	50
85	11
100 (M^+)	10

◀ **Figure 12-16**
Mass spectrum of 2,4-dimethyl-pentane, given both as a bar graph and in tabular form. Notice that abundances are given as percentages of the strongest peak (the base peak). In this example, the base peak is at m/z 41 and the molecular ion peak (parent peak) is at m/z 100.

▶ **Figure 12-17**
Block diagram of a gas chromatograph–mass spectrometer (GC–MS). The gas chromatograph column separates the mixture into its components. The quadrupole mass spectrometer scans mass spectra of the components as they leave the column.

A molecular ion peak (also called the *parent peak*) is observed in most mass spectra, meaning that a detectable number of molecular ions (M^+) reach the detector without fragmenting. These molecular ions are usually the particles of highest mass in the spectrum. The value of m/z for the molecular ion immediately gives the molecular weight of the compound.

12-13C   Mass Spectrometry of Mixtures: The GC–MS

Mass spectrometry is combined with gas chromatography for routine analysis of mixtures of compounds, such as reaction mixtures or environmental samples. Figure 12-17 shows a simplified diagram of a common type of GC–MS. The *gas chromatograph* uses a heated capillary column coated on the inside with silicone rubber (or other *stationary phase*) to separate the components of the mixture. A small amount of sample (about 10^{-6} gram is enough) is injected into a heated injector, where a gentle flow of helium sweeps it into the column. As the sample passes through the column, the more volatile components (that interact less with the stationary phase) move through the column faster than the less volatile components. The separated components leave the column at different times, passing through a transfer line into the ion source of the mass spectrometer, where the molecules are ionized and allowed to fragment.

Most gas chromatograph–mass spectrometer systems use a *quadrupole mass filter* to separate the ions. In a high vacuum, the ions pass down the length of four rods, which have varying voltages applied to them. (Figure 12–17 shows two of the four rods.) The varying electric fields cause the ions to follow complex orbits, and only one mass reaches the detector at any instant. By scanning the voltages, a wide range of masses can be measured in less than 1 second. In this way, many mass spectra are taken and stored on a computer disk as the components of the sample pass from the chromatograph column into the mass spectrometer. This powerful GC–MS combination allows many components of a mixture to be separated by the gas chromatograph and later identified by their mass spectra.

We can establish the identity of a drug by infrared spectroscopy (IR) or gas chromatography-mass spectrometry (GC-MS). IR provides clues to the functional groups, and it can confirm the structure by comparing the spectrum with that of an authentic sample. GC-MS provides a characteristic retention time that suggests possible compounds, and it confirms the structure by comparison of the retention time and the mass spectrum with those of an authentic sample.

12-14 Determination of the Molecular Formula by Mass Spectrometry

12-14A   High-Resolution Mass Spectrometry

Although mass spectra usually show the particle masses rounded to the nearest whole number, the masses are not really integral. The ^{12}C nucleus is *defined* to have a mass of exactly 12 atomic mass units (amu), and all other nuclei have masses based on this standard. For example, a proton has a mass of about 1, but not exactly: Its mass is 1.007825 amu. Table 12-3 shows the atomic masses for the most common isotopes found in organic compounds.

Determination of a molecular formula is possible using a **high-resolution mass spectrometer** (HRMS), one that uses extra stages of electrostatic or magnetic focusing to form a very precise beam and to detect particle masses to an accuracy of about 1 part in 20,000. A mass determined to several significant figures using an HRMS is called an *exact mass*. Although it is not really exact, it is much more accurate than the usual integral mass numbers. Comparing the exact mass with masses calculated by molecular formula makes it possible to identify the correct formula.

Consider a molecular ion with a mass of 44. This approximate molecular weight might correspond to C_3H_8 (propane), C_2H_4O (acetaldehyde), CO_2, or CN_2H_4. Each of these molecular formulas corresponds to a different exact mass:

	C_3H_8		C_2H_4O		CO_2		CN_2H_4
3 C	36.00000	2 C	24.00000	1 C	12.00000	1 C	12.00000
8 H	8.06260	4 H	4.03130			4 H	4.03130
		1 O	15.99490	2 O	31.98983	2 N	28.00610
	44.06260		44.02620		43.98983		44.03740

If the HRMS measured the exact mass of this ion as 44.029 mass units, we would conclude that the compound has a molecular formula of C_2H_4O, because the mass corresponding to this formula is closest to the observed value. Published tables of exact masses are available for comparison with values obtained from the HRMS. Depending on the completeness of the tables, they may include sulfur, halogens, or other elements.

12-14B   Use of Heavier Isotope Peaks

Whether or not a high-resolution mass spectrometer is available, molecular ion peaks often provide information about the molecular formula. Most elements do not consist of a single isotope but contain heavier isotopes in varying amounts. These heavier isotopes give rise to small peaks at higher mass numbers than the major M^+ molecular ion peak. A peak that is one mass unit heavier than the M^+ peak is called the **M+1 peak**; two units heavier, the **M+2 peak**; and so on. Table 12-4 gives the isotopic composition of some common elements, showing how they contribute to M+1 and M+2 peaks.

Ideally, we could use the isotopic compositions in Table 12-4 to determine the entire molecular formula of a compound, by carefully measuring the abundances of the M^+, M+1, and M+2 peaks. In practice, however, there are background peaks at every mass number. These background peaks are often similar in intensity to the M+1 peak, preventing an accurate measurement of the M+1 peak. High-resolution mass spectrometry is much more reliable.

Some elements (particularly S, Cl, Br, I, and N) are recognizable from molecular ion peaks, however. A compound with no sulfur, chlorine, or bromine has a small M+1 peak and an even smaller M+2 peak. If a compound contains sulfur, the M+2 peak is larger than the M+1 peak: about 4% of the M^+ peak. If chlorine is present, the M+2 peak is about a third as large as the M^+ peak. If bromine is present, the M^+ and M+2 ions have about equal abundances; the molecular ion appears as a doublet separated by two mass units, with one mass corresponding to ^{79}Br and one to ^{81}Br.

TABLE 12-3 "Exact" Masses of Common Isotopes	
Isotope	**Atomic Mass (amu)**
^{12}C	12.000000
^1H	1.007825
^{16}O	15.994914
^{14}N	14.003050

Isotope ratios can help to identify banned substances in Olympic athletes. For example, mass spectrometry can distinguish between synthetic testosterone and the naturally occurring hormone by detecting differences in the isotope ratios of ^{13}C and ^{12}C.

TABLE 12-4 Isotopic Composition of Some Common Elements						
Element	**M^+**		**M+1**		**M+2**	
hydrogen	^1H	100.0%				
carbon	^{12}C	98.9%	^{13}C	1.1%		
nitrogen	^{14}N	99.6%	^{15}N	0.4%		
oxygen	^{16}O	99.8%			^{18}O	0.2%
sulfur	^{32}S	95.0%	^{33}S	0.8%	^{34}S	4.2%
chlorine	^{35}Cl	75.5%			^{37}Cl	24.5%
bromine	^{79}Br	50.5%			^{81}Br	49.5%
iodine	^{127}I	100.0%				

Iodine is recognized by the presence of the iodonium ion, I^+, at m/z 127. This clue is combined with a characteristic 127-unit gap in the spectrum corresponding to loss of the iodine radical. Nitrogen (or an odd number of nitrogen atoms) is suggested by an odd molecular weight. Stable compounds containing only carbon, hydrogen, and oxygen have even molecular weights.

Recognizable elements in the mass spectrum

Br	M+2 as large as M^+
Cl	M+2 a third as large as M^+
I	I^+ at 127; large gap
N	odd M^+
S	M+2 larger than usual (4% of M^+)

The following spectra show compounds containing sulfur, chlorine, and bromine.

PROBLEM 12-7

Point out which of these four mass spectra indicate the presence of sulfur, chlorine, bromine, iodine, or nitrogen. Suggest a molecular formula for each.

12-15
Fragmentation Patterns in Mass Spectrometry

In addition to the molecular formula, the mass spectrum provides structural information. An electron with a typical energy of 70 eV (1610 kcal/mol or 6740 kJ/mol) has far more energy than needed to ionize a molecule. The impact forms the radical cation, and it often breaks a bond to give a cation and a radical. The resulting cation is observed by the mass spectrometer, but the uncharged radical is not accelerated or detected. We can infer the mass of the uncharged radical from the amount of mass lost from the molecular ion to give the observed cation fragment.

Ionization

$$R{:}R' \ + \ e^- \ \longrightarrow \ [R{\cdot}R']^{\ddagger} \ + \ 2\,e^-$$
$$\text{radical cation}$$
$$\text{(molecular ion)}$$

Fragmentation

$$[R{\cdot}R']^{\ddagger} \ \longrightarrow \ R^+ \ + \ {\cdot}R'$$
$$\qquad\qquad \text{cation fragment} \quad \text{radical fragment}$$
$$\qquad\qquad \text{(observed)} \qquad \text{(not observed)}$$

This bond breaking does not occur randomly; it tends to form the most stable fragments. By knowing what stable fragments result from different kinds of compounds, we can recognize structural features and use the mass spectrum to confirm a proposed structure.

12-15A   Mass Spectra of Alkanes

The mass spectrum of *n*-hexane (Fig. 12-18) shows several characteristics typical of straight-chain alkanes. The base peak (*m/z* 57) corresponds to loss of an ethyl group, giving an ethyl radical and a butyl cation. The neutral ethyl radical is not detected, because it is not charged and is not accelerated or deflected.

m/z of the charged fragment on this side of the broken bond

$$[CH_3CH_2CH_2CH_2 \overset{57}{+} CH_2CH_3]^{\ddagger} \ \longrightarrow \ CH_3CH_2CH_2CH_2^+ \ + \ {\cdot}CH_2CH_3$$

hexane radical cation	1-butyl cation	ethyl radical (29)
M⁺ 86	detected at *m/z* 57	not detected

▶ **Figure 12-18**
Mass spectrum of *n*-hexane. Groups of ions correspond to loss of one-, two-, three-, and four-carbon fragments.

A similar fragmentation gives an ethyl cation and a butyl radical. In this case, the ethyl fragment (m/z 29) is detected.

$$[CH_3CH_2CH_2CH_2 \overset{\lceil \; 29}{|} CH_2CH_3]^{\ddagger} \longrightarrow CH_3CH_2CH_2CH_2 \cdot \; + \; {}^+CH_2CH_3$$

hexane radical cation   1-butyl radical (57)   ethyl cation
M⁺ 86   not detected   detected at m/z 29

Symmetric cleavage of hexane gives a propyl cation and a propyl radical.

$$[CH_3CH_2CH_2 \overset{43 \; \lceil}{|} CH_2CH_2CH_3]^{\ddagger} \longrightarrow CH_3CH_2CH_2{}^+ \; + \; \cdot CH_2CH_2CH_3$$

hexane radical cation   propyl cation   propyl radical (43)
M⁺ 86   detected at m/z 43   not detected

> **PROBLEM-SOLVING HINT**
>
> Most molecular ions have even mass numbers. Most fragments have odd mass numbers. (With a nitrogen atom, the molecular ion is odd and most fragments containing N are even.)

Cleavage to give a pentyl cation (m/z 71) and a methyl radical is weak because the methyl radical is less stable than a substituted radical. Cleavage to give a methyl cation (m/z 15) and a pentyl radical is not visible because the methyl cation is less stable than a substituted cation. The stability of the cation is apparently more important than the stability of the radical, since a weak peak appears corresponding to loss of a methyl radical, but we see no cleavage to give a methyl cation.

$$[CH_3CH_2CH_2CH_2CH_2 \overset{71 \; \lceil}{|} CH_3]^{\ddagger} \longrightarrow CH_3CH_2CH_2CH_2CH_2{}^+ \; + \; \cdot CH_3$$

hexane radical cation   pentyl cation   methyl radical (15)
M⁺ 86   weak at m/z 71   not detected

$$[CH_3CH_2CH_2CH_2CH_2 \overset{\lceil \; 15 \; \text{(not formed)}}{|} CH_3]^{\ddagger} \overset{X}{\longrightarrow} CH_3CH_2CH_2CH_2CH_2 \cdot \; + \; {}^+CH_3$$

hexane radical cation   pentyl radical (71)   methyl cation
M⁺ 86   not detected   (too unstable)

Cation and radical stabilities help to explain the mass spectra of branched alkanes as well. Figure 12-19 shows the mass spectrum of 2-methylpentane. Fragmentation of a branched alkane commonly occurs at a branch carbon atom to give the most highly substituted cation and radical. Fragmentation of 2-methylpentane at the branched carbon atom can give a secondary carbocation in either of two ways:

$$\begin{bmatrix} & CH_3 \\ & | \\ CH_3CH_2CH_2 \overset{}{+} CH \overset{}{+} CH_3 \end{bmatrix}^{+ \cdot}$$

2-methylpentane radical cation
m/z 86

$$CH_3CH_2CH_2 - \overset{\overset{\displaystyle CH_3}{|}}{CH}{}^+ \; + \; \cdot CH_3$$

2-pentyl cation   methyl radical
m/z 71

$${}^+CH - \overset{\overset{\displaystyle CH_3}{|}}{} CH_3 \; + \; CH_3CH_2CH_2 \cdot$$

isopropyl cation   propyl radical
m/z 43 (base peak)

> **PROBLEM-SOLVING HINT**
>
> The guidelines we used to predict carbocation stability in El and S_N1 reactions are also useful for interpreting mass spectra. Relatively stable carbocations are generally more abundant in the mass spectrum.

Both fragmentations give secondary cations, but the second gives a primary radical instead of a methyl radical. Therefore, the second fragmentation accounts for the base (largest) peak, while the first accounts for another large peak at m/z 71. Other fragmentations (to give primary cations) account for the weaker peaks.

▶ **Figure 12-19**

Mass spectrum of 2-methylpentane. The base peak corresponds to loss of a propyl radical to give an isopropyl cation.

PROBLEM 12-8

Show the fragmentation that accounts for the cation at m/z 57 in the mass spectrum of 2-methylpentane.

PROBLEM 12-9

Show the fragmentations that give rise to the peaks at m/z 43, 57, and 85 in the mass spectrum of 2,4-dimethylpentane (Fig. 12-16).

12-15B   Fragmentation Giving Resonance-Stabilized Cations; Mass Spectra of Alkenes

Fragmentation in the mass spectrometer gives resonance-stabilized cations whenever possible. The most common fragmentation of alkenes is cleavage of an allylic bond to give a resonance-stabilized allylic cation. Figure 12-20 shows how the radical cation of 2-hexene undergoes allylic cleavage to give the resonance-stabilized cation responsible

▲ **Figure 12-20**

The radical cation of 2-hexene cleaves at an allylic bond to give a resonance-stabilized methallyl cation, m/z 55.

for the base peak at m/z 55. We will encounter other types of resonance-stabilized cations in the mass spectra of ethers, amines, and carbonyl compounds in later chapters covering the chemistry of these functional groups.

PROBLEM 12-10

Catalytic hydrogenation of compound X gives 2,6-dimethyloctane as the only product. The mass spectrum of compound X shows a molecular ion at m/z 140 and prominent peaks at m/z 57 and m/z 83. Suggest a structure for compound X, and justify your answer.

12-15C Fragmentation Splitting out a Small Molecule; Mass Spectra of Alcohols

Mass spectral peaks are often seen corresponding to loss of small, stable molecules. Loss of a small molecule is usually indicated by a peak with an even mass number, corresponding to loss of an even mass number. A radical cation may lose water (18), CO (28), CO_2 (44), and even ethene (28) or other alkenes. The most common example is the loss of water from alcohols, which occurs so readily that the molecular ion is often weak or absent. The peak corresponding to loss of water (the M–18 peak) is usually strong, however.

The mass spectrum of 3-methyl-1-butanol (Fig. 12-21) shows a favorable loss of water. The peak at m/z 70 that *appears* to be the molecular ion is actually the intense M–18 peak. The molecular ion (m/z 88) is not observed because it loses water very readily. The base peak at m/z 55 corresponds to loss of water and a methyl group.

In addition to losing water, alcohols commonly fragment next to the carbinol carbon atom to give a resonance-stabilized carbocation.

◀ **Figure 12-21**
The mass spectrum of 3-methyl-1-butanol. The strong peak at m/z 70 is actually the M–18 peak, corresponding to loss of water. The molecular ion is not visible because it loses water easily.

This type of *alpha cleavage* is prominent in the spectrum of 2,6-dimethyl-4-heptanol shown in Problem 12-11.

PROBLEM 12-11

Account for the peaks at *m/z* 87, 111, and 126 in the mass spectrum of 2,6-dimethyl-4-heptanol.

SUMMARY   Common Fragmentation Patterns

This summary is provided for rapid reference to the common fragmentation patterns of simple functional groups. Some of these functional groups are discussed in later chapters, but they are included here so this reference table can be used throughout the course.

1. *Alkanes:* cleavage to give the most stable carbocations (Section 12-15A)

$$\begin{bmatrix} R' \\ | \\ R-C+R'' \\ | \\ H \end{bmatrix}^{\ddagger} \longrightarrow \begin{array}{c} R' \\ | \\ R-C^+ \\ | \\ H \end{array} + \; \cdot R''$$

even   odd

2. *Alcohols:* loss of water (Section 12-15C)

$$\begin{bmatrix} H & OH \\ | & | \\ -C-C- \\ | & | \end{bmatrix}^{\ddagger} \longrightarrow \begin{bmatrix} >C=C< \end{bmatrix}^{\ddagger} + \; H_2O$$

even   even

or α cleavage (Section 12-15C)

$$\begin{bmatrix} OH \\ | \\ -C+C- \\ | & | \end{bmatrix}^{\ddagger} \longrightarrow \begin{bmatrix} OH & {}^+OH \\ | & || \\ -C^+ \longleftrightarrow -C \\ | & | \end{bmatrix} + \; \begin{array}{c} | \\ \cdot C- \\ | \end{array}$$

even   odd

3. *Alkenes and aromatics:* cleavage to give allylic and benzylic carbocations

$$[R-CH=CH-CH_2+R']^{\ddagger} \longrightarrow R-CH=CH-\overset{+}{CH_2} + \; \cdot R'$$
(Section 12-15B)   allylic cation (odd)

(The following fragmentations are covered in later chapters.)

$$\begin{bmatrix} \text{Ph}-CH_2+R \end{bmatrix}^{\ddagger} \longrightarrow \text{Ph}-\overset{+}{CH_2} + \; \cdot R \longrightarrow \text{(tropylium)}^+ + \; \cdot R$$

(Section 16-14)   benzylic cation   tropylium ion
                                      m/z 91   *m/z* 91

4. *Amines:* α cleavage to give stabilized cations (Section 19-9)

$$[R_2N\text{—}CH_2\text{—}R']^{+} \longrightarrow R_2\overset{+}{N}\text{=}CH_2 + \cdot R'$$

odd   iminium ion (even)

5. *Ethers:* loss of an alkyl group (Section 14-4)

$$[R\text{—}CH_2\text{—}O\text{—}R']^{+} \longrightarrow R\text{—}CH\text{=}\overset{+}{O}H + \cdot R'$$

even   stabilized cation (odd)

$$or \quad [R\text{—}CH_2\text{—}O\text{—}R']^{+} \longrightarrow R\text{—}CH_2\text{—}\overset{..}{\underset{..}{O}}\cdot + {}^{+}R$$

even   alkyl cation (odd)

or α cleavage

$$[R\text{—}CH_2\text{—}O\text{—}R']^{+} \xrightarrow{\alpha\ cleavage} H_2C\text{=}\overset{+}{O}\text{—}R' + \cdot R$$

even   stabilized cation (odd)

6. *Ketones and aldehydes:* loss of alkyl groups to give acylium ions (Section 18-5)

$$\begin{bmatrix} \overset{O}{\underset{\|}{R\text{—}C}}\text{—}R' \end{bmatrix}^{+} \longrightarrow R\text{—}C\text{≡}O^{+} + \cdot R'$$

even   acylium ion (odd)

The McLafferty rearrangement splits out alkenes (Section 18-5)

γ-hydrogen

$$\begin{bmatrix} \overset{O}{\underset{\|}{R\text{—}C}}\text{—}\overset{\alpha}{C}\text{—}\overset{\beta}{C}\text{—}\overset{\gamma}{C}\text{—}R' \end{bmatrix}^{+} \longrightarrow \begin{bmatrix} O\text{—}H \\ \underset{R}{C}\overset{\alpha}{\underset{C}{}} \end{bmatrix}^{+} + \overset{\beta}{C}\text{=}\overset{\gamma}{C}\text{—}R'$$

even   even

Chapter 12 Glossary

absorption spectroscopy The measurement of the amount of light absorbed by a compound as a function of the wavelength (p. 490)

base peak The strongest peak in a mass spectrum. (p. 521)

conjugated double bonds Double bonds that alternate with single bonds, so their pi bonding orbitals can overlap with each other. (p. 499)

electromagnetic spectrum The range of all possible electromagnetic frequencies from zero to infinity. In practice, it ranges from radio waves up to gamma rays. (p. 491)

fingerprint region The portion of the infrared spectrum between 600 and 1400 cm^{-1}, where many complex vibrations occur; so named because no two different compounds (except enantiomers) have exactly the same absorptions in this region. (p. 494)

Fourier transform infrared spectrometer (FT–IR) Infrared light passes through both the sample and a scanning interferometer to give an interference pattern (interferogram). The interferogram is digitized, and the Fourier-transformed spectrum is calculated. (p. 497)

fragmentation The breaking apart of a molecular ion upon ionization in a mass spectrometer. (p. 520)

frequency (ν) The number of complete wave cycles that pass a fixed point in a second; or the number of reversals of the electromagnetic field per second. (p. 491)

high-resolution mass spectrometer A mass spectrometer that can measure masses very accurately, usually to 1 part in 20,000. This high precision allows calculation of molecular formulas using the known atomic masses of the elements. (p. 522)

infrared spectrometer A device that measures a compound's absorption of infrared light as a function of frequency or wavelength. (p. 496)

infrared spectrum A graph of the infrared energy absorbed by a sample as a function of the frequency ($\bar{\nu}$, expressed as a wavenumber, cm^{-1}) or the wavelength (λ, expressed in μm). (p. 498)

interferometer The light-measuring portion of an FT–IR spectrometer. The light is split into two beams. One beam is reflected from a stationary mirror, and the other from a moving mirror. The beams are recombined to form an interference pattern called an **interferogram**. Fourier transformation of the interferogram gives the spectrum. (p. 497)

IR active A vibration that changes the dipole moment of the molecule and thus can absorb infrared light. (p. 495)

IR inactive A vibration that does not change the dipole moment of the molecule and thus cannot absorb infrared light. (p. 495)

mass spectrometer An instrument that ionizes molecules, sorts the ions according to their masses, and records the abundance of ions of each mass. (p. 519)

mass spectrum The graph produced by a mass spectrometer, showing the masses along the x axis and their abundance along the y axis. (p. 521)

  m/z (formerly m/e): The mass-to-charge ratio of an ion. Most ions have a charge of $+1$, and m/z simply represents their masses.

molecular ion, M$^+$ (parent ion) In mass spectrometry, the ion with the same mass as the molecular weight of the original compound; no fragmentation has occurred. (p. 520)

  M+1 peak: An isotopic peak that is one mass unit heavier than the major molecular ion peak. (p. 523)

  M+2 peak: An isotopic peak that is two mass units heavier than the major molecular ion peak. (p. 523)

photon A massless packet of electromagnetic energy. (p. 491)

radical cation A positively charged ion with an unpaired electron; commonly formed by electron impact ionization, when the impinging electron knocks out an additional electron. (p. 520)

$$R{:}R + e^- \longrightarrow [R{\cdot}R]^{\ddot{+}} + 2e^-$$
$$\text{radical cation}$$

source (ion source) The part of a mass spectrometer where the sample is ionized and undergoes fragmentation. (p. 520)

wavelength (λ) The distance between any two peaks (or any two troughs) of a wave. (p. 491)

wavenumber ($\bar{\nu}$) The number of wavelengths that fit into one centimeter (cm^{-1} or reciprocal centimeters); proportional to the frequency. The product of the wavenumber (in cm^{-1}) and the wavelength (in μm) is 10,000. (p. 492)

Essential Problem-Solving Skills in Chapter 12

1. Given an IR spectrum, identify the reliable characteristic peaks.
2. Explain why some characteristic peaks are usually strong or weak and why some may be absent.
3. Predict the stretching frequencies of common functional groups.
4. Identify functional groups from IR spectra.
5. Identify conjugated and strained C=O bonds and conjugated and aromatic C=C bonds from their absorptions in the IR spectrum.
6. Determine molecular weights from mass spectra.
7. When possible, use mass spectra to recognize the presence of Br, Cl, I, N, and S atoms.
8. Predict the major ions from fragmentation of alkanes, alkenes, and alcohols.
9. Use the fragmentation pattern to determine whether a proposed structure is consistent with the mass spectrum.

Study Problems

12-12   Define and give an example of each term.
(a) fingerprint region   (b) an IR-active vibration   (c) an IR-inactive vibration
(d) wavelength   (e) conjugated double bonds   (f) a radical cation
(g) a base peak   (h) wavenumber   (i) a molecular ion

12-13   Convert the following infrared wavelengths to cm^{-1}
(a) 6.24 μm, typical for an aromatic $C=C$ (b) 3.38 μm, typical for a saturated $C-H$ bond
(c) 5.85 μm, typical for a ketone carbonyl   (d) 5.75 μm, typical for an ester carbonyl
(e) 4.52 μm, typical for a nitrile   (f) 3.03 μm, typical for an alcohol $O-H$

12-14   All of the following compounds absorb infrared radiation between 1600 and 1800 cm^{-1}. In each case
(1) Show which bonds absorb in this region.
(2) Predict the approximate absorption frequencies.
(3) Predict which compound of each pair absorbs more strongly in this region.

12-15   Describe the characteristic infrared absorption frequencies that would allow you to distinguish between the following pairs of compounds.
(a) 2,3-dimethyl-2-butene and 2,3-dimethyl-1-butene
(b) 1,3-cyclohexadiene and 1,4-cyclohexadiene

(c) $CH_3(CH_2)_3-\overset{O}{\overset{\|}{C}}-H$ and $CH_3(CH_2)_2-\overset{O}{\overset{\|}{C}}-CH_3$
          pentanal   2-pentanone

          cyclohexanol   cyclohexanone

(e) $CH_3(CH_2)_5-C\equiv C-H$ and $CH_3(CH_2)_3-C\equiv C-CH_2CH_3$
          1-octyne   3-octyne

(f) cyclohexanol and cyclohexene

(g) $CH_3CH_2CH_2-\overset{O}{\overset{\|}{C}}-OH$ and $CH_3-\overset{OH}{\overset{|}{CH}}-CH_2-\overset{O}{\overset{\|}{C}}-H$
          butanoic acid   3-hydroxybutanal

(h) $CH_3CH_2CH_2-\overset{O}{\overset{\|}{C}}-NH_2$ and $CH_3CH_2-\overset{O}{\overset{\|}{C}}-CH_2CH_3$
          butanamide   3-pentanone

12-16   Four infrared spectra are shown below, corresponding to four of the following compounds. For each spectrum, determine the structure and explain how the peaks in the spectrum correspond to the structure you have chosen.

12-17   Predict the masses and the structures of the most abundant fragments observed in the mass spectra of the following compounds.

(a) 2-methylpentane   (b) 3-methyl-2-hexene   (c) 4-methyl-2-pentanol

12-18   Give logical fragmentation reactions to account for the following ions observed in these mass spectra.

(a) *n*-octane: 114, 85, 71, 57   (b) methylcyclohexane: 98, 83

(c) 2-methyl-2-pentene: 84, 69   (d) 1-pentanol: 70, 55, 41, 31

12-19   A common lab experiment is the dehydration of cyclohexanol to cyclohexene.

(a) Explain how you could tell from the IR spectrum whether your product was pure cyclohexene, pure cyclohexanol, or a mixture of cyclohexene and cyclohexanol. Give approximate frequencies for distinctive peaks.

(b) Explain why mass spectrometry might not be a good way to distinguish cyclohexene from cyclohexanol.

12-20   (A true story.) While organizing the undergraduate stockroom, a new chemistry professor found a half-gallon jug containing a cloudy liquid (bp 100–105°C), marked only "STUDENT PREP." She ran a quick mass spectrum, which is printed below. As soon as she saw the spectrum (without even checking the actual mass numbers), she said, "I know what it is."

(a) What compound is the "student prep"? Any uncertainty in the structure?

(b) Suggest structures for the fragments at 136, 107, and 93. Why is the base peak (at m/z 57) so strong?

*12-21   A C—D (carbon–deuterium) bond is electronically much like a C—H bond, and it has a similar stiffness, measured by the *spring constant, k*. The deuterium atom has twice the mass of a hydrogen atom, however.

(a) The infrared absorption frequency is proportional to $\sqrt{k/m}$. Use this relationship to calculate the IR absorption frequency of a typical C—D bond.

(b) A chemist dissolves a sample in deuterochloroform ($CDCl_3$), then decides to take the IR spectrum and simply evaporates most of the $CDCl_3$. What functional group will *appear* to be present in this IR spectrum as a result of the $CDCl_3$ impurity?

*12-22   The mass spectrum of *n*-octane shows a prominent molecular ion peak (m/z 114). There is also a large peak at m/z 57, but it is not the base peak. The mass spectrum of 3,4-dimethylhexane shows a smaller molecular ion, and the peak at mass 57 is the base peak. Explain these trends in abundance of the molecular ions and the ions at mass 57, and predict the intensities of the peaks at masses 57 and 114 in the spectrum of 2,2,3,3-tetramethylbutane.

12-23   An unknown, foul-smelling hydrocarbon gives the mass spectrum and infrared spectrum shown below.

(a) Use the mass spectrum to propose a molecular formula. How many elements of unsaturation are there?

(b) Use the IR spectrum to determine the functional group(s), if any.

(c) Propose one or more structures for this compound. What parts of the structure are uncertain? If you knew that hydrogenation of the compound gives *n*-octane, would the structure still be uncertain?

(d) Propose structures for the major fragments at 39, 67, 81, and 95 in the mass spectrum. Explain why the base peak is so strong.

*12-24   Chapter 9 covered a synthesis of alkynes by a double dehydrohalogenation of dihalides. A student tried to convert 2,5-dimethyl-3-hexene to 2,5-dimethyl-3-hexyne by adding bromine across the double bond, then doing a double elimination. The infrared and mass spectra of the major product are shown below.

(a) Do the spectra confirm the right product? If not, what is it?

(b) Explain the important peaks in the IR spectrum.

12-25 Four IR spectra are shown below, corresponding to four of the following compounds. For each spectrum, determine the structure and explain how the peaks in the spectrum correspond to the structure you have chosen.

12-26   A laboratory student added 1-bromobutane to a flask containing dry ether and magnesium turnings. An exothermic reaction resulted, and the ether boiled vigorously for several minutes. Then she added acetone to the reaction mixture, and the ether boiled even more vigorously. She added dilute acid to the mixture, and separated the layers. She evaporated the ether layer, and distilled a liquid that boiled at 143°C. GC–MS analysis of the distillate showed one major product with a few minor impurities. The mass spectrum of the major product is shown below.

(a) Draw out the reactions that took place and show the product that was formed.

(b) Explain why the molecular ion is or is not visible in the mass spectrum, and show what ions are likely to be responsible for the strong peaks at m/z 59 and 101.

CHAPTER 13

Nuclear Magnetic Resonance Spectroscopy

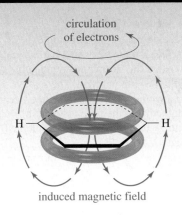

circulation of electrons

H — H

induced magnetic field

Nuclear magnetic resonance spectroscopy (NMR) is the most powerful tool available for organic structure determination. Like IR spectroscopy, NMR can be used with a very small sample, and it does not harm the sample. The NMR spectrum provides a great deal of information about the structure of the compound, and some structures can be determined using only the NMR spectrum. More commonly, however, the NMR spectrum is used in conjunction with other forms of spectroscopy and chemical analysis to determine the structures of complicated organic molecules.

NMR is used to study a wide variety of nuclei, including ^1H, ^{13}C, ^{15}N, ^{19}F, and ^{31}P. Organic chemists find proton (^1H) and carbon-13 (^{13}C) NMR to be most useful because hydrogen and carbon are major components of organic compounds. Historically, NMR was first used to study protons (the nuclei of hydrogen atoms), and proton magnetic resonance (^1H NMR) spectrometers are the most common. "Nuclear magnetic resonance" is assumed to mean "proton magnetic resonance" unless a different nucleus is specified. We begin our study of NMR with proton magnetic resonance and conclude with a discussion of ^{13}C NMR.

A nucleus with an odd atomic number or an odd mass number has a *nuclear spin* that can be observed by the NMR spectrometer. A proton is the simplest nucleus, and its odd atomic number of 1 implies it has a spin. We can visualize a spinning proton as a rotating sphere of positive charge (Figure 13-1). This movement of charge is like an electric current in a loop of wire. It generates a magnetic field (symbolized by B) called the **magnetic moment**, that looks like the field of a small bar magnet.

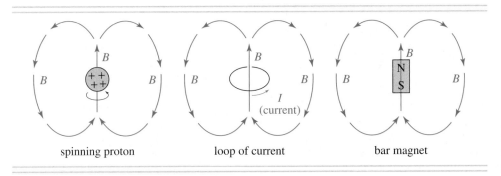

spinning proton   loop of current   bar magnet

◄ Figure 13-1
A spinning proton generates a magnetic field, called its magnetic moment. This magnetic field (B) resembles that of a small bar magnet.

539

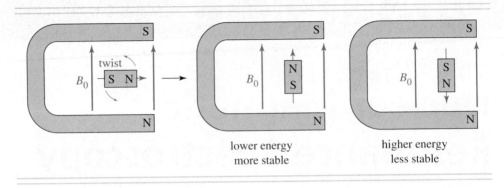

▶ Figure 13-2
▶ **Figure 13-2**
An external magnetic field (B_0) applies a force to a small bar magnet, twisting the bar magnet to align it with the external field. The arrangement of the bar magnet aligned *with* the field is lower in energy than the arrangement aligned *against* the field.

When a small bar magnet is placed in the field of a larger magnet (Figure 13-2), it twists to align itself with the field of the larger magnet—a lower energy arrangement than an orientation against the field. The same effect is seen when a proton is placed in an external magnetic field (B_0), as shown below. Quantum mechanics requires the proton's magnetic moment to be aligned either *with* the external field or *against* the field. The lower-energy state with the proton aligned with the field is called the *alpha-spin* (α-spin) *state*. The higher-energy state with the proton aligned against the external magnetic field is called the *beta-spin* (β-spin) *state*.

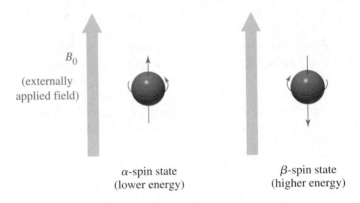

In the absence of an external magnetic field, proton magnetic moments have random orientations. When an external magnetic field is applied, each proton in a sample assumes the α state or the β state. Because the α-spin state is lower in energy, there are more α spins than β spins.

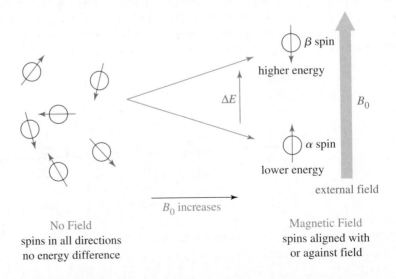

In a strong magnetic field, the energy difference between the two spin states is larger than it is in a weaker field. In fact, the energy difference is proportional to the strength of the magnetic field, as expressed in the equation

$$\Delta E = \gamma \frac{h}{2\pi} B_0, \text{ where } \Delta E = \text{energy difference between } \alpha \text{ and } \beta \text{ states}$$

$$h = \text{Planck's constant}$$

$$B_0 = \text{strength of the external magnetic field}$$

$$\gamma = \text{gyromagnetic ratio, } 26{,}753 \text{ sec}^{-1} \text{ gauss}^{-1} \text{ for a proton}$$

The **gyromagnetic ratio (γ)** is a constant that depends on the magnetic moment of the nucleus under study. Magnetic fields are measured in *gauss*; for example, the strength of the earth's magnetic field is about 0.57 gauss. The new SI unit is the *tesla* (T), which is simply 10,000 gauss.

The energy difference between a proton's two spin states is small. For a strong external magnetic field of 25,000 gauss (2.5 T), it is only about 10^{-5} kcal/mol (4×10^{-5} kJ/mol). Even this small energy difference can be detected by NMR. When a proton interacts with a photon with just the right amount of electromagnetic energy, the proton's spin can flip from α to β or from β to α. A nucleus aligned with the field can absorb the energy needed to flip and become aligned against the field.

When a nucleus is subjected to the right combination of magnetic field and electromagnetic radiation to flip its spin, it is said to be "in resonance" (Figure 13-3), and its absorption of energy is detected by the NMR spectrometer. This is the origin of the term "nuclear magnetic resonance."

A photon's energy is given by $E = h\nu$, showing that the energy E is proportional to ν, the frequency of the electromagnetic wave. This equation can be combined with the equation for the energy difference between the spin states:

$$\Delta E = h\nu = \gamma \frac{h}{2\pi} B_0$$

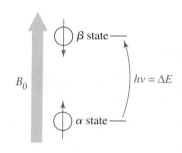

▲ Figure 13-3
A nucleus is "in resonance" when it is irradiated with radio-frequency photons having energy equal to the energy difference between the spin states. Under these conditions, a proton in the α-spin state can absorb a photon and flip to the β-spin state.

Rearranging to solve for ν shows that the resonance frequency ν is proportional to the applied magnetic field (B_0) and the gyromagnetic ratio (γ):

$$\nu = \frac{1}{2\pi} \gamma B_0$$

For a proton, $\gamma = 26{,}753 \text{ sec}^{-1} \text{ gauss}^{-1}$, and

$$\nu = \frac{(26{,}753 \text{ sec}^{-1} \text{ gauss}^{-1})}{2\pi} \times B_0 = (4257.8 \text{ sec}^{-1} \text{ gauss}^{-1}) \times B_0$$

For the fields of currently available magnets, proton resonance frequencies occur in the radio-frequency (RF) region of the spectrum. NMR spectrometers are usually designed for the most powerful magnet that is practical for the price range of the spectrometer (to make ΔE as large and easily detected as possible), and the radio frequency needed for resonance is calculated based on the field. In the past, the most common operating frequency for student spectrometers has been 60 MHz (megahertz; 1 million cycles per second), corresponding to a magnetic field of 14,092 gauss. Higher-resolution instruments commonly operate at frequencies of 100 to 300 MHz (and higher), corresponding to fields of 23,486 to 70,459 gauss.

SOLVED PROBLEM 13-1

Calculate the magnetic fields that correspond to proton resonance frequencies of 60 MHz and 300 MHz.

SOLUTION

We substitute into the equation $\nu - (1/2\pi)\gamma B_0$.

$$60 \text{ MHz} = 60 \times 10^6 \text{ sec}^{-1} = (4257.8 \text{ sec}^{-1} \text{ gauss}^{-1}) \times B_0$$

$$B_0 = 14{,}092 \text{ gauss (1.4092 tesla)}$$

$$300 \text{ MHz} = 300 \times 10^6 \text{ sec}^{-1} = (4257.8 \text{ sec}^{-1} \text{ gauss}^{-1}) \times B_0$$

$$B_0 = 70{,}459 \text{ gauss (7.0459 tesla)}$$

13-3
Magnetic Shielding by Electrons

Up to now, we have considered the resonance of a naked proton in a magnetic field; but real protons in organic compounds are not naked. They are surrounded by electrons that partially shield them from the external magnetic field. The electrons circulate and generate a small "induced" magnetic field that opposes the externally applied field.

A similar effect occurs when a loop of wire is moved into a magnetic field. The electrons in the wire are induced to flow around the loop in the direction shown in Figure 13-4; this is the principle of the electric generator. The induced electric current creates a magnetic field that opposes the external field.

In a molecule, the electron cloud around each nucleus acts like a loop of wire, rotating in response to the external field. This induced rotation is a circular current whose magnetic field opposes the external field. The result is that the magnetic field at the nucleus is weaker than the external field, and we say the nucleus is **shielded**. The effective magnetic field *at the shielded proton* is always weaker than the external field, so the applied field must be increased for resonance to occur at a given frequency (Figure 13-5).

$$B_{\text{effective}} = B_{\text{external}} - B_{\text{shielding}}$$

At 300 MHz, an unshielded naked proton absorbs at 70,459 gauss, but a shielded proton requires a stronger field. For example, if a proton is shielded by 1 gauss when

▶ **Figure 13-4**
When a loop of wire is moved into a magnetic field, a current is induced in the wire. This current produces its own smaller magnetic field, in the direction opposite the applied field. In a molecule, electrons can circulate around a nucleus. The resulting "current" sets up a magnetic field that opposes the external field, so the nucleus feels a slightly weaker field.

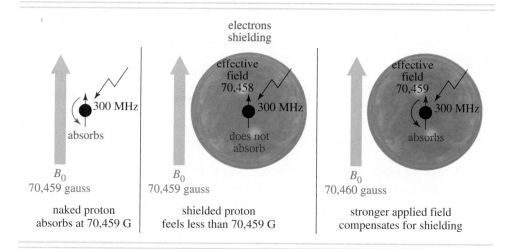

the external field is 70,459 gauss, the effective magnetic field *at the proton* is 70,458 gauss. If the external field is increased to 70,460 gauss, the effective magnetic field at the proton is increased to 70,459 gauss, which brings this proton into resonance.

If all protons were shielded by the same amount, they would all be in resonance at the same combination of frequency and magnetic field. Fortunately, protons in different chemical environments are shielded by different amounts. In methanol, for example, the electronegative oxygen atom withdraws some electron density from around the hydroxyl proton. The hydroxyl proton is not shielded as much as the methyl protons, so the hydroxyl proton absorbs at a lower field than the methyl protons (but still at a higher field than a naked proton). We say that the hydroxyl proton is **deshielded** somewhat by the presence of the electronegative oxygen atom.

more shielded, absorb at a higher field

$$H-C-\overset{\displaystyle H}{\underset{\displaystyle H}{|}}-\ddot{O}:$$

H — less shielded, absorbs at a lower field

Because of the diverse and complex structures of organic molecules, the shielding effects of electrons at various positions are generally different. A careful measurement of the field strengths required for resonance of the various protons in a molecule provides us with two important types of information:

1. The *number of different absorptions* (also called *signals* or *peaks*) implies how many different types of protons are present.

2. The *amount of shielding* shown by these absorptions often implies the electronic structure of the molecule close to each type of proton.

Two other aspects of the NMR spectrum we will consider are the intensities of the signals and their splitting patterns:

3. The *intensities of the signals* imply how many protons of each type are present.

4. The *splitting of the signals* gives information about other nearby protons.

Before discussing the design of spectrometers, let's review what happens in an NMR spectrometer. Protons (in the sample compound) are placed in a magnetic field, where they align either with the field or against it. While still in the magnetic field, the protons are subjected to radiation of a frequency they can absorb by changing their orientation relative to the field. If protons were isolated, they would all absorb at the same frequency, proportional to the magnetic field.

But protons in a molecule are partially shielded from the magnetic field, and this shielding depends on each proton's environment. Thus, protons in different environments within a molecule exposed to a constant frequency absorb the radiation at different magnetic field strengths. The NMR spectrometer must be equipped to vary the magnetic field and plot a graph of energy absorption as a function of the magnetic field strength. Such a graph is called a **nuclear magnetic resonance spectrum**.

13-4
The NMR
Spectrometer

The simplest type of NMR spectrometer (Figure 13-6) consists of four parts:

1. A stable magnet, with a sensitive controller to produce a precise magnetic field
2. A radio-frequency (RF) transmitter, emitting a precise frequency
3. A detector to measure the sample's absorption of RF energy
4. A recorder to plot the output from the detector versus the applied magnetic field

The recorder prints a graph of absorption (on the y axis) as a function of the applied magnetic field (on the x axis). Higher values of the magnetic field are toward the right (**upfield**), and lower values are toward the left (**downfield**). The absorptions of more shielded protons appear upfield, toward the right of the spectrum, and less shielded protons appear downfield, toward the left. The NMR spectrum of methanol is shown in Figure 13-7.

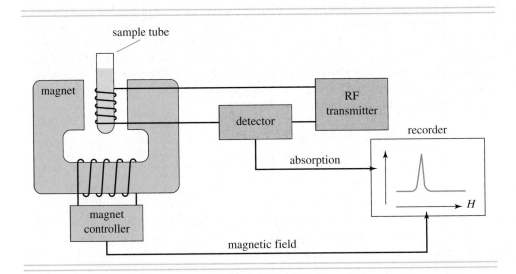

▶ **Figure 13-6**
Simplified block diagram of a nuclear magnetic resonance spectrometer.

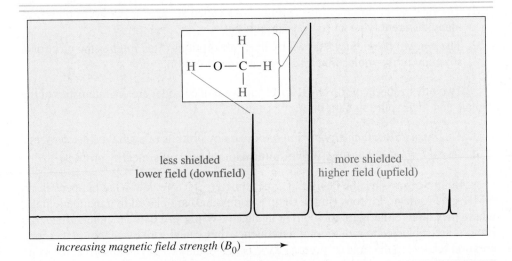

▶ **Figure 13-7**
Proton NMR spectrum of methanol. The more shielded methyl protons appear toward the right of the spectrum (higher field); the less shielded hydroxyl proton appears toward the left (lower field)

13-5A Measurement of Chemical Shifts

13-5
The Chemical Shift

The variations in the positions of NMR absorptions, arising from electronic shielding and deshielding, are called **chemical shifts**.

> **chemical shift** The difference (in parts per million) between the resonance frequency of the proton being observed and that of tetramethylsilane (TMS).

In practice, it is difficult to measure the absolute field where a proton absorbs with enough accuracy to distinguish individual protons, because the signals often differ by only a few thousandths of a gauss at an applied field of 70,459 gauss. A more accurate method for expressing chemical shifts is to determine the value relative to a reference compound added to the sample. The *difference* in the magnetic field strength for resonance of the sample protons and the reference protons can be measured very accurately.

The most common NMR reference compound is *tetramethylsilane* $(CH_3)_4Si$, abbreviated **TMS**. Because silicon is less electronegative than carbon, the methyl groups of TMS are relatively electron rich, and their protons are well shielded. They absorb at a higher field strength than most hydrogens bonded to carbon or other elements, so most NMR signals appear *downfield* (to the left) of the TMS signal. All 12 protons in TMS absorb at exactly the same applied magnetic field, giving one strong absorption.

A small amount of TMS is added to the sample, and the instrument measures the difference in magnetic field between where the protons in the sample absorb and where TMS absorbs. For each type of proton in the sample, the distance downfield of TMS is the chemical shift of those protons.

Chemical shifts are measured in *parts per million* (ppm), a dimensionless fraction of the total applied field. By custom, the difference in field (the chemical shift) between the NMR signal of a proton and that of TMS is not measured in gauss, but in frequency units (hertz or Hz). Remember that frequency units and magnetic field units are always proportional in NMR, with $\nu = \gamma B_0 / 2\pi$. The horizontal axis of the NMR spectrum is calibrated in hertz. A chemical shift in ppm can be calculated by dividing the shift measured in hertz by the spectrometer frequency measured in millions of hertz (megahertz, or MHz).

$$\text{chemical shift (ppm)} = \frac{\text{shift downfield from TMS (Hz)}}{\text{total spectrometer frequency (MHz)}}$$

The chemical shift (in ppm) of a given proton is the same regardless of the operating field and frequency of the spectrometer. The use of chemical shifts to describe absorptions standardizes values for all NMR spectrometers.

The most common scale of chemical shifts is the δ (delta) scale, which we will use (Fig. 13-8). The signal from tetramethylsilane (TMS) is *defined* as 0.00 ppm on the δ scale.

$$CH_3 - \underset{\underset{CH_3}{|}}{\overset{\overset{CH_3}{|}}{Si}} - CH_3$$

tetramethylsilane (TMS)

Photo of a modern 300-MHz NMR spectrometer. The metal container at right contains the superconducting magnet, cooled by a liquid helium bath inside a liquid nitrogen bath. The electronics used to control the spectrometer and to calculate spectra are at left and in the background.

$$\text{chemical shift, ppm } \delta = \frac{\text{shift downfield from TMS (in Hz)}}{\text{spectrometer frequency (in MHz)}}$$

600 Hz	480 Hz	360 Hz	240 Hz	120 Hz	0 Hz
10 9	8 7	6 5	4 3	2 1	0

ppm δ ↑ TMS

60 MHz

3000 Hz	2400 Hz	1800 Hz	1200 Hz	600 Hz	0 Hz
10 9	8 7	6 5	4 3	2 1	0

ppm δ ↑ TMS

300 MHz

◀ **Figure 13-8**
Use of the δ scale with 60- and 300-MHz spectrometers. The absorption of TMS is defined as 0, with the scale increasing from right to left (toward the lower field). Each δ unit is 1 ppm difference from TMS: 60 Hz at 60 MHz and 300 Hz at 300 MHz.

Most protons absorb at lower fields than TMS, so the δ scale increases toward lower field (toward the left of the spectrum). The spectrum is calibrated in both frequency and ppm δ.

SOLVED PROBLEM 13-2

A 300-MHz spectrometer records a proton that absorbs at a frequency 2130 Hz downfield from TMS.
(a) Determine its chemical shift, and express this shift as a magnetic field difference.
(b) Predict this proton's chemical shift at 60 MHz. In a 60-MHz spectrometer, how far downfield (in gauss and in hertz) from TMS would this proton absorb?

SOLUTION

(a) The chemical shift is the fraction

$$\frac{\text{shift downfield (Hz)}}{\text{spectrometer frequency (MHz)}} = \frac{2130 \text{ Hz}}{300 \text{ MHz}} = 7.10 \text{ ppm}$$

The chemical shift of this proton is $\delta 7.10$. The field shift is 70,459 gauss $\times$ $(7.10 \times 10^{-6}) = 0.500$ gauss.
(b) The chemical shift is unchanged at 60 MHz: $\delta 7.10$. The field shift is 14,092 gauss $\times$ $(7.10 \times 10^{-6}) = 0.100$ gauss. The frequency shift is 60 MHz $\times$ $(7.10 \times 10^{-6}) =$ 426 Hz.

PROBLEM 13-1

In a 300-MHz spectrometer, the protons in iodomethane absorb at a position 650 Hz downfield from TMS.
(a) What is the chemical shift of these protons?
(b) Determine the difference in the magnetic field required for resonance of the iodomethane protons compared with the TMS protons.
(c) What is the chemical shift of the iodomethane protons in a 60 MHz spectrometer?
(d) How many hertz downfield from TMS would they absorb at 60 MHz?

The 300-MHz NMR spectrum of methanol (Figure 13-9) shows the two signals from methanol together with the TMS reference peak at $\delta 0.0$. The methyl protons absorb

▶ **Figure 13-9**
A 300-MHz NMR spectrum of methanol. The methyl protons absorb at $\delta 3.4$, and the hydroxyl proton absorbs at $\delta 4.8$.

TABLE 13-1 Variation of Chemical Shift with Electronegativity

	X in CH$_3$—X				
	F	**OH**	**Cl**	**Br**	**I**
electronegativity of X	4.0	3.4	3.2	3.0	2.7
chemical shift of CH$_3$—X	δ4.3	δ3.4	δ3.0	δ2.7	δ2.2

1025 Hz (0.241 gauss) downfield from TMS. Their chemical shift is 3.4 ppm, so we say that the methyl protons absorb at δ3.4. The hydroxyl proton absorbs farther downfield, at a position around 1450 Hz (0.340 gauss) from TMS. Its chemical shift is δ4.8.

Both the hydroxyl proton and the methyl protons of methanol show the deshielding effects of the electronegative oxygen atom. The chemical shift of a methyl group in an alkane is about δ0.9. Therefore, the methanol oxygen deshields the methyl protons by an additional 2.5 ppm. Other electronegative atoms produce similar deshielding effects. Table 13-1 compares the chemical shifts of methanol with those of the methyl halides. Notice that the chemical shift of the methyl protons depends on the electronegativity of the substituent, with more electronegative substituents deshielding more and giving larger chemical shifts.

The effect of an electronegative group on the chemical shift also depends on its distance from the protons. In methanol, the hydroxyl proton is separated from oxygen by one bond, and its chemical shift is δ4.8. The methyl protons are separated from oxygen by two bonds, and their chemical shift is δ3.4. In general, the effect of an electron-withdrawing substituent decreases with increasing distance, and the effects are usually negligible on protons that are separated from the electronegative group by four or more bonds.

This decreasing effect can be seen by comparing the chemical shifts of the various protons in 1-bromobutane with those in butane. The deshielding effect of an electronegative substituent drops off rapidly with distance. In 1-bromobutane, protons on the α-carbon are deshielded by about 2.5 ppm, and the β-protons are deshielded by about 0.4 ppm. Protons that are more distant than the β-protons are deshielded by a negligible amount.

TABLE 13-2 Chemical Shifts of the Chloromethanes

butane

chemical shift: 0.9 1.3 1.3 0.9

1-bromobutane

deshielding resulting from Br, ppm: 0.0 0.0 0.4 2.5

If more than one electron-withdrawing group is present, the deshielding effects are nearly (but not quite) additive. In the chloromethanes (Table 13-2), the addition of the first chlorine atom causes a shift to δ3.0, the second chlorine shifts the absorption further to δ5.3, and the third chlorine moves the chemical shift to δ7.2 for chloroform. The chemical shift *difference* is about 2 to 3 ppm each time another chlorine atom is added, but each additional chlorine moves the peak slightly less than the previous one did.

13-5B Characteristic Values of Chemical Shifts

Since the chemical shift of a proton is determined by its environment, we can construct a table of approximate chemical shifts for many types of compounds. Let's begin with a short table of representative chemical shifts (Table 13-3) and consider the reasons for some of the more interesting and unusual values. A comprehensive table of chemical shifts appears in Appendix 1.

Compound	Chemical Shift	Difference
H—C—H (H, H)	δ0.2	
		2.8 ppm
H—C—Cl (H, H)	δ3.0	
		2.3 ppm
H—C—Cl (Cl, H)	δ5.3	
		1.9 ppm
H—C—Cl (Cl, Cl)	δ7.2	

Note: Each chlorine atom added changes the chemical shift of the remaining methyl protons by 2 to 3 ppm. These changes are nearly additive.

TABLE 13-3 Typical Values of Chemical Shifts

Type of Proton	Approximate δ	Type of Proton	Approximate δ
alkane ($-CH_3$)	0.9	$>C=C<_{CH_3}$	1.7
alkane ($-CH_2-$)	1.3		
alkane $\left(-\underset{\vert}{CH}-\right)$	1.4	Ph$-$H	7.2
		Ph$-CH_3$	2.3
$-\overset{\overset{\displaystyle O}{\|\|}}{C}-CH_3$	2.1	R$-$CHO	9–10
		R$-$COOH	10–12
$-C{\equiv}C-H$	2.5	R$-$OH	variable, about 2–5
R$-CH_2-$X	3–4	Ar$-$OH	variable, about 4–7
(X = halogen, O)		R$-NH_2$	variable, about 1.5–4
$>C=C<_{H}$	5–6		

Note: These values are approximate, as all chemical shifts are affected by neighboring substituents. The numbers given here assume that alkyl groups are the only other substituents present. A more complete table of chemical shifts appears in Appendix 1.

SOLVED PROBLEM 13-3

Using Table 13.3 above, predict the chemical shifts of the protons in the following compounds.

(a) $CH_3-\overset{\overset{\displaystyle O}{\|\|}}{C}-OH$ (b) $Cl-CH_2^a-CH_2^b-CH_3^c$ (c) $(CH_3^a)_3CCH^b{=}CH_2^c$

SOLUTION

(a) The methyl group in acetic acid is next to a carbonyl group; Table 13-3 predicts a chemical shift of about $\delta 2.1$. (The experimental value is $\delta 2.10$). The acid proton ($-$COOH) should absorb between $\delta 10$ and $\delta 12$. (The experimental value is $\delta 11.4$, variable.)

(b) Protons *a* are on the carbon atom bearing the chlorine, and they absorb between $\delta 3$ and $\delta 4$ (experimental: $\delta 3.7$). Protons *b* are one carbon removed, and they are predicted to absorb about $\delta 1.7$, like the β protons in 1-bromobutane (experimental: $\delta 1.8$). The methyl protons *c* will be nearly unaffected, absorbing around $\delta 0.9$ ppm (experimental: $\delta 1.0$).

(c) Methyl protons *a* are expected to absorb around $\delta 0.9$ (experimental: $\delta 1.0$). The vinyl protons *b* and *c* are expected to absorb between $\delta 5$ and $\delta 6$ (experimental $\delta 5.8$ for *b* and $\delta 4.9$ for *c*).

PROBLEM-SOLVING **HINT**

Table 13-3 provides the *numbers* but not the understanding and practice needed to work most NMR problems. Learn the material in this table, then practice doing problems until you feel confident.

Vinyl and Aromatic Protons   Table 13-3 shows that double bonds and aromatic rings produce large deshielding effects on their vinyl and aromatic protons. These deshielding effects result from the same type of circulation of electrons that normally shields nuclei from the magnetic field. In benzene and its derivatives, the aromatic ring of pi-bonding electrons acts as a conductor, and the external magnetic field induces a *ring current* (Figure 13-10). At the center of the ring, the induced field acts to oppose the external field. These induced field lines curve around, however, and on the edge of the ring the induced field *adds to* the external field. As a result, the aromatic protons are actually *deshielded*, resulting in absorption at low values of the applied magnetic field. Benzene absorbs at $\delta 7.2$, and most aromatic protons absorb in the range of $\delta 7$ to $\delta 8$.

We should remember that the benzene molecule is not always lined up in the position shown in Figure 13-10. Because benzene is constantly tumbling in the solution, the

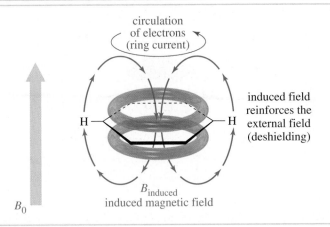

◀ **Figure 13-10**
The induced magnetic field of the circulating aromatic electrons opposes the applied magnetic field along the axis of the ring. The aromatic hydrogens are on the equator of the ring, where the induced field lines curve around and reinforce the applied field.

chemical shift observed for its protons is an average of all the possible orientations. If we could hold a benzene molecule in the position shown in Figure 13-10, its protons would absorb at a field even lower than $\delta 7.2$. Other orientations, such as the one with the benzene ring edge-on to the magnetic field, would be less deshielded and would absorb at a higher field. It is the *average* of all these orientations that is observed by the resonance at $\delta 7.2$.

Figure 13-11 shows the NMR spectrum of toluene (methylbenzene). The aromatic protons absorb around $\delta 7.2$. The methyl protons are deshielded by a smaller amount, absorbing at $\delta 2.3$.

The pi electrons of an alkene deshield the vinyl protons in the same way that an aromatic ring of electrons deshields the aromatic protons. The effect is not as large in the alkene, however, because there is not such a large, effective ring of electrons as there is in benzene. Once again, the motion of the pi electrons generates an induced magnetic field that opposes the applied field at the middle of the double bond. The vinyl protons are on the periphery of this field, however, where the induced field bends around and reinforces the external field (Figure 13-12). As a result of this deshielding effect, most vinyl protons absorb in the range of $\delta 5$ to $\delta 6$.

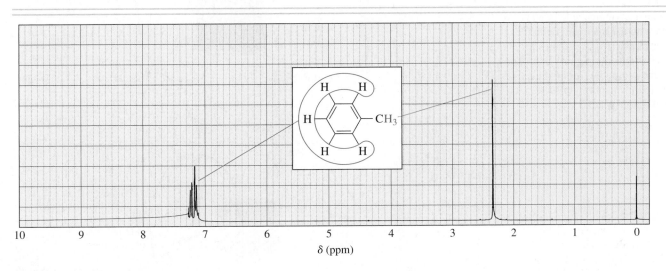

▲ **Figure 13-11**
Proton NMR spectrum of toluene. The aromatic protons absorb at chemical shifts near $\delta 7.2$, and the methyl protons absorb at $\delta 2.3$.

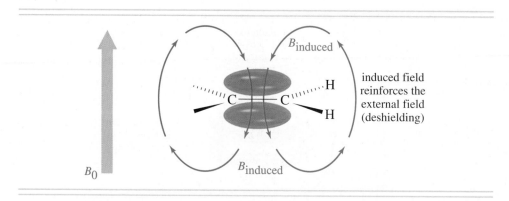

► **Figure 13-12**
Vinyl protons are positioned on the periphery of the induced magnetic field of the pi electrons. In this position, they are deshielded by the induced magnetic field.

Acetylenic Hydrogens   Since the pi bond of an alkene deshields the vinyl protons, we might expect an acetylenic hydrogen ($-C\equiv C-H$) to be even more deshielded by the two pi bonds of the triple bond. The opposite is true: Acetylenic hydrogens absorb around $\delta 2.5$, compared with $\delta 5$ to $\delta 6$ for vinyl protons. Figure 13-13 shows that the triple bond has a cylinder of electron density surrounding the sigma bond. As the molecules tumble in solution, in some orientations this cylinder of electrons can circulate to produce an induced magnetic field. The acetylenic proton lies along the *axis* of this induced field, which is a shielded region. When this shielded orientation is averaged with all other possible orientations, the result is a resonance around $\delta 2.5$.

Aldehyde Protons   Aldehyde protons ($-CHO$) absorb at even lower fields than vinyl protons and aromatic protons: between $\delta 9$ and $\delta 10$. Figure 13-14 shows that the aldehyde proton is deshielded both by the circulation of the electrons in the double bond and by the inductive electron-withdrawing effect of the carbonyl oxygen atom.

► **Figure 13-13**
When the acetylenic triple bond is aligned with the magnetic field, the cylinder of electrons circulates to create an induced magnetic field. The acetylenic proton lies along the axis of this field, which opposes the external field.

► **Figure 13-14**
Like a vinyl proton, the aldehyde proton is deshielded by the circulation of electrons in the pi bond. It is also deshielded by the electron-withdrawing effect of the carbonyl ($C=O$) group, giving a resonance between $\delta 9$ and $\delta 10$.

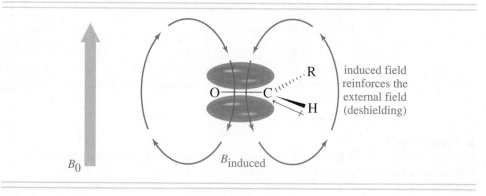

Hydrogen-Bonded Protons   The chemical shifts of O—H protons in alcohols and N—H protons in amines depend on the concentration. In concentrated solutions, these protons are deshielded by hydrogen bonding, and they absorb at a relatively low field: about $\delta3.5$ for an amine N—H and about $\delta4.5$ for an alcohol O—H. When the alcohol or amine is diluted with a non-hydrogen-bonding solvent such as CCl_4, hydrogen bonding becomes less important. In dilute solutions, these signals are observed around $\delta2$.

Hydrogen bonding and the proton exchange that accompanies it may contribute to a broadening of the peak from an O—H or N—H proton. A broad peak appears because protons exchange from one molecule to another during the NMR resonance (see Section 13-12). The protons pass through a variety of environments during this exchange, absorbing over a wider range of frequencies and field strengths.

Carboxylic Acid Protons   Because carboxylic acid protons are bonded to an oxygen next to a carbonyl group, they have considerable positive character. They are strongly deshielded and absorb at chemical shifts greater than $\delta10$. Carboxylic acids frequently exist as hydrogen-bonded dimers (shown at right), with moderate rates of proton exchange that broaden the absorption of the acid proton.

The proton NMR spectrum of acetic acid is shown in Figure 13-15. As we expect, the methyl group next to the carbonyl absorbs at a chemical shift of $\delta2.1$. The acid proton signal appears at a chemical shift that is not scanned in the usual range of the NMR spectrum. It is seen in a second trace with a 2.5 ppm offset, meaning that this trace corresponds to frequencies with chemical shifts 2.5 ppm *larger* than shown on the trace. The acid proton appears around $\delta11.8$: $\delta9.8$ read from the trace, plus the $\delta2.0$ offset.

carboxylic acid dimer

PROBLEM 13-2

Predict the chemical shifts of the protons in the following compounds.

(a) $(CH_3)_3C$, H / $C=C$ / H, $C(CH_3)_3$

(b) [benzene ring with H, CH_3, CH_3, CH_3, CH_3, H]

(c) CH_3O— [benzene ring with H, H, H, H] —OCH_3

(d) $CH_3-\overset{\displaystyle CH_3}{\underset{\displaystyle OH}{C}}-C\equiv C-H$

(e) [benzene ring]$-CH_2-\overset{\displaystyle O}{C}-OH$

(f) $CH_3-\overset{\displaystyle CH_3}{\underset{\displaystyle Br}{C}}-CH_2Br$

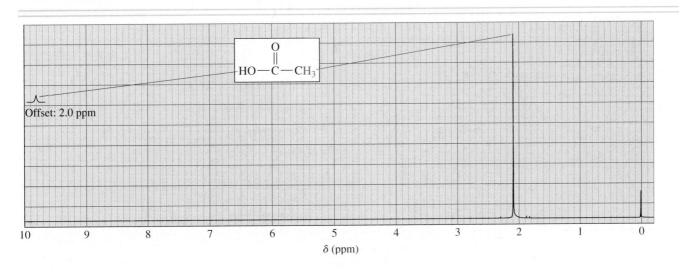

Offset: 2.0 ppm

$$HO-\overset{\displaystyle O}{C}-CH_3$$

10   9   8   7   6   5   4   3   2   1   0
δ (ppm)

▲ **Figure 13-15**
In the NMR spectrum of acetic acid, the methyl protons are deshielded to about $\delta2.1$ by the adjacent carbonyl group. The acid proton appears at $\delta11.8$, shown on an offset trace.

13-6
The Number
of Signals

In general, the number of NMR signals corresponds to the number of different kinds of protons present in the molecule. For example methyl *t*-butyl ether has two types of protons (Figure 13-16). The three methoxyl protons are chemically identical, and they give rise to a single absorption at $\delta 3.2$. The *t*-butyl protons are chemically different from the methyl protons, absorbing at $\delta 1.2$.

Protons in identical chemical environments with the same shielding have the same chemical shift. Such protons are said to be **chemically equivalent**. This is what is meant whenever we use the term *equivalent* in discussing NMR spectroscopy. In methyl *t*-butyl ether, the three methoxyl protons are chemically equivalent, and the nine *t*-butyl protons are chemically equivalent.

The spectrum of *t*-butyl acetoacetate (Figure 13-17) shows three types of protons: the *t*-butyl protons *(a)*, with a chemical shift of $\delta 1.5$; the methyl protons *(b)*, deshielded by an adjacent carbonyl group, with a chemical shift of $\delta 2.25$; and the methylene protons *(c)*, deshielded by two adjacent carbonyl groups, at $\delta 3.35$.

▲ **Figure 13-16**
Methyl *t*-butyl ether has two types of protons, giving two NMR signals.

▲ **Figure 13-17**
t-Butyl acetoacetate has three types of protons, giving three signals in the NMR spectrum.

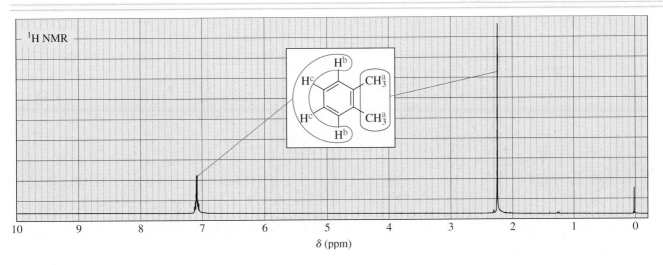

▲ Figure 13-18
There are three types of protons in *o*-xylene, but only two absorptions are seen in the
spectrum. The aromatic protons H^b and H^c are accidentally equivalent, producing a broadened
peak at $\delta 7.1$.

In some cases, there may be fewer signals in the NMR spectrum than there are different types of protons in the molecule. For example, Figure 13-18 shows the structure
and spectrum of *o*-xylene (1,2-dimethylbenzene). There are three different types of protons, labeled *a* for the two equivalent methyl groups, *b* for the protons adjacent to the
methyl groups, and *c* for the protons two carbons removed. The spectrum shows only
two distinct signals, however.

The upfield signal at $\delta 2.3$ corresponds to the six methyl protons, H^a. The absorption at $\delta 7.2$ corresponds to all four of the aromatic protons, H^b and H^c. Although the
two types of aromatic protons are different, the methyl groups do not strongly influence
the electron density of the ring or the amount of shielding felt by any of the substituents
on the ring. The aromatic protons produce two signals, but these signals happen to occur
at nearly the same chemical shift. Protons that are not chemically equivalent but happen to absorb at the same chemical shift are said to be **accidentally equivalent**.

PROBLEM 13-3

Determine the number of different kinds of protons in each compound.
(a) 1-chloropropane
(b) 2-chloropropane
(c) 2,2-dimethylbutane
(d) 1-bromo-2-methylbenzene,

PROBLEM 13-4

The NMR spectrum of toluene (methylbenzene) was shown in Figure 13-11.
(a) How many different kinds of protons are there in toluene?
(b) Explain why the aromatic region around $\delta 7.2$ is broad, with more than one sharp absorption.

The area under a peak is proportional to the number of hydrogens contributing to that peak.
For example, in the methyl *t*-butyl ether spectrum (Figure 13-19), the absorption of the
t-butyl protons is larger and stronger than that of the methoxyl protons because there are
three times as many *t*-butyl protons as methoxyl protons. We cannot simply compare peak
heights, however; the *area* under the peak is proportional to the number of protons.

13-7
Areas of the Peaks

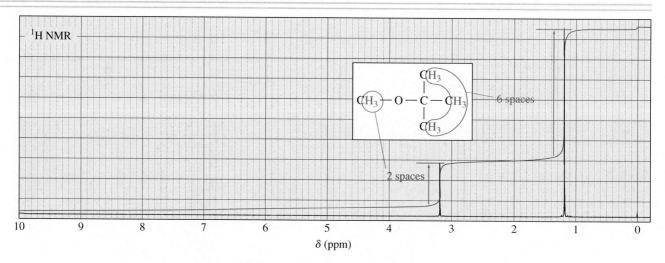

▲ **Figure 13-19**
Integrated NMR spectrum of methyl *t*-butyl ether. In going over a peak, the integrator trace (blue) rises by an amount that is proportional to the area under the peak.

PROBLEM-SOLVING HINT

1. If you are having trouble counting the fractional spaces, use a millimeter ruler to measure the integrals.
2. You don't know the total number of hydrogens, so try setting the smallest integral equal to one hydrogen and the others proportionally. If some of the other integrals are not whole numbers of hydrogens, then set the smallest equal to 2 or 3 as required. For example, 1:1.3:2 would become 3:4:6 and you would look for a compound with this ratio or 6:8:12 or 9:12:18, etc.

NMR spectrometers have **integrators** that compute the relative areas of peaks. The integrator draws a second trace (the integral trace) that rises when it goes over a peak. The amount the integral trace rises is proportional to the area of that peak. Newer digital instruments also print a number representing the area of each peak. These numbers correspond to the heights of the rises in an integral trace.

Neither an integral trace (shown in blue in Figure 13-19) nor a digital integral can specifically indicate that methyl *t*-butyl ether has three methyl hydrogens and nine *t*-butyl hydrogens. Each simply shows that about three times as many hydrogens are represented by the peak at δ1.2 as are represented by the peak of δ3.2. We must interpret what the 3 : 1 ratio means in terms of the structure.

Figure 13-20 shows the integrated spectrum of a compound with molecular formula $C_6H_{12}O_2$. Because we know the molecular formula, we can use the integral trace to determine exactly how many protons are responsible for each peak. The integrator

▲ **Figure 13-20**
Proton NMR spectrum for a compound of molecular formula $C_6H_{12}O_2$.

has moved a total of 6 spaces vertically in integrating the 12 protons in the molecule. Each proton is represented by

$$\frac{6 \text{ spaces}}{12 \text{ hydrogens}} = \text{about } 0.5 \text{ space per hydrogen}$$

The signal at $\delta 3.8$ has an integral of 0.5 space, so it must represent one proton. At $\delta 2.6$, the integrator moves 1 space, corresponding to two protons. The signal at $\delta 2.2$ has an integral of 1.5 spaces, for three protons; and the signal at $\delta 1.2$ (3 spaces) corresponds to six protons. Considering the expected chemical shifts together with the information provided by the integrator leaves no doubt which protons are responsible for which signals in the spectrum.

PROBLEM 13-5

Draw the integral trace expected for the NMR spectrum of *t*-butyl acetoacetate, shown in Figure 13-17.

PROBLEM 13-6

Determine the ratios of the peak areas in the following spectra. Then use this information, together with the chemical shifts, to pair up the compounds with their spectra. Assign the peaks in each spectrum to the protons they represent in the molecular structure. Possible structures:

PROBLEM-SOLVING **HINT**

Oxygen atoms are *σ*-withdrawers and *π*-donors of electron density. They deshield protons on the adjacent carbon atom to $\delta 3–\delta 4$.

σ-withdrawing

When attached to aromatic rings, however, O—H and O—R groups donate electron density into the π system of the ring. Nearby protons absorb *upfield* of the usual $\delta 7.2$ for benzene (often around $\delta 6.8$).

π-donating

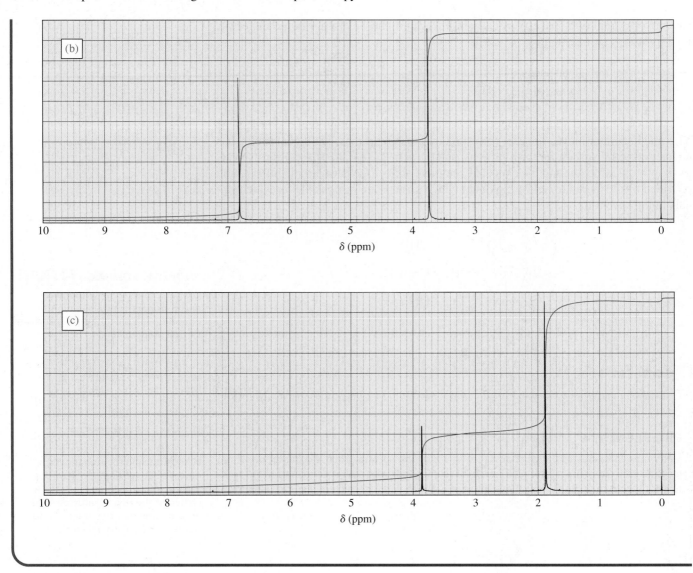

13-8
Spin-Spin Splitting

13-8A Theory of Spin-Spin Splitting

A proton in the NMR spectrometer is subjected to both the external magnetic field and the induced field of the shielding electrons. If there are other protons nearby, their small magnetic fields also affect the absorption frequency of the proton we are observing. Consider the spectrum of 1,1,2-tribromoethane (Fig. 13-21). As expected, there are two signals with areas in the ratio of 1:2. The smaller signal (H^a) appears at $\delta 5.7$, shifted downfield by the two adjacent bromine atoms. The larger signal (H^b) appears at $\delta 4.1$. These signals do not appear as single peaks but as a triplet (three peaks) and a doublet (two peaks), respectively. This splitting of signals into multiplets, called **spin-spin splitting**, results when two different types of protons are close enough that their magnetic fields influence each other. Such protons are said to be **magnetically coupled**.

Spin-spin splitting can be explained by considering the individual spins of the magnetically coupled protons. Assume that our spectrometer is scanning the signal for the H^b protons of 1,1,2-tribromoethane at $\delta 4.1$ (Figure 13-22). These protons are under the influence of the small magnetic field of the adjacent proton, H^a. The orientation of H^a is not the same for every molecule in the sample. In some molecules, H^a is aligned with the external magnetic field, and in others, it is aligned against the field.

◀ Figure 13-21
The proton NMR spectrum of
1,1,2-tribromoethane shows a triplet
of area 1 at $\delta 5.7$ and a doublet of area
2 at $\delta 4.1$.

◀ Figure 13-22
When the nearby H^a proton is aligned
with the external magnetic field,
it deshields H^b; when H^a is aligned
against the field, it shields H^b.

When H^a is aligned with the field, the H^b protons feel a slightly stronger total field: They are effectively deshielded, and they absorb at a lower field. When the H^a proton is aligned against the field, the H^b protons are shielded, and they absorb at a higher field. These are the two absorptions of the doublet seen for the H^b protons. About half of the molecules have H^a aligned with the field and about half against the field, so the two absorptions of the doublet are nearly equal in area.

Spin-spin splitting is a reciprocal property. That is, if one proton splits another, the second proton must split the first.

Proton *a* appears as a triplet (at $\delta 5.7$) because there are four permutations of the two H^b proton spins, with two of them giving the same magnetic field (Figure 13-23). When both H^b spins are aligned with the applied field, proton *a* is deshielded; when both H^b spins are aligned against the field, proton *a* is shielded; and when the two H^b spins are opposite each other (two possible permutations), they cancel each other out. Three signals result, with the middle signal twice as large as the others because it corresponds to two possible spin permutations.

▶ **Figure 13-23**
The H^a absorption is affected by the three combinations of H^b spins. When the H^b spins reinforce the external field, the H^a absorption occurs at a lower field. When the H^b spins oppose the external field, the H^a absorption occurs at a higher field. Two permutations, where the H^b proton spins cancel each other, allow H^a to absorb at its "normal" position. The peak area ratios are 1:2:1.

13-8B   The $N+1$ Rule

The analysis used above for the splitting of 1,1,2-tribromoethane can be extended to more complicated systems. In general, the multiplicity (number of peaks) of a signal is given by the $N+1$ **rule:**

$N+1$ rule: If a signal is split by N equivalent protons, it is split into $N+1$ peaks.

The relative areas of the $N+1$ **multiplet** that results are approximately given by the appropriate line of Pascal's triangle:

Relative Peak Intensities of Symmetric Multiplets

Number of Equivalent Protons Causing Splitting	Number of Peaks (multiplicity)	Area Ratios (Pascal's triangle)
0	1 (singlet)	1
1	2 (doublet)	1 1
2	3 (triplet)	1 2 1
3	4 (quartet)	1 3 3 1
4	5 (quintet)	1 4 6 4 1
5	6 (sextet)	1 5 10 10 5 1
6	7 (septet)	1 6 15 20 15 6 1

Consider the splitting of the signals for the ethyl group in ethylbenzene (Figure 13-24). The methyl protons are split by two adjacent protons, and they appear upfield as a triplet of areas 1:2:1. The methylene (CH_2) protons are split by three protons, appearing farther downfield as a quartet of areas 1:3:3:1. This splitting pattern is typical for an ethyl group. Because ethyl groups are common, you should learn to recognize this familiar pattern. All five aromatic protons absorb close to 7.2 ppm because the alkyl substituent has only a small effect on the chemical shifts of the aromatic protons. The aromatic protons split each other in a complicated manner (Section 13-9).

PROBLEM SOLVING HINT

When you see splitting, look for nonequivalent protons on adjacent carbon atoms.

▲ **Figure 13-24**
Proton NMR spectrum of ethylbenzene. The ethyl group appears as a triplet at $\delta 1.2$ (—CH_3)
and a quartet at $\delta 2.6$ (—CH_2—). The aromatic protons appear as a multiplet near $\delta 7.2$.

13-8C   The Range of Magnetic Coupling

In ethylbenzene there is no spin-spin splitting between the aromatic protons and the
protons of the ethyl group. These protons are not on adjacent carbon atoms, so they are
too far away to be magnetically coupled.

The magnetic coupling that causes spin-spin splitting takes place primarily through the bonds
of the molecule. Most examples of spin-spin splitting involve coupling between protons that
are separated by three bonds, so they are bonded to adjacent carbon atoms (vicinal protons).

  Most spin-spin splitting is between protons on adjacent carbon atoms.

Protons bonded to the same carbon atom (geminal protons) can split each other *only if
they are nonequivalent*. In most cases, protons on the same carbon atom are equivalent,
and equivalent protons cannot split each other.

Bonded to the same carbon: two bonds between protons

spin-spin splitting is normally observed
(if nonequivalent)

Bonded to adjacent carbons: three bonds between protons

spin-spin splitting is normally observed
(this is the most common case)

PROBLEM-SOLVING $\mathbf{HINT}$

In most cases, protons on the same
carbon atom are equivalent and do
not split each other. If they are
nonequivalent, however
(Section 13-10), they will split each
other.

characteristic ethyl group

▲ **Figure 13-25**
A multiplet often "leans" upward toward the protons that are causing the splitting. The ethyl multiplets in ethylbenzene lean toward each other.

Bonded to nonadjacent carbons: four or more bonds between protons

spin-spin splitting is *not* normally observed

Protons separated by more than three bonds usually do not produce observable spin-spin splitting. Occasionally, such "long-range coupling" does occur, but these cases are unusual. For now, we consider only nonequivalent protons on adjacent carbon atoms (or closer) to be magnetically coupled.

You may have noticed that the two multiplets in the upfield part of the ethylbenzene spectrum are not quite symmetrical. In general, a multiplet "leans" upward toward the signal of the protons responsible for the splitting. In the ethyl signal (Figure 13-25) the quartet at lower field leans toward the triplet at a higher field, and vice versa.

Another characteristic splitting pattern is shown in the NMR spectrum of methyl isopropyl ketone (3-methyl-2-butanone) in Figure 13-26.

The three protons *(a)* of the methyl group bonded to the carbonyl appear as a singlet of relative area 3, near $\delta 2.1$. Methyl ketones and acetate esters characteristically give such singlets around $\delta 2.1$, since there are no protons on the adjacent carbon atom.

singlet, $\delta 2.1$ → $CH_3 - \overset{\overset{\displaystyle O}{\|}}{C} - R$
a methyl ketone

singlet, $\delta 2.1$ → $CH_3 - \overset{\overset{\displaystyle O}{\|}}{C} - O - R$
an acetate ester

The six methyl protons *(b)* of the isopropyl group are equivalent. They appear as a doublet of relative area 6 at about $\delta 1.1$, slightly deshielded by the carbonyl group two bonds away. This doublet leans downfield because these protons are magnetically coupled to the methine proton *c*. This doublet is also shown in an insert box with the horizontal scale expanded for clarity. The vertical scale is adjusted so the peaks fit in the box.

The methine proton H^c appears as a multiplet of relative area 1, at $\delta 2.6$. This absorption is a septet (seven peaks) because it is coupled to the six adjacent methyl protons *(b)*. Some small peaks of this septet may not be visible unless the spectrum is amplified, as it is in the

▲ **Figure 13-26**
Proton NMR spectrum of methyl isopropyl ketone. The isopropyl group appears as a characteristic pattern of a strong doublet at a higher field and a weak multiplet (a septet) at a lower field. The methyl group appears as a singlet at $\delta 2.1$.

◀ **Figure 13-27**
Characteristic isopropyl group
pattern

expanded insert box. The pattern seen in this spectrum, summarized in Figure 13-27, is typical for an isopropyl group: The methyl protons give a strong doublet at a higher field, and
the methine proton gives a weak multiplet (usually difficult to count the peaks) at a lower
field. Isopropyl groups are easily recognized from this characteristic pattern.

PROBLEM-SOLVING HINT

Ethyl and isopropyl groups are
common. Learn to recognize them
from their splitting patterns.

PROBLEM-SOLVING STRATEGY
Drawing an NMR Spectrum

In learning about NMR spectra, you have seen that chemical shift values can be assigned to
specific types of protons, that the areas under peaks are proportional to the numbers of protons, and that nearby protons cause spin-spin splitting. By analyzing the structure of a molecule with these principles in mind, you can predict the features of an NMR spectrum. Learning
to draw spectra will help you to recognize the features of actual spectra. The process is not
difficult if a systematic approach is used. A stepwise method is illustrated here, by drawing
the NMR spectrum of the compound shown below.

$$CH_3^a \\ CH^b - O - \overset{O}{\overset{\|}{C}} - CH_2^c - CH_3^d \\ CH_3^a$$

1. **Determine how many types of protons are present, together with their proportions.**
   In the example above, there are four types of protons, labeled *a*, *b*, *c*, and *d*. The area ratios
   should be 6:1:2:3.

2. **Estimate the chemical shifts of the protons. (Table 13.3 and Appendix 1 serve as**
   guides.)
   Proton *b* is on a carbon atom bonded to oxygen; it should absorb around $\delta 3$ to $\delta 4$. Protons *a*
   are less deshielded by the oxygen, probably around $\delta 1$ to $\delta 2$. Protons *c* are on a carbon bonded to a carbonyl group; they should absorb around $\delta 2.1$ to $\delta 2.5$. Protons *d*, one carbon removed
   from a carbonyl, will be deshielded less than protons *c*, and also less than protons *a*, which
   are next to a more strongly deshielded carbon atom. Protons *d* should absorb around $\delta 1.0$.

3. **Determine the splitting patterns.**
   Protons *a* and *b* split each other into a doublet and a septet, respectively (a typical isopropyl group pattern). Protons *c* and *d* split each other into a quartet and a triplet, respectively (a typical ethyl group pattern).

4. **Summarize each absorption in order, from the lowest field to the highest.**

	Proton b	Protons c	Protons a	Protons d
area	1	2	6	3
chemical shift	3–4	2.1–2.5	1–2	1
splitting	septet	quartet	doublet	triplet

5. Draw the spectrum, using the information from your summary.

Work through the following problem to become comfortable with predicting NMR spectra.

PROBLEM 13-7

Draw the NMR spectra you would expect for the following compounds.

(a) $(CH_3)_2CH - O - CH(CH_3)_2$

(b) $Cl - CH_2 - CH_2 - \overset{\overset{\displaystyle O}{\|}}{C} - O - CH_3$

(c) $Ph - CH(CH_3)_2$

(d) $CH_3CH_2O - \langle\ \rangle - OCH_2CH_3$

(e) $\begin{array}{l} CH_2 - COOCH_2CH_3 \\ | \\ CH_2 - COOCH_2CH_3 \end{array}$

13-8D   Coupling Constants

The distances between the peaks of multiplets can provide additional structural information. These distances are all about 7 Hz in the methyl isopropyl ketone spectrum (Figs. 13–26 and 13–27). These splittings are equal because *any two magnetically coupled protons must have equal effects on each other*. The distance between adjacent peaks of the H^c multiplet (split by H^b) must equal the distance between the peaks of the H^b doublet (split by H^c).

The distance between the peaks of a multiplet (measured in hertz) is called the **coupling constant**. Coupling constants are represented by *J*, and the coupling constant between H^a and H^b is represented by J$_{ab}$. In complicated spectra with many types of protons, groups of neighboring protons can sometimes be identified by measuring their coupling constants. Multiplets that have the same coupling constant may arise from adjacent groups of protons that split each other.

The magnetic effect that one proton has on another depends on the bonds connecting the protons, but it does not depend on the strength of the external magnetic field. For this reason, the coupling constant (measured in hertz) does not vary with the field strength of the spectrometer. A spectrometer operating at 300 MHz records the same coupling constants as a 60 MHz instrument.

Figure 13-28 shows typical values of coupling constants. The most commonly observed coupling constant is the 7-Hz splitting of protons on adjacent carbon atoms in freely rotating alkyl groups. Many of the insert boxes have a 7-Hz grid superimposed for easy comparison with this common splitting.

Coupling constants help to distinguish among the possible isomers of a compound, as in the spectrum of *p*-nitrotoluene (Figure 13-29). The methyl protons (*c*) absorb as a singlet near δ2.5, and the aromatic protons appear as a pair of doublets. The doublet centered around δ7.3 corresponds to the two aromatic protons ortho to the methyl group (*a*). The doublet centered around δ8.1 corresponds to the two protons ortho to the electron-withdrawing nitro group (*b*).

PROBLEM-SOLVING **HINT**

Watch for unusually large coupling constants, especially in the vinyl region, where they may indicate stereochemistry about a double bond.

The value of 7 Hz in an alkyl group is averaged for rapid rotation about the carbon–carbon bond. If rotation is hindered by a ring or bulky groups, other splitting constants may be observed.

◀ **Figure 13-28**
Typical values of proton coupling constants.

▲ **Figure 13-29**
Proton NMR spectrum of *p*-nitrotoluene.

Each proton *a* is magnetically coupled to one *b* proton, splitting the H^a absorption into a doublet. Similarly, each proton *b* is magnetically coupled to one proton *a*, splitting the H^b absorption into a doublet. The coupling constant is 8 Hz, suggesting that the magnetically coupled protons H^a and H^b are ortho to each other.

Both the ortho and meta isomers of nitrotoluene have four distinct types of aromatic protons, and the spectra for these isomers are more complex. Figure 13-29 must correspond to the para isomer of nitrotoluene.

Coupling constants also help to distinguish stereoisomers. In Figure 13-30(a), the 9-Hz coupling constant between the two vinyl protons shows that they are cis to one another. In Figure 13-30(b), the 15-Hz coupling constant shows that the two vinyl protons are trans. Notice that the 9-Hz coupling looks too wide for common alkyl group splitting (7 Hz). The 15-Hz coupling looks about double the common 7-Hz alkyl group splitting.

▲ **Figure 13-30**
Proton NMR spectra of 4,4-dimethylcyclohex-2-en-1-one and β-ionone.

PROBLEM 13-8

(a) Assign protons to the peaks in the NMR spectrum of 4,4-dimethylcyclohex-2-en-1-one. Explain the splitting that gives the triplets at δ1.8 and δ2.4.

(b) Assign protons to the peaks in the NMR spectrum of β-ionone. Explain the splitting seen in the three multiplets at δ1.4, δ1.6, and δ2.1. Explain how you know which of these multiplets corresponds to which methylene groups in the molecule.

PROBLEM 13-9

Draw the NMR spectra you expect for the following compounds.

(a)

Ph—C(H)=C(H)—C(CH₃)₃

(b)

CH₃O—C(Cl)=C(CH₃)(H)

(c)

(CH₃)₃C—C(H)=C(H)—C(=O)—OCH₂CH₃

(d)

PROBLEM 13-10

An unknown compound (C_3H_2NCl) shows moderately strong IR absorptions around 1650 cm^{-1} and 2200 cm^{-1}. Its NMR spectrum consists of two doublets ($J = 14$ Hz) at $\delta 5.9$ and $\delta 7.1$. Propose a structure consistent with these data.

PROBLEM 13-11

Two spectra are given below. Propose a structure that corresponds to each spectrum.

There are many cases of **complex splitting**, where signals are split by adjacent protons of more than one type, with different coupling constants. Consider the vinyl proton H^a, adjacent to the phenyl ring of styrene. The chemical shift of H^a is $\delta 6.6$ because it is deshielded by both the vinyl group and the aromatic ring.

13-9
Complex Splitting

styrene

▶ **Figure 13-31**
A splitting tree. The H^a signal in styrene is split (J_{ab} = 17 Hz) by coupling with H^b, and further split (J_{ac} = 11 Hz) by coupling with H^c.

▲ **Figure 13-32**
Proton NMR spectrum of styrene.

▲ **Figure 13-33**
The H^b proton in styrene is split by coupling with H^a (J_{ab} = 17 Hz), and further split by coupling with H^c (J_{bc} = 1.4 Hz).

H^a is coupled to H^b with a typical trans coupling constant J_{ab} = 17 Hz. It is also coupled to proton H^c with J_{ac} = 11 Hz. The H^a signal is split into a doublet of spacing 17 Hz, and each of those peaks is further split into a doublet of spacing 11 Hz, for a total of four peaks. This complex splitting, called a *doublet of doublets*, can be analyzed by a diagram called a *splitting tree*, shown in Figure 13-31.

The proton NMR spectrum of styrene is shown in Figure 13-32. The signal from H^a, with splitting as in Figure 13-31, is centered at $\delta 6.6$. H^b is also split by two nonequivalent protons: It is split by H^a with a trans coupling constant J_{ab} = 17 Hz and further split by H^c with a geminal coupling constant J_{bc} = 1.4 Hz. The H^b doublet of doublets, centered at $\delta 5.65$, is shown in Figure 13-33.

PROBLEM 13-12

Draw a splitting tree, similar to Figures 13-31 and 13-33, for proton H^c in styrene. What is the chemical shift of proton H^c?

▲ **Figure 13-34**
The NMR spectrum of *n*-propyl iodide seems to show the H^b signal split into a sextet by the five hydrogens on the adjacent carbon atoms. On closer inspection, the multiplet is seen to be an imperfect sextet, the result of complex splitting by two sets of protons (*a* and *c*) with similar splitting constants.

Sometimes a signal is split by two or more different kinds of protons with similar coupling constants. Consider *n*-propyl iodide (Fig. 13-34), where the *b* protons on the middle carbon atom are split by two types of protons: the methyl protons (H^c) and the CH$_2$I protons (H^a).

The coupling constants for these two interactions are similar: J_{ab} = 6.8 Hz, and J_{bc} = 7.3 Hz. The spectrum shows the H^b signal as a sextet, almost as though there were five equivalent protons coupled with H^b. The trace in the insert box, enlarged and offset, shows that the pattern is not a perfect sextet. The analysis of the splitting pattern serves as a reminder that the $N+1$ rule works only in a perfect multiplet, when the signal is split by *equivalent* protons.

PROBLEM 13-13

The spectrum of *trans*-2-hexenoic acid is shown below.
(a) Assign peaks to show which protons give rise to which peaks in the spectrum.
(b) Draw a tree to show the complex splitting of the vinyl proton centered around 7 ppm. Estimate the values of the coupling constants.

PROBLEM 13-14

The NMR spectrum of cinnamaldehyde is shown below.

(a) Determine the chemical shifts of H^a, H^b, and H^c. The absorption of one of these protons is difficult to see; look carefully at the integrals.

(b) Estimate the coupling constants J_{ab} and J_{bc}.

(c) Draw a tree to analyze the complex splitting of the proton centered at $\delta6.7$.

PROBLEM 13-15

Consider the proton NMR spectrum of the following ketone.

(a) Predict the approximate chemical shift of each type of proton.

(b) Predict the number of NMR peaks for each type of proton.

(c) Draw a tree to show the splitting predicted for the absorption of the circled proton.

PROBLEM-SOLVING **HINT**

Protons on the β carbon of an α,β-unsaturated carbonyl compound absorb at very low fields (about $\delta7$) because of the electron-withdrawing effect of the carbonyl group.

13-10
Stereochemical Nonequivalence of Protons

Stereochemical differences often result in different chemical shifts for protons on the same carbon atom. For example, the two protons on C_1 of allyl bromide (3-bromopropene) are not equivalent. H^a is cis to the $-CH_2Br$ group, and H^b is trans. H^a absorbs at $\delta5.3$; H^b absorbs at $\delta5.1$. There are four different (by NMR) types of protons in allyl bromide, as shown in the structure:

To determine whether similar-appearing protons are equivalent, mentally substitute another atom for each of the protons in question. *If the same product is formed by imaginary replacement of either of two protons, those protons are chemically equivalent.*

For example, replacing any of the three methyl protons in ethanol by an imaginary Z atom gives the same compound; these hydrogens are chemically equivalent.

different conformations of the same compound

When this imaginary replacement test is applied to the protons on C_1 of allyl bromide, the imaginary products are different. Replacing the cis hydrogen gives the cis diastereomer, and replacing the trans hydrogen gives the trans diastereomer. Because the two imaginary products are diastereomers, these protons on C_1 are called **diastereotopic** protons.

Cyclobutanol shows these stereochemical relationships in a cyclic system. The hydroxyl proton H^a is clearly unique; it absorbs between $\delta 3$ and $\delta 5$, depending on the solvent and concentration. H^b is also unique, absorbing between $\delta 3$ and $\delta 4$. Protons H^e and H^f are diastereotopic (and absorb at different fields) because H^e is cis to the hydroxyl group, while H^f is trans.

To distinguish among the other four protons, notice that cyclobutanol has an internal mirror plane of symmetry. Protons H^c are cis to the hydroxyl group, while protons H^d are trans. Therefore, protons H^c are diastereotopic to protons H^d, and the two sets of protons absorb at different magnetic fields and are capable of splitting each other.

PROBLEM 13-16

Use the imaginary replacement technique to show that protons H^c and H^d in cyclobutanol are diastereotopic.

PROBLEM 13-17*

If the imaginary replacement of either of two protons forms enantiomers, then those protons are said to be *enantiotopic*. The NMR is not a chiral probe, and it cannot distinguish between enantiotopic protons. They are seen to be "equivalent by NMR."

(a) Use the imaginary replacement technique to show that the two allylic protons (those on C_3) of allyl bromide are enantiotopic.

(b) Similarly, show that the two H^c protons in cyclobutanol are enantiotopic.

(c) What other protons in cyclobutanol are enantiotopic?

Diastereomerism also occurs in saturated, acyclic compounds; for example, 1,2-dichloropropane is a simple compound that contains diastereotopic protons. The two protons on the $-CH_2Cl$ group are diastereotopic; their imaginary replacement gives diastereomers.

The most stable conformation of 1,2-dichloropropane (at left) shows that the diastereotopic protons on C^1 exist in different chemical environments. They experience different magnetic fields and are nonequivalent by NMR. The NMR spectrum of 1,2-dichloropropane is shown in Figure 13-35. The methyl protons appear as a doublet at $\delta 1.6$, and the single proton on C^2 appears as a complex multiplet at $\delta 4.15$. The two protons on C^1 appear as distinct absorptions at $\delta 3.60$ and $\delta 3.77$. They are split by the proton on C^2, and they also split each other.

The presence of an asymmetric carbon atom adjacent to the CH_2Cl group gives rise to the different chemical environments of these diastereotopic protons. When a molecule contains an asymmetric carbon atom, the protons on any methylene groups are

▲ **Figure 13-35**
The proton NMR spectrum of 1,2-dichloropropane shows distinct absorptions for the methylene protons on C^1. These hydrogen atoms are diastereotopic and are chemically nonequivalent.

usually diastereotopic. They may or may not be resolved in the NMR, however, depending on the differences in their environments.

PROBLEM 13-18*

Predict the theoretical number of different NMR signals produced by each compound, and give approximate chemical shifts. Point out any diastereotopic relationships.
(a) 2-bromobutane   **(b)** cyclopentanol
(c) Ph—CHBr—CH_2Br **(d)** vinyl chloride

We have already seen evidence that NMR does not provide an instantaneous picture of a molecule. For example, a terminal alkyne does not give a spectrum where the molecules oriented along the field absorb at a high field and those oriented perpendicular to the field absorb at a lower field. What we see is one signal whose position is averaged over the chemical shifts of all the orientations of a rapidly tumbling molecule. In general, any type of movement or change that takes place faster than about a hundredth of a second will produce an averaged NMR spectrum.

13-11
Time Dependence
of NMR
Spectroscopy

13-11A   Conformational Changes

This principle is illustrated by the cyclohexane spectrum. In the chair conformation, there are two kinds of protons: the axial hydrogens and the equatorial hydrogens. The axial hydrogens become equatorial and the equatorial hydrogens become axial by chair–chair interconversions. These interconversions are fast on an NMR time scale at room temperature. The NMR spectrum of cyclohexane shows only one sharp, averaged peak (at $\delta 1.4$) at room temperature.

H_1 (ax)

H_2 (eq) $\rightleftharpoons$ H_1 (eq)

H_2 (ax)

Low temperatures retard the chair–chair interconversion of cyclohexane. The NMR spectrum at $-89°C$ shows two nonequivalent types of protons that split each other, giving two broad bands corresponding to the absorptions of the axial and equatorial protons. The broadening of the bands results from spin-spin splitting between axial and equatorial protons on the same carbon atom and on adjacent carbons. This technique of using low temperatures to stop conformational interconversions is called *freezing out* the conformations.

NMR spectroscopy is useful for determining the conformations of biological molecules such as neurotransmitters. For example, information about the conformations of acetylcholine has been used to design rigid analogs that are used as drugs to treat Alzheimer's disease.

13-11B   Fast Proton Transfers

Hydroxyl Protons   Like conformational interconversions, chemical processes often occur faster than the NMR technique can observe them. Figure 13-36 shows two NMR spectra for ethanol.

Figure 13-36(a) shows coupling between the hydroxyl (—OH) proton and the adjacent methylene (—CH_2—) protons, with a coupling constant of about 5 Hz. This is an ultrapure sample of ethanol with no contamination of acid, base, or water. Part (b) shows a typical sample of ethanol, with some acid or base present to catalyze the interchange of the hydroxyl protons. No splitting is seen between the hydroxyl proton and the methylene protons. During the NMR measurement, each hydroxyl proton becomes attached to a large number of different ethanol molecules and experiences all possible spin arrangements of the methylene group. What we see is a single, unsplit hydroxyl

▲ **Figure 13-36**

Comparison of the NMR spectrum of unusually pure ethanol and the spectrum of ethanol with a trace of an acidic (or basic) impurity. The impurity catalyzes a fast exchange of the —OH proton from one ethanol molecule to another. This rapidly exchanging proton produces a single, unsplit absorption at an averaged field.

absorption corresponding to the averaged field the proton experiences from bonding to many different ethanol molecules.

Proton exchange occurs in most alcohols and carboxylic acids, and in many amines and amides. If the exchange is fast (as it usually is for —OH protons), we see one sharp averaged signal. If the exchange is very slow, we see splitting. If the exchange is moderately slow, we may see a broadened peak that is neither cleanly split nor cleanly averaged.

PROBLEM 13-19

Propose mechanisms to show the interchange of protons between ethanol molecules under
(a) acid catalysis   **(b)** base catalysis

N—H Protons   Protons on nitrogen often show broadened signals in the NMR, both because of moderate rates of exchange and because of the magnetic properties of the nitrogen nucleus. Depending on the rate of exchange and other factors, N—H protons may give absorptions that are sharp and cleanly split, sharp and unsplit (averaged), or broad and shapeless. Figure 13-37 illustrates an NMR spectrum where the —NH$_2$ protons produce a very broad absorption, the shapeless peak centered at $\delta 5.2$.

▲ **Figure 13-37**
Example of a proton NMR spectrum with a broad N—H absorption.

Because the chemical shifts of O—H and N—H protons depend on the concentration and the solvent, it is often difficult to tell whether or not a given peak corresponds to one of these types of protons. We can use proton exchange to identify their NMR signals by shaking the sample with an excess of deuterium oxide, D_2O. Any exchangeable hydrogens are quickly replaced by deuterium atoms, which are invisible in the proton NMR spectrum.

> Complex splitting patterns can often be simplified by replacing a hydrogen with deuterium. Deuterium is invisible in the proton NMR, so the resulting spectrum shows the loss of a signal and simplified signals from the adjacent hydrogens.

$$R—O—H + D—O—D \quad \rightleftarrows \quad R—O—D + D—O—H$$

$$R—NH_2 + 2D—O—D \quad \rightleftarrows \quad R—ND_2 + 2D—O—H$$

When a second NMR spectrum is recorded (after shaking with D_2O), the signals from any exchangeable protons are either absent or much less intense.

PROBLEM 13-20

Draw the NMR spectrum expected from ethanol that has been shaken with a drop of D_2O.

PROBLEM 13-21

Propose chemical structures consistent with the following NMR spectra and molecular formulas. In spectrum (a) explain why the peak at $\delta 1.7$ is not a clean triplet with this very pure sample. In spectrum (b) explain why some of the protons are likely to be missed.

PROBLEM-SOLVING STRATEGY
Interpreting Proton NMR Spectra

Learning to interpret NMR spectra requires practice with a large number of examples and problems. The problems at the end of this chapter should help you gain confidence in your ability to assemble a structure from the NMR spectrum combined with other information. This section provides some hints that can help make spectral analysis a little easier.

When you first look at a spectrum, consider the major features before getting bogged down in the minor details. Here are a few major characteristics you might watch for:

1. If the molecular formula is known, use it to determine the number of elements of unsaturation (see Section 7-3). The elements of unsaturation suggest rings, double bonds, or triple bonds. Matching the integrated peak areas with the number of protons in the formula gives the numbers of protons represented by the individual peaks.

2. Any broadened singlets in the spectrum might be due to —OH or —NH protons. If the broad singlet is deshielded past 10 ppm, an acid —OH group is likely.

$$—O—H \qquad —\overset{|}{N}—H \qquad —\overset{\overset{\displaystyle O}{\|}}{C}—OH$$

frequently broad singlets   broad (or sharp) singlet, $\delta > 10$

3. A signal around $\delta3$ to $\delta4$ suggests protons on a carbon bearing an electronegative element such as oxygen or a halogen. Protons that are more distant from the electronegative atom will be less strongly deshielded.

$$—O—\overset{|}{\underset{|}{C}}—H \qquad Br—\overset{|}{\underset{|}{C}}—H \qquad Cl—\overset{|}{\underset{|}{C}}—H \qquad I—\overset{|}{\underset{|}{C}}—H$$

around $\delta3$–$\delta4$ for hydrogens on carbons bearing oxygen or halogen

4. Signals around $\delta7$ to $\delta8$ suggest the presence of an aromatic ring. If some of the aromatic absorptions are farther downfield than $\delta7.2$, an electron-withdrawing substituent may be attached.

around $\delta7$–$\delta8$ H

5. Signals around $\delta 5$ to $\delta 6$ suggest vinyl protons. Splitting constants can differentiate cis and trans isomers.

around $\delta 5-\delta 6$; $J = 10$ Hz   around $\delta 5-\delta 6$; $J = 15$ Hz

6. Learn to recognize ethyl groups and isopropyl groups (and structures that resemble these groups) by their characteristic splitting patterns.

ethyl group   isopropyl group

7. Signals around $\delta 2.1$ to $\delta 2.5$ may suggest protons adjacent to a carbonyl group or next to an aromatic ring. A singlet at $\delta 2.1$ often results from a methyl group bonded to a carbonyl group.

around $\delta 2.1-\delta 2.5$ singlet, $\delta 2.1$ singlet, $\delta 2.3$

8. Signals in the range $\delta 9$ to $\delta 10$ suggest an aldehyde.

aldehyde, $\delta 9-\delta 10$

9. A sharp singlet around $\delta 2.5$ suggests a terminal alkyne.

$$-C \equiv C-H$$

around $\delta 2.5$

These hints are neither exact nor complete. They are simple methods for making educated guesses about the major features of a compound from its NMR spectrum. The hints can be used to draw partial structures to examine all the possible ways they might be combined to give a molecule that corresponds with the spectrum. Figure 13-38 gives a graphic presentation of some of the most common chemical shifts. A more complete table of chemical shifts appears in Appendix 1.

PROBLEM-SOLVING HINT

Remember to look for structural information based on
1. number of absorptions
2. chemical shifts
3. areas of peaks
4. spin-spin splitting

◀ **Figure 13-38**
Common chemical shifts in the ^{1}H NMR spectrum.

▲ **Figure 13-39**
Proton NMR spectrum for a compound of formula $C_4H_8O_2$.

SAMPLE PROBLEM

Consider how you might approach the NMR spectrum shown in Figure 13-39. The molecular formula is known to be $C_4H_8O_2$, implying one element of unsaturation (the saturated formula would be $C_4H_{10}O_2$). Three types of protons appear in this spectrum. The signals at $\delta 4.1$ and $\delta 1.3$ resemble an ethyl group—confirmed by the 2:3 ratio of the integrals.

partial structure: $-CH_2-CH_3$

The ethyl group is probably bonded to an electronegative element, since its methylene ($-CH_2-$) protons absorb close to $\delta 4$. The molecular formula contains oxygen, so an ethoxy group is suggested.

partial structure: $-O-CH_2-CH_3$

The singlet at $\delta 2.1$ (area = 3) might be a methyl group bonded to a carbonyl group. A carbonyl group would also account for the element of unsaturation.

partial structure: $-\overset{\overset{\displaystyle O}{\|}}{C}-CH_3$

We have accounted for all eight hydrogen atoms in the spectrum. Putting together all the clues, we arrive at a proposed structure.

$CH_3^a-CH_2^b-O-\overset{\overset{\displaystyle O}{\|}}{C}-CH_3^c$
ethyl acetate

At this point, the structure should be rechecked to make sure it is consistent with the molecular formula, the proton ratios given by the integrals, the chemical shifts of the signals, and the spin-spin splitting. In ethyl acetate, the H^a protons give a triplet (split by the adjacent CH_2 group, $J = 7$ Hz) of area 3 at $\delta 1.3$; the H^b protons give a quartet (split by the adjacent CH_3 group, $J = 7$ Hz) of area 2 at $\delta 4.1$; and the H^c protons give a singlet of area 3 at $\delta 2.1$.

PROBLEM 13-22

Draw the expected NMR spectrum of methyl propionate, and point out how it differs from the spectrum of ethyl acetate.

$$CH_3-O-\overset{\overset{\textstyle O}{\|}}{C}-CH_2-CH_3$$
methyl propionate

SOLVED PROBLEM 13-4

Propose a structure for the compound of molecular formula $C_4H_{10}O$ whose proton NMR spectrum appears below.

SOLUTION

The molecular formula $C_4H_{10}O$ indicates no elements of unsaturation. Four types of hydrogens appear in this spectrum, in the ratio $2:1:1:6$. The singlet (one proton) at $\delta2.4$ might be a hydroxyl group, and the signal (two protons) at $\delta3.4$ corresponds to protons on a carbon atom bonded to oxygen. The $\delta3.4$ signal is a doublet, implying that the adjacent carbon atom bears one hydrogen.

partial structure: $H-O-CH_2-\overset{\overset{\textstyle H}{|}}{\underset{|}{C}}-$

(Since we cannot be certain that the $\delta2.4$ absorption is actually a hydroxyl group, we might consider shaking the sample with D_2O. If the 2.4 ppm absorption represents a hydroxyl group, it will shrink or vanish after shaking with D_2O.)

The signals at $\delta1.8$ and $\delta0.9$ resemble the pattern for an isopropyl group. The integral ratio of $1:6$ supports this assumption. Since the methine $(-\overset{|}{C}H-)$ proton of the isopropyl group absorbs at a fairly high field, the isopropyl group must be bonded to a carbon atom rather than an oxygen.

partial structure: $-\overset{|}{\underset{|}{C}}-CH\overset{\displaystyle CH_3}{\underset{\displaystyle CH_3}{}}$

Our two partial structures add to a total of six carbon atoms (compared with the four in the molecular formula) because two of the carbon atoms appear in both partial structures. Drawing the composite of the partial structures, we have isobutyl alcohol:

$$H^a-O-CH_2^b-CH^c\overset{\displaystyle CH_3^d}{\underset{\displaystyle CH_3^d}{}}$$

This structure must be rechecked to make sure that it has the correct molecular formula and that it accounts for all the structural evidence provided by the spectrum (Problem 13-23).

PROBLEM 13-23

Give the spectral assignments for the protons in isobutyl alcohol (Solved Problem 13.4). For example,

$$H^a \text{ is a singlet, area } = 1, \text{ at } \delta 2.4$$

PROBLEM 13-24

Five proton NMR spectra are given here, together with molecular formulas. In each case, propose a structure that is consistent with the spectrum.

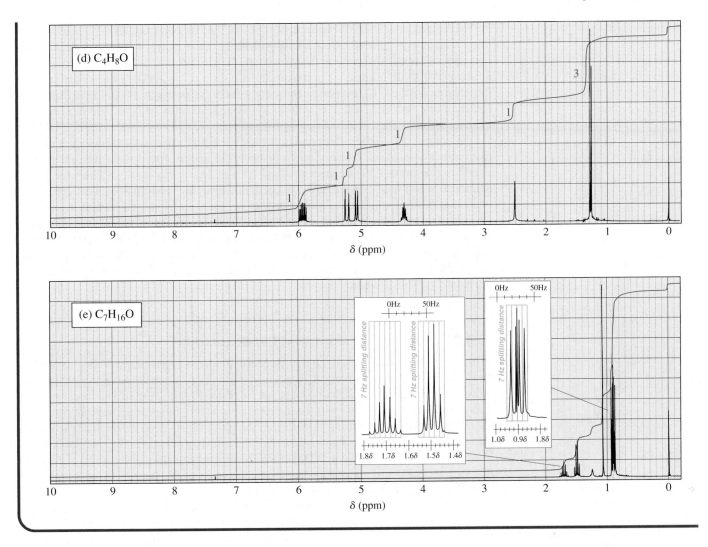

Where does a carbonyl group absorb in the NMR? Where does an internal alkyne absorb? In the proton NMR, both of these groups are invisible. Sometimes we can *infer* their presence: If the carbonyl group has a proton attached (an aldehyde proton), the peak between $\delta 9$ and 10 alerts us to its presence. If the adjacent carbon atom has hydrogens, their signals between $\delta 2.1$ and 2.5 are suggestive, but we still can't see the carbonyl group. An internal alkyne is even more difficult, because there are no distinctive absorptions in the proton NMR and usually none in the IR either.

The development of Fourier transform NMR spectroscopy made carbon NMR (^{13}C NMR or CMR) possible, and high-field superconducting spectrometers allowed it to become nearly as convenient as proton NMR (^{1}H NMR). Carbon NMR determines the magnetic environments of the carbon atoms themselves. Carbonyl carbon atoms, alkyne carbon atoms, and aromatic carbon atoms all have characteristic chemical shifts in the ^{13}C NMR spectrum.

13-12 Carbon-13 NMR Spectroscopy

13-12A   Sensitivity of Carbon NMR

Carbon NMR took longer than proton NMR to become a routine technique because carbon NMR signals are much weaker than proton signals. About 99% of the carbon atoms in a natural sample are the isotope ^{12}C. This isotope has an even number of protons and an even number of neutrons, so it has no magnetic spin and cannot give rise to NMR signals. The less abundant isotope ^{13}C has an odd number of neutrons, giving it a magnetic spin of $\frac{1}{2}$, just like a proton. Because only 1% of the carbon atoms in a sample are the magnetic ^{13}C isotope, the sensitivity of ^{13}C NMR is decreased by a factor of 100. In addition, the gyromagnetic ratio of ^{13}C is only one-fourth that of the proton, so the ^{13}C resonance frequency (at a given magnetic field) is only one-fourth of that for ^1H NMR. The smaller gyromagnetic ratio leads to a further decrease in sensitivity.

Because ^{13}C NMR is less sensitive than ^1H NMR, special techniques are needed to obtain a spectrum. If we simply operate the spectrometer in a normal (called *continuous wave* or CW) manner, the desired signals are very weak and become lost in the noise. When many spectra are averaged, however, the random noise tends to cancel while the desired signals are reinforced. If several spectra are taken and stored in a computer, they can be averaged and the accumulated spectrum plotted by the computer. Since the ^{13}C NMR technique is much less sensitive than the ^1H NMR technique, hundreds of spectra are commonly averaged to produce a usable result. Several minutes are required to scan each CW spectrum, and this averaging procedure is long and tedious. Fortunately, there is a better way.

13-12B   Fourier Transform NMR Spectroscopy

When magnetic nuclei are placed in a magnetic field and irradiated with a pulse of radio frequency close to their resonant frequency, the nuclei absorb some of the energy and precess like little tops at their resonant frequencies (Figure 13-40). This precession of many nuclei at slightly different frequencies produces a complex signal that decays as the nuclei lose the energy they gained from the pulse. This signal is called a **free induction decay** (or **transient**) and it contains all the information needed to calculate a spectrum. The free induction decay (**FID**) can be recorded by a radio receiver and a computer in 1 to 2 seconds, and many FIDs can be averaged in a few minutes. A computer converts the averaged transients into a spectrum.

A *Fourier transform* is the mathematical technique used to compute the spectrum from the free induction decay, and this technique of using pulses and collecting transients is called **Fourier transform spectroscopy**. A Fourier transform spectrometer is usually more expensive than a continuous wave spectrometer, since it must have a fairly sophisticated computer with the capability of storing thousands of complicated transients. A good ^{13}C NMR instrument usually has the capability to do ^1H NMR spectra as well. When used with proton spectroscopy, the Fourier transform technique produces good spectra with very small amounts (less than a milligram) of sample.

pulse B_0 nuclei precess   free induction decay (FID)   spectrum

▲ **Figure 13-40**
The Fourier transform NMR spectrometer delivers a radio-frequency pulse close to the resonance frequency of the nuclei. Each nucleus precesses at its own resonance frequency, generating a free induction decay (FID). Many of these transient FIDs are accumulated and averaged in a short period of time. A computer does a Fourier transform (FT) on the averaged FID, producing the spectrum printed on the recorder.

13-12C Carbon Chemical Shifts

Figure 13-41 gives typical ranges of chemical shifts for carbon atoms in organic molecules. A more detailed table of carbon chemical shifts is provided as Appendix 1C. Carbon chemical shifts are usually about 15 to 20 times larger than comparable proton chemical shifts, which makes sense because the carbon atom is one atom closer to a shielding or deshielding group than its attached hydrogen. For example, an aldehyde proton absorbs around $\delta 9.4$ in the ^{1}H NMR spectrum, and the carbonyl carbon atom absorbs around 180 ppm downfield from TMS in the ^{13}C spectrum. Figure 13-42 compares the proton and carbon spectra of a complex aldehyde to show this relationship between proton and carbon chemical shifts.

The proton (lower) and carbon (upper) spectra in Figure 13-42 are calibrated so the full width of the proton spectrum is 10 ppm, while the width of the ^{13}C spectrum is 200 ppm (20 times as large). Notice how the corresponding peaks in the two spectra almost line up vertically. This proportionality between ^{13}C NMR and ^{1}H NMR chemical shifts is an approximation that allows us to make a first estimate of a carbon atom's chemical shift. For example, the peak for the aldehyde *proton* is at $\delta 9.5$ in the proton spectrum, so we expect the peak for the aldehyde *carbon* to appear at a chemical shift

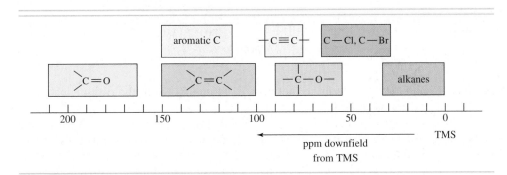

◀ **Figure 13-41**
Table of approximate chemical shift values for ^{13}C NMR. Most of these values for a carbon atom are about 15 to 20 times the chemical shift of a proton if it were bonded to the carbon atom.

▲ **Figure 13-42**
^{1}H NMR (lower) and ^{13}C NMR (upper) spectra of a heterocyclic aldehyde. Notice the correlation between the chemical shifts in the two spectra. The proton spectrum has a sweep width of 10 ppm, and the carbon spectrum has a width of 200 ppm.

► **Figure 13-43**
Proton and ^{13}C NMR spectra of
1,2,2-trichloropropane.

between 15 and 20 times as large (between $\delta 144$ and $\delta 192$) in the carbon spectrum.
The actual position is at $\delta 180$.

Notice also the triplet at $\delta 77$ in the ^{13}C NMR spectrum in Figure 13-42. This is the
carbon signal for deuterated chloroform ($CDCl_3$), split into three equal-sized peaks by coupling with the deuterium atom. Chloroform-d ($CDCl_3$) is a common solvent for ^{13}C NMR
because the spectrometer can "lock" onto the signal from deuterium at a different frequency from carbon. The $CDCl_3$ solvent signal is a common feature of carbon NMR spectra.

Because chemical shift effects are larger in ^{13}C NMR, an electron-withdrawing
group has a substantial effect on the chemical shift of a carbon atom beta (one carbon
removed) to the group. For example, Figure 13-43 shows the ^{1}H NMR and ^{13}C NMR
spectra of 1,2,2-trichloropropane. The methyl (CH_3) carbon absorbs at 33 ppm downfield from TMS because the two chlorine atoms on the adjacent $—CCl_2—$ carbon
have a substantial effect on the methyl carbon. The chemical shift of this methyl carbon is about 15 times that of its attached protons ($\delta 2.1$), in accordance with our prediction. Similarly, the chemical shift of the $—CH_2Cl$ carbon (56 ppm) is about 15
times that of its protons ($\delta 4.0$). Although the CCl_2 carbon has no protons, the proton in
a $—CHCl_2$ group generally absorbs around $\delta 5.8$. The carbon absorption at 87 ppm is
about 15 times this proton shift.

13-12D   Important Differences Between Proton and Carbon Techniques

Most of the characteristics of ^{13}C NMR spectroscopy are similar to those of the ^{1}H
NMR technique. There are some important differences, however.

Operating Frequency   The gyromagnetic ratio for ^{13}C is about one-fourth that of
the proton, so the resonance frequency is also about one-fourth. A spectrometer with a
70,459-gauss magnet needs a 300-MHz transmitter for protons and a 75.6-MHz transmitter for ^{13}C. A spectrometer with a 14,092-gauss magnet needs a 60-MHz transmitter for protons and a 15.1-MHz transmitter for ^{13}C.

Peak Areas   The areas of ^{13}C NMR peaks are not necessarily proportional to the
number of carbons giving rise to the peaks. Carbon atoms with two or three protons
attached usually give the strongest absorptions and carbons with no protons tend to give

weak absorptions. Newer spectrometers have an integrating mode that uses gated decoupling to equalize the absorptions of different carbon atoms. This mode makes peak integrals nearly proportional to the relative numbers of carbon atoms.

13-12E   Spin-Spin Splitting

^{13}C NMR splitting patterns are quite different from those observed in ^{1}H NMR. Only 1% of the carbon atoms in the ^{13}C NMR sample are magnetic, so there is only a small probability that an observed ^{13}C nucleus is adjacent to another ^{13}C nucleus. Therefore, carbon–carbon splitting can be ignored. Carbon–hydrogen coupling is common, however. Most carbon atoms are bonded directly to hydrogen atoms or are sufficiently close to hydrogen atoms for carbon–hydrogen spin-spin coupling to be observed. Extensive carbon–hydrogen coupling produces splitting patterns that can be complicated and difficult to interpret.

Proton Spin Decoupling   To simplify ^{13}C NMR spectra, they are commonly recorded using **proton spin decoupling**, where the protons are continuously irradiated with a broadband ("noise") proton transmitter. As a result, all the protons are continuously in resonance, and they rapidly flip their spins. The carbon nuclei see an *average* of the possible combinations of proton spin states. Each carbon signal appears as a single, unsplit peak because any carbon–hydrogen splitting has been eliminated. The spectra in Figures 13-42 and 13-43 were generated in this manner.

> **PROBLEM 13-25**
>
> Draw the expected broadband-decoupled ^{13}C NMR spectra of the following compounds. Use Figure 13-41 to estimate the chemical shifts.
>
> **(a)**   **(b)**   **(c)**   **(d)**

Off-Resonance Decoupling   Proton spin decoupling produces spectra that are very simple, but some valuable information is lost in the process. **Off-resonance decoupling** simplifies the spectrum but allows some of the splitting information to be retained (Figure 13-44). With off-resonance decoupling, the ^{13}C nuclei are split only by the protons directly

^{13}C NMR

δ (ppm)

◄ **Figure 13-44**
Off-resonance–decoupled ^{13}C NMR spectrum of 1,2,2-trichloropropane. The CCl_2 group appears as a singlet, the CH_3Cl group as a triplet, and the CH_3 group as a quartet. Compare this spectrum with Figure 13-43.

▲ **Figure 13-45**
Off-resonance–decoupled (upper) and broadband-decoupled (lower) ^{13}C NMR spectra of
2-butanone.

bonded to them. The $N+1$ rule applies, so a carbon atom with one proton (a methine) appears as a doublet, a carbon with two attached protons (a methylene) gives a triplet, and
a methyl carbon is split into a quartet. Off-resonance–decoupled spectra are easily recognized by the appearance of TMS as a quartet at 0 ppm, split by the three protons of each
methyl group.

The best procedure for obtaining a ^{13}C NMR spectrum is to run the spectrum
twice: The singlets in the broadband-decoupled spectrum indicate the number of nonequivalent types of carbon atoms and their chemical shifts. The multiplicities of the
signals in the off-resonance–decoupled spectrum indicate the number of hydrogen atoms
bonded to each carbon atom. ^{13}C spectra are often given with two traces, one broadband
decoupled and the other off-resonance decoupled. If just one trace is given, it is usually
broadband decoupled. Figure 13-45 shows both spectra for 2-butanone.

PROBLEM 13-26

Show which carbon atoms correspond with which peaks in the ^{13}C NMR spectrum of
2-butanone (Figure 13-45).

PROBLEM 13-27

Repeat Problem 13-25, sketching the off-resonance—decoupled ^{13}C spectra of the compounds.

PROBLEM 13-28

Draw the proton NMR spectrum you would expect for 2-butanone. How well do the proton chemical shifts predict the carbon chemical shifts, using the "15 to 20 times as large" rule of thumb?

13-13
Interpreting Carbon NMR Spectra

Interpreting ^{13}C NMR spectra uses the same principles as interpreting ^{1}H NMR spectra. In fact, carbon spectra are often easier to interpret. A ^{13}C NMR spectrum provides
the following information:

1. The *number of different signals* implies how many different types of carbons are
present.

Off-resonance–decoupled and
broadband-decoupled spectra of
δ-valerolactone, molecular formula
$C_5H_8O_2$.

2. The *chemical shifts* of those signals suggest what types of functional groups contain those carbon atoms.

3. The *peak areas* (in the integrating mode) imply how many carbons of each type are present.

4. The *splitting of signals* in the off-resonance–decoupled spectrum indicates how many protons are bonded to each carbon atom ($N+1$ rule).

For example, in the ^{13}C NMR spectrum of δ-valerolactone (Figure 13-46), the CH_2 groups in the upper (off-resonance–decoupled) spectrum are split into triplets, but they appear as singlets in the lower (broadband-decoupled) spectrum.

Let's consider how we might solve this structure, given only the ^{13}C NMR spectrum and the molecular formula. As we have seen in Figures 13-41 and 13-42, the signal at 173 ppm is appropriate for a carbonyl carbon. The off-resonance–decoupled spectrum shows a singlet at 173 ppm, implying that no hydrogens are bonded to the carbonyl carbon.

The chemical shift of the next absorption is about 70 ppm. This is about 20 times the chemical shift of a proton on a carbon bonded to an electronegative element. The molecular formula implies that the electronegative element must be oxygen. Since the signal at 70 ppm is a triplet in the off-resonance–decoupled spectrum, this carbon must be a methylene ($—CH_2—$) group.

partial structures: $—CH_2—O—$ $\overset{\displaystyle O}{\overset{\displaystyle \|}{—C—}}$

The signal at 30 ppm corresponds to a carbon atom bonded to a carbonyl group. Remember that a proton on a carbon adjacent to a carbonyl group absorbs around 2.1 ppm, and we expect the carbon to have a chemical shift about 15 to 20 times as large. This carbon atom is a methylene group, as shown by the triplet in the off-resonance–decoupled spectrum.

partial structures: $—CH_2—O—$ $—CH_2—\overset{\displaystyle O}{\overset{\displaystyle \|}{C}}—$

The two signals at 19 and 22 ppm are from carbon atoms that are not directly bonded to any deshielding group, although the carbon at 22 ppm is probably closer to one of the oxygen atoms. These are also triplets in the off-resonance–decoupled spectrum and they correspond to methylene groups. We can propose:

partial structures: $-\overset{|}{\underset{|}{C}}-CH_2-CH_2-CH_2-O-$ $-CH_2-\overset{\overset{O}{\|}}{C}-$

The molecular formula $C_5H_8O_2$ implies the presence of two elements of unsaturation. The carbonyl (C=O) group accounts for one, but there are no more carbonyl groups and no double-bonded alkene carbon atoms. The other element of unsaturation must be a ring. Combining the partial structures into a ring gives the complete structure.

In the following problems, only the broadband-decoupled spectra are provided. In cases where off-resonance–decoupled spectra are available, the off-resonance multiplicity of each peak is indicated: (s) = singlet, (d) = doublet, (t) = triplet, and (q) = quartet.

PROBLEM 13-29

A bottle of allyl bromide was found to contain a large amount of an impurity. A careful distillation separated the impurity, which has the molecular formula C_3H_6O. The following ^{13}C NMR spectrum of the impurity was obtained:

(a) Propose a structure for this impurity.
(b) Assign the peaks in the ^{13}C NMR spectrum to the carbon atoms in the structure.
(c) Suggest how this impurity arose in the allyl bromide sample.

PROBLEM 13-30

An inexperienced graduate student was making some 4-hydroxybutanoic acid. He obtained an excellent yield of a different compound, whose ^{13}C NMR spectrum appears below.

(a) Propose a structure for this product.
(b) Assign the peaks in the ^{13}C NMR spectrum to the carbon atoms in the structure.

PROBLEM 13-31

A laboratory student was converting cyclohexanol to cyclohexyl bromide by using one equivalent of sodium bromide in a large excess of concentrated sulfuric acid. The major product she recovered was not cyclohexyl bromide but a compound of formula C_6H_{10} that gave the following ^{13}C NMR spectrum:

(a) Propose a structure for this product.
(b) Assign the peaks in the ^{13}C NMR spectrum to the carbon atoms in the structure.
(c) Suggest modifications in the reaction to obtain a better yield of cyclohexyl bromide.

13-14
Nuclear Magnetic Resonance Imaging

When chemists use NMR spectroscopy, they take great pains to get the most uniform magnetic field possible (often homogeneous to within one part per billion). They place small tubes of homogeneous solutions in the magnetic field and spin the tubes to average out any remaining variations in the magnetic field. Their goal is to have the sample behave as if it were all at a single point in the magnetic field, with every molecule subjected to exactly the same external magnetic field.

Nuclear magnetic resonance imaging uses the same physical effect, but its goals are almost the opposite of chemical NMR. In NMR imaging, a heterogeneous sample (commonly a living human body) is placed in the magnetic field of a large-bore superconducting magnet. The magnetic field is purposely nonuniform, with a gradient that allows just the protons in one plane of the sample to be in resonance at any one time. By using a combination of field gradients and sophisticated Fourier transform techniques, the instrument can look selectively at one point within the sample, or a line within the sample, or a plane within the sample. The computer generates an image of a two-dimensional slice through the sample. A succession of slices can be accumulated in the computer to give a three-dimensional plot of the proton resonances within the bulk of the sample.

Medical NMR imaging is commonly called **magnetic resonance imaging (MRI)** to avoid the common fear of the word "nuclear" and the misconception that "nuclear" means "radioactive." There is nothing radioactive about an NMR spectrometer. In fact, MRI is the least invasive, least hazardous method available for imaging the interior of the body. The only common side effect is claustrophobia from being confined within the ring of the wide-bore magnet.

The MRI image can easily distinguish watery tissues, fatty tissues, bone, air spaces, blood, etc. by their differences in composition and movement. By using proton **relaxation times**, the technique becomes even more useful. In a strong magnetic field, slightly more proton spins are aligned with the field (the lower-energy state) than against it. A radio-frequency pulse of just the right duration inverts some spins, increasing the number of spins oriented against the magnetic field. The spins gradually relax to their normal state

The ^{31}P signal of phosphates in cells and tissues can be followed by ^{31}P NMR spectroscopy. This technique has been used to study the effects of exercise and oxygen starvation on the metabolism of phosphate esters such as ATP.

(a)

(b)

▲ Figure 13-47
(a) MRI scan of a human brain showing a metastatic tumor in one hemisphere. (b) MRI image of the pelvic region showing severe damage in an arthritic hip.

over a period of a few seconds. By following the free-induction decay, the spectrometer measures how quickly spin relaxation occurs in each pixel of the sample.

Differing relaxation times are coded by color or intensity in the image, giving valuable information about the tissues involved. For example, cancerous tissues tend to have longer relaxation times than the corresponding normal tissues so tumors are readily apparent in the NMR image. Figure 13-47 shows two actual MRI images: The first image is a slice through a patient's head showing a brain tumor. The second image is a slice through another patient's pelvic region showing an arthritic hip.

PROBLEM-SOLVING STRATEGY
Spectroscopy Problems

We have now learned to use both IR and NMR spectroscopy, as well as mass spectrometry, to determine the structures of unknown organic compounds. These techniques usually provide a unique structure with little chance of error. A major part of successful spectral interpretation is using an effective strategy rather than simply looking at the spectra, hoping that something obvious will jump out. A systematic approach should take into account the strengths and weaknesses of each technique. The following table summarizes the information provided by each spectroscopic technique.

Summary of the Information Provided by Each Type of Spectroscopy			
	MS	**IR**	**NMR**
molecular weight	✓		
molecular formula	✓ (HRMS)		
heteroatoms	✓	H	S
functional groups	S	✓	H
alkyl substituents	S		✓

Notes: ✓, usually provides this information.
H, usually provides helpful information.
S, sometimes provides helpful information.

We can summarize how you might go about identifying an unknown compound, but the actual process depends on what you already know about the chemistry of the compound and what you learn from each spectrum. Always go through the process with scratch paper and a pencil, so you can keep track of mass numbers, formulas, possible functional groups, and carbon skeletons.

1. *Mass spectrum.* Look for a molecular ion, and determine a tentative molecular weight. Remember that some compounds (alcohols, for example) may fail to give a visible molecular ion. If the molecular weight is odd, consider a nitrogen atom. If an HRMS is available, compare the "exact" mass with the tables to find a molecular formula with a mass close to the experimental value.

   Look for anything unusual or characteristic about the mass spectrum: Does the M+2 peak of the parent ion look larger than the M+1 peak? It might contain S, Cl, or Br. Is there a large gap and a peak at 127 characteristic of iodine?

   Although you might look at the MS fragmentation patterns to help determine the structure, this is more time-consuming than going on to other spectra. You can verify the fragmentation patterns more easily once you have a proposed structure.

2. *Infrared spectrum.* Look for O—H, N—H, or ≡C—H peaks in the 3300 cm^{-1} region. Are there saturated C—H peaks to the right of 3000 cm^{-1}? Unsaturated =C—H peaks to the left of 3000 cm^{-1}? Also look for C≡C or C≡N stretch around 2200 cm^{-1}, and for C=O, C=C, or C=N stretch between 1600 and 1800 cm^{-1}. The exact position of the peak, plus other characteristics (intensity, broadening), should help to determine the functional groups. For example, a broad O—H band centered over the C—H stretch at 3000 cm^{-1} might imply a carboxylic acid group, —COOH.

   The combination of IR and an odd molecular ion in the mass spectrum should confirm amines, amides, and nitriles. A strong alcohol —OH absorption in the IR might suggest that the apparent molecular ion in the mass spectrum could be low by 18 units from loss of water.

3. *Nuclear magnetic resonance spectrum.* First look for strongly deshielded protons, such as carboxylic acids ($\delta 10$ to $\delta 12$) aldehydes ($\delta 9$ to $\delta 10$), and aromatic protons ($\delta 7$ to $\delta 8$). Moderately deshielded peaks might be vinyl protons ($\delta 5$ to $\delta 6$) or protons on a carbon bonded to an electronegative atom such as oxygen or halogen ($\delta 3$ to $\delta 4$). A peak around $\delta 2.1$ to $\delta 2.5$ might be an acetylenic proton or a proton on a carbon next to a carbonyl group, a benzene ring, or a vinyl group.

These possibilities should be checked to see which are consistent with the IR spectrum. Finally, the spin-spin splitting patterns should be analyzed to suggest the structures of the alkyl groups present.

Once you have considered all the spectra, there should be one or two tentative structures. Each structure should be checked to see whether it accounts for the major characteristics of all the spectra.

- Are the molecular weight and formula of the tentative structure consistent with the appearance (or the absence) of the molecular ion in the mass spectrum? Are there peaks in the mass spectrum corresponding to the expected fragmentation products?
- Does the tentative structure explain each of the characteristic stretching frequencies in the infrared spectrum? Does it account for any shifting of frequencies from their usual positions?
- Does the tentative structure account for each proton (or carbon) in the NMR spectrum? Does it also account for the observed chemical shifts and spin-spin splitting patterns?

If the tentative structure successfully accounts for all these features of the spectra, you can be confident that it is correct.

PROBLEM (Partially Solved) 13-32

Sets of spectra are given for two compounds. For each set,

(1) Look at each spectrum individually, and list the structural characteristics you can determine from that spectrum.
(2) Look at the set of spectra as a group, and propose a tentative structure.
(3) Verify that your proposed structure accounts for the major features of each spectrum. The solution for compound 1 is given after the problem, but go as far as you can before looking at the solution.

SOLUTION TO COMPOUND I

Mass spectrum: The MS shows an odd molecular weight (121), possibly indicating the presence of a nitrogen atom.

Infrared spectrum: The IR shows a sharp peak around 3400 cm^{-1}, possibly the N—H of an amine or the ≡C—H of a terminal alkyne. Because the MS suggests a nitrogen atom, and there is no other evidence for an alkyne (no C≡C stretch around 2200 cm^{-1}), the 3400 cm^{-1} absorption is probably an N—H bond. The unsaturated =C—H absorptions above 3000 cm^{-1}, combined with an aromatic C=C stretch around 1600 cm^{-1}, indicate an aromatic ring.

NMR spectrum: The NMR shows complex splitting in the aromatic region, probably from a benzene ring: the total integral of 5 suggests the ring is monosubstituted. Part of the aromatic absorption is shifted upfield of δ7.2, suggesting that the substituent on the benzene ring is a pi electron-donating group like an amine or an ether. An ethyl group (total area 5) is seen at δ1.2 and δ3.1, appropriate for protons on a carbon atom bonded to nitrogen. A broad singlet of area 1 appears at δ3.5, probably resulting from the N—H seen in the IR spectrum. Combining this information, we propose a nitrogen atom bonded to a hydrogen atom, a benzene ring, and an ethyl group. The total molecular weight for this structure would be 121, in agreement with the molecular ion in the mass spectrum.

Proposed structure for compound 1

The proposed structure shows an aromatic ring with 5 protons, which explains the aromatic signals in the NMR and the C=C at 1600 cm^{-1} and the =C—H above 3000 cm^{-1} in the IR. The aromatic ring is bonded to an electron-donating —NHR group, which explains the odd molecular weight, the N—H absorption in the IR, and the aromatic signals shifted above δ7.2 in the NMR. The ethyl group bonded to nitrogen explains the ethyl signals in the NMR, deshielded to δ3.1 by the nitrogen atom. The base peak in the MS (M − 15 = 106) is explained by the loss of a methyl group to give a resonance-stabilized cation:

m/z 106

Chapter 13 Glossary

accidentally equivalent nuclei Nuclei that are nonequivalent by NMR yet absorb at nearly the same chemical shift and are not resolved. Nuclei that absorb at the same chemical shift cannot split each other, whether they are chemically equivalent or accidentally equivalent. (p. 553)

chemically equivalent atoms Atoms that cannot be distinguished chemically. The replacement test for chemically equivalent atoms gives identical compounds. (p. 552)

chemical shift The difference (in ppm) between the resonance frequency of the proton (or carbon nucleus) being observed and that of tetramethylsilane (TMS). Chemical shifts are usually given on the δ (delta) scale, in parts per million downfield from TMS. (p. 545)

complex splitting Splitting by two or more different kinds of protons with different coupling constants. (p. 565)

coupling constant (J) The distance (in hertz) between two adjacent peaks of a multiplet. (p. 562)

deshielded Bonded to a group that withdraws part of the electron density from around the nucleus; results in a larger chemical shift, moved downfield. (p. 543)

diastereotopic atoms Nuclei that occupy diastereomeric positions. The replacement test for diastereotopic atoms gives diastereomers. Diastereotopic nuclei can be distinguished by NMR, and they can split each other unless they are *accidentally equivalent*. (p. 569)

downfield At a lower value of the applied magnetic field, toward the left on the NMR spectrum. The more deshielded a nucleus is, the farther downfield it absorbs. (p. 544)

Fourier transform spectroscopy Spectroscopy that involves collecting transients (containing all the different resonance frequencies) and converting the averaged transients into a spectrum using the mathematical Fourier transform. (p. 580)

> **transient (free induction decay, or FID):** The signal that results when many nuclei are irradiated by a pulse of energy and precess at their resonance frequencies.

gyromagnetic ratio (γ) A measure of the magnetic properties of a nucleus. The resonance frequency (ν) is given by the equation $\nu = \gamma B_{eff}/2\pi$, where B_{eff} is the effective magnetic field at the nucleus. The gyromagnetic ratio of a proton is 26,753 sec^{-1} gauss^{-1}. The gyromagnetic ratio of a ^{13}C nucleus is 6728 sec^{-1} gauss^{-1}. (p. 541)

induced magnetic field The magnetic field set up by the motion of electrons in a molecule (or in a wire) in response to the application of an external magnetic field. (p. 542)

integration The measurement of the area under a peak, proportional to the number of protons giving rise to that peak. (p. 554)

magnetically coupled Protons that are close enough that their magnetic fields influence each other, resulting in spin-spin splitting. (p. 556)

magnetic moment The magnitude of a nuclear magnetic field, related to the gyromagnetic ratio γ. (p. 539)

magnetic resonance imaging (MRI) The medical term for NMR imaging, avoiding the word "nuclear." Use of field gradients in a large-bore magnet to scan two-dimensional slices of a patient's body. (p. 587)

nuclear magnetic resonance spectroscopy (NMR) A form of spectroscopy that measures the absorption of radio-frequency energy by nuclei in a magnetic field. The energy absorbed causes nuclear spin transitions. (p. 539)

> **carbon magnetic resonance** (^{13}C NMR, CMR): NMR of the ^{13}C isotope of carbon. (p. 579)
> **proton magnetic resonance** (^1H NMR, PMR); NMR of protons. (p. 539)

off-resonance decoupling A technique used with ^{13}C NMR in which only the protons directly bonded to a carbon atom cause spin-spin splitting. (p. 583)

relaxation time A measure of how slowly the nuclear spins return to their normal state after an RF pulse near their resonance frequency. Alternatively, the evening after a chemistry exam. (p. 587)

shielded Surrounded by electrons whose induced magnetic field opposes the externally applied magnetic field. The effective magnetic field at the shielded nucleus is less than the applied magnetic field. (p. 542)

spin decoupling Elimination of spin-spin splitting by constantly irradiating one type of nuclei at its resonance frequency. (p. 583)

spin-spin splitting (magnetic coupling) The interaction of the magnetic fields of two or more nuclei, usually through the bonds connecting them. Spin-spin splitting converts a single signal to a **multiplet**, a set of smaller peaks. (p. 556)

> **multiplet:** A group of peaks resulting from the spin-spin splitting of the signal from a single type of nucleus. A **doublet** has two peaks, a **triplet** has three peaks, a **quartet** has four peaks, and so on. (p. 558)
> **$N+1$ rule:** A signal that is being split by N equivalent protons is split into a multiplet with $N+1$ individual peaks. (p. 558)

TMS Tetramethylsilane, an NMR standard whose absorption is defined as $\delta 0.00$ (p. 545)
upfield At a higher value of the applied magnetic field, toward the right on the NMR spectrum.
The more shielded a nucleus, the farther upfield it absorbs. (p. 544)

Essential Problem-Solving Skills in Chapter 13

1. Given a structure, determine which protons are equivalent and which are nonequivalent; predict the number of signals and their approximate chemical shifts.
2. Given the chemical shifts of absorptions, suggest likely types of protons.
3. Use the integral trace to determine the relative numbers of different types of protons.
4. Predict which protons in a structure will be magnetically coupled, and predict the number of peaks and approximate coupling constants of their multiplets.
5. Use proton spin-spin splitting patterns to determine the structure of alkyl and other groups.
6. Draw the general features of the NMR spectrum of a given compound.
7. Predict the approximate chemical shifts of carbon atoms in a given compound. Given the chemical shifts of ^{13}C absorptions, suggest likely types of carbons.
8. Use the off-resonance–decoupled ^{13}C spectrum to determine the number of hydrogens bonded to a given carbon.
9. Combine the chemical shifts, integrals, and spin-spin splitting patterns in the NMR spectrum with information from infrared and mass spectra to determine the structures of organic compounds.

Study Problems

13-33   An unknown compound has the molecular formula $C_9H_{11}Br$. Its proton NMR spectrum shows the following absorptions.

  singlet, $\delta 7.1$, integral 4.4 cm
  singlet, $\delta 2.3$, integral 13.0 cm
  singlet, $\delta 2.2$, integral 6.7 cm

Propose a structure for this compound.

13-34   Predict the multiplicity (the number of peaks as a result of splitting) for each shaded proton in the following compounds.

(a) $CH_3-CH_2-CCl_2-CH_3$ (b) $CH_3-CH-OH$ (c) $CH_3-CH-CH_3$
                                          | |
                                         CH_3 CH_3

(d) H—⟨benzene ring with H, H, H, H⟩—CH_3 (e) ⟨tetrahydropyran ring⟩—CH_3

13-35   Predict the approximate chemical shifts of the protons in the following compounds.
(a) benzene   (b) cyclohexane   (c) $CH_3-O-CH_2-CH_2-CHCl_2$

(d) $CH_3-CH_2-C\equiv C-H$ (e) $CH_3-CH_2-\overset{\displaystyle O}{\overset{\|}{C}}-CH_3$ (f) $(CH_3)_2CH-CH_2-CH_2-OH$

(g) $CH_3-CH_2-\overset{\displaystyle O}{\overset{\|}{C}}-H$ (h) $CH_3-CH=CH-CHO$ (i) $HOOC-CH_2-CH_2-\overset{\displaystyle O}{\overset{\|}{C}}-O-CH(CH_3)_2$

(j) ⟨cyclohexane ring⟩$=C\overset{\displaystyle H}{\underset{\displaystyle H}{}}$ (k) ⟨indane structure⟩   (l) ⟨γ-butyrolactone structure⟩

   methylenecyclohexane   indane γ-butyrolactone

13-36   The following proton NMR spectrum is of a compound of molecular formula C_3H_8O.
    (a) Propose a structure for this compound.
    (b) Assign peaks to show which protons give rise to which signals in the spectrum.

13-37   Using a 60-MHz spectrometer, a chemist observes the following absorption:

$$\text{doublet, } J = 7 \text{ Hz, at } \delta4.00$$

    (a) How many hertz from the TMS peak is this absorption?
    (b) Where would this peak be located (in ppm and in hertz) in the 300-MHz spectrum of this sample?
    (c) What would J be in the 300-MHz spectrum?

13-38   A compound ($C_{10}H_{12}O_2$) whose spectrum appears below was isolated from a reaction mixture containing 2-phenylethanol and acetic acid.
    (a) Propose a structure for this compound.
    (b) Assign peaks to show which protons give rise to which signals in the spectrum.

13-39   Sketch your predictions of the proton NMR spectra of the following compounds.

    (a) $CH_3-O-CH_2CH_3$

    (b) $(CH_3)_2CH-\overset{\overset{\textstyle O}{\|}}{C}-CH_3$

    (c) $Cl-CH_2-CH_2-CH_2-Cl$

    (d)

    (e)

13-40   Tell precisely how you would use the proton NMR spectra to distinguish between the following pairs of compounds.

   (a) 1-bromopropane and 2-bromopropane

   (b) $CH_3-CH_2-\overset{\overset{\displaystyle O}{\|}}{C}-H$ and $CH_3-\overset{\overset{\displaystyle O}{\|}}{C}-CH_3$

   (c) $CH_3-CH_2-O-\overset{\overset{\displaystyle O}{\|}}{C}-CH_3$ and $CH_3-CH_2-\overset{\overset{\displaystyle O}{\|}}{C}-O-CH_3$

   (d) $CH_3-CH_2-C\equiv C-H$ and $CH_3-C\equiv C-CH_3$

13-41   Give the approximate chemical shift and multiplicity of each absorption band in the off-resonance–decoupled ^{13}C NMR spectra of the following compounds.

   (a) $CH_3-\overset{\overset{\displaystyle O}{\|}}{C}-O-CH_2-CH_3$
   ethyl acetate

   (b) $H_2C=CH-CH_2Cl$
   3-chloropropene

13-42   The following off-resonance–decoupled carbon NMR was obtained from a compound of formula $C_3H_5Cl_3$. Propose a structure for this compound, and show which carbon atoms give rise to which peaks in the spectrum.

13-43   A small pilot plant was adding bromine across the double bond of 2-butene to make 2,3-dibromobutane. A controller malfunction allowed the reaction temperature to rise beyond safe limits. A careful distillation of the product showed that several impurities had formed, including the one whose NMR spectra appear below. Determine its structure, and assign the peaks to the protons in your structure.

13-44   A new chemist moved into an industrial lab where work was being done on oxygenated gasoline additives. Among the additives that had been tested, she found an old bottle containing a clear, pleasant-smelling liquid that was missing its label. She took a quick NMR spectrum (shown below) and was able to determine the identity of the compound without any additional information. Propose a structure, and assign the peaks. (*Hint:* This is a very pure sample.)

*13-45   When 2-chloro-2-methylbutane is treated with a variety of strong bases, the products always seem to contain two isomers (A and B) of formula C_5H_{10}. When sodium hydroxide is used as the base, isomer A predominates. When potassium *t*-butoxide is used as the base, isomer B predominates. The ^1H and ^{13}C NMR spectra of A and B are given below.

(a) Determine the structures of isomers A and B.

(b) Explain why A is the major product when using sodium hydroxide as the base and why B is the major product when using potassium *t*-butoxide as the base.

13-46   (A true story.) A major university was designated as a national nuclear magnetic resonance center by the National Science Foundation. Several large superconducting instruments were being installed when a government safety inspector appeared and demanded to know what provisions were being made to handle the nuclear waste produced by these instruments. Assume you are the manager of the NMR center, and offer an explanation that could be understood by a nonscientist.

13-47   A compound was isolated as a minor constituent in an extract from garden cress. Its spectra appear below.
    (1) Look at each spectrum individually, and list the structural characteristics you can determine from that spectrum.
    (2) Look at the set of spectra as a group, and propose a tentative structure.
    (3) Verify that your proposed structure accounts for the major features of each spectrum.

*13-48   The following spectra are taken from a compound that is an important starting material for organic synthesis. Determine the structure, first by considering each spectrum individually, then by considering all the spectra together. Assign peaks to show that your proposed structure accounts for all the major features of each spectrum.

CHAPTER 14

Ethers, Epoxides, and Sulfides

18-crown-6
with K$^+$ solvated

14-1
Introduction

Ethers are compounds of formula R—O—R′, where R and R′ may be alkyl groups or aryl (benzene ring) groups. Like alcohols, ethers are related to water, with alkyl groups replacing the hydrogen atoms. In an alcohol, one hydrogen atom of water is replaced by an alkyl group. In an ether, both hydrogens are replaced by alkyl groups. The two alkyl groups are the same in a **symmetrical ether** and different in an **unsymmetrical ether**.

H—O—H   R—O—H   R—O—R′
water   alcohol   ether

Examples of ethers

CH$_3$CH$_2$—O—CH$_2$CH$_3$ 〔benzene〕—O—CH$_3$ 〔THF ring〕

diethyl ether
(a symmetrical ether)

methyl phenyl ether
(an unsymmetrical ether)

tetrahydrofuran
(a symmetrical, cyclic ether)

As with other functional groups, we will discuss how ethers are formed and how they react. Ethers (other than epoxides) are relatively unreactive, however, and they are not frequently used as synthetic intermediates. Because they are stable with many types of reagents, ethers are commonly used as solvents for organic reactions. In this chapter, we consider the properties of ethers and how these properties make ethers such valuable solvents for organic reactions.

The most important commercial ether is diethyl ether, often called "ethyl ether," or simply "ether." Ether is a good solvent for reactions and extractions, and it is used as a volatile starting fluid for diesel and gasoline engines. Ether was used as a surgical anesthetic for over a hundred years (starting in 1842), but it is highly flammable, and patients often vomited as they regained consciousness. Several compounds that are less flammable and more easily tolerated are now in use, including nitrous oxide (N$_2$O) and halothane (CF$_3$—CHClBr).

Diethyl ether was found to be a much safer anesthetic than chloroform. Like chloroform, ether is more soluble in fatty tissue than in water, so it passes quickly into the central nervous system and takes effect quickly. Ether is also volatile, making it easy to administer. But ether is much less toxic than chloroform because ether degrades to ethanol, which the body can oxidize.

14-2
Physical Properties of Ethers

14-2A Structure and Polarity of Ethers

Like water, ethers have a bent structure, with an sp^3 hybrid oxygen atom giving a nearly tetrahedral bond angle. In water, the nonbonding electrons compress the H—O—H bond angle to 104.5°, but in a typical ether, the bulk of the alkyl groups enlarges the bond angle. Figure 14-1 shows the structure of dimethyl ether, with a tetrahedral bond angle of 110°.

Although ethers lack the polar hydroxyl group of alcohols, they are still strongly polar compounds. The dipole moment of an ether is the vector sum of two polar C—O bonds, with a substantial contribution from the two lone pairs of electrons. Table 14-1

◀ **Figure 14-1**
Structure of dimethyl ether.

TABLE 14-1 Comparison of the Boiling Points of Ethers, Alkanes, and Alcohols of Similar Molecular Weights

Compound	Formula	MW	bp (°C)	Dipole Moment (D)
water	H_2O	18	100	1.9
ethanol	$CH_3CH_2—OH$	46	78	1.7
dimethyl ether	$CH_3—O—CH_3$	46	−25	1.3
propane	$CH_3CH_2CH_3$	44	−42	0.1
n-butanol	$CH_3CH_2CH_2CH_2—OH$	74	118	1.7
tetrahydrofuran		72	66	1.6
diethyl ether	$CH_3CH_2—O—CH_2CH_3$	74	35	1.2
pentane	$CH_3CH_2CH_2CH_2CH_3$	72	36	0.1

Note: The alcohols are hydrogen bonded, giving them much higher boiling points. The ethers have boiling points that are closer to those of alkanes with similar molecular weights.

compares the dipole moments of dimethyl ether, diethyl ether, and tetrahydrofuran (THF) with those of alkanes and alcohols of similar molecular weights. An ether such as THF provides a strongly polar solvent without the reactivity of a hydroxyl group.

14-2B Boiling Points of Ethers; Hydrogen Bonding

Table 14-1 compares the boiling points of several ethers, alcohols, and alkanes. Notice that the boiling points of dimethyl ether and diethyl ether are nearly 100°C lower than those of alcohols having similar molecular weights. This large difference results mostly from hydrogen bonding in the alcohols. Pure ethers cannot engage in hydrogen bonding because they have no O—H groups. Ethers do have large dipole moments, resulting in dipole–dipole attractions, but these attractions appear to have relatively little effect on their boiling points.

Although pure ethers have no hydroxyl groups to engage in hydrogen bonding, they can hydrogen bond with other compounds that do have O—H or N—H groups. Figure 14-2 shows that a hydrogen bond requires both a hydrogen bond *donor* and a hydrogen

An ether inhaler used by William Morton in the first public demonstration of ether anesthesia on October 16, 1846, at Massachusetts General Hospital.

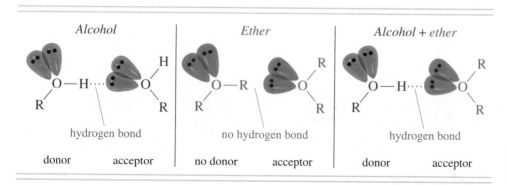

◀ **Figure 14-2**
A molecule of water or an alcohol can serve as both a hydrogen bond donor and acceptor. Ether molecules have no hydroxyl groups, so they are not hydrogen bond donors. If a hydrogen bond donor is present, ethers can serve as hydrogen bond acceptors.

TABLE 14-2   Physical Properties of Some Representative Ethers

Name	Structure	mp (°C)	bp (°C)	Density (g/mL)
dimethyl ether	$CH_3—O—CH_3$	−140	−25	0.66
ethyl methyl ether	$CH_3CH_2—O—CH_3$		8	0.72
diethyl ether	$CH_3CH_2—O—CH_2CH_3$	−116	35	0.71
di-*n*-propyl ether	$CH_3CH_2CH_2—O—CH_2CH_2CH_3$	−122	91	0.74
diisopropyl ether	$(CH_3)_2CH—O—CH(CH_3)_2$	−86	68	0.74
1,2-dimethoxyethane (DME)	$CH_3—O—CH_2CH_2—O—CH_3$	−58	83	0.86
methyl phenyl ether (anisole)	$CH_3—O—\phenyl$	−37	154	0.99
diphenyl ether		27	259	1.07
furan		−86	32	0.94
tetrahydrofuran (THF)		−108	65	0.89
1,4-dioxane		11	101	1.03

bond *acceptor*. The donor is the molecule with an O—H or N—H group. The acceptor is the molecule whose lone pair of electrons forms a weak partial bond to the hydrogen atom provided by the donor. An ether molecule has the lone pair to form a hydrogen bond with an alcohol (or other hydrogen bond donor), but it cannot form a hydrogen bond with another ether molecule. Because ether molecules are not held together by hydrogen bonds, they are more volatile than alcohols having similar molecular weights. Table 14-2 lists the physical properties of a representative group of common ethers.

14-2C   Ethers as Polar Solvents

Ethers are ideally suited as solvents for many organic reactions. They dissolve a wide range of polar and nonpolar substances, and their relatively low boiling points simplify their evaporation from the reaction products. Nonpolar substances tend to be more soluble in ethers than in alcohols because ethers have no hydrogen-bonding network to be broken up by the nonpolar solute.

Polar substances tend to be nearly as soluble in ethers as in alcohols because ethers have large dipole moments as well as the ability to serve as hydrogen bond acceptors. The nonbonding electron pairs of an ether effectively solvate cations, as shown in Figure 14-3.

▶ **Figure 14-3**
An ionic substance such as lithium iodide (LiI) is moderately soluble in ethers because the small lithium cation is strongly solvated by the ether's lone pairs of electrons. Unlike alcohols, ethers cannot serve as hydrogen bond donors, so they do not solvate small anions well.

Ethers do not solvate anions as well as alcohols do, however. Ionic substances with small, "hard" anions requiring strong solvation to overcome their ionic bonding are often insoluble in ether solvents. Substances with large, diffuse anions, such as iodides, acetates, and other organic anions, tend to be more soluble in ether solvents than substances with smaller, harder anions.

Alcohols cannot be used as solvents for reagents that are more strongly basic than the alkoxide ion. The hydroxyl group quickly protonates the base, destroying the basic reagent.

$$ \text{B:}^- \ + \ \text{R}-\overset{..}{\underset{..}{\text{O}}}\text{H} \ \rightleftharpoons \ \text{B}-\text{H} \ + \ \text{R}-\overset{..}{\underset{..}{\text{O}}}\text{:}^- $$

strong base   alcohol   protonated base   alkoxide ion

Ethers are nonhydroxylic (no hydroxyl group), and they are normally unreactive toward strong bases. For this reason, ethers are frequently used as solvents for very strong polar bases (like the Grignard reagent) that require polar solvents. The four ethers shown here are common solvents for organic reactions. DME, THF, and dioxane are miscible with water, while diethyl ether is sparingly soluble in water.

$CH_3CH_2-O-CH_2CH_3$	$CH_3-O-CH_2CH_2-O-CH_3$	tetrahydrofuran	1,4-dioxane
diethyl ether	1,2-dimethoxyethane	THF, oxolane	dioxane
"ether"	DME, "glyme"	bp 65°C	bp 101°C
bp 35°C	bp 82°C		

PROBLEM 14-1

Rank the given solvents in decreasing order of their ability to dissolve each compound.

Solutes

(a) NaOAc   **(b)** naphthalene   **(c)** 2-naphthol

Solvents

ethyl ether
water
ethanol
dichloromethane

14-2D Stable Complexes of Ethers with Reagents

The special properties of ethers (polarity, lone pairs, but relatively unreactive) enhance the formation and use of many reagents. For example, Grignard reagents cannot form unless an ether is present, possibly to share its lone pairs of electrons with the magnesium atom. This sharing of electrons stabilizes the reagent and helps keep it in solution (Figure 14-4).

Complexes with Electrophiles An ether's nonbonding electrons also stabilize borane, BH_3. Pure borane exists as a dimer called diborane, B_2H_6. Diborane is a toxic, flammable, and explosive gas, whose use is both dangerous and inconvenient. Borane forms a stable complex with tetrahydrofuran. The $BH_3 \cdot THF$ complex is commercially available as a 1 M solution, easily measured and transferred like any other air-sensitive liquid reagent. The availability of $BH_3 \cdot THF$ has contributed greatly to the convenience of hydroboration (Section 8-7).

▲ **Figure 14-4**

Complexation of an ether with a Grignard reagent stabilizes the reagent and helps keep it in solution.

diborane   +   2   tetrahydrofuran   ⟶   2 $BH_3 \cdot THF$

Boron trifluoride is used as a Lewis acid catalyst in a wide variety of reactions. Like diborane, BF_3 is a toxic gas, but BF_3 forms a stable complex with ethers, allowing it to be conveniently stored and measured. The complex of BF_3 with diethyl ether is called "boron trifluoride etherate."

boron
trifluoride

diethyl ether

$BF_3 \cdot OEt_2$
"boron trifluoride etherate"

PROBLEM 14-2

Aluminum trichloride ($AlCl_3$) dissolves in ether with the evolution of a large amount of heat. (In fact, this reaction can become rather violent if it gets too warm.) Show the structure of the resulting aluminum chloride etherate complex.

Crown Ether Complexes   In Chapter 6, we encountered the use of **crown ethers**, large cyclic polyethers that specifically solvate metal cations by complexing the metal in the center of the ring. Different crown ethers solvate different cations, depending on the relative sizes of the crown ether and the cation and the number of binding sites around the cation.

12-crown-4
solvates Li^+

15-crown-5
solvates Na^+

18-crown-6
solvates K^+

18-crown-6
with K^+ solvated

Complexation by crown ethers often allows polar inorganic salts to dissolve in nonpolar organic solvents. This enhanced solubility allows polar salts to be used under aprotic conditions, where the uncomplexed anions may have greatly enhanced reactivity. For example, in Section 6-10B, we used 18-crown-6 to dissolve potassium fluoride in acetonitrile (CH_3CN), where the poorly solvated fluoride ion is a moderately strong nucleophile. Many other salts, including carboxylate salts ($RCOO^- {}^+K$), cyanides (KCN), and permanganates ($KMnO_4$), can be dissolved in aprotic (and often nonpolar) organic solvents using crown ethers. In each case, the crown ether complexes only the cation, leaving the anion bare and highly reactive.

PROBLEM 14-3

In the presence of 18-crown-6, potassium permanganate dissolves in benzene to give "purple benzene," a useful reagent for oxidizing alkenes in an aprotic environment. Use a drawing of the complex to show why $KMnO_4$ dissolves in benzene and why the reactivity of the permanganate ion is enhanced.

We have been using the common nomenclature of ethers, which is sometimes called the *alkyl alkyl ether* system. The IUPAC system, generally used with more complicated ethers, is sometimes called the *alkoxy alkane* system. Common names are almost always used for simple ethers.

14-3A   Common Names (Alkyl Alkyl Ether Names)

Common names of ethers are formed by naming the two alkyl groups on oxygen and adding the word *ether*. Under the current system, the alkyl groups should be named in alphabetical order, but many people still use the old system, which named the groups in order of increasing complexity. For example, if one of the alkyl groups is methyl and the other is *t*-butyl, the current common name should be "*t*-butyl methyl ether," but most chemists use the older common name, "methyl *t*-butyl ether" (or MTBE). If both groups are methyl, the name is "dimethyl ether." If just one alkyl group is described in the name, it implies the ether is symmetrical, as in "ethyl ether."

14-3B   IUPAC Names (Alkoxy Alkane Names)

IUPAC names use the more complex alkyl group as the root name, and the rest of the ether as an alkoxy group. For example, cyclohexyl methyl ether is named methoxycyclohexane. This systematic nomenclature is often the only clear way to name complex ethers.

	$CH_3-O-CH_2CH_3$		$Cl-CH_2-O-CH_3$
IUPAC name:	methoxyethane		chloromethoxymethane
common name:	ethyl methyl ether	methoxybenzene	chloromethyl methyl ether
		methyl phenyl ether, anisole	

IUPAC name:	3-ethoxy-1,1-dimethylcyclohexane	*trans*-1-chloro-2-methoxycyclobutane	2-ethoxyethanol

---

PROBLEM 14-4

Give a common name and a systematic name for each compound.

(a) ▷—OCH₃   **(b)** $CH_3CH_2-O-CH(CH_3)_2$ **(c)** $ClCH_2CH_2OCH_3$

(d)   **(e)** $(CH_3)_3C-O-\overset{\displaystyle CH_3}{\underset{\displaystyle CH_2CH_3}{CH}}$ **(f)**

14-3C   Nomenclature of Cyclic Ethers

Cyclic ethers are our first examples of **heterocyclic compounds**, containing a ring in which a ring atom is an element other than carbon. This atom, called the **heteroatom**, is numbered 1 in numbering the ring atoms. Heterocyclic ethers are an especially important and useful class of ethers.

Epoxides (Oxiranes)   We have already encountered some of the chemistry of epoxides (Section 8-12). **Epoxides** are three-membered cyclic ethers, usually formed by peroxyacid oxidation of the corresponding alkenes. The common name of an epoxide is formed by adding "oxide" to the name of the alkene that is oxidized. The following reactions show the synthesis and common names of two simple epoxides.

$$H_2C=CH_2 \ + \ Ph-\overset{O}{\overset{\|}{C}}-OOH \ \longrightarrow \ H_2C-CH_2 \ + \ Ph-\overset{O}{\overset{\|}{C}}-OH$$

ethylene   peroxybenzoic acid   ethylene oxide   benzoic acid

cyclohexene → peroxybenzoic acid → cyclohexene oxide

One systematic method for naming epoxides is to name the rest of the molecule and use the term "epoxy" as a substituent, giving the numbers of the two carbon atoms bonded to the epoxide oxygen.

trans-1,2-epoxy-4-methylcyclohexane   *cis*-2,3-epoxy-4-methoxyhexane

Another systematic method names epoxides as derivatives of the parent compound, ethylene oxide, using "oxirane" as the systematic name for ethylene oxide. In this system, the ring atoms of a heterocyclic compound are numbered starting with the heteroatom and going in the direction to give the lowest substituent numbers.

oxirane   2,2-diethyl-3-isopropyloxirane   *trans*-2-methoxy-3-methyloxirane

Oxetanes   The least common cyclic ethers are the four-membered **oxetanes**. Because these four-membered rings are strained, they are more reactive than larger cyclic ethers and open-chain ethers. They are not as reactive as the highly strained oxiranes (epoxides), however.

oxetane   2-ethyl-3,3-dimethyloxetane

Furans (Oxolanes)   The five-membered cyclic ethers are commonly named after an aromatic member of this group, **furan**. We consider the aromaticity of furan and other heterocycles in Chapter 16. The systematic term **oxolane** is also used for a five-membered ring containing an oxygen atom.

furan   3-methoxyfuran   tetrahydrofuran (THF)
(oxolane)

The saturated five-membered cyclic ether resembles furan but has four additional hydrogen atoms. Therefore, it is called *tetrahydrofuran* (THF). One of the most polar ethers, tetrahydrofuran is an excellent nonhydroxylic organic solvent for polar reagents. Grignard reactions sometimes succeed in THF even when they fail in diethyl ether.

Pyrans (Oxanes)   The six-membered cyclic ethers are commonly named as derivatives of **pyran**, an unsaturated ether. The saturated compound has four more hydrogen atoms, so it is called *tetrahydropyran* (THP). The systematic term **oxane** is also used for a six-membered ring containing an oxygen atom.

pyran   4-methylpyran   tetrahydropyran (THP)
                                       (oxane)

Dioxanes   Heterocyclic ethers with two oxygen atoms in a six-membered ring are called **dioxanes**. The most common form of dioxane is the one with the two oxygen atoms in a 1,4-relationship. 1,4-dioxane is miscible with water, and it is widely used as a polar solvent for organic reactions.

1,4-dioxane   4-methyl-1,3-dioxane   dibenzodioxin (dioxin)

Dioxin is a common name for dibenzodioxin, 1,4-dioxane fused with two benzene rings. The name "dioxin" is often used incorrectly in the news media for 2,3,7,8-tetrachlorodibenzodioxin (TCDD), a toxic contaminant in the synthesis of the herbicide called 2,4,5-T or Agent Orange. Surprisingly, TCDD has been in the environment for many millions of years because it is also formed in forest fires. Most dioxins are toxic and carcinogenic (cause cancer) because they associate with DNA and cause a misreading of the genetic code.

2,4,5-trichlorophenoxyacetic acid   2,3,7,8-tetrachlorodibenzodioxin
(2,4,5-T or Agent Orange)   (TCDD, incorrectly "dioxin")

PROBLEM 14-5

1,4-Dioxane is made commercially by the acid-catalyzed dehydration of an alcohol.
(a) Show what alcohol will dehydrate to give 1,4-dioxane.
(b) Propose a mechanism for this reaction.

PROBLEM 14-6

Name the following heterocyclic ethers.

(a) [structure: dihydropyran ring with O]

(b) [structure: 1,4-dioxane ring with Cl substituent]

(c) [structure: pyran ring with CH(CH₃)₂ substituent]

(d) H, CH₃CH₂ — epoxide with O — CH₂CH₃, H

(e) [furan ring with Br and OCH₂CH₃ substituents]

(f) [oxetane ring with Br, CH₃, CH₃ substituents]

14-4
Spectroscopy of Ethers

Infrared Spectroscopy of Ethers   Infrared spectra do not show obvious or reliable absorptions for ethers. Most ethers give a moderate to strong $C—O$ stretch around 1000 to 1200 cm^{-1} (in the fingerprint region), but many compounds other than ethers give similar absorptions. Nevertheless, the IR spectrum can be useful because it shows the *absence* of carbonyl ($C=O$) groups and hydroxyl ($O—H$) groups. If the molecular formula contains an oxygen atom, the lack of carbonyl or hydroxyl absorptions in the IR suggests an ether.

Mass Spectrometry of Ethers   The most common fragmentation of ethers is cleavage next to one of the carbon atoms bonded to oxygen. Because this carbon is *alpha* to the oxygen atom, this fragmentation is called **α cleavage**. The resulting *oxonium ion* (oxygen with three bonds and a positive charge) is resonance-stabilized by the nonbonding electrons on oxygen.

α Cleavage

$$[R{\dashv}CH_2—O—R']^{\ddagger} \longrightarrow \underset{\text{not observed}}{R\cdot} + \left[\begin{array}{c} H \\ {>}C{\overset{+}{—}}\ddot{\ddot{O}}—R' \\ H \end{array} \longleftrightarrow \begin{array}{c} H \\ {>}C{=}\overset{+}{\ddot{O}}—R' \\ H \end{array} \right]$$

$$\underset{\text{oxonium ion}}{}$$

Another common cleavage is the loss of either of the two alkyl groups to give another oxonium ion or an alkyl cation.

Loss of an alkyl group

$$[R—CH_2—O{\dashv}R']^{\ddagger} \overset{\sim H}{\longrightarrow} \left[R—\overset{+}{C}H—\ddot{\ddot{O}}—H \longleftrightarrow R—CH{=}\overset{+}{\underset{..}{O}}—H \right] + \underset{\text{not observed}}{\cdot R'}$$

$$\underset{\text{oxonium ion}}{}$$

or

$$[R—CH_2—O{\dashv}R']^{\ddagger} \longrightarrow \underset{\text{not observed}}{R—CH_2—O\cdot} + \underset{\text{alkyl cation}}{{}^+R'}$$

The mass spectrum of diethyl ether appears in Figure 14-5. The four most abundant ions correspond to the molecular ion, loss of an ethyl group, α cleavage, and loss of an ethylene molecule combined with α cleavage. All these modes of cleavage form resonance-stabilized oxonium ions.

Loss of an ethyl group

$$[CH_3-CH_2 \overset{\lceil 45}{+} O-CH_2-CH_3]^{\ddagger} \longrightarrow H-\overset{+}{O}=CH-CH_3 + \cdot CH_2CH_3$$

m/z 74 *m/z* 45 loss of 29

α Cleavage

$$[CH_3-CH_2-O-CH_2 \overset{\lceil 59}{+} CH_3]^{\ddagger} \longrightarrow CH_3-CH_2-\overset{+}{O}=CH_2 + \cdot CH_3$$

m/z 74 *m/z* 59 loss of 15

α Cleavage combined with loss of an ethylene molecule

$$CH_3-CH_2-\overset{+}{O}=CH_2 \longrightarrow H-\overset{+}{O}=CH_2 + CH_2=CH_2$$

m/z 59 *m/z* 31 loss of 28

◀ **Figure 14-5**

The mass spectrum of diethyl ether shows major peaks for the molecular ion, loss of an ethyl group, α cleavage, and α cleavage combined with loss of a molecule of ethylene.

PROBLEM 14-7

Propose a fragmentation to account for each numbered peak in the mass spectrum of *n*-butyl isopropyl ether.

NMR Spectroscopy of Ethers   In the ^{13}C NMR spectrum, a carbon atom bonded to oxygen generally absorbs between $\delta 65$ and $\delta 90$. Protons on carbon atoms bonded to oxygen usually absorb at chemical shifts between $\delta 3.5$ and $\delta 4$ in the ^{1}H NMR spectrum. Both alcohols and ethers have resonances in this range. See, for example, the NMR spectra of methyl *t*-butyl ether (page 554) and ethanol (page 572). If a compound containing C, H, and O has resonances in the correct range, and if there is no O—H stretch or C=O stretch in the IR spectrum, an ether is the most likely functional group.

$$-\overset{\overset{\displaystyle H}{|}}{\underset{\underset{\displaystyle H}{|}}{C}}-O-$$

^{13}C $\delta 65-\delta 90$

^{1}H $\delta 3.5-\delta 4$

14-5
The Williamson Ether Synthesis

We have already seen most of the common methods for synthesizing ethers. We review them at this time, looking more closely at the mechanisms to see which methods are most suitable for preparing various kinds of ethers. The **Williamson ether synthesis** (Section 11-14) is the most reliable and versatile ether synthesis. This method involves the S_N2 attack of an alkoxide ion on an unhindered primary alkyl halide or tosylate. Secondary alkyl halides and tosylates are occasionally used in the Williamson synthesis, but elimination competes, and the yields are often poor.

$$R-\ddot{\underset{..}{O}}{:}^- \quad R'\!-\!\ddot{\underset{..}{X}}{:} \quad \longrightarrow \quad R-\ddot{\underset{..}{O}}-R' \;+\; {:}\ddot{\underset{..}{X}}{:}^-$$

The alkoxide is commonly made by adding Na, K, or NaH to the alcohol (Section 11-14).

Examples

cyclohexanol
$\xrightarrow[\text{(2) CH}_3\text{CH}_2-\text{OTs}]{\text{(1) Na}}$
ethoxycyclohexane
(92%)

3,3-dimethyl-2-pentanol
$\xrightarrow[\text{(2) CH}_3-\text{I}]{\text{(1) NaH}}$
2-methoxy-3,3-dimethylpentane
(90%)

SOLVED PROBLEM 14-1

(a) Why is the following reaction a poor method for the synthesis of *t*-butyl propyl ether?
(b) What would be the major product from this reaction?
(c) Propose a better synthesis of *t*-butyl propyl ether.

$$\text{CH}_3\text{CH}_2\text{CH}_2-\ddot{\underset{..}{O}}{:}^-\ ^+\text{Na} \;+\; \text{CH}_3-\overset{\overset{\displaystyle \text{CH}_3}{|}}{\underset{\underset{\displaystyle \text{CH}_3}{|}}{C}}-\text{Br} \quad \xrightarrow[\times]{\text{does } not \text{ give}} \quad \text{CH}_3-\overset{\overset{\displaystyle \text{CH}_3}{|}}{\underset{\underset{\displaystyle \text{CH}_3}{|}}{C}}-\text{O}-\text{CH}_2\text{CH}_2\text{CH}_3$$

sodium propoxide   *t*-butyl bromide   *t*-butyl propyl ether

SOLUTION

(a) The desired S_N2 reaction cannot occur on the tertiary alkyl halide.

(b) The alkoxide ion is a strong base as well as a nucleophile, and elimination prevails.

$$CH_3CH_2CH_2-\overset{..}{\underset{..}{O}}:^-{}^+Na + H-\overset{\overset{\displaystyle H}{|}}{\underset{\underset{\displaystyle H}{|}}{C}}-\overset{\overset{\displaystyle CH_3}{|}}{\underset{\underset{\displaystyle Br}{|}}{C}}-CH_3 \xrightarrow{E2} H_2C=\overset{\overset{\displaystyle CH_3}{}}{\underset{\underset{\displaystyle CH_3}{}}{C}}$$

sodium propoxide   isobutylene

t-butyl bromide

$$+ \quad CH_3CH_2CH_2OH \quad + \quad NaBr$$

(c) A better synthesis would use the less hindered alkyl group as the S_N2 substrate and the alkoxide of the more hindered alkyl group.

$$CH_3-\overset{\overset{\displaystyle CH_3}{|}}{\underset{\underset{\displaystyle CH_3}{|}}{C}}-O^-{}^+Na + CH_3CH_2-\overset{\overset{\displaystyle H}{|}}{\underset{\underset{\displaystyle Br}{|}}{C}}-H \xrightarrow{S_N2} CH_3-\overset{\overset{\displaystyle CH_3}{|}}{\underset{\underset{\displaystyle CH_3}{|}}{C}}-O-CH_2CH_2CH_3$$

sodium *t*-butoxide   1-bromopropane   *t*-butyl propyl ether

> **PROBLEM-SOLVING** $\mathbb{HINT}$
>
> To convert two alcohols to an ether, convert the more hindered alcohol to its alkoxide. Convert the less hindered alcohol to its tosylate (or an alkyl halide). Make sure the tosylate (or halide) is a good S_N2 substrate.

PROBLEM 14-8

Propose a Williamson synthesis of 3-butoxy-1,1-dimethylcyclohexane from 3,3-dimethyl-cyclohexanol and butanol.

Synthesis of Phenyl Ethers   A phenol (aromatic alcohol) can be used as the alkoxide fragment (but not the halide fragment) for the Williamson ether synthesis. Phenols are more acidic than aliphatic alcohols (Section 10-6), and sodium hydroxide is sufficiently basic to form the phenoxide ion. As with other alkoxides, the electrophile should have an unhindered primary alkyl group and a good leaving group.

2-nitrophenol   2-butoxynitrobenzene
                                                (80%)

Reaction conditions: (1) NaOH (2) $CH_3CH_2CH_2CH_2-I$

PROBLEM 14-9

Show how you would use the Williamson ether synthesis to prepare the following ethers. You may use any alcohols or phenols as your organic starting materials.

(a) cyclohexyl propyl ether
(b) isopropyl methyl ether
(c) 1-methoxy-4-nitrobenzene
(d) ethyl *n*-propyl ether (two ways)
(e) benzyl *t*-butyl ether (benzyl = Ph—CH_2—)

14-6
Synthesis of Ethers by Alkoxymercuration–Demercuration

The **alkoxymercuration–demercuration** process adds a molecule of an alcohol across the double bond of an alkene (Section 8-6). The product is an ether, as shown below.

☞ $\underset{}{C}=\underset{}{C}$ $\xrightarrow[\text{ROH}]{\text{Hg(OAc)}_2}$ $-\underset{\underset{\text{AcOHg}}{|}}{C}-\underset{\underset{:\ddot{O}-R}{|}}{C}-$ $\xrightarrow{\text{NaBH}_4}$ $-\underset{\underset{H}{|}}{C}-\underset{\underset{OR}{|}}{C}-$

mercurial ether

Example

$$CH_3(CH_2)_3-CH=CH_2 \xrightarrow[\text{(2) NaBH}_4]{\text{(1) Hg(OAc)}_2, \text{CH}_3\text{OH}} CH_3(CH_2)_3-\underset{\underset{OCH_3}{|}}{CH}-CH_3$$

1-hexene

2-methoxyhexane, 80%
(Markovnikov product)

PROBLEM-SOLVING $\mathbb{HINT}$

Alkoxymercuration adds the —OR group of the alcohol to the more highly substituted carbon atom of the C=C double bond.

PROBLEM 14-10

Show how the following ethers might be synthesized using (1) alkoxymercuration–demercuration and (2) the Williamson synthesis. (When one of these methods cannot be used for the given ether, point out why it will not work.)

(a) 2-methoxybutane   (b) ethyl cyclohexyl ether

(c) 2-methyl-1-methoxycyclopentane   (d) 1-methyl-1-methoxycyclopentane

(e) 1-methyl-1-isopropoxycyclopentane   (f) *t*-butyl phenyl ether

14-7
Industrial Synthesis: Bimolecular Dehydration of Alcohols

The least expensive method for synthesizing simple symmetrical ethers is the acid-catalyzed bimolecular dehydration, discussed in Section 11-10B. Unimolecular dehydration (to give an alkene) competes with bimolecular dehydration. To form an ether, the alcohol must have an unhindered primary alkyl group, and the temperature must be kept low. If the alcohol is hindered or the temperature is too high, the delicate balance between substitution and elimination shifts in favor of elimination, and very little ether is formed. Bimolecular dehydration is used in industry to make symmetrical ethers from primary alcohols. Because the dehydration is so limited in its scope, it finds little use in the laboratory synthesis of ethers.

Bimolecular dehydration

☞ $2\,R-OH \underset{}{\overset{H^+}{\rightleftharpoons}} R-O-R + H_2O$

Examples

$$2\,CH_3OH \xrightarrow{\text{H}_2\text{SO}_4, 140°C} CH_3-O-CH_3 + H_2O$$

methyl alcohol   dimethyl ether
(100%)

$$2\,CH_3CH_2OH \xrightarrow{\text{H}_2\text{SO}_4, 140°C} CH_3CH_2-O-CH_2CH_3 + H_2O$$

ethyl alcohol   diethyl ether
(88%)

$$2\ CH_3CH_2CH_2OH \xrightarrow{H_2SO_4,\ 140°C} CH_3CH_2CH_2\!-\!O\!-\!CH_2CH_2CH_3\ +\ H_2O$$

n-propyl alcohol   *n*-propyl ether
                                                         (75%)

$$CH_3\!-\!\underset{\underset{OH}{|}}{CH}\!-\!CH_3 \xrightarrow{H_2SO_4,\ 140°C} H_2C\!=\!CH\!-\!CH_3\ +\ H_2O$$

isopropyl alcohol   unimolecular dehydration
                                              (no ether is formed)

If the conditions are carefully controlled, bimolecular dehydration is a cheap synthesis of diethyl ether. In fact, this is the industrial method used to produce millions of gallons of diethyl ether each year.

PROBLEM 14-11

Explain why bimolecular dehydration is a poor method for making unsymmetrical ethers such as ethyl methyl ether.

PROBLEM-SOLVING $\mathbf{HINT}$

Bimolecular dehydration of alcohols is generally a poor synthetic method.

PROBLEM 14-12

Propose a mechanism for the acid-catalyzed dehydration of *n*-propyl alcohol to *n*-propyl ether, as shown above. When the temperature is allowed to rise too high, propene is formed. Propose a mechanism for the formation of propene, and explain why it is favored at higher temperatures.

PROBLEM 14-13

Which of the following ethers can be formed in good yield by dehydrating the corresponding alcohols? For those that cannot be formed by dehydration, suggest an alternative method that will work.
(a) dibutyl ether   (b) ethyl *n*-propyl ether   (c) di-*sec*-butyl ether

---

SUMMARY   Syntheses of Ethers

1. *The Williamson ether synthesis* (Sections 11-14 and 14-5)

$$R\!-\!\ddot{O}\!:^- +\ R'\!-\!X \longrightarrow R\!-\!\ddot{O}\!-\!R' +\ X^-$$
X = Cl, Br, I, OTs, etc.   R′ must be primary

2. *Addition of an alcohol across a double bond: alkoxymercuration–demercuration* (Sections 8-6 and 14-6)

$$\underset{}{>}\!C\!=\!C\!\underset{}{<}\ +\ Hg(OAc)_2 \xrightarrow{R-OH} \underset{AcOHg\quad OR}{-\overset{|}{C}\!-\!\overset{|}{C}-} \xrightarrow{NaBH_4} \underset{H\quad OR}{-\overset{|}{C}\!-\!\overset{|}{C}-}$$

Markovnikov orientation

3. *Bimolecular dehydration of alcohols: industrial synthesis* (Section 11-10B and 14-7)

$$2\,R\!-\!OH \underset{}{\overset{H^+}{\rightleftharpoons}} R\!-\!O\!-\!R + H_2O$$
R must be primary

Unlike alcohols, ethers are not commonly used as synthetic intermediates because they do not undergo many reactions. This unreactivity makes ethers attractive as solvents. Even so, ethers do undergo a limited number of characteristic reactions.

Ethers are cleaved by heating with HBr or HI to give alkyl bromides or alkyl iodides.

14-8
Cleavage of Ethers by HBr and HI

$$\text{☞} \quad R\!-\!O\!-\!R' \xrightarrow[\text{(X = Br or I)}]{\text{excess HX}} R\!-\!X + R'\!-\!X$$

Ethers are unreactive toward most bases, but they can react under acidic conditions. A protonated ether can undergo substitution or elimination with the expulsion of an alcohol. Ethers react with concentrated HBr and HI because these reagents are sufficiently acidic to protonate the ether, while bromide and iodide are good nucleophiles for the substitution.

$$R\!-\!\overset{..}{\underset{..}{O}}\!-\!R' + H^+{}^-X \rightleftharpoons R\!-\!\overset{H}{\underset{}{O^+}}\!-\!R' \longrightarrow X\!-\!R + :\!\overset{H}{\underset{}{O}}\!-\!R' \xrightarrow{HX} X\!-\!R + X\!-\!R'$$

ether   (X = Br or I)   protonated ether   alkyl halide   alcohol

In effect, this reaction converts a dialkyl ether into two alkyl halides. The conditions are very strong, however, and the molecule must not contain any acid-sensitive functional groups.

Iodide and bromide ions are good nucleophiles but weak bases, so they are more likely to substitute by the S_N2 mechanism than to eliminate by the E2 mechanism. The reaction of diethyl ether with HBr is an example of this displacement. The ethanol produced by the cleavage reacts with HBr (see Section 11-7), and the final products are two molecules of ethyl bromide.

MECHANISM   Cleavage of an Ether by HBr

Protonation and cleavage of the ether

$$CH_3CH_2\!-\!\overset{..}{\underset{..}{O}}\!-\!CH_2CH_3 + H\!-\!Br \rightleftharpoons CH_3\!-\!\underset{\underset{H-O^+-CH_2CH_3}{|}}{\overset{H}{\underset{|}{C}}}\!-\!H \longrightarrow CH_3\!-\!\underset{\underset{H}{|}}{\overset{Br}{\underset{|}{C}}}\!-\!H + H\!-\!\overset{..}{\underset{..}{O}}\!-\!CH_2CH_3$$

diethyl ether   protonated ether   ethyl bromide   ethanol

Conversion of ethanol to ethyl bromide

$$H\!-\!\overset{..}{\underset{..}{O}}\!-\!CH_2CH_3 + H\!-\!Br \rightleftharpoons H\!-\!\underset{\underset{H}{|}}{\overset{..}{O^+}}\!-\!CH_2\!-\!CH_3 \longrightarrow H_2O + Br\!-\!CH_2CH_3$$

ethanol   ethyl bromide

Overall reaction

$$CH_3CH_2\!-\!O\!-\!CH_2CH_3 \xrightarrow{\text{excess HBr/H}_2\text{O}} 2\,CH_3CH_2\!-\!Br$$

diethyl ether   ethyl bromide

Hydroiodic acid (HI) reacts with ethers the same way HBr does. Aqueous iodide is a stronger nucleophile than aqueous bromide, and iodide reacts at a faster rate. We can rank the hydrohalic acids in order of their reactivity toward the cleavage of ethers:

$$HI > HBr \gg HCl$$

PROBLEM 14-14

Propose a mechanism for the following reaction.

$$\text{tetrahydrofuran} \xrightarrow{\text{excess HBr}} \text{1,4-dibromobutane}$$

Phenyl Ethers Phenyl ethers (one of the groups bonded to oxygen is a benzene ring) react with HBr or HI to give alkyl halides and phenols. Phenols do not react further to give halides because the sp^2 hybridized carbon atom of the phenol cannot undergo the S_N2 (or S_N1) reaction needed for conversion to the halide.

ethyl phenyl ether   protonated ether   phenol   ethyl bromide
(no further reaction)

PROBLEM 14-15

Predict the products of the following reactions. An excess of acid is available in each case.
(a) ethoxycyclohexane + HBr   **(b)** tetrahydropyran + HI

(c) anisole (methoxybenzene) + HBr   **(d)** + HI

(e) Ph—O—CH₂CH₂—CH—CH₂—O—CH₂CH₃ + HBr
                          |
                          CH₃

PROBLEM 14-16

Boron tribromide (BBr_3) cleaves ethers to give alkyl halides and alcohols.

$$R-O-R' + BBr_3 \longrightarrow R-O-BBr_2 + R'Br$$

$$R-O-BBr_2 + 3 H_2O \longrightarrow ROH + B(OH)_3 + 2 HBr$$

The reaction is thought to involve attack by a bromide ion on the Lewis acid–base adduct of the ether with BBr_3 (a strong Lewis acid). Propose a mechanism for the reaction of butyl methyl ether with BBr_3 to give (after hydrolysis) butanol and bromomethane.

When ethers are stored in the presence of atmospheric oxygen, they slowly oxidize to produce hydroperoxides and dialkyl peroxides, both of which are explosive. Such a spontaneous oxidation by atmospheric oxygen is called an **autoxidation**.

14-9 Autoxidation of Ethers

ether   hydroperoxide   dialkyl peroxide

Example

diisopropyl ether   hydroperoxide   diisopropyl peroxide

Organic chemists often buy large containers of ethers and use small quantities over several months. Once a container has been opened, it contains atmospheric oxygen, and the autoxidation process begins. After several months, a large amount of peroxide may be present. Distillation or evaporation concentrates the peroxides, and an explosion may occur.

Such an explosion may be avoided by taking a few simple precautions. Ethers should be bought in small quantities, kept in tightly sealed containers, and used promptly. Any procedure requiring evaporation or distillation should use only peroxide-free ether. Any ether that might be contaminated with peroxides should be discarded or treated to destroy the peroxides.

SUMMARY   Reactions of Ethers

1. *Cleavage by HBr and HI* (Section 14-8)

$$R{-}O{-}R' \xrightarrow[\text{(X = Br, I)}]{\text{excess HX}} R{-}X + R'{-}X$$

$$Ar{-}O{-}R \xrightarrow[\text{(X = Br, I)}]{\text{excess HX}} Ar{-}OH + R{-}X$$

Ar = aromatic ring

Example

$$CH_3CH_2{-}O{-}CH_3 \xrightarrow{\text{excess HI}} CH_3CH_2I + CH_3I$$
ethyl methyl ether   ethyl iodide   methyl iodide

2. *Autoxidation* (Section 14-9)

$$R{-}O{-}CH_2{-}R' \xrightarrow[\text{(slow)}]{\text{excess } O_2} R{-}O{-}\overset{\overset{\displaystyle OOH}{|}}{C}H{-}R' + R{-}O{-}O{-}CH_2{-}R'$$
ether   hydroperoxide   dialkyl peroxide

14-10
Sulfides (Thioethers)

Sulfides are also called **thioethers** because they are the sulfur analogues of ethers. Like thiols, sulfides have strong characteristic odors: The odor of dimethyl sulfide is reminiscent of oysters that have been kept in the refrigerator for too long. Sulfides are named like ethers, with "sulfide" replacing "ether" in the common names. In the IUPAC (alkoxy alkane) names, "alkylthio" replaces "alkoxy."

$$CH_3{-}S{-}CH_3$$
dimethyl sulfide

methyl phenyl sulfide

4-ethylthio-2-methyl-2-pentene

Sulfides are easily synthesized by the Williamson ether synthesis, using a thiolate ion as the nucleophile.

$$CH_3CH_2{-}S^- + CH_3CH_2CH_2{-}Br \longrightarrow CH_3CH_2CH_2{-}S{-}CH_2CH_3 + Br^-$$
ethanethiolate   1-bromopropane   ethyl propyl sulfide

Thiols are more acidic than water. Therefore, thiolate ions are easily generated by treating thiols with aqueous sodium hydroxide.

$$CH_3CH_2{-}SH + Na^+\ ^-OH \rightleftharpoons CH_3CH_2{-}S^-\ ^+Na + H_2O$$
pK_a = 10.5   sodium ethanethiolate   pK_a = 15.7

Because sulfur is larger and more polarizable than oxygen, thiolate ions are even better nucleophiles than alkoxide ions. Thiolates are such effective nucleophiles that secondary alkyl halides often react to give good yields of S_N2 products.

(R)-2-bromobutane $\xrightarrow[\text{CH}_3\text{OH}]{\text{CH}_3\text{S}^-}$ (S)-2-(methylthio) butane

PROBLEM 14-17

Show how you would synthesize butyl isopropyl sulfide using 1-butanol, 2-propanol, and any solvents and reagents you need.

Sulfides are much more reactive than ethers. In a sulfide, sulfur's valence is not necessarily filled: Sulfur can form additional bonds with other atoms. Sulfur forms particularly strong bonds with oxygen, and sulfides are easily oxidized to sulfoxides and sulfones. Sulfoxides and sulfones are drawn using either hypervalent double-bonded structures or formally charged single-bonded structures as shown below.

The hydrogen peroxide/acetic acid combination is a good oxidant for sulfides. One equivalent of peroxide gives the sulfoxide, and a second equivalent further oxidizes the sulfoxide to the sulfone. This reagent combination probably reacts via the peroxyacid, which is formed in equilibrium with hydrogen peroxide.

Because they are easily oxidized, sulfides are often used as mild reducing agents. For example, we have used dimethyl sulfide to reduce the potentially explosive ozonides that result from ozonolysis of alkenes (Section 8-15).

Sulfur compounds are more nucleophilic than the corresponding oxygen compounds, because sulfur is larger and more polarizable and its electrons are less tightly held in orbitals that are farther from the nucleus. Although ethers are weak nucleophiles, sulfides are relatively strong nucleophiles. Sulfides attack unhindered alkyl halides to give **sulfonium salts**.

$$R—\ddot{S}—R' \qquad R'—X \xrightarrow{S_N2} R—\overset{R'}{\underset{+}{S}}—R \qquad X^-$$

sulfide   alkyl halide   sulfonium salt

Example

$$CH_3—\ddot{S}—CH_3 + CH_3I \longrightarrow CH_3—\overset{CH_3}{\underset{+}{S}}—CH_3 \qquad I^-$$

dimethyl sulfide   trimethylsulfonium iodide

Sulfonium salts are good alkylating agents because the leaving group is an uncharged sulfide. Sulfur's polarizability enhances partial bonding in the transition state, lowering its energy.

$$Nuc:^- \qquad CH_3—\overset{+}{\underset{R}{S}}{\overset{R}{:}} \qquad X^- \longrightarrow Nuc—CH_3 \quad + \quad R—\ddot{S}—R \quad + \quad X^-$$

nucleophile   sulfonium salt   sulfide

Example

$$\text{pyridine} \quad N: \quad + \quad CH_3—\overset{CH_3}{\underset{\ddot{S}}{|}}—CH_3 \quad I^- \longrightarrow \quad N^+—CH_3 \quad I^- \quad + \quad CH_3—\ddot{S}—CH_3$$

pyridine   trimethylsulfonium iodide   *N*-methylpyridinium iodide   dimethyl sulfide

Sulfonium salts are common alkylating agents in biological systems. For example, ATP activation of methionine forms the sulfonium salt *S*-adenosylmethionine (SAM), a biological methylating agent.

adenosine triphosphate (ATP)   *S*-adenosylmethionine (SAM)

SAM converts norepinephrine to epinephrine (adrenaline) in the adrenal glands.

norepinephrine

SAM

epinephrine

PROBLEM 14-18

Mustard gas, $Cl-CH_2CH_2-S-CH_2CH_2-Cl$, was used as a poisonous chemical agent in World War I. Mustard gas is much more toxic than a typical primary alkyl chloride. Its toxicity stems from its ability to alkylate amino groups on important metabolic enzymes, rendering the enzymes inactive.
(a) Propose a mechanism to explain why mustard gas is an exceptionally potent alkylating agent.
(b) Bleach (sodium hypochlorite, NaOCl, a strong oxidizing agent) neutralizes and inactivates mustard gas. Bleach is also effective on organic stains because it oxidizes colored compounds to colorless compounds. Propose products that might be formed by the reaction of mustard gas with bleach.

Epoxides are easily made from alkenes, and (unlike other ethers) they undergo a variety of useful synthetic reactions. For these reasons, epoxides are valuable synthetic intermediates. Here we review the epoxidation techniques already covered (Section 8-12) and consider in more detail the useful syntheses and reactions of epoxides.

14-11
Synthesis
of Epoxides

14-11A Peroxyacid Epoxidation

Peroxyacids (sometimes called *peracids*) are used to convert alkenes to epoxides. If the reaction takes place in aqueous acid, the epoxide opens to a glycol. Therefore, to make an epoxide, we use a weakly acidic peroxyacid that is soluble in aprotic solvents such as CH_2Cl_2. Because of its desirable solubility properties, *meta*-chloroperoxybenzoic acid (**MCPBA**) is often used for these epoxidations.

alkene   peroxyacid   epoxide   acid

Example

cyclohexene   epoxycyclohexane
(100%)

MCPBA =   *meta*-chloroperoxybenzoic acid

The epoxidation takes place in a one-step, **concerted reaction** that maintains the stereochemistry of any substituents on the double bond.

The peroxyacid epoxidation is quite general, with electron-rich double bonds reacting fastest. The following reactions are difficult transformations made possible by this selective, stereospecific epoxidation procedure. The second example uses magnesium monoperoxyphthalate (**MMPP**), a relatively stable peroxyacid often used in large-scale epoxidations.

1,2-dimethyl-1,4-cyclohexadiene   *cis*-4,5-epoxy-4,5-dimethylcyclohexene

(*E*)-2-nitro-1-phenylpropene   (*E*)-2-methyl-2-nitro-3-phenyloxirane

14-11B Base-Promoted Cyclization of Halohydrins

A second synthesis of epoxides and other cyclic ethers involves a variation of the Williamson ether synthesis. If an alkoxide ion and a halogen atom are located in the same molecule, the alkoxide may displace a halide ion and form a ring. Treatment of a **halohydrin** with a base leads to an epoxide through this internal S_N2 attack.

(X = Cl, Br, I)

Halohydrins are easily generated by treating alkenes with aqueous solutions of halogens. Bromine water and chlorine water add across double bonds with Markovnikov orientation (Section 8-11). The following reaction shows the reaction of cyclopentene with chlorine water to give the chlorohydrin. Treatment of the chlorohydrin with aqueous sodium hydroxide gives the epoxide.

Formation of the chlorohydrin

cyclopentene   chlorine water   chloronium ion   *trans*-chlorohydrin (mixture of enantiomers)

Displacement of the chlorohydrin

trans-chlorohydrin   alkoxide   epoxide
(50% overall)

This reaction can be used to synthesize cyclic ethers with larger rings. The difficulty lies in preventing the base (added to deprotonate the alcohol) from attacking and displacing the halide. 2,6-Lutidine, a bulky base that cannot easily attack a carbon atom, can be used to deprotonate the hydroxyl group to give a five-membered cyclic ether. Five-, six-, and seven-membered (and occasionally four-membered) cyclic ethers are formed this way.

chloro-alcohol   2,6-lutidine   alkoxide   2-methyltetrahydrofuran
(2,6-dimethylpyridine)   (85%)

PROBLEM 14-19

Show how you would accomplish the following transformations. Some of these examples require more than one step.
(a) 2-methylpropene → 2,2-dimethyloxirane
(b) 1-phenylethanol → 2-phenyloxirane
(c) 5-chloro-1-pentene → tetrahydropyran
(d) 5-chloro-1-pentene → 2-methyltetrahydrofuran
(e) 2-chloro-1-hexanol → 1,2-epoxyhexane

PROBLEM 14-20

The 2001 Nobel Prize in chemistry was awarded to three organic chemists who have developed methods for catalytic asymmetric syntheses. An asymmetric (or enantioselective) synthesis is one that converts an achiral starting material into mostly one enantiomer of a chiral product. K. Barry Sharpless (The Scripps Research Instutute) developed an asymmetric epoxidation of allylic alcohols that gives excellent chemical yields and greater than 90% enantiomeric excess.

The Sharpless epoxidation uses *tert*-butyl hydroperoxide, titanium (IV) isopropoxide, and a dialkyl tartrate ester as the reagents. The following epoxidation of geraniol is typical.

geraniol   80% yield, > 90% ee

Reagents:

tert-butyl hydroperoxide   titanium (IV) isopropoxide   diethyl L-tartrate

(a) Which of these reagents is most likely to be the actual oxidizing agent? That is, which reagent is reduced in the reaction? What is the likely function of the other reagents?

(b) When achiral reagents react to give a chiral product, that product is normally formed as a racemic mixture of enantiomers. How can the Sharpless epoxidation give just one nearly pure enantiomer of the product?

(c) Draw the other enantiomer of the product. What reagents would you use if you wanted to epoxidize geraniol to give this other enantiomer?

SUMMARY Epoxide Syntheses

1. *Peroxyacid epoxidation* (Section 14-11A)

2. *Base-promoted cyclization of halohydrins* (Section 14-11B)

$$X = Cl, Br, I, OTs, etc.$$

Example

2-chloro-1-phenylethanol   2-phenyloxirane

14-12
Acid-Catalyzed Ring Opening of Epoxides

Epoxides are much more reactive than common dialkyl ethers because of the large strain energy (about 25 kcal/mol, or 105 kJ/mol) associated with the three-membered ring. Unlike other ethers, epoxides react under both acidic and basic conditions. The products of acid-catalyzed opening depend primarily on the solvent used.

In Water In Section 8-13 we saw that acid-catalyzed hydrolysis of epoxides gives glycols with anti stereochemistry. The mechanism of this hydrolysis involves protonation of oxygen (forming a good leaving group), followed by S_N2 attack by water. Anti stereochemistry results from the back-side attack of water on the protonated epoxide.

MECHANISM Acid-Catalyzed Opening of an Epoxide

Acid protonates the epoxide.   Water attacks and opens the ring.

1,2-epoxycyclopentane

trans-cyclopentane-1,2-diol
(mixture of enantiomers)

Direct anti hydroxylation of an alkene (without isolation of the epoxide intermediate) is possible by using an acidic aqueous solution of a peroxyacid. As soon as the epoxide is formed, it hydrolyzes to the glycol. Peroxyacetic acid (CH_3CO_3H) and peroxyformic acid (HCO_3H) are often used for the anti hydroxylation of alkenes.

trans-2-butene   →   *meso*-butane-1,2-diol

PROBLEM 14-21

Propose mechanisms for the epoxidation and ring-opening steps of the epoxidation and hydrolysis of *trans*-2-butene shown above. Predict the product of the same reaction with *cis*-2-butene.

In Alcohols   When the acid-catalyzed opening of an epoxide takes place with an alcohol as the solvent, a molecule of alcohol acts as the nucleophile. This reaction produces an alkoxy alcohol with anti stereochemistry. This is an excellent method for making compounds with ether and alcohol functional groups on adjacent carbon atoms. For example, the acid-catalyzed opening of 1,2-epoxycyclopentane in a methanol solution gives *trans*-2-methoxycyclopentanol.

MECHANISM   Acid-Catalyzed Opening of an Epoxide in an Alcohol Solution

Acid protonates the epoxide.   Alcohol attacks the protonated epoxide.

1,2-epoxycyclopentane   *trans*-2-methoxycyclopentanol (82%)
(mixture of enantiomers)

PROBLEM 14-22

Cellosolve® is the trade name for 2-ethoxyethanol, a common industrial solvent. This compound is produced in chemical plants that use ethylene as their only organic feedstock. Show how you would accomplish this industrial process.

Using Hydrohalic Acids   When an epoxide reacts with a hydrohalic acid (HCl, HBr, or HI), a halide ion attacks the protonated epoxide. This reaction is analogous to the cleavage of ethers by HBr or HI. The halohydrin initially formed reacts further with HX to give a 1,2-dihalide. This is rarely a useful synthetic reaction, because the 1,2-dihalide can be made directly from the alkene by electrophilic addition of X_2.

(several steps)

▲ **Figure 14-6**
The biosynthesis of steroids starts with epoxidation of squalene to squalene-2,3-epoxide. The opening of this epoxide promotes cyclization of the carbon skeleton under the control of an enzyme. The cyclized intermediate is converted to lanosterol, then to other steroids.

PROBLEM 14-23

When ethylene oxide is treated with anhydrous HBr gas, the major product is 1,2-dibromo-ethane. When ethylene oxide is treated with concentrated aqueous HBr, the major product is ethylene glycol. Use mechanisms to explain these results.

Inhibitors of squalene epoxidase are used in anti-fungal drugs to treat athlete's foot, jock itch, ring worm, and nail infections. The drug Tinactin™ (tolnaftate) inhibits squalene epoxidase, which blocks the synthesis of the steroids the fungus needs to make its cell membrane. The defective cell membrane kills the fungus.

The Opening of Squalene-2,3-Epoxide   *Steroids* are tetracyclic compounds that serve a wide variety of biological functions, including hormones (sex hormones), emulsifiers (bile acids), and membrane components (cholesterol). The biosynthesis of steroids is believed to involve an acid-catalyzed opening of squalene-2,3-epoxide (Figure 14-6). Squalene is a member of the class of natural products called *terpenes* (see Section 25-8). The enzyme squalene epoxidase oxidizes squalene to the epoxide, which opens and forms a carbocation that cyclizes under the control of another enzyme. The cyclized intermediate rearranges to lanosterol, which is converted to cholesterol and other steroids.

Although cyclization of squalene-2,3-epoxide is controlled by an enzyme, its mechanism is similar to the acid-catalyzed opening of other epoxides. The epoxide oxygen becomes protonated and is attacked by a nucleophile. In this case, the nucleophile is a pi bond. The initial result is a tertiary carbocation (Figure 14-7)

This initial carbocation is attacked by another double bond, leading to the formation of another ring and another tertiary carbocation. A repetition of this process leads to the cyclized intermediate shown in Figure 14-6.

PROBLEM 14-24

Show the rest of the mechanism for formation of the cyclized intermediate in Figure 14-6.

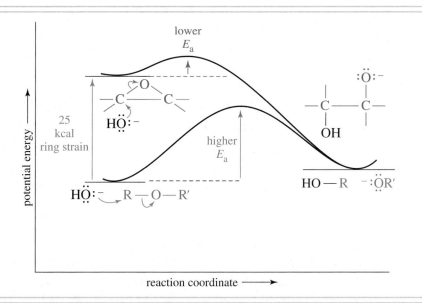

protonated squalene-2,3-epoxide   tertiary carbocation   tertiary carbocation

▲ **Figure 14-7**
Cyclization of squalene epoxide begins with the acid-catalyzed opening of the epoxide. Each additional cyclization step forms another carbocation.

Most ethers do not undergo nucleophilic substitutions or eliminations under basic conditions, because an alkoxide ion is a poor leaving group. Epoxides have about 25 kcal/mol (105 kJ/mol) of ring strain that is released upon ring opening, however, and this strain is enough to compensate for the poor alkoxide leaving group. Figure 14-8 compares the energy profiles for nucleophilic attack on an ether and on an epoxide. The starting epoxide is about 25 kcal/mol (105 kJ/mol) higher in energy than the ether, and its displacement has a lower activation energy.

The reaction of an epoxide with hydroxide ion leads to the same product as the acid-catalyzed opening of the epoxide: a 1,2-diol (glycol), with anti stereochemistry. In fact, either the acid-catalyzed or base-catalyzed reaction may be used to open an epoxide, but the acid-catalyzed reaction takes place under milder conditions. Unless there is an acid-sensitive functional group present, the acid-catalyzed hydrolysis is preferred.

14-13
Base-Catalyzed Ring Opening of Epoxides

◀ **Figure 14-8**
An epoxide is higher in energy than an acyclic ether by about 25 kcal/mol (105 kJ/mol) ring strain. The ring strain is released in the product, giving it an energy similar to the products from the acyclic ether. Release of the ring strain makes the displacement of an epoxide thermodynamically favorable.

MECHANISM   Base-Catalyzed Opening of an Epoxide

Base attacks and opens the ring.   Protonation of the alkoxide.

cyclopentene oxide

trans-cyclopentane-1,2-diol
(mixture of enantiomers)

Like hydroxide, alkoxide ions react with epoxides to form ring-opened products. For example, cyclopentene oxide reacts with sodium methoxide in methanol to give the same *trans*-2-methoxycyclopentanol produced in the acid-catalyzed opening in methanol.

cyclopentene oxide

trans-2-methoxycyclopentanol
(mixture of enantiomers)

Amines can also open epoxides. Ethylene oxide reacts with aqueous ammonia to give the important industrial reagent ethanolamine. The nitrogen atom in ethanolamine is still nucleophilic, and ethanolamine can react further to give diethanolamine and triethanolamine. Good yields of ethanolamine are achieved by using excess ammonia.

H_2C—CH_2 + $:NH_3$ $\xrightarrow{H_2O}$ CH_2—CH_2 $\longrightarrow$ $HOCH_2CH_2\ddot{N}H_2$

ethylene oxide   ammonia   ethanolamine

$HOCH_2CH_2\ddot{N}H_2$ $\longrightarrow$ $(HOCH_2CH_2)_2\ddot{N}H$ $\longrightarrow$ $(HOCH_2CH_2)_3N:$

ethanolamine   diethanolamine   triethanolamine

PROBLEM 14-25

Propose a complete mechanism for the reaction of cyclopentene oxide with sodium methoxide in methanol.

PROBLEM 14-26

Predict the major product when each reagent reacts with ethylene oxide.
(a) sodium ethoxide   (b) NaNH$_2$ (c) Ph—ONa   (d) Ph—NH$_2$
                     sodium amide   sodium phenoxide   aniline

Symmetrically substituted epoxides (such as cyclopentene oxide, on p. 626) give the same product in both the acid-catalyzed and base-catalyzed ring openings. An unsymmetrical epoxide gives different products under acid-catalyzed and base-catalyzed conditions, however.

14-14

14-14
Orientation of Epoxide Ring Opening

2,2-dimethyloxirane

H^+, CH_3CH_2—OH →

H₃C   OH
H₃C—C—CH₂
CH_3CH_2—O
2-ethoxy-2-methyl-1-propanol
acid-catalyzed product

CH_3CH_2—Ö:⁻, EtOH →

OH
H₃C—C—CH₂
H₃C   O—CH₂CH₃
1-ethoxy-2-methyl-2-propanol
base-catalyzed product

Under basic conditions, the alkoxide ion simply attacks the less hindered carbon atom in an S_N2 displacement.

Under acidic conditions, the alcohol attacks the protonated epoxide. It might seem that the alcohol would attack at the less hindered oxirane carbon, but this is not the case. In the protonated epoxide, there is a balancing act between ring strain and the energy it costs to put some of the positive charge on the carbon atoms. We can represent this sharing of positive charge by drawing the following three resonance structures:

structure I   structure II   structure III

Structure I is the conventional structure for the protonated epoxide, while structures II and III show that the oxirane carbon atoms share part of the positive charge. The tertiary carbon bears a larger part of the positive charge, and it is more strongly electrophilic; that is, structure II is more important than structure III. The bond between the tertiary carbon and oxygen is weaker, implying a lower transition state energy for attack at the tertiary carbon. Attack by the weak nucleophile (ethanol in this case) is

PROBLEM-SOLVING HINT

In proposing mechanisms for acid-catalyzed opening of epoxides, imagine that the protonated epoxide opens to the more stable (more highly substituted) carbocation.

sensitive to the strength of the electrophile, and it occurs at the more electrophilic tertiary carbon.

This ring opening is similar to the opening of a bromonium ion in the formation of a bromohydrin (Section 8-11) and the opening of the mercurinium ion during oxymercuration (Section 8-5). All three reactions involve the opening of an electrophilic three-membered ring by a weak nucleophile. Attack takes place at the more electrophilic carbon atom, which is usually the more highly substituted carbon because it can better support the positive charge. Most base-catalyzed epoxide openings, on the other hand, involve attack by a strong nucleophile at the less hindered carbon atom.

SOLVED PROBLEM 14-2

Predict the major products for the reaction of 1-methyl-1,2-epoxycyclopentane with
(a) sodium ethoxide in ethanol
(b) H_2SO_4 in ethanol

SOLUTION

(a) Sodium ethoxide attacks the less hindered secondary carbon to give (E)-2-ethoxy-1-methylcyclopentanol.

(b) Under acidic conditions, the alcohol attacks the more electrophilic tertiary carbon atom of the protonated epoxide. The product is (E)-2-ethoxy-2-methylcyclopentanol.

PROBLEM 14-27

Predict the major products of the following reactions, including stereochemistry where appropriate.
(a) 2,2-dimethyloxirane + $H^+/H_2^{18}O$ (oxygen-labeled water)
(b) 2,2-dimethyloxirane + $H^{18}O^-/H_2^{18}O$
(c) (Z)-2-ethyl-2,3-dimethyloxirane + CH_3O^-/CH_3OH
(d) (Z)-2-ethyl-2,3-dimethyloxirane + H^+/CH_3OH

Like other strong nucleophiles, Grignard and organolithium reagents attack epoxides to give (after hydrolysis) ring-opened alcohols.

☞ $R—MgX$ + $CH_2—CH—R'$ $\xrightarrow[\text{(2) } H_3O^+]{\text{(1) ether}}$ $CH_2—CH—R'$
   or $R—Li$ R

For example, ethylmagnesium bromide reacts with oxirane (ethylene oxide) to form the magnesium salt of 1-butanol.

$H_2C—CH_2$ $\longrightarrow$ $H_2C—CH_2$ $\xrightarrow{H_3O^+}$ $H_2C—CH_2$
$CH_3CH_2—MgBr$ CH_3CH_2 CH_3CH_2
                                                   1-butanol

   Substituted epoxides can be used in this reaction, with the carbanion attacking the less hindered epoxide carbon atom. This reaction works best if one of the oxirane carbons is unsubstituted, to allow an unhindered nucleophilic attack. Organolithium reagents (RLi) are more selective than Grignard reagents in attacking the less hindered epoxide carbon atom. Unless one carbon atom is very strongly hindered, Grignard reagents may give mixtures of products.

$H_2C—CH—CH_3$ + cyclohexyllithium (Li) $\xrightarrow[\text{ether}]{}$ $\xrightarrow{H_3O^+}$ $CH_2—CH—CH_3$ (OH)

methyloxirane   cyclohexyllithium   1-cyclohexyl-2-propanol

2-cyclohexyl-2-ethyloxirane + phenylmagnesium bromide (MgBr) $\xrightarrow[\text{ether}]{}$ $\xrightarrow{H_3O^+}$ $CH_2—C—CH_2CH_3$ (OH)

2-cyclohexyl-2-ethyloxirane   phenylmagnesium bromide   2-cyclohexyl-1-phenyl-2-butanol

PROBLEM 14-28

Give the expected products of the following reactions. Include a hydrolysis step where necessary.
(a) ethylene oxide + isopropylmagnesium bromide
(b) 2,2-dimethyloxirane + methyllithium
(c) cyclopentyloxirane + ethyllithium

The earliest glues were made of carbohydrates and proteins. Wheat paste uses the gluten in wheat, the sticky carbohydrate that holds bread together. Hide glue is a collagen-containing protein extract of animal hides, hooves, and tendons. Hide glue was used for wood and paper gluing for hundreds of years, and it is still used for fine musical instruments and other articles that must be readily taken apart without damaging the wood. Hide glue is water soluble, however, and the bond quickly fails in a damp environment.

14-16

Epoxy Resins: The Advent of Modern Glues

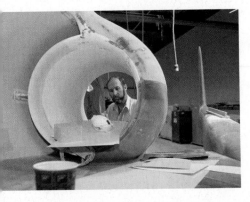

Epoxy resins bind graphite and polymer fibers into laminates used in the aerospace industry.

It does not fill gaps because it shrinks to a fraction of its wet volume as it dries. Glues based on casein (a milk protein) were developed to give a stronger, water-resistant bond. A casein glue (such as Elmer's®) gives a bond as strong as most woods, and it resists water for hours before it softens. But it does not fill gaps well, and it works poorly with metals and plastics.

Imagine a glue that does not shrink at all as it hardens; it fills gaps perfectly so that pieces don't need to be fitted closely. It holds forever in water, is at least as strong as wood and plastic, and sticks to anything: wood, metal, plastic, etc. It lasts forever on the shelf without hardening, yet hardens quickly once the pieces are in place. It can be made runny so it fills tiny voids, or thick and pasty so it stays in place while it hardens.

This ideal glue was only a dream until the development of epoxy adhesives. Epoxies polymerize in place, so they match the shape of the joint perfectly and adhere to microscopic irregularities in the surfaces. There is no solvent to evaporate, so there is no shrinkage. Epoxies are bonded by ether linkages, so they are unaffected by water. Epoxies use a prepolymer that can be made as runny or as gummy as desired, and they use a hardening agent that can be modified to control the curing time. In the absence of the hardening agent, they have a long shelf life.

The most common epoxy resins use a prepolymer made from *bisphenol A* and epichlorohydrin.

Under base-catalyzed conditions, the anion of bisphenol A opens the epoxide of epichlorohydrin to give an alkoxide that snaps shut on the other end, forming another epoxide.

This second epoxide reacts with another molecule of bisphenol A. Each molecule of bisphenol A can also react with two molecules of epichlorohydrin.

With exactly equal amounts of bisphenol A and epichlorohydrin, this polymerization would continue until the polymer chains were very long and the material would be a solid polymer. In making epoxy resins, however, excess epichlorohydrin is added to form short chains with epichlorohydrins on both ends. More epichlorohydrin gives shorter chains and a runny prepolymer. Less epichlorohydrin gives longer chains (containing up to 25 epichlorohydrin/bisphenol A units) and a more viscous prepolymer.

When you buy epoxy glues, they come in two parts: the resin (prepolymer) and the hardener. The hardener can be any of a wide variety of compounds having basic or nucleophilic properties. Polyamines are the most common hardeners. The hardener can attack a terminal epoxide group, initiating a polymerization of the chain ends.

Or the hardener can deprotonate a hydroxyl group from the interior of a chain, cross-linking one chain with another. The final polymer is an intricate three-dimensional network that is strong and resistant to chemical attack.

middle of chain 1

SUMMARY   Reactions of Epoxides

1. *Acid-catalyzed opening* (Sections 8-13 and 14-12)
  a. *In water*

anti stereochemistry

  b. *In alcohols*

The alkoxy group bonds to the more highly substituted carbon.

Example

$$\underset{\substack{\text{methyl oxirane} \\ \text{(propylene oxide)}}}{\overset{\displaystyle \overset{H}{\underset{H_3C}{\diagdown}} \overset{\displaystyle C - CH_2}{\underset{O}{\diagup}}}{}} \quad \xrightarrow[\text{CH}_3\text{OH}]{\text{H}^+} \quad \underset{\substack{\text{2-methoxy-1-propanol}}}{CH_3 - \underset{\underset{OCH_3}{|}}{CH} - CH_2 - OH}$$

c. *Using hydrohalic acids* (X = Cl, Br, I)

$$\underset{O}{\overset{| \quad |}{-C - C-}} \quad \xrightarrow{H-X} \quad \underset{X}{\overset{OH}{-C - C-}} \quad \xrightarrow{H-X} \quad \underset{X}{\overset{X}{-C - C-}}$$

2. *Base-catalyzed opening*
a. *With alkoxides* (Section 14-13)

$$\underset{O}{\overset{|}{-C - CH_2}} \quad \xrightarrow[R-OH]{R - \ddot{\underset{\cdot\cdot}{O}}{:}^-} \quad \underset{OH}{\overset{|}{-C - CH_2 - OR}}$$

The alkoxy group bonds to the less highly substituted carbon.

Example

$$\underset{\substack{\text{propylene oxide}}}{\overset{\displaystyle \overset{H}{\underset{H_3C}{\diagdown}} \overset{\displaystyle C - CH_2}{\underset{O}{\diagup}}}{}} \quad \xrightarrow[\text{CH}_3\text{OH}]{CH_3\ddot{\underset{\cdot\cdot}{O}}{:}^-\ ^+Na} \quad \underset{\substack{\text{1-methoxy-2-propanol}}}{CH_3 - \underset{\underset{OH}{|}}{CH} - CH_2 - OCH_3}$$

b. *With organometallics* (Section 14-15)

$$\underset{O}{\overset{|}{-C - CH_2}} \quad \xrightarrow[\text{(2) H}_3\text{O}^+]{\text{(1) R}-M} \quad \underset{OH}{\overset{|}{-C - CH_2 - R}}$$

M = Li or MgX   R bonds to the less substituted carbon

Example

$$\underset{\substack{\text{propylene oxide}}}{\overset{\displaystyle \overset{H}{\underset{H_3C}{\diagdown}} \overset{\displaystyle C - CH_2}{\underset{O}{\diagup}}}{}} \quad \xrightarrow[\text{(2) H}_3\text{O}^+]{\text{(1)} \bigcirc - MgBr} \quad \underset{\substack{\text{1-cyclohexyl-2-propanol}}}{CH_3 - \underset{\underset{OH}{|}}{CH} - CH_2 - \bigcirc}$$

Chapter 14
Glossary

alkoxy group (alkoxyl group) A substituent consisting of an alkyl group bonded through an oxygen atom, —O—R. (p. 605)

alkoxymercuration Addition of mercury and an alkoxy group to a double bond, usually by a solution of mercuric acetate in an alcohol; usually followed by sodium borohydride reduction (demercuration) to give an ether. (p. 612)

$$\begin{array}{c} >\!\!C\!\!=\!\!C\!\!< \xrightarrow[\text{R}-\text{O}-\text{H}]{\text{Hg(OAc)}_2} \quad \begin{array}{c} \text{R}-\text{O} \quad \text{Hg(OAc)} \\ | \qquad | \\ -\text{C}-\text{C}- \\ | \qquad | \end{array} \xrightarrow{\text{NaBH}_4} \begin{array}{c} \text{R}-\text{O} \quad \text{H} \\ | \qquad | \\ -\text{C}-\text{C}- \\ | \qquad | \end{array} \end{array}$$

(alkoxymercuration)   (reduction)

α cleavage The breaking of a bond between the first and second carbon atoms adjacent to the ether oxygen atom (or other functional group). (p. 608)

autoxidation Any oxidation that proceeds spontaneously using the oxygen in the air. Autoxidation of ethers gives hydroperoxides and dialkyl peroxides. (p. 615)

concerted reaction A reaction that takes place in one step, with simultaneous bond breaking and bond forming. (p. 619)

crown ether A large cyclic polyether used to complex and solvate cations in nonpolar solvents. (p. 604)

dioxane A heterocyclic ether with two oxygen atoms in a six-membered ring. (p. 607)

epoxidation Oxidation of an alkene to an epoxide. Usually accomplished by treating the alkene with a peroxyacid. (p. 619)

epoxide (oxirane) A compound containing a three-membered heterocyclic ether. (p. 605)

epoxy resins Polymers formed by condensing epichlorohydrin with a dihydroxy compound, most frequently bisphenol A. (p. 629)

ether A compound with two alkyl (or aryl) groups bonded to an oxygen atom, R—O—R′. (p. 600)
  symmetrical ether: An ether with two identical alkyl groups.
  unsymmetrical ether: An ether with two different alkyl groups.

furan The five-membered heterocyclic ether with two carbon–carbon double bonds; or a derivative of furan. (p. 606)

halohydrin A compound containing a halogen atom and a hydroxyl group on adjacent carbon atoms. Chlorohydrins, bromohydrins, and iodohydrins are most common. (p. 620)

heterocyclic compound (heterocycle) A compound containing a ring in which one or more of the ring atoms are elements other than carbon. The noncarbon ring atoms are called **heteroatoms**. (p. 605)

heterocyclic ethers:

| epoxide (oxirane) | oxetane | furan | THF (oxolane) | pyran | tetrahydropyran oxane | 1,4-dioxane |

MCPBA An abbreviation for *meta*-chloroperoxybenzoic acid, an epoxidizing agent. (p. 619)

MMPP An abbreviation for magnesium monoperoxyphthalate, a relatively stable peroxyacid often used in large-scale epoxidations. (p. 620)

oxane The systematic name for a six-membered cyclic ether (a tetrahydropyran). (p. 607)

oxetane A compound containing a four-membered heterocyclic ether. (p. 606)

oxirane The systematic name for an epoxide, or for ethylene oxide specifically. (p. 605)

oxolane The systematic name for a five-membered cyclic ether (a tetrahydrofuran). (p. 606)

peroxide Any compound containing the —O—O— linkage. The oxygen–oxygen bond is easily cleaved, and organic peroxides are prone to explosions. (p. 615)

H—O—O—H   R—O—O—H   R—O—O—R′
hydrogen peroxide   an alkyl hydroperoxide   a dialkyl peroxide

peroxyacid (peracid) A carboxylic acid with an extra oxygen in the hydroxyl group. (p. 619)

$$\begin{array}{c} \quad\quad \text{O} \\ \quad\quad \| \\ \text{R}-\text{C}-\text{O}-\text{O}-\text{H} \end{array}$$

pyran The six-membered heterocyclic ether with two carbon–carbon double bonds; or a derivative of pyran. (p. 607)

sulfide (thioether) A compound with two alkyl (or aryl) groups bonded to a sulfur atom R—S—R′. (p. 616)

sulfone A compound of the form R—SO$_2$—R′ (see next page). (p. 617)

sulfonium salt A salt containing a sulfur atom bonded to three alkyl groups, R$_3$S$^+$, and a counterion (see next page). (p. 618)

sulfoxide A compound of the form R—SO—R′ (see below). (p. 617)

$$R—S—R' \qquad R—\overset{R''}{\underset{X^-}{\overset{|}{S^+}}}—R' \qquad R—\overset{O}{\overset{\|}{S}}—R' \qquad R—\overset{O}{\underset{O}{\overset{\|}{\underset{\|}{S}}}}—R'$$

sulfide   sulfonium salt   sulfoxide   sulfone

Williamson ether synthesis Formation of an ether by the S_N2 reaction of an alkoxide ion with an alkyl halide or tosylate. In general, the electrophile must be primary, or occasionally secondary. (p. 610)

$$R—\ddot{\underset{\cdot\cdot}{O}}{:}^- \quad R'{-}X \longrightarrow R—O—R' + X^-$$

Essential Problem-Solving Skills in Chapter 14

1. Draw and name ethers and heterocyclic ethers, including epoxides.
2. Predict relative boiling points and solubilities of ethers.
3. Explain how ether solvents stabilize electrophilic reagents and why they are compatible with organometallic reagents.
4. Determine the structures of ethers from their spectra, and explain their characteristic absorptions and fragmentations.
5. Devise efficient laboratory syntheses of ethers and epoxides, including:
   (a) The Williamson ether synthesis
   (b) Alkoxymercuration–demercuration
   (c) Peroxyacid epoxidation
   (d) Base-promoted cyclization of halohydrins.
6. Predict the products of reactions of ethers and epoxides, including:
   (a) Cleavage and autoxidation of ethers
   (b) Acid- and base-promoted opening of epoxides
   (c) Reactions of epoxides with organometallic reagents.
7. Use your knowledge of the mechanisms of ether and epoxide reactions to propose mechanisms and products for similar reactions you have never seen before.

Study Problems

14-29 Briefly define each term, and give an example.
   (a) autoxidation   (b) Williamson ether synthesis   (c) alkoxymercuration–demercuration
   (d) heterocyclic compound   (e) epoxidation   (f) concerted reaction
   (g) unsymmetrical ether   (h) crown ether

14-30 Write structural formulas for the following compounds.
   (a) ethyl isopropyl ether   (b) di-*n*-butyl ether   (c) 2-ethoxyoctane
   (d) divinyl ether   (e) allyl methyl ether   (f) cyclohexene oxide
   (g) *cis*-2,3-epoxyhexane   (h) (2R, 3S)-2-methoxy-3-pentanol

14-31 Give common names for the following compounds.
   (a) $(CH_3)_2CH—O—CH(CH_3)CH_2CH_3$ (b) $(CH_3)_3C—O—CH_2CH(CH_3)_2$
   (c) $Ph—O—CH_2CH_3$ (d) $Cl—CH_2—O—CH_2CH_2CH_3$

   (e)   (f)   (g)   (h)

14-32   Give IUPAC names for the following compounds.

(a) $CH_3-O-CH(CH_3)CH_2OH$ (b) $Ph-O-CH_2CH_3$ (c)   (d)

(e)   (f)   (g)   (h)   (i)

14-33   Predict the products of the following reactions.

(a) *sec*-butyl isopropyl ether + concd. HBr, heat
(b) *t*-butyl ethyl ether + concd. HBr, heat
(c) di-*n*-butyl ether + hot concd. NaOH
(d) di-*n*-butyl ether + Na metal
(e) ethoxybenzene + concd. HI, heat
(f) 1,2-epoxyhexane + H^+, CH_3OH
(g) *trans*-2,3-epoxyoctane + H^+, H_2O
(h) propylene oxide + methylamine (CH_3NH_2)

(i) potassium *t*-butoxide + *n*-butyl bromide

(j) $\xrightarrow{\text{(1) phenyllithium}}{\text{(2) hydrolysis}}$

(k) $\xrightarrow{\text{MCPBA, } CH_2Cl_2}$

(l) $\xrightarrow{\text{HBr}}$

(m) $\xrightarrow{CH_3O^-,\ CH_3OH}$

(n) $\xrightarrow{CH_3OH,\ H^+}$

14-34   (A true story.) An inexperienced graduate student moved into a laboratory and began work. He needed some diethyl ether for a reaction, so he opened an old, rusty 1-gallon can marked "ethyl ether" and found there was half a gallon left. To purify the ether, the student set up a distillation apparatus, started a careful distillation, and went to the stockroom for the other reagents he needed. While he was at the stockroom, the student heard a muffled "boom." He quickly returned to his lab to find a worker from another laboratory putting out a fire. Most of the distillation apparatus was embedded in the ceiling.

(a) Explain what probably happened.   (b) Explain how this near-disaster might have been prevented.

14-35   (a) Show how you would synthesize the pure (R) enantiomer of 2-butyl methyl sulfide starting with pure (R)-2-butanol and any reagents you need.

(b) Show how you would synthesize the pure (S) enantiomer of the product.

14-36   (a) Predict the values of m/z and the structures of the most abundant fragments you would observe in the mass spectrum of di-*n*-propyl ether.

(b) Give logical fragmentations to account for the following ions observed in the mass spectrum of 2-methoxypentane: 102, 87, 71, 59, 31.

14-37   The following reaction resembles the acid-catalyzed cyclization of squalene oxide. Propose a mechanism for this reaction.

14-38   Show how you would accomplish the following synthetic transformations in good yield.

(a) 1-hexene → 1-phenyl-2-hexanol   (b) 1-hexene → 1-methoxy-2-hexanol   (c) 1-hexene → 2-methoxy-1-hexanol

14-39   Give the structures of intermediates **A** through **H** in the following synthesis of *trans*-1-cyclohexyl-2-methoxycyclohexane.

14-40   (Another true story.) An organic lab student carried out the reaction of methylmagnesium iodide with acetone (CH_3COCH_3), followed by hydrolysis. During the distillation to isolate the product, she forgot to mark the vials she used to collect the fractions. She turned in a product of formula $C_4H_{10}O$ that boiled at 35°C. The IR spectrum showed only a weak O—H stretch around 3300 cm^{-1}, and the mass spectrum showed a base peak at m/z 59. The NMR spectrum showed a quartet ($J = 7$ Hz) of area 2 at $\delta 3.5$ and a triplet ($J = 7$ Hz) of area 3 at $\delta 1.3$. Propose a structure for this product, explain how it corresponds to the observed spectra, and suggest how the student isolated this compound.

14-41   Show how you would synthesize the following ethers in good yield from the indicated starting materials and any additional reagents needed.

(a) cyclopentyl *n*-propyl ether from cyclopentanol and 1-propanol   (b) *n*-butyl phenyl ether from phenol and 1-butanol
(c) 2-methoxydecane from a decene   (d) 1-methoxydecane from a decene
(e) 1-ethoxy-1-methylcyclohexane from 1-methylcyclohexene   (f) *trans*-2,3-epoxyoctane from *trans*-2-octene

14-42   There are two different ways of making 2-ethoxyoctane from 2-octanol using the Williamson ether synthesis. When pure (−)-2-octanol of specific rotation −8.24° is treated with sodium metal and then ethyl iodide, the product is 2-ethoxyoctane with a specific rotation of −15.6°. When pure (−)-2-octanol is treated with tosyl chloride and pyridine and then with sodium ethoxide, the product is also 2-ethoxyoctane. Predict the rotation of the 2-ethoxyoctane made using the tosylation/sodium ethoxide procedure, and propose a detailed mechanism to support your prediction.

14-43   Under base-catalyzed conditions, several molecules of propylene oxide can react to give short polymers. Propose a mechanism for the base-catalyzed formation of the following trimer.

14-44   Under the right conditions, the following acid-catalyzed double cyclization proceeds in remarkably good yields. Propose a mechanism. Does this reaction resemble a biological process you have seen?

*14-45   Propylene oxide is a chiral molecule. Hydrolysis of propylene oxide gives propylene glycol, another chiral molecule.
(a) Draw the enantiomers of propylene oxide.
(b) Propose a mechanism for the acid-catalyzed hydrolysis of pure (*R*)-propylene oxide.
(c) Propose a mechanism for the base-catalyzed hydrolysis of pure (*R*)-propylene oxide.
(d) Explain why the acid-catalyzed hydrolysis of optically active propylene oxide gives a product with a rotation opposite that of the product of the base-catalyzed hydrolysis.

*14-46   An acid-catalyzed reaction was carried out using methyl cellosolve (2-methoxyethanol) as the solvent. When the 2-methoxyethanol was redistilled, a higher-boiling fraction (bp 162°C) was also recovered. The mass spectrum of this fraction showed the molecular weight to be 134. The IR and NMR spectra appear below. Determine the structure of this compound, and propose a mechanism for its formation.

***14-47**   A compound of molecular formula C_8H_8O gives the IR and NMR spectra shown below. Propose a structure, and show how it is consistent with the observed absorptions.

CHAPTER 15

Conjugated Systems, Orbital Symmetry, and Ultraviolet Spectroscopy

15-1
Introduction

Double bonds can interact with each other if they are separated by just one single bond. Such interacting double bonds are said to be **conjugated**. Double bonds with two or more single bonds separating them have little interaction and are called **isolated double bonds**. For example, 1,3-pentadiene has conjugated double bonds, while 1,4-pentadiene has isolated double bonds.

conjugated double bonds
(more stable than isolated double bonds)

1,3-pentadiene

isolated double bonds

1,4-pentadiene

Because of the interaction between the double bonds, systems containing conjugated double bonds tend to be more stable than similar systems with isolated double bonds. In this chapter, we consider the unique properties of conjugated systems, the theoretical reasons for this extra stability, and some of the characteristic reactions of molecules containing conjugated double bonds. We also study ultraviolet spectroscopy, a tool for determining the structures of conjugated systems.

15-2
Stabilities of Dienes

In Chapter 7, we used **heats of hydrogenation** to compare the relative stabilities of alkenes. For example, the heats of hydrogenation of 1-pentene and *trans*-2-pentene show that the disubstituted double bond in *trans*-2-pentene is 2.6 kcal/mol (10 kJ/mol) more stable than the monosubstituted double bond in 1-pentene.

1-pentene $\xrightarrow[\text{Pt}]{\text{H}_2}$ $\Delta H° = -30.0 \text{ kcal} (-125 \text{ kJ})$

trans-2-pentene $\xrightarrow[\text{Pt}]{\text{H}_2}$ $\Delta H° = -27.4 \text{ kcal} (-115 \text{ kJ})$

When a molecule has two isolated double bonds, the heat of hydrogenation is close to the sum of the heats of hydrogenation for the individual double bonds. For example, the heat of hydrogenation of 1,4-pentadiene is -60.2 kcal (-252 kJ), about twice that of 1-pentene.

$\Delta H° = -60.2$ kcal (-252 kJ)

1,4-pentadiene

For conjugated dienes, the heat of hydrogenation is less than the sum for the individual double bonds. For example, *trans*-1,3-pentadiene has a monosubstituted double bond like the one in 1-pentene and a disubstituted double bond like the one in 2-pentene. The sum of the heats of hydrogenation of 1-pentene and 2-pentene is -57.4 kcal (-240 kJ), but the heat of hydrogenation of *trans*-1,3-pentadiene is -53.7 kcal (-225 kJ), showing that the conjugated diene has about 3.7 kcal (15 kJ) extra stability.

1-pentene	2-pentene	predicted
(-30.0 kcal) + (-27.4 kcal) =		-57.4 kcal (-240 kJ)

actual value -53.7 kcal (-225 kJ)
more stable by   3.7 kcal   (15 kJ)

trans-1,3-pentadiene

What happens if two double bonds are even closer together than in the conjugated case? Successive double bonds with no intervening single bonds are called **cumulated double bonds**. Consider 1,2-pentadiene, which contains cumulated double bonds. Such 1,2-diene systems are also called **allenes**, after the simplest member of the class, 1,2-propadiene or "allene", $H_2C=C=CH_2$. The heat of hydrogenation of 1,2-pentadiene is -69.8 kcal/mol (-292 kJ/mol).

$CH_3CH_2CH_2CH_2CH_3$ $\Delta H° = -69.8$ kcal (-292 kJ)
pentane

1,2-pentadiene (ethylallene)

sum of   1-pentene + 2-pentene $\Delta H° = -57.4$ kcal (-240 kJ)

1,2-pentadiene is *less* stable by 12.4 kcal (52 kJ)

Because 1,2-pentadiene has a larger heat of hydrogenation than 1,4-pentadiene, we conclude that the cumulated double bonds of allenes are less stable than isolated double bonds and much less stable than conjugated double bonds. Figure 15-1 summarizes the relative stability of isolated, conjugated, and cumulated dienes and compares them with alkynes.

PROBLEM 15-1

Rank each group of compounds in order of increasing heat of hydrogenation.
(a) 1,2-hexadiene; 1,3,5-hexatriene; 1,3-hexadiene; 1,4-hexadiene; 1,5-hexadiene; 2,4-hexadiene

(b)

PROBLEM 15-2

In a strongly acidic solution, 1,4-cyclohexadiene tautomerizes to 1,3-cyclohexadiene. Propose a mechanism for this rearrangement, and explain why it is energetically favorable.

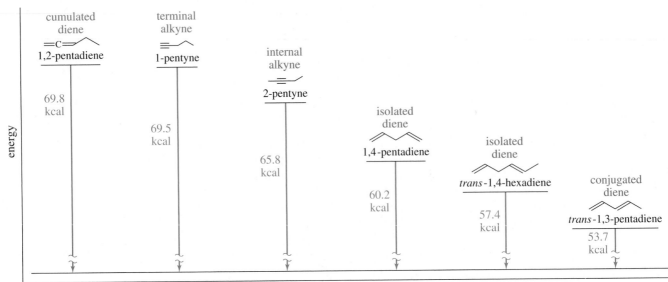

▲ **Figure 15-1**
Relative energies of conjugated, isolated, and cumulated dienes compared with alkynes, based on heats of hydrogenation (kcal/mol).

PROBLEM 15-3

(Review) The central carbon atom of an allene is a member of two double bonds, and it has an interesting orbital arrangement that holds the two ends of the molecule at right angles to each other.
(a) Draw an orbital diagram of allene, showing why the two ends are perpendicular.
(b) Draw the two enantiomers of 1,3-dichloroallene. A model may be helpful.

15-3
Molecular Orbital Picture of a Conjugated System

Figure 15-1 shows that the compound with conjugated double bonds is 3.7 kcal/mol (15 kJ/mol) more stable than a similar compound with isolated double bonds. This 3.7 kcal of extra stability in the conjugated molecule is called the **resonance energy** of the system. (Other terms favored by some chemists are *conjugation energy, delocalization energy,* and *stabilization energy.*) We can best explain this extra stability of conjugated systems by examining their **molecular orbitals**. Let's begin with the molecular orbitals of the simplest conjugated diene, 1,3-butadiene.

15-3A   Structure and Bonding of 1,3-Butadiene

The heat of hydrogenation of 1,3-butadiene is about 3.6 kcal (15 kJ) less than twice that of 1-butene, showing that 1,3-butadiene has a resonance energy of 3.6 kcal.

$$H_2C=CH-CH=CH_2 \xrightarrow{H_2,\,Pt} CH_3-CH_2-CH_2-CH_3 \qquad \Delta H° = -56.6\,kcal\,(-237\,kJ)$$

1,3-butadiene

$$H_2C=CH-CH_2-CH_3 \xrightarrow{H_2,\,Pt} CH_3-CH_2-CH_2-CH_3 \qquad \Delta H° = -30.1\,kcal\,(-126\,kJ)$$

1-butene

$$\times 2 = -60.2\,kcal\,(-252\,kJ)$$

resonance energy of 1,3-butadiene = 60.2 kcal − 56.6 kcal   = 3.6 kcal (15 kJ)

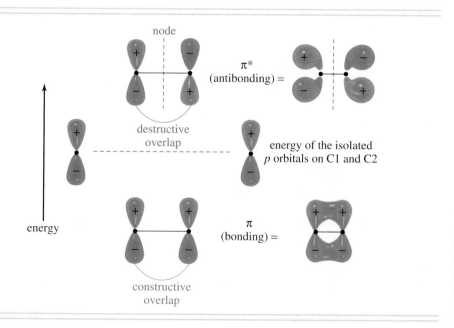

Structure of 1,3-butadiene in its most stable conformation. The 1.48 Å central carbon–carbon single bond is shorter than the 1.54 Å bonds typical of alkanes because of its partial double-bond character.

Figure 15-2 shows the most stable conformation of 1,3-butadiene. Note that this conformation is planar, with the p orbitals on the two pi bonds aligned.

The C2—C3 bond in 1,3-butadiene (1.48 Å) is considerably shorter than a carbon–carbon single bond in an alkane (1.54 Å). This bond is shortened slightly by the increased s character of the sp^2 hybrid orbitals; but the most important cause of this bond shortening is its pi bonding overlap and partial double-bond character. The planar conformation, with the p orbitals of the two double bonds aligned, allows overlap between the pi bonds. In effect, the electrons in the double bonds are **delocalized** over the entire molecule, creating some pi overlap and pi bonding in the C2—C3 bond. The length of this bond is intermediate between the normal length of a single bond and that of a double bond.

Lewis structures are not adequate to represent delocalized molecules such as 1,3-butadiene. To represent the bonding in conjugated systems accurately, we must consider molecular orbitals that represent the entire conjugated pi system, and not just one bond at a time.

15-3B   Constructing the Molecular Orbitals of 1,3-Butadiene

All four carbon atoms of 1,3-butadiene are sp^2 hybridized, and (in the planar conformation) they all have overlapping p orbitals. Let's review how we constructed the pi molecular orbitals (MOs) of ethylene from the p atomic orbitals of the two carbon atoms (Figure 15-3). Each p orbital consists of two lobes, with opposite phases of the wave

The pi bonding orbital of ethylene is formed by constructive overlap of unhybridized p orbitals on the sp^2 hybrid carbon atoms. Destructive overlap of these two orbitals forms the antibonding pi orbital. Combination of two p orbitals must give exactly two molecular orbitals.

function in the two lobes. The plus and minus signs used in drawing these orbitals indicate the *phase of the wave function*, **not electrical charges**. To minimize confusion, we will color the lobes of the *p* orbitals to emphasize the phase difference.

In the pi bonding orbital of ethylene, there is overlap of lobes with the same sign (+ with + and − with −) in the bonding region between the nuclei. We call this reinforcement **constructive overlap**. In the antibonding orbital (marked by an *), there is canceling of opposite signs (+ with −) in the bonding region. This canceling of the wave function is called **destructive overlap**. Midway between the nuclei is a **node**: a region of zero electron density where the positive and negative phases exactly cancel. Electrons have lower potential energy in the **bonding MO** than in the original *p* orbitals, and higher potential energy in the **antibonding MO**. In the ground state of ethylene, two electrons are in the bonding MO, but the antibonding MO is vacant. Stable molecules tend to have filled bonding MOs and empty antibonding MOs.

When viewing Figure 15-3, there are several important principles to keep in mind. Constructive overlap results in a bonding interaction; destructive overlap results in an antibonding interaction. Also, the number of pi molecular orbitals is always the same as the number of *p* orbitals used to form the MOs. These molecular orbitals have energies that are symmetrically distributed above and below the energy of the starting *p* orbitals. Half are bonding MOs, and half are antibonding MOs.

Now we are ready to construct the molecular orbitals of 1,3-butadiene. The *p* orbitals on C1 through C4 overlap, giving an extended system of four *p* orbitals that form four pi molecular orbitals. Two MOs are bonding, and two are antibonding. To represent the four *p* orbitals, we draw four *p* orbitals in a line. Although 1,3-butadiene is not linear, this simple straight-line representation makes it easier to draw and visualize the molecular orbitals.

represented by

The lowest-energy molecular orbital always consists entirely of bonding interactions. We indicate such an orbital by drawing all the positive phases of the *p* orbitals overlapping constructively on one face of the molecule, and the negative phases overlapping constructively on the other face. Figure 15-4 shows the lowest-energy MO for 1,3-butadiene. This MO places electron density on all four *p* orbitals, with slightly more on C2 and C3. (In these figures, larger and smaller *p* orbitals are used to show which atoms bear more of the electron density in a particular MO.)

This lowest-energy orbital is exceptionally stable for two reasons: There are three bonding interactions, and the electrons are delocalized over four nuclei. This orbital helps to illustrate why the conjugated system is more stable than two isolated double bonds. It also shows some pi bond character between C2 and C3, which lowers the energy of the planar conformation and helps to explain the short C2—C3 bond length.

As with ethylene, the second molecular orbital (π_2) of butadiene (Figure 15-5) has one node in the center of the molecule. This MO represents the classic picture of a diene. There are bonding interactions at the C1—C2 and C3—C4 bonds, and there is a (weaker) antibonding interaction between C2 and C3.

The π_2 orbital has two bonding interactions and one antibonding interaction, so we expect it to be a bonding orbital (two bonding − one antibonding = one bonding). It is not as strongly bonding nor as low in energy as the all-bonding π_1 orbital. Adding and subtracting bonding and antibonding interactions is not a reliable method for calculating energies of molecular orbitals, but it is useful for predicting whether a given orbital is bonding or antibonding and for ranking orbitals in order of their energy.

The third butadiene MO (π_3^*) has two nodes (Figure 15-6). There is a bonding interaction at the C2—C3 bond, and there are two antibonding interactions, one between

▲ **Figure 15-4**
The lowest-energy orbital of 1,3-butadiene has bonding interactions between all adjacent carbon atoms. This orbital is labeled π_1 because it is a pi bonding orbital and it has the lowest energy.

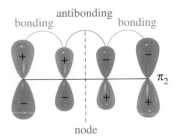

▲ **Figure 15-5**
The second MO of 1,3-butadiene has one node in the center of the molecule. There are bonding interactions at the C1—C2 and C3—C4 bonds, and there is a (weaker) antibonding interaction between C2 and C3. This π_2 orbital is bonding, but is not as strongly bonding as π_1.

C1 and C2 and the other between C3 and C4. This is an antibonding orbital (*), and it is vacant in the ground state.

The fourth, and last, molecular orbital (π_4^*) of 1,3-butadiene has three nodes and is totally antibonding (Figure 15-7). This MO has the highest energy and is unoccupied in the molecule's ground state. This highest-energy MO (π_4^*) is typical: For most systems, the highest-energy MO has antibonding interactions between all pairs of adjacent atoms.

Butadiene has four pi electrons (two electrons in each of the two double bonds in the Lewis structure) to be placed in the four MOs described above. Each MO can accommodate two electrons, and the lowest-energy MOs are filled first. Therefore, the four pi electrons go into π_1 and π_2. Figure 15-8 shows the electronic configuration of 1,3-butadiene. Both bonding MOs are filled, and both antibonding MOs are empty. Most stable molecules have this arrangement of filled bonding orbitals and vacant antibonding orbitals. Figure 15-8 also compares the relative energies of the ethylene MOs with the butadiene MOs to show that the conjugated butadiene system is slightly more stable than two ethylene double bonds.

The partial double-bond character between C2 and C3 in 1,3-butadiene explains why the molecule is most stable in a planar conformation. There are actually two planar conformations that allow overlap between C2 and C3. These conformations arise by rotation about the C2—C3 bond, and they are considered single-bond analogues of trans and cis isomers about a double bond. Thus, they are named **s-trans** ("single"-trans) and **s-cis** ("single"-cis) **conformations**.

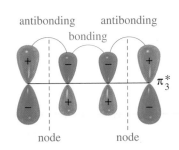

▲ **Figure 15-6**
The third butadiene MO is an antibonding orbital, and it is vacant in the ground state.

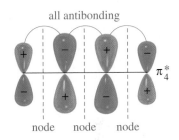

▲ **Figure 15-7**
The highest-energy MO of 1,3-butadiene has three nodes and three antibonding interactions. It is strongly antibonding, and its energy is very high.

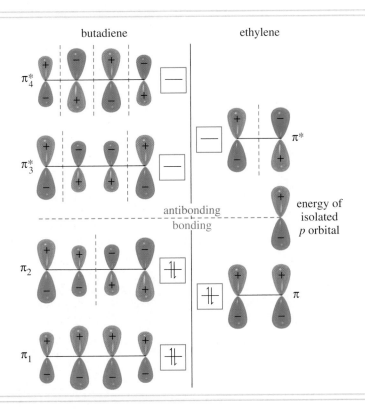

◀ **Figure 15-8**
In both 1,3-butadiene and ethylene, the bonding MOs are filled and the antibonding MOs are vacant. The average energy of the electrons is slightly lower in butadiene. This lower energy is the resonance stabilization of the conjugated diene.

The *s*-trans conformation is 2.3 kcal (9.6 kJ) more stable than the *s*-cis conformation, which shows interference between the two nearby hydrogen atoms. The rotational barrier for these conformers (rotation about the C2—C3 bond) is only about 4.9 kcal/mol (20.5 kJ/mol), compared with about 60 kcal/mol (250 kJ/mol) for rotation of a double bond in an alkene. The *s*-cis and *s*-trans conformers of butadiene (and all the skew conformations in between) easily interconvert at room temperature.

15-4
Allylic Cations

Conjugated compounds undergo a variety of reactions, many of which involve intermediates that retain some of the resonance stabilization of the conjugated system. Common intermediates include allylic systems, particularly allylic cations and radicals. Allylic cations and radicals are stabilized by delocalization. First, we consider some reactions involving allylic cations and radicals, then (Section 15-8) we derive the molecular orbital picture of their bonding.

In Chapter 7, we saw that the —CH$_2$—CH=CH$_2$ group is called the **allyl group**. Many common names use this terminology.

allylic position   allyl bromide   allyl alcohol   allylbenzene

When allyl bromide is heated with a good ionizing solvent, it ionizes to the **allyl cation**, an allyl group with a positive charge. More highly substituted analogues are called **allylic cations**. All allylic cations are stabilized by resonance with the adjacent double bond, which delocalizes the positive charge over two carbon atoms.

allyl bromide   allyl cation

substituted allylic cations

PROBLEM 15-4

Draw another resonance form for each of the substituted allylic cations shown above, showing how the positive charge is shared by another carbon atom. In each case, state whether your second resonance form is a more important or less important resonance contributor than the first structure. (Which structure places the positive charge on the more highly substituted carbon atom?)

PROBLEM 15-5

When 3-bromo-1-methylcyclohexene undergoes solvolysis in hot ethanol, two products are formed. Propose a mechanism that accounts for both of these products.

We can represent a delocalized ion such as the allyl cation either by resonance forms, as shown on the left below, or by a combined structure, as shown on the right. Although the combined structure is more concise, it is sometimes confusing because it attempts to convey all the information provided by two or more resonance forms.

$$\left[\underset{1}{H_2C} = \underset{2}{\overset{H}{\underset{|}{C}}} - \overset{+}{\underset{3}{CH_2}} \longleftrightarrow \overset{+}{\underset{1}{H_2C}} - \underset{2}{\overset{H}{\underset{|}{C}}} = \underset{3}{CH_2} \right] \quad \text{or} \quad \underset{1}{\overset{\frac{1}{2}+}{H_2C}} \cdots \underset{2}{\overset{H}{\underset{|}{C}}} \cdots \underset{3}{\overset{\frac{1}{2}+}{CH_2}}$$

resonance forms   combined representation

Because of its resonance stabilization, the (primary) allyl cation is about as stable as a simple secondary carbocation, such as the isopropyl cation. Substituted allylic cations generally have at least one secondary carbon atom bearing part of the positive charge; they are about as stable as simple tertiary carbocations such as the *t*-butyl cation.

Stability of carbocations

$$H_3C^+ < 1° < 2°, \text{allyl} < 3°, \text{substituted allylic}$$

$$\overset{\frac{1}{2}+}{H_2C} = CH = \overset{\frac{1}{2}+}{CH_2} \quad \text{is about as stable as} \quad CH_3 - \overset{+}{CH} - CH_3$$

$$CH_3 - \overset{\delta+}{CH} = CH = \overset{\delta+}{CH_2} \quad \text{is about as stable as} \quad CH_3 - \overset{+}{C} \overset{\displaystyle CH_3}{\underset{\displaystyle CH_3}{\big\langle}}$$

15-5
1,2- and 1,4-Addition to Conjugated Dienes

Electrophilic additions to conjugated dienes usually involve allylic cations as intermediates. Unlike simple carbocations, an allylic cation can react with a nucleophile at either of its positive centers. Let's consider the addition of HBr to 1,3-butadiene, an electrophilic addition that produces a mixture of two constitutional isomers. One product, 3-bromo-1-butene, results from Markovnikov addition across one of the double bonds. In the other product, 1-bromo-2-butene, the double bond shifts to the C2—C3 position.

$$H_2C = CH - CH = CH_2 + HBr \longrightarrow \underset{\substack{\text{3-bromo-1-butene}\\\text{1,2-addition}}}{H_2C - \overset{H}{\underset{|}{\underset{|}{C}}}\overset{Br}{\underset{|}{H}} - CH = CH_2} + \underset{\substack{\text{1-bromo-2-butene}\\\text{1,4-addition}}}{\overset{H}{\underset{|}{H_2C}} - CH = CH - \overset{Br}{\underset{|}{CH_2}}}$$

The first product results from electrophilic addition of HBr across a double bond. This process is called a **1,2-addition** whether or not these two carbon atoms are numbered 1 and 2 in naming the compound. In the second product, the proton and bromide ion add at the ends of the conjugated system, to carbon atoms with a 1,4-relationship. Such an addition is called a **1,4-addition** whether or not these carbon atoms are numbered 1 and 4 in naming the compound.

$$\underset{}{\overset{}{C}} = \overset{}{\underset{}{C}} - \overset{}{\underset{}{C}} = \overset{}{\underset{}{C}} \quad \overset{A-B}{\longrightarrow} \quad \underset{\underset{\text{1,2-addition}}{A \quad B}}{-\overset{1}{C} - \overset{2}{\underset{}{C}} - \overset{}{C} = \overset{}{C}} \quad + \quad \underset{\underset{\text{1,4-addition}}{A \qquad\qquad B}}{-\overset{1}{C} - \overset{2}{C} = \overset{3}{C} - \overset{4}{C} -}$$

The mechanism is similar to other electrophilic additions to alkenes. The proton is the electrophile, adding to the alkene to give the most stable carbocation. Protonation of 1,3-butadiene gives an allylic cation, which is stabilized by resonance delocalization of the positive charge over two carbon atoms.

Bromide can attack this resonance-stabilized intermediate at either of the two carbon atoms sharing the positive charge. Attack at the secondary carbon gives 1,2-addition, while attack at the primary carbon gives 1,4-addition.

MECHANISM   1,2- and 1,4-Addition to a Conjugated Diene

Step 1: Protonation forms a resonance-stabilized allylic cation.

allylic cation

Step 2: A nucleophile attacks at either electrophilic carbon atom.

1,2-addition   and   1,4-addition

The key to formation of these two products is the presence of a double bond in position to form a stabilized allylic cation. Molecules having such double bonds are likely to react via resonance-stabilized intermediates.

PROBLEM 15-6

Treatment of an alkyl halide with alcoholic $AgNO_3$ often promotes ionization.

$$Ag^+ + R\text{—}Cl \longrightarrow AgCl + R^+$$

When 3-chloro-1-methylcyclopentene reacts with $AgNO_3$ in ethanol, two isomeric ethers are formed. Suggest structures, and propose a mechanism for their formation.

PROBLEM 15-7

Propose a mechanism for each reaction, showing explicitly how the observed mixtures of products are formed.
(a) 3-methyl-2-buten-1-ol + HBr $\rightarrow$
                         1-bromo-3-methyl-2-butene + 3-bromo-3-methyl-1-butene
(b) 2-methyl-3-buten-2-ol + HBr $\rightarrow$
                         1-bromo-3-methyl-2-butene + 3-bromo-3-methyl-1-butene
(c) 1,3-butadiene + Br_2 $\rightarrow$ 3,4-dibromo-1-butene + 1,4-dibromo-2-butene
(d) 1-chloro-2-butene + $AgNO_3$, H_2O $\rightarrow$ 2-buten-1-ol + 3-buten-2-ol
(e) 3-chloro-1-butene + $AgNO_3$, H_2O $\rightarrow$ 2-buten-1-ol + 3-buten-2-ol

One of the interesting peculiarities of the reaction of 1,3-butadiene with HBr is the effect of temperature on the products. If the reagents are allowed to react briefly at $-80°C$, the 1,2-addition product predominates. If this reaction mixture is later allowed to warm to $40°C$, however, or if the reaction itself is carried out at $40°C$, the composition favors the 1,4-addition product.

15-6
Kinetic versus Thermodynamic Control in the Addition of HBr to 1,3-Butadiene

$$
\begin{array}{c}
HBr \\
+ \\
H_2C{=}CH{-}CH{=}CH_2
\end{array}
$$

$-80°C$

(80%) $H_2C{-}CH{-}CH{=}CH_2$ (1,2-product)
$\qquad\quad |\quad\ \ |$
$\qquad\quad H\ \ \ Br$

(20%) $H_2C{-}CH{=}CH{-}CH_2$ (1,4-product)
$\qquad\quad |\qquad\qquad\ \ |$
$\qquad\quad H\qquad\qquad\ Br$

$\downarrow 40°C$

$40°C$

(15%) $H_2C{-}CH{-}CH{=}CH_2$ (1,2-product)
$\qquad\quad |\quad\ \ |$
$\qquad\quad H\ \ \ Br$

(85%) $H_2C{-}CH{=}CH{-}CH_2$ (1,4-product)
$\qquad\quad |\qquad\qquad\ \ |$
$\qquad\quad H\qquad\qquad\ Br$

This variation in product composition with temperature reminds us that the most stable product is not always the major product. Of the two products, we expect 1-bromo-2-butene (the 1,4-product) to be more stable, since it has the more highly substituted double bond. This prediction is supported by the fact that this isomer predominates when the reaction mixture is warmed to $40°C$ and allowed to equilibrate.

A reaction-energy diagram for the second step of this reaction (Figure 15-9) helps to show why one product is favored at low temperatures and another at higher temperatures. The allylic cation is in the center of the diagram; it can react toward the left to give the 1,2-product or toward the right to give the 1,4-product. The initial product depends on where bromide attacks the resonance-stabilized allylic cation. Bromide can attack at either of the two carbon atoms that share the positive charge.

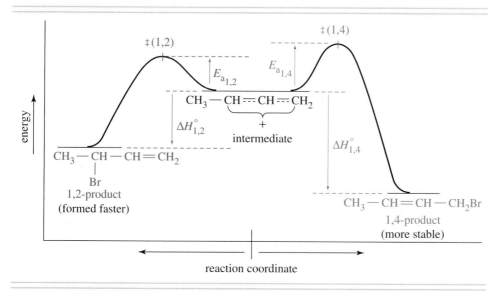

◀ **Figure 15-9**
The allylic carbocation (center) formed in the addition of HBr to 1,3-butadiene can react at either of its electrophilic carbon atoms. The transition state (‡) leading to 1,2-addition has a lower energy than that leading to the 1,4 product, so the 1,2-product is formed faster (kinetic product). The 1,2-product is not as stable as the 1,4-product, however. If equilibrium is reached, the 1,4-product predominates (thermodynamic product).

Attack at the secondary carbon gives 1,2-addition, and attack at the primary carbon gives 1,4-addition.

Kinetic Control at −80°C   The transition state for 1,2-addition has a lower energy than the transition state for 1,4-addition, giving the 1,2-addition a lower activation energy (E_a). This is not surprising, because 1,2-addition results from bromide attack at the more highly substituted secondary carbon, which bears more of the positive charge because it is better stabilized than the primary carbon. Because the 1,2-addition has a lower activation energy than the 1,4-addition, the 1,2-addition takes place faster (at *all* temperatures).

Attack by bromide on the allylic cation is a strongly exothermic process, so the reverse reaction has a large activation energy. At −80°C, few collisions take place with this much energy, and the rate of the reverse reaction is practically zero. Under these conditions, the product that is formed faster predominates. Because the kinetics of the reaction determine the results, this situation is called **kinetic control** of the reaction. The 1,2-product, favored under these conditions, is called the **kinetic product**.

Thermodynamic Control at 40°C   At 40°C, a significant fraction of molecular collisions have enough energy for reverse reactions to occur. Notice that the activation energy for the reverse of the 1,2-addition is less than that for the reverse of the 1,4-addition. Although the 1,2-product is still formed faster, it also reverts to the allylic cation faster than the 1,4-product does. At 40°C, an equilibrium is set up, and the relative energy of each species determines its concentration. The 1,4-product is the most stable species, and it predominates. Since thermodynamics determine the results, this situation is called **thermodynamic control** (or **equilibrium control**) of the reaction. The 1,4-product, favored under these conditions, is called the **thermodynamic product**.

We will see many additional reactions whose products may be determined by kinetic control or by thermodynamic control, depending on the conditions. In general, reactions that do not reverse easily are kinetically controlled because an equilibrium is rarely established. In kinetically controlled reactions, the product with the lowest-energy transition state predominates. Reactions that are easily reversible are thermodynamically controlled unless something happens to prevent equilibrium from being attained. In thermodynamically controlled reactions, the lowest energy product predominates.

PROBLEM 15-8

When Br_2 is added to 1,3-butadiene at −15°C, the product mixture contains 60% of product A and 40% of product B. When the same reaction takes place at 60°C, the product ratio is 10% A and 90% B.
 (a) Propose structures for products A and B. (*Hint:* In many cases, an allylic carbocation is more stable than a bromonium ion.)
 (b) Propose a mechanism to account for formation of both A and B.
 (c) Show why A predominates at −15°C, but B predominates at 60°C.
 (d) If you had a solution of pure A, and its temperature was raised to 60°C, what would you expect to happen? Propose a mechanism to support your prediction.

Like allylic cations, allylic radicals are stabilized by resonance delocalization. For example, the mechanism of free-radical bromination of cyclohexene is shown below. Substitution occurs entirely at the allylic position, where abstraction of a hydrogen gives a resonance-stabilized allylic radical as the intermediate.

15-7
Allylic Radicals

MECHANISM Free-Radical Allylic Bromination

Initiation step $Br_2 \xrightarrow{h\nu} 2\ Br\cdot$

Propagation steps

cyclohexene   an allylic radical

allylic radical   allylic bromide

continues the chain

In larger concentrations, bromine *adds* across double bonds (via a bromonium ion) to give saturated dibromides (Section 8-10). In the reaction shown above, bromine *substitutes* for a hydrogen atom. The key to getting substitution is to have only a low concentration of bromine available, together with light or free radicals to initiate the reaction. Because free radicals are highly reactive, even a small concentration of radicals can produce a fast chain reaction. At higher concentrations of bromine, the ionic (bromonium ion) addition becomes faster than the free-radical substitution.

Stability of Allylic Radicals Why is it that (in the first propagation step) a bromine radical abstracts only an allylic hydrogen atom, not one from another secondary site? Abstraction of allylic hydrogens is preferred because the allylic free radical that results is resonance-stabilized. The bond-dissociation energies required to generate several free radicals are compared below. Notice that the allyl radical (a primary free radical) is actually 3 kcal/mol (13 kJ/mol) *more* stable than the tertiary butyl radical.

$$CH_3CH_2-H \longrightarrow CH_3CH_2\cdot + H\cdot \qquad \Delta H = +98\ \text{kcal}\ (+410\ \text{kJ})\ \text{(primary)}$$
$$(CH_3)_2CH-H \longrightarrow (CH_3)_2CH\cdot + H\cdot \qquad \Delta H = +94\ \text{kcal}\ (+393\ \text{kJ})\ \text{(secondary)}$$
$$(CH_3)_3C-H \longrightarrow (CH_3)_3C\cdot + H\cdot \qquad \Delta H = +91\ \text{kcal}\ (+381\ \text{kJ})\ \text{(tertiary)}$$
$$H_2C=CH-CH_2-H \longrightarrow H_2C=CH-CH_2\cdot + H\cdot \qquad \Delta H = +88\ \text{kcal}\ (+368\ \text{kJ})\ \text{(allyl)}$$

The allylic 2-cyclohexenyl radical in the free-radical bromination of cyclohexene has its odd electron delocalized over two secondary carbon atoms, so it is even more stable than the unsubstituted allyl radical. The second propagation step may occur at either of the radical carbons, but in this symmetrical case, either position gives 3-bromocyclohexene as the product. Less symmetrical compounds often give mixtures of products resulting from an **allylic shift**: In the product, the double bond can appear at either of the positions it occupies in the resonance forms of the allylic radical. An allylic shift in a radical reaction is similar to the 1,4-addition of an electrophilic reagent such as HBr to a diene (Section 15-5).

The following propagation steps show how a mixture of products results from the free-radical allylic bromination of 1-butene.

$$CH_3-CH-CH=CH_2 + Br\cdot \longrightarrow \left[CH_3-\dot{C}H-CH=CH_2 \longleftrightarrow CH_3-CH=CH-\dot{C}H_2\right] + HBr$$

$$\text{resonance-stabilized allylic radical}$$

$$\downarrow Br_2$$

$$CH_3-CH-CH=CH_2 + CH_3-CH=CH-CH_2 + Br\cdot$$

$$\overset{|}{Br} \qquad\qquad\qquad\qquad \overset{|}{Br}$$

(mixture)

PROBLEM 15-9

When methylenecyclohexane is treated with a low concentration of bromine under irradiation by a sunlamp, two substitution products are formed.

methylenecyclohexane   + Br_2 $\longrightarrow$ two substitution products   +   HBr

(a) Propose structures for these two products.
(b) Propose a mechanism to account for their formation.

Bromination Using NBS   Effective use of free-radical allylic bromination requires a low concentration of bromine in the reaction mixture to enhance allylic substitution over ionic addition. Simply adding bromine would make the concentration too high, and ionic addition of bromine to the double bond would result. A convenient bromine source for allylic bromination is *N*-bromosuccinimide (NBS), a brominated derivative of succinimide. Succinimide is a cyclic amide of the four-carbon diacid succinic acid.

succinic acid   succinimide   *N*-bromosuccinimide (NBS)

NBS provides a fairly constant, low concentration of Br_2 because it reacts with HBr liberated in the substitution, converting it back into Br_2. This reaction also removes the HBr by-product, preventing it from adding across the double bond by its own free-radical chain reaction.

Step 1:   Free-radical allylic substitution (mechanism on p. 649)

$$R-H + Br_2 \xrightarrow{h\nu} R-Br + HBr$$

Step 2:   NBS converts the HBr by-product back into Br_2

N—Br   +   HBr $\longrightarrow$ N—H   + Br_2

NBS   succinimide

The NBS reaction is carried out in a clever way. The allylic compound is dissolved in carbon tetrachloride, and one equivalent of NBS is added. NBS is denser than CCl_4 and not very soluble in it, so it sinks to the bottom of the CCl_4 solution. The reaction is initiated using a sunlamp for illumination or a radical initiator such as a peroxide. The NBS gradually *appears* to rise to the top of the CCl_4 layer. It is actually converted to succinimide, which is less dense than CCl_4. Once all the solid succinimide has risen to the top, the sunlamp is turned off, the solution is filtered to remove the succinimide, and the CCl_4 is evaporated to recover the product.

PROBLEM 15-10

Devise a complete mechanism for the light-initiated reaction of 1-hexene with NBS in carbon tetrachloride solution.

PROBLEM 15-11

Predict the product(s) of light-initiated reaction with NBS in CCl_4 for the following starting materials.

(a) cyclopentene   **(b)** *trans*-2-pentene   **(c)** ⟨benzene ring⟩—CH_3

toluene

15-8
Molecular Orbitals of the Allylic System

Let's take a closer look at the electronic structure of allylic systems, using the allyl radical as our example. One resonance form shows a pi bond between C2 and C3 with the radical electron on C1, and the other shows a pi bond between C1 and C2 with the radical electron on C3. These two resonance forms imply that there is half a pi bond between C1 and C2 and half a pi bond between C2 and C3, with the radical electron half on C1 and half on C3.

⟨resonance structures: two resonance forms linked by a double-headed arrow, labeled "resonance forms", equal sign, and "combined representation"⟩

Remember that no resonance form has an independent existence: a compound has characteristics of all its resonance forms at the same time, but it does not "resonate" among them. To have pi bonding overlap simultaneously between C1 and C2 and between C2 and C3, the *p* orbitals of all three carbon atoms must be parallel. The geometric structure of the allyl system is shown in Figure 15-10. The allyl cation, the allyl

π bonding

π bonding

◀ **Figure 15-10**
Geometric structure of the allyl cation, allyl radical, and allyl anion.

▶ **Figure 15-11**
The three molecular orbitals of the allyl system. The lowest-energy MO (π_1) has no nodes and is entirely bonding. The intermediate orbital (π_2) is nonbonding, having one symmetrical node that coincides with the center carbon atom. The highest-energy MO (π_3^*) has two nodes and is entirely antibonding. In the allyl radical, π_1 is filled. The odd electron is in π_2, having its electron density entirely on C1 and C3.

PROBLEM-SOLVING $\mathbf{HINT}$

In drawing pi MO's, several p orbitals combine to give the same number of MOs: half bonding and half antibonding. If there are an odd number of MOs, the middle one is nonbonding.

The lowest-energy MO has no nodes; each higher MO has one more node.

The highest-energy MO is entirely antibonding, with a node at each overlap.

In a stable system, the bonding MOs are filled, and the antibonding MOs are empty.

radical, and the allyl anion all have this same geometric structure, differing only in the number of pi electrons.

Just as the four p orbitals of 1,3-butadiene overlap to form four molecular orbitals, the three atomic p orbitals of the allyl system overlap to form three molecular orbitals, shown in Figure 15-11. These three MOs share several important features with the MOs of the butadiene system. The first MO is entirely bonding, the second has one node, and the third has two nodes and (because it is the highest-energy MO) is entirely antibonding.

As with butadiene, we expect that half of the MOs will be bonding, and half antibonding; but with an odd number of MOs, they cannot be symmetrically divided. One of the MOs must appear at the middle of the energy levels, neither bonding nor antibonding: It is a **nonbonding molecular orbital**. Electrons in a nonbonding orbital have the same energy as in an isolated p orbital.

The structure of the nonbonding orbital (π_2) may seem strange because there is zero electron density on the center p orbital (C2). This is the case because π_2 must have one node, and the only symmetrical position for one node is in the center of the molecule, crossing C2. We can tell from its structure that π_2 must be nonbonding, because C1 and C3 both have zero overlap with C2. The total is zero bonding, or a nonbonding orbital.

15-9
Electronic Configurations of the Allyl Radical, Cation, and Anion

The right-hand column of Figure 15-11 shows the electronic structure for the allyl radical, with three pi electrons in the lowest available molecular orbitals. Two electrons are in the all-bonding MO (π_1), representing the pi bond shared between the C1—C2 bond and the C2—C3 bond. The odd electron goes into π_2 with zero electron density on the center carbon atom (C2). This MO representation agrees with the resonance picture showing the radical electron shared equally by C1 and C3, but not C2. Both the resonance and MO pictures successfully predict that the radical will react at either of the end carbon atoms, C1 or C3.

The electronic configuration of the allyl cation (Figure 15-12) differs from that of the allyl radical; it lacks the odd electron in π_2, which has half of its electron density on C1 and half on C3. In effect, we have removed half an electron from each of C1 and C3, while C2 remains unchanged. This MO picture is consistent with the resonance picture showing the positive charge shared by C1 and C3.

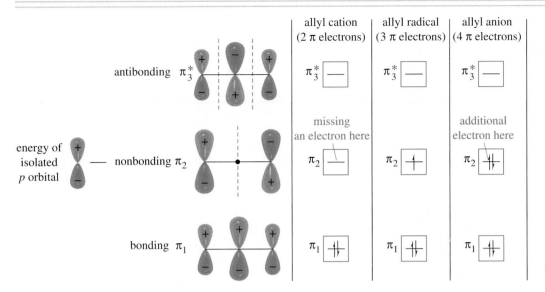

▲ Figure 15-12
Comparison of the electronic structure of the allyl cation and allyl anion with the allyl radical. The allyl cation has no electron in π_2, leaving half a positive charge on each of C1 and C3. The allyl anion has two electrons in π_2, giving half a negative charge to each of C1 and C3.

Figure 15-12 also shows the electronic configuration of the allyl anion, which differs from the allyl radical in having an additional electron in π_2, the nonbonding orbital with its electron density divided between C1 and C3. In agreement with the resonance picture, this electron's negative charge is divided equally between C1 and C3.

The molecular orbital representation shows the allyl anion with a pair of electrons in π_2, a nonbonding orbital. This picture is also consistent with the resonance forms shown above, with a lone pair of nonbonding electrons evenly divided between C1 and C3.

PROBLEM 15-12

When 1-bromo-2-butene is added to magnesium metal in dry ether, a Grignard reagent is formed. Addition of water to this Grignard reagent gives a mixture of 1-butene and 2-butene (cis and trans). When the Grignard reagent is made using 3-bromo-1-butene, addition of water produces exactly the same mixture of products in the same ratios. Explain this curious result.

15-10
S$_N$2 Displacement Reactions of Allylic Halides and Tosylates

Allylic halides and tosylates show enhanced reactivity toward nucleophilic displacement reactions by the S$_N$2 mechanism. For example, allyl bromide reacts with nucleophiles by the S$_N$2 mechanism about 40 times faster than *n*-propyl bromide.

Figure 15-13 shows how this rate enhancement can be explained by allylic delocalization of electrons in the transition state. The transition state for the S$_N$2 reaction looks like a trigonal carbon atom with a *p* orbital perpendicular to the three substituents. The electrons of the attacking nucleophile are forming a bond using one lobe of the *p* orbital while the leaving group's electrons are leaving from the other lobe.

When the substrate is allylic, the transition state receives resonance stabilization through conjugation with the *p* orbitals of the pi bond. This stabilization lowers the energy of the transition state, resulting in a lower activation energy and an enhanced rate.

The enhanced reactivity of allylic halides and tosylates makes them particularly attractive as electrophiles for S$_N$2 reactions. Allylic halides are so reactive that they couple with Grignard and organolithium reagents, a reaction that does not work well with unactivated halides.

$$H_2C{=}CH{-}CH_2Br + CH_3{-}(CH_2)_3{-}Li \longrightarrow H_2C{=}CH{-}CH_2{-}(CH_2)_3{-}CH_3 + LiBr$$

allyl bromide   *n*-butyllithium   1-heptene (85%)

PROBLEM 15-13

Show how you might synthesize the following compounds starting with alkyl or alkenyl halides containing four carbon atoms or fewer.
(a) 1-heptene   **(b)** 5-methyl-2-hexene

S$_N$2 reaction on *n*-propyl bromide

S$_N$2 reaction on allyl bromide

transition state

▲ Figure 15-13
In the transition state for the S$_N$2 reaction of allyl bromide with a nucleophile, the double bond is conjugated with the *p* orbital that is momentarily present on the reacting carbon atom. The resulting overlap lowers the energy of the transition state, increasing the reaction rate.

In 1928, German chemists Otto Diels and Kurt Alder discovered that alkynes and alkenes with electron-withdrawing groups add to conjugated dienes to form six-membered rings. The **Diels–Alder reaction** has proven to be a useful synthetic tool, providing one of the best ways to make six-membered rings with diverse functionality and controlled stereochemistry. In 1950, Diels and Alder were awarded the Nobel Prize for their work.

The Diels–Alder reaction is also called a **[4 + 2] cycloaddition** because a ring is formed by the interaction of four pi electrons in the diene with two pi electrons of the alkene or alkyne. Since the electron-poor alkene or alkyne is prone to react with a diene, it is called a **dienophile** (lover of dienes). In effect, the Diels–Alder reaction converts two pi bonds into two sigma bonds. We can symbolize the Diels–Alder reaction by using three arrows to show the movement of three pairs of electrons. This electron movement is **concerted**, with three pairs of electrons moving simultaneously. The electron-withdrawing groups ($-W$) are usually carbonyl-containing ($C=O$) groups or cyano ($-C\equiv N$) groups.

The Diels–Alder reaction is like a nucleophile-electrophile reaction. The diene is electron-rich, while the dienophile is electron-poor. Simple dienes such as 1,3-butadiene are sufficiently electron-rich to be effective dienes for the Diels–Alder reaction. The presence of electron-releasing groups such as alkyl groups or alkoxy ($-OR$) groups may further enhance the reactivity of the diene.

Simple alkenes and alkynes such as ethene and ethyne are poor dienophiles, however. A good dienophile generally has one or more electron-withdrawing groups ($-W$) pulling electron density away from the pi bond. Dienophiles commonly have carbonyl-containing ($C=O$) groups or cyano ($-C\equiv N$) groups to enhance their Diels–Alder reactivity. Figure 15-14 shows some representative Diels–Alder reactions involving a variety of different dienes and dienophiles.

15-11
The Diels–Alder Reaction

The precursor to vitamin D undergoes a reverse cycloaddition reaction (that is, the ring opens) in the skin after exposure to the sun's ultraviolet rays. In cold climates, children who have little exposure to sunlight often cannot synthesize or consume enough vitamin D, and as a result they develop rickets. Fish liver oils are rich sources of vitamin D.

diene *dienophile* *Diels–Alder adduct*

◀ **Figure 15-14**
Examples of the Diels–Alder reaction. Electron-releasing substituents activate the diene; electron-withdrawing substituents activate the dienophile.

KEY MECHANISM   The Diels–Alder Reaction

The Diels–Alder is a one-step, concerted mechanism.
A diene reacts with an electron-poor alkene to give a cyclohexene.

diene   dienophile
electron-rich   electron-poor

$\xrightarrow{\Delta \text{ (heat)}}$

A diene reacts with an electron-poor alkyne to give a cyclohexadiene.

diene   dienophile

$\xrightarrow{\Delta}$

PROBLEM-SOLVING HINT

A Diels–Alder product always contains one more ring than the reactants. The two ends of the diene form new bonds to the ends of the dienophile. The center (formerly single) bond of the diene becomes a double bond. The dienophile's double bond becomes a single bond (or its triple bond becomes a double bond).

PROBLEM 15-14

Predict the products of the following proposed Diels–Alder reactions.

(a)

(b)

(c)

(d)

(e)

(f)

PROBLEM-SOLVING HINT

To deconstruct a Diels–Alder product, look for the double bond at the center of what was the diene. Directly across the ring is the dienophile bond, usually with electron-withdrawing groups. (If a single bond, the dienophile had a double bond; if double, the dienophile had a triple bond.) Break the two bonds that join the diene and dienophile, and restore the two double bonds of the diene and the double (or triple) bond of the dienophile.

PROBLEM 15-15

What dienes and dienophiles would react to give the following Diels–Alder products?

(a)

(b)

(c)

(d)

(e)

(f)

▲ **Figure 15-15**

The Diels–Alder reaction has a concerted mechanism, with all the bond making and bond breaking occurring in a single step. Three pairs of electrons move simultaneously, requiring a transition state with overlap between the end *p* orbitals of the diene and those of the dienophile.

15-11A   Stereochemical Requirements of the Diels–Alder Transition State

The mechanism of the Diels–Alder reaction is a concerted cyclic movement of six electrons: four in the diene and two in the dienophile. For the three pairs of electrons to move simultaneously, the transition state must have a geometry that allows overlap of the two end *p* orbitals of the diene with those of the dienophile. Figure 15-15 shows the required geometry of the transition state. The geometry of the Diels–Alder transition state explains why some isomers react differently from others, and it enables us to predict the stereochemistry of the products. Three stereochemical features of the Diels–Alder reaction are controlled by the requirements of the transition state:

s-cis Conformation of the Diene   The diene must be in the *s*-cis conformation to react. When the diene is in the *s*-trans conformation, the end *p* orbitals are too far apart to overlap with the *p* orbitals of the dienophile. The *s*-trans conformation usually has a lower energy than the *s*-cis, but this energy difference is not enough to prevent most dienes from undergoing Diels–Alder reactions. For example, the *s*-trans conformation of butadiene is only 2.3 kcal/mol (9.6 kJ/mol) lower in energy than the *s*-cis conformation.

s-cis

s-trans
2.3 kcal (9.6 kJ) more stable

Structural features that aid or hinder the diene in achieving the *s*-cis conformation affect its ability to participate in Diels–Alder reactions. Figure 15-16 shows that dienes with functional groups that hinder the *s*-cis conformation react more slowly than butadiene. Dienes with functional groups that hinder the *s*-trans conformation react faster than butadiene.

Diels–Alder rate compared with that of 1,3-butadiene

← *slower*   similar to butadiene   *faster* →

(no Diels–Alder)

▶ **Figure 15-16**
Dienes that easily adopt the *s*-cis conformation undergo the Diels–Alder reaction more readily.

Because cyclopentadiene is fixed in the *s*-cis conformation, it is highly reactive in the Diels–Alder reaction. It is so reactive, in fact, that at room temperature, cyclopentadiene slowly reacts with itself to form dicyclopentadiene. Cyclopentadiene is regenerated by heating the dimer above 200°C. Above 200°C, the Diels–Alder reaction reverses, and the more volatile cyclopentadiene monomer distills over into a cold flask. The monomer can be stored indefinitely at dry-ice temperatures.

syn Stereochemistry The Diels–Alder reaction is a syn addition with respect to both the diene and the dienophile. The dienophile adds to one face of the diene, and the diene adds to one face of the dienophile. As you can see from the transition state in Figure 15-15, there is no opportunity for any of the substituents to change their stereochemical positions during the course of the reaction. Substituents that are on the same side of the diene or dienophile will be cis on the newly formed ring. The following examples show the results of this syn addition.

cis	cis (meso)
trans	trans (racemic)
cis (meso)	trans (racemic)

The Endo Rule When the dienophile has a pi bond in its electron-withdrawing group (as in a carbonyl group or a cyano group), the *p* orbitals in that electron-withdrawing group approach the central carbon atoms (C2 and C3) of the diene. This proximity results in

transition state

▲ Figure 15-17
In most Diels–Alder reactions, there is *secondary overlap* between the *p* orbitals of the
electron-withdrawing group and those of the central carbon atoms of the diene. Secondary
overlap stabilizes the transition state, and it favors products having the electron-withdrawing
groups in *endo* positions.

secondary overlap: an overlap of the *p* orbitals of the electron-withdrawing group with
the *p* orbitals of C2 and C3 of the diene (Fig. 15-17). Secondary overlap helps to stabilize the transition state.

The influence of secondary overlap was first observed in reactions using cyclopentadiene to form bicyclic ring systems. In the bicyclic product (called *norbornene*), the
electron-withdrawing substituent occupies the stereochemical position closest to the central atoms of the diene. This position is called the **endo position** because the substituent
seems to be inside the pocket formed by the six-membered ring of norbornene. This
stereochemical preference for the electron-withdrawing substituent to go in the endo
position is called the **endo rule**.

stereochemical positions
of norbornene

The endo rule is useful for predicting the products of many types of Diels–Alder
reactions, regardless of whether they use cyclopentadiene to form norbornene systems.
The following examples show the use of the endo rule with other types of Diels–Alder
reactions.

SOLVED PROBLEM 15-1

Use the endo rule to predict the product of the following cycloaddition.

SOLUTION

Imagine this diene to be a substituted cyclopentadiene; the endo product will be formed.

In the imaginary reaction, we replaced the two inside hydrogens with the rest of the cyclopentadiene ring. Now we put them back and have the actual product.

PROBLEM 15-16

For each proposed Diels–Alder reaction, predict the major product. Include stereochemistry where appropriate.

(a)

(b)

(c)

15-11B Diels–Alder Reactions Using Unsymmetrical Reagents

Even when the diene and dienophile are both unsymmetrically substituted, the Diels–Alder reaction usually gives a single product (or a major product) rather than a random mixture. We can usually predict the major product by considering how the substituents polarize the diene and the dienophile in their charge-separated resonance forms. If we then arrange the reactants to connect the most negatively charged carbon in the (electron-rich) diene with the most positively charged carbon in the (electron-poor) dienophile, we can usually predict the correct orientation. The following examples show that an electron-donating substituent on the diene (D) and an electron-withdrawing substituent on the dienophile (W) usually show either a 1,2- or 1,4-relationship in the product.

Formation of 1,4-product

1,4-product   1,3-product

Predicting this product

diene   dienophile   charge-separated resonance forms   1,4-product

Formation of 1,2-product

1,2-product   1,3-product

Predicting this product

minus reacts with plus

diene   dienophile   charge-separated resonance forms   1,2-product

Stable compounds that mimic these charge-separated structures are used to generate protein catalysts or catalytic antibodies to carry out the Diels–Alder reaction. These catalysts accelerate the corresponding transformation, and they can be designed to produce the disfavored isomer.

In most cases, we don't even need to draw the charge-separated resonance forms to determine which orientation of the reactants is preferred. We can predict the major products of unsymmetrical Diels–Alder reactions simply by remembering that the electron-donating groups of the diene and the electron-withdrawing groups of the dienophile usually bear either a 1,2-relationship or a 1,4-relationship in the products, but not a 1,3-relationship.

SOLVED PROBLEM 15-2

Predict the products of the following proposed Diels–Alder reactions.

(a)

(b)

SOLUTION

(a) The methyl group is weakly electron-donating to the diene, and the carbonyl group is electron-withdrawing from the dienophile. The two possible orientations place these groups in a 1,4-relationship or a 1,3-relationship. We select the 1,4-relationship for our predicted product. (Experimental results show a 70:30 preference for the 1,4-product.)

1,4-relationship (major)   1,3-relationship (minor)
(70%)   (30%)

(b) The methoxyl group ($-OCH_3$) is strongly electron-donating to the diene, and the cyano group ($-C \equiv N$) is electron-withdrawing from the dienophile. Depending on the orientation of addition, the product has either a 1,2- or a 1,3-relationship of these two groups. We select the 1,2-relationship, and the endo rule predicts cis stereochemistry of the two substituents.

1,2-relationship (product)   1,3-relationship (not formed)

PROBLEM 15-17

In Solved Problem 15-2, we simply predicted that the products would have a 1,2- or 1,4-relationship of the proper substituents. Draw the charge-separated resonance forms of the reactants to support these predictions.

PROBLEM 15-18

Predict the products of the following Diels–Alder reactions.

15-12
The Diels–Alder as an Example of a Pericyclic Reaction

To understand why the Diels–Alder reaction takes place, we must consider the molecular orbitals involved. The Diels–Alder is a **cycloaddition**: Two molecules combine in a one-step, **concerted** (simultaneous) **reaction** to form a new ring. Cycloadditions such as the Diels–Alder are one class of **pericyclic reactions**, which involve the concerted forming and breaking of bonds within a closed ring of interacting orbitals. Figure 15-15 (page 657) shows the closed loop of interacting orbitals in the Diels–Alder transition state. Each carbon atom of the new ring has one orbital involved in this closed loop.

A concerted pericyclic reaction has a single transition state, whose activation energy may be supplied by heat (thermal induction) or by ultraviolet light (photochemical induction). Some pericyclic reactions proceed only under thermal induction, while others proceed only under photochemical induction. Some pericyclic reactions take place under both thermal and photochemical conditions, but the two sets of conditions give different products.

For many years, pericyclic reactions were poorly understood and unpredictable. Around 1965, Robert B. Woodward and Roald Hoffmann developed a theory for predicting the results of pericyclic reactions by considering the symmetry of the molecular orbitals of the reactants and products. Their theory, called **conservation of orbital symmetry**, says that the MOs of the reactants must flow smoothly into the MOs of the products without any drastic changes in symmetry. In that case, there will be bonding interactions to help stabilize the transition state. Without these bonding interactions in the transition state, the concerted cyclic reaction cannot occur. Conservation of symmetry has been used to develop "rules" to predict which pericyclic reactions are feasible and what products will result. These rules are often called the **Woodward–Hoffmann rules**.

Pericyclic reactions are used frequently in organic synthesis, but are rarely used in nature. One exception is the well-characterized enzyme chorismate mutase. This enzyme plays a central role in the formation of aromatic compounds in bacteria, fungi, and plants.

15-12A Conservation of Orbital Symmetry in the Diels–Alder Reaction

We will not develop all the Woodward–Hoffmann rules, but we will show how the molecular orbitals can indicate whether a cycloaddition will take place. The simple Diels–Alder reaction of butadiene with ethylene serves as our first example. The molecular orbitals of butadiene and ethylene are represented in Figure 15-18. Butadiene, with four atomic p orbitals, has four molecular orbitals: two bonding MOs (filled) and two antibonding MOs (vacant). Ethylene, with two atomic p orbitals, has two MOs: a bonding MO (filled) and an antibonding MO (vacant).

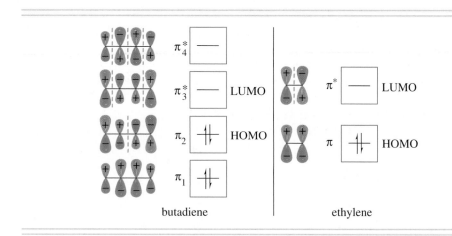

butadiene   ethylene

▶ **Figure 15-18**
Molecular orbitals of butadiene and ethylene.

butadiene HOMO

bonding interactions

ethylene LUMO

▲ **Figure 15-19**
The HOMO of butadiene forms a bonding overlap with the LUMO of ethylene because the orbitals have similar symmetry. This reaction is therefore symmetry-allowed.

In the Diels–Alder reaction, the diene acts as the electron-rich nucleophile, and the dienophile acts as the electron-poor electrophile. If we imagine the diene contributing a pair of electrons to the dienophile, the highest-energy electrons of the diene require the least activation energy for such a donation. The electrons in the highest-energy occupied orbital, called the **Highest Occupied Molecular Orbital (HOMO)**, are the important ones because they are the most weakly held. The HOMO of butadiene is π_2, and its symmetry determines the course of the reaction.

The orbital in ethylene that receives these electrons is the lowest-energy orbital available, the **Lowest Unoccupied Molecular Orbital (LUMO)**. In ethylene, the LUMO is the π^* antibonding orbital. If the electrons in the HOMO of butadiene can flow smoothly into the LUMO of ethylene, a concerted reaction can take place.

Figure 15-19 shows that the HOMO of butadiene has the correct symmetry to overlap in phase with the LUMO of ethylene. Having the correct symmetry means the orbitals that form the new bonds can overlap constructively: plus with plus and minus with minus. These bonding interactions stabilize the transition state and promote the concerted reaction. This favorable result shows the reaction is **symmetry-allowed**. The Diels–Alder reaction is common, and this theory correctly predicts a favorable transition state.

15-12B The "Forbidden" [2 + 2] Cycloaddition

If a cycloaddition produces an overlap of positive-phase orbitals with negative-phase orbitals (destructive overlap), antibonding interactions are generated. Antibonding interactions raise the activation energy; thus the reaction is classified as **symmetry-forbidden**. The [2 + 2] cycloaddition of two ethylenes to give cyclobutane is a symmetry-forbidden reaction.

two ethylenes   (transition state)   cyclobutane

This [2 + 2] cycloaddition requires the HOMO of one of the ethylenes to overlap with the LUMO of the other. Figure 15-20 shows that an antibonding interaction results from this overlap, raising the activation energy. For a cyclobutane molecule to result, one of the MOs would have to change its symmetry. Orbital symmetry would not be conserved, so the reaction is symmetry-forbidden. Such a symmetry-forbidden reaction can occasionally be made to occur, but it cannot occur in the concerted pericyclic manner shown above.

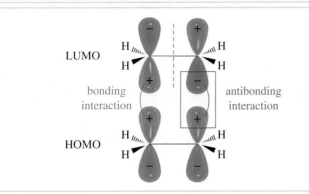

◀ **Figure 15-20**
The HOMO and LUMO of two ethylene molecules have different symmetries, and they overlap to form an antibonding interaction. The concerted $[2 + 2]$ cycloaddition is therefore symmetry-forbidden.

15-12C Photochemical Induction of Cycloadditions

When ultraviolet light rather than heat is used to induce pericyclic reactions, our predictions are generally reversed. For example, the $[2 + 2]$ cycloaddition of two ethylenes is photochemically "allowed." When a photon with the correct energy strikes ethylene, one of the pi electrons is excited to the next higher molecular orbital (Figure 15-21). This higher orbital, formerly the LUMO, is now occupied: It is the new HOMO*, the HOMO of the excited molecule.

Exposure of DNA to ultraviolet light induces a [2 + 2] cyclcoaddition reaction between the double bonds of adjacent thymines. The resulting cyclobutane ring, or thymine dimer, prevents DNA reproduction, and can lead to the development of skin cancer.

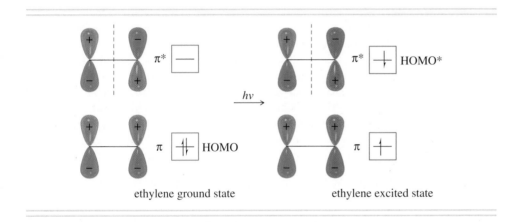

ethylene ground state   ethylene excited state

◀ **Figure 15-21**
Ultraviolet light excites one of the ethylene pi electrons into the antibonding orbital. The antibonding orbital is now occupied, so it is the new HOMO*.

The HOMO* of the excited ethylene molecule has the same symmetry as the LUMO of a ground-state ethylene. An excited molecule can react with a ground-state molecule to give cyclobutane (Figure 15-22). The $[2 + 2]$ cycloaddition is therefore *photochemically allowed* but *thermally forbidden*. In most cases, photochemically allowed reactions are thermally forbidden, and thermally allowed reactions are photochemically forbidden.

PROBLEM 15-19

Show that the $[4 + 2]$ Diels–Alder reaction is photochemically forbidden.

PROBLEM 15-20

(a) Show that the $[4 + 4]$ cycloaddition of two butadiene molecules to give cycloocta-1,5-diene is thermally forbidden but photochemically allowed.
(b) There is a different, thermally allowed cycloaddition of two butadiene molecules. Show this reaction, and explain why it is thermally allowed. (*Hint:* Consider the dimerization of cyclopentadiene.)

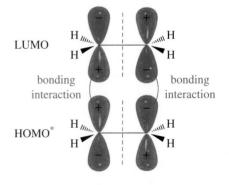

▲ **Figure 15-22**
Photochemical $[2 + 2]$ cycloaddition. The HOMO* of the excited ethylene overlaps favorably with the LUMO of an unexcited (ground-state) molecule.

15-13
Ultraviolet Absorption Spectroscopy

We have already encountered three powerful analytical techniques used by organic chemists. Infrared spectroscopy (IR, Chapter 12) observes the vibrations of molecular bonds, providing information about the nature of the bonding and the functional groups in a molecule. Nuclear magnetic resonance spectroscopy (NMR, Chapter 13) detects nuclear transitions, providing information about the electronic and molecular environment of the nuclei. From the NMR spectrum we can determine the structure of the alkyl groups present and often infer the functional groups. A mass spectrometer (MS, Chapter 12) bombards molecules with electrons, causing them to break apart in predictable ways. The masses of the molecular ion and the fragments provide a molecular weight (and perhaps a molecular formula) as well as structural information about the original compound.

We now study **ultraviolet (UV) spectroscopy**, which detects the electronic transitions of conjugated systems and provides information about the length and structure of the conjugated part of a molecule. UV spectroscopy gives more specialized information than do IR or NMR, and it is less commonly used than the other techniques.

15-13A   Spectral Region

In our study of infrared spectroscopy, we saw that an organic molecule can absorb electromagnetic radiation if the frequency of the waves corresponds to the frequency of the motions of bonds in the molecule. Common IR spectrometers operate at wavelengths between 2.5×10^{-4} and 25×10^{-4} cm, corresponding to energies of about 1.1 to 11 kcal/mol (4.6 to 46 kJ/mol).

Ultraviolet frequencies correspond to shorter wavelengths and much higher energies than infrared (Table 15-1). The UV region is a range of frequencies just beyond the visible: *ultra*, meaning beyond, and *violet*, the highest-frequency visible light. Wavelengths of the UV region are given in units of nanometers (nm; 10^{-9} m). Common UV spectrometers operate in the range of 200 to 400 nm (2×10^{-5} to 4×10^{-5} cm), corresponding to photon energies of about 70 to 140 kcal/mol (300 to 600 kJ/mol). These spectrometers often extend into the visible region (longer wavelength, lower energy) and are called **UV–visible spectrometers**. UV–visible energies correspond to electronic transitions: the energy needed to excite an electron from one molecular orbital to another.

TABLE 15-1   Comparison of Infrared and Ultraviolet Wavelengths

Spectral Region	Wavelength, λ	Energy Range, kcal/mol (kJ/mol)
ultraviolet	200–400 nm ($2–4 \times 10^{-5}$ cm)	70–140 (300–600)
infrared	2.5–25 μm ($2.5–25 \times 10^{-4}$ cm)	1.1–11 (4.6–46)

15-13B   Ultraviolet Light and Electronic Transitions

The wavelengths of UV light absorbed by a molecule are determined by the electronic energy differences between orbitals in the molecule. Sigma bonds are very stable, and the electrons in sigma bonds are usually unaffected by wavelengths of light above 200 nm. Pi bonds have electrons that are more easily excited into higher energy orbitals. Conjugated systems are particularly likely to have low-lying vacant orbitals, and electronic transitions into these orbitals produce characteristic ultraviolet absorptions.

Ethylene, for example, has two pi orbitals: the bonding orbital (π, the HOMO) and the antibonding orbital (π^*, the LUMO). The ground state has two electrons in the bonding orbital and none in the antibonding orbital. A photon with the right amount of energy can excite an electron from the bonding orbital (π) to the antibonding orbital (π^*). This transition from a π bonding orbital to a π^* antibonding orbital is called a $\pi \rightarrow \pi^*$ *transition* (Figure 15-23).

◀ **Figure 15-23**
The absorption of a 171-nm photon excites an electron from the π bonding MO of ethylene to the π^* antibonding MO. This absorption requires light of greater energy (shorter wavelength) than the range covered by a typical UV spectrometer.

The $\pi \rightarrow \pi^*$ transition of ethylene requires absorption of light at 171 nm (164 kcal/mol, or 686 kJ/mol). Most UV spectrometers cannot detect this absorption because it is obscured by the absorption caused by oxygen in the air. In conjugated systems, however, there are electronic transitions with lower energies that correspond to wavelengths longer than 200 nm. Figure 15-24 compares the MO energies of ethylene with those of butadiene to show that the HOMO and LUMO of butadiene are closer in energy than those of ethylene. The HOMO of butadiene is higher in energy than the HOMO of ethylene, and the LUMO of butadiene is lower in energy than the LUMO of ethylene. Both differences reduce the relative energy of the $\pi_2 \rightarrow \pi_3^*$ transition. The resulting absorption is at 217 nm (129 kcal/mol, or 540 kJ/mol), which can be measured using a standard UV spectrometer.

Just as conjugated dienes absorb at longer wavelengths than simple alkenes, conjugated trienes absorb at even longer wavelengths. In general, the energy difference between HOMO and LUMO decreases as the length of conjugation increases.

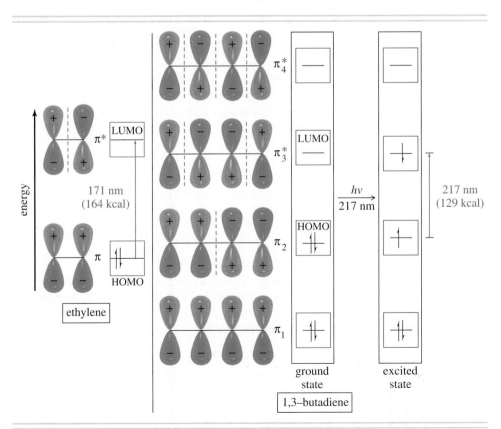

◀ **Figure 15-24**
In 1,3-butadiene, the $\pi \rightarrow \pi^*$ transition absorbs at a wavelength of 217 nm (129 kcal/mol), compared with 171 nm (164 kcal/mol), for ethylene. This longer wavelength (lower-energy) absorption results from a smaller energy difference between the HOMO and LUMO in butadiene than in ethylene.

▶ **Figure 15-25**
1,3,5-Hexatriene has a smaller energy difference (108 kcal/mol) between its HOMO and LUMO than does 1,3-butadiene (129 kcal/mol). The $\pi \rightarrow \pi^*$ transition corresponding to this energy difference absorbs at a longer wavelength: 258 nm, compared with 217 nm for 1,3-butadiene.

Carotene derivatives absorb different wavelengths of light, depending on the length of the conjugated system and the presence of other functional groups.

In 1,3,5-hexatriene, for example (Fig. 15-25), the HOMO is π_3, and the LUMO is π_4^*. The HOMO in 1,3,5-hexatriene is slightly higher in energy than that for 1,3-butadiene, and the LUMO is slightly lower in energy. Once again, the narrowing of the energy between the HOMO and the LUMO gives a lower energy, longer wavelength absorption. The principal $\pi \rightarrow \pi^*$ transition in 1,3,5-hexatriene occurs at 258 nm (108 kcal/mol, or 452 kJ/mol).

We can summarize the effects of conjugation on the wavelength of UV absorption by stating a general rule: *A compound that contains a longer chain of conjugated double bonds absorbs light at a longer wavelength.* β-Carotene, which has 11 conjugated double bonds in its pi system, absorbs at 454 nm, well into the visible region of the spectrum, corresponding to absorption of blue light. White light from which blue has been removed appears orange. β-Carotene is the principal compound responsible for giving carrots their orange color. Carotene derivatives provide many of the colors we see in fruits, vegetables, and autumn leaves.

β-carotene

Because they have no interaction with each other, isolated double bonds do not contribute to shifting the UV absorption to longer wavelengths. Both their reactions and their UV absorptions are like those of simple alkenes. For example, 1,4-pentadiene absorbs at 178 nm, a value that is typical of simple alkenes rather than conjugated dienes.

	isolated	conjugated
1-pentene, 176 nm	1,4-pentadiene, 178 nm	1,3-pentadiene, 223 nm

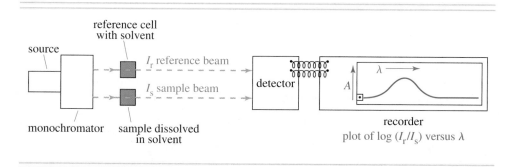

15-13C   Obtaining an Ultraviolet Spectrum

To measure the ultraviolet (or UV–visible) spectrum of a compound, the sample is dissolved in a solvent (often ethanol) that does not absorb above 200 nm. The sample solution is placed in a quartz cell, and some of the solvent is placed in a **reference cell**. An ultraviolet spectrometer operates by comparing the amount of light transmitted through the sample (the **sample beam**) with the amount of light in the **reference beam**. The reference beam passes through the reference cell to compensate for any absorption of light by the cell and the solvent.

The spectrometer (Figure 15-26) has a *source* that emits all frequencies of UV light (above 200 nm). This light passes through a *monochromator*, which uses a diffraction grating or a prism to spread the light into a spectrum and select one wavelength. This single wavelength of light is split into two beams, with one beam passing through the sample cell and the other passing through the reference (solvent) cell. The detector continuously measures the intensity ratio of the reference beam (I_r) compared with the sample beam (I_s). As the spectrometer scans the wavelengths in the UV region, a chart recorder draws a graph (called a *spectrum*) of the absorbance of the sample as a function of the wavelength.

The *absorbance, A,* of the sample at a particular wavelength is governed by *Beer's law*.

$$A = \log\left(\frac{I_r}{I_s}\right) = \varepsilon c l$$

where

c = sample concentration in moles per liter
l = path length of light through the cell in centimeters
ε = the **molar absorptivity** (or **molar extinction coefficient**) of the sample.
Molar absorptivity (ε) is a measure of how strongly the sample absorbs light at that wavelength.

If the sample absorbs light at a particular wavelength, the sample beam (I_s) is less intense than the reference beam (I_r), and the ratio I_r/I_s is greater than 1. The ratio is equal to 1 when there is no absorption. The absorbance (the logarithm of the ratio) is therefore greater than zero when the sample absorbs and equal to zero when it does not. A UV spectrum is a plot of A, the absorbance of the sample, as a function of the wavelength.

UV–visible spectra tend to show broad peaks and valleys. The spectral data that are most characteristic of a sample are as follows:

1. The wavelength(s) of maximum absorbance, called λ_{max}
2. The value of the molar absorptivity ε at each maximum

Since UV–visible spectra are broad and lacking in detail, they are rarely printed as actual spectra. The spectral information is given as a list of the value or values of λ_{max} together with the molar absorptivity for each value of λ_{max}.

The molar extinction coefficient ε, associated with a wavelength of maximum absorbance λ_{max}, is particularly useful for determining drug concentrations. For example, the concentration of tetracycline is measured at 380 nm where the molar absorptivity value is 16,200.

▶ **Figure 15-27**
The UV spectrum of isoprene dissolved in methanol shows $\lambda_{max} = 222$ nm, $\varepsilon = 20,000$.

The UV spectrum of isoprene (2-methyl-1,3-butadiene) is shown in Figure 15-27. This spectrum could be summarized as follows:

$$\lambda_{max} = 222 \text{ nm} \qquad \varepsilon = 20,000$$

The value of λ_{max} is read directly from the spectrum, but the molar absorptivity ε must be calculated from the concentration of the solution and the path length of the cell. For an isoprene concentration of 4×10^{-5} M and a 1-cm cell, the molar absorptivity is found by rearranging Beer's law ($A = \varepsilon cl$).

$$\varepsilon = \frac{A}{cl} = \frac{0.8}{4 \times 10^{-5}} = 20,000$$

Molar absorptivities in the range of 5000 to 30,000 are typical for the $\pi \longrightarrow \pi^*$ transitions of conjugated polyene systems. Such large molar absorptivities are helpful, since spectra may be obtained with very small amounts of sample. On the other hand, samples and solvents for UV spectroscopy must be extremely pure. A minute impurity with a large molar absorptivity can easily obscure the spectrum of the desired compound.

PROBLEM 15-21

One milligram of a compound of molecular weight 160 is dissolved in 10 ml of ethanol, and the solution is poured into a 1-cm UV cell. The UV spectrum is taken, and there is an absorption at $\lambda_{max} = 247$ nm. The maximum absorbance at 247 nm is 0.50. Calculate the value of ε for this absorption.

15-13D   Interpreting UV–Visible Spectra

The values of λ_{max} and ε for conjugated molecules depend on the exact nature of the conjugated system and its substituents. R. B. Woodward and L. F. Fieser developed an extensive set of correlations between molecular structures and absorption maxima, called the **Woodward–Fieser rules**. These rules are summarized in Appendix 3. For most purposes, however, we can use some simple generalizations for estimating approximate values of λ_{max} for common types of systems. Table 15-2 gives the values of λ_{max} for several types of isolated alkenes, conjugated dienes, conjugated trienes, and a conjugated tetraene.

TABLE 15-2 Ultraviolet Absorption Maxima of Some Representative Molecules

Isolated

$$CH_2=CH_2$$

ethylene	cyclohexene	1,4-hexadiene
λ_{max}: 171 nm	182 nm	180 nm

Conjugated dienes

1,3-butadiene	2,4-hexadiene	1,3-cyclohexadiene	3-methylenecyclohexene
λ_{max}: 217 nm	227 nm	256 nm	232 nm

Conjugated trienes

1,3,5-hexatriene	a steroid triene
λ_{max}: 258 nm	304 nm

Conjugated tetraene

1,3,5,7-octatetraene
290 nm

The examples in Table 15-2 show that the addition of another conjugated double bond to a conjugated system has a large effect on λ_{max}. In going from ethylene (171 nm) to 1,3-butadiene (217 nm) to 1,3,5-hexatriene (258 nm) to 1,3,5,7-octatetraene (290 nm), the values of λ_{max} increase by about 30 to 40 nm for each double bond extending the conjugated system. Alkyl groups also increase the value of λ_{max} by about 5 nm per alkyl group. For example, 2,4-dimethyl-1,3-pentadiene has the same conjugated system as 1,3-butadiene, but with three additional alkyl groups (circled below). Its absorption maximum is at 232 nm, a wavelength 15 nm longer than λ_{max} for 1,3-butadiene (217 nm).

1,3-butadiene	2,4-dimethyl-1,3-pentadiene
$\lambda_{max} = 217$ nm	3 additional alkyl groups, $\lambda_{max} = 232$ nm

Structural Difference	*Effect on* λ_{max}
additional conjugated $C=C$	30–40 nm longer
additional alkyl substituent	about 5 nm longer

PROBLEM-SOLVING HINT

Some good rules of thumb: An additional conjugated $C=C$ increases λ_{max} about 30 to 40 nm; an additional alkyl group increases it about 5 nm. Useful base values:

217 nm   256 nm   232 nm

SOLVED PROBLEM 15-3

Rank the following dienes in order of increasing values of λ_{max}. (Their actual absorption maxima are 185 nm, 235 nm, 273 nm, and 300 nm).

SOLUTION

λ_{max}:   185 nm   235 nm   273 nm   300 nm

These compounds are an isolated diene, two conjugated dienes, and a conjugated triene. The isolated diene will have the shortest value of λ_{max} (185 nm), close to that of cyclohexene (182 nm).

The second compound looks like 3-methylenecyclohexene (232 nm) with an additional alkyl substituent (circled). Its absorption maximum should be around (232 + 5) nm, and 235 nm must be the correct value.

The third compound looks like 1,3-cyclohexadiene (256 nm), but with an additional alkyl substituent (circled) raising the value of λ_{max}, so 273 nm must be the correct value.

The fourth compound looks like 1,3-cyclohexadiene (256 nm), but with an additional conjugated double bond (circled) and another alkyl group (circled). We predict a value of λ_{max} about 35 nm longer than for 1,3-cyclohexadiene, so 300 nm must be the correct value.

PROBLEM 15-22

Using the examples in Table 15-2 to guide you, match the following UV absorption maxima (λ_{max}) with the corresponding compounds: (1) 232 nm; (2) 237 nm; (3) 273 nm; (4) 283 nm; (5) 313 nm; (6) 353 nm.

(a)   (b)   (c)   (d)   (e)

Chapter 15 Glossary

1,2-addition An addition in which two atoms or groups add to adjacent atoms. (p. 645)

a 1,2-addition

1,4-addition An addition in which two atoms or groups add to atoms that bear a 1,4-relationship. (p. 645)

a 1,4-addition

allyl group The common name for the 2-propenyl group, $-CH_2-CH=CH_2$ (p. 644)
allylic position The carbon atom next to a carbon–carbon double bond. The term is used in naming compounds, such as an **allylic halide**, or in referring to reactive intermediates, such as an **allylic cation**, an **allylic radical**, or an **allylic anion**. (p. 644)

allylic position

$$H_2C=CH-CHBr-CH_3 \qquad (CH_3)_2C=CH-\overset{+}{C}(CH_3)_2$$

an allylic halide   an allylic cation

allylic shift The isomerization of a double bond that occurs through the delocalization of an allylic intermediate. (p. 649)

$$H_2C{=}CH{-}CH_2{-}CH_3 \xrightarrow[h\nu]{NBS} H_2C{=}CH{-}CHBr{-}CH_3 + BrCH_2{-}CH{=}CH{-}CH_3$$
allylic shift product

concerted reaction A reaction in which all bond making and bond breaking occurs in the same step. (p. 655)

conjugated double bonds Double bonds that alternate with single bonds, with interaction by overlap of the p orbitals in the pi bonds. (p. 638)

conjugated   isolated   cumulated

    isolated double bonds: Double bonds separated by two or more single bonds. Isolated double bonds react independently, as they do in a simple alkene.
    cumulated double bonds: Successive double bonds with no intervening single bonds.
    allene (cumulene): A compound containing cumulated carbon–carbon double bonds.

conservation of orbital symmetry A theory of pericyclic reactions stating that the MOs of the reactants must flow smoothly into the MOs of the products without any drastic changes in symmetry. That is, there must be bonding interactions to help stabilize the transition state. (p. 663)

constructive overlap An overlap of orbitals that contributes to bonding. Overlap of lobes with similar phases (+ phase with + phase, or − phase with − phase) is generally constructive overlap. (p. 642)

cycloaddition A reaction of two alkenes or polyenes to form a cyclic product. Cycloadditions often take place through concerted interaction of the pi electrons in two unsaturated molecules. (p. 663)

delocalized orbital A molecular orbital that results from the combination of three or more atomic orbitals. When filled, these orbitals spread electron density over all the atoms involved. (p. 641)

destructive overlap An overlap of orbitals that contributes to antibonding. Overlap of lobes with opposite phases (+ phase with − phase) is generally destructive overlap. (p. 642)

Diels–Alder reaction A synthesis of six-membered rings by a **[4 + 2] cycloaddition**. This notation means that four pi electrons in one molecule interact with two pi electrons in the other molecule to form a new ring. (p. 655)

    dienophile: The component with two pi electrons that reacts with a diene in the Diels–Alder reaction.

    endo rule: The stereochemical preference for electron-poor substituents on the dienophile to assume endo positions in a bicyclic Diels–Alder product.

    secondary overlap: Overlap of the p orbitals of the electron-withdrawing group of the dienophile with those of the central atoms (C2 and C3) of the diene. This overlap helps stabilize the transition state. With cyclic dienes, it favors endo products. (p. 659)

cyclopentadiene   acrylonitrile   Diels–Alder adduct
a diene   a dienophile   *endo* stereochemistry

heat of hydrogenation The enthalpy of reaction that accompanies the addition of hydrogen to a mole of an unsaturated compound. (p. 638)

$$\Delta H° = \text{heat of hydrogenation}$$

HOMO An acronym for **highest occupied molecular orbital**. In a photochemically excited state, this orbital is represented as HOMO*. (p. 664)

kinetic control Product distribution is governed by the rates at which the various products are formed. (p. 648)

> **kinetic product:** The product that is formed fastest; the major product under kinetic control.

LUMO An acronym for **lowest unoccupied molecular orbital**. (p. 664)

molar absorptivity, ε (molar extinction coefficient) A measure of how strongly a compound absorbs light at a particular wavelength. It is defined by *Beer's law*,

$$A = \log\left(\frac{I_r}{I_s}\right) = \varepsilon c l$$

where A is the absorbance, I_r and I_s are the amounts of light passing through the reference and sample beams, c is the sample concentration in moles per liter, and l is the path length of light through the cell. (p. 669)

molecular orbitals (MOs) Orbitals that include more than one atom in a molecule. Molecular orbitals can be bonding, antibonding, or nonbonding. (p. 640)

> **bonding molecular orbitals:** MOs that are lower in energy than the isolated atomic orbitals from which they are made. Electrons in these orbitals serve to hold the atoms together.
>
> **antibonding molecular orbitals:** MOs that are higher in energy than the isolated atomic orbitals from which they are made. Electrons in these orbitals tend to push the atoms apart.
>
> **nonbonding molecular orbitals:** MOs with the same energy as the isolated atomic orbitals from which they are made. Electrons in these orbitals have no effect on the bonding of the atoms. (p. 652)

node A region of a molecular orbital with zero electron density. (p. 642)

pericyclic reaction A reaction involving concerted reorganization of electrons within a closed loop of interacting orbitals. Cycloadditions are one class of pericyclic reactions. (p. 663)

reference beam A second beam in the spectrometer that passes through a **reference cell** containing only the solvent. The **sample beam** is compared with this beam to compensate for any absorption by the cell or the solvent. (p. 669)

resonance energy The extra stabilization provided by delocalization, compared with a localized structure. For dienes and polyenes, the resonance energy is the extra stability of the conjugated system compared with the energy of a compound with an equivalent number of isolated double bonds. (p. 640)

s-cis conformation A cis-like conformation of a single bond in a conjugated diene or polyene. (p. 643)

s-trans conformation A trans-like conformation of a single bond in a conjugated diene or polyene. (p. 643)

s-cis conformation   *s*-trans conformation

symmetry-allowed The MOs of the reactants can flow into the MOs of the products in one concerted step according to the rules of conservation of orbital symmetry. In a symmetry-allowed cycloaddition there is constructive overlap (+ phase with + phase, − phase with − phase) between the HOMO of one molecule and the LUMO of the other. (p. 664)

symmetry-forbidden The MOs of the reactants are of incorrect symmetries to flow into those of the products in one concerted step. (p. 664)

thermodynamic control (equilibrium control) Product distribution is governed by the stabilities of the products. Thermodynamic control operates when the reaction mixture is allowed to come to equilibrium. (p. 648)

> **thermodynamic product:** The most stable product; the major product under thermodynamic control.

UV–visible spectroscopy The measurement of the absorption of ultraviolet and visible light as a function of wavelength. Ultraviolet light consists of wavelengths from about 100 to 400 nm. Visible light is from about 400 nm (violet) to 750 nm (red). (p. 666)

Woodward–Fieser rules A set of rules that correlate values of λ_{max} in the UV–visible spectrum with structures of conjugated systems. (p. 670 and Appendix 3)

Woodward–Hoffman rules A set of symmetry rules that predict whether a particular pericyclic reaction is symmetry-allowed or symmetry-forbidden. (p. 663)

Essential Problem-Solving Skills in Chapter 15

1. Show how to construct the molecular orbitals of ethylene, butadiene, and the allylic system. Show the electronic configurations of ethylene, butadiene, and the allyl cation, radical, and anion.

2. Recognize reactions that are enhanced by resonance stabilization of the intermediates, such as free-radical reactions and cationic reactions. Develop mechanisms to explain the enhanced rates and observed products, and draw resonance forms of the stabilized intermediates.

3. Predict the products of Diels–Alder reactions, including the orientation of cycloaddition with unsymmetrical reagents and the stereochemistry of the products.

4. By comparing the molecular orbitals of the reactants, predict which cycloadditions will be thermally allowed and which will be photochemically allowed.

5. Use values of λ_{max} from UV–visible spectra to estimate the length of conjugated systems, and compare compounds with similar structures.

Study Problems

15-23   Briefly define each term and give an example.

(a) λ_{max} (b) molar absorptivity   (c) an allylic alcohol
(d) an endo product   (e) conjugated double bonds   (f) cumulated double bonds
(g) isolated double bonds   (h) a substituted allene   (i) a molecular orbital
(j) an antibonding MO   (k) an allylic radical   (l) an s-trans conformation
(m) a 1,2-addition   (n) a 1,4-addition   (o) a cycloaddition
(p) kinetic control of a reaction (q) a Diels–Alder reaction   (r) thermodynamic control
(s) a dienophile   (t) a concerted reaction   (u) a HOMO, a HOMO*, and a LUMO
(v) a symmetry-forbidden reaction

15-24   Classify the following dienes and polyenes as isolated, conjugated, cumulated, or some combination of these classifications.

(a) 1,5-cyclooctadiene   (b) 1,3-cyclooctadiene   (c) 1,2-cyclodecadiene
(d) 1,3,6-cyclooctatriene   (e) 1,3,5-cyclohexatriene (benzene)

15-25   Predict the products of the following reactions.

(a) allyl bromide + cyclohexyl magnesium bromide   (b) cyclopentadiene + anhydrous HCl
(c) 2-methylpropene + NBS, light   (d) 1-pentene + NBS, light
(e) 1,3-butadiene + bromine water   (f) 1,3,5-hexatriene + bromine in CCl_4
(g) 1-(bromomethyl)-2-methylcyclopentene, heated in methanol
(h) cyclopentadiene + methyl acrylate, $CH_2\!=\!CH\!-\!COOCH_3$
(i) 1,3-cyclohexadiene + $CH_3OOC\!-\!C\!\equiv\!C\!-\!COOCH_3$

15-26   Show how the reaction of an allylic halide with a Grignard reagent might be used to synthesize the following hydrocarbons.

(a) 5-methyl-1-hexene   (b) 2,5,5-trimethyl-2-heptene

15-27   Draw the important resonance contributors for the following cations, anion, and radical.

(a)   (b)   (c)   (d)

15-28   A solution was prepared using 0.0010 g of an unknown steroid (of molecular weight around 255) in 100 ml of ethanol. Some of this solution was placed in a 1-cm cell, and the UV spectrum was measured. This solution was found to have $\lambda_{max} = 235$ nm, with $A = 0.74$.
(a) Compute the value of the molar absorptivity at 235 nm.
(b) Which of the following compounds might give this spectrum?

15-29   When *N*-bromosuccinimide is added to 1-hexene and a sunlamp is shone on the mixture, three products result.
(a) Give the structures of these three products.
(b) Propose a mechanism that accounts for the formation of these three products.

15-30   Predict the products of the following Diels–Alder reactions. Include stereochemistry where appropriate.

15-31   For each structure
(1) Draw all the important resonance contributors.
(2) Evaluate the significance of each resonance contributor.

15-32   A graduate student was following a procedure to make 3-propyl-1,4-cyclohexadiene. During the workup procedure, his research adviser called him into her office. By the time the student returned to his bench, the product had warmed to a higher temperature than recommended. He isolated the product, which gave the appropriate $=$C—H stretch in the IR, but the C$=$C stretch appeared around 1630 cm^{-1} as opposed to the literature value of 1650 cm^{-1} for the desired product. The mass spectrum showed the correct molecular weight, but the base peak was at M − 29 rather than at M − 43 as expected.
(a) Should he have his IR recalibrated or should he repeat the experiment, watching the temperature more carefully? What does the 1630 cm^{-1} absorption suggest?
(b) Draw the structure of the desired product, and propose a structure for the actual product.
(c) Show why he expected the MS base peak to be at M − 43, and show how your proposed structure would give an intense peak at M − 29.

15-33   Show how Diels–Alder reactions might be used to synthesize the following compounds.

(d)

(e)

(f)

(g)

chlordane

(h)

aldrin

(i)

*15-34 Furan and maleimide undergo a Diels–Alder reaction at 25°C to give the *endo* isomer of the product. When the reaction takes place at 90°C, however, the major product is the *exo* isomer. Further study shows that the *endo* isomer of the product isomerizes to the *exo* isomer at 90°C.

furan: O   maleimide: N—H

(a) Draw and label the *endo* and *exo* isomers of the Diels–Alder adduct of furan and maleimide.
(b) Which isomer of the product would you usually expect from this reaction? Explain why this isomer is usually favored.
(c) Examine your answer to (b) and determine whether this answer applies to a reaction that is kinetically controlled or one that is thermodynamically controlled or both.
(d) Explain why the *endo* isomer predominates when the reaction takes place at 25°C and why the *exo* isomer predominates at 90°C.

*15-35 **(a)** Sketch the pi molecular orbitals of 1,3,5-hexatriene (Figure 15-25).
(b) Show the electronic configuration of the ground state of 1,3,5-hexatriene.
(c) Show what product would result from the $[6 + 2]$ cycloaddition of 1,3,5-hexatriene with maleic anhydride.

1,3,5-hexatriene   maleic anhydride

(d) Show that the $[6 + 2]$ cyclization of 1,3,5-hexatriene with maleic anhydride is thermally forbidden but photochemically allowed.
(e) Show the Diels–Alder product that would actually result from heating 1,3,5-hexatriene with maleic anhydride.

*15-36 The pentadienyl radical, $H_2C{=}CH{-}CH{=}CH{-}CH_2\cdot$, has its odd electron delocalized over three carbon atoms.
(a) Use resonance forms to show which three carbon atoms bear the odd electron.
(b) How many MOs are there in the molecular orbital picture of the pentadienyl radical?
(c) How many nodes are there in the lowest-energy MO of the pentadienyl system? How many in the highest-energy MO?
(d) Draw the MOs of the pentadienyl system in order of increasing energy.
(e) Show how many electrons are in each MO for the pentadienyl radical (ground state).
(f) Show how your molecular orbital picture agrees with the resonance picture showing delocalization of the odd electron onto three carbon atoms.
(g) Remove the highest-energy electron from the pentadienyl radical to give the pentadienyl cation. Which carbon atoms share the positive charge? Does this picture agree with the resonance picture?
(h) Add an electron to the pentadienyl radical to give the pentadienyl anion. Which carbon atoms share the negative charge? Does this picture agree with the resonance picture?

*15-37   A student was studying terpene synthesis, and she wanted to make the compound shown below. First she converted
3-bromo-6-methylcyclohexene to alcohol **A**, shown below. She heated alcohol **A** with sulfuric acid and purified one of the
components (compound **B**) from the resulting mixture. Compound **B** has the correct molecular formula for the desired
product.

(a) Suggest how 3-bromo-6-methylcyclohexene might be converted to alcohol **A**.

(b) The UV spectrum of compound **B** shows λ_{max} at 273 nm. Is Compound **B** the correct product? If not, suggest a structure for compound **B** consistent with these UV data.

(c) Propose a mechanism for the dehydration of alcohol **A** to compound **B**.

*15-38   Part of a synthesis by E. J. Corey and David Watt (Harvard University) involves the Diels–Alder cycloaddition of the
following pyrone and cyclohexenone. The initial reaction gives the *endo* product, which loses carbon dioxide in a retro-
Diels–Alder to generate a diene with predictable stereochemistry and functionality. IR and UV spectroscopy of the final
product shows that it contains a diene conjugated with an ester, and an unconjugated ketone. Determine the structures of
the intermediate and the final product, with particular attention to their stereochemistry.

Aromatic Compounds

In 1825, Michael Faraday isolated a pure compound of boiling point 80°C from the oily mixture that condensed from illuminating gas, the fuel burned in gaslights. Elemental analysis showed an unusually small hydrogen-to-carbon ratio of 1 : 1, corresponding to an empirical formula of CH. Faraday named the new compound "bicarburet of hydrogen." Eilhard Mitscherlich synthesized the same compound in 1834 by heating benzoic acid, isolated from gum benzoin, in the presence of lime. Like Faraday, Mitscherlich found that the empirical formula was CH. He also used a vapor-density measurement to determine a molecular weight of about 78, for a molecular formula of C_6H_6. Since the new compound was derived from gum benzoin, he named it benzin, now called *benzene*.

Many other compounds discovered in the nineteenth century seemed to be related to benzene. These compounds also had low hydrogen-to-carbon ratios as well as pleasant aromas, and they could be converted to benzene or related compounds. This group of compounds was called **aromatic** because of their pleasant odors. Other organic compounds, without these properties, were called **aliphatic**, meaning "fatlike." As the unusual stability of aromatic compounds was investigated, the term "aromatic" came to be applied to compounds with this stability, regardless of their odors.

16-1
Introduction; The Discovery of Benzene

The Kekulé Structure   In 1866, Friedrich Kekulé proposed a cyclic structure for benzene with three double bonds. Considering that multiple bonds had been proposed only recently (1859), the cyclic structure with alternating single and double bonds was considered somewhat bizarre.

16-2
The Structure and Properties of Benzene

The **Kekulé structure** has its shortcomings, however. For example, it predicts two different 1,2-dichlorobenzenes, but only one is known to exist. Kekulé suggested (incorrectly) that a fast equilibrium interconverts the two isomers of 1,2-dichlorobenzene.

1,2-dichlorobenzene

The Resonance Representation   The resonance picture of benzene is a natural extension of Kekulé's hypothesis. In a Kekulé structure, the C—C single bonds would be longer than the double bonds. Spectroscopic methods have shown that the benzene ring is planar and all the bonds are the same length (1.397 Å). Because the ring is planar and the carbon nuclei are positioned at equal distances, the two Kekulé structures differ only in the positioning of the pi electrons.

Benzene is actually a resonance hybrid of the two Kekulé structures. This representation implies that the pi electrons are delocalized, with a bond order of $1\frac{1}{2}$ between adjacent carbon atoms. The carbon–carbon bond lengths in benzene are shorter than typical single-bond lengths, yet longer than typical double-bond lengths.

all C—C bond
lengths 1.397 Å

double bond
1.34 Å

single bond
1.48 Å

resonance representation

bond order = $1\frac{1}{2}$

butadiene

combined representation

The resonance-delocalized picture explains most of the structural properties of benzene and its derivatives—the *benzenoid* aromatic compounds. Because the pi bonds are delocalized over the ring, we often inscribe a circle in the hexagon rather than drawing three localized double bonds. This representation helps us remember there are no localized single or double bonds, and it prevents us from trying to draw supposedly different isomers that differ only in the placement of double bonds in the ring. We often use Kekulé structures in drawing reaction mechanisms, however, to show the movement of individual pairs of electrons.

PROBLEM 16-1
Write Lewis structures for the Kekulé representations of benzene. Show all the valence electrons.

Using this resonance picture, we can draw a more realistic representation of benzene (Figure 16-1). Benzene is a ring of six sp^2 hybrid carbon atoms, each bonded to one hydrogen atom. All the carbon–carbon bonds are the same length, and all of the bond angles are exactly 120°. Each sp^2 carbon atom has an unhybridized p orbital perpendicular to the plane of the ring, and six electrons occupy this circle of p orbitals.

At this point, we can define an **aromatic compound** to be a cyclic compound containing some number of conjugated double bonds and having an unusually large resonance energy. Using benzene as the example, we will consider how aromatic compounds differ from aliphatic compounds. Then we will discuss why an aromatic structure confers extra stability and how we can predict aromaticity in some interesting and unusual compounds.

Friedrich August Kekulé von
Stradoniz (1829–96), pictured on a
Belgian postage stamp.

▶ **Figure 16-1**
Benzene is a flat ring of sp^2 hybrid
carbon atoms with their unhybridized
p orbitals all aligned and overlapping.
The ring of p orbitals contains six
electrons. The carbon–carbon bond
lengths are all 1.397 Å, and all the
bond angles are exactly 120°.

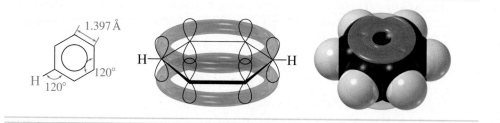

1.397 Å

120°

120°

The Unusual Reactions of Benzene Benzene is actually much more stable than we would expect from the simple resonance-delocalized picture. Both the Kekulé structure and the resonance-delocalized picture show that benzene is a cyclic conjugated triene. We might expect benzene to undergo the typical reactions of polyenes. In fact, its reactions are quite unusual. For example, an alkene decolorizes potassium permanganate by reacting to form a glycol (Section 8-14). The purple permanganate color disappears, and a precipitate of manganese dioxide forms. When permanganate is added to benzene, however, no reaction occurs.

Most alkenes decolorize solutions of bromine in carbon tetrachloride (Section 8-10). The red bromine color disappears as bromine adds across the double bond. When bromine is added to benzene, no reaction occurs, and the red bromine color remains.

Addition of a catalyst such as ferric bromide to the mixture of bromine and benzene causes the bromine color to disappear slowly. HBr gas is evolved as a by-product, but the expected *addition* of Br_2 does not take place. Instead, the organic product results from *substitution* of a bromine atom for a hydrogen, and all three double bonds are retained.

is not formed

The Unusual Stability of Benzene Benzene's reluctance to undergo typical alkene reactions suggests that it must be unusually stable. By comparing molar heats of hydrogenation, we can get a quantitative idea of its stability. Benzene, cyclohexene, and the cyclohexadienes all hydrogenate to cyclohexane. Figure 16-2 shows how the experimentally determined heats of hydrogenation are used to compute the resonance energies of 1,3-cyclohexadiene and benzene, based on the following reasoning.

1. Hydrogenation of cyclohexene is exothermic by 28.6 kcal/mol (120 kJ/mol).
2. Hydrogenation of 1,4-cyclohexadiene is exothermic by 57.4 kcal/mol (240 kJ/mol), about twice the heat of hydrogenation of cyclohexene. The resonance energy of the isolated double bonds in 1,4-cyclohexadiene is about zero.

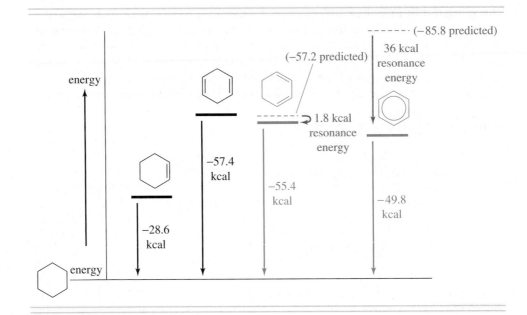

▶ **Figure 16-2**
The molar heats of hydrogenation and the relative energies of cyclohexene, 1,4-cyclohexadiene, 1,3-cyclohexadiene, and benzene. The dashed lines represent the energies that would be predicted if every double bond had the same energy as the double bond in cyclohexene.

3. Hydrogenation of 1,3-cyclohexadiene is exothermic by 55.4 kcal/mol (232 kJ/mol), about 1.8 kcal (8 kJ) less than twice the value for cyclohexene. A resonance energy of 1.8 kcal (8 kJ) is typical for a conjugated diene.

4. Hydrogenation of benzene requires higher pressures of hydrogen and a more active catalyst. This hydrogenation is exothermic by 49.8 kcal/mol (208 kJ/mol), about 36.0 kcal (152 kJ) less than three times the value for cyclohexene.

$$
\begin{aligned}
\Delta H^\circ &= -49.8 \text{ kcal} \\
3 \times \text{cyclohexene} &= -85.8 \text{ kcal} \\
\hline
\text{resonance energy} &= 36.0 \text{ kcal}
\end{aligned}
$$

The huge 36 kcal/mol (152 kJ/mol) resonance energy of benzene cannot be explained by conjugation effects alone. The heat of hydrogenation for benzene is actually smaller than that for 1,3-cyclohexadiene. The hydrogenation of the *first* double bond of benzene is endothermic, the first endothermic hydrogenation we have encountered. In practice, this reaction is difficult to stop after the addition of 1 mole of H_2 because the product (1,3-cyclohexadiene) hydrogenates more easily than benzene itself. Clearly, the benzene ring is exceptionally unreactive.

$$
\begin{aligned}
&\Delta H^\circ_{\text{hydrogenation}} \\
\text{benzene:} \quad &-49.8 \text{ kcal } (-208 \text{ kJ}) \\
\text{1,3-cyclohexadiene:} \quad &-55.4 \text{ kcal } (-232 \text{ kJ}) \\
\hline
\Delta H^\circ = \ &+5.6 \text{ kcal } (+24 \text{ kJ})
\end{aligned}
$$

PROBLEM 16-2

Using the information in Figure 16-2, calculate the values of ΔH° for the following reactions:

(a) [benzene] + H_2 $\xrightarrow{\text{catalyst}}$ [1,3-cyclohexadiene]   (b) [benzene] + 2 H_2 $\xrightarrow{\text{catalyst}}$ [cyclohexene]

(c) [1,3-cyclohexadiene] + H_2 $\xrightarrow{\text{catalyst}}$ [cyclohexane]

Failures of the Resonance Picture For many years, chemists assumed that benzene's large resonance energy results from having two identical, stable resonance structures. It was thought that other hydrocarbons with analogous conjugated systems of alternating single and double bonds would show similar stability. These cyclic hydrocarbons with alternating single and double bonds are called **annulenes**. For example, benzene is the six-membered annulene, so it can be named [6]annulene. Cyclobutadiene is [4]annulene, cyclooctatetraene is [8]annulene, and larger annulenes are named similarly.

cyclobutadiene [4]annulene	benzene [6]annulene	cyclooctatetraene [8]annulene	cyclodecapentaene [10]annulene

For the double bonds to be completely conjugated, the annulene must be planar so the p orbitals of the pi bonds can overlap. As long as an annulene is assumed to be planar, we can draw two Kekulé-like structures that seem to show a benzene-like resonance. Figure 16-3 shows proposed benzene-like resonance forms for cyclobutadiene and cyclooctatetraene. Although these resonance structures imply that the [4] and [8]annulenes should be unusually stable (like benzene), experiments have shown that cyclobutadiene and cyclooctatetraene are not unusually stable. These results imply that the simple resonance picture is incorrect.

Cyclobutadiene has never been isolated and purified. It undergoes an extremely fast Diels–Alder dimerization. To avoid the Diels–Alder reaction, cyclobutadiene has been prepared at low concentrations in the gas phase and as individual molecules trapped in frozen argon at low temperatures. This is not the behavior we expect from a molecule with exceptional stability!

In 1911, Richard Willstätter synthesized cyclooctatetraene and found that it reacts like a normal polyene. Bromine adds readily to cyclooctatetraene, and permanganate oxidizes its double bonds. This evidence shows that cyclooctatetraene is much less stable than benzene. In fact, structural studies have shown that cyclooctatetraene is not planar. It is most stable in a "tub" conformation, with poor overlap between adjacent pi bonds.

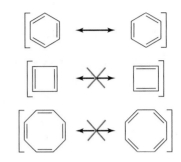

▲ **Figure 16-3**
Cyclobutadiene and cyclooctatetraene have alternating single and double bonds similar to those of benzene. These compounds were mistakenly expected to be aromatic.

"tub" conformation of cyclooctatetraene

PROBLEM 16-3

(a) Draw the resonance structures of benzene, cyclobutadiene, and cyclooctatetraene, showing all the carbon and hydrogen atoms.
(b) Assuming that these molecules are all planar, show how the p orbitals on the sp^2 hybrid carbon atoms form continuous rings of overlapping orbitals above and below the plane of the carbon atoms.

PROBLEM 16-4

Show the product of the Diels–Alder dimerization of cyclobutadiene. (This reaction is similar to the dimerization of cyclopentadiene, discussed in Section 15-11.)

Visualizing benzene as a resonance hybrid of two Kekulé structures cannot fully explain the unusual stability of the aromatic ring. As we have seen with other conjugated systems, molecular orbital theory provides the key to understanding aromaticity and predicting which compounds will have the stability of an aromatic system.

16-3
The Molecular Orbitals of Benzene

Benzene has a planar ring of six sp^2 hybrid carbon atoms, each with an unhybridized p orbital that overlaps with the p orbitals of its neighbors to form a continuous ring of orbitals above and below the plane of the carbon atoms. Six pi electrons are contained in this ring of overlapping p orbitals.

The six overlapping p orbitals create a cyclic system of molecular orbitals. Cyclic systems of molecular orbitals differ from linear systems such as 1,3-butadiene and the allyl system. A two-dimensional cyclic system requires two-dimensional MOs, with the possibility of two distinct MOs having the same energy. We can still follow the same principles in developing a molecular orbital representation for benzene, however.

PROBLEM-SOLVING HINT

These principles, used in drawing the MOs of benzene, are applicable to many MO problems.

1. There are six atomic p orbitals that overlap to form the benzene pi system. Therefore, there must be six molecular orbitals.

2. The lowest-energy molecular orbital is entirely bonding, with constructive overlap between all pairs of adjacent p orbitals. There are no nodes in this lowest lying MO.

3. The number of nodes increases as the MOs increase in energy.

4. The MOs should be evenly divided between bonding and antibonding MOs, with the possibility of nonbonding MOs in some cases.

Figure 16-4 shows the six π molecular orbitals of benzene as viewed from above, showing the sign of the top lobe of each p orbital. The first MO (π_1) is entirely bonding, with no nodes. It is very low in energy because it has six bonding interactions and the electrons are delocalized over all six carbon atoms. The top lobes of the p orbitals

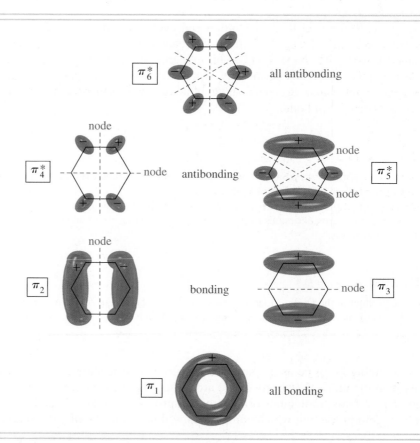

▶ **Figure 16-4**
The six π molecular orbitals of benzene, viewed from above. The number of nodal planes increases with energy, and there are two degenerate MOs at each intermediate energy level.

all have the same sign, as do the bottom lobes. The six p orbitals overlap to form a continuously bonding ring of electron density.

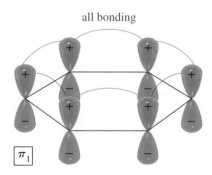

In a cyclic system of overlapping p orbitals, the intermediate energy levels are **degenerate** (equal in energy), with two orbitals at each energy level. Both π_2 and π_3 have one nodal plane, as we expect at the second energy level.

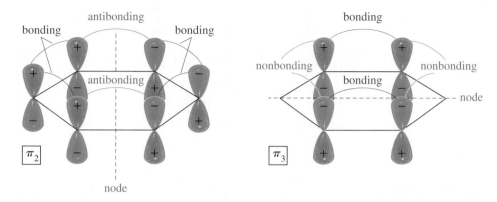

Notice that π_2 has four bonding interactions and two antibonding interactions, for a total of two net bonding interactions. Similarly, π_3 has two bonding interactions and four nonbonding interactions, also totaling two net bonding interactions. Although we cannot use the number of bonding and antibonding interactions as a quantitative measure of an orbital's energy, it is clear that π_2 and π_3 are bonding MOs, but not as strongly bonding as π_1.

The next orbitals, π_4^* and π_5^*, are also degenerate, with two nodal planes in each. The π_4^* orbital has two antibonding interactions and four nonbonding interactions; it is an antibonding (*) orbital. Its degenerate partner, π_5^*, has four antibonding interactions and two bonding interactions, for a total of two antibonding interactions. This degenerate pair of MOs, π_4^* and π_5^*, are about as strongly antibonding as π_2 and π_3 are bonding.

People and animals cannot make the benzene nucleus, a key component of the essential amino acid phenylalanine. Thus, they must obtain phenylalanine in their diet. Bacteria can synthesize the aromatic ring, however, and they synthesize their own phenylalanine. Compounds that block these pathways are being developed as potential antibiotics.

$$H_2N-CH-COOH$$
$$|$$
$$CH_2-\bigcirc$$

phenylalanine

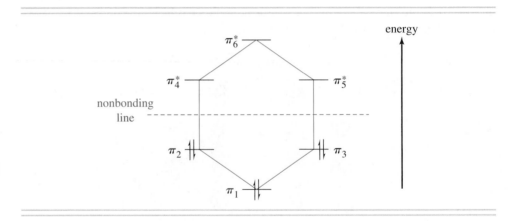

▶ **Figure 16-5**
Energy diagram of the molecular orbitals of benzene. Benzene's six pi electrons fill the three bonding orbitals, leaving the antibonding orbitals vacant.

The all-antibonding π_6^* has three nodal planes. Each pair of adjacent p orbitals is out of phase and interacts destructively.

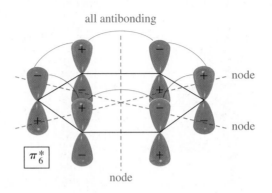

The Energy Diagram of Benzene   The energy diagram of the benzene MOs (Figure 16-5) shows them to be symmetrically distributed above and below the non-bonding line (the energy of an isolated p orbital). The all-bonding and all-antibonding orbitals (π_1 and π_6^*) are lowest and highest in energy, respectively. The degenerate bonding orbitals (π_2 and π_3) are higher in energy than π_1, but still bonding. The degenerate pair π_4^* and π_5^* are antibonding, yet not as high in energy as the all-antibonding π_6^* orbital.

The Kekulé structure for benzene shows three pi bonds, representing six electrons (three pairs) involved in pi bonding. Six electrons fill the three bonding MOs of the benzene system. This electronic configuration explains the unusual stability of benzene. The first MO is all-bonding and is particularly stable. The second and third (degenerate) MOs are still strongly bonding, and all three of these bonding MOs delocalize the electrons over several nuclei. This configuration, with all the bonding MOs filled (a "closed bonding shell"), is energetically very favorable.

16-4
The Molecular Orbital Picture of Cyclobutadiene

Although we can draw benzene-like resonance structures (Figure 16-3) for cyclobuta-diurene, experimental evidence shows that cyclobutadiene is unstable. Its instability is explained by the molecular orbitals, shown in Figure 16-6. Four sp^2 hybrid carbon atoms form the cyclobutadiene ring, and their four p orbitals overlap to form four molecular orbitals. The lowest energy MO is π_1, the all-bonding MO with no nodes.

The next two orbitals, π_2 and π_3, are degenerate (equal energy), each having one symmetrically situated nodal plane. Each of these MOs has two bonding interactions and two antibonding interactions. The net bonding order is zero, so these two MOs are non-bonding. The final MO, π_4^*, has two nodal planes and is entirely antibonding.

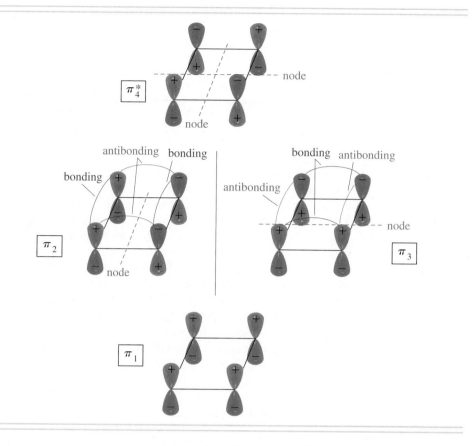

◀ **Figure 16-6**
The pi molecular orbitals of cyclobutadiene. There are four MOs: the lowest-energy bonding orbital, the highest energy antibonding orbital, and two degenerate nonbonding orbitals.

Figure 16-7 is an energy diagram of the four cyclobutadiene MOs. The lowest-lying MO (π_1) is strongly bonding, and the highest-lying MO (π_4^*) is equally antibonding. The two degenerate nonbonding orbitals are intermediate in energy, falling on the nonbonding line (the energy of an isolated p orbital).

The localized structure of cyclobutadiene shows two double bonds, implying four pi electrons. Two electrons fill π_1, the lowest-lying orbital. Once π_1 is filled, there are two orbitals of equal energy available for the remaining two electrons. If the two electrons go into the same orbital, they must have paired spins and they must share the same region of space. Since electrons repel each other, less energy is required for the electrons to occupy different degenerate orbitals, with unpaired spins. This principle is another application of Hund's rule (Section 1-2).

The electronic configuration in Figure 16-7 indicates that cyclobutadiene should be unstable. Its highest-lying electrons are in nonbonding orbitals (π_2 and π_3) and are therefore very reactive. According to Hund's rule, the compound exists as a diradical (two unpaired electrons) in its ground state. Thus, molecular orbital theory successfully predicts the dramatic stability difference between benzene and cyclobutadiene.

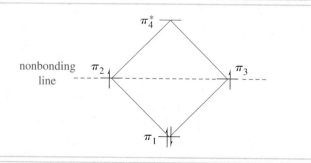

◀ **Figure 16-7**
An electronic energy diagram of cyclobutadiene shows that two electrons are unpaired in separate nonbonding molecular orbitals.

► **Figure 16-8**
The polygon rule predicts that the MO energy diagrams for these annulenes will resemble the polygonal shapes of the annulenes.

nonbonding line

benzene   cyclobutadiene   cyclooctatetraene

The Polygon Rule   The patterns of molecular orbitals in benzene (Figure 16-5) and in cyclobutadiene (Figure 16-7) are similar to the patterns in other annulenes: The lowest-lying MO is the unique one with no nodes; thereafter, the molecular orbitals occur in degenerate (equal-energy) pairs until only one highest-lying MO remains. In benzene, the energy diagram looks like the hexagon of a benzene ring. In cyclobutadiene, the pattern looks like the diamond of the cyclobutadiene ring.

The **polygon rule** states that the molecular orbital energy diagram of a regular, completely conjugated cyclic system has the same polygonal shape as the compound, with one vertex (the all-bonding MO) at the bottom. The nonbonding line cuts horizontally through the center of the polygon. Figure 16-8 shows how the polygon rule predicts the MO energy diagrams for benzene, cyclobutadiene, and cyclooctatetraene. The pi electrons are filled into the orbitals in accordance with the aufbau principle (lowest-energy orbitals are filled first) and Hund's rule.

PROBLEM 16-5

Does the MO energy diagram of cyclooctatetraene (Figure 16-8) appear to be a particularly stable or unstable configuration? Explain.

16-5
Aromatic, Antiaromatic, and Nonaromatic Compounds

Our working definition of aromatic compounds has included cyclic compounds containing conjugated double bonds, with unusually large resonance energies. At this point we can be more specific about the properties that are required for a compound (or an ion) to be aromatic.

Aromatic compounds are those that meet the following criteria:

1. The structure must be cyclic, containing some number of conjugated pi bonds.
2. Each atom in the ring must have an unhybridized *p* orbital. (The ring atoms are usually sp^2 hybridized or occasionally sp hybridized.)
3. The unhybridized *p* orbitals must overlap to form a continuous ring of parallel orbitals. In most cases, the structure must be planar (or nearly planar) for effective overlap to occur.
4. Delocalization of the pi electrons over the ring must *lower* the electronic energy.

An **antiaromatic** compound is one that meets the first three criteria, but delocalization of the pi electrons over the ring *increases* the electronic energy.

Aromatic structures are more stable than their open-chain counterparts. For example, benzene is more stable than 1,3,5-hexatriene.

more stable (aromatic)   less stable

Cyclobutadiene meets the first three criteria for a continuous ring of overlapping *p* orbitals, but delocalization of the pi electrons *increases* the electronic energy. Cyclobutadiene is less stable than its open-chain counterpart (1,3-butadiene), and it is antiaromatic.

less stable (antiaromatic)   more stable

A cyclic compound that does not have a continuous, overlapping ring of *p* orbitals cannot be aromatic or antiaromatic. It is said to be **nonaromatic**, or aliphatic. Its electronic energy is similar to that of its open-chain counterpart. For example, 1,3-cyclohexadiene is about as stable as *cis,cis*-2,4-hexadiene.

⟵ similar stabilities ⟶

(nonaromatic)

16-6 Hückel's Rule

Erich Hückel developed a shortcut for predicting which of the annulenes and related compounds are aromatic and which are antiaromatic. In using Hückel's rule, we must be certain that the compound under consideration meets the criteria for an aromatic or antiaromatic system.

It must have a continuous ring of overlapping *p* orbitals, usually in a planar conformation.

Once these criteria are met, **Hückel's rule** applies:

Hückel's Rule: If the number of pi electrons in the cyclic system is:

$(4N + 2)$, the system is aromatic;
$(4N)$, the system is antiaromatic,

where N is an integer.

Common aromatic systems have 2, 6, and 10 pi electrons, for $N = 0, 1,$ and 2. Antiaromatic systems might have 4, 8, or 12 pi electrons, for $N = 1, 2,$ or 3.

Benzene is [6]annulene, cyclic, with a continuous ring of overlapping *p* orbitals. There are six pi electrons in benzene (three double bonds in the classical structure), so it is a $(4N + 2)$ system, with $N = 1$. Hückel's rule predicts benzene to be aromatic.

Like benzene, cyclobutadiene ([4]annulene) has a continuous ring of overlapping *p* orbitals, but it has four pi electrons (two double bonds in the classical structure). Hückel's rule predicts cyclobutadiene to be antiaromatic.

Cyclooctatetraene is [8]annulene, with eight pi electrons (four double bonds) in the classical structure. It is a $4N$ system, with $N = 2$. If Hückel's rule were applied to cyclooctatetraene, it would predict antiaromaticity. However, cyclooctatetraene is a stable hydrocarbon with a boiling point of 152°C. It does not show the high reactivity associated with antiaromaticity, yet it is not aromatic either. Its reactions are typical of alkenes.

Cyclooctatetraene would be antiaromatic if Hückel's rule applied, so the conjugation of its double bonds is energetically unfavorable. Remember that Hückel's rule applies to a compound *only* if there is a continuous ring of overlapping *p* orbitals, usually in a planar system. Cyclooctatetraene is more flexible than cyclobutadiene, and it assumes a nonplanar "tub" conformation that avoids most of the overlap between adjacent pi bonds. Hückel's rule simply does not apply.

PROBLEM-SOLVING HINT

Hückel's rule is commonly used to determine aromaticity and antiaromaticity. A continuous, planar ring of overlapping *p* orbitals is required for the rule to apply. Otherwise, the system is nonaromatic.

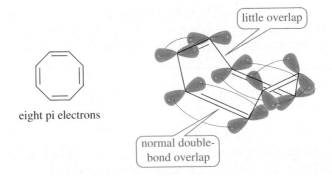

eight pi electrons

little overlap

normal double-
bond overlap

PROBLEM 16-6

Make a model of cyclooctatetraene in the tub conformation. Draw this conformation, and estimate the angle between the *p* orbitals of adjacent pi bonds.

Large-Ring Annulenes   Like cyclooctatetraene, larger annulenes with $4N$ systems do not show antiaromaticity because they have the flexibility to adopt nonplanar conformations. Even though [12]annulene, [16]annulene, and [20]annulene are $4N$
systems (with $N = 3$, 4, and 5, respectively), they all react as partially conjugated
polyenes.

[12]annulene   [16]annulene   [20]annulene

Aromaticity in the larger $4N + 2$ annulenes depends on whether the molecule
can adopt the necessary planar conformation. In the all-cis [10]annulene, the planar
conformation requires an excessive amount of angle strain. The [10]annulene isomer
with two trans double bonds cannot adopt a planar conformation either, because two
hydrogen atoms interfere with each other. Neither of these [10]annulene isomers is
aromatic, even though each has $(4N + 2)$ pi electrons, with $N = 2$. If the interfering hydrogen atoms in the partially trans isomer are removed, the molecule can be planar. When these hydrogen atoms are replaced with a bond, the aromatic compound
naphthalene results.

all-cis   two trans   naphthalene
nonaromatic   nonaromatic   aromatic

Some of the larger annulenes with $(4N + 2)$ pi electrons can achieve planar conformations. For example, the following [14]annulene and [18]annulene have aromatic
properties.

[14]annulene (aromatic)   [18]annulene (aromatic)

PROBLEM 16-7

Classify the following compounds as aromatic, antiaromatic, or nonaromatic:

(a)   (b)   (c)   (d)

PROBLEM 16-8

One of the following compounds is much more stable than the other two. Classify each as aromatic, antiaromatic, or nonaromatic.

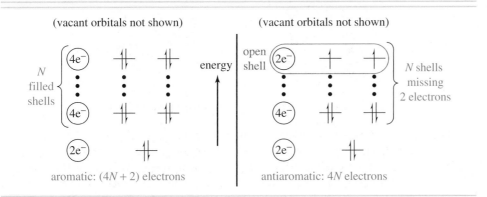

heptalene   azulene   pentalene

Benzene is aromatic because it has a filled shell of equal-energy orbitals. The degenerate orbitals π_2 and π_3 are filled, and all the electrons are paired. Cyclobutadiene, by contrast, has an open shell of electrons. There are two half-filled orbitals easily capable of donating or accepting electrons. To derive Hückel's rule, we must show under what general conditions there is a filled shell of orbitals.

Recall the pattern of MOs in a cyclic conjugated system. There is one all-bonding, lowest-lying MO, followed by degenerate pairs of bonding MOs. (There is no need to worry about the antibonding MOs because they are vacant in the ground state.) The lowest-lying MO is always filled (two electrons). Each additional shell consists of two degenerate MOs, requiring four electrons to fill a shell. Figure 16-9 shows this pattern of two electrons for the lowest orbital, and then four electrons for each additional shell.

16-7
Molecular Orbital Derivation of Hückel's Rule

◄ Figure 16-9
In a cyclic conjugated system, the lowest-lying MO is filled with two electrons. Each of the additional shells consists of two degenerate MOs, with space for four electrons. If a molecule has $(4N + 2)$ pi electrons, it will have a filled shell. If it has $4N$ electrons, there will be two unpaired electrons in two degenerate orbitals.

(vacant orbitals not shown)   (vacant orbitals not shown)

aromatic: $(4N + 2)$ electrons   antiaromatic: $4N$ electrons

A compound has a filled shell of orbitals if it has two electrons for the lowest-lying orbital, plus $4N$ electrons, where N is the number of filled pairs of degenerate orbitals. The total number of pi electrons in this case is $(4N + 2)$. If the system has a total of only $4N$ electrons, it is two electrons short of filling N pairs of degenerate orbitals. There are only two electrons in the Nth pair of degenerate orbitals. This is a half-filled shell, and Hund's rule predicts these electrons will be unpaired (a diradical).

PROBLEM 16-9

(a) Use the polygon rule to draw an energy diagram (as in Figures 16-5 and 16-7) for the MOs of a planar cyclooctatetraenyl system.
(b) Fill in the eight pi electrons for cyclooctatetraene. Is this electronic configuration aromatic or antiaromatic?
(c) Draw pictorial representations (like Figures 16-4 and 16-6) for the three bonding MOs and the two nonbonding MOs of cyclooctatetraene. The antibonding MOs are difficult to draw, except for the all-antibonding MO,

---

16-8
Aromatic Ions

Up to this point, we have discussed aromaticity using the annulenes as examples. Annulenes are uncharged molecules having even numbers of carbon atoms with alternating single and double bonds. Hückel's rule also applies to systems having odd numbers of carbon atoms and bearing positive or negative charges. We now consider some common aromatic ions and their antiaromatic counterparts.

16-8A   The Cyclopentadienyl Ions

We can draw a five-membered ring of sp^2 hybrid carbon atoms with all the unhybridized p orbitals lined up to form a continuous ring. With five pi electrons, this system would be neutral, but it would be a radical because an odd number of electrons cannot all be paired. With four pi electrons (a cation), Hückel's rule predicts this system to be antiaromatic. With six pi electrons (an anion), Hückel's rule predicts aromaticity.

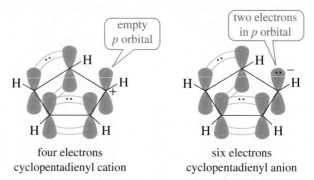

four electrons
cyclopentadienyl cation

six electrons
cyclopentadienyl anion

Because the cyclopentadienyl anion (six pi electrons) is aromatic, it is unusually stable compared with other carbanions. It can be formed by abstracting a proton from cyclopentadiene, which is unusually acidic for an alkene. Cyclopentadiene has a pK_a of 16, compared with a pK_a of 46 for cyclohexene. In fact, cyclopentadiene is nearly as acidic as water and more acidic than many alcohols. It is entirely ionized by potassium t-butoxide:

the cyclopentadienyl anion
(six pi electrons)

pKₐ = 18

(on left reactant) pKₐ = 16

Cyclopentadiene is unusually acidic because loss of a proton converts the nonaromatic diene to the aromatic cyclopentadienyl anion. Cyclopentadiene contains an sp^3 hybrid ($-CH_2-$) carbon atom without an unhybridized p orbital, so there can be no continuous ring of p orbitals. Deprotonation of the $-CH_2-$ group leaves an orbital occupied by a pair of electrons. This orbital can rehybridize to a p orbital, completing a ring of p orbitals containing six pi electrons: the two electrons on the deprotonated carbon, plus the four electrons in the original double bonds.

cyclopentadiene
nonaromatic

cyclopentadienyl anion
aromatic

When we say the cyclopentadienyl anion is aromatic, this does not necessarily imply that it is as stable as benzene. As a carbanion, the cyclopentadienyl anion reacts readily with electrophiles. Because this ion is aromatic, it is more stable than the corresponding open-chain ion.

more stable
(aromatic)

less stable

Hückel's rule predicts that the cyclopentadienyl cation, with four pi electrons, is antiaromatic. In agreement with this prediction, the cyclopentadienyl cation is not easily formed. Protonated 2,4-cyclopentadienol does not lose water (to give the cyclopentadienyl cation), even in concentrated sulfuric acid. The antiaromatic cation is simply too unstable.

2,4-cyclopentadienol

(does not occur)

not formed
(four pi electrons)

Using a simple resonance approach, we might incorrectly expect both of the cyclopentadienyl ions to be unusually stable. Shown below are resonance structures that spread the negative charge of the anion and the positive charge of the cation over all five carbon atoms of the ring. With conjugated cyclic systems such as these, the resonance approach is a poor predictor of stability. The Hückel rule, based on molecular orbital theory, is a much better predictor of stability for these aromatic and antiaromatic systems.

$$\left[\begin{array}{ccccccccc} & \longleftrightarrow & & \longleftrightarrow & & \longleftrightarrow & & \longleftrightarrow & \end{array} \right]$$

cyclopentadienyl anion: six pi electrons, aromatic

$$\left[\begin{array}{ccccccccc} & \longleftrightarrow & & \longleftrightarrow & & \longleftrightarrow & & \longleftrightarrow & \end{array} \right]$$

cyclopentadienyl cation: four pi electrons, antiaromatic
The resonance picture gives a misleading suggestion of stability.

PROBLEM 16-10

(a) Draw the molecular orbitals for the cyclopropenyl case.

(Since there are three *p* orbitals, there must be three MOs: one all-bonding MO and one degenerate pair of MOs.)

(b) Draw an energy diagram for the cyclopropenyl MOs. (The polygon rule may be helpful.) Label each MO as bonding, nonbonding, or antibonding, and add the nonbonding line. Notice that it goes through the approximate average of the MOs.

(c) Add electrons to your energy diagram to show the configuration of the cyclopropenyl cation and the cyclopropenyl anion. Which is aromatic and which is antiaromatic?

PROBLEM 16-11*

Repeat Problem 16-10 for the cyclopentadienyl ions. Draw one all-bonding MO, then a pair of degenerate MOs, and then a final pair of degenerate MOs. Draw the energy diagram, fill in the electrons, and determine the electronic configurations of the cyclopentadienyl cation and anion.

16-8B   The Cycloheptatrienyl Ions

As with the five-membered ring, we can imagine a flat seven-membered ring with seven *p* orbitals aligned. The cation has six pi electrons, and the anion has eight pi electrons. Once again, we can draw resonance forms that seem to show either the positive charge of the cation or the negative charge of the anion delocalized over all seven atoms of the ring. By now, however, we know that the six-electron system is aromatic and the eight-electron system is antiaromatic (if it remains planar).

$$\left[\begin{array}{ccccccccccccc} & \longleftrightarrow & & \longleftrightarrow & & \longleftrightarrow & & \longleftrightarrow & & \longleftrightarrow & & \longleftrightarrow & \end{array} \right]$$

cycloheptatrienyl cation (tropylium ion): six pi electrons, aromatic

$$\left[\begin{array}{ccccccccccccc} & \longleftrightarrow & & \longleftrightarrow & & \longleftrightarrow & & \longleftrightarrow & & \longleftrightarrow & & \longleftrightarrow & \end{array} \right]$$

cycloheptatrienyl anion: eight pi electrons, antiaromatic (if planar)
The resonance picture gives a misleading suggestion of stability.

The cycloheptatrienyl cation is easily formed by treating the corresponding alcohol with dilute (0.01N) aqueous sulfuric acid. This is our first example of a hydrocarbon cation that is stable in aqueous solution.

tropylium ion, six pi electrons

The cycloheptatrienyl cation is called the **tropylium ion**. This aromatic ion is much less reactive than most carbocations. Some tropylium salts can be isolated and stored for months without decomposing. Nevertheless, the tropylium ion is not necessarily as stable as benzene. Its aromaticity implies that the cyclic ion is more stable than the corresponding open-chain ion.

more stable (aromatic)   less stable

Although the tropylium ion forms easily, the corresponding anion is difficult to form because it is antiaromatic. Cycloheptatriene ($pK_a = 39$) is barely more acidic than propene ($pK_a = 43$), and the anion is very reactive. This result agrees with the prediction of Hückel's rule that the cycloheptatrienyl anion is antiaromatic.

cycloheptatriene
$pK_a = 39$ cycloheptatrienyl anion
eight pi electrons

16-8C The Cyclooctatetraene Dianion

We have seen that aromatic stabilization leads to unusually stable hydrocarbon anions such as the cyclopentadienyl anion. Dianions of hydrocarbons are rare and are usually much more difficult to form. Cyclooctatetraene reacts with potassium metal, however, to form an aromatic dianion.

ten pi electrons

The cyclooctatetraene dianion has a planar, regular octagonal structure, with C—C bond lengths of 1.40 Å, close to the 1.397 Å bond lengths in benzene. Cyclooctatetraene itself has eight pi electrons, so the dianion has ten: $(4N + 2)$, with $N = 2$. The cyclooctatetraene dianion is easily prepared because it is aromatic.

PROBLEM-SOLVING HINT

Use Hückel's rule (and the criteria for its application), rather than resonance, to determine which annulenes and ions are aromatic, antiaromatic, and nonaromatic.

PROBLEM 16-12

Explain why each compound or ion should be aromatic, antiaromatic, or nonaromatic.

(a) the cyclonona-tetraene cation

(b) the cyclonona-tetraene anion

(c) the [16]annulene dianion

(d) the [18]annulene dianion

(e)

(f) the [20]annulene dication

PROBLEM 16-13

The following hydrocarbon has an unusually large dipole moment. Explain how a large dipole moment might arise.

PROBLEM 16-14

When 3-chlorocyclopropene is treated with AgBF₄, AgCl precipitates. The organic product can be obtained as a crystalline material, soluble in polar solvents such as nitromethane but insoluble in hexane. When the crystalline material is dissolved in nitromethane containing KCl, the original 3-chlorocyclopropene is regenerated. Determine the structure of the crystalline material, and write equations for its formation and its reaction with chloride ion.

PROBLEM 16-15

The polarization of a carbonyl group can be represented by a pair of resonance structures:

$$\left[\begin{array}{c} \diagup \\ C=\ddot{O} \end{array} \quad \longleftrightarrow \quad \begin{array}{c} \diagup \\ \overset{+}{C}-\ddot{O}{:}^{-} \end{array} \right]$$

Cyclopropenone and cycloheptatrienone are more stable than anticipated. Cyclopentadienone, however, is relatively unstable and rapidly undergoes a Diels–Alder dimerization. Explain.

cyclopropenone   cycloheptatrienone   cyclopentadienone

16-8D   Summary of Annulenes and Their Ions

Applications of Hückel's rule to a variety of cyclic pi systems are summarized below. These systems are classified according to the number of pi electrons: The 2, 6, and 10 pi-electron systems are aromatic, while the 4 and 8 pi-electron systems are antiaromatic if they are planar.

Two pi-electron systems (aromatic)

 cyclopropenyl cation (cyclopropenium ion)

Four pi-electron systems (antiaromatic)

cyclobutadiene   cyclopropenyl anion   cyclopentadienyl cation

Six pi-electron systems (aromatic)

benzene   cyclopentadienyl anion   cycloheptatrienyl cation   pyridine   pyrrole   furan
               (cyclopentadienide ion)   (tropylium ion)

Eight pi-electron systems (antiaromatic if planar)

cyclooctatetraene   cycloheptatrienyl   cyclononatetraenyl   pentalene
(not planar)   anion   cation

Ten pi-electron systems (aromatic)

naphthalene   azulene   cyclononatetraenyl   cyclooctatetraenyl   indole
                                         anion   dianion

(Naphthalene can also be considered as two fused benzenes.)

Twelve pi-electron systems (antiaromatic if planar)

[12]annulene   heptalene
(not planar)

The criteria for Hückel's rule require a ring of atoms, all with unhybridized *p* orbitals overlapping in a continuous ring. In discussing aromaticity, we have considered only compounds composed of rings of sp^2 hybrid carbon atoms. **Heterocyclic compounds**, with rings containing sp^2 hybridized atoms of other elements, can also be aromatic. Nitrogen, oxygen, and sulfur are the most common heteroatoms in heterocyclic aromatic compounds.

**16-9
Heterocyclic
Aromatic
Compounds**

▲ Figure 16-10
Pyridine has six delocalized electrons in its cyclic pi system. The two nonbonding electrons on nitrogen are in an sp^2 orbital, and they do not interact with the pi electrons of the ring.

16-9A   Pyridine

Pyridine is an aromatic nitrogen analogue of benzene. It has a six-membered heterocyclic ring with six pi electrons. Pyridine has a nitrogen atom in place of one of the six C—H units of benzene, and the nonbonding pair of electrons on nitrogen replaces the bond to a hydrogen atom. These nonbonding electrons are in an sp^2 hybrid orbital in the plane of the ring (Figure 16-10). They are perpendicular to the pi system and do not overlap with it.

Pyridine shows all the characteristics of aromatic compounds. It has a resonance energy of 27 kcal/mol (113 kJ/mol), and it usually gives substitution rather than addition. Because it has an available pair of nonbonding electrons, pyridine is basic (Figure 16-11). In an acidic solution, pyridine protonates to give the pyridinium ion. The pyridinium ion is still aromatic because the additional proton has no effect on the electrons of the aromatic sextet: It simply bonds to pyridine's nonbonding pair of electrons.

▶ Figure 16-11
Pyridine is basic, with nonbonding electrons available to abstract a proton. The protonated pyridine (a pyridinium ion) is still aromatic.

pyridine, pK_b = 8.8

pyridinium ion, pK_a = 5.2

16-9B   Pyrrole

Pyrrole is an aromatic five-membered heterocycle, with one nitrogen atom and two double bonds (Figure 16-12). Although it may seem that pyrrole has only four pi electrons, the nitrogen atom has a lone pair of electrons. The pyrrole nitrogen atom is sp^2 hybridized, and its unhybridized p orbital overlaps with the p orbitals of the carbon atoms to form a continuous ring. The lone pair on nitrogen occupies the p orbital, and (unlike the lone pair of pyridine) these electrons take part in the pi bonding system. These two electrons, added to the four pi electrons of the two double bonds, complete an aromatic sextet. Pyrrole has a resonance energy of 22 kcal/mol (92 kJ/mol).

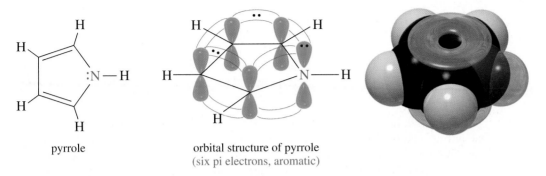

▲ Figure 16-12
The pyrrole nitrogen atom is sp^2 hybridized, with a lone pair of electrons in the p orbital. This p orbital overlaps with the p orbitals of the carbon atoms to form a continuous ring. Counting the four electrons of the double bonds and the two electrons in the nitrogen p orbital, there are six pi electrons.

pyrrole, $pK_b = 13.6$
(weak base)

N-protonated pyrrole, $pK_a = 0.4$
(strong acid)

pyrrole
(aromatic)

N-protonated pyrrole
(nonaromatic)

◀ Figure 16-13
The pyrrole nitrogen atom must become sp^3 hybridized to abstract a proton. This eliminates the unhybridized p orbital needed for aromaticity.

Pyrrole ($pK_b = 13.6$) is a much weaker base than pyridine ($pK_b = 8.8$). This difference is due to the structure of the protonated pyrrole (Figure 16-13). To form a bond to a proton requires the use of one of the electron pairs in the aromatic sextet. In the protonated pyrrole, the nitrogen atom is bonded to four different atoms (two carbon atoms and two hydrogen atoms), requiring sp^3 hybridization and leaving no unhybridized p orbital. The protonated pyrrole is nonaromatic. In fact, a sufficiently strong acid actually protonates pyrrole at the 2-position, on one of the carbon atoms of the ring, rather than on nitrogen.

Porphobilinogen, a substituted pyrrole, is the building block of the heme ring, which has many physiological functions, such as the transport and storage of oxygen.

porphobilinogen

16-9C   Pyrimidine and Imidazole

Pyrimidine is a six-membered heterocycle with two nitrogen atoms situated in a 1,3-arrangement. Both nitrogen atoms are like the pyridine nitrogen. Each has its lone pair of electrons in the sp^2 hybrid orbital in the plane of the aromatic ring. These lone pairs are not needed for the aromatic sextet, and they are basic, like the lone pair of pyridine.

PROBLEM-SOLVING $\mathbf{HINT}$

Practice spotting basic and nonbasic nitrogen atoms. Most nonbasic (pyrrole-like) nitrogens have three single bonds, and most basic (pyridine-like) nitrogens have a double bond in the ring.

pyrimidine   imidazole   purine

Imidazole is an aromatic five-membered heterocycle with two nitrogen atoms. One nitrogen atom (the one not bonded to a hydrogen) has its lone pair in an sp^2 orbital that is not involved in the aromatic system; this lone pair is basic. The other nitrogen uses its third sp^2 orbital to bond to hydrogen, and its lone pair is part of the aromatic sextet. Like the pyrrole nitrogen atom, this imidazole N—H nitrogen is not very basic. Once imidazole is protonated, the two nitrogens become chemically equivalent. Either nitrogen can lose a proton and return to an imidazole molecule.

imidazole   protonated imidazole

imidazole

Purine has an imidazole ring fused to a pyrimidine ring. Purine has three basic nitrogen atoms and one pyrrole-like nitrogen.

Pyrimidine and purine derivatives serve in DNA and RNA to specify the genetic code. Imidazole derivatives enhance the catalytic activity of enzymes. We will consider these important heterocyclic derivatives in more detail in Chapters 23 and 24.

PROBLEM 16-16

Show which of the nitrogen atoms in purine are basic, and which one is not basic. For the nonbasic nitrogen, explain why its nonbonding electrons are not easily available to become protonated.

PROBLEM 16-17

The proton NMR spectrum of 2-pyridone gives the chemical shifts shown below:

2-pyridone   thymine

(a) Is 2-pyridone aromatic?

(b) Use resonance forms to explain your answer to (a). Also explain why the protons at $\delta 7.31$ and 7.26 are more deshielded than the other two ($\delta 6.15$ and 6.57).

(c) Thymine is one of the heterocyclic bases found in DNA. Do you expect thymine to be aromatic? Explain.

A number of purine and pyrimidine analogs are used as anti-cancer drugs. For example, 5-fluorouracil blocks the enzyme that produces thymidine, a key base in DNA, and results in the death of many cancerous as well as some healthy cells.

5-fluorouracil

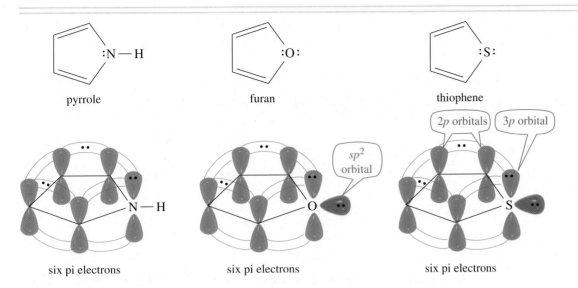

▲ Figure 16-14
Pyrrole, furan, and thiophene are isoelectronic. In furan and thiophene, the pyrrole N—H bond is replaced by a nonbonding pair of electrons in the sp^2 hybrid orbital.

16-9D   Furan and Thiophene

Furan is an aromatic five-membered heterocycle-like pyrrole, but the heteroatom is oxygen instead of nitrogen. The classical structure for furan (Figure 16-14) shows that the oxygen atom has two lone pairs of electrons. The oxygen atom is sp^2 hybridized, and one of the lone pairs occupies an sp^2 hybrid orbital. The other lone pair occupies the unhybridized p orbital, combining with the four electrons in the double bonds to give an aromatic sextet. Furan has a resonance energy of 16 kcal/mol (67 kJ/mol).

   Thiophene is similar to furan, with a sulfur atom in place of the furan oxygen. The bonding in thiophene is similar to that in furan, except that the sulfur atom uses an unhybridized $3p$ orbital to overlap with the $2p$ orbitals on the carbon atoms. The resonance energy of thiophene is 29 kcal/mol (121 kJ/mol).

> The removal of inorganic and organic sulfur from coal is beneficial because combustion of coal releases the sulfur into the atmosphere where it contributes to acid rain. Microorganisms having the capability to use thiophene as a food source are being exploited for the desulfurization of coal.

PROBLEM 16-18

Explain why each compound is aromatic, antiaromatic, or nonaromatic.

(a) isoxazole

(b) 1,3-thiazole

(c) pyran

(d) pyrylium ion

(e) γ-pyrone

(f) 1,2-dihydropyridine

(g) cytosine

PROBLEM 16-19

Borazole, $B_3N_3H_6$, is an unusually stable cyclic compound. Propose a structure for borazole, and explain why it is aromatic.

16-10
Polynuclear Aromatic Hydrocarbons

The **polynuclear aromatic hydrocarbons** (abbreviated PAHs or PNAs) are composed of two or more fused benzene rings. **Fused rings** share two carbon atoms and the bond between them.

Naphthalene   Naphthalene ($C_{10}H_8$) is the simplest fused aromatic compound, consisting of two fused benzene rings. We represent naphthalene by using one of the three Kekulé resonance structures or using the circle notation for the aromatic rings.

naphthalene

The two aromatic rings in naphthalene contain a total of 10 pi electrons. Two isolated aromatic rings would contain 6 pi electrons in each aromatic system, for a total of 12. The smaller amount of electron density gives naphthalene less than twice the resonance energy of benzene: 60 kcal/mol (252 kJ/mol), or 30 kcal (126 kJ) per aromatic ring, compared with benzene's resonance energy of 36 kcal/mol (151 kJ/mol).

Anthracene and Phenanthrene   As the number of fused aromatic rings increases, the resonance energy per ring continues to decrease and the compounds become more reactive. Tricyclic anthracene has a resonance energy of 84 kcal/mol (351 kJ/mol), or 28 kcal (117 kJ) per aromatic ring. Phenanthrene has a slightly higher resonance energy of 91 kcal/mol (381 kJ/mol), or about 30.3 kcal (127 kJ) per aromatic ring. Each of these compounds has only 14 pi electrons in its three aromatic rings, compared with 18 electrons for three separate benzene rings.

anthracene

phenanthrene

(Only one Kekulé structure is shown for each compound.)

Because they are not as strongly stabilized as benzene, anthracene and phenanthrene can undergo addition reactions that are more characteristic of their nonaromatic polyene relatives. Anthracene undergoes 1,4-addition at the 9- and 10-positions to give a product with two isolated, fully aromatic benzene rings. Similarly, phenanthrene undergoes 1,2-addition at the 9- and 10-positions to give a product with two fully aromatic rings. (Because they are less likely to be substituted, the bridgehead carbon atoms of fused aromatics are often left unnumbered.)

anthracene

$\xrightarrow[\text{CCl}_4]{\text{Br}_2}$

(mixture of cis and trans)

phenanthrene

$\xrightarrow[\text{CCl}_4]{\text{Br}_2}$

(mixture of cis and trans)

PROBLEM 16-20

(a) Draw all the Kekulé structures of anthracene and phenanthrene.

(b) Propose mechanisms for the two additions shown above.

(c) In Chapter 8, most of the additions of bromine to double bonds gave entirely *anti* stereochemistry. Explain why the addition to phenanthrene gives a mixture of syn and anti stereochemistry.

(d) When the product from (c) is heated, HBr is evolved and 9-bromophenanthrene results. Propose a mechanism for this dehydrohalogenation.

Larger Polynuclear Aromatic Hydrocarbons   There is a high level of interest in the larger PAHs because they are formed in most combustion processes and many of them are carcinogenic (capable of causing cancer). The following three compounds, for example, are present in tobacco smoke. These compounds are so hazardous that laboratories must install special containment facilities to work with them, yet smokers expose their lung tissues to them.

The black material in diesel exhaust consists of small particles that are rich in polynuclear aromatic hydrocarbons.

pyrene   benzo[*a*]pyrene   dibenzopyrene

Benzo[*a*]pyrene, one of the most thoroughly studied carcinogens, is formed whenever organic compounds undergo incomplete combustion. For example, benzo[*a*]pyrene is found in chimney soot, in broiled steaks, and in cigarette smoke. Long before our ancestors learned to use fire, they were exposed to benzo[*a*]pyrene in the smoke and ash from forest fires. Its carcinogenic effects appear to result from its epoxidation to arene oxides, which can be attacked by nucleophilic sites of DNA. The resulting DNA derivatives cannot be properly transcribed; on replication, they produce mutations.

Benzo[*a*]pyrene in soot was the culprit for a large number of skin cancers in young boys who cleaned chimneys in the 1700s. The body transforms this compound to a highly reactive epoxide, which forms a covalent bond with DNA.

benzo[*a*]pyrene   4,5-benzo[*a*]pyrene oxide   7,8-benzo[*a*]pyrene oxide

cytidine
(a DNA base)

DNA derivative

16-11
Aromatic Allotropes of Carbon

What do you get when you make an extremely large polynuclear aromatic hydrocarbon, with millions or billions of benzene rings joined together? You get graphite, one of the oldest-known forms of pure elemental carbon. Let's consider how aromaticity plays a role in the stability of both the old and the new forms of carbon.

16-11A   Allotropes of Carbon; Diamond

We don't normally think of elemental carbon as an organic compound. Historically, carbon was known to exist as three allotropes (elemental forms with different properties): amorphous carbon, diamond, and graphite.

"Amorphous carbon" refers to charcoal, soot, coal, and carbon black. These materials are mostly microcrystalline forms of graphite. They are characterized by small particle sizes and large surface areas with partially saturated valences. These small particles readily absorb gases and solutes from solution, and they form strong, stable dispersions in polymers, such as the dispersion of carbon black in tires.

Diamond is the hardest naturally occurring substance known. Diamond has a crystalline structure containing tetrahedral carbon atoms linked together in a three-dimensional lattice (Figure 16-15). This lattice extends throughout each crystal, so that a diamond is actually one giant molecule. Diamond is an electrical insulator because the electrons are all tightly bound in sigma bonds (length 1.54 Å, typical of C—C single bonds), and they are unavailable to carry a current.

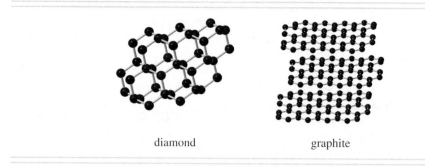

diamond   graphite

◀ **Figure 16-15**
Structures of diamond and graphite. Diamond is a lattice of tetrahedral carbon atoms linked in a rigid three-dimensional array. Graphite consists of planar layers of fused aromatic rings.

16-11B   Graphite

Graphite has the layered planar structure shown in Figure 16-15. Within a layer, the C—C bond lengths are all 1.415 Å, which is fairly close to the C—C bond length in benzene (1.397 Å). Between layers, the distance is 3.35 Å, which is about twice the van der Waals radius for carbon, and suggests there is little or no bonding between layers. The layers can easily cleave and slide across each other, making graphite a good lubricant. This layered picture also helps to explain graphite's unusual electrical properties: It is a good electrical conductor parallel to the layers, but it resists electrical currents perpendicular to the layers.

We picture each layer of graphite as a nearly infinite lattice of fused aromatic rings. All the valences are satisfied (except at the edges), and no bonds are needed between layers. Only van der Waals forces hold the layers together, consistent with their ability to slide easily over one another. The pi electrons within a layer can conduct electrical currents parallel to the layer, but electrons cannot easily jump between layers, so graphite is resistive perpendicular to the layers.

Because of its aromaticity, graphite is slightly more stable than diamond, and the transition from diamond to graphite is slightly exothermic ($\Delta H° = -0.7$ kcal/mol or -2.9 kJ/mol). Fortunately for those who have invested in diamonds, the favorable conversion of diamond to graphite is exceedingly slow. Diamond (3.51 g/cm³) has a higher density than graphite (2.25 g/cm³), implying that graphite might be converted to diamond under very high pressures. Indeed, small industrial diamonds can be synthesized by subjecting graphite to pressures over 125,000 atm and temperatures around 3000°C, using catalysts such as Cr and Fe.

16-11C   Fullerenes

Around 1985, Kroto, Smalley, and Curl (Rice University) isolated a molecule of formula C_{60} from the soot produced by using a laser (or an electrical arc) to vaporize graphite. Molecular spectra showed that C_{60} is unusually symmetrical: It has only one type of carbon by ^{13}C NMR ($\delta 143$ ppm), and there are only two types of bonds (1.39 Å and 1.45 Å). Figure 16-16 shows the structure of C_{60}, which was named **buckminsterfullerene** in

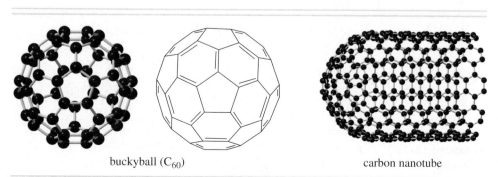

buckyball (C_{60})   carbon nanotube

◀ **Figure 16-16**
Structure of C_{60} and a carbon nanotube. Each carbon in C_{60} is a bridgehead carbon for a five-membered ring and two six-membered rings. A nanotube is a cylinder composed of aromatic six-membered rings similar to graphite. The end of the tube is half of a C_{60} sphere. Notice how the five-membered rings cause the structure to curve at the end of the tube.

honor of the American architect R. Buckminster Fuller, whose geodesic domes used similar five- and six-membered rings to form a curved roof. The C_{60} molecules are sometimes called "buckyballs," and these types of compounds (C_{60} and similar carbon clusters) are called **fullerenes**.

A soccer ball has the same structure as C_{60}, with each vertex representing a carbon atom. All the carbon atoms are chemically the same. Each carbon serves as a bridgehead for two six-membered rings and one five-membered ring. There are only two types of bonds: The bonds that are shared by a five-membered ring and a six-membered ring (1.45 Å) and the bonds shared between two six-membered rings (1.39 Å). Compare these bond lengths with a typical double bond (1.33 Å), a typical aromatic bond (1.40 Å), and a typical single bond (1.48 Å between sp^2 carbons). It appears that the double bonds are somewhat localized between the six-membered rings as shown in Figure 16-16. These double bonds are less reactive than typical alkene double bonds, yet they do undergo some of the addition reactions of alkenes.

Nanotubes (Figure 16-16) were discovered around 1991. These structures begin with half of the C_{60} sphere, fused to a cylinder composed entirely of fused six-membered rings (as in a layer of graphite). Nanotubes have aroused interest because they are electrically conductive only along the length of the tube, and they have an enormous strength-to-weight ratio.

16-12
Fused Heterocyclic Compounds

Purine is one of many fused heterocyclic compounds whose rings share two atoms and the bond between them. For example, the following compounds all contain fused heterocyclic aromatic rings:

purine   indole   benzimidazole   quinoline   benzofuran   benzothiophene

The properties of fused-ring heterocycles are generally similar to those of the simple heterocycles. Fused heterocyclic compounds are common in nature, and they are also used as drugs to treat a wide variety of illnesses. Figure 16-17 shows some fused heterocycles that occur naturally or are synthesized for use as drugs.

L-tryptophan, an amino acid   benziodarone, a vasodilator   LSD, a hallucinogen   quinine, an antimalarial drug

▲ **Figure 16-17**
Examples of biologically active fused heterocycles.

Benzene derivatives have been isolated and used as industrial reagents for well over 100 years. Many of their names are rooted in the historical traditions of chemistry. The following compounds are usually called by their historical common names, and almost never by the systematic IUPAC names.

16-13
Nomenclature of Benzene Derivatives

common name: phenol   toluene   aniline   anisole
                (benzenol)   (methylbenzene) (benzenamine)   (methoxybenzene)

common name: styrene   acetophenone   benzaldehyde   benzoic acid
                (vinylbenzene) (methyl phenyl ketone)

Many compounds are named as derivatives of benzene, with their substituents named just as though they were attached to an alkane.

tert-butylbenzene   nitrobenzene   ethynylbenzene   benzenesulfonic acid
                                       (phenylacetylene)

Disubstituted benzenes are named using the prefixes ***ortho-***, ***meta-***, and ***para-*** to specify the substitution patterns. These terms are abbreviated *o*-, *m*-, and *p*-. Numbers can also be used to specify the substitution in disubstituted benzenes.

1,2 or ortho   1,3 or meta   1,4 or para

common name: *o*-dichlorobenzene   *m*-chloroperoxybenzoic acid   *p*-nitrophenol
IUPAC name: 1,2-dichlorobenzene   3-chloroperoxybenzoic acid   4-nitrophenol

With three or more substituents on the benzene ring, numbers are used to indicate their positions. Assign the numbers as you would with a substituted cyclohexane, to give the lowest possible numbers to the substituents. The carbon atom bearing the functional group that defines the base name (as in phenol or benzoic acid) is assumed to be C1.

1,3,5-trinitrobenzene   2,4-dinitrophenol   3,5-dihydroxybenzoic acid

Many disubstituted benzenes (and polysubstituted benzenes) have historical names. Some of these are obscure, with no obvious connection to the structure of the molecule.

common name:	*m*-xylene	mesitylene	*o*-toluic acid	*p*-cresol
IUPAC name:	1,3-dimethylbenzene	1,3,5-trimethylbenzene	2-methylbenzoic acid	4-methylphenol

When the benzene ring is named as a substituent on another molecule, it is called a **phenyl group**. The phenyl group is used in the name just like the name of an alkyl group, and it is often abbreviated **Ph** (or ϕ) in drawing a complex structure.

or Ph—CH$_2$—C≡C—CH$_3$ or Ph$_2$O   or PhCH$_2$CH$_2$OH
1-phenyl-2-butyne   diphenyl ether   3-phenoxycyclohexene   2-phenylethanol

The seven-carbon unit consisting of a benzene ring and a methylene (—CH$_2$—) group is often named as a **benzyl group**. Be careful not to confuse the *benzyl group* (seven carbons) with the *phenyl group* (six carbons).

PROBLEM-SOLVING HINT

A benzene ring substituent is a *phenyl group* (6 carbons).
A *benzyl group* contains an additional CH$_2$ group (7 carbons).

a phenyl group   a benzyl group   benzyl bromide   benzyl alcohol
                                       (α-bromotoluene)

Aromatic hydrocarbons are sometimes called **arenes**. An **aryl group**, abbreviated **Ar**, is the aromatic group that remains after the removal of a hydrogen atom from an aromatic ring. The phenyl group, **Ph**, is the simplest aryl group. The generic aryl group (**Ar**) is the aromatic relative of the generic alkyl group, which we symbolize by **R**.

Examples of aryl groups

the phenyl group   the *o*-nitrophenyl group   the *p*-tolyl group   the 3-pyridyl group

Examples of the use of a generic aryl group

Ar—MgBr   Ar$_2$O or Ar—O—Ar′   Ar—NH$_2$ Ar—SO$_3$H
an arylmagnesium   a diaryl ether   an arylamine   an arylsulfonic acid
bromide

PROBLEM 16-21

Draw and name all the chlorinated benzenes having from one to six chlorine atoms.

PROBLEM 16-22

Name the following compounds:

(a) [benzene ring with F]

(b) [benzene ring with CH$_2$CH$_2$C≡CH]

(c) [benzene ring with OH and CH$_3$]

(d) [benzene ring with CH=CH$_2$ and NO$_2$]

(e) [benzene ring with COOH and Br]

(f) [benzene ring with OCH(CH$_3$)$_2$]

(g) [benzene ring with OH, NO$_2$, NO$_2$]

(h) [benzene ring with CH$_2$OCH$_2$CH$_3$]

PROBLEM 16-23

Draw and name a specific example of each class of compounds.
(a) an alkyl aryl ether, Ar—O—R
(b) an arylsulfonic acid, Ar—SO$_3$H
(c) an aryllithium reagent
(d) an aryl alcohol (What is a better generic name for this class of compounds?)
(e) a diaryl methanol
(f) an arylbenzene
(g) a substituted benzyl alcohol

The melting points, boiling points, and densities of benzene and some derivatives are given in Table 16-1. Benzene derivatives tend to be more symmetrical than similar aliphatic compounds, so they pack better into crystals and have higher melting points. For example, benzene melts at 6°C, while hexane melts at −95°C. Similarly, *para*-disubstituted benzenes are more symmetrical than the ortho and meta isomers, and they pack better into crystals and have higher melting points.

The relative boiling points of many benzene derivatives are related to their dipole moments. For example, the dichlorobenzenes have boiling points that follow their dipole moments. Symmetrical *p*-dichlorobenzene has zero dipole moment and the lowest boiling point. *m*-Dichlorobenzene has a small dipole moment and a slightly higher boiling point. *o*-Dichlorobenzene has the largest dipole moment and the highest boiling point. Even though *p*-dichlorobenzene has the lowest boiling point, it packs best into a crystal, and it has the highest melting point of the dichlorobenzenes.

16-14
Physical Properties of Benzene and Its Derivatives

Mothballs are composed of either *p*-dichlorobenzene or naphthalene.

o-dichlorobenzene
bp 181°C
mp −17°C

m-dichlorobenzene
bp 173°C
mp −25°C

p-dichlorobenzene
bp 170°C
mp 54°C

TABLE 16-1   Physical Properties of Benzene and Its Derivatives

Compound	mp (°C)	bp (°C)	Density (g/mL)	Compound	mp (°C)	bp (°C)	Density (g/mL)
benzene	6	80	0.88	*o*-xylene	−26	144	0.88
toluene	−95	111	0.87	*m*-xylene	−48	139	0.86
ethylbenzene	−95	136	0.87	*p*-xylene	13	138	0.86
styrene	−31	146	0.91	*o*-chlorotoluene	−35	159	1.08
ethynylbenzene	−45	142	0.93	*m*-chlorotoluene	−48	162	1.07
fluorobenzene	−41	85	1.02	*p*-chlorotoluene	8	162	1.07
chlorobenzene	−46	132	1.11	*o*-dichlorobenzene	−17	181	1.31
bromobenzene	−31	156	1.49	*m*-dichlorobenzene	−25	173	1.29
iodobenzene	−31	188	1.83	*p*-dichlorobenzene	54	170	1.07
benzyl bromide	−4	199	1.44	*o*-dibromobenzene	7	225	1.62
nitrobenzene	6	211	1.20	*m*-dibromobenzene	−7	218	1.61
phenol	43	182	1.07	*p*-dibromobenzene	87	218	1.57
anisole	37	156	0.98	*o*-toluic acid	106	263	1.06
benzoic acid	122	249	1.31	*m*-toluic acid	111	263	1.05
benzyl alcohol	−15	205	1.04	*p*-toluic acid	180	275	1.06
aniline	−6	186	1.02	*o*-cresol	30	192	1.03
diphenyl ether	28	259	1.08	*m*-cresol	12	202	1.03
mesitylene	−45	165	0.87	*p*-cresol	36	202	1.03

Benzene and other aromatic hydrocarbons are slightly denser than the nonaromatic analogues, but they are still less dense than water. The halogenated benzenes are denser than water. Aromatic hydrocarbons and halogenated aromatics are generally insoluble in water, although some derivatives with strongly polar functional groups (phenol, benzoic acid, etc.) are moderately soluble in water.

16-15
Spectroscopy of Aromatic Compounds

Infrared Spectroscopy (Review)   Aromatic compounds are readily identified by their infrared spectra because they show a characteristic C=C stretch around 1600 cm^{-1}. This is a lower C=C stretching frequency than for isolated alkenes (1640 to 1680 cm^{-1}) or conjugated dienes (1620 to 1640 cm^{-1}) because the aromatic bond order is only about $1\frac{1}{2}$. The aromatic bond is therefore less stiff than a normal double bond, and it vibrates at a lower frequency.

Like alkenes, aromatic compounds show unsaturated =C—H stretching just above 3000 cm^{-1} (usually around 3030 cm^{-1}). The combination of the aromatic C=C stretch around 1600 cm^{-1} and the =C—H stretch just above 3000 cm^{-1} leaves little doubt of the presence of an aromatic ring. The sample spectra labeled Compounds 4, 5, and 7 in Chapter 12 (pages 516–517) show compounds containing aromatic rings.

NMR Spectroscopy (Review)   Aromatic compounds give readily identifiable ^{1}H NMR signals around $\delta 7$ to $\delta 8$, strongly deshielded by the aromatic ring current (Section 13-5B). In benzene, the aromatic protons absorb around $\delta 7.2$. The signals may be moved farther downfield by electron-withdrawing groups such as carbonyl, nitro, or cyano groups, or upfield by electron-donating groups such as hydroxyl, alkoxy, or amino groups.

◀ **Figure 16-18**
The mass spectrum of
n-butylbenzene has its base peak at
m/z 91, corresponding to cleavage of
a benzylic bond. The fragments are a
benzyl cation and a propyl radical.
The benzyl cation rearranges to the
tropylium ion, detected at m/z 91.

Nonequivalent aromatic protons that are ortho or meta usually split each other. The spin-spin splitting constants are about 8 Hz for ortho protons and 2 Hz for meta protons. Figures 13-11, 13-18, 13-24, 13-29, and 13-32 show proton NMR spectra of aromatic compounds.

Aromatic carbon atoms absorb around $\delta 120$ to $\delta 150$ in the ^{13}C NMR spectrum. Alkene carbon atoms can also absorb in this spectral region, but the combination of ^{13}C NMR with ^{1}H NMR or IR spectroscopy usually leaves no doubt about the presence of an aromatic ring.

Mass Spectrometry The most common mass spectral fragmentation of alkylbenzene derivatives is the cleavage of a benzylic bond to give a resonance-stabilized benzylic cation. For example, in the mass spectrum of n-butylbenzene (Figure 16-18), the base peak is at m/z 91, from the benzyl cation. The benzyl cation may rearrange to give the aromatic tropylium ion. Alkylbenzenes frequently give ions corresponding to the tropylium ion at m/z 91.

PROBLEM 16-24

Draw three more resonance forms for the benzyl cation shown in Figure 16-18.

Ultraviolet Spectroscopy The ultraviolet spectra of aromatic compounds are quite different from those of nonaromatic polyenes. For example, benzene has three absorptions in the ultraviolet region: an intense band at $\lambda_{max} = 184$ nm ($\varepsilon = 68,000$), a moderate band at $\lambda_{max} = 204$ nm ($\varepsilon = 8800$), and a characteristic low-intensity band of multiple absorptions centered around 254 nm ($\varepsilon = 200$ to 300). In the UV spectrum of benzene in Figure 16-19, the absorption at 184 nm does not appear because wavelengths shorter than 200 nm are not accessible by standard UV–visible spectrometers.

All three major bands in the benzene spectrum correspond to $\pi \rightarrow \pi^*$ transitions. The absorption at 184 nm corresponds to the energy of the transition from one of the two HOMOs to one of the two LUMOs. The weaker band at 204 nm corresponds to a "forbidden" transition that would be impossible to observe if benzene were always an unperturbed, perfectly hexagonal structure.

▶ **Figure 16-19**
Ultraviolet spectra of benzene and
styrene.

The most characteristic part of the spectrum is the band centered at 254 nm, called
the **benzenoid band**. About three to six small, sharp peaks (called *fine structure*) usually appear in this band. Their molar absorptivities are weak, usually 200 to 300. These
benzenoid absorptions correspond to additional forbidden transitions.

Simple benzene derivatives show most of the characteristics of benzene, including
the moderate band in the 210-nm region and the benzenoid band in the 260-nm region.
Alkyl and halogen substituents increase the values of λ_{max} by about 5 nm, as shown by
the examples in Table 16-2. An additional conjugated double bond can increase the value
of λ_{max} by about 30 nm, as shown by the UV spectrum of styrene in Figure 16-19.

TABLE 16-2 Ultraviolet Spectra of Benzene and Some Derivatives

Compound	Structure	Moderate Band λ_{max}(nm)	ε	Benzenoid Band λ_{max}(nm)	ε
benzene		204	8,800	254	250
ethylbenzene	CH_2CH_3	208	7,800	260	220
m-xylene	CH_3 CH_3	212	7,300	264	300
bromobenzene	Br	210	7,500	258	170
styrene		248	15,000	282	740

PROBLEM 16-25

The UV spectrum of 1-phenyl-2-propen-1-ol shows an intense absorption at 220 nm and a weaker absorption at 258 nm. When this compound is treated with dilute sulfuric acid, it rearranges to an isomer with an intense absorption at 250 nm and a weaker absorption at 290 nm. Suggest a structure for the isomeric product and propose a mechanism for its formation.

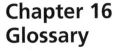

Chapter 16
Glossary

aliphatic compound An organic compound that is not aromatic. (p. 679)

allotropes Different forms of an element with different properties. For example, diamond, graphite, and fullerenes are different allotropic forms of elemental carbon. (p. 704)

annulenes Cyclic hydrocarbons with alternating single and double bonds. (p. 683)

[6]annulene (benzene)   [10]annulene (cyclodecapentaene)

aromatic compound A cyclic compound containing some number of conjugated double bonds, characterized by an unusually large resonance energy. (pp. 679, 680, 688)

To be aromatic, all its ring atoms must have unhybridized p orbitals that overlap to form a continuous ring. In most cases, the structure must be planar and have $(4N + 2)$ pi electrons, with N an integer. Delocalization of the pi electrons over the ring results in a lowering of the electronic energy.

antiaromatic compound A compound that has a continuous ring of p orbitals as in an aromatic compound, but delocalization of the pi electrons over the ring increases the electronic energy. (p. 688)

In most cases, the structure must be planar and have $(4N)$ pi electrons, with N an integer.

arenes Aromatic hydrocarbons, usually based on the benzene ring as a structural unit. (p. 708)

aryl group (abbreviated **Ar**) The aromatic group that remains after taking a hydrogen atom off an aromatic ring; the aromatic equivalent of the generic alkyl group (**R**). (p. 708)

benzenoid band The weak band around 250 to 270 nm in the UV spectra of benzenoid aromatics. This band is characterized by multiple sharp absorptions (fine structure). (p. 714)

benzyl group ($PhCH_2$—) The seven-carbon unit consisting of a benzene ring and a methylene group. (p. 708)

buckminsterfullerene ("buckyballs") A common name for the C_{60} molecule with the same symmetry as a soccer ball. The arrangement of five-membered and six-membered rings is similar to that in a geodesic dome. (p. 706)

degenerate orbitals Orbitals having the same energy. (p. 685)

fullerenes A common generic term for carbon clusters similar to C_{60} (buckminsterfullerene) and compounds related to them.

fused rings Rings that share a common carbon–carbon bond and its two carbon atoms. (p. 702)

heterocyclic compound (heterocycle) A cyclic compound in which one or more of the ring atoms is not carbon. (p. 697)

    aromatic heterocycle: A heterocyclic compound that fulfills the criteria for aromaticity and has a substantial resonance energy.

Hückel's rule A cyclic molecule or ion that has a continuous ring of overlapping p orbitals will be

  1. aromatic if the number of pi electrons is $(4N + 2)$, with N an integer.

  2. antiaromatic if the number of pi electrons is $(4N)$, with N an integer. (p. 689)

Kekulé structure A classical structural formula for an aromatic compound, showing localized double bonds. (p. 679)

nanotubes A common term for carbon tubes consisting of a cylinder of fused graphite-like six-membered rings and ending with half of a C_{60} sphere. (p. 706)

nonaromatic compound Neither aromatic nor antiaromatic; lacking the continuous ring of overlapping p orbitals required for aromaticity or antiaromaticity. (p. 689)

ortho Having a 1,2-relationship on a benzene ring. (p. 707)

meta Having a 1,3-relationship on a benzene ring. (p. 707)

para Having a 1,4-relationship on a benzene ring. (p. 707)

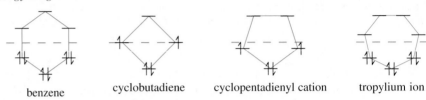

ortho (1,2)   meta (1,3)   para (1,4)

phenyl group (**Ph** or **φ**) The benzene ring, minus one hydrogen atom, when named as a substituent on another molecule. (p. 708)

polygon rule The energy diagram of the MOs of a regular, completely conjugated cyclic system has the same polygonal shape as the compound, with one vertex (the all-bonding MO) at the bottom. The nonbonding line cuts horizontally through the center of the polygon. (p. 688)

Energy diagrams

benzene   cyclobutadiene   cyclopentadienyl cation   tropylium ion

polynuclear aromatic compounds Aromatic compounds with two or more fused aromatic rings. Naphthalene is a **polynuclear aromatic hydrocarbon** (PAH or PNA). Indole is a polynuclear aromatic heterocycle. (p. 702)

naphthalene   indole

resonance energy The extra stabilization provided by delocalization, compared with a localized structure. For aromatic compounds, the resonance energy is the extra stabilization provided by the delocalization of the electrons in the aromatic ring. (p. 682)

tropylium ion The cycloheptatrienyl cation. (p. 695)

Essential Problem-Solving Skills in Chapter 16

1. Be able to construct the molecular orbitals of a cyclic system of *p* orbitals similar to benzene and cyclobutadiene.

2. Use the polygon rule to draw the energy diagram for a cyclic system of *p* orbitals, and fill in the electrons to show whether a given compound or ion is aromatic or antiaromatic.

3. Use Hückel's rule to predict whether a given annulene, heterocycle, or ion will be aromatic, antiaromatic, or nonaromatic.

4. For heterocycles containing nitrogen atoms, determine whether the lone pairs are used in the aromatic system, and predict whether the nitrogen atom is strongly or weakly basic.

5. Recognize fused aromatic systems such as polynuclear aromatic hydrocarbons and fused heterocyclic compounds, and use the theory of aromatic compounds to explain their properties.

6. Name aromatic compounds, and draw their structures from the names.

7. Use IR, NMR, UV, and mass spectra to determine the structures of aromatic compounds. Given an aromatic compound, predict the important features of its spectra.

Study Problems

16-26 Define each term and give an example.
   (a) a heterocyclic aromatic compound   **(b)** an antiaromatic compound   **(c)** a Kekulé structure
   (d) an annulene   **(e)** degenerate orbitals   **(f)** the polygon rule
   (g) a polynuclear aromatic heterocycle   **(h)** fused rings   **(i)** a polynuclear aromatic hydrocarbon
   (j) the benzenoid UV band   **(k)** a filled shell of MOs   **(l)** Hückel's rule
   (m) resonance energy   **(n)** an aryl group

16-27 Draw the structure of each compound.
   (a) *o*-nitroanisole   **(b)** 2,4-dimethoxyphenol   **(c)** *p*-aminobenzoic acid
   (d) 4-nitroaniline   **(e)** *m*-chlorotoluene   **(f)** *p*-divinylbenzene
   (g) *p*-bromostyrene   **(h)** 3,5-dimethoxybenzaldehyde   **(i)** tropylium chloride
   (j) sodium cyclopentadienide **(k)** 2-phenylpropan-1-ol   **(l)** benzyl methyl ether

16-28 Name the following compounds:

(a)   **(b)**   **(c)**   **(d)**

(e)   **(f)**   **(g)**   **(h)**

16-29 Draw and name all the methyl, dimethyl, and trimethylbenzenes.

16-30 One of the following hydrocarbons is much more acidic than the others. Indicate which one, and explain why it is unusually acidic.

16-31 In Kekulé's time, cyclohexane was unknown, and there was no proof that benzene must be a six-membered ring. Determination of the structure relied largely on the known numbers of monosubstituted and disubstituted benzenes, together with the knowledge that benzene did not react like a normal alkene. The following structures were the likely candidates.

(localized double bonds)

   (a) Show where the six hydrogen atoms are in each structure.
   (b) For each structure, draw all the possible monobrominated derivatives (C_6H_5Br) that would result from randomly substituting one hydrogen with a bromine. Benzene was known to have only one monobromo derivative.
   (c) For each of the structures that had only one monobromo derivative in part (b), draw all the possible dibromo derivatives. Benzene was known to have three dibromo derivatives, but resonance theory was unknown at the time.
   (d) Determine which structure was most consistent with what was known about benzene at that time: Benzene gives one monobrominated derivative and three dibrominated derivatives, and it gives negative chemical tests for an alkene.
   (e) The structure that was considered the most likely structure for benzene is called *Ladenburg benzene*, after the chemist who proposed it. What factors would make Ladenburg benzene relatively unstable, in contrast with the stability observed with real benzene?

16-32   The following molecules and ions are grouped by similar structures. Classify each as aromatic, antiaromatic, or nonaromatic. For the aromatic and antiaromatic species, give the number of pi electrons in the ring.

16-33   Azulene is a deep blue hydrocarbon with resonance energy of 49 kcal/mol (205 kJ/mol). Azulene has ten pi electrons, so it might be considered one large aromatic ring. Azulene has an unusually large dipole moment (1.0 D) for a hydrocarbon, indicating significant charge separation. Show how this charge separation might arise.

azulene

16-34   Each of the following heterocycles includes one or more nitrogen atoms. Classify each nitrogen atom as strongly basic or weakly basic, according to the availability of its lone pair of electrons.

*16-35   Some of the following compounds show aromatic properties, while others do not.
(1) Predict which ones are likely to be aromatic, and explain why they are aromatic.
(2) Predict which nitrogen atoms are more basic than water, and which are less basic.

*16-36   The benzene ring alters the reactivity of a neighboring group in the *benzylic position* much like a double bond alters the reactivity of groups in the allylic position.

$$H_2C=CH-CH_2-R$$

allylic position   benzyl group   benzylic position   benzyl radical

Benzylic cations, anions, and radicals are all more stable than simple alkyl intermediates.

(a) Use resonance structures to show the delocalization (over four carbon atoms) of the positive charge, odd electron, and negative charge of the benzyl cation, radical, and anion.

(b) Toluene reacts with bromine in the presence of light to give benzyl bromide. Propose a mechanism for this reaction.

$$\text{toluene} + Br_2 \xrightarrow{h\nu} \text{benzyl bromide} + HBr$$

toluene   benzyl bromide

(c) Which of the following reactions will have the faster rate and give the better yield? Use a drawing of the transition state to explain your answer.

$$\text{—CH}_2\text{Br} \xrightarrow[CH_3OH]{NaOCH_3} \text{—CH}_2\text{OCH}_3$$

$$\text{—CH}_2\text{Br} \xrightarrow[CH_3OH]{NaOCH_3} \text{—CH}_2\text{OCH}_3$$

16-37   Before spectroscopy was invented, *Körner's absolute method* was used to determine whether a disubstituted benzene derivative was the ortho, meta, or para isomer. Körner's method involves adding a third group (often a nitro group) and determining how many isomers are formed. For example, when *o*-xylene is nitrated (by a method shown in Chapter 17), two isomers are formed.

$$\text{o-xylene} \xrightarrow{\text{nitrate}} \text{(NO}_2 \text{ isomer)} + \text{(O}_2\text{N isomer)}$$

(a) How many isomers are formed by nitration of *m*-xylene?

(b) How many isomers are formed by nitration of *p*-xylene?

(c) A turn-of-the-century chemist isolated an aromatic compound of molecular formula $C_6H_4Br_2$. He carefully nitrated this compound and purified three isomers of formula $C_6H_3Br_2NO_2$. Propose structures for the original compound and the three nitrated derivatives.

16-38   For each NMR spectrum, propose a structure consistent with the spectrum and the additional information provided.
  (a) Elemental analysis shows the molecular formula to be C_8H_7OCl. The IR spectrum shows a moderate absorption at 1602 cm^{-1} and a strong absorption at 1690 cm^{-1}.

  (b) The mass spectrum shows a double molecular ion of ratio $1:1$ at m/z 184 and 186.

16-39   Recall (Section 16-10) that two positions of anthracene sometimes react more like polyenes than like aromatic compounds.
  (a) Draw a Kekulé structure that shows how the reactive positions of anthracene are the ends of a diene, appropriate for a Diels–Alder reaction.
  (b) The Diels–Alder reaction of anthracene with maleic anhydride is a common organic lab experiment. Predict the product of this Diels–Alder reaction.

maleic anhydride

16-40   Biphenyl has the following structure.

biphenyl

(a) Is biphenyl a (fused) polynuclear aromatic hydrocarbon?

(b) How many pi electrons are there in the two aromatic rings of biphenyl? How does this number compare with that for naphthalene?

(c) The heat of hydrogenation for biphenyl is about 100 kcal/mol (418 kJ/mol). Calculate the resonance energy of biphenyl.

(d) Compare the resonance energy of biphenyl with that of naphthalene and with that of two benzene rings. Explain the difference in the resonance energies of naphthalene and biphenyl.

16-41   The following hydrocarbon reacts with two equivalents of butyllithium to form a dianion of formula $[C_8H_6]^{2-}$. Propose a structure for this dianion, and suggest why it forms so readily.

$$+ \ 2 \ C_4H_9Li \ \longrightarrow \ [C_8H_6]^{2-} \ (Li^+)_2 \ + \ 2 \ C_4H_{10}\uparrow$$

16-42   How would you convert 1,3,5,7-cyclononatetraene to an aromatic compound?

*__16-43__   The ribonucleosides that make up ribonucleic acid (RNA) are composed of D-ribose (a sugar) and four heterocyclic "bases." The general structure of a ribonucleoside is

$$HO-CH_2 \quad O \quad \boxed{base}$$

OH   OH

a ribonucleoside

The four heterocyclic bases are cytosine, uracil, guanine, and adenine. Cytosine and uracil are called *pyrimidine bases* because their structures resemble pyrimidine. Guanine and adenine are called *purine bases* because their structures resemble purine.

pyrimidine   cytosine   uracil   purine   guanine   adenine

(a) Determine which rings of these bases are aromatic.

(b) Predict which nitrogen atoms are basic.

(c) Do any of these bases have easily formed tautomers that are aromatic? (Consider moving a proton from nitrogen to a carbonyl group to form a phenolic derivative.)

*__16-44__   Consider the following compound, which has been synthesized and characterized:

$$(CH_3)_3C \qquad C(CH_3)_3$$
$$N$$
$$(CH_3)_3C$$

(a) Assuming this molecule is entirely conjugated, do you expect it to be aromatic, antiaromatic, or nonaromatic?

(b) Why was this molecule synthesized with three *t*-butyl substituents? Why not make the unsubstituted compound and study it instead?

(c) Do you expect the nitrogen atom to be basic? Explain.

(d) At room temperature, the proton NMR spectrum shows only two singlets of ratio 1 : 2. The smaller signal remains unchanged at all temperatures. As the temperature is lowered to −110°C, the larger signal broadens and separates into two new singlets, one on either side of the original chemical shift. At −110°C, the spectrum consists of three separate singlets of areas 1 : 1 : 1. Explain what these NMR data indicate about the bonding in this molecule. How does your conclusion based on the NMR data agree with your prediction in part (a)?

16-45   A student found an old bottle labeled "thymol" on the stockroom shelf. After noticing a pleasant odor, she obtained the
following mass, IR, and NMR spectra. The NMR peak at δ4.8 disappears on shaking with D_2O. Propose a structure for
thymol, and show how your structure is consistent with the spectra. Propose a fragmentation to explain the MS peak at
m/z 135, and show why the resulting ion is relatively stable.

*16-46   An unknown compound gives the following mass, IR, and NMR spectra. Propose a structure, and show how it is consistent with the spectra. Show the fragmentations that give the prominent peaks at m/z 127 and 155 in the mass spectrum.

*16-47   Hexahelicene (below) seems a poor candidate for optical activity because all its carbon atoms are sp^2 hybrids and presumably flat. Nevertheless, hexahelicene has been synthesized and separated into enantiomers. Its optical rotation is enormous: $[\alpha]_D = 3700°$. Explain why hexahelicene is optically active, and speculate as to why the rotation is so large.

hexahelicene

CHAPTER 17

Reactions of Aromatic Compounds

EPM of anisole

With an understanding of the properties that make a compound aromatic, we now consider the reactions of aromatic compounds. A large part of this chapter is devoted to *electrophilic aromatic substitution*, the most important mechanism involved in the reactions of aromatic compounds. Many reactions of benzene and its derivatives are explained by minor variations of electrophilic aromatic substitution. We will study several of these reactions and then consider how substituents on the ring influence its reactivity toward electrophilic aromatic substitution and the regiochemistry seen in the products. We will also study other reactions of aromatic compounds, including nucleophilic aromatic substitution, addition reactions, reactions of side chains, and special reactions of phenols.

17-1
Electrophilic Aromatic Substitution

Like an alkene, benzene has clouds of pi electrons above and below its sigma bond framework. Although benzene's pi electrons are in a stable aromatic system, they are available to attack a strong electrophile to give a carbocation. This resonance-stabilized carbocation is called a **sigma complex** because the electrophile is joined to the benzene ring by a new sigma bond.

sigma complex   substituted

The sigma complex (also called an *arenium ion*) is not aromatic because the sp^3 hybrid carbon atom interrupts the ring of *p* orbitals. This loss of aromaticity contributes to the highly endothermic nature of this first step. The sigma complex regains aromaticity either by a reversal of the first step (returning to the reactants) or by loss of the proton on the tetrahedral carbon atom, leading to the substitution product.

The overall reaction is the *substitution* of an electrophile (E^+) for a proton (H^+) on the aromatic ring: **electrophilic aromatic substitution**. This class of reactions includes substitutions by a wide variety of electrophilic reagents. Because it enables us to introduce functional groups directly onto the aromatic ring, electrophilic aromatic substitution is the most important method for synthesis of substituted aromatic compounds.

PROBLEM-SOLVING HINT

Note that the three resonance forms of the sigma complex show the positive charge on the three carbon atoms *ortho* and *para* to the site of substitution.

KEY MECHANISM   Electrophilic Aromatic Substitution

Step 1: Attack on the electrophile forms the sigma complex.

sigma complex (arenium ion)

Step 2: Loss of a proton gives the substitution product.

Bromination of Benzene   Bromination follows the general mechanism for electrophilic aromatic substitution. Bromine itself is not sufficiently electrophilic to react with benzene, but a strong Lewis acid such as $FeBr_3$ catalyzes the reaction. Bromine donates a pair of electrons to $FeBr_3$, forming a stronger electrophile with a weakened Br—Br bond and a partial positive charge on one of the bromine atoms. Attack by benzene forms the sigma complex. Bromide ion from $FeBr_4^-$ acts as a weak base to remove a proton from the sigma complex, giving the aromatic product and HBr, and regenerating the catalyst.

17-2
Halogenation
of Benzene

MECHANISM   Bromination of Benzene

Step 1: Formation of a stronger electrophile.

$Br_2 \cdot FeBr_3$ intermediate
(a stronger electrophile than Br_2)

Step 2: Electrophilic attack and formation of the sigma complex.

sigma complex

Step 3: Loss of a proton gives the products.

bromobenzene

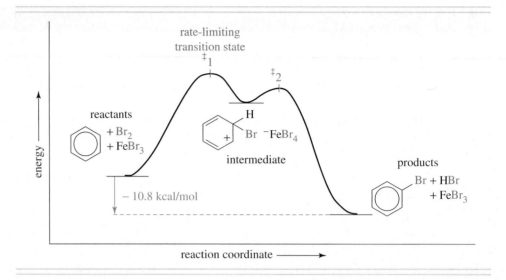

▶ **Figure 17-1**
The energy diagram for the bromination of benzene shows that the first step is endothermic and rate limiting, and the second step is strongly exothermic.

Formation of the sigma complex is rate limiting, and the transition state leading to it occupies the highest-energy point on the energy diagram (Figure 17-1). This step is strongly endothermic because it forms a nonaromatic carbocation. The second step is exothermic because aromaticity is regained and a molecule of HBr is evolved. The overall reaction is exothermic by 10.8 kcal/mol (45 kJ/mol).

Comparison with Alkenes   Benzene is not as reactive as alkenes, which react rapidly with bromine at room temperature to give addition products (Section 8-10). For example, cyclohexene reacts to give *trans*-1,2-dibromocyclohexane. This reaction is exothermic by about 29 kcal/mol (121 kJ/mol).

$$\Delta H° = -29 \text{ kcal}$$
$$(-121 \text{ kJ})$$

The analogous addition of bromine to benzene is *endothermic* because it requires the loss of aromatic stability. The addition is not seen under normal circumstances. The *substitution* of bromine for a hydrogen atom gives an aromatic product. The substitution is exothermic, but it requires a Lewis acid catalyst to convert bromine to a stronger electrophile.

$$\Delta H° = +2 \text{ kcal}$$
$$(+8 \text{ kJ})$$

$$\Delta H° = -10.8 \text{ kcal}$$
$$(-45 \text{ kJ})$$

bromobenzene
(80%)

Chlorination of Benzene

Chlorination of benzene works much like bromination, except that aluminum chloride ($AlCl_3$) is most often used as the Lewis acid catalyst.

benzene $+$ Cl_2 $\xrightarrow{AlCl_3}$ chlorobenzene (85%) $+$ HCl

---

PROBLEM 17-1

Propose a detailed mechanism for the aluminum chloride–catalyzed reaction of benzene with chlorine.

Iodination of Benzene

Iodination of benzene requires an acidic oxidizing agent, such as nitric acid. Nitric acid is consumed in the reaction, so it is a reagent (an oxidant) rather than a catalyst.

benzene $+$ $\frac{1}{2}I_2$ $+$ HNO_3 $\longrightarrow$ iodobenzene (85%) $+$ NO_2 $+$ H_2O

Iodination probably involves an electrophilic aromatic substitution with iodonium ion (I^+) acting as the electrophile. The iodonium ion results from oxidation of iodine by nitric acid.

$$H^+ \;+\; HNO_3 \;+\; \tfrac{1}{2}I_2 \;\longrightarrow\; I^+ \;+\; NO_2 \;+\; H_2O$$

iodonium
ion

PROBLEM 17-2

Propose a mechanism for the reaction of benzene with the iodonium ion.

PROBLEM 17-3

Controlled fluorination of benzene is difficult, but it can be accomplished by a two-step thallation procedure. Benzene reacts with thallium tris(trifluoroacetate), $Tl(OCOCF_3)_3$, to give an organothallium intermediate. Further reaction with potassium fluoride and boron trifluoride gives the aryl fluoride. Propose a mechanism for the first step, the thallation of benzene.

benzene $+$ $Tl(OCOCF_3)_3$ $\longrightarrow$ organothallium intermediate —$Tl(OCOCF_3)_2$ $\xrightarrow{KF, BF_3}$ fluorobenzene —F

| benzene | thallium tris(trifluoroacetate) | organothallium intermediate | fluorobenzene |

Hint: The ionization of mercuric acetate gives the electrophile that oxymercurates an alkene (Section 8-5); a similar ionization of thallium tris(trifluoroacetate) gives an electrophile that substitutes onto an aromatic ring.

Thallation reactions are potentially useful, but organothallium compounds are highly toxic and easily absorbed through the skin.

17-3
Nitration of Benzene

Benzene reacts with hot, concentrated nitric acid to give nitrobenzene. This sluggish reaction is not convenient because a hot mixture of concentrated nitric acid with any oxidizable material might explode. A safer and more convenient procedure uses a mixture of nitric acid and sulfuric acid. Sulfuric acid is a catalyst, allowing **nitration** to take place more rapidly and at lower temperatures.

nitrobenzene (85%)

Aromatic nitro compounds are components of many drugs, such as the antibiotic chloramphenicol. The body reduces aromatic nitro drugs to the corresponding aromatic amines.

The mechanism is shown below. Sulfuric acid reacts with nitric acid to form the **nitronium ion** ($^+NO_2$), a powerful electrophile. The mechanism is similar to other sulfuric acid—catalyzed dehydrations. Sulfuric acid protonates the hydroxyl group of nitric acid, allowing it to leave as water, forming a nitronium ion. The nitronium ion reacts with benzene to form a sigma complex. Loss of a proton from the sigma complex gives nitrobenzene.

MECHANISM   Nitration of Benzene

Formation of the nitronium ion.

nitronium ion

Step 1: Electrophilic attack.

Step 2: Loss of a proton.

benzene   nitronium
                    ion

sigma complex
(resonance-delocalized)

nitrobenzene

PROBLEM 17-4

p-Xylene undergoes nitration much faster than benzene. Use resonance forms of the sigma complex to explain this accelerated rate.

17-4
Sulfonation
of Benzene

We have already used esters of *p*-toluenesulfonic acid as activated derivatives of alcohols with a good leaving group, the tosylate group (Section 11-5). *p*-Toluenesulfonic acid is an example of an *arylsulfonic acid* (general formula Ar—SO_3H), which are often used as strong acid catalysts that are soluble in nonpolar organic solvents. Arylsulfonic acids are easily synthesized by **sulfonation** of benzene derivatives, an electrophilic aromatic substitution using sulfur trioxide (SO_3) as the electrophile.

benzene · · · · · sulfur trioxide · · · · · · · · · · · · · · · · · · benzenesulfonic acid (95%)

"Fuming sulfuric acid" is the common name for a solution of 7% SO_3 in H_2SO_4. Sulfur trioxide is the *anhydride* of sulfuric acid, meaning that the addition of water to SO_3 gives H_2SO_4. Although it is uncharged, sulfur trioxide is a strong electrophile, with three sulfonyl (S=O) bonds drawing electron density away from the sulfur atom. Benzene attacks sulfur trioxide, forming a sigma complex. Loss of a proton on the tetrahedral carbon and reprotonation on oxygen gives benzenesulfonic acid.

sulfur trioxide, a powerful electrophile

MECHANISM   Sulfonation of Benzene

Step 1: Electrophilic attack.   *Step 2:* Loss of a proton.

benzene   sulfur trioxide   sigma complex   benzenesulfonic acid
                                        (resonance-delocalized)

PROBLEM 17-5

Use resonance forms to show that the dipolar sigma complex shown above has its positive charge delocalized over three carbon atoms and its negative charge delocalized over three oxygen atoms.

Desulfonation   Sulfonation is reversible, and a sulfonic acid group may be removed from an aromatic ring by heating in dilute sulfuric acid. In practice, steam is often used as a source of both water and heat for **desulfonation**.

benzenesulfonic acid · benzene (95%)

Desulfonation follows the same mechanistic path as sulfonation, except in the opposite order. A proton adds to a ring carbon to form a sigma complex, then loss of sulfur trioxide gives the unsubstituted aromatic ring.

(resonance-delocalized) · · · · · · · (SO_3 + H_2O ⇌ H_2SO_4)

Substantial quantities of sulfonated aromatic compounds are released into the environment as a result of industrial and domestic uses. The major domestic source is laundry detergent, which contains alkylbenzenesulfonates.

Protonation of the Aromatic Ring; Hydrogen–Deuterium Exchange Desulfonation involves protonation of an aromatic ring to form a sigma complex. Similarly, if a proton attacks benzene, the sigma complex can lose either of the two protons at the tetrahedral carbon. We can prove that a reaction has occurred by using a deuterium ion (D^+) rather than a proton and by showing that the product contains a deuterium atom in place of hydrogen. This experiment is easily accomplished by adding SO_3 to some D_2O (heavy water) to generate D_2SO_4. Benzene reacts to give a deuterated product.

The reaction is reversible, but at equilibrium, the final products reflect the D/H ratio of the solution. A large excess of deuterium gives a product with all six of the benzene hydrogens replaced by deuterium. This reaction serves as a synthesis of benzene-d_6 (C_6D_6), a common NMR solvent.

benzene   benzene-d_6

17-5
Nitration of Toluene: The Effect of Alkyl Substitution

Up to now, we have considered only benzene as the substrate for electrophilic aromatic substitution. To synthesize more complicated aromatic compounds, we need to consider the effects other substituents might have on further substitutions. For example, toluene (methylbenzene) reacts with a mixture of nitric and sulfuric acids much like benzene does, but with some interesting differences:

1. Toluene reacts about 25 times faster than benzene under the same conditions. We say that toluene is **activated** toward electrophilic aromatic substitution and that the methyl group is an **activating group**.

2. Nitration of toluene gives a mixture of products, primarily those resulting from substitution at the ortho and para positions. Because of this preference, we say that the methyl group of toluene is an **ortho, para-director**.

toluene   o-nitrotoluene   m-nitrotoluene   p-nitrotoluene
                        (60%)   (4%)   (36%)

These product ratios show that the orientation of substitution is not random. If each C—H position were equally reactive, there would be equal amounts of ortho and meta substitution and half as much para substitution: 40% ortho, 40% meta, and 20% para. This is the statistical prediction based on the two ortho positions, two meta positions, and just one para position available for substitution.

The rate-limiting step (the highest-energy transition state) for electrophilic aromatic substitution is the first step, formation of the sigma complex. This step is where the electrophile bonds to the ring, determining the substitution pattern. We can explain both the enhanced reaction rate and the preference for ortho and para substitution by considering the structures of the intermediate sigma complexes. In this endothermic reaction, the structure of the transition state leading to the sigma complex resembles the product, the sigma complex (Hammond postulate, Section 4-14). We are justified in using the stabilities of the sigma complexes to indicate the relative energies of the transition states leading to them.

When benzene reacts with the nitronium ion, the resulting sigma complex has the positive charge distributed over three secondary (2°) carbon atoms.

Benzene

In ortho or para substitution of toluene, the positive charge is spread over two secondary carbons and one tertiary (3°) carbon (bearing the CH_3 group).

Ortho attack

Para attack

p-Nitrotoluene is the starting point for the syntheses of benzocaine and procaine, two compounds used as local anesthetics. (See Section 19-19C3)

benzocaine (R=H)
procaine (R=NEt$_2$)

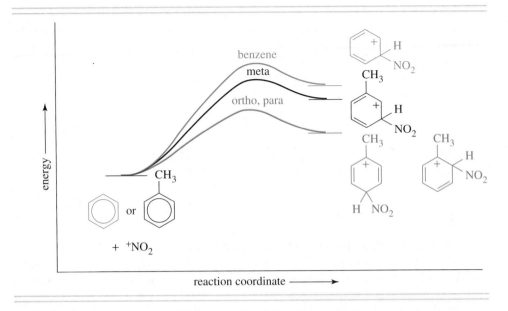

▶ **Figure 17-2**
The methyl group of toluene stabilizes the sigma complexes and the transition states leading to them. This stabilization is most effective when the methyl group is ortho or para to the site of substitution.

Because the sigma complexes for ortho and para attack have resonance forms with tertiary carbocations, they are more stable than the sigma complex for nitration of benzene. Therefore, the ortho and para positions of toluene react faster than benzene.

The sigma complex for meta substitution has its positive charge spread over three 2° carbons; this intermediate is similar in energy to the intermediate for substitution of benzene. Therefore, meta substitution of toluene does not show the large rate enhancement seen with ortho and para substitution.

Meta attack

The methyl group in toluene is electron donating; it stabilizes the intermediate sigma complex and the rate limiting transition state leading to its formation. This stabilizing effect is large when it is situated ortho or para to the site of substitution and the positive charge is delocalized onto the tertiary carbon atom. When substitution occurs at the meta position, the positive charge is not delocalized onto the tertiary carbon, and the methyl group has a smaller effect on the stability of the sigma complex. Figure 17-2 compares the energy profiles for nitration of benzene and toluene at the ortho, meta, and para positions.

17-6
Activating, Ortho, Para-Directing Substituents

17-6A Alkyl Groups

The results observed with toluene are general for any alkylbenzene undergoing electrophilic aromatic substitution. Substitution ortho or para to the alkyl group gives an intermediate (and a transition state) with the positive charge shared by the tertiary carbon atom. As a result, alkylbenzenes undergo electrophilic aromatic substitution faster than benzene, and the products are predominantly ortho- and para-substituted. An alkyl group is therefore an activating substituent, and it is **ortho, para-directing**. This effect is called **inductive stabilization** because the alkyl group donates electron density through the sigma bond joining it with the benzene ring.

Shown below is the reaction of ethylbenzene with bromine, catalyzed by ferric bromide. As with toluene, the rates of formation of the ortho- and para-substituted isomers are greatly enhanced with respect to the meta isomer.

ethylbenzene   o-bromo   m-bromo   p-bromo
                   (38%)   (<1%)   (62%)

PROBLEM 17-6

(a) Draw a detailed mechanism for the FeBr₃-catalyzed reaction of ethylbenzene with bromine, and show why the sigma complex (and the transition state leading to it) is lower in energy for substitution at the ortho and para positions than it is for substitution at the meta position.

(b) Explain why m-xylene undergoes nitration 100 times faster than p-xylene.

PROBLEM 17-7

Styrene (vinylbenzene) undergoes electrophilic aromatic substitution much faster than benzene, and the products are found to be primary ortho- and para-substituted styrenes. Use resonance forms of the intermediates to explain these results.

17-6B   Substituents with Nonbonding Electrons

The Methoxyl Group   Anisole (methoxybenzene) undergoes nitration about 10,000 times faster than benzene and about 400 times faster than toluene. This result seems curious because oxygen is a strongly electronegative group, yet it donates electron density to stabilize the transition state and the sigma complex. Recall that the nonbonding electrons of an oxygen atom adjacent to a carbocation stabilize the positive charge through resonance.

only six   each atom has eight
valence electrons   valence electrons

The second resonance form puts the positive charge on the electronegative oxygen atom, but it has more covalent bonds, and it provides each atom with an octet in its valence shell. This type of stabilization is called **resonance stabilization**, and the oxygen atom is called **resonance-donating** or **pi-donating** because it donates electron density through a pi bond in one of the resonance structures. Like alkyl groups, the methoxyl group of anisole preferentially activates the ortho and para positions.

anisole   o-nitroanisole   m-nitroanisole   p-nitroanisole
               (31%)   (2%)   (67%)

Resonance forms show that the methoxyl group effectively stabilizes the sigma complex if it is ortho or para to the site of substitution, but not if it is meta. Resonance stabilization is provided by a pi bond between the —OCH$_3$ substituent and the ring.

Ortho attack

Meta attack

Para attack

A methoxyl group is so strongly activating that anisole quickly brominates in water without a catalyst. In the presence of excess bromine, this reaction proceeds to the tribromide.

anisole

2,4,6-tribromoanisole
(100%)

EPM of anisole

PROBLEM 17-8

Propose a detailed mechanism for the bromination of ethoxybenzene to give *o*- and *p*-bromo-ethoxybenzene.

The Amino Group   Like an alkoxyl group, a nitrogen atom with a nonbonding pair of electrons serves as a powerful activating group. For example, aniline undergoes a fast bromination (without a catalyst) in bromine water to give the tribromide. Sodium bicarbonate is added to neutralize the HBr formed and to prevent protonation of the basic amino (—NH$_2$) group (see Problem 17-11).

aniline → 2,4,6-tribromoaniline (100%) + HBr

Reagents: 3 Br$_2$, H$_2$O, NaHCO$_3$ (to neutralize HBr)

EPM of aniline

Nitrogen's nonbonding electrons provide resonance stabilization to the sigma complex if attack takes place ortho or para to the position of the nitrogen atom.

Ortho attack

Para attack

(plus other resonance forms)

(plus other resonance forms)

PROBLEM 17-9

Draw all the resonance forms for the sigma complexes corresponding to bromination of aniline at the ortho, meta, and para positions.

Thus, any substituent with a lone pair of electrons on the atom bonded to the ring can provide resonance stabilization to a sigma complex. Several examples are illustrated below in decreasing order of their activation of an aromatic ring. All these substituents are strongly activating, and they are all ortho, para-directing.

SUMMARY   Activating, Ortho, Para-Directors

Groups

$$-\ddot{\text{O}}:^- \;>\; -\overset{R}{\underset{..}{\text{N}}}-R \;>\; -\ddot{\text{O}}-H \;>\; -\ddot{\text{O}}-R \;>\; -\overset{H}{\underset{..}{\text{N}}}-\overset{O}{\overset{\|}{\text{C}}}-R \;>\; -R$$

(no lone pairs)

Compounds

phenoxides > anilines > phenols > phenyl ethers > anilides > alkylbenzenes

---

PROBLEM 17-10

When bromine is added to two beakers, one containing phenyl isopropyl ether and the other containing cyclohexene, the bromine color in both beakers disappears. What observation could you make while performing this test that would allow you to distinguish the alkene from the aryl ether?

17-7
Deactivating, Meta-Directing Substituents

EPM of nitrobenzene

Nitrobenzene is about 100,000 times *less* reactive than benzene toward electrophilic aromatic substitution. For example, nitration of nitrobenzene requires concentrated nitric and sulfuric acids at temperatures above 100°C. Nitration proceeds slowly, giving the meta isomer as the major product.

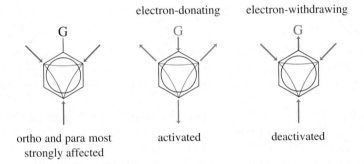

These results should not be surprising. We have already seen that a substituent on a benzene ring has its greatest effect on the carbon atoms ortho and para to the substituent. An electron-donating substituent activates primarily the ortho and para positions, and an electron-withdrawing substituent (such as a nitro group) deactivates primarily the ortho and para positions.

electron-donating   electron-withdrawing

ortho and para most   activated   deactivated
strongly affected

This selective deactivation leaves the meta positions the most reactive, and meta substitution is seen in the products. **Meta-directors**, often called **meta-allowing** substituents, deactivate the meta position less than the ortho and para positions, allowing meta substitution.

We can show why the nitro group is a strong **deactivating group** by considering its resonance forms. No matter how we position the electrons in a Lewis dot diagram, the nitrogen atom always has a formal positive charge.

The positively charged nitrogen inductively withdraws electron density from the aromatic ring. This aromatic ring is less electron-rich than benzene, so it is deactivated toward reactions with electrophiles.

The following reactions show why this deactivating effect is strongest at the ortho and para positions. Each sigma complex has its positive charge spread over three carbon atoms. In ortho and para substitution, one of the carbon atoms bearing this positive charge is the carbon attached to the positively charged nitrogen atom of the nitro group. Since like charges repel, this close proximity of two positive charges is especially unstable.

Ortho attack

especially unstable

Meta attack

Para attack

especially unstable

In the sigma complex for meta substitution, the carbon bonded to the nitro group does not share the positive charge of the ring. This is a more stable situation because the positive charges are farther apart. As a result, nitrobenzene reacts primarily at the meta position. We can summarize by saying that the nitro group is a deactivating group, and that it is a meta-director (or meta-allower).

The energy diagram in Figure 17-3 compares the energies of the transition states and intermediates leading to ortho, meta, and para substitution of nitrobenzene with those for benzene. Notice that a higher activation energy is involved for substitution of nitrobenzene at any position, resulting in slower reaction rates than for benzene.

Just as activating substituents are all ortho, para-directors, most deactivating substituents are meta-directors. In general, deactivating substituents are groups with a positive charge (or a partial positive charge) on the atom bonded to the aromatic ring. As we saw with the nitro group, this positively charged atom repels any positive charge on the adjacent carbon atom of the ring. Of the possible sigma complexes, only the one corresponding to meta substitution avoids putting a positive charge on this ring carbon.

▶ **Figure 17-3**
Nitrobenzene is deactivated toward electrophilic aromatic substitution at *any* position, but deactivation is strongest at the ortho and para positions. Reaction occurs at the meta position, but it is slower than the reaction of benzene.

For example, the partial positive charge on a carbonyl carbon allows substitution primarily at the meta position:

Ortho attack

+ charges adjacent (unfavorable)

acetophenone

(+ charge here in other resonance forms)

Meta attack

(+ charge here in other resonance forms)

This sigma complex does not place the positive charge on the ring carbon bearing the carbonyl group.

The following summary table lists some common substituents that are deactivating and meta-directing. Resonance forms are also given to show how a positive charge arises on the atom bonded to the aromatic ring.

PROBLEM 17-11

In an aqueous solution containing sodium bicarbonate, aniline reacts quickly with bromine to give 2,4,6-tribromoaniline. Nitration of aniline requires very strong conditions, however, and the yields (mostly *m*-nitroaniline) are poor.
(a) What conditions are used for nitration, and what form of aniline is present under these conditions?

(b) Explain why nitration of aniline is so sluggish and why it gives mostly meta substitution.

*(c) Although nitration of aniline is slow and gives mostly meta substitution, nitration of acetanilide (PhNHCOCH$_3$) goes quickly and gives mostly para substitution. Use resonance forms to explain this difference in reactivity.

SUMMARY **Deactivating, Meta-Directors**

Group	Resonance Forms	Example
—NO$_2$ nitro		nitrobenzene
—SO$_3$H sulfonic acid		benzenesulfonic acid
—C≡N: cyano		benzonitrile
ketone or aldehyde		acetophenone
ester		methyl benzoate
—NR$_3^+$ quaternary ammonium		trimethylanilinium iodide

The halobenzenes are exceptions to the general rules. Halogens are deactivating groups, yet they are ortho, para-directors. We can explain this unusual combination of properties by considering that

1. The halogens are strongly electronegative, withdrawing electron density from a carbon atom through the sigma bond. (inductive withdrawal)
2. The halogens have nonbonding electrons that can donate electron density through pi bonding. (resonance donation)

These inductive and resonance effects oppose each other. The carbon–halogen bond (shown at right) is strongly polarized, with the carbon atom at the positive end of the dipole. This polarization draws electron density away from the benzene ring, making it less reactive toward electrophilic substitution.

17-8
Halogen Substituents: Deactivating, but Ortho, Para-Directing

less electron-rich

If an electrophile reacts at the ortho or para position, however, the positive charge of the sigma complex is shared by the carbon atom bearing the halogen. The nonbonding electrons of the halogen can further delocalize the charge onto the halogen, giving a **halonium ion** structure. This resonance stabilization allows a halogen to be pi-donating, even though it is sigma-withdrawing.

Ortho attack   *Para attack*   *Meta attack*

Reaction at the meta position gives a sigma complex whose positive charge is not delocalized onto the halogen-bearing carbon atom. Therefore, the meta intermediate is not stabilized by the halonium ion structure. The following reaction illustrates the preference for ortho and para substitution in the nitration of chlorobenzene.

Figure 17-4 shows the effect of the halogen atom graphically, with an energy diagram comparing energies of the transition states and intermediates for electrophilic

▶ **Figure 17-4**
The energies of the intermediates and transition states are higher for chlorobenzene than for benzene. The highest energy results from substitution at the meta position, while the energies for ortho and para substitution are slightly lower because of stabilization by the halonium ion structure.

attack on chlorobenzene and benzene. Higher energies are required for the reactions of chlorobenzene, especially for attack at the meta position.

PROBLEM 17-12

Draw all the resonance forms of the sigma complex for nitration of bromobenzene at the ortho, meta, and para positions. Point out why the intermediate for meta substitution is less stable than the other two.

PROBLEM 17-13

(a) Predict the structure of the product formed when HCl adds to 1-bromocyclohexene.
(b) Propose a mechanism with resonance forms to support your prediction.
(c) Explain how this prediction is in accord with the ortho, para-directing effect of bromine on an aromatic ring.

PROBLEM-SOLVING **HINT**

Remember which substituents are activating and which are deactivating. Activators are ortho, para-directing, and deactivators are meta-directing, except for the halogens.

SUMMARY **Directing Effects of Substituents**

π Donors	σ Donors	Halogens	Carbonyls	Other
—$\ddot{\text{N}}\text{H}_2$ —$\ddot{\text{O}}\text{H}$ —$\ddot{\text{O}}\text{R}$ —$\ddot{\text{N}}\text{HCOCH}_3$	—R (alkyl) (aryl)	—F —Cl —Br —I	$\overset{\text{O}}{\overset{\|}{-\text{C}-\text{R}}}$ $\overset{\text{O}}{\overset{\|}{-\text{C}-\text{OH}}}$ $\overset{\text{O}}{\overset{\|}{-\text{C}-\text{OR}}}$	—SO$_3$H —C≡N —NO$_2$ —$\overset{+}{\text{N}}\text{R}_3$
	ortho, para-directing		meta-directing	
◄━━━━━ ACTIVATING		DEACTIVATING		━━━━━►

17-9
Effects of Multiple Substituents on Electrophilic Aromatic Substitution

Two or more substituents exert a combined effect on the reactivity of an aromatic ring. If the groups reinforce each other, the result is easy to predict. For example, we can predict that all the xylenes (dimethylbenzenes) are activated toward electrophilic substitution because the two methyl groups are both activating. In the case of a nitrobenzoic acid, both substituents are deactivating, so we predict that a nitrobenzoic acid is deactivated toward attack by an electrophile.

o-xylene
activated

m-nitrobenzoic acid
deactivated

m-toluic acid
not obvious

The orientation of addition is easily predicted in many cases. For example, in *m*-xylene there are two positions ortho to one of the methyl groups and para to the other. Electrophilic substitution occurs primarily at these two equivalent positions. There may be some substitution at the position between the two methyl groups (ortho to both), but this position is sterically hindered, and it is less reactive than the other two activated positions. In *p*-nitrotoluene, the methyl group directs an electrophile toward its ortho positions. The nitro group directs toward the same locations because they are its meta positions.

m-xylene → HNO₃ / H₂SO₄ → major product (65%)

each is *ortho* to one CH₃, *para* to the other

ortho to both CH₃'s, but hindered

p-nitrotoluene → HNO₃ / H₂SO₄ → major product (99%)

ortho to CH₃ *meta* to NO₂

PROBLEM 17-14

Predict the mononitration products of the following compounds.
(a) *o*-nitrotoluene
(b) *m*-chlorotoluene
(c) *o*-bromobenzoic acid
(d) *p*-methoxybenzoic acid
(e) *m*-cresol (*m*-methylphenol)

When the directing effects of two or more substituents conflict, it is more difficult to predict where an electrophile will react. In many cases, mixtures result. For example, *o*-xylene is activated at all the positions, so it gives mixtures of substitution products.

o-xylene → HNO₃ / H₂SO₄ → (58%) + (42%)

When there is a conflict between an activating group and a deactivating group, the activating group usually directs the substitution. We can make an important generalization:

Activating groups are usually stronger directors than deactivating groups.

In fact, it is helpful to separate substituents into three classes, from strongest to weakest.

1. Powerful ortho, para-directors that stabilize the sigma complexes through resonance. Examples are —OH, —OR, and —NR₂ groups.
2. Moderate ortho, para-directors, such as alkyl groups and halogens.
3. All meta-directors.

$$-OH, -OR, -NR_2 \;>\; -R, -X \;>\; \overset{\overset{\displaystyle O}{\|}}{-C}-R, -SO_3H, -NO_2$$

If two substituents direct an incoming electrophile toward different reaction sites, the substituent in the stronger class predominates. If both are in the same class, mixtures are likely. In the following reaction, the stronger group predominates and directs the incoming substituent. The methoxyl group is a stronger director than the nitro group, and substitution occurs ortho and para to the methoxyl group. Steric effects prevent much substitution at the crowded position ortho to both the methoxyl group and the nitro group.

m-nitroanisole   major products

SOLVED PROBLEM 17-1

Predict the major product(s) of bromination of *p*-chloroacetanilide.

PROBLEM-SOLVING **HINT**
To predict products of compounds with multiple substituents, look for the most strongly activating substituent(s).

SOLUTION

The amide group (—NHCOCH$_3$) is a strong activating and directing group because the nitrogen atom with its nonbonding pair of electrons is bonded to the aromatic ring. The amide group is a stronger director than the chlorine atom, and substitution occurs mostly at the positions ortho to the amide. Like an alkoxyl group, the amide is a particularly strong activating group, and the reaction gives some of the dibrominated product.

p-chloroacetanilide

PROBLEM 17-15

Predict the mononitration products of the following aromatic compounds.
(a) *p*-methylanisole
(b) *m*-nitrochlorobenzene
(c) *p*-chlorophenol
(d) *m*-nitroanisole

(e) *o*-methylacetanilide

(f) (Consider the structures of these groups. One is activating, and the other is deactivating.)

PROBLEM 17-16

Biphenyl is two benzene rings joined by a single bond. The site of substitution for a biphenyl is determined by (1) which phenyl ring is more activated (or less deactivated), and (2) which position on that ring is most reactive, using the fact that a phenyl substituent is ortho, para-directing.

PROBLEM-SOLVING **HINT**
When predicting substitution products for compounds with more than one ring, first decide which ring is more activated (or less deactivated), then consider only that ring, and decide which position is most reactive.

(a) Use resonance forms of a sigma complex to show why a phenyl substituent should be ortho, para-directing.

(b) Predict the mononitration products of the following compounds.

(i) [structure: biphenyl]

biphenyl

(ii) [structure with OH]

(iii) [structure with $\overset{O}{\overset{\|}{C}}$—CH$_3$]

(iv) [structure with NO$_2$]

(v) [structure: three rings]

17-10
The Friedel–Crafts Alkylation

Carbocations are perhaps the most important electrophiles capable of substituting onto aromatic rings, because this substitution forms a new carbon–carbon bond. Reactions of carbocations with aromatic compounds were first studied in 1877 by the French alkaloid chemist Charles Friedel and his American partner, James Crafts. In the presence of Lewis acid catalysts such as aluminum chloride ($AlCl_3$) or ferric chloride ($FeCl_3$), alkyl halides were found to alkylate benzene to give alkylbenzenes. This useful reaction is called the **Friedel–Crafts alkylation**.

Friedel–Crafts alkylation

☞ [benzene with H] + R—X $\xrightarrow{\text{Lewis acid} \atop (AlCl_3, \text{FeBr}_3, \text{etc.})}$ [benzene with R] + H—X

(X = Cl, Br, I)

For example, aluminum chloride catalyzes the alkylation of benzene by *t*-butyl chloride. HCl gas is evolved.

[benzene] + CH_3—$\overset{CH_3}{\underset{CH_3}{C}}$—Cl $\xrightarrow{AlCl_3}$ [benzene with $\overset{CH_3}{\underset{CH_3}{C}}$—$CH_3$] + HCl

benzene   *t*-butyl chloride   *t*-butylbenzene
                                                        (90%)

This alkylation is a typical electrophilic aromatic substitution, with the *t*-butyl cation acting as the electrophile. The *t*-butyl cation is formed by reaction of *t*-butyl chloride with the catalyst, aluminum chloride. The *t*-butyl cation reacts with benzene to form a sigma complex. Loss of a proton gives the product, *t*-butylbenzene. The aluminum chloride catalyst is regenerated in the final step.

Friedel–Crafts alkylations are used with a wide variety of primary, secondary, and tertiary alkyl halides. With secondary and tertiary halides, the reacting electrophile is probably the carbocation.

R—X + $AlCl_3$ $\rightleftharpoons$ R$^+$ + $^-AlCl_3X$

(R is secondary   reacting
or tertiary)   electrophile

MECHANISM Friedel–Crafts Alkylation

Step 1: Formation of a carbocation.

t-butyl chloride

t-butyl cation

Step 2: Electrophilic attack.

Step 3: Loss of a proton.

+ AlCl$_3$
+ HCl

With primary alkyl halides, the free primary carbocation is too unstable. The actual electrophile is a complex of aluminum chloride with the alkyl halide. In this complex, the carbon–halogen bond is weakened (as indicated by dashed lines) and there is considerable positive charge on the carbon atom. The mechanism for the aluminum chloride–catalyzed reaction of ethyl chloride with benzene is as follows:

$$CH_3\!-\!CH_2\!-\!Cl \ + \ AlCl_3 \ \rightleftharpoons \ \overset{\delta+}{CH_3\!-\!CH_2}\!\cdots\!Cl\!\cdots\!\overset{\delta-}{AlCl_3}$$

sigma complex

CH_2CH_3 + H—Cl + AlCl$_3$

PROBLEM 17-17

Propose products (if any) and mechanisms for the following AlCl$_3$-catalyzed reactions:
(a) chlorocyclohexane with benzene   **(b)** methyl chloride with anisole
(c) 3-chloro-2,2-dimethylbutane with isopropylbenzene

Friedel–Crafts Alkylation Using Other Carbocation Sources We have seen several ways of generating carbocations, and most of these can be used for Friedel–Crafts alkylations. Two common methods are protonation of alkenes and treatment of alcohols with BF$_3$.

Alkenes are protonated by HF to give carbocations. Fluoride ion is a weak nucleophile and does not immediately attack the carbocation. If benzene (or an activated benzene derivative) is present, electrophilic substitution occurs. The protonation step follows Markovnikov's rule, forming the more stable carbocation, which alkylates the aromatic ring.

Alcohols are another source of carbocations for Friedel–Crafts alkylations. Alcohols commonly form carbocations when treated with Lewis acids such as boron trifluoride (BF$_3$). If benzene (or an activated benzene derivative) is present, substitution may occur.

Formation of the cation

Electrophilic substitution on benzene

The BF$_3$ used in this reaction is consumed and not regenerated. A full equivalent of the Lewis acid is needed, so we say that the reaction is *promoted* by BF$_3$ rather than *catalyzed* by BF$_3$.

PROBLEM 17-18

For each reaction, show the generation of the electrophile and predict the products.
(a) benzene + cyclohexene + HF
(b) *t*-butyl alcohol + benzene + BF$_3$
(c) *t*-butylbenzene + 2-methylpropene + HF
(d) 2-propanol + toluene + BF$_3$

Limitations of the Friedel–Crafts Alkylation   Although the Friedel–Crafts alkylation looks good in principle, it has three major limitations that severely restrict its use.

Friedel–Crafts reactions fail with strongly deactivated systems.

Limitation 1   Friedel–Crafts reactions work only with benzene, halobenzenes, and activated benzene derivatives; they fail with strongly deactivated systems such as nitrobenzene, benzenesulfonic acid, and phenyl ketones. In some cases, we can get around this limitation by adding the deactivating group or changing an activating group into a deactivating group *after* the Friedel–Crafts step.

SOLVED PROBLEM 17-2

Devise a synthesis of *p*-nitro-*t*-butylbenzene from benzene.

SOLUTION

To make *p*-nitro-*t*-butylbenzene, we would first use a Friedel–Crafts reaction to make *t*-butylbenzene. Nitration gives the correct product. If we were to make nitrobenzene first, the Friedel–Crafts reaction to add the *t*-butyl group would fail.

Good

Bad

Limitation 2 Like other carbocation reactions, the Friedel–Crafts alkylation is susceptible to carbocation rearrangements. As a result, only certain alkylbenzenes can be made using the Friedel–Crafts alkylation. *t*-Butylbenzene, isopropylbenzene, and ethylbenzene can be synthesized using the Friedel–Crafts alkylation because the corresponding cations are not prone to rearrangement. Consider what happens, however, when we try to make *n*-propylbenzene by the Friedel–Crafts alkylation.

Alkyl carbocations for Friedel–Crafts alkylations are prone to rearrangements.

Ionization with rearrangement gives isopropyl cation

Reaction with benzene gives isopropylbenzene

Limitation 3 Because alkyl groups are activating substituents, the product of the Friedel–Crafts alkylation is *more reactive* than the starting material. Multiple alkylations are hard to avoid. This limitation can be severe. If we need to make ethylbenzene, we might try adding some $AlCl_3$ to a mixture of 1 mole of ethyl chloride and 1 mole of benzene. As some ethylbenzene is formed, however, it is activated, reacting even faster than benzene itself. The product is a mixture of some (ortho and para) diethylbenzenes, some triethylbenzenes, a small amount of ethylbenzene, and some leftover benzene.

Friedel–Crafts alkylations are prone to multiple alkylation.

The problem of dialkylation can be avoided by using a large excess of benzene. For example, if 1 mole of ethyl chloride is used with 50 moles of benzene, the concentration of ethylbenzene is always low, and the electrophile is more likely to react with benzene than with ethylbenzene. Distillation separates the product from excess benzene. This is a common industrial approach, since a continuous distillation can recycle the unreacted benzene.

In the laboratory, we must often alkylate aromatic compounds that are more expensive than benzene. Because we cannot afford to use a large excess of the starting material, a more selective method is needed. Fortunately, the Friedel–Crafts acylation, discussed in Section 17-11, introduces just one group without danger of polyalkylation or rearrangement.

PROBLEM 17-19

Predict the products (if any) of the following reactions.
(a) (excess) benzene + isobutyl chloride + $AlCl_3$
(b) (excess) toluene + 1-butanol + BF_3
(c) (excess) nitrobenzene + 2-chloropropane + $AlCl_3$
(d) (excess) benzene + 3, 3-dimethyl-1-butene + HF

PROBLEM 17-20

Which reactions will produce the desired product in good yield? You may assume that aluminum chloride is added as a catalyst in each case. For the reactions that will not give a good yield of the desired product, predict the major products.

Reagents	Desired Product
(a) benzene + *n*-butyl bromide	*n*-butylbenzene
(b) ethylbenzene + *t*-butyl chloride	*p*-ethyl-*t*-butylbenzene
(c) bromobenzene + ethyl chloride	*p*-bromoethylbenzene
(d) ethylbenzene + bromine	*p*-bromoethylbenzene
(e) anisole + methyl iodide (3 moles)	2,4,6-trimethylanisole

PROBLEM 17-21

Show how you would synthesize the following aromatic derivatives from benzene:
(a) *p*-*t*-butylnitrobenzene   (b) *p*-toluenesulfonic acid   (c) *p*-chlorotoluene

17-11
The Friedel–Crafts Acylation

An **acyl group** is a carbonyl group with an alkyl group attached. Acyl groups are named systematically by dropping the final *-e* from the alkane name and adding the *-oyl* suffix. Historical names are often used for the *formyl group*, the *acetyl group*, and the *propionyl group*, however.

An **acyl chloride** is an acyl group bonded to a chlorine atom. Acyl chlorides are made by reaction of the corresponding carboxylic acids with thionyl chloride. Therefore, acyl chlorides are also called **acid chlorides**. We consider acyl chlorides in more detail when we study acid derivatives in Chapter 21.

In the presence of aluminum chloride, an acyl chloride reacts with benzene (or an activated benzene derivative) to give a phenyl ketone: an *acylbenzene*. The **Friedel–Crafts acylation** is analogous to the Friedel–Crafts alkylation, except that the reagent is an acyl chloride instead of an alkyl halide and the product is an acylbenzene (a "phenone") instead of an alkylbenzene.

Friedel–Crafts acylation

Example

17-11A   Mechanism of Acylation

The mechanism of Friedel–Crafts acylation (shown below) resembles that for alkylation, except that the carbonyl group helps to stabilize the cationic intermediate. The acyl halide forms a complex with aluminum chloride; loss of the tetrachloroaluminate ion ($^-$AlCl$_4$) gives a resonance-stabilized **acylium ion**. The acylium ion is a strong electrophile. It reacts with benzene or an activated benzene derivative to form an acylbenzene.

MECHANISM   Friedel–Crafts Acylation

Step 1: Formation of an acylium ion.

Step 2: Electrophilic attack.

Step 3: Loss of a proton. Complexation of the product.

The product of acylation (the acylbenzene) is a ketone. The ketone's carbonyl group has nonbonding electrons that complex with the Lewis acid catalyst ($AlCl_3$), requiring a full equivalent of $AlCl_3$ in the acylation. The initial product is the aluminum chloride complex of the acylbenzene. Addition of water hydrolyzes this complex, giving the free acylbenzene.

The electrophile in the Friedel–Crafts acylation appears to be a large, bulky complex, such as $R—\overset{+}{C}{=}O \; {}^-AlCl_4$. Para substitution usually prevails when the aromatic substrate has an ortho, para-directing group, possibly because the electrophile is too bulky for effective attack at the ortho position. For example, when ethylbenzene reacts with acetyl chloride, the major product is *p*-ethylacetophenone.

One of the most attractive features of the Friedel–Crafts acylation is the deactivation of the product toward further substitution. The acylbenzene has a carbonyl group (a deactivating group) bonded to the aromatic ring. Since Friedel–Crafts reactions do not occur on strongly deactivated rings, the acylation stops after one substitution.

Thus, Friedel–Crafts acylation overcomes two of the three limitations of the alkylation: The acylium ion is resonance-stabilized, so that no rearrangements occur; and the acylbenzene product is deactivated, so that no further reaction occurs. Like the alkylation, however, the acylation fails with strongly deactivated aromatic rings.

SUMMARY   **Comparison of Friedel–Crafts Alkylation and Acylation**

Alkylation	*Acylation*
The alkylation cannot be used with strongly deactivated derivatives.	Also true: Only benzene, halobenzenes, and activated derivatives are suitable.
The carbocations involved in the alkylation may rearrange.	Resonance-stabilized acylium ions are not prone to rearrangement.
Polyalkylation is commonly a problem.	The acylation forms a deactivated acylbenzene, which does not react further.

17-11B   The Clemmensen Reduction: Synthesis of Alkylbenzenes

How do we synthesize alkylbenzenes that cannot be made by Friedel–Crafts alkylation? We use the Friedel–Crafts acylation to make the acylbenzene, then we reduce the acylbenzene to the alkylbenzene using the **Clemmensen reduction**: treatment with aqueous HCl and amalgamated zinc (zinc treated with mercury salts).

This two-step sequence can synthesize many alkylbenzenes that are impossible to make by direct alkylation. For example, we saw earlier that *n*-propylbenzene cannot be made by Friedel–Crafts alkylation. Benzene reacts with *n*-propyl chloride and AlCl$_3$ to give isopropylbenzene, together with some diisopropylbenzene. In the acylation, however, benzene reacts with propanoyl chloride and AlCl$_3$ to give ethyl phenyl ketone (propiophenone), which is easily reduced to *n*-propylbenzene.

Carboxylic acids and acid anhydrides also serve as acylating agents in Friedel–Crafts reactions. We consider these acylating agents in Chapters 20 and 21 when we study the reactions of carboxylic acids and their derivatives.

17-11C   The Gatterman–Koch Formylation: Synthesis of Benzaldehydes

We cannot add a formyl group to benzene by Friedel–Crafts acylation in the usual manner. The problem lies with the necessary reagent, formyl chloride, which is unstable and cannot be bought or stored.

Formylation can be accomplished by using a high-pressure mixture of carbon monoxide and HCl together with a catalyst consisting of a mixture of cuprous chloride (CuCl) and aluminum chloride. This mixture generates the formyl cation, possibly through a small concentration of formyl chloride. The reaction with benzene gives formylbenzene, better known as benzaldehyde. This reaction, called the **Gatterman–Koch synthesis**, is widely used in industry to synthesize aryl aldehydes.

benzaldehyde

PROBLEM 17-22

Show how you would use the Friedel–Crafts acylation, Clemmensen reduction, and/or Gatterman–Koch synthesis to prepare the following compounds:

(a) $Ph-\overset{O}{\overset{\|}{C}}-CH_2CH(CH_3)_2$
    isobutyl phenyl ketone

(b) $Ph-\overset{O}{\overset{\|}{C}}-C(CH_3)_3$
    t-butyl phenyl ketone

(c) $Ph-\overset{O}{\overset{\|}{C}}-Ph$
    diphenyl ketone

(d) *p*-methoxybenzaldehyde

(e) 1-phenyl-2,2-dimethylpropane

(f) *n*-butylbenzene

17-12
Nucleophilic Aromatic Substitution

Nucleophiles can displace halide ions from aryl halides, particularly if there are strong electron-withdrawing groups ortho or para to the halide. Because a nucleophile substitutes for a leaving group on an aromatic ring, this class of reactions is called **nucleophilic aromatic substitution**. The following examples show that both ammonia and hydroxide ion can displace chloride from 2,4-dinitrochlorobenzene:

Electrophilic aromatic substitution is the most important reaction of aromatic compounds because it has broad applications for a wide variety of aromatic compounds. In contrast, *nucleophilic* aromatic substitution has restricted applications. In nucleophilic aromatic substitution, a strong nucleophile replaces a leaving group, such as a halide. What is the mechanism of nucleophilic aromatic substitution? It cannot be the S_N2 mechanism because aryl halides cannot achieve the correct geometry for back-side displacement. The aromatic ring blocks approach of the nucleophile to the back of the carbon bearing the halogen.

The S_N1 mechanism cannot be involved either. Strong nucleophiles are required for nucleophilic aromatic substitution, and the reaction rate is proportional to the concentration of the nucleophile. Thus, the nucleophile must be involved in the transition state.

Electron-withdrawing substituents (such as nitro groups) *activate* the ring toward nucleophilic aromatic substitution, suggesting that the transition state is developing a negative charge on the ring. In fact, nucleophilic aromatic substitutions are difficult without at least one powerful electron-withdrawing group. (This effect is the opposite of that for *electrophilic* aromatic substitution, where electron-withdrawing substituents slow or stop the reaction.)

Nucleophilic aromatic substitutions have been studied in detail. Either of two mechanisms may be involved, depending on the reactants. One mechanism is similar to the electrophilic aromatic substitution mechanism, except that nucleophiles and carbanions are involved rather than electrophiles and carbocations. The other mechanism involves "benzyne," an interesting and unusual reactive intermediate.

17-12A The Addition–Elimination Mechanism

Consider the reaction of 2,4-dinitrochlorobenzene with sodium hydroxide (shown below). When hydroxide (the nucleophile) attacks the carbon bearing the chlorine, a negatively charged sigma complex results. The negative charge is delocalized over the ortho and para carbons of the ring and further delocalized into the electron-withdrawing nitro groups. Loss of chloride from the sigma complex gives 2,4-dinitrophenol, which is deprotonated in this basic solution.

MECHANISM Nucleophilic Aromatic Substitution (Addition–Elimination)

Step 1: Attack by hydroxide gives a resonance-stabilized sigma complex.

Step 2: Loss of chloride gives the product.   *Step 3:* Excess base deprotonates the product.

The resonance forms shown above illustrate how nitro groups ortho and para to the halogen help to stabilize the intermediate (and the transition state leading to it). Without strong electron-withdrawing groups in these positions, formation of the negatively charged sigma complex is unlikely.

activates positions ortho and para

activated

not activated

PROBLEM 17-23

Fluoride ion is usually a poor leaving group because it is not very polarizable. Fluoride serves as the leaving group in the Sanger reagent (2,4-dinitrofluorobenzene), used in the determination of peptide structures (Chapter 24). Explain why fluoride works as a leaving group in this nucleophilic aromatic substitution, even though it is a poor leaving group in the S_N1 and S_N2 mechanisms.

2,4-dinitrofluorobenzene   +   amine   2,4-dinitrophenyl derivative
(Sanger reagent)

17-12B   The Benzyne Mechanism: Elimination–Addition

The addition–elimination mechanism for nucleophilic aromatic substitution requires strong electron-withdrawing substituents on the aromatic ring. Under extreme conditions, however, unactivated halobenzenes react with strong bases. For example, a commercial synthesis of phenol (the "Dow process") involves treatment of chlorobenzene with sodium hydroxide and a small amount of water in a pressurized reactor at 350°C:

chlorobenzene   sodium phenoxide   +   NaCl   phenol

Similarly, chlorobenzene reacts with sodium amide ($NaNH_2$, an extremely strong base) to give aniline, $Ph\!-\!NH_2$. This reaction does not require high temperatures, taking place in liquid ammonia at $-33°C$.

Nucleophilic substitution of unactivated benzene derivatives occurs by a mechanism different from the addition-elimination we saw with the nitro-substituted halobenzenes. A clue to the mechanism is provided by the reaction of *p*-bromotoluene with sodium amide. The products are a 50:50 mixture of *m*- and *p*-toluidine.

p-bromotoluene   *p*-toluidine   *m*-toluidine
                                          (50%)   (50%)

These two products can be explained by an elimination–addition mechanism, called the **benzyne mechanism** because of the unusual intermediate. Sodium amide (or sodium hydroxide in the Dow process) reacts as a *base*, abstracting a proton. The product is a carbanion with a negative charge and a nonbonding pair of electrons localized in the sp^2 orbital that once formed the C—H bond.

The carbanion can expel bromide ion to become a neutral species. As bromide leaves with its bonding electrons, an empty sp^2 orbital remains. This orbital overlaps with the filled orbital adjacent to it, giving additional bonding between these two carbon atoms. The two sp^2 orbitals are directed 60° away from each other, so their overlap is not very effective. This reactive intermediate is called a **benzyne** because it can be symbolized by drawing a triple bond between these two carbon atoms. Triple bonds are usually linear, however, so this is a very reactive, highly strained triple bond.

Amide ion is a strong nucleophile, attacking at either end of the weak, reactive benzyne triple bond. Subsequent protonation gives toluidine. About half of the product results from attack by the amide ion at the meta carbon, and about half from attack at the para carbon.

In summary, the benzyne mechanism operates when the halobenzene is unactivated toward nucleophilic aromatic substitution, and forcing conditions are used with a strong base. A two-step elimination forms a reactive benzyne intermediate. Nucleophilic attack, followed by protonation, gives the substituted product.

MECHANISM Nucleophilic Aromatic Substitution

Addition–elimination (with strong withdrawing groups)

Step 1: Addition of the nucleophile.   *Step 2:* Elimination of the leaving group.

(continues on next page)

(continued)

Elimination–addition (without strong withdrawing groups)

Step 1: Deprotonation.   *Step 2:* Elimination of the leaving group.   *Step 3:* Addition of the nucleophile.   *Step 4:* Protonation.

"benzyne"

PROBLEM 17-24

Propose a mechanism that shows why *p*-chlorotoluene reacts with sodium hydroxide at 350°C to give a mixture of *p*-cresol and *m*-cresol.

PROBLEM 17-25

Propose mechanisms and show the expected products of the following reactions:
(a) 2,4-dinitrochlorobenzene + sodium methoxide (NaOCH$_3$)
(b) 2,4-dimethylchlorobenzene + sodium hydroxide, 100°C
(c) *p*-nitrobromobenzene + methylamine (CH$_3$—NH$_2$)
(d) 2,4-dinitrochlorobenzene + excess hydrazine (H$_2$N—NH$_2$)

PROBLEM 17-26

Nucleophilic aromatic substitution provides one of the common methods for making phenols. (Another method is discussed in Section 19-18.) Show how you would synthesize the following phenols, using benzene or toluene as your aromatic starting material, and explain why mixtures of products would be obtained in some cases:
(a) *p*-nitrophenol   (b) 2,4,6-tribromophenol   (c) *p*-chlorophenol
(d) *m*-cresol   (e) *p*-*n*-butylphenol

PROBLEM 17-27

The highly reactive triple bond of benzyne is a powerful dienophile. Predict the product of the Diels–Alder reaction of benzyne (from chlorobenzene and NaOH, heated) with cyclopentadiene.

17-13
Addition Reactions of Benzene Derivatives

17-13A   Chlorination

Although substitution is more common, aromatic compounds may undergo addition if forcing conditions are used. When benzene is treated with an excess of chlorine under heat and pressure (or with irradiation by light), six chlorine atoms add to form 1,2,3,4,5,6-hexachlorocyclohexane. This product is often called *benzene hexachloride* (BHC) because it is synthesized by direct chlorination of benzene.

benzene   benzene hexachloride, BHC
                       (eight isomers)

The addition of chlorine to benzene, believed to involve a free-radical mechanism, is normally impossible to stop at an intermediate stage. The first addition destroys the ring's aromaticity, and the next 2 moles of Cl_2 add very rapidly. All eight possible stereoisomers are produced in various amounts. The most important isomer for commercial purposes is the insecticide *lindane*, which is used in a shampoo to kill head lice.

lindane

17-13B Catalytic Hydrogenation of Aromatic Rings

Catalytic hydrogenation of benzene to cyclohexane takes place at elevated temperatures and pressures, often catalyzed by ruthenium or rhodium. Substituted benzenes react to give substituted cyclohexanes; disubsited benzenes usually give mixtures of cis and trans isomers.

benzene $\xrightarrow[\text{Pt, Pd, Ni, Ru, or Rh}]{3\ H_2,\ 1000\ \text{psi}}$ cyclohexane (100%)

m-xylene $\xrightarrow[\substack{\text{Ru or Rh catalyst} \\ 100°C}]{3\ H_2,\ 1000\ \text{psi}}$ 1,3-dimethylcyclohexane (100%) (mixture of cis and trans)

Catalytic hydrogenation of benzene is the commercial method for producing cyclohexane and substituted cyclohexane derivatives. The reduction cannot be stopped at an intermediate stage (cyclohexene or cyclohexadiene) because these alkenes are reduced faster than benzene.

17-13C Birch Reduction

In 1944, the Australian chemist A. J. Birch found that benzene derivatives are reduced to nonconjugated 1,4-cyclohexadienes by treatment with sodium or lithium in a mixture of liquid ammonia and an alcohol. The **Birch reduction** provides a convenient method for making a wide variety of interesting and useful cyclic dienes.

benzene $\xrightarrow[\text{NH}_3(l),\ \text{ROH}]{\text{Na or Li}}$ 1,4-cyclohexadiene (90%)

The mechanism of the Birch reduction (shown below) is similar to the sodium/liquid ammonia reduction of alkynes to *trans*-alkenes (Section 9-9C). A solution of sodium in liquid ammonia contains solvated electrons that can add to benzene, forming a radical anion. The strongly basic radical anion abstracts a proton from the alcohol in the solvent, giving a cyclohexadienyl radical. The radical quickly adds another solvated electron to form a cyclohexadienyl anion. Protonation of this anion gives the reduced product.

MECHANISM   The Birch Reduction

Formation of solvated electrons.

$$NH_3 + Na \rightleftharpoons NH_3 \cdot e^- \text{ (deep blue solution)} + Na^+$$
$$\text{solvated electron}$$

Addition of an electron, followed by a proton, forms a radical.

Addition of a second electron, followed by a proton, gives the product.

The two carbon atoms that are reduced go through anionic intermediates; electron-withdrawing substituents stabilize the carbanions, while electron-donating substituents destabilize them. Therefore, reduction takes place on carbon atoms bearing electron-withdrawing substituents (such as those containing carbonyl groups) and not on carbon atoms bearing electron-releasing substituents (such as alkyl and alkoxyl groups).

A carbon bearing an electron-withdrawing carbonyl group is reduced

A carbon bearing an electron-releasing alkoxyl group is not reduced

(85%)

Substituents that are strongly electron-releasing ($-OCH_3$ for example) deactivate the aromatic ring toward Birch reduction. Lithium is often used with these deactivated systems, together with a cosolvent (often THF) and a weaker proton source (*t*-butyl alcohol). The stronger reducing agent, combined with a weaker proton source, enhances the reduction.

PROBLEM 17-28

Propose mechanisms for the Birch reductions of benzoic acid and anisole shown above. Show why the observed orientation of reduction is favored in each case.

PROBLEM 17-29

Predict the major products of the following reactions:
(a) toluene + excess Cl_2 (heat, pressure)
(b) toluene + Na (liquid NH_3, CH_3CH_2OH)
(c) *o*-xylene + H_2 (1000 psi, 100°C, Rh catalyst)
(d) *p*-xylene + Na (liquid NH_3, CH_3CH_2OH)

(e) 2,7-dimethoxynaphthalene

(f) benzamide

Many reactions are not affected by the presence of a nearby benzene ring; yet others depend on the aromatic ring to promote the reaction. For example, the Clemmensen reduction is occasionally used to reduce aliphatic ketones to alkanes, but it works best reducing aryl ketones to alkylbenzenes. Several additional side-chain reactions show the effects of a nearby aromatic ring.

17-14 Side-Chain Reactions of Benzene Derivatives

17-14A   Permanganate Oxidation

An aromatic ring imparts extra stability to the nearest carbon atom of its side chains. The aromatic ring and *one* carbon atom of a side chain can survive a vigorous permanganate oxidation. The product is a carboxylate salt of benzoic acid. This oxidation is occasionally useful for making benzoic acid derivatives, as long as any other functional groups are resistant to oxidation. (Hot chromic acid can also be used for this oxidation.)

$$\text{(1) KMnO}_4\text{, H}_2\text{O, 100°C} \quad \text{(2) H}^+$$

(or Na$_2$Cr$_2$O$_7$, H$_2$SO$_4$, heat)

PROBLEM 17-30

Predict the major products of treating the following compounds with hot, concentrated potassium permanganate, followed by acidification with dilute HCl:

(a) isopropylbenzene

(b) *p*-xylene

(c) (tetralin)

17-14B   Side-Chain Halogenation

Alkylbenzenes undergo free-radical halogenation much more easily than alkanes because abstraction of a hydrogen atom at a **benzylic position** gives a resonance-stabilized benzylic radical. For example, ethylbenzene reacts with chlorine in the presence of light to give α-chloroethylbenzene. Further chlorination can occur to give a dichlorinated product.

resonance-stabilized benzylic radical

benzylic radical   α-chloroethylbenzene   chlorine radical continues the chain   dichlorinated

---

PROBLEM 17-31

Propose a mechanism for the formation of 1,1-dichloro-1-phenylethane from 1-chloro-1-phenylethane. The IUPAC names of these compounds are (1,1-dichloroethyl)benzene and (1-chloroethyl)benzene.

   Although chlorination shows a preference for α substitution (the α position is the benzylic carbon bonded to the benzene ring), the chlorine radical is too reactive to give entirely benzylic substitution. Mixtures of isomers are often produced. In the chlorination of ethylbenzene, for example, there is a significant amount of substitution at the β carbon.

ethylbenzene   →(Cl₂, hv)   α-chloroethylbenzene (56%)   +   β-chloroethylbenzene (44%)   +   dichlorinated products

Bromine radicals are not as reactive as chlorine radicals, so bromination is more selective than chlorination (Section 4-13C). Bromine reacts exclusively at the benzylic position.

ethylbenzene   →(Br₂ or NBS, hv)   α-bromoethylbenzene   +   (trace)   α,α-dibromoethylbenzene

Either elemental bromine (much cheaper) or *N*-bromosuccinimide may be used as the reagent for benzylic bromination. *N*-Bromosuccinimide is preferred for allylic bromination (Section 15-7) because Br₂ can add to the double bond. This is not a problem with the relatively unreactive benzene ring unless it has powerful activating substituents.

PROBLEM 17-32

Propose a mechanism for the bromination of ethylbenzene shown above.

PROBLEM 17-33

What would be the ratio of products in the reaction of chlorine with ethylbenzene if chlorine *randomly* abstracted a methyl or methylene proton? What is the reactivity ratio for the benzylic hydrogens compared with the methyl hydrogens?

PROBLEM 17-34

Predict the major products when the following compounds are irradiated by light and treated with (1) 1 mole of Br₂ and (2) excess Br₂:

(a) isopropylbenzene

(b)   (tetralin)

PROBLEM-SOLVING HINT

In predicting reactions on side chains of aromatic rings, consider resonance forms that delocalize a charge or a radical electron onto the ring.

17-14C Nucleophilic Substitution at the Benzylic Position

In Chapter 15, we saw that allylic halides are more reactive than most alkyl halides in both S_N1 and S_N2 reactions. Benzylic halides are also more reactive in these substitutions, for reasons similar to those for allylic halides.

First-Order Reactions   First-order nucleophilic substitution requires ionization of the halide to give a carbocation. In the case of a benzylic halide, the carbocation is resonance-stabilized. For example, the 1-phenylethyl cation (2°) is about as stable as a 3° alkyl cation.

1-phenylethyl cation (2°)   is about as stable as   *t*-butyl cation (3°)

Because they form relatively stable carbocations, benzyl halides undergo S_N1 reactions relatively easily.

benzyl bromide $\xrightarrow{\text{CH}_3\text{CH}_2\text{OH, } \Delta \text{ (heat)}}$ benzyl ethyl ether

triphenylmethyl fluoroborate

$^-BF_4$

If a benzylic cation is bonded to more than one phenyl group, the stabilizing effects are additive. An extreme example is the triphenylmethyl cation. This cation is exceptionally stable, with three phenyl groups to stabilize the positive charge. In fact, triphenylmethyl fluoroborate can be stored for years as a stable ionic solid.

Second-Order Reactions   Like allylic halides, benzylic halides are about 100 times more reactive than primary alkyl halides in S_N2 displacement reactions. The explanation for this enhanced reactivity is similar to the one used for the reactivity of allylic halides.

During S_N2 displacement on a benzylic halide, the *p* orbital that partially bonds with the nucleophile and the leaving group also overlaps with the pi electrons of the ring (Figure 17-5). This stabilizing conjugation lowers the energy of the transition state, increasing the reaction rate.

S_N2 reactions of benzyl halides efficiently convert aromatic methyl groups to functional groups. Halogenation, followed by substitution, gives the functionalized product.

CH_3 / NO_2 $\xrightarrow[hv]{\text{Br}_2}$ CH_2Br / NO_2 $\xrightarrow[\text{CH}_3\text{OH}]{\text{NaOCH}_3}$ CH_2OCH_3 / NO_2

CH_3, Br $\xrightarrow[hv]{\text{Br}_2}$ CH_2Br, Br $\xrightarrow[\text{acetone}]{\text{NaCN}}$ CH_2CN, Br

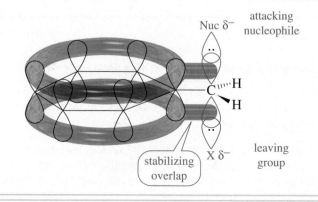

▶ Figure 17-5
The transition state for S_N2 displacement of a benzylic halide is stabilized by conjugation with the pi electrons in the ring.

Nuc δ−   attacking nucleophile

stabilizing overlap

X δ−   leaving group

PROBLEM 17-35

Propose a mechanism for the reaction of benzyl bromide with ethanol to give benzyl ethyl ether (shown on page 760).

PROBLEM 17-36

(a) Based on what you know about the relative stabilities of alkyl cations and benzylic cations, predict the product of addition of HBr to 1-phenylpropene.
(b) Propose a mechanism for this reaction.

PROBLEM 17-37

(a) Based on what you know about the relative stabilities of alkyl radicals and benzylic radicals, predict the product of addition of HBr to 1-phenylpropene in the presence of a free-radical initiator.
(b) Propose a mechanism for this reaction.

PROBLEM 17-38

Show how you would synthesize the following compounds, using the indicated starting materials:
(a) 3-phenyl-1-butanol from styrene

(b) [structure: CH$_3$—CH—OCH$_3$ attached to benzene ring with OCH$_3$] from anisole

(c) [structure: benzene ring with CH$_2$CN and O$_2$N substituents] from toluene

---

17-15
Reactions of Phenols

Much of the chemistry of phenols is like that of aliphatic alcohols. For example, phenols can be acylated to give esters, and phenoxide ions can serve as nucleophiles in the Williamson ether synthesis (Section 14-5). Formation of phenoxide ions is particularly easy because phenols are more acidic than water; aqueous sodium hydroxide deprotonates phenols to give phenoxide ions.

Aspirin inactivates two cyclooxygenases, named COX-1 and COX-2, and prevents the formation of critical prostaglandins in the body. Inhibition of COX-2 relieves then pain, while inhibition of COX-1 results in stomach irritation and upset.

salicylic acid   acetic acid   acetylsalicylic acid
(aspirin)

anethole
(licorice flavoring)

All the alcohol-like reactions shown above involve breaking of the phenolic O—H bond. This is a common way for phenols to react. It is far more difficult to break the C—O bond of a phenol, however. Most alcohol reactions in which the C—O bond breaks are not possible with phenols. For example, phenols do not undergo acid-catalyzed elimination or S$_N$2 back-side attack.

Phenols also undergo reactions that are not possible with aliphatic alcohols. Let's consider some reactions that are peculiar to phenols.

17-15A   Oxidation of Phenols to Quinones

Phenols undergo oxidation, but they give different types of products from those seen with aliphatic alcohols. Chromic acid oxidation of a phenol gives a conjugated 1,4-diketone called a **quinone**. In the presence of air, many phenols slowly autoxidize to dark mixtures containing quinones.

m-cresol $\xrightarrow[\text{H}_2\text{SO}_4]{\text{Na}_2\text{Cr}_2\text{O}_7}$ 2-methyl-1,4-benzoquinone

Hydroquinone (1,4-benzenediol) is easily oxidized because it already has two oxygen atoms bonded to the ring. Even very weak oxidants like silver bromide (AgBr) can oxidize hydroquinone. Silver bromide is reduced to black metallic silver in a light-sensitive reaction: Any grains of silver bromide that have been exposed to light (AgBr*) react faster than unexposed grains.

hydroquinone (1,4-benzenediol) + 2 AgBr* ⟶ quinone (1,4-benzoquinone) + 2 Ag↓ + 2 HBr

Black-and-white photography is based on this reaction. A film containing small grains of silver bromide is exposed by a focused image. Where light strikes the film, the grains are activated. The film is then treated with a hydroquinone solution (the *developer*) to reduce the activated silver bromide grains, leaving black silver deposits where the film was exposed to light. The result is a negative image, with dark areas where light struck the film.

PROBLEM 17-39

The bombardier beetle defends itself by spraying a hot quinone solution from its abdomen (see photo). This solution is formed by the enzyme-catalyzed oxidation of hydroquinone by hydrogen peroxide. Write a balanced equation for this oxidation.

Quinones occur widely in nature, where they serve as biological oxidation–reduction reagents. The quinone *coenzyme Q* (CoQ) is also called *ubiquinone* because it seems ubiquitous (found everywhere) in oxygen-consuming organisms. Coenzyme Q serves as an oxidizing agent within the mitochondria of cells. The following reaction shows the reduction of coenzyme Q by NADH (the reduced form of nicotinamide adenine dinucleotide), which becomes oxidized to NAD^+.

Large quantities of acetaminophen can be toxic because the body transforms it to a benzoquinoneimine (a nitrogen replaces one oxygen of benzoquinone). This highly reactive metabolite causes extensive liver damage and can be deadly.

acetaminophen $\xrightarrow{\text{[O]}}$ benzoquinoneimine

When threatened, the bombardier beetle mixes hydroquinone and H_2O_2 with enzymes. Peroxide oxidizes hydroquinone to quinone, and the strongly exothermic reaction brings the solution to boiling. The hot, irritating liquid sprays from the tip of the insect's abdomen.

coenzyme Q
oxidized form

NADH
reduced form

coenzyme Q
reduced form

NAD⁺
oxidized form

$$R = -(CH_2-CH=\overset{\overset{\textstyle CH_3}{|}}{C}-CH_2)_{\overline{10}}H$$

17-15B Electrophilic Aromatic Substitution of Phenols

Phenols are highly reactive substrates for electrophilic aromatic substitution because the nonbonding electrons of the hydroxyl group stabilize the sigma complex formed by attack at the ortho or para position (Section 17-6B). Therefore, the hydroxyl group is strongly activating and ortho, para-directing. Phenols are excellent substrates for halogenation, nitration, sulfonation, and some Friedel–Crafts reactions. Because they are highly reactive, phenols are usually alkylated or acylated using relatively weak Friedel–Crafts catalysts (such as HF) to avoid overalkylation or overacylation.

Phenoxide ions, easily generated by treating a phenol with sodium hydroxide, are even more reactive than phenols toward electrophilic aromatic substitution. Because they are negatively charged, phenoxide ions react with positively charged electrophiles to give neutral sigma complexes whose structures resemble quinones.

phenoxide ion

sigma complex

Phenoxide ions are so strongly activated that they undergo electrophilic aromatic substitution with carbon dioxide, a weak electrophile. The carboxylation of phenoxide ion is an industrial synthesis of salicylic acid, which is then converted to aspirin as shown on p. 761.

salicylic acid

PROBLEM 17-40

Predict the products formed when *m*-cresol (*m*-methylphenol) reacts with

(a) NaOH and then ethyl bromide

(b) acetyl chloride, $CH_3-\overset{\displaystyle O}{\overset{\|}{C}}-Cl$

(c) bromine in CCl_4 in the dark

(d) excess bromine in CCl_4 in the light

(e) sodium dichromate in H_2SO_4

(f) 2 equivalents of *t*-butyl chloride and $AlCl_3$

PROBLEM 17-41

Benzoquinone is a good Diels–Alder dienophile. Predict the products of its reaction with
(a) 1,3-butadiene　　　(b) 1,3-cyclohexadiene　　　(c) furan

PROBLEM 17-42

Phenol reacts with 3 equivalents of bromine in CCl_4 (in the dark) to give a product of formula $C_6H_3OBr_3$. When this product is added to bromine water, a yellow solid of molecular formula $C_6H_2OBr_4$ precipitates out of the solution. The IR spectrum of the yellow precipitate shows a strong absorption (much like that of a quinone) around 1680 cm^{-1}. Propose structures for the two products.

SUMMARY　**Reactions of Aromatic Compounds**

1. *Electrophilic aromatic substitution*
   a. *Halogenation* (Section 17-2)

bromobenzene

   b. *Nitration* (Section 17-3)

nitrobenzene

   c. *Sulfonation* (Section 17-4)

benzenesulfonic acid

   d. *Friedel–Crafts alkylation* (Section 17-10)

t-butylbenzene

e. *Friedel–Crafts acylation* (Section 17-11)

propiophenone

f. *Gatterman–Koch synthesis* (Section 17-11C)

benzaldehyde

g. *Substituent effects* (Sections 17-5 through 17-9)

Activating, ortho, para-directing: —R, —ÖR, —ÖH, —Ö:⁻, —N̈R₂ (amines, amides)

Deactivating, ortho, para-directing: —Cl, —Br, —I

Deactivating, meta-allowing: —NO₂, —SO₃H, —NR₃⁺, —C̈=O, —C≡N

2. *Nucleophilic aromatic substitution* (Section 17-12)

a halobenzene
(G = strong withdrawing group)

strong
nucleophile

Example

2,4-dinitrochlorobenzene

2,4-dinitroaniline

If G is not a strong electron-withdrawing group, severe conditions are required, and a benzyne mechanism is involved.

3. *Addition reactions*
   a. *Chlorination* (Section 17-13A)

benzene

benzene hexachloride (BHC)

b. *Catalytic hydrogenation* (Section 17-13B)

o-diethylbenzene

1,2-diethylcyclohexane
(mixture of cis and trans)

c. *Birch reduction* (Section 17-13C)

ethylbenzene   1-ethyl-1,4-cyclohexadiene

4. *Side-chain reactions*

a. *The Clemmensen reduction* (converts acylbenzenes to alkylbenzenes, Section 17-11B)

an acylbenzene   an alkylbenzene

b. *Permanganate oxidation* (Section 17-14A)

an alkylbenzene   a benzoic acid salt

c. *Side-chain halogenation* (Section 17-14B)

an alkylbenzene   an α-bromo alkylbenzene

d. *Nucleophilic substitution at the benzylic position* (Section 17-14C)

The benzylic position is activated toward both S_N1 and S_N2 displacements.

an α-halo alkylbenzene

5. *Oxidation of phenols to quinones* (Section 17-15A)

o-chlorophenol   2-chloro-1,4-benzoquinone

Chapter 17 Glossary

activating group A substituent that makes the aromatic ring more reactive (usually toward electrophilic aromatic substitution) than benzene. (p. 728)

acyl group ($R-\overset{\overset{\textstyle O}{\|}}{C}-$) A carbonyl group with an alkyl group attached. (p. 746)

   acyl chloride (acid chloride): An acyl group bonded to a chlorine atom, RCOCl.

acylium ion ($R-C\equiv O^+$) An acyl group fragment with a positive charge. (p. 747)

benzylic position The carbon atom of an alkyl group that is directly bonded to a benzene ring; the position α to a benzene ring. (p. 758)

The benzylic positions are circled in red.

benzyne A reactive intermediate in some nucleophilic aromatic substitutions, benzyne is benzene with two hydrogen atoms removed. It can be drawn with a highly strained triple bond in the six-membered ring. (p. 752)

Birch reduction The partial reduction of a benzene ring by sodium or lithium in liquid ammonia. The products are usually 1,4-cyclohexadienes. (p. 755)

Clemmensen reduction The reduction of a carbonyl group to a methylene group by zinc amalgam, Zn(Hg), in dilute hydrochloric acid. (p. 749)

   amalgam: An alloy of a metal with mercury.

deactivating group A substituent that makes the aromatic ring less reactive (usually toward electrophilic aromatic substitution) than benzene. (p. 734)

electrophilic aromatic substitution (EAS) Replacement of a hydrogen on an aromatic ring by a strong electrophile. (p. 722)

electrophilic attack   sigma complex   substituted product

Friedel–Crafts acylation Formation of an acylbenzene by substitution of an acylium ion on an aromatic ring. (p. 746)

acylium ion

an acylbenzene

Friedel–Crafts alkylation Formation of an alkyl-substituted benzene derivative by substitution of an alkyl carbocation or carbocation-like species on an aromatic ring. (p. 742)

Gatterman–Koch synthesis The synthesis of benzaldehydes by treating a benzene derivative with CO and HCl using an $AlCl_3/CuCl$ catalyst. (p. 749)

halonium ion Any positively charged ion that has a positive charge (or partial positive charge) on a halogen atom. The positive charge is usually delocalized onto the halogen through resonance. (Specific: chloronium ion, bromonium ion, etc.) (p. 738)

inductive stabilization Stabilization of a reactive intermediate by donation or withdrawal of electron density through sigma bonds. (p. 730)

meta-director (meta allower) A substituent that deactivates primarily the ortho and para positions, leaving the meta position the least deactivated and most reactive. (p. 734)

nitration Replacement of a hydrogen atom by a nitro group, $-NO_2$. (p. 726)

nitronium ion The NO_2^+ ion, $O{=}\overset{+}{N}{=}O$. (p. 726)

nucleophilic aromatic substitution (NAS) Replacement of a leaving group on an aromatic ring by a strong nucleophile. (p. 750)

ortho, para-director A substituent that activates primarily the ortho and para positions toward attack. (p. 728)

quinone A derivative of a cyclohexadiene-dione. Common quinones are the 1,4-quinones (*para*-quinones); the less stable 1,2-quinones (*ortho*-quinones) are relatively uncommon. (p. 762)

p-quinone   *o*-quinone

resonance stabilization Stabilization of a reactive intermediate by donation or withdrawal of electron density through pi bonds.

  resonance-donating (pi-donating): Capable of donating electrons through resonance involving pi bonds. (p. 731)

  resonance-withdrawing (pi-withdrawing): Capable of withdrawing electron density through resonance involving pi bonds. (p. 734)

alkoxyl groups are pi-donating

nitro groups are pi-withdrawing

sigma complex An intermediate in electrophilic aromatic substitution or nucleophilic aromatic substitution with a sigma bond between the electrophile or nucleophile and the former aromatic ring. The sigma complex bears a delocalized positive charge in electrophilic aromatic substitution and a delocalized negative charge in nucleophilic aromatic substitution. (p. 722)

sulfonation Replacement of a hydrogen atom by a sulfonic acid group, $-SO_3H$. (p. 726)

  desulfonation: Replacement of the $-SO_3H$ group by a hydrogen. In benzene derivatives, this is done by heating with water or steam. (p. 727)

Essential Problem-Solving Skills in Chapter 17

1. Predict products and propose mechanisms for the common electrophilic aromatic substitutions: halogenation, nitration, sulfonation, and Friedel–Crafts alkylation and acylation.

2. Draw resonance structures for the sigma complexes resulting from electrophilic attack on substituted aromatic rings. Explain which substituents are activating and which are deactivating, and show why they are ortho, para-directing or meta-allowing.

3. Predict the position(s) of electrophilic aromatic substitution on molecules containing substituents on one or more aromatic rings.

4. Design syntheses that use the influence of substituents to generate the correct isomers of multisubstituted aromatic compounds.

5. Determine which nucleophilic aromatic substitutions are likely, and propose mechanisms of the addition–elimination type and of the benzyne type.

6. Predict the products of Birch reduction, hydrogenation, and chlorination of aromatic compounds, and use these reactions in syntheses.

7. Explain how the reactions of side chains are affected by the presence of the aromatic ring, predict the products of side-chain reactions, and use these reactions in syntheses.

8. Predict the products of oxidation and substitution of phenols, and use these reactions in syntheses.

Study Problems

17-43 Define each term and give an example.
   (a) activating group   **(b)** deactivating group   **(c)** sigma complex
   (d) sulfonation   **(e)** desulfonation   **(f)** nitration
   (g) ortho, para-director   **(h)** meta-director   **(i)** resonance stabilization
   (j) Friedel–Crafts acylation   **(k)** Friedel–Crafts alkylation **(l)** Clemmensen reduction
   (m) Gatterman–Koch synthesis   **(n)** benzyne mechanism   **(o)** Birch reduction
   (p) quinone   **(q)** benzylic position

17-44 Predict the major products formed when benzene reacts with the following reagents:
   (a) t-butyl bromide, $AlCl_3$ **(b)** 1-chlorobutane, $AlCl_3$
   (c) isobutyl alcohol + BF_3 **(d)** bromine + a nail
   (e) isobutylene + HF   **(f)** fuming sulfuric acid
   (g) 1,2-dichloroethane + $AlCl_3$ **(h)** benzoyl chloride + $AlCl_3$
   (i) iodine + HNO_3 **(j)** nitric acid + sulfuric acid
   (k) carbon monoxide, HCl, and $AlCl_3/CuCl$ **(l)** 1-chloro-2,2-dimethylpropane + $AlCl_3$

17-45 Predict the major products formed when isopropylbenzene reacts with the following reagents:
   (a) 1 equivalent of Br_2 and light   **(b)** Br_2 and $FeBr_3$ **(c)** SO_3 and H_2SO_4
   (d) hot, concd. $KMnO_4$ **(e)** acetyl chloride and $AlCl_3$ **(f)** n-propyl chloride and $AlCl_3$

17-46 Show how you would synthesize the following compounds, starting with benzene or toluene and any necessary acyclic reagents:
   (a) 1-phenyl-1-bromobutane   **(b)** 1-phenyl-1-methoxybutane   **(c)** 3-phenyl-1-propanol
   (d) ethoxybenzene   **(e)** 1,2-dichloro-4-nitrobenzene **(f)** 1-phenyl-2-propanol
   (g) 3,4-dibromobenzoic acid

17-47 Predict the major products of the following reactions:
   (a) 2,4-dinitrochlorobenzene + $NaOCH_3$ **(b)** isopropoxybenzene + t-butyl chloride + $AlCl_3$
   (c) nitrobenzene + fuming sulfuric acid   **(d)** nitrobenzene + acetyl chloride + $AlCl_3$
   (e) p-methylanisole + acetyl chloride + $AlCl_3$ **(f)** p-methylanisole + Br_2, light

   (g) 1,2-dichloro-4-nitrobenzene + $NaNH_2$ **(h)** $Ph-\overset{\overset{O}{\|}}{C}-NHPh + CH_3CH_2-\overset{\overset{O}{\|}}{C}-Cl, AlCl_3$

(i)   *p*-ethylbenzenesulfonic acid + HNO$_3$, H$_2$SO$_4$ **(j)**   *p*-ethylbenzenesulfonic acid + steam

(k)   [structure: indane] + hot, concd. KMnO$_4$

indane

(l)   [structure: *p*-methylacetanilide with NH—C(=O)—CH$_3$ and CH$_3$] + acetyl chloride, AlCl$_3$

p-methylacetanilide

17-48   Predict the major products of bromination of the following compounds, using Br$_2$ and FeBr$_3$ in the dark:

(a) [biphenyl with NO$_2$ and OCH$_3$ substituents]

(b) [phenyl benzoate structure with C(=O)—O]

(c) [biphenyl with O=C—OCH$_3$ and OCH$_3$ substituents]

17-49   Give the structures of compounds **A** through **H** in the following series of reactions:

[reaction scheme: benzene + isovaleryl chloride (C(=O)Cl)/AlCl$_3$ → **A** → HNO$_3$/H$_2$SO$_4$ → **B**; **A** → Zn(Hg)/HCl → **C** → KMnO$_4$ (hot, concd.) → **D**; **C** → Br$_2$/hν → **E**; **E** → NH$_3$ (excess) → 1-amino-3-methylbutylbenzene (NH$_2$ structure); **E** → (CH$_3$)$_3$CO$^-$ K$^+$ → **G** → HBr → **H**; **E** → NaOCH$_3$ → **F**]

17-50   A student added 3-phenylpropanoic acid (PhCH$_2$CH$_2$COOH) to a molten salt consisting of a 1 : 1 mixture of NaCl and AlCl$_3$ maintained at 170°C. After 5 minutes, he poured the molten mixture into water and extracted it into dichloromethane. Evaporation of the dichloromethane gave a 96% yield of the product whose spectra appear below. The mass spectrum of the product shows a molecular ion at *m/z* 132. What is the product?

17-51   The compound shown below reacts with HBr to give a product of molecular formula $C_{10}H_{11}Br$.

(a) Propose a mechanism for this reaction, and predict the structure of the product. Carefully show the resonance stabilization of the intermediate.

(b) When this reaction takes place in the presence of a free-radical initiator, the product is a different isomer of formula $C_{10}H_{11}Br$. Propose a structure for this second product, and give a mechanism to account for its formation.

17-52   The following compound reacts with a hot, concentrated solution of NaOH (in a sealed tube) to give a mixture of two products. Propose structures for these products, and give a mechanism to account for their formation.

$$\text{Cl}$$

$$\xrightarrow[\text{350°C}]{\text{NaOH, H}_2\text{O}}\quad\text{2 products}$$

17-53 α-Tetralone undergoes Birch reduction to give an excellent yield of a single product. Predict the structure of the product, and propose a mechanism for its formation.

$$\text{O}$$

$$\xrightarrow[\text{CH}_3\text{CH}_2\text{OH}]{\text{Na, NH}_3(l)}$$

α -tetralone

17-54   Electrophilic aromatic substitution usually occurs at the 1-position of naphthalene, also called the α position. Predict the major products of the reactions of naphthalene with the following reagents:

α position

(a) HNO_3, H_2SO_4 **(b)** Br_2, $FeBr_3$ **(c)** CH_3CH_2COCl, $AlCl_3$
(d) isobutylene and HF   **(e)** cyclohexanol and BF_3 **(f)** fuming sulfuric acid

17-55   Triphenylmethanol is insoluble in water, but when it is treated with concentrated sulfuric acid, a bright yellow solution results. As this yellow solution is diluted with water, its color disappears and a precipitate of triphenylmethanol reappears. Suggest a structure for the bright yellow species, and explain this unusual behavior.

17-56   The most common selective herbicide for killing broadleaf weeds is 2,4-dichlorophenoxyacetic acid (2,4-D). Show how you would synthesize 2,4-D from benzene, chloroacetic acid ($ClCH_2COOH$), and any necessary reagents and solvents.

2,4-dichlorophenoxyacetic acid (2,4-D)

17-57   Furan undergoes electrophilic aromatic substitution more readily than benzene; mild reagents and conditions are sufficient. For example, furan reacts with bromine to give 2-bromofuran.

furan   2-bromofuran

(a) Propose mechanisms for the bromination of furan at the 2-position and at the 3-position. Draw the resonance structures of each sigma complex, and compare their stabilities.

(b) Explain why furan undergoes bromination (and other electrophilic aromatic substitutions) primarily at the 2-position.

17-58   In Section 17-12A, we saw that the Sanger reagent, 2,4-dinitrofluorobenzene, is a good substrate for nucleophilic aromatic substitution. Show the product, and propose a mechanism for the reaction of the Sanger reagent with the amino acid phenylalanine.

phenylalanine

*__17-59__   *Bisphenol A* is an important component of many polymers, including polycarbonates, polyurethanes, and epoxy resins. It is synthesized from phenol and acetone with HCl as a catalyst. Propose a mechanism for this reaction.

phenol   acetone   bisphenol A

17-60   Unlike most other electrophilic aromatic substitutions, sulfonation is often reversible. (See Section 17-4.) When one sample of toluene is sulfonated at 0°C and another sample is sulfonated at 100°C, the following ratios of substitution products result:

Isomer of the Product	Reaction Temperature	
	0°C	*100°C*
o-toluenesulfonic acid	43%	13%
m-toluenesulfonic acid	4%	8%
p-toluenesulfonic acid	53%	79%

(a) Explain the change in the product ratios when the temperature is increased.

(b) Predict what will happen when the product mixture from the reaction at 0°C is heated to 100°C.

17-61   Devise a synthesis of the following compound, starting from toluene and using any necessary reagents:

*__17-62__   When anthracene is added to the reaction of chlorobenzene with concentrated NaOH at 350°C, an interesting Diels–Alder adduct of formula $C_{20}H_{14}$ results. The proton NMR spectrum of the product shows a singlet of area 2 around $\delta 3$ and a broad singlet of area 12 around $\delta 7$. Propose a structure for the product, and explain why one of the aromatic rings of anthracene reacted as a diene.

*17-63   In Chapter 14, we saw that Agent Orange contains (2,4,5-trichlorophenoxy) acetic acid, called 2,4,5-T. This compound is synthesized by the partial reaction of 1,2,4,5-tetrachlorobenzene with sodium hydroxide, followed by reaction with sodium chloroacetate, $ClCH_2CO_2Na$.
   (a) Draw the structures of these compounds, and write equations for these reactions.
   (b) One of the impurities in the Agent Orange used in Vietnam was 2,3,7,8-tetrachlorodibenzodioxin (2,3,7,8-TCDD), often incorrectly called "dioxin." Propose a mechanism to show how 2,3,7,8-TCDD is formed in the synthesis of 2,4,5-T.
   (c) Show how the TCDD contamination might be eliminated, both after the first step and on completion of the synthesis.

2,4,5-T   2,3,7,8-tetrachlorodibenzodioxin (TCDD)

*17-64   Phenolphthalein, a common nonprescription laxative, is also an acid–base indicator that is colorless in acid and red in base. Phenolphthalein is synthesized by the acid-catalyzed reaction of phthalic anhydride with 2 equivalents of phenol.

phthalic anhydride   phenolphthalein   red dianion

   (a) Propose a mechanism for the synthesis of phenolphthalein.
   (b) Propose a mechanism for the conversion of phenolphthalein to its red dianion in base.
   (c) Use resonance structures to show that the two phenolic oxygen atoms are equivalent (each with half a negative charge) in the red phenolphthalein dianion.

*17-65   Benzene reacts with propylene oxide in the presence of aluminum chloride to give an alcohol of formula $C_9H_{12}O$. Propose a mechanism for this reaction, and predict the structure of the product. (*Hint:* Consider how epoxides usually open under acidic conditions.)

*17-66   A graduate student tried to make o-fluorophenylmagnesium bromide by adding magnesium to an ether solution of o-fluorobromobenzene. After obtaining puzzling results with this reaction, she repeated the reaction by using as solvent some tetrahydrofuran that contained a small amount of furan. From this reaction, she isolated a fair yield of the compound that follows. Propose a mechanism for its formation.

*17-67   A common illicit synthesis of methamphetamine involves an interesting variation of the Birch reduction. A solution of ephedrine in alcohol is added to liquid ammonia, followed by several pieces of lithium metal. The Birch reduction usually reduces the aromatic ring (Section 17-13C), but in this case it eliminates the hydroxyl group of ephedrine to give methamphetamine. Propose a mechanism, similar to that for the Birch reduction, to explain this unusual course of the reaction.

ephedrine   methamphetamine

CHAPTER 18

Ketones and Aldehydes

18-1
Carbonyl
Compounds

We will study compounds containing the **carbonyl group** (C═O) in detail because they are of central importance to organic chemistry, biochemistry, and biology. Some of the common types of carbonyl compounds are listed in Table 18-1.

Carbonyl compounds are everywhere. In addition to their uses as reagents and solvents, they are constituents of fabrics, flavorings, plastics, and drugs. Naturally occurring carbonyl compounds include proteins, carbohydrates, and nucleic acids that make up all plants and animals. In the next few chapters, we will discuss the properties and reactions of simple carbonyl compounds. Then, in Chapters 23 and 24, we apply this carbonyl chemistry to carbohydrates, nucleic acids, and proteins.

The simplest carbonyl compounds are ketones and aldehydes. A **ketone** has two alkyl (or aryl) groups bonded to the carbonyl carbon atom. An **aldehyde** has one alkyl (or aryl) group and one hydrogen atom bonded to the carbonyl carbon atom.

ketone	aldehyde	carbonyl group

condensed structures: RCOR′   RCHO

Ketone: Two alkyl groups bonded to a carbonyl group.
Aldehyde: One alkyl group and one hydrogen bonded to a carbonyl group.

Ketones and aldehydes are similar in structure, and they have similar properties. There are some differences, however, particularly in their reactions with oxidizing agents and with nucleophiles. In most cases, aldehydes are more reactive than ketones, for reasons we discuss shortly.

TABLE 18-1 Some Common Classes of Carbonyl Compounds

Class	General Formula	Class	General Formula
ketones	R—C(═O)—R′	aldehydes	R—C(═O)—H
carboxylic acids	R—C(═O)—OH	acid chlorides	R—C(═O)—Cl
esters	R—C(═O)—O—R′	amides	R—C(═O)—NH$_2$

The carbonyl carbon atom is sp^2 hybridized and bonded to three other atoms through coplanar sigma bonds oriented about 120° apart. The unhybridized p orbital overlaps with a p orbital of oxygen to form a pi bond. The double bond between carbon and oxygen is similar to an alkene C=C double bond, except that the carbonyl double bond is shorter, stronger, and polarized.

18-2 Structure of the Carbonyl Group

	length	energy
ketone C=O bond	1.23 Å	178 kcal/mol (745 kJ/mol)
alkene C=C bond	1.34 Å	146 kcal/mol (611 kJ/mol)

The double bond of the carbonyl group has a large dipole moment because oxygen is more electronegative than carbon, and the bonding electrons are not shared equally. In particular, the less tightly held pi electrons are pulled more strongly toward the oxygen atom, giving ketones and aldehydes larger dipole moments than most alkyl halides and ethers. We can use resonance forms to symbolize this unequal sharing of the pi electrons.

major   minor

The first resonance form is more important because it involves more bonds and less charge separation. The contribution of the second structure is evidenced by the large dipole moments of the ketones and aldehydes shown below.

Compare with:

$\mu = 2.7$ D
acetaldehyde

$\mu = 2.9$ D
acetone

$\mu = 1.9$ D
chloromethane

$\mu = 1.30$ D
dimethyl ether

This polarization of the carbonyl group contributes to the reactivity of ketones and aldehydes: The positively polarized carbon atom acts as an electrophile (Lewis acid), and the negatively polarized oxygen acts as a nucleophile (Lewis base).

18-3 Nomenclature of Ketones and Aldehydes

IUPAC Names   Systematic names of ketones are derived by replacing the final *-e* in the alkane name with *-one*. The "alkane" name becomes "alkanone." In open-chain ketones, we number the longest chain that includes the carbonyl carbon from the end closest to the carbonyl group, and we indicate the position of the carbonyl group by a number. In cyclic ketones, the carbonyl carbon atom is assigned the number 1.

2-butanone
butan-2-one

2,4-dimethyl-3-pentanone
2,4-dimethylpentan-3-one

1-phenyl-1-propanone
1-phenylpropan-1-one

3-methylcyclopentanone

2-cyclohexenone
cyclohex-2-en-1-one

4-hydroxy-4-methyl-2-pentanone
4-hydroxy-4-methylpentan-2-one

Systematic names for aldehydes are derived by replacing the final *-e* of the alkane name with *-al*. An aldehyde carbon is at the end of a chain, so it is number 1. If the aldehyde group is attached to a large unit (usually a ring), the suffix *-carbaldehyde* is used.

ethanal

4-bromo-3-methylheptanal

3-hydroxybutanal

2-pentenal
pent-2-enal

cyclohexanecarbaldehyde

2-hydroxycyclopentane-1-carbaldehyde

A ketone or aldehyde group can also be named as a substituent on a molecule with another functional group as its root. The ketone carbonyl is designated by the prefix *oxo-*, and the —CHO group is named as a *formyl* group. Carboxylic acids frequently contain ketone or aldehyde groups named as substituents.

3-oxopentanal

2-formylbenzoic acid

3-oxobutanoic acid

Common Names   As with other classes of compounds, ketones and aldehydes are often called by common names instead of their systematic IUPAC names. Ketone common names are formed by naming the two alkyl groups bonded to the carbonyl group. Substituent locations are given using Greek letters, beginning with the carbon *next to* the carbonyl group.

methyl ethyl ketone

di-*sec*-butyl ketone

β-bromoethyl isopropyl ketone

t-butyl α-methoxypropyl ketone

Some ketones have historical common names. Dimethyl ketone is always called *acetone*, and alkyl phenyl ketones are usually named as the acyl group followed by the suffix *-phenone*.

acetone   acetophenone   propiophenone   benzophenone

Common names of aldehydes are derived from the common names of carboxylic acids (Table 18-2). These names often reflect the Latin or Greek term for the original source of the acid or the aldehyde.

Greek letters are used with common names of aldehydes to give the locations of substituents. The first letter (α) is given to the carbon atom *next to* the carbonyl group.

Common name: β-bromobutyraldehyde α-methoxypropionaldehyde
IUPAC name: 3-bromobutanal   2-methoxypropanal

TABLE 18-2 Common Names of Aldehydes

Carboxylic Acid	Derivation	Aldehyde
formic acid	*formica*, "ants"	formaldehyde (methanal)
acetic acid	*acetum*, "sour"	acetaldehyde (ethanal)
propionic acid	*protos pion*, "first fat"	propionaldehyde (propanal)
butyric acid	*butyrum*, "butter"	butyraldehyde (butanal)
benzoic acid	*gum benzoin*, "blending"	benzaldehyde

PROBLEM 18-1

Give the IUPAC name and (if possible) a common name for each compound.

(a)   (b)   (c)   (d)

18-4
Physical Properties of Ketones and Aldehydes

Polarization of the carbonyl group creates dipole–dipole attractions between the molecules of ketones and aldehydes, resulting in higher boiling points than for hydrocarbons and ethers of similar molecular weights. Ketones and aldehydes have no O—H or N—H bonds, however, so their molecules cannot form hydrogen bonds with each other. Their boiling points are therefore lower than those of alcohols of similar molecular weight. The following compounds of molecular weight 58 or 60 are ranked in order of increasing boiling points. The ketone and the aldehyde are more polar and higher boiling than the ether and the alkane, but lower boiling than the hydrogen-bonded alcohol.

$CH_3CH_2CH_2CH_3$ $CH_3-O-CH_2CH_3$

butane   methoxyethane
bp 0°C   bp 8°C

$$CH_3CH_2-\overset{\overset{\displaystyle O}{\|}}{C}-H \qquad CH_3-\overset{\overset{\displaystyle O}{\|}}{C}-CH_3 \qquad CH_3CH_2CH_2-OH$$

propanal   acetone   1-propanol
bp 49°C   bp 56°C   bp 97°C

The melting points, boiling points, and water solubilities of some representative ketones and aldehydes are given in Table 18-3.

Although pure ketones and aldehydes cannot engage in hydrogen bonding with each other, they have lone pairs of electrons and can act as hydrogen bond acceptors with other compounds having O—H or N—H bonds. For example, the —OH hydrogen of water or an alcohol can form a hydrogen bond with the unshared electrons on a carbonyl oxygen atom.

TABLE 18-3 Physical Properties of Some Representative Ketones and Aldehydes

IUPAC Name	Common Name	Structure	mp (°C)	bp (°C)	Density (g/cm^3)	H$_2$O Solubility (%)
		Ketones				
2-propanone	acetone	CH_3COCH_3	−95	56	0.79	∞
2-butanone	methyl ethyl ketone (MEK)	$CH_3COCH_2CH_3$	−86	80	0.81	25.6
2-pentanone	methyl *n*-propyl ketone	$CH_3COCH_2CH_2CH_3$	−78	102	0.81	5.5
3-pentanone	diethyl ketone	$CH_3CH_2COCH_2CH_3$	−41	101	0.81	4.8
2-hexanone		$CH_3CO(CH_2)_3CH_3$	−57	127	0.83	1.6
3-hexanone		$CH_3CH_2COCH_2CH_2CH_3$		124	0.82	
2-heptanone		$CH_3CO(CH_2)_4CH_3$	−36	151	0.81	1.4
3-heptanone		$CH_3CH_2CO(CH_2)_3CH_3$	−39	147	0.82	0.4
4-heptanone	di-*n*-propyl ketone	$(CH_3CH_2CH_2)_2CO$	−34	144	0.82	
4-methyl-3-penten-2-one	mesityl oxide	$(CH_3)_2C{=}CHCOCH_3$	−59	131	0.86	
3-buten-2-one	methyl vinyl ketone (MVK)	$CH_2{=}CHCOCH_3$	−6	80	0.86	
cyclohexanone			−16	157	0.94	15
acetophenone	methyl phenyl ketone	$C_6H_5COCH_3$	21	202	1.02	0.5
propiophenone	ethyl phenyl ketone	$C_6H_5COCH_2CH_3$	21	218		
benzophenone	diphenyl ketone	$C_6H_5COC_6H_5$	48	305	1.08	
		Aldehydes				
methanal	formaldehyde	HCHO or CH_2O	−92	−21	0.82	55
ethanal	acetaldehyde	CH_3CHO	−123	21	0.78	∞
propanal	propionaldehyde	CH_3CH_2CHO	−81	49	0.81	20
butanal	*n*-butyraldehyde	$CH_3(CH_2)_2CHO$	−97	75	0.82	7.1
2-methylpropanal	isobutyraldehyde	$(CH_3)_2CHCHO$	−66	61	0.79	11
pentanal	*n*-valeraldehyde	$CH_3(CH_2)_3CHO$	−91	103	0.82	
3-methylbutanal	isovaleraldehyde	$(CH_3)_2CHCH_2CHO$	−51	93	0.80	
hexanal	caproaldehyde	$CH_3(CH_2)_4CHO$	−56	129	0.83	0.1
heptanal	*n*-heptaldehyde	$CH_3(CH_2)_5CHO$	−45	155	0.85	0.02
propenal	acrolein	$CH_2{=}CH-CHO$	−88	53	0.84	30
2-butenal	crotonaldehyde	$CH_3-CH{=}CH-CHO$	−77	104	0.86	18
benzaldehyde		C_6H_5CHO	−56	179	1.05	0.3

Because of this hydrogen bonding, ketones and aldehydes are good solvents for polar hydroxylic substances such as alcohols. They are also remarkably soluble in water. Table 18-3 shows that acetaldehyde and acetone are miscible (soluble in all proportions) with water. Other ketones and aldehydes with up to four carbon atoms are fairly soluble in water. These solubility properties are similar to those of ethers and alcohols, which also engage in hydrogen bonding with water.

Formaldehyde and acetaldehyde are the most common aldehydes. Formaldehyde is a gas at room temperature, so it is often stored and used as a 40% aqueous solution called *formalin*. When dry formaldehyde is needed, it can be generated by heating one of its solid derivatives, usually *trioxane* or *paraformaldehyde*. Trioxane is a cyclic *trimer*, containing three formaldehyde units. Paraformaldehyde is a linear *polymer*, containing many formaldehyde units. These solid derivatives form spontaneously when a small amount of acid catalyst is added to pure formaldehyde.

One symptom of untreated diabetes is the characteristic fruity smell of acetone in the patient's breath. Because diabetics cannot use carbohydrates properly, the body goes into a state called ketosis, in which it produces acetone and other ketones.

Acetaldehyde boils near room temperature, and it can be handled as a liquid. Acetaldehyde is also used as a trimer (*paraldehyde*) and a tetramer (*metaldehyde*), formed from acetaldehyde under acid catalysis. Heating either of these compounds provides dry acetaldehyde. Paraldehyde is used in medicines as a sedative, and metaldehyde is used as a bait and poison for snails and slugs.

18-5
Spectroscopy of Ketones and Aldehydes

18-5A Infrared Spectra of Ketones and Aldehydes

The carbonyl ($C=O$) stretching vibrations of simple ketones and aldehydes occur around 1710 cm^{-1}. Because the carbonyl group has a large dipole moment, these absorptions are very strong. In addition to the carbonyl absorption, an aldehyde shows a set of two low-frequency C—H stretching absorptions around 2710 and 2810 cm^{-1}.

$$\underset{\text{ketone}}{R-\overset{\displaystyle O}{\overset{\|}{C}}-R'} \qquad \underset{\text{aldehyde}}{R-\overset{\displaystyle O}{\overset{\|}{C}}-H}$$

1710 cm^{-1} 1710 cm^{-1} 2710, 2810 cm^{-1}

Figure 12-11 (page 506) compares the IR spectra of a simple ketone and aldehyde.

Conjugation lowers the carbonyl stretching frequencies of ketones and aldehydes because the partial pi bonding character of the single bond between the conjugated double bonds reduces the electron density of the carbonyl pi bond. The stretching frequency of this weakened carbonyl bond is lowered to about 1685 cm^{-1}. Ring strain has the opposite effect, raising the carbonyl stretching frequency in ketones with three-, four-, and five-membered rings.

acetophenone 2-butenal cyclopentanone cyclopropanone

18-5B Proton NMR Spectra of Ketones and Aldehydes

When considering the proton NMR spectra of ketones and aldehydes, we are interested primarily in the protons bonded to the carbonyl group (aldehyde protons) and the protons bonded to the adjacent carbon atom (the α-carbon atom). Aldehyde protons normally absorb at chemical shifts between $\delta 9$ and $\delta 10$. The aldehyde proton's absorption may be split ($J = 1$ to 5 Hz) if there are protons on the α-carbon atom. Protons on the α-carbon atom of a ketone or aldehyde usually absorb at a chemical shift between $\delta 2.1$ and $\delta 2.4$ if there are no other electron-withdrawing groups nearby. Methyl ketones are characterized by a singlet at about $\delta 2.1$.

Figure 18-1 shows the proton NMR spectrum of butanal. The aldehyde proton appears at $\delta 9.75$, split by the protons on the α-carbon atom with a small ($J = 1$ Hz) coupling constant. The α protons appear at $\delta 2.4$, and the β and γ protons appear at increasing magnetic fields, as they are located farther from the deshielding effects of the carbonyl group.

▲ Figure 18-1

The proton NMR spectrum of butanal (butyraldehyde) shows the aldehyde proton at $\delta 9.8$, split into a triplet ($J = 1$ Hz) by the two α protons. The α, β, and γ protons appear at values of δ that decrease with increasing distance from the carbonyl group.

18-5C Carbon NMR Spectra of Ketones and Aldehydes

Carbonyl carbon atoms have chemical shifts around 200 ppm in the carbon NMR spectrum. Because they have no hydrogens attached, ketone carbonyl carbon atoms usually give weak absorptions. The α-carbon atoms usually absorb at chemical shifts of about 30 to 40 ppm. Figure 18-2 shows the spin-decoupled carbon NMR spectrum of 2-heptanone, in which the carbonyl carbon absorbs at 208 ppm, and the α-carbon atoms absorb at 30 ppm (methyl) and 44 ppm (methylene).

▲ Figure 18-2

This spin-decoupled carbon NMR spectrum of 2-heptanone shows the carbonyl carbon at 208 ppm and the α carbons at 30 ppm (methyl) and 44 ppm (methylene).

PROBLEM 18-2

NMR spectra for two compounds are given below, together with the molecular formulas. Each compound is a ketone or an aldehyde. In each case, show what characteristics of the spectrum imply the presence of a ketone or an aldehyde, and propose a structure for the compound.

18-5D   Mass Spectra of Ketones and Aldehydes

In the mass spectrometer, a ketone or an aldehyde may lose an alkyl group to give a resonance-stabilized acylium ion, like the acylium ion that serves as the electrophile in the Friedel–Crafts acylation (Section 17-11).

$$\left[\begin{array}{c} O \\ \parallel \\ R-C-R' \end{array}\right]^{\ddagger} \longrightarrow \quad [R-\overset{+}{C}=O \quad \longleftrightarrow \quad R-C\equiv O^{+}] \;+\; \cdot R'$$

$$\text{acylium ion}$$

Figure 18-3 shows the mass spectrum of methyl ethyl ketone (2-butanone). The molecular ion is prominent at m/z 72. The base peak at m/z 43 corresponds to loss of

◀ Figure 18-3
The mass spectrum of 2-butanone shows a prominent molecular ion and a base peak from loss of an ethyl radical to give an acylium ion.

the ethyl group. Because a methyl radical is less stable than an ethyl radical, the peak corresponding to loss of the methyl group (m/z 57) is much weaker than the base peak from loss of the ethyl group.

McLafferty Rearrangement of Ketones and Aldehydes   The mass spectrum of butyraldehyde (Fig. 18-4) shows the peaks we expect at m/z 72 (molecular ion), m/z 57 (loss of a methyl group), and m/z 29 (loss of a propyl group). The peak at m/z 57 is from cleavage between the β and γ carbons to give a resonance-stabilized carbocation. This is also a common fragmentation with carbonyl compounds; like the other odd-numbered peaks, it results from loss of a radical.

The base peak is at m/z 44, from loss of a fragment of mass 28. Loss of a fragment with an even mass number corresponds to loss of a stable, neutral molecule (as when water, mass 18, is lost from an alcohol). A fragment of mass 28 corresponds to a molecule of ethylene (C_2H_4). This fragment is lost through a process called the **McLafferty rearrangement**, involving a cyclic intramolecular transfer of a hydrogen atom from the γ (gamma) carbon to the carbonyl oxygen (shown in Figure 18-5).

The McLafferty rearrangement is a characteristic fragmentation of ketones and aldehydes as long as they have γ hydrogens. It is equivalent to a cleavage between the α and β carbon atoms, plus one mass unit for the hydrogen that is transferred.

▶ **Figure 18-4**

The mass spectrum of butyraldehyde shows the expected ions of masses 72, 57, and 29. The base peak at m/z 44 results from the loss of ethylene via McLafferty rearrangement.

▲ **Figure 18-5**

Mechanism of the McLafferty rearrangement. This rearrangement may be concerted, as shown here, or the γ hydrogen may be transferred first, followed by fragmentation.

PROBLEM 18-3

Why were no products from McLafferty rearrangement observed in the spectrum of 2-butanone (Fig. 18-3)?

PROBLEM 18-4

Use equations to show the fragmentation leading to each numbered peak in the mass spectrum of 2-octanone.

18-5E   Ultraviolet Spectra of Ketones and Aldehydes

The $\pi \rightarrow \pi^*$ Transition   The strongest absorptions in the ultraviolet spectra of aldehydes and ketones are the ones resulting from $\pi \rightarrow \pi^*$ electronic transitions. As with alkenes, these absorptions are observable ($\lambda_{max} > 200$ nm) only if the carbonyl double bond is conjugated with another double bond. The simplest conjugated carbonyl system is propenal, shown below. The $\pi \rightarrow \pi^*$ transition of propenal occurs at λ_{max} of 210 nm ($\varepsilon = 11,000$). Alkyl substitution increases the value of λ_{max} by about 10 nm per alkyl group. An additional conjugated double bond increases the value of λ_{max} by about 30 nm. Notice the large values of the molar absorptivities ($\varepsilon > 5000$), as we also observed for the $\pi \rightarrow \pi^*$ transitions of conjugated dienes.

propenal
$\lambda_{max} = 210$ nm, $\varepsilon = 11,000$

three alkyl groups
$\lambda_{max} = 237$ nm, $\varepsilon = 12,000$

three alkyl groups
$\lambda_{max} = 244$ nm, $\varepsilon = 12,500$

The $n \rightarrow \pi^*$ Transition   An additional band of absorptions in the ultraviolet spectra of ketones and aldehydes results from promoting one of the nonbonding electrons on oxygen to a π^* antibonding orbital. This transition involves a smaller amount of energy than the $\pi \rightarrow \pi^*$ transition because the promoted electron leaves a nonbonding (n) orbital that is higher in energy than the π bonding orbital (Fig. 18-6).

Because the $n \rightarrow \pi^*$ transition requires less energy than a $\pi \rightarrow \pi^*$ transition, it gives a lower frequency (longer wavelength) absorption. The $n \rightarrow \pi^*$ transitions of simple, unconjugated ketones and aldehydes give absorptions with values of λ_{max} between 280 and 300 nm. Each double bond added in conjugation with the carbonyl group increases the value of λ_{max} by about 30 nm. For example, the $n \rightarrow \pi^*$ transition of acetone occurs at λ_{max} of 280 nm ($\varepsilon = 15$). Figure 18-7 shows the UV spectrum of a ketone conjugated with one double bond, having λ_{max} around 315 to 330 nm ($\varepsilon = 110$).

Figures 18-6 and 18-7 show that $n \rightarrow \pi^*$ transitions have small molar absorptivities, generally about 10 to 200. These absorptions are around 1000 times weaker than $\pi \rightarrow \pi^*$ transitions because the $n \rightarrow \pi^*$ transition corresponds to a "forbidden"

PROBLEM-SOLVING $HINT$

Conjugated carbonyl compounds have characteristic $\pi \rightarrow \pi^*$ absorptions in the UV spectrum.

Base value:

210 nm

An additional conjugated C=C increases λ_{max} about 30 nm; an additional alkyl group increases it about 10 nm.

► **Figure 18-6**

Comparison of the $\pi \rightarrow \pi^*$ and the $n \rightarrow \pi^*$ transitions. The $n \rightarrow \pi^*$ transition requires less energy because the nonbonding (n) electrons are higher in energy than the bonding π electrons.

► **Figure 18-7**

UV spectrum of 4-methyl-3-penten-2-one. This spectrum would be listed as λ_{max} 237, $\varepsilon = 12,000$; λ_{max} 315, $\varepsilon = 110$.

PROBLEM-SOLVING **HINT**

Carbonyl $n \rightarrow \pi^*$ absorptions are very weak, and they are not as useful as the $\pi \rightarrow \pi^*$ absorptions.
Base value: $\lambda_{max} = 280$–300 nm.
The value of λ_{max} increases about 30 nm for a conjugated C=C double bond.

electronic transition with a low probability of occurrence. The nonbonding orbitals on oxygen are perpendicular to the π^* antibonding orbitals, and there is zero overlap between these orbitals (see Figure 18-6). This forbidden transition occurs occasionally, but much less frequently than the "allowed" $\pi \rightarrow \pi^*$ transition.

Notice that the y axis of the spectrum in Figure 18-7 is logarithmic, allowing both the $\pi \rightarrow \pi^*$ and the much weaker $n \rightarrow \pi^*$ absorptions to be charted on the same spectrum. It is often necessary to run the spectrum twice, using different concentrations of the sample, to observe both absorptions.

More complete information for predicting UV spectra is given in Appendix 3.

PROBLEM 18-5

Predict the approximate values of λ_{max} for the $\pi \rightarrow \pi^*$ transition and the $n \rightarrow \pi^*$ transition in each compound.

(a)   (b)   (c)   (d)

In the chemical industry, ketones and aldehydes are used as solvents, starting materials, and reagents for the synthesis of other products. Although formaldehyde is well known as the formalin solution used to preserve biological specimens, most of the 4 billion kilograms of formaldehyde produced each year is used to make Bakelite®, phenol–formaldehyde resins, urea–formaldehyde glues, and other polymeric products. Acetaldehyde is used primarily as a starting material in the manufacture of acetic acid, polymers, and drugs.

Acetone is the most important commercial ketone, with over 3 billion kilograms used each year. Both acetone and methyl ethyl ketone (2-butanone) are common industrial solvents. These ketones dissolve a wide range of organic materials, have convenient boiling points for easy distillation, and have low toxicities.

Many other ketones are used as flavorings and additives to foods, drugs, and other products. Table 18-4 lists some simple ketones and aldehydes with well-known odors and flavors. *Pyrethrin*, isolated from pyrethrum flowers, is commercially extracted for use as a "natural" insecticide. "Natural" or not, pyrethrin can cause severe allergic reactions, nausea, vomiting, and other toxic effects in animals.

18-6 Industrial Importance of Ketones and Aldehydes

TABLE 18-4 Ketones and Aldehydes Used in Household Products

	butyraldehyde	vanillin	acetophenone	*trans*-cinnamaldehyde
Odor:	buttery	vanilla	pistachio	cinnamon
Uses:	margarine, foods	foods, perfumes	ice cream	candy, foods, drugs

	camphor	pyrethrin	carvone	muscone
Odor:	"camphoraceous"	floral	(−) enantiomer: spearmint (+) enantiomer: caraway seed	musky aroma
Uses:	liniments, inhalants	plant insecticide	candy, toothpaste, etc.	perfumes

In studying reactions of other functional groups, we have already encountered some of the best methods for making ketones and aldehydes. Let's review and summarize these reactions, and then consider some additional synthetic methods. A summary table of syntheses of ketones and aldehydes begins on page 795.

18-7 Review of Syntheses of Ketones and Aldehydes

18-7A Ketones and Aldehydes from Oxidation of Alcohols (Section 11-2)

Because there are so many ways of making alcohols, they are the most important intermediates for synthesis of ketones and aldehydes. For example, a Grignard reaction can assemble a complicated alcohol, which can be oxidized to a ketone or an aldehyde.

$$\text{R}-\text{MgX} + \text{R}'-\overset{\overset{\displaystyle O}{\|}}{\text{C}}-\text{H} \xrightarrow[\text{ether}]{} \xrightarrow{\text{H}_3\text{O}^+} \text{R}-\overset{\overset{\displaystyle OH}{|}}{\text{CH}}-\text{R}' \xrightarrow[\text{H}_2\text{SO}_4]{\text{Na}_2\text{Cr}_2\text{O}_7} \text{R}-\overset{\overset{\displaystyle O}{\|}}{\text{C}}-\text{R}'$$

Grignard   aldehyde   secondary   ketone

alcohol

Secondary alcohols → ketones

Secondary alcohols are readily oxidized to ketones by sodium dichromate in sulfuric acid ("chromic acid") or by potassium permanganate ($KMnO_4$).

borneol   camphor
(88%)

Primary alcohols → aldehydes

$$\text{R}-\overset{\overset{\displaystyle OH}{|}}{\text{CH}_2} \xrightarrow[-2\,\text{H}]{\text{[oxidizing agent]}} \text{R}-\overset{\overset{\displaystyle O}{\|}}{\text{C}}-\text{H} \xrightarrow[\text{[O]}]{\text{[overoxidation]}} \text{R}-\overset{\overset{\displaystyle O}{\|}}{\text{C}}-\text{OH}$$

primary alcohol   aldehyde   carboxylic acid

Oxidation of a primary alcohol to an aldehyde requires careful selection of an oxidizing agent. Because aldehydes are easily oxidized to carboxylic acids, strong oxidants like chromic acid often give overoxidation. *Pyridinium chlorochromate* (PCC), a complex of chromium trioxide with pyridine and HCl, provides good yields of aldehydes without overoxidation.

cyclohexylmethanol   cyclohexanecarbaldehyde
(90%)

18-7B   Ketones and Aldehydes from Ozonolysis of Alkenes (Section 8-15B)

Ozonolysis, followed by a mild reduction, cleaves alkenes to give ketones and aldehydes.

Ozonolysis is useful as a synthetic method or as an analytical technique. Yields are generally good.

1-methylcyclohexene   6-oxoheptanal
(65%)

18-7C Phenyl Ketones and Aldehydes: Friedel–Crafts Acylation (Section 17-11)

Friedel–Crafts acylation is an excellent method for making alkyl aryl ketones or diaryl ketones. It cannot be used on strongly deactivated aromatic systems, however.

R is alkyl or aryl; G is hydrogen, a halogen, or an activating group.

p-nitrobenzoyl chloride

p-nitrobenzophenone
(90%)

The Gatterman–Koch synthesis is a variant of the Friedel–Crafts acylation in which carbon monoxide and HCl generate an intermediate that reacts like formyl chloride. Like Friedel–Crafts reactions, the Gatterman–Koch formylation succeeds only with benzene and activated benzene derivatives.

toluene

p-methylbenzaldehyde (major)
(50%)

18-7D Ketones and Aldehydes from Hydration of Alkynes (Section 9-9F)

Catalyzed by Acid and Mercuric Salts   Hydration of a terminal alkyne is a convenient way of making methyl ketones. This reaction is catalyzed by a combination of sulfuric acid and mercuric ion. The initial product of Markovnikov hydration is an enol, which quickly tautomerizes to its keto form. Internal alkynes can be hydrated, but mixtures of ketones often result.

alkyne

enol (not isolated)

methyl ketone

Example

ethynylcyclohexane

enol

cyclohexyl methyl ketone
(90%)

Hydroboration–Oxidation of Alkynes   Hydroboration–oxidation of an alkyne gives anti-Markovnikov addition of water across the triple bond. Di (secondary isoamyl) borane, called *disiamylborane*, is used, since this bulky borane cannot add twice across the triple bond. On oxidation of the borane, the unstable enol quickly tautomerizes to an aldehyde.

$$R{-}C{\equiv}C{-}H \xrightarrow[\text{(2) } H_2O_2, \text{ NaOH}]{\text{(1) Sia}_2\text{BH}} \left[\begin{array}{c} R \quad\quad H \\ C{=}C \\ H \quad\quad OH \end{array}\right] \xrightarrow{\text{-OH}} R{-}CH_2{-}\overset{\displaystyle O}{\overset{\|}{C}}{-}H$$

alkyne   enol (not isolated)   aldehyde

Example

ethynylcyclohexane   cyclohexylethanal
                                                                   (65%)

In the following sections, we consider additional syntheses of ketones and aldehydes that we have not covered before. These syntheses form ketones and aldehydes from carboxylic acids, nitriles, acid chlorides, and alkyl halides (used to alkylate 1,3-dithiane).

SOLVED PROBLEM 18-1

Show how you would synthesize each compound from starting materials containing no more than six carbon atoms.

(a)   (b)

SOLUTION

(a) This compound is a ketone with 12 carbon atoms. The carbon skeleton might be assembled from two six-carbon fragments using a Grignard reaction, which gives an alcohol that is easily oxidized to the target compound.

An alternative route to the target compound involves Friedel–Crafts acylation.

(b) This compound is an aldehyde with eight carbon atoms. An aldehyde might come from oxidation of an alcohol (possibly a Grignard product) or hydroboration of an alkyne. If we use a Grignard, the restriction to six-carbon starting materials means we need to add two carbons to a methylcyclopentyl fragment, ending in a primary alcohol. Grignard addition to an epoxide does this.

target (b) $\xleftarrow{\text{PCC}}$ [cyclopentane]CH$_2$—CH$_2$—$\overset{\text{OH}}{\underset{}{\text{CH}}}_2$ $\xleftarrow[\text{(2) H}_3\text{O}^+]{\text{(1)} \triangle\text{O, ether}}$ [cyclopentane]CH$_2$MgBr

Alternatively, we could construct the carbon skeleton using acetylene as the two-carbon fragment. The resulting terminal alkyne undergoes hydroboration to the correct aldehyde.

target (b) $\xleftarrow[\text{(2) H}_2\text{O}_2,\ ^-\text{OH}]{\text{(1) Sia}_2\text{BH}}$ [cyclopentane]CH$_2$—C≡C—H $\xleftarrow{\text{H—C≡C:}^- \text{Na}^+}$ [cyclopentane]CH$_2$Br

PROBLEM 18-6

Show how you would synthesize each compound from starting materials containing no more than six carbon atoms.

(a) [cyclohexane ring]$\overset{\text{O}}{\underset{}{\text{C}}}$—CH$_2CH_3$ (b) [benzene ring]$\overset{\text{O}}{\underset{}{\text{C}}}$—CH$_2CH_3$ (c) [cyclopentane ring]CH$_2$—$\overset{\text{O}}{\underset{}{\text{C}}}$—CH$_3$

1,3-Dithiane is a weak proton acid (pK_a = 32) that can be deprotonated by strong bases such as *n*-butyllithium. The resulting carbanion is stabilized by the electron-withdrawing effect of two highly polarizable sulfur atoms.

[1,3-dithiane ring with S, S and H H] + C$_4$H$_9$—Li ⟶ [dithiane anion ring with S, S and H] + C$_4$H$_{10}$↑

1,3-dithiane, pK_a = 32   *n*-butyllithium   dithiane anion   butane

Alkylation of the dithiane anion by a primary alkyl halide or tosylate gives a thioacetal (sulfur acetal) that can be hydrolyzed using an acidic solution of mercuric chloride. The product is an aldehyde bearing the alkyl group that was added by the alkylating agent. This is a useful synthesis of aldehydes bearing primary alkyl groups.

[dithiane anion] + R—X ⟶ [thioacetal] $\xrightarrow[\text{H}_2\text{O}]{\text{H}^+,\ \text{HgCl}_2}$ [aldehyde]

dithiane anion   alkylating agent (primary alkyl halide)   thioacetal   aldehyde

Alternatively, the thioacetal can be alkylated once more to give a thioketal. Hydrolysis of the thioketal gives a ketone. (Acetals and ketals are discussed in more detail in Section 18-18.)

[thioacetal H R] $\xrightarrow[\text{(2) 1° R'—X}]{\text{(1) BuLi}}$ [thioketal R' R] $\xrightarrow[\text{H}_2\text{O}]{\text{H}^+,\ \text{HgCl}_2}$ [ketone R' R]

thioacetal   thioketal   ketone

For example, 1-phenyl-2-pentanone may be synthesized as shown below:

1,3-dithiane → thioacetal → thioketal → ketone

In each of these sequences, dithiane is alkylated once or twice, then hydrolyzed to give a carbonyl group bearing the alkyl group(s) used in the alkylation. We often consider dithiane to be a synthetic equivalent of a carbonyl group that can be made nucleophilic and alkylated.

PROBLEM-SOLVING $HINT$

You can think of dithiane as a "masked" carbonyl group. To make an aldehyde or ketone, add to dithiane whatever alkyl groups are on the carbonyl group of the target compound.

PROBLEM 18-7

Show how you would use the dithiane method to make the following ketones and aldehydes:
(a) 3-phenylpropanal   (b) 4-phenyl-2-hexanone
(c) dibenzyl ketone   (d) 1-cyclohexyl-4-phenyl-2-butanone

18-9
Synthesis of Ketones from Carboxylic Acids

Organolithium reagents can be used to synthesize ketones from carboxylic acids. Organolithiums are so reactive toward carbonyls that they attack the lithium salts of carboxylate anions to give dianions. Protonation of the dianion forms the hydrate of a ketone, which quickly loses water to give the ketone (see Section 18-14).

carboxylic acid → lithium carboxylate → dianion → hydrate → ketone

If the organolithium reagent is inexpensive, we can simply add two equivalents to the carboxylic acid. The first equivalent generates the carboxylate salt, and the second attacks the carbonyl group. Subsequent protonation gives the ketone.

cyclohexane carboxylic acid → dianion → hydrate → cyclohexyl phenyl ketone

PROBLEM 18-8

Predict the products of the following reactions.

(a) [benzoic acid structure] $\xrightarrow[\text{(2) H}_3\text{O}^+]{\text{(1) excess CH}_3\text{Li}}$

(b) CH_3COOH $\xrightarrow[\text{(2) H}_3\text{O}^+]{\text{(1) 2 [cyclohexyllithium]}}$

(c) $CH_3(CH_2)_3COOH$ $\xrightarrow[\text{(2) H}_3\text{O}^+]{\text{(1) excess CH}_3\text{CH}_2\text{Li}}$

Nitriles can also be used as starting materials for the synthesis of ketones. Discussed in Chapter 21, nitriles are compounds containing the cyano ($-C\equiv N$) functional group. Since nitrogen is more electronegative than carbon, the $-C\equiv N$ triple bond is polarized like the $C=O$ bond of the carbonyl group. Nucleophiles can add to the $-C\equiv N$ triple bond by attacking the electrophilic carbon atom.

A Grignard or organolithium reagent attacks a nitrile to give the magnesium salt of an imine. Acidic hydrolysis of the imine leads to the ketone. The mechanism of this acid hydrolysis is the reverse of acid-catalyzed imine formation, covered in Section 18-16. Note that the ketone is formed during the hydrolysis after any excess Grignard reagent has been destroyed; thus, the ketone is not attacked.

18-10
Synthesis of Ketones from Nitriles

nucleophilic attack   Mg salt of imine   imine   ketone

Example

benzonitrile   phenylmagnesium bromide   benzophenone imine (magnesium salt)   benzophenone (80%)

PROBLEM 18-9

Predict the products of the following reactions.
(a) $CH_3CH_2CH_2CH_2-C\equiv N + CH_3CH_2-MgBr$, then H_3O^+
(b) benzyl bromide + sodium cyanide
(c) product of (b) + cyclopentylmagnesium bromide, then acidic hydrolysis.

PROBLEM 18-10

Show how the following transformations may be accomplished in good yield. You may use any additional reagents that are needed.
(a) bromobenzene $\rightarrow$ propiophenone
(b) $CH_3CH_2CN \rightarrow$ 3-heptanone
(c) pentanoic acid $\rightarrow$ 3-heptanone
(d) toluene $\rightarrow$ benzyl cyclopentyl ketone

Because aldehydes are easily oxidized to acids, one might wonder whether acids are easily reduced back to aldehydes. Aldehydes tend to be more reactive than acids, however, and reducing agents that are strong enough to reduce acids also reduce aldehydes even faster.

18-11
Synthesis of Aldehydes and Ketones from Acid Chlorides

acid   aldehyde (not isolable)   alkoxide

Acids can be reduced to aldehydes by first converting them to a functional group that is easier to reduce than an aldehyde: the acid chloride. Acid chlorides (acyl chlorides) are reactive derivatives of carboxylic acids in which the acidic hydroxyl group is replaced by a chlorine atom. Acid chlorides are often synthesized by treatment of carboxylic acids with thionyl chloride, $SOCl_2$.

$$\underset{\text{acid}}{R-\overset{\overset{\displaystyle O}{\|}}{C}-OH} \;+\; \underset{\text{thionyl chloride}}{Cl-\overset{\overset{\displaystyle O}{\|}}{S}-Cl} \;\longrightarrow\; \underset{\text{acid chloride}}{R-\overset{\overset{\displaystyle O}{\|}}{C}-Cl} \;+\; HCl \;+\; SO_2\uparrow$$

Strong reducing agents like $LiAlH_4$ reduce acid chlorides all the way to primary alcohols. Lithium aluminum tri(*t*-butoxy)hydride is a milder reducing agent that reacts faster with acid chlorides than with aldehydes. Reduction of acid chlorides with lithium aluminum tri(*t*-butoxy)hydride gives good yields of aldehydes.

$$\underset{\text{acid chloride}}{R-\overset{\overset{\displaystyle O}{\|}}{C}-Cl} \;\xrightarrow[\text{lithium aluminum tri(\textit{t}-butoxy)hydride}]{Li^+\;{}^-AlH(O\text{-}t\text{-Bu})_3}\; \underset{\text{aldehyde}}{R-\overset{\overset{\displaystyle O}{\|}}{C}-H}$$

Example

$$\underset{\text{isovaleric acid}}{\overset{\overset{\displaystyle CH_3}{|}}{CH_3CHCH_2}-\overset{\overset{\displaystyle O}{\|}}{C}-OH} \;\xrightarrow{SOCl_2}\; \underset{\text{isovaleroyl chloride}}{\overset{\overset{\displaystyle CH_3}{|}}{CH_3CHCH_2}-\overset{\overset{\displaystyle O}{\|}}{C}-Cl} \;\xrightarrow{Li^+\;{}^-AlH(O\text{-}t\text{-Bu})_3}\; \underset{\text{isovaleraldehyde (65\%)}}{\overset{\overset{\displaystyle CH_3}{|}}{CH_3CHCH_2}-\overset{\overset{\displaystyle O}{\|}}{C}-H}$$

Synthesis of Ketones Grignard and organolithium reagents react with acid chlorides much like hydride reagents: They add R^- where a hydride reagent would add H^-. As we saw in Section 10-9, Grignard and organolithium reagents add to acid chlorides to give ketones, but they add again to the ketones to give tertiary alcohols.

$$\underset{\text{acid chloride}}{R'-\overset{\overset{\displaystyle O}{\|}}{C}-Cl} \;\xrightarrow[\text{(fast)}]{RMgX}\; \underset{\text{ketone}}{\left[R'-\overset{\overset{\displaystyle O}{\|}}{C}-R\right]} \;\xrightarrow[\text{(fast)}]{RMgX}\; \underset{\text{alkoxide}}{R'-\overset{\overset{\displaystyle O^-\;{}^+MgX}{|}}{\underset{\underset{\displaystyle R}{|}}{C}}-R}$$

To stop at the ketone stage, a weaker organometallic reagent is needed: one that reacts faster with acid chlorides than with ketones. A **lithium dialkylcuprate** (*Gilman reagent*) is such a reagent.

$$\underset{\substack{\text{a lithium dialkylcuprate}\\ \text{(Gilman reagent)}}}{R_2CuLi} \;+\; R'-\overset{\overset{\displaystyle O}{\|}}{C}-Cl \;\longrightarrow\; R'-\overset{\overset{\displaystyle O}{\|}}{C}-R \;+\; R-Cu \;+\; LiCl$$

The lithium dialkylcuprate is formed by the reaction of two equivalents of the corresponding organolithium reagent (Section 10-8B) with cuprous iodide.

$$2\;\;R-Li \;+\; CuI \;\longrightarrow\; R_2CuLi \;+\; LiI$$

Example

80%

PROBLEM 18-11

Predict the products of the following reactions:

(a)

$\dfrac{(1)\ LiAlH_4}{(2)\ H_3O^+}$

(b)

$LiAlH(O\text{-}t\text{-}Bu)_3$

(c)

$\left(\diagdown\diagup\right)_2 CuLi$

SUMMARY **Syntheses of Ketones and Aldehydes**

1. *Oxidation of alcohols* (Section 11-2)

   a. *Secondary alcohols* → *ketones*

$$\underset{\text{secondary alcohol}}{R-\overset{\overset{\displaystyle OH}{|}}{C}H-R'} \xrightarrow{Na_2Cr_2O_7/H_2SO_4} \underset{\text{ketone}}{R-\overset{\overset{\displaystyle O}{\|}}{C}-R'}$$

   b. *Primary alcohols* → *aldehydes*

$$\underset{\text{primary alcohol}}{R-CH_2OH} \xrightarrow{C_5H_5NH^+ CrO_3Cl^-(PCC)} \underset{\text{aldehyde}}{R-\overset{\overset{\displaystyle O}{\|}}{C}-H}$$

2. *Ozonolysis of alkenes* (Section 8-15B)

$$\underset{\text{alkene}}{\overset{R}{\underset{H}{\diagdown}}C=C\overset{R'}{\underset{R''}{\diagup}}} \xrightarrow[(2)\ (CH_3)_2S]{(1)\ O_3} \underset{\text{aldehyde}}{\overset{R}{\underset{H}{\diagdown}}C=O} + \underset{\text{ketone}}{O=C\overset{R'}{\underset{R''}{\diagup}}}$$

(gives aldehydes or ketones, depending on the starting alkene)

3. *Friedel–Crafts acylation* (Section 17-11)

$$R-\overset{\overset{\displaystyle O}{\|}}{C}-Cl + \underset{}{\overset{G}{\bigcirc}} \xrightarrow{AlCl_3} G-\overset{}{\bigcirc}-\underset{\text{aryl ketone}}{\overset{\overset{\displaystyle O}{\|}}{C}-R} \quad (+\ \ ortho)$$

R = alkyl or aryl;
G = hydrogen, halogen, or an activating group.

The Gatterman–Koch formylation (Section 17-11C)

$$HCl + CO + \overset{G}{\bigcirc} \xrightarrow{AlCl_3,\ CuCl} G-\bigcirc-\overset{\overset{\displaystyle O}{\|}}{C}-H$$

G = hydrogen or an activating group.

benzaldehyde derivative

4. *Hydration of alkynes (Section 9-9F)*

   a. *Catalyzed by acid and mercuric salts (Markovnikov orientation)*

$$R-C\equiv C-H \xrightarrow[\text{H}_2\text{O}]{\text{Hg}^{2+},\ \text{H}_2\text{SO}_4} \left[\begin{matrix} R \\ C=C \\ HO \end{matrix} \begin{matrix} H \\ \\ H \end{matrix}\right] \longrightarrow R-\overset{\overset{\displaystyle O}{\|}}{C}-CH_3$$

alkyne   enol (not isolated)   methyl ketone

   b. *Hydroboration–oxidation (anti-Markovnikov orientation)*

$$R-C\equiv C-H \xrightarrow[\text{(2) H}_2\text{O}_2,\ \text{NaOH}]{\text{(1) Sia}_2\text{BH}} \left[\begin{matrix} R \\ C=C \\ H \end{matrix} \begin{matrix} H \\ \\ OH \end{matrix}\right] \longrightarrow R-CH_2-\overset{\overset{\displaystyle O}{\|}}{C}-H$$

alkyne   enol (not isolated)   aldehyde

5. *Alkylation of 1,3-dithianes (Section 18-8)*

6. *Synthesis of ketones using organolithium reagents with carboxylic acids (Section 18-9)*

$$R-\overset{\overset{\displaystyle O}{\|}}{C}-OH \xrightarrow{2\ R'-Li} R-\overset{\overset{\displaystyle OLi}{|}}{\underset{\underset{\displaystyle R'}{|}}{C}}-OLi \xrightarrow{\text{H}_3\text{O}^+} R-\overset{\overset{\displaystyle O}{\|}}{C}-R'$$

carboxylic acid   dianion   ketone

Example

7. *Synthesis of ketones from nitriles* (Section 18-10)

$$R-C{\equiv}N \ + \ R'-Mg-X \ \longrightarrow \ \underset{\substack{\text{Mg salt}\\\text{of imine}}}{R-\overset{\overset{\displaystyle N-MgX}{\|}}{C}-R'} \ \xrightarrow{\ H_3O^+\ } \ \underset{\text{ketone}}{R-\overset{\overset{\displaystyle O}{\|}}{C}-R'}$$

(or R'—Li)

Example

benzonitrile $\xrightarrow[\text{(2) }H_3O^+]{\text{(1) }CH_3CH_2CH_2-MgBr}$ butyrophenone

8. *Aldehyde synthesis by reduction of acid chlorides* (Section 18-11)

$$\underset{\text{acid chloride}}{R-\overset{\overset{\displaystyle O}{\|}}{C}-Cl} \ \xrightarrow[\text{(or } H_2,\ Pd,\ BaSO_4,\ S)]{Li^+\ {}^-AlH(O\text{-}t\text{-Bu})_3} \ \underset{\text{aldehyde}}{R-\overset{\overset{\displaystyle O}{\|}}{C}-H}$$

Example

$$\underset{\text{3-phenylbutanoyl chloride}}{CH_3-\overset{\overset{\displaystyle Ph}{|}}{CH}-CH_2-\overset{\overset{\displaystyle O}{\|}}{C}-Cl} \ \xrightarrow{Li^+\ {}^-AlH(O\text{-}t\text{-Bu})_3} \ \underset{\text{3-phenylbutanal}}{CH_3-\overset{\overset{\displaystyle Ph}{|}}{CH}-CH_2-\overset{\overset{\displaystyle O}{\|}}{C}-H}$$

9. *Ketone synthesis from acid chlorides* (Section 18-11)

$$\underset{\text{acid chloride}}{R'-\overset{\overset{\displaystyle O}{\|}}{C}-Cl} \ + \ R_2CuLi \ \longrightarrow \ \underset{\text{ketone}}{R'-\overset{\overset{\displaystyle O}{\|}}{C}-R}$$

Example

$$CH_3(CH_2)_4\overset{\overset{\displaystyle O}{\|}}{C}-Cl \ + \ \left(\diagdown\diagup\diagdown\diagup\right)_2CuLi \ \longrightarrow \ CH_3(CH_2)_4\overset{\overset{\displaystyle O}{\|}}{C}-(CH_2)_3CH_3$$

Ketones and aldehydes undergo many reactions to give a wide variety of useful derivatives. Their most common reaction is **nucleophilic addition**, addition of a nucleophile and a proton across the C=O double bond. The reactivity of the carbonyl group arises from the electronegativity of the oxygen atom and the resulting polarization of the carbon–oxygen double bond. The electrophilic carbonyl carbon atom is sp^2 hybridized and flat, leaving it relatively unhindered and open to attack from either face of the double bond.

As a nucleophile attacks the carbonyl group, the carbon atom changes hybridization from sp^2 to sp^3. The electrons of the pi bond are forced out to the oxygen atom to form an alkoxide anion, which protonates to give the product of nucleophilic addition.

18-12
Reactions of Ketones and Aldehydes: Nucleophilic Addition

$$\underset{\text{nucleophilic attack}}{\overset{Nuc{:}^-}{\underset{R}{\overset{R}{\diagdown}}}C\overset{\delta+\quad\delta-}{=}\overset{\cdot\cdot}{O}\overset{\cdot\cdot}{}} \ \rightleftharpoons \ \underset{\text{alkoxide}}{\overset{Nuc}{\underset{R}{\overset{R}{\diagdown}}}C-\overset{\cdot\cdot}{\underset{\cdot\cdot}{O}}{:}^-} \ \underset{}{\overset{H-Nuc}{\rightleftharpoons}} \ \underset{\text{product}}{\overset{Nuc}{\underset{R}{\overset{R}{\diagdown}}}C-\overset{H}{\underset{\cdot\cdot}{O}}{:}} \ + \ Nuc{:}^-$$

We have seen at least two examples of nucleophilic addition to ketones and aldehydes. A Grignard reagent (a strong nucleophile resembling a carbanion, $R:^-$) attacks the electrophilic carbonyl carbon atom to give an alkoxide intermediate. Subsequent protonation gives an alcohol.

Hydride reduction of a ketone or aldehyde is another example of nucleophilic addition, with hydride ion ($H:^-$) serving as the nucleophile. Attack by hydride gives an alkoxide that protonates to form an alcohol.

Weak nucleophiles, such as water and alcohols, can add to activated carbonyl groups under acidic conditions. A carbonyl group is a weak base, and it can become protonated in an acidic solution. A carbonyl group that is protonated (or bonded to some other electrophile) is strongly electrophilic, inviting attack by a weak nucleophile.

activated carbonyl

The following reaction is the acid-catalyzed nucleophilic addition of water across the carbonyl group of acetone. This hydration of a ketone or aldehyde is discussed in Section 18-14.

In effect, the base-catalyzed addition to a carbonyl group results from nucleophilic attack of a strong nucleophile followed by protonation. Acid-catalyzed addition begins with protonation, followed by the attack of a weaker nucleophile. Many additions are reversible, with the position of the equilibrium depending on the relative stabilities of the reactants and products.

In most cases, aldehydes are more reactive than ketones toward nucleophilic additions. They usually react more quickly than ketones, and the position of the equilibrium usually lies more toward the products than with ketones. We explain aldehydes' enhanced reactivity by noticing that an aldehyde has only one electron-donating alkyl group, making the aldehyde carbonyl group slightly more electron-poor and electrophilic (an *electronic effect*). Also, an aldehyde has only one bulky alkyl group (compared with two in a ketone), leaving the carbonyl group more exposed toward nucleophilic attack. Especially with a bulky nucleophile, the product of attack on the aldehyde is less hindered than the product from the ketone (a *steric effect*).

PROBLEM 18-12 (review)

Show how you would accomplish the following synthetic conversions. You may use any additional reagents and solvents you need.

(a) Ph—CHO ⟶ Ph—C(=O)—Ph

(b) Ph—C(=O)—Ph ⟶ Ph₃C—OH

(c) Ph—C(=O)—Ph ⟶ Ph—CH(OH)—Ph

(d) Ph—CHO ⟶ Ph—CH(OH)—C≡C—CH₂CH₃

PROBLEM-SOLVING HINT

Please become familiar with these simple mechanisms. You will see many examples in the next few pages. Also, most of the important multistep mechanisms in this chapter are combinations of these simple steps.

PROBLEM 18-13

Sodium triacetoxyborohydride, $Na(CH_3COO)_3BH$, is a mild reducing agent that reduces aldehydes much more quickly than ketones. It can be used to reduce aldehydes in the presence of ketones, such as in the following reaction:

$$CH_3-\overset{O}{\overset{||}{C}}-CH_2-\overset{O}{\overset{||}{C}}-H \xrightarrow[CH_3COOH]{Na(CH_3COO)_3BH} CH_3-\overset{O}{\overset{||}{C}}-CH_2-\overset{OH}{\overset{|}{CH_2}}$$

(a) Draw a complete Lewis structure for sodium triacetoxyborohydride.
(b) Propose a mechanism for the reduction of an aldehyde by sodium triacetoxyborohydride.

The following table summarizes the base-catalyzed and acid-catalyzed mechanisms for nucleophilic addition, together with their reverse reactions.

KEY MECHANISM   Nucleophilic Additions to Carbonyl Groups

Basic conditions (strong nucleophile)

Step 1: Addition of the nucleophile.   **Step 2:** Protonation.

Reverse reaction:

Acidic conditions (weak nucleophile, activated carbonyl)

Step 1: Protonation.   **Step 2:** Addition of the nucleophile.

Reverse reaction:

18-13
The Wittig Reaction

We have seen carbonyl groups undergoing addition by a variety of carbanion-like reagents, including Grignard reagents, organolithium reagents, and acetylide ions. In 1954, Georg Wittig discovered a way of adding a phosphorus-stabilized carbanion to a ketone or aldehyde. The product is not an alcohol, however, because the intermediate undergoes elimination to an alkene. In effect, the **Wittig reaction** converts the carbonyl group of a ketone or an aldehyde into a new double bond where no bond existed before. This reaction proved so useful that Wittig received the Nobel Prize in chemistry in 1979 for this discovery.

The Wittig reaction

ketone or aldehyde   phosphorus ylide

alkene

The phosphorus-stabilized carbanion is an **ylide** (pronounced "ill'-id")—a molecule that bears no overall charge but has a negatively charged carbon atom bonded to a positively charged heteroatom. Phosphorus ylides are prepared from triphenylphosphine and alkyl halides in a two-step process. The first step is nucleophilic attack by triphenylphosphine on an unhindered (usually primary) alkyl halide. The product is an alkyltriphenylphosphonium salt. The phosphonium salt is treated with a strong base (usually butyllithium) to abstract a proton from the carbon atom bonded to phosphorus.

Examples

The phosphorus ylide has two resonance forms: one with a double bond between carbon and phosphorus, and another with charges on carbon and phosphorus. The double-bonded resonance form requires ten electrons in the valence shell of phosphorus, using a *d* orbital. The pi bond between carbon and phosphorus is weak, and the charged structure is the major contributor. The carbon atom actually bears a partial negative charge, balanced by a corresponding positive charge on phosphorus.

PROBLEM 18-14

Trimethylphosphine is much less expensive than triphenylphosphine. Why is trimethylphosphine unsuitable for making most phosphorus ylides?

Because of its carbanion character, the ylide carbon atom is strongly nucleophilic. It attacks a carbonyl group to give a charge-separated intermediate called a *betaine* (pronounced "bay'-tuh-ene"). A betaine is an unusual compound because it contains a negatively charged oxygen and a positively charged phosphorus on adjacent carbon atoms. Phosphorus and oxygen form strong bonds, and the attraction of opposite charges promotes the fast formation of a four-membered *oxaphosphetane* ring. (In some cases, the oxaphosphetane may be formed directly by a cycloaddition, rather than via a betaine.)

The four-membered ring quickly collapses to give the alkene and triphenylphosphine oxide. Triphenylphosphine oxide is exceptionally stable, and the conversion of triphenylphosphine to triphenylphosphine oxide provides the driving force for the Wittig reaction.

MECHANISM The Wittig Reaction

Step 1: The ylide attacks the carbonyl.

ylide   ketone or aldehyde

Step 2: The betaine closes to a 4-membered ring.

a betaine   oxaphosphetane

Step 3: The ring collapses to products.

four-membered ring   triphenylphosphine oxide + alkene

The following examples show the formation of carbon–carbon double bonds using the Wittig reaction. Mixtures of cis and trans isomers often result when geometric isomerism is possible.

85%

PROBLEM 18-15

Like other strong nucleophiles, triphenylphosphine attacks and opens epoxides. The initial product (a betaine) quickly cyclizes to an oxaphosphetane that collapses to an alkene and triphenylphosphine oxide.

(a) Show each step in the reaction of *trans*-2,3-epoxybutane with triphenylphosphine to give 2-butene. What is the stereochemistry of the double bond in the product?

(b) Show how this reaction might be used to convert *cis*-cyclooctene to *trans*-cyclooctene.

Planning a Wittig Synthesis   The Wittig reaction is a valuable synthetic tool that converts a carbonyl group to a carbon–carbon double bond. A wide variety of alkenes may be synthesized by the Wittig reaction. To determine the necessary reagents, mentally divide the target molecule at the double bond and decide which of the two components should come from the carbonyl compound and which should come from the ylide.

In general, the ylide should come from an unhindered alkyl halide. Triphenyl-phosphine is a bulky reagent, reacting best with unhindered primary and methyl halides. It occasionally reacts with unhindered secondary halides, but these reactions are sluggish and often give poor yields. The following example and Solved Problem show the planning of some Wittig syntheses.

Analysis

$$\text{(CH}_3)_2\text{C}=\text{C(H)CH}_2\text{CH}_3 \quad \xrightarrow{\text{could come from}}$$

$$\begin{cases} \text{(CH}_3)_2\text{C}=\text{O} \; + \; \text{Ph}_3\overset{+}{\text{P}}-\overset{-}{\text{C}}(\text{H})\text{CH}_2\text{CH}_3 \\ \text{(preferred)} \\ \text{or} \\ \text{(CH}_3)_2\overset{+}{\text{C}}-\text{PPh}_3 \; + \; \text{O}=\text{C(H)CH}_2\text{CH}_3 \end{cases}$$

Synthesis

$$\text{Br}-\text{CH(H)CH}_2\text{CH}_3 \quad \xrightarrow[\text{(2) BuLi}]{\text{(1) Ph}_3\text{P}} \quad \text{Ph}_3\overset{+}{\text{P}}-\overset{-}{\text{C}}(\text{H})\text{CH}_2\text{CH}_3 \quad \xrightarrow{\text{(CH}_3)_2\text{C}=\text{O}} \quad \text{(CH}_3)_2\text{C}=\text{C(H)CH}_2\text{CH}_3$$

SOLVED PROBLEM 18-2

Show how you would use a Wittig reaction to synthesize 1-phenyl-1,3-butadiene.

1-phenyl-1,3-butadiene

SOLUTION

This molecule has two double bonds that might be formed by Wittig reactions. The central double bond could be formed in either of two ways. Both of these syntheses will probably work, and both will produce a mixture of cis and trans isomers.

Analysis

You should complete this solution by drawing out the syntheses indicated by this analysis (Problem 18-16).

PROBLEM-SOLVING $\mathbb{HINT}$

Plan a Wittig synthesis so that the less hindered end of the double bond comes from the ylide. Remember that the ylide is made by S$_N$2 attack of triphenylphosphine on an unhindered alkyl halide, followed by deprotonation.

PROBLEM 18-16

(a) Outline the syntheses indicated in Solved Problem 18-2, beginning with aldehydes and alkyl halides.

(b) Both of these syntheses of 1-phenyl-1,3-butadiene form the central double bond. Show how you would synthesize this target molecule by forming the terminal double bond.

PROBLEM 18-17

Show how Wittig reactions might be used to synthesize the following compounds. In each case, start with an alkyl halide and a ketone or an aldehyde.

(a) $Ph-CH=C(CH_3)_2$ (b) $Ph-C(CH_3)=CH_2$

(c) $Ph-CH=CH-CH=CH-Ph$

(d)

(e)

18-14
Hydration of Ketones and Aldehydes

In an aqueous solution, a ketone or an aldehyde is in equilibrium with its hydrate, a geminal diol. With most ketones, the equilibrium favors the unhydrated keto form of the carbonyl.

$$K = \frac{[\text{hydrate}]}{[\text{ketone}][\text{H}_2\text{O}]}$$

keto form   hydrate (a geminal diol)

Example

$K = 0.002$

acetone + H_2O ⇌ acetone hydrate

Hydration occurs through the nucleophilic addition mechanism, with water (in acid) or hydroxide ion (in base) serving as the nucleophile.

MECHANISM   Hydration of Ketones and Aldehydes

In acid

Step 1: Protonation.   *Step 2:* Water adds.   *Step 3:* Deprotonation.

(continued)

In base

Step 1: Hydroxide adds.   *Step 2:* Protonation.

These stability effects are apparent in the equilibrium constants for hydration of ketones and aldehydes. Ketones have values of K_{eq} of about 10^{-4} to 10^{-2}. For most aldehydes, the equilibrium constant for hydration is close to 1. Formaldehyde, with no alkyl groups bonded to the carbonyl carbon, has a hydration equilibrium constant of about 40. Strongly electron-withdrawing substituents on the alkyl group of a ketone or aldehyde also destabilize the carbonyl group and favor the hydrate. Chloral (trichloroacetaldehyde) has an electron-withdrawing trichloromethyl group that favors the hydrate. Chloral forms a stable, crystalline hydrate that became famous in the movies as knockout drops or a Mickey Finn.

Aldehydes are more likely than ketones to form stable hydrates. The electrophilic carbonyl group of a ketone is stabilized by its two electron-donating alkyl groups, but an aldehyde carbonyl has only one stabilizing alkyl group; its partial positive charge is not as well stabilized. Aldehydes are thus more electrophilic and less stable than ketones. Formaldehyde, with no electron-donating groups, is even less stable than other aldehydes.

The body rapidly reduces chloral (trichloroacetaldehye) to trichloroethanol, which is responsible for the drug's sleep-inducing effect.

PROBLEM 18-18

Propose mechanisms for
(a) The acid-catalyzed hydration of chloral to form chloral hydrate.
(b) The base-catalyzed hydration of acetone to form acetone hydrate.

PROBLEM-SOLVING HINT

Don't be surprised to see some O—H stretch, from the hydrate, in the IR spectra of many aldehydes.

PROBLEM 18-19

Rank the following compounds in order of increasing amount of hydrate present at equilibrium:

18-15
Formation
of Cyanohydrins

Hydrogen cyanide (H—C≡N) is a toxic, water-soluble liquid that boils at 26°C. Because it is mildly acidic, HCN is sometimes called hydrocyanic acid.

$$H—C≡N: + H_2O \;\rightleftarrows\; H_3O^+ + \;^-:C≡N: \qquad pK_a = 9.2$$

The conjugate base of hydrogen cyanide is the cyanide ion ($^-:C≡N:$). Cyanide ion is a strong base and a strong nucleophile. It attacks ketones and aldehydes to give addition products called **cyanohydrins**. The mechanism is a base-catalyzed nucleophilic addition: attack by cyanide ion on the carbonyl group, followed by protonation of the intermediate.

MECHANISM   Formation of Cyanohydrins

Step 1: Cyanide adds.   *Step 2:* Protonation.

Cyanohydrins may be formed using liquid HCN with a catalytic amount of sodium cyanide or potassium cyanide. HCN is highly toxic and volatile, however, and therefore dangerous to handle. Many procedures use a full equivalent of sodium or potassium cyanide (rather than HCN), dissolved in some other proton-donating solvent.

Cyanohydrin formation is reversible, and the equilibrium constant may or may not favor the cyanohydrin. These equilibrium constants follow the general reactivity trend of ketones and aldehydes,

formaldehyde > other aldehydes > ketones

Formaldehyde reacts quickly and quantitatively with HCN. Most other aldehydes have equilibrium constants that favor cyanohydrin formation. Reactions of HCN with ketones have equilibrium constants that may favor either the ketones or the cyanohydrins, depending on the structure. Ketones that are hindered by large alkyl groups react slowly with HCN and give poor yields of cyanohydrins.

2-butanone   2-butanone cyanohydrin
(95%)

di-*t*-butyl ketone   slow reaction, poor yields

The failure with bulky ketones is largely due to steric effects. Cyanohydrin formation involves rehybridizing of the sp^2 carbonyl carbon to sp^3, narrowing the angle between the alkyl groups from about 120° to about 109.5°, increasing their steric interference.

PROBLEM 18-20

Propose a mechanism for each cyanohydrin synthesis shown above.

Organic compounds containing the cyano group ($-C\equiv N$) are called **nitriles**. A cyanohydrin is therefore an α-hydroxynitrile. Nitriles hydrolyze to carboxylic acids under acidic conditions (discussed in Section 21-7D), so cyanohydrins hydrolyze to α-hydroxy acids. This is the most convenient method for making many α-hydroxy acids:

PROBLEM 18-21

Show how you would accomplish the following syntheses.
(a) acetophenone $\rightarrow$ acetophenone cyanohydrin   **(b)** 1-hexanol $\rightarrow$ hexanal cyanohydrin

(c)

Under the proper conditions, either ammonia or a primary amine reacts with a ketone or an aldehyde to form an **imine**. Imines are nitrogen analogues of ketones and aldehydes, with a carbon–nitrogen double bond in place of the carbonyl group. Like amines, imines are basic; a substituted imine is also called a **Schiff base**. Imine formation is an example of a large class of reactions called **condensations**, reactions in which two (or more) organic compounds are joined, often with the loss of water or another small molecule.

18-16
Formation of Imines

ketone or aldehyde   primary amine   carbinolamine   imine (Schiff base)

The mechanism of imine formation (shown below) begins with a basic nucleophilic addition of the amine to the carbonyl group. Attack by the amine, followed by protonation of the oxygen atom (and deprotonation of the nitrogen atom), gives an unstable intermediate called a **carbinolamine**.

A carbinolamine converts to an imine by losing water and forming a double bond: a dehydration. This dehydration follows the same mechanism as the acid-catalyzed dehydration of an alcohol (Section 11-10). Protonation of the hydroxyl group converts it to a good leaving group, and it leaves as water. The resulting cation is stabilized by a resonance structure with all octets filled and the positive charge on nitrogen. Loss of a proton gives the imine.

KEY MECHANISM Formation of Imines

Step 1: The amine adds to the carbonyl.   *Step 2:* Protonation and deprotonation.

nucleophilic attack   fast proton transfer   carbinolamine

The second half of this mechanism is an acid-catalyzed dehydration.

Step 3: Protonation of hydroxyl.   *Step 4:* Loss of water.   *Step 5:* Deprotonation.

carbinolamine   protonated   intermediate (all octets filled)   imine

---

The proper pH is crucial to imine formation. The second half is acid-catalyzed, so the solution must be somewhat acidic. If the solution is too acidic, however, the amine becomes protonated and nonnucleophilic, inhibiting the first step. Figure 18-8 shows that the rate of imine formation is fastest around pH 4.5.

Some typical imine-forming reactions are shown on the following page. In each case, notice that the C=O group of the ketone or aldehyde is replaced by the C=N—R group of the imine.

▶ **Figure 18-8**
Although dehydration of the carbinolamine is acid-catalyzed, too much acid stops the first step by protonating the amine. Formation of the imine is fastest around pH 4.5.

nucleophilic   nonnucleophilic

cyclohexanone + $\ddot{N}H_3$ (ammonia) $\underset{}{\overset{H^+}{\rightleftharpoons}}$ cyclohexanone imine + H_2O

cyclopentanone + aniline ($\ddot{N}H_2$) $\underset{}{\overset{H^+}{\rightleftharpoons}}$ cyclopentanone phenyl imine + H_2O

benzaldehyde + $CH_3-\ddot{N}H_2$ (methylamine) $\underset{}{\overset{H^+}{\rightleftharpoons}}$ benzaldehyde methyl imine + H_2O

PROBLEM 18-22

Propose a mechanism for each of the imine-forming reactions above.

PROBLEM 18-23

Depending on the reaction conditions, two different imines of formula C_8H_9N might be formed by the reaction of benzaldehyde with methylamine. Explain, and give the structures of the two imines.

PROBLEM 18-24

Give the structures of the carbonyl compound and the amine used to form the following imines:

(a), (b), (c), (d)

PROBLEM-SOLVING HINT

Imine formation is one of the important mechanisms in this chapter. If you know the simple mechanisms, you can remember this mechanism as having two parts:
1. base-catalyzed nucleophilic addition to the carbonyl, and
2. acid-catalyzed dehydration (as with an alcohol).

Imine formation is reversible, and most imines can be hydrolyzed back to the amine and the ketone or aldehyde. The principle of microscopic reversibility (Section 8-4A) states that the reverse reaction taking place under the same conditions should follow the same pathway but in reverse order. Therefore, the mechanism for hydrolysis of an imine is simply the reverse of the mechanism for its formation.

benzaldehyde methyl imine $\underset{}{\overset{H^+, \text{excess } H_2O}{\rightleftharpoons}}$ benzaldehyde + $CH_3-\overset{+}{N}H_3$

PROBLEM 18-25

Propose a mechanism for the hydrolysis of benzaldehyde methyl imine shown above.

18-17
Condensations with Hydroxylamine and Hydrazines

Ketones and aldehydes also condense with other ammonia derivatives, such as hydroxylamine and substituted hydrazines, to give imine derivatives. The equilibrium constants for these reactions are usually more favorable than for reactions with simple amines. Hydroxylamine reacts with ketones and aldehydes to form **oximes**; hydrazine derivatives react to form **hydrazones**; and semicarbazide reacts to form **semicarbazones**. The mechanisms of these reactions are similar to the mechanism of imine formation.

Anhydrous hydrazine is a common rocket fuel. In most cases, liquid oxygen serves as the oxidizer. The fuel and oxidizer are sprayed into the combustion chamber, where they react to generate heat and pressure, forcing the reaction products out through the rocket nozzle.

phenyl-2-propanone + hydroxylamine → phenyl-2-propanone oxime + H₂O

benzaldehyde + hydrazine → benzaldehyde hydrazone + H₂O

cyclohexanone + phenylhydrazine → cyclohexanone phenylhydrazone + H₂O

2-butanone + semicarbazide → 2-butanone semicarbazone + H₂O

These derivatives are useful both as starting materials for further reactions (see Section 19-19) and for characterization and identification of the original carbonyl compounds. Oximes, semicarbazones, and phenylhydrazones are often solid compounds with characteristic melting points. Standard tables give the melting points of these derivatives for thousands of different ketones and aldehydes.

If an unknown compound forms one of these derivatives, the melting point can be compared with that in the table. If the compound's physical properties match those of a known compound and the melting point of its oxime, semicarbazide, or phenylhydrazone matches as well, we can be fairly certain of a correct identification.

PROBLEM-SOLVING HINT

Please learn these common derivatives. You will see many examples, especially in the laboratory

PROBLEM 18-26

2,4-Dinitrophenylhydrazine is frequently used for making derivatives of ketones and aldehydes because the products (2,4-dinitrophenylhydrazones, called **2,4-DNP derivatives**) are even more likely than the phenylhydrazones to be solids with sharp melting points. Propose a mechanism for the reaction of acetone with 2,4-dinitrophenylhydrazine in a mildly acidic solution.

SUMMARY Condensations of Amines with Ketones and Aldehydes

$$\text{☞} \quad {>}C{=}O \ + \ H_2\ddot{N}{-}\boxed{Z} \quad \xrightarrow{H^+} \quad {>}C{=}\ddot{N}{-}\boxed{Z} \ + \ H_2O$$

Z in Z—NH$_2$	Reagent	Product
—H	$H_2\ddot{N}{-}\boxed{H}$ ammonia	${>}C{=}\ddot{N}{-}\boxed{H}$ an imine
—R	$H_2\ddot{N}{-}\boxed{R}$ primary amine	${>}C{=}\ddot{N}{-}\boxed{R}$ an imine (Schiff base)
—OH	$H_2\ddot{N}{-}\boxed{OH}$ hydroxylamine	${>}C{=}\ddot{N}{-}\boxed{OH}$ an oxime
—NH$_2$	$H_2\ddot{N}{-}\boxed{NH_2}$ hydrazine	${>}C{=}\ddot{N}{-}\boxed{NH_2}$ a hydrazone
—NHPh	$H_2\ddot{N}{-}\boxed{NHPh}$ phenylhydrazine	${>}C{=}\ddot{N}{-}\boxed{NHPh}$ a phenylhydrazone
$-NH\overset{\displaystyle O}{\overset{\|}{C}}NH_2$	$H_2\ddot{N}{-}\boxed{NH{-}\overset{\displaystyle O}{\overset{\|}{C}}{-}NH_2}$ semicarbazide	${>}C{=}\ddot{N}{-}\boxed{NH{-}\overset{\displaystyle O}{\overset{\|}{C}}{-}NH_2}$ a semicarbazone

PROBLEM 18-27

Predict the products of the following reactions.

(a) [cyclopentanone] $+ \ HO{-}NH_2 \ \xrightarrow{H^+}$

(b) [α-tetralone] $+ \ H_2N{-}NH_2 \ \xrightarrow{H^+}$

(c) $Ph{-}CH{=}CH{-}CHO \ + \ H_2N{-}\overset{\displaystyle O}{\overset{\|}{C}}{-}NH{-}NH_2 \ \xrightarrow{H^+}$

(d) $Ph{-}\overset{\displaystyle O}{\overset{\|}{C}}{-}Ph \ + \ Ph{-}NH{-}NH_2 \ \xrightarrow{H^+}$

PROBLEM 18-28

Show what amines and carbonyl compounds combine to give the following derivatives.

(a) $Ph{-}CH{=}N{-}NH{-}\overset{\displaystyle O}{\overset{\|}{C}}{-}NH_2$

(b) [camphor N—OH (oxime) structure]

(c) [tetralone N—NHPh structure]

(d) [cyclohexanone =N—NH—dinitrophenyl structure with O$_2$N and NO$_2$]

(e) [quinoline N, CH$_3$ structure]

Just as ketones and aldehydes react with water to form hydrates, they also react with alcohols to form **acetals**. Acetals formed from ketones are often called **ketals**, although this term has been dropped from the IUPAC nomenclature. In the formation of an acetal, two molecules of alcohol add to the carbonyl group, and one molecule of water is eliminated.

18-18

Formation of Acetals

$$\underset{\text{aldehyde}}{\overset{\displaystyle \overset{O}{\underset{}{\parallel}}}{R-C-H}} \;+\; 2\,R'-OH \;\;\underset{}{\overset{H^+}{\rightleftharpoons}}\;\; \underset{\text{acetal}}{\overset{R'O\quad OR'}{\underset{R\quad\;\; H}{\diagdown C \diagup}}} \;+\; H_2O$$

$$\underset{\text{ketone}}{\overset{\displaystyle \overset{O}{\underset{}{\parallel}}}{R-C-R'}} \;+\; 2\,R''-OH \;\;\underset{}{\overset{H^+}{\rightleftharpoons}}\;\; \underset{\substack{\text{acetal (IUPAC)}\\\text{ketal (common)}}}{\overset{R''O\quad OR''}{\underset{R\quad\;\; R'}{\diagdown C \diagup}}} \;+\; H_2O$$

Although hydration is catalyzed by either acid or base, acetal formation must be acid-catalyzed. For example, consider the reaction of cyclohexanone with methanol, catalyzed by *p*-toluenesulfonic acid.

Overall reaction

PROBLEM-SOLVING HINT

In basic conditions, a strong nucleophile usually adds directly to the carbonyl group. In acidic conditions, strong nucleophiles are rarely present; an acid (or Lewis acid) usually protonates the carbonyl to activate it toward attack by a weak nucleophile.

The mechanism for this reaction is shown below. The first step is a typical acid-catalyzed addition to the carbonyl group. The acid catalyst protonates the carbonyl group, and the alcohol (a weak nucleophile) attacks the protonated, activated carbonyl. Loss of a proton from the positively charged intermediate gives a **hemiacetal**. The hemiacetal gets its name from the Greek prefix *hemi-*, meaning "half." Having added one molecule of the alcohol, the hemiacetal is halfway to becoming a "full" acetal. Like the hydrates of ketones and aldehydes, most hemiacetals are too unstable to be isolated and purified.

The second half of the mechanism converts the hemiacetal to the more stable acetal. Protonation of the hydroxyl group, followed by loss of water, gives a resonance-stabilized carbocation. Attack on the carbocation by methanol, followed by loss of a proton, gives the acetal.

KEY MECHANISM   Formation of Acetals

The first half is an acid-catalyzed addition to the carbonyl

Step 1: Protonation.   **Step 2:** Alcohol adds.   **Step 3:** Deprotonation.

(continued)

The second half begins like an acid-catalyzed dehydration

Step 4: Protonation.   *Step 5:* Loss of water.

hemiacetal   protonation, loss of water   resonance-stabilized carbocation

Step 6: Second alcohol adds.   *Step 7:* Deprotonation.

attack by methanol   acetal

PROBLEM 18-29

Propose a mechanism for the acid-catalyzed reaction of acetaldehyde with ethanol to give acetaldehyde diethyl acetal.

PROBLEM-SOLVING HINT

Acetal formation is one of the important mechanisms in this chapter. If you know the simple mechanisms, you can remember this mechanism as having two parts:
1. acid-catalyzed nucleophilic addition to the carbonyl, and
2. S_N1 by protonation and loss of the OH group, then attack by the alcohol.

Since hydration is catalyzed by either acid or base, you might wonder why acetal formation is catalyzed only by acid. In fact, the first step (formation of the hemiacetal) can be base-catalyzed, involving attack by alkoxide ion and protonation of the alkoxide. The second step requires replacement of the hemiacetal —OH group by the alcohol —OR″ group. Hydroxide ion is a poor leaving group for the S_N2 reaction, so alkoxide cannot displace the —OH group. This replacement occurs under acidic conditions, however, because protonation of the —OH group and loss of water gives a resonance-stabilized cation.

Attempted base-catalyzed acetal formation

attack on ketone
(or aldehyde)   hemiacetal   (no S_N2 displacement)

Equilibrium of Acetal Formation   Acetal formation is reversible, and the equilibrium constant determines the proportions of reactants and products present at equilibrium. For simple aldehydes, the equilibrium constants generally favor the acetal products. For example, the acid-catalyzed reaction of acetaldehyde with ethanol gives a good yield of the acetal.

With hindered aldehydes and with most ketones, the equilibrium constants favor the carbonyl compounds rather than the acetals. To enhance these reactions, the alcohol is often used as the solvent to assure a large excess. The water formed as a by-product is removed by distillation to force the equilibrium toward the right.

Conversely, most acetals are hydrolyzed simply by shaking them with dilute acid in water. The large excess of water drives the equilibrium toward the ketone or aldehyde. The mechanism is simply the reverse of acetal formation. For example, cyclohexanone dimethyl acetal is quantitatively hydrolyzed to cyclohexanone by brief treatment with dilute aqueous acid.

$$CH_3-O \quad O-CH_3 \qquad \xrightarrow[\text{H}^+, \text{ excess H}_2\text{O}]{} \qquad + \quad 2\ CH_3OH$$

PROBLEM 18-30

Propose a mechanism for the acid-catalyzed hydrolysis of cyclohexanone dimethyl acetal.

Fluocinolone acetonide is a steroid acetal used to treat skin conditions such as eczema and psoriasis. The acetal group decreases the water solubility of the parent steroid, thereby enhancing its potency and allowing for a longer duration of action.

fluocinolone acetonide

Cyclic Acetals   Formation of an acetal using a diol as the alcohol gives a cyclic acetal. Cyclic acetals often have more favorable equilibrium constants, since there is a smaller entropy loss when two molecules (a ketone and a diol) condense than when three molecules (a ketone and two molecules of an alcohol) condense. Ethylene glycol is often used to make cyclic acetals; its acetals are called **ethylene acetals** (or **ethylene ketals**). Dithiane (Section 18-8) and its alkylated derivatives are examples of cyclic thioacetals (sulfur acetals).

benzaldehyde + ethylene glycol $\xrightarrow{\text{H}^+}$ benzaldehyde ethylene acetal + H_2O

formaldehyde + propane-1,3-dithiol $\xrightarrow{\text{H}^+}$ dithiane + H_2O

Carbohydrates   Sugars and other carbohydrates most commonly exist as cyclic acetals and hemiacetals. For example, glucose is a six-carbon sugar that is most stable as a hemiacetal. Lactose is a disaccharide (composed of two sugar units) that has one acetal and one hemiacetal. We discuss the structures of carbohydrates in detail in Chapter 23.

glucose (open chain)   glucose (cyclic hemiacetal)   lactose

PROBLEM 18-31

Show what alcohols and carbonyl compounds give the following derivatives:

(a) CH₃CH₂O OCH₂CH₃

(b)

(c)

(d)

(e)

(f)

PROBLEM-SOLVING STRATEGY
Proposing Reaction Mechanisms

Here we apply the general principles for proposing reaction mechanisms to the hydrolysis of an acetal. These principles were introduced in Chapters 7 and 11 and are summarized in Appendix 4. Remember that you should draw all the bonds and substituents of each carbon atom involved in a mechanism. Show each step separately, using curved arrows to show the movement of electron pairs (from the nucleophile to the electrophile).

Our problem is to propose a mechanism for the acid-catalyzed hydrolysis of the following acetal:

The type of mechanism is stated to be acid-catalyzed. Therefore, we assume it involves strong electrophiles and cationic intermediates (possibly carbocations), but no strong nucleophiles or strong bases and certainly no carbanions or free radicals.

1. **Consider the carbon skeletons of the reactants and products, and decide which carbon atoms in the products are likely derived from which carbon atoms in the reactants.**
   First you must decide what products are formed by hydrolysis of the acetal. In dealing with acetals and hemiacetals, any carbon atom with *two* bonds to oxygen is derived from a carbonyl group. Draw an equation showing all the affected atoms. The equation shows that water must somehow add (probably by a nucleophilic attack), and the ring must be cleaved.

2. **Consider whether any of the reactants is a strong enough electrophile to react without being activated. If not, consider how one of the reactants might be converted to a strong electrophile by protonation of a Lewis basic site (or complexation with a Lewis acid).**
   The reactant probably will not react with water until it is activated, most likely by protonation. It can become protonated at either oxygen atom. We will arbitrarily choose the

ring oxygen for protonation. The protonated compound is well suited for ring cleavage to form a stabilized (and strongly electrophilic) cation.

protonation   cleavage   resonance-stabilized cation

3. Consider how a nucleophilic site on another reactant can attack the strong electrophile to form a bond needed in the product. Draw the product of this bond formation.
Attack by water on the cation gives a protonated hemiacetal.

attack by water   hemiacetal
deprotonation

4. Consider how the product of nucleophilic attack might be converted to the final product (if it has the right carbon skeleton) or reactivated to form another bond needed in the product.
Just as an —OH group can be lost by protonation and loss of water, the —OCH_3 group can be lost by protonating it and losing methanol. A protonated version of the product results.

protonation

resonance-stabilized   deprotonation
intermediate   products

5. Draw all the steps of the mechanism, using curved arrows to show the movement of electrons.
The complete mechanism is given by combining the equations written immediately above. You should write out the mechanism to review the steps involved.

    As further practice in proposing reaction mechanisms, do Problems 18-32 and 18-33 by completing the five steps listed in this section.

PROBLEM 18-32

In the mechanism for acetal hydrolysis shown above, the ring oxygen atom was protonated first, the ring was cleaved, and then the methoxyl group was lost. The mechanism could also be written to show the methoxyl oxygen protonating and cleaving first, followed by ring cleavage. Draw this alternative mechanism.

PROBLEM 18-33

(a) Propose a mechanism for the acid-catalyzed reaction of cyclohexanone with ethylene glycol to give cyclohexanone ethylene acetal.
(b) Propose a mechanism for the acid-catalyzed hydrolysis of cyclohexanone ethylene acetal.

PROBLEM-SOLVING HINT
The mechanism of a reverse reaction is normally the reverse of the mechanism of the forward reaction, as long as they take place under similar conditions. If you know the mechanism for formation of an acetal, you can always write the mechanism for its hydrolysis, using the same intermediates in reverse order.

(c) Compare the mechanisms you drew in parts (a) and (b). How similar are these mechanisms, comparing them in reverse order?

(d) Propose a mechanism for the acid-catalyzed hydrolysis of the compound given in Problem 18-31(f).

Acetals hydrolyze under acidic conditions, but they are stable to strong bases and nucleophiles. Acetals are easily made from the corresponding ketones and aldehydes and easily converted back to the parent carbonyl compounds. This easy interconversion makes acetals attractive as **protecting groups** to prevent ketones and aldehydes from reacting with strong bases and nucleophiles.

As an example, consider the proposed synthesis below. The necessary Grignard reagent could not be made because the carbonyl group would react with the nucleophilic organometallic group.

18-19
Use of Acetals as Protecting Groups

Proposed synthesis

If the aldehyde is protected as an acetal, however, it is unreactive toward a Grignard reagent. The "masked" aldehyde is converted to the Grignard reagent, which is allowed to react with cyclohexanone. Dilute aqueous acid both protonates the alkoxide to give the alcohol and hydrolyzes the acetal to give the deprotected aldehyde.

Actual synthesis

Selective Acetal Formation   Because aldehydes form acetals more readily than ketones do, an aldehyde can be protected selectively in the presence of a ketone. This selective protection leaves the ketone available for modification under neutral or basic conditions without disturbing the more reactive aldehyde group. The following example shows the reduction of a ketone in the presence of a more reactive aldehyde:

PROBLEM 18-34

Show how you would accomplish the following syntheses. You may use whatever additional reagents you need.

(a)

(b)

(c)

(d)

(e)

(f) $BrCH_2CH_2CCH_3$ ⟶ $HC{\equiv}CCH_2CH_2CCH_3$

18-20
Oxidation of Aldehydes

Unlike ketones, aldehydes are easily oxidized to carboxylic acids by common oxidants such as chromic acid, chromium trioxide, permanganate, and peroxy acids. Aldehydes oxidize so easily that air must be excluded from their containers to avoid slow oxidation by atmospheric oxygen. Because aldehydes oxidize so easily, mild reagents such as Ag_2O can oxidize them selectively in the presence of other oxidizable functional groups.

A Tollens test (p. 819) is usually done on a small scale, but it can also create a silver mirror on a large object.

$$R{-}\overset{O}{\overset{\|}{C}}{-}H \xrightarrow[\text{(oxidizing agent)}]{[O]} R{-}\overset{O}{\overset{\|}{C}}{-}OH$$

Examples

$$CH_3{-}\underset{\underset{CH_3}{|}}{CH}{-}\overset{O}{\overset{\|}{C}}{-}H \xrightarrow[\text{dil } H_2SO_4]{Na_2Cr_2O_7} CH_3{-}\underset{\underset{CH_3}{|}}{CH}{-}\overset{O}{\overset{\|}{C}}{-}OH$$

isobutyraldehyde   isobutyric acid (90%)

$$\xrightarrow[\text{THF/H}_2\text{O}]{Ag_2O}$$

(97%)

Silver ion, Ag^+, oxidizes aldehydes selectively in a convenient functional group test for aldehydes. The **Tollens test** involves adding a solution of silver–ammonia complex (the **Tollens reagent**) to the unknown compound. If an aldehyde is present, its oxidation reduces silver ion to metallic silver in the form of a black suspension or a silver mirror deposited on the inside of the container. Simple hydrocarbons, ethers, ketones, and even alcohols do not react with the Tollens reagent.

$$R\overset{O}{\underset{}{-\!\!\overset{\|}{C}\!\!-}}H \;+\; 2\,Ag(NH_3)_2^+ \;+\; 3\,^-OH \quad\xrightarrow{H_2O}\quad 2\,Ag\downarrow \;+\; R\overset{O}{\underset{}{-\!\!\overset{\|}{C}\!\!-}}O^- \;+\; 4\,NH_3 \;+\; 2\,H_2O$$

aldehyde   Tollens reagent   silver   carboxylate

PROBLEM 18-35

Predict the major products of the following reactions:

(a) [structure: 4-hydroxycyclohexanecarbaldehyde, HO on one position, CHO on the para position] + Ag_2O

(b) [structure: 4-hydroxycyclohexanecarbaldehyde, HO and CHO] + $K_2Cr_2O_7/H_2SO_4$

(c) [structure: cyclohexanone ring with CHO substituent and ketone O] + $Ag(NH_3)_2^+\,^-OH$

(d) [structure: cyclohexene ring with CHO substituent] + $KMnO_4$ (cold, dilute)

18-21A Catalytic Hydrogenation

Ketones and aldehydes are most commonly reduced by sodium borohydride (see Sections 10-11 and 18-12). Sodium borohydride ($NaBH_4$) reduces ketones to secondary alcohols and aldehydes to primary alcohols. Lithium aluminum hydride ($LiAlH_4$) also accomplishes these reductions, but it is a more powerful reducing agent and more difficult to work with.

Like alkene double bonds, carbonyl double bonds can be reduced by catalytic hydrogenation. Catalytic hydrogenation is slower with carbonyl groups than with olefinic double bonds, however. Before sodium borohydride was available, catalytic hydrogenation was often used to reduce aldehydes and ketones, but any olefinic double bonds were reduced as well. In the laboratory, we prefer sodium borohydride over catalytic reduction because it reduces ketones and aldehydes faster than olefins, and no gas-handling equipment is required. Catalytic hydrogenation is still widely used in industry, however, because H_2 is much cheaper than $NaBH_4$, and pressure equipment is more readily available there.

The most common catalyst for catalytic hydrogenation of ketones and aldehydes is **Raney nickel**. Raney nickel is a finely divided hydrogen-bearing form of nickel made by treating a nickel–aluminum alloy with a strong sodium hydroxide solution. The aluminum in the alloy reacts to form hydrogen, leaving behind a finely divided nickel powder saturated with hydrogen. Pt and Rh catalysts are also used for hydrogenation of ketones and aldehydes.

18-21
Other Reductions of Ketones and Aldehydes

$$\xrightarrow[\text{(Raney nickel)}]{\text{Ni-H}_2} \qquad \xrightarrow{\text{Ni-H}_2} \qquad$$

(90%)

18-21B   Deoxygenation of Ketones and Aldehydes

A *deoxygenation* replaces the carbonyl oxygen atom of a ketone or aldehyde with two hydrogen atoms, reducing the carbonyl group past the alcohol stage all the way to a methylene group. The following equation compares deoxygenation with the common hydride reductions that give alcohols.

Clemmensen Reduction (Review)

The **Clemmensen reduction** is most commonly used to convert acylbenzenes (from Friedel–Crafts acylation, Section 17-11B) to alkylbenzenes, but it also works with other ketones and aldehydes that are not sensitive to acid. The carbonyl compound is heated with an excess of amalgamated zinc (zinc treated with mercury) and hydrochloric acid. The actual reduction occurs by a complex mechanism on the surface of the zinc.

$$CH_3-(CH_2)_5-CHO \xrightarrow[\text{HCl, H}_2\text{O}]{\text{Zn(Hg)}} CH_3-(CH_2)_5-CH_3$$

heptanal   *n*-heptane (72%)

Wolff–Kishner Reduction

Compounds that cannot survive treatment with hot acid can be deoxygenated using the **Wolff–Kishner reduction**. The ketone or aldehyde is converted to its hydrazone, which is heated with a strong base such as KOH or potassium *t*-butoxide. Ethylene glycol, diethylene glycol, or another high-boiling solvent is used to facilitate the high temperature needed in the second step.

Examples

The mechanism for formation of the hydrazone is the same as the mechanism for imine formation (Section 18-16). The actual reduction step involves two tautomeric proton transfers from nitrogen to carbon. In this strongly basic solution, we expect a proton transfer from N to C to occur by loss of a proton from nitrogen, followed by reprotonation on carbon. A second deprotonation sets up the intermediate for loss of nitrogen to form a carbanion. This carbanion is quickly reprotonated to give the product.

MECHANISM   Wolff–Kishner Reduction

Step 1: Deprotonation from N.

Step 2: Reprotonation on C.

Step 3: Remove second proton from N.

Step 4: Lose N_2.

Step 5: Protonate.

PROBLEM 18-36

Propose a mechanism for both parts of the Wolff–Kishner reduction of cyclohexanone: the formation of the hydrazone, then the base-catalyzed reduction with evolution of nitrogen gas.

PROBLEM 18-37

Predict the major products of the following reactions:

(a)

$$\xrightarrow[\text{HCl, H}_2\text{O}]{\text{Zn(Hg)}}$$

(b)

$$\xrightarrow[\text{(2) KOH, heat}]{\text{(1) H}_2\text{NNH}_2}$$

(c)

$$\xrightarrow[\text{(2) KOH, heat}]{\text{(1) N}_2\text{H}_4}$$

(d)

$$\xrightarrow[\text{HCl, H}_2\text{O}]{\text{Zn(Hg)}}$$

SUMMARY   Reactions of Ketones and Aldehydes

1. *Addition of organometallic reagents* (Sections 9-7B and 10-9)

$$\underset{\text{(metal = MgX, Li, Na, etc.)}}{R-\overset{\overset{\displaystyle O}{\|}}{C}-R' \ + \ R''-\text{metal}} \ \longrightarrow \ \underset{\text{alkoxide}}{R-\overset{\overset{\displaystyle O^{-+}\text{metal}}{|}}{\underset{\underset{\displaystyle R''}{|}}{C}}-R'} \ \xrightarrow{\text{H}_3\text{O}^+} \ \underset{\text{alcohol}}{R-\overset{\overset{\displaystyle OH}{|}}{\underset{\underset{\displaystyle R''}{|}}{C}}-R'}$$

2. *Reduction* (Sections 10-12 and 18-21A)

$$R-\overset{\overset{\displaystyle O}{\|}}{C}-R' + NaBH_4 \text{ (or LiAlH}_4) \longrightarrow R-\overset{\overset{\displaystyle O^-}{|}}{\underset{\underset{\displaystyle H}{|}}{C}}-R' \xrightarrow{H^+} R-\overset{\overset{\displaystyle OH}{|}}{\underset{\underset{\displaystyle H}{|}}{C}}-R$$

ketone or aldehyde   (or H$_2$/Raney nickel)   alkoxide   alcohol

3. *The Wittig reaction* (Section 18-13)

$$Ph_3\overset{+}{P}-\overset{\overset{\displaystyle R}{|}}{\underset{\underset{\displaystyle R}{|}}{C}\!:^-} + \overset{\underset{\displaystyle R'}{}}{\underset{\displaystyle R'}{}}C=O \longrightarrow \overset{\overset{\displaystyle R}{|}}{\underset{\underset{\displaystyle R}{|}}{C}}=\overset{\overset{\displaystyle R'}{|}}{\underset{\underset{\displaystyle R'}{|}}{C}} + Ph_3P=O$$

phosphorus ylide   ketone or aldehyde   alkene

Example

$$Ph_3\overset{+}{P}-\overset{\overset{\displaystyle H}{|}}{\underset{\underset{\displaystyle CH_3}{|}}{C}\!:^-} + \text{(cyclopentanone)} \longrightarrow \text{(cyclopentane)}=\overset{\overset{\displaystyle H}{|}}{\underset{\underset{\displaystyle CH_3}{|}}{C}} + Ph_3P=O$$

4. *Hydration* (Section 18-14)

$$R-\overset{\overset{\displaystyle O}{\|}}{C}-R' + H_2O \rightleftharpoons R-\overset{\overset{\displaystyle HO}{|}}{\underset{}{C}}\overset{OH}{}-R'$$

ketone or aldehyde   hydrate

Example

$$Cl_3C-\overset{\overset{\displaystyle O}{\|}}{C}-H + H_2O \rightleftharpoons Cl_3C-\overset{HOOH}{\underset{}{C}}-H$$

chloral   chloral hydrate

5. *Formation of cyanohydrins* (Section 18-15)

$$R-\overset{\overset{\displaystyle O}{\|}}{C}-R' + HCN \underset{^-CN}{\rightleftharpoons} R-\overset{HOCN}{\underset{}{C}}-R'$$

ketone or aldehyde   cyanohydrin

Example

$$CH_3CH_2CH_2-\overset{\overset{\displaystyle O}{\|}}{C}-H \xrightarrow[^-CN]{HCN} CH_3CH_2CH_2-\overset{HOCN}{\underset{}{C}}-H$$

butanal   butanal cyanohydrin

6. *Formation of imines* (Section 18-16)

$$R-\overset{\overset{\displaystyle O}{\|}}{C}-R' + R''-NH_2 \underset{}{\overset{H^+}{\rightleftharpoons}} R-\overset{\overset{\displaystyle N-R''}{\|}}{C}-R' + H_2O$$

ketone or aldehyde   primary amine   imine (Schiff base)

Example

cyclopentanone   + CH_3-NH_2 $\overset{H^+}{\rightleftharpoons}$ cyclopentanone methyl imine

methylamine

7. *Formation of oximes and hydrazones* (Section 18-17)

$$R-\overset{\overset{\displaystyle O}{\|}}{C}-R' \quad + \quad H_2N-OH \quad \xrightarrow{H^+} \quad R-\overset{\overset{\displaystyle N-OH}{\|}}{C}-R'$$

ketone or aldehyde   hydroxylamine   oxime

$$R-\overset{\overset{\displaystyle O}{\|}}{C}-R' \quad + \quad H_2N-NH-R'' \quad \xrightarrow{H^+} \quad R-\overset{\overset{\displaystyle N-NH-R''}{\|}}{C}-R'$$

ketone or aldehyde   hydrazine reagent   hydrazone derivative

$R'' =$	Reagent Name	Derivative Name
—H	hydrazine	hydrazone
—Ph	phenylhydrazine	phenylhydrazone
$-\overset{\overset{\displaystyle O}{\|}}{C}-NH_2$	semicarbazide	semicarbazone

Example

benzaldehyde   phenylhydrazine   benzaldehyde phenylhydrazone

$$+ \quad H_2N-NHPh \quad \xrightarrow{H^+} \quad \cdots \quad + \quad H_2O$$

8. *Formation of acetals* (Section 18-18)

$$R-\overset{\overset{\displaystyle O}{\|}}{C}-R' \quad + \quad 2\,R''-OH \quad \rightleftharpoons \quad R-\overset{\overset{\displaystyle R''O\quad OR''}{\diagdown\,\diagup}}{C}-R' \quad + \quad H_2O$$

ketone (aldehyde)   alcohol   acetal

Example

benzaldehyde   ethylene glycol   benzaldehyde ethylene acetal

$$+ \quad \begin{matrix} CH_2-CH_2 \\ | \qquad | \\ OH \quad\; OH \end{matrix} \quad \rightleftharpoons \quad \cdots \quad + \quad H_2O$$

9. *Oxidation of aldehydes* (Section 18-20)

$$R-\overset{\overset{\displaystyle O}{\|}}{C}-H \quad \xrightarrow{\text{chromic acid, permanganate, Ag}^+,\text{ etc.}} \quad R-\overset{\overset{\displaystyle O}{\|}}{C}-OH$$

aldehyde   acid

Example

benzaldehyde   chromic acid   benzoic acid

$$\text{CHO} \quad + \quad Na_2Cr_2O_7/H_2SO_4 \quad \longrightarrow \quad \text{COOH}$$

Tollens test

$$R-\overset{\overset{\displaystyle O}{\|}}{C}-H \; + \; 2\,Ag(NH_3)_2^+ \; + \; 3\;^-OH \xrightarrow{H_2O} 2\,Ag\downarrow \; + \; R-\overset{\overset{\displaystyle O}{\|}}{C}-O^- \; + \; 4\,NH_3 \; + \; 2\,H_2O$$

aldehyde   Tollens reagent   silver   carboxylate

10. *Deoxygenation reactions*
   a. *Clemmensen reduction* (Sections 17-11B and 18-21B)

$$R-\overset{\overset{\displaystyle O}{\|}}{C}-R' \;+\; Zn(Hg) \;\xrightarrow{\;HCl\;}\; R-\overset{\overset{\displaystyle H}{|}}{\underset{\underset{\displaystyle H}{|}}{C}}-R'$$

ketone or aldehyde

   b. *Wolff–Kishner reduction* (Section 18-21B)

$$R-\overset{\overset{\displaystyle O}{\|}}{C}-R' \;+\; H_2N-NH_2 \;\longrightarrow\; R-\overset{\overset{\displaystyle N-NH_2}{\|}}{C}-R' \;\xrightarrow[\text{heat}]{\;KOH\;}\; R-\overset{\overset{\displaystyle H}{|}}{\underset{\underset{\displaystyle H}{|}}{C}}-R' \;+\; H_2O \;+\; N\equiv N\uparrow$$

ketone or aldehyde   hydrazine   hydrazone

Example

cyclohexanone

$$\xrightarrow[\text{(2) KOH, heat}]{\text{(1) } H_2N-NH_2}$$

cyclohexane

11. *Reactions of ketones and aldehydes at their α positions*
This large group of reactions is covered in Chapter 22.

Example

Aldol condensation

$$2\;CH_3-\overset{\overset{\displaystyle O}{\|}}{C}-H \;\xrightarrow{\;\text{base}\;}\; CH_3-\overset{\overset{\displaystyle OH}{|}}{\underset{\underset{\displaystyle H}{|}}{C}}-CH_2-\overset{\overset{\displaystyle O}{\|}}{C}-H$$

Chapter 18
Glossary

acetal A derivative of an aldehyde or ketone having two alkoxy groups in place of the carbonyl group. The acetal of a ketone is sometimes called a **ketal**. (p. 812)

$$CH_3-\overset{\overset{\displaystyle O}{\|}}{C}-H \;+\; 2\;CH_3OH \;\underset{}{\overset{H^+}{\rightleftharpoons}}\; CH_3-\overset{\overset{\displaystyle CH_3O\;\;\;OCH_3}{\diagdown\;\diagup}}{C}-H \;+\; H_2O$$

acetaldehyde   acetaldehyde
dimethyl acetal

ethylene acetal (ethylene ketal): A cyclic acetal using ethylene glycol as the alcohol. (p. 814)
aldehyde A compound containing a carbonyl group bonded to an alkyl (or aryl) group and a hydrogen atom. (p. 774)
carbinolamine An intermediate in the formation of an imine, having an amine and a hydroxyl group bonded to the same carbon atom. (p. 808)

$$R-\overset{\overset{\displaystyle O}{\|}}{C}-R \;+\; R'-NH_2 \;\rightleftharpoons\; \left[\;R-\overset{\overset{\displaystyle HO\;\;\;NH-R'}{\diagdown\;\diagup}}{C}-R\;\right] \;\rightleftharpoons\; R-\overset{\overset{\displaystyle N-R'}{\|}}{C}-R \;+\; H_2O$$

carbinolamine   imine

carbonyl group The C=O functional group. (p. 774)
Clemmensen reduction The deoxygenation of a ketone or aldehyde by treatment with zinc amalgam and dilute HCl. (p. 820)
condensation A reaction in which two or more organic compounds are joined, often with the loss of a small molecule such as water. (p. 807)

cyanohydrin A compound with a hydroxyl group and a cyano group on the same carbon atom. Cyanohydrins are generally made by the reaction of a ketone or aldehyde with HCN. (p. 806)

$$CH_3-\underset{\underset{\text{acetone}}{}}{\overset{\overset{O}{\|}}{C}}-CH_3 \quad + \quad HCN \quad \rightleftharpoons \quad CH_3-\underset{\underset{\text{acetone cyanohydrin}}{}}{\overset{\overset{HO\quad CN}{\diagdown\,\diagup}}{C}}-CH_3$$

dithiane (1,3-dithiane) A thioacetal of formaldehyde that is sufficiently acidic to be deprotonated by exceptionally strong bases. See Section 18-8. (p. 791)

propane-1,3-thiol   formaldehyde   1,3-dithiane   dithiane anion

enol A vinyl alcohol. Simple enols generally tautomerize to their keto forms. (p. 790)

enol   keto

hemiacetal A derivative of an aldehyde or ketone similar to an acetal, but with just one alkoxy group and one hydroxyl group on the former carbonyl carbon atom. (p. 812)
hydrate (of an aldehyde or ketone) The geminal diol formed by addition of water across the carbonyl double bond. (p. 804)

$$Cl_3C-\underset{\underset{\text{chloral}}{}}{\overset{\overset{O}{\|}}{C}}-H \quad + \quad H_2O \quad \underset{\longleftarrow}{\overset{H^+ \text{ or } {}^-OH}{\longrightarrow}} \quad Cl_3C-\underset{\underset{\text{chloral hydrate}}{}}{\overset{\overset{HO\quad OH}{\diagdown\,\diagup}}{C}}-H$$

hydrazone A compound containing the $C{=}N{-}NH_2$ group, formed by the reaction of a ketone or aldehyde with hydrazine. (p. 810)
  2,4-DNP derivative: A hydrazone made using 2,4-dinitrophenylhydrazine. (p. 810)

cyclopentanone   2,4-DNP derivative of cyclopentanone

imine A compound with a carbon–nitrogen double bond, formed by the reaction of a ketone or aldehyde with a primary amine. A substituted imine is often called a **Schiff base**. (p. 807)

ketal A common name for the acetal of a ketone. The term *ketal* was recently eliminated from the IUPAC nomenclature. (p. 812)
ketone A compound containing a carbonyl group bonded to two alkyl or aryl groups. (p. 774)
lithium dialkylcuprate (Gilman reagent) An organometallic reagent that couples with alkyl halides and acyl halides (acid chlorides). (p. 794)

$$R_2CuLi \quad + \quad R'-\overset{\overset{O}{\|}}{C}-Cl \quad \longrightarrow \quad R'-\overset{\overset{O}{\|}}{C}-R \quad + \quad R-Cu \quad + \quad LiCl$$

McLafferty rearrangement In mass spectrometry, the loss of an alkene fragment by a cyclic rearrangement of a carbonyl compound having γ hydrogens. (p. 783)

nitrile A compound containing the cyano group, $C\equiv N$. (p. 793)

nucleophilic addition Addition of a reagent across a multiple bond by attack of a nucleophile at the electrophilic end of the multiple bond. As used in this chapter, the addition of a nucleophile and a proton across the $C=O$ bond. (p. 797)

oxime A compound containing the $C=N-OH$ group, formed by the reaction of a ketone or aldehyde with hydroxylamine. (p. 810)

protecting group A group used to prevent a sensitive functional group from reacting while another part of the molecule is being modified. The protecting group is later removed. (p. 817)

Raney nickel A finely divided, hydrogen-bearing form of nickel made by treating a nickel–aluminum alloy with strong sodium hydroxide. The aluminum in the alloy reacts to form hydrogen, leaving a finely divided nickel powder saturated with hydrogen. (p. 819)

semicarbazone A compound containing the $C=N-NH-CONH_2$ group, formed by the reaction of a ketone or aldehyde with semicarbazide. (p. 810)

Tollens test A test for aldehydes: Adding the **Tollens reagent**, a silver–ammonia complex $[Ag(NH_3)_2{}^+\ {}^-OH]$, gives a carboxylate salt and a silver mirror on the inside of a glass container. (p. 819)

Wittig reaction Reaction of an aldehyde or ketone with a phosphorus ylide to form an alkene. One of the most versatile syntheses of alkenes. (p. 800)

$$\underset{\text{ketone or aldehyde}}{\overset{R}{\underset{R}{>}}C=O} \quad + \quad \underset{\text{phosphorus ylide}}{\overset{R'}{\underset{R'}{>}}\overset{-}{C}\overset{+}{-}\overset{Ph}{\underset{Ph}{P}}-Ph} \quad \longrightarrow\longrightarrow \quad \underset{\text{alkene}}{\overset{R}{\underset{R}{>}}C=C\overset{R'}{\underset{R'}{<}}} \quad + \quad Ph_3P=O$$

ylide An uncharged molecule containing a carbon atom with a negative charge bonded to a heteroatom with a positive charge. A phosphorus ylide is the nucleophilic species in the Wittig reaction. (p. 801)

Wolff–Kishner reduction Deoxygenation of a ketone or aldehyde by conversion to the hydrazone, followed by treatment with a strong base. (p. 820)

Essential Problem-Solving Skills in Chapter 18

1. Name ketones and aldehydes, and draw the structures from their names.

2. Interpret the IR, NMR, UV, and mass spectra of ketones and aldehydes, and use spectral information to determine the structures.

3. Write equations for syntheses of ketones and aldehydes from alcohols, alkenes, alkynes, carboxylic acids, nitriles, acid chlorides, dithianes, and aromatic compounds.

4. Propose effective single-step and multistep syntheses of ketones and aldehydes.

5. Predict the products of reactions of ketones and aldehydes with the following types of compounds (give mechanisms where appropriate):
   (a) hydride reducing agents; Clemmensen and Wolff–Kishner reagents
   (b) Grignard and organolithium reagents
   (c) phosphorus ylides
   (d) water
   (e) hydrogen cyanide
   (f) ammonia and primary amines
   (g) hydroxylamine and hydrazine derivatives
   (h) alcohols
   (i) oxidizing agents

6. Use your knowledge of the mechanisms of ketone and aldehyde reactions to propose mechanisms and products of similar reactions you have never seen before.

7. Show how to convert ketones and aldehydes to other functional groups.

8. Use retrosynthetic analysis to propose effective multistep syntheses using ketones and aldehydes as intermediates and protecting the carbonyl group if necessary.

Study Problems

18-38   Define each term and give an example.
- **(a)** ketone
- **(b)** aldehyde
- **(c)** enol form
- **(d)** cyanohydrin
- **(e)** imine
- **(f)** hydrate
- **(g)** carbinolamine
- **(h)** oxime
- **(i)** phenylhydrazone
- **(j)** 2,4-DNP derivative
- **(k)** semicarbazone
- **(l)** acetal
- **(m)** Wittig reaction
- **(n)** 1,3-dithiane
- **(o)** hemiacetal
- **(p)** Tollens test
- **(q)** Wolff–Kishner reduction
- **(r)** Clemmensen reduction
- **(s)** ketal
- **(t)** ethylene acetal

18-39   Name the following ketones and aldehydes. When possible, give both a common name and an IUPAC name.
- **(a)** $CH_3CO(CH_2)_4CH_3$
- **(b)** $CH_3(CH_2)_2CO(CH_2)_2CH_3$
- **(c)** $CH_3(CH_2)_5CHO$
- **(d)** PhCOPh
- **(e)** $CH_3CH_2CH_2CHO$
- **(f)** CH_3COCH_3
- **(g)** $CH_3CH_2CHBrCH_2CH(CH_3)CHO$
- **(h)** $Ph-CH{=}CH-CHO$
- **(i)** $CH_3CH{=}CH-CH{=}CH-CHO$
- **(j)** $CH_3CH_2COCH_2CHO$

(k)

(l)

18-40   Rank the following carbonyl compounds in order of increasing equilibrium constant for hydration:

$$CH_3COCH_2Cl \quad ClCH_2CHO \quad CH_2O \quad CH_3COCH_3 \quad CH_3CHO$$

18-41   Sketch the expected proton NMR spectrum of 3,3-dimethylbutanal.

18-42   Predict the values of λ_{max} for the $\pi \rightarrow \pi^*$ and $n \rightarrow \pi^*$ transitions in the UV spectrum of 3-methylcyclohex-2-enone.

18-43   A compound of formula $C_6H_{10}O_2$ shows only two absorptions in the proton NMR: a singlet at 2.67 ppm and a singlet at 2.15 ppm. These absorptions have areas in the ratio 2:3. The IR spectrum shows a strong absorption at 1708 cm^{-1}. Propose a structure for this compound.

18-44   The proton NMR spectrum of a compound of formula $C_{10}H_{12}O$ appears below. This compound reacts with an acidic solution of 2,4-dinitrophenylhydrazine to give a crystalline derivative, but it gives a negative Tollens test. Propose a structure for this compound and give peak assignments to account for the absorptions in the spectrum.

18-45   The following compounds undergo McLafferty rearrangement in the mass spectrometer. Predict the masses of the resulting charged fragments.
- **(a)** pentanal
- **(b)** 3-methyl-2-pentanone
- **(c)** 3-methylpentanal
- **(d)** $CH_3CH_2CH_2-\overset{\overset{\displaystyle O}{\|}}{C}-OCH_3$ (methyl butyrate)

18-46   An unknown compound gives a positive 2,4-dinitrophenylhydrazine test and a negative Tollens test. Its mass spectrum shows prominent ions at m/z 128, 100, 86, 85, and 71. Show which of the following structures is most likely: 2-octanone; 4-octanone; 2-octen-3-ol; 5-propoxy-1-pentanol. Also show the fragmentations that lead to the observed ions.

18-47   An unknown compound gives a molecular ion of m/z 70 in the mass spectrum. It reacts with semicarbazide hydrochloride to give a crystalline derivative, but it gives a negative Tollens test. The NMR and IR spectra appear below. Propose a structure for this compound, and give peak assignments to account for the absorptions in the spectra. Explain why the signal at 1790 cm^{-1} in the IR spectrum appears at an unusual frequency.

18-48   The following compounds are easily differentiated by their UV spectra:

(a) Predict values of λ_{max} for these compounds.
(b) Explain how the UV spectra would be used to distinguish between these compounds.

18-49   For each compound,
(1) name the functional group
(2) show what compound(s) result from complete hydrolysis

(a)

(b) CH₃CH₂CH₂—C—CH₃ with CH₃O and OCH₃

(c) [structure]   **(d)** [structure]

(e) [cyclopentyl-O-CH(OCH₃)—CH₃]   **(f)** [dithiane structure]   **(g)** [pyrazine structure]   **(h)**

18-50   Propose mechanisms for the following reactions:

(a) CH₃—C(=O)—H →(PhNHNH₂, H⁺)→ CH₃—C(=N—NHPh)—H

(b) CH₃—C(=O)—H →(CH₃OH, H⁺)→ CH₃—C(OCH₃)(OCH₃... CH₃O OCH₃)—H

(c) [cyclohexanone] →(Ph₃P=CH₂)→ [methylenecyclohexane, =CH₂]

(d) [spiro ketal] →(H⁺, H₂O)→ [cyclopentanone] =O + HO—CH₂—CH₂—OH

(e) [pyrroline with N—H] →(H⁺, H₂O)→ [CHO with ⁺NH₃]

18-51   Show how you would accomplish the following syntheses efficiently and in good yield. You may use any necessary reagents.
(a)   acetaldehyde → lactic acid, CH₃CH(OH)COOH

(b) [cyclopentanone] → [benzylidene cyclopentane, CHPh]

(c) [cyclopentanone with CHO] → [cyclopentanone with CH₂OH]

(d) [cyclopentanone with CHO] → [cyclopentanol with CHO, OH]

(e) [cyclopentanone with CHO] → [cyclopentanone with CHCH₂CH₃, CHO]

(f) →

(g) → [decalin with OH, H]

(h) [octalone with O] → [decalin with OH, H]

18-52   Show how you would synthesize the following derivatives from appropriate carbonyl compounds:

(a) [cyclobutanone N—OH structure]

(b) [benzaldehyde N-cyclopentyl imine]

(c) [benzyl N-cyclopentyl imine]

(d) [indane spiro dioxolane]   **(e)** [cyclohexyl N= isopropyl imine]   **(f)** [cyclopentane with CH₃O and OCH₃]

18-53   Draw structures of the following derivatives.
(a)   the 2,4-dinitrophenylhydrazone of benzaldehyde
(b)   the semicarbazone of cyclobutanone
(c)   cyclopropanone oxime
(d)   the ethylene ketal of 3-hexanone
(e)   acetaldehyde dimethyl acetal
(f)   the methyl hemiacetal of formaldehyde
(g)   the (E) isomer of the ethyl imine of propiophenone
(h)   the dithiane thioacetal of propanal

18-54   Section 18-8 covers the synthesis of aldehydes and ketones using 1,3-dithiane as a masked carbonyl group (the thioacetal of a carbonyl group). Like (oxygen) acetals, thioacetals hydrolyze in dilute acid. Thioacetals are somewhat more stable, however, and mercuric (Hg^{2+}) salts are often added as a specific Lewis acid to promote the hydrolysis.

(a) Show how you would make 2-methyl-1,3-dithiane from 1,3-dithiane.
(b) Propose a mechanism for the acid-catalyzed hydrolysis of 2-methyl-1,3-dithiane.
(c) Propose a mechanism for how Hg^{+2} might assist the hydrolysis.

2-methyl-1,3-dithiane

18-55   There are three dioxane isomers: 1,2-dioxane, 1,3-dioxane, and 1,4-dioxane. One of these acts like an ether and is an excellent solvent for Grignard reactions. Another one is potentially explosive when heated. The third one quickly hydrolyzes in dilute acid. Show which isomer acts like a simple ether, and explain why one of them is potentially explosive. Propose a mechanism for the acid hydrolysis of the third isomer.

1,2-dioxane   1,3-dioxane   1,4-dioxane

18-56   Predict the products formed when cyclohexanone reacts with the following reagents:
(a) CH_3NH_2, H^+ (b) excess CH_3OH, H^+
(c) hydroxylamine and weak acid   (d) ethylene glycol and *p*-toluenesulfonic acid
(e) phenylhydrazine and weak acid   (f) PhMgBr and then mild H_3O^+
(g) Tollens reagent   (h) sodium acetylide, then mild H_3O^+
(i) sodium cyanide   (j) acidic hydrolysis of the product from (i)
(k) hydrazine, then hot, fused KOH   (l) $Ph_3P{=}CH_2$

18-57   Both $NaBH_4$ and $NaBD_4$ are commercially available, and D_2O is common and inexpensive. Show how you would synthesize the following labeled compounds, starting with 2-butanone:

(a) $CH_3{-}\overset{\overset{\displaystyle OH}{|}}{\underset{\underset{\displaystyle D}{|}}{C}}{-}CH_2{-}CH_3$ (b) $CH_3{-}\overset{\overset{\displaystyle OD}{|}}{\underset{\underset{\displaystyle D}{|}}{C}}{-}CH_2{-}CH_3$ (c) $CH_3{-}\overset{\overset{\displaystyle OD}{|}}{\underset{\underset{\displaystyle H}{|}}{C}}{-}CH_2{-}CH_3$

18-58   When $LiAlH_4$ reduces 3-methylcyclopentanone, the product mixture contains 60% *cis*-3-methylcyclopentanol and 40% *trans*-3-methylcyclopentanol. Use your models and make three-dimensional drawings to explain this preference for the cis isomer.

18-59   Some Grignard reagents react with ethyl orthoformate, followed by acidic hydrolysis, to give aldehydes. Propose mechanisms for the two steps in this synthesis.

ethyl orthoformate   acetal

18-60   Show how you would accomplish the following syntheses:
(a) benzene $\longrightarrow$ *n*-butylbenzene
(b) benzonitrile $\longrightarrow$ propiophenone
(c) benzene $\longrightarrow$ *p*-methoxybenzaldehyde
(d) Ph$-$(CH$_2$)$_4$$-$OH $\longrightarrow$

tetralone

18-61   Predict the products formed when cyclohexanecarbaldehyde reacts with the following reagents:
(a) PhMgBr, then H_3O^+ (b) Tollens reagent   (c) semicarbazide and weak acid
(d) excess ethanol and acid   (e) 1,3-propanedithiol, H^+ (f) zinc amalgam and dilute hydrochloric acid

18-62   Show how you would synthesize 2-octanone from each compound. You may use any necessary reagents.
(a) heptanal   (b) 1-octyne   (c) 1,3-dithiane   (d) 2-octanol
(e) heptanoic acid   (f) $CH_3(CH_2)_5CN$ (g) 2,3-dimethyl-2-nonene

18-63   Show how you would synthesize octanal from each compound. You may use any necessary reagents.
(a) 1-octanol   (b) 1-nonene   (c) 1-octyne
(d) 1,3-dithiane   (e) 1,1-dichlorooctane   (f) octanoic acid

18-64   Hydration of alkynes (via oxymercuration) gives good yields of single compounds only with symmetrical or terminal alkynes. Show what the products would be from hydration of each compound.
(a) 3-hexyne   (b) 2-hexyne   (c) 1-hexyne   (d) cyclodecyne   (e) 3-methylcyclodecyne

18-65   Which of the following compounds would give a positive Tollens test? (Remember that the Tollens test involves mild basic aqueous conditions.)

(a) $CH_3CH_2CH_2COCH_3$

(b) $CH_3CH_2CH_2CH_2CHO$

(c) $CH_3CH{=}CH{-}CH{=}CH{-}OH$

(d) $CH_3CH_2CH_2CH_2{-}CH(OH)OCH_3$

(e) $CH_3CH_2CH_2CH_2{-}CH(OCH_3)_2$

(f)

18-66   Solving the following road-map problem depends on determining the structure of **A**, the key intermediate. Give structures for compounds **A** through **K**:

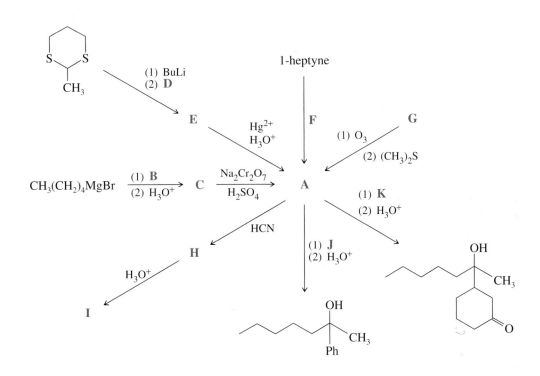

18-67   The UV spectrum of an unknown compound shows values of λ_{max} at 225 nm ($\varepsilon = 10,000$) and at 318 nm ($\varepsilon = 40$). The mass spectrum shows a molecular ion at m/z 96 and a prominent base peak at m/z 68. The IR and NMR spectra appear below. Propose a structure, and show how your structure corresponds to the observed absorptions. Propose a favorable fragmentation to account for the MS base peak at m/z 68 (loss of C_2H_4).

18-68   Two structures for the sugar **glucose** are shown on page 814. Interconversion of the open-chain and cyclic hemiacetal forms is catalyzed by either acid or base.

(a) Propose a mechanism for the cyclization, assuming a trace of acid is present.

(b) The cyclic hemiacetal is more stable than the open-chain form, so very little of the open-chain form is present at equilibrium. Will an aqueous solution of glucose reduce Tollens reagent and give a positive Tollens test? Explain.

18-69   Two structures of the sugar **fructose** are shown below. The cyclic structure predominates in aqueous solution.

(a) Number the carbon atoms in the cyclic structure. What is the functional group at C2 in the cyclic form?

(b) Propose a mechanism for the cyclization, assuming a trace of acid is present.

18-70   The following road-map problem centers on the structure and properties of **A**, a key intermediate in these reactions. Give structures for compounds **A** through **J**.

*18-71   A dithiane synthesis can convert an aldehyde to a ketone. The aldehyde is first converted to its dithiane derivative, which is deprotonated and alkylated. A mercuric chloride–assisted hydrolysis gives the ketone. Show how this technique might be used to convert benzaldehyde to benzyl phenyl ketone.

*18-72   Under acid catalysis, an alcohol reacts with dihydropyran to give the tetrahydropyranyl derivative (called a "THP ether") of the alcohol.

dihydropyran

tetrahydropyranyl derivative
R—O—THP, a "THP ether"

(a) Propose a mechanism for this reaction.
(b) The "THP ether" is not an ether. What functional group does it actually contain? How will it react under basic conditions? Under acidic conditions?
(c) Propose a mechanism for hydrolysis of the THP derivative in dilute aqueous acid, and predict the products.

*18-73   The mass spectrum of unknown compound **A** shows a molecular ion at m/z 116 and prominent peaks at m/z 87 and m/z 101. Its UV spectrum shows no maximum above 200 nm. The IR and NMR spectra of **A** are shown below. When **A** is washed with dilute aqueous acid, extracted into dichloromethane, and the solvent evaporated, the product **B** shows a strong carbonyl signal at 1715 cm^{-1} in the IR spectrum and a weak maximum at 274 nm ($\varepsilon = 16$) in the UV spectrum. The mass spectrum of **B** shows a molecular ion of m/z 72.
(a) Determine the structures of **A** and **B**, and show the fragmentation to account for the peaks at m/z 87 and 101.
(b) Propose a mechanism for the acid-catalyzed hydrolysis of **A** to **B**.

*18-74   (A true story.) The chemistry department custodian was cleaning the organic lab when an unmarked bottle fell off a shelf and smashed on the floor, leaving a puddle of volatile liquid. The custodian began to wipe up the puddle, but he was overcome with burning in his eyes and a feeling of having an electric drill thrust up his nose. He left the room and called the fire department, who used breathing equipment to go in to clean up the chemical. Three students were asked to identify the chemical quickly so the custodian could be treated and the chemical could be handled properly. The students took IR and NMR spectra, which appear below. The UV spectrum showed values of λ_{max} at 220 nm ($\varepsilon = 16{,}000$) and at 314 nm ($\varepsilon = 65$). The mass spectrometer was down, so no molecular weight was available. Determine the structure of this nasty compound, and show how your structure fits the spectra.

*18-75   Because of its great utility, the Wittig reaction has inspired interesting and useful variations, some with colorful names. The *Horner–Emmons modification* uses a phosphonate-stabilized carbanion in a Wittig reaction. This reagent is an anion that is more nucleophilic than the usual Wittig ylide. The by-product $(RO)_2PO_2^-$ is water-soluble and easily separated from the product. These phosphonate-stabilized carbanions are compatible with carbonyl groups (especially esters) in the reagent.

The reagent is prepared by the *Arbusov reaction:*

The Arbusov reaction can be carried out with an α-bromo ketone or α-bromo ester, in which case it is called the *Perkow reaction.*

(a) Propose a mechanism for the Horner–Emmons reaction.
(b) Propose a mechanism for the Arbusov reaction.
(c) Show how you would use these reactions for the following conversions:

(i)

(ii)

*18-76   (a) Simple aminoacetals hydrolyze quickly and easily in dilute acid. Propose a mechanism for hydrolysis of the following aminoacetal:

(b) The nucleosides that make up DNA have heterocyclic rings linked to deoxyribose by an aminoacetal functional group. Point out the aminoacetal linkages in deoxycytidine and deoxyadenosine.

deoxycytidine   deoxyadenosine

(c) Our genetic code depends on the stability of DNA. We are fortunate that the aminoacetal linkages of DNA are not easily cleaved. Show why your mechanism for part (a) does not work so well with deoxycytidine and deoxyadenosine.

CHAPTER 19

Amines

electrostatic potential
map of trimethylamine

19-1
Introduction

Amines are organic derivatives of ammonia with one or more alkyl or aryl groups bonded to the nitrogen atom. As a class, amines include some of the most important biological compounds. Amines serve many functions in living organisms, such as bioregulation, neuro-transmission, and defense against predators. Because of their high degree of biological activity, many amines are used as drugs and medicines. The structures and uses of some important biologically active amines are shown in Figure 19-1.

The *alkaloids* are an important group of biologically active amines, mostly synthesized by plants to protect them from being eaten by insects and other animals. The structures of some representative alkaloids are shown in Figure 19-2. Although some alkaloids are used medicinally (chiefly as painkillers), all alkaloids are toxic and cause death if taken in large quantities. The Greeks chose the alkaloid coniine to kill Socrates, although morphine, nicotine, or cocaine would have served equally well. Mild cases of alkaloid poisoning can produce psychological effects that resemble peacefulness, euphoria, or hallucinations. People seeking these effects often become addicted to alkaloids. Alkaloid addiction often ends in death; current estimates are over 400,000 deaths in the United States per year, including both natural alkaloids like nicotine and cocaine, and synthetic alkaloids like amphetamine.

(*S*)-coniine

dopamine
a neurotransmitter

epinephrine
an adrenal hormone

L-tryptophan
an amino acid

piperazine
kills intestinal worms

nicotinic acid
niacin, a vitamin

pyridoxine
vitamin B_6

histamine
dilates blood vessels

▲ **Figure 19-1**
Examples of some biologically active amines.

▲ Figure 19-2
Some representative alkaloids.

Amines are classified as **primary** (1°), **secondary** (2°), or **tertiary** (3°), corresponding to one, two, or three alkyl or aryl groups bonded to nitrogen.

19-2
Nomenclature of Amines

Primary (1°) amines   *Secondary (2°) amines*   *Tertiary (3°) amines*

cyclohexylamine (1°)   *tert*-butylamine (1°)   *N*-ethylaniline (2°)   piperidine (2°)   *N,N*-diethylaniline (3°)   quinuclidine (3°)

Quaternary ammonium salts have four alkyl or aryl bonds to a nitrogen atom. The nitrogen atom bears a positive charge, just as it does in simple ammonium salts such as ammonium chloride. The following are examples of quaternary (4°) ammonium salts:

tetraethylammonium iodide   *N*-butylpyridinium bromide   acetylcholine, a neurotransmitter

19-2A   Common Names

Common names of amines are formed from the names of the alkyl groups bonded to nitrogen, followed by the suffix *-amine*. The prefixes *di-*, *tri-*, and *tetra-* are used to describe two, three, or four identical substituents.

$CH_3CH_2\ddot{N}H_2$ $(CH_3\overset{CH_3}{\underset{}{CH}}CH_2CH_2)_2\ddot{N}H$ $(CH_3CH_2)_2\ddot{N}CH_3$ $(CH_3CH_2CH_2CH_2)_4N^{+-}Cl$

ethylamine   diisopentylamine   diethylmethylamine   tetrabutylammonium chloride

cyclohexyldimethylamine   benzylamine   diphenylamine

In naming amines with more complicated structures, the —NH_2 group is called the **amino** group. It is treated like any other substituent, with a number or other symbol indicating its position on the ring or carbon chain.

3-aminocyclopentene

γ-aminobutyric acid
(4-aminobutanoic acid)

trans-3-aminocyclohexanol

p-aminobenzoic acid (PABA)

Using this system, secondary and tertiary amines are named by classifying the nitrogen atom (together with its alkyl groups) as an alkylamino group. The largest or most complicated alkyl group is taken to be the parent molecule.

3-dimethylamino-1-hexanol

4-(ethylmethylamino)cyclohexanone

Aromatic and heterocyclic amines are generally known by historical names. Phenylamine is called *aniline*, for example, and its derivatives are named as derivatives of aniline.

aniline

3-ethylaniline

N,N-diethylaniline

4-methylaniline
or p-toluidine

Mitomycin C, an anticancer agent used to treat stomach and colon cancer, contains an aziridine ring. The aziridine functional group participates in the drug's degradation of DNA, resulting in the death of cancerous cells.

mitomycin C

We first considered nitrogen heterocycles in Section 16-9. The names and structures of some common ones are shown below. The heteroatom is usually assigned position number 1.

aziridine

pyrrole

pyrrolidine

1-methylpyrrolidine
(*N*-methylpyrrolidine)

imidazole

indole

pyridine

2-methylpyridine

piperidine

pyrimidine

purine

PROBLEM 19-1

Determine which of the heterocyclic amines shown above are aromatic. Give the reasons for your conclusions.

19-2B IUPAC Names

The IUPAC nomenclature for amines is similar to that for alcohols. The longest continuous chain of carbon atoms determines the root name. The *-e* ending in the alkane name is changed to *-amine*, and a number shows the position of the amino group along the chain. Other substituents on the carbon chain are given numbers, and the prefix *N-* is used for each substituent on nitrogen.

$$\overset{\overset{\displaystyle \ddot{N}H_2}{|}}{CH_3CH_2CHCH_3} \qquad \overset{\overset{\displaystyle CH_3 \;\; \ddot{N}H_2}{| \quad\;\; |}}{CH_3CHCH_2CH_2} \qquad \overset{\overset{\displaystyle \ddot{N}HCH_3}{|}}{CH_3CH_2CHCH_3} \qquad \overset{\overset{\displaystyle CH_3 \;\; CH_3}{| \quad\;\; |}}{CH_3CH_2CHCHCHCH_3}$$
$$\qquad\qquad\qquad\qquad\qquad\qquad\qquad\qquad\qquad\qquad\qquad\qquad\qquad\qquad \underset{\displaystyle :N(CH_3)_2}{|}$$

2-butanamine   3-methyl-1-butanamine   *N*-methyl-2-butanamine   2,4,*N*,*N*-tetramethyl-3-hexanamine

PROBLEM 19-2

Draw the structures of the following compounds:
(a) *t*-butylamine   (b) α-aminopropionaldehyde
(c) 4-(dimethylamino)pyridine   (d) 2-methylaziridine
(e) *N*-methyl-*N*-ethyl-3-hexanamine   (f) *m*-chloroaniline

PROBLEM 19-3

Give correct names for the following amines:

(a) $CH_3-CH_2-CH_2-\underset{\underset{\displaystyle NH_2}{|}}{CH}-CH_3$ (b) $CH_3-CH_2-\underset{\underset{\displaystyle NHCH_3}{|}}{CH}-CH_3$

(c) [structure: *m*-aminophenol — benzene ring with NH₂ and OH] (d) [structure: 3-methylpyrrole] (e) [structure: cyclopentane-1,2-diamine] (f) [structure: cyclohexane with NH₂ and CHO]

In Chapter 2, we saw that ammonia has a slightly distorted tetrahedral shape. A lone pair of nonbonding electrons occupies one of the tetrahedral positions. This geometry is represented by sp^3 hybridization of nitrogen, with the bulky lone pair compressing the H—N—H bond angles to 107° from the "ideal" sp^3 bond angle of 109.5°. Trimethylamine shows less angle compression because the bulky methyl groups open the angle slightly.

19-3
Structure of Amines

ammonia   trimethylamine

electrostatic potential
map for trimethylamine

The electrostatic potential map for trimethylamine shows how the nonbonding electrons give rise to a red region (high negative potential) above the pyramidal nitrogen atom.

► **Figure 19-3**
Nitrogen inversion interconverts the two enantiomers of a simple chiral amine. The transition state is a planar, sp^2 hybrid structure with the lone pair in a p orbital.

A tetrahedral amine with three different substituents (and a lone pair) is non-superimposable on its mirror image. In most cases we cannot resolve such an amine into two enantiomers, however, because the enantiomers interconvert rapidly (see Figure 19-3). This interconversion takes place by **nitrogen inversion**, in which the lone pair moves from one face of the molecule to the other. The nitrogen atom is sp^2 hybridized in the transition state, and the non-bonding electrons occupy a p orbital. This is a fairly stable transition state, as reflected by the small activation energy of about 6 kcal/mol (25 kJ/mol). Interconversion of (R)- and (S)-ethylmethylamine is shown in Figure 19-3. In naming the enantiomers of chiral amines, the Cahn–Ingold–Prelog convention is used, with the nonbonding electron pair having the lowest priority.

Although most simple amines cannot be resolved into enantiomers, several types of chiral amines can be resolved:

1. *Amines whose chirality stems from the presence of asymmetric carbon atoms.* For example, 2-butanamine can be resolved into enantiomers because the 2-butyl group is chiral.

(S)-2-butanamine   (R)-2-butanamine

2. *Quaternary ammonium salts with asymmetric nitrogen atoms.* Inversion of configuration is not possible because there is no lone pair to undergo nitrogen inversion. For example, the methyl ethyl isopropyl anilinium salts can be resolved into enantiomers.

3. *Amines that cannot attain the sp^2 hybrid transition state for nitrogen inversion.* If the nitrogen atom is contained in a small ring, for example, it is prevented from attaining the 120° bond angles that facilitate inversion. Such a compound has a higher activation energy for inversion, the inversion is slow, and the enantiomers may be resolved. Chiral aziridines (three-membered rings containing a nitrogen) often may be resolved into enantiomers.

(R)-1,2,2-trimethylaziridine   (S)-1,2,2-trimethylaziridine

19-4
Physical Properties of Amines

Amines are strongly polar because the large dipole moment of the lone pair of electrons adds to the dipole moments of the C $\leftrightarrow$ N and H $\leftrightarrow$ N bonds. Primary and secondary amines have N—H bonds, allowing them to form hydrogen bonds. Having no N—H bonds, pure tertiary amines cannot engage in hydrogen bonding. They can accept hydrogen bonds from molecules having O—H or N—H bonds, however.

overall dipole moment

1° or 2° amine:
hydrogen bond donor and acceptor

3° amine:
hydrogen bond acceptor only

Because nitrogen is less electronegative than oxygen, the N—H bond is less polar than the O—H bond. Therefore, amines form weaker hydrogen bonds than do alcohols of similar molecular weights. Primary and secondary amines have boiling points that are lower than those of alcohols, yet higher than those of ethers of similar molecular weights. With no hydrogen bonding, tertiary amines have lower boiling points than primary and secondary amines of similar molecular weights. Table 19-1 compares the boiling points of an ether, an alcohol, and amines of similar molecular weights.

All amines, even tertiary ones, form hydrogen bonds with hydroxylic solvents such as water and alcohols. Therefore, amines tend to be soluble in alcohols, and the lower-molecular-weight amines (up to about six carbon atoms) are relatively soluble in water. Table 19-2 lists the melting points, boiling points, and water solubilities of some simple aliphatic and aromatic amines.

Perhaps the most obvious property of amines is their characteristic odor of rotting fish. Some of the diamines are particularly pungent; the following diamines have common names that describe their odors:

$$CH_2CH_2CH_2CH_2$$
$$|\qquad\qquad |$$
$$NH_2\qquad\quad NH_2$$

putrescine
(1,4-butanediamine)

$$CH_2CH_2CH_2CH_2CH_2$$
$$|\qquad\qquad\qquad |$$
$$NH_2\qquad\qquad NH_2$$

cadaverine
(1,5-pentanediamine)

TABLE 19-1 Comparison of the Boiling Points of an Ether, an Alcohol, and Amines of Similar Molecular Weights

Compound	bp (°C)	Type	Molecular Weight
$(CH_3)_3N$:	3	tertiary amine	59
CH_3—O—CH_2—CH_3	8	ether	60
CH_3—NH—CH_2—CH_3	37	secondary amine	59
$CH_3CH_2CH_2$—NH_2	48	primary amine	59
$CH_3CH_2CH_2$—OH	97	alcohol	60

TABLE 19-2 Melting Points, Boiling Points, and Water Solubilities of Some Simple Amines

Name	Structure	Molecular Weight	mp (°C)	bp (°C)	H₂O Solubility (g/100 g H₂O)
		Primary amines			
methylamine	CH_3NH_2	31	−93	−7	very soluble
ethylamine	$CH_3CH_2NH_2$	45	−81	17	∞
n-propylamine	$CH_3CH_2CH_2NH_2$	59	−83	48	∞
isopropylamine	$(CH_3)_2CHNH_2$	59	−101	33	∞
n-butylamine	$CH_3CH_2CH_2CH_2NH_2$	73	−50	77	∞
isobutylamine	$(CH_3)_2CHCH_2NH_2$	73	−86	68	∞
sec-butylamine	$CH_3CH_2CH(NH_2)CH_3$	73	−104	63	∞
tert-butylamine	$(CH_3)_3CNH_2$	73	−68	45	∞
cyclohexylamine	*cyclo*-$C_6H_{11}NH_2$	99		134	slightly soluble
benzylamine	$C_6H_5CH_2NH_2$	107		185	∞
allylamine	$CH_2{=}CH{-}CH_2NH_2$	57		53	very soluble
aniline	$C_6H_5NH_2$	93	−6	184	3.7
		Secondary amines			
dimethylamine	$(CH_3)_2NH$	45	−96	7	very soluble
ethylmethylamine	$CH_3CH_2NHCH_3$	59		37	very soluble
diethylamine	$(CH_3CH_2)_2NH$	73	−42	56	very soluble
di-*n*-propylamine	$(CH_3CH_2CH_2)_2NH$	101	−40	111	slightly soluble
diisopropylamine	$[(CH_3)_2CH]_2NH$	101	−61	84	slightly soluble
di-*n*-butylamine	$(CH_3CH_2CH_2CH_2)_2NH$	129	−59	159	slightly soluble
N-methylaniline	$C_6H_5NHCH_3$	107	−57	196	slightly soluble
diphenylamine	$(C_6H_5)_2NH$	169	54	302	insoluble
		Tertiary amines			
trimethylamine	$(CH_3)_3N$	59	−117	3.5	91
triethylamine ·	$(CH_3CH_2)_3N$	101	−115	90	14
tri-*n*-propylamine	$(CH_3CH_2CH_2)_3N$	143	−94	156	slightly soluble
N,N-dimethylaniline	$C_6H_5N(CH_3)_2$	121	2	194	1.4
triphenylamine	$(C_6H_5)_3N$	251	126	365	insoluble

PROBLEM 19-5

Rank each set of compounds in order of increasing boiling points.
(a) triethylamine, di-*n*-propylamine, *n*-propyl ether
(b) ethanol, dimethylamine, dimethyl ether
(c) trimethylamine, diethylamine, diisopropylamine

19-5
Basicity of Amines

An amine is a nucleophile (a Lewis base) because its lone pair of nonbonding electrons can form a bond with an electrophile. An amine can also act as a Brønsted–Lowry base by accepting a proton from a proton acid.

Reaction of an amine as a nucleophile

Reaction of an amine as a proton base

Because amines are fairly strong bases, their aqueous solutions are basic. An amine can abstract a proton from water, giving an ammonium ion and a hydroxide ion. The equilibrium constant for this reaction is called the **base-dissociation constant** for the amine, symbolized by K_b.

$$R-\overset{\overset{\displaystyle H}{|}}{\underset{\underset{\displaystyle H}{|}}{N}}: \; + \; H-O-H \; \underset{}{\overset{K_b}{\rightleftharpoons}} \; R-\overset{\overset{\displaystyle H}{|}}{\underset{\underset{\displaystyle H}{|}}{N}}\overset{+}{}H \; + \; {}^-OH$$

$$K_b = \frac{[RNH_3{}^+][{}^-OH]}{[RNH_2]} \qquad pK_b = -\log_{10}K_b$$

Values of K_b for most amines are fairly small (about 10^{-3} or smaller), and the equilibrium for this dissociation lies toward the left. Nevertheless, aqueous solutions of amines are distinctly basic, and they turn litmus paper blue.

Because they vary by many orders of magnitude, base-dissociation constants are usually listed as their negative logarithms, or pK_b values. For example, if a certain amine has $K_b = 10^{-3}$, then $pK_b = 3$. Just as we used pK_a values to indicate acid strengths (stronger acids have smaller pK_a values), we use pK_b values to compare the relative strengths of amines as proton bases.

Stronger bases have smaller values of pK_b.

The values of pK_b for some representative amines are listed in Table 19-3.

TABLE 19-3 Values of pK_b for Some Representative Amines

Amine	K_b	pK_b	pK_a of $R_3\overset{+}{N}H$	Amine	K_b	pK_b	pK_a of $R_3\overset{+}{N}H$
ammonia	1.8×10^{-5}	4.74	9.26	*Aryl amines*			
				aniline	4.0×10^{-10}	9.40	4.60
Primary alkyl amines				N-methylaniline	6.1×10^{-10}	9.21	4.79
methylamine	4.3×10^{-4}	3.36	10.64	N,N-dimethylaniline	11.6×10^{-10}	8.94	5.06
ethylamine	4.4×10^{-4}	3.36	10.64	p-toluidine	1.2×10^{-9}	8.92	5.08
n-propylamine	4.7×10^{-4}	3.32	10.68	p-fluoroaniline	4.4×10^{-10}	9.36	4.64
isopropylamine	4.0×10^{-4}	3.40	10.60	p-chloroaniline	1×10^{-10}	10.00	4.00
n-butylamine	4.8×10^{-4}	3.32	10.68	p-bromoaniline	7×10^{-11}	10.15	3.85
cyclohexylamine	4.7×10^{-4}	3.33	10.67	p-iodoaniline	6×10^{-11}	10.22	3.78
benzylamine	2.0×10^{-5}	4.67	9.33	p-methoxyaniline	2×10^{-9}	8.70	5.30
				p-nitroaniline	1×10^{-13}	13.00	1.00
Secondary amines							
dimethylamine	5.3×10^{-4}	3.28	10.72	*Heterocyclic amines*			
diethylamine	9.8×10^{-4}	3.01	10.99	pyrrole	1×10^{-15}	~15	~−1
di-n-propylamine	10.0×10^{-4}	3.00	11.00	pyrrolidine	1.9×10^{-3}	2.73	11.27
				imidazole	8.9×10^{-8}	7.05	6.95
Tertiary amines				pyridine	1.8×10^{-9}	8.75	5.25
trimethylamine	5.5×10^{-5}	4.26	9.74	piperidine	1.3×10^{-3}	2.88	11.12
triethylamine	5.7×10^{-4}	3.24	10.76				
tri-n-propylamine	4.5×10^{-4}	3.35	10.65				

Some references do not list values of K_b or pK_b for amines. Instead, they list values of K_a or pK_a for the conjugate acid, which is the ammonium ion. We can show that the product of K_a for the ammonium ion and K_b for the amine is K_W, the ion product for water, which is 10^{-14} at room temperature. This is true for any conjugate acid–base pair. (See Section 1-13B.)

$$R-\overset{+}{N}H_3 + H_2O \; \underset{}{\overset{K_a}{\rightleftharpoons}} \; R-\overset{..}{N}H_2 + H_3O^+$$

$$\underset{\text{ammonium ion}}{\phantom{R-\overset{+}{N}H_3}} \qquad\qquad \underset{\text{amine}}{\phantom{R-\overset{..}{N}H_2}}$$

$$K_a = \frac{[RNH_2][H_3O^+]}{[RNH_3^+]} \qquad K_b = \frac{[RNH_3^+][^-OH]}{[RNH_2]}$$

$$K_a \times K_b = [H_3O^+][^-OH] = K_W = 1.0 \times 10^{-14}$$

$$pK_a + pK_b = 14 \qquad pK_b = 14 - pK_a$$

PROBLEM-SOLVING HINT

The pK_a of RNH_3^+ is the pH at which half of the molecules are protonated. At lower (more acidic) pH, the amine is mostly protonated (RNH_3^+). At higher (more basic) pH, the amine is mostly deprotonated (RNH_2).

These relationships allow us to convert values of K_a (or pK_a) for the ammonium ion and K_b (or pK_b) for the amine. They also remind us that a strongly basic amine has a weakly acidic ammonium ion and a weakly basic amine has a strongly acidic ammonium ion.

19-6
Effects on Amine Basicity

Figure 19-4 shows an energy diagram for the reaction of an amine with water. On the left are the reactants: the free amine and water. On the right are the products: the ammonium ion and hydroxide ion.

Any structural feature that stabilizes the ammonium ion (relative to the free amine) shifts the reaction toward the right, making the amine a stronger base. Any feature that stabilizes the free amine (relative to the ammonium ion) shifts the reaction toward the left, making the amine a weaker base.

Substitution by Alkyl Groups   As an example, consider the relative basicities of ammonia and methylamine. Alkyl groups are electron-donating toward cations, and methylamine has a methyl group to help stabilize the positive charge on nitrogen. This stabilization lowers the potential energy of the methylammonium cation, making methylamine a stronger base than ammonia. The simple alkylamines are generally stronger bases than ammonia.

$$H-\overset{H}{\underset{H}{N:}} \;+\; H_2O \;\rightleftharpoons\; H-\overset{H}{\underset{H}{\overset{|}{N}{}^+}}-H \;+\; {}^-OH \qquad \begin{array}{l} pK_b = 4.74 \\ \text{(weaker base)} \end{array}$$

$$H_3C-\overset{H}{\underset{H}{N:}} \;+\; H_2O \;\rightleftharpoons\; H_3C\rightarrow\overset{H}{\underset{H}{\overset{|}{N}{}^+}}-H \;+\; {}^-OH \qquad \begin{array}{l} pK_b = 3.36 \\ \text{(stronger base)} \end{array}$$

stabilized by the alkyl group

We might expect secondary amines to be stronger bases than primary amines and tertiary amines to be the strongest bases of all. The actual situation is more complicated

▶ **Figure 19-4**
Potential-energy diagram of the base-dissociation reaction of an amine.

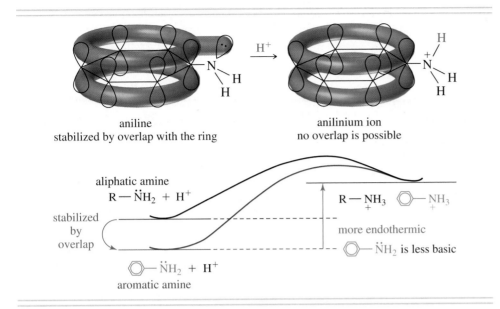

aniline
stabilized by overlap with the ring

anilinium ion
no overlap is possible

◀ **Figure 19-5**
Aniline is stabilized by overlap of the lone pair with the aromatic ring. No such overlap is possible in the anilinium ion.

because of solvation effects. Because ammonium ions are charged, they are strongly solvated by water and the energy of solvation contributes to their stability. The additional alkyl groups around the ammonium ions of secondary and tertiary amines decrease the number of water molecules that can approach closely and solvate the ions. The opposing trends of inductive stabilization and steric hindrance of solvation tend to cancel out in most cases; as a result, primary, secondary, and tertiary amines show similar ranges of basicity.

Resonance Effects on Basicity Aromatic amines (anilines and their derivatives) are weaker bases than simple aliphatic amines (Table 19-3). This reduced basicity is due to resonance delocalization of the nonbonding electrons in the free amine. Figure 19-5 shows how stabilization of the reactant (the free amine) makes the amine less basic. In aniline, overlap between the aromatic ring and the orbital containing nitrogen's lone pair stabilizes the lone pair and makes it less reactive. This overlap is lost in the anilinium ion, so the reactant (aniline) is stabilized compared with the product. The reaction is shifted toward the left, and aniline is less basic than most aliphatic amines.

Resonance effects also influence the basicity of pyrrole. Pyrrole is a very weak base, with a pK_b of about 15. As we saw in Chapter 15, pyrrole is aromatic because the nonbonding electrons on nitrogen are located in a p orbital, where they contribute to the aromatic sextet. When the pyrrole nitrogen is protonated, pyrrole loses its aromatic stabilization. Therefore, protonation on nitrogen is unfavorable, and pyrrole is a very weak base.

pyrrole
(aromatic)

$K_b = 10^{-15}$

protonated
(not aromatic)

Hybridization Effects Our study of terminal alkynes (Section 9-6) showed that electrons are held more tightly by orbitals with more s character. This principle helps to explain why unsaturated amines tend to be weaker bases than simple aliphatic amines. In pyridine, for example, the nonbonding electrons occupy an sp^2 orbital, with greater s character and more tightly held electrons than those in the sp^3 orbital of an aliphatic amine. Pyridine's nonbonding electrons are less available for bonding to a proton. Pyridine does not lose its aromaticity on protonation, however, and it is a much stronger base than pyrrole.

pyridine, $pK_b = 8.75$

sp^2 hybridized (less basic)

piperidine, $pK_b = 2.88$

sp^3 hybridized (more basic)

The effect of increased *s* character on basicity is even more pronounced in nitriles with *sp* hybridization. For example, acetonitrile has a pK_b of 24, showing that it is a very weak base. In fact, a concentrated mineral acid is required to protonate acetonitrile.

sp hybridized

$$CH_3 - C \equiv N :$$

$pK_b = 24$

very weakly basic

PROBLEM 19-6

Rank each set of compounds in order of increasing basicity.
(a) $NaOH, NH_3, CH_3NH_2, Ph - NH_2$ **(b)** aniline, *p*-methylaniline, *p*-nitroaniline
(c) aniline, pyrrole, pyridine   **(d)** pyrrole, imidazole, 3-nitropyrrole

---

19-7
Salts of Amines

Protonation of an amine gives an **amine salt**. The amine salt is composed of two types of ions: the protonated amine cation (an ammonium ion) and the anion derived from the acid. Simple amine salts are named as the substituted **ammonium salts**. Salts of complex amines use the names of the amine and the acid that make up the salt.

$$CH_3CH_2CH_2 - \ddot{N}H_2 \quad + \quad HCl \quad \rightleftharpoons \quad CH_3CH_2CH_2 - NH_3^+ \ ^-Cl$$

n-propylamine   hydrochloric acid   *n*-propylammonium chloride

$$(CH_3CH_2)_3N: \quad + \quad H_2SO_4 \quad \rightleftharpoons \quad (CH_3CH_2)_3NH^+ HSO_4^-$$

triethylamine   sulfuric acid   triethylammonium hydrogen sulfate

pyridine   acetic acid   pyridinium acetate

Amine salts are ionic, high-melting, nonvolatile solids. They are much more soluble in water than the parent amines, and they are only slightly soluble in nonpolar organic solvents.

Formation of amine salts can be used to isolate and characterize amines. Most amines containing more than six carbon atoms are relatively insoluble in water. In dilute aqueous acid, these amines form their corresponding ammonium salts, and they dissolve. Formation of a soluble salt is one of the characteristic functional group tests for amines.

$R_3N:$ Aq. HCl $R_3\overset{+}{N}H \ ^-Cl$
"free" amine   Aq. NaOH   amine salt
(water insoluble)   (water soluble)

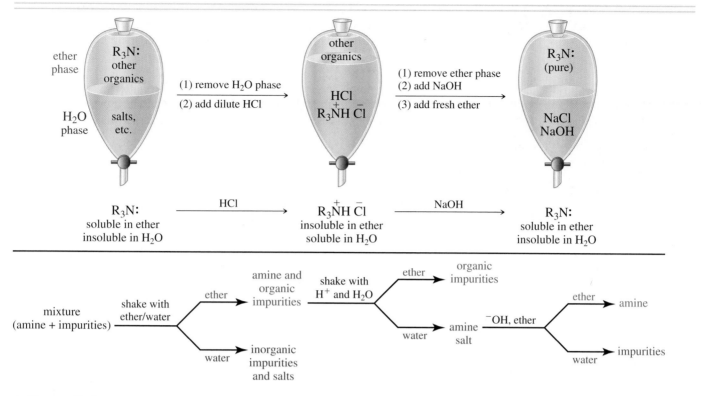

▲ Figure 19-6
The basicity of an amine can be used for purification. The amine is initially more soluble in ether than in water. Addition of dilute HCl converts it to the water-soluble hydrochloride salt. Neutralization with NaOH regenerates the free amine.

The formation of amine salts is also used to separate amines from less basic compounds (Fig. 19-6). In dilute acid, the amine forms a salt and dissolves. When the solution is made alkaline (by the addition of NaOH), the free amine is regenerated. The purified free amine either separates out of the aqueous solution or is extracted into an organic solvent.

Many drugs and other biologically important amines are commonly stored and used as their salts. Amine salts are less prone to decomposition by oxidation and other reactions, and they have virtually no fishy odor. The salts are soluble in water, and they are easily converted to solutions for syrups and injectables.

As an example, the drug ephedrine is widely used in cold and allergy medications. Ephedrine melts at 79°C, has an unpleasant fishy odor, and is oxidized by air to undesirable products. Ephedrine hydrochloride melts at 217°C, does not oxidize easily, and has virtually no odor. Obviously, the hydrochloride salt is preferable for compounding medications.

> Ephedrine is the principal ingredient in Ma Huang, an herbal remedy originally used to treat asthma. While the remedy is now advertised as a "natural" way to enhance mood, reduce fatigue, and lose weight, its stimulating effects on the heart can be fatal.

ephedrine
mp 79°C, foul-smelling
easily air oxidized

ephedrine hydrochloride
mp 217°C, no odor
stable

The chemistry of amine salts plays a large role in drug addiction. Cocaine, for example, is usually smuggled and "snorted" as the hydrochloride salt, which is

Cocaine hydrochloride is often divided into "lines" on a mirror and then snorted. "Crack" cocaine is sold as "rocks," commonly smoked in a crude pipe.

more stable and gives off less odor to alert the authorities. Smoking cocaine gives a more intense rush (and stronger addiction) because of fast absorption by lung tissues. But cocaine hydrochloride is not volatile; it tends to decompose before it vaporizes. Treating cocaine hydrochloride with sodium hydroxide and extracting it into ether converts it back to the volatile "free base" for smoking. "Free-basing" cocaine is hazardous because it involves large amounts of ether. A simpler alternative is to mix a paste of cocaine hydrochloride with sodium bicarbonate and let it dry into "rocks." This mixture is called "crack cocaine" because it makes a crackling sound when it is heated.

cocaine hydrochloride

cocaine "free base"

19-8
Amine Salts as Phase-Transfer Catalysts

Quaternary ammonium salts ($R_4N^+ \ ^-X$) are especially useful because they are somewhat soluble in both water and nonpolar organic solvents. They work as **phase-transfer catalysts** to move ionic reagents into organic solvents in which they would not otherwise dissolve. As an example, consider the reaction shown in Figure 19-7. This reaction generates dichlorocarbene ($:CCl_2$) by alpha elimination using sodium hydroxide (see Section 8-9B). Dichlorocarbene adds to cyclohexene to form a new cyclopropane ring.

The flask pictured on the left in Figure 19-7 shows the reaction before adding a phase-transfer catalyst. Water forms the upper layer, and chloroform forms the lower layer. Sodium hydroxide dissolves in water but not in chloroform, so it is entirely in the water layer. Cyclohexene dissolves in chloroform but not in water, so it is entirely in the chloroform layer. Very little reaction occurs under these conditions, even when using a high-speed stirrer.

The flask pictured on the right shows the reaction after tetrabutylammonium chloride, a phase-transfer catalyst, is added. The tetrabutylammonium ion forms an ion pair with hydroxide ion. This ion pair has large alkyl groups that enhance its solubility in organic solvents, so it can migrate (aided by high-speed stirring) into the chloroform layer. In the organic phase, hydroxide ion is more reactive than in the aqueous phase because it is stripped of its solvating water molecules. Hydroxide reacts with chloroform to give dichlorocarbene, which reacts with cyclohexene to give the cyclopropanated product.

Other anions may be transferred into organic phases by tetra-alkylammonium phase-transfer catalysts. For example, sodium cyanide (NaCN) is not soluble in most organic solvents, but the cyanide ion (^-CN) can be used as a nucleophile in organic solvents under phase-transfer conditions, as shown here. Like the hydroxide ion, cyanide is a stronger nucleophile in the organic phase because it is not solvated by water molecules.

$$Bu_4N^+ \ ^-Cl + Na^+ \ ^-CN \ \rightleftharpoons \ Bu_4N^+ \ ^-CN + Na^+ \ ^-Cl \qquad \text{(aqueous phase)}$$

organic insoluble   ion pair

$$Bu_4N^+ \ ^-CN + R-CH_2-Cl \ \longrightarrow \ R-CH_2-C\equiv N + Bu_4N^+ \ ^-Cl \quad \text{(organic phase)}$$

organic soluble   a nitrile

Overall reaction

(80%)

Mechanism

1. Aqueous phase

$$Bu_4N^+\ {}^-Cl + Na^+\ {}^-OH \rightleftharpoons Bu_4N^+\ {}^-OH + Na^+\ {}^-Cl$$

tetrabutylammonium   (ion pair)
chloride

2. Organic phase

$$Bu_4N^+\ {}^-OH + CHCl_3 \rightleftharpoons Bu_4N^+\ {}^-CCl_3 + H_2O$$
organic soluble

$$Bu_4N^+\ {}^-CCl_3 \rightleftharpoons Bu_4N^+\ {}^-Cl\ +\ :CCl_2$$

catalyst regenerated   dichlorocarbene

▲ **Figure 19-7**
Use of a phase-transfer catalyst. This example shows the reaction of cyclohexene and chloroform, both insoluble in water, with a 50% aqueous solution of sodium hydroxide.

19-9A   Infrared Spectroscopy

The most reliable IR absorption of primary and secondary amines is the N—H stretch whose frequency appears between 3200 and 3500 cm^{-1}. Since this absorption is often broad, it is easily confused with the O—H absorption of an alcohol. In most cases, however, there are one or more spikes visible in the N—H stretching region of an amine spectrum. Primary amines (R—NH_2) usually give two N—H spikes, while secondary

19-9
Spectroscopy
of Amines

▲ **Figure 19-8**
Infrared spectrum of 1-propanamine. Notice the characteristic N—H stretching absorptions at 3300 and 3400 cm^{-1}.

amines (R$_2$N—H) usually give just one. Tertiary amines (R$_3$N) give no N—H absorptions. Notice in Figure 19-8 the characteristic N—H absorptions in the IR spectrum of 1-propanamine, a primary amine. The IR spectrum of a secondary amine appears in Figure 12-10, and that of an alcohol appears in Figure 12-9 for comparison.

Although an amine IR spectrum also contains absorptions resulting from vibrations of C—N bonds, these vibrations appear around 1000 to 1200 cm^{-1}, in the same region as C—C and C—O vibrations. Therefore, they are not very useful for identifying an amine.

PROBLEM 19-7

The following partial IR spectra correspond to a primary amine, a secondary amine, and an alcohol: Give the functional group for each spectrum.

19-9B   Proton NMR Spectroscopy

Like the O—H protons of alcohols, the N—H protons of amines absorb at chemical shifts that depend on the extent of hydrogen bonding. The solvent and the sample concentration influence hydrogen bonding and therefore the chemical shift. Typical N—H chemical shifts appear in the range $\delta 1$ to $\delta 4$.

Another similarity between O—H and N—H protons is their failure, in many cases, to show spin–spin splitting. In some samples, N—H protons exchange from one molecule to another at a rate that is faster than the time scale of the NMR experiment, and the N—H protons fail to show magnetic coupling. Sometimes the N—H protons of a very pure amine will show clean splitting, but these cases are rare. More commonly the N—H protons appear as broad peaks. A broad peak should arouse suspicion of N—H protons. As with O—H protons, an absorption of N—H protons decreases or disappears after shaking the sample with D_2O.

Nitrogen is not as electronegative as oxygen and the halogens, so the protons on the α-carbon atoms of amines are not as strongly deshielded. Protons on an amine's α-carbon atom generally absorb between $\delta2$ and $\delta3$, but the exact position depends on the structure and substitution of the amine.

$$\overset{\downarrow}{CH_3}\!-\!NR_2 \qquad R\!-\!\overset{\downarrow}{CH_2}\!-\!NR_2 \qquad R_2\overset{\downarrow}{CH}\!-\!NR_2$$

methyl $\delta2.3$ methylene $\delta2.7$ methine $\delta2.9$

Protons that are beta to a nitrogen atom show a much smaller effect, usually absorbing in the range $\delta1.1$ to $\delta1.8$. These chemical shifts show a downfield movement of about 0.2 ppm resulting from the beta relationship. The NMR spectrum of 1-propanamine (Figure 19-9) shows these characteristic chemical shifts.

γ protons β protons α protons

$$CH_3\!-\!CH_2\!-\!CH_2\!-\!NH_2$$
$\delta0.9$ $\delta1.4$ $\delta2.6$ variable ($\delta1.7$ in this spectrum)

▲ **Figure 19-9**
^{13}C and proton NMR spectra of 1-propanamine.

PROBLEM 19-8

The proton NMR spectrum of a compound of formula $C_4H_{11}N$ is shown below. Determine the structure of this amine, and give peak assignments for all of the protons in the structure.

19-9C Carbon NMR Spectroscopy

The α-carbon atom bonded to the nitrogen of an amine usually shows a chemical shift of about 30 to 50 ppm. This range agrees with our general rule that a carbon atom shows a chemical shift about 20 times as great as the protons bonded to it. In propanamine (Figure 19-9), for example, the α-carbon atom absorbs at 45 ppm, while its protons absorb at 2.7 ppm. The β carbon is less deshielded, absorbing at 27 ppm, compared with its protons' absorption at 1.5 ppm. The γ-carbon atom shows little effect from the presence of the nitrogen atom, absorbing at 11 ppm. Table 19-4 shows the carbon NMR chemical shifts of some representative amines.

TABLE 19-4 Carbon NMR Chemical Shifts of Some Representative Amines

δ	γ	β	α	
			CH_3-NH_2	methanamine
			26.9	
		$CH_3-CH_2-NH_2$		ethanamine
		17.7	35.9	
	$CH_3-CH_2-CH_2-NH_2$			1-propanamine
	11.2	27.3	44.9	
$CH_3-CH_2-CH_2-CH_2-NH_2$				1-butanamine
14.0	20.4	36.7	42.3	

PROBLEM 19-9

The carbon NMR chemical shifts of diethylmethylamine, piperidine, 1-propanol, and propanal are given below. Determine which spectrum corresponds to each structure, and show which carbon atom(s) are responsible for each absorption.
(a) 25.9, 27.8, 47.9   **(b)** 13.8, 47.5, 58.2   **(c)** 7.9, 44.7, 201.9   **(d)** 10.0, 25.8, 63.6

19-9D   Mass Spectrometry

The most obvious piece of information provided by the mass spectrum is the molecular weight. Stable compounds containing only carbon, hydrogen, oxygen, chlorine, bromine, and iodine give molecular ions with even mass numbers. Most of their fragments have odd mass numbers. This is because carbon and oxygen have even valences and even mass numbers, while hydrogen, chlorine, bromine, and iodine have odd valences and odd mass numbers.

Nitrogen has an odd valence and an even mass number. When a nitrogen atom is present in a stable molecule, the molecular weight is odd. In fact, whenever an odd number of nitrogen atoms are present in a molecule, the molecular ion has an odd mass number. Most of the fragments have even mass numbers.

The most common fragmentation of amines is α cleavage to give a resonance-stabilized cation: an *iminium* ion. This ion is simply a protonated version of an imine (Section 18-16).

Figure 19-10 shows the mass spectrum of butyl propyl amine. The base peak (m/z 72) corresponds to α cleavage with loss of a propyl radical to give a resonance-stabilized iminium ion. A similar α cleavage, with loss of an ethyl radical, gives the peak at m/z 86.

▲ **Figure 19-10**
Mass spectrum of butyl propyl amine. Note the odd mass number of the molecular ion and the even mass numbers of most fragments. The base peak corresponds to α cleavage in the butyl group, giving a propyl radical and a resonance-stabilized iminium ion.

PROBLEM 19-10

(a) Show how fragmentation occurs to give the base peak at *m/z* 58 in the mass spectrum of ethyl propyl amine, shown below.
(b) Show how a similar cleavage in the ethyl group gives an ion of *m/z* 72.
(c) Explain why the peak at *m/z* 72 is much weaker than the one at *m/z* 58.

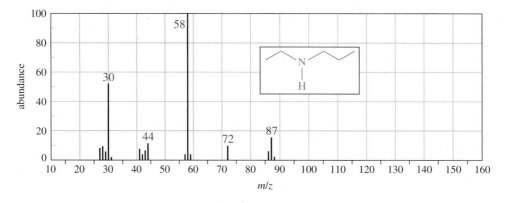

19-10
Reactions of Amines with Ketones and Aldehydes (Review)

In contrast to other functional groups, we will study the reactions of amines before we study their syntheses. This organization is better because most amine syntheses begin with an amine (or ammonia) and add groups to make more highly substituted amines. In other words, most amine syntheses apply the reactions of amines. By studying the reactions first, we can readily understand how to use these reactions to convert simpler amines to more complicated amines.

In Section 18-16, we saw that amines attack ketones and aldehydes. When this nucleophilic attack is followed by dehydration, an imine (Schiff base) results. The analogous reaction of a hydrazine derivative gives a hydrazone, and the reaction with hydroxylamine gives an oxime. In Section 19-19, we will use these reactions to synthesize amines.

$$
\underset{\substack{\text{ketone or aldehyde}}}{\overset{\overset{\displaystyle O}{\|}}{\underset{R \quad R'}{C}}} + Y-\ddot{N}H_2 \;\overset{H^+}{\rightleftharpoons}\; \underset{\substack{\text{carbinolamine}}}{\overset{\overset{\displaystyle HO \quad :N-H}{\underset{|}{|}}}{\underset{R \quad R'}{C}}} \;\overset{H^+}{\rightleftharpoons}\; \underset{\substack{\text{derivative}}}{\overset{\overset{\displaystyle :N}{\|}}{\underset{R \quad R'}{C}}} + H_2O
$$

Y = H or alkyl	gives an imine (Schiff base)
Y = OH	gives an oxime
Y = NHR	gives a hydrazone

19-11
Aromatic Substitution of Arylamines and Pyridine (Review)

19-11A   Electrophilic Aromatic Substitution of Arylamines

In an arylamine, the nonbonding electrons on nitrogen help stabilize intermediates resulting from electrophilic attack at the positions ortho or para to the amine substituent. As a result, amino groups are strong activating groups and ortho, para-directors. Figure 19-11 shows the sigma complexes involved in ortho and para substitution of aniline.

The following reactions show halogenation of aniline derivatives, which occurs readily without a catalyst. If an excess of the reagent is used, all the unsubstituted positions ortho and para to the amino group become substituted.

◀ **Figure 19-11**
The amino group is a strong activator and ortho, para-director. The nonbonding electrons on nitrogen stabilize the σ complex when attack occurs at the ortho or para positions.

Care must be exercised in reactions with aniline derivatives, however. Strongly acidic reagents protonate the amino group, giving an ammonium salt that bears a full positive charge. The $—NH_3^+$ group is strongly deactivating (and meta-allowing). Therefore, strongly acidic reagents are unsuitable for substitution of anilines. Oxidizing acids (such as nitric and sulfuric acids) may oxidize the amino group, leading to decomposition and occasional violent reactions. In Section 19-13, we see how the amino group may be acylated to decrease its basicity and permit substitution by a wide variety of electrophiles.

19-11B Electrophilic Aromatic Substitution of Pyridine

In its aromatic substitution reactions, pyridine resembles a strongly deactivated benzene. Friedel–Crafts reactions fail completely, and other substitutions require unusually strong conditions. Deactivation results from the electron-withdrawing effect of the electronegative nitrogen atom. Its nonbonding electrons are perpendicular to the π system, and they cannot stabilize the positively charged intermediate. When pyridine does react, it gives substitution at the 3-position, analogous to the meta substitution shown by deactivated benzene derivatives. The following reactions compare the intermediates formed by nitration of pyridine at the 2-position and at the 3-position.

MECHANISM Electrophilic Aromatic Substitution of Pyridine

Attack at the 3-position is observed

Attack at the 3-position gives the most stable intermediate

Loss of a proton gives the product

3-nitropyridine
(observed)

Attack at the 2-position (not observed)

no octet
unfavorable

2-nitropyridine
(not observed)

Electrophilic attack on pyridine at the 2-position gives an unstable intermediate, with one of the resonance structures showing a positive charge and only six electrons on nitrogen. In contrast, all three resonance forms of the intermediate from attack at the 3-position place the positive charge on the less electronegative carbon atoms.

Electrophilic substitution of pyridine is further hindered by the tendency of the nitrogen atom to attack electrophiles and take on a positive charge. The positively charged pyridinium ion is even more resistant than pyridine to electrophilic substitution.

pyridine   electrophile   pyridinium ion
(less reactive)

PROBLEM 19-11

Propose a mechanism for nitration of pyridine at the 4-position, and explain why this orientation is not observed.

Two electrophilic substitutions of pyridine are shown below. Notice that these reactions require severe conditions, and the yields are poor to fair.

pyridine → Br$_2$, 300°C / NaHCO$_3$ → 3-bromopyridine (30%)

pyridine → fuming H$_2$SO$_4$, HgSO$_4$ / 230°C → pyridine-3-sulfonic acid (protonated) (70%)

PROBLEM 19-12

Propose a mechanism for the sulfonation of pyridine, pointing out why sulfonation occurs at the 3-position.

19-11C Nucleophilic Aromatic Substitution of Pyridine

Pyridine is deactivated toward electrophilic attack, but it is activated toward attack by electron-rich nucleophiles; that is, it is activated toward nucleophilic aromatic substitution. If there is a good leaving group at either the 2-position or the 4-position, a nucleophile can attack and displace the leaving group. The reaction below shows nucleophilic attack at the 2-position. The intermediate is stabilized by delocalization of the negative charge onto the electronegative nitrogen atom. This stabilization is not possible if attack occurs at the 3-position.

MECHANISM   Nucleophilic Aromatic Substitution of Pyridine

Nucleophilic attack at the 2-position is observed

negative charge on electronegative nitrogen (favorable)

expulsion of the leaving group gives the product

+ Cl$^-$

Nucleophilic attack at the 3-position (not observed)

+ Cl$^-$

(no delocalization of negative charge onto N)

19-12
Alkylation of Amines by Alkyl Halides

Amines react with primary alkyl halides to give alkylated ammonium halides. Alkylation proceeds by the S_N2 mechanism, so it is not feasible with tertiary halides because they are too hindered. Secondary halides often give poor yields, with elimination predominating over substitution.

$$R-\ddot{N}H_2 \ + \ R'-CH_2-Br \ \longrightarrow \ R-\overset{+}{N}H_2-CH_2-R' \ \ ^-Br$$

primary amine   primary halide   salt of a secondary amine

Unfortunately, the initially formed salt may become deprotonated. The resulting secondary amine is nucleophilic, and it can react with another molecule of the halide.

$$R-\overset{+}{N}H_2-CH_2-R'\,^-Br \ + \ R-\ddot{N}H_2 \ \rightleftharpoons \ R-\ddot{N}H-CH_2-R' \ + \ R-\overset{+}{N}H_3 \ ^-Br$$

2° amine

$$R-\ddot{N}H-CH_2-R' \ + \ R'-CH_2-Br \ \longrightarrow \ R-\underset{+}{\overset{CH_2-R'}{\underset{|}{N}}}H-CH_2-R' \ ^-Br$$

2° amine   salt of a tertiary amine

The disadvantage of direct alkylation lies in stopping it at the desired stage. Even if just one equivalent of the halide is added, some amine molecules will react once, some will react twice, and some will react three times (to give the tetraalkylammonium salt). Others will not react at all. A complex mixture results.

Alkylation of amines can give good yields of the desired alkylated products in two types of reactions:

1. *"Exhaustive" alkylation to the tetraalkylammonium salt.* Mixtures of different alkylated products are avoided if enough alkyl halide is added to alkylate the amine as many times as possible. This **exhaustive alkylation** gives a tetraalkylammonium salt. A mild base (often $NaHCO_3$ or dilute $NaOH$) is added to deprotonate the intermediate alkylated amines and to neutralize the large quantities of HX formed.

$$CH_3CH_2CH_2-NH_2 \ + \ 3\,CH_3-I \ \xrightarrow{\ NaHCO_3\ } \ CH_3CH_2CH_2-N^+(CH_3)_3 \ ^-I$$

(90%)

2. *Reaction with a large excess of ammonia.* Because ammonia is inexpensive and has a low molecular weight, it is convenient to use a very large excess. Addition of a primary alkyl halide to a large excess of ammonia forms the primary amine, and the probability of dialkylation is small. Excess ammonia is simply allowed to evaporate.

$$\ddot{N}H_3 \; + \; R{-}CH_2{-}X \; \longrightarrow \; R{-}CH_2{-}\overset{+}{N}H_3 \; {}^-X$$

10 moles   1 mole

PROBLEM 19-16

Show how you would use direct alkylation to synthesize the following compounds.
(a) benzyltrimethylammonium iodide   **(b)** 1-pentanamine   **(c)** benzylamine

---

Primary and secondary amines react with acid halides to form amides. This reaction is a *nucleophilic acyl substitution*: the replacement of a leaving group on a carbonyl carbon by a nucleophile. We will study nucleophilic acyl substitution in detail in Chapters 20 and 21. In this case, the amine replaces chloride ion.

19-13
Acylation of Amines by Acid Chlorides

$$R'{-}\ddot{N}H_2 + R'{-}\overset{\overset{O}{\|}}{C}{-}Cl \; \longrightarrow \; R{-}\overset{\overset{O}{\|}}{C}{-}\ddot{N}H{-}R' + HCl$$

The amine attacks the carbonyl group of an acid chloride much like it attacks the carbonyl group of a ketone or aldehyde. The acid chloride is more reactive than a ketone or an aldehyde because the electronegative chlorine atom draws electron density away from the carbonyl carbon, making it more electrophilic. The chlorine atom in the tetrahedral intermediate is a good leaving group. The tetrahedral intermediate expels chloride to give the amide. A base such as pyridine or NaOH is often added to neutralize the HCl produced.

MECHANISM   Acylation of an Amine by an Acid Chloride

Example

The amide produced in this reaction usually does not undergo further acylation. Amides are stabilized by a resonance structure that involves nitrogen's non-bonding electrons and places a positive charge on nitrogen. As a result, amides are much less basic and less nucleophilic than amines.

The diminished basicity of amides can be used to advantage in electrophilic aromatic substitutions. For example, if the amino group of aniline is acetylated to give acetanilide, the resulting amide is still activating and ortho, para-directing. Unlike aniline, however, acetanilide may be treated with acidic (and mild oxidizing) reagents, as shown below. Aryl amino groups are frequently acylated before further substitutions are attempted on the ring, and the acyl group is removed later by acidic or basic hydrolysis (Section 21-7C).

SOLVED PROBLEM 19-1

Show how you would accomplish the following synthetic conversion in good yield:

SOLUTION

An attempted Friedel–Crafts acylation on aniline would likely meet with disaster. The free amino group would attack both the acid chloride and the Lewis acid catalyst.

We can control the nucleophilicity of aniline's amino group by converting it to an amide, which is still activating and ortho, para-directing for the Friedel–Crafts reaction. Acylation, followed by hydrolysis of the amide, gives the desired product.

PROBLEM 19-17

Give the products expected from the following reactions:
(a) acetyl chloride + ethylamine

(b) benzoyl chloride + (CH$_3$)$_2$NH dimethylamine

(c) CH$_3$—(CH$_2$)$_4$—C—Cl + piperidine hexanoyl chloride

19-14
Formation of Sulfonamides

Sulfonyl chlorides are the acid chlorides of sulfonic acids. Like acyl chlorides, sulfonyl chlorides are strongly electrophilic.

R—C—OH a carboxylic acid

R—C—Cl an acyl chloride (acid chloride)

R—C—OH a sulfonic acid

R—C—Cl a sulfonyl chloride

A primary or secondary amine attacks a sulfonyl chloride and displaces chloride ion to give an amide. Amides of sulfonic acids are called **sulfonamides**. This reaction is similar to the formation of a sulfonate ester from a sulfonyl chloride (such as tosyl chloride) and an alcohol (Section 11-5).

R'—NH$_2$ amine, R—S—Cl sulfonyl chloride $\xrightarrow{\text{NaOH}}$ R—S—NH—R' $\longrightarrow$ R—S—NHR' sulfonamide, H$_2$O, Cl$^-$

The *sulfa drugs* are a class of sulfonamides used as antibacterial agents. In 1936, sulfanilamide was found to be effective against streptococcal infections. Sulfanilamide is synthesized from acetanilide (having the amino group protected as an amide) by chlorosulfonation followed by treatment with ammonia. The final reaction is hydrolysis of the protecting group to give sulfanilamide.

acetanilide $\xrightarrow{\text{Cl—S—OH}}$ (sulfonyl chloride product) $\xrightarrow[\text{H}_2\text{O}]{\text{:NH}_3}$ (sulfonamide product) $\xrightarrow[\text{heat}]{\text{dil HCl}}$ sulfanilamide

PROBLEM 19-18

What would happen in the synthesis of sulfanilamide if the amino group were not protected as an amide in the chlorosulfonation step?

The biological activity of sulfanilamide has been studied in detail. It appears that sulfanilamide is an analogue of *p*-aminobenzoic acid. Streptococci use *p*-aminobenzoic acid to synthesize folic acid, an essential compound for growth and reproduction.

p-aminobenzoic acid   folic acid

The effectiveness of sulfa drugs is now limited due to bacterial resistance. One method used by resistant strains is to overproduce *p*-aminobenzoic acid, effectively diluting the drug's concentration.

Sulfanilamide cannot be used to make folic acid, but the bacterial enzymes cannot distinguish between sulfanilamide and *p*-aminobenzoic acid. The production of active folic acid is inhibited, and the organism stops growing. Sulfanilamide does not kill the bacteria, but it inhibits their growth and reproduction, allowing the body's own defense mechanisms to destroy the infection.

PROBLEM 19-19

Show how you would use the same sulfonyl chloride as used in the sulfanilamide synthesis to make sulfathiazole and sulfapyridine.

sulfathiazole   sulfapyridine

19-15
Amines as Leaving Groups: The Hofmann Elimination

Amines can be converted to alkenes by elimination reactions, much like alcohols and alkyl halides undergo elimination to give alkenes (Sections 11-10 and 7-9). An amine cannot undergo elimination directly, however, because the leaving group would be an amide ion ($^-NH_2$ or ^-NHR), which is a very strong base and a poor leaving group. An amino group can be converted to a good leaving group by exhaustive methylation, which converts it to a quaternary ammonium salt that can leave as a neutral amine. Exhaustive methylation is usually accomplished using methyl iodide.

Exhaustive methylation of an amine

$$R-\ddot{N}H_2 + 3\,CH_3-I \longrightarrow R-\overset{+}{N}(CH_3)_3\ ^-I + 2\,HI$$

poor leaving group   good leaving group

Elimination of the quaternary ammonium salt generally takes place by the E2 mechanism, which requires a strong base. To provide the base, the quaternary ammonium iodide is converted to the hydroxide salt by treatment with silver oxide. Heating of the quaternary ammonium hydroxide results in E2 elimination and formation of an alkene. This elimination of a quaternary ammonium hydroxide is called the **Hofmann elimination**.

Conversion to the hydroxide salt

$$R-\overset{+}{N}(CH_3)_3\ ^-I\ +\ \tfrac{1}{2}Ag_2O\ +\ H_2O\ \longrightarrow\ R-\overset{+}{N}(CH_3)_3\ ^-OH\ +\ AgI\downarrow$$

quaternary ammonium iodide   quaternary ammonium hydroxide

MECHANISM   Hofmann Elimination

Hofmann elimination

For example, when 2-butanamine is exhaustively methylated, converted to the hydroxide salt, and heated, elimination takes place to form a mixture of 1-butene and 2-butene.

Exhaustive methylation and conversion to the hydroxide salt

Heating and Hofmann elimination

In Chapter 7, we saw that eliminations of alkyl halides usually follow the Saytzeff rule; that is, the most highly substituted product predominates. This rule applies because the most highly substituted alkene is usually the most stable. In the Hofmann elimination, however, the product is commonly the *least* highly substituted alkene. We often classify an elimination as giving mostly the *Saytzeff product* (the most highly substituted alkene) or the *Hofmann product* (the least highly substituted alkene).

Saytzeff elimination

The Hofmann elimination's preference for the least highly substituted alkene stems from several factors, but one of the most compelling involves the sheer bulk of the leaving group. Remember that the E2 mechanism requires an anti-coplanar arrangement of the proton and the leaving group (Section 7–9). The extremely large trialkylamine leaving group in the Hofmann elimination often interferes with this coplanar arrangement.

Figure 19-12 shows the stereochemistry of the Hofmann elimination of 2-butanamine. The methylated ammonium salt eliminates by losing trimethylamine and a proton on either

PROBLEM-SOLVING HINT

Some of the stereochemical features of the Hofmann elimination are best studied using your models. Models are essential for working problems involving this elimination, such as Problem 19–20.

Looking along the C2—C3 bond

needed for E2 (less stable)

The most stable C2—C3 conformation

more stable (E2 is impossible in this conformation)

Looking along the C1—C2 bond

(any of the three staggered conformations is suitable for the E2)

▲ **Figure 19-12**
Hofmann elimination of 2-butanamine. The most stable conformation of the C2—C3 bond has no proton on C3 in an anti relationship to the leaving group. Along the C1—C2 bond, however, any staggered conformation has an anti relationship between a proton and the leaving group. Abstraction of a proton from C1 gives the Hofmann product.

C1 or C3. The possible conformations along the C2—C3 bond are shown at the top of Figure 19-12. An anti-coplanar arrangement between a C3 proton and the leaving group requires an unfavorable gauche interaction between the C4 methyl group and the bulky trimethylammonium group. The most stable conformation about the C2—C3 bond has a methyl group in the anti-coplanar position, preventing elimination along the C2—C3 bond.

The bottom half of Figure 19-12 shows the conformations along the C1—C2 bond. *Any* of the three staggered conformations of the C1—C2 bond provides an anti relationship between one of the protons and the leaving group. The Hofmann product predominates because elimination of one of the C1 protons involves a lower-energy, more probable transition state than the hindered transition state required for Saytzeff (C2—C3) elimination.

The Hofmann elimination is frequently used to determine the structures of complex amines by converting them to simpler amines. The direction of elimination is usually predictable, giving the least substituted alkene. Figure 19-13 shows two examples using the Hofmann elimination to simplify complex amines.

▶ **Figure 19-13**
Examples of the Hofmann elimination. The least substituted alkene is usually the favored product.

SOLVED PROBLEM 19-2

Predict the major product(s) formed when the following amine is treated with excess iodomethane, followed by heating with silver oxide.

SOLUTION

Solving this type of problem requires finding every possible elimination of the methylated salt. In this case, the salt has the following structure:

$$\xrightarrow{\text{excess CH}_3\text{I}} \xrightarrow[\text{heat}]{\text{Ag}_2\text{O}}$$

The green, blue, and red arrows show the three possible elimination routes. The corresponding products are:

The first (green) alkene has a disubstituted double bond. The second (blue) alkene is monosubstituted, and the red alkene (ethylene) has an unsubstituted double bond. We predict that the red products will be favored.

PROBLEM 19-20

Predict the major products formed when the following amines undergo exhaustive methylation, treatment with Ag$_2$O, and heating:

(a) 2-hexanamine   **(b)** 2-methylpiperidine   **(c)** N-ethylpiperidine

(d)   **(e)**   **(f)**

PROBLEM-SOLVING HINT

The key to solving Hofmann elimination problems is to find all possible ways the compound can eliminate.
Then, the one that gives the least substituted alkene is probably favored.

Amines are notoriously easy to oxidize, and oxidation is often a side reaction in amine syntheses. Amines also oxidize during storage in contact with the air. Preventing air oxidation is one of the reasons for converting amines to their salts for storage or use as medicines.

19-16
Oxidation of Amines; The Cope Elimination

The following partial structures show some of the bonding and oxidation states of amines:

Depending on their specific structures, these states are generally more oxidized as you go from left to right. (Note the increasing number of bonds to oxygen.)

Most amines are oxidized by common oxidants such as H_2O_2, permanganate, and peroxyacids. Primary amines oxidize easily, but complex mixtures of products often result. The following sequence shows increasingly oxidized products of a primary amine, as it becomes more oxidized from left to right. [O] is a generic oxidizing agent.

Secondary amines are easily oxidized to **hydroxylamines**. Side products are often formed, however, and the yields may be low. The mechanisms of amine oxidations are not well characterized, partly because many reaction paths (especially those involving free radicals) are available.

Tertiary amines are oxidized to **amine oxides**, often in good yields. Either H_2O_2 or a peroxyacid may be used for this oxidation. Notice that an amine oxide must be drawn with a full positive charge on nitrogen and a negative charge on oxygen, as in nitro compounds. Because the N—O bond of the amine oxide is formed by donation of the electrons on nitrogen, this bond is often written as an arrow (N → O) in older literature.

Because of the positive charge on nitrogen, the amine oxide may undergo a **Cope elimination** much like the Hofmann elimination of a quaternary ammonium salt. The amine oxide acts as its own base through a cyclic transition state, so a strong base is not needed. The Cope elimination generally gives the same orientation as Hofmann elimination, resulting in the least substituted alkene.

Primary amines are oxidized in the body by monoamine oxidases (MAO). MAO converts the amine to an imine, which is hydrolyzed to yield an aldehyde and ammonia. One function of MAO is to regulate the levels of the neurotransmitters serotonin and norepinephrine. Monoamine oxidase inhibitors prevent the oxidation (and inactivation) of these neurotransmitters, thereby elevating mood. MAO inhibitors were the first antidepressants, but they are used sparingly now because of numerous side effects.

serotonin

norepinephrine

MECHANISM   Cope Elimination of an Amine Oxide

[transition state]

Cope elimination occurs under milder conditions than Hofmann elimination. It is particularly useful when a sensitive or reactive alkene must be synthesized by the elimination of an amine. Because the Cope elimination involves a cyclic transition state, it occurs with syn stereochemistry.

SOLVED PROBLEM 19-3

Predict the products expected when the following compound is treated with H_2O_2 and heated:

SOLUTION

Oxidation converts the tertiary amine to an amine oxide. Cope elimination can give either of two alkenes. We expect the less hindered elimination to be favored, giving the Hofmann product.

PROBLEM 19-21

Give the products expected when the following tertiary amines are treated with a peroxyacid and heated:
(a) *N,N*-dimethyl-2-hexanamine
(b) *N,N*-diethyl-2-hexanamine
(c) cyclohexyldimethylamine
(d) *N*-ethylpiperidine

PROBLEM 19-22

When the (*R*,*R*) isomer of the amine shown below is treated with an excess of methyl iodide, then silver oxide, then heated, the major product is the Hofmann product.

(a) Draw the structure of the major (Hofmann) product.

(b) Some Saytzeff product is also formed. It has the (*E*) configuration. When the same amine is treated with MCPBA and heated, the Saytzeff product has the (*Z*) configuration. Use stereochemical drawings of the transition states to explain these observations.

19-17
Reactions of Amines with Nitrous Acid

Reactions of amines with nitrous acid (H—O—N=O) are particularly useful for synthesis. Because nitrous acid is unstable, it is generated in situ (in the reaction mixture) by mixing sodium nitrite ($NaNO_2$) with cold, dilute hydrochloric acid.

In an acidic solution, nitrous acid may protonate and lose water to give the nitrosonium ion, $^+N{=}O$. The nitrosonium ion appears to be the reactive intermediate in most reactions of amines with nitrous acid.

Reaction with Primary Amines: Formation of Diazonium Salts   Primary amines react with nitrous acid, via the nitrosonium ion, to give diazonium cations of the form $R{-}\overset{+}{N}{\equiv}N$. This procedure is called **diazotization of an amine**. Diazonium salts are the most useful products obtained from the reactions of amines with nitrous acid. The mechanism for diazonium salt formation begins with a nucleophilic attack on the nitrosonium ion to form an *N*-nitrosoamine.

MECHANISM   Diazotization of an Amine

Formation of the *N*-nitrosamine

(continued)

A proton transfer (a tautomerism) from nitrogen to oxygen forms a hydroxyl group and a second N—N bond.

$$R-\overset{H}{\underset{}{N}}-\ddot{N}=\overset{..}{\underset{..}{O}} + H_3O^+ \rightleftharpoons \left[R-\overset{H}{\underset{..}{N}}-\ddot{N}=\overset{..}{\underset{..}{O}}^+-H \longleftrightarrow R-\overset{H}{\underset{..}{N}}=N-\overset{..}{O}H \right] + H_2\ddot{O}: \rightleftharpoons$$

N-nitrosoamine   protonated *N*-nitrosoamine

$$R-\ddot{N}=N-\ddot{O}H + H_3O^+$$
second N—N bond formed

Protonation of the hydroxyl group, followed by loss of water, gives the diazonium cation.

$$R-\ddot{N}=\ddot{N}-\ddot{O}H \xrightarrow{H_3O^+} R-\ddot{N}=\ddot{N}-\overset{+}{O}H_2 \longrightarrow R-\overset{+}{N}\equiv N: + H_2\ddot{O}:$$
diazonium ion

The overall diazotization reaction is

$$R-\ddot{N}H_2 + NaNO_2 + 2HCl \longrightarrow R-\overset{+}{N}\equiv N\ Cl^- + H_2O + NaCl$$
primary amine   sodium nitrite   diazonium salt

Alkanediazonium salts are unstable. They decompose to give nitrogen and carbocations.

$$R-\overset{+}{N}\equiv N: \longrightarrow R^+ + :N\equiv N:$$
alkanediazonium cation   carbocation   nitrogen

The driving force for this reaction is the formation of N_2, an exceptionally stable molecule. The carbocations generated in this manner react like others we have seen: by nucleophilic attack to give substitution, by proton loss to give elimination, and by rearrangement. Because of the many competing reaction pathways, alkanediazonium salts usually decompose to give complex mixtures of products. Therefore, the diazotization of primary alkylamines is not widely used for synthesis.

Arenediazonium salts (formed from arylamines) are relatively stable, however, and they serve as intermediates in a variety of important synthetic reactions. These reactions are discussed in Section 19-18.

Reaction with Secondary Amines: Formation of *N*-Nitrosoamines   Secondary amines react with the nitrosonium ion to form secondary *N*-**nitrosoamines**, sometimes called *nitrosamines*.

$$R-\overset{H}{\underset{R}{N}}: + {}^+\overset{..}{N}=\overset{..}{O}: \rightleftharpoons R-\overset{H}{\underset{R}{N}}-\ddot{N}=\overset{..}{O}: \xrightarrow{H_2\ddot{O}:} R-\overset{..}{\underset{R}{N}}-\ddot{N}=\overset{..}{O}: + H_3O^+$$
2° amine   nitrosonium   2° *N*-nitrosoamine
              ion

Secondary *N*-nitrosoamines are stable under the reaction conditions because they do not have the N—H proton needed for the tautomerism (shown above with a primary amine) to form a diazonium ion. The secondary *N*-nitrosoamine usually separates from the reaction mixture as an oily liquid.

Small quantities of *N*-nitrosoamines have been shown to cause cancer in laboratory animals. These findings have generated concern about the common practice of using sodium nitrite to preserve meats such as bacon, ham, and hot dogs. When the meat is eaten, sodium nitrite combines with stomach acid to form nitrous acid, which can convert amines in the food to *N*-nitrosoamines. Because nitrites are naturally present in many other foods, it is unclear just how much additional risk is involved in using sodium nitrite to preserve meats. More research is being done in this area to evaluate the risk.

The most useful reaction of amines with nitrous acid is the reaction of arylamines to form arenediazonium salts. We consider next how these diazonium salts may be used as synthetic intermediates.

> **PROBLEM 19-23**
>
> Predict the products from the reactions of the following amines with sodium nitrite in dilute HCl:
> (a) cyclohexanamine (b) *N*-ethyl-2-hexanamine (c) piperidine (d) aniline

19-18 Reactions of Arenediazonium Salts

In contrast with alkanediazonium salts, arenediazonium salts are relatively stable in aqueous solutions around 0 to 10°C. Above these temperatures, they decompose, and they may explode if they are isolated and allowed to dry. The diazonium ($-\overset{+}{N}\equiv N$) group can be replaced by many different functional groups, including $-H$, $-OH$, $-CN$, and halogens.

Arenediazonium salts are formed by diazotizing a primary aromatic amine. Primary aromatic amines are commonly prepared by nitrating an aromatic ring, then reducing the nitro group to an amino ($-NH_2$) group. In effect, by forming and diazotizing an amine, an activated aromatic position can be converted into a wide variety of functional groups. For example, toluene might be converted to a variety of substituted derivatives by using this procedure:

The following flowchart shows some of the functional groups that can be introduced via arenediazonium salts:

PROBLEM-SOLVING HINT
These reactions of diazonium salts are extremely useful for solving aromatic synthesis problems.

	Products
Ar—OH	phenols
Ar—Cl (Br)	aryl halides
Ar—C≡N	benzonitriles
Ar—F (I)	aryl halides
Ar—H	(deamination)
Ar—N=N—Ar′	azo dyes

Replacement of the Diazonium Group by Hydroxide: Hydrolysis Hydrolysis takes place when a solution of an arenediazonium salt is strongly acidified (usually by adding H_2SO_4) and warmed. The hydroxyl group of water replaces N_2, forming a phenol. This is a useful laboratory synthesis of phenols because (unlike nucleophilic aromatic substitution) it does not require strong electron-withdrawing substituents or powerful bases and nucleophiles.

$$\text{Ar}\overset{+}{-}\text{N}\equiv\text{N} \quad \text{Cl}^- \xrightarrow[\text{H}_2\text{O}]{\text{H}_2\text{SO}_4,\ \text{heat}} \text{Ar}-\text{OH} + \text{N}_2\uparrow + \text{H}^+$$

Example

(75%)

Replacement of the Diazonium Group by Chloride, Bromide, and Cyanide: The Sandmeyer Reaction

Copper (I) salts (cuprous salts) have a special affinity for diazonium salts. Cuprous chloride, cuprous bromide, and cuprous cyanide react with arenediazonium salts to give aryl chlorides, aryl bromides, and aryl cyanides. It is often necessary to heat the reaction mixture to drive these reactions to completion. The use of cuprous salts to replace arenediazonium groups is called the **Sandmeyer reaction**. The Sandmeyer reaction (using cuprous cyanide) is an excellent method for attaching another carbon substituent to an aromatic ring.

The Sandmeyer reaction

$$\text{Ar}\overset{+}{-}\text{N}\equiv\text{N} \quad \text{Cl}^- \xrightarrow[(X = Cl,\ Br,\ C\equiv N)]{\text{CuX}} \text{Ar}-\text{X} + \text{N}_2\uparrow$$

Examples

(75%)

(90%)

(70%)

Replacement of the Diazonium Group by Fluoride and Iodide

When an arenediazonium salt is treated with fluoroboric acid (HBF_4), the diazonium fluoroborate precipitates out of solution. If this precipitated salt is filtered and then heated, it decomposes to give the aryl fluoride. Although this reaction requires the isolation and heating of a potentially explosive diazonium salt, it may be carried out safely if it is done carefully with the proper equipment. There are few other methods for making aryl fluorides.

$$Ar\overset{+}{-N}\equiv N \quad Cl^- \xrightarrow{HBF_4} Ar\overset{+}{-N}\equiv N \quad ^-BF_4 \xrightarrow{heat} Ar-F + N_2\uparrow + BF_3$$
diazonium fluoroborate

Example

Aryl iodides are formed by treating arenediazonium salts with potassium iodide. This is one of the best methods for making iodobenzene derivatives.

$$Ar\overset{+}{-N}\equiv N \quad Cl^- \xrightarrow{KI} Ar-I + N_2\uparrow$$

Example

Reduction of the Diazonium Group to Hydrogen: Deamination of Anilines

Hypophosphorus acid (H_3PO_2) reacts with arenediazonium salts, replacing the diazonium group with a hydrogen. In effect, this is a reduction of the arenediazonium ion.

$$Ar\overset{+}{-N}\equiv N \quad Cl^- \xrightarrow{H_3PO_2} Ar-H + N_2\uparrow$$

Example

This reaction is sometimes used to remove an amino group that was added to activate the ring. Solved Problem 19-4 shows how one might use this technique.

SOLVED PROBLEM 19-4

Show how you would convert toluene to 3,5-dibromotoluene in good yield.

SOLUTION

Direct bromination of toluene cannot give 3,5-dibromotoluene because the methyl group activates the ortho and para positions.

Starting with *p*-toluidine (*p*-methylaniline), however, the strongly activating amino group directs bromination to its ortho positions. Removal of the amino group (deamination) gives the desired product.

Diazonium Salts as Electrophiles: Diazo Coupling   Arenediazonium ions act as weak electrophiles in electrophilic aromatic substitutions. The products have the structure Ar—N=N—Ar, containing the —N=N— **azo** linkage. For this reason, the products are called azo compounds, and the reaction is called **diazo coupling**. Because they are weak electrophiles, diazonium salts react only with strongly activated rings (such as derivatives of aniline and phenol).

$$Ar—\overset{+}{N}≡N + H—Ar' \longrightarrow Ar—N=N—Ar' + H^+$$

diazonium ion   (activated)   an azo compound

Example

methyl orange (an indicator)

PROBLEM 19-24

Propose a mechanism for the synthesis of methyl orange.

Azo compounds bring two substituted aromatic rings into conjugation with an azo group, which is a strong chromophore. Therefore, most azo compounds are strongly colored, and they make excellent dyes, known as *azo dyes*. The diazo coupling syntheses of some common azo dyes follow.

para red

alizarin yellow

Diazo coupling often takes place in basic solutions because deprotonation of the phenolic —OH groups and the sulfonic acid and carboxylic acid groups helps to activate the aromatic rings toward electrophilic aromatic substitution. Many of the common azo dyes have one or more sulfonate (—SO_3^-) or carboxylate (—COO^-) groups on the molecule to promote solubility in water and to help bind the dye to the polar surfaces of common fibers such as cotton and wool.

PROBLEM 19-25

Show how you would convert aniline to the following compounds:
(a) fluorobenzene **(b)** chlorobenzene **(c)** 1,3,5-trimethylbenzene
(d) bromobenzene **(e)** iodobenzene **(f)** benzonitrile

(g) phenol **(h)**

(use aniline and resorcinol)

SUMMARY   Reactions of Amines

1. *Reaction as a proton base* (Section 19-5)

2. *Reactions with ketones and aldehydes* (Sections 18-16 and 18-17)

Y=H or alkyl gives an imine (Schiff base)

Y=OH gives an oxime

Y=NHR gives a hydrazone

3. *Alkylation* (Section 19-12)

$$R—\ddot{N}H_2 + R'—CH_2—Br \longrightarrow R—\overset{+}{N}H_2—CH_2—R' \ ^-Br$$
amine   primary halide   salt of alkylated amine

(Overalkylation is common.)

Examples

$$CH_3—CH_2—CH_2—NH_2 + 3\ CH_3—I \xrightarrow{NaHCO_3} CH_3—CH_2—CH_2—\overset{+}{N}(CH_3)_3 \ ^-I$$

$$excess\ NH_3 + CH_3CH_2CH_2CH_2CH_2—Br \longrightarrow CH_3CH_2CH_2CH_2CH_2—NH_2$$

4. *Acylation to form amides* (Section 19-13)

Example

$$H_2N—Ph + CH_3—\overset{\overset{O}{\|}}{C}—Cl \xrightarrow{\text{pyridine}} CH_3—\overset{\overset{O}{\|}}{C}—NH—Ph$$

aniline   acetyl chloride   acetanilide

5. *Reaction with sulfonyl chlorides to give sulfonamides* (Section 19-14)

$$R—\ddot{N}H_2 + Cl—\overset{\overset{O}{\|}}{\underset{\underset{O}{\|}}{S}}—R' \longrightarrow R—\ddot{N}H—\overset{\overset{O}{\|}}{\underset{\underset{O}{\|}}{S}}—R' + HCl$$

amine

sulfonyl chloride   sulfonamide

Example

$$CH_3(CH_2)_3—NH_2 + Cl—\overset{\overset{O}{\|}}{\underset{\underset{O}{\|}}{S}}—Ph \longrightarrow CH_3(CH_2)_3NH—\overset{\overset{O}{\|}}{\underset{\underset{O}{\|}}{S}}—Ph + HCl$$

1-butanamine

benzenesulfonyl   *N*-butyl benzenesulfonamide
chloride

6. *Hofmann and Cope eliminations*

a. Hofmann elimination (Section 19-15)

Conversion to quaternary ammonium hydroxide

$$R—CH_2—CH_2—\ddot{N}H_2 \xrightarrow{3\ CH_3I} R—CH_2—CH_2—\overset{+}{N}(CH_3)_3\ ^-I$$

$$\xrightarrow{Ag_2O} R—CH_2—CH_2—\overset{+}{N}(CH_3)_3\ ^-OH$$

Elimination

$$HO^- \quad \overset{H}{\underset{R—\overset{|}{\underset{H}{C}}—\overset{|}{\underset{\overset{+}{N}(CH_3)_3}{C}}—H}{\,}} \xrightarrow{\text{heat}} H—O—H \quad \overset{R}{\underset{H}{\,}}C{=}C\overset{H}{\underset{H}{\,}} \quad :N(CH_3)_3$$

Hofmann elimination usually gives the least substituted alkene.

Example

$$\overset{1}{C}H_3—\overset{2}{C}H—\overset{3}{C}H_2—\overset{4}{C}H_3 \xrightarrow{150°C} CH_3—CH{=}CH—CH_3 + H_2C{=}CH—CH_2—CH_3$$
$$\underset{^+N(CH_3)_3\ ^-OH}{|}$$

(Saytzeff product)   (Hofmann product)
(5%)   (95%)

b. Cope elimination of a tertiary amine oxide (Section 19-16)

$$\overset{H\quad :N(CH_3)_2}{\underset{R—\overset{|}{\underset{H}{C}}—\overset{|}{\underset{H}{C}}—R'}{}} \xrightarrow[\text{or } H_2O_2]{\text{peracid}} \overset{H\quad \overset{-O}{\overset{|}{\underset{N(CH_3)_2}{+}}}}{\underset{R—\overset{|}{\underset{H}{C}}—\overset{|}{\underset{H}{C}}—R'}{}} \xrightarrow{\text{heat}} \overset{HO—\ddot{N}(CH_3)_2}{\underset{H}{\overset{R}{\,}}C{=}C\overset{R'}{\underset{H}{\,}}}$$

Cope elimination also gives the least highly substituted alkene.

7. *Oxidation* (Section 19-16)

a. Secondary amines

$$R_2\ddot{N}—H + H_2O_2 \longrightarrow R_2\ddot{N}—OH + H_2O$$

2° amine   a 2° hydroxyamine

b. Tertiary amines

$$R_3N: + H_2O_2 \longrightarrow R_3N^{\pm}O^- + H_2O$$

3° amine   (or ArCO_3H)   3° amine oxide   (or ArCOOH)

8. *Diazotization* (Section 19-17)

$$R-\overset{\cdot\cdot}{N}H_2 \xrightarrow{\text{NaNO}_2,\ \text{HCl}} R-\overset{+}{N}\equiv N\text{: Cl}^-$$

primary alkylamine   alkanediazonium salt

$$Ar-\overset{\cdot\cdot}{N}H_2 \xrightarrow{\text{NaNO}_2,\ \text{HCl}} Ar-\overset{+}{N}\equiv N\text{: Cl}^-$$

primary arylamine   arenediazonium salt

a. Reactions of diazonium salts (Section 19-18)

  (I) Hydrolysis

$$Ar-\overset{+}{N}\equiv N\text{: Cl}^- \xrightarrow[\text{H}_2\text{O}]{\text{H}_2\text{SO}_4,\ \text{heat}} Ar-OH + N_2\uparrow + HCl$$

Example

$$Ph-\overset{+}{N}\equiv N\text{: Cl}^- \xrightarrow[\text{H}_2\text{O}]{\text{H}_2\text{SO}_2,\ \text{heat}} Ph-OH + N_2\uparrow + HCl$$

benzenediazonium chloride   phenol

  (II) The Sandmeyer reaction

$$Ar-\overset{+}{N}\equiv N\text{: Cl}^- \xrightarrow[\text{X=Cl, Br, C}\equiv\text{N}]{\text{CuX}} Ar-X + N_2\uparrow$$

Examples

$$Ph-\overset{+}{N}\equiv N\text{: Cl}^- \xrightarrow{\text{CuCl}} Ph-Cl + N_2\uparrow$$

benzenediazonium chloride   chlorobenzene

p-nitrobenzenediazonium chloride   *p*-nitrobenzonitrile

  (III) Replacement by fluoride or iodide

$$Ar-\overset{+}{N}\equiv N\text{: Cl}^- \xrightarrow{\text{HBF}_4} Ar-\overset{+}{N}\equiv N\text{: BF}_4^- \xrightarrow{\text{heat}} Ar-F + N_2\uparrow + BF_3$$

$$Ar-\overset{+}{N}\equiv N\text{: Cl}^- \xrightarrow{\text{KI}} Ar-I + N_2\uparrow + KCl$$

Example

2-iodonaphthalene

(IV) Reduction to hydrogen

$$Ar-\overset{+}{N}\equiv N\text{: Cl}^- \xrightarrow{\text{H}_3\text{PO}_2} Ar-H + N_2\uparrow$$

Example

p-ethylaniline   ethylbenzene

(V) Diazo coupling

$$Ar \overset{+}{-N} \equiv N: \; + \; H-Ar' \longrightarrow Ar-\overset{..}{N} = \overset{..}{N} - Ar' \; + \; H^+$$
diazonium ion   (activated)   an azo compound

Example

methyl orange
(an indicator)

+ HCl

19-19
Synthesis of Amines

Many methods are available for making amines. Most of these methods derive from the reactions of amines covered in the preceding sections. The most common amine syntheses start with ammonia or an amine and add another alkyl group. Such a process converts ammonia to a primary amine, or a primary amine to a secondary amine, or a secondary amine to a tertiary amine.

$$:NH_3 \longrightarrow \longrightarrow R-\overset{..}{N}H_2$$
ammonia   1° amine

$$\overset{\backslash}{\underset{/}{N}}-H \longrightarrow \longrightarrow \overset{\backslash}{\underset{/}{N}}-R$$
1° or 2° amine   2° or 3° amine

Reductive amination is the most general technique, capable of adding a primary or secondary alkyl group to an amine. Acylation–reduction can only add a primary alkyl group, and (like reductive amination) it can form an amine with one, two, or three alkyl substituents. Finally, we will consider some specialized syntheses that are limited to making primary amines.

19-19A   Reductive Amination

Reductive amination is the most general method for synthesizing amines. First we form an imine or oxime derivative of a ketone or aldehyde, and then reduce it to the amine. In effect, reductive amination adds one alkyl group to the nitrogen atom. The product can be a primary, secondary, or tertiary amine, depending on whether the amine used as the starting material had zero, one, or two alkyl groups.

Primary Amines   Primary amines result from condensation of hydroxylamine (zero alkyl groups) with a ketone or an aldehyde, followed by reduction of the oxime.

Hydroxylamine is used in place of ammonia because most oximes are stable, easily isolated compounds. The oxime is reduced using catalytic reduction, lithium aluminum hydride, or zinc and HCl.

Examples

Secondary Amines   Condensation of a ketone or an aldehyde with a primary amine forms an *N*-substituted imine (a Schiff base). Reduction of the imine gives a secondary amine.

Example

Tertiary Amines   Condensation of a ketone or an aldehyde with a secondary amine gives an iminium salt. Iminium salts are frequently unstable, so they are rarely isolated. A reducing agent in the solution reduces the iminium salt to a tertiary amine. The reducing agent must be capable of reducing the iminium salt, but it must not reduce the carbonyl group of the ketone or aldehyde. Sodium triacetoxyborohydride [$Na(CH_3COO)_3BH$ or $Na(AcO)_3BH$] works well for this reduction because it is less reactive than sodium borohydride and it does not reduce the carbonyl group. For these reductions of iminium salts, sodium triacetoxyborohydride has largely replaced an older reagent, sodium cyanoborohydride ($NaBH_3CN$), which is more toxic and not as effective.

Example

cyclohexanone   iminium salt   *N,N*-dimethycyclohexylamine

(85%)

SOLVED PROBLEM 19-5

Show how to synthesize the following amines from the indicated starting materials:
(a) *N*-cyclopentylaniline from aniline   (b) *N*-ethylpyrrolidine from pyrrolidine

SOLUTION

(a) This synthesis requires adding a cyclopentyl group to aniline (primary) to make a secondary amine. Cyclopentanone is the carbonyl compound.

aniline   cyclopentanone

PROBLEM-SOLVING $\mathbf{HINT}$

Reductive amination is the most useful amine synthesis: It adds a 1° or 2° alkyl group to nitrogen. Use an aldehyde to add a 1° group, and a ketone to add a 2° group.

(b) This synthesis requires adding an ethyl group to a secondary amine to make a tertiary amine. The carbonyl compound is acetaldehyde. Formation of a tertiary amine by reductive amination involves an iminium intermediate, which is reduced by $Na(AcO)_3BH$.

pyrrolidine   acetaldehyde

PROBLEM 19-26

Show how to synthesize the following amines from the indicated starting materials by reductive amination:
(a) benzylmethylamine from benzaldehyde

(b) $Ph—CH_2—\overset{:NH_2}{\underset{|}{CH}}—CH_3$ from $Ph—CH_2—\overset{O}{\overset{||}{C}}—CH_3$
     (±)-amphetamine   1-phenyl-2-propanone
(c) *N*-benzylpiperidine from piperidine
(d) *N*-cyclohexylaniline from cyclohexanone
(e) cyclohexylamine from cyclohexanone

(f) from compounds containing no more than five carbon atoms

hydroxylamine ⟶ primary amine
primary amine ⟶ secondary amine
secondary amine ⟶ tertiary amine

19-19B   Acylation–Reduction

The second general synthesis of amines is by **acylation–reduction**. Like reductive amination, acylation–reduction adds one alkyl group to the nitrogen atom of the starting amine. Acylation of the starting amine by an acid chloride gives an amide, with no tendency toward overacylation (Section 19-13). Reduction of the amide by lithium aluminum hydride (LiAlH$_4$) gives the corresponding amine.

$$R-\ddot{N}H_2 + Cl-\overset{\overset{\displaystyle O}{\|}}{C}-R' \xrightarrow[\substack{\text{pyridine} \\ \text{or NaOH}}]{\text{acylation}} R-\ddot{N}H-\overset{\overset{\displaystyle O}{\|}}{C}-R' \xrightarrow[\substack{(2)\ H_2O}]{\substack{\text{reduction} \\ (1)\ LiAlH_4}} R-\ddot{N}H-CH_2-R'$$

amine · acid chloride · amide · alkylated amine

Acylation–reduction converts ammonia to a primary amine, a primary amine to a secondary amine, or a secondary amine to a tertiary amine. These reactions are quite general, with one restriction: The added alkyl group is always 1° because the carbon bonded to nitrogen is derived from the carbonyl group of the amide, reduced to a methylene (—CH$_2$—) group.

Primary amines

$$R-\overset{\overset{\displaystyle O}{\|}}{C}-Cl + \ddot{N}H_3 \longrightarrow R-\overset{\overset{\displaystyle O}{\|}}{C}-\ddot{N}H_2 \xrightarrow[(2)\ H_2O]{(1)\ LiAlH_4} R-CH_2-\ddot{N}H_2$$

acid chloride · ammonia · 1° amide · 1° amine

Example

$$CH_3-\underset{\underset{\displaystyle CH_3}{|}}{CH}-CH_2-\overset{\overset{\displaystyle O}{\|}}{C}-Cl + \ddot{N}H_3 \longrightarrow CH_3-\underset{\underset{\displaystyle CH_3}{|}}{CH}-CH_2-\overset{\overset{\displaystyle O}{\|}}{C}-\ddot{N}H_2 \xrightarrow[(2)\ H_2O]{(1)\ LiAlH_4}$$

3-methylbutanoyl chloride · 3-methylbutanamide

$$CH_3-\underset{\underset{\displaystyle CH_3}{|}}{CH}-CH_2-CH_2-\ddot{N}H_2$$

3-methyl-1-butanamine

Secondary amines

$$R-\overset{\overset{\displaystyle O}{\|}}{C}-Cl + R'-\ddot{N}H_2 \longrightarrow R-\overset{\overset{\displaystyle O}{\|}}{C}-\ddot{N}H-R' \xrightarrow[(2)\ H_2O]{(1)\ LiAlH_4} R-CH_2-\ddot{N}H-R'$$

acid chloride · primary amine · *N*-substituted amide · 2° amine

Example

$$CH_3CH_2CH_2-\overset{\overset{\displaystyle O}{\|}}{C}-Cl + \ddot{N}H_2\text{—}\bigcirc \longrightarrow CH_3CH_2CH_2-\underset{\underset{\displaystyle O}{\|}}{C}-\ddot{N}H\text{—}\bigcirc \xrightarrow[(2)\ H_2O]{(1)\ LiAlH_4} CH_3CH_2CH_2-CH_2-\ddot{N}H\text{—}\bigcirc$$

butanoyl chloride · aniline · *N*-phenylbutanamide · *N*-butylaniline

Tertiary amines

$$R-\overset{\overset{\displaystyle O}{\|}}{C}-Cl + R'_2\ddot{N}H \longrightarrow R-\overset{\overset{\displaystyle O}{\|}}{C}-\ddot{N}R'_2 \xrightarrow[(2)\ H_2O]{(1)\ LiAlH_4} R-CH_2-\ddot{N}R'_2$$

acid chloride · secondary amine · *N,N*-disubstituted amide · 3° amine

Example

$$\underset{\text{benzoyl chloride}}{\overset{\overset{\displaystyle O}{\underset{\displaystyle C}{\|}}\text{—}Cl}{\bigcirc}} + H-\ddot{N}(CH_2CH_3)_2 \longrightarrow \underset{\text{N,N-diethylbenzamide}}{\overset{(CH_3CH_2)_2\ddot{N}\overset{\displaystyle O}{\underset{\displaystyle C}{\diagdown}}}{\bigcirc}} \xrightarrow[(2)\ H_2O]{(1)\ LiAlH_4} \underset{\text{benzyldiethylamine}}{\overset{(CH_3CH_2)_2\ddot{N}\diagdown CH_2}{\bigcirc}}$$

benzoyl chloride · diethylamine · *N,N*-diethylbenzamide · benzyldiethylamine

SOLVED PROBLEM 19-6

Show how to synthesize *N*-ethylpyrrolidine from pyrrolidine using acylation–reduction.

SOLUTION

This synthesis requires adding an ethyl group to pyrrolidine to make a tertiary amine. The acid chloride needed will be acetyl chloride (ethanoyl chloride). Reduction of the amide gives *N*-ethylpyrrolidine.

Compare this synthesis with Solved Problem 19-5(b) to show how reductive amination and acylation–reduction can accomplish the same result.

PROBLEM 19-27

Show how to synthesize the following amines from the indicated starting materials by acylation–reduction:

(a) *N*-propylpiperidine from piperidine   **(b)** *N*-benzylaniline from aniline

---

PROBLEM-SOLVING HINT

Like reductive amination, acylation–reduction adds an alkyl group to nitrogen. It is more restrictive, though, because the group added is always 1°.

1° group added

ammonia → primary amine
primary amine → secondary amine
secondary amine → tertiary amine

19-19C   Syntheses Limited to Primary Amines

Primary amines are the most common class of amines, and they are also needed as starting materials for synthesis of secondary and tertiary amines. Many methods have been developed for making primary amines, ranging from simple alkylation of ammonia to sophisticated multistep syntheses. We will consider some of the more common syntheses.

Direct Alkylation and Gabriel Synthesis   The S_N2 reaction of amines with alkyl halides is complicated by a tendency for over-alkylation to form a mixture of monoalkylated and polyalkylated products (Section 19-12). Simple primary amines can be synthesized, however, by adding a large excess of ammonia to a halide or tosylate that is a good S_N2 substrate. Because there is a large excess of ammonia present, the probability that a molecule of the halide will alkylate ammonia is much larger than the probability that it will overalkylate the amine product.

$$\text{R—CH}_2\text{—X} + \text{excess NH}_3 \longrightarrow \text{R—CH}_2\text{—NH}_2 + \text{NH}_4\text{X}$$

Example

$$\text{CH}_3\text{CH}_2\text{CH}_2\text{CH}_2\text{CH}_2\text{—Br} + \text{excess NH}_3 \longrightarrow \text{CH}_3\text{CH}_2\text{CH}_2\text{CH}_2\text{CH}_2\text{—NH}_2 + \text{NH}_4\text{Br}$$

1-bromopentane   1-pentanamine

PROBLEM 19-28

Show how you would use direct alkylation of ammonia to synthesize 1-heptanamine.

In 1887, Siegmund Gabriel (at the University of Berlin) developed the **Gabriel amine synthesis** for making primary amines without danger of overalkylation. He used the phthalimide anion as a protected form of ammonia that cannot alkylate more than once.

Phthalimide has one acidic N—H proton (pK_a 8.3) that is abstracted by potassium hydroxide to give the phthalimide anion.

phthalimide $\xrightarrow[\text{H}_2\text{O}]{\text{KOH}}$ resonance-stabilized phthalimide anion

The phthalimide anion is a strong nucleophile, displacing a halide or tosylate ion from a good S$_N$2 substrate. Heating the *N*-alkyl phthalimide with hydrazine displaces the primary amine, giving the very stable hydrazide of phthalimide.

phthalimide anion $\xrightarrow[\substack{\text{alkyl halide} \\ \text{(usually 1°)}}]{\text{R—X}}$ *N*-alkyl phthalimide $\xrightarrow[\text{heat}]{\text{H}_2\ddot{\text{N}}-\ddot{\text{N}}\text{H}_2}$ phthalimide hydrazide + R—$\ddot{\text{N}}$H$_2$ primary amine

Example

Br—CH$_2$CH$_2$CHCH$_3$ (CH$_3$) isopentyl bromide $\xrightarrow[\text{(phthalimide anion)}]{}$ *N*-isopentylphthalimide :N—CH$_2$CH$_2$CHCH$_3$ $\xrightarrow[\text{hydrazine}]{\text{H}_2\text{N—NH}_2}$ H$_2$$\ddot{\text{N}}$—CH$_2CH_2$CHCH$_3$ isopentylamine (95%)

PROBLEM 19-29

Show how Gabriel syntheses might be used to prepare the following amines:
(a) benzylamine **(b)** 1-hexanamine **(c)** γ-aminobutyric acid

Reduction of Azides and Nitriles Just as Gabriel used the anion of phthalimide to put the nitrogen atom into a primary amine, we can use other nucleophiles as well. We need a good nucleophile that can alkylate only once, and that is easily converted to an amino group. Two good nucleophiles for introducing a nitrogen atom are the azide ion and the cyanide ion. Azide ion introduces (after reduction) an —NH$_2$ group, and cyanide ion introduces a —CH$_2$—NH$_2$ group.

Formation and Reduction of Azides **Azide** ion ($^-$N$_3$) is an excellent nucleophile that displaces leaving groups from unhindered primary and secondary alkyl halides and tosylates. The products are alkyl azides (RN$_3$), which have no tendency to react further. Azides are easily reduced to primary amines, either by LiAlH$_4$ or by catalytic hydrogenation. Alkyl azides can be explosive, so they are reduced without purification.

R—X + Na$^+$:$\ddot{\text{N}}$=$\overset{+}{\text{N}}$=$\ddot{\text{N}}$: $\xrightarrow{\text{S}_N2}$ [R—N=$\overset{+}{\text{N}}$=$\ddot{\text{N}}$: $\longleftrightarrow$ R—$\ddot{\text{N}}$—$\overset{+}{\text{N}}$≡N:] $\xrightarrow[\text{or H}_2/\text{Pd}]{\text{LiAlH}_4}$ R—$\ddot{\text{N}}$H$_2$

halide or tosylate (must be 1° or 2°)   sodium azide   an alkyl azide   1° amine

Examples

1-bromo-2-phenylethane → 2-phenylethyl azide → 2-phenylethylamine (89%)

cyclohexyl bromide → cyclohexyl azide → cyclohexylamine (54%)

Azide ion also reacts with a variety of other electrophiles. The following example shows how an azide ion opens an epoxide, (the product can be reduced to an amino alcohol):

epoxycyclohexane

Formation and Reduction of Nitriles Like the azide ion, cyanide ion ($^-$:C≡N:) is a good S_N2 nucleophile; it displaces leaving groups from unhindered primary and secondary alkyl halides and tosylates. The product is a **nitrile** (R—C≡N), which has no tendency to react further. Nitriles are reduced to primary amines by lithium aluminum hydride or by catalytic hydrogenation.

R—X + $^-$:C≡N: ⟶ R—C≡N: $\xrightarrow[\text{or } H_2/\text{catalyst}]{\text{LiAlH}_4}$ R—CH$_2$—N̈H$_2$

halide or tosylate (must be 1° or 2°) nitrile amine (one carbon added)

Example

1-bromopropane → butanenitrile → 1-butanamine (70%)

When the cyano (—C≡N) group is added and reduced, the resulting amine has an additional carbon atom. In effect, the cyanide substitution–reduction process is like adding —CH$_2$—NH$_2$. The following synthesis makes 2-phenylethylamine, which we also made by the azide synthesis:

benzyl bromide phenylacetonitrile 2-phenylethylamine

Notice that the starting material in this case has one less carbon atom because the cyanide synthesis adds both a carbon and a nitrogen.

We have seen (Section 18-15) that cyanide ion adds to ketones and aldehydes to form cyanohydrins. Reduction of the —C≡N group of the cyanohydrin provides a way to synthesize β-hydroxy amines.

cyclopentanone   cyclopentanone cyanohydrin   1-(methylamino)cyclopentanol

PROBLEM-SOLVING $\mathbf{HINT}$

To convert an alkyl halide (or alcohol, via the tosylate) to an amine, form the azide and reduce. To convert it to an amine with an additional carbon atom, form the nitrile and reduce. In either case, the alkyl group must be suitable for S_N2 displacement.

PROBLEM 19-30

Show how you would accomplish the following synthetic conversions:
(a) benzyl bromide → benzylamine
(b) 1-bromo-2-phenylethane → 3-phenylpropanamine
(c) pentanoic acid → 1-pentanamine
(d) pentanoic acid → 1-hexanamine
(e) (R)-2-bromobutane → (S)-2-butanamine
(f) (R)-2-bromobutane → (S)-2-methyl-1-butanamine
(g) 2-hexanone → 1-amino-2-methyl-2-hexanol

Reduction of Nitro Compounds Both aromatic and aliphatic nitro groups are easily reduced to amino groups. The most common methods are catalytic hydrogenation and acidic reduction by an active metal.

$$R-NO_2 \xrightarrow[\textit{or active metal and } H^+]{H_2/\text{catalyst}} R-NH_2$$

catalyst = Ni, Pd, or Pt
active metal = Fe, Zn, or Sn

Examples

o-nitrotoluene   o-toluidine (90%)

2-nitropentane   2-aminopentane (85%)

The most common reason for reducing aromatic nitro compounds is to make substituted anilines. Much of this chemistry was developed by the dye industry, which uses aniline derivatives for azo coupling reactions (Section 19-18) to make aniline dyes. Nitration of an aromatic ring (by electrophilic aromatic substitution) gives a nitro compound, which is reduced to the aromatic amine.

$$Ar-H \xrightarrow{HNO_3, H_2SO_4} Ar-NO_2 \xrightarrow{\text{reduction}} Ar-NH_2$$

For example, nitration followed by reduction is used in the synthesis of benzocaine (a topical anesthetic), shown on the next page. Notice that the stable nitro group is retained through an oxidation and esterification. The final step reduces the nitro group to the relatively sensitive amine (which could not survive the oxidation step).

The reaction scheme showing:

CH₃ (toluene) →[HNO₃, H₂SO₄, nitration]→ CH₃ with NO₂ (p-nitrotoluene) →[(1) KMnO₄, ⁻OH; (2) H⁺, oxidation]→ C(=O)—OH with NO₂ →[CH₃CH₂OH, H⁺ (see Section 11-12), esterification]→ C(=O)—OCH₂CH₃ with NO₂

C(=O)—OCH₂CH₃ with NO₂ →[Zn, HCl / CH₃CH₂OH, reduction]→ C(=O)—OCH₂CH₃ with :NH₂ · HCl

benzocaine · HCl

PROBLEM 19-31

Show how you would prepare the following aromatic amines by aromatic nitration, followed by reduction. You may use benzene and toluene as your aromatic starting materials.

(a) aniline   (b) *p*-bromoaniline
(c) *m*-bromoaniline   (d) *m*-aminobenzoic acid

The Hofmann Rearrangement of Amides   In the presence of a strong base, primary amides react with chlorine or bromine to form shortened amines with loss of the carbonyl carbon atom. This reaction, called the **Hofmann rearrangement**, is used to synthesize primary alkyl and aryl amines.

The Hofmann rearrangement

$$R—\overset{\overset{\text{O}}{\|}}{C}—\ddot{N}H_2 + X_2 + 4\,NaOH \longrightarrow R—\ddot{N}H_2 + 2\,NaX + Na_2CO_3 + 2\,H_2O$$

primary amide   (X₂ = Cl₂ or Br₂)   amine

Although we have several other methods for making primary amines, most of them depend on S_N2 displacements, which fail with 3° alkyl groups. The Hofmann rearrangement can produce primary amines with 1°, 2°, or 3° alkyl groups, or aryl amines. The following examples suggest the wide variety of amines that are accessible by the Hofmann rearrangement:

Examples

$$CH_3CH_2CH_2CH_2CH_2—\overset{\overset{\text{O}}{\|}}{C}—NH_2 \xrightarrow[H_2O]{Cl_2,\ ^-OH} CH_3CH_2CH_2CH_2CH_2—NH_2$$

hexanamide   1-pentanamine (90%)

2-methyl-2-phenylpropanamide →[Cl₂, ⁻OH / H₂O]→ 2-phenyl-2-propanamine

p-nitrobenzamide →[Br₂, ⁻OH / H₂O]→ *p*-nitroaniline

The mechanism of the Hofmann rearrangement is particularly interesting because it involves some intermediates that we have not encountered before. The first step is the replacement of one of the hydrogens on nitrogen by a halogen. This step is possible because the amide N—H protons are slightly acidic, and a strong base deprotonates a small fraction of the amide molecules. The depronated amide is a strong nucleophile, and it attacks bromine to form the *N*-bromo amide.

MECHANISM   Hofmann Rearrangement

Deprotonation of the amide and nucleophilic attack on bromine

Because of the electronegative nature of bromine, the *N*-bromo amide is more easily deprotonated than the original primary amide. Deprotonation of the *N*-bromo amide gives another resonance-stabilized anion. The deprotonated amide has a bromine atom present as a potential leaving group. In order for bromide to leave, however, the alkyl (R—) group must migrate to nitrogen. This is the actual rearrangement step, giving an isocyanate intermediate.

Second deprotonation and rearrangement with loss of bromide

Isocyanates react rapidly with water to give carbamic acids.

Decarboxylation of the carbamic acid gives the amine and carbon dioxide.

PROBLEM-SOLVING HINT

The Hofmann rearrangement mechanism is long and complicated, but it can be broken down:
1. Deprotonation and bromination to form the bromoamide, then another deprotonation.
2. Rearrangement to an isocyanate, with bromide as the leaving group.
3. Hydroxide attack on the carbonyl of the isocyanate.
4. Decarboxylation of the carbamic acid.
(Step 2, the actual rearrangement, may be easier to understand if you compare it with the Curtius rearrangement, in Problem 19-34.)

PROBLEM 19-32

Propose a mechanism for the following Hofmann rearrangement used in the synthesis of phentermine, an appetite suppressant.

PROBLEM 19-33

When (*R*)-2-methylbutanamide reacts with bromine in a strong aqueous solution of sodium hydroxide, the product is an optically active amine. Give the structure of the expected product, and use your knowledge of the reaction mechanism to predict its stereochemistry.

*PROBLEM 19-34

The *Curtius rearrangement* accomplishes the same synthetic goal as the Hofmann rearrangement, and it takes place by a similar mechanism. An acid chloride reacts with azide ion to give an acyl azide, which undergoes Curtius rearrangement when heated.

acyl azide

R—N=C=O: $\xrightarrow{\text{H}_2\text{O}}$ R—NH₂

isocyanate   amine

(a) The Curtius rearrangement takes place through a much shorter mechanism than the Hofmann rearrangement. Which step(s) of the Hofmann rearrangement resemble the Curtius rearrangement?
(b) Bromide serves as the leaving group in the Hofmann rearrangement. What is the leaving group in the Curtius rearrangement?
(c) Propose a mechanism for the following reaction:

SUMMARY Synthesis of Amines

1. *Reductive amination* (Section 19-19A)
   a. Primary amines

ketone or aldehyde   oxime   1° amine

Example

cyclopentanone   cyclopentanone oxime   cyclopentylamine

   b. Secondary amines

ketone or aldehyde   *N*-substituted imine   2° amine

Example

acetone   phenylisopropylamine

c. Tertiary amines

Example

cyclohexanone   iminium salt   *N,N*-dimethylcyclohexylamine

2. *Acylation–reduction of amines* (Section 19-19B)

Example

aniline   butanoyl chloride   *N*-phenylbutanamide

N-butylaniline (2°)

3. *Alkylation of ammonia* (Section 19-19C)

$$R—CH_2—X + excess \ \ddot{N}H_3 \longrightarrow R—CH_2—\ddot{N}H_2 + HX$$

Example

benzyl bromide   benzylamine

4. *The Gabriel synthesis of primary amines* (Section 19-19C)

N-alkyl phthalimide

5. *Reduction of azides* (Section 19-19C)

$$R-\ddot{N}=\overset{+}{N}=\ddot{N}:^{-} \xrightarrow[\text{or } H_2/Pd]{LiAlH_4} R-\ddot{N}H_2$$

alkyl azide   1° amine

Example

cyclohexyl bromide → NaN₃ → cyclohexyl azide → (1) LiAlH₄ (2) H₂O → cyclohexylamine

6. *Reduction of nitriles* (Section 19-19C)

$$R-C\equiv N: \xrightarrow{H_2/\text{catalyst or } LiAlH_4} R-CH_2-\ddot{N}H_2$$

nitrile   1° amine

Example

benzyl bromide → ⁻:C≡N: → phenylacetonitrile → H₂ / Ni → β-phenylethylamine

7. *Reduction of nitro compounds* (Section 19-19C)

$$R-NO_2 \xrightarrow[\text{or active metal and } H^+]{H_2/\text{catalyst}} R-\ddot{N}H_2$$

catalyst = Ni, Pd, or Pt
active metal = Fe, Zn, or Sn

Example

nitrobenzene → Sn, H₂SO₄ → → ⁻OH → aniline

8. *The Hofmann rearrangement* (Section 19-19C)

$$R-\overset{\overset{O}{\|}}{C}-\ddot{N}H_2 + X_2 + 4\,NaOH \longrightarrow R-\ddot{N}H_2 + 2\,NaX + Na_2CO_3 + 2\,H_2O$$

1° amide   (X₂ = Cl₂ or Br₂)   amine

Example

$$CH_3CH_2CH_2CH_2CH_2-\overset{\overset{O}{\|}}{C}-NH_2 \xrightarrow[H_2O]{Cl_2,\ ^-OH} CH_3CH_2CH_2CH_2CH_2-NH_2$$

hexanamide   pentanamine

9. *Nucleophilic aromatic substitution* (Section 17-12)

$$R-\ddot{N}H_2 + Ar-X \longrightarrow R-\ddot{N}H-Ar + HX$$

(The aromatic ring should be activated toward nucleophilic attack.)

Example

CH₃CH₂—NH₂ + 2,4-dinitrofluorobenzene → N-ethyl-2,4-dinitroaniline

ethylamine   2,4-dinitrofluorobenzene   N-ethyl-2,4-dinitroaniline

Chapter 19 Glossary

acylation Addition of an **acyl group** ($R - \overset{\overset{\displaystyle O}{\|}}{C} -$), usually replacing a hydrogen atom. Acylation of an amine gives an amide. (p. 859)

$$R - NH_2 + Cl - \overset{\overset{\displaystyle O}{\|}}{C} - R' \longrightarrow R - NH - \overset{\overset{\displaystyle O}{\|}}{C} - R' + HCl$$

amine   acid chloride   amide

    acetylation: Acylation by an acetyl group ($CH_3 - \overset{\overset{\displaystyle O}{\|}}{C} -$).

acylation–reduction A method for synthesizing amines by acylating ammonia or an amine, then reducing the amide. (p. 879)

$$R - \ddot{N}H_2 + R' - \overset{\overset{\displaystyle O}{\|}}{C} - Cl \longrightarrow R - \ddot{N}H - \overset{\overset{\displaystyle O}{\|}}{C} - R' \xrightarrow[\text{(2) } H_2O]{\text{(1) LiAlH}_4} R - \ddot{N}H - CH_2 - R'$$

amine   acid chloride   amide   alkylated amine

amine A derivative of ammonia with one or more alkyl or aryl groups bonded to the nitrogen atom. (p. 836)
    A **primary amine** (1° amine) has one alkyl group bonded to nitrogen.
    A **secondary amine** (2° amine) has two alkyl groups bonded to nitrogen.
    A **tertiary amine** (3° amine) has three alkyl groups bonded to nitrogen.

$$R - \underset{\ddot{}}{N} - H \qquad R - \underset{\ddot{}}{N} - R' \qquad R - \underset{\ddot{}}{N} - R'$$

primary amine   secondary amine   tertiary amine

amino group The $- NH_2$ group. If alkylated, it becomes an **alkylamino** group, $-NHR$ or a **dialkylamino** group. $-NR_2$. (p. 838)

amine oxide An amine with a fourth bond to an oxygen atom. In the amine oxide, the nitrogen atom bears a positive charge, and the oxygen atom bears a negative charge. (p. 866)

$$R - \overset{\overset{\displaystyle O^-}{|}}{\underset{\underset{\displaystyle R''}{|}}{N^+}} - R' \qquad R - NH_3^+ \ X^- \qquad R - \overset{\overset{\displaystyle R}{|}}{\underset{\underset{\displaystyle R}{|}}{N^+}} - R \ \ X^-$$

an amine oxide   an ammonium salt   a quaternary ammonium salt

ammonium salt (amine salt) A derivative of an amine with a positively charged nitrogen atom having four bonds. An amine is protonated by an acid to give an ammonium salt. (p. 846) A **quaternary ammonium salt** has a nitrogen atom bonded to four alkyl or aryl groups. (p. 837)

azide A compound having the azido group, $-N_3$. (p. 882)

$$\left[CH_3CH_2 - \overset{-}{\underset{\ddot{}}{N}} - \overset{+}{N} \equiv N: \longleftrightarrow CH_3CH_2 - \ddot{N} = \overset{+}{N} = \overset{-}{\ddot{N}:} \right]$$

ethyl azide

base-dissociation constant (K_b) A measure of the basicity of a compound such as an amine, defined as the equilibrium constant for the following reaction. The negative $\log_{10}$ of K_b is given as pK_b. (p. 843)

$$R - \underset{\underset{\displaystyle H}{|}}{N} \overset{\displaystyle H}{\diagdown} + H - O - H \underset{}{\overset{K_b}{\rightleftharpoons}} R - \overset{\overset{\displaystyle H}{|}}{\underset{\underset{\displaystyle H}{|}}{N^+}} - H + {}^-OH$$

Cope elimination A variation of the Hofmann elimination, where a tertiary amine oxide eliminates to an alkene with a hydroxylamine serving as the leaving group. (p. 865)

diazo coupling The use of a diazonium salt as an electrophile in electrophilic aromatic substitution. (p. 873)

$$Ar\overset{+}{-}N\equiv N \ + \ H-\langle\bigcirc\rangle-Y \ \longrightarrow \ Ar-N=N-\langle\bigcirc\rangle-Y \ + \ H^+$$

diazonium ion   (activated)   an **azo** compound

diazotization of an amine The reaction of a primary amine with nitrous acid to form a diazonium salt. (p. 868)

exhaustive alkylation Treatment of an amine with an excess of an alkylating agent (often methyl iodide) to form the quaternary ammonium salt. (p. 858)

$$R-NH_2 \xrightarrow{\text{excess } CH_3I} R-\overset{+}{N}(CH_3)_3 \ \ I^-$$

exhaustive methylation of a primary amine

Gabriel amine synthesis Synthesis of primary amines by alkylation of the potassium salt of phthalimide, followed by displacement of the amine by hydrazine. (p. 881)

Hofmann elimination Elimination of a quaternary ammonium hydroxide with an amine as the leaving group. The Hofmann elimination usually gives the least highly substituted alkene. (p. 862)

Hofmann rearrangement of amides (Hofmann degradation) Treatment of a primary amide with sodium hydroxide and bromine or chlorine gives a primary amine. (p. 885)

$$\underset{\text{1° amide}}{R-\overset{\displaystyle O}{\overset{\|}{C}}-NH_2} \ + \ \underset{(X_2 = Cl_2 \text{ or } Br_2)}{X_2} \ + \ 4\,NaOH \ \longrightarrow \ \underset{\text{amine}}{R-NH_2} \ + \ 2\,NaX \ + \ Na_2CO_3 \ + \ 2\,H_2O$$

hydroxylamine The compound H_2NOH; or generically, an amine in which a hydroxyl group is one of the three substituents bonded to nitrogen. (p. 866)

$$R-\overset{\displaystyle R'}{\underset{\displaystyle \ddot{N}}{|}}-OH$$

nitrile A compound of formula $R-C\equiv N$, containing the *cyano group,* $-C\equiv N$. (p. 883)

nitrogen inversion (pyramidal inversion) Inversion of configuration of a nitrogen atom in which the lone pair moves from one face of the molecule to the other. The transition state is planar, with the lone pair in a *p* orbital. (p. 840)

N-nitrosoamine (nitrosamine) An amine with a nitroso group ($-N=O$) bonded to the amine nitrogen atom. The reaction of secondary amines with nitrous acid gives secondary N-nitrosoamines. (p. 869)

phase-transfer catalyst A compound (such as a quaternary ammonium halide) that is soluble in both water and organic solvents and that helps reagents move between organic and aqueous phases. (p. 848)

reductive amination The reduction of an imine or oxime derivative of a ketone or aldehyde. One of the most general methods for synthesis of amines. (p. 877)

$$\underset{\text{ketone or aldehyde}}{R-\overset{\displaystyle O}{\overset{\|}{C}}-R'} \xrightarrow[H^+]{R''-NH_2} \underset{N\text{-substituted imine}}{R-\overset{\displaystyle \ddot{N}-R''}{\overset{\|}{C}}-R'} \xrightarrow{\text{reduction}} \underset{\text{2° amine}}{R-\overset{\displaystyle \ddot{N}HR''}{\underset{|}{C}H}-R'}$$

Sandmeyer reaction Replacement of the $-\overset{+}{N}\equiv N$ group in an arenediazonium salt by a cuprous salt; usually cuprous chloride, bromide, or cyanide. (p. 871)

$$Ar-\overset{+}{N}\equiv N \ \ ^-Cl \xrightarrow[(X = Cl, Br, C\equiv N)]{CuX} Ar-X \ + \ N_2\uparrow$$

sulfonamide An amide of a sulfonic acid. The nitrogen analogue of a sulfonate ester. (p. 861)

a sulfonamide   a *p*-toluenesulfonamide (a tosylamide)

Essential Problem-Solving Skills in Chapter 19

1. Name amines and draw the structures from their names.
2. Interpret the IR, NMR, and mass spectra of amines, and use the spectral information to determine the structures.
3. Explain how the basicity of amines varies with hybridization and aromaticity.
4. Contrast the physical properties of amines with those of their salts.
5. Predict the products of reactions of amines with the following types of compounds; propose mechanisms where appropriate.
   (a) ketones and aldehydes
   (b) alkyl halides and tosylates
   (c) acid chlorides
   (d) sulfonyl chlorides
   (e) nitrous acid
   (f) oxidizing agents
   (g) arylamines with electrophiles
6. Give examples of the use of arenediazonium salts in diazo coupling reactions and in the synthesis of aryl chlorides, bromides, iodides, fluorides, and nitriles.
7. Illustrate the uses and mechanisms of the Hofmann and Cope eliminations, and predict the major products.
8. Use your knowledge of the mechanisms of amine reactions to propose mechanisms and products of similar reactions you have never seen before.
9. Show how to synthesize amines from other amines, ketones and aldehydes, acid chlorides, nitro compounds, alkyl halides, nitriles, and amides.
10. Use retrosynthetic analysis to propose effective single-step and multistep syntheses of compounds with amines as intermediates or products, protecting the amine as an amide if necessary.

Study Problems

19-35 Define each term and give an example.
   (a) acylation of an amine
   (b) a 1° amine
   (c) a 2° amine
   (d) a 3° amine
   (e) an aromatic heterocyclic amine
   (f) a tertiary amine oxide
   (g) an aliphatic heterocyclic amine
   (h) a quaternary ammonium salt
   (i) diazotization of an amine
   (j) a diazo coupling reaction
   (k) exhaustive methylation
   (l) a sulfa drug
   (m) Gabriel synthesis of an amine
   (n) the Hofmann elimination
   (o) the Hofmann rearrangement
   (p) an *N*-nitrosoamine
   (q) reductive amination
   (r) the Sandmeyer reaction
   (s) a sulfonamide.
   (t) an azide
   (u) a nitrile

19-36 For each compound,
   (1) classify the nitrogen-containing functional groups
   (2) provide an acceptable name

(e) $CH_3-\overset{+}{\underset{\underset{\text{Ph}}{|}}{N}}-CH_2CH_3$ (with O⁻ on N)

(f) $Ph-\underset{\underset{CH_3}{|}}{N}-CH_2CH_3$

(g) pyridinium Cl^-

(h)

19-37   Rank the amines in each set in order of increasing basicity.

(a) aniline (NH_2), cyclohexylamine (NH_2), diphenylamine ($\overset{H}{N}$)

(b) piperidine ($N-H$), pyridine (N—H), pyrrole ($\overset{H}{N}$)

(c) imidazole ($\overset{H}{N}$—N), pyrrole ($\overset{H}{N}$), pyrrolidine ($\overset{H}{N}$)

(d) aniline (NH_2), H_3C-aniline (NH_2), O_2N-aniline (NH_2)

19-38   Which of the following compounds are capable of being resolved into enantiomers?
(a) *N*-ethyl-*N*-methylaniline   **(b)** 2-methylpiperidine   **(c)** 1-methylpiperidine   **(d)** 1,2,2-trimethylaziridine

(e) pyrrolidinium ($\overset{+}{N}$, CH_3, CH_2CH_3) Cl^-

(f) dihydropyrrolium ($\overset{+}{N}$, CH_3, CH_2CH_3) Cl^-

(g) spiro compound ($N-H$)

(h) spiro compound ($N-CH_3$)

19-39   Complete the following proposed acid–base reactions whether the reactants or products are favored and predict:

(a) pyridine + CH_3COOH (acetic acid) $\longrightarrow$

(b) pyrrole + CH_3COOH (acetic acid) $\longrightarrow$

(c) pyridinium chloride ($\overset{+}{N}$, H) Cl^- + piperidine ($N-H$) $\longrightarrow$

(d) anilinium chloride ($\overset{+}{N}H_3$ Cl^-) + pyrrolidine ($N-H$) $\longrightarrow$

19-40   Predict the organic products formed when the following amides are treated with alkaline bromine water:

(a) $Ph-CH_2CH_2-\overset{\overset{O}{\|}}{C}-NH_2$ **(b)** $H_2N-\overset{\overset{O}{\|}}{C}-(CH_2)_4-\overset{\overset{O}{\|}}{C}-NH_2$ **(c)** phenyl cyclopropane with $\overset{\overset{O}{\|}}{C}NH_2$ (H, H)

19-41   Predict the products of the following reactions:
(a)   excess NH_3 + $Ph-CH_2CH_2CH_2Br$ $\longrightarrow$ **(b)** 1-bromopentane $\xrightarrow[\text{(2) LiAlH}_4]{\text{(1) NaN}_3}$

(c) N-methylpiperidine + H_2O_2 $\longrightarrow$ **(d)** product from part (c) $\xrightarrow{\text{heat}}$

(e) quinolizidine $\xrightarrow[\text{(3) heat}]{\overset{\text{(1) excess CH}_3\text{I}}{\text{(2) Ag}_2\text{O}}}$ **(f)** product from part (e) $\xrightarrow[\text{(3) heat}]{\overset{\text{(1) excess CH}_3\text{I}}{\text{(2) Ag}_2\text{O}}}$

(g) piperidine (NH) + $NaNO_2$ + HCl $\longrightarrow$ **(h)** nitrobenzene (NO_2) $\xrightarrow{\text{Zn, HCl}}$

(i) CH_3NH_2 + cyclohexyl$-CH_2-\overset{\overset{O}{\|}}{C}-Cl$ $\longrightarrow$ **(j)** product from part(i) $\xrightarrow{\text{LiAlH}_4}$

(k) $CH_3—(CH_2)_3—\overset{\displaystyle NCH_3}{\overset{\|}{C}}—CH_2CH_3 \xrightarrow{LiAlH_4}$

(m) 2-butanone + diethylamine $\xrightarrow{Na(CH_3COO)_3BH}$

(o) 3-nitroaniline $\xrightarrow[\text{(2) } H_3PO_2]{\text{(1) HCl, NaNO}_2}$

(l) $Ph—CH_2—\overset{\displaystyle CN}{\underset{\displaystyle |}{CH}}—CH_3 \xrightarrow{LiAlH_4}$

(n) 4-fluoropyridine $\xrightarrow{NaOCH_2CH_3}$

(p) 2-butanone $\xrightarrow[\text{(2) } LiAlH_4]{\text{(1) KCN, HCN}}$

19-42   Show how *m*-toluidine can be converted to the following compounds, using any necessary reagents:

(a) *m*-toluidine (CH_3 ... NH_2)   *m*-toluonitrile (CH_3 ... $C≡N$)   **(b)** (CH_3 ... CH_2NH_2)   **(c)** *m*-iodotoluene (CH_3 ... I)

(d) *m*-cresol (CH_3 ... OH)   **(e)** 3-methyl-4-nitroaniline (CH_3 ... NH_2, O_2N)   **(f)** (CH_3 ... $\overset{\displaystyle H}{N}$—cyclopentyl)

19-43   The mass spectrum of *t*-butylamine is given below. Use a diagram to show the cleavage that accounts for the base peak. Suggest why no molecular ion is visible in this spectrum.

$(CH_3)_3CNH_2$

mass spectrum: abundance (y-axis, 0 to 100) vs. m/z (x-axis, 10 to 160); base peak near 58

19-44   Using any necessary reagents, show how you would accomplish the following syntheses:

(a) benzene $\rightarrow$ $Ph—NH—\overset{\displaystyle O}{\overset{\|}{C}}$—cyclopentyl

(b) aniline (NH_2) $\rightarrow$ $—NH—\overset{\displaystyle O}{\underset{\displaystyle O}{\overset{\|}{\underset{\|}{S}}}}—C_6H_4—CH_3$

(c) piperidine (N—H) $\rightarrow$ $N(CH_3)_2$

(d) cycloheptanone $\rightarrow$ cycloheptylamine (NH_2)

(e) N-methylpyrrolidine (N—CH_3) $\rightarrow$ $\overset{+}{N}\overset{O^-}{\underset{CH_3}{}}$

(f) N-methylpyrrolidine (N—CH_3) $\rightarrow$ $N\overset{OH}{\underset{CH_3}{}}$

(g) CH_3 ...—COOH $\rightarrow$ CH_3 ...—$\overset{\displaystyle O}{\overset{\|}{C}}$—$N\overset{CH_2CH_3}{\underset{CH_2CH_3}{}}$

(mosquito repellent)

19-45 The following drugs are synthesized using the methods in this chapter and in previous chapter. Devise a synthesis for each, starting with any compound containing no more than six carbon atoms.
- **(a)** Phenacetin, used with aspirin and caffeine in pain-relief medications.
- **(b)** Methamphetamine, once considered a safe diet pill, but now known to be addictive and destructive of brain tissue.
- **(c)** Dopamine, one of the neurotransmitters in the brain. Parkinson's disease is thought to be the result of a dopamine deficiency.

phenacetin   methamphetamine   dopamine

19-46 Propose mechanisms for the following reactions:

19-47 The two most general amine syntheses are the reduction of amides and the reductive amination of carbonyl compounds. Show how these techniques can be used to accomplish the following syntheses:
- **(a)** benzoic acid → benzylamine
- **(b)** benzaldehyde → benzylamine
- **(c)** pyrrolidine → N-ethylpyrrolidine
- **(d)** cyclohexanone → N-cyclohexylpyrrolidine.
- **(e)** HOOC—(CH₂)₃—COOH → 1,5-pentanediamine (cadaverine)

19-48 Several additional amine syntheses are effectively limited to making primary amines. The reduction of azides and nitro compounds and the Gabriel synthesis leave the carbon chain unchanged. Formation and reduction of a nitrile adds one carbon atom and the Hofmann rearrangement eliminates one carbon atom. Show how these amine syntheses can be used for the following conversions:
- **(a)** allyl bromide → allylamine
- **(b)** ethylbenzene → p-ethylaniline
- **(c)** 3-phenylheptanoic acid → 2-phenyl-1-hexanamine
- **(d)** 1-bromo-3-phenylheptane → 3-phenyl-1-heptanamine
- **(e)** 1-bromo-3-phenylheptane → 4-phenyl-1-octanamine

19-49
- **(a)** Guanidine (below) is about as strong a base as hydroxide ion. Explain why guanidine is a much stronger base than most other amines.
- **(b)** Show why p-nitroaniline is a much weaker base (3 pK_b units weaker) than aniline.
- *****(c)** Explain why N,N,2,6-tetramethylaniline (below) is a much stronger base than N,N-dimethylaniline.

guanidine   N,N,2,6-tetramethylaniline   N,N-dimethylaniline

19-50 Using toluene and alcohols containing no more than four carbon atoms as your organic starting materials, show how you would synthesize the following compounds in good yields:
- **(a)** 1-pentanamine
- **(b)** N-methyl-1-butanamine
- **(c)** N-ethyl-N-propyl-2-butanamine
- **(d)** benzyl-n-propylamine
- **(e)**

19-51 Using any necessary reagents, show how you would accomplish the following multistep syntheses:

19-52   The alkaloid *coniine* has been isolated from hemlock and purified. Its molecular formula is $C_8H_{17}N$. Treatment of coniine with excess methyl iodide, followed by silver oxide and heating, gives the pure (S)-enantiomer of N,N-dimethyl-7-octene-4-amine. Propose a complete structure for coniine, and show how this reaction gives the observed product.

19-53   A chemist is summoned to an abandoned waste-disposal site to determine the contents of a leaking, corroded barrel. The barrel reeks of an overpowering fishy odor. The chemist dons a respirator to approach the barrel and collect a sample, which she takes to her laboratory for analysis.

　　　The mass spectrum shows a molecular ion at m/z 101, and the most abundant fragment is at m/z 86. The IR spectrum shows no absorptions above 3000 cm^{-1}, many absorptions between 2800 and 3000 cm^{-1}, no absorptions between 1500 and 2800 cm^{-1}, and a strong absorption at 1200 cm^{-1}. The proton NMR spectrum shows a triplet (J = 7 Hz) at $\delta 1.0$ and a quartet (J = 7 Hz) at $\delta 2.4$, with integrals of 17 spaces and 11 spaces, respectively.

(a)   Show what structural information is implied by each spectrum, and propose a structure for the unknown toxic waste.

(b)   Current EPA regulations prohibit the disposal of liquid wastes because they tend to leak out of their containers. Propose an inexpensive method for converting this waste to a solid, relatively odorless form for reburial.

(c)   Suggest how the chemist might remove the fishy smell from her clothing.

*19-54   Pyrrole undergoes electrophilic aromatic substitution more readily than benzene, and mild reagents and conditions are sufficient. These reactions normally occur at the 2-position rather than the 3-position, as shown in the following example:

(a)   Propose a mechanism for the acetylation of pyrrole as shown above. You may begin with pyrrole and the acylium ion, $CH_3—C\equiv O^+$. Be careful to draw all the resonance structures of the intermediate.

(b)   Explain why pyrrole reacts more readily than benzene, and also why substitution occurs primarily at the 2-position rather than the 3-position.

19-55   Section 17-12 showed how nucleophilic aromatic substitution can give aryl amines, if there is a strong electron-withdrawing group ortho or para to the site of substitution. Consider the following example:

(a)   Propose a mechanism for this reaction.

(b)   We usually think of fluoride ion as a poor leaving group. Explain why this reaction readily displaces fluoride as the leaving group.

(c)   Explain why this reaction stops with the desired product, rather than reacting with another dinitrofluorobenzene.

19-56   The following spectra for **A** and **B** correspond to two structural isomers. The NMR singlet at $\delta 1.16$ in spectrum **A** disappears when the sample is shaken with D_2O. The singlet at $\delta 0.6$ ppm in the spectrum of **B** disappears on shaking with D_2O. Propose structures for these isomers, and show how your structures correspond to the spectra. Show what cleavage is responsible for the base peak at m/z 44 in the mass spectrum of **A** and the prominent peak at m/z 58 in the mass spectrum of **B**.

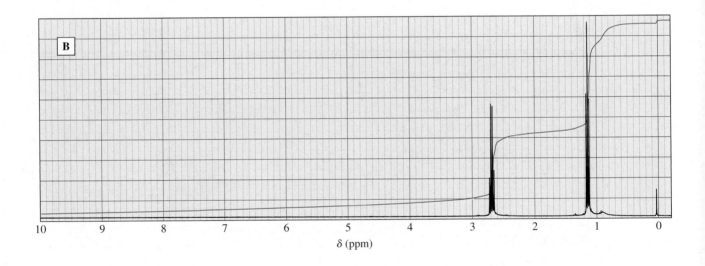

*19-57   (A true story.) A drug user responded to an ad placed by a DEA informant in a drug-culture magazine. He later flew from Colorado to Maryland, where he bought some 1-phenyl-2-propanone (P2P) from the informant. The police waited nearly a month for the suspect to synthesize something, then obtained a search warrant and searched the residence. They found the unopened bottle of P2P; apparently the suspect was not a good chemist and was unable to follow the instructions the informant gave him. They also found pipes and bongs with residues of marijuana and cocaine, plus a bottle of methylamine hydrochloride, some muriatic acid (dilute HCl), zinc strips, flasks, and other equipment.

(a) Assume you are consulting for the police. Show what synthesis the suspect was prepared to carry out, to provide probable cause for the charge of attempting to manufacture a controlled substance.

(b) Assume you are a member of the jury. Would you convict the defendant of attempting to manufacture a controlled substance?

19-58   An unknown compound shows a weak molecular ion at m/z 87 in the mass spectrum, and the only large peak is at m/z 30. The IR spectrum follows. The NMR spectrum shows only three singlets: one of area 9 at $\delta0.9$, one of area 2 at $\delta1.0$, and one of area 2 at $\delta2.4$. The singlet at $\delta1.0$ disappears on shaking with D_2O. Determine the structure of the compound, and show the favorable fragmentation that accounts for the ion at m/z 30.

*19-59   A compound of formula $C_{11}H_{16}N_2$ gives the IR, ^1H NMR, and ^{13}C NMR spectra shown below. The proton NMR peak at $\delta 2.0$ disappears on shaking with D_2O. Propose a structure for this compound, and show how your structure accounts for the observed absorptions.

CHAPTER 20

Carboxylic Acids

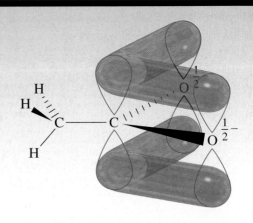

20-1 Introduction

The combination of a **carbo**nyl group and a hyd**roxyl** on the same carbon atom is called a **carboxyl group**. Compounds containing the carboxyl group are distinctly acidic and are called **carboxylic acids**.

$$-\overset{\displaystyle O}{\overset{\|}{C}}-O-H \qquad R-\overset{\displaystyle O}{\overset{\|}{C}}-O-H \qquad R-COOH \quad R-CO_2H$$

carboxyl group   carboxylic acid   condensed structures

Carboxylic acids are classified according to the substituent bonded to the carboxyl group. An **aliphatic acid** has an alkyl group bonded to the carboxyl group, while an **aromatic acid** has an aryl group. The simplest acid is *formic acid*, with a proton bonded to the carboxyl group. **Fatty acids** are long-chain aliphatic acids derived from the hydrolysis of fats and oils (Section 20-6).

$$H-\overset{\displaystyle O}{\overset{\|}{C}}-O-H \qquad CH_3-CH_2-\overset{\displaystyle O}{\overset{\|}{C}}-O-H \qquad \underset{\bigcirc}{}\overset{\displaystyle O}{\overset{\|}{C}}-O-H \qquad CH_3(CH_2)_{16}-\overset{\displaystyle O}{\overset{\|}{C}}-O-H$$

formic acid   propionic acid (an aliphatic acid)   benzoic acid (an aromatic acid)   stearic acid (a fatty acid)

A carboxylic acid donates protons by heterolytic cleavage of the acidic O—H bond to give a proton and a **carboxylate ion**. We consider the ranges of acidity and the factors affecting the acidity of carboxylic acids in Section 20-4.

$$R-\overset{\displaystyle O}{\overset{\|}{C}}-O-H \ + \ H_2O \ \rightleftharpoons \ R-\overset{\displaystyle O}{\overset{\|}{C}}-O^- \ + \ H_3O^+$$

carboxylic acid   carboxylate ion

20-2 Nomenclature of Carboxylic Acids

20-2A   Common Names

Several aliphatic carboxylic acids have been known for hundreds of years, and their common names reflect their historical sources. *Formic acid* was extracted from ants: *formica* in Latin. Acetic acid was isolated from vinegar, called *acetum* ("sour") in Latin. Propionic acid was considered to be the first fatty acid, and the name is derived from the Greek *protos pion* ("first fat"). Butyric acid results from the oxidation of butyraldehyde, which

TABLE 20-1   Names and Physical Properties of Some Carboxylic Acids

IUPAC Name	Common Name	Formula	mp (°C)	bp (°C)	Solubility (g/100 g H$_2$O)
methanoic[a]	formic	HCOOH	8	101	∞ (miscible)
ethanoic[a]	acetic	CH$_3$COOH	17	118	∞
propanoic	propionic	CH$_3$CH$_2$COOH	−21	141	∞
2-propenoic[a]	acrylic	H$_2$C=CH—COOH	14	141	∞
butanoic	butyric	CH$_3$(CH$_2$)$_2$COOH	−6	163	∞
2-methylpropanoic	isobutyric	(CH$_3$)$_2$CHCOOH	−46	155	23
trans-2-butenoic[a]	crotonic	CH$_3$—CH=CH—COOH	71	185	8.6
pentanoic	valeric	CH$_3$(CH$_2$)$_3$COOH	−34	186	3.7
3-methylbutanoic	isovaleric	(CH$_3$)$_2$CHCH$_2$COOH	−29	177	5
2,2-dimethylpropanoic	pivalic	(CH$_3$)$_3$C—COOH	35	164	2.5
hexanoic	caproic	CH$_3$(CH$_2$)$_4$COOH	−4	206	1.0
octanoic	caprylic	CH$_3$(CH$_2$)$_6$COOH	16	240	0.7
decanoic	capric	CH$_3$(CH$_2$)$_8$COOH	31	269	0.2
dodecanoic	lauric	CH$_3$(CH$_2$)$_{10}$COOH	44		i
tetradecanoic	myristic	CH$_3$(CH$_2$)$_{12}$COOH	54		i
hexadecanoic	palmitic	CH$_3$(CH$_2$)$_{14}$COOH	63		i
octadecanoic	stearic	CH$_3$(CH$_2$)$_{16}$COOH	72		i
cis-9-octadecenoic[a]	oleic	CH$_3$(CH$_2$)$_7$CH=CH(CH$_2$)$_7$COOH	16		i
cis,cis-9,12-octadecadienoic[a]	linoleic	CH$_3$(CH$_2$)$_4$CH=CHCH$_2$CH=CH(CH$_2$)$_7$COOH	−5		i
cyclohexanecarboxylic		*c*-C$_6$H$_{11}$COOH	31	233	0.2
benzoic	benzoic	C$_6$H$_5$—COOH	122	249	0.3
2-methylbenzoic	*o*-toluic	*o*-CH$_3$C$_6$H$_4$COOH	106	259	0.1
3-methylbenzoic	*m*-toluic	*m*-CH$_3$C$_6$H$_4$COOH	112	263	0.1
4-methylbenzoic	*p*-toluic	*p*-CH$_3$C$_6$H$_4$COOH	180	275	0.03

[a] IUPAC name is rarely used.

is found in butter: *butyrum* in Latin. Caproic, caprylic, and capric acids are found in the skin secretions of goats: *caper* in Latin. The names and physical properties of some carboxylic acids are shown in Table 20-1.

In common names, the positions of substituents are named using Greek letters. Notice that the lettering begins with the carbon atom *adjacent* to the carboxyl carbon, the α carbon. With common names, the prefix *iso-* is sometimes used for acids ending in the —CH(CH$_3$)$_2$ grouping.

α-chloropropionic acid   γ-aminobutyric acid   isovaleric acid (β-methylbutyric acid)

20-2B   IUPAC Names

The IUPAC nomenclature for carboxylic acids uses the name of the alkane that corresponds to the longest continuous chain of carbon atoms. The final *-e* in the alkane name is replaced by the suffix *-oic acid*. The chain is numbered, *starting with the carboxyl* *carbon atom*, to give positions of substituents along the chain. In naming, the carboxyl group takes priority over any of the functional groups discussed previously.

Unsaturated acids are named using the name of the corresponding alkene, with the final *-e* replaced by *-oic acid*. The carbon chain is numbered starting with the carboxyl carbon, and a number gives the location of the double bond. The stereochemical terms *cis* and *trans* (and *Z* and *E*) are used as they are with other alkenes. Cycloalkanes with —COOH substituents are generally named as *cycloalkanecarboxylic acids*.

IUPAC name: (*E*)-4-methyl-3-hexenoic acid   *trans*-3-phenyl-2-propenoic acid   3,3-dimethylcyclohexanecarboxylic acid
new IUPAC name: (*E*)-4-methylhex-3-enoic acid   (*E*)-3-phenylprop-2-enoic acid
                                                                (cinnamic acid)

Aromatic acids of the form Ar—COOH are named as derivatives of *benzoic acid*, Ph—COOH. As with other aromatic compounds, the prefixes *ortho-*, *meta-*, and *para-* may be used to give the positions of additional substituents. Numbers are used if there are more than two substituents on the aromatic ring. Many aromatic acids have historical names that are unrelated to their structures.

benzoic acid   *p*-aminobenzoic acid   *o*-hydroxybenzoic acid   *p*-methylbenzoic acid   α-naphthoic acid
                                                                (salicyclic acid)   (*p*-toluic acid)

20-2C   Nomenclature of Dicarboxylic Acids

Common Names of Dicarboxylic Acids   A **dicarboxylic acid** (also called a *diacid*) is a compound with two carboxyl groups. The common names of simple dicarboxylic acids are used more frequently than their systematic names. A common

TABLE 20-2   Names and Physical Properties of Dicarboxylic Acids

IUPAC Name	Common Name	Formula	mp (°C)	Solubility (g/100 g H_2O)
ethanedioic	oxalic	HOOC—COOH	189	14
propanedioic	malonic	$HOOCCH_2COOH$	136	74
butanedioic	succinic	$HOOC(CH_2)_2COOH$	185	8
pentanedioic	glutaric	$HOOC(CH_2)_3COOH$	98	64
hexanedioic	adipic	$HOOC(CH_2)_4COOH$	151	2
heptanedioic	pimelic	$HOOC(CH_2)_5COOH$	106	5
cis-2-butenedioic	maleic	cis-HOOCCH=CHCOOH	130.5	79
trans-2-butenedioic	fumaric	trans-HOOCCH=CHCOOH	302	0.7
benzene-1,2-dicarboxylic	phthalic	$1,2\text{-}C_6H_4(COOH)_2$	231	0.7
benzene-1,3-dicarboxylic	isophthalic	$1,3\text{-}C_6H_4(COOH)_2$	348	
benzene-1,4-dicarboxylic	terephthalic	$1,4\text{-}C_6H_4(COOH)_2$	300 subl.	0.002

mnemonic for these names is "*Oh my, such good apple pie*," standing for *o*xalic, *m*alonic, *s*uccinic, *g*lutaric, *a*dipic, and *p*imelic acids. The names and physical properties of some dicarboxylic acids are given in Table 20-2

Substituted dicarboxylic acids are given common names using Greek letters, as with the simple carboxylic acids. Greek letters are assigned beginning with the carbon atom adjacent to the carboxyl group that is closer to the substituents.

β-bromoadipic acid

α-methyl-β-phenylglutaric acid

Benzenoid compounds with two carboxyl groups are named **phthalic acids**. *Phthalic acid* itself is the ortho isomer. The meta isomer is called *isophthalic acid*, and the para isomer is called *terephthalic acid*.

o-phthalic acid
phthalic acid

m-phthalic acid
isophthalic acid

p-phthalic acid
terephthalic acid

IUPAC Names of Dicarboxylic Acids   Aliphatic dicarboxylic acids are named simply by adding the suffix *-dioic acid* to the name of the parent alkane. For straight-chain dicarboxylic acids, the parent alkane name is determined by using the longest continuous chain that contains both carboxyl groups. The chain is numbered beginning with the carboxyl carbon atom that is closer to the substituents, and these numbers are used to give the positions of the substituents.

3-bromohexanedioic acid

2-methyl-3-phenylpentanedioic acid

The system for naming cyclic dicarboxylic acids treats the carboxyl groups as substituents on the cyclic structure.

trans-1,3-cyclopentanedicarboxylic acid   1,3-benzenedicarboxylic acid

PROBLEM 20-1

Draw the structures of the following carboxylic acids.
(a) α-methylbutyric acid
(b) 2-bromobutanoic acid
(c) 4-aminopentanoic acid
(d) *cis*-4-phenyl-2-butenoic acid
(e) *trans*-2-methylcyclohexanecarboxylic acid
(f) 2,3-dimethylfumaric acid
(g) *m*-chlorobenzoic acid
(h) 3-methylphthalic acid
(i) β-aminoadipic acid
(j) 3-chloroheptanedioic acid

PROBLEM 20-2

Name the following carboxylic acids (when possible, give both a common name and a systematic name).

(a)   (b)   (c)

(d)   (e)   (f)

20-3
Structure and Physical Properties of Carboxylic Acids

Structure of the Carboxyl Group   The structure of the most stable conformation of formic acid is shown below. The entire molecule is approximately planar. The sp^2 hybrid carbonyl carbon atom is planar, with nearly trigonal bond angles. The O—H bond also lies in this plane, eclipsed with the C=O bond.

bond angles   bond lengths

It seems surprising that an eclipsed conformation is most stable. It appears that one of the unshared electron pairs on the hydroxyl oxygen atom is delocalized into the electrophilic pi system of the carbonyl group. We can draw the following resonance forms to represent this delocalization:

major   very minor   minor

Boiling Points   Carboxylic acids boil at considerably higher temperatures than do alcohols, ketones, or aldehydes of similar molecular weights. For example, acetic acid (MW 60) boils at 118°C, 1-propanol (MW 60) boils at 97°C, and propionaldehyde (MW 58) boils at 49°C.

$$CH_3-\overset{\overset{\textstyle O}{\|}}{C}-OH \qquad CH_3-CH_2-CH_2-OH \qquad CH_3-CH_2-\overset{\overset{\textstyle O}{\|}}{C}-H$$

acetic acid, bp 118°C   1-propanol, bp 97°C   propionaldehyde, bp 49°C

The high boiling points of carboxylic acids result from formation of a stable, hydrogen-bonded dimer. This dimer contains an eight-membered ring joined by two hydrogen bonds, effectively doubling the molecular weight of the molecules leaving the liquid phase.

hydrogen-bonded acid dimer

Melting Points   The melting points of some common carboxylic acids are given in Table 20-1. Acids containing more than eight carbon atoms are generally solids, unless they contain double bonds. The presence of double bonds (especially cis double bonds) in a long chain impedes formation of a stable crystal lattice, resulting in a lower melting point. For example, both stearic acid (octadecanoic acid) and linoleic acid (*cis,cis*-9,12-octadecadienoic acid) have 18 carbon atoms, but stearic acid melts at 70°C and linoleic acid melts at −5°C.

$$CH_3-(CH_2)_{16}-\overset{\overset{\textstyle O}{\|}}{C}-OH$$

stearic acid, mp 70°C

linoleic acid, mp −5°C

The melting points of dicarboxylic acids (Table 20-2) are relatively high. With two carboxyl groups per molecule, the forces of hydrogen bonding are particularly strong in diacids; a high temperature is required to break the lattice of hydrogen bonds in the crystal and melt the diacid.

Solubilities   Carboxylic acids form hydrogen bonds with water, and the lower-molecular-weight carboxylic acids (up through four carbon atoms) are miscible with water. As the length of the hydrocarbon chain increases, water solubility decreases until acids with more than 10 carbon atoms are essentially insoluble in water. The water solubilities of some simple carboxylic acids and diacids are given in Tables 20-1 and 20-2.

Carboxylic acids are very soluble in alcohols because acids form hydrogen bonds with alcohols. Also, alcohols are not as polar as water, so the longer-chain acids are more soluble in alcohols than they are in water. Most carboxylic acids are quite soluble in relatively nonpolar solvents such as chloroform because the acid continues to exist in its dimeric form in the nonpolar solvent. Thus, the hydrogen bonds of the cyclic dimer are not disrupted when the acid dissolves in a nonpolar solvent.

20-4
Acidity
of Carboxylic Acids

PROBLEM-SOLVING HINT

In an aqueous solution, an acid will be mostly dissociated if the pH is above (more basic than) the acid's pK_a, and mostly undissociated if the pH is below (more acidic than) the acid's pK_a.

20-4A   Measurement of Acidity

A carboxylic acid may dissociate in water to give a proton and a carboxylate ion. The equilibrium constant K_a for this reaction is called the *acid-dissociation constant*. The pK_a of an acid is the negative logarithm of K_a, and we commonly use pK_a as an indication of the relative acidities of different acids (Table 20-3).

$$R-\overset{\overset{\displaystyle O}{\|}}{C}-O-H \;+\; H_2O \;\rightleftharpoons\; R-\overset{\overset{\displaystyle O}{\|}}{C}-O^- \;+\; H_3O^+$$

$$K_a = \frac{[R-CO_2^-][H_3O^+]}{[R-CO_2H]}$$

$$pK_a = -\log_{10} K_a$$

Values of pK_a are about 5 $(K_a = 10^{-5})$ for simple carboxylic acids. For example, acetic acid has a pK_a of 4.7 $(K_a = 1.8 \times 10^{-5})$. Although carboxylic acids are not as strong as most mineral acids, they are still much more acidic than other functional groups we have studied. For example, alcohols have pK_a values in the range 16 to 18. Acetic acid $(pK_a = 4.74)$ is about 10^{11} times as acidic as the most acidic alcohols! In fact, concentrated acetic acid causes acid burns when it comes into contact with the skin.

Dissociation of either an acid or an alcohol involves breaking an O—H bond, but dissociation of a carboxylic acid gives a carboxylate ion with the negative charge spread out equally over *two* oxygen atoms, compared with just one oxygen in an alkoxide ion (Figure 20-1). The delocalized charge makes the carboxylate ion more stable than

---

TABLE 20-3   Values of K_a and pK_a for Carboxylic Acids and Dicarboxylic Acids

Formula	Name	Values			
		Simple carboxylic acids			
		K_a(at 25°C)			pK_a
HCOOH	methanoic acid	1.77×10^{-4}			3.75
CH_3COOH	ethanoic acid	1.76×10^{-5}			4.74
CH_3CH_2COOH	propanoic acid	1.34×10^{-5}			4.87
$CH_3(CH_2)_2COOH$	butanoic acid	1.54×10^{-5}			4.82
$CH_3(CH_2)_3COOH$	pentanoic acid	1.52×10^{-5}			4.81
$CH_3(CH_2)_4COOH$	hexanoic acid	1.31×10^{-5}			4.88
$CH_3(CH_2)_6COOH$	octanoic acid	1.28×10^{-5}			4.89
$CH_3(CH_2)_8COOH$	decanoic acid	1.43×10^{-5}			4.84
C_6H_5COOH	benzoic acid	6.46×10^{-5}			4.19
$p\text{-}CH_3C_6H_4COOH$	*p*-toluic acid	4.33×10^{-5}			4.36
$p\text{-}ClC_6H_4COOH$	*p*-chlorobenzoic acid	1.04×10^{-4}			3.98
$p\text{-}NO_2C_6H_4COOH$	*p*-nitrobenzoic acid	3.93×10^{-4}			3.41
		Dicarboxylic acids			
		K_{a1}	pK_{a1}	K_{a2}	pK_{a2}
HOOC—COOH	oxalic	5.4×10^{-2}	1.27	5.2×10^{-5}	4.28
$HOOCCH_2COOH$	malonic	1.4×10^{-3}	2.85	2.0×10^{-6}	5.70
$HOOC(CH_2)_2COOH$	succinic	6.4×10^{-5}	4.19	2.3×10^{-6}	5.64
$HOOC(CH_2)_3COOH$	glutaric	4.5×10^{-5}	4.35	3.8×10^{-6}	5.42
$HOOC(CH_2)_4COOH$	adipic	3.7×10^{-5}	4.43	3.9×10^{-6}	5.41
cis-HOOCCH=CHCOOH	maleic	1.0×10^{-2}	2.00	5.5×10^{-7}	6.26
trans-HOOCCH=CHCOOH	fumaric	9.6×10^{-4}	3.02	4.1×10^{-5}	4.39
$1,2\text{-}C_6H_4(COOH)_2$	phthalic	1.1×10^{-3}	2.96	4.0×10^{-6}	5.40
$1,3\text{-}C_6H_4(COOH)_2$	isophthalic	2.4×10^{-4}	3.62	2.5×10^{-5}	4.60
$1,4\text{-}C_6H_4(COOH)_2$	terephthalic	2.9×10^{-4}	3.54	3.5×10^{-5}	4.46

R—Ö—H + H₂Ö: ⇌   R—Ö:⁻   + H₃O⁺ $pK_a \cong 16$
alcohol   alkoxide $(K_a \cong 10^{-16})$

$$R-\overset{\overset{\ddot{O}}{\parallel}}{C}-\ddot{O}-H + H_2\ddot{O}: \rightleftharpoons \left[R-C\overset{\ddot{O}:}{\underset{\ddot{O}:^-}{}} \longleftrightarrow R-C\overset{\ddot{O}:^-}{\underset{\ddot{O}:}{}} \right] + H_3O^+$$

acid   carboxylate $pK_a \cong 5$
$(K_a \cong 10^{-5})$

Figure 20-1 energy diagram: R—OH + H₂O and R—COOH + H₂O reacting to give R—O⁻ + H₃O⁺ and R—C(O^½⁻)(O^½⁻) + H₃O⁺, with "stabilization of carboxylate"

▲ Figure 20-1
Carboxylic acids are more acidic than alcohols because carboxylate ions are more stable than alkoxide ions.

the alkoxide ion; therefore, dissociation of a carboxylic acid to a carboxylate ion is less endothermic than dissociation of an alcohol to an alkoxide ion.

The carboxylate ion can be visualized either as a resonance hybrid (as in Figure 20-1) or as a conjugated system of three p orbitals containing four electrons. The carbon atom and the two oxygen atoms are sp^2 hybridized, and each has an unhybridized p orbital. Overlap of these three p orbitals gives a three-center π molecular orbital system. There is half a π bond between the carbon and each oxygen atom, and there is half a negative charge on each oxygen atom (Figure 20-2).

Table 20-3 gives pK_a values for dicarboxylic acids in addition to those for simple carboxylic acids. Diacids have two dissociation constants: K_{a1} is for the first dissociation, and K_{a2} is for the second dissociation, to give a dianion. The second carboxyl group is always less acidic than the first $(K_{a2} \ll K_{a1})$ because extra energy is required to create a second negative charge close to another, mutually repulsive, negative charge.

malonic acid reaction: malonic acid $\overset{K_{a1}}{\rightleftharpoons}$ H₃O⁺ + anion $\overset{K_{a2}}{\rightleftharpoons}$ 2 H₃O⁺ + dianion

malonic acid $K_{a1} = 1.4 \times 10^{-3}$ anion $K_{a2} = 2.0 \times 10^{-6}$ dianion

$$CH_3-C\overset{O^{\frac{1}{2}-}}{\underset{O^{\frac{1}{2}-}}{}}$$

◀ Figure 20-2
Structure of the acetate ion. Each C—O bond has a bond order of $\frac{3}{2}$ from one σ bond and half a π bond. Each oxygen atom bears half of the negative charge.

20-4B Substituent Effects on Acidity

A substituent that stabilizes the negatively charged carboxylate ion enhances dissociation and results in a stronger acid. Electronegative atoms enhance the strength of an acid in this manner. This inductive effect can be quite large if one or more strongly electron-withdrawing groups are present on the α-carbon atom. For example, chloroacetic acid ($ClCH_2$—COOH) has a pK_a of 2.86, indicating that it is a stronger acid than acetic acid ($pK_a = 4.74$). Dichloroacetic acid (Cl_2CH—COOH) is stronger yet, with a pK_a of 1.26. Trichloroacetic acid (Cl_3C—COOH) has a pK_a of 0.64, comparable in strength to some mineral acids.

The magnitude of a substituent effect depends on its distance from the carboxyl group. Substituents on the α-carbon atom are most effective in increasing acid strength. More distant substituents have smaller effects on acidity, showing that inductive effects decrease rapidly with distance.

acetic acid
$pK_a = 4.74$

chloroacetic acid
$pK_a = 2.86$

dichloroacetic acid
$pK_a = 1.26$

trichloroacetic acid
$pK_a = 0.64$

stronger acids →

4-chlorobutanoic acid
$pK_a = 4.52$

3-chlorobutanoic acid
$pK_a = 4.05$

2-chlorobutanoic acid
$pK_a = 2.86$

Substituted benzoic acids show similar trends in acidity, with electron-withdrawing groups enhancing the acid strength and electron-donating groups decreasing the acid strength. These effects are strongest for substituents in the ortho and para positions. In the examples shown below, notice that a nitro substituent (electron-withdrawing) increases the strength of the acid, while a methoxyl substituent (electron-donating) decreases the acid strength. The nitro group has a larger effect in the ortho and para positions than in the meta position.

	p-methoxy	benzoic acid	m-nitro	p-nitro	o-nitro
$pK_a =$	4.46	4.19	3.47	3.41	2.16

stronger acids →

Table 20-4 lists values of K_a and pK_a for some substituted carboxylic acids, showing how electron-withdrawing groups enhance the strength of an acid.

TABLE 20-4 Values of K_a and pK_a for Substituted Carboxylic Acids

Acid	K_a	pK_a	
F_3CCOOH	5.9×10^{-1}	0.23	
Cl_3CCOOH	2.3×10^{-1}	0.64	stronger acids
$Cl_2CHCOOH$	5.5×10^{-2}	1.26	
$O_2N—CH_2COOH$	2.1×10^{-2}	1.68	
$NCCH_2COOH$	3.4×10^{-3}	2.46	
FCH_2COOH	2.6×10^{-3}	2.59	
$ClCH_2COOH$	1.4×10^{-3}	2.86	
$CH_3CH_2CHClCOOH$	1.4×10^{-3}	2.86	
$BrCH_2COOH$	1.3×10^{-3}	2.90	
ICH_2COOH	6.7×10^{-4}	3.18	
$HC{\equiv}CCH_2COOH$	4.8×10^{-4}	3.32	
CH_3OCH_2COOH	2.9×10^{-4}	3.54	
$HOCH_2COOH$	1.5×10^{-4}	3.83	
$CH_3CHClCH_2COOH$	8.9×10^{-5}	4.05	
$C_6H_5CH_2COOH$	4.9×10^{-5}	4.31	
$CH_2{=}CHCH_2COOH$	4.5×10^{-5}	4.35	
$ClCH_2CH_2CH_2COOH$	3.0×10^{-5}	4.52	
CH_3COOH	1.8×10^{-5}	4.74	
$CH_3CH_2CH_2COOH$	1.5×10^{-5}	4.82	
CH_3CH_2COOH	1.3×10^{-5}	4.87	

PROBLEM 20-3

Rank the compounds in each set in order of increasing acid strength.
(a) CH_3CH_2COOH $CH_3CHBrCOOH$ CH_3CBr_2COOH

(b) $CH_3CH_2CH_2CHBrCOOH$ $CH_3CH_2CHBrCH_2COOH$ $CH_3CHBrCH_2CH_2COOH$

(c) $\underset{\underset{NO_2}{|}}{CH_3CHCOOH}$ $\underset{\underset{Cl}{|}}{CH_3CHCOOH}$ CH_3CH_2COOH $\underset{\underset{C{\equiv}N}{|}}{CH_3CHCOOH}$

The absorption of many orally administered drugs containing carboxylic acids depends on their pK_a values. Aspirin, for example, is largely absorbed from the acidic environment of the stomach because it is present as the acid, which readily passes through the membranes into the blood.

aspirin

20-5
Salts of Carboxylic Acids

A strong base can completely deprotonate a carboxylic acid. The products are a carboxylate ion, the cation remaining from the base, and water. The combination of a carboxylate ion and a cation is a **salt of a carboxylic acid**.

$$R{-}\underset{\underset{\displaystyle carboxylic\ acid}{}}{\overset{\overset{\displaystyle O}{\|}}{C}}{-}O{-}H \ + \ \underset{strong\ base}{M^+\ {}^-OH} \ \rightleftharpoons \ R{-}\underset{\underset{\displaystyle acid\ salt}{}}{\overset{\overset{\displaystyle O}{\|}}{C}}{-}O^-\ {}^+M \ + \ \underset{water}{H_2O}$$

For example, sodium hydroxide deprotonates acetic acid to form sodium acetate, the sodium salt of acetic acid.

$$CH_3{-}\underset{\underset{\displaystyle acetic\ acid}{}}{\overset{\overset{\displaystyle O}{\|}}{C}}{-}O{-}H \ + \ \underset{sodium\ hydroxide}{Na^+\ {}^-OH} \ \rightleftharpoons \ CH_3{-}\underset{\underset{\displaystyle sodium\ acetate}{}}{\overset{\overset{\displaystyle O}{\|}}{C}}{-}O^-\ {}^+Na \ + \ H_2O$$

Because mineral acids are stronger than carboxylic acids, addition of a mineral acid converts a carboxylic acid salt back to the original carboxylic acid.

$$\text{☞} \quad R\overset{\overset{\displaystyle O}{\|}}{-C}-O^- \text{ } ^+M \quad + \quad H^+ \quad \rightleftharpoons \quad R\overset{\overset{\displaystyle O}{\|}}{-C}-O-H \quad + \quad M^+$$

acid salt   regenerated acid

Example

$$CH_3\overset{\overset{\displaystyle O}{\|}}{-C}-O^- \text{ } ^+Na \quad + \quad H^+ \text{ } Cl^- \quad \rightleftharpoons \quad CH_3\overset{\overset{\displaystyle O}{\|}}{-C}-O-H \quad + \quad Na^+ \text{ } Cl^-$$

sodium acetate   acetic acid

Carboxylic acid salts have very different properties from the acids, including enhanced solubility in water and less odor. Because acids and their salts are easily inter-converted, these salts serve as useful derivatives of carboxylic acids.

Nomenclature of Carboxylic Acid Salts   Salts of carboxylic acids are named simply by naming the cation, then naming the carboxylate ion by replacing the *-ic acid* part of the acid name with *-ate*. The example above shows that sodium hydroxide reacts with acetic acid to form sodium acetate. The following examples show the formation and nomenclature of some other salts:

$$CH_3CH_2CH_2CH_2\overset{\overset{\displaystyle O}{\|}}{-C}-OH \quad + \quad LiOH \quad \longrightarrow \quad CH_3CH_2CH_2CH_2\overset{\overset{\displaystyle O}{\|}}{-C}-O^- \text{ } ^+Li$$

IUPAC name: pentanoic acid   lithium hydroxide   lithium pentanoate
common name: valeric acid   lithium valerate

$$CH_3CH_2CH_2\overset{\overset{\displaystyle O}{\|}}{-C}-OH \quad + \quad :NH_3 \quad \longrightarrow \quad CH_3CH_2CH_2\overset{\overset{\displaystyle O}{\|}}{-C}-O^- \text{ } ^+NH_4$$

IUPAC name: butanoic acid   ammonia   ammonium butanoate
common name: butyric acid   ammonium butyrate

Properties of Acid Salts   Like the salts of amines (Section 19-7), carboxylic acid salts are solids with little odor. They generally melt at high temperatures, and they often decompose before reaching their melting points. Carboxylate salts of the alkali metals (Li^+, Na^+, K^+) and ammonium (NH_4^+) carboxylates are generally soluble in water but relatively insoluble in nonpolar organic solvents. *Soap* is a common example of carboxylate salts, consisting of the soluble sodium salts of long-chain fatty acids (Chapter 25). Carboxylate salts of most other metal ions are insoluble in water. For example, when soap is used in "hard" water containing calcium, magnesium, or iron ions, the insoluble carboxylate salts precipitate out as "hard-water scum."

$$2 \text{ } CH_3(CH_2)_{16}\overset{\overset{\displaystyle O}{\|}}{-C}-O^- \text{ } ^+Na \quad + \quad Ca^{2+} \quad \longrightarrow \quad [CH_3(CH_2)_{16}\overset{\overset{\displaystyle O}{\|}}{-C}-O]_2Ca\downarrow \quad + \quad 2 \text{ } Na^+$$

a soap   "hard-water scum"

Salt formation can be used to identify and purify acids. Carboxylic acids are deprotonated by the weak base sodium bicarbonate, forming the sodium salt of the acid, carbon dioxide, and water. An unknown compound that is insoluble in water, but dissolves in a sodium bicarbonate solution with a release of bubbles of carbon dioxide, is almost certainly a carboxylic acid.

$$R\overset{\overset{\displaystyle O}{\|}}{-C}-O-H \quad + \quad NaHCO_3 \quad \rightleftharpoons \quad R\overset{\overset{\displaystyle O}{\|}}{-C}-O^- \text{ } ^+Na \quad + \quad H_2O \quad + \quad CO_2\uparrow$$

insoluble in water   water soluble

▲ **Figure 20-3**

The solubility properties of acids and their salts may be used to remove nonacidic impurities. A carboxylic acid is more soluble in the organic phase, but its salt is more soluble in the aqueous phase. Acid–base extractions can move the acid from the ether phase into the aqueous phase and back into the ether phase, leaving impurities behind.

Some purification methods take advantage of the different solubilities of acids and their salts. Nonacidic (or weakly acidic) impurities can be removed from a carboxylic acid using acid–base extractions (Figure 20-3). First, the acid is dissolved in an organic solvent such as ether and shaken with water. The acid remains in the organic phase while any water-soluble impurities are washed out. Next, the acid is washed with aqueous sodium bicarbonate, forming a salt that dissolves in the aqueous phase. Nonacidic impurities (and weakly acidic impurities such as phenols) remain in the ether phase. The phases are separated, and acidification of the aqueous phase regenerates the acid, which is insoluble in water but dissolves in a fresh portion of ether. Evaporation of the final ether layer gives the purified acid.

PROBLEM 20-4

Suppose you have just synthesized heptanoic acid from 1-heptanol. The product is contaminated by sodium dichromate, sulfuric acid, 1-heptanol, and possibly heptanal. Explain how you would use acid–base extractions to purify the heptanoic acid. Use a chart like that in Figure 20-3 to show where the impurities are at each stage.

PROBLEM 20-5

Phenols are less acidic than carboxylic acids, with values of pK_a around 10. Phenols are deprotonated by (and therefore soluble in) solutions of sodium hydroxide but not by solutions of sodium bicarbonate. Explain how you would use extractions to isolate the three pure compounds from a mixture of *p*-cresol (*p*-methylphenol), cyclohexanone, and benzoic acid.

PROBLEM 20-6

Oxidation of a primary alcohol to an aldehyde usually gives some overoxidation to the carboxylic acid. Assume you have used PCC to oxidize 1-pentanol to pentanal.
(a) Show how you would use acid-base extraction to purify the pentanal.
(b) Which of the expected impurities cannot be removed from pentanal by acid–base extractions? How would you remove this impurity?

Thimerosal has been used as a topical antiseptic for many years. Its antimicrobial effect is mostly due to the toxicity of the mercury atom that is bound and stabilized by the thiol group of ortho-mercaptobenzoic acid. The carboxylate salt of the acid is used to enhance solubility of this organomercurial compound. Thimerosal is also used as a preservative in some contact lens solutions. Sensitivity to thimerosal, and concerns about chronic exposure to mercury, have encouraged the development of thimerosal-free antiseptics and contact-lens solutions.

COONa

SHgCH₂CH₃

thimerosal

20-6
Commercial Sources of Carboxylic Acids

The most important commercial aliphatic acid is acetic acid. *Vinegar* is a 5% aqueous solution of acetic acid used in cooking and in prepared foods such as pickles, ketchup, and salad dressings. Vinegar is produced by fermentation of sugars and starches. An intermediate in this fermentation is ethyl alcohol. When fermented alcoholic beverages such as wine and cider are exposed to air, the alcohol oxidizes to acetic acid. This is the source of "wine vinegar" and "cider vinegar."

$$\text{sugar and starches} \xrightarrow{\text{fermentation}} \underset{\text{ethyl alcohol}}{CH_3-CH_2-OH} \xrightarrow[\text{O}_2]{\text{fermentation}} \underset{\text{vinegar}}{CH_3-\overset{\overset{\displaystyle O}{\|}}{C}-OH}$$

Acetic acid is also an industrial chemical. It serves as a solvent, a starting material for synthesis, and a catalyst for a wide variety of reactions. Some industrial acetic acid is produced from ethylene, using a catalytic oxidation to form acetaldehyde, followed by another catalytic oxidation to acetic acid.

$$\underset{\text{ethylene}}{\underset{H}{\overset{H}{_{>}}}C=C\underset{H}{\overset{H}{_{<}}}} \xrightarrow[\substack{PdCl_2/CuCl_2 \\ \text{(catalyst)}}]{O_2} \underset{\text{acetaldehyde}}{CH_3-\overset{\overset{\displaystyle O}{\|}}{C}-H} \xrightarrow[\substack{\text{cobalt acetate} \\ \text{(catalyst)}}]{O_2} \underset{\text{acetic acid}}{CH_3-\overset{\overset{\displaystyle O}{\|}}{C}-O-H}$$

Methanol can also serve as the feedstock for an industrial synthesis of acetic acid. The rhodium-catalyzed reaction of methanol with carbon monoxide requires high pressures, so it is not suitable for a laboratory synthesis.

$$\underset{\text{methanol}}{CH_3OH} + CO \xrightarrow[\text{heat, pressure}]{\text{Rh catalyst}} \underset{\text{acetic acid}}{CH_3COOH}$$

> Moths, flies, and other insects often use fatty acids and derivatives as pheromones to attract mates. There is much interest in exploiting these compounds as a means of insect control.

Figure 20-4 shows how long-chain aliphatic acids are obtained from the hydrolysis of fats and oils, a reaction discussed in Chapter 25. The **fatty acids** found in fats and oils are generally straight-chain acids with even numbers of carbon atoms ranging between about C_6 and C_{18}. The hydrolysis of animal fat gives mostly saturated fatty acids, while plant oils give large amounts of unsaturated fatty acids with one or more olefinic double bonds.

Some aromatic carboxylic acids are also commercially important. Benzoic acid is used as an ingredient in medications, a preservative in foods, and a starting material

▶ **Figure 20-4**
Hydrolysis of a fat or an oil gives a mixture of the salts of straight-chain fatty acids. Animal fats contain primarily saturated fatty acids, while most vegetable oils are polyunsaturated.

for synthesis. Benzoic acid can be produced by the oxidation of toluene with potassium permanganate, nitric acid, or other strong oxidants.

toluene $\xrightarrow{HNO_3}$ benzoic acid

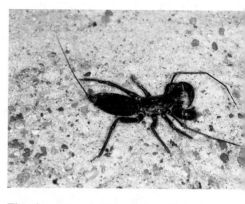

The vinegaroon (whip-tail scorpion) expels a spray of acetic acid to repel predators.

Two important commercial diacids are adipic acid (hexanedioic acid) and phthalic acid (1,2-benzenedicarboxylic acid). Adipic acid is used in the manufacture of nylon 66, and phthalic acid is used to make polyesters. The industrial synthesis of adipic acid uses benzene as the starting material. Benzene is hydrogenated to cyclohexane, whose oxidation (using a cobalt/acetic acid catalyst) gives adipic acid. Phthalic acid is produced by the direct oxidation of naphthalene or xylene using a vanadium pentoxide catalyst.

benzene $\xrightarrow[\text{high pressure}]{H_2,\ Ni}$ cyclohexane $\xrightarrow[\text{Co(OCOCH}_3)_3,\ CH_3COOH]{O_2}$ adipic acid

xylene or naphthalene $\xrightarrow[\text{heat}]{O_2,\ V_2O_5}$ phthalic acid

> The microbial transformation of toluene to adipic acid is being developed as an alternative source of adipic acid. If the process can be made economically feasible, it might produce less environmental impact than the chemical synthesis from benzene.

20-7A   Infrared Spectroscopy

The most obvious feature in the infrared spectrum of a carboxylic acid is the intense carbonyl stretching absorption. In a saturated acid, this vibration occurs around $1710\ \text{cm}^{-1}$, often broadened by hydrogen bonding involving the carbonyl group. In conjugated acids, the carbonyl stretching frequency is lowered to about $1690\ \text{cm}^{-1}$.

20-7
Spectroscopy of Carboxylic Acids

The O—H stretching vibration of a carboxylic acid absorbs in a broad band around $2500-3500\ \text{cm}^{-1}$. This frequency range is lower than the hydroxyl stretching frequencies of water and alcohols, whose O—H groups absorb in a band centered around $3300\ \text{cm}^{-1}$. In the spectrum of a carboxylic acid, the broad hydroxyl band appears right on top of the C—H stretching region. This overlapping of absorptions gives the $3000\ \text{cm}^{-1}$ region a characteristic appearance of a broad peak (the O—H stretching) with sharp peaks (C—H stretching) superimposed on it.

The IR spectrum of 2-methylpropenoic acid (methacrylic acid) is shown in Figure 20-5. Compare this conjugated example with the spectrum of hexanoic acid (Figure 12-12, p. 507). Notice the shift in the position of the carbonyl absorptions, and notice that the conjugated, unsaturated acid has a fairly strong C=C stretching absorption around $1630\ \text{cm}^{-1}$, just to the right of the carbonyl absorption.

▲ **Figure 20-5**
IR spectrum of 2-methyl-propenoic acid.

PROBLEM 20-7

The IR spectrum of *trans-2*-octenoic acid appears below. Point out the spectral characteristics that allow you to tell that this is a carboxylic acid, and show which features lead you to conclude that the acid is unsaturated and conjugated.

20-7B NMR Spectroscopy

Carboxylic acid protons are the most deshielded protons we have encountered, absorbing between $\delta 10$ and $\delta 13$. Depending on the solvent and the concentration, this acid proton peak may be sharp or broad, but it is always unsplit because of proton exchange.

▲ **Figure 20-6**
Proton NMR spectrum of butanoic acid.

The protons on the α-carbon atom absorb between δ2.0 and δ2.5, in about the same position as the protons on a carbon atom alpha to a ketone or an aldehyde. The proton NMR spectrum of butanoic acid is shown in Figure 20-6.

The carbon NMR chemical shifts of carboxylic acids resemble those of ketones and aldehydes. The carbonyl carbon atom absorbs around 180 ppm, and the α-carbon atom absorbs around 30 to 40 ppm. The chemical shifts of the carbon atoms in hexanoic acid are the following:

PROBLEM 20-8

(a) Determine the structure of the carboxylic acid whose proton NMR spectrum appears on the following page.
(b) Draw the NMR spectrum you would expect from the corresponding aldehyde whose oxidation would give this carboxylic acid.
(c) Point out two distinctive differences in the spectra of the aldehyde and the acid.

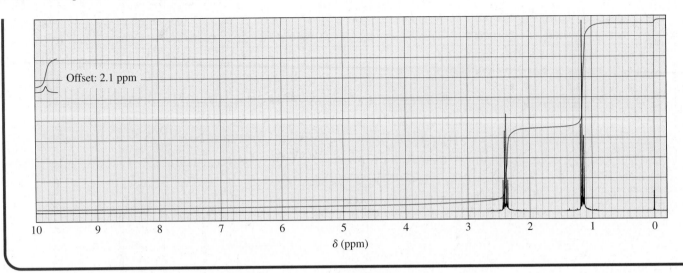

Offset: 2.1 ppm

δ (ppm)

20-7C Ultraviolet Spectroscopy

Saturated carboxylic acids have a weak $n \rightarrow \pi^*$ transition that absorbs around 200 to 215 nm. This absorption corresponds to the weak transition around 270 to 300 nm in the spectra of ketones and aldehydes. The molar absorptivity is very small (about 30 to 100), and the absorption often goes unnoticed.

Conjugated acids show much stronger absorptions. One C=C double bond conjugated with the carboxyl group results in a spectrum with λ_{max} still around 200 nm, but with molar absorptivity of about 10,000. A second conjugated double bond raises the value of λ_{max} to about 250 nm, as illustrated by the following examples:

$$CH_2\text{=}CH\text{—}\overset{\displaystyle O}{\overset{\|}{C}}\text{—}OH \qquad \lambda_{max} = 200 \text{ nm} \qquad \varepsilon = 10,000$$

$$CH_3\text{—}CH\text{=}CH\text{—}CH\text{=}CH\text{—}\overset{\displaystyle O}{\overset{\|}{C}}\text{—}OH \qquad \lambda_{max} = 254 \text{ nm} \qquad \varepsilon = 25,000$$

20-7D Mass Spectrometry

The molecular ion peak of a carboxylic acid is usually small because favorable modes of fragmentation are available. The most common fragmentation is loss of a molecule of an alkene (the McLafferty rearrangement, discussed in Section 18-5D). Another common fragmentation is loss of an alkyl radical to give a resonance-stabilized cation with the positive charge delocalized over an allylic system and two oxygen atoms.

McLafferty rearrangement   *(m/z* is even)

loss of an alkyl group   resonance-stabilized cation
                                       (m/z is odd)

◄ **Figure 20-7**
The mass spectrum of pentanoic acid
shows a weak parent peak, a base
peak from the McLafferty
rearrangement, and another strong
peak from loss of an ethyl radical.

The mass spectrum of pentanoic acid is given in Figure 20-7. The base peak at m/z 60 corresponds to the fragment from loss of propene via the McLafferty rearrangement. The strong peak at m/z 73 corresponds to loss of an ethyl radical with rearrangement to give a resonance-stabilized cation.

PROBLEM 20-9

Draw all four resonance forms of the fragment at m/z 73 in the mass spectrum of pentanoic acid.

PROBLEM 20-10

Use equations to explain the prominent peaks at m/z 74 and m/z 87 in the mass spectrum of 2-methylpentanoic acid.

20-8
Synthesis of Carboxylic Acids

20-8A   Review of Previous Syntheses

We have already encountered three methods for preparing carboxylic acids: (1) oxidation of alcohols and aldehydes, (2) oxidative cleavage of alkenes and alkynes, and (3) severe side-chain oxidation of alkylbenzenes.

1. Primary alcohols and aldehydes are commonly oxidized to acids by chromic acid (H_2CrO_4, formed from $Na_2Cr_2O_7$ and H_2SO_4). Potassium permanganate is occasionally used, but the yields are often lower (Sections 11-2B and 18-20).

☞
$$R-CH_2-OH \xrightarrow[\text{(or KMnO}_4)]{H_2CrO_4} R-\overset{\overset{\displaystyle O}{\|}}{C}-H \xrightarrow[\text{(or KMnO}_4)]{H_2CrO_4} R-\overset{\overset{\displaystyle O}{\|}}{C}-OH$$

primary alcohol   aldehyde   carboxylic acid

Example

$$Ph-CH_2-CH_2-CH_2-OH \xrightarrow{Na_2Cr_2O_7, H_2SO_4} Ph-CH_2-CH_2-\overset{\overset{\displaystyle O}{\|}}{C}-OH$$

3-phenylpropanol   3-phenylpropanoic acid

2. Cold, dilute potassium permanganate reacts with alkenes to give glycols. Warm, concentrated permanganate solutions oxidize the glycols further, cleaving the central carbon–carbon bond. Depending on the substitution of the original double bond, ketones or acids may result (Section 8-15A).

☞
$$\overset{R}{\underset{H}{>}}C=C\overset{R'}{\underset{R''}{<}} \xrightarrow{\text{concd. KMnO}_4} \left[\begin{matrix} R & R' \\ | & | \\ H-C-C-R'' \\ | & | \\ HO & OH \end{matrix} \right] \longrightarrow R-COOH + O=C\overset{R'}{\underset{R''}{<}}$$

alkene   glycol (not isolated)   acid   ketone

Examples

$$\overset{Ph}{\underset{H}{>}}C=C\overset{H}{\underset{CH_2-CH_3}{<}} \xrightarrow{\text{concd. KMnO}_4} Ph-COOH + CH_3-CH_2-COOH$$

$$\bigcirc\!\!\!= \xrightarrow{\text{concd. KMnO}_4} \bigcirc\!\!\!\overset{COOH}{\underset{COOH}{}}$$

cyclohexene   adipic acid

With alkynes, either ozonolysis or a vigorous permanganate oxidation cleaves the triple bond to give carboxylic acids (Section 9-10).

☞
$$R-C\equiv C-R' \xrightarrow[\substack{\text{or (1) O}_3 \\ \text{(2) H}_2O}]{\text{concd. KMnO}_4} \left[R-\overset{\overset{\displaystyle O}{\|}}{C}-\overset{\overset{\displaystyle O}{\|}}{C}-R' \right] \longrightarrow R-COOH + HOOC-R'$$

alkyne   (not isolated)   carboxylic acids

Example

$$CH_3CH_2CH_2-C\equiv C-Ph \xrightarrow[\text{(2) H}_2O]{\text{(1) O}_3} CH_3CH_2CH_2-COOH + Ph-COOH$$

3. Side chains of alkylbenzenes are oxidized to benzoic acid derivatives by treatment with hot potassium permanganate or hot chromic acid. Because this oxidation requires severe reaction conditions, it is useful only for making benzoic acid derivatives with no oxidizable functional groups. Oxidation-resistant functional groups such as $-Cl$, $-NO_2$, $-SO_3H$, and $-COOH$ may be present (Section 17-14A).

an alkylbenzene
(Z must be oxidation-resistant)

a benzoic acid

Example

p-chloroisopropylbenzene

p-chlorobenzoic acid

20-8B Carboxylation of Grignard Reagents

We have seen how Grignard reagents act as strong nucleophiles, adding to the carbonyl groups of ketones and aldehydes (Section 10-9). Similarly, Grignard reagents add to carbon dioxide to form magnesium salts of carboxylic acids. Addition of dilute acid protonates these magnesium salts to give carboxylic acids. This method is useful because it converts a halide functional group to a carboxylic acid functional group with an additional carbon atom.

R—X
(alkyl or
aryl halide)

Example

bromocyclohexane

cyclohexanecarboxylic acid

> Many blood-clotting proteins are carboxylated at specific locations to produce a modified protein that binds calcium, which is essential for the clotting process. In the absence of this carboxylation, blood clotting will not take place.

20-8C Formation and Hydrolysis of Nitriles

Another way to convert an alkyl halide (or tosylate) to a carboxylic acid with an additional carbon atom is to displace the halide with sodium cyanide. The product is a nitrile with one additional carbon atom. Acidic or basic hydrolysis of the nitrile gives a carboxylic acid by a mechanism discussed in Chapter 21. This method is limited to halides and tosylates that are good S_N2 electrophiles: usually primary and unhindered.

PROBLEM-SOLVING HINT

Oxidation of alcohols does not change the number of carbon atoms. Oxidative cleavages of alkenes and alkynes decrease the number of carbon atoms (except in cyclic cases). Carboxylation of Grignard reagents and formation and hydrolysis of nitriles increase the number of carbon atoms by one.

Example

benzyl bromide → phenylacetonitrile

NaCN, acetone

phenylacetonitrile → phenylacetic acid + NH_4^+

H^+, H_2O

PROBLEM 20-11

Show how you would synthesize the following carboxylic acids, using the indicated starting materials:

(a) 4-octyne → butanoic acid
(b) *trans*-cyclodecene → decanedioic acid
(c) bromobenzene → phenylacetic acid
(d) 2-butanol → 2-methylbutanoic acid
(e) *p*-xylene → terephthalic acid
(f) allyl iodide → 3-butenoic acid

SUMMARY **Syntheses of Carboxylic Acids**

1. *Oxidation of primary alcohols and aldehydes* (Section 11-2B and 18-20)

$$R-CH_2-OH \xrightarrow[\text{or } KMnO_4]{H_2CrO_4} R-\overset{\overset{\displaystyle O}{\|}}{C}-H \xrightarrow[\text{or } KMnO_4]{H_2CrO_4} R-\overset{\overset{\displaystyle O}{\|}}{C}-OH$$

primary alcohol → aldehyde → carboxylic acid

2. *Oxidation cleavage of alkenes and alkynes* (Section 8-15A and 9-10)

alkene → acid + ketone

concd. $KMnO_4$

$$R-C\equiv C-R' \xrightarrow[\substack{\text{or (1) } O_3 \\ (2) H_2O}]{\text{concd. } KMnO_4} R-COOH + HOOC-R'$$

alkyne → carboxylic acids

3. *Oxidation of alkylbenzenes* (Section 17-14A)

an alkylbenzene → a benzoic acid
(Z must be oxidation-resistant)

$Na_2Cr_2O_7$, H_2SO_4 or $KMnO_4$, H_2O

4. *Carboxylation of Grignard reagents* (Section 20-8B)

$$R-X \xrightarrow[\text{ether}]{Mg} R-MgX \xrightarrow{O=C=O} R-\overset{\overset{\displaystyle O}{\|}}{C}-O^- \ ^+MgX \xrightarrow{H^+} R-\overset{\overset{\displaystyle O}{\|}}{C}-OH$$

alkyl or aryl halide → → → acid

Example

$$CH_3-\underset{\underset{CH_2Br}{|}}{CH}-CH_3 \xrightarrow[\text{ether}]{Mg} CH_3-\underset{\underset{CH_2MgBr}{|}}{CH}-CH_3 \xrightarrow[\text{(2) H}^+]{\text{(1) CO}_2} CH_3-\underset{\underset{CH_2-COOH}{|}}{CH}-CH_3$$

isobutyl bromide   isovaleric acid

5. *Formation and hydrolysis of nitriles* (Section 20-8C)

$$R-CH_2-X \xrightarrow[\text{acetone}]{NaCN} R-CH_2-C\equiv N\colon \xrightarrow[\text{or } ^-OH, H_2O]{H^+, H_2O} R-CH_2-\overset{O}{\overset{\|}{C}}-OH$$

Example

benzyl bromide $\xrightarrow[\text{(2) H}^+, \text{H}_2\text{O}]{\text{(1) NaCN, acetone}}$ phenylacetic acid

6. *The haloform reaction* (converts methyl ketones to acids and iodoform; Chapter 22)

$$R-\overset{O}{\overset{\|}{C}}-CH_3 \xrightarrow[^-OH]{X_2} R-\overset{O}{\overset{\|}{C}}-O^- + HCX_3$$

X = Cl, Br, I

Example

$$Ph-\overset{O}{\overset{\|}{C}}-CH_3 \xrightarrow[\text{(2) H}^+]{\text{(1) X}_2, ^-OH} Ph-\overset{O}{\overset{\|}{C}}-OH$$

acetophenone   benzoic acid

7. *Malonic ester synthesis* (makes substituted acetic acids; Chapter 22)

$$\underset{\underset{COOEt}{|}}{\overset{\overset{COOEt}{|}}{CH_2}} \xrightarrow[\text{(2) R}-X]{\text{(1) Na}^+ \ ^-OCH_2CH_3} \underset{\underset{COOEt}{|}}{\overset{\overset{COOEt}{|}}{R-CH}} \xrightarrow[\text{(2) H}^+, \text{heat}]{\text{(1) }^-OH} R-CH_2-\overset{O}{\overset{\|}{C}}-OH$$

Example

$$\underset{\underset{COOEt}{|}}{\overset{\overset{COOEt}{|}}{CH_2}} \xrightarrow[\text{(2) CH}_3\text{CH}_2\text{CH}_2\text{CH}_2-Br]{\text{(1) Na}^+ \ ^-OCH_2CH_3} \underset{\underset{COOEt}{|}}{\overset{\overset{COOEt}{|}}{n\text{-Bu}-CH}} \xrightarrow[\text{(2) H}^+, \text{heat}]{\text{(1) }^-OH} n\text{-Bu}-CH_2-\overset{O}{\overset{\|}{C}}-OH$$

hexanoic acid

Ketones, aldehydes, and carboxylic acids all contain the carbonyl group, yet the reactions of acids are quite different from those of ketones and aldehydes. Ketones and aldehydes commonly react by nucleophilic addition to the carbonyl group; but carboxylic acids (and their derivatives) more commonly react by **nucleophilic acyl substitution**, where one nucleophile replaces another on the acyl (C=O) carbon atom.

Nucleophilic acyl substitution

$$R-\overset{\overset{\cdot\cdot}{O}\cdot}{\overset{\|}{C}}-X + Nuc\colon^- \rightleftharpoons R-\overset{\overset{\cdot\cdot}{O}\cdot}{\overset{\|}{C}}-Nuc + \colon X^-$$

20-9
Reactions of Carboxylic Acids and Derivatives; Nucleophilic Acyl Substitution

carboxylic acid   acyl halide   anhydride   ester   amide

Acid derivatives differ in the nature of the nucleophile bonded to the acyl carbon: —OH in the acid, —Cl in the acid chloride, —OR' in the ester, and —NH$_2$ (or an amine) in the amide. Nucleophilic acyl substitution is the most common method for interconverting these derivatives. The mechanisms of these substitutions depend on whether the reaction takes place in acid or base.

Under basic conditions, a strong nucleophile can add to the carbonyl group to give a tetrahedral intermediate. This intermediate then expels the leaving group. The base-catalyzed hydrolysis of an ester to an acid is an example of this mechanism. More examples are discussed in Chapter 21 (Carboxylic Acid Derivatives), since nucleophilic acyl substitution is a common method for interconverting acid derivatives.

MECHANISM   Nucleophilic Acyl Substitution of an Ester

Step 1: Addition of hydroxide.

Step 2: Elimination of alkoxide.

Step 3: Proton transfer.

ester + ⁻OH   tetrahedral intermediate   acid + alkoxide   carboxylate + alcohol

---

Under acidic conditions, the carbonyl group becomes protonated, activating it toward nucleophilic acyl substitution. Attack by a (weak) nucleophile gives a tetrahedral intermediate. In many cases, the leaving group becomes protonated before it leaves, so it leaves as a neutral molecule. We now cover the most useful example of acid-catalyzed nucleophilic acyl substitution: the Fischer esterification.

20-10
Condensation of Acids with Alcohols: The Fischer Esterification

The **Fischer esterification** converts carboxylic acids and alcohols directly to esters by an acid-catalyzed nucleophilic acyl substitution. The net reaction is replacement of the acid —OH group by the —OR group of the alcohol.

R—C—OH + R'—OH $\underset{}{\overset{H^+}{\rightleftharpoons}}$ R—C—O—R' + H$_2$O

acid   alcohol   ester

Examples

CH$_3$—C—OH + CH$_3$CH$_2$—OH $\underset{K_{eq} = 3.38}{\overset{H_2SO_4}{\rightleftharpoons}}$ CH$_3$—C—O—CH$_2$CH$_3$ + H$_2$O

phthalic acid $\underset{}{\overset{excess\ CH_3OH,\ H^+}{\rightleftharpoons}}$ dimethyl phthalate

The Fischer esterification mechanism (facing page) is an acid-catalyzed nucleophilic acyl substitution. The carbonyl group of a carboxylic acid is not sufficiently electrophilic to be attacked by an alcohol. The acid catalyst protonates the carbonyl group and activates it toward nucleophilic attack. Loss of a proton gives the hydrate of an ester.

Loss of water from the hydrate of the ester occurs by the same mechanism as loss of water from the hydrate of a ketone (Section 18-14). Protonation of either one of the hydroxyl groups allows it to leave as water, forming a resonance-stabilized cation. Loss of a proton from the second hydroxyl group gives the ester.

The mechanism of the Fischer esterification may seem long and complicated at first, but it is simplified by breaking it down into two stages: (1) acid-catalyzed addition of the alcohol to the carbonyl and (2) acid-catalyzed dehydration. If you have learned these mechanistic components as we have encountered them, you can write the entire mechanism without having to memorize it.

KEY MECHANISM   Fischer Esterification

(1) Acid-catalyzed addition of the alcohol to the carbonyl.

(2) Acid-catalyzed dehydration.

protonated ester   ester

PROBLEM 20-12

(a) The Key Mechanism shown above for Fischer Esterification omitted some important resonance forms of the intermediates shown in brackets. Complete the mechanism by drawing all the resonance forms of these two intermediates.

(b) Propose a mechanism for the acid-catalyzed reaction of acetic acid with ethanol to give ethyl acetate.

PROBLEM 20-13

Most of the Fischer esterification mechanism is identical with the mechanism for acetal formation. The difference is in the final step, where a carbocation loses a proton to give the ester. Write mechanisms for the following reactions, with the comparable steps directly above and below each other. Explain why the final step of the esterification (proton loss) cannot occur in acetal formation, and show what happens instead.

PROBLEM-SOLVING HINT

The Fischer esterification mechanism is a perfect example of acid-catalyzed nucleophilic acyl substitution. You should understand this mechanism well.

PROBLEM 20-14

A carboxylic acid has two oxygen atoms, each with two nonbonding pairs of electrons.
(a) Draw the resonance forms of a carboxylic acid that is protonated on the hydroxyl oxygen atom.
(b) Compare the resonance forms with those given previously for an acid protonated on the carbonyl oxygen atom.
(c) Explain why the carbonyl oxygen atom of a carboxylic acid is more basic than the hydroxyl oxygen.

Fischer esterification is an equilibrium, and typical equilibrium constants for esterification are not very large. For example, if 1 mole of acetic acid is mixed with 1 mole of ethanol, the equilibrium mixture contains 0.65 mole each of ethyl acetate and water, and 0.35 mole each of acetic acid and ethanol. Esterification using secondary and tertiary alcohols gives even smaller equilibrium constants.

Equilibrium mixture

$$CH_3\overset{\overset{\displaystyle O}{\|}}{-C}-OH + CH_3CH_2OH \underset{}{\overset{K_{eq} = 3.38}{\rightleftharpoons}} CH_3\overset{\overset{\displaystyle O}{\|}}{-C}-OCH_2CH_3 + H_2O$$

0.35 mole   0.35 mole   0.65 mole   0.65 mole

Esterification may be driven to the right either by using an excess of one of the reactants or by removing one of the products. For example, in forming ethyl esters, excess ethanol is often used to drive the equilibrium as far as possible toward the ester. Alternatively, water may be removed either by distilling it out or by adding a dehydrating agent such as magnesium sulfate or molecular sieves (dehydrated zeolite crystals that adsorb water).

Because of the inconvenience of driving the Fischer esterification to completion, we might prefer the reaction of an acid chloride with an alcohol for the laboratory synthesis of esters. The Fischer esterification is preferred in industry, however, where the techniques mentioned above give good yields of products and avoid the expensive step of converting the acid to its acid chloride.

PROBLEM-SOLVING HINT

In equilibrium reactions, look for ways to use an excess of a reagent or else to remove a product as it forms. Is it possible to use one of the reagents as a solvent? Can we distill off a product or drive off water?

PROBLEM 20-15

Show how Fischer esterification might be used to form the following esters (in each case, suggest a method for driving the reaction to completion):
(a) methyl salicylate   **(b)** methyl formate (bp 32°C)   **(c)** ethyl benzoate

PROBLEM 20-16

The mechanism of the Fischer esterification was controversial until 1938, when Irving Roberts and Harold Urey of Columbia University used isotopic labeling to follow the alcohol oxygen atom through the reaction.

A catalytic amount of sulfuric acid was added to a mixture of 1 mole of acetic acid and 1 mole of special methanol containing the heavy ^{18}O isotope of oxygen. After a short period, the acid was neutralized to stop the reaction, and the components of the mixture were separated.

$$CH_3\overset{\overset{\displaystyle O}{\|}}{-C}-O-H + CH_3-^{18}O-H \overset{H_2SO_4}{\rightleftharpoons} CH_3\overset{\overset{\displaystyle O}{\|}}{-C}-O-CH_3 + H_2O$$

(a) Propose a mechanism for this reaction.
(b) Follow the labeled ^{18}O atom through your mechanism, and show where it will be found in the products.
(c) The ^{18}O isotope is not radioactive. Suggest how you could experimentally determine the amounts of ^{18}O in the separated components of the mixture.

SOLVED PROBLEM 20-1

Ethyl orthoformate hydrolyzes easily in dilute acid to give formic acid and three equivalents of ethanol. Propose a mechanism for the hydrolysis of ethyl orthoformate.

ethyl orthoformate   formic acid   ethanol

SOLUTION

Ethyl orthoformate resembles an acetal with an extra alkoxy group, so this mechanism should resemble the hydrolysis of an acetal (Section 18-18). There are three equivalent basic sites: the three oxygen atoms. Protonation of one of these sites allows ethanol to leave, giving a resonance-stabilized cation. Attack by water gives an intermediate that resembles a hemiacetal with an extra alkoxy group.

Protonation and loss of a second ethoxyl group gives an intermediate that is simply a protonated ester.

protonated ester   ethyl formate

Hydrolysis of ethyl formate follows the reverse path of the Fischer esterification. This part of the mechanism is left to you as an exercise.

PROBLEM 20-17

(a) The solution given for Solved Problem 20-1 was missing some important resonance forms of the intermediates shown in brackets. Complete this mechanism by drawing all the resonance forms of these intermediates. Do your resonance forms help to explain why this reaction occurs under very mild conditions (water with a tiny trace of acid)?

(b) Finish the solution for Solved Problem 20-1 by providing a mechanism for the acid-catalyzed hydrolysis of ethyl formate. The principle of microscopic reversibility suggests that the Fischer esterification mechanism should give you a pathway to follow, since this hydrolysis is the reverse of the Fischer esterification.

20-11
Synthesis and Use of Acid Chlorides

Halide ions are excellent leaving groups for nucleophilic acyl substitution; therefore, acyl halides are particularly useful intermediates for making acid derivatives. In particular, acid chlorides (acyl chlorides) are easily made and are commonly used as an activated form of a carboxylic acid. Both the carbonyl oxygen and the chlorine atom withdraw electron density from the acyl carbon atom, making it strongly electrophilic. Acid chlorides react with a wide range of nucleophiles, generally through the addition–elimination mechanism of nucleophilic acyl substitution.

an acid chloride (acyl chloride)   acid chloride   tetrahedral intermediate   acid derivative

The best reagents for converting carboxylic acids to acid chlorides are thionyl chloride ($SOCl_2$) and oxalyl chloride ($COCl)_2$ because they form gaseous by-products that do not contaminate the product. Oxalyl chloride is particularly easy to use because it boils at 62°C and any excess is easily evaporated from the reaction mixture.

Examples

oleic acid

oleoyl chloride
(95%)

3-phenylpropanoic acid

3-phenylpropanoyl chloride
(95%)

The mechanisms of these reactions begin like the reaction of an alcohol with thionyl chloride. Either oxygen atom of the acid can attack sulfur, replacing chloride by a mechanism that looks like sulfur's version of nucleophilic acyl substitution. The product is an interesting, reactive chlorosulfite anhydride.

a chlorosulfite anhydride

This reactive anhydride undergoes nucleophilic acyl substitution by chloride ion to give the acid chloride.

$$R-\overset{\overset{\ddot{O}}{\|}}{C}-\ddot{O}-\overset{}{S}-Cl \longrightarrow R-\overset{\overset{:\ddot{O}:^-}{|}}{\underset{Cl}{C}}-\ddot{O}-\overset{}{S}-Cl \longrightarrow R-\overset{\overset{\ddot{O}}{\|}}{C}_{Cl} + SO_2 + Cl^-$$

*PROBLEM 20-18

Propose a mechanism for the reaction of benzoic acid with oxalyl chloride. This mechanism begins like the thionyl chloride reaction, to give a reactive mixed anhydride. Nucleophilic acyl substitution by chloride ion gives a tetrahedral intermediate that eliminates a leaving group, which then fragments into carbon dioxide, carbon monoxide, and chloride ion.

Acid chlorides react with alcohols to give esters through a nucleophilic acyl substitution by the addition–elimination mechanism discussed on the previous page. Attack by the alcohol at the electrophilic carbonyl group gives a tetrahedral intermediate. Loss of chloride and deprotonation give the ester.

$$R-\overset{\overset{\ddot{O}}{\|}}{C}-Cl + R'-\ddot{O}H \rightleftharpoons \left[R-\overset{\overset{:\ddot{O}:^-}{|}}{\underset{R'-\overset{+}{O}-H}{C}}-Cl \right] \longrightarrow R-\overset{\overset{\ddot{O}:^-}{\diagup}}{C}\underset{R'}{\overset{+}{O}-H} + Cl^- \longrightarrow R-\overset{\overset{\ddot{O}}{\|}}{C}-\ddot{O}-R' + HCl$$
ester

This reaction provides an efficient two-step method for converting a carboxylic acid to an ester. The acid is converted to the acid chloride, which reacts with an alcohol to give the ester.

☞ $$R-\overset{\overset{O}{\|}}{C}-OH \xrightarrow[\text{or } SOCl_2]{(COCl)_2} R-\overset{\overset{O}{\|}}{C}-Cl \xrightarrow[\text{alcohol}]{R'-O-H} R-\overset{\overset{O}{\|}}{C}-O-R' + HCl$$
acid   acid chloride   ester

Example

$$Ph-\overset{\overset{O}{\|}}{C}-OH \xrightarrow{SOCl_2} Ph-\overset{\overset{O}{\|}}{C}-Cl \xrightarrow[\text{ethanol}]{CH_3CH_2-OH} Ph-\overset{\overset{O}{\|}}{C}-O-CH_2CH_3 + HCl$$
benzoic acid   benzoyl chloride   ethyl benzoate

Ammonia and amines react with acid chlorides to give amides, also through the addition–elimination mechanism of nucleophilic acyl substitution. A carboxylic acid is efficiently converted to an amide by forming the acid chloride, which reacts with an amine to give the amide. A base such as pyridine or NaOH is often added to prevent HCl from protonating the amine.

☞ $$R-\overset{\overset{O}{\|}}{C}-Cl + R'-N\ddot{H}_2 \longrightarrow R-\overset{\overset{O}{\|}}{C}-\ddot{N}H-R' + HCl$$
acid chloride   amine   amide

Example

$$CH_3-\overset{\overset{O}{\|}}{C}-Cl + CH_3-N\ddot{H}_2 \xrightarrow{NaOH} CH_3-\overset{\overset{O}{\|}}{C}-\overset{\overset{H}{|}}{\underset{..}{N}}-CH_3 + NaCl + H_2O$$
acetyl chloride   methylamine   N-methylacetamide

PROBLEM 20-19

Propose mechanisms for the nucleophilic acyl substitutions to form ethyl benzoate and *N*-methylacetamide as shown on the previous page.

PROBLEM 20-20

Show how you would use an acid chloride as an intermediate to synthesize
(a) *N*-phenylbenzamide (PhCONHPh) from benzoic acid and aniline
(b) phenyl propionate (CH₃CH₂COOPh) from propionic acid and phenol

20-12
Esterification Using Diazomethane

Carboxylic acids are converted to their methyl esters very simply by adding an ether solution of diazomethane. The only by-product is nitrogen gas, and any excess diazomethane also evaporates. Purification of the ester usually involves only evaporation of the solvent. Yields are nearly quantitative in most cases.

$$R-\overset{\overset{\displaystyle O}{\|}}{C}-OH \ + \ CH_2N_2 \ \longrightarrow \ R-\overset{\overset{\displaystyle O}{\|}}{C}-O-CH_3 \ + \ N_2 \uparrow$$

acid   diazomethane   methyl ester

Example

cyclobutanecarboxylic acid $\xrightarrow{CH_2N_2}$ methyl cyclobutanecarboxylate (100%)   +   N₂ ↑

Diazomethane is often used to esterify polar or reactive compounds for mass spectrometry analysis. For example, a urine test for cocaine might involve treating the sample with diazomethane to convert benzoylecgonine, the major urinary metabolite of cocaine, to its volatile methyl ester for MS analysis.

Diazomethane is a toxic, explosive yellow gas that dissolves in ether and is fairly safe to use in ether solutions. The reaction of diazomethane with carboxylic acids probably involves transfer of the acid proton, giving a methyldiazonium salt. This diazonium salt is an excellent methylating agent, with nitrogen gas as a leaving group.

MECHANISM   Esterification Using Diazomethane

Step 1: Proton transfer.

Step 2: Nucleophilic attack on the methyl group.

Because diazomethane is hazardous in large quantities, it is rarely used industrially or in large-scale laboratory reactions. The yields of methyl esters are excellent, however, so diazomethane is often used for small-scale esterifications of valuable and delicate carboxylic acids.

20-13
Condensation of Acids with Amines: Direct Synthesis of Amides

Amides can be synthesized directly from carboxylic acids, although the acid chloride procedure uses milder conditions and often gives better yields. The initial reaction of a carboxylic acid with an amine gives an ammonium carboxylate salt. The carboxylate ion is a poor electrophile, and the ammonium ion is not nucleophilic, so the reaction stops at this point. Heating this salt to well above 100°C drives off steam and forms an amide. This direct synthesis is an important industrial process because it avoids the expense and hazardous reagents required to make the acid chloride.

$$R-\overset{\overset{O}{\|}}{C}-OH + R'-\ddot{N}H_2 \rightleftharpoons R-\overset{\overset{O}{\|}}{C}-O^- \ H_3\overset{+}{N}-R' \xrightarrow{\text{heat}} R-\overset{\overset{O}{\|}}{C}-\ddot{N}H-R' + H_2O \uparrow$$

acid   amine   an ammonium carboxylate salt   amide

Example

benzoic acid   ethylamine   ethylammonium benzoate   *N*-ethylbenzamide

COOH + $CH_3CH_2\ddot{N}H_2$ ⟶ COO⁻ $H_3\overset{+}{N}CH_2CH_3$ $\xrightarrow{\text{heat}}$ $\overset{\overset{O}{\|}}{C}-\ddot{N}HCH_2CH_3$ + H_2O

PROBLEM 20-21

Show how to synthesize the following compounds:
(1) using benzoyl chloride and any other necessary reagents
(2) using benzoic acid and any other necessary reagents

(a) $\overset{\overset{O}{\|}}{C}-N(CH_3)_2$ (b) $\overset{\overset{O}{\|}}{C}-O-CH(CH_3)_2$ (c) $CH_2-O-\overset{\overset{O}{\|}}{C}-CH_3$

N,N-dimethylbenzamide   isopropyl benzoate   benzyl acetate

Lithium aluminum hydride (LiAlH$_4$ or LAH) reduces carboxylic acids to primary alcohols. The aldehyde is an intermediate in this reduction, but it cannot be isolated because it is reduced more easily than the original acid.

20-14
Reduction of Carboxylic Acids

$$R-\overset{\overset{O}{\|}}{C}-OH \xrightarrow[\text{(2) } H_3O^+]{\text{(1) LiAlH}_4} R-CH_2-OH$$

acid   primary alcohol

Example

$$\text{C}_6\text{H}_5-CH_2-\overset{\overset{O}{\|}}{C}-OH \xrightarrow[\text{(2) } H_3O^+]{\text{(1) LiAlH}_4} \text{C}_6\text{H}_5-CH_2-CH_2OH$$

phenylacetic acid   (75%)
                    2-phenylethanol

Lithium aluminum hydride is a strong base, and the first step is deprotonation of the acid. Hydrogen gas is evolved, and the lithium salt results.

$$R-\overset{\overset{O}{\|}}{C}-O-H + Li^+ \ H-\overset{\overset{H}{|}}{\underset{H}{Al}}-H \longrightarrow H_2 \uparrow + R-\overset{\overset{O}{\|}}{C}-O^- \ ^+Li + AlH_3$$

Several paths are possible for the rest of the mechanism. In one likely path, AlH$_3$ adds to the carbonyl group of the lithium carboxylate salt.

$$R-\overset{\overset{\cdot\cdot}{\ddot{O}}}{\underset{:\ddot{O}:^- Li^+}{C}} + \overset{H \quad H}{\underset{H}{Al}} \longrightarrow R-\overset{\ddot{O}-AlH_2}{\underset{:\ddot{O}:^- Li^+}{C}-H}$$

Elimination gives an aldehyde, which is quickly reduced to a lithium alkoxide.

Water added in the second step protonates the alkoxide to the primary alcohol.

$$R—CH_2—O^- Li^+ + H_2O \longrightarrow R—CH_2—OH + LiOH$$

Borane also reduces carboxylic acids to primary alcohols. Borane (complex with THF; see Section 8-7) reacts with the carboxyl group faster than with any other carbonyl function. It often gives excellent selectivity, as shown by the following example, where a carboxylic acid is reduced while a ketone is unaffected. (LiAlH$_4$ would also reduce the ketone.)

(80%)

Reduction to Aldehydes   Reduction of carboxylic acids to aldehydes is difficult because aldehydes are more reactive than carboxylic acids toward most reducing agents. Almost any reagent that reduces acids to aldehydes also reduces aldehydes to primary alcohols. What is needed is a derivative of the acid that is more reactive than the aldehyde. As you might guess, the reactive acid derivative is the acid chloride.

Lithium aluminum tri(*t*-butoxy) hydride, LiAl[OC(CH$_3$)$_3$]$_3$H, is a weaker reducing agent than lithium aluminum hydride. It reduces acid chlorides because they are strongly activated toward nucleophilic addition of a hydride ion. Under these conditions, the aldehyde reduces more slowly and is easily isolated.

Example

PROBLEM 20-22

Show how you would synthesize the following compounds from the appropriate carboxylic acids or acid derivatives.

A general method of making ketones involves the reaction of a carboxylic acid with two equivalents of an organolithium reagent. This reaction was discussed in Section 18-9.

**20-15
Alkylation of Carboxylic Acids to Form Ketones**

$$\underset{}{R-\overset{\overset{\textstyle O}{\|}}{C}-O-H} \quad \xrightarrow[\text{(2) } H_2O]{\text{(1) 2 R'}-\text{Li}} \quad R-\overset{\overset{\textstyle O}{\|}}{C}-R' \; + \; R'-H$$

Example

benzoic acid   propiophenone

The first equivalent of the organolithium reagent simply deprotonates the acid. The second equivalent adds to the carbonyl to give a stable dianion. Hydrolysis of the dianion (by adding water) gives the hydrate of a ketone. Because the ketone is formed in a separate hydrolysis step (rather than in the presence of the organolithium reagent), overalkylation is not observed.

carboxylic acid   dianion   hydrate of ketone   ketone

PROBLEM 20-23

Propose a mechanism for conversion of the dianion to the ketone under mildly acidic conditions.

PROBLEM 20-24

Show how the following ketones might be synthesized from the indicated acids, using any necessary reagents:
(a) propiophenone from propionic acid (two ways, using alkylation of the acid and using Friedel–Crafts acylation)
(b) methyl cyclohexyl ketone from cyclohexanecarboxylic acid

| SUMMARY | **Reactions of Carboxylic Acids** |

General types of reactions

	$R-\overset{\overset{\textstyle O}{\|}}{C}-O^-$	deprotonation
$R-\overset{\overset{\textstyle O}{\|}}{C}-OH$	$R-\overset{\overset{\textstyle O}{\|}}{C}-Y$	nucleophilic acyl substitution
	$R-CH_2-OH$	reduction
	$R-Y \; + \; CO_2$	decarboxylation

1. *Salt formation* (Section 20-5)

$$\underset{\text{acid}}{R-\overset{\overset{\displaystyle O}{\|}}{C}-OH} + \underset{\text{strong base}}{M^+\;{}^-OH} \;\rightleftharpoons\; \underset{\text{salt}}{R-\overset{\overset{\displaystyle O}{\|}}{C}-O^-\;{}^+M} + H_2O$$

Example

$$2\,\underset{\text{propionic acid}}{CH_3CH_2-\overset{\overset{\displaystyle O}{\|}}{C}-OH} + Ca(OH)_2 \;\longrightarrow\; \underset{\text{calcium propionate}}{(CH_3CH_2-\overset{\overset{\displaystyle O}{\|}}{C}-O^-)_2Ca^{2+}}$$

2. *Conversion to acid chlorides* (Section 20-11)

$$\underset{\text{acid}}{R-\overset{\overset{\displaystyle O}{\|}}{C}-OH} + \underset{\text{thionyl chloride}}{Cl-\overset{\overset{\displaystyle O}{\|}}{S}-Cl} \;\longrightarrow\; \underset{\text{acid chloride}}{R-\overset{\overset{\displaystyle O}{\|}}{C}-Cl} + SO_2\uparrow + HCl\uparrow$$

Example

$$\underset{\text{butanoic acid}}{CH_3-CH_2-CH_2-\overset{\overset{\displaystyle O}{\|}}{C}-OH} + \underset{\text{thionyl chloride}}{SOCl_2} \;\longrightarrow\; \underset{\text{butanoyl chloride}}{CH_3-CH_2-CH_2-\overset{\overset{\displaystyle O}{\|}}{C}-Cl}$$

3. *Conversion to esters* (Sections 20-10, 20-11, and 20-12)

$$\underset{\text{acid}}{R-\overset{\overset{\displaystyle O}{\|}}{C}-OH} + \underset{\text{alcohol}}{R'-OH} \;\underset{}{\overset{H^+}{\rightleftharpoons}}\; \underset{\text{ester}}{R-\overset{\overset{\displaystyle O}{\|}}{C}-O-R'} + H_2O$$

$$\underset{\text{acid chloride}}{R-\overset{\overset{\displaystyle O}{\|}}{C}-Cl} + \underset{\text{alcohol}}{R'-OH} \;\longrightarrow\; \underset{\text{ester}}{R-\overset{\overset{\displaystyle O}{\|}}{C}-O-R'} + HCl\uparrow$$

$$\underset{\text{acid}}{R-\overset{\overset{\displaystyle O}{\|}}{C}-OH} + \underset{\text{diazomethane}}{CH_2N_2} \;\longrightarrow\; \underset{\text{methyl ester}}{R-\overset{\overset{\displaystyle O}{\|}}{C}-O-CH_3} + N_2\uparrow$$

Example

$$\underset{\text{benzoic acid}}{C_6H_5-\overset{\overset{\displaystyle O}{\|}}{C}-OH} + \underset{\text{ethanol}}{CH_3-CH_2-OH} \;\overset{H^+}{\rightleftharpoons}\; \underset{\text{ethyl benzoate}}{C_6H_5-\overset{\overset{\displaystyle O}{\|}}{C}-OCH_2CH_3} + H_2O$$

4. *Conversion to amides* (Sections 20-11 and 20-13)

$$\underset{\text{acid}}{R-\overset{\overset{\displaystyle O}{\|}}{C}-OH} + \underset{\text{amine}}{R'-NH_2} \rightleftharpoons \underset{\text{salt}}{R-\overset{\overset{\displaystyle O}{\|}}{C}-O^-\,H_3N^+-R'} \overset{\text{heat}}{\longrightarrow} \underset{\text{amide}}{R-\overset{\overset{\displaystyle O}{\|}}{C}-NH-R'} + H_2O$$

$$\underset{\text{acid chloride}}{R-\overset{\overset{\displaystyle O}{\|}}{C}-Cl} + \underset{\text{amine}}{R'-NH_2} \overset{\text{NaOH}}{\longrightarrow} \underset{\text{amide}}{R-\overset{\overset{\displaystyle O}{\|}}{C}-NH-R'} + NaCl + H_2O$$

Example

$$CH_3-\overset{\overset{\displaystyle O}{\|}}{C}-OH \; + \; CH_3-NH-CH_3 \; \xrightarrow{\text{heat}} \; CH_3-\overset{\overset{\displaystyle O}{\|}}{C}-N(CH_3)_2 \; + \; H_2O$$

acetic acid   dimethylamine   *N,N*-dimethylacetamide

5. *Conversion to anhydrides* (Section 21-5)

$$R-\overset{\overset{\displaystyle O}{\|}}{C}-Cl \; + \; HO-\overset{\overset{\displaystyle O}{\|}}{C}-R' \; \longrightarrow \; R-\overset{\overset{\displaystyle O}{\|}}{C}-O-\overset{\overset{\displaystyle O}{\|}}{C}-R' \; + \; HCl$$

acid chloride   acid   acid anhydride

Example

$$CH_3-\overset{\overset{\displaystyle O}{\|}}{C}-Cl \; + \; HO-\overset{\overset{\displaystyle O}{\|}}{C}-Ph \; \longrightarrow \; CH_3-\overset{\overset{\displaystyle O}{\|}}{C}-O-\overset{\overset{\displaystyle O}{\|}}{C}-Ph \; + \; HCl$$

acetyl chloride   benzoic acid   a mixed anhydride
(acetic benzoic anhydride)

6. *Reduction to primary alcohols* (Sections 10-11 and 20-14)

$$R-\overset{\overset{\displaystyle O}{\|}}{C}-OH \; \xrightarrow[\substack{\text{(2) } H_3O^+ \\ \text{(or use } BH_3 \cdot THF)}]{\text{(1) } LiAlH_4} \; R-CH_2-OH$$

acid   primary alcohol

7. *Reduction to aldehydes* (Sections 18-11 and 20-14)

$$R-\overset{\overset{\displaystyle O}{\|}}{C}-Cl \; \xrightarrow[\text{lithium aluminum tri(}t\text{-butoxy)hydride}]{LiAl[OC(CH_3)_3]_3H} \; R-\overset{\overset{\displaystyle O}{\|}}{C}-H$$

acid chloride   aldehyde

8. *Alkylation to form ketones* (Sections 18-9 and 20-15)

$$R-\overset{\overset{\displaystyle O}{\|}}{C}-O^- \, {}^+Li \; \xrightarrow[\text{(2) } H_2O]{\substack{\text{(1) } R'-Li \\ \text{alkyllithium}}} \; R-\overset{\overset{\displaystyle O}{\|}}{C}-R'$$

lithium carboxylate   ketone

9. *Side-chain halogenation* (Hell–Volhard–Zelinsky reaction; Section 22-4)

$$R-CH_2-\overset{\overset{\displaystyle O}{\|}}{C}-OH \; \xrightarrow{Br_2/PBr_3} \; R-\overset{\overset{\displaystyle Br}{|}}{CH}-\overset{\overset{\displaystyle O}{\|}}{C}-Br \; \xrightarrow{H_2O} \; R-\overset{\overset{\displaystyle Br}{|}}{CH}-\overset{\overset{\displaystyle O}{\|}}{C}-OH \; + \; HBr$$

α-bromo acyl bromide   *α*-bromoacid

---

acid chloride (acyl chloride) An activated acid derivative in which the hydroxyl group of the acid is replaced by a chlorine atom. (p. 925)

anhydride A composite of two acid molecules, with loss of water. A *mixed anhydride* contains two different acids. (p. 922)

Chapter 20
Glossary

$$2 \; CH_3-\overset{\overset{\displaystyle O}{\|}}{C}-OH \; \rightleftharpoons \; CH_3-\overset{\overset{\displaystyle O}{\|}}{C}-O-\overset{\overset{\displaystyle O}{\|}}{C}-CH_3 \; + \; H_2O \qquad\qquad Ph-\overset{\overset{\displaystyle O}{\|}}{C}-O-\overset{\overset{\displaystyle O}{\|}}{C}-CH_3$$

acetic acid   acetic anhydride   acetic benzoic anhydride

carboxyl group The —COOH functional group of a carboxylic acid. (p. 900)
carboxylate ion The anion resulting from deprotonation of a carboxylic acid. (p. 900)

carboxylation A reaction in which a compound (usually a carboxylic acid) is formed by the addition of CO_2 to an intermediate. The addition of CO_2 to a Grignard reagent is an example of a carboxylation. (p. 919)

carboxylic acid Any compound containing the *carboxyl group*, —COOH. (p. 900)

    An **aliphatic acid** has an alkyl group bonded to the carboxyl group.

    An **aromatic acid** has an aryl group bonded to the carboxyl group.

    A **dicarboxylic acid** (a diacid) has two carboxyl groups.

fatty acid A long-chain linear carboxylic acid. Some fatty acids are saturated, while others are unsaturated. (pp. 900, 912)

Fischer esterification The acid-catalyzed reaction of a carboxylic acid with an alcohol to form an ester. (p. 922)

$$R-\overset{\overset{\textstyle O}{\|}}{C}-O-H \ + \ R'-OH \ \underset{}{\overset{H^+}{\rightleftharpoons}} \ R-\overset{\overset{\textstyle O}{\|}}{C}-O-R' \ + \ H_2O$$

      acid   alcohol   ester

molecular seives Dehydrated zeolite crystals with well-defined pore sizes to admit molecules smaller than the pores. Often used to adsorb water from solvents or reactions. (p. 924)

nucleophilic acyl substitution A reaction in which a nucleophile substitutes for a leaving group on a carbonyl carbon atom. Nucleophilic acyl substitution usually takes place through the following addition–elimination mechanism. (p. 921)

$$R-\overset{\overset{\textstyle \cdot\overset{\cdot\cdot}{O}}{\|}}{C}-X \ + \ Nuc\!:^- \ \rightleftharpoons \ R-\overset{\overset{\textstyle :\overset{\cdot\cdot}{O}:^-}{|}}{\underset{\underset{\textstyle Nuc}{|}}{C}}-X \ \longrightarrow \ R-\overset{\overset{\textstyle \cdot\overset{\cdot\cdot}{O}}{\|}}{C}-Nuc \ + \ :X^-$$

the addition–elimination mechanism of nucleophilic acyl substitution

phthalic acids Benzenedicarboxylic acids. *Phthalic acid* itself is the ortho isomer. The meta isomer is *isophthalic acid*, and the para isomer is *terephthalic acid*. (p. 903)

salt of a carboxylic acid An ionic compound containing the deprotonated anion of a carboxylic acid, called the *carboxylate ion:* R—COO⁻. An acid salt is formed by the reaction of an acid with a base. (p. 909)

Essential Problem-Solving Skills in Chapter 20

1. Name carboxylic acids, and draw the structures from their names.
2. Show how the acidity of acids varies with their substitution.
3. Contrast the physical properties of carboxylic acids and their salts.
4. Interpret the IR, UV, NMR, and mass spectra of carboxylic acids, and use the spectral information to determine the structures.
5. Show how to synthesize carboxylic acids from oxidation of alcohols and aldehydes, carboxylation of Grignard reagents, hydrolysis of nitriles, and oxidation of alkylbenzenes.
6. Show how acids are converted to esters and amides using acid chlorides as intermediates. Propose mechanisms for these nucleophilic acyl substitutions.
7. Give the mechanism of the Fischer esterification, and show how the equilibrium can be driven toward the products or the reactants.
8. Predict the products of reactions of carboxylic acids with the following reagents and give mechanisms where appropriate:

       (a) diazomethane   (b) amines, followed by heating

       (c) lithium aluminum hydride   (d) excess alkyllithium reagents

---

Study Problems

20-25   Define each term, and give an example.

        (a) carboxylic acid   (b) carboxylate ion

        (c) carboxylation of a Grignard reagent   (d) acid-dissociation constant

(e) ester
(g) fatty acid
(i) dicarboxylic acid
(k) acid chloride

(f) Fischer esterification
(h) nucleophilic acyl substitution
(j) salt of a carboxylic acid

20-26   Give the IUPAC names of the following compounds:

(a) $PhCH_2CH_2COOH$
(d) $HOOCCH_2CH(CH_3)CH_2CO_2H$

(b) $CH_3CH_2CH(CH_3)CO_2H$
(e) $CH_3CH_2CH(CH_3)COONa$

(c) $CH_3CH(CH_3)CHBrCOOH$
(f) $(CH_3)_2C{=}CHCOOH$

(g)

(h)

20-27   Give the common names of the following compounds:

(a) $PhCH_2CH_2COOH$
(d) $HOOCCH_2CH(CH_3)CH_2CO_2H$

(b) $CH_3CH_2CH(CH_3)CO_2H$
(e) $(CH_3)_2CHCH_2COONa$

(c) $(CH_3)_2CHCHBrCOOH$
(f) $CH_3CH(NH_2)CH_2COOH$

(g)

(h) $\begin{array}{c} COO^- \\ | \\ COO^- \end{array}$ Mg^{2+}

20-28   Draw the structures of the following compounds:

(a) ethanoic acid   (b) phthalic acid   (c) magnesium formate
(d) malonic acid   (e) chloroacetic acid   (f) acetyl chloride
(g) zinc undecanoate (athlete's-foot powder)   (h) sodium benzoate (a food preservative)
(i) sodium fluoroacetate (Compound 1080, a controversial coyote poison)

20-29   In each pair of compounds, which is the stronger base?

(a) CH_3COO^- or $ClCH_2COO^-$ (b) sodium acetylide or sodium acetate
(c) sodium acetate or sodium ethoxide

20-30   Predict the products (if any) of the following acid-base reactions:

(a) acetic acid + ammonia
(c) *p*-toluic acid + potassium trifluoroacetate
(e) benzoic acid + sodium phenoxide

(b) phthalic acid + excess NaOH
(d) α-bromopropionic acid + sodium propionate

20-31   Rank the following isomers in order of increasing boiling point, the reasons for your order of ranking, and explain:

3-hydroxytetrahydrofuran   ethyl acetate   butyric acid

$$CH_3{-}\overset{\displaystyle O}{\overset{\|}{C}}{-}OCH_2CH_3 \qquad CH_3CH_2CH_2{-}\overset{\displaystyle O}{\overset{\|}{C}}{-}OH$$

20-32   Arrange each group of compounds in order of increasing acidity.

(a) phenol, ethanol, acetic acid   (b) *p*-toluenesulfonic acid, acetic acid, chloroacetic acid
(c) benzoic acid, *o*-nitrobenzoic acid, *m*-nitrobenzoic acid
(d) butyric acid, α-bromobutyric acid, β-bromobutyric acid

20-33   What conclusions can you draw from the pK_a values shown for the following compounds?

$$\underset{NO_2}{CH_2COOH}\ 1.68 \qquad \underset{CN}{CH_2COOH}\ 2.46 \qquad \underset{Cl}{CH_2COOH}\ 2.86 \qquad \underset{OH}{CH_2COOH}\ 3.83 \qquad \underset{H}{CH_2COOH}\ 4.74$$

20-34   Given the structure of ascorbic acid (vitamin C):

ascorbic acid

(a) Is ascorbic acid a carboxylic acid?
(b) Compare the acid strength of ascorbic acid (pK_a = 4.71) with acetic acid.
(c) Predict which proton in ascorbic acid is the most acidic.
(d) Draw the form of ascorbic acid that is present in the body (aqueous solution, pH = 7.4).

20-35   Predict the products, if any, of the following reactions:

(a) $\xrightarrow[\text{(2) } H_3O^+]{\text{(1) LiAlH}_4}$

(b) —CH$_2$Br $\xrightarrow[\text{(2) } H_3O^+]{\text{(1) NaCN}}$

(c) $\xrightarrow[\text{(2) AlCl}_3]{\text{(1) SOCl}_2}$

(d) 4-octyne $\xrightarrow[\text{(warm, concd.)}]{\text{KMnO}_4, H_2O}$

(e) —CH$_2$OH $\xrightarrow{\text{Na}_2\text{Cr}_2\text{O}_7, H_2\text{SO}_4}$

(f) $\underset{\displaystyle \overset{\textstyle Ph}{|}}{CH_3CH_2-CH-COOH}$ $\xrightarrow{B_2H_6}$

(g) —CH$_2$OH $\xrightarrow[\text{(warm, concd.)}]{\text{KMnO}_4, H_2O}$

(h) $\xrightarrow[\text{(hot, concd.)}]{\text{KMnO}_4, H_2O}$

(i) $\xrightarrow[\text{(3) } H_3O^+]{\begin{array}{l}\text{(1) Mg, ether}\\\text{(2) CO}_2\end{array}}$

(j) $\xrightleftharpoons{\text{2-butanol, H}^+}$

(k) $HOCH_2CH_2CH_2-\overset{\displaystyle O}{\overset{\|}{C}}-OH$ $\xrightarrow{H^+}$ (cyclic ester)

20-36   Show how you would accomplish the following syntheses efficiently (you may use any necessary reagents):
(a) *trans*-1-bromo-2-butene → *trans*-3-pentenoic acid (two ways)   **(b)** 3-hexene → propanoic acid
(c) 2-butenal → 2-butenoic acid   **(d)** hexanoic acid → hexanal

(e) $CH_3(CH_2)_3COOH$ ⟶ $CH_3(CH_2)_3-\overset{\displaystyle O}{\overset{\|}{C}}-OCH_3$ (two ways)
          valeric acid   methyl valerate

(f)

(g)

(h)

20-37   Show how you would use extractions with a separatory funnel to separate a mixture of the following compounds.
          benzoic acid   phenol   benzyl alcohol   aniline

20-38   When pure (*S*)-lactic acid is esterified by racemic 2-butanol, the product is 2-butyl lactate, with the following structure:

(a) Draw three-dimensional structures of the two stereoisomers formed, specifying the configuration at each asymmetric carbon atom. (Using your models may be helpful.)
(b) Determine the relationship between the two stereoisomers you have drawn.

20-39   Show how you would accomplish the following multistep syntheses:
(a) Ph—CH$_2$—CH$_2$—OH ⟶ Ph—CH$_2$—CH$_2$—COOH

(b)

(c)

(d)

(e)

*20-40   The IR, NMR, and mass spectra are provided for an organic compound.
   (a) Consider each spectrum individually, and tell what characteristics of the molecule are apparent from that spectrum.
   (b) Propose a structure for the compound, and show how your structure fits the spectral data.
  *(c)** Explain why an important signal is missing from the proton NMR spectrum.

20-41   In the presence of a trace of acid, δ-hydroxyvaleric acid forms a cyclic ester (lactone).

$$HO-CH_2CH_2CH_2CH_2-COOH$$

δ-hydroxyvaleric acid

   (a) Give the structure of the lactone, called δ-valerolactone.
   (b) Propose a mechanism for the formation of δ-valerolactone.

20-42   We have seen that an acid chloride reacts with an alcohol to form an ester.

$$R-\underset{\underset{\displaystyle O}{\|}}{C}-Cl \ + \ R'-OH \ \longrightarrow \ R-\underset{\underset{\displaystyle O}{\|}}{C}-O-R' \ + \ HCl$$

An acid chloride also reacts with another carboxylic acid molecule. The product is an acid anhydride.

$$R-\underset{\underset{\displaystyle O}{\|}}{C}-Cl \ + \ R'-\underset{\underset{\displaystyle O}{\|}}{C}-OH \ \longrightarrow \ \underset{\text{acid anhydride}}{R-\underset{\underset{\displaystyle O}{\|}}{C}-O-\underset{\underset{\displaystyle O}{\|}}{C}-R'} \ + \ HCl$$

Propose a mechanism for the reaction of benzoyl chloride (PhCOCl) with acetic acid, and show the structure of the resulting anhydride.

***20-43**   A student synthesized Compound 1 shown below. To purify the compound, he extracted it into aqueous base and then acidified the solution to protonate the acid so that he could extract it back into ether. When he evaporated the ether, he found that his product had been converted entirely into Compound 2.

Compound 1   →   Compound 2

(a) What is the functional group that forms the ring in Compound 1? In Compound 2?
(b) How many carbon atoms are there in Compound 1? In Compound 2? Where did the other carbon atoms go?
(c) When did the reaction occur: When the student added the base, or when he added the acid?
(d) Propose a mechanism for the conversion of Compound 1 to Compound 2.

***20-44**   The relative acidities of carboxylic acids (and, by inference, the stabilities of their carboxylate ions) have been used to compare the electron-donating and electron-withdrawing properties of substituents. These studies are particularly valuable to distinguish between inductive and resonance effects on the stabilities of compounds and ions. Some examples:
(a) The phenyl group is a mild ortho, para-director in electrophilic aromatic substitution. Is the phenyl group electron-donating or electron-withdrawing in EAS?
(b) The pK_a of phenylacetic acid is 4.31, showing that phenylacetic acid is a stronger acid than acetic acid. Is the phenyl group electron-donating or electron-withdrawing in the ionization of phenylacetic acid? How can you resolve the apparent contradiction?
(c) 4-Methoxybenzoic acid is a weaker acid than benzoic acid, but methoxyacetic acid is a stronger acid than acetic acid. Explain this apparent contradiction.
(d) Methyl groups are usually electron-donating, and propanoic acid is a weaker acid than acetic acid. Yet 2,6-dimethylbenzoic acid is a *stronger* acid than benzoic acid, while 2,6-dimethylphenol is a weaker acid than phenol. Explain these confusing experimental results.

20-45   (A true story) The manager of an organic chemistry stockroom prepared unknowns for a "Ketones and Aldehydes" experiment by placing two drops of the liquid unknowns in test tubes and storing the test tubes for several days until they were needed. One of the unknowns was misidentified by every student, however. This unknown was taken from a bottle marked "Heptaldehyde." The stockroom manager took an IR spectrum of the liquid in the bottle and found a sharp carbonyl stretch around 1710 cm^{-1} and small, sharp peaks around 2710 and 2810 cm^{-1}.

The students complained that their spectra showed no peaks at 2710 or 2810 cm^{-1}, but a broad absorption centered over the 3000 cm^{-1} region and a carbonyl peak around 1715 cm^{-1}. They also maintained that their samples are soluble in dilute aqueous sodium hydroxide.
(a) Identify the compound in the stockroom manager's bottle and the compound in the students' test tubes.
(b) Explain the discrepancy between the stockroom manager's spectrum and the students' results.
(c) Suggest how this misunderstanding might be prevented in the future.

3000 cm^{-1} 1710 cm^{-1} 3000 cm^{-1} 1715 cm^{-1}

stockroom managers' spectrum   students' spectrum

***20-46**   The antidepressant drug *tranylcypromine* is a primary amine with the amino group on a cyclopropane ring. Show how you would convert *trans*-cinnamic acid to tranylcypromine. (*Hint:* The cyclopropyl group is a poor S$_N$2 substrate, like a tertiary group. Consider reactions that can make primary amines with tertiary alkyl groups.)

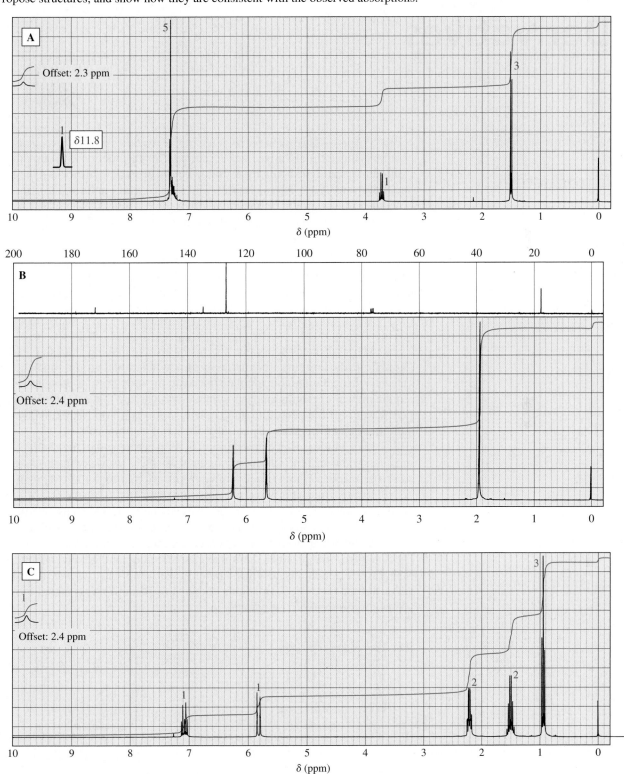

trans-cinnamic acid   tranylcypromine

20-47   The following NMR spectra correspond to compounds of formulas (**A**) $C_9H_{10}O_2$, (**B**) $C_4H_6O_2$, and (**C**) $C_6H_{10}O_2$, respectively. Propose structures, and show how they are consistent with the observed absorptions.

Carboxylic Acid Derivatives

$$
\text{PhOCH}_2\overset{\displaystyle O}{\overset{\displaystyle \|}{\text{C}}}\text{—NH}
$$

acylated, inactive enzyme

21-1 Introduction

Carboxylic acid derivatives are compounds with functional groups that can be converted to carboxylic acids by a simple acidic or basic hydrolysis. The most important acid derivatives are esters, amides, and nitriles. Acid halides and anhydrides are also included in this group, although we often think of them as activated forms of the parent acids rather than completely different compounds.

$$
\underset{\text{acid halide}}{R-\overset{\displaystyle O}{\overset{\displaystyle \|}{C}}-X} \qquad \underset{\text{anhydride}}{R-\overset{\displaystyle O}{\overset{\displaystyle \|}{C}}-O-\overset{\displaystyle O}{\overset{\displaystyle \|}{C}}-R} \qquad \underset{\text{ester}}{R-\overset{\displaystyle O}{\overset{\displaystyle \|}{C}}-O-R'} \qquad \underset{\text{amide}}{R-\overset{\displaystyle O}{\overset{\displaystyle \|}{C}}-NH_2} \qquad \underset{\text{nitrile}}{R-C\equiv N}
$$

Condensed structure: RCOX   (RCO)$_2$O   RCO$_2$R′   RCONH$_2$ RCN

Many advances in organic chemistry involve making and using derivatives of carboxylic acids. Proteins are bonded by amide functional groups, and chemists have created synthetic amides that emulate the desirable properties of proteins. For example, the nylon in your climbing rope is a synthetic polyamide that emulates the protein in a spider's web. All the penicillin and cephalosporin antibiotics are amides that extend the antimicrobial properties of naturally occurring antibiotics.

Like amides, esters are common both in nature and in the chemical industry. Animal fats and vegetable oils are mixtures of esters, as are waxy materials such as beeswax and spermaceti. Plants often synthesize esters that give the characteristic tastes and odors to their fruits and flowers. In addition to making synthetic esters for flavors, odors, and lubricants, chemists have made synthetic polyesters such as Dacron polyester fiber used in clothing and Mylar polyester film used in magnetic recording tapes.

Some examples of naturally occurring esters and amides are shown below. Isoamyl acetate gives ripe bananas their characteristic odor, and geranyl acetate is found in the oil of roses, geraniums, and many other flowers. *N,N*-Diethyl-*meta*-toluamide (DEET®) is one of the best insect repellents known, and penicillin G is one of the antibiotics that revolutionized modern medicine.

isoamyl acetate
(banana oil)

geranyl acetate
(geranium oil)

N,N-diethyl-*meta*-toluamide

penicillin G

21-2A   Esters of Carboxylic Acids

Esters are carboxylic acid derivatives in which the hydroxyl group ($-$OH) is replaced by an alkoxy group ($-$OR). An ester is a combination of a carboxylic acid and an alcohol, with loss of a molecule of water. We have seen that esters can be formed by the Fischer esterification of an acid with an alcohol (Section 20-10).

$$ R-\overset{O}{\overset{\|}{C}}-OH \ + \ R'-OH \ \underset{}{\overset{H^+}{\rightleftharpoons}} \ R-\overset{O}{\overset{\|}{C}}-O-R' \ + \ H_2O $$

acid   alcohol   ester

The names of esters consist of two words that reflect their composite structure. The first word is derived from the *alkyl* group of the alcohol, and the second word from the *carboxylate* group of the carboxylic acid. The IUPAC name is derived from the IUPAC names of the alkyl group and the carboxylate, while the common name is derived from the common names of each. The following examples show both the IUPAC names and the common names of some esters:

$$ CH_3CH_2-OH \ + \ HO-\overset{O}{\overset{\|}{C}}-CH_3 \ \overset{H^+}{\rightleftharpoons} \ CH_3CH_2-O-\overset{O}{\overset{\|}{C}}-CH_3 \ + \ H_2O $$

IUPAC name: ethanol   ethanoic acid   ethyl ethanoate
common name: ethyl alcohol   acetic acid   ethyl acetate

$$ (CH_3)_2CH-O-\overset{O}{\overset{\|}{C}}-H $$

IUPAC name: 1-methylethyl methanoate
common name: isopropyl formate

phenyl benzoate
phenyl benzoate

methyl 2-phenylethanoate
methyl phenylacetate

$$ Ph-CH_2-O-\overset{O}{\overset{\|}{C}}-\overset{CH_3}{\overset{|}{CH}}-CH_3 $$

IUPAC name: benzyl 2-methylpropanoate
common name: benzyl isobutyrate

$$ CH_3-O-\overset{O}{\overset{\|}{C}}- $$

methyl cyclopentanecarboxylate
methyl cyclopentanecarboxylate

cyclohexyl methanoate
cyclohexyl formate

Lactones   Cyclic esters are called **lactones**. A lactone is formed from an open-chain hydroxy acid in which the hydroxyl group has reacted with the acid group to form an ester.

IUPAC name: 4-hydroxybutanoic acid   4-hydroxybutanoic acid lactone
common name: γ-hydroxybutyric acid   γ-butyrolactone

The IUPAC names of lactones are derived by adding the term *lactone* at the end of the name of the parent acid. The common names of lactones, used more often than IUPAC names, are formed by changing the *-ic acid* ending of the hydroxy acid to *-olactone*. A Greek letter designates the carbon atom that bears the hydroxy group to close the ring. Substituents are named just as they are on the parent acid.

IUPAC name: 5-hydroxypentanoic acid lactone   4-hydroxy-2-methylpentanoic acid lactone
common name: δ-valerolactone   α-methyl-γ-valerolactone

21-2B   Amides

An **amide** is a composite of a carboxylic acid and ammonia or an amine. An acid reacts with an amine to form an ammonium carboxylate salt. When this salt is heated to well above 100°C, water is driven off and an amide results.

The simple amide structure shows a nonbonding pair of electrons on the nitrogen atom. Unlike amines, however, amides are only weakly basic, and we consider the amide functional group to be neutral. A concentrated strong acid is required to protonate an amide, and protonation occurs on the carbonyl oxygen atom rather than on nitrogen. This lack of basicity can be explained by picturing the amide as a resonance hybrid of the conventional structure and a structure with a double bond between carbon and nitrogen.

very weakly basic   protonation on oxygen

This resonance representation correctly predicts a planar amide nitrogen atom that is sp^2 hybridized to allow pi bonding with the carbonyl carbon atom. For example, formamide has a planar structure like an alkene. The C—N bond has partial double-bond character, with a rotational barrier of 18 kcal/mol (75 kJ/mol).

formamide

An amide of the form R—CO—NH$_2$ is called a **primary amide** because there is only one carbon atom bonded to the amide nitrogen. An amide with an alkyl group on nitrogen (R—CO—NHR′) is called a **secondary amide** or an **N-substituted amide**. Amides with two alkyl groups on the amide nitrogen (R—CO—NR$_2'$) are called **tertiary amides** or **N,N-disubstituted amides**.

Many local anesthetics are amides. Lidocaine, the prototype for this group of drugs, is the most widely used.

lidocaine

primary amide

secondary amide
(*N*-substituted amide)

tertiary amide
(*N,N*-disubstituted amide)

To name a primary amide, first name the corresponding acid. Drop the *-ic acid* or *-oic acid* suffix, and add the suffix *-amide*. For secondary and tertiary amides, treat the alkyl groups on nitrogen as substituents, and specify their position by the prefix *N-*.

IUPAC name:	*N*-ethylethanamide	*N,N*-dimethylmethanamide	*N*-ethyl-*N*,2-dimethylpropanamide
common name:	*N*-ethylacetamide	*N,N*-dimethylformamide	*N*-ethyl-*N*-methylisobutyramide

For acids that are named as alkanecarboxylic acids, the amides are named by using the suffix *-carboxamide*. Some amides, such as acetanilide, have historical names that are still commonly used.

cyclopentanecarboxamide

N,N-dimethylcyclopropanecarboxamide

acetanilide

Lactams   Cyclic amides are called **lactams**. Lactams are formed from amino acids, where the amino group and the carboxyl group have joined to form an amide. Lactams are named like lactones, by adding the term *lactam* at the end of the IUPAC name of the parent acid. Common names of lactams are formed by changing the *-ic acid* ending of the amino acid to *-olactam*.

IUPAC name:	4-aminobutanoic acid	4-aminobutanoic acid lactam
common name:	γ-aminobutyric acid	γ-butyrolactam

IUPAC name: 3-aminopropanoic acid lactam   6-aminohexanoic acid lactam   4-amino-2-methylpentanoic acid lactam
common name: β-propiolactam   ε-caprolactam   α-methyl-γ-valerolactam

21-2C   Nitriles

Nitriles contain the **cyano group**, —C≡N. Although nitriles lack the carbonyl group of carboxylic acids, they are classified as acid derivatives because they hydrolyze to give carboxylic acids and can be synthesized by dehydration of amides.

Hydrolysis to an acid

$$R-C\equiv N \xrightarrow[\text{H}^+ \text{ or } ^-\text{OH}]{\text{H}_2\text{O}} \underset{\text{primary amide}}{R-\overset{\displaystyle O}{\overset{\|}{C}}-NH_2} \xrightarrow[\text{H}^+]{\text{H}_2\text{O}} \underset{\text{acid}}{R-\overset{\displaystyle O}{\overset{\|}{C}}-OH}$$

nitrile

Synthesis from an acid

$$\underset{\text{acid}}{R-\overset{\displaystyle O}{\overset{\|}{C}}-OH} \xrightarrow[\text{heat}]{\text{NH}_3} \underset{\text{primary amide}}{R-\overset{\displaystyle O}{\overset{\|}{C}}-NH_2} \xrightarrow{\text{POCl}_3} \underset{\text{nitrile}}{R-C\equiv N}$$

Both the carbon atom and the nitrogen atom of the cyano group are *sp* hybridized, and the R—C≡N bond angle is 180° (linear). The structure of a nitrile is similar to that of a terminal alkyne, except that the nitrogen atom of the nitrile has a lone pair of electrons in place of the acetylenic hydrogen of the alkyne. Figure 21-1 compares the structures of acetonitrile and propyne.

Although a nitrile has a lone pair of electrons on nitrogen, it is not very basic. A typical nitrile has a pK_b of about 24, requiring a concentrated solution of mineral acid to protonate the nitrile. We explain this lack of basicity by noting that the nitrile's lone pair resides in an *sp*-hybrid orbital, with 50% *s* character. This orbital is close to the nucleus, and these electrons are tightly bound and relatively unreactive.

Nitrile nomenclature is derived from that of carboxylic acids. The IUPAC name is constructed from the alkane name, with the suffix *-nitrile* added. For common names, the suffix *-ic acid* is replaced by the suffix *-onitrile*.

$$CH_3-C\equiv N \qquad CH_3-\overset{\overset{\displaystyle Br}{|}}{CH}-CH_2-C\equiv N \qquad CH_3-\overset{\overset{\displaystyle OCH_3}{|}}{CH}-CH_2CH_2CH_2-C\equiv N$$

IUPAC name: ethanenitrile   3-bromobutanenitrile   5-methoxyhexanenitrile
common name: acetonitrile   β-bromobutyronitrile   δ-methoxycapronitrile

▶ **Figure 21-1**
Comparison of the electronic structures of acetonitrile and propyne (methylacetylene). In both compounds, the atoms at the ends of the triple bonds are *sp* hybridized, and the bond angles are 180°. In place of the acetylenic hydrogen atom, the nitrile has a lone pair of electrons in the *sp* orbital of nitrogen.

acetonitrile   propyne

For acids that are named as alkanecarboxylic acids, the corresponding nitriles are named by using the suffix -*carbonitrile*. The —C≡N group can also be named as a substituent, the *cyano group*.

$$\triangleright\!-CN$$
cyclopropanecarbonitrile

$$\overset{5}{CH_3}-\overset{4}{CH_2}-\overset{3}{\underset{\underset{CN}{|}}{CH}}-\overset{2}{CH_2}-\overset{1}{COOH}$$
3-cyanopentanoic acid

21-2D   Acid Halides

Acid halides, also called **acyl halides**, are activated derivatives used for the synthesis of other acyl compounds such as esters, amides, and acylbenzenes (in the Friedel–Crafts acylation). The most common acyl halides are the acyl chlorides (acid chlorides), and we will generally use acid chlorides as examples.

$$\underset{\substack{\text{an acid halide}\\ \text{(acyl halide)}}}{R-\overset{\overset{\displaystyle O}{\|}}{C}-halogen} \qquad \underset{\substack{\text{acid chloride}\\ \text{(acyl chloride)}}}{R-\overset{\overset{\displaystyle O}{\|}}{C}-Cl} \qquad \underset{\substack{\text{acid bromide}\\ \text{(acyl bromide)}}}{R-\overset{\overset{\displaystyle O}{\|}}{C}-Br}$$

The halogen atom of an acyl halide inductively withdraws electron density from the carbonyl carbon, enhancing its electrophilic nature and making acyl halides particularly reactive toward nucleophilic acyl substitution. The halide ion also serves as a good leaving group.

$$\underset{Nuc:^-}{R-\overset{\overset{\displaystyle \cdot\ddot{O}\cdot}{\|}}{C}-\ddot{C}l:} \;\rightleftharpoons\; \underset{Nuc}{R-\overset{\overset{\displaystyle :\ddot{O}:^-}{|}}{C}-\ddot{C}l:} \;\longrightarrow\; R-C\overset{\displaystyle \ddot{O}:}{\underset{Nuc}{\diagdown}} \qquad \underset{\text{leaving group}}{:\ddot{C}l:^-}$$

An acid halide is named by replacing the -*ic acid* suffix of the acid name with -*yl* and the halide name. For acids that are named as alkanecarboxylic acids, the acid chlorides are named by using the suffix -*carbonyl chloride*.

$$\underset{\substack{\text{ethanoyl fluoride}\\ \text{acetyl fluoride}}}{CH_3-\overset{\overset{\displaystyle O}{\|}}{C}-F} \qquad \underset{\substack{\text{propanoyl chloride}\\ \text{propionyl chloride}}}{CH_3-CH_2-\overset{\overset{\displaystyle O}{\|}}{C}-Cl} \qquad \underset{\substack{\text{3-bromobutanoyl bromide}\\ \beta\text{-bromobutyryl bromide}}}{CH_3-\underset{\underset{Br}{|}}{CH}-CH_2-\overset{\overset{\displaystyle O}{\|}}{C}-Br} \qquad \underset{\text{cyclopentanecarbonyl chloride}}{\overset{\displaystyle O}{\overset{\|}{\text{⬠}-C-Cl}}}$$

21-2E   Acid Anhydrides

The word **anhydride** means "without water." An acid anhydride contains two molecules of an acid, with loss of a molecule of water. Addition of water to an anhydride regenerates two molecules of the carboxylic acid.

$$\underset{\text{two molecules of acid}}{R-\overset{\overset{\displaystyle O}{\|}}{C}-OH \;+\; HO-\overset{\overset{\displaystyle O}{\|}}{C}-R} \;\rightleftharpoons\; \underset{\text{acid anhydride}}{R-\overset{\overset{\displaystyle O}{\|}}{C}-O-\overset{\overset{\displaystyle O}{\|}}{C}-R} \;+\; \underset{\text{water}}{H_2O}$$

Like acid halides, anhydrides are activated derivatives of carboxylic acids, although anhydrides are not as reactive as acid halides. In an acid chloride, the chlorine atom activates the carbonyl group and serves as a leaving group. In an anhydride, the carboxylate group serves these functions:

The reaction scheme at the top:

$$R\!-\!\overset{\overset{\displaystyle O}{\|}}{C}\!-\!\overset{..}{\underset{..}{O}}\!-\!\overset{\overset{\displaystyle O}{\|}}{C}\!-\!R \quad \rightleftharpoons \quad R\!-\!\overset{\overset{\displaystyle :\overset{..}{O}:^-}{|}}{\underset{\underset{\displaystyle Nuc}{}}{C}}\!-\!\overset{..}{\underset{..}{O}}\!-\!\overset{\overset{\displaystyle O}{\|}}{C}\!-\!R \quad \longrightarrow \quad R\!-\!\overset{\displaystyle C}{\underset{\displaystyle Nuc}{}} \quad \overset{\displaystyle O}{\underset{..}{\overset{..}{O}}}\!-\!\overset{\overset{\displaystyle O}{\|}}{C}\!-\!R$$

Nuc: Nuc

carboxylate
leaving group

Half of an anhydride's acid units are lost as leaving groups. If the acid is expensive, we would not use the anhydride as an activated form to make a derivative; the acid chloride is a more efficient alternative, using chloride as the leaving group. Anhydrides are used primarily when the necessary anhydride is cheap and readily available. Acetic anhydride, phthalic anhydride, succinic anhydride, and maleic anhydride are the ones most often used. Diacids commonly form cyclic anhydrides, especially if a five- or six-membered ring results.

Anhydride nomenclature is very simple; the word *acid* is changed to *anhydride* in both the common name and the IUPAC name (rarely used). The following examples show the names of some common anhydrides:

$$CH_3\!-\!\overset{\overset{\displaystyle O}{\|}}{C}\!-\!O\!-\!\overset{\overset{\displaystyle O}{\|}}{C}\!-\!CH_3 \qquad CF_3\!-\!\overset{\overset{\displaystyle O}{\|}}{C}\!-\!O\!-\!\overset{\overset{\displaystyle O}{\|}}{C}\!-\!CF_3$$

(abbreviated Ac₂O)   (abbreviated TFAA)
ethanoic anhydride   trifluoroethanoic anhydride
acetic anhydride   trifluoroacetic anhydride

1,2-benzenedicarboxylic anhydride
phthalic anhydride

but-2-enedioic anhydride
maleic anhydride

Anhydrides composed of two different acids are called **mixed anhydrides** and are named by using the names of the individual acids.

$$CH_3\!-\!\overset{\overset{\displaystyle O}{\|}}{C}\!-\!O\!-\!\overset{\overset{\displaystyle O}{\|}}{C}\!-\!H \qquad CH_3CH_2\!-\!\overset{\overset{\displaystyle O}{\|}}{C}\!-\!O\!-\!\overset{\overset{\displaystyle O}{\|}}{C}\!-\!CF_3$$

IUPAC name: ethanoic methanoic anhydride   trifluoroethanoic propanoic anhydride
common name: acetic formic anhydride   trifluoroacetic propionic anhydride

21-2F Nomenclature of Multifunctional Compounds

With all the different functional groups we have studied, it is not always obvious which functional group of a multifunctional compound is the "main" one, and which groups should be named as substituents. In choosing the principal group for the root name, we use the following priorities:

acid > ester > amide > nitrile > aldehyde > ketone > alcohol > amine > alkene > alkyne

Table 21-1 summarizes these priorities, together with the suffixes used for main groups and the prefixes used for substituents. The following examples illustrate these priorities in naming multifunctional compounds:

ethyl *o*-cyanobenzoate

2-formylcyclohexanecarboxamide

$$CH_3\!-\!CH_2\!-\!\overset{\overset{\displaystyle OH}{|}}{CH}\!-\!C\!\equiv\!N$$

2-hydroxybutanenitrile

TABLE 21-1 Summary of Functional Group Nomenclature

Functional Group	Name as Main Group	Name as Substituent
Main groups in order of decreasing priority:		
carboxylic acids	-oic acid	carboxy[a]
esters	-oate	alkoxycarbonyl[a]
amides	-amide	amido[a]
nitriles	-nitrile	cyano
aldehydes	-al	formyl
ketones	-one	oxo
alcohols	-ol	hydroxy
amines	-amine	amino
alkenes	-ene	alkenyl
alkynes	-yne	alkynyl
alkanes	-ane	alkyl
ethers		alkoxy
halides		halo

[a] Denotes rare usage.

PROBLEM 21-1

Name the following carboxylic acid derivatives, giving both a common name and an IUPAC name where possible:

(a) $PhCOOCH_2CH(CH_3)_2$

(b) $PhOCHO$

(c) $PhCH(CH_3)COOCH_3$

(d) $PhNHCOCH_2CH(CH_3)_2$

(e) $CH_3CONHCH_2Ph$

(f) $CH_3CH(OH)CH_2CN$

(g) $(CH_3)_2CHCH_2COBr$

(h) $Cl_2CHCOCl$

(i) $(CH_3)_2CHCOOCHO$

(j)

(k)

(l) $PhCONH-$

(m)

(n)

(o)

(p)

(q)

(r)

(*Hint:* Named as a piperidine derivative.)

21-3A Boiling Points and Melting Points

Figure 21-2 is a graph of the boiling points of simple acid derivatives plotted against their molecular weights. The *n*-alkanes are included for comparison. Notice that esters and acid chlorides have boiling points near those of the unbranched alkanes with similar molecular weights. These acid derivatives contain highly polar carbonyl groups, but the polarity of the carbonyl group has only a small effect on the boiling points (Section 18-4).

Carboxylic acids are strongly hydrogen bonded in the liquid phase, resulting in elevated boiling points. The stable hydrogen-bonded dimer has a higher effective

21-3

Physical Properties of Carboxylic Acid Derivatives

Examples (MW 55–60)	bp(°C)
CH₃—C(=O)—NH₂	222
CH₃—C(=O)—OH	118
CH₃CH₂CH₂OH	97
CH₃CH₂—C≡N	97
H—C(=O)—OCH₃	32
CH₃CH₂CH₂CH₃	0

▲ **Figure 21-2**
Boiling points of acid derivatives plotted against their molecular weights. Alcohols and unbranched alkanes are included for comparison.

molecular weight and boils at a higher temperature. Nitriles also have higher boiling points than esters and acid chlorides of similar molecular weight. This effect results from a strong dipolar association between adjacent cyano groups.

carboxylic acid dimer   dipolar association of nitriles

Amides have surprisingly high boiling points and melting points compared with other compounds of similar molecular weight. Primary and secondary amides participate in strong hydrogen bonding, shown in Figure 21-3. The resonance picture shows a partial negative charge on oxygen and a partial positive charge on nitrogen. The positively charged nitrogen polarizes the N—H bond, making the hydrogen strongly electrophilic. The negatively charged oxygen's lone pairs are particularly effective in forming hydrogen bonds to these polarized N—H hydrogens.

Pure tertiary amides lack N—H bonds, so they cannot participate in hydrogen bonding (although they are good hydrogen bond acceptors). Still, they have high boil-hydrogen bonding   intermolecular attraction

◀ **Figure 21-3**
The resonance picture of an amide shows its strongly polar nature. Hydrogen bonds and dipolar attractions stabilize the liquid phase, resulting in higher boiling points.

ing points, close to those of carboxylic acids of similar molecular weights. Figure 21-3 shows how a pairing of two molecules is strongly attractive, helping to stabilize the liquid phase. Vaporization disrupts this arrangement, so a higher temperature is needed for boiling.

Strong hydrogen bonding between molecules of primary and secondary amides also results in unusually high melting points. For example, N-methylacetamide (secondary, one N—H bond) has a melting point of 28°C, which is 89° higher than the melting point (−61°C) of its isomer dimethylformamide (tertiary, no N—H bond). With two N—H bonds to engage in hydrogen bonding, the primary amide propionamide melts at 79°C, about 50° higher than its secondary isomer N-methylacetamide.

dimethylformamide   N-methylacetamide   propionamide
mp −61°C   mp 28°C   mp 79°C

21-3B   Solubility

Acid derivatives (esters, acid chlorides, anhydrides, nitriles, and amides) are soluble in common organic solvents such as alcohols, ethers, chlorinated alkanes, and aromatic hydrocarbons. Acid chlorides and anhydrides cannot be used in nucleophilic solvents such as water and alcohols, however, because they react with these solvents. Many of the smaller esters, amides, and nitriles are relatively soluble in water (Table 21-2) because of their high polarity and their ability to form hydrogen bonds with water.

Esters, tertiary amides, and nitriles are frequently used as solvents for organic reactions because they provide a polar reaction medium without O—H or N—H groups that can donate protons or act as nucleophiles. Ethyl acetate is a moderately polar solvent with a boiling point of 77°C, convenient for easy evaporation from a reaction mixture. Acetonitrile, dimethylformamide (DMF), and dimethylacetamide (DMA) are highly polar solvents that solvate ions almost as well as water, but without the reactivity of O—H or N—H groups. These three solvents are miscible with water and are often used as solvent mixtures with water.

TABLE 21-2 Esters, Amides, and Nitriles Commonly Used as Solvents for Organic Reactions

Compound	Name	mp (°C)	bp (°C)	Water Solubility
$CH_3-\overset{\overset{\text{O}}{\|\|}}{C}-OCH_2CH_3$	ethyl acetate	−83	77	10%
$H-\overset{\overset{\text{O}}{\|\|}}{C}-N(CH_3)_2$	dimethylformamide (DMF)	−61	153	miscible
$CH_3-\overset{\overset{\text{O}}{\|\|}}{C}-N(CH_3)_2$	dimethylacetamide (DMA)	−20	165	miscible
$CH_3-C\equiv N$	acetonitrile	−45	82	miscible

21-4
Spectroscopy of Carboxylic Acid Derivatives

21-4A   Infrared Spectroscopy

Different types of carbonyl groups give characteristic strong absorptions at different positions in the infrared spectrum. As a result, infrared spectroscopy is often the best method to detect and differentiate these carboxylic acid derivatives. Table 21-3 summarizes the characteristic IR absorptions of carbonyl functional groups. As in Chapter 12, we are using about 1710 cm^{-1} for simple ketones, aldehydes, and acids as a standard for comparison. Appendix 2 gives a more complete table of characteristic IR frequencies.

Esters   Ester carbonyl groups absorb at relatively high frequencies, about 1735 cm^{-1}. Except for strained cyclic ketones, few other functional groups absorb strongly in this region. Esters also have a C—O single-bond stretching absorption between 1000 and

TABLE 21-3 Characteristic IR Stretching Absorptions of Acid Derivatives

Functional Group		Frequency	Comments
ketone	$R-\overset{\overset{\text{O}}{\|\|}}{C}-R$	C=O, 1710 cm^{-1}	lower if conjugated, higher if strained
acid	$R-\overset{\overset{\text{O}}{\|\|}}{C}-OH$	C=O, 1710 cm^{-1} O—H, $2500-3500 \text{ cm}^{-1}$	lower if conjugated broad, on top of C—H stretch
ester	$R-\overset{\overset{\text{O}}{\|\|}}{C}-O-R'$	C=O, 1735 cm^{-1}	lower if conjugated, higher if strained
amide	$R-\overset{\overset{\text{O}}{\|\|}}{C}-\underset{\underset{\text{H}}{\|}}{N}-R'$	C=O, $1640-1680 \text{ cm}^{-1}$ N—H, $3200-3500 \text{ cm}^{-1}$	two peaks for R—CO—NH$_2$, one peak for R—CO—NHR'
acid chloride	$R-\overset{\overset{\text{O}}{\|\|}}{C}-Cl$	C=O, 1800 cm^{-1}	very high frequency
acid anhydride	$R-\overset{\overset{\text{O}}{\|\|}}{C}-O-\overset{\overset{\text{O}}{\|\|}}{C}-R$	C=O, 1800 and 1750 cm^{-1}	two peaks
nitrile	$R-C\equiv N$	C≡N, 2200 cm^{-1}	just above 2200 cm^{-1}

▲ **Figure 21-4**
Infrared spectra of (a) ethyl octanoate and (b) methyl benzoate. The carbonyl stretching frequency of simple esters is around 1735 cm^{-1} and that of conjugated esters is around 1710–1725 cm^{-1}.

1200 cm^{-1}, but many other molecules also absorb in this region. We do not consider this absorption to be diagnostic for an ester, but we may look for it in uncertain cases.

Conjugation lowers the carbonyl stretching frequency of an ester. Conjugated esters absorb around 1710 to 1725 cm^{-1} and might easily be confused with simple ketones. The presence of *both* a strong carbonyl absorption in this region and a conjugated C=C absorption around 1620 to 1640 cm^{-1} suggests a conjugated ester. Compare the spectra of ethyl octanoate and methyl benzoate in Figure 21-4 to see these differences.

PROBLEM 21-2

What characteristics of the methyl benzoate spectrum rule out an aldehyde or carboxylic acid functional group giving the absorption at 1723 cm^{-1}?

PROBLEM 21-3

Give the frequencies of the C—O single-bond stretching absorptions in the IR spectra of ethyl octanoate and methyl benzoate.

less than a
full double bond

Amides Simple amides have much lower carbonyl stretching frequencies than the other carboxylic acid derivatives, absorbing around 1640 to 1680 cm^{-1} (often a close doublet). This low-frequency absorption agrees with the resonance picture of the amide. The C=O bond of the amide carbonyl is somewhat less than a full double bond. Because it is not as strong as the C=O bond in a simple ketone or carboxylic acid, the amide C=O has a lower stretching frequency.

Primary and secondary amides have N—H bonds that give infrared stretching absorptions in the region 3200 to 3500 cm^{-1}. These absorptions fall in the same region as the broad O—H absorption of an alcohol, but the amide N—H absorptions are usually sharper. In primary amides (R—CO—NH$_2$), there are two N—H bonds, and two sharp peaks occur in the region 3200 to 3500 cm^{-1}. Secondary amides (R—CO—NHR′) have only one N—H bond, and only one peak is observed in the N—H region of the spectrum. Tertiary amides (R—CO—NR$_2'$) have no N—H bonds, so there is no N—H absorption.

The infrared spectrum of butyramide appears in Figure 12–13a (page 510). Notice the strong carbonyl stretching absorption around 1630–1660 cm^{-1} and two N—H stretching absorptions at 3350 and 3180 cm^{-1}.

Lactones and Lactams Unstrained lactones (cyclic esters) and lactams (cyclic amides) absorb at typical frequencies for esters and amides. Ring strain raises the carbonyl absorption frequency, however. Recall that cyclic ketones with five-membered or smaller rings show a similar increase in carbonyl stretching frequency (Section 18-5A). Figure 21-5 shows the effect of ring strain on the C=O stretching frequencies of lactones and lactams.

δ-valerolactone	γ-butyrolactone	β-propiolactone	δ-valerolactam	γ-butyrolactam	β-propiolactam
1735 cm^{-1}	1770 cm^{-1}	1800 cm^{-1}	1670 cm^{-1}	1700 cm^{-1}	1745 cm^{-1}
no strain	moderate strain	highly strained	no strain	moderate strain	highly strained

▲ **Figure 21-5**
Ring strain in a lactone or lactam increases the carbonyl stretching frequency.

Nitriles Nitriles show a characteristic C≡N stretching absorption around 2200 cm^{-1} in the infrared spectrum. This absorption can be distinguished from the alkyne C≡C absorption by two characteristics: Nitriles usually absorb at frequencies slightly *higher* than 2200 cm^{-1} (to the left of 2200 cm^{-1}), while alkynes usually absorb at frequencies slightly *lower* than 2200 cm^{-1}; and nitrile absorptions are usually stronger because the C≡N triple bond is more polar than the alkyne C≡C triple bond.

The IR spectrum of butyronitrile appears in Figure 12-14 (page 511). Notice the strong triple-bond stretching absorption at 2249 cm^{-1}. The IR spectrum of hexanenitrile (Compound 3, p. 516) shows C≡N stretching at 2246 cm^{-1}.

Acid Halides and Anhydrides Acid halides and anhydrides are rarely isolated as unknown compounds; but they are commonly used as reagents and intermediates, and infrared spectroscopy can confirm that an acid has been converted to a pure acid chloride or anhydride. The carbonyl stretching vibration of an acid chloride occurs at a high frequency, around 1800 cm^{-1}.

Anhydrides give *two* carbonyl stretching absorptions, one around 1800 cm^{-1} and another around 1750 cm^{-1}. Figure 21-6 shows the spectrum of propionic anhydride, with carbonyl absorptions at 1818 and 1751 cm^{-1}.

PROBLEM-SOLVING HINT
The absorptions listed in Table 21-3 are often the best spectroscopic information available for determining the functional group in an unknown acid derivative.

▲ **Figure 21-6**

Infrared spectrum of propionic anhydride, showing C=O stretching absorptions at 1818 and 1751 cm^{-1}.

PROBLEM 21-4

The IR spectra shown next may include a carboxylic acid, an ester, an amide, a nitrile, an acid chloride, or an acid anhydride. Determine the functional group suggested by each spectrum, and list the specific frequencies you used to make your decision.

21-4B   NMR Spectroscopy

NMR spectroscopy of acid derivatives is complementary to IR spectroscopy. For the most part, IR gives information about the functional groups, while NMR gives information about the alkyl groups. In many cases, the combination of IR and NMR provides enough information to determine the structure.

Proton NMR   The proton chemical shifts found in acid derivatives are close to those of similar protons in ketones, aldehydes, alcohols, and amines (Figure 21-7). For example, protons alpha to a carbonyl group absorb between $\delta 2.0$ and $\delta 2.5$, whether the carbonyl group is part of a ketone, aldehyde, acid, ester, or amide. The protons of the alcohol-derived group of an ester or the amine-derived group of an amide give absorptions similar to those in the spectrum of the parent alcohol or amine.

The N—H protons of an amide may be broad, appearing between $\delta 5$ and $\delta 8$, depending on concentration and solvent. Figure 13-37 (page 573) shows the NMR spectrum of an amide with a broad N—H absorption. The *formyl* proton bonded to the carbonyl group of a formate ester or formamide resembles an aldehyde proton, but it is slightly more shielded and appears around $\delta 8$. In a nitrile, the protons on the α-carbon atom absorb around $\delta 2.5$, similar to the α protons of a carbonyl group.

▶ **Figure 21-7**

Typical absorptions of acid derivatives in the proton NMR spectrum.

▲ Figure 21-8
The proton and carbon NMR spectra of *N,N*-dimethylformamide show two methyl singlets resulting from hindered rotation about the amide bond. In both spectra the transoid methyl group is farther downfield than the cisoid methyl group.

The NMR spectrum of *N,N*-dimethylformamide (Figure 21-8) shows the formyl proton (H—C=O) around δ8. The two methyl groups appear as two singlets (not a spin-spin splitting doublet) near δ2.9 and δ3.0. The two singlets result from hindered rotation about the amide bond. The cisoid and transoid methyl groups interconvert slowly with respect to the NMR time scale.

Carbon NMR   The carbonyl carbons of acid derivatives appear at shifts around 170 to 180 ppm, slightly more shielded than the carbonyl carbons of ketones and aldehydes. The α-carbon atoms absorb around 30 to 40 ppm. The sp^3 hybridized carbons bonded to oxygen in esters absorb around 60 to 80 ppm, and those bonded to nitrogen in amides absorb around 40 to 60 ppm. The cyano carbon of a nitrile absorbs around 120 ppm.

Figure 21-8 also shows the carbon NMR spectrum of *N,N*-dimethylformamide (DMF). Note the carbonyl carbon atom at 162 ppm and the two distinct cisoid and transoid carbons at 31 ppm and 36 ppm, respectively.

PROBLEM 21-5

For each set of IR and NMR spectra, determine the structure of the unknown compound. Explain how your proposed structure fits the spectra. **(a)** C_3H_5NO **(b)** $C_5H_8O_2$

Preview   Acid derivatives react with a wide variety of nucleophilic reagents under both basic and acidic conditions. Most of these reactions involve **nucleophilic acyl substitutions**, following similar reaction mechanisms. In each case, the nucleophilic reagent adds to the carbonyl group to produce a tetrahedral intermediate, which expels the leaving group to regenerate the carbonyl group. Through this addition–elimination process, the nucleophilic reagent substitutes for the leaving group. In the sections that follow, we consider several examples of these reactions, first under basic conditions and then under acidic conditions. In each case, we will note the similarities with other reactions that follow this same addition–elimination pathway.

Nucleophilic acyl substitutions are also called **acyl transfer** reactions because they transfer the acyl group from the leaving group to the attacking nucleophile. The following is a generalized **addition–elimination mechanism** for nucleophilic acyl substitution under basic conditions.

21-5
Interconversion of Acid Derivatives by Nucleophilic Acyl Substitution

---

KEY MECHANISM   Addition–Elimination Mechanism of Nucleophilic Acyl Substitution

Step 1: Addition of the nucleophile.   *Step 2:* Elimination of the leaving group.

Nuc:⁻ +

nucleophilic attack   tetrahedral intermediate   products   leaving group

---

Depending on the nature of Nuc:⁻ and Y:⁻, we can imagine converting any acid derivative into almost any other. Reactions that actually occur generally convert a more reactive acid derivative to a less reactive one. Predicting these reactions requires a knowledge of the relative reactivity of acid derivatives.

PROBLEM-SOLVING HINT

This mechanism applies to most of the reactions in this chapter.

21-5A   Reactivity of Acid Derivatives

Acid derivatives differ greatly in their reactivity toward nucleophilic acyl substitution. For example, water hydrolyzes acetyl chloride in a violently exothermic reaction, while acetamide is stable in boiling water. Acetamide is hydrolyzed only by boiling it in strong acid or base for several hours.

$$CH_3-\overset{\displaystyle O}{\overset{\|}{C}}-Cl \ + \ H_2O \ \xrightarrow{\text{(very fast)}} \ CH_3-\overset{\displaystyle O}{\overset{\|}{C}}-OH \ + \ HCl$$

$$CH_3-\overset{\displaystyle O}{\overset{\|}{C}}-NH_2 + Na^+\ ^-OH \ \xrightarrow[\text{(slow)}]{H_2O,\ boil} \ CH_3-\overset{\displaystyle O}{\overset{\|}{C}}-O^-\ ^+Na \ + \ NH_3$$

The reactivity of acid derivatives toward nucleophilic attack depends on their structure and on the nature of the attacking nucleophile. In general, reactivity follows this order:

Reactivity	Derivative		Leaving group	Basicity
more reactive				less basic
↑	acid chloride	$R-\overset{O}{\overset{\|}{C}}-Cl$	Cl^-	↓
	anhydride	$R-\overset{O}{\overset{\|}{C}}-O-\overset{O}{\overset{\|}{C}}-R$	$^-O-\overset{O}{\overset{\|}{C}}-R$	
	ester	$R-\overset{O}{\overset{\|}{C}}-O-R'$	$^-O-R'$	
	amide	$R-\overset{O}{\overset{\|}{C}}-NH_2$	$^-NH_2$	
less reactive	carboxylate	$R-\overset{O}{\overset{\|}{C}}-O^-$	—	more basic

Such as order of reactivity stems partly from the basicity of the leaving groups. Strong bases are not good leaving groups, and the reactivity of the derivatives decreases as the leaving group becomes more basic.

Resonance stabilization also affects the reactivity of acid derivatives. In amides, for example, resonance stabilization is lost when a nucleophile attacks.

$$\left[R-\overset{\overset{\displaystyle \ddot{O}}{\|}}{C}-\ddot{N}H_2 \ \longleftrightarrow \ R-\overset{\overset{\displaystyle :\ddot{O}:^-}{|}}{C}=\overset{+}{N}H_2 \right] \ \underset{}{\overset{Nuc:^-}{\rightleftharpoons}} \ R-\overset{\overset{\displaystyle :\ddot{O}:^-}{|}}{\underset{Nuc}{C}}-\ddot{N}H_2$$

resonance-stabilized   no resonance stabilization

A smaller amount of stabilization is present in esters.

$$\left[R-\overset{\overset{\displaystyle \ddot{O}}{\|}}{C}-\ddot{O}-R' \ \longleftrightarrow \ R-\overset{\overset{\displaystyle :\ddot{O}:^-}{|}}{C}=\overset{+}{\ddot{O}}-R' \right] \ \underset{}{\overset{Nuc:^-}{\rightleftharpoons}} \ R-\overset{\overset{\displaystyle :\ddot{O}:^-}{|}}{\underset{Nuc}{C}}-\ddot{O}-R'$$

weak resonance stabilization   no resonance stabilization

Resonance stabilization of an anhydride is like that in an ester, but the stabilization is shared between two carbonyl groups. Each carbonyl group receives less stabilization than an ester carbonyl.

$$\left[R-\overset{\overset{\displaystyle \ddot{O}}{\|}}{C}-\ddot{O}-\overset{\overset{\displaystyle \ddot{O}}{\|}}{C}-R \ \longleftrightarrow \ R-\overset{\overset{\displaystyle :\ddot{O}:^-}{|}}{C}=\overset{+}{\ddot{O}}-\overset{\overset{\displaystyle \ddot{O}}{\|}}{C}-R \right] \ \underset{}{\overset{Nuc:^-}{\rightleftharpoons}} \ R-\overset{\overset{\displaystyle :\ddot{O}:^-}{|}}{\underset{Nuc}{C}}-\ddot{O}-\overset{\overset{\displaystyle \ddot{O}}{\|}}{C}-R$$

shared, weak resonance stabilization

There is little resonance stabilization of an acid chloride, and it is quite reactive.

Interconversions of acid derivatives

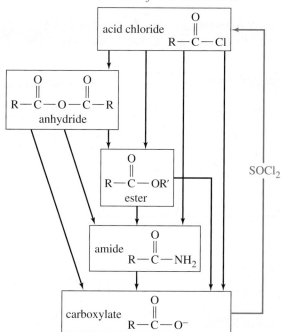

◀ **Figure 21-9**
More reactive acid derivatives are easily converted to less reactive derivatives. A "downhill" reaction from

$$\underset{\substack{\|\\ \text{R—C—W}}}{\overset{O}{}} \text{ to } \underset{\substack{\|\\ \text{R—C—Z}}}{\overset{O}{}}$$

generally requires Z^- or $H—Z$ as the nucleophile for nucleophilic acyl substitution.

PROBLEM-SOLVING **HINT**

All the mechanisms in this section are variations of the addition–elimination mechanism of nucleophilic acyl substitution. You should be able to draw any of these mechanisms without having to memorize them.

In general, we can easily accomplish nucleophilic acyl substitutions that convert more reactive derivatives to less reactive ones. Thus, an acid chloride is easily converted to an anhydride, ester, or amide. An anhydride is easily converted to an ester or an amide. An ester is easily converted to an amide, but an amide can be hydrolyzed only to the acid or the carboxylate ion (in basic conditions). Figure 21-9 graphically summarizes these conversions. Notice that thionyl chloride ($SOCl_2$) converts an acid to its most reactive derivative, the acid chloride (Section 20-11).

Acid chlorides are the most reactive acid derivatives, so they are easily converted to any of the other acid derivatives. Acid chlorides react with carboxylic acids (or their carboxylate salts) to form anhydrides. Either oxygen atom of the acid can attack the strongly electrophilic carbonyl of the acid chloride to form a tetrahedral intermediate that loses chloride ion and a proton to give the anhydride.

MECHANISM Conversion of an Acid Chloride to an Anhydride

Step 1: Addition of the nucleophile.

Step 2: Elimination of the leaving group.

Step 3: Loss of a proton.

$$\underset{\text{acid chloride}}{\overset{O}{\underset{\|}{R—C—Cl}}} + \underset{\text{acid}}{\overset{O}{\underset{\|}{H\ddot{O}—C—R'}}} \rightleftharpoons \underset{\text{tetrahedral intermediate}}{R—C—Cl} \longrightarrow R—C \quad Cl^- \rightleftharpoons \underset{\substack{\text{anhydride}\\ + \text{ H—Cl}}}{\overset{O \quad O}{\underset{\|\quad\|}{R—C—O—C—R'}}}$$

Example

$$\underset{\text{heptanoyl chloride}}{\overset{O}{\underset{\|}{CH_3(CH_2)_5—C—Cl}}} + \underset{\text{heptanoic acid}}{\overset{O}{\underset{\|}{CH_3(CH_2)_5—C—OH}}} \longrightarrow \underset{\text{heptanoic anhydride}}{\overset{O \quad\quad O}{\underset{\|\quad\quad\|}{CH_3(CH_2)_5—C—O—C—(CH_2)_5CH_3}}}$$

Acid chlorides react rapidly with alcohols to give esters in a strongly exothermic reaction. Care must be taken to keep the temperature low to avoid dehydration of the alcohol.

MECHANISM   Conversion of an Acid Chloride to an Ester

Step 1: Addition of the nucleophile.

Step 2: Elimination of the leaving group.

Step 3: Loss of a proton.

$$R-\underset{\overset{\displaystyle O}{\|}}{C}-Cl \ + \ R'-\ddot{O}H \ \rightleftharpoons \ R-\underset{\underset{R'}{\overset{\displaystyle |}{\underset{\displaystyle H}{\overset{+}{O}}}}}{\overset{\overset{\displaystyle :\ddot{O}:^-}{|}}{C}}-Cl \ \longrightarrow \ R-\underset{\underset{R'}{\overset{|}{\underset{}{\overset{+}{O}-H}}}}{\overset{\overset{\displaystyle \ddot{O}}{\|}}{C}} \ + \ Cl^- \ \longrightarrow \ R-\underset{}{\overset{\overset{\displaystyle \ddot{O}}{\|}}{C}}-\ddot{O}-R'$$

acid chloride   alcohol

tetrahedral intermediate

ester

+ HCl

Example

cyclopentanecarbonyl chloride   2-propanol   2-propyl cyclopentanecarboxylate   + HCl

Acid chlorides react rapidly with ammonia and amines to give amides. The HCl generated by the reaction can protonate the amine starting material, so a two-fold excess of the amine is required. Alternatively, a base such as pyridine or NaOH may be added with the amine to neutralize the HCl and avoid having to use a large excess of the amine.

MECHANISM   Conversion of an Acid Chloride to an Amide

Step 1: Addition of the nucleophile.

Step 2: Elimination of the leaving group.

Step 3: Loss of a proton.

$$R-\underset{\overset{\displaystyle O}{\|}}{C}-\ddot{Cl}: \ 2\, R_2'\overset{+}{N}-H \ \rightleftharpoons \ R-\underset{\underset{R_2'N-H}{\overset{|+}{}}}{\overset{\overset{\displaystyle :\ddot{O}:^-}{|}}{C}}-\ddot{Cl}: \ \longrightarrow \ R-\underset{R_2'\overset{+}{N}-H}{\overset{\overset{\displaystyle \ddot{O}}{\diagup\diagdown}}{C}} \ \longrightarrow \ R-\underset{:NR_2'}{\overset{\overset{\displaystyle \ddot{O}}{\diagup\diagdown}}{C}}$$

acid chloride   amine

tetrahedral intermediate

$R_2'NH$

$+ \ R_2'\overset{+}{N}H_2 \ \ Cl^-$

Reaction of an acid chloride with ammonia gives a primary amide. With a primary amine, this reaction gives a secondary amide; and with a secondary amine, it gives a tertiary amide.

Example

$$CH_3-(CH_2)_4-\underset{\overset{\displaystyle O}{\|}}{C}-Cl \ + \ \bigcirc\!\!-NH_2 \ \longrightarrow \ CH_3-(CH_2)_4-\underset{\overset{\displaystyle O}{\|}}{C}-NH-\bigcirc \ + \ HCl$$

hexanoyl chloride   cyclohexylamine (primary amine)   *N*-cyclohexylhexanamide (secondary amide)

Acid anhydrides are not as reactive as acid chlorides, but they are still considered to be activated toward nucleophilic acyl substitution. Reaction with an alcohol gives an ester. Notice that one of the two acid units from the anhydride is expelled as the leaving group.

MECHANISM   Conversion of an Acid Anhydride to an Ester

Step 1: Addition of the nucleophile.

Step 2: Elimination of the leaving group.

Step 3: Loss of a proton.

$$R-\overset{\overset{\displaystyle \ddot{O}}{\|}}{C}-O-\overset{\overset{\displaystyle O}{\|}}{C}-R \ + \ R'-\ddot{O}H \ \rightleftharpoons \ R-\overset{\overset{\displaystyle :\ddot{O}:^-}{|}}{\underset{\overset{\displaystyle O^+}{\underset{R'\diagdown H}{|}}}{C}}-O-\overset{\overset{\displaystyle O}{\|}}{C}-R \ \longrightarrow \ R-\overset{\overset{\displaystyle \ddot{O}\cdot}{|}}{\underset{\overset{\displaystyle +\ddot{O}}{\underset{R'\diagdown H}{|}}}{C}}=O-\overset{\overset{\displaystyle O}{\|}}{C}-R$$

tetrahedral intermediate

$$\rightleftharpoons \ R-\overset{\overset{\displaystyle O}{\|}}{C}-O-R' \ + \ R-\overset{\overset{\displaystyle O}{\|}}{C}-OH$$

ester

Example

cyclopentanol   + $CH_3-\overset{\overset{\displaystyle O}{\|}}{C}-O-\overset{\overset{\displaystyle O}{\|}}{C}-CH_3$ $\longrightarrow$ cyclopentyl acetate   +   AcOH

cyclopentanol   acetic anhydride   cyclopentyl acetate   acetic acid

Anhydrides react quickly with ammonia and amines. Reaction of an anhydride with ammonia gives a primary amide. With a primary amine, the reaction gives a secondary amide; and with a secondary amine, it gives a tertiary amide.

MECHANISM   Conversion of an Acid Anhydride to an Amide

Step 1: Addition of the nucleophile.

Step 2: Elimination of the leaving group.

Step 3: Loss of a proton.

$$R-\overset{\overset{\displaystyle \ddot{O}}{\|}}{C}-O-\overset{\overset{\displaystyle O}{\|}}{C}-R \ + \ R'_2\ddot{N}H \ \rightleftharpoons \ R-\overset{\overset{\displaystyle :\ddot{O}:^-}{|}}{\underset{\overset{\displaystyle +N-H}{\underset{R'\diagdown R'}{|}}}{C}}-O-\overset{\overset{\displaystyle O}{\|}}{C}-R \ \longrightarrow \ R-\overset{\overset{\displaystyle \ddot{O}\cdot}{|}}{\underset{\overset{\displaystyle +N-H}{\underset{R'\diagdown R'}{|}}}{C}}=O-\overset{\overset{\displaystyle O}{\|}}{C}-R$$

anhydride   amine   tetrahedral intermediate

$$\rightleftharpoons \ R-\overset{\overset{\displaystyle O}{\|}}{C}-NR'_2 \ + \ R-\overset{\overset{\displaystyle O}{\|}}{C}-OH$$

amide

Example

aniline   + $CH_3-\overset{\overset{\displaystyle O}{\|}}{C}-O-\overset{\overset{\displaystyle O}{\|}}{C}-CH_3$ $\longrightarrow$ acetanilide   +   AcOH

aniline   acetic anhydride   acetanilide   acetic acid

Esters are converted to amides by heating with ammonia or an amine. This reaction is called **ammonolysis**, meaning "lysis (cleavage) by an amine." Esters are not as reactive as acid chlorides and anhydrides, and prolonged heating is often required.

Ammonolysis using ammonia gives primary amides. Primary amines react to give secondary amides, and secondary amines react (often slowly) to give tertiary amides. In each case, the acyl group of the ester is transferred from the oxygen atom of the alcohol to the nitrogen atom of the amine.

MECHANISM   Conversion of an Ester to an Amide (Ammonolysis of an Ester)

Step 1: Addition of the nucleophile.

Step 2: Elimination of the leaving group.

Step 3: Loss of a proton.

tetrahedral intermediate

$$R-C(=O)-NHR'' + R'-OH$$

Example

ethyl formate   cyclohexylamine   *N*-cyclohexylformamide
(90%)

$$+ \quad CH_3CH_2-OH$$

PROBLEM 21-6

(a) Propose a mechanism for the reaction of benzyl alcohol with acetyl chloride to give benzyl acetate.

(b) Propose a mechanism for the reaction of benzoic acid with acetyl chloride to give acetic benzoic anhydride.

(c) Propose a *second* mechanism for the reaction of benzoic acid with acetyl chloride to give acetic benzoic anhydride. This time, let the *other* oxygen of benzoic acid serve as the nucleophile to attack the carbonyl group of acetyl chloride. Because proton transfers are fast between these oxygen atoms, it is difficult to differentiate between these two mechanisms experimentally.

(d) Propose a mechanism for the reaction of aniline with acetic anhydride to give acetanilide.

(e) Propose a mechanism for the reaction of aniline with ethyl acetate to give acetanilide. What is the leaving group in your proposed mechanism? Would this be a suitable leaving group for an S_N2 reaction?

21-5B   Leaving Groups in Nucleophilic Acyl Substitutions

Loss of an alkoxide ion as a leaving group in the second step of the conversion of an ester to an amide should surprise you.

tetrahedral intermediate

alkoxide
(strong base)

S$_N$2

bond to methoxide is about half broken

transition state

Acyl substitution

$^-$OCH$_3$ leaves in an
exothermic step

transition state

bond to methoxide
has just begun
to break

◀ **Figure 21-10**
Comparison of S$_N$2 and acyl
addition–elimination reactions with
methoxide as the leaving group. In
the concerted S$_N$2, methoxide leaves
in a slightly endothermic step, and
the bond to methoxide is largely
broken in the transition state.
In the acyl substitution, methoxide
leaves in an exothermic second step
with a reactant-like transition state:
The bond to methoxide has just
begun to break in the transition state.

In our study of alkyl substitution and elimination reactions (S$_N$1, S$_N$2, E1, E2), we saw
that strong bases such as hydroxide and alkoxide are poor leaving groups for these reactions. Figure 21-10 compares the acyl addition–elimination mechanism with the S$_N$2
mechanism. The differences in the mechanisms explain why strong bases may serve as
leaving groups in acyl substitution, even though they cannot in alkyl substitution.

The S$_N$2 reaction's one-step mechanism is not strongly endothermic or exothermic. The bond to the leaving group is about half broken in the transition state, so the
reaction rate is sensitive to the nature of the leaving group. With a poor leaving group
such as alkoxide, this reaction is quite slow.

In the acyl substitution, the leaving group leaves in a separate second step. This
second step is highly exothermic, and the Hammond postulate (Section 4-14) predicts
that the transition state resembles the reactant: the tetrahedral intermediate. In this transition state, the bond to the leaving group has barely begun to break. The energy of the
transition state (and therefore the reaction rate) is not very sensitive to the nature of the
leaving group.

Nucleophilic acyl substitution is our first example of a reaction with strong bases
as leaving groups. We will see many additional examples of such reactions. In general,
a strong base may serve as a leaving group if it leaves in a highly exothermic step, usually converting an unstable, negatively charged intermediate to a stable molecule.

PROBLEM-SOLVING **HINT**

A strong base may serve as a leaving
group if it leaves in a highly
exothermic step, usually converting an
unstable, negatively charged
intermediate to a stable molecule.

PROBLEM 21-7

Which of the following proposed reactions would take place quickly under mild conditions?

(a) CH$_3$—C—NH$_2$ + NaCl $\longrightarrow$ CH$_3$—C—Cl + NaNH$_2$

(b) $Ph-\overset{\overset{\displaystyle O}{\|}}{C}-Cl + CH_3NH_2 \longrightarrow Ph-\overset{\overset{\displaystyle O}{\|}}{C}-NHCH_3 + HCl$

(c) $(CH_3)_2CH-\overset{\overset{\displaystyle O}{\|}}{C}-NH_2 + CH_3OH \longrightarrow (CH_3)_2CH-\overset{\overset{\displaystyle O}{\|}}{C}-OCH_3 + NH_3$

(d) $CH_3CH_2-\overset{\overset{\displaystyle O}{\|}}{C}-Cl + CH_3-\overset{\overset{\displaystyle O}{\|}}{C}-OH \longrightarrow CH_3CH_2-\overset{\overset{\displaystyle O}{\|}}{C}-O-\overset{\overset{\displaystyle O}{\|}}{C}-CH_3 + HCl$

(e) $CH_3-\overset{\overset{\displaystyle O}{\|}}{C}-O-\overset{\overset{\displaystyle O}{\|}}{C}-CH_3 + CH_3NH_2 \longrightarrow CH_3-\overset{\overset{\displaystyle O}{\|}}{C}-NHCH_3 + CH_3COOH$

PROBLEM 21-8

Show how you would synthesize the following esters from appropriate acyl chlorides and alcohols:
(a) ethyl propionate
(b) phenyl 3-methylhexanoate
(c) benzyl benzoate
(d) cyclopropyl cyclohexanecarboxylate

PROBLEM 21-9

Show how you would use appropriate acyl chlorides and amines to synthesize the following amides:
(a) *N,N*-dimethylacetamide
(b) acetanilide (PhNHCOCH$_3$)

(c) cyclohexane carboxamide
(d)

PROBLEM 21-10

(a) Show how you would use acetic anhydride and an appropriate alcohol or amine to synthesize (i) benzyl acetate (ii) *N,N*-diethylacetamide.
(b) Propose a mechanism for each synthesis in part (a).

PROBLEM 21-11

Propose a mechanism for the reaction of benzyl acetate with methylamine. Label the attacking nucleophile and the leaving group, and draw the transition state in which the leaving group leaves.

21-6
Acid-Catalyzed Nucleophilic Acyl Substitution

In each substitution discussed in Section 21-5, a nucleophile attacks the carbonyl group to form a tetrahedral intermediate. Some nucleophiles are too weak, however, to attack an unactivated carbonyl group. For example, an alcohol attacks the carbonyl group of an acid chloride, but it does not attack an acid. If a strong acid protonates the carbonyl group of the carboxylic acid, however, the carbonyl group is activated toward attack by the alcohol, resulting in Fischer esterification (Section 20-10).

A similar mechanism is responsible for the acid-catalyzed **transesterification** of an ester, in which one alkoxy group substitutes for another. When an ester is treated with a different alcohol in the presence of an acid catalyst, the two alcohol groups can interchange. An equilibrium results, and the equilibrium can be driven toward the desired ester by using a large excess of the desired alcohol or by removing the other alcohol.

Transesterification also occurs under basic conditions, catalyzed by a small amount of alkoxide ion. Once again, a large excess of the desired alcohol helps to achieve a good conversion.

Transesterification

Example

ethyl benzoate   (large excess)   methyl benzoate

PROBLEM-SOLVING STRATEGY
Proposing Reaction Mechanisms

Rather than simply showing the mechanisms for acid-catalyzed and base-catalyzed transesterification, let's consider how one might work out these mechanisms as in a problem.

Base-Catalyzed Transesterification

First consider the base-catalyzed transesterification of ethyl benzoate with methanol. This is a classic example of nucleophilic acyl substitution by the addition–elimination mechanism. Methoxide ion is sufficiently nucleophilic to attack the ester carbonyl group. Ethoxide ion serves as a leaving group in a strongly exothermic second step.

nucleophilic attack   tetrahedral intermediate

Now try a base-catalyzed mechanism on your own in Problem 21-12.

PROBLEM 21-12

When ethyl 4-hydroxybutyrate is heated in the presence of a trace of a basic catalyst (sodium acetate), one of the products is a lactone. Propose a mechanism for formation of this lactone.

Acid-Catalyzed Transesterification

The acid-catalyzed reaction follows a similar mechanism, but it is more complicated because of additional proton transfers. We use the stepwise procedure to propose a mechanism for the following reaction, in which methanol replaces ethanol:

1. **Consider the carbon skeletons of the reactants and products, and identify which carbon atoms in the products are likely derived from which carbon atoms in the reactants.**

   In this case, an ethoxyl group is replaced by a methoxyl group.

2. **Consider whether any of the reactants is a strong enough electrophile to react without being activated. If not, consider how one of the reactants might be converted to a strong electrophile by protonation of a Lewis basic site.**

   The ester carbonyl group is not a strong enough electrophile to react with methanol. Protonation converts it to a strong electrophile (shown in step 3).

3. **Consider how a nucleophilic site on another reactant can attack the strong electrophile to form a bond needed in the product.**

   Methanol has a nucleophilic oxygen atom that can attack the activated carbonyl group to form the new C—O bond needed in the product.

activation of C=O   (resonance-stabilized)

tetrahedral intermediate

4. **Consider how the product of nucleophilic attack might be converted to the final product or reactivated to form another bond needed in the product.**

   The task here is to break bonds, not form them. The ethoxyl group (OCH₂CH₃) must be lost. The most common mechanism for losing a group under acidic conditions is to protonate it (to make it a good leaving group), then lose it. In fact, losing the ethoxyl group is exactly the reverse of the mechanism used above to gain the methoxyl group.

   Protonation prepares the ethoxyl group to leave. When it leaves, the product is simply a protonated version of the final product.

5. **Draw out all steps of the mechanism, using curved arrows to show the movement of electrons.**

   Once again, this summary is left to you to help you review the mechanism.

PROBLEM 21-13

Complete the mechanism for this acid-catalyzed transesterification by drawing out all the individual steps. Draw the important resonance contributors for each resonance-stabilized intermediate.

PROBLEM 21-14

Propose a mechanism for the following ring-opening transesterification, (use the mechanism in Problem 21-13 as a model):

MECHANISM Transesterification

Base-catalyzed

Step 1: Addition of the nucleophile.   *Step 2:* Elimination of the leaving group.

tetrahedral intermediate

Acid-catalyzed

First half: Acid-catalyzed addition of the nucleophile.

Step 1: Protonation   *Step 2:* Attack by   *Step 3:* Deprotonation.
of the carbonyl.   the nucleophile.

Second half: Acid-catalyzed elimination of the leaving group.

Step 1: Protonation of   *Step 2:* Elimination of   *Step 3:* Deprotonation.
the leaving group.   the leaving group.

Some reactions that can go as basic nucleophilic acyl substitutions actually work much better with an acid catalyst. For example, aspirin is made from salicylic acid and acetic anhydride. When these reagents are mixed, the reaction goes slowly. Addition of a drop of sulfuric acid accelerates the reaction, and it goes to completion in a minute or two.

salicylic acid   +   acetic anhydride $\xrightarrow[\text{fast}]{H_2SO_4}$ aspirin (acetylsalicylic acid)   + CH_3COOH

PROBLEM 21-15

Propose a mechanism for the acid-catalyzed reaction of salicylic acid with acetic anhydride.

21-7
Hydrolysis of Carboxylic Acid Derivatives

All acid derivatives hydrolyze to give carboxylic acids. In most cases, hydrolysis occurs under either acidic or basic conditions. The reactivity of acid derivatives toward hydrolysis varies from highly reactive acyl halides to relatively unreactive amides.

21-7A Hydrolysis of Acid Halides and Anhydrides

Acid halides and anhydrides are so reactive that they hydrolyze under neutral conditions. Hydrolysis of an acid halide or anhydride is usually an annoying side reaction that takes place on exposure to moist air. Hydrolysis can be avoided by storing acid halides and anhydrides under dry nitrogen and by using dry solvents and reagents.

21-7B Hydrolysis of Esters

Acid-catalyzed hydrolysis of an ester is simply the reverse of the Fischer esterification equilibrium. Addition of excess water drives the equilibrium toward the acid and the alcohol.

Basic hydrolysis of esters, called **saponification**, avoids the equilibrium of the Fischer esterification. Hydroxide ion attacks the carbonyl group to give a tetrahedral intermediate. Expulsion of alkoxide ion gives the acid, and a fast proton transfer gives the carboxylate ion and the alcohol. This strongly exothermic proton transfer drives the saponification to completion. A full mole of base is consumed to deprotonate the acid.

MECHANISM   Saponification of an Ester

Step 1: Addition of the nucleophile.   *Step 2:* Elimination of the leaving group.   *Step 3:* Proton transfer.

ester   tetrahedral intermediate   acid   alkoxide   carboxylate   alcohol

Example

$$CH_3CH_2-\overset{\overset{\textstyle O}{\|}}{C}-O-CH_2CH_3 \quad + \quad Na^+ \, {}^-OH \quad \longrightarrow \quad CH_3CH_2-\overset{\overset{\textstyle O}{\|}}{C}-O^- \, {}^+Na \quad + \quad CH_3CH_2-OH$$

ethyl propionate   sodium propionate   ethanol

The term *saponification* (Latin, *saponis*, "soap") literally means "the making of soap." Soap is made by the basic hydrolysis of fats, which are esters of long-chain carboxylic acids (*fatty acids*) with the triol glycerol. When sodium hydroxide hydrolyzes a fat, the resulting long-chain sodium carboxylate salts are what we know as soap. Soaps and detergents are discussed in more detail in Chapter 25.

a fat (triester of glycerol)   glycerol   soap (salts of fatty acids)

PROBLEM 21-16

Suppose we have some optically pure (R)-2-butyl acetate that has been "labeled" with the heavy ^{18}O isotope at one oxygen atom as shown below.

$$CH_3-\overset{\overset{\textstyle O}{\|}}{C}-\overset{18}{O}-\underset{\underset{\textstyle CH_3}{}}{\overset{\overset{\textstyle CH_2CH_3}{}}{C}}\cdots H$$

(a) Draw a mechanism for the hydrolysis of this compound under basic conditions. Predict which of the products will contain the ^{18}O label. Also predict whether the 2-butanol product will be pure (R), pure (S), or racemized.
(b) Repeat part (a) for the acid-catalyzed hydrolysis of this compound.
(c) Explain how you would prove experimentally where the ^{18}O label appears in the products. (^{18}O is not radioactive.)

PROBLEM 21-17

(a) Explain why we speak of acidic hydrolysis of an ester as *acid-catalyzed*, but of basic hydrolysis as *base-promoted*.
(b) Soap manufacturers always use base to hydrolyze fats, and never acid. Suggest two reasons basic hydrolysis is preferred.

PROBLEM 21-18

Propose a mechanism for the base-promoted hydrolysis of γ-butyrolactone:

21-7C Hydrolysis of Amides

Amides hydrolyze to carboxylic acids under both acidic and basic conditions. Amides are the most stable acid derivatives, and stronger conditions are required for their hydrolysis than for hydrolysis of an ester. Typical hydrolysis conditions involve prolonged heating in 6 M HCl or 40% aqueous NaOH.

Basic hydrolysis

$$\text{R}-\overset{\overset{\displaystyle O}{\|}}{\text{C}}-\text{NHR}' \;+\; \text{Na}^+\text{ }^-\text{OH} \;\xrightarrow{\text{H}_2\text{O}}\; \text{R}-\overset{\overset{\displaystyle O}{\|}}{\text{C}}-\text{O}^-\text{ }^+\text{Na} \;+\; \text{R}'\text{NH}_2$$

Example

N,N-diethylbenzamide + NaOH $\xrightarrow{\text{H}_2\text{O}}$ sodium benzoate + (CH$_3$CH$_2$)$_2$NH   diethylamine

Acid hydrolysis

$$\text{R}-\overset{\overset{\displaystyle O}{\|}}{\text{C}}-\text{NHR}' \;+\; \text{H}_3\text{O}^+ \;\longrightarrow\; \text{R}-\overset{\overset{\displaystyle O}{\|}}{\text{C}}-\text{OH} \;+\; \text{R}'\text{NH}_3^+$$

Example

N-methyl-2-phenylacetamide + H$_2$SO$_4$ $\xrightarrow{\text{H}_2\text{O}}$ phenylacetic acid + CH$_3\overset{+}{\text{N}}$H$_3$ HSO$_4^-$ methylammonium sulfate

The basic hydrolysis mechanism (shown below for a primary amide) is similar to that for hydrolysis of an ester. Hydroxide attacks the carbonyl to give a tetrahedral intermediate. Expulsion of an amide ion gives a carboxylic acid, which is quickly deprotonated to give the salt of the acid and ammonia.

MECHANISM Basic Hydrolysis of an Amide

Step 1: Addition of the nucleophile.

Step 2: Elimination of the leaving group.

Step 3: Proton transfer.

tetrahedral intermediate

Under acidic conditions, the mechanism of amide hydrolysis resembles the acid-catalyzed hydrolysis of an ester. Protonation of the carbonyl group activates it toward nucleophilic attack by water to give a tetrahedral intermediate. Protonation of the amino group enables it to leave as the amine. A fast exothermic proton transfer gives the acid and the protonated amine.

MECHANISM Acidic Hydrolysis of an Amide

First half: Acid-catalyzed addition of the nucleophile (water).

protonation
of the carbonyl

addition
of the nucleophile

loss of a proton

Second half: Acid-catalyzed elimination of the leaving group.

protonation of
the leaving group

elimination of the leaving group

deprotonation

PROBLEM 21-19

Draw the important resonance contributors for both resonance-stabilized cations in the mechanism for acid-catalyzed hydrolysis of an amide.

PROBLEM 21-20

Propose a mechanism for the hydrolysis of *N,N*-dimethylacetamide
(a) under basic conditions   (b) under acidic conditions

PROBLEM 21-21

The equilibrium for hydrolysis of amides, under both acidic and basic conditions, favors the products. Use your mechanisms for the hydrolysis of *N,N*-dimethylacetamide to show which steps are sufficiently exothermic to drive the reactions to completion.

21-7D   Hydrolysis of Nitriles

Nitriles are hydrolyzed to amides, and further to carboxylic acids, by heating with aqueous acid or base. Mild conditions can hydrolyze a nitrile only as far as the amide; stronger conditions can hydrolyze it all the way to the carboxylic acid.

Basic hydrolysis

Example

nicotinonitrile

nicotinamide

Acidic hydrolysis

Example

The mechanism for basic hydrolysis begins with attack by hydroxide on the electrophilic carbon of the cyano group. Protonation gives the unstable enol tautomer of an amide. Removal of a proton from oxygen and reprotonation on nitrogen gives the amide. Further hydrolysis of the amide to the carboxylate salt involves the same base-promoted mechanism as that discussed above.

MECHANISM   Base-Catalyzed Hydrolysis of a Nitrile

Attack by hydroxide ion and reprotonation

Removal and replacement of a proton (tautomerization)

PROBLEM 21-22

Propose a mechanism for the basic hydrolysis of benzonitrile to the benzoate ion and ammonia.

PROBLEM 21-23

The mechanism for acidic hydrolysis of a nitrile resembles the basic hydrolysis, except that the nitrile is first protonated, activating it toward attack by a weak nucleophile (water). Under acidic conditions, the proton transfer (tautomerism) involves protonation on nitrogen followed by deprotonation on oxygen. Propose a mechanism for the acid-catalyzed hydrolysis of benzonitrile to benzamide.

21-8
Reduction of Acid Derivatives

Carboxylic acids and their derivatives can be reduced to alcohols, aldehydes, and amines. Because they are relatively difficult to reduce, acid derivatives generally require a strong reducing agent such as lithium aluminum hydride ($LiAlH_4$).

21-8A   Reduction to Alcohols

Lithium aluminum hydride reduces acids, acid chlorides, and esters to primary alcohols. (The reduction of acids was covered in Section 20-14.)

Example

Both esters and acid chlorides react through an addition-elimination mechanism to give an aldehyde, which is quickly reduced to an alkoxide. Dilute acid is added in a second step to protonate the alkoxide.

MECHANISM   Hydride Reduction of an Ester

Step 1: Addition of the nucleophile (hydride).   *Step 2:* Elimination of alkoxide.

Step 3: Addition of a second hydride ion.

PROBLEM 21-24

Propose a mechanism for the reduction of octanoyl chloride by lithium aluminum hydride.

21-8B   Reduction to Aldehydes

Acid chlorides are more reactive than other acid derivatives, and they are reduced to aldehydes by mild reducing agents such as lithium aluminum tri(*t*-butoxy)hydride. This reduction was covered in Sections 18-11 and 20-14.

Example

$$CH_3(CH_2)_6-\overset{\displaystyle O}{\overset{\|}{C}}-Cl \xrightarrow{\text{Li}(t\text{-BuO})_3\text{AlH}} CH_3(CH_2)_6-\overset{\displaystyle O}{\overset{\|}{C}}-H$$

octanoyl chloride   octanal

21-8C   Reduction to Amines

Lithium aluminum hydride reduces amides and nitriles to amines, providing some of the best synthetic routes to amines (Sections 19-19B and 19-19C). Primary amides and nitriles are reduced to primary amines. Secondary amides are reduced to secondary amines, and tertiary amides are reduced to tertiary amines.

Amides
{
☞ $R-\overset{\displaystyle O}{\overset{\|}{C}}-NH_2$ $\xrightarrow[\text{(2) H}_2\text{O}]{\text{(1) LiAlH}_4}$ $R-CH_2-NH_2$
primary amide   primary amine

☞ $R-\overset{\displaystyle O}{\overset{\|}{C}}-NHR'$ $\xrightarrow[\text{(2) H}_2\text{O}]{\text{(1) LiAlH}_4}$ $R-CH_2-NHR'$
secondary amide   secondary amine

☞ $R-\overset{\displaystyle O}{\overset{\|}{C}}-NR'_2$ $\xrightarrow[\text{(2) H}_2\text{O}]{\text{(1) LiAlH}_4}$ $R-CH_2-NR'_2$
tertiary amide   tertiary amine

Example

$$CH_3-\overset{\displaystyle O}{\overset{\|}{C}}-NH-Ph \xrightarrow[\text{(2) H}_2\text{O}]{\text{(1) LiAlH}_4} CH_3-CH_2-NH-Ph$$

acetanilide   *N*-ethylaniline

Nitriles   ☞ $R-C\equiv N\colon$ $\xrightarrow[\text{or (1) LiAlH}_4; \text{(2) H}_2\text{O}]{\text{H}_2/\text{Pt}}$ $R-\overset{\displaystyle H}{\underset{\displaystyle H}{\overset{|}{\underset{|}{C}}}}-\overset{\displaystyle H}{\underset{\displaystyle H}{N\colon}}$

Example

Ph$-CH_2-C\equiv N\colon$ $\xrightarrow[\text{(2) H}_2\text{O}]{\text{(1) LiAlH}_4}$ Ph$-CH_2-CH_2-\ddot{N}H_2$

PROBLEM 21-25

Give the expected products of lithium aluminum hydride reduction of the following compounds (followed by hydrolysis):

(a) butyronitrile   **(b)** *N*-cyclohexylacetamide   **(c)** ε-caprolactam

(d)   **(e)**   **(f)**

Esters and Acid Chlorides   Grignard and organolithium reagents add twice to acid chlorides and esters to give alkoxides (Section 10-9D). Protonation of the alkoxides gives alcohols.

21-9
Reactions of Acid Derivatives with Organometallic Reagents

$$R-\overset{\overset{\displaystyle O}{\|}}{C}-OR' \xrightarrow[\text{(or 2 R''Li)}]{2\ R''MgX} R-\overset{\overset{\displaystyle OMgX}{|}}{\underset{\underset{\displaystyle R''}{|}}{C}}-R'' \ +\ R'OMgX \xrightarrow{H_3O^+} R-\overset{\overset{\displaystyle OH}{|}}{\underset{\underset{\displaystyle R''}{|}}{C}}-R''$$

ester   alkoxide salt   tertiary alcohol

Examples

$$Ph-\overset{\overset{\displaystyle O}{\|}}{C}-OEt\ +\ 2\ PhMgBr \longrightarrow Ph-\overset{\overset{\displaystyle OMgBr}{|}}{\underset{\underset{\displaystyle Ph}{|}}{C}}-Ph \xrightarrow{H_3O^+} Ph-\overset{\overset{\displaystyle OH}{|}}{\underset{\underset{\displaystyle Ph}{|}}{C}}-Ph$$

an ester   a 3° alcohol

$$H-\overset{\overset{\displaystyle O}{\|}}{C}-OEt\ +\ 2\ C_4H_9Li \longrightarrow H-\overset{\overset{\displaystyle OLi}{|}}{\underset{\underset{\displaystyle C_4H_9}{|}}{C}}-C_4H_9 \xrightarrow{H_3O^+} H-\overset{\overset{\displaystyle OH}{|}}{\underset{\underset{\displaystyle C_4H_9}{|}}{C}}-C_4H_9$$

a formate ester   a 2° alcohol

$$CH_3CH_2-\overset{\overset{\displaystyle O}{\|}}{C}-Cl\ +\ 2\ PhMgBr \longrightarrow CH_3CH_2-\overset{\overset{\displaystyle OMgBr}{|}}{\underset{\underset{\displaystyle Ph}{|}}{C}}-Ph \xrightarrow{H_3O^+} CH_3CH_2-\overset{\overset{\displaystyle OH}{|}}{\underset{\underset{\displaystyle Ph}{|}}{C}}-Ph$$

an acid chloride   a 3° alcohol

The mechanism involves nucleophilic substitution at the acyl carbon atom. Attack by the carbanion-like organometallic reagent, followed by elimination of alkoxide (from an ester) or chloride (from an acid chloride), gives a ketone. A second equivalent of the organometallic reagent adds to the ketone to give the alkoxide. Hydrolysis gives tertiary alcohols, unless the original ester is a formate (R = H), which gives a secondary alcohol. In each case, two of the groups on the product are the same, derived from the organometallic reagent.

MECHANISM   Reaction of an Ester with 2 Moles of a Grignard Reagent

Step 1: Addition of the Grignard.   *Step 2:* Elimination of alkoxide.   *Step 3:* Addition of another Grignard.

ester   Grignard   tetrahedral intermediate   ketone   alkoxide

Nitriles   A Grignard or organolithium reagent attacks the electrophilic cyano group to form the salt of an imine. Acidic hydrolysis of the salt (in a second step) gives the imine, which is further hydrolyzed to a ketone (Section 18-10).

$$\overset{\delta^-}{R'}-\overset{\delta^+}{Mg}-X$$

$$R-C\equiv N: \longrightarrow \underset{R}{\overset{R'}{>}}C=\underset{}{N}\overset{MgX}{} \xrightarrow{H_3O^+} \underset{R}{\overset{R'}{>}}C=\underset{}{N}\overset{H}{} \xrightarrow{H_3O^+} \underset{R}{\overset{R'}{>}}C=O:$$

salt of imine   imine   ketone

Example

$$Ph-C\equiv N: \;+\; CH_3MgI \longrightarrow \underset{CH_3}{\overset{Ph}{>}}C=\underset{}{N}\overset{MgI}{} \xrightarrow{H_3O^+} \underset{CH_3}{\overset{Ph}{>}}C=O:$$

benzonitrile   methylmagnesium   magnesium salt   acetophenone
                iodide

PROBLEM-SOLVING **HINT**

Grignards add to esters and acid chlorides to give tertiary alcohols, with one group from the ester or acid chloride, and two identical groups from the Grignard. *Formate* esters give secondary alcohols, with a hydrogen from the ester and two identical groups from the Grignard.

PROBLEM 21-26

Draw a mechanism for the acidic hydrolysis of the above magnesium salt to acetophenone.

PROBLEM 21-27

Draw a mechanism for the reaction of propanoyl chloride with 2 moles of phenylmagnesium bromide.

PROBLEM 21-28

Show how you would add a Grignard reagent to an ester or a nitrile to synthesize
(a) 4-phenyl-4-heptanol
(b) 4-heptanol
(c) 2-pentanone

21-10
Summary of the Chemistry of Acid Chlorides

Having discussed the reactions and mechanisms characteristic of all the common acid derivatives, we now review the syntheses and reactions of each type of compound. In addition, these sections cover any reactions that are peculiar to a specific class of acid derivatives.

Synthesis of Acid Chlorides   Acid chlorides (acyl chlorides) are synthesized from the corresponding carboxylic acids using a variety of reagents. Thionyl chloride ($SOCl_2$) and oxalyl chloride ($COCl)_2$ are the most convenient reagents because they produce only gaseous side products (Section 20-11).

$$\begin{array}{c} O \\ \parallel \\ R-C-OH \end{array} \xrightarrow[or\ (COCl)_2]{SOCl_2} \begin{array}{c} O \\ \parallel \\ R-C-Cl \end{array} + SO_2\uparrow + HCl\uparrow$$

Reactions of Acid Chlorides   Acid chlorides react quickly with water and other nucleophiles and are therefore not found in nature. Because they are the most reactive acid derivatives, acid chlorides are easily converted to other acid derivatives. Often, the best synthetic route to an ester, anhydride, or amide may involve using the acyl chloride as an intermediate.

$$
R-\overset{\overset{\displaystyle O}{\|}}{C}-Cl \quad \text{acid chloride (acyl chloride)}
$$

$\xrightarrow{\text{H}_2\text{O}}$ $R-\overset{\overset{\displaystyle O}{\|}}{C}-OH$ + HCl   (Section 21-7A)
acid

$\xrightarrow{\text{R'OH}}$ $R-\overset{\overset{\displaystyle O}{\|}}{C}-OR'$ + HCl   (Sections 20-11 and 21-5)
ester

$\xrightarrow{\text{R'NH}_2}$ $R-\overset{\overset{\displaystyle O}{\|}}{C}-NHR'$ + HCl   (Sections 20-11 and 21-5)
amide

$\xrightarrow{\text{R'COOH}}$ $R-\overset{\overset{\displaystyle O}{\|}}{C}-O-\overset{\overset{\displaystyle O}{\|}}{C}-R'$ + HCl   (Section 21-5)
anhydride

Grignard and organolithium reagents add twice to acid chlorides to give 3° alcohols (after hydrolysis). Lithium dialkylcuprates add just once to give ketones. Lithium aluminum hydride adds hydride twice to acid chlorides, reducing them to 1° alcohols (after hydrolysis). Acid chlorides react with the weaker reducing agent lithium tri-*t*-butoxyaluminum hydride to give aldehydes.

$$
R-\overset{\overset{\displaystyle O}{\|}}{C}-Cl \quad \text{acid chloride (acyl chloride)}
$$

$\xrightarrow[\text{(2) H}_2\text{O}]{\text{(1) 2 R'MgX}}$ $R-\overset{\overset{\displaystyle OH}{|}}{\underset{\underset{\displaystyle R'}{|}}{C}}-R'$ (Sections 10-9 and 21-9)
3° alcohol

$\xrightarrow{\text{R'}_2\text{CuLi}}$ $R-\overset{\overset{\displaystyle O}{\|}}{C}-R'$ (Section 18-11)
ketone

$\xrightarrow[\text{(2) H}_2\text{O}]{\text{(1) LiAlH}_4}$ $R-CH_2OH$ (Sections 10-11 and 21-8A)
1° alcohol

$\xrightarrow{\text{Li}(t\text{-BuO})_3\text{AlH}}$ $R-\overset{\overset{\displaystyle O}{\|}}{C}-H$ (Sections 18-11 and 21-8B)
aldehyde

Friedel–Crafts Acylation of Aromatic Rings In the presence of aluminum chloride, acyl halides acylate benzene, halobenzenes, and activated benzene derivatives. Friedel–Crafts acylation is discussed in detail in Section 17-11.

☞ $R-\overset{\overset{\displaystyle O}{\|}}{C}-Cl$ + (benzene ring with Z) $\xrightarrow{\text{AlCl}_3}$ (benzene ring with Z)$-\overset{\overset{\displaystyle O}{\|}}{C}-R$

(Z = H, halogen, or an activating group)   an acylbenzene

Example

$$CH_3-CH_2-\overset{\overset{\displaystyle O}{\|}}{C}-Cl \ + \ CH_3O-\langle\bigcirc\rangle \ \xrightarrow{AlCl_3} \ CH_3O-\langle\bigcirc\rangle-\overset{\overset{\displaystyle O}{\|}}{C}-CH_2CH_3$$

propionyl chloride   anisole   *p*-methoxypropiophenone
(major product)

PROBLEM 21-29

Draw a mechanism for the acylation of anisole by propionyl chloride. Recall that Friedel-Crafts acylation involves an acylium ion as the electrophile in electrophilic aromatic substitution.

PROBLEM 21-30

Show how Friedel–Crafts acylation might be used to synthesize the following compounds:
(a) acetophenone   **(b)** benzophenone   **(c)** *n*-butylbenzene

21-11
Summary of the Chemistry of Anhydrides

cantharidin

Like acid chlorides, anhydrides are activated acid derivatives, and they are often used for the same types of acylations. Anhydrides are not as reactive as acid chlorides, and they are occasionally found in nature. For example, cantharidin is a toxic ingredient of "Spanish fly," which is used as a vesicant (causing burning and blistering) to destroy warts on the skin.

Because anhydrides are not as reactive as acid chlorides, they are often more selective in their reactions. Anhydrides are valuable when the appropriate acid chloride is too reactive, does not exist, or is more expensive than the corresponding anhydride.

Acetic Anhydride Acetic anhydride is the most important carboxylic acid anhydride. It is produced at the rate of about 4 billion pounds per year, primarily for synthesis of plastics, fibers, and drugs. (See the synthesis of aspirin on p. 968.) Acetic anhydride consists of two molecules of acetic acid, minus a molecule of water. The most common industrial synthesis begins by dehydrating acetic acid to give ketene.

$$CH_3-\overset{\overset{\displaystyle O}{\|}}{C}-OH \ \underset{\xleftarrow{\hspace{1cm}}}{\overset{750°C}{\xrightarrow{(EtO)_3P=O}}} \ \overset{H}{\underset{H}{>}}C=C=O \ + \ H_2O$$

acetic acid   ketene

This dehydration is highly endothermic ($\Delta H = +35\ \text{kcal/mol} = +147\ \text{kJ/mol}$), but a large increase in entropy results from breaking one molecule into two. Thus, at a sufficiently high temperature (750°C is typical), the equilibrium favors the products. Triethyl phosphate is added as a catalyst to improve the rate of the reaction.

Ketene (a gas at room temperature) is fed directly into acetic acid, where it reacts quickly and quantitatively to give acetic anhydride. This inexpensive large-scale manufacture makes acetic anhydride a convenient and inexpensive acylating reagent.

$$CH_3-\overset{\overset{\displaystyle O}{\|}}{C}-OH \ + \ \overset{H}{\underset{H}{>}}C=C=O \ \longrightarrow \ CH_3-\overset{\overset{\displaystyle O}{\|}}{C}-O-\overset{\overset{\displaystyle O}{\|}}{C}-CH_3$$

acetic acid   ketene   acetic anhydride

General Anhydride Synthesis. Other anhydrides must be made by less specialized methods. The most general method for making anhydrides is the reaction of an acid chloride with a carboxylic acid or a carboxylate salt.

$$R\!-\!\overset{O}{\underset{\|}{C}}\!-\!Cl \ + \ {}^-O\!-\!\overset{O}{\underset{\|}{C}}\!-\!R' \ \longrightarrow \ R\!-\!\overset{O}{\underset{\|}{C}}\!-\!O\!-\!\overset{O}{\underset{\|}{C}}\!-\!R' \ + \ Cl^-$$

acid chloride   carboxylate   acid anhydride
                  (or acid)

Examples

$$CH_3\!-\!\overset{O}{\underset{\|}{C}}\!-\!Cl \ + \ HO\!-\!\overset{O}{\underset{\|}{C}}\!-\!Ph \ \xrightarrow{\text{pyridine}} \ CH_3\!-\!\overset{O}{\underset{\|}{C}}\!-\!O\!-\!\overset{O}{\underset{\|}{C}}\!-\!Ph \ + \ pyr \cdot HCl$$

acetyl chloride   benzoic acid   acetic benzoic anhydride

$$CH_3\!-\!\overset{O}{\underset{\|}{C}}\!-\!Cl \ + \ H\!-\!\overset{O}{\underset{\|}{C}}\!-\!O^- \ {}^+Na \ \longrightarrow \ CH_3\!-\!\overset{O}{\underset{\|}{C}}\!-\!O\!-\!\overset{O}{\underset{\|}{C}}\!-\!H \ + \ NaCl$$

acetyl chloride   sodium formate   acetic formic anhydride

Some cyclic anhydrides are made simply by heating the corresponding diacid. A dehydrating agent, such as acetyl chloride or acetic anhydride, is occasionally added to accelerate this reaction. Because five- and six-membered cyclic anhydrides are particularly stable, the equilibrium favors the cyclic products.

phthalic acid $\xrightarrow{\text{heat}}$ phthalic anhydride   + H_2O (steam)

succinic acid   + CH_3CCl $\longrightarrow$ succinic anhydride   + CH_3COH +   HCl

Blister beetles secrete cantharidin, a powerful vesicant. Crushing a blister beetle between the fingers results in severe blistering of the skin. When horses eat hay containing blister beetles, they often die from gastroenteritis and kidney failure caused by cantharidin poisoning.

Reactions of Anhydrides   Anhydrides undergo many of the same reactions as acid chlorides. Like acid chlorides, anhydrides are easily converted to less reactive acid derivatives.

$$R\!-\!\overset{O}{\underset{\|}{C}}\!-\!O\!-\!\overset{O}{\underset{\|}{C}}\!-\!R$$

anhydride

$\xrightarrow{H_2O}$ $R\!-\!\overset{O}{\underset{\|}{C}}\!-\!OH$ (acid)   + $R\!-\!COOH$ (Section 21-7A)

$\xrightarrow[H^+]{R'OH}$ $R\!-\!\overset{O}{\underset{\|}{C}}\!-\!OR'$ (ester)   + $R\!-\!COOH$ (Section 21-5)

$\xrightarrow{R'NH_2}$ $R\!-\!\overset{O}{\underset{\|}{C}}\!-\!NHR'$ (amide)   + $R\!-\!COOH$ (Section 21-5)

Like acid chlorides, anhydrides participate in the Friedel–Crafts acylation. The catalyst may be aluminum chloride, polyphosphoric acid (PPA), or other acidic reagents. Cyclic anhydrides can provide additional functionality on the side chain of the aromatic product.

(Z = H, halogen, or an activating group)

an acylbenzene

Example

benzene   succinic anhydride   4-oxo-4-phenylbutanoic acid

Most reactions of anhydrides involve loss of one of the two acid molecules as a leaving group. If a precious acid needed to be activated, converting it to the anhydride would allow only half of the acid groups to react. Converting the acid to an acid chloride would be easier, and it would allow all the acid groups to react. In most cases, it is easier and more efficient to make and use acid chlorides rather than anhydrides. There are three specific instances when anhydrides are preferred, however.

1. *Use of acetic anhydride*. Acetic anhydride is inexpensive and convenient to use, and it often gives better yields than acetyl chloride for acetylation of alcohols (to make acetate esters) and amines (to make acetamides).

2. *Use of acetic formic anhydride*. Formyl chloride (the acid chloride of formic acid) cannot be used for formylation because it quickly decomposes to CO and HCl. Acetic formic anhydride, made from sodium formate and acetyl chloride, reacts primarily at the formyl group. Lacking a bulky, electron-donating alkyl group, the formyl group is both less hindered and more electrophilic than the acetyl group. Alcohols and amines are formylated by acetic formic anhydride to give formate esters and formamides, respectively.

3. *Use of cyclic anhydrides to give difunctional compounds*. It is often necessary to convert just one acid group of a diacid to an ester or an amide. This conversion is easily accomplished using a cyclic anhydride.

When an alcohol or an amine reacts with a cyclic anhydride, only one of the carboxyl groups in the anhydride is converted to an ester or an amide. The other is expelled as a carboxylate ion, and a monofunctionalized derivative results.

glutaric anhydride   monoethyl ester

PROBLEM 21-31

(a) Give the products expected when acetic formic anhydride reacts with (i) aniline and (ii) benzyl alcohol.

(b) Propose mechanisms for these reactions.

PROBLEM 21-32

Show how you would use anhydrides to synthesize the following compounds. In each case, explain why an anhydride might be preferable to an acid chloride.

(a) *n*-octyl formate   (b) *n*-octyl acetate

(c) phthalic acid monoamide   (d) succinic acid monomethyl ester

21-12
Summary of the Chemistry of Esters

Esters are among the most common acid derivatives. They are found in plant oils, where they give the fruity aromas we associate with ripeness. For example, the odor of ripe bananas comes mostly from isoamyl acetate. Oil of wintergreen contains methyl salicylate, which has also been used as a medicine. Lavender oil and sweet clover contain small amounts of coumarin, which gives depth and longevity to their odors. Sperm whales use spermaceti, a waxy ester, to regulate their buoyancy in the water and possibly as a resonating chamber for communicating underwater.

isoamyl acetate
(isopentyl acetate)

methyl salicylate
oil of wintergreen

coumarin

spermaceti
(cetyl palmitate)

Esters are widely used in industry as solvents. Ethyl acetate is a good solvent for a wide variety of compounds, and its toxicity is low compared with other solvents. Ethyl acetate is also found in household products such as cleaners, polishes, glues, and spray finishes. Ethyl butyrate and butyl butyrate were once widely used as solvents for paints and finishes, including the "butyrate dope" that was sprayed on the fabric covering of aircraft wings to make them tight and stiff. Polyesters (covered below and in Chapter 26) are among the most common polymers, used in fabrics (Dacron®), films (VCR tapes), and solid plastics (soft-drink bottles).

> Coumarin is an anticoagulant that slows clotting of the blood, leading to hemorrhaging. Its aroma is similar to vanilla, so it is sometimes found as an adulterant in imported vanilla flavoring. Ingestion can be fatal.

Synthesis of Esters   Esters are usually synthesized by the Fischer esterification of an acid with an alcohol or by the reaction of an acid chloride (or anhydride) with an alcohol. Methyl esters can be made by treating the acid with diazomethane. The alcohol group in an ester can be changed by transesterification, which can be catalyzed by either acid or base.

$$\underset{\text{ester}}{R-\overset{\overset{\textstyle O}{\|}}{C}-OR''} \quad + \quad \underset{\text{alcohol}}{R'-OH} \quad \underset{\longleftarrow}{\overset{H^+ \text{ or } ^-OR'}{\rightleftharpoons}} \quad \underset{\text{ester}}{R-\overset{\overset{\textstyle O}{\|}}{C}-OR'} \quad + \quad R''OH \qquad \text{(Section 21-6)}$$

$$\underset{\text{acid}}{R-\overset{\overset{\textstyle O}{\|}}{C}-OH} \quad + \quad \underset{\text{diazomethane}}{CH_2N_2} \quad \longrightarrow \quad \underset{\text{methyl ester}}{R-\overset{\overset{\textstyle O}{\|}}{C}-OCH_3} \quad + \quad N_2\uparrow \qquad \text{(Section 20-12)}$$

Reactions of Esters   Esters are much more stable than acid chlorides and anhydrides. For example, most esters do not react with water under neutral conditions. They hydrolyze under acidic or basic conditions, however, and an amine can displace the alkoxyl group to form an amide. Lithium aluminum hydride reduces esters to primary alcohols, and Grignard and organolithium reagents add twice to give alcohols (after hydrolysis).

$$\underset{\text{ester}}{R-\overset{\overset{\textstyle O}{\|}}{C}-OR'}$$

$$\xrightarrow[H^+ \text{ or } ^-OH]{H_2O} \quad \underset{\text{acid}}{R-\overset{\overset{\textstyle O}{\|}}{C}-OH} \;+\; R'OH \qquad \text{(Section 21-7B)}$$

$$\xrightarrow[H^+ \text{ or } ^-OR'']{R''OH} \quad \underset{\text{ester}}{R-\overset{\overset{\textstyle O}{\|}}{C}-OR''} \;+\; R'OH \qquad \text{(Section 21-6)}$$

$$\xrightarrow{R''NH_2} \quad \underset{\text{amide}}{R-\overset{\overset{\textstyle O}{\|}}{C}-NHR''} \;+\; R'OH \qquad \text{(Section 21-5)}$$

$$\xrightarrow[(2)\ H_2O]{(1)\ LiAlH_4} \quad \underset{1°\ \text{alcohol}}{R-CH_2OH} \;+\; R'OH \qquad \text{(Sections 10-11 and 21-8A)}$$

$$\xrightarrow[(2)\ H_2O]{(1)\ 2\ R''MgX} \quad \underset{3°\ \text{alcohol}}{R-\overset{\overset{\textstyle OH}{|}}{\underset{\underset{\textstyle R''}{|}}{C}}-R''} \;+\; R'OH \qquad \text{(Sections 10-9D and 21-9)}$$

Formation of Lactones   Simple lactones containing five- and six-membered rings are often more stable than the open-chain hydroxy acids. Such lactones form spontaneously under acidic conditions (via the Fischer esterification).

27%   73%

Lactones that are not energetically favored may be synthesized by driving the equilibrium toward the products. For example, the 10-membered 9-hydroxynonanoic acid lactone is formed in a dilute benzene solution containing a trace of *p*-toluenesulfonic acid. The reaction is driven to completion by distilling the benzene/water azeotrope to remove water and shift the equilibrium to the right.

9-hydroxynonanoic acid   9-hydroxynonanoic acid lactone   (removed)
                                                          (95%)

Lactones are common among natural products. For example, L-ascorbic acid (vitamin C) is necessary in the human diet to avoid the connective tissue disease known as scurvy. In acid solutions, ascorbic acid is an equilibrium mixture of the cyclic and acyclic forms, but the cyclic form predominates. Erythromycin is a member of the macrolide (large-ring lactone) group of antibiotics, which is isolated from *Streptomyces erythraeus*. It inhibits bacterial protein synthesis, thus arresting bacterial growth and development. Erythromycin is effective against a wide range of diseases, including staphylococcus, streptococcus, chlamydia, and Legionnaires' disease.

L-ascorbic acid (vitamin C)   erythromycin

PROBLEM 21-33

Propose a mechanism for the formation of 9-hydroxynonanoic acid lactone, as shown above.

PROBLEM 21-34

Suggest the most appropriate reagent for each synthesis, and explain your choice:

PROBLEM 21-35

Show how you would synthesize each compound, starting with an ester containing no more than eight carbon atoms. Any other necessary reagents may be used.

(a) $Ph_3C—OH$ (b) $(PhCH_2)_2CHOH$ (c) $PhCONHCH_2CH_3$
(d) Ph_2CHOH (e) $PhCH_2OH$ (f) $PhCOOH$
(g) $PhCH_2COOCH(CH_3)_2$ (h) $PhCH_2—C(CH_2CH_3)_2$
                                                        $|$
                                                       OH

Polyesters   Right now, you are probably using at least five things that are made from polyester. Your clothes probably have some Dacron® polyester fiber in them, and they are almost certainly sewn with Dacron® thread. Your computer is using floppy disks made of Mylar®, and your VCR is playing a tape made of Mylar®. Some of the electronics in your computer are probably "potted" (covered and insulated from shock) in Glyptal® polyester resin. The soft drink in your hand probably came in a plastic bottle that was blow-molded from poly(ethylene terephthalate) resin, better known as PET.

All these plastics are essentially the same compound, composed of terephthalic acid (*para*-phthalic acid) esterified with ethylene glycol. This polyester is made by a base-catalyzed transesterification of dimethyl terephthalate with ethylene glycol at a temperature around 150°. At this temperature, methanol escapes as a gas, driving the reaction to completion. We will study polyesters and other polymers in more detail in Chapter 26.

$$CH_3O-\overset{\overset{\displaystyle O}{\|}}{C}-\!\!\left\langle\bigcirc\right\rangle\!\!-\overset{\overset{\displaystyle O}{\|}}{C}-OCH_3 \;+\; HO-CH_2CH_2-OH \xrightarrow[\text{NaOCH}_3]{\text{heat, loss of CH}_3\text{OH}}$$

dimethyl terephthalate   ethylene glycol

$$----\overset{\overset{\displaystyle O}{\|}}{C}-\!\!\left\langle\bigcirc\right\rangle\!\!-\overset{\overset{\displaystyle O}{\|}}{C}-\!\!\left[O-CH_2CH_2-O-\overset{\overset{\displaystyle O}{\|}}{C}-\!\!\left\langle\bigcirc\right\rangle\!\!-\overset{\overset{\displaystyle O}{\|}}{C}\right]_n\!\!-O-CH_2CH_2-O----$$

poly(ethylene terephthalate) or PET, also called Dacron® polyester or Mylar® film

21-13
Summary of the Chemistry of Amides

Synthesis of Amides   Amides are the least reactive acid derivatives, and they can be made from any of the others. In the laboratory, amides are commonly synthesized by the reaction of an acid chloride (or anhydride) with an amine. The most common industrial synthesis involves heating an acid with an amine to drive off water and promote condensation. Esters react with amines and ammonia to give amides, and the partial hydrolysis of nitriles also gives amides.

$$R-\overset{\overset{\displaystyle O}{\|}}{C}-OH \;+\; R'-NH_2 \xrightarrow{\text{heat}} R-\overset{\overset{\displaystyle O}{\|}}{C}-NHR' \;+\; H_2O \qquad \text{(Section 20-13)}$$
acid   amine   amide

$$R-\overset{\overset{\displaystyle O}{\|}}{C}-Cl \;+\; 2\,R_2'NH \longrightarrow R-\overset{\overset{\displaystyle O}{\|}}{C}-NR_2' \;+\; R_2'\overset{+}{N}H_2\;Cl^- \qquad \text{(Sections 20-11 and 21-5)}$$
acid chloride   amine   amide

$$R-\overset{\overset{\displaystyle O}{\|}}{C}-O-\overset{\overset{\displaystyle O}{\|}}{C}-R \;+\; R_2'NH \longrightarrow R-\overset{\overset{\displaystyle O}{\|}}{C}-NR_2' \;+\; RCOOH \qquad \text{(Section 21-5)}$$
anhydride   amine   amide

$$R-\overset{\overset{\displaystyle O}{\|}}{C}-OR'' \;+\; R'-NH_2 \longrightarrow R-\overset{\overset{\displaystyle O}{\|}}{C}-NHR' \;+\; R''OH \qquad \text{(Section 21-5)}$$
ester   amine   amide

$$R-C\equiv N \;+\; H_2O \xrightarrow{\text{H}^+ \text{ or } ^-\text{OH}} R-\overset{\overset{\displaystyle O}{\|}}{C}-NH_2 \qquad 1° \text{ amide} \qquad \text{(Section 21-7D)}$$
nitrile

Reactions of Amides   Because amides are the most stable acid derivatives, they are not easily converted to other derivatives by nucleophilic acyl substitution. From a synthetic standpoint, their most important reaction is the reduction to amines, which is one of the best methods for synthesizing amines. The Hofmann rearrangement (Section 19-19C) also converts amides to amines, with loss of one carbon atom. Amides are hydrolyzed by strong acid or strong base. Just as nitriles can be hydrolyzed to amides, amides can be dehydrated to nitriles.

$$
R\text{—}\overset{\displaystyle O}{\overset{\|}{C}}\text{—}NHR'
$$
amide

$\xrightarrow[\text{H}^+ \text{ or } ^-\text{OH}]{\text{H}_2\text{O}}$ $R\text{—}\overset{\displaystyle O}{\overset{\|}{C}}\text{—OH}$ + $R'NH_2$ (Section 21-7C)
acid

$\xrightarrow[\text{(2) H}_2\text{O}]{\text{(1) LiAlH}_4}$ $R\text{—CH}_2NHR'$ (Sections 19-19B and 21-8C)
amine

$\xrightarrow[\text{(Hofmann rearr.)}]{\text{Br}_2, \ ^-\text{OH}}$ $R\text{—NH}_2$ + CO_3^{2-} (Section 19-19C4)
1° amine

$\xrightarrow[\text{(or P}_2\text{O}_5)]{\text{POCl}_3}$ $R\text{—C}\equiv N$ (Section 21-7D)
nitrile

Dehydration of Amides to Nitriles

Strong dehydrating agents can remove the elements of water from a primary amide to give a nitrile. Dehydration of amides is one of the most common methods for synthesis of nitriles. Phosphorus pentoxide (P_2O_5) is the traditional reagent for this dehydration, but phosphorus oxychloride ($POCl_3$) sometimes gives better yields.

☞ $R\text{—}\overset{\displaystyle O}{\overset{\|}{C}}\text{—}\ddot{N}H_2$ $\xrightarrow[\text{(or P}_2\text{O}_5)]{\text{POCl}_3}$ $R\text{—C}\equiv N\text{:}$
primary amide   nitrile

Example

$CH_3CH_2CH_2CH_2\text{—}\overset{\displaystyle CH_3CH_2}{\overset{|}{CH}}\text{—}\overset{\displaystyle O}{\overset{\|}{C}}\text{—}\ddot{N}H_2$ $\xrightarrow{\text{P}_2\text{O}_5}$ $CH_3CH_2CH_2CH_2\text{—}\overset{\displaystyle CH_3CH_2}{\overset{|}{CH}}\text{—C}\equiv N\text{:}$
2-ethylhexanamide   2-ethylhexanenitrile (90%)

Formation of Lactams

Five-membered lactams (γ-lactams) and six-membered lactams (δ-lactams) often form upon heating or adding a dehydrating agent to the appropriate γ-amino acids and δ-amino acids. Lactams containing smaller or larger rings do not form readily under these conditions.

γ-aminobutyric acid $\xrightarrow{\text{heat}}$ γ-butyrolactam + H_2O

δ-aminovaleric acid $\xrightarrow{\text{heat}}$ δ-valerolactam + H_2O

Biological Reactivity of β-Lactams

β-Lactams are unusually reactive amides, and are capable of acylating a variety of nucleophiles. The considerable strain in the four-membered ring appears to be the driving force behind the unusual reactivity of

β-lactams. When a β-lactam acylates a nucleophile, the ring opens and the ring strain is relieved.

β-propiolactam

The β-lactam ring is found in three important classes of antibiotics, all isolated from fungi. *Penicillins* have a β-lactam ring fused to a five-membered ring containing a sulfur atom. *Cephalosporins* have a β-lactam ring fused to an unsaturated six-membered ring containing a sulfur atom. *Carbapenems* have a β-lactam ring fused to an unsaturated five-membered ring with a sulfur atom bonded to the ring. The structures of penicillin V, cephalexin, and imipenem illustrate these three classes of antibiotics.

penicillin V
a penicillin

cephalexin (Keflex®)
a cephalosporin

imipenem (Primaxin®)
a carbapenem

Resistant bacteria inactivate β-lactam antibiotics by hydrolysis of the lactam ring. Augmentin™ consists of a β-lactam antibiotic and a compound that blocks the enzyme responsible for hydrolysis. The combination enables the antibiotic to reach the cell wall intact.

These β-lactam antibiotics apparently work by interfering with the synthesis of bacterial cell walls. Figure 21-11 shows how the carbonyl group of the β-lactam acylates a hydroxyl group (from a serine residue) on one of the enzymes involved in making the cell wall. The acylated enzyme is inactive for synthesis of the cell wall protein. This acylation step is unusual because it converts an amide to an ester, an "uphill" reaction that we would assume to be endothermic. With this β-lactam, however, the strain of the four-membered ring activates the amide enough for it to acylate an alcohol to form an ester in an exothermic step.

▶ **Figure 21-11**
β-Lactam antibiotics function by acylating and inactivating one of the enzymes needed to make the bacterial cell wall.

PROBLEM 21-36

Show how you would accomplish the following syntheses. You may use any necessary reagents.
(a) benzoic acid → benzyldimethylamine
(b) pyrrolidine → *N*-methylpyrrolidine

PROBLEM 21-37

Nitriles are considered carboxylic acid derivatives because they are often prepared from acids through the primary amides. Show how you would convert the following carboxylic acids to nitriles.
(a) butyric acid $\rightarrow$ butyronitrile   **(b)** benzoic acid $\rightarrow$ benzonitrile
(c) cyclopentanecarboxylic acid $\rightarrow$ cyclopentanecarbonitrile

Polyamides: Nylon   The discovery of nylon in 1938 made possible a wide range of high-strength fibers, fabrics, and plastics that we take for granted today. The most common form of nylon is called nylon 6,6 because it consists of a 6-carbon diacid and a 6-carbon diamine in repeating blocks. Nylon 6,6 is made by mixing adipic acid and 1,6-hexanediamine (common name, *hexamethylene diamine*) to form a *nylon salt*, then heating the salt to drive off water and form amide bonds. The molten product is extruded in continuous filaments and stretched to align the polymer chains. The combination of polymer chains aligned with the fiber, plus the strong amide hydrogen bonding between the chains, gives nylon fibers great strength. We consider nylon chemistry in more detail in Chapter 26.

Production of continuous-filament nylon tire cord.

$$HO-\underset{\underset{O}{\|}}{C}-(CH_2)_4-\underset{\underset{O}{\|}}{C}-OH \;+\; H_2N-(CH_2)_6-NH_2 \;\longrightarrow\;$$

adipic acid   hexamethylene diamine

$$^-O-\underset{\underset{O}{\|}}{C}-(CH_2)_4-\underset{\underset{O}{\|}}{C}-O^-$$
$$\overset{+}{H_3N}-(CH_2)_6-\overset{+}{NH_3}$$

nylon salt

$$\Big\downarrow \text{heat}, -H_2O$$

$$---\underset{\underset{O}{\|}}{C}-(CH_2)_4-\underset{\underset{O}{\|}}{C}\Big[-NH-(CH_2)_6-NH-\underset{\underset{O}{\|}}{C}-(CH_2)_4-\underset{\underset{O}{\|}}{C}\Big]_n-NH-(CH_2)_6-NH---$$

poly(hexamethylene adipamide), called nylon 6,6

Although nitriles lack an acyl group, they are considered acid derivatives because they hydrolyze to carboxylic acids. Nitriles are frequently made from carboxylic acids (with the same number of carbons) by conversion to primary amides followed by dehydration. They are also made from primary alkyl halides and tosylates (adding one carbon) by nucleophilic substitution with cyanide ion. Aryl cyanides can be made by the Sandmeyer reaction of an aryldiazonium salt with cuprous cyanide. α-Hydroxynitriles (cyanohydrins) are made by the reaction of ketones and aldehydes with HCN.

21-14 Summary of the Chemistry of Nitriles

$$R-\underset{\underset{O}{\|}}{C}-NH_2 \xrightarrow{POCl_3} R-C\equiv N \qquad \text{(Section 21-13)}$$

primary amide   nitrile

$$R-X\ (1°) \xrightarrow{NaCN} R-C\equiv N \;+\; Na^+\,X^- \qquad \text{(Section 6-9)}$$

alkyl halide   nitrile

$$Ar-\overset{+}{N}\equiv N \xrightarrow{CuCN} Ar-C\equiv N \;+\; N_2\!\uparrow \qquad \text{(Section 19-18)}$$

diazonium salt   aryl nitrile

$$R-\underset{\underset{O}{\|}}{C}-R' \xrightarrow[KCN]{HCN} R-\overset{HO\;\;\;C\equiv N}{\underset{|}{C}}-R' \qquad \text{(Section 18-15)}$$

ketone or aldehyde   cyanohydrin

Reactions of Nitriles Nitriles undergo acidic or basic hydrolysis to amides, which may be further hydrolyzed to carboxylic acids. Reduction of a nitrile by lithium aluminum hydride gives a primary amine, and the reaction with a Grignard reagent gives an imine that hydrolyzes to a ketone.

$$
\underset{\text{nitrile}}{R-C\equiv N}
$$

$$
\xrightarrow[\text{H}^+ \text{ or } ^-\text{OH}]{\text{H}_2\text{O}} \quad \underset{\text{amide}}{R-\overset{\overset{\displaystyle O}{\|}}{C}-NH_2} \quad \xrightarrow[\text{H}^+ \text{ or } ^-\text{OH}]{\text{H}_2\text{O}} \quad \underset{\text{acid}}{R-\overset{\overset{\displaystyle O}{\|}}{C}-OH} \qquad \text{(Section 21-7D)}
$$

$$
\xrightarrow[\text{(2) H}_2\text{O}]{\text{(1) LiAlH}_4} \quad \underset{\text{amine}}{R-CH_2NH_2} \qquad \text{(Sections 19-19C2 and 21-8C)}
$$

$$
\xrightarrow{R'\text{MgX}} \quad \underset{\text{imine salt}}{R-\overset{\overset{\displaystyle N^{\diagup MgX}}{\|}}{C}-R'} \quad \xrightarrow{\text{H}_3\text{O}^+} \quad \underset{\text{ketone}}{R-\overset{\overset{\displaystyle O}{\|}}{C}-R'} \qquad \text{(Sections 18-10 and 21-9)}
$$

The presence of nitriles in the atmospheres of other planets is significant because they may be precursors to critical biological molecules. For example, nitriles can give rise to amino acids (Section 24-5D), the building blocks of proteins.

PROBLEM 21-38

Show how you would convert the following starting materials to the indicated nitriles:
(a) phenylacetic acid → phenylacetonitrile
(b) phenylacetic acid → 3-phenylpropionitrile
(c) *p*-chloronitrobenzene → *p*-chlorobenzonitrile

PROBLEM 21-39

Show how each transformation may be accomplished by using a nitrile as an intermediate. You may use any necessary reagents.
(a) 1-hexanol → 1-heptanamine
(b) cyclohexanecarboxamide → cyclohexyl ethyl ketone
(c) 1-octanol → 2-decanone

21-15 Thioesters

Most carboxylic esters are composites of carboxylic acids and alcohols. A **thioester** is formed from a carboxylic acid and a thiol. Thioesters are also called *thiol esters* to emphasize the fact that they are derivatives of thiols.

$$
\underset{\text{acid}}{R-\overset{\overset{\displaystyle O}{\|}}{C}-OH} + \underset{\text{alcohol}}{R'-OH} \rightleftharpoons \underset{\text{ester}}{R-\overset{\overset{\displaystyle O}{\|}}{C}-O-R'} + \text{H}_2\text{O}
$$

$$
\underset{\text{acid}}{R-\overset{\overset{\displaystyle O}{\|}}{C}-OH} + \underset{\text{thiol}}{R'-SH} \rightleftharpoons \underset{\text{thioester}}{R-\overset{\overset{\displaystyle O}{\|}}{C}-S-R'} + \text{H}_2\text{O}
$$

Thioesters are more reactive toward nucleophilic acyl substitution than normal esters, but less reactive than acid chlorides and anhydrides. If we add thioesters to the order of reactivity, we have the following sequence:

Relative reactivity

$$
\underset{\text{acid chloride}}{R-\overset{\overset{\displaystyle O}{\|}}{C}-Cl} > \underset{\text{anhydride}}{R-\overset{\overset{\displaystyle O}{\|}}{C}-O-\overset{\overset{\displaystyle O}{\|}}{C}-R} > \underset{\text{thioester}}{R-\overset{\overset{\displaystyle O}{\|}}{C}-S-R'} > \underset{\text{ester}}{R-\overset{\overset{\displaystyle O}{\|}}{C}-O-R'} > \underset{\text{amide}}{R-\overset{\overset{\displaystyle O}{\|}}{C}-NH_2}
$$

The enhanced reactivity of thioesters results from two major differences. First, the resonance stabilization of a thioester is less than that of an ester. In the thioester, the

◀ **Figure 21-12**
The resonance overlap in a thioester is not as effective as that in an ester.

second resonance form involves overlap between a 2*p* orbital on carbon and a 3*p* orbital on sulfur (Figure 21-12). These orbitals are of different sizes and are located at different distances from the nuclei. The overlap is weak and relatively ineffective, leaving the C—S bond of a thioester weaker than the C—O bond of an ester.

The second difference is in the leaving groups: An alkyl sulfide anion ($^-$:S̈—R) is a better leaving group than an alkoxide anion ($^-$:Ö—R) because the sulfide is less basic than an alkoxide, and the larger sulfur atom carries the negative charge spread over a larger volume of space. Sulfur is also more polarizable than oxygen, allowing more bonding as the alkyl sulfide anion is leaving (Section 6-11A).

Living systems need acylating reagents, but acid halides and anhydrides are too reactive for selective acylation. Also, they would hydrolyze under the aqueous conditions found in living organisms. Thioesters are not so prone to hydrolysis, yet they are excellent selective acylating reagents; therefore, thioesters are common acylating agents in living systems. Many biochemical acylations involve transfer of acyl groups from thioesters of coenzyme A (CoA). Figure 21-13 shows the structure of acetyl coenzyme A, together with the mechanism for transfer of the acetyl group to a nucleophile. In effect, acetyl CoA serves as a water-stable equivalent of acetyl chloride (or acetic anhydride) in living systems.

▲ **Figure 21-13**
Coenzyme A (CoA) is a thiol whose thioesters serve as biochemical acyl transfer reagents. Acetyl CoA transfers an acetyl group to a nucleophile, with coenzyme A serving as the leaving group.

21-16
Esters and Amides of Carbonic Acid

Carbonic acid (H_2CO_3) is formed when carbon dioxide dissolves in water. Although carbonic acid itself is constantly in equilibrium with carbon dioxide and water, it has several important stable derivatives. **Carbonate esters** are diesters of carbonic acid, with two alkoxy groups replacing the hydroxyl groups of carbonic acid. **Ureas** are diamides of carbonic acid, with two nitrogen atoms bonded to the carbonyl group. The unsubstituted urea, simply called *urea*, is the waste product excreted by mammals from the metabolism of excess protein. **Carbamate esters (urethanes)** are the stable esters of the unstable **carbamic acid**, the monoamide of carbonic acid.

$$O=C=O \; + \; H_2O \; \rightleftharpoons \; \left[H-O-\overset{\overset{\displaystyle O}{\|}}{C}-O-H \right]$$
carbonic acid (unstable)

$$R-O-\overset{\overset{\displaystyle O}{\|}}{C}-O-R$$
a carbonate ester

$$CH_3CH_2-O-\overset{\overset{\displaystyle O}{\|}}{C}-O-CH_2CH_3$$
diethyl carbonate

cyclohexyl ethyl carbonate

$$R-NH-\overset{\overset{\displaystyle O}{\|}}{C}-NH-R$$
a substituted urea

$$H_2N-\overset{\overset{\displaystyle O}{\|}}{C}-NH_2$$
urea

$$(CH_3)_2N-\overset{\overset{\displaystyle O}{\|}}{C}-N(CH_3)_2$$
tetramethylurea

$$R-NH-\overset{\overset{\displaystyle O}{\|}}{C}-O-R$$
a carbamate or urethane

$$\left[H_2N-\overset{\overset{\displaystyle O}{\|}}{C}-OH \right]$$
carbamic acid (unstable)

$$H_2N-\overset{\overset{\displaystyle O}{\|}}{C}-OEt$$
ethyl carbamate

1-naphthyl-*N*-methylcarbamate (Sevin® insecticide)

Most of these derivatives are synthesized by nucleophilic acyl substitution from phosgene, the acid chloride of carbonic acid.

$$Cl-\overset{\overset{\displaystyle O}{\|}}{C}-Cl \; + \; 2\,CH_3CH_2-OH \; \longrightarrow \; CH_3CH_2-O-\overset{\overset{\displaystyle O}{\|}}{C}-O-CH_2CH_3 \; + \; 2\,HCl$$
phosgene   diethyl carbonate

$$Cl-\overset{\overset{\displaystyle O}{\|}}{C}-Cl \; \xrightarrow{CH_3CH_2OH} \; Cl-\overset{\overset{\displaystyle O}{\|}}{C}-OCH_2CH_3 \; \xrightarrow{} \; $$
ethyl *N*-cyclohexyl carbamate

$$Cl-\overset{\overset{\displaystyle O}{\|}}{C}-Cl \; + \; 2\,(CH_3)_2NH \; \longrightarrow \; (CH_3)_2N-\overset{\overset{\displaystyle O}{\|}}{C}-N(CH_3)_2 \; + \; 2\,HCl$$
tetramethylurea

Another way of making urethanes is to treat an alcohol or a phenol with an **isocyanate**, which is an anhydride of a carbamic acid. Although the carbamic acid is unstable, the urethane is stable. This is the way Sevin® insecticide is made.

$$R-N=C=O \quad + \quad H_2O \quad \longrightarrow \quad \left[R-NH-\overset{\overset{\displaystyle O}{\|}}{C}-OH \right] \quad \longrightarrow \quad R-NH_2 \quad + \quad CO_2$$

an isocyanate   a carbamic acid   an amine

$$☞ \quad R-N=C=O \quad + \quad HO-R' \quad \longrightarrow \quad R-NH-\overset{\overset{\displaystyle O}{\|}}{C}-O-R'$$

an isocyanate   alcohol   a carbamate ester
(urethane)

Example

$CH_3-N=C=O$ +
methyl isocyanate

1-naphthol

Sevin® insecticide

PROBLEM 21-40

Propose a mechanism for the reaction of methyl isocyanate with 1-naphthol to give Sevin®
insecticide.

PROBLEM 21-41

For each heterocyclic compound
(i)   Explain what type of acid derivative is present.
(ii) Show what compounds would result from complete hydrolysis.
(iii) Are any of the rings aromatic? Explain.

(a)   (b)   (c)

(d) HN   NH   (e)   (f)

PROBLEM 21-42

Biosynthesis of the pyrimidine bases used in DNA occurs via *N*-carbamoylaspartate, which is
formed as follows:

carbamoyl phosphate   aspartate   *N*-carbamoylaspartate

N-Carbamoylaspartate cyclizes under the control of an enzyme, giving dihydroorotate, which
is dehydrogenated to orotate, a direct precursor to the pyrimidine bases. (The structures of the
pyrimidine bases are shown in Section 23-21.)

N-carbamoylaspartate   dihydroorotate   orotate

The development of Sevin™ and
related insecticides resulted from
studies on the alkaloid physostigmine, which has a methyl carbamate portion. These studies also led
to the development of potent nerve
gases such as Sarin™.

physostigmine

sarin

(a) What kind of compound is carbamoyl phosphate? Would you expect such a compound to react with an amine to give an amide?

(b) What special kind of amide is *N*-carbamoylaspartate?

(c) What kind of reaction cyclizes *N*-carbamoylaspartate to dihydroorotate?

(d) Is orotate aromatic? Draw the structure of pyrimidine. Why is orotate called a "pyrimidine base"? (*Hint:* Consider tautomers.)

Before the development of tough, resilient polyurethane wheels, street roller skates used steel wheels that stopped dead when they hit the smallest pebble or crack in the pavement. Rollerblades would not exist without polymer technology, both in the wheels and in the strong ABS plastic used for the uppers. The helmet is molded from Lexan® polycarbonate.

Polycarbonates and Polyurethanes The chemistry of carbonic acid derivatives is particularly important because two large classes of polymers are bonded by linkages containing these functional groups: the *polycarbonates* and the *polyurethanes*. Polycarbonates are polymers bonded by the carbonate ester linkage, and polyurethanes are polymers bonded by the carbamate ester linkage. Lexan® polycarbonate is a strong, clear polymer used in bulletproof windows and crash helmets. The diol used to make Lexan® is a phenol called *bisphenol A*, a common intermediate in polyester and polyurethane synthesis.

Lexan® polycarbonate

A polyurethane results when a diol reacts with a diisocyanate, a compound with two isocyanate groups. A common form of polyurethane is made by the reaction of ethylene glycol with *toluene diisocyanate*.

a polyurethane

Chapter 21 Glossary

acid halide (acyl halide) An activated acid derivative in which the hydroxyl group of the acid is replaced by a halogen, usually chlorine. (p. 945)

acyl transfer Another term for a *nucleophilic acyl substitution*. The term *acyl transfer* emphasizes the "transfer" of the acyl group from the leaving group to the attacking nucleophile. (p. 957)

amide An acid derivative in which the hydroxyl group of the acid is replaced by a nitrogen atom and its attached hydrogens or alkyl groups. An amide is a composite of a carboxylic acid and an amine. (p. 942)

$$R-\overset{\overset{\displaystyle O}{\|}}{C}-NH_2 \qquad R-\overset{\overset{\displaystyle O}{\|}}{C}-\overset{\overset{\displaystyle H}{|}}{N}-R' \qquad R-\overset{\overset{\displaystyle O}{\|}}{C}-\overset{\overset{\displaystyle R'}{|}}{N}-R'$$

primary amide   secondary amide   tertiary amide

(*N*-substituted amide)   (*N,N*-disubstituted amide)

ammonolysis of an ester Cleavage of an ester by ammonia (or an amine) to give an amide and an alcohol. (p. 961)

anhydride (carboxylic acid anhydride) An activated acid derivative formed from two acid molecules with loss of a molecule of water. A **mixed anhydride** is an anhydride derived from two different acid molecules. (p. 945)

$$2\ R-\overset{\overset{\displaystyle O}{\|}}{C}-OH \quad \rightleftharpoons \quad R-\overset{\overset{\displaystyle O}{\|}}{C}-O-\overset{\overset{\displaystyle O}{\|}}{C}-R \ + \ H_2O$$

carbamate ester See **urethane**. (p. 990)

carbonate ester A diester of carbonic acid. (p. 990)

carbonic acid The one-carbon dicarboxylic acid, $HOCOOH$. Carbonic acid is constantly in equilibrium with carbon dioxide and water. Its esters and amides are stable, however. (p. 990)

ester An acid derivative in which the hydroxyl group of the acid is replaced by an alkoxyl group. An ester is a composite of a carboxylic acid and an alcohol. (p. 941)

Fischer esterification (p. 964)

$$R-\overset{\overset{\displaystyle O}{\|}}{C}-OH \ + \ R'-OH \ \underset{}{\overset{H^+}{\rightleftharpoons}} \ R-\overset{\overset{\displaystyle O}{\|}}{C}-O-R' \ + \ H_2O$$

acid   alcohol   ester

Hofmann rearrangement of amides Conversion of a primary amide to an amine (with one less carbon) by treatment with a basic bromine solution. The $C{=}O$ group is lost as CO_2. (pp. 885, 985)

isocyanate A compound of formula $R-N{=}C{=}O$. (p. 990)

lactam A cyclic amide. (p. 943)

lactone A cyclic ester. (p. 941)

nitrile An organic compound containing the **cyano group**, $C{\equiv}N$. (p. 944)

nucleophilic acyl substitution A nucleophile substitutes for a leaving group on a carbonyl carbon atom. Nucleophilic acyl substitution usually takes place through the following **addition–elimination mechanism**. (p. 957)

$$R-\overset{\overset{\displaystyle \ddot{O}}{\|}}{C}-X \ + \ Nuc{:}^- \ \rightleftharpoons \ \left[R-\overset{\overset{\displaystyle :\ddot{O}:^-}{|}}{\underset{\underset{\displaystyle Nuc}{|}}{C}}-X \right] \ \rightleftharpoons \ R-\overset{\overset{\displaystyle \ddot{O}}{\|}}{C}-Nuc \ + \ {:}X^-$$

addition–elimination mechanism of nucleophilic acyl substitution

polymer A large molecule composed of many smaller units (monomers) bonded together. (p. 983)

 polyamide (nylon): A polymer in which the monomer units are bonded by amide linkages. (p. 987)

 polycarbonate: A polymer in which the monomer units are bonded together by carbonate ester linkages. (p. 992)

 polyester: A polymer in which the monomer units are bonded by ester linkages. (p. 983)

 polyurethane: A polymer in which the monomer units are bonded together by carbamate ester (urethane) linkages. (p. 992)

saponification Basic hydrolysis of an ester to an alcohol and a carboxylate salt. (p. 968)

thioester An acid derivative in which the hydroxyl group of the acid is replaced by a sulfur atom and its attached alkyl or aryl group. A thioester is a composite of a carboxylic acid and a thiol. (p. 988)

transesterification Substitution of one alkoxy group for another in an ester. Transesterification can take place under either acidic or basic conditions. (p. 964)

urea A diamide of carbonic acid. (p. 990)

urethane (carbamate ester) An ester of **carbamic acid**, $H_2N-COOH$; a monoester, monoamide of carbonic acid. (p. 990)

Essential Problem-Solving Skills in Chapter 21

1. Name carboxylic acid derivatives, and draw the structures from their names.
2. Compare the physical properties of acid derivatives, and explain the unusually high boiling points and melting points of amides.
3. Interpret the spectra of acid derivatives, and use the spectral information to determine the structures. Show how the carbonyl stretching frequency in the IR depends on the structure of the acid derivative.
4. Show how acid derivatives are easily interconverted by nucleophilic acyl substitution from more reactive derivatives to less reactive derivatives. Show how acid chlorides serve as activated intermediates to convert acids to acid derivatives.
5. Show how acid catalysis is used to synthesize acid derivatives, as in the Fischer esterification and in transesterification. Propose mechanisms for these reactions.
6. Show how acid derivatives hydrolyze to carboxylic acids, and explain why either acid or base is a suitable catalyst for the hydrolysis. Propose mechanisms for these hydrolyses.
7. Show what reagents are used to reduce acid derivatives, and show the products of reduction.
8. Show what products result from the addition of Grignard and organolithium reagents to acid derivatives, and propose mechanisms for these reactions.
9. Summarize the importance, uses, and special reactions of each type of acid derivative.

Study Problems

21-43 Define each term and give an example:

(a) ester
(b) lactone
(c) Friedel–Crafts acylation
(d) lactam
(e) tertiary amide
(f) primary amide
(g) nitrile
(h) acid chloride
(i) anhydride
(j) mixed anhydride
(k) Fischer esterification
(l) nucleophilic acyl substitution
(m) polyester
(n) ammonolysis of an ester
(o) carbonate ester
(p) Hofmann rearrangement
(q) urethane
(r) saponification
(s) thioester
(t) transesterification
(u) urea

21-44 Give appropriate names for the following compounds:

(a) $CH_3CH_2CHCH_2-\overset{\overset{\displaystyle O}{\|}}{C}-Cl$ with CH_3 on the CH

(b) $Ph-\overset{\overset{\displaystyle O}{\|}}{C}-O-\overset{\overset{\displaystyle O}{\|}}{C}-H$

(c) $CH_3-\overset{\overset{\displaystyle O}{\|}}{C}-NH-Ph$

(d) $CH_3-NH-\overset{\overset{\displaystyle O}{\|}}{C}-Ph$

(e) $Ph-O-\overset{\overset{\displaystyle O}{\|}}{C}-CH_3$

(f) $Ph-\overset{\overset{\displaystyle O}{\|}}{C}-O-CH_3$

(g) (benzene ring)—C≡N

(h) (benzene ring with propyl chain)—CN

(i) $CH_3O-\overset{\overset{\displaystyle O}{\|}}{C}$ and $\overset{\overset{\displaystyle O}{\|}}{C}-OCH_3$ on benzene ring

(j) (toluene ring with CH₃)—$\overset{\overset{\displaystyle O}{\|}}{C}-N(CH_2CH_3)_2$

(k) H_3C—(lactone ring with O and =O)

(l) (four-membered ring with CH_3CH_2, N—H, and =O; a β-lactam)

21-45 Predict the major products formed when benzoyl chloride (PhCOCl) reacts with the following reagents:

(a) ethanol
(b) sodium acetate
(c) aniline
(d) anisole and aluminum chloride
(e) excess phenylmagnesium bromide, then dilute acid

21-46 Acid-catalyzed transesterification and Fischer esterification take place by nearly identical mechanisms. Transesterification can also take place by a base-catalyzed mechanism, but all attempts at base-catalyzed Fischer esterification (using ⁻OR″, for example) seem doomed to failure. Explain why Fischer esterification cannot be catalyzed by base.

21-47   Predict the products of the following reactions:

(a) phenol + acetic anhydride

(b) phenol + acetic formic anhydride

(c) aniline + phthalic anhydride

(d) anisole + succinic anhydride and aluminum chloride

(e) Ph—CH—CH₂—NH₂ + 1 equivalent of acetic anhydride
       |
      OH

(f) Ph—CH—CH₂—NH₂ + excess acetic anhydride
       |
      OH

21-48   Show how you would accomplish the following syntheses in good yields:

(a)

(b)

(c)

(d)

(e)

(f)

21-49   Propose mechanisms for the following reactions:

(a)

benzoyl chloride   isopropyl alcohol   isopropyl benzoate

(b)

ethyl benzoate   sodium benzoate   ethanol

NaOH / H₂O

+ CH₃CH₂OH

(c)

ethyl benzoate   benzoic acid   ethanol

H⁺ / H₂O

+ CH₃CH₂OH

(d)

EtO⁻ / EtOH

(e) CH₃—*CH—CH₂CH₃   →   CH₃—*CH—CH₂CH₃
      | |
     OH   Ac₂O   OAc
  (R)-2-butanol   (acetic anhydride)   2-butyl acetate

Does this reaction proceed with retention, inversion, or racemization of the asymmetric carbon atom?

21-50   Predict the products of the following reactions:

(a) [structure: benzoyl chloride] + [structure: cyclohexanol] ⟶

(b) [structure: cyclohexanecarboxylic acid methyl ester] $\xrightarrow[\text{heat}]{CH_3NH_2}$

(c) Ph—C(=O)—Cl + [structure: pyrrolidine N—H] ⟶

(d) [structure: glutaric anhydride] + [structure: cyclohexylamine] ⟶

(e) Ph—C(=O)—OCH$_2$CH$_3$ $\xrightarrow[\text{(2) H}_2\text{O}]{\text{(1) LiAlH}_4}$

(f) [structure: 2-piperidinone] $\xrightarrow[\text{(2) H}_2\text{O}]{\text{(1) LiAlH}_4}$

(g) [structure: 7-membered lactone] $\xrightarrow[\text{CH}_3\text{OH}]{^-\text{OCH}_3}$

(h) [structure: δ-valerolactone] $\xrightarrow[\text{(2) H}_2\text{O}]{\text{(1) excess PhMgBr}}$

(i) [structure: cyclopentyl—C≡N] $\xrightarrow[\text{(2) H}_3\text{O}^+]{\text{(1) CH}_3\text{MgI}}$

(j) [structure: 7-membered lactam N—H] $\xrightarrow[\text{H}_2\text{O}]{\text{NaOH}}$

(k) PhCH$_2$—CH(CH$_3$)—CH$_2$—C(=O)—NH$_2$ $\xrightarrow{\text{Br}_2,\ \text{NaOH}}$

21-51   Predict the products of saponification of the following esters:

(a) H—C(=O)—O—Ph   (b) CH$_3$CH$_2$—C(=O)—OCH$_2$CH$_3$ (c) [structure: dihydrocoumarin]   (d) [structure: 1,4-dioxane-2,3-dione type]

21-52   An ether extraction of nutmeg gives large quantities of *trimyristin*, a waxy crystalline solid of melting point 57°C. The IR spectrum of trimyristin shows a very strong absorption at 1733 cm^{-1}. Basic hydrolysis of trimyristin gives 1 equivalent of glycerol and 3 equivalents of myristic acid (tetradecanoic acid).
  (a) Draw the structure of trimyristin.
  (b) Predict the products formed when trimyristin is treated with lithium aluminum hydride, followed by aqueous hydrolysis of the aluminum salts.

21-53   Two widely used pain relievers are aspirin and acetaminophen. Show how you would synthesize these drugs from phenol:

[structure: aspirin — O—C(=O)—CH$_3$ and COOH on benzene ring]   [structure: acetaminophen — NH—C(=O)—CH$_3$ and HO on benzene ring]

aspirin   acetaminophen

21-54   Show how you would accomplish the following syntheses. Some of these conversions may require more than one step.
  (a) isopentyl alcohol → isopentyl acetate (banana oil)
  (b) 3-ethylpentanoic acid → 3-ethylpentanenitrile
  (c) isobutylamine → N-isobutylformamide
  (d) ethyl acetate → 3-methyl-3-pentanol
  (e) pyrrolidine → N-methylpyrrolidine
  (f) cyclohexylamine → N-cyclohexylacetamide
  (g) bromocyclohexane → dicyclohexylmethanol

  (h)   dimethyl oxalate   ⟶   [structure: 2,3-dioxopiperazine with two N—H]

  (i) [structure: cyclopentyl—CH$_2$OH] ⟶ [structure: cyclopentyl—CN]

(j)

21-55 Grignard reagents add to carbonate esters as they add to other esters.
(a) Predict the major product of the following reaction:

$$CH_3CH_2-O-\overset{\overset{\displaystyle O}{\|}}{C}-O-CH_2CH_3 \quad \xrightarrow[\text{(2) } H_2O]{\text{(1) excess PhMgBr}}$$

diethyl carbonate

(b) Show how you would synthesize 3-ethyl-3-pentanol using diethyl carbonate and ethyl bromide as your only organic reagents.

*__21-56__ One mole of acetyl chloride is added to a liter of triethylamine, resulting in a vigorous exothermic reaction. Once the reaction mixture has cooled, 1 mole of ethanol is added. Another vigorous exothermic reaction results. The mixture is analyzed and found to contain triethylamine, ethyl acetate, and triethylammonium chloride. Propose mechanisms for the two exothermic reactions.

21-57 Show how you would accomplish the following multistep syntheses, using the indicated starting material and any necessary reagents:
(a) 6-hepten-1-ol $\longrightarrow$ ε-caprolactone   **(b)** methoxybenzene $\longrightarrow$ p-methoxybenzamide

(c)

(d)

gallic acid   mescaline

21-58 Phosgene is the acid chloride of carbonic acid. Although phosgene was used as a war gas in World War I, it is now used as a reagent for the synthesis of many useful products. Phosgene reacts like other acid chlorides, but it can react twice.

carbonic acid   phosgene

Predict the products formed when phosgene reacts with
(a) excess ethanol
(b) excess methylamine
(c) 1 equivalent of methanol, followed by 1 equivalent of aniline
(d) ethylene glycol
(e) t-Butyloxycarbonyl chloride is an important reagent for the synthesis of peptides and proteins (Chapter 24). Show how you would use phosgene to synthesize t-butyloxycarbonyl chloride.

t-butyloxycarbonyl chloride   Alar®

21-59 Alar® was once used as a growth retardant for apples to keep them on the tree longer, resulting in sweeter, jucier apples. A controversy arose in 1989, when a group suggested that Alar® residues in apple juice might be harmful to children because the hydrolysis of Alar® produces a toxic hydrazine derivative. Apple juice is slightly acidic, and it may be stored for a long time, possibly allowing hydrolysis to take place. Many people stopped eating apples, and a large part of that year's apple crop was wasted. This result was unnecessary because there was never any serious concern about fresh apples, and most of the apples had never been treated with Alar®. Eventually it was shown that the amounts of Alar® in apple juice were not hazardous, but apple growers stopped using Alar® anyway. The most unfortunate outcome of this controversy was the enactment of "veggie libel laws" that expose people to lawsuits if they express concern about food products.
(a) Show what products are formed by the hydrolysis of Alar®.
(b) Propose a mechanism for this hydrolysis under mildly acidic conditions.

21-60 In Section 21-16, we saw that Sevin® insecticide is made by the reaction of 1-naphthol with methyl isocyanate. A Union Carbide plant in Bhopal, India, once used this process to make Sevin® for use as an agricultural insecticide. On December

3, 1984, either by accident or by sabotage, a valve was opened that admitted water to a large tank of methyl isocyanate. The pressure and temperature within the tank rose dramatically, and pressure-relief valves opened to keep the tank from bursting. A large quantity of methyl isocyanate rushed out through the pressure-relief valves, and the vapors flowed with the breeze into populated areas, killing about 2500 people and injuring many more.

(a) Write an equation for the reaction that took place in the tank. Explain why the pressure and temperature rose dramatically.

(b) Propose a mechanism for the reaction you wrote in part (a).

(c) Propose an alternative synthesis of Sevin®. Unfortunately, the best alternative synthesis uses phosgene, a gas that is even more toxic than methyl isocyanate.

21-61   The structures of five useful polymers are shown below, together with some of their best-known products. In each case,

(i) Determine the kind of polymer (polyamide, polyester, etc.).

(ii) Draw the structures of the monomers that would be released by complete hydrolysis.

(iii) Suggest what monomers or stable derivatives of the monomers might be used to make these polymers.

(a)

soft, sheer fabrics; synthetic silk

(b)

climbing ropes, violin strings

(c)

crash helmets, bulletproof "glass"

(d)

high-strength fabrics; bulletproof vests

(e)

skateboard wheels, foam mattresses

21-62   The following two compounds, which exhibit powerful antibacterial activity, were isolated from the fungus *Cephalosporium acremonium* in 1948 from the sea near a sewage outlet on the Sardinian coast:

cephalosporin N

cephalosporin C

(a) Name the class of antibiotics represented by each compound.

(b) After its structure had been determined, one of these compounds came to be known by a more appropriate name. Determine which compound is poorly named, and replace the inappropriate part of the name.

*21-63   Methyl *p*-nitrobenzoate has been found to undergo saponification faster than methyl benzoate.

(a) Consider the mechanism of saponification, and explain the reasons for this rate enhancement.

(b) Would you expect methyl *p*-methoxybenzoate to undergo saponification faster or slower than methyl benzoate?

21-64   A student has just added ammonia to hexanoic acid and has begun to heat the mixture when he is called away to the telephone. After a long telephone conversation, he returns to find that the mixture has overheated and turned black. He distills the volatile components and recrystallizes the solid residue. Among the components he isolates are compounds A (a liquid, molecular formula $C_6H_{11}N$) and B (a solid, molecular formula $C_6H_{13}NO$). The infrared spectrum of A shows a strong, sharp absorption at 2247 cm^{-1}. The infrared spectrum of B shows absorptions at 3390, 3200, and 1665 cm^{-1}. Determine the structures of compounds A and B.

21-65   A chemist was called to an abandoned aspirin factory to determine the contents of a badly corroded vat. Knowing that two salvage workers had become ill from breathing the fumes, she put on her breathing apparatus as soon as she noticed an overpowering odor like that of vinegar but much more pungent. She entered the building and took a sample of the contents of the vat. The mass spectrum showed a molecular weight of 102, and the NMR spectrum showed only a singlet at $\delta2.15$. The IR spectrum, which appears below, left no doubt about the identity of the compound. Identify the compound, and suggest a method for its safe disposal.

21-66   An unknown compound gives a mass spectrum with a weak molecular ion at m/z 113 and a prominent ion at m/z 68. Its NMR and IR spectra are shown below. Determine the structure, and show how it is consistent with the observed absorptions. Propose a favorable fragmentation to explain the prominent MS peak at m/z 68.

*21-67 An unknown compound of molecular formula C_5H_9NO gives the IR and NMR spectra shown below. The broad NMR peak at $\delta 7.55$ disappears when the sample is shaken with D_2O. Propose a structure, and show how it is consistent with the absorptions in the spectra.

21-68   An unknown compound gives the NMR, IR, and the mass spectra shown next. Propose a structure, and show how it is consistent with the observed absorptions. Show fragmentations that account for the prominent ion at m/z 69 and the smaller peak at m/z 99.

*21-69   The ^{1}H NMR spectrum, ^{13}C NMR spectrum, and IR spectrum of an unknown compound ($C_6H_8O_3$) appear next. Determine the structure, and show how it is consistent with the spectra.

Alpha Substitutions and Condensations of Enols and Enolate Ions

Up to now, we have studied two of the main types of carbonyl reactions: nucleophilic addition and nucleophilic acyl substitution. In these reactions, the carbonyl group serves as an *electrophile* by accepting electrons from an attacking nucleophile. In this chapter, we consider two more types of reactions: substitution at the carbon atom next to the carbonyl group (called alpha substitution) and carbonyl condensations. **Alpha (α) substitutions** involve the replacement of a hydrogen atom at the **α carbon atom** (the carbon next to the carbonyl) by some other group. Alpha substitution generally takes place when the carbonyl compound is converted to its enolate ion or enol tautomer. Both of these have lost a hydrogen atom at the alpha position, and both are *nucleophilic*. Attack on an electrophile completes the substitution.

22-1
Introduction

MECHANISM   Alpha Substitution

Step 1: Deprotonation.

Step 2: Attack on electrophile.

enolate ion

Carbonyl **condensations** are alpha substitutions where the electrophile is another carbonyl compound. From the electrophile's point of view, the condensation is either a nucleophilic addition or a nucleophilic acyl substitution. With ketones and aldehydes, protonation of the alkoxide gives the product of nucleophilic addition. With esters, loss of alkoxide gives the product of nucleophilic acyl substitution.

MECHANISM   Addition of an Enolate to Ketones and Aldehydes (a Condensation)

Step 1: Addition of the enolate.

Step 2: Protonation.

enolate   ketone

addition product

$+ \ RO^-$

MECHANISM   Substitution of an Enolate on an Ester (a Condensation)

Step 1: Addition of the enolate.

Step 2: Elimination of the alkoxide.

enolate   ester   substitution product

Alpha substitutions and condensations of carbonyl compounds are some of the most common methods for forming carbon–carbon bonds. A wide variety of compounds can participate as nucleophiles or electrophiles (or both) in these reactions, and many useful products can be synthesized. We begin our study of these reactions by considering the structure and formation of enols and enolate ions.

22-2
Enols
and Enolate Ions

22-2A   Keto–Enol Tautomerism

In the presence of strong bases, ketones and aldehydes act as weak proton acids. A proton on the α carbon is abstracted to form a resonance-stabilized **enolate ion** with the negative charge spread over a carbon atom and an oxygen atom. Reprotonation can occur either on the α carbon (returning to the **keto** form) or on the oxygen atom, giving a vinyl alcohol, the **enol** form.

MECHANISM   Base-Catalyzed Keto–Enol Tautomerism

Step 1: Deprotonation on C.   *Step 2:* Reprotonation on O.

keto form   enolate ion   enol form
                                                                        (vinyl alcohol)

In this way, base catalyzes an equilibrium between isomeric keto and enol forms of a carbonyl compound. For simple ketones and aldehydes, the keto form predominates. Therefore, a vinyl alcohol (an enol) is best described as an alternative isomeric form of a ketone or aldehyde. In Section 9-9F, we saw that an enol intermediate, formed by hydrolysis of an alkyne, quickly isomerizes to its keto form.

keto form   enol form   keto form   enol form
(99.99%)   (0.01%)   (99.95%)   (0.05%)

This type of isomerization, occurring by the migration of a proton and the movement of a double bond, is called **tautomerism**, and the isomers that interconvert are called **tautomers**. Don't confuse tautomers with resonance forms. Tautomers are true isomers (different compounds) with their atoms arranged differently. Under the right circumstances, with no catalyst present, either individual tautomeric form may be isolated. Resonance forms are different representations of the *same* structure, with all the atoms in the same places, showing how the electrons are delocalized.

Keto–enol tautomerism is also catalyzed by acid. In acid, a proton is moved from the α carbon to oxygen by first protonating oxygen and then removing a proton from carbon.

MECHANISM Acid-Catalyzed Keto–Enol Tautomerism

Step 1: Protonation on O. *Step 2:* Deprotonation on C.

keto form   protonated carbonyl   enol form

Compare the base-catalyzed and acid-catalyzed mechanisms shown above for keto–enol tautomerism. In base, the proton is removed from carbon, then replaced on oxygen. In acid, oxygen is protonated first, then carbon is deprotonated. Most proton-transfer mechanisms work this way. In base, the proton is removed from the old location, then replaced at the new location. In acid, protonation occurs at the new location, followed by deprotonation at the old location.

In addition to its mechanistic importance, keto–enol tautomerism affects the stereochemistry of ketones and aldehydes. A hydrogen atom on an α carbon may be lost and regained through keto–enol tautomerism; such a hydrogen is said to be **enolizable**. If an asymmetric carbon atom has an enolizable hydrogen atom, a trace of acid or base allows that carbon to invert its configuration, with the enol serving as the intermediate. A racemic mixture (or an equilibrium mixture of diastereomers) is the result.

enolizable hydrogens

(R) configuration   enol (achiral)   (S) configuration

α carbons

PROBLEM 22-1

Phenylacetone can form two different enols.
(a) Show the structures of these enols.
(b) Predict which enol will be present in the larger concentration at equilibrium.
(c) Propose mechanisms for the formation of the two enols in acid and in base.

PROBLEM 22-2

Show each step in the mechanism of the acid-catalyzed interconversion of (R)- and (S)-2-methylcyclohexanone.

PROBLEM 22-3

When *cis*-2,4-dimethylcyclohexanone is dissolved in aqueous ethanol containing a trace of NaOH, a mixture of cis and trans isomers results. Propose a mechanism for this isomerization.

PROBLEM-SOLVING HINT

In acid, proton transfers usually occur by adding a proton in the new position, then deprotonating the old position.

In base, proton transfers occur by deprotonating the old position, then reprotonating at the new position.

22-2B   Formation and Stability of Enolate Ions

A carbonyl group dramatically increases the acidity of the protons on the α carbon atom because most of the enolate ion's negative charge resides on the electronegative oxygen atom. The pK_a for removal of an α proton from a typical ketone or aldehyde is about 20, showing that a typical ketone or aldehyde is much more acidic than an alkane or an alkene ($pK_a > 40$), or even an alkyne ($pK_a = 25$). Still, a ketone or aldehyde is less acidic than water ($pK_a = 15.7$) or an alcohol ($pK_a = 16$ to 18). When a simple ketone or aldehyde is treated with hydroxide ion or an alkoxide ion, the equilibrium mixture contains only a small fraction of the deprotonated, enolate form.

Example

(equilibrium lies to the left)

Even though the equilibrium concentration of the enolate ion may be small, it serves as a useful, reactive nucleophile. When an enolate reacts with an electrophile (other than a proton), the enolate concentration decreases, and the equilibrium shifts to the right (Fig. 22-1). Eventually, all the carbonyl compound reacts via a low concentration of the enolate ion.

> **PROBLEM 22-4**
>
> Give the important resonance forms for the enolate ion of
> **(a)** acetone   **(b)** cyclopentanone   **(c)** 2,4-pentanedione

Sometimes this equilibrium mixture of enolate and base won't work, usually because the base (hydroxide or alkoxide) reacts with the electrophile faster than the enolate does. In these cases, we need a base that reacts completely to convert the carbonyl

▶ **Figure 22-1**
Reaction of the enolate ion with an electrophile removes it from the equilibrium, shifting the equilibrium to the right.

compound to its enolate before adding the electrophile. Although sodium hydroxide and alkoxides are not sufficiently basic, powerful bases are available to convert a carbonyl compound completely to its enolate. The most effective and useful base for this purpose is lithium diisopropylamide (LDA), the lithium salt of diisopropylamine. LDA is made by using an alkyllithium reagent to deprotonate diisopropylamine.

$$CH_3-CH(CH_3)-\overset{..}{N}-H \quad + \quad C_4H_9Li \quad \longrightarrow \quad C_4H_{10} \quad + \quad CH_3-CH(CH_3)-\overset{..}{N}:^- \quad Li^+$$

diisopropylamine   *n*-butyllithium   butane   lithium diisopropylamide (LDA)

Diisopropylamine has a pK_a of about 40, showing that it is much *less* acidic than a typical ketone or aldehyde. By virtue of its two isopropyl groups, LDA is a bulky reagent; it does not easily attack a carbon atom or add to a carbonyl group. Thus, it is a powerful base, but not a strong nucleophile. When LDA reacts with a ketone, it abstracts the α proton to form the lithium salt of the enolate. We will see that these lithium enolate salts are very useful in synthesis.

$$R-\overset{O}{\overset{\|}{C}}-\overset{H}{\overset{|}{C}}- \quad + \quad (i\text{-}C_3H_7)_2N^-\ Li^+ \quad \rightleftarrows \quad R-\overset{O^-\ ^+Li}{\overset{|}{C}}=C\diagdown \quad + \quad (i\text{-}C_3H_7)_2N-H$$

ketone ($pK_a \cong 20$)   LDA   lithium salt of enolate   diisopropylamine ($pK_a = 40$)

(equilibrium lies to the right)

Example

cyclohexanone ($pK_a = 19$)   +   LDA $\rightleftarrows$ lithium enolate of cyclohexanone (100%)   + $(i\text{-}C_3H_7)_2N-H$ ($pK_a = 40$)

22-3A Base-Promoted α Halogenation

When a ketone is treated with a halogen and base, an α-halogenation reaction occurs.

☞ $$-\overset{O}{\overset{\|}{C}}-\overset{H}{\overset{|}{C}}- \quad + \quad ^-OH \quad + \quad X_2 \quad \longrightarrow \quad -\overset{O}{\overset{\|}{C}}-\overset{X}{\overset{|}{C}}- \quad + \quad X^- \quad + \quad H_2O$$

ketone   (X_2 = Cl_2, Br_2, or I_2)

Example

cyclohexanone $\xrightarrow[^-OH, H_2O]{Cl_2}$ 2-chlorocyclohexanone   + Cl^- + H_2O

The base-promoted halogenation takes place by a nucleophilic attack of an enolate ion on the electrophilic halogen molecule. The products are the halogenated ketone and a halide ion.

MECHANISM Base-Promoted Halogenation

Step 1: Reprotonation.

Step 2: Attack on the electrophile.

enolate ion   + H_2O

This reaction is called *base-promoted*, rather than base-catalyzed, because a full equivalent of the base is consumed in the reaction.

PROBLEM-SOLVING HINT

In drawing mechanisms, you can show either resonance form of an enolate attacking the electrophile. It is often more intuitive to show the carbanion form attacking.

SOLVED PROBLEM 22-1

Propose a mechanism for the reaction of 3-pentanone with sodium hydroxide and bromine to give 2-bromo-3-pentanone.

SOLUTION

In the presence of sodium hydroxide, a small amount of 3-pentanone is present as its enolate.

The enolate reacts with bromine to give the observed product.

enolate

α-haloketone

PROBLEM 22-5

Propose a mechanism for the example showing the formation of 2-chlorocyclohexanone above.

Multiple Halogenation   In many cases, base-promoted halogenation does not stop with replacement of just one hydrogen. The product (the α-haloketone) is more reactive toward further halogenation than is the starting material, because the electron-withdrawing halogen stabilizes the enolate ion, enhancing its formation.

(enolate stabilized by X)

For example, bromination of 3-pentanone gives mostly 2,2-dibromo-3-pentanone. After one hydrogen is replaced by bromine, the enolate ion is stabilized by both the carbonyl group and the bromine atom. A second bromination takes place faster than the first. Notice that the second substitution takes place at the same carbon atom as the first, because that carbon atom bears the enolate-stabilizing halogen.

Because of this tendency for multiple halogenation, base-promoted halogenation is rarely used for the preparation of monohalo ketones. The acid-catalyzed procedure (discussed in Section 22-3C) is preferred.

> **PROBLEM 22-6**
>
> Propose a mechanism to show how acetophenone undergoes base-promoted chlorination to give trichloroacetophenone.

22-3B The Haloform Reaction

With most ketones, base-promoted halogenation continues until the α-carbon atom is completely halogenated. Methyl ketones have three α protons on the methyl carbon, and they undergo halogenation three times to give trihalomethyl ketones.

With three electron-withdrawing halogen atoms, the trihalomethyl group can serve as a reluctant leaving group for nucleophilic acyl substitution. The trihalomethyl ketone reacts with hydroxide ion to give a carboxylic acid. A fast proton exchange gives a carboxylate ion and a haloform (chloroform, $CHCl_3$; bromoform, $CHBr_3$; or iodoform, CHI_3). The overall reaction is called the **haloform reaction**.

The overall haloform reaction is summarized below. A methyl ketone reacts with a halogen under strongly basic conditions to give a carboxylate ion and a haloform.

Example

$$CH_3CH_2-\overset{\overset{\displaystyle O}{\|}}{C}-CH_3 \quad \xrightarrow[^-OH]{\text{excess } Br_2} \quad CH_3CH_2-\overset{\overset{\displaystyle O}{\|}}{C}-CBr_3$$

2-butanone

$$CH_3CH_2-\overset{\overset{\displaystyle O}{\|}}{C}-CBr_3 \quad \xrightarrow{^-OH} \quad CH_3CH_2-\overset{\overset{\displaystyle O}{\|}}{C}-O^- \quad + \quad HCBr_3$$

propionate   bromoform

When the halogen is iodine, the haloform product (iodoform) is a solid that separates out as a yellow precipitate. This **iodoform test** identifies methyl ketones, which halogenate three times, then lose $^-CI_3$ to give iodoform.

$$Ph-\overset{\overset{\displaystyle O}{\|}}{C}-CH_3 \quad \xrightarrow[^-OH]{\text{excess } I_2} \quad Ph-\overset{\overset{\displaystyle O}{\|}}{C}-CI_3 \quad \xrightarrow{^-OH} \quad Ph-\overset{\overset{\displaystyle O}{\|}}{C}-O^- \quad + \quad HCI_3 \downarrow$$

acetophenone α,α,α-triiodoacetophenone   benzoate   iodoform

Iodine is an oxidizing agent, and an alcohol can give a positive iodoform test if it oxidizes to a methyl ketone. The iodoform reaction can convert such an alcohol to a carboxylic acid with one less carbon atom.

$$☞ \quad R-\overset{\overset{\displaystyle OH}{|}}{CH}-CH_3 + I_2 \quad \longrightarrow \quad R-\overset{\overset{\displaystyle O}{\|}}{C}-CH_3 + 2\,HI \quad \xrightarrow[^-OH]{\text{excess } I_2} \quad R-\overset{\overset{\displaystyle O}{\|}}{C}-O^- + HCI_3 \downarrow$$

(one less carbon)

Example

$$CH_3(CH_2)_3-\overset{\overset{\displaystyle OH}{|}}{CH}-CH_3 \quad \xrightarrow[^-OH]{I_2} \quad CH_3(CH_2)_3-\overset{\overset{\displaystyle O}{\|}}{C}-CH_3 \quad \xrightarrow[^-OH]{I_2} \quad CH_3(CH_2)_3-\overset{\overset{\displaystyle O}{\|}}{C}-O^- + HCI_3 \downarrow$$

2-hexanol   2-hexanone   pentanoate

PROBLEM 22-7

Propose a mechanism for the reaction of methyl cyclohexyl ketone with excess bromine in the presence of sodium hydroxide.

PROBLEM 22-8

Predict the products of the following reactions.
(a) methyl cyclopentyl ketone + excess Cl_2 + excess NaOH
(b) 1-cyclopentylethanol + excess I_2 + excess NaOH
(c) cyclohexanone + excess I_2 + excess NaOH
(d) propiophenone + excess Br_2 + excess NaOH

PROBLEM 22-9

Which compounds will give a positive iodoform test?
(a) 1-phenylethanol   **(b)** 2-pentanone   **(c)** 2-pentanol
(d) 3-pentanone   **(e)** acetone   **(f)** isopropyl alcohol

22-3C   Acid-Catalyzed α Halogenation

Ketones also undergo acid-catalyzed α halogenation. One of the most effective procedures is to dissolve the ketone in acetic acid, which serves as both the solvent and the acid catalyst. In contrast with basic halogenation, acidic halogenation can selectively replace just one hydrogen or more than one if appropriate amounts of the halogen are used.

acetophenone $+$ Br_2 $\xrightarrow{CH_3COOH}$ α-bromoacetophenone (70%) $+$ HBr

acetophenone $+$ 2 Cl_2 $\xrightarrow{CH_3COOH}$ α,α-dichloroacetophenone $+$ 2 HCl

The mechanism of acid-catalyzed halogenation involves attack of the enol form on the electrophilic halogen molecule. Loss of a proton gives the α-haloketone and the hydrogen halide.

MECHANISM   Acid-Catalyzed Halogenation

Step 1: The enol attacks the halogen.   *Step 2:* Deprotonation.

enol   halogen   carbocation intermediate α-haloketone

This reaction is similar to the attack of an alkene on a halogen, resulting in addition of the halogen across the double bond. The pi bond of an enol is more reactive toward halogens, however, because the carbocation that results is stabilized by resonance with the enol —OH group. Loss of the enol proton converts the intermediate to the product, an α-haloketone.

Unlike ketones, aldehydes are easily oxidized, and halogens are strong oxidizing agents. Attempted halogenation of aldehydes usually results in oxidation to carboxylic acids.

PROBLEM-SOLVING $HINT$

Under acidic conditions, reactions at positions α to carbonyl groups often involve the enol tautomer acting as a nucleophile.

aldehyde $+$ X_2 $+$ H_2O $\longrightarrow$ acid $+$ 2 H—X

SOLVED PROBLEM 22-2

Propose a mechanism for the acid-catalyzed conversion of cyclohexanone to 2-chlorocyclohexanone.

cyclohexanone $\xrightarrow[CH_3COOH]{Cl_2}$ 2-chlorocyclohexanone (65%)

SOLUTION

Under acid catalysis, the ketone is in equilibrium with its enol form.

keto form   stabilized intermediate   enol form

The enol acts as a weak nucleophile, attacking chlorine to give a resonance-stabilized intermediate. Loss of a proton gives the product.

PROBLEM 22-10

Propose a mechanism for the acid-catalyzed bromination of 3-pentanone.

PROBLEM 22-11

Acid-catalyzed halogenation is synthetically useful for converting ketones to α,β-unsaturated ketones, which are useful in Michael reactions (Section 22-18). Propose a method for converting cyclohexanone to 2-cyclohexenone (newer name cyclohex-2-en-1-one), an important synthetic starting material.

ketone α,β-unsaturated   cyclohexanone   2-cyclohexenone

22-4
α Bromination of Acids: The HVZ Reaction

The **Hell–Volhard–Zelinsky (HVZ) reaction** replaces a hydrogen atom with a bromine atom on the α carbon of a carboxylic acid. The carboxylic acid is treated with bromine and phosphorous tribromide, followed by water to hydrolyze the intermediate α-bromo acyl bromide.

The HVZ reaction

Example

$$CH_3CH_2CH_2-\overset{\overset{\displaystyle O}{\|}}{C}-OH \xrightarrow{Br_2/PBr_3} CH_3CH_2\overset{\overset{\displaystyle Br}{|}}{C}H-\overset{\overset{\displaystyle O}{\|}}{C}-Br$$

butanoic acid   2-bromobutanoyl bromide

$$CH_3CH_2\overset{\overset{\displaystyle Br}{|}}{C}H-\overset{\overset{\displaystyle O}{\|}}{C}-Br \xrightarrow{H_2O} CH_3CH_2\overset{\overset{\displaystyle Br}{|}}{C}H-\overset{\overset{\displaystyle O}{\|}}{C}-OH + HBr$$

2-bromobutanoyl bromide   2-bromobutanoic acid

The mechanism is similar to other acid-catalyzed α halogenations; the enol form of the acyl bromide serves as a nucleophilic intermediate. The first step is formation of acyl bromide, which enolizes more easily than does the acid.

acid   keto form   enol form

The enol is nucleophilic, attacking bromine to give the α-brominated acyl bromide.

enol

α-bromo acyl bromide

If a derivative of the α-bromoacid is desired, the α-bromo acyl bromide serves as an activated intermediate (similar to an acid chloride) for the synthesis of an ester, amide, or other derivative. If the α-bromoacid itself is needed, a water hydrolysis completes the synthesis.

PROBLEM 22-12

Show the products of the reactions of these carboxylic acids with PBr_3/Br_2 followed by hydrolysis.

(a) propanoic acid   (b) benzoic acid   (c) succinic acid   (d) oxalic acid

---

We have seen many reactions where nucleophiles attack unhindered alkyl halides and tosylates by the S_N2 mechanism. An enolate ion can serve as the nucleophile, becoming alkylated in the process. Because the enolate has two nucleophilic sites (the oxygen and the α carbon), it can react at either of these sites. The reaction usually takes place primarily at the α carbon, forming a new C—C bond. In effect, this is another type of α substitution, with an alkyl group substituting for an α hydrogen.

22-5
Alkylation of Enolate Ions

C-alkylation product
(more common)

O-alkylation product
(less common)

Typical bases such as sodium hydroxide or an alkoxide ion cannot be used to form enolates for alkylation because at equilibrium, a large quantity of the hydroxide or alkoxide base is still present. These strongly nucleophilic bases give side reactions with the alkyl halide or tosylate. Lithium diisopropylamide (LDA) avoids these side reactions. Because it is a much stronger base, LDA converts the ketone entirely to its enolate. All the LDA is consumed in forming the enolate, leaving the enolate to react without interference from the LDA. Also, LDA is a very bulky base and thus a poor nucleophile, so it generally does not react with the alkyl halide or tosylate.

Example

Direct alkylation of enolates (using LDA) gives the best yields when only one kind of α hydrogen can be replaced by an alkyl group. If there are two different kinds of α protons that may be abstracted to give enolates, mixtures of products alkylated at the different α carbons may result. Aldehydes are not suitable for direct alkylation because they undergo side reactions when treated with LDA.

PROBLEM 22-13

Predict the major products of the following reactions.

(a) acetone $\dfrac{\text{(1) LDA}}{\text{(2) CH}_2\text{=CHCH}_2\text{Br}}$

(b) $\dfrac{\text{(1) LDA}}{\text{(2) CH}_3\text{CH}_2\text{I}}$

(c) $\dfrac{\text{(1) LDA}}{\text{(2) CH}_3\text{I}}$

A milder alternative to direct alkylation of enolate ions is the formation and alkylation of an enamine derivative. An **enamine** (a **vinyl amine**) is the nitrogen analogue of an enol. The resonance picture of an enamine shows that it has some carbanion character.

22-6
Formation and Alkylation of Enamines

The electrostatic potential map (EPM) of a simple enamine shows a high negative electrostatic potential (red) near the α carbon atom of the double bond. This is the nucleophilic carbon atom of the enamine.

pyrrolidine enamine
of cyclohexanone

electrostatic potential map

An enamine is a stronger nucleophile than an enol, but still quite selective in its alkylation reactions. The nucleophilic carbon atom attacks an electrophile to give a resonance-stabilized cationic intermediate (an iminium ion).

an enamine   electrophile   minor   major

An enamine results from the reaction of a ketone or aldehyde with a secondary amine. Recall that a ketone or aldehyde reacts with a *primary* amine (Section 18-16) to form a carbinolamine, which dehydrates to give the $C{=}N$ double bond of an imine. But a carbinolamine from a *secondary* amine does not form a $C{=}N$ double bond because there is no proton on nitrogen to eliminate. A proton is lost from the α carbon, forming the $C{=}C$ double bond of an enamine.

2° amine   2° carbinolamine   no proton on N

removes α proton   enamine

Example

cyclohexanone   pyrrolidine   pyrrolidine enamine
                                           of cyclohexanone

PROBLEM 22-14

Propose a mechanism for the acid-catalyzed reaction of cyclohexanone with pyrrolidine.

Enamines displace halides from reactive alkyl halides, giving alkylated iminium salts. The iminium ions are unreactive toward further alkylation or acylation. The following example shows benzyl bromide reacting with the pyrrolidine enamine of cyclohexanone.

benzyl bromide   alkylated iminium salt   alkylated ketone

The alkylated iminium salt hydrolyzes to the alkylated ketone. The mechanism of this hydrolysis is similar to the mechanism for acid-catalyzed hydrolysis of an imine (Section 18-16).

PROBLEM 22-15

Without looking back, propose a mechanism for the hydrolysis of this iminium salt to the alkylated ketone. The first step is attack by water, followed by loss of a proton to give an amino alcohol. Protonation on nitrogen allows pyrrolidine to leave, giving the protonated ketone.

Overall reaction

enamine   iminium salt

The enamine alkylation procedure is sometimes called the **Stork reaction**, after its inventor, Gilbert Stork of Columbia University. The Stork reaction is often the best method for alkylating or acylating ketones, using a variety of reactive alkyl and acyl halides. Some halides that react well with enamines to give alkylated and acylated ketone derivatives are the following:

Ph—CH$_2$—X $>$C=C—CH$_2$—X   CH$_3$—X   R—C—Cl

benzyl halides   allylic halides   methyl halides   acyl halides

The following sequence shows the acylation of an enamine to synthesize a β-diketone. The initial acylation gives an acyl iminium salt, which hydrolyzes to the β-diketone product. As we will see in Section 22-15, β-dicarbonyl compounds are easily alkylated, and they serve as useful intermediates in the synthesis of more complicated molecules.

| enamine | acyl chloride | intermediate | acyl iminium salt | β-diketone |

PROBLEM 22-16

Give the expected products of the following acid-catalyzed reactions.
(a) acetophenone + methylamine (b) acetophenone + dimethylamine
(c) cyclohexanone + aniline (d) cyclohexanone + piperidine

PROBLEM 22-17

Show how you would accomplish each conversion using an enamine synthesis.
(a) cyclopentanone ⟶ 2-allylcyclopentanone
(b) 3-pentanone ⟶ 2-methyl-1-phenyl-3-pentanone

(c) acetophenone ⟶ Ph—C(=O)—CH$_2$—C(=O)—Ph

Condensations are some of the most important enolate reactions of carbonyl compounds. Condensations combine two or more molecules, often with the loss of a small molecule such as water or an alcohol. Under basic conditions, the **aldol condensation** involves the nucleophilic addition of an enolate ion to another carbonyl group. The product, a β-hydroxy ketone or aldehyde, is called an **aldol** because it contains both an *ald*ehyde group and the hydroxy group of an alco*hol*. The aldol product may dehydrate to an α,β-unsaturated carbonyl compound.

22-7
The Aldol Condensation of Ketones and Aldehydes

The aldol condensation

| ketone or aldehyde | aldol product | α,β-unsaturated ketone or aldehyde |

22-7A Base-Catalyzed Aldol Condensations

Under basic conditions, the aldol condensation occurs by a nucleophilic addition of the enolate ion (a strong nucleophile) to a carbonyl group. Protonation gives the aldol product. Note that the carbonyl group serves as the electrophile that is attacked by the nucleophilic enolate ion. From the electrophile's viewpoint, the reaction is a nucleophilic addition across the carbonyl double bond. From the viewpoint of the enolate ion, the reaction is an alpha substitution: The other carbonyl compound replaces an alpha hydrogen.

KEY MECHANISM   Base-Catalyzed Aldol Condensation

Step 1: Addition of the enolate.   *Step 2:* Protonation.

Consider the aldol condensation of acetaldehyde, shown below. Deprotonation of acetaldehyde gives an enolate ion, which acts as a strong nucleophile. Attack on the carbonyl group of another acetaldehyde molecule gives addition across the carbonyl double bond, forming the aldol product.

Step 1: Formation of the enolate ion

Step 2: Nucleophilic attack at the carbonyl

The aldol condensation is reversible, establishing an equilibrium between reactants and products. For acetaldehyde, conversion to the aldol product is about 50%. Ketones also undergo aldol condensation, but equilibrium concentrations of the products are generally small. Aldol condensations are sometimes accomplished by clever experimental methods. For example, Figure 22-2 shows how a good yield of the acetone aldol product ("diacetone alcohol") is obtained, even though the equilibrium concentration of the product is only about 1%. Acetone is boiled so it condenses into a chamber containing an insoluble basic catalyst. The reaction can take place only in the catalyst chamber. When the solution returns to the boiling flask, it contains about 1% diacetone alcohol. Diacetone alcohol is less volatile than acetone, remaining in the boiling flask while acetone boils and condenses (refluxes) in contact with the catalyst. After several hours, nearly all the acetone is converted to diacetone alcohol.

▲ Figure 22-2
The aldol condensation of acetone gives only 1% product at equilibrium, yet a clever technique gives a good yield. Acetone refluxes onto a basic catalyst such as Ba(OH)$_2$. The nonvolatile diacetone alcohol does not reflux, so its equilibrium concentration gradually increases until all the acetone is converted to diacetone alcohol.

SOLVED PROBLEM 22-3

Propose a mechanism for the base-catalyzed aldol condensation of acetone (Fig. 22-2).

SOLUTION

The first step is formation of the enolate to serve as a nucleophile.

The second step is a nucleophilic attack by the enolate on another molecule of acetone. Protonation gives the aldol product.

PROBLEM 22-18

Propose a mechanism for the aldol condensation of cyclohexanone. Do you expect the equilibrium to favor the reactant on the product?

PROBLEM 22-19

Give the expected products for the aldol condensations of
(a) propanal   **(b)** phenylacetaldehyde.

PROBLEM 22-20

A student wanted to dry some diacetone alcohol, and allowed it to stand over anhydrous potassium carbonate for a week. At the end of the week, the sample was found to contain nearly pure acetone. Propose a mechanism for the reaction that took place.

22-7B   Acid-Catalyzed Aldol Condensations

Aldol condensations also take place under acidic conditions. The enol serves as a weak nucleophile to attack an activated (protonated) carbonyl group. As an example, consider the acid-catalyzed aldol condensation of acetaldehyde. The first step is formation of the enol by the acid-catalyzed keto–enol tautomerism, as discussed earlier. The enol attacks the protonated carbonyl of another acetaldehyde molecule. Loss of the enol proton gives the aldol product.

MECHANISM   Acid-Catalyzed Aldol Condensation

Step 1: Addition of the enol to the protonated carbonyl.   *Step 2:* Deprotonation.

attack by enol   resonance-stabilized intermediate   aldol product

---

PROBLEM 22-21

Propose a complete mechanism for the acid-catalyzed aldol condensation of acetone.

22-8
Dehydration
of Aldol Products

Heating a basic or acidic mixture of an aldol product leads to dehydration of the alcohol functional group. The product is a conjugated α,β-unsaturated aldehyde or ketone. Thus, an aldol condensation, followed by dehydration, forms a *new carbon–carbon double bond*. Before the Wittig reaction was discovered (Section 18-13), the aldol with dehydration was probably the best method for joining two molecules with a double bond. It is still often the cheapest and easiest method.

diacetone alcohol   4-methyl-3-penten-2-one
                                       (mesityl oxide)

Under acidic conditions, dehydration follows a mechanism similar to those of other acid-catalyzed alcohol dehydrations (Section 11-10). We have not previously seen a base-catalyzed dehydration, however. Base-catalyzed dehydration depends on the acidity of the α proton of the aldol product. Abstraction of an α proton gives an enolate that can expel hydroxide ion to give a more stable product. Hydroxide is not a good leaving group in an E2 elimination, but it can serve as a leaving group in a strongly exothermic step like this one, which stabilizes a negatively charged intermediate. The following mechanism shows the base-catalyzed dehydration of 3-hydroxybutanal:

KEY MECHANISM   Dehydration of an Aldol

Step 1: Formation of the enolate.

Step 2: Elimination of hydroxide.

removal of α proton

loss of hydroxide

conjugated system

Even when the aldol equilibrium is unfavorable for formation of a β-hydroxy ketone or aldehyde, the dehydration product may be obtained in good yield by heating the reaction mixture. Dehydration is usually exothermic because it leads to a conjugated system. In effect, the exothermic dehydration drives the aldol equilibrium to the right.

PROBLEM 22-22

Propose a mechanism for the dehydration of diacetone alcohol to mesityl oxide.
(a) in acid   (b) in base

PROBLEM 22-23

When propionaldehyde is warmed with sodium hydroxide, one of the products is 2-methyl-2-pentenal. Propose a mechanism for this reaction.

PROBLEM 22-24

Predict the products of aldol condensation, followed by dehydration, of the following ketones and aldehydes.
(a) butyraldehyde   (b) acetophenone   (c) cyclohexanone

> Aldolases are enzymes that form aldol products, most commonly in the metabolism of carbohydrates or sugars. In contrast to the chemical reaction, aldolases generate just one product stereospecifically. Hence, they are sometimes used in organic synthesis for key transformations.

22-9
Crossed Aldol Condensations

When the enolate of one aldehyde (or ketone) adds to the carbonyl group of another, the result is called a **crossed aldol condensation**. The compounds used in the reaction must be selected carefully, or a mixture of several products will be formed.

Consider the aldol condensation between acetaldehyde (ethanal) and propanal shown on page 1022. Either of these reagents can form an enolate ion. Attack by the enolate of ethanal on propanal gives a product different from the one formed by attack of the enolate of propanal on ethanal. Also, self-condensations of ethanal and propanal continue to take place. Depending on the reaction conditions, various proportions of the four possible products result.

Enolate of ethanal adds to propanal

$$CH_3CH_2—\overset{\overset{O}{\|}}{C}—H \;\;\rightleftharpoons\;\; CH_3CH_2—\overset{\overset{OH}{|}}{\underset{\underset{CH_2—CHO}{|}}{C}}—H$$

$$\underset{\;\;^-:CH_2—CHO}{}$$

Enolate of propanal adds to ethanal

$$CH_3—\overset{\overset{O}{\|}}{C}—H \;\;\rightleftharpoons\;\; CH_3—\overset{\overset{OH}{|}}{\underset{\underset{CH_3—CH—CHO}{}}{C}}—H$$

$$\underset{CH_3—\overset{..}{C}H—CHO}{}$$

Self-condensation of ethanal

$$CH_3—\overset{\overset{O}{\|}}{C}—H \;\;\rightleftharpoons\;\; CH_3—\overset{\overset{OH}{|}}{\underset{\underset{CH_2—CHO}{|}}{C}}—H$$

$$\underset{\;\;^-:CH_2—CHO}{}$$

Self-condensation of propanal

$$CH_3CH_2—\overset{\overset{O}{\|}}{C}—H \;\;\rightleftharpoons\;\; CH_3CH_2—\overset{\overset{OH}{|}}{\underset{\underset{CH_3—CH—CHO}{}}{C}}—H$$

$$\underset{CH_3—\overset{..}{C}H—CHO}{}$$

A crossed aldol condensation can be effective if it is planned so that only one of the reactants can form an enolate ion and so that the other compound is more likely to react with the enolate. If only one of the reactants has an α hydrogen, only one enolate will be present in the solution. If the other reactant is present in excess or contains a particularly electrophilic carbonyl group, it is more likely to be attacked by the enolate ion.

The following two reactions are successful crossed aldol condensations. The aldol products may or may not undergo dehydration, depending on the reaction conditions and the structure of the products.

To carry out these reactions, slowly add the compound with α protons to a basic solution of the compound with no α protons. This way, the enolate ion is formed in the presence of a large excess of the other component, and the desired reaction is favored.

PROBLEM-SOLVING STRATEGY
Proposing Reaction Mechanisms

The general principles for proposing reaction mechanisms, first introduced in Chapter 4 and summarized in Appendix 4, are applied here to a crossed aldol condensation. This example emphasizes a base-catalyzed reaction involving strong nucleophiles. In drawing mechanisms, be careful to draw all the bonds and substituents of each carbon atom involved. Show each step separately, and draw curved arrows to show the movement of electrons from the nucleophile to the electrophile.

Our problem is to propose a mechanism for the base-catalyzed reaction of methylcyclohexanone with benzaldehyde:

First, we must determine the type of mechanism. Sodium ethoxide, a strong base and a strong nucleophile, implies the reaction involves strong nucleophiles as intermediates. We expect to see strong nucleophiles and anionic intermediates (possibly stabilized carbanions), but no strong electrophiles or strong acids, and certainly no carbocations or free radicals.

1. **Consider the carbon skeletons of the reactants and products, and decide which carbon atoms in the products are likely derived from which carbon atoms in the reactants.**
   Because one of the rings is aromatic, it is clear which ring in the product is derived from which ring in the reactants. The carbon atom that bridges the two rings in the products must be derived from the carbonyl group of benzaldehyde. The two α protons from methylcyclohexanone and the carbonyl oxygen are lost as water.

2. **Consider whether any of the reactants is a strong enough nucleophile to react without being activated. If not, consider how one of the reactants might be converted to a strong nucleophile by deprotonation of an acidic site or by attack on an electrophilic site.**
   Neither of these reactants is a strong enough nucleophile to attack the other. If ethoxide removes an α proton from methylcyclohexanone, however, a strongly nucleophilic enolate ion results.

3. **Consider how an electrophilic site on another reactant (or, in a cyclization, another part of the same molecule) can undergo attack by the strong nucleophile to form a bond needed in the product. Draw the product of this bond formation.**
   Attack at the electrophilic carbonyl group of benzaldehyde, followed by protonation, gives a β-hydroxy ketone (an aldol).

aldol

4. Consider how the product of nucleophilic attack might be converted to the final product (if it has the right carbon skeleton) or reactivated to form another bond needed in the product.

The β-hydroxy ketone must be dehydrated to give the final product. Under these basic conditions, the usual alcohol dehydration mechanism (protonation of hydroxyl, followed by loss of water) cannot occur. Removal of another proton gives an enolate ion that can lose hydroxide in a strongly exothermic step to give the final product.

aldol   enolate   dehydrated

5. Draw out all the steps using curved arrows to show the movement of electrons. Be careful to show only one step at a time.

The complete mechanism is given by combining the equations shown above. We suggest you write out the mechanism as a review of the steps involved.

As further practice in proposing mechanisms for base-catalyzed reactions, do Problems 22-25 and 22-26B, using the steps shown above.

PROBLEM 22-25

Propose mechanisms for the following base-catalyzed condensations:
(a) 2,2-dimethylpropanal with acetaldehyde
(b) benzaldehyde with propionaldehyde

PROBLEM 22-26

When acetone is treated with excess benzaldehyde in the presence of base, the crossed condensation adds two equivalents of benzaldehyde and expels two equivalents of water. Propose a structure for the condensation product of acetone with two molecules of benzaldehyde.

PROBLEM-SOLVING HINT

Practice predicting the structures of aldol products (before and after dehydration) and drawing the mechanisms. These reactions are among the most important in this chapter.

PROBLEM 22-27

In the problem-solving feature above, methylcyclohexanone was seen to react at its *unsubstituted* α carbon. Try to write a mechanism for the same reaction at the methyl-substituted carbon atom, and explain why this regiochemistry is not observed.

PROBLEM 22-28

Predict the major products of the following base-catalyzed aldol condensations with dehydration.
(a) benzophenone (PhCOPh) + propionaldehyde
(b) 2,2-dimethylpropanal + acetophenone

PROBLEM 22-29

Cinnamaldehyde is used as a flavoring agent in cinnamon candies. Show how cinnamaldehyde is synthesized by a crossed aldol condensation followed by dehydration.

cinnamaldehyde:

Intramolecular aldol reactions of diketones are often useful for making five- and six-membered rings. Aldol cyclizations of rings larger than six and smaller than five are less common because larger and smaller rings are less favored by their energy and entropy. The following reactions show how a 1,4-diketone can condense and dehydrate to give a cyclopentenone and how a 1,5-diketone gives a cyclohexenone.

22-10
Aldol Cyclizations

enolate of a1,4-diketone   aldol product   a cyclopentenone   + H_2O

Example

cis-8-undecene-2,5-dione   aldol product   *cis*-jasmone (a perfume)   + H_2O
(90%)

enolate of a 1,5-diketone   aldol product   a cyclohexenone   + H_2O

Example

2,6-heptanedione   aldol product   3-methylcyclohex-2-enone   + H_2O
(a 1,5-diketone)

The following example shows how the carbonyl group of the product may be outside the ring in some cases.

2,7-octanedione   aldol product   1-acetyl-2-methylcyclopentene   + H_2O

PROBLEM 22-30

Show how 2,7-octanedione might cyclize to a cycloheptenone. Explain why ring closure to the cycloheptenone is not observed.

PROBLEM 22-31

When 1,6-cyclodecanedione is treated with sodium carbonate, the product gives a UV spectrum similar to that of 1-acetyl-2-methylcyclopentene. Propose a structure for the product, and give a mechanism for its formation.

1,6-cyclodecanedione

22-11
Planning Syntheses Using Aldol Condensations

As long as we remember their limitations, aldol condensations can serve as useful synthetic reactions for making a variety of organic compounds. In particular, aldol condensations (with dehydration) form new carbon–carbon double bonds. We can use some general principles to decide whether a compound might be an aldol product and which reagents to use as starting materials.

Aldol condensations produce β-hydroxy aldehydes and ketones (aldols) and α,β-unsaturated aldehydes and ketones. If a target molecule has one of these functionalities, an aldol should be considered. To determine the starting materials, divide the structure at the α,β bond. In the case of the dehydrated product, the α,β bond is the double bond. Figure 22-3 shows the division of some aldol products into their starting materials.

PROBLEM 22-32

Show how each compound can be dissected into reagents joined by an aldol condensation, then decide whether the necessary aldol condensation is feasible.

PROBLEM 22-33

The following compound results from base-catalyzed aldol cyclization of a 2-substituted cyclohexanone.
(a) Show the diketone that would cyclize to give this product.
(b) Propose a mechanism for the cyclization.

▲ **Figure 22-3**

Aldol products are β-hydroxyl aldehydes and ketones, or α,β-unsaturated aldehydes and ketones. An aldol product is dissected into its starting materials by mentally breaking the α,β bond.

The α hydrogens of esters are weakly acidic, and they can be deprotonated to give enolate ions. Esters are less acidic than ketones and aldehydes because the ester carbonyl group is stabilized by resonance with the other oxygen atom. This resonance makes the carbonyl group less capable of stabilizing the negative charge of an enolate ion.

22-12
The Claisen Ester Condensation

A typical pK_a for an α proton of an ester is about 24, compared with a pK_a of about 20 for a ketone or aldehyde. Even so, strong bases do deprotonate esters.

Ester enolates are strong nucleophiles, and they undergo a wide range of interesting and useful reactions. Most of these reactions are related to the Claisen condensation, the most important of all ester condensations.

The **Claisen condensation** results when an ester molecule undergoes nucleophilic acyl substitution by an enolate. The intermediate has an alkoxy (—OR) group that acts as a leaving group, leaving a β-keto ester. The overall reaction combines two ester molecules to give a β-keto ester.

KEY MECHANISM   Claisen Condensation

Step 1: Addition of the enolate.   *Step 2:* Elimination of the alkoxide leaving group.

ester enolate   intermediate   a β-keto ester

Notice that one molecule of the ester (deprotonated, reacting as the enolate) serves as the nucleophile to attack another molecule of the ester, which serves as the acylating reagent in this nucleophilic acyl substitution.

β-Keto esters are more acidic than simple ketones, aldehydes, and esters because the negative charge of the enolate is delocalized over both carbonyl groups. β-Keto esters have pK_a values around 11, showing they are stronger acids than water. In strong base such as ethoxide ion or hydroxide ion, the β-keto ester is rapidly and completely deprotonated.

a β-keto ester
(pK_a = 11)

resonance-stabilized enolate ion

Deprotonation of the β-keto ester provides a driving force for the Claisen condensation. The deprotonation is strongly exothermic, making the overall reaction exothermic and driving the reaction to completion. Because the base is consumed in the deprotonation step, a full equivalent of base must be used, and the Claisen condensation is said to be *base-promoted* rather than *base-catalyzed*. After the reaction is complete, addition of dilute acid converts the enolate back to the β-keto ester.

The following example shows the self-condensation of ethyl acetate to give ethyl acetoacetate (ethyl 3-oxobutanoate). Ethoxide is used as the base to avoid transesterification or hydrolysis of the ethyl ester (see Problem 22-34). The initial product is the enolate of ethyl acetoacetate, which is reprotonated in the final step.

2 CH_3—C—OCH_2CH_3 $\xrightarrow{Na^+ \ ^-OCH_2CH_3}$ $Na^+ \ ^-:CH$—C—OCH_2CH_3 $\xrightarrow{H_3O^+}$ CH_2—C—OCH_2CH_3

ethyl acetate   sodium ethoxide   keto ester enolate   ethyl acetoacetate (75%)

SOLVED PROBLEM 22-4

Propose a mechanism for the self-condensation of ethyl acetate to give ethyl acetoacetate.

SOLUTION

The first step is formation of the ester enolate. The equilibrium for this step lies far to the left; ethoxide deprotonates only a small fraction of the ester.

$$
\underset{(pK_a = 24)}{\overset{H \quad \ddot{O}}{\underset{\vert \quad \Vert}{CH_2-C-OCH_2CH_3}}} + \ {}^-\!\!:\ddot{O}CH_2CH_3 \ \rightleftharpoons \ \underset{\text{enolate}}{{}^-\!\!:CH_2-\overset{O}{\overset{\Vert}{C}}-OCH_2CH_3} + \underset{(pK_a = 16)}{H-\ddot{O}CH_2CH_3}
$$

The enolate ion attacks another molecule of the ester; expulsion of ethoxide ion gives ethyl acetoacetate.

$$
\underset{\substack{{}^-\!\!:CH_2-\overset{O}{\overset{\Vert}{C}}-OCH_2CH_3 \\ \text{nucleophilic attack}}}{CH_3-\overset{\ddot{O}}{\overset{\Vert}{C}}-OCH_2CH_3} \ \rightleftharpoons \ \underset{\substack{CH_2-\overset{O}{\overset{\Vert}{C}}-OCH_2CH_3 \\ \text{expulsion of ethoxide}}}{CH_3-\overset{:\ddot{O}:^-}{\overset{\vert}{C}}-\ddot{O}CH_2CH_3} \ \rightleftharpoons \ \underset{\substack{CH_2-\overset{O}{\overset{\Vert}{C}}-OCH_2CH_3 \\ \text{ethyl acetoacetate}}}{CH_3-C\overset{\ddot{O}^-}{\diagup}\ \ {}^-\!\!:\ddot{O}CH_2CH_3}
$$

In the presence of ethoxide ion, ethyl acetoacetate is deprotonated to give its enolate. This exothermic deprotonation helps to drive the reaction to completion.

$$
\underset{(pK_a = 11)}{\overset{O \quad H \quad O}{\underset{\Vert \quad \vert \quad \Vert}{CH_3-C-CH-C-OCH_2CH_3}}} + \ {}^-\!\!:\ddot{O}CH_2CH_3 \ \rightleftharpoons \ \underset{\text{enolate}}{\overset{O \quad \quad O}{\underset{\Vert \quad \quad \Vert}{CH_3-C-\ddot{C}H-C-OCH_2CH_3}}} + \underset{(pK_a = 16)}{H-\ddot{O}CH_2CH_3}
$$

When the reaction is complete, the enolate ion is reprotonated to give ethyl acetoacetate.

$$
\underset{\text{enolate}}{\overset{O \quad \quad O}{\underset{\Vert \quad \quad \Vert}{CH_3-C-\ddot{C}H-C-OCH_2CH_3}}} \ \xrightarrow{\ H_3O^+\ } \ \underset{\text{ethyl acetoacetate}}{\overset{O \quad H \quad O}{\underset{\Vert \quad \vert \quad \Vert}{CH_3-C-CH-C-OCH_2CH_3}}}
$$

> Enzymes called polyketide synthases catalyze a series of Claisen-type reactions to generate many useful natural products such as the antibiotic erythromycin (page 983). These enzymes use thioesters instead of the oxygen esters.

PROBLEM 22-34

Ethoxide is used as the base in the condensation of ethyl acetate to avoid some unwanted side reactions. Show what side reactions would occur if the following bases were used.
(a) sodium methoxide   (b) sodium hydroxide

PROBLEM 22-35

Esters with only one α hydrogen generally give poor yields in the Claisen condensation. Propose a mechanism for the Claisen condensation of ethyl isobutyrate, and explain why a poor yield is obtained.

PROBLEM 22-36

Predict the products of self-condensation of the following esters.
(a) methyl propanoate + NaOCH$_3$
(b) ethyl phenylacetate + NaOCH$_2$CH$_3$

(c) [cyclopentyl]$-CH_2-\overset{O}{\overset{\Vert}{C}}-OCH_3$ + NaOCH$_3$

SOLVED PROBLEM 22-5

Show what ester would undergo Claisen condensation to give the following β-keto ester.

$$Ph-CH_2-CH_2-\overset{\overset{\displaystyle O}{\|}}{C}-\underset{\underset{\displaystyle CH_2-Ph}{|}}{CH}-\overset{\overset{\displaystyle O}{\|}}{C}-OCH_3$$

SOLUTION

First, break the structure apart at the α,β bond (α,β to the ester carbonyl). This is the bond formed in the Claisen condensation.

$$Ph-CH_2-CH_2-\overset{\overset{\displaystyle O}{\|}}{\underset{\beta}{C}}-\quad\quad\underset{\underset{\displaystyle CH_2-Ph}{|}}{\overset{\alpha}{CH}}-\overset{\overset{\displaystyle O}{\|}}{C}-OCH_3$$

Next, replace the α proton that was lost, and replace the alkoxy group that was lost from the carbonyl. Two molecules of methyl 3-phenylpropionate result.

$$Ph-CH_2-CH_2-\overset{\overset{\displaystyle O}{\|}}{C}-OCH_3 \quad\quad H-\underset{\underset{\displaystyle CH_2-Ph}{|}}{CH}-\overset{\overset{\displaystyle O}{\|}}{C}-OCH_3$$

Now draw out the reaction. Sodium methoxide is used as the base because the reactants are methyl esters.

$$2\ Ph-CH_2-CH_2-\overset{\overset{\displaystyle O}{\|}}{C}-OCH_3 \xrightarrow[\text{(2) H}_3O^+]{\text{(1) Na}^+\ ^-OCH_3} Ph-CH_2-CH_2-\overset{\overset{\displaystyle O}{\|}}{C}-\underset{\underset{\displaystyle CH_2-Ph}{|}}{CH}-\overset{\overset{\displaystyle O}{\|}}{C}-OCH_3$$

PROBLEM-SOLVING HINT

The Claisen condensation occurs by a nucleophilic acyl substitution, with different forms of the ester acting as both the nucleophile (the enolate) and the electrophile (the ester carbonyl).

PROBLEM 22-37

Propose a mechanism for the self-condensation of methyl 3-phenylpropionate catalyzed by sodium methoxide.

PROBLEM 22-38

Show what esters would undergo Claisen condensation to give the following β-keto esters.

(a) $CH_3CH_2CH_2-\overset{\overset{\displaystyle O}{\diagup}}{C}$ ... $CH_3CH_2-\underset{}{CH}-\overset{\overset{\displaystyle O}{\|}}{C}-OCH_2CH_3$

(b) $Ph-CH_2-\overset{\overset{\displaystyle O}{\diagup}}{C}$ ... $Ph-CH-\overset{\overset{\displaystyle O}{\|}}{C}-OCH_3$

(c) ... $CH_2CH_2-\overset{\overset{\displaystyle O}{\diagup}}{C}$ ... $CH-\overset{\overset{\displaystyle O}{\|}}{C}-OCH_3$ ... CH_2

(d) ...

An internal Claisen condensation of a diester forms a ring. Such an internal Claisen cyclization is called a **Dieckmann condensation** or a **Dieckmann cyclization**. Five- and six-membered rings are easily formed by Dieckmann condensations. Rings larger than six carbons or smaller than five carbons are rarely formed by this method.

The following examples of the Dieckmann condensation show that a 1,6-diester gives a five-membered ring, and a 1,7-diester gives a six-membered ring.

22-13
The Dieckmann Condensation: A Claisen Cyclization

diethyl adipate
(a 1,6-diester)

cyclic β-keto ester
(80%)

dimethyl pimelate
(a 1,7-diester)

cyclic β-keto ester

PROBLEM 22-39

Propose mechanisms for the two Dieckmann condensations shown above.

PROBLEM 22-40

Some of the following keto esters can be formed by Dieckmann condensations, but others cannot. Determine which ones are possible, and draw the starting diesters.

(a)

(b)

(c)

(d)

(*Hint:* Consider using a protecting group.)

Claisen condensations can take place between different esters, particularly when only one of the esters has the α hydrogens needed to form an enolate. In a **crossed Claisen condensation**, an ester without α hydrogens serves as the electrophilic component. Some useful esters without α hydrogens are benzoate, formate, carbonate, and oxalate esters.

22-14
Crossed Claisen Condensations

$$\underset{\text{methyl benzoate}}{\bigcirc\!\!-\!\!\overset{\overset{\displaystyle O}{\|}}{C}\!\!-\!\!OCH_3} \qquad \underset{\text{methyl formate}}{H\!\!-\!\!\overset{\overset{\displaystyle O}{\|}}{C}\!\!-\!\!OCH_3} \qquad \underset{\text{dimethyl carbonate}}{CH_3O\!\!-\!\!\overset{\overset{\displaystyle O}{\|}}{C}\!\!-\!\!OCH_3} \qquad \underset{\text{dimethyl oxalate}}{CH_3O\!\!-\!\!\overset{\overset{\displaystyle O}{\|}}{C}\!\!-\!\!\overset{\overset{\displaystyle O}{\|}}{C}\!\!-\!\!OCH_3}$$

A crossed Claisen condensation is carried out by first adding the ester without α hydrogens to a solution of the alkoxide base. The ester with α hydrogens is slowly added to this solution, where it forms an enolate and condenses. The condensation of ethyl acetate with ethyl benzoate is an example of a crossed Claisen condensation.

$$\underset{\substack{\text{ethyl benzoate}\\(\text{no }\alpha\text{ hydrogens})}}{\bigcirc\!\!-\!\!\overset{\overset{\displaystyle O}{\|}}{C}\!\!-\!\!OC_2H_5} + \underset{\substack{\text{ethyl acetate}\\(\text{forms enolate})}}{\overset{\alpha}{CH_3}\!\!-\!\!\overset{\overset{\displaystyle O}{\|}}{C}\!\!-\!\!OC_2H_5} \xrightarrow[\text{(2) H}_3\text{O}^+]{\text{(1) Na}^+\,^-\text{OCH}_2\text{CH}_3} \underset{\text{ethyl benzoylacetate}}{\bigcirc\!\!-\!\!\overset{\overset{\displaystyle O}{\|}}{C}\!\!-\!\!\underset{\beta}{\overset{\alpha}{CH_2}}\!\!-\!\!\overset{\overset{\displaystyle O}{\|}}{C}\!\!-\!\!OC_2H_5}$$

PROBLEM 22-41

Propose a mechanism for the crossed Claisen condensation between ethyl acetate and ethyl benzoate.

PROBLEM 22-42

Predict the products from crossed Claisen condensation of the following pairs of esters. Indicate which combinations are poor choices for crossed Claisen condensations.

(a) $Ph\!\!-\!\!CH_2\!\!-\!\!\overset{\overset{\displaystyle O}{\|}}{C}\!\!-\!\!OCH_3$ + $Ph\!\!-\!\!\overset{\overset{\displaystyle O}{\|}}{C}\!\!-\!\!OCH_3 \longrightarrow$

(b) $Ph\!\!-\!\!CH_2\!\!-\!\!\overset{\overset{\displaystyle O}{\|}}{C}\!\!-\!\!OCH_3$ + $CH_3\!\!-\!\!\overset{\overset{\displaystyle O}{\|}}{C}\!\!-\!\!OCH_3 \longrightarrow$

(c) $CH_3\!\!-\!\!\overset{\overset{\displaystyle O}{\|}}{C}\!\!-\!\!OC_2H_5$ + $C_2H_5O\!\!-\!\!\overset{\overset{\displaystyle O}{\|}}{C}\!\!-\!\!\overset{\overset{\displaystyle O}{\|}}{C}\!\!-\!\!OC_2H_5 \longrightarrow$

(d) $CH_3\!\!-\!\!CH_2\!\!-\!\!\overset{\overset{\displaystyle O}{\|}}{C}\!\!-\!\!OC_2H_5$ + $C_2H_5O\!\!-\!\!\overset{\overset{\displaystyle O}{\|}}{C}\!\!-\!\!OC_2H_5 \longrightarrow$

> Fatty acids are made in the body by a series of series of Claisen-type reactions catalyzed by an enzyme called a fatty acid synthase. The enzyme uses the thioesters of malonate and acetate as building blocks (see page 1038).

SOLVED PROBLEM 22-6

Show how a crossed Claisen condensation might be used to prepare

$$H\!\!-\!\!\overset{\overset{\displaystyle O}{\|}}{C}\!\!-\!\!\underset{\underset{\displaystyle Ph}{|}}{CH}\!\!-\!\!\overset{\overset{\displaystyle O}{\|}}{C}\!\!-\!\!OCH_3$$

SOLUTION

Break the α,β bond of this β-keto ester, since that is the bond formed in the Claisen condensation.

$$H\!\!-\!\!\underset{\beta}{\overset{\overset{\displaystyle O}{\|}}{C}}\!\!-\!\!\Big| \qquad \Big|\!\!-\!\!\underset{\underset{\displaystyle Ph}{|}}{\overset{\alpha}{CH}}\!\!-\!\!\overset{\overset{\displaystyle O}{\|}}{C}\!\!-\!\!OCH_3$$

Now add the alkoxy group to the carbonyl and replace the proton on the α carbon.

$$H-\overset{\overset{\displaystyle O}{\|}}{C}-OCH_3 \qquad H-\overset{\overset{\displaystyle O}{\|}}{\underset{\underset{\displaystyle Ph}{|}}{C}}H-\overset{\overset{\displaystyle O}{\|}}{C}-OCH_3$$

Write out the reaction, making sure that one of the components has α hydrogens and the other does not.

$$H-\overset{\overset{\displaystyle O}{\|}}{C}-OCH_3 + H-\underset{\underset{\displaystyle Ph}{|}}{C}H-\overset{\overset{\displaystyle O}{\|}}{C}-OCH_3 \xrightarrow[\text{(2) } H_3O^+]{\text{(1) } Na^+ \ ^-OCH_3} H-\overset{\overset{\displaystyle O}{\|}}{C}-\underset{\underset{\displaystyle Ph}{|}}{C}H-\overset{\overset{\displaystyle O}{\|}}{C}-OCH_3$$

no α hydrogens   forms enolate

PROBLEM 22-43

Show how crossed Claisen condensations could be used to prepare the following esters.

(a) $Ph-\overset{\overset{\displaystyle O}{\|}}{C}-\underset{\underset{\displaystyle CH_3}{|}}{C}H-\overset{\overset{\displaystyle O}{\|}}{C}-OCH_2CH_3$

(b) $Ph-\underset{\underset{\displaystyle \overset{\displaystyle O}{\|}}{\underset{\displaystyle C}{\|}}-\overset{\overset{\displaystyle O}{\|}}{C}-OCH_3}{C}H-\overset{\overset{\displaystyle O}{\|}}{C}-OCH_3$

(c) $EtO-\overset{\overset{\displaystyle O}{\|}}{C}-\underset{\underset{\displaystyle Ph}{|}}{C}H-\overset{\overset{\displaystyle O}{\|}}{C}-OCH_2CH_3$

(d) $(CH_3)_3C-\overset{\overset{\displaystyle O}{\|}}{C}-\underset{\underset{\displaystyle CH_2CH_2CH_3}{|}}{C}H-\overset{\overset{\displaystyle O}{\|}}{C}-OCH_3$

Crossed Claisen condensations between ketones and esters are also possible. Ketones are more acidic than esters, and the ketone component is more likely to deprotonate and serve as the enolate component in the condensation. The ketone enolate attacks the ester, which undergoes nucleophilic acyl substitution and thereby acylates the ketone.

$$R-CH_2-\overset{\overset{\displaystyle O}{\|}}{C}-R' \qquad R-CH_2-\overset{\overset{\displaystyle O}{\|}}{C}-OR'$$
ketone, $pK_a = 20$ ester, $pK_a = 24$
more acidic   less acidic

$$-\overset{\overset{\displaystyle O}{\|}}{C}-\overset{\alpha}{C}: \quad R-\overset{\overset{\displaystyle O}{\|}}{C}-OR' \rightleftharpoons -\overset{\overset{\displaystyle O}{\|}}{C}-\overset{\alpha}{\underset{|}{C}}\overset{\beta}{\underset{\underset{\displaystyle OR'}{|}}{C}}-O^- \longrightarrow -\overset{\overset{\displaystyle O}{\|}}{C}-\overset{\alpha}{\underset{|}{C}}-\overset{\beta}{C}\overset{R}{\underset{\displaystyle O}{\diagdown}}$$

ketone enolate   ester   tetrahedral intermediate   acylated ketone

This condensation works best if the ester has no α hydrogens, so that it cannot form an enolate. Because of the difference in acidities, however, the reaction is often successful between ketones and esters even when both have α hydrogens. The following examples show some crossed Claisen condensations between ketones and esters. Notice the variety of difunctional and trifunctional compounds that can be produced by appropriate choices of esters.

$$CH_3-\overset{\overset{\displaystyle O}{\|}}{C}-CH_3 \;+\; \text{(phenyl)}-\overset{\overset{\displaystyle O}{\|}}{C}-OCH_3 \quad\xrightarrow[\text{(2) } H_3O^+]{\text{(1) Na}^+ \text{ }^-OCH_3}\quad \text{(phenyl)}-\overset{\overset{\displaystyle O}{\|}}{C}-\overset{\beta}{CH_2}-\overset{\overset{\displaystyle O}{\|}}{C}-CH_3$$

| acetone | methyl benzoate | | a β-diketone |

| cyclohexanone | diethyl carbonate | a β-keto ester |

| cyclopentanone | diethyl oxalate | a diketo ester |

PROBLEM-SOLVING HINT

Claisen and crossed Claisen condensations are important synthetic tools and interesting mechanistic examples. Practice predicting product structures and drawing mechanisms until you gain confidence.

PROBLEM 22-44

Predict the major products of the following crossed Claisen condensations.

(a)

(b) $CH_3-\overset{\overset{\displaystyle O}{\|}}{C}-CH_3 \;+\; CH_3-\overset{\overset{\displaystyle O}{\|}}{C}-OCH_3 \quad\xrightarrow{\text{NaOCH}_3}$

(c) $CH_3-\overset{\overset{\displaystyle O}{\|}}{C}-CH_2CH_2-\overset{\overset{\displaystyle O}{\|}}{C}-OCH_2CH_3 \quad\xrightarrow{\text{NaOCH}_2CH_3}$

PROBLEM 22-45

Show how Claisen condensations could be used to make the following compounds.

(a)

(b)

(c)

(d)

22-15

Syntheses Using β-Dicarbonyl Compounds

Many alkylation and acylation reactions are most effective using anions of β-dicarbonyl compounds that can be completely deprotonated and converted to their enolate ions by common bases such as methoxide ion and ethoxide ion. The *malonic ester synthesis* and the *acetoacetic ester synthesis* use the enhanced acidity of the α protons in malonic ester and acetoacetic ester to accomplish alkylations and acylations that are difficult or impossible with simple esters. We have seen that most ester condensations use alkoxides to form enolate ions. With simple esters, only a small amount of enolate is formed; the equilibrium favors the alkoxide and the ester. The alkoxide often interferes with the desired reaction. For example, if we want an alkyl halide to alkylate an enolate, alkoxide ion in the solution will attack the alkyl halide and form an ether.

$$R-\overset{..}{\underset{..}{O}}:^- \;+\; \underset{\substack{| \\ }}{\overset{\substack{H \quad O \\ | \quad \parallel}}{-C-C-OR}} \;\rightleftharpoons\; ROH \;+\; \underset{\substack{| \\ }}{\overset{\substack{O \\ \parallel}}{-\overset{..}{C}-C}}-OR$$

>99%   <1%

add alkylating agent $R'\!-\!X \;\longrightarrow\; R-\overset{..}{\underset{..}{O}}-R' \;+\; :X^-$

In contrast, β-dicarbonyl compounds, such as malonic ester and acetoacetic ester, are more acidic than alcohols. They are completely deprotonated by alkoxides, and the resulting enolates are easily alkylated and acylated. At the end of the synthesis, one of the carbonyl groups can be removed by decarboxylation, leaving a compound that is difficult or impossible to make by direct alkylation or acylation of a simple ester.

$$CH_3CH_2O-\overset{\substack{O \\ \parallel}}{C}-CH_2-\overset{\substack{O \\ \parallel}}{C}-OCH_2CH_3 \qquad CH_3-\overset{\substack{O \\ \parallel}}{C}-CH_2-\overset{\substack{O \\ \parallel}}{C}-OCH_2CH_3$$

diethyl malonate (malonic ester)   ethyl acetoacetate (acetoacetic ester)

First we compare the acidity advantages of β-dicarbonyl compounds, and then we consider how these compounds are used in synthesis.

Acidities of β-Dicarbonyl Compounds   Table 22-1 compares the acidities of some carbonyl compounds with the acidities of alcohols and water. Notice the large increase in

TABLE 22-1 Typical Acidities of Carbonyl Compounds

Conjugate Acid	Conjugate Base	pK_a
Commonly used bases		
H—O—H water	^-OH	15.7
CH_3O-H methanol	CH_3O^-	15.5
CH_3CH_2O-H ethanol	$CH_3CH_2O^-$	15.9
Simple ketones and esters		
$\overset{\alpha}{CH_3}-\overset{\substack{O \\ \parallel}}{C}-CH_3$ acetone	$^-:CH_2-\overset{\substack{O \\ \parallel}}{C}-CH_3$	20
$\overset{\alpha}{CH_3}-\overset{\substack{O \\ \parallel}}{C}-OCH_2CH_3$ ethyl acetate	$^-:CH_2-\overset{\substack{O \\ \parallel}}{C}-OCH_2CH_3$	24
β-dicarbonyl compounds		
$CH_3CH_2O-\overset{\substack{O \\ \parallel \\ \beta}}{C}-\overset{\alpha}{CH_2}-\overset{\substack{O \\ \parallel}}{C}-OCH_2CH_3$ diethyl malonate (malonic ester)	$CH_3CH_2O-\overset{\substack{O \\ \parallel}}{C}-\overset{..}{\underset{}{C}H}-\overset{\substack{O \\ \parallel}}{C}-OCH_2CH_3$	13
$CH_3-\overset{\substack{O \\ \parallel \\ \beta}}{C}-\overset{\alpha}{CH_2}-\overset{\substack{O \\ \parallel}}{C}-OCH_2CH_3$ ethyl acetoacetate (acetoacetic ester)	$CH_3-\overset{\substack{O \\ \parallel}}{C}-\overset{..}{\underset{}{C}H}-\overset{\substack{O \\ \parallel}}{C}-OCH_2CH_3$	11

acidity for compounds with two carbonyl groups beta to each other. The α protons of the β-dicarbonyl compounds are more acidic than the hydroxyl protons of water and alcohols. This enhanced acidity results from increased stability of the enolate ion. The negative charge is delocalized over two carbonyl groups rather than just one, as shown by the resonance forms for the enolate ion of diethyl malonate (also called *malonic ester*).

$$CH_3CH_2O-\overset{\overset{O}{\|}}{C}\underset{\underset{H \quad H}{C}}{}\overset{\overset{O}{\|}}{C}-OCH_2CH_3 \quad + \quad {}^-{:}\ddot{O}CH_2CH_3 \quad \rightleftharpoons$$

diethyl malonate (malonic ester)
($pK_a = 13$)

$$\left[CH_3CH_2O-\overset{\overset{:\ddot{O}{:}^-}{\|}}{C}\underset{\underset{H}{\overset{\cdot\cdot}{C}}}{}\overset{\overset{:\ddot{O}{:}}{\|}}{C}-OCH_2CH_3 \longleftrightarrow CH_3CH_2O-\overset{\overset{:\ddot{O}{:}^-}{\|}}{C}\underset{\underset{H}{C}}{}\overset{\overset{:\ddot{O}{:}}{\|}}{C}-OCH_2CH_3 \longleftrightarrow CH_3CH_2O-\overset{\overset{:\ddot{O}{:}}{\|}}{C}\underset{\underset{H}{C}}{}\overset{\overset{:\ddot{O}{:}^-}{\|}}{C}-OCH_2CH_3 \right]$$

resonance-stabilized enolate ion

PROBLEM 22-46

Show the resonance forms for the enolate ions that result when the following compounds are treated with a strong base.
(a) ethyl acetoacetate   (b) 2,4-pentanedione
(c) ethyl α-cyanoacetate   (d) nitroacetone

22-16
The Malonic Ester Synthesis

The **malonic ester synthesis** makes substituted derivatives of acetic acid. Malonic ester (diethyl malonate) is alkylated or acylated on the carbon that is α to both carbonyl groups, and the resulting derivative is hydrolyzed and allowed to decarboxylate.

Malonic ester synthesis

$$\overset{\overset{O}{\|}}{\underset{\overset{\|}{O}}{\underset{C-OC_2H_5}{\overset{C-OC_2H_5}{H-C-H}}}} \xrightarrow[\text{(2) } R-X]{\text{(1) NaOC}_2\text{H}_5} \overset{\overset{O}{\|}}{\underset{\overset{\|}{O}}{\underset{C-OC_2H_5}{\overset{C-OC_2H_5}{R-C-H}}}} \xrightarrow[\text{heat}]{\text{H}_3\text{O}^+} \overset{\overset{CO_2}{}}{\underset{\overset{\|}{O}}{\underset{C-OH}{\overset{H}{R-C-H}}}} + 2\ C_2H_5OH$$

malonic ester   alkylated malonic ester   substituted acetic acid

Malonic ester is completely deprotonated by sodium ethoxide. The resulting enolate ion is alkylated by an unhindered alkyl halide, tosylate, or other electrophilic reagent. This step is an S_N2 displacement, requiring a good S_N2 substrate.

$$CH_3CH_2O\overset{\overset{O}{\|}}{C}-CH_2-\overset{\overset{O}{\|}}{C}OCH_2CH_3 \xrightarrow{\text{Na}^+ {}^-\text{OCH}_2\text{CH}_3} CH_3CH_2O\overset{\overset{O}{\|}}{C}-\overset{-}{C}H-\overset{\overset{O}{\|}}{C}OCH_2CH_3 \longrightarrow CH_3CH_2O\overset{\overset{O}{\|}}{C}-\underset{\underset{R}{C}H}{}-\overset{\overset{O}{\|}}{C}OCH_2CH_3$$
$$\underset{R-X}{\curvearrowright}$$

malonic ester   alkylated diethyl malonate

Hydrolysis of the alkylated diethyl malonate (a diethyl alkylmalonic ester) gives a malonic acid derivative.

a diethyl alkylmalonate   →   an alkylmalonic acid   (H⁺, heat, H_2O)

Any carboxylic acid with a carbonyl group in the β position is prone to decarboxylate. At the temperature of the hydrolysis, the alkylmalonic acid loses CO_2 to give a substituted derivative of acetic acid. Decarboxylation takes place through a cyclic transition state, initially giving an enol from that quickly tautomerizes to the product.

alkylmalonic acid CO_2 + enol   substituted acetic acid + $CO_2\uparrow$

The product of a malonic ester synthesis is a substituted acetic acid, with the substituent being the group used to alkylate malonic ester. In effect, the second carboxyl group is temporary, allowing the ester to be easily deprotonated and alkylated. Hydrolysis with decarboxylation removes the temporary carboxyl group, leaving the substituted acetic acid.

malonic ester →(1) ⁻OC_2H_5 (2) R—X→ alkylmalonic ester

alkylmalonic ester →(H⁺, heat, H_2O)→ substituted acetic acid + $CO_2\uparrow$

The alkylmalonic ester has a second acidic proton that can be removed by a base. Removing this proton and alkylating the enolate with another alkyl halide gives a dialkylated malonic ester. Hydrolysis and decarboxylation leads to a disubstituted derivative of acetic acid.

alkylmalonic ester →(1) $NaOCH_2CH_3$ (2) R′—X→ dialkylmalonic ester

dialkylmalonic ester →(H⁺, heat, H_2O)→ disubstituted acetic acid + $CO_2\uparrow$

▲ **Figure 22-4**
Fatty acid biosynthesis. Activated as its coenzyme A thioester, the growing fatty acid (acyl-CoA) acylates malonyl-CoA in a malonic ester synthesis. Two carbon atoms are added, with the third lost as CO_2. Enzymatic reduction, dehydration, and further reduction gives a fatty acid that has been lengthened by two carbon atoms.

The malonic ester synthesis is useful for making cycloalkanecarboxylic acids, some of which are not easily made by any other method. The ring is formed from a dihalide, using a double alkylation of malonic ester. The following synthesis of cyclobutanecarboxylic acid shows that a strained four-membered ring system can be generated by this ester alkylation, even though most other condensations cannot form four-membered rings.

The malonic ester synthesis might seem like an arcane technique that only an organic chemist would use. Still, it is much like the method that cells use to synthesize the long-chain fatty acids found in fats, oils, waxes, and cell membranes. Figure 22-4 outlines the steps that take place in the lengthening of a fatty acid chain by two carbon atoms at a time. The growing acid (acyl-CoA) is activated as its thioester with coenzyme A (structure on page 989). A malonic ester acylation adds two of the three carbons of malonic acid (as malonyl-CoA), with the third carbon lost in the decarboxylation. A β-ketoester results. Reduction of the ketone, followed by dehydration and reduction of the double bond, gives an acyl group that has been lengthened by two carbon atoms. The cycle is repeated until the acid has reached the necessary length, always with an even number of carbon atoms.

PROBLEM-SOLVING HINT

A malonic ester synthesis goes through alkylation of the enolate, hydrolysis, and decarboxylation. To design a synthesis, look at the product and see what groups are added to acetic acid. Use those groups to alkylate malonic ester, then hydrolyze and decarboxylate.

SOLVED PROBLEM 22-7

Show how the malonic ester synthesis is used to prepare 2-benzylbutanoic acid.

SOLUTION

2-Benzylbutanoic acid is a substituted acetic acid having the substituents $Ph-CH_2-$ and CH_3CH_2-.

$$CH_3-CH_2-CH-C-OH$$

substituent | acetic acid

CH$_2$—Ph
substituent

Adding these substituents to the enolate of malonic ester eventually gives the correct product.

COOC$_2$H$_5$

$$CH_2-C-OC_2H_5$$

malonic ester

$$\xrightarrow[\text{(2) PhCH}_2\text{Br}]{\text{(1) NaOCH}_2\text{CH}_3}$$

COOC$_2$H$_5$

$$CH-C-OC_2H_5$$

CH$_2$Ph

$$\xrightarrow[\text{(2) CH}_3\text{CH}_2\text{Br}]{\text{(1) NaOCH}_2\text{CH}_3}$$

COOC$_2$H$_5$

$$CH_3CH_2-C-C-OC_2H_5$$

CH$_2$Ph

dialkylmalonic ester

$$\xrightarrow[\text{H}_2\text{O}]{\text{H}^+, \text{ heat}}$$

CO$_2$ ↑

$$CH_3CH_2-CH-C-OH$$

CH$_2$Ph

disubstituted acetic acid

PROBLEM 22-47

Show how the following compounds can be made using the malonic ester synthesis.
(a) 3-phenylpropanoic acid
(b) 2-methylpropanoic acid
(c) 4-phenylbutanoic acid
(d) cyclopentanecarboxylic acid

PROBLEM 22-48

(a) Explain why the following substituted acetic acid cannot be formed by the malonic ester synthesis.

(b) Sections 22-2B and 22-5 showed the use of lithium diisopropylamide (LDA) to deprotonate a ketone quantitatively. Draw the acid–base reaction between LDA and the following ester, and use estimated pK_a values to decide whether the reaction favors the reactants or products at equilibrium.

(c) Show how you might use a modern alternative to the malonic ester synthesis to make the acid shown in part (a). You may use the ester shown in part (b) as your starting material.

The **acetoacetic ester synthesis** is similar to the malonic ester synthesis, but the final products are ketones: specifically, substituted derivatives of acetone. In the acetoacetic ester synthesis, substituents are added to ethyl acetoacetate (acetoacetic ester), followed by hydrolysis and decarboxylation to produce an alkylated derivative of acetone.

22-17
The Acetoacetic Ester Synthesis

$$CH_3-\overset{\displaystyle O}{\overset{\|}{C}}-CH_2-\overset{\displaystyle O}{\overset{\|}{C}}-OC_2H_5 \xrightarrow[\text{(2) R—X}]{\text{(1) }^-OC_2H_5}$$

ethyl acetoacetate
(acetoacetic ester)

$$CH_3-\overset{\displaystyle O}{\overset{\|}{C}}-\overset{\displaystyle R}{\overset{|}{C}H}-\overset{\displaystyle O}{\overset{\|}{C}}-OC_2H_5 \xrightarrow[\text{heat}]{H_3O^+} CH_3-\overset{\displaystyle O}{\overset{\|}{C}}-\overset{\displaystyle R}{\overset{|}{C}H_2}$$

alkylated ester   substituted acetone

$$+ \; C_2H_5OH \; + \; CO_2 \uparrow$$

Acetoacetic ester is like a molecule of acetone with a temporary ester group attached to enhance its acidity. Ethoxide ion completely deprotonates acetoacetic ester. The resulting enolate is alkylated by an unhindered alkyl halide or tosylate to give an alkylacetoacetic ester. Once again, the alkylating agent must be a good S$_N$2 substrate.

temporary ester group

ethyl acetoacetate
(pK_a = 11)

enolate ion

alkylacetoacetic ester

Acidic hydrolysis of the alkylacetoacetic ester initially gives an alkylacetoacetic acid. The alkylacetoacetic acid is a β-keto acid; the keto group in the β position allows decarboxylation to occur.

alkylacetoacetic ester
(a β-keto ester)

alkylacetoacetic acid
(a β-keto acid)

$CO_2 \uparrow$

a substituted acetone

The β-keto acid decarboxylates by the same mechanism as the alkylmalonic acid in the malonic ester synthesis. A six-membered cyclic transition state splits out carbon dioxide to give the enol form of the substituted acetone. This decarboxylation usually takes place spontaneously at the temperature of the hydrolysis.

the β-keto acid → CO$_2$ + enol → tautomerization → a substituted acetone

Disubstituted acetones are formed simply by alkylating acetoacetic ester a second time before the hydrolysis and decarboxylation steps, as shown in the following general synthesis.

temporary ester group

$$R—CH(COOC_2H_5)—\overset{O}{\underset{\|}{C}}—CH_3 \xrightarrow[\text{(2) } R'—X]{\text{(1) } ^-OC_2H_5} R—\underset{R'}{\overset{COOC_2H_5}{C}}—\overset{O}{\underset{\|}{C}}—CH_3 \xrightarrow[H_2O]{H^+,\text{ heat}} R—\underset{R'}{\overset{H}{C}}—\overset{O}{\underset{\|}{C}}—CH_3 \quad CO_2\uparrow$$

dialkylacetoacetic ester → disubstituted acetone

SOLVED PROBLEM 22-8

Show how the acetoacetic ester synthesis is used to make 3-propyl-5-hexen-2-one.

SOLUTION

The target compound is acetone with an *n*-propyl group and an allyl group as substituents:

$$(CH_3CH_2CH_2)—CH—\overset{O}{\underset{\|}{C}}—CH_3 \quad \text{acetone}$$

n-propyl group   (CH$_2$—CH=CH$_2$)   allyl group

With an *n*-propyl halide and an allyl halide as the alkylating agents, the acetoacetic ester synthesis should produce 3-propyl-5-hexen-2-one.

Two alkylation steps give the required substitution:

$$CH_2(COOC_2H_5)—\overset{O}{\underset{\|}{C}}—CH_3 \xrightarrow[\text{(2) } CH_3CH_2CH_2Br]{\text{(1) } ^-OC_2H_5} CH_3CH_2CH_2—CH(COOC_2H_5)—\overset{O}{\underset{\|}{C}}—CH_3 \xrightarrow[\text{(2) } CH_2=CH—CH_2Br]{\text{(1) } ^-OC_2H_5} CH_3CH_2CH_2—\underset{H_2C=CH—CH_2}{\overset{COOC_2H_5}{C}}—\overset{O}{\underset{\|}{C}}—CH_3$$

Hydrolysis proceeds with decarboxylation to give the disubstituted acetone product.

$$CH_3CH_2CH_2—\underset{H_2C=CH—CH_2}{\overset{COOC_2H_5}{C}}—\overset{O}{\underset{\|}{C}}—CH_3 \xrightarrow[H_2O]{H^+,\text{ heat}} \left[CH_3CH_2CH_2—\underset{H_2C=CH—CH_2}{\overset{COOH}{C}}—\overset{O}{\underset{\|}{C}}—CH_3 \right] \longrightarrow CH_3CH_2CH_2—\underset{H_2C=CH—CH_2}{\overset{H}{C}}—\overset{O}{\underset{\|}{C}}—CH_3 \quad CO_2\uparrow$$

β-keto acid   3-propyl-5-hexen-2-one

PROBLEM 22-49

Show the ketones that would result from hydrolysis–decarboxylation of the following β-keto esters.

(a) PhCH₂—CH—C—CH₃ with COOC₂H₅ (b) CH₂—C—C—CH₃ (c) cyclopentanone with C—OCH₂CH₃

PROBLEM 22-50

Show how the following ketones might be synthesized by using the acetoacetic ester synthesis.

(a) PhCH₂CH₂—C—CH₃ (b) (c) H₂C=CHCH₂CH—CCH₃ with Ph—CH₂

PROBLEM 22-51

(a) Although the following compound is a substituted acetone derivative, it cannot be made by the acetoacetic ester synthesis. Explain why (two reasons).

(b) The use of LDA to make enolate ions (Sections 22-2B and 22-5) has provided alternatives to the acetoacetic ester synthesis. Show how you might make the compound shown in part (a), beginning with 1,3-diphenylacetone.

(c) Enamine reactions (Section 22-6) occur under relatively mild conditions, and they often give excellent yields of compounds like the one shown in part (a). Show how you might use an enamine reaction for this synthesis, beginning with 1,3-diphenylacetone.

22-18
Conjugate Additions: The Michael Reaction

α,β-Unsaturated carbonyl compounds have unusually electrophilic double bonds. The β carbon is electrophilic because it shares the partial positive charge of the carbonyl carbon through resonance.

A nucleophile can attack an α,β-unsaturated carbonyl compound either at the carbonyl group itself or at the β position. When attack occurs at the carbonyl group protonation of the oxygen leads to a **1,2-addition** product in which the nucleophile and the proton have added to adjacent atoms. When attack occurs at the β position, the oxygen atom is the fourth atom counting from the nucleophile, and the addition is called a **1,4-addition**. The net result of 1,4-addition is addition of the nucleophile and a hydrogen

KEY MECHANISM 1,2-Addition and 1,4-Addition

1,2-addition

Step 1: Addition of the nucleophile to C=O.

attack at carbonyl

Step 2: Protonation.

protonation of alkoxide

1,4-addition (Michael addition)

Step 1: Conjugate addition of the nucleophile.

attack at β carbon

Step 2: Protonation.

protonation of enolate   (enol)   tautomerism   (keto)

atom across a double bond that was conjugated with a carbonyl group. For this reason, 1,4-addition is often called **conjugate addition**.

Addition of a stabilized enolate ion to the double bond of an α,β-unsaturated carbonyl compound is called a **Michael addition**. The electrophile (the α,β-unsaturated carbonyl compound) accepts a pair of electrons; it is called the **Michael acceptor**. The attacking nucleophile donates a pair of electrons; it is called the **Michael donor**. A wide variety of compounds can serve as Michael donors and acceptors. Some of the most common ones are shown in Table 22-2. Common donors are enolate ions that are stabilized by two strong electron-withdrawing groups such as carbonyl groups, cyano groups, or nitro groups. These enolates are formed quantitatively by common bases, without extra base around to attack the Michael acceptor. Common acceptors contain a double bond conjugated with a carbonyl group, a cyano group, or a nitro group.

TABLE 22-2 Some Common Michael Donors and Michael Acceptors

Michael Donors		Michael Acceptors	
$R-\overset{O}{\overset{\|}{C}}-\overset{..}{\underset{}{C}}H-\overset{O}{\overset{\|}{C}}-R'$	β-diketone	$H_2C=CH-\overset{O}{\overset{\|}{C}}-H$	conjugated aldehyde
$R-\overset{O}{\overset{\|}{C}}-\overset{..}{\underset{}{C}}H-\overset{O}{\overset{\|}{C}}-OR'$	β-keto ester	$H_2C=CH-\overset{O}{\overset{\|}{C}}-R$	conjugated ketone
R_2CuLi	dialkyl cuprate	$H_2C=CH-\overset{O}{\overset{\|}{C}}-OR$	conjugated ester
$\overset{N:}{>}C=C<$	enamine	$H_2C=CH-\overset{O}{\overset{\|}{C}}-NH_2$	conjugated amide
$R-\overset{O}{\overset{\|}{C}}-\overset{..}{\underset{}{C}}H-C\equiv N$	β-keto nitrile	$H_2C=CH-C\equiv N$	conjugated nitrile
$R-\overset{O}{\overset{\|}{C}}-\overset{..}{\underset{}{C}}H-NO_2$	α-nitro ketone	$H_2C=CH-NO_2$	nitroethylene

As an example, let's consider the addition of the malonic ester enolate to methyl vinyl ketone (MVK). The crucial step is the nucleophilic attack by the enolate at the carbon. The resulting enolate is strongly basic, and it is quickly protonated.

The product of this Michael addition may be treated like any other substituted malonic ester in the malonic ester synthesis. Hydrolysis and decarboxylation lead to a δ-keto acid. It is not easy to imagine other ways to synthesize this interesting keto acid.

SOLVED PROBLEM 22-9

Show how the following diketone might be synthesized using a Michael addition.

SOLUTION

A Michael addition would have formed a new bond at the β carbon of the acceptor. Therefore, we break this molecule apart at the β,γ bond.

The top fragment, where we broke the β bond, must have come from a conjugated ketone, and it must have been the Michael acceptor. The bottom fragment is a simple ketone. It is unlikely that this ketone was used without some sort of additional stabilizing group. We can add a temporary ester group to the ketone (making a substituted acetoacetic ester) and use the acetoacetic ester synthesis to give the correct product.

PROBLEM 22-52

In Solved Problem 22-9, the target molecule was synthesized using a Michael addition to form the bond that is β,γ to the upper carbonyl group. Another approach is to use a Michael addition to form the bond that is β,γ to the other (lower) carbonyl group. Show how you would accomplish this alternative synthesis.

PROBLEM 22-53

Show how cyclohexanone might be converted to the following δ-diketone, (*Hint:* Stork).

PROBLEM 22-54

Show how an acetoacetic ester synthesis might be used to form a δ-diketone.

PROBLEM 22-55

Propose a mechanism for the conjugate addition of a nucleophile (Nuc:⁻) to acrylonitrile (H₂C=CHCN) and to nitroethylene. Use resonance forms to show how the cyano and nitro groups activate the double bond toward conjugate addition.

PROBLEM 22-56

Show how the following products might be synthesized from suitable Michael donors and acceptors.

PROBLEM-SOLVING HINT

Claisen condensations usually give 1,3-dicarbonyl products, with one saturated carbon between two carbonyl groups. Michael additions commonly give 1,5-dicarbonyl products, with three saturated carbons between two carbonyl groups. When you need a compound with three carbons between two carbonyl groups, consider a Michael addition.

22-19
The Robinson Annulation

British chemist Sir Robert Robinson (1886–1975) invented the Robinson annulation to form complicated ring systems.

We have seen that Michael addition of a ketone enolate (or its enamine) to an α,β-unsaturated ketone gives a δ-diketone. If the conjugate addition takes place under strongly basic or acidic conditions, the δ-diketone undergoes a spontaneous intramolecular aldol condensation, usually with dehydration, to give a new six-membered ring: a conjugated cyclohexenone. This synthesis is called the **Robinson annulation** (ring-forming) reaction. Consider an example using a substituted cyclohexanone as the Michael donor and methyl vinyl ketone (MVK) as the Michael acceptor.

The Robinson annulation

new cyclohexenone
(65%)

The mechanism begins with addition of the enolate of the cyclohexanone to MVK, forming a δ-diketone.

Step 1: Michael addition

a δ-diketone

This δ-diketone might take part in several different aldol condensations, but it is ideally suited for a particularly favorable one: formation of a six-membered ring. To form a six-membered ring, the enolate of the methyl ketone attacks the cyclohexanone carbonyl. The aldol product dehydrates to give a cyclohexenone.

Step 2: cyclic aldol

Step 3: dehydration

enolate

It is not difficult to predict the products of the Robinson annulation and to draw the mechanisms if you remember that the Michael addition is first, followed by an aldol condensation with dehydration to give a cyclohexenone.

PROBLEM-SOLVING STRATEGY
Proposing Reaction Mechanisms

This problem-solving example addresses a complicated base-catalyzed reaction, using the system for proposing mechanisms summarized in Appendix 4. The problem is to propose a mechanism for the base-catalyzed reaction of ethyl acetoacetate with methyl vinyl ketone.

First, we must determine the type of mechanism. The use of a basic catalyst suggests the reaction involves strong nucleophiles as intermediates. We expect to see anionic intermediates (possibly stabilized carbanions), but no strong electrophiles or strong acids, and no carbocations or free radicals.

1. **Consider the carbon skeletons of the reactants and products, and decide which carbon atoms in the products are likely derived from which carbon atoms in the reactants.**

   The ester group in the product must be derived from ethyl acetoacetate. The β carbon from the ester (now part of the C=C double bond) should be derived from the ketone of ethyl acetoacetate. The structure of MVK can be seen in the remaining four carbons.

2. **Consider whether one of the reactants is a strong enough nucleophile to react without being activated. If not, consider how one of the reactants might be converted to a strong nucleophile by deprotonation of an acidic site or by attack on an electrophilic site.**

   Neither reactant is a strong enough nucleophile to attack the other. Ethyl acetoacetate is more acidic than ethanol, so ethoxide ion quickly removes a proton to give the enolate ion.

3. **Consider how an electrophilic site on another reactant (or, in a cyclization, another part of the same molecule) can undergo attack by the strong nucleophile to form a bond needed in the product. Draw the product of this bond formation.**

   The enolate of acetoacetic ester might attack either the electrophilic double bond (Michael addition) or the carbonyl group of MVK. A Michael addition forms one of the bonds needed in the product.

4. **Consider how the product of nucleophilic attack might be converted to the final product (if it has the right carbon skeleton) or reactivated to form another bond needed in the product.**

The carbonyl group of ethyl acetoacetate must be converted to a C=C double bond in the α,β position of the other ketone. This conversion corresponds to an aldol condensation with dehydration.

5. **Draw out all the steps using curved arrows to show the movement of electrons. Be careful to show only one step at a time.**

The complete mechanism is given by combining the equations shown above. We suggest you write out the mechanism as a review of the steps.

As further practice in proposing mechanisms for multistep condensations, try Problems 22-57 and 22-58 by using the approach shown.

PROBLEM 22-57

Propose a mechanism for the following reaction.

PROBLEM 22-58

The base-catalyzed reaction of an aldehyde (having no α hydrogens) with an anhydride is called the *Perkin condensation*. Propose a mechanism for the following example of the Perkin condensation. (Sodium acetate serves as the base.)

$$CH_3-\overset{\overset{\displaystyle O}{\|}}{C}-O-\overset{\overset{\displaystyle O}{\|}}{C}-CH_3$$

+

$$\overset{\overset{\displaystyle O}{\|}}{C}-H$$

(1) CH_3CO_2Na, Δ
(2) H_3O^+

$$CH=CH-\overset{\overset{\displaystyle O}{\|}}{C}-OH$$

cinnamic acid

+ CH_3COOH

PROBLEM 22-59

Show how you would use Robinson annulations to synthesize the following compounds. Work backward, remembering that the cyclohexenone is the new ring and that the double bond of the cyclohexenone is formed by the aldol with dehydration. Take apart the double bond, then see what structures the Michael donor and acceptor must have.

(a)

(b)

PROBLEM-SOLVING HINT

You can usually spot a product of Robinson annulation because it has a new cyclohexenone ring. The mechanism is not difficult if you remember "Michael goes first," followed by an aldol with dehydration.

SUMMARY **Enolate Additions and Condensations**

A complete summary of additions and condensations would be long and involved. This summary covers the major classes of condensations and related reactions.

1. *α Halogenation* (Sections 22-3 and 22-4)

$$R-\overset{\overset{\displaystyle O}{\|}}{C}-\overset{\overset{\displaystyle H}{|}\alpha}{\underset{|}{C}}- \ + \ X_2 \quad \xrightarrow{H^+ \text{ or } ^-OH} \quad R-\overset{\overset{\displaystyle O}{\|}}{C}-\overset{\overset{\displaystyle X}{|}}{\underset{|}{C}}-$$

a. *The iodoform (or haloform) reaction*

$$R-\overset{\overset{\displaystyle O}{\|}}{C}-CH_3 \ + \ \text{excess } I_2 \quad \xrightarrow{^-OH} \quad R-\overset{\overset{\displaystyle O}{\|}}{C}-O^- \ + \ HCI_3 \downarrow$$

methyl ketone

b. *The Hell–Volhard–Zelinsky (HVZ) reaction*

$$R-CH_2-\overset{\overset{\displaystyle O}{\|}}{C}-OH \quad \xrightarrow{Br_2/PBr_3} \quad R-\overset{\overset{\displaystyle Br}{|}}{\underset{}{CH}}-\overset{\overset{\displaystyle O}{\|}}{C}-Br \quad \xrightarrow{H_2O} \quad R-\overset{\overset{\displaystyle Br}{|}}{\underset{}{CH}}-\overset{\overset{\displaystyle O}{\|}}{C}-OH$$

α-bromo acid

2. *Alkylation of lithium enolates* (Section 22-5)

$$R-\overset{\overset{\displaystyle O}{\|}}{C}-CH_2-R \quad \xrightarrow[\text{(2) } R'-X]{\text{(1) LDA}} \quad R-\overset{\overset{\displaystyle O}{\|}}{C}-\overset{\overset{\displaystyle R'}{|}}{\underset{}{CH}}-R$$

(LDA = lithium diisopropylamide; $R'-X$ = unhindered 1° halide or tosylate)

3. *Alkylation of enamines (Stork reaction) (Section 22-6)*

enamine   alkylated enamine   alkylated ketone

4. *The aldol condensation and subsequent dehydration (Sections 22-7 through 22-11)*

ketone or aldehyde   aldol product α,β-unsaturated ketone or aldehyde

5. *The Claisen ester condensation (Sections 22-12 through 22-14)*
   (Cyclizations are the Dieckmann condensation.)

The product is initially formed as its anion.

6. *The malonic ester synthesis (Section 22-16)*

malonic ester   substituted malonic ester   substituted acetic acid

7. *The acetoacetic ester synthesis (Section 22-17)*

acetoacetic ester   substituted acetoacetic ester   substituted acetone

8. *The Michael addition (conjugate addition) (Sections 22-18 and 22-19)*

(Y and Z are carbonyl or other electron-withdrawing groups.)

Example: The Robinson annulation

cyclohexanone   MVK   Michael adduct   annulated product

Chapter 22 Glossary

acetoacetic ester synthesis Alkylation or acylation of acetoacetic ester (ethyl acetoacetate), followed by hydrolysis and decarboxylation, to give substituted acetone derivatives. (p. 1040)

aldol condensation An acid- or base-catalyzed conversion of two ketone or aldehyde molecules to a β-hydroxy ketone or aldehyde (called an **aldol**). Aldol condensations often take place with subsequent dehydration to give α,β-unsaturated ketones and aldehydes. (p. 1017)

   crossed aldol condensation: An aldol condensation between two different ketones or aldehydes. (p. 1021)

alpha (α) carbon atom The carbon atom next to a carbonyl group. The hydrogen atoms on the α carbon are called α hydrogens or **α protons**. (p. 1003)

alpha (α) substitution Replacement of a hydrogen atom at the α carbon atom by some other group. (p. 1003)

Claisen condensation The base-catalyzed conversion of two ester molecules to a β-keto ester. (p. 1027)

   crossed Claisen condensation: A Claisen condensation between two different esters or between a ketone and an ester. (p. 1032)

condensation A reaction that bonds two or more molecules, often with the loss of a small molecule such as water or an alcohol. (p. 1003)

conjugate addition (1,4-addition) Another term for Michael addition. (p. 1042)

Dieckmann condensation (Dieckmann cyclization) A Claisen condensation that forms a ring. (p. 1031)

enamine A **vinyl amine**, usually generated by the acid-catalyzed reaction of a secondary amine with a ketone or an aldehyde. (p. 1015)

enol A vinyl alcohol. Simple enols usually tautomerize to their **keto** forms. (p. 1004)

enolate ion The resonance-stabilized anion formed by deprotonating the carbon atom next to a carbonyl group. (p. 1004)

enolate ion

enolizable hydrogen (α hydrogen) A hydrogen atom on a carbon adjacent to a carbonyl group. Such a hydrogen may be lost and regained through keto–enol tautomerism, losing its stereochemistry in the process. (p. 1005)

haloform reaction The conversion of a methyl ketone to a carboxylate ion and a haloform (CHX_3) by treatment with a halogen and base. The **iodoform reaction** uses iodine to give a precipitate of solid iodoform. (p. 1009)

Hell–Volhard–Zelinsky (HVZ) reaction Reaction of a carboxylic acid with Br_2 and PBr_3 to give an α-bromo acyl bromide, often hydrolyzed to an α-bromo acid. (p. 1012)

malonic ester synthesis Alkylation or acylation of malonic ester (diethyl malonate), followed by hydrolysis and decarboxylation, to give substituted acetic acids. (p. 1036)

Michael addition (conjugate addition) A **1,4-addition** of a nucleophile (the **Michael donor**, usually a resonance-stabilized carbanion) to a conjugated double bond such as an α,β-unsaturated ketone or ester (the **Michael acceptor**). (p. 1042)

Robinson annulation Formation of a cyclohexenone ring by condensation of methyl vinyl ketone (MVK) or a substituted MVK derivative with a ketone. Robinson annulation proceeds by Michael addition to MVK, followed by an aldol condensation with dehydration. (p. 1046)

Stork reaction Alkylation or acylation of a ketone or aldehyde using its enamine derivative as the nucleophile. Acidic hydrolysis regenerates the alkylated or acylated ketone or aldehyde. (p. 1016)

tautomerism An isomerism involving the migration of a proton and the corresponding movement of a double bond. An example is the **keto–enol tautomerism** of a ketone or aldehyde with its enol form. (p. 1004)

    tautomers: The isomers related by a tautomerism.

keto tautomer   enol tautomer

keto–enol tautomerism

Essential Problem-Solving Skills in Chapter 22

This is a difficult chapter because condensations take on a wide variety of forms. You should try to *understand* the reactions and their mechanisms so you can generalize and predict related reactions. Work enough problems to get a feel for the standard reactions (aldol, Claisen, Michael) and to gain confidence in working out new variations of the standard mechanisms. Make sure you feel comfortable with condensations that form new rings.

1. Show how enols and enolate ions act as nucleophiles. Give mechanisms for acid-catalyzed and base-catalyzed keto–enol tautomerisms.

2. Give mechanisms for acid-catalyzed and base-promoted alpha halogenation of ketones and acid-catalyzed halogenation of acids (the HVZ reaction). Explain why multiple halogenation is common with basic catalysis, and give a mechanism for the haloform reaction.

3. Show how alkylation and acylation of enamines and lithium enolates are used synthetically. Give mechanisms for these reactions.

4. Predict the products of aldol and crossed aldol reactions before and after dehydration of the aldol products. Give mechanisms for the acid- and base-catalyzed reactions. (Aldols are reversible, so be sure you can write these mechanisms *backward* as well.) Show how aldols are used to make β-hydroxy carbonyl compounds and α,β-unsaturated carbonyl compounds.

5. Predict the products of Claisen and crossed Claisen condensations, and propose mechanisms. Show how a Claisen condensation constructs the carbon skeleton of a target compound.

6. Show how the malonic ester synthesis and the acetoacetic ester synthesis are used to make substituted acetic acids and substituted acetones. Give mechanisms for these reactions.

7. Predict the products of Michael additions, and show how to use these reactions in syntheses. Show the general mechanism of the Robinson annulation, and use it to form cyclohexenone ring systems.

Study Problems

22-60 Define each term and give an example.
    (a) haloform reaction   **(b)** condensation   **(c)** aldol condensation
    (d) crossed aldol condensation   **(e)** alpha substitution   **(f)** α,β-unsaturated compound
    (g) HVZ reaction   **(h)** Claisen condensation   **(i)** Dieckmann condensation
    (j) crossed Claisen condensation   **(k)** enamine   **(l)** Stork reaction
    (m) tautomerism   **(n)** enolizable hydrogen   **(o)** malonic ester synthesis
    (p) acetoacetic ester synthesis   **(q)** Michael addition   **(r)** Robinson annulation

22-61 For each molecule shown below,
    (1) Indicate the most acidic hydrogens.
    (2) Draw the important resonance contributors of the anion that results from removal of the most acidic hydrogen.

(f)

(g) $CH_3—CH=CH—\overset{\overset{\displaystyle O}{\|}}{C}—H$

(h) $CH_2=CH—CH_2—\overset{\overset{\displaystyle O}{\|}}{C}—H$

22-62 (1) Rank the following compounds in order of increasing acidity.

(2) Indicate which compounds would be more than 99% deprotonated by a solution of sodium ethoxide in ethanol.

(a) **(b)** **(c)** **(d)** ...COOH

(e) **(f)** ...CH$_2$OH **(g)**

22-63 Pentane-2,4-dione (acetylacetone) exists as a tautomeric mixture of 8% keto and 92% enol forms. Draw the stable enol tautomer, and explain its unusual stability.

$$CH_3—\overset{\overset{\displaystyle O}{\|}}{C}—CH_2—\overset{\overset{\displaystyle O}{\|}}{C}—CH_3$$
acetylacetone

22-64 Predict the products of the following aldol condensations. Show the products both before and after dehydration.

(a) $\begin{matrix} CH_3 \\ \diagdown \\ CH—CH_2—\overset{\overset{\displaystyle O}{\|}}{C}—H \\ \diagup \\ CH_3 \end{matrix} \xrightarrow{\ ^-OH\ }$

(b) $\xrightarrow{\ H^+\ }$

(c) $2\ Ph—CHO + CH_3—\overset{\overset{\displaystyle O}{\|}}{C}—CH_3 \xrightarrow{\ ^-OH\ }$

(d) $Ph—CH_2—\overset{\overset{\displaystyle O}{\|}}{C}—H + H_2C=O \xrightarrow{\ ^-OH\ }$

22-65 Predict the products of the following Claisen condensations.

(a) $\begin{matrix} CH_3 \\ \diagdown \\ CH—CH_2—\overset{\overset{\displaystyle O}{\|}}{C}—OCH_3 \\ \diagup \\ CH_3 \end{matrix} \xrightarrow[CH_3OH]{\ ^-OCH_3\ }$

(b) $\xrightarrow[CH_3OH]{\ ^-OCH_3\ }$ ...COOCH$_3$

(c) $CH_3CH_2—\overset{\overset{\displaystyle O}{\|}}{C}—CH_2CH_2CH_2CH_2—\overset{\overset{\displaystyle O}{\|}}{C}—OCH_3 \xrightarrow[CH_3OH]{\ ^-OCH_3\ }$ (Dieckmann)

(d) $+ CH_3O—\overset{\overset{\displaystyle O}{\|}}{C}—\overset{\overset{\displaystyle O}{\|}}{C}—OCH_3 \xrightarrow[CH_3OH]{\ NaOCH_3\ }$

(e) $\begin{matrix} CH_2—\overset{\overset{\displaystyle O}{\|}}{C}—OCH_3 \\ \\ CH_2—\overset{\underset{\displaystyle O}{\|}}{C}—CH_3 \end{matrix} \xrightarrow[CH_3OH]{\ NaOCH_3\ }$

22-66 Propose mechanisms for the reactions shown in Problems 22-64 (a) and (b) and 22-65 (a) and (b).

22-67 Show how you would use an aldol, Claisen, or other type of condensation to make each compound.

(a) CHO

(b) COOEt

(c) CH$_3$ / CH$_2$CH$_3$

(d)   **(e)**   **(f)**

22-68 Predict the products of the following reactions.

(a) $CH_3CH_2-\overset{O}{\underset{\|}{C}}-CH_2-\overset{O}{\underset{\|}{C}}-OCH_3$
$\xrightarrow{\substack{(1)\ NaOCH_3 \\ (2)\ CH_3I \\ (3)\ H_3O^+,\ heat}}$

(b) $CH_3-\overset{O}{\underset{\|}{C}}-CH_2-\overset{O}{\underset{\|}{C}}-OCH_2CH_3$ +
$\xrightarrow{\substack{(1)\ NaOCH_2CH_3 \\ (2)\ H_3O^+,\ heat}}$

(c)
$\xrightarrow{\substack{(1)\ LDA \\ (2)\ CH_3CH_2CH_2Br}}$

(d) + $\xrightarrow{^-OH}$

(e)
$\xrightarrow{\substack{(1)\ H_2C=CH-CH_2Br \\ (2)\ H_3O^+}}$

(f)
$\xrightarrow{\substack{(1)\ NaOCH_3 \\ (2)\ CH_3CH_2CH_2CH_2Br}}$

(g) product from part (f) $\xrightarrow[heat]{H_3O^+}$ (decarboxylation)

22-69 Show how you would accomplish the following conversions in good yields. You may use any necessary reagents.

(a) $CH_3-\overset{CH_3}{\underset{CH_3}{\overset{|}{\underset{|}{C}}}}-\overset{O}{\underset{\|}{C}}-CH_3 \longrightarrow CH_3-\overset{CH_3}{\underset{CH_3}{\overset{|}{\underset{|}{C}}}}-\overset{O}{\underset{\|}{C}}-CH_2Br$

(b) $\longrightarrow$

(c) $CH_3-\overset{CH_3}{\underset{CH_3}{\overset{|}{\underset{|}{C}}}}-\overset{O}{\underset{\|}{C}}-CH_3 \longrightarrow CH_3-\overset{CH_3}{\underset{CH_3}{\overset{|}{\underset{|}{C}}}}-\overset{O}{\underset{\|}{C}}-O^-$

(d) $Ph-\overset{O}{\underset{\|}{C}}-H \longrightarrow Ph-CH=CH-CH_3$

(e) $\longrightarrow$ (*Hint:* aldol)

22-70 Show how you would use the malonic ester synthesis to make the following compounds.

(a)   **(b)**   **(c)**

22-71 Show how you would use the acetoacetic ester synthesis to make the following compounds.

(a)   **(b)**   **(c)** (*Hint:* Consider using 2,6-heptanedione as an intermediate.)

*22-72   The following compounds can be synthesized by aldol condensations, followed by further reactions. (In each case, work backward from the target molecule to an aldol product, and show what compounds are needed for the condensation.)

(a) Ph—CH$_2$—CH$_2$—CH(OH)—Ph

(b)

(c)

22-73   Propose mechanisms for the following reactions.

(a)   + PhCHO $\xrightarrow{\ ^-OH\ }$

(b)   + Ph—C(=O)—OCH$_3$ $\xrightarrow{\ ^-OCH_3\ }$

(c)   + H$_2$C=CH—C(=O)—CH$_2$CH$_3$ $\xrightarrow{\ ^-OH\ }$

(d) $\xrightarrow[\text{(2) H}_3\text{O}^+]{\text{(1) MVK}}$

22-74   Write equations showing the expected products of the following enamine alkylation and acylation reactions. Then give the final products expected after hydrolysis of the iminium salts.
(a) pyrrolidine enamine of 3-pentanone + allyl chloride
(b) pyrrolidine enamine of acetophenone + butanoyl chloride
(c) piperidine enamine of cyclopentanone + methyl iodide
(d) piperidine enamine of cyclopentanone + methyl vinyl ketone

*22-75   Show how you would accomplish the following multistep conversions. You may use any additional reagents you need.

(a) dimethyl adipate and allyl bromide $\longrightarrow$

(b) cyclopentanone $\longrightarrow$

(c) $\longrightarrow$

*22-76   Many of the condensations we have studied are reversible. The reverse reactions are often given the prefix *retro-*, the Latin word meaning "backward." Propose mechanisms to account for the following reactions.

(a) $\xrightarrow{\text{H}^+}$ CH$_3$—C(=O)—(CH$_2$)$_5$—CHO

(retro-aldol)

(b) $\xrightarrow{\text{H}^+}$

(retro-aldol and further condensation)

(c) $\xrightleftharpoons{\ ^-OH\ }$ + H$_2$C=CH—CN

(retro-Michael)

22-77   Show how you would use the Robinson annulation to synthesize the following compounds.

(a)   (b)   (c)

CH₃

22-78   Propose a mechanism for the following reaction. Show the structure of the compound that results from hydrolysis and decarboxylation of the product.

CHO

$+$ $CH_2(COOCH_2CH_3)_2$ $\xrightarrow{NaOCH_2CH_3}$ $CH{=}C(COOCH_2CH_3)_2$ $\xrightarrow[\text{decarboxylation}]{\text{hydrolysis,}}$ H_3O^+, Δ

benzaldehyde   malonic ester

22-79   A reaction involved in the metabolism of sugars is the splitting of fructose 1,6-diphosphate to give glyceraldehyde 3-phosphate and dihydroxyacetone phosphate. In the living system, this retro-aldol is catalyzed by an enzyme called *aldolase*; however, it can also be catalyzed by a mild base. Propose a mechanism for the base-catalyzed reaction.

$\xrightleftharpoons{\text{aldolase or } ^-OH}$

fructose 1,6-diphosphate

dihydroxyacetone phosphate

glyceraldehyde 3-phosphate

22-80   Biochemists studying the structure of collagen (a fibrous protein in connective tissue) found cross-links containing α,β-unsaturated aldehydes between protein chains. Show the structures of the side chains that react to form these cross-links, and propose a mechanism for their formation in a weakly acidic solution.

protein chain   protein chain

*22-81   Show reaction sequences (not detailed mechanisms) that explain these transformations:

(a) CH_2O $+$ $2\ CH_2(COOEt)_2$ $\xrightarrow[\text{(2) H}^+]{\text{(1) NaOEt}}$

EtOOC   COOEt

EtOOC   COOEt

(b) $+$ $CH_2(COOEt)_2$ $\xrightarrow[\text{(2) H}_3O^+]{\text{(1) NaOEt}}$

(c) CH_2O $+$ 2 $\xrightarrow[\text{(2) H}^+]{\text{(1) NaOEt}}$

COOH

Carbohydrates and Nucleic Acids

Carbohydrates are the most abundant organic compounds in nature. Nearly all plants and animals synthesize and metabolize carbohydrates, using them to store energy and deliver it to their cells. Plants synthesize carbohydrates through *photosynthesis*, a complex series of reactions that use sunlight as the energy source to convert carbon dioxide and water into glucose and oxygen. Many molecules of glucose can be linked together to form either *starch* for energy storage or *cellulose* to support the plant.

$$6\,CO_2 + 6\,H_2O \xrightarrow{\text{light}} 6\,O_2 + \underset{\text{glucose}}{C_6H_{12}O_6} \longrightarrow \text{starch, cellulose} + H_2O$$

Most living organisms oxidize glucose to carbon dioxide and water to provide the energy needed by their cells. Plants can retrieve the glucose units from starch when needed; in effect, starch is a plant's storage unit for solar energy for later use. Animals can also store glucose energy by linking many molecules together to form *glycogen*, another form of starch. *Cellulose* makes up the cell walls of plants and forms their structural framework. Cellulose is the major component of *wood*, a strong yet supple material that supports the great weight of the oak, yet allows the willow to bend with the wind.

Almost every aspect of human life involves carbohydrates in one form or another. Like other animals, we use the energy content of carbohydrates in our food to produce and store energy in our cells. Clothing is made from cotton and linen, two forms of cellulose. Other fabrics are made by manipulating cellulose to convert it to the semisynthetic fibers *rayon* and *cellulose acetate*. In the form of wood, we use cellulose to construct our houses and as a fuel to heat them. Even this page is made from cellulose fibers.

Carbohydrate chemistry is one of the most interesting areas of organic chemistry. Many chemists are employed by companies that use carbohydrates to make foods, building materials, and other consumer products. All biologists need to understand carbohydrates, which play pivotal roles throughout the plant and animal kingdoms. At first glance, the structures and reactions of carbohydrates may seem complicated. We will learn how these structures and reactions are consistent and predictable, however, and we can study carbohydrates as easily as we study the simplest organic compounds.

23-1 Introduction

Low carbohydrate diets restrict the intake of carbohydrates, sometimes resulting in rapid weight loss. The weight is lost because glycogen and fatty acids are burned to maintain blood glucose levels.

The term **carbohydrate** arose because most **sugars** have molecular formulas $C_n(H_2O)_m$, suggesting that carbon atoms are combined in some way with water. In fact, the empirical formula of most simple sugars is $C(H_2O)$. Chemists named these compounds "hydrates of carbon" or "carbohydrates" because of these molecular formulas. Our modern definition of carbohydrates includes polyhydroxyaldehydes, polyhydroxyketones, and compounds that are easily hydrolyzed to them.

23-2 Classification of Carbohydrates

◄ Figure 23-1
Glucose and fructose are monosaccharides. Glucose is an aldose (a sugar with an aldehyde group), and fructose is a ketose (a sugar with a ketone group). Carbohydrate structures are commonly drawn using Fischer projections.

glucose   fructose

Monosaccharides, or *simple sugars*, are carbohydrates that cannot be hydrolyzed to simpler compounds. Figure 23-1 shows the Fischer projections of the monosaccharides *glucose* and *fructose*. Glucose is a polyhydroxyaldehyde, while fructose is a polyhydroxyketone. Polyhydroxyaldehydes are called **aldoses** (*ald-* is for *ald*ehyde and *-ose* is the suffix for a sugar), and polyhydroxyketones are called **ketoses** (*ket-* for *ket*one, and *-ose* for sugar).

We have used Fischer projections to draw the structures of glucose and fructose because Fischer projections conveniently show the stereochemistry at all the asymmetric carbon atoms. The Fischer projection was originally developed by Emil Fischer, a carbohydrate chemist who received the Nobel Prize for his proof of the structure of glucose. Fischer used this shorthand notation for drawing and comparing sugar structures quickly and easily. We will use Fischer projections extensively in our work with carbohydrates, so you may want to review them (Section 5–10) and make models of the structures in Figure 23-1 to review the stereochemistry implied by these structures. In aldoses, the aldehyde carbon is the most highly oxidized (and numbered 1 in the IUPAC name), so it is always at the top of the Fischer projection. In ketoses, the carbonyl group is usually the second carbon from the top.

PROBLEM-SOLVING $\mathcal{HINT}$

The Fischer projection represents each asymmetric carbon atom by a cross with the horizontal bonds projecting toward the viewer and the vertical bonds projecting away. The carbon chain is arranged along the vertical bonds, with the most oxidized end (or carbon #1 in the IUPAC name) at the top.

For more than one asymmetric carbon atom, the Fischer projection represents a totally eclipsed conformation: not the most stable conformation, but usually the most symmetric conformation, helpful for comparing stereochemistry.

PROBLEM 23-1

Draw the mirror images of glucose and fructose. Are glucose and fructose chiral? Do you expect them to be optically active?

A **disaccharide** is a sugar that can be hydrolyzed to two monosaccharides. For example, sucrose ("table sugar") is a disaccharide that can be hydrolyzed to one molecule of glucose and one molecule of fructose.

$$1 \text{ sucrose} \xrightarrow[\text{heat}]{H_3O^+} 1 \text{ glucose} + 1 \text{ fructose}$$

Both monosaccharides and disaccharides are highly soluble in water, and most have the characteristic sweet taste we associate with sugars.

Polysaccharides are carbohydrates that can be hydrolyzed to many monosaccharide units. Polysaccharides are naturally occurring polymers (*biopolymers*) of carbohydrates. They include starch and cellulose, both biopolymers of glucose. **Starch** is a polysaccharide whose carbohydrate units are easily added to store energy, or removed to provide energy to cells. The polysaccharide **cellulose** is a major structural component of plants. Hydrolysis of either starch or cellulose gives many molecules of glucose.

$$\text{starch} \xrightarrow[\text{heat}]{H_3O^+} \text{over 1000 glucose molecules}$$

$$\text{cellulose} \xrightarrow[\text{heat}]{H_3O^+} \text{over 1000 glucose molecules}$$

Before we can understand the chemistry of these more complex carbohydrates, we must first learn the principles of carbohydrate structure and reactions, using the simplest monosaccharides as examples. Then we will apply these principles to more complex disaccharides and polysaccharides. The chemistry of carbohydrates is understood by applying the chemistry of alcohols, aldehydes, and ketones. In general, the chemistry of biomolecules can be predicted by applying the chemistry of simple organic molecules with similar functional groups.

23-3
Monosaccharides

23-3A   Classification of Monosaccharides

Most sugars have their own specific common names, such as glucose, fructose, galactose, and mannose. These names are not systematic, although there are simple ways to remember the common structures. We simplify the study of monosaccharides by grouping similar structures together. This classification is done according to three criteria:

1. Whether the sugar contains a ketone or an aldehyde group
2. The number of carbon atoms in the carbon chain
3. The stereochemical configuration of the asymmetric carbon atom farthest from the carbonyl group

As we have seen, sugars with aldehyde groups are called **aldoses**, and those with ketone groups are called **ketoses**. The number of carbon atoms in the sugar generally ranges from three to seven, designated by the terms *triose* (three carbons), *tetrose* (four carbons), *pentose* (five carbons), *hexose* (six carbons), and *heptose* (seven carbons). These two groups of terms are often combined. For example, glucose has an aldehyde and contains six carbon atoms, so it is an aldohexose. Fructose also contains six carbon atoms, but it is a ketose, so it is called a ketohexose. Most ketoses have the ketone on C2, the second carbon atom of the chain. The most common naturally occurring sugars are aldohexoses and aldopentoses.

$$
\begin{array}{llll}
^1\text{CHO} & ^1\text{CH}_2\text{OH} & & \\
\mid & \mid & & \\
^2\text{CHOH} & ^2\text{C}{=}\text{O} & & \\
\mid & \mid & & \\
^3\text{CHOH} & ^3\text{CHOH} & ^1\text{CHO} & ^1\text{CH}_2\text{OH} \\
\mid & \mid & \mid & \mid \\
^4\text{CHOH} & ^4\text{CHOH} & ^2\text{CHOH} & ^2\text{C}{=}\text{O} \\
\mid & \mid & \mid & \mid \\
^5\text{CHOH} & ^5\text{CHOH} & ^3\text{CHOH} & ^3\text{CHOH} \\
\mid & \mid & \mid & \mid \\
^6\text{CH}_2\text{OH} & ^6\text{CH}_2\text{OH} & ^4\text{CH}_2\text{OH} & ^4\text{CH}_2\text{OH} \\
\text{an aldohexose} & \text{a ketohexose} & \text{an aldotetrose} & \text{a ketotetrose}
\end{array}
$$

PROBLEM 23-2

(a) How many asymmetric carbon atoms are there in an aldotetrose? Draw all the aldotetrose stereoisomers.
(b) How many asymmetric carbons are there in a ketotetrose? Draw all the ketotetrose stereoisomers.
(c) How many asymmetric carbons and stereoisomers are there for an aldohexose? For a ketohexose?

PROBLEM 23-3

(a) There is only one ketotriose, called *dihydroxyacetone*. Draw its structure.
(b) There is only one aldotriose, called *glyceraldehyde*. Draw the two enantiomers of glyceraldehyde.

23-3B The D and L Configurations of Sugars

Around 1880–1900, carbohydrate chemists made great strides in determining the structures of natural and synthetic sugars. They found ways to build larger sugars out of smaller ones, adding a carbon atom to convert a tetrose to a pentose and a pentose to a hexose. The opposite conversion, removing one carbon atom at a time (called a *degradation*), was also developed. A degradation could convert a hexose to a pentose, a pentose to a tetrose, and a tetrose to a triose. There is only one aldotriose, glyceraldehyde.

These chemists noticed they could start with any of the naturally occurring sugars, and degradation to glyceraldehyde always gave the dextrorotatory (+) enantiomer of glyceraldehyde. Some synthetic sugars, on the other hand, degraded to the levorotatory (−) enantiomer of glyceraldehyde. Carbohydrate chemists started using the *Fischer–Rosanoff convention*, which uses a D to designate the sugars that degrade to (+) glyceraldehyde and an L for those that degrade to (−) glyceraldehyde. Although these chemists did not know the absolute configurations of any of these sugars, the D and L relative configurations were useful to distinguish the naturally occurring D sugars from their unnatural L enantiomers.

We now know the absolute configurations of (+) and (−) glyceraldehyde. These structures serve as the configurational standards for all monosaccharides.

(+)-glyceraldehyde
D series of sugars

(−)-glyceraldehyde
L series of sugars

Figure 23-2 shows that degradation (covered in Section 23-14) removes the aldehyde carbon atom, and it is the *bottom asymmetric carbon* in the Fischer projection (the asymmetric carbon farthest removed from the carbonyl group) that determines which enantiomer of glyceraldehyde is formed by successive degradation.

The (+) enantiomer of glyceraldehyde has its OH group on the right in the Fischer projection, as shown in Figure 23-2. Therefore, sugars of the D series have the OH group of the bottom asymmetric carbon on the right in the Fischer projection; sugars of the L series have the OH group of the bottom asymmetric carbon on the left. In the examples at the top of page 1051, notice that the D or L configuration is determined by the bottom asymmetric carbon, and the enantiomer of a D sugar is always an L sugar.

D-(+)-glucose   D-(−)-arabinose   D-(−)-erythrose   D-(+)-glyceraldehyde

▲ **Figure 23-2**
Degradation of an aldose removes the aldehyde carbon atom to give a smaller sugar. Sugars of the D series give (+)-glyceraldehyde on degradation to the triose. Therefore, the OH group of the bottom asymmetric carbon atom of the D sugars must be on the right in the Fischer projection.

Most naturally occurring sugars have the D configuration, and most members of the D family of aldoses (up through six carbon atoms) are found in nature. Figure 23-3 shows the D family of aldoses. Notice that the D or L configuration does not tell us which way a sugar rotates the plane of polarized light. This must be determined by experiment. Some D sugars have (+) rotations, while others have (−) rotations.

On paper, the family tree of D aldoses (Fig. 23-3) can be generated by starting with D-(+)-glyceraldehyde and adding another carbon at the top to generate two aldotetroses: erythrose with the OH group of the new asymmetric carbon on the right, and threose with the new OH group on the left. Adding another carbon to these aldotetroses gives four aldopentoses, and adding a sixth carbon gives eight

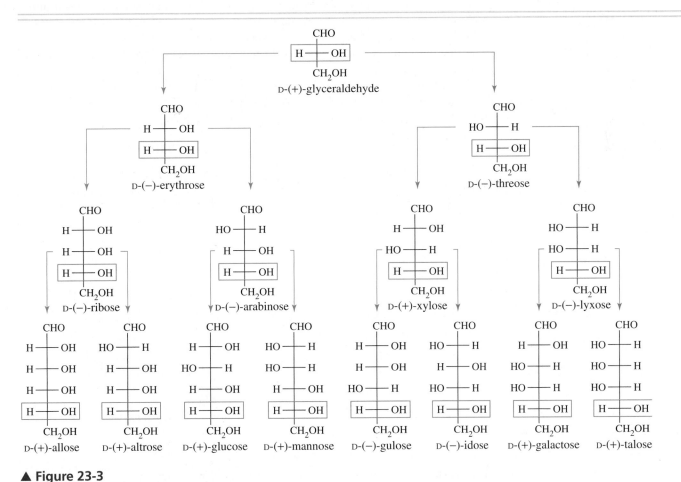

▲ **Figure 23-3**
The D family of aldoses. All these sugars occur naturally except for threose, lyxose, allose, and gulose.

aldohexoses.* In Section 23-15, we describe the Kiliani–Fischer synthesis, which actually adds a carbon atom and generates the pairs of elongated sugars just as we have drawn them in this family tree.

At the time the D and L system of relative configurations was introduced, chemists could not determine the absolute configurations of chiral compounds. They decided to draw the D series with the glyceraldehyde OH group on the right, and the L series with it on the left. This guess later proved to be correct, so it was not necessary to revise all the old structures.

PROBLEM 23-4

Draw and name the enantiomers of the sugars shown in Figure 23-2. Give the relative configuration (D or L) and the sign of the rotation in each case.

PROBLEM 23-5

Which configuration (*R* or *S*) does the bottom asymmetric carbon have for the D series of sugars? Which configuration for the L series?

PROBLEM-SOLVING HINT

Most naturally occurring sugars are of the D series, with the OH group of the bottom asymmetric carbon on the right in the Fischer projection.

23-4
Erythro and Threo Diastereomers

Erythrose is the aldotetrose with the OH groups of its two asymmetric carbons situated on the same side of the Fischer projection, and *threose* is the diastereomer with the OH groups on opposite sites of the Fischer projection. These names have evolved into a shorthand way of naming diastereomers with two adjacent asymmetric carbon atoms. A diastereomer is called **erythro** if its Fischer projection shows similar groups on the same side of the molecule. It is called **threo** if similar groups are on opposite sides of the Fischer projection.

For example, syn hydroxylation of *trans*-crotonic acid gives two enantiomers of the threo diastereomer of 2,3-dihydroxybutanoic acid. The same reaction with *cis*-crotonic acid gives the erythro diastereomer of the product.

trans-crotonic acid $\xrightarrow{OsO_4/H_2O_2}$ (2R,3S) | (2S,3R)
threo-2,3-dihydroxybutanoic acid

cis-crotonic acid $\xrightarrow{OsO_4/H_2O_2}$ (2R,3R) | (2S,3S)
erythro-2,3-dihydroxybutanoic acid

---

*Drawn in this order, the names of the four aldopentoses (ribose, arabinose, xylose, and lyxose) are remembered by the mnemonic "*Rib*s *ar*e *ex*tra *l*ean." The mnemonic for the eight aldohexoses (allose, altrose, glucose, mannose, gulose, idose, galactose, and talose) is "*All altr*uists *gl*adly *m*ake *gu*m *in* *gal*lon *t*anks."

◀ Figure 23-4
The terms *erythro* and *threo* are used with dissymmetric molecules whose ends are different. The erythro diastereomer is the one with similar groups on the same side of the Fischer projection, while the threo diastereomer has similar groups on opposite sides of the Fischer projection. The terms *meso* and (±) [or (*d, l*)] are preferred with symmetric molecules.

The terms *erythro* and *threo* are generally used only with molecules that do not have symmetric ends. In symmetric molecules such as 2,3-dibromobutane and tartaric acid, the terms *meso* and (*d, l*) are preferred because these terms indicate the diastereomer and tell whether or not it has an enantiomer. Figure 23-4, shows the proper use of the terms *erythro* and *threo* for dissymmetric molecules, as well as the terms *meso* and (*d, l*) for symmetric molecules.

PROBLEM 23-6

Draw Fischer projections for the enantiomers of *threo*-1,2,3-hexanetriol.

$$HOCH_2 - CH(OH) - CH(OH) - CH_2CH_2CH_3$$

PROBLEM 23-7

The bronchodilator *ephedrine* is *erythro*-2-(methylamino)-1-phenylpropan-1-ol. The decongestant *pseudoephedrine* is *threo*-2-(methylamino)-1-phenylpropan-1-ol.
(a) Draw the four stereoisomers of 2-(methylamino)-1-phenylpropan-1-ol, first as Fischer projections, and then as three-dimensional representations (dotted lines and wedges).
(b) Label ephedrine and pseudoephedrine. What is the relationship between them?
(c) Label the D and L isomers of ephedrine and pseudoephedrine using the Fischer–Rosanoff convention.
(d) Both ephedrine and pseudoephedrine are commonly used as racemic mixtures. Ephedrine is also available as the pure levorotatory (−) isomer (Biophedrine®), and pseudoephedrine is also available as the more active (+) isomer (Sudafed®). Can you label the (−) isomer of ephedrine and the (+) isomer of pseudoephedrine?

23-5
Epimers

Many common sugars are closely related, differing only by the stereochemistry at a single carbon atom. For example, glucose and mannose differ only at C2, the first asymmetric carbon atom. Sugars that differ only by the stereochemistry at a single carbon are called **epimers**, and the carbon atom where they differ is generally stated. If the number of a carbon atom is not specified, it is assumed to be C2. Therefore, glucose and mannose are "C2 epimers" or simply "epimers." The C4 epimer of glucose is galactose, and the C2 epimer of erythrose is threose.

D-mannose   D-glucose   D-galactose   D-erythrose   D-threose

---

PROBLEM 23-8

(a) Draw D-allose, the C3 epimer of glucose.

(b) Draw D-talose, the C2 epimer of D-galactose.

(c) Draw D-idose, the C3 epimer of D-talose. Now compare your answers with Figure 23-3.

(d) Draw the C4 "epimer" of D-xylose. Notice that this "epimer" is actually an L-series sugar, and we have seen its enantiomer. Give the correct name for this L-series sugar.

23-6
Cyclic Structures of Monosaccharides

Cyclic Hemiacetals   In Chapter 18, we saw that an aldehyde reacts with one molecule of an alcohol to give a hemiacetal, and with a second molecule of the alcohol to give an acetal. The hemiacetal is not as stable as the acetal, and most hemiacetals decompose spontaneously to the aldehyde and the alcohol. Therefore, hemiacetals are rarely isolated.

If the aldehyde group and the hydroxyl group are part of the same molecule, a cyclic hemiacetal results. Cyclic hemiacetals are particularly stable if they result in five- or six-membered rings. In fact, five- and six-membered cyclic hemiacetals are often more stable than their open-chain forms.

MECHANISM   Formation of a Cyclic Hemiacetal

The Cyclic Hemiacetal Form of Glucose   Aldoses contain an aldehyde group and several hydroxyl groups. The solid, crystalline form of an aldose is normally a cyclic hemiacetal. In solution, the aldose exists as an equilibrium mixture of the cyclic hemiacetal and the open-chain form. For most sugars, the equilibrium favors the cyclic hemiacetal.

Aldohexoses such as glucose can form cyclic hemiacetals containing either five-membered or six-membered rings. For most common aldohexoses, the equilibrium favors six-membered rings with a hemiacetal linkage between the aldehyde carbon and the hydroxyl group on C5. Figure 23-5 shows formation of the cyclic hemiacetal of glucose. Notice that the hemiacetal has a new asymmetric carbon atom at C1. Figure 23-5 shows the C1 hydroxyl group up, but another possible stereoisomer would have this hydroxyl group directed down. We discuss the stereochemistry at C1 in more detail in Section 23-7.

▲ **Figure 23-5**
Glucose exists almost entirely as its cyclic hemiacetal form.

The cyclic structure is often drawn initially in the **Haworth projection**, which depicts the ring as being flat (of course, it is not). The Haworth projection is widely used in biology texts, although most chemists prefer to use the more realistic chair conformation. Figure 23-5 shows the cyclic form of glucose both as a Haworth projection and as a chair conformation.

Drawing Cyclic Monosaccharides   Cyclic hemiacetal structures may seem complicated at first glance, but they can be drawn and recognized by following the process illustrated in Figure 23-5:

1. Mentally lay the Fischer projection over on its right side. The groups that were on the right in the Fischer projection are down in the cyclic structure, and the groups that were on the left are up.
2. C5 and C6 curl back away from you. The C4—C5 bond must be rotated so that the C5 hydroxyl group can form a part of the ring. For a sugar of the D series, this rotation puts the terminal —CH$_2$OH (C6 in glucose) upward.
3. Close the ring, and draw the result. Always draw the Haworth projection or chair conformation with the oxygen at the back, right-hand corner, with C1 at the far right. C1 is easily identified because it is the hemiacetal carbon—the only carbon bonded to two oxygens. The hydroxyl group on C1 can be either up or down, as discussed in Section 23-7. Sometimes the ambiguous stereochemistry is symbolized by a wavy line.

Chair conformations can be drawn by recognizing the differences between the sugar in question and glucose. The following procedure is useful for drawing D-aldohexoses.

1. Draw the chair conformation puckered, as shown in Figure 23-5. The hemiacetal carbon (C1) is the footrest.
2. Glucose has its substituents on alternating sides of the ring. In drawing the chair conformation, just put all the ring substituents in equatorial positions. (In the Haworth projection, the —OH on C4 is opposite the —CH$_2$OH on C5, and the —OH on C3 is opposite that on C4.)
3. To draw or recognize other common sugars, notice how they differ from glucose and make the appropriate changes.

SOLVED PROBLEM 23-1

Draw the cyclic hemiacetal forms of D-mannose and D-galactose both as chair conformations and as Haworth projections. Mannose is the C2 epimer of glucose, and galactose is the C4 epimer of glucose.

SOLUTION

The chair conformations are easier to draw, so we will do them first. Draw the rings, and number the carbon atoms, starting with the hemiacetal carbon. Mannose is the C2 epimer of glucose, so the substituent on C2 is axial, while all the others are equatorial as in glucose. Galactose is the C4 epimer of glucose, so its substituent on C4 is axial.

D-mannose   D-galactose

The simplest way to draw Haworth structures for these two sugars is to draw the chair conformations and then draw the flat rings with the same substituents in the up and down positions. For practice, however, let's lay down the Fischer projection for galactose. You should follow along with your molecular models.

(1) Lay down the Fischer projection: right → down and left → up.

D-galactose

(2) Rotate the C4—C5 bond to put the —OH in place. (For a D sugar, the —CH$_2$OH goes up.)

(3) Close the ring, and draw the final hemiacetal. The hydroxyl group on C1 can be either up or down, as discussed in Section 23-7. Sometimes this ambiguous stereochemistry is symbolized by a wavy line.

PROBLEM 23-9

Draw the Haworth projection for the cyclic structure of D-mannose by laying down the Fischer projection.

PROBLEM 23-10

Allose is the C3 epimer of glucose. Draw the cyclic hemiacetal form of D-allose, first in the chair conformation and then in the Haworth projection.

The Five-Membered Cyclic Hemiacetal Form of Fructose   Not all sugars exist as six-membered rings in their hemiacetal forms. Many aldopentoses and ketohexoses form five-membered rings. The five-membered ring of fructose is shown in Figure 23-6. Five-membered rings are not puckered as much as six-membered rings, so they are usually depicted as flat Haworth projections. The five-membered ring is customarily drawn with the ring oxygen in back and the hemiacetal carbon (the one bonded to two oxygens) on the right. The —CH₂OH at the back left (C6) is in the up position for D-series ketohexoses.

▲ **Figure 23-6**
Fructose forms a five-membered cyclic hemiacetal. Five-membered rings are usually represented as flat Haworth structures.

Pyranose and Furanose Names   Cyclic structures of monosaccharides are named according to their five- or six-membered rings. A six-membered cyclic hemiacetal is called a **pyranose**, derived from the name of the six-membered cyclic ether *pyran*. A five-membered cyclic hemiacetal is called a **furanose**, derived from the name of the five-membered cyclic ether *furan*. The ring is still numbered as it is in the sugar, not beginning with the heteroatom as it would be in the heterocyclic nomenclature. These structural names are incorporated into the systematic names of sugars.

PROBLEM 23-11

Talose is the C4 epimer of mannose. Draw the chair conformation of D-talopyranose.

PROBLEM 23-12

(a) Figure 23-3 shows that the degradation of D-glucose gives D-arabinose, an aldopentose. Arabinose is most stable in its furanose form. Draw D-arabinofuranose.

(b) Ribose, the C2 epimer of arabinose, is most stable in its furanose form. Draw D-ribofuranose.

PROBLEM 23-13

The carbonyl group in D-galactose may be isomerized from C1 to C2 by brief treatment with dilute base (by the enediol rearrangement, Section 23-8). The product is the C4 epimer of fructose. Draw the furanose structure of the product.

23-7
Anomers
of Monosaccharides;
Mutarotation

When a pyranose or furanose ring closes, the hemiacetal carbon atom is converted from a flat carbonyl group to an asymmetric carbon. Depending on which face of the (protonated) carbonyl group is attacked, the hemiacetal —OH group can be directed either up or down. These two orientations of the hemiacetal —OH group give diastereomeric products called **anomers**. Figure 23-7 shows the anomers of glucose.

The hemiacetal carbon atom is called the **anomeric carbon**, easily identified as the only carbon atom bonded to two oxygens. Its —OH group is called the anomeric hydroxyl group. Notice in Figure 23-7 that the anomer with the anomeric —OH group down (axial) is called the α (alpha) anomer, while the one with the anomeric —OH group up (equatorial) is called the β (beta) anomer. We can draw the α and β anomers of most aldohexoses by remembering that the β form of glucose (β-D-glucopyranose) has all its substituents in equatorial positions. To draw an α anomer, simply move the anomeric —OH group to the axial position.

Another way to remember the anomers is to notice that the α anomer has its anomeric hydroxyl group trans to the terminal —CH$_2$OH group, while it is cis in the β anomer. This rule works for all sugars, from both the D and L series, as well as for furanoses. Figure 23-8 shows the two anomers of fructose, whose anomeric carbon is C2. The α anomer has the anomeric —OH group down, trans to the terminal —CH$_2$OH group, while the β anomer has it up, cis to the terminal —CH$_2$OH.

PROBLEM 23-14

Draw the following monosaccharides, using chair conformations for the pyranoses and Haworth projections for the furanoses.
(a) α-D-mannopyranose (C2 epimer of glucose)
(b) β-D-galactopyranose (C4 epimer of glucose)
(c) β-D-allopyranose (C3 epimer of glucose)
(d) α-D-arabinofuranose
(e) β-D-ribofuranose (C2 epimer of arabinose)

▲ **Figure 23-7**
The anomers of glucose. The hydroxyl group on the anomeric (hemiacetal) carbon is down (axial) in the α anomer and up (equatorial) in the β anomer. The β anomer of glucose has all its substituents in equatorial positions.

▲ **Figure 23-8**

The α anomer of fructose has the anomeric —OH group down, trans to the terminal —CH$_2$OH group.

The β anomer has the anomeric hydroxyl group up, cis to the terminal —CH$_2$OH.

Properties of Anomers: Mutarotation   Because anomers are diastereomers, they generally have different properties. For example, α-D-glucopyranose has a melting point of 146°C and a specific rotation of +112.2°, while β-D-glucopyranose has a melting point of 150°C and a specific rotation of +18.7°. When glucose is crystallized from water at room temperature, pure crystalline α-D-glucopyranose results. If glucose is crystallized from water by letting the water evaporate at a temperature above 98°C, crystals of pure β-D-glucopyranose are formed (Figure 23-9).

In each of these cases, *all* the glucose in the solution crystallizes as the favored anomer. In the solution, the two anomers are in equilibrium through a small amount of the open-chain form, and this equilibrium continues to supply more of the anomer that is crystallizing out of solution.

When one of the glucose anomers dissolves in water, an interesting change in the specific rotation is observed. When the α anomer dissolves, its specific rotation gradually decreases from an initial value of +112.2° to +52.6°. When the pure β anomer

> Many diabetics have long-standing elevated blood glucose levels. In the open-chain form, glucose condenses with the amino groups of proteins. This glycosylation of proteins may cause some of the chronic effects of diabetes.

▲ **Figure 23-9**

An aqueous solution of D-glucose contains an equilibrium mixture of α-D-glucopyranose, β-D-glucopyranose, and the intermediate open-chain form. Crystallization below 98°C gives the α anomer, and crystallization above 98°C gives the β anomer.

dissolves, its specific rotation gradually increases from +18.7° to the same value of +52.6°. This change ("mutation") in the specific rotation is called **mutarotation**. Mutarotation occurs because the two anomers interconvert in solution. When either of the pure anomers dissolves in water, its rotation gradually changes to an intermediate rotation that results from equilibrium concentrations of the anomers. The specific rotation of glucose is usually listed as +52.6°, the value for the equilibrium mixture of anomers. The positive sign of this rotation is the source of the name *dextrose*, an old common name for glucose.

SOLVED PROBLEM 23-2

Calculate how much of the α anomer and how much of the β anomer are present in an equilibrium mixture with a specific rotation of +52.6°.

SOLUTION

If the fraction of glucose present as the α anomer ($[\alpha] = +112.2°$) is a, the fraction present as the β anomer ($[\alpha] = +18.7°$) is b, and the rotation of the mixture is +52.6°, we have

$$a(+112.2°) + b(+18.7°) = +52.6°$$

There is very little of the open-chain form present, so the fraction present as the α anomer (a) plus the fraction present as the β anomer (b) should account for all the glucose:

$$a + b = 1 \quad \text{or} \quad b = 1 - a$$

Substituting $(1 - a)$ for b in the first equation, we have

$$a(112.2°) + (1 - a)(18.7°) = 52.6°$$

Solving this equation for a, we have $a = 0.36$, or 36%. Thus, b must be $(1 - 0.36) = 0.64$, or 64%. The amounts of the two anomers present at equilibrium are

$$\alpha \text{ anomer, 36%} \qquad \beta \text{ anomer, 64%}$$

When we remember that the anomeric hydroxyl group is axial in the α anomer and equatorial in the β anomer, it is reasonable that the more stable β anomer should predominate.

PROBLEM 23-15

Like glucose, galactose shows mutarotation when it dissolves in water. The specific rotation of α-D-galactopyranose is +150.7°, while that of the β anomer is +52.8°. When either of the pure anomers dissolves in water, the specific rotation gradually changes to +80.2°. Determine the percentages of the two anomers present at equilibrium.

23-8
Reactions
of Monosaccharides:
Side Reactions
in Base

Sugars are multifunctional compounds that can undergo reactions typical of any of their functional groups. Most sugars exist as cyclic hemiacetals, yet in solution they are in equilibrium with their open-chain aldehyde or ketone forms. As a result, sugars undergo most of the usual reactions of ketones, aldehydes, and alcohols. Reagents commonly used with monofunctional compounds often give unwanted side reactions with sugars, however. Carbohydrate chemists have developed reactions that work well with sugars while avoiding the undesired side reactions. As we learn about the unique reactions of

MECHANISM Base-Catalyzed Epimerization of Glucose

Abstraction of the α proton   Reprotonation

▲ **Figure 23-10**
Under basic conditions, stereochemistry is lost at the carbon atom next to the carbonyl group.

simple sugars, we will often draw them as their open-chain forms because it is often the small equilibrium amount of the open-chain form that reacts.

Epimerization and the Enediol Rearrangement   One of the most important aspects of sugar chemistry is the inability, in most cases, to use basic reagents because they cause unwanted side reactions. Two common base-catalyzed side reactions are epimerization and enediol rearrangement.

Under basic conditions, the proton alpha to the aldehyde (or ketone) carbonyl group is reversibly removed (Figure 23-10). In the resulting enolate ion, C2 is no longer asymmetric, and its stereochemistry is lost. Reprotonation can occur on either face of the enolate, giving either configuration. The result is an equilibrium mixture of the original sugar and its C2 epimer. Because a mixture of epimers results, this stereochemical change is called **epimerization**. Figure 23-10 shows the rapid base-catalyzed epimerization of glucose to a mixture of glucose and its C2 epimer, mannose.

PROBLEM 23-16

Propose a mechanism for the base-catalyzed epimerization of erythrose to a mixture of erythrose and threose.

Another base-catalyzed side reaction is the **enediol rearrangement**, which moves the carbonyl group up and down the chain (Figure 23-11). If the enolate ion formed by removal of a proton on C2 reprotonates on the C1 oxygen, an **enediol** intermediate results. Removal of a proton from the C2 oxygen of the enediol and reprotonation on C1 gives fructose, a ketose.

Under strongly basic conditions, the combination of enediol rearrangements and epimerization leads to a complex mixture of sugars. Except when using protected sugars, most chemists doing sugar chemistry employ neutral or acidic reagents to avoid these annoying side reactions.

MECHANISM   Base-Catalyzed Enediol Rearrangement

Step 1: Remove the α proton.   *Step 2:* Reprotonate on O.   *Step 3:* Deprotonate the O on C2.   *Step 4:* Reprotonate on C1.

D-glucose ⇌ enolate ⇌ enediol ⇌ enolate ⇌ D-fructose

▲ **Figure 23-11**
Under basic conditions, the carbonyl group can isomerize to other carbon atoms. Aldoses equilibrate with ketoses via enediol intermediates.

PROBLEM 23-17

Show how C3 of fructose can epimerize under basic conditions.

PROBLEM 23-18

Show how another enediol rearrangement can move the carbonyl group from C2 in fructose to C3.

23-9
Reduction of Monosaccharides

Like other aldehydes and ketones, aldoses and ketoses can be reduced to the corresponding polyalcohols, called **sugar alcohols** or **alditols**. The most common reagents are sodium borohydride or catalytic hydrogenation using a nickel catalyst. Alditols are named by adding the suffix *-itol* to the root name of the sugar. The following equation shows the reduction of glucose to glucitol, sometimes called *sorbitol*.

β-D-glucopyranose ⇌ open-chain aldehyde $\xrightarrow{H_2, Ni}$ D-glucitol (D-sorbitol) an alditol

Reduction of a ketose creates a new asymmetric carbon atom, formed in either of two configurations, resulting in two epimers. For example, reduction of fructose gives a mixture of glucitol and mannitol.

α-D-fructofuranose $\quad\rightleftharpoons\quad$ open-chain ketone $\quad\xrightarrow{\text{NaBH}_4}\quad$ D-glucitol + D-mannitol — a mixture of alditols

Sugar alcohols are widely used in industry, primarily as food additives and sugar substitutes. Glucitol has the common name *sorbitol* because it was first isolated from the berries of the mountain ash, *Sorbus aucuparia*. Industrially, sorbitol is made by catalytic hydrogenation of glucose. Sorbitol is used as a sugar substitute, a moistening agent, and a starting material for making vitamin C. Mannitol was first isolated from plant exudates known as *mannas* (of Biblical fame), the origin of the names *mannose* and *mannitol*. Mannitol is derived commercially from seaweed, or it can be made by catalytic hydrogenation of mannose. Galactitol (*dulcitol*) also can be obtained from many plants, or it can be made by catalytic hydrogenation of galactose.

PROBLEM 23-19

When D-glucose is treated with sodium borohydride, optically active glucitol results. When optically active D-galactose is reduced, however, the product is optically inactive. Explain this loss of optical activity.

PROBLEM 23-20

Emil Fischer synthesized L-gulose, an unusual aldohexose that reduces to give D-glucitol. Suggest a structure for this L sugar, and show how L-gulose gives the same alditol as D-glucose. (*Hint:* D-glucitol has —CH$_2$OH groups at both ends. *Either* of these primary alcohol groups might have come from reduction of an aldehyde.)

Monosaccharides are oxidized by a variety of reagents. The aldehyde group of an aldose oxidizes easily. Some reagents also selectively oxidize the terminal —CH$_2$OH group at the far end of the molecule. Oxidation is used to identify the functional groups of a sugar, to help to determine its stereochemistry, and as part of a synthesis to convert one sugar into another.

Bromine Water Bromine water oxidizes the aldehyde group of an aldose to a carboxylic acid. Bromine water is used for this oxidation because it does not oxidize the alcohol groups of the sugar and it does not oxidize ketoses. Also, bromine water is acidic and does not cause epimerization or rearrangement of the carbonyl group. Because bromine water oxidizes aldoses but not ketoses, it serves as a useful test to distinguish aldoses from ketoses. The product of bromine water oxidation is an **aldonic acid** (older term: **glyconic acid**). For example, bromine water oxidizes glucose to gluconic acid.

23-10
Oxidation of Monosaccharides; Reducing Sugars

Example

aldehyde

$$O\!\!=\!\!\overset{H}{\underset{(CHOH)_n}{\overset{|}{C}}}$$

CH$_2$OH

aldose

$\xrightarrow[\text{H}_2\text{O}]{\text{Br}_2}$

acid

$$O\!\!=\!\!\overset{OH}{\underset{(CHOH)_n}{\overset{|}{C}}}$$

CH$_2$OH

aldonic acid
(glyconic acid)

CHO
H——OH
HO——H
H——OH
H——OH
CH$_2$OH
glucose

$\xrightarrow[\text{H}_2\text{O}]{\text{Br}_2}$

COOH
H——OH
HO——H
H——OH
H——OH
CH$_2$OH
gluconic acid

PROBLEM 23-21

Draw and name the products of bromine water oxidation of
(a) D-mannose   (b) D-galactose   (c) D-fructose

Nitric Acid   Nitric acid is a stronger oxidizing agent than bromine water, oxidizing both the aldehyde group and the terminal —CH$_2$OH group of an aldose to carboxylic acid groups. The resulting dicarboxylic acid is called an **aldaric acid** (older terms: **glycaric acid** or **saccharic acid**). For example, nitric acid oxidizes glucose to glucaric acid.

Example

aldehyde [CHO]
|
(CHOH)$_n$
|
alcohol [CH$_2$OH]

aldose

$\xrightarrow{\text{HNO}_3}$

acid [COOH]
|
(CHOH)$_n$
|
acid [COOH]

aldaric acid
(glycaric acid)

CHO
H——OH
HO——H
H——OH
H——OH
CH$_2$OH
glucose

$\xrightarrow{\text{HNO}_3}$

COOH
H——OH
HO——H
H——OH
H——OH
COOH
glucaric acid

Diabetics must monitor their blood glucose levels several times a day. A common diagnostic kit uses the enzymatic oxidation of glucose to determine its concentration. A paper strip is treated with glucose oxidase, which specifically catalyzes oxidation of the aldehyde group by oxygen in the air. The products are gluconic acid and hydrogen peroxide. Peroxide oxidizes a dye on the same strip of paper to produce a color change.

PROBLEM 23-22

Draw and name the products of nitric acid oxidation of
(a) D-mannose   (b) D-galactose

PROBLEM 23-23

Two sugars, **A** and **B**, are known to be glucose and galactose, but it is not certain which one is which. On treatment with nitric acid, **A** gives an optically inactive aldaric acid, while **B** gives an optically active aldaric acid. Which sugar is glucose, and which is galactose?

Tollens Test   **Tollens test** detects aldehydes, which react with Tollens reagent to give carboxylate ions and metallic silver, often in the form of a silver mirror on the inside of the container.

$$R\!-\!\overset{O}{\overset{\|}{C}}\!-\!H + 2\ Ag(NH_3)_2^+\ {}^-OH + {}^-OH \longrightarrow R\!-\!\overset{O}{\overset{\|}{C}}\!-\!O^- + 2\ Ag\downarrow + 4\ NH_3 + 2\ H_2O$$

aldehyde   Tollens reagent   oxidized   reduced
                                                          acid anion   silver mirror

In its open-chain form, an aldose has an aldehyde group, which reacts with Tollens reagent to give an aldonic acid and a silver mirror. This oxidation is not a good synthesis of the aldonic acid, however, because Tollens reagent is strongly basic and promotes epimerization and enediol rearrangements. Sugars that reduce Tollens reagent to give a silver mirror are called **reducing sugars**.

β-D-glucose ⇌ open-chain form →(Ag(NH₃)₂⁺ ⁻OH, Tollens reagent) gluconic acid (+ side products) + Ag↓

Tollens test cannot distinguish between aldoses and ketoses because the basic Tollens reagent promotes enediol rearrangements. Under basic conditions, the open-chain form of a ketose can isomerize to an aldose, which reacts to give a positive Tollens test.

a ketose →(⁻OH) enediol intermediate →(⁻OH) an aldose →(Ag(NH₃)₂⁺ ⁻OH) positive Tollens test + Ag↓

What good is Tollens test if it doesn't distinguish between aldoses and ketoses? The answer lies in the fact that Tollens reagent must react with the open-chain form of the sugar, which has a free aldehyde or ketone. If the cyclic form cannot open to the free carbonyl compound, the sugar does not react with Tollens reagent. Hemiacetals are easily opened, but an acetal is stable under neutral or basic conditions (Section 18-18). If the carbonyl group is in the form of a cyclic acetal, the cyclic form cannot open to the free carbonyl compound, and the sugar gives a negative Tollens test (Figure 23-12).

Sugars in the form of acetals are called **glycosides**, and their names end in the -oside suffix. For example, a glycoside of glucose would be a **glucoside**, and if it were a six-membered ring, it would be a *glucopyranoside*. Similarly, a glycoside of ribose would be a *riboside*, and if it were a five-membered ring it would be a *ribofuranoside*. In general, a sugar whose name ends with the suffix -ose is a reducing sugar, and one whose name ends with -oside is nonreducing. Because they exist as stable acetals rather than hemiacetals, glycosides cannot spontaneously open to their open-chain forms, and they do not mutarotate. They are locked in a particular anomeric form.

We can summarize by saying that Tollens test distinguishes between reducing sugars and nonreducing sugars: Reducing sugars (aldoses and ketoses) are hemiacetals, and they mutarotate. Nonreducing sugars (glycosides) are acetals, and they do not mutarotate.

23-11
Nonreducing Sugars: Formation of Glycosides

a glycoside

Examples of nonreducing sugars

methyl β-D-glucopyranoside
(or methyl β-D-glucoside)

ethyl α-D-fructofuranoside
(or ethyl α-D-fructoside)

▶ **Figure 23-12**
Sugars that are full acetals are stable to Tollens reagent and are nonreducing sugars. Such sugars are called glycosides.

PROBLEM 23-24

Which of the following are reducing sugars? Comment on the common name *sucrose* for table sugar.

(a) methyl α-D-galactopyranoside
(b) β-L-idopyranose (an aldohexose)
(c) α-D-allopyranose
(d) ethyl β-D-ribofuranoside

(e)

(f)

sucrose

PROBLEM 23-25

Draw the structures of the compounds named in Problem 23-24 parts (a), (c), and (d). Allose is the C3 epimer of glucose, and ribose is the C2 epimer of arabinose.

Formation of Glycosides   Recall that aldehydes and ketones are converted to acetals by treatment with an alcohol and a trace of acid catalyst (Section 18-18). These conditions also convert aldoses and ketoses to the acetals we call glycosides. Regardless of the anomer used as the starting material, both anomers of the glycoside are formed (as an equilibrium mixture) under these acidic conditions. The more stable anomer predominates. For example, the acid-catalyzed reaction of glucose with methanol gives a mixture of methyl glucosides.

methyl α-D-glucopyranoside

methyl β-D-glucopyranoside

Like other acetals, glycosides are stable to basic conditions but are hydrolyzed to a free sugar and an alcohol by aqueous acid. Glycosides may be used with basic reagents and in basic solutions.

An **aglycone** is the group bonded to the anomeric carbon atom of a glycoside. For example, methanol is the aglycone in a methyl glycoside. Many aglycones are bonded by an oxygen atom, while others are bonded through a nitrogen atom or some other heteroatom. Figure 23-13 shows the structures of some glycosides with interesting aglycones.

Disaccharides and polysaccharides are glycosides in which the alcohol forming the glycoside bond is an —OH group of another monosaccharide. We will consider disaccharides and polysaccharides in Sections 23-18 and 23-19.

ethyl α-D-glucopyranoside

cytidine, a nucleoside (Section 23-21)

salicin, from willow bark

amygdalin
a component of laetrile, a controversial cancer drug

a glycoprotein N-glycoside
(showing the linkage from carbohydrate to protein)

▲ **Figure 23-13**
The group bonded to the anomeric carbon of a glycoside is called an aglycone. Some aglycones are bonded through an oxygen atom (a true acetal), and others are bonded through other atoms such as nitrogen.

PROBLEM 23-26

The mechanism of glycoside formation is the same as the second part of the mechanism for acetal formation. Propose a mechanism for the formation of methyl β-D-glucopyranoside.

PROBLEM 23-27

Show the products that result from hydrolysis of amygdalin in dilute acid. Can you suggest why amygdalin might be toxic to tumor (and possibly other) cells?

PROBLEM 23-28

Treatment of either anomer of fructose with excess ethanol in the presence of a trace of HCl gives a mixture of the α and β anomers of ethyl-D-fructofuranoside. Draw the starting materials, reagents, and products for this reaction. Circle the aglycone in each product.

23-12
Ether and Ester Formation

Because they contain several hydroxyl groups, sugars are very soluble in water and rather insoluble in organic solvents. Sugars are difficult to recrystallize from water because they often form supersaturated syrups like honey and molasses. If the hydroxyl groups are alkylated to form ethers, sugars behave like simpler organic compounds. The ethers are soluble in organic solvents, and they are more easily purified by recrystallization and simple chromatographic methods.

Treating a sugar with methyl iodide and silver oxide converts its hydroxyl groups to methyl ethers. Silver oxide polarizes the H_3C—I bond, making the methyl carbon strongly electrophilic. Attack by the carbohydrate —OH group, followed by deprotonation, gives the ether. Figure 23-14 shows that the anomeric hydroxyl group is also converted to an ether. If the conditions are carefully controlled, the hemiacetal C—O bond is not broken, and the configuration at the anomeric carbon is preserved.

The Williamson ether synthesis is the most common method for forming simple ethers, but it involves a strongly basic alkoxide ion. Under these basic conditions, a simple sugar would isomerize and decompose. A modified Williamson method may be used if the sugar is first converted to a glycoside (by treatment with an alcohol and an acid catalyst). The glycoside is an acetal, stable to base. Treatment of a glycoside with sodium hydroxide and methyl iodide or dimethyl sulfate gives the methylated carbohydrate.

▲ Figure 23-14
Treatment of an aldose or a ketose with methyl iodide and silver oxide gives the totally methylated ether. If the conditions are carefully controlled, the stereochemistry at the anomeric carbon is usually preserved.

methyl α-D-glucopyranoside
(stable to base)

(dimethyl sulfate)

methyl 2,3,4,6-tetra-*O*-methyl-α-D-glucopyranoside

PROBLEM 23-29

Propose a mechanism for methylation of any one of the hydroxyl groups of methyl α-D-glucopyranoside, using NaOH and dimethyl sulfate.

PROBLEM 23-30

Draw the expected product of the reaction of the following sugars with methyl iodide and silver oxide.

(a) α-D-fructofuranose   **(b)** β-D-galactopyranose

Ester Formation   Another way to convert sugars to easily handled derivatives is to acylate the hydroxyl groups to form esters. Sugar esters are readily crystallized and purified, and they dissolve in common organic solvents. Treatment with acetic anhydride and pyridine (as a mild basic catalyst) converts sugar hydroxyl groups to acetate esters, as shown in Figure 23-15. This reaction acetylates all the hydroxyl groups, including that of the hemiacetal on the anomeric carbon. The anomeric C—O bond is not broken in the acylation, and the stereochemistry of the anomeric carbon atom is usually preserved. If we start with a pure α anomer or a pure β anomer, the product is the corresponding anomer of the acetate.

Oxidation of glucose at C-6 produces glucuronic acid. This sugar is attached to many drugs to make a glucuronide derivative, a water-soluble metabolite that is readily excreted.

glucuronic acid

sugar

acetate ester

Example

β-D-fructofuranose

penta-*O*-acetyl-β-D-fructofuranoside

▲ **Figure 23-15**
Acetic anhydride and pyridine convert all the hydroxyl groups on a sugar to acetate esters. The stereochemistry at the anomeric carbon is usually preserved.

PROBLEM 23-31

Predict the products formed when the following sugars react with acetic anhydride and pyridine.

(a) α-D-glucopyranose   (b) β-D-ribofuranose

23-13
Reactions with Phenylhydrazine: Osazone Formation

One of the best methods for derivatizing ketones and aldehydes is conversion to hydrazones, especially phenylhydrazones and 2,4-dinitrophenylhydrazones (Section 18-17). In his exploratory work on sugar structures, Emil Fischer often made and used phenylhydrazone derivatives. In fact, his constant use of phenylhydrazine ultimately led to Fischer's death from chronic phenylhydrazine poisoning in 1919.

Sugars do not form the simple phenylhydrazone derivatives we might expect, however. Two molecules of phenylhydrazine condense with each molecule of the sugar to give an **osazone**, in which both C1 and C2 have been converted to phenylhydrazones. The term *osazone* is derived from the *-ose* suffix of a sugar and the last half of the word hydr*azone*. Most osazones are easily crystallized, with sharp melting points. Melting points of osazone derivatives provide valuable clues for the identification and comparison of sugars.

In the formation of an osazone, both C1 and C2 are converted to phenylhydrazones. Therefore, a ketose gives the same osazone as its related aldose. Also notice that the stereochemistry at C2 is lost in the phenylhydrazone. Thus, C2 epimers give the same osazone.

PROBLEM-SOLVING $HINT$

If two aldoses form the same osazone, they are C2 epimers. If an aldose and a ketose form the same osazone, they have the same structure at all carbons except C1 and C2.

PROBLEM 23-32

(a) Show that D-glucose, D-mannose, and D-fructose all give the same osazone. Show the structure and stereochemistry of this osazone.

(b) D-Talose is an aldohexose that gives the same osazone as D-galactose. Give the structure of D-talose, and give the structure of its osazone.

In our discussion of D and L sugars, we briefly mentioned a method for shortening the chain of an aldose by removing the aldehyde carbon at the top of the Fischer projection. Such a reaction, removing one of the carbon atoms, is called a **degradation**.

The most common method used to shorten sugar chains is the **Ruff degradation**, developed by Otto Ruff, a prominent German chemist around the turn of the 20th century. The Ruff degradation is a two-step process that begins with a bromine-water oxidation of the aldose to its aldonic acid. Treatment of the aldonic acid with hydrogen peroxide and ferric sulfate oxidizes the carboxyl group to CO_2 and gives an aldose with one less carbon atom. The Ruff degradation is used mainly for structure determination and synthesis of new sugars.

23-14
Chain Shortening: The Ruff Degradation

Ruff degradation

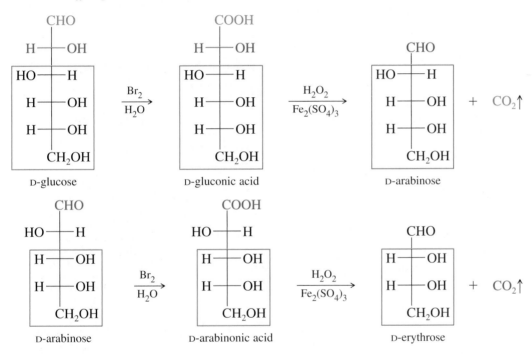

PROBLEM 23-33

Show that Ruff degradation of D-mannose gives the same aldopentose (D-arabinose) as does D-glucose.

PROBLEM 23-34

D-Lyxose is formed by Ruff degradation of galactose. Give the structure of D-lyxose. Ruff degradation of D-lyxose gives D-threose. Give the structure of D-threose.

PROBLEM 23-35

D-Altrose is an aldohexose. Ruff degradation of D-altrose gives the same aldopentose as does degradation of D-allose, the C3 epimer of glucose. Give the structure of D-altrose.

The **Kiliani–Fischer synthesis** lengthens an aldose carbon chain by adding one carbon atom to the aldehyde end of the aldose. The result of this process is a chain-lengthened sugar with a new carbon atom at C1 and the former aldehyde group (the former C1) now at C2. This synthesis is useful both for determining the structure of existing sugars and for synthesizing new sugars.

23-15
Chain Lengthening: The Kiliani–Fischer Synthesis

The Kiliani–Fischer synthesis

The aldehyde carbon atom is made asymmetric in the first step, formation of the cyanohydrin. Two epimeric cyanohydrins result. For example, D-arabinose reacts with HCN to give the following cyanohydrins.

Aqueous hydrogenation of these cyanohydrins gives two imines, which quickly hydrolyze to aldehydes. A poisoned catalyst of palladium on barium sulfate is used for the hydrogenation, to avoid over-reduction.

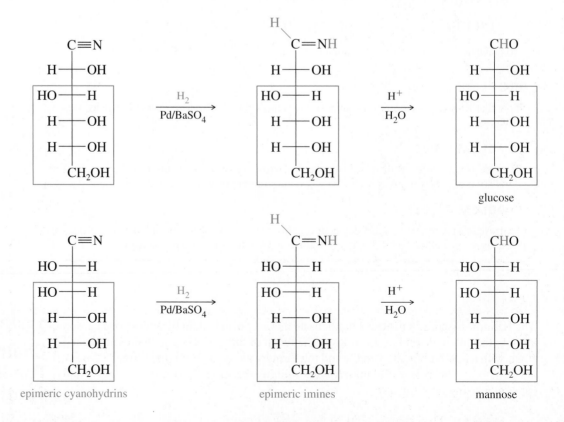

The Kiliani–Fischer synthesis accomplishes the opposite of the Ruff degradation. Ruff degradation of either of two C2 epimers gives the same shortened aldose, and the Kiliani–Fischer synthesis converts this shortened aldose back into a mixture of the same two C2 epimers. For example, glucose and mannose are degraded to arabinose, and the Kiliani–Fischer synthesis converts arabinose into a mixture of glucose and mannose.

PROBLEM 23-36

Ruff degradation of D-arabinose gives D-erythrose. The Kiliani–Fischer synthesis converts D-erythrose to a mixture of D-arabinose and D-ribose. Draw out these reactions, and give the structure of D-ribose.

PROBLEM 23-37

The *Wohl degradation*, an alternative to the Ruff degradation, is nearly the reverse of the Kiliani–Fischer synthesis. The aldose carbonyl group is converted to the oxime, which is dehydrated by acetic anhydride to the nitrile (a cyanohydrin). Cyanohydrin formation is reversible, and a basic hydrolysis allows the cyanohydrin to lose HCN. Using the following sequence of reagents, give equations for the individual reactions in the Wohl degradation of D-arabinose to D-erythrose. Mechanisms are not required.

(1) hydroxylamine hydrochloride
(2) acetic anhydride
(3) $^{-}OH, H_2O$

SUMMARY   Reactions of Sugars

1. *Undesirable rearrangements catalyzed by base* (Section 23-8)
Because of these side reactions, basic reagents are rarely used with sugars.
a. *Epimerization of the alpha carbon*

b. *Enediol rearrangements*

2. *Reduction* (Section 23-9)

3. *Oxidation* (Section 23-10)

   a. *To aldonic acids (glyconic acids) by bromine water*

$$\underset{\text{aldose}}{\begin{array}{c} \text{CHO} \\ | \\ (\text{CHOH})_n \\ | \\ \text{CH}_2\text{OH} \end{array}} \quad \xrightarrow[\text{H}_2\text{O}]{\text{Br}_2} \quad \underset{\text{aldonic acid}}{\begin{array}{c} \text{COOH} \\ | \\ (\text{CHOH})_n \\ | \\ \text{CH}_2\text{OH} \end{array}}$$

   b. *To aldaric acids (glycaric acids) by nitric acid*

$$\underset{\text{aldose}}{\begin{array}{c} \text{CHO} \\ | \\ (\text{CHOH})_n \\ | \\ \text{CH}_2\text{OH} \end{array}} \quad \xrightarrow{\text{HNO}_3} \quad \underset{\text{aldaric acid}}{\begin{array}{c} \text{COOH} \\ | \\ (\text{CHOH})_n \\ | \\ \text{COOH} \end{array}}$$

   c. *Tollens test for reducing sugars*

$$\underset{\text{aldose}}{\begin{array}{c} \text{CHO} \\ | \\ \text{CHOH} \\ | \\ (\text{CHOH})_n \\ | \\ \text{CH}_2\text{OH} \end{array}} \quad \text{or} \quad \underset{\text{ketose}}{\begin{array}{c} \text{CH}_2\text{OH} \\ | \\ \text{C}=\text{O} \\ | \\ (\text{CHOH})_n \\ | \\ \text{CH}_2\text{OH} \end{array}} \quad \xrightarrow{\text{Ag(NH}_3)_2\text{OH}} \quad \begin{array}{c} \text{COO}^- \\ | \\ \text{CHOH} \\ | \\ (\text{CHOH})_n \\ | \\ \text{CH}_2\text{OH} \end{array} \quad \begin{array}{l} + \quad \text{rearrangement} \\ \\ + \quad \text{Ag (silver mirror)} \end{array}$$

4. *Glycoside formation (conversion to an acetal)* (Section 23-11)

(either anomer) $\xrightarrow[\text{H}^+]{\text{CH}_3\text{OH}}$ a methyl glycoside
(more stable anomer predominates)

5. *Alkylation to give ethers* (Section 23-12)

$\xrightarrow[\text{Ag}_2\text{O}]{\text{excess CH}_3\text{I}}$ (gives the same anomer as the starting material)

6. *Acylation to give esters* (Section 23-12)

$\xrightarrow[\text{pyridine}]{\overset{\text{O}\quad\quad\text{O}}{\text{CH}_3\text{C}-\text{O}-\text{CCH}_3}}$ (gives the same anomer as the starting materials)

7. *Osazone formation* (Section 23-13)

either aldose epimer or ketose

osazone

8. *Ruff degradation* (Section 23-14)

aldose

shortened aldose

9. *Kiliani–Fischer synthesis* (Section 23-15)

aldose

epimers of lengthened aldose

Considering the stereochemical complexity of sugars, it is amazing that Emil Fischer determined the structures of glucose and the other aldohexoses in 1891, only 14 years after the tetrahedral structure of carbon had been proposed. Fischer received the Nobel Prize for this work in 1902. Much of Fischer's proof used the carbohydrate reactions we have studied, together with some clever reasoning about the symmetry and dissymmetry of the resulting products. We will use Fischer's work with glucose as an elegant example of these reactions, showing the determination of complex stereochemistry by clever use of simple methods.

In 1891, there were no methods for determining the absolute configuration of molecules, so Fischer could not know which enantiomer was the naturally occurring one. He made the assumption that the —OH group on C2 of (+)-glyceraldehyde (and on the bottom asymmetric carbon of the D family of sugars) is on the right in the Fischer projection. Eventually, this was shown to be a correct guess, but all his reasoning would have applied to the other enantiomers if the guess had been wrong.

Fischer had done many chemical tests on glucose, and he had used Ruff degradations to degrade it to D-(+)-glyceraldehyde. He knew that glucose is an aldose and that it has six carbon atoms; therefore, the eight members of the D family of aldohexoses (Figure 23-3) are the possible structures. Fischer used four major clues to determine which of these structures corresponds to glucose. We will consider the four clues individually and study the information obtained from each.

23-16
Fischer's Proof of the Configuration of Glucose

CLUE 1:   On Ruff degradation, glucose and mannose give the same aldopentose: D-(−)-arabinose.

This clue suggests that glucose and mannose are C2 epimers, a hypothesis he confirmed by treating them with phenylhydrazine and showing that glucose and mannose give the same osazone. Even more important, this information relates the structure of glucose to the simpler structure of arabinose, the chain-shortened aldopentose.

CLUE 2:   On Ruff degradation, D-(−)-arabinose gives the aldotetrose D-(−)-erythrose. Upon treatment with nitric acid, erythrose gives an *optically inactive* aldaric acid, *meso*-tartaric acid.

D-Erythrose, obtained from Ruff degradation of a D-aldopentose, must be a D-aldotetrose. There are only two D-aldotetroses, labeled below as structures I and II. Nitric acid oxidation of structure I gives a symmetrical meso product, but structure II gives an optically active product.

Because oxidation of D-erythrose gives an optically inactive aldaric acid, erythrose must correspond to structure I. Structure I would give *meso*-tartaric acid when it is oxidized to an aldaric acid. D-Arabinose must be one of the two epimeric structures that would degrade to this structure for D-erythrose.

CLUE 3:   On oxidation with nitric acid, D-(−)-arabinose gives an *optically active* aldaric acid.

Of the two possible structures for D-arabinose (below), only the second would oxidize to give an optically active aldaric acid. Structure B must be arabinose.

Since glucose and mannose degrade to arabinose, structures X and Y shown below must be glucose and mannose. At this point, however, it is impossible to tell which structure is glucose and which is mannose.

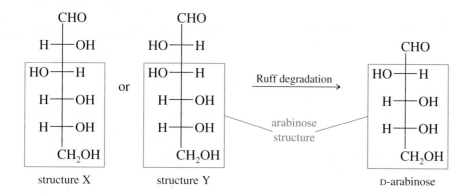

CLUE 4:   When the —CHO and —CH$_2$OH groups of D-mannose are interchanged, the product is still D-mannose. When the —CHO and —CH$_2$OH groups of D-glucose are interchanged, the product is an unnatural L sugar.

Fischer had developed a clever method for converting the aldehyde group of an aldose to an alcohol while converting the terminal alcohol group to an aldehyde. In effect, this synthesis interchanges the two end groups of the aldose chain.

$$
\begin{array}{ccc}
\text{CHO} & & \text{CH}_2\text{OH} \\
| & \xrightarrow{\text{several steps}} & | \\
(\text{CHOH})_n & & (\text{CHOH})_n \\
| & & | \\
\text{CH}_2\text{OH} & & \text{CHO} \\
\text{original aldose} & & \text{end groups interchanged}
\end{array}
$$

If the two end groups of structure X are interchanged, the product looks strange indeed. Remember that we can rotate a Fischer projection by 180°; when we do that, it becomes clear that the product is an unusual sugar of the L series (L-gulose). Structure X must be D-glucose.

structure X
(D-glucose)

end groups interchanged

rotate 180°

an L sugar
(L-gulose)

Structure Y gives a D sugar when its end groups are interchanged. In fact, a 180° rotation shows that the product of end-group interchange in structure Y gives back the original structure! Structure Y must be D-mannose.

structure Y
(D-mannose)

end groups interchanged

structure Y
(D-mannose)

This type of reasoning can be used to determine the structures of all the other aldoses. Problems 23-38 and 23-39 will give you some practice determining sugar structures.

PROBLEM-SOLVING HINT

In working this type of problem, it is often easier to start with the smallest structure mentioned (often glyceraldehyde) and work backward to larger structures. Write out all possible structures and use the clues to eliminate the wrong ones.

PROBLEM 23-38

On treatment with phenylhydrazine, aldohexoses **A** and **B** give the same osazone. On treatment with warm nitric acid, **A** gives an optically inactive aldaric acid, but sugar **B** gives an optically active aldaric acid. Sugars **A** and **B** are both degraded to aldopentose **C**, which gives an optically active aldaric acid on treatment with nitric acid. Aldopentose **C** is degraded to aldotetrose **D**, which gives optically active tartaric acid when it is treated with nitric acid. Aldotetrose **D** is degraded to (+)-glyceraldehyde. Deduce the structures of sugars **A**, **B**, **C**, and **D**, and use Figure 23-3 to determine the correct names of these sugars.

PROBLEM 23-39

Aldose **E** is optically active, but treatment with sodium borohydride converts it to an optically inactive alditol. Ruff degradation of **E** gives **F**, whose alditol is optically inactive. Ruff degradation of **F** gives optically active D-glyceraldehyde. Give the structures and names of **E** and **F** and their optically inactive alditols.

23-17
Determination of Ring Size; Periodic Acid Cleavage of Sugars

Using methods similar to Fischer's, the straight-chain form of any monosaccharide can be worked out. As we have seen, however, monosaccharides exist mostly as cyclic pyranose or furanose hemiacetals. These hemiacetals are in equilibrium with the open-chain forms, so sugars can react like hemiacetals or like ketones and aldehydes. How can we freeze this equilibrium and determine the optimum ring size for any given sugar? Sir Walter Haworth (inventor of the Haworth projection) used some simple chemistry to determine the pyranose structure of glucose in 1926.

Glucose is converted to a pentamethyl derivative by treatment with methyl iodide and silver oxide (Section 23-12). The five methyl groups are not the same, however. Four are methyl ethers, but one is the glycosidic methyl group of an acetal.

β-D-glucose

methyl 2,3,4,6-tetra-O-methyl-β-D-glucoside

Acetals are easily hydrolyzed by dilute acid, but ethers are stable under these conditions. Treatment of the pentamethyl glucose derivative with dilute acid hydrolyzes only the acetal methyl group. Haworth determined that the free hydroxyl group is on C5 of the hydrolyzed ether, showing that the cyclic form of glucose is a pyranose.

pentamethyl derivative

free hemiacetal

open-chain form
2,3,4,6,-*O*-tetramethyl-D-glucose

PROBLEM 23-40

(a) Show the product that results when fructose is treated with an excess of methyl iodide and silver oxide.
(b) Show what happens when the product of part (a) is hydrolyzed using dilute acid.
(c) Show what the results of parts (a) and (b) imply about the hemiacetal structure of fructose.

Periodic Acid Cleavage of Carbohydrates Another method used to determine the size of carbohydrate rings is cleavage by periodic acid. Recall that periodic acid cleaves vicinal diols to give two carbonyl compounds, either ketones or aldehydes, depending on the substitution of the reactant (Section 11-11B).

vicinal diol

periodic acid

ketones and aldehydes

Because ether and acetal groups are not affected, periodic acid cleavage of a glycoside can help to determine the size of the ring. For example, periodic acid oxidation of methyl β-D-glucopyranoside gives the products shown below. The structure of the fragment containing C4, C5, and C6 implies that the original glycoside was a six-membered ring bonded through the C5 oxygen atom.

methyl β-D-glucopyranoside

D-glyceraldehyde

On the other hand, if glucose were a furanose (five-membered ring), the periodic acid cleavage would give an entirely different set of products. Because glucose actually exists as a pyranose (six-membered ring), these products are not observed.

methyl β-D-glucofuranoside

(not observed)

PROBLEM-SOLVING HINT

Cleavage occurs only between two carbon atoms that bear hydroxyl groups.

PROBLEM 23-41

(a) Draw the reaction of methyl β-D-fructofuranoside with periodic acid, and predict the products.

(b) Draw the structure of a hypothetical methyl β-D-fructopyranoside, and predict the products from periodic acid oxidation.

(c) The reaction of methyl β-D-glucopyranoside with periodic acid (shown above) gives only the D-(+) enantiomer of glyceraldehyde (among other products). If you oxidized an aldohexose glycoside with periodic acid and one of the products was the L-(−) enantiomer of glyceraldehyde, what would that tell you about the sugar?

23-18
Disaccharides

As we have seen, the anomeric carbon of a sugar can react with the hydroxyl group of an alcohol to give an acetal called a *glycoside*. If the hydroxyl group is part of another sugar molecule, then the glycoside product is a **disaccharide**, a sugar composed of two monosaccharide units (Figure 23-16).

▲ **Figure 23-16**
A sugar reacts with an alcohol to give an acetal called a glycoside. When the alcohol is part of another sugar, the product is a disaccharide.

In principle, the anomeric carbon can react with *any* of the hydroxyl groups of another sugar to form a disaccharide. In naturally occurring disaccharides, however, there are three common glycosidic bonding arrangements.

1. A 1,4′ link. The anomeric carbon is bonded to the oxygen atom on C4 of the second sugar. The prime symbol (′) in 1,4′ indicates that C4 is on the second sugar.
2. A 1,6′ link. The anomeric carbon is bonded to the oxygen atom on C6 of the second sugar.
3. A 1,1′ link. The anomeric carbon of the first sugar is bonded through an oxygen atom to the anomeric carbon of the second sugar.

We will consider some naturally occurring disaccharides with these common glycosidic linkages.

23-18A   The 1,4′ Linkage: Cellobiose, Maltose, and Lactose

The most common glycosidic linkage is the 1,4′ link. The anomeric carbon of one sugar is bonded to the oxygen atom on C4 of the second ring.

Cellobiose: A β-1,4′ Glucosidic Linkage   *Cellobiose*, the disaccharide obtained by partial hydrolysis of cellulose, contains a 1,4′ linkage. In cellobiose, the anomeric carbon of one glucose unit is linked through an equatorial (β) carbon–oxygen bond to C4 of another glucose unit. This β-1,4′ linkage from a *glucose* acetal is called a β-1,4′ **gluco**sidic linkage.

Cellobiose, 4-O-(β-D-glucopyranosyl)-β-D-glucopyranose or 4-O-(β-D-glucopyranosyl)-D-glucopyranose

Two alternative ways of drawing and naming cellobiose

The complete name for cellobiose, 4-*O*-(β-D-glucopyranosyl)-β-D-glucopyranose, gives its structure. This name says that a β-D-glucopyranose ring (the right-hand ring) is substituted in its 4-position by an oxygen attached to a (β-D-glucopyranosyl) ring, drawn on the left. The name in parentheses says the substituent is a β-glucose, and the -*syl* ending indicates that this ring is a glycoside. The left ring with the -*syl* ending is an acetal and cannot mutarotate, while the right ring with the -*ose* ending is a hemiacetal and can mutarotate. Because cellobiose has a glucose unit in the hemiacetal form (and therefore is in equilibrium with its open-chain aldehyde form), it is a reducing sugar. Once again, the -*ose* ending indicates a mutarotating, reducing sugar.

Mutarotating sugars are often shown with a wavy line to the free anomeric hydroxyl group, signifying that they can exist as an equilibrium mixture of the two anomers. Their names are often given without specifying the stereochemistry of this mutarotating hydroxyl group, as in 4-*O*-(β-D-glucopyranosyl)-D-glucopyranose.

Maltose: An α-1,4′ Glucosidic Linkage   *Maltose* is a disaccharide formed when starch is treated with sprouted barley, called *malt*. This malting process is the first step in brewing beer, converting polysaccharides to disaccharides and monosaccharides that ferment more easily. Like cellobiose, maltose contains a 1,4′ glycosidic linkage between two glucose units. The difference in maltose is that the stereochemistry of the glucosidic linkage is α rather than β.

Maltose, 4-O-(α-D-glucopyranosyl)-D-glucopyranose

α-1,4′ glucosidic linkage

Like cellobiose, maltose has a free hemiacetal ring (on the right). This hemiacetal is in equilibrium with its open-chain form, and it mutarotates and can exist in either the α or β anomeric form. Because maltose exists in equilibrium with an open-chain aldehyde, it reduces Tollens reagent, and maltose is a reducing sugar.

PROBLEM 23-42

Draw the structure of the individual mutarotating α and β anomers of maltose.

PROBLEM 23-43

Give an equation to show the reduction of Tollens reagent by maltose.

Lactose: A β-1,4′ Galactosidic Linkage   *Lactose* is similar to cellobiose, except that the glycoside (the left ring) in lactose is galactose rather than glucose. Lactose is composed of one galactose unit and one glucose unit. The two rings are linked by a β-glycosidic bond of the galactose acetal to the 4-position on the glucose ring: a β-1,4′ *galacto*sidic linkage.

Lactose, 4-O-(β-D-galactopyranosyl)-D-glucopyranose

axial 4-hydroxyl group of galactose

β-galactosidic linkage

Lactose occurs naturally in the milk of mammals, including cows and humans. Hydrolysis of lactose requires a β-galactosidase enzyme (sometimes called lactase). Some humans synthesize a β-galactosidase, but others do not. This enzyme is present in the digestive fluids of normal infants to hydrolyze their mother's milk. Once the child stops drinking milk, production of the enzyme gradually stops. In most parts of the world, people do not use milk products after early childhood, and the adult population can no longer digest lactose. Consumption of milk or milk products can cause digestive discomfort in *lactose-intolerant* people who lack the β-galactosidase enzyme. Lactose-intolerant infants must drink soybean milk or another lactose-free formula.

PROBLEM 23-44

Does lactose mutarotate? Is it a reducing sugar? Explain. Draw the two anomeric forms of lactose.

23-18B   The 1,6′ Linkage: Gentiobiose

In addition to the common 1,4′ glycosidic linkage, the 1,6′ linkage is also found in naturally occurring carbohydrates. In a 1,6′ linkage, the anomeric carbon of one sugar is linked to the oxygen of the terminal carbon (C6) of another. This linkage gives a different sort of stereochemical arrangement, because the hydroxyl group on C6 is one carbon atom removed from the ring. Gentiobiose is a sugar with two glucose units joined by a β-1,6′ glucosidic linkage.

Gentiobiose, 6-O-(β-D-glucopyranosyl)-D-glucopyranose

Although the 1,6′ linkage is rare in disaccharides, it is commonly found as a branch point in polysaccharides. For example, branching in amylopectin (insoluble starch) occurs at 1,6′ linkages, as discussed in Section 23-19B.

PROBLEM 23-45

Is gentiobiose a reducing sugar? Does it mutarotate? Explain your reasoning.

23-18C   Linkage of Two Anomeric Carbons: Sucrose

Some sugars are joined by a direct glycosidic linkage between their anomeric carbon atoms: a 1,1′ linkage. *Sucrose* (common table sugar), for example, is composed of one glucose unit and one fructose unit bonded by an oxygen atom linking their anomeric carbon atoms. (Because fructose is a ketose and its anomeric carbon is C2, this is actually a 1,2′ linkage.) Notice that the linkage is in the α position with respect to the glucose ring and in the β position with respect to the fructose ring.

Sucrose, α-D-glucopyranosyl-β-D-fructofuranoside
(or β-D-fructofuranosyl-α-D-glucopyranoside)

Sucrose (a nonreducing sugar) is not as easily oxidized as a reducing sugar, so it is much more useful for preserving foods such as jams and jellies. A reducing sugar like glucose would oxidize and spoil.

Both monosaccharide units in sucrose are present as acetals, or glycosides. Neither ring is in equilibrium with its open-chain aldehyde or ketone form, so sucrose does not reduce Tollens reagent and it cannot mutarotate. Because both units are glycosides, the systematic name for sucrose can list either of the two glycosides as being a substituent on the other. Both systematic names end in the *-oside* suffix, indicating a nonmutarotating, nonreducing sugar. Like many other common names, *sucrose* ends in the *-ose* ending even though it is a nonreducing sugar. Common names are not reliable indicators of the properties of sugars.

Sucrose is hydrolyzed by enzymes called *invertases*, found in honeybees and yeasts, that specifically hydrolyze the β-D-fructofuranoside linkage. The resulting mixture of glucose and fructose is called *invert sugar* because hydrolysis converts the positive rotation [+66.5°] of sucrose to a negative rotation that is the average of glucose [+52.7°] and fructose [−92.4°]. The most common form of invert sugar is honey, a supersaturated mixture of glucose and fructose hydrolyzed from sucrose by the invertase enzyme of honeybees. Glucose and fructose were once called *dextrose* and *levulose*, respectively, according to their opposite signs of rotation.

SOLVED PROBLEM 23-3

An unknown carbohydrate of formula $C_{12}H_{22}O_{11}$ reacts with Tollens reagent to form a silver mirror. An α-glycosidase has no effect on the carbohydrate, but a β-galactosidase hydrolyzes it to D-galactose and D-mannose. When the carbohydrate is methylated (using methyl iodide and silver oxide) and then hydrolyzed with dilute HCl, the products are 2,3,4,6-tetra-*O*-methylgalactose and 2,3,4-tri-*O*-methylmannose. Propose a structure for this unknown carbohydrate.

SOLUTION

The formula shows this is a disaccharide composed of two hexoses. Hydrolysis gives D-galactose and D-mannose, identifying the two hexoses. Hydrolysis requires a β-galactosidase, showing that galactose and mannose are linked by a β-galactosyl linkage. Since the original carbohydrate is a reducing sugar, one of the hexoses must be in a free hemiacetal form. Galactose is present as a glycoside; thus mannose must be present in its hemiacetal form. The unknown carbohydrate must be a (β-galactosyl)-mannose.

The methylation/hydrolysis procedure shows the point of attachment of the glycosidic bond to mannose and also confirms the size of the six-membered rings. In galactose, all the hydroxyl groups are methylated except C1 and C5. C1 is the anomeric carbon, and the C5 oxygen is used to form the hemiacetal of the pyranose ring. In mannose, all the hydroxyl groups are methylated except C1, C5, and C6. The C5 oxygen is used to form the pyranose ring (the C6 oxygen would form a less stable seven-membered ring); therefore, the oxygen on C6 must be involved in the glycosidic linkage. The structure and systematic name are shown below.

6-*O*-(β-D-galactopyranosyl)-D-mannopyranose

PROBLEM 23-46

Trehalose is a nonreducing disaccharide ($C_{12}H_{22}O_{11}$) isolated from the poisonous mushroom *Amanita muscaria*. Treatment with an α-glucosidase converts trehalose to two molecules of glucose, but no reaction occurs when trehalose is treated with a β-glucosidase. When trehalose is methylated by dimethyl sulfate in mild base and then hydrolyzed, the only product is 2,3,4,6-tetra-*O*-methylglucose. Propose a complete structure and systematic name for trehalose.

Raffinose is a trisaccharide ($C_{18}H_{32}O_{16}$) isolated from cottonseed meal. Raffinose does not reduce Tollens reagent, and it does not mutarotate. Complete hydrolysis of raffinose gives D-glucose, D-fructose, and D-galactose. When raffinose is treated with invertase, the products are D-fructose and a reducing disaccharide called *melibiose*. Raffinose is unaffected by treatment with a β-galactosidase, but an α-galactosidase hydrolyzes it to D-galactose and sucrose. When raffinose is treated with dimethyl sulfate and base followed by hydrolysis, the products are 2,3,4-tri-*O*-methylglucose, 1,3,4,6-tetra-*O*-methylfructose, and 2,3,4,6-tetra-*O*-methylgalactose. Determine the complete structures of raffinose and melibiose, and give a systematic name for melibiose.

Polysaccharides are carbohydrates that contain many monosaccharide units joined by glycosidic bonds. They are one class of *biopolymers*, or naturally occurring polymers. Smaller polysaccharides, containing about three to ten monosaccharide units, are sometimes called **oligosaccharides**. Most polysaccharides have hundreds or thousands of simple sugar units linked together into long polymer chains. Except for units at the ends of chains, all the anomeric carbon atoms of polysaccharides are involved in acetal glycosidic links. Therefore, polysaccharides give no noticeable reaction with Tollens reagent, and they do not mutarotate.

23-19
Polysaccharides

23-19A   Cellulose

Cellulose, a polymer of D-glucose, is the most abundant organic material. Cellulose is synthesized by plants as a structural material to support the weight of the plant. Long cellulose molecules, called *microfibrils*, are held in bundles by hydrogen bonding between the many —OH groups of the glucose rings. About 50% of dry wood and about 90% of cotton fiber is cellulose.

Cellulose is composed of D-glucose units linked by β-1,4′ glycosidic bonds. This bonding arrangement (like that in cellobiose) is rather rigid and very stable, giving cellulose desirable properties for a structural material. Figure 23-17 shows a partial structure of cellulose.

Humans and other mammals lack the β-glucosidase enzyme needed to hydrolyze cellulose, so they cannot use it directly for food. Several groups of bacteria and protozoa can hydrolyze cellulose, however. Termites and ruminants maintain colonies of these bacteria in their digestive tracts. When a cow eats hay, these bacteria convert about 20 to 30% of the cellulose to digestible carbohydrates.

Rayon is a fiber made from cellulose that has been converted to a soluble derivative, and then regenerated. In the common *viscose process*, wood pulp is treated with carbon disulfide and sodium hydroxide to convert the free hydroxyl groups to xanthates, which are soluble in water. The viscous solution (called *viscose*) is forced through a spinneret into an aqueous sodium bisulfate solution, where a fiber of insoluble cellulose is regenerated.

The acoustic properties of cellulose have never been surpassed by other substances. Here, a luthier carves maple for use in a violin.

β-glucosidic linkage

◀ **Figure 23-17**
Cellulose is a β-1,4′ polymer of D-glucose, systematically named poly(1,4′-*O*-β-D-glucopyranoside).

Alternatively, the viscose solution can be extruded in sheets to give *cellophane* film. Rayon and cotton are both cellulose, yet rayon thread can be much stronger because it consists of long, continuously extruded fibers, rather than short cotton fibers spun together.

$$\text{ROH} \;+\; \text{CS}_2 \;+\; \text{NaOH} \;\longrightarrow\; \left[\text{RO}-\overset{\displaystyle S}{\underset{\displaystyle \|}{C}}-\text{S}^-\right]\text{Na}^+ \;+\; \text{H}_2\text{O}$$

cellulose   xanthate derivatives
(viscose)

$$\left[\text{RO}-\overset{\displaystyle S}{\underset{\displaystyle \|}{C}}-\text{S}^-\right]\text{Na}^+ \;+\; \text{NaHSO}_4 \;\xrightarrow[\text{H}_2\text{O}]{}\; \text{ROH} \;+\; \text{CS}_2 \;+\; \text{Na}_2\text{SO}_4$$

extruded into solution   rayon
(regenerated cellulose)

PROBLEM 23-48

Cellulose is converted to *cellulose acetate* by treatment with acetic anhydride and pyridine. Cellulose acetate is soluble in common organic solvents, and it is easily dissolved and spun into fibers. Show the structure of cellulose acetate.

23-19B   Starches: Amylose, Amylopectin, and Glycogen

Plants use starch granules for storing energy. When the granules are dried and ground up, different types of starches can be separated by mixing them with hot water. About 20% of the starch is water-soluble *amylose*, and the remaining 80% is water-insoluble *amylopectin*. When starch is treated with dilute acid or appropriate enzymes, it is progressively hydrolyzed to maltose and then to glucose.

Amylose   Like cellulose, **amylose** is a linear polymer of glucose with 1,4' glycosidic linkages. The difference is in the stereochemistry of the linkage. Amylose has α-1,4' links, while cellulose has β-1,4' links. A partial structure of amylose is shown in Figure 23-18.

The subtle stereochemical difference between cellulose and amylose results in some striking physical and chemical differences. The α linkage in amylose kinks the polymer chain into a helical structure. This kinking increases hydrogen bonding with water and lends additional solubility. As a result, amylose is soluble in water, while cellulose is not. Cellulose is stiff and sturdy, but amylose is not. Unlike cellulose, amylose is an excellent food source. The α-1,4' glucosidic linkage is easily hydrolyzed by an α-glucosidase enzyme, present in all animals.

▶ **Figure 23-18**
Amylose is an α-1,4' polymer of glucose, systematically named poly (1,4'-*O*-α-D-glucopyranoside). Amylose differs from cellulose only in the stereochemistry of the glycosidic linkage.

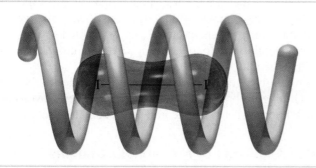

◀ **Figure 23-19**
The amylose helix forms a blue charge-transfer complex with molecular iodine.

The helical structure of amylose also serves as the basis for an interesting and useful reaction. The inside of the helix is just the right size and polarity to accept an iodine (I_2) molecule. When iodine is lodged within this helix, a deep blue starch-iodine complex results (Figure 23-19). This is the basis of the *starch-iodide* test for oxidizers. The material to be tested is added to an aqueous solution of amylose and potassium iodide. If the material is an oxidizer, some of the iodide (I^-) is oxidized to iodine (I_2), which forms the blue complex with amylose.

Amylopectin **Amylopectin**, the insoluble fraction of starch, is also primarily an α-1,4′ polymer of glucose. The difference between amylose and amylopectin lies in the branched nature of amylopectin, with a branch point about every 20 to 30 glucose units. Another chain starts at each branch point, connected to the main chain by an α-1,6′ glycosidic linkage. A partial structure of amylopectin, including one branch point, is shown in Figure 23-20.

α-1,6′ glucosidic linkage branch point

▲ **Figure 23-20**
Amylopectin is a branched α-1,4′ polymer of glucose. At the branch points, there is a single α-1,6′ linkage that provides the attachment point for another chain. Glycogen has a similar structure, except that its branching is more extensive.

Glycogen   **Glycogen** is the carbohydrate that animals use to store glucose for readily available energy. A large amount of glycogen is stored in the muscles themselves, ready for immediate hydrolysis and metabolism. Additional glycogen is stored in the liver, where it can be hydrolyzed to glucose for secretion into the bloodstream, providing an athlete with a "second wind."

The structure of glycogen is similar to that of amylopectin, but with more extensive branching. The highly branched structure of glycogen leaves many end groups available for quick hydrolysis to provide glucose needed for metabolism.

23-19C   Chitin: A Polymer of *N*-Acetylglucosamine

Chitin (pronounced *kī'-t'n*, rhymes with Titan) forms the exoskeletons of insects. In crustaceans, chitin forms a matrix that binds calcium carbonate crystals into the exoskeleton. Chitin is different from the other carbohydrates we have studied. It is a polymer of *N*-acetylglucosamine, an amino sugar (actually an amide) that is common in living organisms. In *N*-acetylglucosamine, the hydroxyl group on C2 of glucose is replaced by an amino group (forming glucosamine), and that amino group is acetylated.

This cicada is shedding its nymphal exoskeleton. Chitin lends strength and rigidity to the exoskeletons of insects, but it cannot grow and change with the insect.

N-*Acetylglucosamine, or 2-acetamido-2-deoxy-D-glucose*

Chitin synthase inhibitors are used commercially to kill termite colonies. These compounds prevent the formation of a new exoskeleton, trapping the termites in an old exoskeleton that cannot grow with them.

Chitin is bonded like cellulose, except using *N*-acetylglucosamine instead of glucose. Like other amides, *N*-acetylglucosamine forms exceptionally strong hydrogen bonds between the amide carbonyl groups and N—H protons. The glycosidic bonds are β-1,4′ links, giving chitin structural rigidity, strength, and stability that exceed even that of cellulose. Unfortunately, this strong, rigid polymer cannot easily expand, so it must be shed periodically by molting as the animal grows.

Chitin, or poly (1,4′-O-β-2-acetamido-2-deoxy-D-glucopyranoside),
a β-1,4-linked polymer of N-*acetylglucosamine*

23-20
Nucleic Acids: Introduction

Nucleic acids are substituted polymers of the aldopentose ribose that carry an organism's genetic information. A tiny amount of DNA in a fertilized egg cell determines the physical characteristics of the fully developed animal. The difference between a frog and a human being is encoded in a relatively small part of this DNA. Each cell carries a complete set of genetic instructions that determine the type of cell, what its function will be, when it will grow and divide, and how it will synthesize all the proteins, enzymes, carbohydrates, and other substances the cell and the organism need to survive.

The two major classes of nucleic acids are **ribonucleic acids (RNA)** and **deoxyribonucleic acids (DNA)**. In a typical cell, DNA is found primarily in the nucleus, where it carries the permanent genetic code. The molecules of DNA are huge, with molecular weights up to 50 billion. When the cell divides, DNA replicates to form two copies for the daughter cells. DNA is relatively stable, providing a medium for transmission of genetic information from one generation to the next.

RNA molecules are typically much smaller than DNA, and they are more easily hydrolyzed and broken down. RNA commonly serves as a working copy of the nuclear DNA being decoded. Nuclear DNA directs the synthesis of *messenger RNA*, which leaves the nucleus to serve as a template for the construction of protein molecules in the ribosomes. The messenger RNA is then enzymatically cleaved to its component parts, which become available for assembly into new RNA molecules to direct other syntheses.

The backbone of a nucleic acid is a polymer of ribofuranoside rings (five-membered rings of the sugar ribose) linked by phosphate ester groups. Each ribose unit carries a heterocyclic *base* that provides part of the information needed to specify a particular amino acid in protein synthesis. Figure 23-21 shows the ribose-phosphate backbone of RNA.

DNA and RNA each contain four monomers, called **nucleotides**, that differ in the structure of the bases bonded to the ribose units. Yet this deceptively simple structure encodes complex information just as the 0 and 1 bits used by a computer encode complex programs. First we consider the structure of individual nucleotides, then the bonding of these monomers into single-stranded nucleic acids, and finally the base pairing that binds two strands into the double helix of nuclear DNA.

HIV (the AIDS virus) is shown here attacking a T-4 lymphocyte. HIV is an RNA virus whose genetic material must be translated to DNA before inserting itself into the host cell's DNA. Several of the anti-AIDS drugs are directed toward stopping this reverse transcription of RNA to DNA. (Magnification 1000X)

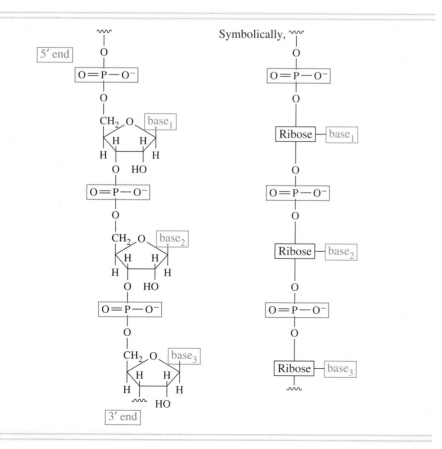

◀ **Figure 23-21**
A short segment of the RNA polymer. Nucleic acids are assembled on a backbone made up of ribofuranoside units linked by phosphate esters.

23-21
Ribonucleosides and Ribonucleotides

Ribonucleosides are components of RNA based on glycosides of the furanose form of D-ribose. We have seen (Section 23-11) that a glycoside may have an aglycone (the substituent on the anomeric carbon) bonded by a nitrogen atom. A ribonucleoside is a β-D-ribofuranoside (a β-glycoside of D-ribofuranose) whose aglycone is a heterocyclic nitrogen base. The following structures show the open-chain and furanose forms of ribose, and a ribonucleoside with a generic base bonded through a nitrogen atom.

CHO

H—OH

H—OH

H—OH

CH₂OH

D-ribose

β-D-ribofuranose

a ribonucleoside

The four bases commonly found in RNA are divided into two classes: The monocyclic compounds cytosine and uracil are called *pyrimidine bases* because they resemble substituted pyrimidines, and the bicyclic compounds adenine and guanine are called *purine bases* because they resemble the bicyclic heterocycle purine (Section 19-3).

pyrimidine   cytosine (C)   uracil (U)

pyrimidine bases

adenine (A)   guanine (G)   purine

purine bases

When bonded to ribose through the nitrogen atoms circled above, the four heterocyclic bases make up the four ribonucleosides cytidine, uridine, adenosine, and guanosine (Figure 23-22). Notice that the two ring systems (the base and the sugar) are numbered separately, and the carbons of the sugar are given primed numbers. For example, the 3′ carbon of cytidine is C3 of the ribose ring.

cytidine (C)   uridine (U)   adenosine (A)   guanosine (G)

▲ **Figure 23-22**
The four common ribonucleosides are cytidine, uridine, adenosine, and guanosine.

PROBLEM 23-49

Cytosine, uracil, and guanine have tautomeric forms with phenolic hydroxyl groups. Draw these tautomeric forms.

PROBLEM 23-50

(a) An aliphatic aminoglycoside is relatively stable to base, but it is quickly hydrolyzed by dilute acid. Propose a mechanism for the acid-catalyzed hydrolysis.

$$\text{HO—CH}_2 \quad \xrightarrow{\text{H}_3\text{O}^+} \quad \text{R}_2\overset{+}{\text{NH}}_2 \; + \; \text{sugar}$$

an aliphatic riboside

(b) Ribonucleosides are not so easily hydrolyzed, requiring relatively strong acid. Using your mechanism for part (a), show why cytidine and adenosine (for example) are not so readily hydrolyzed. Explain why this stability is important for living organisms.

Uric acid is one of the principal end products of purine metabolism. Gout is caused by elevated levels of uric acid in the body, causing uric acid crystals to precipitate in the joints.

uric acid

Ribonucleotides Ribonucleic acid consists of ribonucleosides bonded together into a polymer. This polymer cannot be bonded by glycosidic linkages like those of other polysaccharides because the glycosidic bonds are already used to attach the heterocyclic bases. Instead, the ribonucleoside units are linked by phosphate esters. The 5'-hydroxyl group of each ribofuranoside is esterified to phosphoric acid. A ribonucleoside that is phosphorylated at its 5' carbon is called **a ribonucleotide** ("tied" to phosphate). The four common ribonucleotides, shown in Figure 23-23, are simply phosphorylated versions of the four common ribonucleosides.

The phosphate groups of these ribonucleotides can exist in any of three ionization states, depending on the pH of the solution. At the nearly neutral pH of most organisms (pH = 7.4), there is one proton on the phosphate group. By convention, however, these groups are usually written completely ionized.

$$\underset{\text{in acid}}{\text{HO—P—O—ribose}} \qquad \underset{\text{nearly neutral}}{^-\text{O—P—O—ribose}} \qquad \underset{\substack{\text{in base}\\ \text{(usually written)}}}{^-\text{O—P—O—ribose}}$$

cytidine monophosphate, CMP (cytidylic acid)

uridine monophosphate, UMP (uridylic acid)

adenosine monophosphate, AMP (adenylic acid)

guanosine monophosphate, GMP (guanidylic acid)

▲ Figure 23-23

Four common ribonucleotides. These are ribonucleosides esterified by phosphoric acid at their 5'-position, the —CH_2OH at the end of the ribose chain.

23-22
The Structure of Ribonucleic Acid

Now that we recognize the individual ribonucleotides, we can consider how these units are bonded into the RNA polymer. Each nucleotide has a phosphate group on its 5′ carbon (the end carbon of ribose) and a hydroxyl group on the 3′ carbon. Two nucleotides are joined by a phosphate ester linkage between the 5′-phosphate group of one nucleotide and the 3′-hydroxyl group of another (Figure 23-24).

The RNA polymer consists of many nucleotide units bonded this way, with a phosphate ester linking the 5′ end of one nucleoside to the 3′ end of another. A molecule of RNA always has two ends (unless it is in the form of a large ring); one end has a free 3′ group, and the other end has a free 5′ group. We refer to the ends as the *3′ end* and the *5′ end*, and we refer to directions of replication as the *3′ → 5′ direction* and the *5′ → 3′ direction*. Figures 23-21 and 23-24 show short segments of RNA with the 3′ end and the 5′ end labeled.

▶ **Figure 23-24**
Two nucleotides are joined by a phosphate linkage between the 5′-phosphate group of one and the 3′-hydroxyl group of the other.

23-23
Deoxyribose and the Structure of Deoxyribonucleic Acid

All our descriptions of ribonucleosides, ribonucleotides, and ribonucleic acid also apply to the components of DNA. The principal difference between RNA and DNA is the presence of D-2-deoxyribose as the sugar in DNA instead of the D-ribose found in RNA. The prefix *deoxy-* means that an oxygen atom is missing, and the number 2 means it is missing from C2.

D-2-deoxyribose   β-D-2-deoxyribofuranose   a deoxyribonucleoside

Another key difference between RNA and DNA is the presence of thymine in DNA instead of the uracil in RNA. Thymine is simply uracil with an additional methyl group. The four common bases of DNA are cytosine, thymine, adenine, and guanine.

cytosine (C)   thymine (T)   adenine (A)   guanine (G)

pyrimidine bases   purine bases

These four bases are incorporated into deoxyribonucleosides and deoxyribonucleotides similar to the bases in ribonucleosides and ribonucleotides. The following structures show the common nucleosides that make up DNA. The corresponding nucleotides are simply the same structures with phosphate groups at the 5′ positions.

The structure of the DNA polymer is similar to that of RNA, except there are no hydroxyl groups on the 2′ carbon atoms of the ribose rings. The alternating deoxyribose rings and phosphates act as the backbone, while the bases attached to the ribose units carry the genetic information. The sequence of nucleotides is called the **primary structure** of the DNA strand.

Four common deoxyribonucleosides that make up DNA

deoxycytidine   deoxythymidine   deoxyadenosine   deoxyguanosine

23-23A   Base Pairing

Having discussed the primary structure of DNA and RNA, we now consider how the nucleotide sequence is reproduced or transcribed into another molecule. This information transfer takes place by an interesting hydrogen-bonding interaction between specific pairs of bases.

Each pyrimidine base forms a stable hydrogen-bonded pair with only one of the two purine bases (Figure 23-25). Cytosine forms a base pair, joined by three hydrogen bonds, with guanine. Thymine (or uracil in RNA) forms a base pair with adenine, joined by two hydrogen bonds. Guanine is said to be *complementary* to cytosine, and adenine is complementary to thymine. This base pairing was first suspected in 1950, when Erwin Chargaff of Columbia University noticed that various DNAs, taken from a wide variety of species, had about equal amounts of adenine and thymine, and about equal amounts of guanine and cytosine.

PROBLEM 23-51

All the rings of the four heterocyclic bases are aromatic. This is more apparent when the polar resonance forms of the amide groups are drawn, as is done for thymine at the right. Redraw the hydrogen-bonded guanine-cytosine and adenine-thymine pairs shown in Figure 23-25, using the polar resonance forms of the amides. Show how these forms help to explain why the hydrogen bonds involved in these pairings are particularly strong. Remember that a hydrogen bond arises between an electron-deficient hydrogen atom and an electron-rich pair of nonbonding electrons.

thymine

guanine   G ⦂⦂⦂ C   cytosine   **adenine**   A ⦂⦂⦂ T   thymine

▲ **Figure 23-25**

Base pairing in DNA and RNA. Each purine base forms a stable hydrogen-bonded pair with a specific pyrimidine base. Guanine forms a base pair with three hydrogen bonds to cytosine, and adenine forms a base pair with two hydrogen bonds to thymine (or uracil in RNA). The electrostatic potential maps show that hydrogen bonding takes place between electron-poor hydrogen atoms (blue regions) and electron-rich nitrogen or oxygen atoms (red regions). (In these drawings, "ribose" means β-D-2-deoxyribofuranoside in DNA and β-D-ribofuranoside in RNA.)

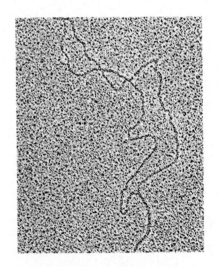

Electron micrograph of double-stranded DNA that has partially uncoiled to show the individual strands. (Magnification 13,000X)

23-23B   The Double Helix of DNA

In 1953, James D. Watson and Francis C. Crick used X-ray diffraction patterns of DNA fibers to determine the molecular structure and conformation of DNA. They found that DNA contains two complementary polynucleotide chains held together by hydrogen bonds between the paired bases. Figure 23-26 shows a portion of the double strand of DNA, with each base paired with its complement. The two strands are *antiparallel*: One strand is arranged $3' \rightarrow 5'$ from left to right, while the other runs in the opposite direction, $5' \rightarrow 3'$ from left to right.

Watson and Crick also found that the two complementary strands of DNA are coiled into a helical conformation about 20 Å in diameter, with both chains coiled around the same axis. The helix makes a complete turn for every ten residues, or about one turn in every 34 Å of length. Figure 23-27 shows the double helix of DNA. In this drawing, the two sugar-phosphate backbones form the vertical double helix with the heterocyclic bases stacked horizontally in the center. Attractive stacking forces between the pi clouds of the aromatic pyrimidine and purine bases are substantial, further helping to stabilize the helical arrangement.

When DNA undergoes replication (in preparation for cell division), an enzyme uncoils part of the double strand. Individual nucleotides naturally hydrogen-bond to their complements on the uncoiled part of the original strand, and a *DNA polymerase* enzyme couples the nucleotides to form a new strand. This process is depicted

▲ Figure 23-26
DNA usually consists of two complementary strands, with all the base pairs hydrogen-bonded together. The two strands are antiparallel, running in opposite directions.

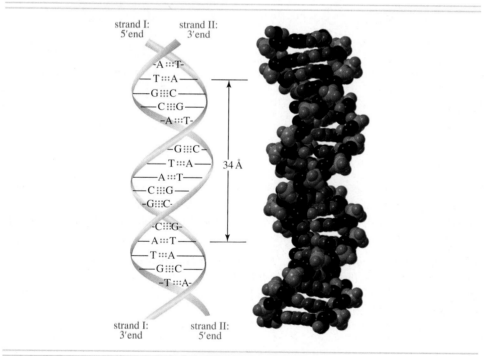

◀ Figure 23-27
Double helix of DNA. Two complementary strands are joined by hydrogen bonds between the base pairs. This double strand coils into a helical arrangement.

schematically in Figure 23-28. A similar process transcribes DNA into a complementary molecule of messenger RNA for use by ribosomes as a template for protein synthesis.

A great deal is known about replication of DNA and translation of the DNA/RNA sequence of bases into proteins. These exciting aspects of nucleic acid chemistry are part of the field of *molecular biology*, and they are covered in detail in biochemistry courses.

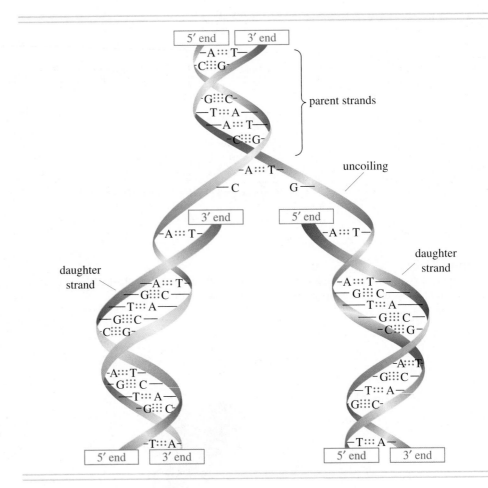

▶ **Figure 23-28**
Replication of the double strand of DNA. A new strand is assembled on each of the original strands, with the DNA polymerase enzyme forming the phosphate ester bonds of the backbone.

23-24
Additional Functions of Nucleotides

We generally think of nucleotides as the monomers that form DNA and RNA, yet these versatile biomolecules serve a variety of additional functions. Here we briefly consider a few additional functions.

AMP: A Regulatory Hormone Adenosine monophosphate (AMP) also occurs in a cyclic form, where the 3'- and 5'-hydroxyl groups are both esterified by the same phosphate group. This *cyclic AMP* is involved in transmitting and amplifying the chemical signals of other hormones.

Adenosine deaminase replaces the C-6 amino group with a hydroxyl group, an important step in purine metabolism. A genetic deficiency of the enzyme causes a severe immunodeficiency, called "baby in a bubble syndrome," because the child must live in a sterile environment.

adenosine monophosphate (AMP)

cyclic AMP

NAD: A Coenzyme   Nicotinamide adenine dinucleotide (NAD) is one of the principal oxidation–reduction reagents in biological systems. This nucleotide has the structure of two D-ribose rings (a *di*nucleotide) linked by their 5′ phosphates. The aglycone of one ribose is nicotinamide, and the aglycone of the other is adenine. A dietary deficiency of nicotinic acid (niacin) leads to the disease called *pellagra*, caused by the inability to synthesize enough nicotinamide adenine dinucleotide.

nicotinic acid (niacin)

nicotinamide

adenine

NAD⁺
nicotinamide adenine dinucleotide

The following equation shows how NAD^+ serves as the oxidizing agent in the biological oxidation of an alcohol. Just the nicotinamide portion of NAD shown takes part in the reaction. The enzyme that catalyzes this reaction is called alcohol dehydrogenase (ADH).

ethanol

NAD⁺

acetaldehyde

NADH (reduced)

ATP: An Energy Source   When glucose is oxidized in the living cell, the energy released is used to synthesize *adenosine triphosphate* (ATP), an anhydride of phosphoric acid. As with most anhydrides, hydrolysis of ATP is highly exothermic. The hydrolysis products are adenosine diphosphate (ADP) and inorganic phosphate.

adenosine triphosphate (ATP)   + H_2O ⟶

adenosine diphosphate (ADP)   phosphate

$\Delta H° = -7.3$ kcal/mol (-31 kJ/mol)

The highly exothermic nature of ATP hydrolysis is largely explained by the heats of hydration of the products. ADP is hydrated about as well as ATP, but inorganic phosphate has a large heat of hydration. Hydrolysis also reduces the electrostatic repulsion of the three negatively charged phosphate groups in ATP. Hydrolysis of adenosine triphosphate (ATP) liberates 7.3 kcal (31 kJ) of energy per mole of ATP. This is the energy that muscle cells use to contract and all cells use to drive their endothermic chemical processes.

Chapter 23 Glossary

aglycone A nonsugar residue bonded to the anomeric carbon of a glycoside (the acetal form of a sugar). Aglycones are commonly bonded to the sugar through oxygen or nitrogen. (p. 1077)

aldaric acid (glycaric acid, saccharic acid) A dicarboxylic acid formed by oxidation of both end carbon atoms of a monosaccharide. (p. 1074)

alditol (sugar alcohol) A polyalcohol formed by reduction of the carbonyl group of a monosaccharide. (p. 1072)

aldonic acid (glyconic acid) A monocarboxylic acid formed by oxidation of the aldehyde group of an aldose. (p. 1073)

aldose A monosaccharide containing an aldehyde carbonyl group. (p. 1059)

amino sugar A sugar (such as glucosamine) in which a hydroxyl group is replaced by an amino group. (p. 1098)

anomeric carbon The hemiacetal carbon in the cyclic form of a sugar (the carbonyl carbon in the open-chain form). The anomeric carbon is easily identified because it is the only carbon with two bonds to oxygen atoms. (p. 1068)

anomers Sugar stereoisomers that differ in configuration only at the anomeric carbon. Anomers are classified as α or β depending on whether the anomeric hydroxyl group (or the aglycone in a glycoside) is trans (α) or cis (β) to the terminal —CH_2OH. (p. 1068)

carbohydrates (sugars) A class of polyhydroxy aldehydes and ketones, many of which have formula $C_n(H_2O)_m$ from which they received the name "hydrates of carbon" or "carbohydrates." (p. 1057)

cellulose A linear β-1,4' polymer of D-glucopyranose. Cellulose forms the cell walls of plants and is the major constituent of wood and cotton. (p. 1095)

chitin A β-1,4' polymer of N-acetylglucosamine that lends strength and rigidity to the exoskeletons of insects and crustaceans. (p. 1098)

degradation A reaction that causes loss of a carbon atom. (p. 1081)

deoxyribonucleic acid (DNA) A biopolymer of deoxyribonucleotides that serves as a template for the synthesis of ribonucleic acid. DNA is also the template for its own replication, through uncoiling and the pairing and enzymatic linking of complementary bases. (p. 1102)

deoxy sugar A sugar in which a hydroxyl group is replaced by a hydrogen. Deoxy sugars are recognized by the presence of a methylene group or a methyl group. (p. 1102)

dextrose The common dextrorotatory isomer of glucose, D-(+)-glucose. (p. 1070)

D series of sugars All sugars whose asymmetric carbon atom farthest from the carbonyl group has the same configuration as the asymmetric carbon atom in D-(+)-glyceraldehyde. Most naturally occurring sugars are members of the D series. (p. 1060)

disaccharide A carbohydrate whose hydrolysis gives two monosaccharide molecules. (p. 1090)

enediol rearrangement (Lobry de Bruyn–Alberta van Ekenstein reaction) A base-catalyzed tantomerization that interconverts aldoses and ketoses with an enediol as an intermediate. This enolization also epimerizes C2 and other carbon atoms. (p. 1071)

epimers Two diastereomeric sugars differing only in the configuration at a single asymmetric carbon atom. The epimeric carbon atom is usually specified, as in "C4 epimers." If no epimeric carbon is specified, it is assumed to be C2. The interconversion of epimers is called **epimerization**. (pp. 1063)

erythro and **threo** Diastereomers having similar groups on the same side (erythro) or on opposite sides (threo) of the Fischer projection. This terminology was adapted from the names of the aldotetroses *erythrose* and *threose*. (p. 1062)

D-erythrose   *erythro*-2,3-dibromopentanoic acid   D-threose   *threo*-3-chloro-2-butanol

furanose A five-membered cyclic hemiacetal form of a sugar. (p. 1067)

furanoside A five-membered cyclic glycoside. (p. 1075)

glucoside A glycoside derived from glucose. (p. 1075)

glycoside A cyclic acetal form of a sugar. Glycosides are stable to base, and they are nonreducing sugars. Glycosides are generally **furanosides** (five-membered) or **pyranosides** (six-membered), and they exist in anomeric α and β forms. (p. 1075)

glycosidic linkage A general term for an acetal bond from an anomeric carbon joining two monosaccharide units. (p. 1091)

> **glucosidic linkage:** A glycosidic linkage using an acetal bond from the anomeric carbon of glucose.

> **galactosidic linkage:** A glycosidic linkage using an acetal bond from the anomeric carbon of galactose.

Haworth projection A flat-ring representation of a cyclic sugar. The Haworth projection does not show the axial and equatorial positions of a pyranose, but it does show the cis and trans relationships. (p. 1065)

ketose A monosaccharide containing a ketone carbonyl group. (p. 1059)

Kiliani–Fischer synthesis A method for elongating an aldose at the aldehyde end. The aldose is converted into two epimeric aldoses with an additional carbon atom. For example, Kiliani–Fischer synthesis converts D-arabinose to a mixture of D-glucose and D-mannose. (p. 1081)

L series of sugars All sugars whose asymmetric carbon atom farthest from the carbonyl group has the same configuration as the asymmetric carbon atom in L-(−)-glyceraldehyde. Sugars of the L series are not common in nature. (p. 1060)

monosaccharide A carbohydrate that does not undergo hydrolysis of glycosidic bonds to give smaller sugar molecules. (p. 1069)

mutarotation A spontaneous change in optical rotation that occurs when a pure anomer of a sugar in its hemiacetal form equilibrates with the other anomer to give an equilibrium mixture with an averaged value of the optical rotation. (p. 1069)

nucleoside An *N*-glycoside of β-D-ribofuranose or β-D-deoxyribofuranose, where the aglycone is one of several derivatives of pyrimidine or purine. (p. 1100)

nucleotide A 5′-phosphate ester of a nucleoside. (p. 1101)

oligosaccharide A carbohydrate whose hydrolysis gives about two to ten monosaccharide units, but not as many as a polysaccharide. (p. 1095)

osazone The product, containing two phenylhydrazone residues, that results from reaction of a reducing sugar with phenylhydrazine. (p. 1080)

polysaccharide A carbohydrate whose hydrolysis gives many monosaccharide molecules. (p. 1095)

primary structure The primary structure of a nucleic acid is the sequence of nucleotides forming the polymer. This sequence determines the genetic characteristics of the nucleic acid. (p. 1103)

pyranose A six-membered cyclic hemiacetal form of a sugar. (p. 1067)

pyranoside A six-membered cyclic glycoside. (p. 1075)

rayon A commercial fiber made from regenerated cellulose. (p. 1095)

reducing sugar Any sugar that gives a positive Tollens test. Both ketoses and aldoses (in their hemiacetal forms) give positive Tollens tests. (p. 1075)

ribonucleic acid (RNA) A biopolymer of ribonucleotides that controls the synthesis of proteins. The synthesis of RNA is generally controlled by and patterned after DNA in the cell. (p. 1102)

ribonucleotide The 5′-phosphate ester of a **ribonucleoside**, a component of RNA based on β-D-ribofuranose and containing one of four heterocyclic bases as the aglycone. (p. 1101)

Ruff degradation A method for shortening the chain of an aldose by one carbon atom by treatment with bromine water, followed by hydrogen peroxide and $Fe_2(SO_4)_3$. (p. 1081)

starches A class of α-1,4′ polymers of glucose used for carbohydrate storage in plants and animals. (p. 1096)

> **amylose:** A linear α-1,4′ polymer of D-glucopyranose used for carbohydrate storage in plants.
>
> **amylopectin:** A branched α-1,4′ polymer of D-glucopyranose used for carbohydrate storage in plants. Branching occurs at α-1,6′ glycosidic linkages.
>
> **glycogen:** An extensively branched α-1,4′ polymer of D-glucopyranose used for carbohydrate storage in animals. Branching occurs at α-1,6′ glycosidic linkages.

sugar (saccharide) Any carbohydrate, regardless of structure, complexity, or taste. A simple sugar is a monosaccharide. (p. 1057)

Tollens test A test for reducing sugars, employing the same silver–ammonia complex used as a test for aldehydes. A positive test gives a silver precipitate, often in the form of a silver mirror. Tollens reagent is basic, and it promotes enediol rearrangements that interconvert ketoses and aldoses. Therefore, both aldoses and ketoses give positive Tollens tests if they are in their hemiacetal forms, in equilibrium with open-chain carbonyl structures. (p. 1074)

Essential Problem-Solving Skills in Chapter 23

1. Draw the Fischer projection of glucose and the chair conformation of the β anomer of glucose (all substituents equatorial) from memory.

2. Recognize the structures of other anomers and epimers of glucose, drawn as either Fischer projections or chair structures, by noticing the differences from the glucose structure.

3. Correctly name monosaccharides and disaccharides, and draw their structures from their names.

4. Predict which carbohydrates mutarotate, which reduce Tollens reagent, and which undergo epimerization and isomerization under basic conditions. (Those with free hemiacetals will, but glycosides with full acetals will not.)

5. Predict the products of the following reactions of carbohydrates:

bromine in water	NaOH and dimethyl sulfate
nitric acid	acetic anhydride and pyridine
$NaBH_4$ or H_2/Ni	phenylhydrazine
alcohols and H^+	Ruff degradation
CH_3I and Ag_2O	Kiliani–Fischer synthesis

6. Use the information gained from these reactions to determine the structure of an unknown carbohydrate. Use the information gained from methylation and from periodic acid cleavage to determine the ring size.

7. Draw the common types of glycosidic linkages, and recognize these linkages in disaccharides and polysaccharides.

8. Recognize the structures of DNA and RNA, and draw the structures of a ribonucleotide and a deoxyribonucleotide.

Study Problems

23-52   Define each term and give an example.

  (a) aldose   (b) ketose   (c) aldonic acid
  (d) aldaric acid   (e) glycoside   (f) aglycone
  (g) sugar   (h) anomers   (i) erythro and threo
  (j) epimers   (k) furanose   (l) pyranose
  (m) Haworth projection   (n) monosaccharide   (o) polysaccharide
  (p) disaccharide   (q) ribonucleoside   (r) ribonucleotide
  (s) deoxyribonucleotide   (t) osazone   (u) reducing sugar
  (v) amino sugar   (w) glycosidic linkage

23-53   Glucose is the most abundant monosaccharide. From memory, draw glucose in
  (a) the Fischer projection of the open chain
  (b) the most stable chair conformation of the most stable pyranose anomer
  (c) the Haworth projection of the most stable pyranose anomer

23-54   Without referring to the chapter, draw the chair conformations of
  (a) β-D-mannopyranose (the C2 epimer of glucose)
  (b) α-D-allopyranose (the C3 epimer of glucose)
  (c) β-D-galactopyranose (the C4 epimer of glucose)
  (d) *N*-acetylglucosamine, glucose with the C2 oxygen atom replaced by an acetylated amino group

23-55   Classify the following monosaccharides. (*Examples:* D-aldohexose, L-ketotetrose.)

  (a) (+)-glucose   (b) (−)-arabinose   (c) L-fructose

(d)
```
        CHO
  HO ── H
  HO ── H
   H ── OH
  HO ── H
       CH2OH
```
   (+)-gulose

(e)
```
       CH2OH
         |
        C=O
   H ── OH
   H ── OH
       CH2OH
```
   (−)-ribulose

(f)
```
        CHO
   H ── OH
  HO ── H
       CH2OH
```
   (+)-threose

23-56   Fructose is the ketose that results from the enediol rearrangement of glucose that shifts the carbonyl group to C2.
  (a) Propose a mechanism for the enediol rearrangement that converts D-glucose to D-fructose.
  (b) Draw the α and β anomers of D-fructofuranose. How can you tell which anomer is α and which is β regardless of the ring size?

23-57   The relative configurations of the stereoisomers of tartaric acid were established by the following syntheses.
  (1) D-(+)-glyceraldehyde $\xrightarrow{\text{HCN}}$ diastereomers A and B (separated).
  (2) Hydrolysis of A and B using aqueous $Ba(OH)_2$ gave C and D, respectively.
  (3) HNO_3 oxidation of C and D gave (−)-tartaric acid and *meso*-tartaric acid, respectively.
  (a) You know the absolute configuration of D-(+)-glyceraldehyde. Use Fischer projections to show the absolute configurations of products A, B, C, and D.
  (b) Show the absolute configurations of the three stereoisomers of tartaric acid: (+)-tartaric acid, (−)-tartaric acid, and *meso*-tartaric acid.

23-58   Use Figure 23-3 (the D family of aldoses) to name the following aldoses.
  (a) the C2 epimer of D-arabinose   (b) the C3 epimer of D-mannose   (c) the C3 epimer of D-threose
  (d) the enantiomer of D-galactose   (e) the C5 epimer of D-glucose

23-59   Draw the following sugar derivatives.
  (a) methyl β-D-glucopyranoside   (b) 2,3,4,6-tetra-*O*-methyl-D-galactopyranose
  (c) 1,3,6-tri-*O*-methyl-D-fructofuranose   (d) methyl 1,3,6-tri-*O*-methyl-α-D-fructofuranoside

23-60   Draw the structures (using chair conformations of pyranoses) of the following disaccharides.
  (a) 4-*O*-(α-D-glucopyranosyl)-D-galactopyranose   (b) α-D-fructofuranosyl-β-D-mannopyranoside
  (c) 6-*O*-(β-D-galactopyranosyl)-D-glucopyranose

23-61   Give the complete systematic name for each structure.

(a)

(b)

(c)

(d)

23-62   Which of the sugars mentioned in Problems 23-59, 23-60, and 23-61 are reducing sugars? Which ones would undergo mutarotation?

23-63   Predict the products obtained when D-galactose reacts with each reagent.
(a) Br_2 and H_2O (b) $NaOH$, H_2O (c) CH_3OH, H^+ (d) $Ag(NH_3)_2^+$ ^-OH
(e) H_2, Ni   (f) Ac_2O (g) excess CH_3I, Ag_2O (h) $NaBH_4$
(i) Br_2, H_2O, then H_2O_2 and $Fe_2(SO_4)_3$ (j) HCN, then H_3O^+, then Na(Hg)   (k) excess HIO_4

23-64   Draw the structures of the products expected when the following carbohydrates are subjected to methylation followed by acidic hydrolysis. In each case, suggest what reagent would be most appropriate for the methylation step.
(a) D-fructose   (b) ethyl α-D-glucopyranoside   (c) sucrose   (d) lactose   (e) gentiobiose   (f) chitin

23-65   (a) Which of the D-aldopentoses will give optically active aldaric acids on oxidation with HNO_3?
(b) Which of the D-aldotetroses will give optically active aldaric acids on oxidation with HNO_3?
(c) Sugar X is known to be a D-aldohexose. On oxidation with HNO_3, X gives an optically inactive aldaric acid. When X is degraded to an aldopentose, oxidation of the aldopentose gives an optically active aldaric acid. Determine the structure of X.
(d) Even though sugar X gives an optically inactive aldaric acid, the pentose formed by degradation gives an optically active aldaric acid. Does this finding contradict the principle that optically inactive reagents cannot form optically active products?
(e) Show what product results if the aldopentose formed from degradation of X is further degraded to an aldotetrose. Does HNO_3 oxidize this aldotetrose to an optically active aldaric acid?

23-66   (a) Give the products expected when (+)-glyceraldehyde reacts with HCN.
(b) What is the relationship between the products? How might they be separated?
(c) Are the products optically active? Explain.

23-67   When fructose reacts with Tollens reagent, the major products are the carboxylate ions of mannonic acid and gluconic acid.
(a) Propose a mechanism to show how fructose isomerizes to a mixture of glucose and mannose in the presence of Tollens reagent.
(b) Explain why bromine water is a better reagent than Tollens reagent for the oxidation of aldoses to aldonic acids.

23-68   When the gum of the shrub *Sterculia setigera* is subjected to acidic hydrolysis, one of the water-soluble components of the hydrolysate is found to be tagatose. The following information is known about tagatose.
(1) Molecular formula $C_6H_{12}O_6$
(2) Undergoes mutarotation
(3) Does not react with bromine water
(4) Reduces Tollens reagent to give D-galactonic acid and D-talonic acid
(5) Methylation of tagatose (using CH_3I and Ag_2O) followed by acidic hydrolysis gives 1,3,4,5-tetra-*O*-methyltagatose
(a) Draw a Fischer projection structure for the open-chain form of tagatose.
(b) Draw the most stable conformation of the most stable cyclic hemiacetal form of tagatose.

23-69   After a series of Kiliani–Fischer syntheses on (+)-glyceraldehyde, an unknown sugar is isolated from the reaction mixture. The following experimental information is obtained:

(1) Molecular formula $C_6H_{12}O_6$

(2) Undergoes mutarotation

(3) Reacts with bromine water to give an aldonic acid

(4) Reacts with phenylhydrazine to give an osazone, mp 178°C

(5) Reacts with HNO_3 to give an optically active aldaric acid

(6) Ruff degradation followed by HNO_3 oxidation gives an optically inactive aldaric acid

(7) Two Ruff degradations followed by HNO_3 oxidation give *meso*-tartaric acid

(8) Formation of the methyl glycoside (using CH_3OH and HCl), followed by periodic acid oxidation, gives a mixture of products that includes (+)-glyceraldehyde.

(a) Draw a Fischer projection structure for the open-chain form of this unknown sugar. Use Figure 23-3 to name the sugar.

(b) Draw the most stable conformation of the most stable cyclic hemiacetal form of this sugar, and give the structure a complete systematic name.

23-70   An unknown reducing disaccharide is found to be unaffected by invertase enzymes. Treatment with an α-galactosidase cleaves the disaccharide to give one molecule of D-fructose and one molecule of D-galactose. When the disaccharide is treated with iodomethane and silver oxide and then hydrolyzed in dilute acid, the products are 2,3,4,6-tetra-*O*-methylgalactose and 1,3,4-tri-*O*-methylfructose. Propose a structure for this disaccharide, and give its complete systematic name.

23-71   Draw the structures of the following nucleotides.

(a) guanosine triphosphate (GTP)

(b) deoxycytidine monophosphate (dCMP)

(c) cyclic guanosine monophosphate (cGMP)

23-72   Draw the structure of a four-residue segment of DNA with the following sequence.

(3′ end)   G-T-A-C   (5′ end)

23-73   Erwin Chargaff's discovery that DNA contains equimolar amounts of guanine and cytosine and also equimolar amounts of adenine and thymine has come to be known as *Chargaff's rule*.

G = C   and   A = T

(a) Does Chargaff's rule imply that equal amounts of guanine and adenine are present in DNA? That is, does G = A?

(b) Does Chargaff's rule imply that the sum of the purine residues equals the sum of the pyrimidine residues? That is, does A + G = C + T?

(c) Does Chargaff's rule apply only to double-stranded DNA, or would it also apply to each individual strand if the double helical strand were separated into its two complementary strands?

23-74   Retroviruses like HIV, the pathogen responsible for AIDS, incorporate an RNA template that is copied into DNA during infection. The *reverse transcriptase* enzyme that copies RNA into DNA is relatively nonselective and error-prone, leading to a high mutation rate. Its lack of selectivity is exploited by the anti-HIV drug AZT (3′-azido-2′,3′-dideoxythymidine), which becomes phosphorylated and is incorporated by reverse transcriptase into DNA, where it acts as a chain terminator. Mammalian DNA polymerases are more selective, having a low affinity for AZT, so its toxicity is relatively low.

Draw the structures of AZT and natural deoxythymidine. Draw the structure of AZT 5′-triphosphate, the derivative that inhibits reverse transcriptase.

*23-75   Exposure to nitrous acid (see Section 19-17), sometimes found in cells, can convert cytosine to uracil.

(a) Propose a mechanism for this conversion.

(b) Explain how this conversion would be mutagenic upon replication.

(c) DNA generally includes thymine, rather than uracil (found in RNA). Based on this fact, explain why the nitrous acid–induced mutation of cytosine to uracil is more easily repaired in DNA than it is in RNA.

CHAPTER 24

Amino Acids, Peptides, and Proteins

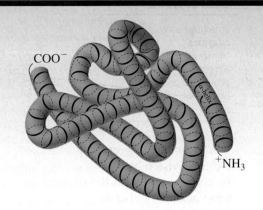

24-1
Introduction

Proteins are the most abundant organic molecules in animals, playing important roles in all aspects of cell structure and function. Proteins are biopolymers of **α-amino acids,** so named because the amino group is bonded to the α carbon atom, next to the carbonyl group. The physical and chemical properties of a protein are determined by its constituent amino acids. The individual amino acid subunits are joined by amide linkages called **peptide bonds**. Figure 24-1 shows the general structure of an α-amino acid and a protein.

Proteins have an amazing range of structural and catalytic properties as a result of their varying amino acid composition. Because of this versatility, proteins serve an astonishing variety of functions in living organisms. Some of the functions of the major classes of proteins are outlined in Table 24-1.

α carbon atom

$$H_2N-CH-C-OH$$
with O double bonded to C above

α-amino group — R — side chain

an α-amino acid

peptide bonds

$$-NH-CH-C-NH-CH-C-NH-CH-C-NH-CH-C-NH-CH-C-$$
with side chains CH_3, CH_2OH, H, CH_2SH, $CH(CH_3)_2$

a short section of a protein

$H_2N-CH-C-OH$ (CH_3) alanine
$H_2N-CH-C-OH$ (CH_2OH) serine
$H_2N-CH-C-OH$ (H) glycine
$H_2N-CH-C-OH$ (CH_2SH) cysteine
$H_2N-CH-C-OH$ ($CH(CH_3)_2$) valine

individual amino acids

▲ **Figure 24-1**
Structure of a general protein and its constituent amino acids. The amino acids are joined by amide linkages called peptide bonds.

TABLE 24-1 Examples of Protein Functions

Class of Protein	Example	Function of Example
structural proteins	collagen, keratin	tendons, skin, hair, nails
enzymes	DNA polymerase	replicate and repair DNA
transport proteins	hemoglobin	transport O_2 to the cells
contractile proteins	actin, myosin	cause contraction of muscles
protective proteins	antibodies	complex with foreign proteins
hormones	insulin	regulate glucose metabolism
toxins	snake venoms	incapacitate prey

The study of proteins is one of the major branches of biochemistry, and there is no clear division between the organic chemistry of proteins and their biochemistry. In this chapter, we begin the study of proteins by learning about their constituents, the amino acids. We also discuss how amino acid monomers are linked into the protein polymer, and how the properties of a protein depend on those of its constituent amino acids. These essentials are needed for the further study of protein structure and function in a biochemistry course.

24-2
Structure and Stereochemistry of the α-Amino Acids

The term **amino acid** might mean any molecule containing both an amino group and any type of acid group; however, the term is almost always used to refer to an α-amino carboxylic acid. The simplest α-amino acid is aminoacetic acid, called *glycine*. Other common amino acids have side chains (symbolized by R) substituted on the α carbon atom. For example, alanine is the amino acid with a methyl side chain.

glycine   a substituted amino acid   alanine (R = CH₃)

Except for glycine, the α-amino acids are all chiral. In every case (except glycine), the chirality center is the asymmetric α carbon atom. Nearly all the naturally occurring amino acids are found to have the *(S)* configuration at the α carbon atom. Figure 24-2 shows the *(S)* enantiomer of alanine arranged in the Fischer projection, with the carbon chain along the vertical and the carbonyl carbon at the top. Notice that *(S)*-alanine has a configuration similar to that of L-(−)-glyceraldehyde, with the amino group on the left in the Fischer projection. Because their stereochemistry is similar to that of L-(−)-glyceraldehyde, the naturally occurring *(S)*-amino acids are classified as **L-amino acids**.

(S)-alanine   L-(−)-glyceraldehyde   an L-amino acid
(L-alanine)   (S) configuration

◀ **Figure 24-2**
Almost all the naturally occurring amino acids have the *(S)* configuration, with stereochemistry resembling that of L-(−)-glyceraldehyde. They are therefore called L-amino acids.

Bacteria require specific enzymes, called racemases, to interconvert D and L amino acids. Mammals do not use D amino acids, so compounds that block racemases are being developed as potential antibiotics.

Although D-amino acids are occasionally found in nature, we usually assume the amino acids under discussion are the common L-amino acids. Remember once again that the D and L nomenclature, like the R and S designation, gives the configuration of the asymmetric carbon atom. It does not imply the sign of the optical rotation, $(+)$ or $(-)$, which must be determined experimentally.

Amino acids combine many of the properties and reactions of both amines and carboxylic acids. The combination of a basic amino group and an acidic carboxyl group in the same molecule also results in some unique properties and reactions. The side chains of some amino acids also have functional groups that lend interesting properties and undergo reactions of their own.

24-2A   The Standard Amino Acids of Proteins

There are 20 α-amino acids, called the **standard amino acids**, that are found in nearly all proteins. The standard amino acids differ from each other in the structure of the side chains bonded to their α carbon atoms. All the standard amino acids are L-amino acids. Table 24-2 shows the 20 standard amino acids, grouped according to the chemical properties of their side chains. Each amino acid is given a three-letter abbreviation and a one-letter symbol [green] for use in writing protein structures.

TABLE 24-2   The Standard Amino Acids

Name	Symbol	Abbreviation	Structure	Functional Group in Side Chain	Isoelectric Point
side chain is H or alkyl (nonpolar)					
glycine	G	Gly	H_2N—CH—COOH, H	none	6.0
alanine	A	Ala	H_2N—CH—COOH, CH_3	alkyl group	6.0
*valine	V	Val	H_2N—CH—COOH, CH, CH_3 CH_3	alkyl group	6.0
*leucine	L	Leu	H_2N—CH—COOH, CH_2—CH—CH_3, CH_3	alkyl group	6.0
*isoleucine	I	Ile	H_2N—CH—COOH, CH_3—CH—CH_2CH_3	alkyl group	6.0
*phenylalanine	F	Phe	H_2N—CH—COOH, CH_2⬡	aromatic group	5.5
proline	P	Pro	HN—CH—COOH, H_2C CH_2 CH_2	rigid cyclic structure	6.3
side chain contains an —OH					
serine	S	Ser	H_2N—CH—COOH, CH_2—OH	hydroxyl group	5.7
*threonine	T	Thr	H_2N—CH—COOH, HO—CH—CH_3	hydroxyl group	5.6

Continues on next page

TABLE 24-2 *(continued)*

Name	Symbol	Abbreviation	Structure	Functional Group in Side Chain	Isoelectric Point
tyrosine	Y	Tyr	H_2N—CH—COOH, CH_2——OH	phenolic—OH group	5.7

side chain contains sulfur

Name	Symbol	Abbreviation	Structure	Functional Group in Side Chain	Isoelectric Point
cysteine	C	Cys	H_2N—CH—COOH, CH_2—SH	thiol	5.0
*methionine	M	Met	H_2N—CH—COOH, CH_2—CH_2—S—CH_3	sulfide	5.7

side chain contains nonbasic nitrogen

Name	Symbol	Abbreviation	Structure	Functional Group in Side Chain	Isoelectric Point
asparagine	N	Asn	H_2N—CH—COOH, CH_2—C—NH_2 (∥O)	amide	5.4
glutamine	Q	Gln	H_2N—CH—COOH, CH_2—CH_2—C—NH_2 (∥O)	amide	5.7
*tryptophan	W	Trp	H_2N—CH—COOH, CH_2 indole ring (N–H)	indole	5.9

side chain is acidic

Name	Symbol	Abbreviation	Structure	Functional Group in Side Chain	Isoelectric Point
aspartic acid	D	Asp	H_2N—CH—COOH, CH_2—COOH	carboxylic acid	2.8
glutamic acid	E	Glu	H_2N—CH—COOH, CH_2—CH_2—COOH	carboxylic acid	3.2

side chain is basic

Name	Symbol	Abbreviation	Structure	Functional Group in Side Chain	Isoelectric Point
*lysine	K	Lys	H_2N—CH—COOH, CH_2—CH_2—CH_2—CH_2—NH_2	amino group	9.7
*arginine	R	Arg	H_2N—CH—COOH, CH_2—CH_2—CH_2—NH—C—NH_2 (∥NH)	guanidino group	10.8
*histidine	H	His	H_2N—CH—COOH, CH_2 imidazole ring (N, NH)	imidazole ring	7.6

*essential amino acid

Notice in Table 24-2 that proline is different from the other standard amino acids because its amino group is fixed in a ring with its α carbon atom. This cyclic structure lends additional strength and rigidity to proline-containing peptides.

PROBLEM 24-1

Draw three-dimensional representations of the following amino acids.
(a) L-phenylalanine   (b) L-arginine   (c) D-serine   (d) L-tryptophan

PROBLEM 24-2

Most naturally occurring amino acids have chirality centers (the asymmetric α carbon atoms) that are named *(S)* by the Cahn–Ingold Prelog convention (Section 5-3). The common naturally occurring form of cysteine has a chirality center that is named *(R)*, however.

(a) What is the relationship between *(R)*-cysteine and *(S)*-alanine? Do they have the opposite three-dimensional configuration (as the names might suggest) or the same configuration?

(b) *(S)*-alanine is an L-amino acid (Fig. 24-2). Is *(R)*-cysteine a D-amino acid or an L-amino acid?

24-2B   Essential Amino Acids

Human beings can synthesize about half of the amino acids needed to make proteins. Other amino acids, called the **essential amino acids**, must be provided in the diet. The ten essential amino acids, starred (*) in Table 24-2, are the following.

arginine (Arg)	valine (Val)	methionine (Met)	leucine (Leu)
threonine (Thr)	phenylalanine (Phe)	histidine (His)	isoleucine (Ile)
lysine (Lys)	tryptophan (Trp)		

Proteins that provide all the essential amino acids in about the right proportions for human nutrition are called **complete proteins**. Examples of complete proteins are those in meat, fish, milk, and eggs. About 50 g of complete protein per day is adequate for adult humans.

Proteins that are severely deficient in one or more of the essential amino acids are called **incomplete proteins**. If the protein in a person's diet comes mostly from one incomplete source, the amount of human protein that can be synthesized is limited by the amounts of the deficient amino acids. Plant proteins are generally incomplete. Rice, corn, and wheat are all deficient in lysine. Rice also lacks threonine, and corn also lacks tryptophan. Beans, peas, and other legumes have the most complete proteins among the common plants, but they are deficient in methionine.

Vegetarians can achieve an adequate intake of the essential amino acids if they eat many different plant foods. Plant proteins can be chosen to be complementary, with some foods supplying amino acids that others lack. An alternative is to supplement the vegetarian diet with a rich source of complete protein such as milk or eggs.

> Gelatin is made from collagen, which is a structural protein composed primarily of glycine, proline, and hydroxyproline. As a result, gelatin has low nutritional value because it lacks many of the essential amino acids.

PROBLEM 24-3

The herbicide *glyphosate* (Roundup®) kills plants by inhibiting an enzyme needed for synthesis of phenylalanine. Deprived of phenylalanine, the plant cannot make the proteins it needs, and it gradually weakens and dies. Although a small amount of glyphosate is deadly to a plant, its human toxicity is quite low. Suggest why this powerful herbicide has little effect on humans.

24-2C   Rare and Unusual Amino Acids

In addition to the standard amino acids, other amino acids are found in protein in smaller quantities. For example, 4-hydroxyproline and 5-hydroxylysine are hydroxylated versions of standard amino acids. These are called *rare* amino acids, even though they are commonly found in collagen.

$$\underset{\text{4-hydroxyproline}}{\text{structure}} \qquad \underset{\text{5-hydroxylysine}}{\underset{\overset{|}{\text{OH}}\qquad\qquad\overset{|}{\text{NH}_2}}{\text{H}_2\text{N}-\overset{6}{\text{CH}_2}-\overset{5}{\text{CH}}-\overset{4}{\text{CH}_2}-\overset{3}{\text{CH}_2}-\overset{2}{\text{CH}}-\overset{1}{\text{COOH}}}}$$

Some of the less common D enantiomers of amino acids are also found in nature. For example, D-glutamic acid is found in the cell walls of many bacteria, and D-serine is found in earthworms. Some naturally occurring amino acids are not α-amino acids: γ-Aminobutyric acid (GABA) is one of the neurotransmitters in the brain, and β-alanine is a constituent of the vitamin pantothenic acid.

D-glutamic acid   D-serine γ-aminobutyric acid β-alanine

Although we commonly write amino acids with an intact carboxyl ($-$COOH) group and amino ($-$NH$_2$) group, their actual structure is ionic and depends on the pH. The carboxyl group loses a proton, giving a carboxylate ion, and the amino group is protonated to an ammonium ion. This structure is called a **dipolar ion** or a **zwitterion** (German for "dipolar ion").

24-3
Acid–Base Properties of Amino Acids

$$\underset{\substack{\text{uncharged structure}\\ \text{(minor component)}}}{\text{H}_2\text{N}-\underset{\overset{|}{\text{R}}}{\text{CH}}-\overset{\overset{\text{O}}{\|}}{\text{C}}-\text{OH}} \quad \rightleftharpoons \quad \underset{\substack{\text{dipolar ion, or zwitterion}\\ \text{(major component)}}}{\overset{+}{\text{H}_3}\text{N}-\underset{\overset{|}{\text{R}}}{\text{CH}}-\overset{\overset{\text{O}}{\|}}{\text{C}}-\text{O}^-}$$

The dipolar nature of amino acids gives them some unusual properties:

1. Amino acids have **high melting points**, generally over 200°C.

$$\overset{+}{\text{H}_3}\text{N}-\text{CH}_2-\text{COO}^-$$

glycine, mp 262°C

2. Amino acids are more **soluble in water** than they are in ether, dichloromethane, and other common organic solvents.

3. Amino acids have much **larger dipole moments (μ)** than simple amines or simple acids.

$$\underset{\text{glycine, } \mu = 14\,\text{D}}{\overset{+}{\text{H}_3}\text{N}-\text{CH}_2-\text{COO}^-} \qquad \underset{\text{propylamine, } \mu = 1.4\,\text{D}}{\text{CH}_3-\text{CH}_2-\text{CH}_2-\text{NH}_2} \qquad \underset{\text{propionic acid, } \mu = 1.7\,\text{D}}{\text{CH}_3-\text{CH}_2-\text{COOH}}$$

4. Amino acids are **less acidic than most carboxylic acids** and **less basic than most amines**. In fact, the acidic part of the amino acid molecule is the $-NH_3^+$ group, not a $-COOH$ group. The basic part is the $-COO^-$ group, and not a free $-NH_2$ group.

$$R-COOH \qquad R-NH_2 \qquad \overset{R}{\underset{}{\overset{|}{H_3\overset{+}{N}-CH-COO^-}}}$$

$$pK_a = 5 \qquad pK_b = 4 \qquad \begin{matrix} pK_a = 10 \\ pK_b = 12 \end{matrix}$$

Because amino acids contain both acidic ($-NH_3^+$) and basic ($-COO^-$) groups, they are *amphoteric* (having both acidic and basic properties). The predominant form of the amino acid depends on the pH of the solution. In an acidic solution, the $-COO^-$ group is protonated to a free $-COOH$ group, and the molecule has an overall positive charge. As the pH is raised, the $-COOH$ loses its proton at about pH 2. This point is called pK_{a1}, the first acid-dissociation constant. As the pH is raised further, the $-NH_3^+$ group loses its proton at about pH 9 or 10. This point is called pK_{a2}, the second acid-dissociation constant. Above this pH, the molecule has an overall negative charge.

$$\underset{\text{cationic in acid}}{\overset{+}{H_3N}-\underset{\underset{R}{|}}{CH}-COOH} \quad \underset{\underset{pK_{a1} \cong 2}{H^+}}{\overset{^-OH}{\rightleftharpoons}} \quad \underset{\text{neutral}}{\overset{+}{H_3N}-\underset{\underset{R}{|}}{CH}-COO^-} \quad \underset{\underset{pK_{a2} \cong 9-10}{H^+}}{\overset{^-OH}{\rightleftharpoons}} \quad \underset{\text{anionic in base}}{H_2N-\underset{\underset{R}{|}}{CH}-COO^-}$$

Figure 24-3 shows a titration curve for glycine. The curve starts at the bottom left, where glycine is entirely in its cationic form. Base is slowly added, and the pH is recorded. At pH 2.3, half of the cationic form has been converted to the zwitterionic form.

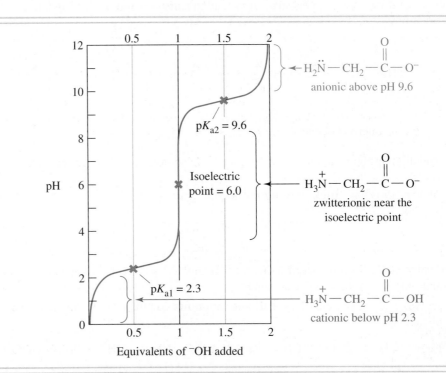

▶ **Figure 24-3**
A titration curve for glycine. The pH controls the charge on glycine: cationic below pH 2.3; anionic above pH 9.6; and zwitterionic between pH 2.3 and 9.6. The isoelectric pH is 6.0.

At pH 6.0, essentially all the glycine is in the zwitterionic form. At pH 9.6, half of the zwitterionic form has been converted to the basic form. From this graph, we can see that glycine is mostly in the cationic form at pH values below 2.3, mostly in the anionic form at pH values above 9.6, and mostly in the zwitterionic form at pH values between 2.3 and 9.6. By varying the pH of the solution, we can control the charge on the molecule. This ability to control the charge of an amino acid is useful for separating and identifying amino acids by electrophoresis, as described in Section 24-4.

24-4
Isoelectric Points and Electrophoresis

An amino acid bears a positive charge in acidic solution (low pH) and a negative charge in basic solution (high pH). There must be an intermediate pH where the amino acid is evenly balanced between the two forms, as the dipolar zwitterion with a net charge of zero. This pH is called the **isoelectric pH** or the **isoelectric point**.

$$H_2N-CH-COO^- \underset{^-OH}{\overset{H^+}{\rightleftarrows}} \overset{+}{H_3N}-CH-COO^- \underset{^-OH}{\overset{H^+}{\rightleftarrows}} \overset{+}{H_3N}-CH-COOH$$
$$\qquad\quad | \qquad\qquad\qquad\qquad | \qquad\qquad\qquad\qquad |$$
$$\qquad\quad R \qquad\qquad\qquad\qquad R \qquad\qquad\qquad\qquad R$$

high pH   isoelectric pH   low pH
(anionic in base)   (neutral)   (cationic in acid)

The isoelectric points of the standard amino acids are given in Table 24-2. Notice that the isoelectric pH depends on the amino acid structure in a predictable way.

acidic amino acids: aspartic acid (2.8), glutamic acid (3.2)

neutral amino acids: (5.0 to 6.3)

basic amino acids: lysine (9.7), arginine (10.8), histidine (7.6)

Aspartic acid and glutamic acid have side chains containing acidic carboxyl groups. These amino acids have acidic isoelectric points around pH 3. An acidic solution is needed to prevent deprotonation of the second carboxylic acid group and keep the amino acid in its neutral isoelectric state.

Basic amino acids (histidine, lysine, and arginine) have isoelectric points at pH values of 7.6, 9.7, and 10.8, respectively. These values reflect the weak basicity of the imidazole ring, the intermediate basicity of an amino group, and the strong basicity of the guanidino group. A basic solution is needed in each case to prevent protonation of the basic side chain to keep the amino acid electrically neutral.

The other amino acids are considered neutral, with no strongly acidic or basic side chains. Their isoelectric points are slightly acidic (from about 5 to 6) because the $-NH_3^+$ group is slightly more acidic than the $-COO^-$ group is basic.

PROBLEM 24-4

Draw the structure of the predominant form of
(a) valine at pH 11   (b) proline at pH 2
(c) arginine at pH 7   (d) glutamic acid at pH 7
(e) a mixture of alanine, lysine, and aspartic acid at (i) pH 6; (ii) pH 11; (iii) pH 2

PROBLEM 24-5

Draw the resonance forms of a protonated guanidino group, and explain why arginine has such a strongly basic isoelectric point.

PROBLEM 24-6

Although tryptophan contains a heterocyclic amine, it is considered a neutral amino acid. Explain why the indole nitrogen of tryptophan is more weakly basic than one of the imidazole nitrogens of histidine.

PROBLEM-SOLVING HINT

At its isoelectric point (IEP), an amino acid has a net charge of zero, with NH_3^+ and COO^- balancing each other. In more acidic solution (lower pH), the carboxyl group becomes protonated and the net charge is positive. In more basic solution (higher pH), the amino group loses its proton and the net charge is negative.

Differences in isoelectric points can be used to separate mixtures of amino acids by **electrophoresis**. A streak of the amino acid mixture is placed in the center of a layer of acrylamide gel or a piece of filter paper wet with a buffer solution. Two electrodes are placed in contact with the edges of the gel or paper, and a potential of several thousand volts is applied across the electrodes. Positively charged (cationic) amino acids are attracted to the negative electrode (the cathode), and negatively charged (anionic) amino acids are attracted to the positive electrode (the anode). An amino acid at its isoelectric point has no net charge, so it does not move.

As an example, consider a mixture of alanine, lysine, and aspartic acid in a buffer solution at pH 6. Alanine is at its isoelectric point, in its dipolar zwitterionic form with a net charge of zero. A pH of 6 is more acidic than the isoelectric pH for lysine (9.7), so lysine is in the cationic form. Aspartic acid has an isoelectric pH of 2.8, so it is in the anionic form.

Structure at pH 6

$$\overset{+}{H_3}N-CH-COO^- \qquad \overset{+}{H_3}N-CH-COO^- \qquad \overset{+}{H_3}N-CH-COO^-$$
$$\overset{|}{CH_3} \qquad\qquad (CH_2)_4-NH_3^+ \qquad\qquad CH_2-COO^-$$

alanine (charge 0)   lysine (charge +1)   aspartic acid (charge −1)

When a voltage is applied to a mixture of alanine, lysine, and aspartic acid at pH 6, alanine does not move. Lysine moves toward the cathode, and aspartic acid moves toward the anode (Fig. 24-4). After a period of time, the separated amino acids are recovered by cutting the paper or scraping the bands out of the gel. If electrophoresis is being used as an analytical technique (to determine the amino acids present in the mixture), the paper or gel is treated with a reagent such as ninhydrin (Section 24-9) to make the bands visible. The amino acids are then identified by comparing their positions with those of standards.

PROBLEM 24-7

Draw the electrophoretic separation of Ala, Lys, and Asp at pH 9.7.

PROBLEM 24-8

Draw the electrophoretic separation of Trp, Cys, and His at pH 6.0.

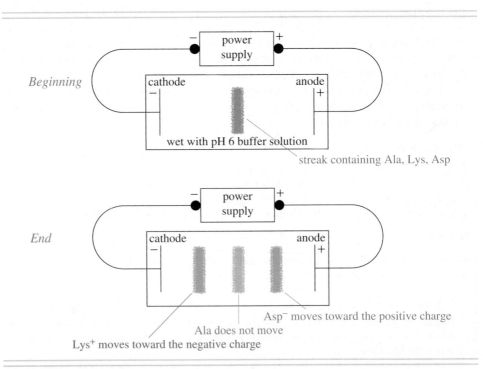

▶ **Figure 24-4**
A simplified picture of the electrophoretic separation of alanine, lysine, and aspartic acid at pH 6. Cationic lysine is attracted to the cathode; anionic aspartic acid is attracted to the anode. Alanine is at its isoelectric point, so it does not move.

Naturally occurring amino acids can be obtained by hydrolyzing proteins and separating the amino acid mixture. Even so, it is often less expensive to synthesize the pure amino acid. In some cases, an unusual amino acid or an unnatural enantiomer is needed, and it must be synthesized. In this chapter, we consider four methods for making amino acids. All these methods are extensions of reactions we have already studied.

24-5
Synthesis
of Amino Acids

24-5A   Reductive Amination

Reductive amination of ketones and aldehydes is one of the best methods for synthesizing amines (Section 19-19). It also forms amino acids. When an α-ketoacid is treated with ammonia, the ketone reacts to form an imine. The imine is reduced to an amine by hydrogen and a palladium catalyst. Under these conditions, the carboxylic acid is not reduced.

This entire synthesis is accomplished in one step by treating the α-ketoacid with ammonia and hydrogen in the presence of a palladium catalyst. The product is a racemic α-amino acid. The following reaction shows the synthesis of racemic phenylalanine from 3-phenyl-2-oxopropanoic acid.

Reductive amination resembles the biological synthesis of amino acids. We can call it a **biomimetic** ("mimicking the biological process") synthesis. The biosynthesis begins with reductive amination of α-ketoglutaric acid (an intermediate in the metabolism of carbohydrates), using ammonium ion as the aminating agent and NADH as the reducing agent. The product of this enzyme-catalyzed reaction is the pure L enantiomer of glutamic acid.

Biosynthesis of other amino acids uses L-glutamic acid as the source of the amino group. Such a reaction, moving an amino group from one molecule to another, is called a **transamination**, and the enzymes that catalyze these reactions are called *transaminases*. For example, the following reaction shows the biosynthesis of aspartic acid using glutamic acid as the nitrogen source. Once again, the enzyme-catalyzed biosynthesis gives the pure L enantiomer of the product.

$$
\left.\begin{array}{c}
\underset{\text{L-glutamic acid}}{\text{HOOC}-\text{CH}_2\text{CH}_2-\overset{\overset{+\text{NH}_3}{|}}{\text{CH}}-\text{COO}^-} \\[2em]
+ \\[2em]
\underset{\text{oxaloacetic acid}}{\text{HOOC}-\text{CH}_2-\overset{\overset{\text{O}}{\|}}{\text{C}}-\text{COO}^-}
\end{array}\right\}
\xrightarrow{\text{transaminase}}
\left\{\begin{array}{c}
\underset{\alpha\text{-ketoglutaric acid}}{\text{HOOC}-\text{CH}_2\text{CH}_2-\overset{\overset{\text{O}}{\|}}{\text{C}}-\text{COO}^-} \\[2em]
+ \\[2em]
\underset{\text{L-aspartic acid}}{\text{HOOC}-\text{CH}_2-\overset{\overset{+\text{NH}_3}{|}}{\text{CH}}-\text{COO}^-}
\end{array}\right.
$$

PROBLEM 24-9

Show how the following amino acids might be formed in the laboratory by reductive amination of the appropriate α-ketoacid:
(a) alanine   (b) leucine   (c) serine   (d) glutamine

24-5B   Amination of an α-Halo Acid

The Hell–Volhard–Zelinsky reaction (Section 22-4) is an effective method for introducing bromine at the α position of a carboxylic acid. The α-bromo acid is converted to a racemic α-amino acid by direct amination, using a large excess of ammonia.

$$
\underset{\text{carboxylic acid}}{\text{R}-\text{CH}_2-\overset{\overset{\text{O}}{\|}}{\text{C}}-\text{OH}}
\xrightarrow[\text{(2) H}_2\text{O}]{\text{(1) Br}_2/\text{PBr}_3}
\underset{\alpha\text{-bromo acid}}{\text{R}-\overset{\overset{\text{Br}}{|}}{\text{CH}}-\overset{\overset{\text{O}}{\|}}{\text{C}}-\text{OH}}
\xrightarrow[\text{(large excess)}]{\text{NH}_3}
\underset{\substack{\text{(D,L)-}\alpha\text{-amino acid}\\ \text{(ammonium salt)}}}{\text{R}-\overset{\overset{\text{NH}_2}{|}}{\text{CH}}-\overset{\overset{\text{O}}{\|}}{\text{C}}-\text{O}^- {}^+\text{NH}_4}
$$

In Section 19-19, we saw that direct amination is often a poor synthesis of amines, giving large amounts of overalkylated products. In this case, however, the reaction gives acceptable yields because a large excess of ammonia is used, making ammonia the nucleophile that is most likely to displace bromine. Also, the adjacent carboxylate ion in the product reduces the nucleophilicity of the amino group. The following sequence shows bromination of 3-phenylpropanoic acid, followed by displacement of bromide ion, to form the ammonium salt of racemic phenylalanine.

$$
\underset{\text{3-phenylpropanoic acid}}{\text{Ph}-\text{CH}_2-\text{CH}_2-\text{COOH}}
\xrightarrow[\text{(2) H}_2\text{O}]{\text{(1) Br}_2/\text{PBr}_3}
\text{Ph}-\text{CH}_2-\overset{\overset{\text{Br}}{|}}{\text{CH}}-\text{COOH}
$$

$$
\text{Ph}-\text{CH}_2-\overset{\overset{\text{Br}}{|}}{\text{CH}}-\text{COOH}
\xrightarrow{\text{excess NH}_3}
\underset{\substack{\text{(D,L)-phenylalanine (salt)}\\ \text{(30–50\%)}}}{\text{Ph}-\text{CH}_2-\overset{\overset{\text{NH}_2}{|}}{\text{CH}}-\text{COO}^- {}^+\text{NH}_4}
$$

PROBLEM 24-10

Show how you would use bromination followed by amination to synthesize the following amino acids.
(a) glycine   (b) leucine   (c) valine   (d) glutamic acid

24-5C   The Gabriel–Malonic Ester Synthesis

One of the best methods of amino acid synthesis is a combination of the Gabriel synthesis of amines (Section 19-19C) with the malonic ester synthesis of carboxylic acids (Section 22-16). The conventional malonic ester synthesis involves alkylation of diethyl malonate, followed by hydrolysis and decarboxylation to give an alkylated acetic acid.

temporary ester group

$$H-\underset{\underset{H}{|}}{\overset{\overset{COOEt}{|}}{C}}-\overset{\overset{O}{||}}{C}-OEt \quad \xrightarrow[(2)\ RX]{(1)\ ^-OEt} \quad H-\underset{\underset{R}{|}}{\overset{\overset{COOEt}{|}}{C}}-\overset{\overset{O}{||}}{C}-OEt \quad \xrightarrow{H_3O^+,\ heat} \quad H-\underset{\underset{R}{|}}{\overset{\overset{H}{|}}{C}}-\overset{\overset{O}{||}}{C}-OH \quad CO_2 \uparrow$$

malonic ester   alkylated acetic acid

To adapt this synthesis to making amino acids, we should begin with a malonic ester that contains an α-amino group. The amino group should be protected as a non-nucleophilic amide to prevent it from attacking the alkylating agent (RX).

The Gabriel–malonic ester synthesis begins with *N*-phthalimidomalonic ester. We can think of *N*-phthalimidomalonic ester as a molecule of glycine (aminoacetic acid) with the amino group protected as an amide (a phthalimide in this case) to keep it from acting as a nucleophile. The acid is protected as an ethyl ester, and the α position is further activated by the additional (temporary) ester group of diethyl malonate.

temporary ester group

N-phthalimidomalonic ester   glycine

protected amine   protected acid

Just as the malonic ester synthesis gives substituted acetic acids, the *N*-phthalimidomalonic ester synthesis gives substituted aminoacetic acids: α-amino acids. *N*-Phthalimidomalonic ester is alkylated in the same way as malonic ester. When the alkylated *N*-phthalimidomalonic ester is hydrolyzed, the phthalimido group is hydrolyzed along with the ester groups. The product is an alkylated aminomalonic acid. Decarboxylation gives a racemic α-amino acid.

The Gabriel–malonic ester synthesis

temporary ester group

N-phthalimidomalonic ester   alkylated   hydrolyzed α-amino acid

$$CO_2 \uparrow$$

The Gabriel–malonic ester synthesis is used to synthesize many amino acids that cannot be formed by direct amination of haloacids. The following example shows the synthesis of methionine, which is formed in very poor yield by direct amination.

(D, L)-methionine
(50%)

PROBLEM 24-11

Show how the Gabriel-malonic ester synthesis could be used to make
(a) valine   **(b)** phenylalanine   **(c)** glutamic acid   **(d)** leucine

*PROBLEM 24-12

The Gabriel–malonic ester synthesis uses an aminomalonic ester with the amino group protected as a phthalimide. A variation has the amino group protected as an acetamido group. Propose how you might use an *acetamidomalonic ester* synthesis to make phenylalanine.

acetamidomalonic ester

24-5D   The Strecker Synthesis

The first known synthesis of an amino acid occurred in 1850 in the laboratory of Adolph Strecker in Tübingen, Germany. Strecker added acetaldehyde to an aqueous solution of ammonia and HCN. The product was α-amino propionitrile, which Strecker hydrolyzed to racemic alanine.

The Strecker synthesis of alanine

acetaldehyde α-amino propionitrile   (D,L)-alanine
                                                                                       (60%)

The **Strecker synthesis** can form a large number of amino acids from appropriate aldehydes. The mechanism is shown below. First, the aldehyde reacts with ammonia to give an imine. The imine is a nitrogen analogue of a carbonyl group, and it is electrophilic when protonated. Attack of cyanide ion on the protonated imine gives the α-amino nitrile. This mechanism is similar to that for formation of a cyanohydrin, except that in the Strecker synthesis cyanide ion attacks an imine rather than the aldehyde itself.

Step 1: formation of the imine (mechanism in Section 18-16)

aldehyde   imine

Step 2: attack by cyanide

In a separate step, hydrolysis of the α-amino nitrile (Section 21-7D) gives an α-amino acid.

SOLVED PROBLEM 24-1

Show how you would use a Strecker synthesis to make isoleucine.

SOLUTION

Isoleucine has a *sec*-butyl group for its side chain. Remember that CH_3—CHO undergoes Strecker synthesis to give alanine, with CH_3 as the side chain. Therefore, *sec*-butyl—CHO should give isoleucine.

PROBLEM 24-13

(a) Show how you would use a Strecker synthesis to make phenylalanine.
(b) Propose a mechanism for each step in the synthesis in part (a).

PROBLEM 24-14

Show how you would use a Strecker synthesis to make
(a) leucine   (b) glycine   (c) valine

PROBLEM-SOLVING HINT

In the malonic ester synthesis, use the side chain of the desired amino acid (must be a good S_N2 substrate) to alkylate the ester. In the Strecker synthesis, the aldehyde carbon becomes the α carbon of the amino acid: Begin with [side chain]—CHO.

SUMMARY   Syntheses of Amino Acids

1. *Reductive amination* (Section 24-5A)

2. *Amination of an α-haloacid* (Section 24-5B)

3. *The Gabriel–malonic ester synthesis* (Section 24-5C)

4. *The Strecker synthesis* (Section 24-5D)

24-6
Resolution
of Amino Acids

All the laboratory syntheses of amino acids described in Section 24-5 produce racemic products. In most cases, only the L enantiomers are biologically active; the D enantiomers may even be toxic. Pure L enantiomers are needed for peptide synthesis if the product is to have the activity of the natural material. Therefore, we must be able to resolve a racemic amino acid into its enantiomers.

In many cases, amino acids can be resolved by the methods we have already discussed (Section 5-16). If a racemic amino acid is converted to a salt with an optically pure chiral acid or base, two diastereomeric salts are formed. These salts can be separated by physical means such as selective crystallization or chromatography. Pure enantiomers are then regenerated from the separated diastereomeric salts. Strychnine and brucine are naturally occurring optically active bases, and tartaric acid is used as an optically active acid for resolving racemic mixtures.

Enzymatic resolution is also used to separate the enantiomers of amino acids. Enzymes are chiral molecules with specific catalytic activities. For example, when an acylated amino acid is treated with an enzyme like hog kidney acylase or carboxypeptidase, the enzyme cleaves the acyl group from just the molecules having the natural (L) configuration. The enzyme does not recognize D-amino acids, so they are unaffected. The resulting mixture of acylated D-amino acid and deacylated L-amino acid is easily separated. Figure 24-5 shows how this selective enzymatic deacylation is accomplished.

> **PROBLEM 24-15**
>
> Suggest how you would separate the free L-amino acid from its acylated D enantiomer in Figure 24-5.

▲ Figure 24-5
An acylase enzyme (such as hog kidney acylase or carboxypeptidase) deacylates only the
natural L-amino acid.

Amino acids undergo many of the standard reactions of both amines and carboxylic
acids. Conditions for some of these reactions must be carefully selected, however, so
that the amino group does not interfere with a carboxyl group reaction, and vice versa.
We will consider two of the most useful reactions, esterification of the carboxyl group
and acylation of the amino group. These reactions are often used to protect either the
carboxyl group or the amino group while the other group is being modified or coupled
to another amino acid. Amino acids also undergo reactions that are specific to the
α-amino acid structure. One of these unique amino acid reactions is the formation of a
colored product on treatment with ninhydrin, discussed in Section 24-7C.

24-7
Reactions
of Amino Acids

24-7A Esterification of the Carboxyl Group

Like monofunctional carboxylic acids, amino acids are esterified by treatment with a large
excess of an alcohol and an acidic catalyst (often gaseous HCl). Under these acidic conditions, the amino group is present in its protonated ($-NH_3^+$) form, so it does not interfere with esterification. The following example illustrates esterification of an amino acid.

proline → (Ph—CH₂—OH / HCl) → proline benzyl ester (90%)

Esters of amino acids are often used as protected derivatives to prevent the carboxyl group from reacting in some undesired manner. Methyl, ethyl, and benzyl esters
are the most common protecting groups. Aqueous acid hydrolyzes the ester and regenerates the free amino acid.

phenylalanine ethyl ester → (H_3O^+) → phenylalanine + CH_3CH_2—OH

Benzyl esters are particularly useful as protecting groups because they can be removed either by acidic hydrolysis or by neutral **hydrogenolysis** ("breaking apart by addition of hydrogen"). Catalytic hydrogenation cleaves the benzyl ester, converting the benzyl group to toluene and leaving the deprotected amino acid. Although the mechanism of this hydrogenolysis is not well known, it apparently hinges on the ease of formation of benzylic intermediates.

phenylalanine benzyl ester $\xrightarrow{H_2,\ Pd}$ phenylalanine   +   toluene

PROBLEM 24-16

Propose a mechanism for the acid-catalyzed hydrolysis of phenylalanine ethyl ester.

PROBLEM 24-17

Give equations for the formation and hydrogenolysis of glutamine benzyl ester.

Decarboxylation is an important reaction of amino acids in many biological processes. Histamine, which causes runny noses and itchy eyes, is synthesized in the body by decarboxylation of histidine. The enzyme that catalyzes this reaction is called histidine decarboxylase.

histamine

24-7B   Acylation of the Amino Group: Formation of Amides

Just as an alcohol esterifies the carboxyl group of an amino acid, an acylating agent converts the amino group to an amide. Acylation of the amino group is often done to protect it from unwanted nucleophilic reactions. A wide variety of acid chlorides and anhydrides are used for acylation. Benzyl chloroformate acylates the amino group to give a benzyloxycarbonyl derivative, often used as a protecting group in peptide synthesis (Section 24-10).

histidine $\xrightarrow{\text{(acetic anhydride)}}$ N-acetylhistidine

leucine $\xrightarrow[\text{(benzyl chloroformate)}]{\text{PhCH}_2\text{OC—Cl}}$ N-benzyloxycarbonyl leucine (90%)

The amino group of the N-benzyloxycarbonyl derivative is protected as the amide half of a carbamate ester (a urethane, Section 21-16), which is more easily hydrolyzed than most other amides. In addition, the ester half of this urethane is a benzyl ester that undergoes hydrogenolysis. Catalytic hydrogenolysis of the N-benzyloxycarbonyl amino acid gives an unstable carbamic acid that quickly decarboxylates to give the deprotected amino acid.

N-benzyloxycarbonyl leucine   toluene   a carbamic acid   leucine

PROBLEM 24-18

Give equations for the formation and hydrogenolysis of N-benzyloxycarbonyl methionine.

24-7C   Reaction with Ninhydrin

Ninhydrin is a common reagent for visualizing spots or bands of amino acids that have been separated by chromatography or electrophoresis. When ninhydrin reacts with an amino acid, one of the products is a deep violet, resonance-stabilized anion called *Ruhemann's purple*. Ninhydrin produces this same purple dye regardless of the structure of the original amino acid. The side chain of the amino acid is lost as an aldehyde.

Reaction of an amino acid with ninhydrin

amino acid   ninhydrin   Ruhemann's purple

The reaction of amino acids with ninhydrin can detect amino acids on a wide variety of substrates. For example, if a kidnapper touches a ransom note with his fingers, the dermal ridges on his fingers leave traces of amino acids from skin secretions. Treatment of the paper with ninhydrin and pyridine causes these secretions to turn purple, forming a visible fingerprint.

PROBLEM 24-19

Use resonance forms to show delocalization of the negative charge in the Ruhemann's purple anion.

SUMMARY   Reactions of Amino Acids

1. *Esterification of the carboxyl group* (Section 24-7A)

amino acid   alcohol   amino ester

2. *Acylation of the amino group: formation of amides* (Section 24-7B)

amino acid   acylating agent   acylated amino acid

3. *Reaction with ninhydrin* (Section 24-7C)

$$H_2N-CH-COOH \quad + \quad 2 \text{ [ninhydrin]} \xrightarrow{\text{pyridine}} \text{[Ruhemann's purple]} \quad + \quad R-CHO$$
$$\qquad | \qquad\qquad\qquad\qquad\qquad\qquad\qquad\qquad\qquad\qquad\qquad\qquad + \quad CO_2\uparrow$$
$$\qquad R$$

amino acid   ninhydrin   Ruhemann's purple

4. *Formation of peptide bonds* (Sections 24-10 and 24-11)

peptide bond

$$H_3\overset{+}{N}-CH-\overset{O}{\underset{|}{C}}-O^- \quad + \quad H_3\overset{+}{N}-CH-\overset{O}{\underset{|}{C}}-O^- \xrightarrow{\text{loss of } H_2O} H_3\overset{+}{N}-CH-\overset{O}{\underset{|}{C}}-NH-CH-\overset{O}{\underset{|}{C}}-O^-$$
$$\qquad\quad | \qquad\qquad\qquad\qquad | \qquad\qquad\qquad\qquad\qquad\qquad | \qquad\qquad\quad |$$
$$\qquad\quad R^1 \qquad\qquad\qquad\qquad R^2 \qquad\qquad\qquad\qquad\qquad\qquad R^1 \qquad\qquad\quad R^2$$

Amino acids also undergo many other common reactions of amines and acids.

24-8

Structure and Nomenclature of Peptides and Proteins

24-8A   Peptide Structure

The most important reaction of amino acids is the formation of peptide bonds. Amines and acids can condense, with the loss of water, to form amides. In the laboratory, we might make an amide simply by mixing the acid and the amine, then heating the mixture to drive off water.

$$R-\overset{O}{\underset{}{\overset{||}{C}}}-OH \quad + \quad H_2\ddot{N}-R' \longrightarrow R-\overset{O}{\underset{}{\overset{||}{C}}}-O^- \ H_3\overset{+}{N}-R' \xrightarrow{\text{heat}} R-\overset{O}{\underset{}{\overset{||}{C}}}-\ddot{N}H-R' \ + \ H_2O$$

acid   amine   salt   amide

Recall from Section 21-13 that amides are the most stable acid derivatives. This stability is partly due to the strong resonance interaction between the nonbonding electrons on nitrogen and the carbonyl group. The amide nitrogen is no longer a strong base, and the C—N bond has restricted rotation because of its partial double-bond character. Figure 24-6 shows the resonance forms we use to explain the partial double-bond character and restricted rotation of an amide bond. In a peptide, this partial double-bond character results in six atoms being held rather rigidly in a plane.

Having both an amino group and a carboxyl group, an amino acid is ideally suited to form an amide linkage. Under the proper conditions, the amino group of one

▶ **Figure 24-6**
Resonance stabilization of an amide accounts for its enhanced stability, the diminished basicity of the nitrogen atom, and the restricted rotation (partial double-bond character) of the C—N bond. In a peptide, the amide bond is called a peptide bond. It holds six atoms in a plane: The C and O of the carbonyl, the N and its H, and the two associated α carbon atoms.

▲ Figure 24-7
The human hormone bradykinin is a nonapeptide with a free $-NH_3^+$ at its N terminus and a free $-COO^-$ at its C terminus.

molecule condenses with the carboxyl group of another. The product is an amide called a *dipeptide* because it consists of two amino acids. The amide linkage between the amino acids is called a **peptide bond**. Although it has a special name, a peptide bond is just like other amide bonds we have studied.

In this manner, any number of amino acids can be bonded in a continuous chain. A **peptide** is any polymer of amino acids linked by amide bonds between the amino group of each amino acid and the carboxyl group of the neighboring amino acid. Each amino acid unit in the peptide is called a **residue**. A **polypeptide** is a peptide containing many amino acid residues but usually having a molecular weight of less than about 5000. **Proteins** contain more amino acid units, with molecular weights ranging from about 6000 to about 40,000,000. The term **oligopeptide** is occasionally used for peptides containing about four to ten amino acid residues. Figure 24-7 shows the structure of the nonapeptide bradykinin, a human hormone that helps to control blood pressure.

The end of the peptide with the free amino group ($-NH_3^+$) is called the **N-terminal end** or the **N terminus**, and the end with the free carboxyl group ($-COO^-$) is called the **C-terminal end** or the **C terminus**. Peptide structures are generally drawn with the N terminus at the left and the C terminus at the right, as bradykinin is drawn in Figure 24-7.

Oxerin-A is a recently discovered neuropeptide containing 33 amino acids. One function of the peptide may be to regulate the sleep-wake cycle. Oxerin-A induces wakefulness, while its absence results in sleepiness.

24-8B Peptide Nomenclature

Peptides are named beginning at the N terminus, and the names of amino acid residues involved in amide linkages (all except the last) are given the *-yl* suffix of acyl groups. For example, the following dipeptide is named alanylserine. The alanine residue has the *-yl* suffix because it has acylated the nitrogen of serine.

Bradykinin (Fig. 24-7) is named as follows (without any spaces):

arginyl prolyl prolyl glycyl phenylalanyl seryl prolyl phenylalanyl arginine

This is a cumbersome and awkward name. A shorthand system is more convenient, representing each amino acid by its three-letter abbreviation. These abbreviations, given in Table 24-2, are generally the first three letters of the name. Once again, the amino acids are arranged from the N terminus at the left to the C terminus at the right. Bradykinin has the following abbreviated name:

Arg-Pro-Pro-Gly-Phe-Ser-Pro-Phe-Arg

Single-letter symbols (also given in Table 24-2) are becoming widely used as well. Using single letters, we symbolize bradykinin by

RPPGFSPFR

> **PROBLEM 24-20**
>
> Draw the complete structures of the following peptides:
> **(a)** Thr-Phe-Met   **(b)** serylarginylglycylphenylalanine   **(c)** IMQDK   **(d)** ELVIS

24-8C   Disulfide Linkages

Amide linkages (peptide bonds) form the backbone of the amino acid chains we call peptides and proteins. A second kind of covalent bond is possible between any cysteine residues present. Cysteine residues can form **disulfide bridges** (also called **disulfide linkages**) which can join two chains or link a single chain into a ring.

Mild oxidation joins two molecules of a thiol into a disulfide, forming a disulfide linkage between the two thiol molecules. This reaction is reversible, and a mild reduction cleaves the disulfide.

$$R-SH \; + \; HS-R \; \underset{[\text{reduction}]}{\overset{[\text{oxidation}]}{\rightleftharpoons}} \; R-S-S-R \; + \; H_2O$$

two molecules of thiol   disulfide

Similarly, two cysteine sulfhydryl (—SH) groups are oxidized to give a disulfide-linked pair of amino acids. This disulfide-linked dimer of cysteine is called *cystine*. Figure 24-8 shows formation of a cystine disulfide bridge linking two peptide chains.

▶ **Figure 24-8**
Cystine, a dimer of cysteine, results when two cysteine residues undergo oxidation to form a disulfide bridge.

▲ Figure 24-9
Structure of human oxytocin. A disulfide linkage holds part of the molecule in a large ring.

Two cysteine residues may form a disulfide bridge within a single peptide chain, making a ring. Figure 24-9 shows the structure of human oxytocin, a peptide hormone that causes contraction of uterine smooth muscle and induces labor. Oxytocin is a non-apeptide with two cysteine residues (at positions 1 and 6) linking part of the molecule in a large ring. In drawing the structure of a complicated peptide, arrows are often used to connect the amino acids, showing the direction from N terminus to C terminus. Notice that the C terminus of oxytocin is a primary amide (Gly · NH$_2$) rather than a free carboxyl group.

Figure 24-10 shows the structure of bovine insulin, a more complex peptide hormone that regulates glucose metabolism. Insulin is composed of two separate peptide chains, the *A chain*, containing 21 amino acid residues, and the *B chain*, containing 30. The A and B chains are joined at two positions by disulfide bridges, and the A chain has an additional disulfide bond that holds six amino acid residues in a ring. The C-terminal amino acids of both chains occur as primary amides.

Disulfide bridges are commonly manipulated in the process of giving hair a *permanent wave*. Hair is composed of protein, which is made rigid and tough partly by disulfide bonds. When hair is treated with a solution of a thiol such as 2-mercaptoethanol (HS—CH$_2$—CH$_2$—OH), the disulfide bridges are reduced and cleaved. The hair is wrapped around curlers, and the disulfide bonds are allowed to re-form, either by air oxidation or by application of a *neutralizer*. The disulfide bonds reform in new positions, holding the hair in the bent conformation enforced by the curlers.

▲ **Figure 24-10**
Structure of bovine insulin. Two chains are joined at two positions by disulfide bridges, and a third disulfide bond holds the A chain in a ring.

24-9
Peptide Structure Determination

Insulin is a relatively simple protein, yet it is a complicated organic structure. How is it possible to determine the complete structure of a protein with hundreds of amino acid residues and a molecular weight of many thousands? Chemists have developed clever ways to determine the exact sequence of amino acids in a protein. We will consider some of the most common methods.

24-9A   Cleavage of Disulfide Linkages

The first step in structure determination is to break all the disulfide bonds, separating the individual peptide chains and opening any disulfide-linked rings. The individual peptide chains are then purified and analyzed separately.

Cystine bridges are easily cleaved by reducing them to the thiol (cysteine) form. These reduced cysteine residues have a tendency to reoxidize and re-form disulfide bridges, however. A more permanent cleavage involves oxidizing the disulfide linkages with per-oxyformic acid (Figure 24-11). This oxidation converts the disulfide bridges to sulfonic acid ($-SO_3H$) groups. The oxidized cysteine units are called *cysteic acid* residues.

24-9B   Determination of the Amino Acid Composition

Once the disulfide bridges have been broken and the individual peptide chains have been separated and purified, the structure of each chain must be determined. The first step is to determine which amino acids are present and in what proportions. To analyze the amino acid composition, the peptide chain is completely hydrolyzed by boiling it for 24 hours in 6 M HCl. The resulting mixture of amino acids (the *hydrolysate*) is placed on the column of an *amino acid analyzer*, diagrammed in Figure 24-12.

In the amino acid analyzer, the components of the hydrolysate are dissolved in an aqueous buffer solution and separated by passing them down an ion-exchange column. The solution emerging from the column is mixed with ninhydrin, which reacts with amino acids to give the purple ninhydrin color. The absorption of light is recorded on a strip chart as a function of time.

The time required for each amino acid to pass through the column (its *retention time*) depends on how strongly that amino acid interacts with the ion-exchange resin. The retention time of each amino acid is known from standardization with pure amino acids. The amino acids present in the sample are identified by comparing their retention

◀ **Figure 24-11**
Oxidation of a protein by peroxyformic acid cleaves all the disulfide linkages by oxidizing cystine to cysteic acid.

◀ **Figure 24-12**
In an amino acid analyzer, the hydrolysate passes through an ion-exchange column. The solution emerging from the column is treated with ninhydrin, and its absorbance is recorded as a function of time. Each amino acid is identified by the retention time required to pass through the column.

times with the known values. The area under each peak is nearly proportional to the amount of the amino acid producing that peak, so we can determine the relative amounts of amino acids present.

Figure 24-13 shows a standard trace of an equimolar mixture of amino acids, followed by the trace produced by the hydrolysate from human bradykinin (Arg-Pro-Pro-Gly-Phe-Ser-Pro-Phe-Arg).

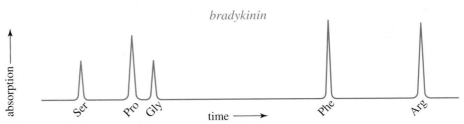

▶ Figure 24-13
Use of an amino acid analyzer to determine the composition of human bradykinin. The bradykinin peaks for Pro, Arg, and Phe are larger than those in the standard equimolar mixture because bradykinin has three Pro residues, two Arg residues, and two Phe residues.

Sequencing the Peptide: Terminal Residue Analysis   The amino acid analyzer determines the amino acids present in a peptide, but it does not reveal their **sequence**: the order in which they are linked together. The peptide sequence is destroyed in the hydrolysis step. To determine the amino acid sequence, we must cleave just one amino acid from the chain and leave the rest of the chain intact. The cleaved amino acid can be separated and identified, and the process can be repeated on the rest of the chain. The amino acid may be cleaved from either end of the peptide (either the N terminus or the C terminus), and we will consider one method used for each end. This general method for peptide sequencing is called **terminal residue analysis**.

24-9C   Sequencing from the N Terminus: The Edman Degradation

The most efficient method for sequencing peptides is the **Edman degradation**. A peptide is treated with phenyl isothiocyanate, followed by mild acid hydrolysis. The products are the shortened peptide chain and a heterocyclic derivative of the N-terminal amino acid called a *phenylthiohydantoin*.

Step 1: Attack by the free amino group on phenyl isothiocyanate.
The product is a phenylthiourea.

Step 2: Treatment with HCl results in cyclization and expulsion of the shortened peptide chain.

Step 1: Cleavage and determination of the N-terminal amino acid

cysteic acid → cysteic acid phenylthiohydantoin

Step 2: Cleavage and determination of the second amino acid (the new N-terminal amino acid)

tyrosine phenylthiohydantoin

▲ **Figure 24-14**
The first two steps in sequencing bovine oxytocin. Each Edman degradation cleaves the N-terminal amino acid and forms its phenylthiohydantoin derivative. The shortened peptide is available for the next step.

The phenylthiohydantoin derivative is identified by chromatography, by comparing it with phenylthiohydantoin derivatives of the standard amino acids. This gives the identity of the original N-terminal amino acid. The rest of the peptide is cleaved intact, and further Edman degradations are used to identify additional amino acids in the chain. This process is well suited to automation, and several types of automatic sequencers have been developed.

Figure 24-14 shows the first two steps in the sequencing of bovine oxytocin. Before sequencing, the oxytocin sample is treated with peroxyformic acid to convert the disulfide bridge to cysteic acid residues.

In theory, Edman degradations could sequence a peptide of any length. In practice, however, the repeated cycles of degradation cause some internal hydrolysis of the peptide, with loss of sample and accumulation of by-products. After about 30 cycles of degradation, further accurate analysis becomes impossible. A small peptide such as bradykinin can be completely determined by Edman degradation, but larger proteins must be broken into smaller fragments (Section 24-9E) before they can be completely sequenced.

PROBLEM 24-21

Draw the structure of the phenylthiohydantoin derivatives of
(a) alanine   **(b)** valine   **(c)** lysine   **(d)** proline

PROBLEM 24-22

Show the third and fourth steps in the sequencing of bovine oxytocin. Use Figure 24-14 as a guide.

PROBLEM 24-23

The **Sanger method** for N-terminus determination is a less common alternative to the Edman degradation. In the Sanger method, the peptide is treated with the Sanger reagent, 2,4-dinitrofluorobenzene, and then hydrolyzed by reaction with 6 M aqueous HCl. The N-terminal amino acid is recovered as its 2,4-dinitrophenyl derivative and identified.

The Sanger method

(a) Propose a mechanism for the reaction of the N terminus of the peptide with 2,4-dinitrofluorobenzene.
(b) Explain why the Edman degradation is usually preferred over the Sanger method.

24-9D C-Terminal Residue Analysis

The selective enzymatic cleavage of proteins is critical to many biological processes. For example, the clotting of blood depends on the enzyme thrombin cleaving fibrinogen at specific points to produce fibrin, the protein that forms a clot.

There is no efficient method for sequencing several amino acids of a peptide starting from the C terminus. In many cases, however, the C-terminal amino acid can be identified using the enzyme *carboxypeptidase*, which cleaves the C-terminal peptide bond. The products are the free C-terminal amino acid and a shortened peptide. Further reaction cleaves the second amino acid that has now become the new C terminus of the shortened peptide. Eventually, the entire peptide is hydrolyzed to its individual amino acids.

A peptide is incubated with the carboxypeptidase enzyme, and the appearance of free amino acids is monitored. In theory, the amino acid whose concentration increases first should be the C terminus, and the next amino acid to appear should be the second residue from the end. In practice, different amino acids are cleaved at different rates, making it difficult to determine amino acids past the C terminus and occasionally the second residue in the chain.

24-9E   Breaking the Peptide into Shorter Chains: Partial Hydrolysis

Before a large protein can be sequenced, it must be broken into smaller chains, not longer than about 30 amino acids. Each of these shortened chains is sequenced, and then the entire structure of the protein is deduced by fitting the short chains together like pieces of a jigsaw puzzle.

Partial cleavage can be accomplished either by using dilute acid with a short reaction time or by using enzymes such as *trypsin* and *chymotrypsin*. The acid-catalyzed cleavage is not very selective, leading to a mixture of short fragments resulting from cleavage at various positions. Enzymes are more selective, giving cleavage at predictable points in the chain.

TRYPSIN: Cleaves the chain at the carboxyl groups of the basic amino acids lysine and arginine.

CHYMOTRYPSIN: Cleaves the chain at the carboxyl groups of the aromatic amino acids phenylalanine, tyrosine, and tryptophan.

Let's use oxytocin (Figure 24-9) as an example to illustrate the use of partial hydrolysis. Oxytocin could be sequenced directly by C-terminal analysis and a series of Edman degradations, but it provides a simple example of how a structure can be pieced together. Acid-catalyzed partial hydrolysis of oxytocin (after cleavage of the disulfide bridge) gives a mixture that includes the following peptides:

Ile-Gln-Asn-Cys   Gln-Asn-Cys-Pro   Pro-Leu-Gly $\cdot$ NH$_2$ Cys-Tyr-Ile-Gln-Asn   Cys-Pro-Leu-Gly

When we match the overlapping regions of these fragments, the complete sequence of oxytocin appears:

Ile-Gln-Asn-Cys

Gln-Asn-Cys-Pro

Cys-Tyr-Ile-Gln-Asn

Cys-Pro-Leu-Gly

Pro-Leu-Gly $\cdot$ NH$_2$

Complete structure

Cys-Tyr-Ile-Gln-Asn-Cys-Pro-Leu-Gly $\cdot$ NH$_2$

> The enzymatic hydrolysis of proteins has many applications. For example, papain (from papaya extract) is used as a meat tenderizer. It cleaves the fibrous proteins at several locations, making the meat less tough.

The two Cys residues in oxytocin may be involved in disulfide bridges, either linking two of these peptide units or forming a ring. By measuring the molecular weight of oxytocin, we can show that it contains just one of these peptide units; therefore, the Cys residues must link the molecule in a ring.

PROBLEM 24-24

Show where trypsin and chymotrypsin would cleave the following peptide.

Tyr-Ile-Gln-Arg-Leu-Gly-Phe-Lys-Asn-Trp-Phe-Gly-Ala-Lys-Gly-Gln-Gln $\cdot$ NH$_2$

PROBLEM 24-25

After treatment with peroxyformic acid, the peptide hormone vasopressin is partially hydrolyzed. The following fragments are recovered. Propose a structure for vasopressin.

Phe-Gln-Asn   Pro-Arg-Gly $\cdot$ NH$_2$ Cys-Tyr-Phe

Asn-Cys-Pro-Arg   Tyr-Phe-Gln-Asn

24-10
Solution-Phase Peptide Synthesis

24-10A Introduction

Total synthesis of peptides is rarely an economical method for their commercial production. Important peptides are usually derived from biological sources. For example, insulin for diabetics was originally taken from pork pancreas. Now, recombinant DNA techniques are improving the quality and availability of peptide pharmaceuticals. It is possible to extract the piece of DNA that contains the code for a particular protein, insert it into a bacterium, and induce the bacterium to produce the protein. Strains of *Escherichia coli* have been developed to produce human insulin that avoids dangerous reactions in people who are allergic to pork products.

Laboratory peptide synthesis is still an important area of chemistry, however. When the structure of a new peptide is determined, a synthesis is usually attempted. The purpose of the synthesis is twofold: If the synthetic material is the same as the natural material, it proves that the proposed structure is correct; and the synthesis provides a larger amount of the material for further biological testing.

Peptide synthesis requires the formation of amide bonds between the proper amino acids in the proper sequence. With simple acids and amines, we would form an amide bond simply by converting the acid to an activated derivative (such as an acyl halide or anhydride) and adding the amine.

$$R-\overset{\overset{\displaystyle O}{\|}}{C}-X \;+\; H_2\ddot{N}-R' \;\longrightarrow\; R-\overset{\overset{\displaystyle O}{\|}}{C}-NH-R' \;+\; H-X$$

(X is a good leaving group, preferably electron-withdrawing)

Amide formation is not so easy with amino acids, however. Each amino acid has both an amino group and a carboxyl group. If we activate the carboxyl group, it reacts with its own amino group. If we mix some amino acids and add a reagent to make them couple, they form every conceivable sequence. Also, some amino acids have side chains that might interfere with peptide formation. For example, glutamic acid has an extra carboxyl group, and lysine has an extra amino group. As a result, peptide synthesis always involves both activating reagents to form the correct peptide bonds and protecting reagents to block formation of incorrect bonds.

Chemists have developed two types of methods for synthesizing peptides. The classical *solution-phase method* involves adding reagents to solutions of growing peptide chains, and the *solid-phase method* involves adding reagents to growing peptide chains bonded to solid particles. Although a variety of reagents and procedures can be used with each method, we will consider only one set of reagents for the solution-phase method and one set for the solid-phase method.

24-10B Solution-Phase Method

Consider the structure of alanylvalylphenylalanine, a simple tripeptide:

$$H_2N-\underset{\underset{\displaystyle CH_3}{|}}{CH}-\overset{\overset{\displaystyle O}{\|}}{C}-NH-\underset{\underset{\displaystyle CH(CH_3)_2}{|}}{CH}-\overset{\overset{\displaystyle O}{\|}}{C}-NH-\underset{\underset{\displaystyle CH_2Ph}{|}}{CH}-\overset{\overset{\displaystyle O}{\|}}{C}-OH$$

alanyl valyl phenylalanine

Ala-Val-Phe

Solution-phase peptide synthesis begins at the N terminus and ends at the C terminus, or left to right as we draw the peptide. The first major step is to couple the carboxyl group of alanine to the amino group of valine. This cannot be done simply by activating the carboxyl group of alanine and adding valine. If we activated the carboxyl group of alanine, it would react with another molecule of alanine.

To prevent side reactions, the amino group of alanine must be protected to make it non-nucleophilic. Treating the free amino acid with benzyl chloroformate forms a urethane, or carbamate ester, that is easily removed at the end of the synthesis. This protecting group has been used for many years, and it has acquired several names. It is called the *benzyloxycarbonyl group*, the *carbobenzoxy group* (Cbz), or simply the *Z group* (abbreviated Z).

Preliminary step: Protect the amino group with Z

The amino group in Z-Ala is protected as the non-nucleophilic amide half of a carbamate ester. The carboxyl group can be activated without reacting with the protected amino group. In the solution-phase synthesis, the carboxyl group is activated by treatment with ethyl chloroformate. The product is a mixed anhydride of the amino acid and carbonic acid. It is strongly activated toward nucleophilic attack.

Step 1: Activate the carboxyl group

When the second amino acid (valine) is added to the protected, activated alanine, the nucleophilic amino group of valine attacks the activated carbonyl of alanine, displacing the anhydride and forming a peptide bond.

Step 2: Couple the next amino acid

$$Z-NHCH-\overset{\overset{O}{\|}}{C}-O-\overset{\overset{O}{\|}}{C}-OCH_2CH_3 \;+\; H_2\ddot{N}-CH-\overset{\overset{O}{\|}}{C}-OH \longrightarrow$$

$$\underset{CH_3}{|} \qquad\qquad\qquad\qquad \underset{CH(CH_3)_2}{|}$$

protected, activated alanine   valine

$$Z-NHCH-\overset{\overset{O}{\|}}{C}-NH-CH-\overset{\overset{O}{\|}}{C}-OH \quad \begin{array}{l} + \; CO_2\uparrow \\ + \; CH_3CH_2OH \end{array}$$

$$\underset{CH_3}{|} \qquad\qquad \underset{CH(CH_3)_2}{|}$$

Z-Ala-Val

PROBLEM 24-26

Give complete mechanisms for the formation of Z-Ala, its activation by ethyl chloroformate, and the coupling with valine.

At this point, we have the N-protected dipeptide Z-Ala-Val. Phenylalanine must be added to the C terminus to complete the Ala-Val-Phe tripeptide. Activation of the valine carboxyl group, followed by addition of phenylalanine, gives the protected tripeptide.

Step 1: Activate the carboxyl group

$$Z-NHCHCNHCH-\overset{\overset{O}{\|}}{C}\boxed{OH} \;+\; Cl-\overset{\overset{O}{\|}}{C}-OEt \longrightarrow Z-NHCHCNHCH-\overset{\overset{O}{\|}}{C}\boxed{O-\overset{\overset{O}{\|}}{C}-OEt} \;+\; HCl$$

$$\underset{CH_3}{|}\;\underset{CH(CH_3)_2}{|} \qquad\qquad\qquad\qquad \underset{CH_3}{|}\;\underset{CH(CH_3)_2}{|}$$

Step 2: Couple the next amino acid

$$Z-Ala-NHCH-\overset{\overset{O}{\|}}{C}\boxed{O-\overset{\overset{O}{\|}}{C}-OEt} \;+\; H_2N-CH-\overset{\overset{O}{\|}}{C}-OH \longrightarrow Z-Ala-NHCH-\overset{\overset{O}{\|}}{C}\boxed{NH-CH-\overset{\overset{O}{\|}}{C}-OH} \;+\; CO_2\uparrow + EtOH$$

$$\underset{CH(CH_3)_2}{|} \qquad\qquad \underset{CH_2-Ph}{|} \qquad\qquad H_3C-\underset{}{CH}-CH_3 \; \underset{CH_2-Ph}{|}$$

phenylalanine   Z-Ala-Val-Phe

To make a larger peptide, repeat these two steps for the addition of each amino acid residue:

1. Activate the C terminus of the growing peptide by reaction with ethyl chloroformate.
2. Couple the next amino acid.

The final step in the solution-phase synthesis is to deprotect the N terminus of the completed peptide. The N-terminal amide bond must be cleaved without breaking any of the peptide bonds in the product. Fortunately, the benzyloxycarbonyl group is partly an amide and partly a benzyl ester, and hydrogenolysis of the benzyl ester takes place under mild conditions that do not cleave the peptide bonds. This mild cleavage is the reason for using the benzyloxycarbonyl group (as opposed to some other acyl group) to protect the N terminus.

Final step: Remove the protecting group

Z-Ala-Val-Phe

Ala-Val-Phe

PROBLEM 24-27

Show how you would synthesize Ala-Val-Phe-Gly-Leu starting with Z-Ala-Val-Phe.

PROBLEM 24-28

Show how the solution-phase synthesis would be used to synthesize Ile-Gly-Asn.

PROBLEM-SOLVING HINT

Remember that classical (solution-phase) peptide synthesis:
1. Goes N → C. Protect N terminus (Z group) first, deprotect last.
2. Couple each AA by activating C terminus (ethyl chloroformate), adding new AA.

The solution-phase method works well for small peptides, and many peptides have been synthesized by this process. Larger proteins are not easily synthesized by the solution-phase method, however. A large number of chemical reactions and purifications are required even for a small peptide. Although the individual yields are excellent, with a large peptide, the overall yield becomes so small as to be unusable, and several months (or years) are required to complete so many steps. The large amounts of time required and the low overall yields are due largely to the purification steps. For larger peptides and proteins, solid-phase peptide synthesis is usually preferred.

In 1962, Robert Bruce Merrifield of Rockefeller University developed a method for synthesizing peptides without having to purify the intermediates. He did this by attaching the growing peptide chains to solid polystyrene beads. After each amino acid is added, the excess reagents are washed away by rinsing the beads with solvent. This ingenious method lends itself to automation, and Merrifield built a machine that can add several amino acid units while running unattended. Using this machine, Merrifield synthesized ribonuclease (124 amino acids) in just six weeks, obtaining an overall yield of 17%. Merrifield's work in **solid-phase peptide synthesis** won the Nobel Prize in chemistry in 1984.

24-11
Solid-Phase
Peptide Synthesis

24-11A The Individual Reactions

There are three important reactions to consider before we use solid-phase peptide synthesis. We must first learn how an amino acid is attached to the solid support, how and why the amino group is protected, and how peptide bonds are formed.

Attaching the Peptide to the Solid Support   The greatest difference between solution-phase and solid-phase peptide synthesis is that solid-phase synthesis is done in the opposite direction: starting with the C terminus and going toward the N terminus, right to left as we write the peptide. The first step is to attach the *last* amino acid (the C terminus) to the solid support.

The solid support is a special polystyrene bead in which some of the aromatic rings have chloromethyl groups. This polymer, often called the *Merrifield resin*, is made by copolymerizing styrene with a few percent of *p*-(chloromethyl)styrene.

Formation of the Merrifield resin

Like other benzyl halides, the chloromethyl groups on the polymer are reactive toward S_N2 attack. The carboxyl group of an N-protected amino acid displaces chloride, giving an amino acid ester of the polymer. In effect, the polymer serves as the alcohol part of an ester protecting group for the carboxyl end of the C-terminal amino acid.

Attachment of the C-terminal amino acid

Once the C-terminal amino acid is fixed to the polymer, the chain is built on the amino group of this amino acid. At the completion of the synthesis, the ester bond to the polymer is cleaved by anhydrous HF. Because this is an ester bond, it is more easily cleaved than the amide bonds of the peptide.

Cleavage of the finished peptide

Use of the *t*-Butyloxycarbonyl (Boc) Protecting Group

The benzyloxycarbonyl group (the Z group) cannot be used with the solid-phase process because the Z group is removed by hydrogenolysis in contact with a solid catalyst. A polymer-bound peptide cannot achieve the intimate contact with a solid catalyst required for hydrogenolysis. The N-protecting group used in the Merrifield procedure is the *t*-butyloxycarbonyl group, abbreviated Boc or *t*-Boc. The Boc group is similar to the Z group, except that it has a *t*-butyl group in place of the benzyl group. Like other *t*-butyl esters, the Boc protecting group is easily removed under acidic conditions.

The acid chloride of the Boc group is unstable, so we use the anhydride, di-*t*-butyldicarbonate, to attach the group to the amino acid.

Protection of the amino group as its Boc derivative

$$CH_3-\underset{\underset{CH_3}{|}}{\overset{\overset{CH_3}{|}}{C}}-O-\overset{\overset{O}{||}}{C}-O-\overset{\overset{O}{||}}{C}-O-\underset{\underset{CH_3}{|}}{\overset{\overset{CH_3}{|}}{C}}-CH_3 \;+\; H_2N-\underset{\underset{R}{|}}{CH}-COOH \;\longrightarrow$$

di-*t*-butyldicarbonate   amino acid

$$CH_3-\underset{\underset{CH_3}{|}}{\overset{\overset{CH_3}{|}}{C}}-O-\overset{\overset{O}{||}}{C}-NH-\underset{\underset{R}{|}}{CH}-COOH \;+\; CO_2 \;+\; CH_3-\underset{\underset{CH_3}{|}}{\overset{\overset{CH_3}{|}}{C}}-OH$$

Boc-amino acid

The Boc group is easily cleaved by brief treatment with trifluoroacetic acid (TFA), CF_3COOH. Loss of a relatively stable *t*-butyl cation from the protonated ester gives an unstable carbamic acid. Decarboxylation of the carbamic acid gives the deprotected amine group of the amino acid. Loss of a proton from the *t*-butyl cation gives isobutylene.

$$CH_3-\underset{\underset{CH_3}{|}}{\overset{\overset{CH_3}{|}}{C}}-O-\overset{\overset{\ddot{O}}{||}}{C}-NH-\underset{\underset{R}{|}}{CH}-COOH \;\xrightarrow{CF_3COOH}\; CH_3-\underset{\underset{CH_3}{|}}{\overset{\overset{CH_3}{|}}{C}}-\overset{..}{\underset{}{O}}-\overset{\overset{\overset{..}{O}^{\pm}-H}{||}}{C}-NH-\underset{\underset{R}{|}}{CH}-COOH \;\longrightarrow$$

Boc-amino acid   protonated

$$CH_3-\underset{\underset{CH_3}{|}}{\overset{\overset{CH_3}{|}}{C^+}} \;+\; \left[\ddot{O}=\overset{\overset{:\ddot{O}-H}{|}}{C}-NH-\underset{\underset{R}{|}}{CH}-COOH \right] \;\longrightarrow$$

a carbamic acid

$$CH_2=C\underset{CH_3}{\overset{CH_3}{<}} \;+\; \overset{+}{H_3N}-\underset{\underset{R}{|}}{CH}-COOH \;+\; CO_2\uparrow$$

isobutylene   free amino acid

People who synthesize peptides generally do not make their own Boc-protected amino acids. Because they use all their amino acids in protected form, they buy and use commercially available Boc-amino acids.

Use of DCC as a Peptide Coupling Agent

The final reaction needed for the Merrifield procedure is the peptide bond-forming condensation. When a mixture of an amine and an acid is treated with *N,N'*-dicyclohexylcarbodiimide (abbreviated DCC), the amine and the acid couple to form an amide. The molecule of water lost in this condensation converts DCC to *N,N'*-dicyclohexyl urea (DCU).

acid   amine   *N,N'*-dicyclohexylcarbodiimide (DCC)

amide   *N,N'*-dicyclohexyl urea (DCU)

The mechanism for DCC coupling is not as complicated as it may seem. The carboxylate ion adds to the strongly electrophilic carbon of the diimide, giving an activated acyl derivative of the acid. This activated derivative reacts readily with the amine to give the amide. In the final step, DCU serves as an excellent leaving group.

Formation of an activated acyl derivative

activated

Coupling with the amine and loss of DCU

amide

Now we consider an example to illustrate how these procedures are combined in the Merrifield solid-phase peptide synthesis.

24-11B   An Example of Solid-Phase Peptide Synthesis

For easy comparison of the solution-phase and solid-phase methods, we will consider the synthesis of the same tripeptide we made using the solution-phase method.

Ala-Val-Phe

The solid-phase synthesis is carried out in the direction opposite that of the solution-phase synthesis. The first step is attachment of the N-protected C-terminal amino acid (Boc-phenylalanine) to the polymer.

Trifluoroacetic acid (TFA) cleaves the Boc protecting group of phenylalanine so that its amino group can be coupled with the next amino acid.

The second amino acid (valine) is added in its N-protected Boc form so that it cannot couple with itself. Addition of DCC couples the valine carboxyl group with the free —NH$_2$ group of phenylalanine.

To couple the final amino acid (alanine), the chain is first deprotected by treatment with trifluoroacetic acid. The N-protected Boc-alanine and DCC are then added.

Step 1: Deprotection

Step 2: Coupling

If we were making a longer peptide, the addition of each subsequent amino acid would require the repetition of two steps:

1. Use trifluoroacetic acid to deprotect the amino group at the end of the growing chain.
2. Add the next Boc-amino acid, using DCC as a coupling agent.

Once the peptide is completed, the final Boc protecting group must be removed, and the peptide must be cleaved from the polymer. Anhydrous HF cleaves the ester linkage that bonds the peptide to the polymer, and it also removes the Boc protecting group. In our example, the following reaction occurs.

PROBLEM 24-30

Show how you would synthesize Leu-Gly-Ala-Val-Phe starting with Boc-Ala-Val-Phe—(P).

PROBLEM 24-31

Show how solid-phase peptide synthesis would be used to make Ile-Gly-Asn.

PROBLEM-SOLVING HINT

Remember that solid-phase peptide synthesis:

1. Goes C → N. Attach the Boc-protected C terminus to the bead first.
2. Couple each AA by removing (TFA) the Boc group from the N terminus, then add the next Boc-protected AA with DCC.
3. Cleave (HF) the finished peptide from the bead.

24-12 Classification of Proteins

Proteins may be classified according to their chemical composition, their shape, or their function. Protein composition and function are treated in detail in a biochemistry course. For now, we briefly survey the types of proteins and their general classifications.

Proteins are grouped into *simple* and *conjugated* proteins according to their chemical composition. **Simple proteins** are those that hydrolyze to give only amino acids. All the protein structures we have considered so far are simple proteins. Examples are insulin, ribonuclease, oxytocin, and bradykinin. **Conjugated proteins** are bonded to a nonprotein **prosthetic group** such as a sugar, a nucleic acid, a lipid, or some other group. Table 24-3 lists some examples of conjugated proteins.

TABLE 24-3 Classes of Conjugated Proteins

Class	Prosthetic Group	Examples
glycoproteins	carbohydrates	γ-globulin, interferon
nucleoproteins	nucleic acids	ribosomes, viruses
lipoproteins	fats, cholesterol	high-density lipoprotein
metalloproteins	a complexed metal	hemoglobin, cytochromes

Proteins are classified as *fibrous* or *globular* depending on whether they form long filaments or coil up on themselves. **Fibrous proteins** are stringy, tough, and usually insoluble in water. They function primarily as structural parts of the organism. Examples of fibrous proteins are α-keratin in hooves and fingernails, and collagen in tendons. **Globular proteins** are folded into roughly spherical shapes. They usually function as enzymes, hormones, or transport proteins. Examples of globular proteins are insulin, ribonuclease, and hemoglobin.

24-13 Levels of Protein Structure

24-13A Primary Structure

Up to now, we have discussed the *primary structure* of proteins. The **primary structure** is the covalently bonded structure of the molecule. This definition includes the sequence of amino acids, together with any disulfide bridges. All the properties of the protein are determined, directly or indirectly, by the primary structure. Any folding, hydrogen bonding, or catalytic activity depends on the proper primary structure.

24-13B Secondary Structure

Peptide chains tend to form orderly hydrogen-bonded arrangements. In particular, the carbonyl oxygen atoms form hydrogen bonds with the amide (N — H) hydrogens. Two arrangements allow an orderly arrangement of hydrogen bonds: the **α-helix** and the **pleated sheet**. These hydrogen-bonded arrangements, if present, are called the **secondary structure** of the protein.

C = gray
N = blue
O = red
R = green

▲ **Figure 24-15**
The α-helical arrangement. Each peptide carbonyl group is hydrogen-bonded to an N—H hydrogen on the next turn of the helix. Side chains are symbolized by green atoms in the space-filling structure.

Spider web is composed mostly of fibroin, a protein with pleated-sheet secondary structure. The pleated-sheet arrangement allows for multiple hydrogen bonds between molecules, conferring great strength.

If the molecule winds into a helical coil, each carbonyl oxygen can hydrogen-bond with an N—H hydrogen on the next turn of the coil. Many proteins wind into an α-helix (a helix that looks like the thread on a right-handed screw) with the side chains positioned on the outside of the helix. For example, the fibrous protein α-keratin is arranged in the α-helical structure, and most globular proteins contain segments of α-helix. Figure 24-15 shows the α-helical arrangement.

Segments of peptides can also form orderly arrangements of hydrogen bonds by lining up side by side. In this arrangement, each carbonyl group on one chain forms a hydrogen bond with an N—H hydrogen on an adjacent chain. This arrangement may involve many peptide molecules lined up side by side, resulting in a two-dimensional *sheet*. The bond angles between amino acid units are such that the sheet is *pleated* (creased), with the amino acid side chains arranged on alternating sides of the sheet. Silk fibroin, the principal fibrous protein in the silks of insects and arachnids, has a pleated sheet secondary structure. Figure 24-16 shows the pleated sheet structure.

A protein may or may not have the same secondary structure throughout its length. Some parts may be curled into an α-helix, while other parts are lined up in a pleated

▶ **Figure 24-16**
The pleated sheet arrangement. Each peptide carbonyl group is hydrogen-bonded to an N—H hydrogen on an adjacent peptide chain.

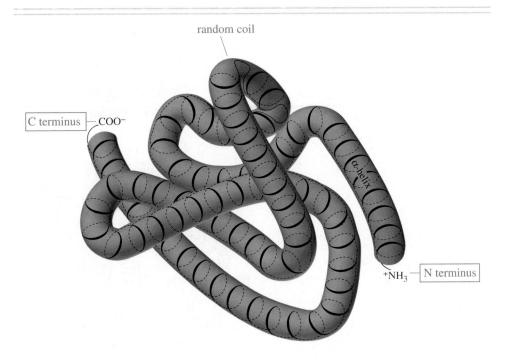

▲ Figure 24-17
The tertiary structure of a typical globular protein includes segments of α-helix with segments of random coil at the points where the helix is folded.

sheet. Parts of the chain may have no secondary structure at all. Such a structureless part is called a **random coil**. Most globular proteins, for example, contain segments of α-helix or pleated sheet separated by kinks of random coil, allowing the molecule to fold into its globular shape.

24-13C Tertiary Structure

The **tertiary structure** of a protein is its complete three-dimensional conformation. Remember that the secondary structure is a local structure. Parts of the protein may have the α-helical structure, while other parts may have the pleated-sheet structure, and still other parts may be random coils. The tertiary structure includes all the secondary structure and all the kinks and folds in between. The tertiary structure of a typical globular protein is represented in Figure 24-17.

Coiling of an enzyme can produce important catalytic effects. Polar, *hydrophilic* (water-loving) side chains are oriented toward the outside of the globule. Nonpolar, *hydrophobic* (water-hating) groups are arranged toward the interior. Coiling in the proper conformation creates an enzyme's **active site**, the region that binds the substrate and catalyzes the reaction. A reaction taking place at the active site in the interior of an enzyme may occur under essentially anhydrous, nonpolar conditions—while the whole system is dissolved in water!

24-13D Quaternary Structure

Quaternary structure refers to the association of two or more peptide chains in the complete protein. For example, hemoglobin, the oxygen carrier in mammalian blood, consists of four peptide chains fitted together to form a globular protein. Figure 24-18 summarizes the four levels of protein structure.

Tertiary structures of proteins are determined by X-ray crystallography. A single crystal of the protein is bombarded with X rays, whose wavelengths are appropriate to be diffracted by the regular atomic spacings in the crystal. A computer then determines the locations of the atoms in the crystal.

primary structure

secondary structure

▶ **Figure 24-18**
A schematic comparison of the levels of protein structure. Primary structure is the covalently bonded structure, including the amino acid sequence and any disulfide bridges. Secondary structure refers to the areas of α-helix, pleated sheet, or random coil. Tertiary structure refers to the overall conformation of the molecule. Quaternary structure refers to the association of two or more peptide chains in the active protein.

tertiary structure

quaternary structure

24-14
Protein Denaturation

For a protein to be biologically active, it must have the correct structure at all levels. The sequence of amino acids must be right, with the correct disulfide bridges linking the cysteines on the chains. The secondary and tertiary structures are important, as well. The protein must be folded into its natural conformation, with the appropriate areas of α-helix and pleated sheet. For an enzyme, the active site must have the right conformation, with the necessary side-chain functional groups in the correct positions. Conjugated proteins must have the right prosthetic groups, and multichain proteins must have the right combination of individual peptides.

With the exception of the covalent primary structure, all these levels of structure are maintained by weak solvation and hydrogen-bonding forces. Small changes in the environment can cause a chemical or conformational change resulting in **denaturation**: disruption of the normal structure and loss of biological activity. Many factors can cause denaturation, but the most common ones are heat and pH.

24-14A   Reversible and Irreversible Denaturation

The cooking of egg white is an example of protein denaturation by high temperature. Egg white contains soluble globular proteins called *albumins*. When egg white is heated, the albumins unfold and coagulate to produce a solid rubbery mass. Different proteins have different abilities to resist the denaturing effect of heat. Egg albumin is quite sensitive to heat, but bacteria that live in geothermal hot springs have developed proteins that retain their activity in boiling water.

When a protein is subjected to an acidic pH, some of the side-chain carboxyl groups become protonated and lose their ionic charge. Conformational changes result, leading to denaturation. In a basic solution, amino groups become deprotonated, similarly losing their ionic charge, causing conformational changes and denaturation.

Milk turns sour because of the bacterial conversion of carbohydrates to lactic acid. When the pH becomes strongly acidic, soluble proteins in milk are denatured and

Irreversible denaturation of egg albumin. The egg white does not become clear and runny again when it is cooled.

precipitate. This process is called *curdling*. Some proteins are more resistant to acidic and basic conditions than others. For example, most digestive enzymes such as amylase and trypsin remain active under acidic conditions in the stomach, even at a pH of about 1.

In many cases, denaturation is irreversible. When cooked egg white is cooled, it does not become uncooked. Curdled milk does not uncurdle when it is neutralized. Denaturation may be reversible, however, if the protein has undergone only mild denaturing conditions. For example, a protein can be *salted out* of solution by a high salt concentration, which denatures and precipitates the protein. When the precipitated protein is redissolved in a solution with a lower salt concentration, it usually regains its activity together with its natural conformation.

24-14B   Prion Diseases

Up through 1980, people thought that all infectious diseases were caused by microbes of some sort. They knew about diseases caused by viruses, bacteria, protozoa, and fungi. There were some strange diseases, however, for which no one had isolated and cultured the pathogen. *Creutzfeldt-Jakob Disease* (CJD) in humans, *scrapie* in sheep, and *transmissible encephalopathy* in mink (TME) all involved a slow, gradual loss of mental function and eventual death. The brains of the victims all showed unusual plaques of amyloid protein surrounded by spongelike tissue.

Workers studying these diseases thought there was an infectious agent involved (as opposed to genetic or environmental causes) because they knew that scrapie and TME could be spread by feeding healthy animals the ground-up remains of sick animals. They had also studied *kuru*, a disease much like CJD among tribes where family members showed their respect for the dead by eating their brains. These diseases were generally attributed to "slow viruses" that were yet to be isolated.

In the 1980s, neurologist Stanley B. Prusiner (of the University of California at San Francisco) made a homogenate of scrapie-infected sheep brains and systematically separated out all the cell fragments, bacteria, and viruses, and found that the remaining material was still infectious. He separated out the proteins and found a protein fraction that was still infectious. He suggested that scrapie (and presumably similar diseases) is caused by a protein infectious agent that he called **prion protein**. This conclusion contradicted the established principle that contagious diseases require a living pathogen. Many skeptical workers repeated Prusiner's work in hopes of finding viral contaminants in the infectious fractions, and most of them finally came to the same conclusion. Prusiner received the 1998 Nobel Prize in medicine for this work.

Micrograph of normal human brain tissue. The nuclei of neurons appear as dark spots.

Since Prusiner's work, prion diseases have become more important because of their threat to humans. Beginning in 1996, some cows in the United Kingdom developed "mad cow disease," and would threaten other animals, wave their heads, fall down, and eventually die. The disease, called *bovine spongiform encephalopathy*, or BSE, was probably transmitted to cattle by feeding them the remains of scrapie-infected sheep. The most frightening aspect of the BSE outbreak was that people could contract a fatal disease, called *new-variant Creutzfeldt-Jakob Disease* (vCJD) from eating the infected meat. Since that time, a similar disease, called *chronic wasting disease*, or CWD, has been found in wild deer and elk in the Rocky Mountains. All of these (presumed) prion diseases are now classified as *transmissible spongiform encephalopathies*, or TSEs.

The most widely accepted theory of prion diseases suggests that the infectious prion protein has the same primary structure as a normal protein found in nerve cells, but it differs in its tertiary structure. In effect, it is a misfolded, denatured version of a normal protein that polymerizes to form the amyloid protein plaques seen in the brains of infected animals. When animal ingests infected food, the polymerized protein resists digestion. Because it is simply a misfolded version of a normal protein, the infectious prion does not provoke the host's immune system to attack the pathogen.

When the abnormal prion interacts with normal prion protein on the membranes of nerve cells, the abnormal protein somehow induces the normal molecules to change their shape. (This is the part of the process we know the least about.) These newly

Brain tissue of a patient infected with vCJD. Note the formation of (white) vacuole spaces and (dark, irregular) plaques of prion protein. (Magnification 200X)

misfolded protein molecules then induce more molecules to change shape. The polymerized abnormal protein cannot be broken down by the usual protease enzymes, so it builds up in the brain and causes the plaques and spongy tissue associated with TSEs.

We once thought that a protein with the correct primary structure, placed in the right physiological solution, would naturally fold into the correct tertiary structure and stay that way. We were wrong. We now know that protein folding is a carefully controlled process in which enzymes and *chaperone proteins* promote correct folding as the protein is being synthesized. Prion diseases have taught us that there are many factors that cause proteins to fold into natural or unnatural conformations, and that the folding of the protein can have major effects on its biological properties within an organism.

Chapter 24 Glossary

active site The region of an enzyme that binds the substrate and catalyzes the reaction. (p. 1153)

amino acid Literally, any molecule containing both an amino group ($-NH_2$) and a carboxyl group ($-COOH$). The term usually means an **α-amino acid,** with the amino group on the carbon atom next to the carboxyl group. (p. 1115)

biomimetic synthesis A laboratory synthesis that is patterned after a biological synthesis. For example, the synthesis of amino acids by reductive amination resembles the biosynthesis of glutamic acid. (p. 1123)

complete proteins Proteins that provide all the essential amino acids in about the right proportions for human nutrition. Examples include those in meat, fish, milk, and eggs. **Incomplete proteins** are severely deficient in one or more of the essential amino acids. Most plant proteins are incomplete. (p. 1118)

conjugated protein A protein that contains a nonprotein prosthetic group such as a sugar, nucleic acid, lipid, or metal ion. (p. 1151)

C terminus (C-terminal end) The end of the peptide chain with a free or derivatized carboxyl group. As the peptide is written, the C terminus is usually on the right. The amino group of the C-terminal amino acid links it to the rest of the peptide. (p. 1133)

denaturation An unnatural alteration of the conformation or the ionic state of a protein. Denaturation generally results in precipitation of the protein and loss of its biological activity. Denaturation may be reversible, as in salting out a protein, or irreversible, as in cooking egg white. (p. 1154)

dipolar ion (zwitterion) A structure with an overall charge of zero but having a positively charged substituent and a negatively charged substituent. Most amino acids exist in dipolar ionic forms. (p. 1119)

disulfide linkage (disulfide bridge) A bond between two cysteine residues formed by mild oxidation of their thiol groups to a disulfide. (p. 1134)

Edman degradation A method for removing and identifying the N-terminal amino acid from a peptide without destroying the rest of the peptide chain. The peptide is treated with phenylisothiocyanate, followed by a mild acid hydrolysis to convert the N-terminal amino acid to its phenylthiohydantoin derivative. The Edman degradation can be used repeatedly to determine the sequence of many residues beginning at the N terminus. (p. 1138)

electrophoresis A procedure for separating charged molecules by their migration in a strong electric field. The direction and rate of migration are governed largely by the average charge on the molecules. (p. 1122)

enzymatic resolution The use of enzymes to separate enantiomers. For example, the enantiomers of an amino acid can be acylated and then treated with hog kidney acylase. The enzyme hydrolyzes the acyl group from the natural L-amino acid, but it does not react with the D-amino acid. The resulting mixture of the free L-amino acid and the acylated D-amino acid is easily separated. (p. 1128)

enzyme A protein-containing biological catalyst. Many enzymes also include *prosthetic groups*, nonprotein constituents that are essential to the enzyme's catalytic activity. (p. 1151)

essential amino acids Ten standard amino acids that are not biosynthesized by humans and must be provided in the diet. (p. 1118)

fibrous proteins A class of proteins that are stringy, tough, threadlike, and usually insoluble in water. (p. 1151)

globular proteins A class of proteins that are relatively spherical in shape. Globular proteins generally have lower molecular weights and are more soluble in water than fibrous proteins. (p. 1151)

α-helix A helical peptide conformation in which the carbonyl groups on one turn of the helix are hydrogen-bonded to N—H hydrogens on the next turn. Extensive hydrogen bonding makes this helical arrangement quite stable. (p. 1151)

hydrogenolysis Cleavage of a bond by the addition of hydrogen. For example, catalytic hydrogenolysis cleaves benzyl esters. (p. 1130)

benzyl ester   acid   toluene

isoelectric point (isoelectric pH) The pH at which an amino acid (or protein) does not move under electrophoresis. This is the pH where the average charge on its molecules is zero, with most of the molecules in their zwitterionic form. (p. 1121)

L-amino acid An amino acid having a stereochemical configuration similar to that of L-(−)-glyceraldehyde. Most naturally occurring amino acids have the L configuration. (p. 1115)

N terminus (N-terminal end) The end of the peptide chain with a free or derivatized amino group. As the peptide is written, the N terminus is usually on the left. The carboxyl group of the N-terminal amino acid links it to the rest of the peptide. (p. 1133)

oligopeptide A small polypeptide, containing about four to ten amino acid residues. (p. 1133)

peptide Any polymer of amino acids linked by amide bonds between the amino group of each amino acid and the carboxyl group of the neighboring amino acid. The terms *dipeptide, tripeptide,* etc. may specify the number of amino acids in the peptide. (p. 1133)

peptide bonds Amide linkages between amino acids. (p. 1133)

pleated sheet A two-dimensional peptide conformation with the peptide chains lined up side by side. The carbonyl groups on each peptide chain are hydrogen-bonded to N—H hydrogens on the adjacent chain, and the side chains are arranged on alternating sides of the sheet. (p. 1151)

polypeptide A peptide containing many amino acid residues. Although proteins are polypeptides, the term *polypeptide* is commonly used for molecules with lower molecular weights than proteins. (p. 1133)

primary structure The covalently bonded structure of a protein; the sequence of amino acids, together with any disulfide bridges. (p. 1151)

prion protein A protein infectious agent that is thought to promote misfolding and polymerization of normal protein molecules, leading to amyloid plaques and destruction of nerve tissue. (p. 1155)

prosthetic group The nonprotein part of a conjugated protein. Examples of prosthetic groups are sugars, lipids, nucleic acids, and metal complexes. (p. 1153)

protein A biopolymer of amino acids. Proteins are polypeptides with molecular weights higher than about 6000 amu. (pp. 1133)

quaternary structure The association of two or more peptide chains into a composite protein. (p. 1153)

random coil A type of protein secondary structure where the chain is neither curled into an α-helix nor lined up in a pleated sheet. In a globular protein, the kinks that fold the molecule into its globular shape are usually segments of random coil. (p. 1153)

residue An amino acid unit of a peptide. (p. 1133)

Sanger method A method for determining the N-terminal amino acid of a peptide. The peptide is treated with 2,4-dinitrofluorobenzene (Sanger's reagent), then completely hydrolyzed. The derivatized amino acid is easily identified, but the rest of the peptide is destroyed in the hydrolysis. (p. 1140)

secondary structure The local hydrogen-bonded arrangement of a protein. The secondary structure is generally the α-helix, pleated sheet, or random coil. (p. 1151)

sequence As a noun, the order in which amino acids are linked together in a peptide. As a verb, to determine the sequence of a peptide. (p. 1138)

simple proteins Proteins composed of only amino acids (having no prosthetic groups). (p. 1151)

solid-phase peptide synthesis A method in which the C-terminal amino acid is attached to a solid support (polystyrene beads) and the peptide is synthesized in the C → N direction by successive coupling of protected amino acids. When the peptide is complete, it is cleaved from the solid support. (p. 1145)

solution-phase peptide synthesis (classical peptide synthesis) Any of several methods in which protected amino acids are coupled in solution in the correct sequence to give a desired peptide. Most of these methods proceed in the N → C direction (p. 1142)

standard amino acids The 20 α-amino acids found in nearly all naturally occurring proteins. (p. 1116)

Strecker synthesis Synthesis of α-amino acids by reaction of an aldehyde with ammonia and cyanide ion, followed by hydrolysis of the intermediate α-amino nitrile. (p. 1126)

terminal residue analysis Sequencing a peptide by removing and identifying the residue at the N terminus or at the C terminus. (p. 1138)

tertiary structure The complete three-dimensional conformation of a protein. (p. 1153)

transamination Transfer of an amino group from one molecule to another. Transamination is a common method for the biosynthesis of amino acids, often involving glutamic acid as the source of the amino group. (p. 1123)

zwitterion (dipolar ion) A structure with an overall charge of zero but having a positively charged substituent and a negatively charged substituent. Most amino acids exist in zwitterionic forms. (p. 1119)

Essential Problem-Solving Skills in Chapter 24

1. Correctly name amino acids and peptides, and draw the structures from their names.

2. Use perspective drawings and Fischer projections to show the stereochemistry of D- and L-amino acids.

3. Explain which amino acids are acidic, which are basic, and which are neutral. Use the isoelectric point to predict whether a given amino acid will be positively charged, negatively charged, or neutral at a given pH.

4. Show how one of the following syntheses might be used to make a given amino acid:

   Reductive amination

   HVZ followed by ammonia

   Gabriel–malonic ester synthesis

   Strecker synthesis

5. Predict products of the following reactions of amino acids: esterification, acylation, reaction with ninhydrin.

6. Use information from terminal residue analysis and partial hydrolysis to determine the structure of an unknown peptide.

7. Show how solution-phase peptide synthesis or solid-phase peptide synthesis would be used to make a given peptide. Use appropriate protecting groups to prevent unwanted couplings.

8. Discuss and identify the four levels of protein structure (primary, secondary, tertiary, quaternary). Explain how the structure of a protein affects its properties and how denaturation changes the structure.

Study Problems

24-32　Define each term and give an example.

(a) α-amino acid	(b) conjugated protein	(c) protein denaturation	(d) dipolar ion
(e) disulfide bridge	(f) Edman degradation	(g) enzymatic resolution	(h) essential amino acid
(i) hydrogenolysis	(j) isoelectric point	(k) L-amino acid	(l) peptide
(m) prosthetic group	(n) primary structure	(o) secondary structure	(p) tertiary structure
(q) quaternary structure	(r) solid-phase peptide synthesis	(s) Strecker synthesis	(t) zwitterion
(u) oligopeptide			

24-33　Draw the complete structure of the following peptide.

$$\text{Val-Gln-Met} \cdot \text{NH}_2$$

24-34　Predict the products of the following reactions.

(a) Phe + [ninhydrin structure] $\xrightarrow[\text{heat}]{\text{pyridine}}$

(b) $\text{Ph}-\text{CH}_2-\text{O}-\overset{\overset{\text{O}}{\|}}{\text{C}}-\text{NH}-\underset{\underset{\text{CH}_3}{|}}{\text{CH}}-\text{COOH} \xrightarrow{\text{H}_2, \text{Pd}}$

(c) Lys + excess $(CH_3CO)_2O \longrightarrow$

(d) (D,L)-proline $\xrightarrow[\text{(2) hog kidney acylase, } H_2O]{\text{(1) excess } Ac_2O}$

(e)
$$CH_3CH_2-\underset{\underset{CHO}{|}}{CH}-CH_3 \xrightarrow[H_2O]{NH_3, HCN}$$

(f) product from part (e) $\xrightarrow{H_3O^+}$

(g) isovaleric acid + $Br_2/PBr_3 \longrightarrow$

(h) product from part (g) + excess $NH_3 \longrightarrow$

24-35 Show how you would synthesize any of the standard amino acids from each starting material. You may use any necessary reagents.

(a) $(CH_3)_2CH-\overset{\overset{O}{\|}}{C}-COOH$

(b) $CH_3-\underset{\underset{CH_2CH_3}{|}}{CH}-CH_2-COOH$

(c) $(CH_3)_2CH-CH_2-CHO$

(d) ⬡$-CH_2Br$

24-36 Show how you would convert alanine to the following derivatives. Show the structure of the product in each case.
(a) alanine isopropyl ester   **(b)** *N*-benzoylalanine
(c) *N*-benzyloxycarbonyl alanine   **(d)** *t*-butyloxycarbonyl alanine

24-37 Suggest a method for the synthesis of the unnatural D enantiomer of alanine from the readily available L enantiomer of lactic acid.

$$CH_3-CHOH-COOH$$
lactic acid

24-38 Show how you would use the Gabriel-malonic ester synthesis to make histidine. What stereochemistry would you expect in your synthetic product?

24-39 Show how you would use the Strecker synthesis to make tryptophan. What stereochemistry would you expect in your synthetic product?

24-40 Write the complete structures for the following peptides. Tell whether each peptide is acidic, basic, or neutral.
(a) methionylthreonine   **(b)** threonylmethionine   **(c)** arginylleucyllysine   **(d)** Glu-Cys-Gln

24-41 The following structure is drawn in an unconventional manner.

$$CH_3CH_2-\underset{\underset{CONH_2}{|}}{CH}-\underset{\underset{CH_3}{|}}{CH}-NH-\overset{\overset{O}{\|}}{C}-\underset{\underset{NH-CO-CH_2NH_2}{|}}{CH}-CH_2CH_2-\overset{\overset{O}{\|}}{C}-NH_2$$

(a) Label the N terminus and the C terminus.   **(b)** Label the peptide bonds.
(c) Identify and label each amino acid present.   **(d)** Give the full name and the abbreviated name.

24-42 *Aspartame* (Nutrasweet®) is a remarkably sweet-tasting dipeptide ester. Complete hydrolysis of aspartame gives phenylalanine, aspartic acid, and methanol. Mild incubation with carboxypeptidase has no effect on aspartame. Treatment of aspartame with phenyl isothiocyanate followed by mild hydrolysis gives the phenylthiohydantoin of aspartic acid. Propose a structure for aspartame.

24-43 A molecular weight determination has shown that an unknown peptide is a pentapeptide, and an amino acid analysis shows that it contains the following residues: one Gly, two Ala, one Met, one Phe.

Treatment of the original pentapeptide with carboxypeptidase gives alanine as the first free amino acid released. Sequential treatment of the pentapeptide with phenyl isothiocyanate followed by mild hydrolysis gives the following derivatives:

first time   second time   third time

Propose a structure for the unknown pentapeptide.

24-44 Show the steps and intermediates in the synthesis of Ile-Leu-Phe
(a) by the solution-phase process   **(b)** by the solid-phase process

24-45   Using classical solution-phase techniques, show how you would synthesize Ala-Val and then combine it with Ile-Leu-Phe to give Ile-Leu-Phe-Ala-Val.

24-46   Peptides often have functional groups other than free amino groups at the N terminus and other than carboxyl groups at the C terminus.

(a)   A tetrapeptide is hydrolyzed by heating with 6 M HCl, and the hydrolysate is found to contain Ala, Phe, Val, and Glu. When the hydrolysate is neutralized, the odor of ammonia is detected. Explain where this ammonia might have been incorporated in the original peptide.

(b)   The tripeptide *thyrotropic hormone releasing factor* (TRF) has the full name pyroglutamylhistidylprolinamide. The structure appears below. Explain the functional groups at the N terminus and at the C terminus.

(c)   On acidic hydrolysis, an unknown pentapeptide gives glycine, alanine, valine, leucine, and isoleucine. No odor of ammonia is detected when the hydrolysate is neutralized. Reaction with phenyl isothiocyanate followed by mild hydrolysis gives *no* phenylthiohydantoin derivative. Incubation with carboxypeptidase has no effect. Explain these findings.

*24-47   Lipoic acid is often found near the active sites of enzymes, usually bound to the peptide by a long, flexible amide linkage with a lysine residue.

lipoic acid   bound to lysine residue

(a)   Is lipoic acid a mild oxidizing agent or a mild reducing agent? Draw it in both its oxidized and reduced forms.

(b)   Show how lipoic acid might react with two Cys residues to form a disulfide bridge.

(c)   Give a balanced equation for the hypothetical oxidation or reduction, as you predicted in part (a), of an aldehyde by lipoic acid.

24-48   Histidine is an important catalytic residue found at the active sites of many enzymes. In many cases, histidine appears to remove protons or to transfer protons from one location to another.

(a)   Show which nitrogen atom of the histidine heterocycle is basic and which is not.

(b)   Use resonance forms to show why the protonated form of histidine is a particularly stable cation.

(c)   Show the structure that results when histidine accepts a proton on the basic nitrogen of the heterocycle and then is deprotonated on the other heterocyclic nitrogen. Explain how histidine might function as a pipeline to transfer protons between sites within an enzyme and its substrate.

24-49   Metabolism of arginine produces urea and the rare amino acid *ornithine*. Ornithine has an isoelectric point close to 10. Propose a structure for ornithine.

24-50   Glutathione (GSH) is a tripeptide that serves as a mild reducing agent to detoxify peroxides and maintain the cysteine residues of hemoglobin and other red blood cell proteins in the reduced state. Complete hydrolysis of glutathione gives Gly, Glu, and Cys. Treatment of glutathione with carboxypeptidase gives glycine as the first free amino acid released. Treatment of glutathione with phenyl isothiocyanate gives the phenylthiohydantoin of glutamic acid.

(a)   Propose a structure for glutathione consistent with this information.

(b)   Oxidation of glutathione forms glutathione disulfide (GSSG). Propose a structure for glutathione disulfide, and write a balanced equation for the reaction of glutathione with hydrogen peroxide.

24-51   Complete hydrolysis of an unknown basic decapeptide gives Gly, Ala, Leu, Ile, Phe, Tyr, Glu, Arg, Lys, and Ser. Terminal residue analysis shows that the N terminus is Ala and the C terminus is Ile. Incubation of the decapeptide with chymotrypsin gives two tripeptides, **A** and **B**, and a tetrapeptide, **C**. Amino acid analysis shows that peptide **A** contains Gly, Glu, Tyr, and NH_3; peptide **B** contains Ala, Phe, and Lys; and peptide **C** contains Leu, Ile, Ser, and Arg. Terminal residue analysis gives the following results.

	N terminus	C terminus
A	Glu	Tyr
B	Ala	Phe
C	Arg	Ile

Incubation of the decapeptide with trypsin gives a dipeptide **D**, a pentapeptide **E**, and a tripeptide **F**. Terminal residue analysis of **F** shows that the N terminus is Ser, and the C terminus is Ile. Propose a structure for the decapeptide and for fragments **A** through **F**.

CHAPTER 25

Lipids

25-1
Introduction

What do the following occurrences all have in common? An athlete is disqualified from the Olympics for illegal use of anabolic steroids. You coat a bread pan with lard to keep the bread from sticking. Your mother is rushed to surgery to remove a gallbladder packed with cholesterol. You wax your shiny new car with carnauba wax. Your father is treated with a prostaglandin to lower his blood pressure. An artist uses turpentine to thin her brushes after painting the brilliant autumn colors.

All these actions involve the use, misuse, or manipulation of lipids. Steroids, prostaglandins, fats, oil, waxes, terpenes, and even the colorful carotenes in the falling leaves are all lipids. In our study of organic chemistry, we have usually classified compounds according to their functional groups. Lipids, however, are classified by their solubility: **Lipids** are substances that can be extracted from cells and tissues by nonpolar organic solvents.

Lipids include many types of compounds containing a wide variety of functional groups. You could easily prepare a solution of lipids by grinding a T-bone steak in a blender and then extracting the puree with chloroform or diethyl ether. The resulting solution of lipids would contain a multitude of compounds, many with complex structures. To facilitate the study of lipids, chemists have divided this large family into two major classes: complex lipids and simple lipids.

Complex lipids are those that are easily hydrolyzed to simpler constituents. Most complex lipids are esters of long-chain carboxylic acids called *fatty acids*. The two major groups of fatty acid esters are *waxes* and *glycerides*. Waxes are esters of long-chain alcohols, and glycerides are esters of glycerol.

Simple lipids are those that are not easily hydrolyzed by aqueous acid or base. This term often seems inappropriate, because many so-called "simple" lipids are quite complex molecules. We will consider three important groups of simple lipids: steroids, prostaglandins, and terpenes. Figure 25-1 shows some examples of complex and simple lipids.

25-2
Waxes

Waxes are esters of long-chain fatty acids with long-chain alcohols. They occur widely in nature and serve a number of purposes in plants and animals. *Spermaceti* (Fig. 25-1) is found in the head of the sperm whale and probably helps to regulate the animal's buoyancy for deep diving. It may also serve to amplify high-frequency sounds for locating prey. *Beeswax* is a mixture of waxes, hydrocarbons, and alcohols that bees use to form their honeycomb. *Carnauba wax* is a mixture of waxes of very high molecular weights. The carnauba plant secretes this waxy material to coat its leaves to prevent excessive loss of water by evaporation. Waxes are also found in the protective coatings of insects' exoskeletons, mammals' fur, and birds' feathers. In contrast to these waxes,

Examples of complex lipids | Examples of simple lipids

$$CH_2-O-\overset{\overset{\displaystyle O}{\|}}{C}-(CH_2)_{16}CH_3$$

$$CH-O-\overset{\overset{\displaystyle O}{\|}}{C}-(CH_2)_{16}CH_3$$

$$CH_2-O-\overset{\overset{\displaystyle O}{\|}}{C}-(CH_2)_{16}CH_3$$

tristearin, a fat

cholesterol, a steroid

$$CH_3(CH_2)_{15}-O-\overset{\overset{\displaystyle O}{\|}}{C}-(CH_2)_{14}CH_3$$

spermaceti (cetyl palmitate), a wax

α-pinene, a terpene

◀ **Figure 25-1**
Complex lipids contain ester functional groups that can be hydrolyzed to acids and alcohols. Simple lipids are not easily hydrolyzed.

Plant leaves often have a wax coating to prevent excessive loss of water.

the "paraffin wax" used to seal preserves is not a true wax; rather, it is a mixture of high molecular weight alkanes.

$$CH_3(CH_2)_{29}-O-\overset{\overset{\displaystyle O}{\|}}{C}-(CH_2)_{24}CH_3 \qquad CH_3(CH_2)_{33}-O-\overset{\overset{\displaystyle O}{\|}}{C}-(CH_2)_{26}CH_3$$

a component of beeswax | a component of carnauba wax

For many years, natural waxes were used in making cosmetics, adhesives, varnishes, and waterproofing materials. Synthetic materials have now replaced natural waxes for most of these uses.

25-3 Triglycerides

Glycerides are simply fatty acid esters of the triol *glycerol*. The most common glycerides are **triglycerides (triacylglycerols)**, in which all three of the glycerol —OH groups have been esterified by fatty acids. For example, tristearin (Fig. 25-1) is a component of beef fat in which all three —OH groups of glycerol are esterified by stearic acid, $CH_3(CH_2)_{16}COOH$.

Triglycerides are commonly called **fats** if they are solid at room temperature and **oils** if they are liquid at room temperature. Most triglycerides derived from mammals are fats, such as beef fat or lard. Although these fats are solid at room temperature, the warm body temperature of the living animal keeps them somewhat fluid, allowing for body movement. In plants and cold-blooded animals, triglycerides are generally oils, such as corn oil, peanut oil, or fish oil. A fish requires liquid oils rather than solid fats because it would have difficulty moving if its triglycerides solidified whenever it swam in a cold stream.

Fats and oils are commonly used for long-term energy storage in plants and animals. Fat is a more efficient source of long-term energy than carbohydrates because metabolism of a gram of fat releases over twice as much energy as a gram of sugar or starch. An average 70-kg adult male stores about 1000 kcal of readily available energy as glycogen (0.2 kg), and about 140,000 kcal of long-term energy as fat (15 kg): enough to supply his resting metabolic needs for nearly three months!

The **fatty acids** of common triglycerides are long, unbranched carboxylic acids with about 12 to 20 carbon atoms. Most fatty acids contain even numbers of carbon atoms because they are derived from two-carbon acetic acid units. Some of the common

TABLE 25-1 Structures and Melting Points of Some Common Fatty Acids

Name	Carbons	Structure	Melting Point (°C)
Saturated acids			
lauric acid	12	⋀⋁⋀⋁⋀⋁COOH	44
myristic acid	14	⋀⋁⋀⋁⋀⋁COOH	59
palmitic acid	16	⋀⋁⋀⋁⋀⋁⋀COOH	64
stearic acid	18	⋀⋁⋀⋁⋀⋁⋀COOH	70
arachidic acid	20	⋀⋁⋀⋁⋀⋁⋀⋁COOH	76
Unsaturated acids			
oleic acid	18	⋀⋁⋀⋁═⋁COOH	4
linoleic acid	18	⋀⋁═⋀═⋁COOH	−5
linolenic acid	18	⋀═⋀═⋀═⋁COOH	−11
eleostearic acid	18	⋀⋁══⋁COOH	49
arachidonic acid	20	⋀⋁═⋀═⋀═⋀═⋁COOH	−49

fatty acids have saturated carbon chains, while others have one or more carbon–carbon double bonds. Table 25-1 shows the structures of some common fatty acids derived from fats and oils.

PROBLEM 25-1

Trimyristin, a solid fat present in nutmeg, is hydrolyzed to give one equivalent of glycerol and three equivalents of myristic acid. Give the structure of trimyristin.

Table 25-1 shows that saturated fatty acids have melting points that increase gradually with their molecular weights. The presence of a cis double bond lowers the melting point, however. Notice that the 18-carbon saturated acid (stearic acid) has a melting point of 70°C, while the 18-carbon acid with a cis double bond (oleic acid) has a melting point of 4°C. This lowering of the melting point results from the unsaturated acid's "kink" at the position of the double bond (Figure 25-2). Kinked molecules cannot pack as tightly together in a solid as the uniform zigzag chains of a saturated acid.

stearic acid, mp 70°C

oleic acid, mp 4°C

▲ Figure 25-2
The cis double bond in oleic acid lowers the melting point by 66°C.

tristearin, mp 72°C

triolein, mp −4°C

◀ **Figure 25-3**
Unsaturated triglycerides have lower melting points because their unsaturated fatty acids do not pack as well in a solid lattice.

A second double bond lowers the melting point further (linoleic acid, mp −5°C), and a third double bond lowers it still further (linolenic acid, mp −11°C). The trans double bonds in eleostearic acid (mp 49°C) have a smaller effect on the melting point than the cis double bonds of linolenic acid. The geometry of a trans double bond is similar to the zigzag conformation of a saturated acid, so it does not kink the chain as much as a cis double bond.

Fats and oils also have melting points that depend on the unsaturation (especially cis double bonds) in their fatty acids. A triglyceride derived from saturated fatty acids has a higher melting point because it packs more easily into a solid lattice than a triglyceride derived from kinked, unsaturated fatty acids. Figure 25-3 shows typical conformations of triglycerides containing saturated and unsaturated fatty acids. Tristearin (mp 72°C) is a saturated fat that packs well in a solid lattice. Triolein (mp −4°C) has the same number of carbon atoms as tristearin, but triolein has three cis double bonds, whose kinked conformations prevent optimum packing in the solid.

Most saturated triglycerides are *fats* because they are solid at room temperature. Most triglycerides with several unsaturations are *oils* because they are liquid at room temperature. The term *polyunsaturated* simply means there are several double bonds in the fatty acids of the triglyceride.

Most naturally occurring fats and oils are mixtures of triglycerides containing a variety of saturated and unsaturated fatty acids. In general, oils from plants and cold-blooded animals contain more unsaturations than fats from warm-blooded animals. Table 25-2

TABLE 25-2 Fatty Acid Composition of Some Fats and Oils, Percent by Weight

Source	Saturated Fatty Acids				Unsaturated Fatty Acids		
	Lauric	Myristic	Palmitic	Stearic	Oleic	Linoleic	Linolenic
beef fat	0	6	27	14	49	2	0
lard	0	1	24	9	47	10	0
human fat	1	3	27	8	48	10	0
herring oil	0	5	14	3	0	0	30[a]
corn oil	0	1	10	3	50	34	0
olive oil	0	0.1	7	2	84	5	2
soybean oil	0.2	0.1	10	2	29	51	7

[a]Contains large amounts of even more highly unsaturated fatty acids.

gives the approximate composition of the fatty acids obtained from hydrolysis of some common fats and oils.

For many years, *lard* (a soft, white solid obtained by rendering animal fat) was commonly used for cooking and baking. Although vegetable oil could be produced more cheaply and in greater quantities, consumers were unwilling to use vegetable oils because they were accustomed to using white, creamy lard. Then vegetable oils were treated with hydrogen gas and a nickel catalyst, reducing some of the double bonds to give a creamy, white *vegetable shortening* that resembles lard. This "partially hydrogenated vegetable oil" largely replaced lard for cooking and baking. *Margarine* is a similar material flavored with butyraldehyde to give it a taste like that of butter. More recently, consumers have learned that "polyunsaturated" vegetable oils are more easily digested, prompting a switch to natural vegetable oils.

> **PROBLEM 25-2**
>
> Give an equation for the complete hydrogenation of triolein using an excess of hydrogen. What is the name of the product, and what are the melting points of the starting material and the product?

25-4
Saponification of Fats and Oils; Soaps and Detergents

Saponification is the base-promoted hydrolysis of the ester linkages in fats and oils (review Section 21-7B). One of the products is soap, and the word *saponification* is derived from the Latin word *saponis*, meaning "soap." Saponification was discovered (before 500 B.C.), when it was found that a curdy material resulted when animal fat was heated with wood ashes. Alkaline substances in the ashes promote hydrolysis of the ester linkages of the fat. Soap is currently made by boiling animal fat or vegetable oil with a solution of sodium hydroxide. The following reaction shows formation of soap from tristearin, a component of beef fat.

Chemically, a **soap** is the sodium or potassium salt of a fatty acid. The negatively charged carboxylate group is **hydrophilic** (attracted to water), and the long hydrocarbon chain is **hydrophobic** (repelled by water) and lipophilic (attracted to oils). The electrostatic potential map of the stearate ion is shown in Figure 25-4. Notice the high electron density (red) around the negatively charged carboxylate end of the molecule. The carboxylate oxygen atoms share the negative charge and participate in strong hydrogen bonding with water molecules. The rest of the molecule (green) is hydrocarbon chain that cannot participate in hydrogen bonding with water.

In water, soap forms a cloudy solution of **micelles**: clusters of about 100 to 200 soap molecules with their polar "heads" (the carboxylate groups) on the surface of the cluster and their hydrophobic "tails" (the hydrocarbon chains) enclosed within. The micelle (Figure 25-4) is an energetically stable particle because the hydrophilic groups are hydrogen-bonded to the surrounding water, while the hydrophobic groups are shielded within the interior of the micelle, interacting with other hydrophobic groups.

Soaps are useful cleaning agents because of the different affinities of a soap molecule's two ends. Greasy dirt is not easily removed by pure water because grease is

Ionic head

Hydrocarbon tail

Water

boundary of water into structure

▲ **Figure 25-4**
The electrostatic potential map of a soap molecule shows high electron density in the negatively charged head and medium electron density (green) in the hydrocarbon tail. In water, soap forms a cloudy solution of micelles, with the hydrophilic heads in contact with water and the hydrophobic tails clustered in the interior. The Na^+ ions (not shown) are dissolved in the water surrounding the micelle.

hydrophobic and insoluble in water. Soap, however, has a long hydrocarbon chain that dissolves in the grease, with its hydrophilic head at the surface of the grease droplet. Once the surface of the grease droplet is covered by many soap molecules, a micelle can form with a tiny grease droplet at its center. This grease droplet is easily suspended in water because it is covered by the hydrophilic carboxylate groups of the soap (Figure 25-5). The resulting mixture of two insoluble phases (grease and water), with one phase dispersed throughout the other in small droplets, is called an **emulsion**. We say the grease has been **emulsified** by the soapy solution. When the wash water is rinsed away, the grease goes with it.

The usefulness of soaps is limited by their tendency to precipitate out of solution in hard water. **Hard water** is water that is acidic or that contains ions of calcium, magnesium, or iron. In acidic water (such as the "acid rain" of environmental concern), soap molecules are protonated to the free fatty acids. Without the ionized carboxylate group, the fatty acid floats to the top as a greasy "acid scum" precipitate.

$$CH_3(CH_2)_n - \overset{\overset{\displaystyle O}{\|}}{C} - O^- \, {}^+Na \quad + \quad H^+ \quad \longrightarrow \quad CH_3(CH_2)_n - \overset{\overset{\displaystyle O}{\|}}{C} - OH \downarrow \quad + \quad Na^+$$

a soap $\qquad\qquad\qquad\qquad\qquad\qquad\qquad\qquad\qquad$ acid scum

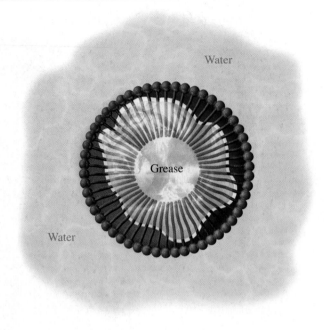

▶ Figure 25-5
In a soapy solution, grease is emulsified by forming micelles coated by the hydrophilic carboxylate groups of the soap.

Many areas have household water containing calcium, magnesium, and iron ions. Although these mineral-rich waters can be healthful for drinking, the ions react with soaps to form insoluble salts called *hard-water scum*. The following equation shows the reaction of a soap with calcium, common in areas where water comes in contact with limestone rocks.

$$2\ CH_3(CH_2)_n\!-\!\overset{\displaystyle O}{\overset{\|}{C}}\!-\!O^-\ {}^+Na \quad + \quad Ca^{+2} \quad \longrightarrow \quad [CH_3(CH_2)_n\!-\!\overset{\displaystyle O}{\overset{\|}{C}}\!-\!O]_2Ca \downarrow \quad + \quad 2\ Na^+$$

$$\underset{\text{a soap}}{} \qquad\qquad\qquad\qquad\qquad \underset{\text{hard-water scum}}{}$$

PROBLEM 25-3

Give equations to show the reactions of sodium stearate with
(a) Ca^{2+} **(b)** Mg^{2+} **(c)** Fe^{3+}

PROBLEM 25-4

Several commercial laundry soaps contain water-softening agents, usually sodium carbonate (Na_2CO_3) or sodium phosphate (Na_3PO_4 or Na_2HPO_4). Explain how these water-softening agents allow soaps to be used in water that is hard by virtue of its
(a) low pH   **(b)** dissolved Ca^{2+}, Mg^{2+}, and Fe^{3+} salts

Soaps precipitate in hard water because of the chemical properties of the carboxylic acid group. **Synthetic detergents** avoid precipitation by using other functional groups in place of carboxylic acid salts. Sodium salts of sulfonic acids are the most widely used class of synthetic detergents. Sulfonic acids are more acidic than carboxylic acids, so their salts are not protonated, even in strongly acidic wash water. Calcium, magnesium, and iron salts of sulfonic acids are soluble in water, so sulfonate salts can be used in hard water without forming a scum (Figure 25-6).

Like soaps, synthetic detergents combine hydrophilic and hydrophobic regions in the same molecule. Hydrophobic regions are generally alkyl groups or aromatic rings. Hydrophilic regions may contain anionic groups, cationic groups, or nonionic groups

An alkylbenzenesulfonate detergent

Examples of other types of detergents

benzylcetyldimethylammonium chloride
(benzalkonium chloride)

$CH_3-(CH_2)_8-\bigcirc-(OCH_2CH_2)_9-OH$

Nonoxynol®, Ortho Pharmaceuticals

sodium dodecyl sulfate
(sodium lauryl sulfate)

N-lauroyl-N-methylglycine, sodium salt
Gardol®, Colgate-Palmolive Co.

▲ **Figure 25-6**
Synthetic detergents may have anionic, cationic, or nonionic hydrophilic functional groups. Of these detergents, only Gardol® is a carboxylate salt and forms a precipitate in hard water.

containing several oxygen atoms or other hydrogen-bonding atoms. Figure 25-6 shows the electrostatic potential map of a sulfonate detergent, with red (electron-rich) regions around the hydrophilic sulfonate group. Figure 25-6 also shows more examples of anionic, cationic, and nonionic detergents.

PROBLEM 25-5

Draw a diagram, similar to Figure 25-5, of an oil droplet emulsified by the alkylbenzenesulfonate detergent shown in Figure 25-6.

PROBLEM 25-6

Point out the hydrophilic and hydrophobic regions in the structures of benzalkonium chloride, Nonoxynol®, and Gardol® (Fig. 25-6).

PROBLEM 25-7

The synthesis of the alkylbenzenesulfonate detergent shown in Figure 25-6 begins with the partial polymerization of propylene to give a pentamer.

$$5\ H_2C{=}CH{-}CH_3 \xrightarrow{\text{acidic catalyst}}$$

a pentamer

Show how Friedel–Crafts reactions can convert this pentamer to the final synthetic detergent.

25-5
Phospholipids

Phospholipids are lipids that contain groups derived from phosphoric acid. The most common phospholipids are **phosphoglycerides**, which are closely related to common fats and oils. A phosphoglyceride generally has a phosphoric acid group in place of one of the fatty acids of a triglyceride. The simplest class of phosphoglycerides are **phosphatidic acids**, which consist of glycerol esterified by two fatty acids and one phosphoric acid group. Although it is often drawn in its acid form, a phosphatidic acid is actually deprotonated at neutral pH.

a phosphatidic acid   ⇌   2 H⁺ +   ionized form   =   schematic representation

nonpolar, hydrocarbon tails

polar head

PROBLEM 25-8

Draw the important resonance forms for a phosphatidic acid that has lost
(a) one proton   (b) two protons

Many phospholipids contain an additional alcohol esterified to the phosphoric acid group. **Cephalins** are esters of ethanolamine, and **lecithins** are esters of choline. Both cephalins and lecithins are widely found in plant and animal tissues.

$$HO-CH_2CH_2-NH_2$$
ethanolamine

$$HO-CH_2CH_2-\overset{+}{N}(CH_3)_3$$
choline

a cephalin,
or phosphatidyl ethanolamine

a lecithin,
or phosphatidyl choline

Like phosphatidic acids, lecithins and cephalins contain a polar "head" and two long, nonpolar hydrocarbon "tails." This soaplike structure gives phospholipids some

▲ **Figure 25-7**
Phosphoglycerides can aggregate into a bilayer membrane with their polar heads exposed to the aqueous solution and the hydrocarbon tails protected within. This lipid bilayer is an important part of the cell membrane.

interesting properties. Like soaps, they form micelles and other aggregations with their polar heads on the outside and their nonpolar tails protected on the inside.

Another stable form of aggregation is a **lipid bilayer**, with the heads on the two surfaces of a membrane and the tails protected within. Cell membranes contain phosphoglycerides oriented in a lipid bilayer, forming a barrier that restricts the flow of water and dissolved substances. Figure 25-7 shows the arrangement of phospholipids in a bilayer membrane.

25-6 Steroids

Steroids are complex polycyclic molecules found in all plants and animals. They are classified as *simple lipids* because they do not undergo hydrolysis like fats, oils, and waxes do. Steroids encompass a wide variety of compounds, including hormones, emulsifiers, and components of membranes. **Steroids** are compounds whose structures are based on the tetracyclic androstane ring system, shown below. The four rings are designated A, B, C, and D, beginning with the ring at lower left, and the carbon atoms are numbered beginning with the A ring and ending with the two "angular" (axial) methyl groups.

androstane

We have seen (Section 3-16B) that fused ring systems such as androstane can have either trans or cis stereochemistry at each ring junction. A simple example is the

▶ **Figure 25-8**
Cis-trans isomers of decalin. In
trans-decalin, the two bonds to the
second ring are trans to one another,
and the hydrogens on the junction are
also trans. In *cis*-decalin, the bonds to
the second ring are cis, and the
junction hydrogens are also cis.

geometric isomerism of *trans-* and *cis*-decalin shown in Figure 25-8. If you make models of these isomers, you will find that the trans isomer is quite rigid and flat (aside
from the ring puckering). In contrast, the cis isomer is relatively flexible, and it has the
two rings situated at a sharp angle to each other.

Each of the ring junctions is trans in the androstane structure shown on the previous
page. Most steroids have this all-trans structure, which results in a stiff, nearly flat molecule with the two axial methyl groups perpendicular to the plane. In some steroids, the
junction between rings A and B is cis, requiring the A ring to fold down below the rest of
the ring system. Figure 25-9 shows the androstane ring system with both trans and cis A-B
ring junctions. The B-C and C-D ring junctions are nearly always trans in natural steroids.

Most steroids have an oxygen functional group ($=$O or $-$OH) at C3, and some
kind of side chain or other functional group at C17. Many also have a double bond from
C5 to either C4 or C6. The structures of androsterone and cholesterol serve as examples.
Androsterone, a male sex hormone, is based on the simple androstane ring system. Cholesterol is a common biological intermediate and is believed to be the biosynthetic precursor to other steroids. It has a side chain at C17 and a double bond between C5 and C6.

▲ **Figure 25-9**
Common steroids may have either a cis or a trans A-B ring junction. The other ring junctions
are normally trans.

androsterone

cholesterol

These gallstones, shown here within the gallbladder, are composed mostly of cholesterol.

The principal sex hormones have been characterized and studied extensively. Testosterone is the most potent of the natural male sex hormones, and estradiol is the most potent natural female hormone. Notice that the female sex hormone differs from the male hormone by its aromatic A ring. For the A ring to be aromatic, the C19 methyl group must be lost. In mammals, testosterone is converted to estradiol in the female's ovaries, where enzymes remove C19 and two hydrogen atoms to give the aromatic A ring.

testosterone

estradiol

PROBLEM 25-9

How would you use a simple extraction to separate a mixture of testosterone and estradiol?

When steroid hormones were first isolated, people believed that no synthetic hormone could rival the astonishing potency of natural steroids. In the past 20 years, however, many synthetic steroids have been developed. Some of these synthetic hormones are hundreds or thousands of times more potent than natural steroids. One example is ethynyl estradiol, a synthetic female hormone that is more potent than estradiol. Ethynyl estradiol is a common ingredient in oral contraceptives.

ethynyl estradiol

The biological activities of steroid hormones result from interactions with specific receptors. For example, estradiol interacts with the estrogen receptor. Some industrial chemicals, such as DDT and polychlorinated biphenyls, may possibly interact with these receptors and cause hormonal effects on people and wildlife.

Some of the most important physiological steroids are the adrenocortical hormones, synthesized by the adrenal cortex. Most of these hormones have either a carbonyl group or a hydroxyl group at C11 of the steroid skeleton. The principal adrenocortical hormone is cortisol, used for the treatment of inflammatory diseases of the skin (psoriasis), the joints (rheumatoid arthritis), and the lungs (asthma). Figure 25-10 compares the structure of natural cortisol with two synthetic corticoids: fluocinolone acetonide, a fluorinated synthetic hormone that is more potent than cortisol for treating skin inflammation; and beclomethasone, a chlorinated synthetic hormone that is more potent than cortisol for treating asthma.

cortisol

fluocinolone acetonide

beclomethasone

▲ **Figure 25-10**
Cortisol is the major natural hormone of the adrenal cortex. Fluocinolone acetonide is more potent for treating skin inflammation, and beclomethasone is more potent for treating asthma.

PROBLEM 25-10

Draw each molecule in a stable chair conformation, and tell whether each red group is axial or equatorial.

(a)

(b)

(c)

androsterone

(d)

digitoxigenin, a cardiac stimulant

25-7
Prostaglandins

Prostaglandins are fatty acid derivatives that are even more powerful biochemical regulators than steroids. They are called prostaglandins because they were first isolated from secretions of the prostate gland. They were later found to be present in all body tissues and fluids, usually in minute quantities. Prostaglandins affect many different body systems, including the nervous system, smooth muscle, blood, and the reproductive system. They play important roles in regulating such diverse functions as blood pressure, blood clotting, the allergic inflammatory response, activity of the digestive system, and the onset of labor.

Prostaglandins have a cyclopentane ring with two long side chains trans to each other, with one side chain ending in a carboxylic acid. Most prostaglandins have 20 carbon atoms, numbered as follows:

▲ **Figure 25-11**
Biosynthesis of prostaglandins begins with an enzyme-catalyzed oxidative cyclization of arachidonic acid.

Many prostaglandins have hydroxyl groups on C11 and C15, and a trans double bond between C13 and C14. They also have a carbonyl group or a hydroxyl group on C9. If there is a carbonyl group at C9, the prostaglandin is a member of the *E series*. If there is a hydroxyl group at C9, it is a member of the *F series*, and the symbol α means the hydroxyl group is directed down. Many prostaglandins have a cis double bond between C5 and C6. The number of double bonds is also given in the name, as shown below for two common prostaglandins.

PGE₁
(PG means prostaglandin;
E means ketone at C9;
1 means one C=C double bond)

PGF₂α
(PG means prostaglandin;
F means hydroxyl at C9, and α means down;
2 means two C=C double bonds)

Prostaglandins are derived from arachidonic acid, a 20-carbon fatty acid with four cis double bonds. Figure 25-11 shows schematically how an enzymatic cyclo-oxidation converts arachidonic acid to the prostaglandin skeleton.

Despite their range of activities, many naturally occurring prostaglandins would make poor drugs because they are rapidly converted to inactive products. Synthetic modifications can prolong their activity. Misoprostol is a stable synthetic derivative of PGE₁ that is used for the treatment of ulcers.

misoprostol

25-8 Terpenes

Terpenes are a diverse family of compounds with carbon skeletons composed of five-carbon isopentyl (isoprene) units. Terpenes are commonly isolated from the **essential oils** of plants: the fragrant oils that are concentrated from plant material, usually by steam distillation. The term *essential oils* literally means "oils derived from the essence" of plants. They often have pleasant tastes or aromas, and they are widely used as flavorings, deodorants, and medicines. Table 25-3 lists several common types of essential oils and their principal components.

25-8A Characteristics and Nomenclature of Terpenes

Hundreds of essential oils were used as perfumes, flavorings, and medicines for centuries before chemistry was capable of studying the mixtures. In 1818, it was found that oil of turpentine has a C:H ratio of 5:8, and many other essential oils have similar C:H ratios. This group of piney-smelling natural products with similar C:H ratios came to be known as **terpenes**.

TABLE 25-3 Some Useful Essential Oils

Essential Oil	Source	Major Components
perfume	flowers	mixtures of terpenes and terpenoids
oil of turpentine	evergreens	mixtures of terpenes and terpenoids
oil of celery	celery	β-selinene, a terpene
oil of anise	anise seed	anethole
oil of bay	bay leaves	myrcene, a terpene
cedar leaf oil	leaves of the "white cedar" (actually a pine)	α-pinene, a terpene

In 1887, German chemist Otto Wallach determined the structures of several terpenes and discovered that all of them are composed of two or more five-carbon units of **isoprene** (2-methyl-1,3-butadiene). The isoprene unit maintains its isopentyl structure in a terpene, usually with modification of the isoprene double bonds.

Isoprene

$$H_2C{=}C{-}CH{=}CH_2 \quad \text{or}$$

with CH_3

An isoprene unit

$$C{-}C{-}C{-}C \quad \text{or}$$

head   tail   (may have double bonds)

The isoprene molecule and the isoprene unit are said to have a "head" (the branched end) and a "tail" (the unbranched ethyl group). Myrcene can be divided into two isoprene units, with the head of one unit bonded to the tail of the other.

myrcene

β-selinene

β-Selinene has a more complicated structure, with two rings and a total of 15 carbon atoms. Nevertheless, β-selinene is composed of three isoprene units. Once again, these three units are bonded head to tail, although the additional bonds used to form the rings make the head-to-tail arrangement more difficult to see.

Many terpenes contain additional functional groups, especially carboxyl groups and hydroxyl groups. A terpene aldehyde, a terpene alcohol, a terpene ketone, and a terpene acid are shown below.

geranial   menthol   camphor   abietic acid

PROBLEM 25-11

Circle the isoprene units in geranial, menthol, camphor, and abietic acid.

25-8B   Classification of Terpenes

Terpenes are classified according to the number of carbon atoms, in units of ten. A terpene with 10 carbon atoms (two isoprene units) is called a **monoterpene**, one with 20 carbon atoms (four isoprene units) is a **diterpene**, and so on. Terpenes with 15 carbon atoms (three isoprene units) are called **sesquiterpenes**, meaning that they have $1\frac{1}{2}$ times 10 carbon atoms. Myrcene, geranial, menthol, and camphor are monoterpenes, β-selinene is a sesquiterpene, abietic acid is a diterpene, and squalene (Figure 25-12) is a **triterpene**.

Class Name	Carbons
monoterpenes	10
sesquiterpenes	15
diterpenes	20
triterpenes	30
tetraterpenes	40

Carotenes, with 40 carbon atoms, are tetraterpenes. Their extended system of conjugated double bonds moves the intense $\pi \rightarrow \pi^*$ ultraviolet absorption into the visible region, making them brightly colored. Carotenes are responsible for the pigmentation of carrots, tomatoes, and squash, and they give a fiery color to tree leaves in autumn. β-Carotene is the most common carotene isomer. It can be divided into two head-to-tail diterpenes, linked tail to tail.

β-carotene: λ_{max} = 454 nm, ε = 140,000

PROBLEM 25-12

Circle the eight isoprene units in β-carotene.

Carotenes are believed to be biological precursors of retinol, commonly known as vitamin A. If a molecule of β-carotene is split in half at the tail-to-tail linkage, each of the diterpene fragments may be converted to retinol.

$$\text{carotene} \quad \xrightarrow[\text{enzyme}]{2\ H_2O} \quad 2$$

retinol (vitamin A)

PROBLEM 25-13

(a) Circle the isoprene units in the following terpenes.
(b) Classify each of these as a monoterpene, diterpene, etc.

(1) α-farnesene
(from oil of citronella)

(2) limonene
(from oil of lemon)

(3) α-pinene
(from turpentine)

(4) zingiberene
(from oil of ginger)

25-8C   Terpenoids

Many natural products are derived from terpenes, even though they do not have carbon skeletons composed exclusively of C_5 isoprene units. These terpene-like compounds are called **terpenoids**. They may have been altered through rearrangements, loss of carbon atoms, or introduction of additional carbon atoms. Cholesterol is an example of a terpenoid that has lost some of the isoprenoid carbon atoms.

Figure 25-12 shows that cholesterol is a triterpenoid, formed from six isoprene units with loss of three carbon atoms. The six isoprene units are bonded head to tail, with

squalene

intermediate

loss of three carbons

cholesterol

▶ **Figure 25-12**
Cholesterol is a triterpenoid that has lost three (blue) carbon atoms from the original six isoprene units of squalene. Another carbon atom has migrated (red arrow) to form the axial methyl group between rings C and D.

the exception of one tail-to-tail linkage. The triterpene precursor of cholesterol is believed to be squalene. We can envision an acid-catalyzed cyclization of squalene to give an intermediate that is converted to cholesterol with loss of three carbon atoms. Possible mechanisms are outlined in Figures 14-6 and 14-7 (pages 624 and 625).

Chapter 25 Glossary

detergent (synthetic detergent) A synthesized compound that acts as an emulsifying agent. Some of the common classes of synthetic detergents are alkylbenzenesulfonate salts, alkyl sulfate salts, alkylammonium salts, and nonionic detergents containing several hydroxyl groups or ether linkages. (p. 1168)

emulsify To promote formation of an emulsion. (p. 1167)

emulsion A mixture of two immiscible liquids, one dispersed throughout the other in small droplets. (p. 1167)

essential oils Fragrant oils (*essences*) that are concentrated from plant material, usually by steam distillation. (p. 1175)

fat A fatty acid triester of glycerol (a triglyceride) that is solid at room temperature. (p. 1163)

fatty acid A long-chain carboxylic acid. Most naturally occurring fatty acids contain even numbers of carbon atoms between 12 and 20. (p. 1163)

glyceride A fatty acid ester of glycerol. (p. 1163)

hard water Water that contains acids or ions (such as Ca^{2+}, Mg^{2+}, or Fe^{3+}) that react with soaps to form precipitates. (p. 1167)

hydrophilic Attracted to water; polar. (p. 1166)

hydrophobic Repelled by water; usually nonpolar and lipophilic (soluble in oils and in nonpolar solvents). (p. 1166)

isoprene The common name for 2-methyl-1,3-butadiene, the structural building block for terpenes. (p. 1176)

lipid bilayer A form of aggregation of phosphoglycerides with the hydrophilic heads forming the two surfaces of a planar structure and the hydrophobic tails protected within. A lipid bilayer forms part of the cell membrane. (p. 1171)

lipids Substances that can be extracted from cells and tissues by nonpolar organic solvents. (p. 1162)

   complex lipids: Lipids that are easily hydrolyzed to simpler constituents, usually by saponification of an ester.

   simple lipids: Lipids that are not easily hydrolyzed to simpler constituents.

micelle A cluster of molecules of a soap, phospholipid, or other emulsifying agent suspended in a solvent, usually water. The hydrophilic heads of the molecules are in contact with the solvent, and the hydrophobic tails are enclosed within the cluster. The micelle may or may not contain an oil droplet. (p. 1166)

oil A fatty acid triester of glycerol (a triglyceride) that is liquid at room temperature. (p. 1163)

phosphoglyceride An ester of glycerol in which the three hydroxyl groups are esterified by two fatty acids and a phosphoric acid derivative. (p. 1170)

   phosphatidic acids: A variety of phosphoglycerides consisting of glycerol esterified by two fatty acids and one free phosphoric acid group.

   cephalins (phosphatidyl ethanolamines): A variety of phosphoglycerides with ethanolamine esterified to the phosphoric acid group.

   lecithins (phosphatidyl cholines): A variety of phosphoglycerides with choline esterified to the phosphoric acid group.

phospholipid Any lipid that contains one or more groups derived from phosphoric acid. (p. 1170)

prostaglandins A class of biochemical regulators consisting of a 20-carbon carboxylic acid containing a cyclopentane ring and various other functional groups. (p. 1174)

saponification Base-promoted hydrolysis of an ester. Originally used to describe the hydrolysis of fats to make soap. (p. 1166)

soap The alkali metal salt of a fatty acid. (p. 1166)

steroid A compound whose structure is based on the tetracyclic androstane ring system. (p. 1171)

terpenes A diverse family of compounds with carbon skeletons composed of two or more 5-carbon isoprene units. **Monoterpenes** contain 10 carbon atoms, **sesquiterpenes** contain 15, **diterpenes** contain 20, and **triterpenes** contain 30. (p. 1175)

terpenoids A family of compounds including both terpenes and compounds of terpene origin whose carbon skeletons have been altered or rearranged. (p. 1178)

triglyceride (triacylglycerol) A fatty acid triester of glycerol. Triglycerides that are solid at room temperature are *fats*, and those that are liquid are *oils*. (p. 1163)

wax An ester of a long-chain fatty acid with a long-chain alcohol. (p. 1162)

Essential Problem-Solving Skills In Chapter 25

1. Classify lipids both into the large classifications (such as simple lipids, complex lipids, phospholipids, etc.) and into the more specific classifications (such as waxes, triglycerides, cephalins, lecithins, steroids, prostaglandins, terpenes, etc.).
2. Predict the physical properties of fats and oils from their structures.
3. Identify the isoprene units in terpenes, and classify terpenes according to the number of carbon atoms.
4. Predict the products of reactions of lipids with standard organic reagents. In particular, consider the reactions of the ester and olefinic groups of glycerides and the carboxyl groups of fatty acids.
5. Explain how soaps and detergents work, with particular attention to their similarities and differences.

Study Problems

25-14 Define each term and give an example.
    (a) lipid   **(b)** fat   **(c)** oil   **(d)** fatty acid
    (e) wax   **(f)** soap   **(g)** detergent   **(h)** hard water
    (i) micelle   **(j)** phospholipid   **(k)** triglyceride   **(l)** simple lipid
    (m) complex lipid   **(n)** prostaglandin   **(o)** steroid   **(p)** terpene

25-15 Give the general classification of each compound.
    (a) glyceryl tripalmitate

(b) $CH_3-(CH_2)_{10}-CH_2-O-\overset{\overset{\displaystyle O}{\|}}{\underset{\underset{\displaystyle O}{\|}}{S}}-O^-\ ^+Na$

sodium lauryl sulfate (in shampoo)

(c) $CH_3-(CH_2)_{13}-O-\overset{\overset{\displaystyle O}{\|}}{C}-(CH_2)_{16}-CH_3$

tetradecyl octadecanoate

(d)

caryophyllene (from cloves)

(e)

norethindrone
(a synthetic hormone)

25-16 Predict the products obtained from the reaction of triolein with the following reagents.
    (a) NaOH in water   **(b)** H_2 and a nickel catalyst   **(c)** Br_2 in CCl_4
    (d) ozone, then dimethyl sulfide   **(e)** warm $KMnO_4$ in water   **(f)** $CH_2I_2/Zn(Cu)$
    (g) saponification, then $LiAlH_4$

25-17 Show how you would convert oleic acid to the following fatty acid derivatives.
    (a) 1-octadecanol   **(b)** stearic acid   **(c)** octadecyl stearate
    (d) nonanal   **(e)** nonanedioic acid   **(f)** 2,9,10-tribromostearic acid

25-18 Phospholipids undergo saponification much like triglycerides. Draw the structure of a phospholipid meeting the following criteria; then draw the products that would result from its saponification.
    (a) a cephalin containing stearic acid and oleic acid   **(b)** a lecithin containing palmitic acid

25-19 Some of the earliest synthetic detergents were the sodium alkyl sulfates.

$$CH_3(CH_2)_nCH_2-O-\overset{\overset{\displaystyle O}{\|}}{\underset{\underset{\displaystyle O}{\|}}{S}}-O^-\ ^+Na$$

Show how you would make sodium octadecylsulfate using tristearin as your organic starting material.

25-20 Which of the following chemical reactions could be used to distinguish between a polyunsaturated vegetable oil and a petroleum oil containing a mixture of saturated and unsaturated hydrocarbons? Explain your reasoning.
    (a) addition of bromine in CCl_4 **(b)** hydrogenation   **(c)** saponification   **(d)** ozonolysis

25-21 How would you use simple chemical tests to distinguish between the following pairs of compounds?
(a) sodium stearate and *p*-dodecylbenzenesulfonate   (b) beeswax and "paraffin wax"
(c) trimyristin and myristic acid   (d) trimyristin and triolein

25-22 A triglyceride can be optically active if it contains two or more different fatty acids.
(a) Draw the structure of an optically active triglyceride containing one equivalent of myristic acid and two equivalents of oleic acid.
(b) Draw the structure of an optically inactive triglyceride with the same fatty acid composition.

25-23 Draw the structure of an optically active triglyceride containing one equivalent of stearic acid and two equivalents of oleic acid. Draw the products expected when this triglyceride reacts with the following reagents. In each case, predict whether the products will be optically active.
(a) H_2 and a nickel catalyst   (b) Br_2 in CCl_4 (c) hot aqueous NaOH   (d) ozone followed by $(CH_3)_2S$

25-24 The structure of limonene appears in Problem 25-13. Predict the products formed when limonene reacts with the following reagents.
(a) excess HBr   (b) excess HBr, peroxides   (c) excess Br_2 in CCl_4
(d) ozone, followed by dimethyl sulfide   (e) warm, concentrated $KMnO_4$ (f) $BH_3 \cdot THF$, followed by basic H_2O_2

25-25 *Olestra*® is a fat-based fat substitute that became available in snack foods such as potato chips in 1998. Previous fat substitutes were carbohydrate-based or protein-based mixtures that do not give as good a sensation in the mouth, and are not suitable for frying. With Olestra®, the glycerol molecule of a fat is replaced by sucrose (p. 1093). In Olestra®, the sucrose molecule has six, seven, or (most commonly) eight fatty acids esterified to its hydroxyl groups. The fatty acids come from hydrolysis of vegetable oils such as soybean, corn, palm, coconut, and cottonseed oils. This unnaturally bulky, fat-like molecule does not pass through the intestinal walls, and digestive enzymes cannot get close to the sucrose center to bind it to their active sites. Olestra® passes through the digestive system unchanged, and it provides zero calories. Draw a typical Olestra® molecule, using any fatty acids that are commonly found in vegetable oils.

25-26 Cholic acid, a major constituent of bile, has the following structure.
(a) Draw the structure of cholic acid showing the rings in their chair conformations, and label each methyl group and hydroxyl group as axial or equatorial. (Making a model may be helpful.)
(b) Cholic acid is secreted in bile as an amide linked to the amino group of glycine. This cholic acid–amino acid combination acts as an emulsifying agent to disperse lipids in the intestines for easier digestion. Draw the structure of the cholic acid–glycine combination, and explain why it is a good emulsifying agent.

25-27 Carefully circle the isoprene units in the following terpenes, and label each compound as a monoterpene, sesquiterpene, or diterpene.

(a) γ-bisabolene

(b) carvone

(c) patchouli alcohol

(d) cedrene

25-28 When an extract of parsley seed is saponified and acidified, one of the fatty acids isolated is *petroselenic acid*, formula $C_{18}H_{34}O_2$. Hydrogenation of petroselenic acid gives pure stearic acid. When petroselenic acid is treated with warm potassium permanganate followed by acidification, the only organic products are dodecanoic acid and adipic acid. The NMR spectrum shows absorptions of vinyl protons split by coupling constants of 7 Hz and 10 Hz. Propose a structure for petroselenic acid, and show how your structure is consistent with these observations.

25-29 The long-term health effects of eating partially hydrogenated vegetable oils concern some nutritionists because many unnatural fatty acids are produced. Consider the partial hydrogenation of linolenic acid by the addition of one or two equivalents of hydrogen. Show how this partial hydrogenation can produce at least three different fatty acids we have not seen before.

25-30 Fatty alcohols can react with reducing sugars to give glycosides such as the cetyl glucoside shown below. Predict the solubility properties and the most obvious uses of this cetyl glucoside.

a cetyl glucoside

CHAPTER 26

Synthetic Polymers

Introduction

People have always used polymers. Prehistoric tools and shelters were made from wood and straw. Both of these building materials contain cellulose, a biopolymer of glucose. Clothing was made from the hides and hair of animals, which contain protein, a biopolymer of amino acids. After people learned to use fire, they made ceramic pottery and glass, using naturally occurring inorganic polymers.

A **polymer** is a large molecule composed of many smaller repeating units (the **monomers**) bonded together. Today when we speak of polymers, we generally mean *synthetic organic polymers* rather than natural organic biopolymers such as DNA, cellulose, and protein, or inorganic polymers such as glass and concrete. The first synthetic organic polymer was made in 1838, when vinyl chloride was accidentally polymerized. Polystyrene was discovered in 1839, shortly after styrene was synthesized and purified. The discovery of polystyrene was inevitable, since styrene polymerizes spontaneously unless a stabilizer is added.

Also in 1839, Charles Goodyear (of tire and blimp fame) discovered how to convert the gummy polymeric sap of the rubber tree to a strong, stretchy material by heating it with sulfur. *Vulcanized rubber* quickly revolutionized the making of boots, tires, and rainwear. This was the first time that someone had artificially cross-linked a natural biopolymer to give it more strength and stability.

In fewer than 150 years, we have become literally surrounded by synthetic polymers. We wear clothes of nylon and polyester, we walk on polypropylene carpets, we drive cars with plastic fenders and synthetic rubber tires, and we use artificial hearts and other organs made of silicone polymers. Our pens and computers, our toys and our televisions are made largely of plastics.

Articles that are not made from polymers are often held together or coated with polymers. A bookcase may be made from wood, but the wood is bonded by a phenol-formaldehyde polymer and painted with a latex polymer. Each year, about *250 billion* pounds of synthetic organic polymers are produced worldwide, mostly for use in consumer products. Large numbers of organic chemists are employed to develop and produce these polymers.

In this chapter, we discuss some of the fundamental principles of polymer chemistry. We begin with a survey of the different kinds of polymers, then consider the reactions used to induce **polymerization**. Finally, we discuss some of the structural characteristics that determine the physical properties of a polymer.

Classes of Synthetic Polymers   There are two major classes of polymers: addition polymers and condensation polymers. **Addition polymers** result from the rapid addition of one molecule at a time to a growing polymer chain, usually with a reactive intermediate (cation, radical, or anion) at the growing end of the chain. Addition polymers are sometimes called **chain-growth polymers**, because growth usually occurs at the end of a chain. The monomers are usually alkenes, and polymerization involves successive additions across the double bonds. Poly(vinyl chloride), widely used as a synthetic leather, is an addition polymer.

vinyl chloride → poly(vinyl chloride)

Condensation polymers result from condensation (bond formation with loss of a small molecule) between the monomers. The most common condensations involve the formation of amides and esters. In a condensation polymerization, any two molecules can condense; they do not need to be at the end of a chain. Condensation polymers are sometimes called **step-growth polymers** because any pair of monomer molecules can react to give a step in the condensation. Dacron polyester is a condensation polymer.

dimethyl terephthalate + ethylene glycol

Dacron® polyester + $2n\ CH_3OH$

Many alkenes undergo chain-growth polymerization when treated with small amounts of suitable initiators. Table 26-1 shows some of the most common addition polymers, all made from substituted alkenes. The chain-growth mechanism involves addition of the reactive end of the growing chain across the double bond of the monomer.

26-2
Addition Polymers

TABLE 26-1 Some of the Most Important Addition Polymers

Polymer	Polymer Uses	Monomer Formula	Polymer Repeating Unit
polyethylene	bottles, bags, films	$H_2C{=}CH_2$	$-(CH_2-CH_2)_n$
polypropylene	plastics, olefin fibers		
polystyrene	plastics, foam insulation		
poly(isobutylene)	specialized rubbers		

(continues on next page)

TABLE 26-1 *continued*

Polymer	Polymer Uses	Monomer Formula	Polymer Repeating Unit
poly(vinyl chloride)	vinyl plastics, films, water pipes	H Cl $C=C$ H H	$\left[CH_2-CH \right]_n$ with Cl
poly(acrylonitrile)	Orlon®, Acrilan® fibers	H $C\equiv N$ $C=C$ H H	$\left[CH_2-CH \right]_n$ with CN
poly(methyl α-methacrylate)	acrylic fibers, Plexiglas®, Lucite® paints	H CH_3 $C=C$ H $C-OCH_3$ O	$\left[CH_2-C \right]_n$ with CH_3 and $COOCH_3$
poly(methyl α-cyanoacrylate)	"super" glues	H $C\equiv N$ $C=C$ H $C-OCH_3$ O	$\left[CH_2-C \right]_n$ with CN and $COOCH_3$
poly(tetrafluoroethylene)	Teflon® coatings, PTFE plastics	F F $C=C$ F F	$\left(CF_2-CF_2 \right)_n$

Depending on the monomer and the initiator used, the reactive intermediates may be free radicals, carbocations, or carbanions. Although these three types of chain-growth polymerizations are similar, we consider them individually.

26-2A   Free-Radical Polymerization

Free-radical polymerization results when a suitable alkene is heated with a radical initiator. For example, styrene polymerizes to polystyrene when it is heated to 100°C in the presence of benzoyl peroxide. This chain-growth polymerization is a free-radical chain reaction. Benzoyl peroxide cleaves when heated to give two carboxyl radicals, which quickly decarboxylate to give phenyl radicals.

A phenyl radical adds to styrene to give a resonance-stabilized benzylic radical. This reaction starts the growth of the polymer chain. Each propagation step adds another molecule of styrene to the growing chain. This addition takes place with the orientation that gives another resonance-stabilized benzylic radical.

Chain growth may continue with addition of several hundred or several thousand styrene units. The length of a polymer chain depends on the number of additions of monomers that occur before a termination step stops the process. Strong polymers with high molecular weights result from conditions that favor fast chain growth and minimize termination steps. Eventually the chain reaction stops, either by the coupling of two chains or by reaction with an impurity (such as oxygen) or simply by running out of monomer.

MECHANISM Free-Radical Polymerization

Initiation step: The initiator reacts with the monomer to start the chain.

phenyl radicals   styrene   benzylic radical

Propagation step: Another molecule of monomer adds to the chain.

growing chain   styrene   elongated chain   polystyrene
n = about 100 to 10,000

PROBLEM 26-1

Show the intermediate that would result if the growing chain added to the other end of the styrene double bond. Explain why the final polymer has phenyl groups substituted on alternating carbon atoms rather than randomly distributed.

Ethylene and propylene are also polymerized by free-radical chain-growth polymerization. With ethylene, the free-radical intermediates are less stable, so stronger reaction conditions are required. Ethylene is commonly polymerized by free-radical initiators at pressures around 3000 atm and temperatures of about 200°C. The product, called *low-density polyethylene*, is the material commonly used in polyethylene bags.

> In the presence of limited nutrients, bacteria can be induced to produce polyhydroxybutyrates and valerates, which are processed into a copolymer known as Biopol™. Biopol™ has properties similar to polypropylene, but it is biodegradable and obtained from nonpetroleum sources.

PROBLEM 26-2

Propose a mechanism for reaction of the first three propylene units in the polymerization of propylene in the presence of benzoyl peroxide.

$$n \ H_2C{=}CH{-}CH_3 \xrightarrow[\text{high pressure}]{\text{benzoyl peroxide}} \left[\begin{array}{cc} H & CH_3 \\ | & | \\ C & C \\ | & | \\ H & H \end{array} \right]_n$$

propylene   polypropylene

Chain Branching by Hydrogen Abstraction   Low-density polyethylene is soft and flimsy because it has a highly branched, amorphous structure. (High-density polyethylene, discussed in Section 26-4, is much stronger because of the orderly structure of unbranched linear polymer chains.) Chain branching in low-density polyethylene results from abstraction of a hydrogen atom in the middle of a chain by the free radical at the end of a chain. A new chain grows from the point of the free radical in the middle of the chain. Figure 26-1 shows abstraction of a hydrogen from a polyethylene chain and the first step in the growth of a branch chain at that point.

▲ Figure 26-1
Chain branching occurs when the growing end of a chain abstracts a hydrogen atom from the middle of a chain. A new branch grows off the chain at that point.

PROBLEM 26-3

Give a mechanism, using Figure 26-1 as a guide, showing chain branching during the free-radical polymerization of styrene. There are two types of aliphatic hydrogens in the polystyrene chain. Which type is more likely to be abstracted?

26-2B   Cationic Polymerization

Cationic polymerization occurs by a mechanism similar to the free-radical process, except that it involves carbocation intermediates. Strongly acidic catalysts are used to initiate cationic polymerization. BF_3 is a particularly effective catalyst, requiring a trace of water or methanol as a co-catalyst. Even when the reagents are carefully dried, there is enough water present for the first initiation step of the following mechanism:

MECHANISM Cationic Polymerization

Initiation steps: The catalyst protonates the monomer, starting the chain.

Propagation step: Another molecule of monomer adds to the chain.

Good monomers for cationic polymerization

R^+ + styrene $\longrightarrow$ benzylic carbocation (good)

growing chain

R^+ + isobutylene $\longrightarrow$ tertiary carbocation (good)

growing chain

Poor monomers for cationic polymerization

R^+ + ethylene $\longrightarrow$ primary carbocation (poor)

growing chain

R^+ + acrylonitrile $\longrightarrow$ destabilized carbocation (poor)

growing chain

◀ **Figure 26-2**
Cationic polymerization requires formation of a relatively stable carbocation intermediate.

A major difference between cationic and free-radical polymerization is that the cationic process needs a monomer that forms a relatively stable carbocation when it reacts with the cationic end of the growing chain. Some monomers form more stable intermediates than others. For example, styrene and isobutylene undergo cationic polymerization easily, while ethylene and acrylonitrile do not polymerize well under these conditions. Figure 26-2 compares the intermediates involved in these cationic polymerizations.

PROBLEM 26-4

The mechanism given for cationic polymerization of isobutylene shows that all the monomer molecules add with the same orientation, giving a polymer with methyl groups on alternate carbon atoms of the chain. Explain why no isobutylene molecules add with the opposite orientation.

PROBLEM 26-5

Suggest which of the following monomers might polymerize well on treatment with BF_3.
(a) vinyl chloride   (b) propylene   (c) methyl α-cyanoacrylate

PROBLEM 26-6

Chain branching occurs in cationic polymerization much as it does in free-radical polymerization. Propose a mechanism to show how branching occurs in the cationic polymerization of styrene. Suggest why isobutylene might be a better monomer for cationic polymerization than styrene.

26-2C Anionic Polymerization

Anionic polymerization occurs through carbanion intermediates. Effective anionic polymerization requires a monomer that gives a stabilized carbanion when it reacts with the anionic end of the growing chain. A good monomer for anionic polymerization should contain at least one strong electron-withdrawing group such as a carbonyl group, a cyano group, or a nitro group. The following reaction shows the chain-lengthening step in the polymerization of methyl acrylate. Notice that the chain-growth step of an anionic polymerization is simply a conjugate addition to a Michael acceptor (Section 22-18).

Chain-growth step in anionic polymerization

growing chain   methyl acrylate   stabilized anion   polymer

PROBLEM 26-7

Draw the important resonance forms of the stabilized anion formed in the anionic polymerization of methyl acrylate.

Anionic polymerization is usually initiated by a strong carbanion-like reagent such as an organolithium or Grignard reagent. Conjugate addition of the initiator to a monomer molecule starts the growth of the chain. Under the polymerization conditions, there is no good proton source available, and many monomer units react before the carbanion is protonated. The following reactions show a butyllithium-initiated anionic polymerization of acrylonitrile to give Orlon®.

MECHANISM Anionic Polymerization

Initiation step: The initiator adds to the monomer to form an anion.

butyllithium   acrylonitrile   stabilized anion

Propagation step: Another molecule of monomer adds to the chain.

growing chain   acrylonitrile   elongated chain   polymer

PROBLEM 26-8

Methyl α-cyanoacrylate (Super Glue) is easily polymerized, even by weak bases. Draw a mechanism for its base-catalyzed polymerization, and explain why this polymerization goes so quickly and easily.

methyl α-cyanoacrylate

PROBLEM 26-9

Chain branching is not as common with anionic polymerization as it is with free-radical poly-merization and cationic polymerization.

(a) Propose a mechanism for chain branching in the polymerization of acrylonitrile.

(b) Compare the relative stabilities of the intermediates in this mechanism with those you drew for chain branching in the cationic polymerization of styrene (Problem 26-6). Explain why chain branching is less common in this anionic polymerization.

Chain-growth polymerization of alkenes usually gives a head-to-tail bonding arrange-ment, with any substituent(s) appearing on alternate carbons of the polymer chain. This bonding arrangement is shown here for a generic polyalkene. Although the polymer backbone is joined by single bonds (and can undergo conformational changes), it is shown in the most stable all-anti conformation.

26-3
Stereochemistry of Polymers

The stereochemistry of the side groups (R) in the polymer has a major effect on the polymer's properties. The polymer has many chirality centers, raising the possibil-ity of millions of stereoisomers. Polymers are grouped into three classes, according to their predominant stereochemistry. If the side groups are generally on the same side of the polymer backbone, the polymer is called **isotactic** (Greek, *iso*, meaning "same," and *tactic*, meaning "order"). If the side groups generally alternate from one side to the other, the polymer is called **syndiotactic** (Greek, meaning "alternating order"). If the side groups occur randomly on either side of the polymer backbone, the polymer is called **atactic** (Greek, meaning "no order"). In most cases, isotactic and syndiotactic polymers have enhanced strength, clarity, and thermal properties over the atactic form of the polymer. Figure 26-3 shows these three types of polymers.

An isotactic polymer (side groups on the same side of the backbone)

A syndiotactic polymer (side groups on alternating sides of the backbone)

An atactic polymer (side groups on random sides of the backbone)

▲ **Figure 26-3**
Three stereochemical types of addition polymers.

> **PROBLEM 26-10**
>
> Draw the structures of isotactic poly(acrylonitrile) and syndiotactic polystyrene.

26-4
Stereochemical Control of Polymerization; Ziegler–Natta Catalysts

For any particular polymer, the three stereochemical forms have distinct properties. In most cases, the stereoregular isotactic and syndiotactic polymers are stronger and stiffer because of their greater crystallinity (a regular packing arrangement). The conditions used for polymerization often control the stereochemistry of the polymer. Anionic polymerizations are the most stereospecific; they usually give isotactic or syndiotactic polymers, depending on the nature of the side group. Cationic polymerizations are often stereospecific, depending on the catalysts and conditions used. Free-radical polymerization is nearly random, resulting in branched, atactic polymers.

In 1953, Karl Ziegler and Giulio Natta discovered that aluminum–titanium initiators catalyze the polymerization of alkenes, with two major advantages over other catalysts:

1. The polymerization is completely stereospecific. Either the isotactic form or the syndiotactic form may be made, by selecting the proper Ziegler–Natta catalyst.
2. Because the intermediates are stabilized by the catalyst, very little hydrogen abstraction occurs. The resulting polymers are linear with almost no branching.

A **Ziegler–Natta catalyst** is an organometallic complex, often containing titanium and aluminum. A typical catalyst is formed by adding a solution of $TiCl_4$ (titanium tetrachloride) to a solution of $(CH_3CH_2)_3Al$ (triethyl aluminum). This mixture is then "aged" by heating it for about an hour. The precise structure of the active catalyst is not known, but the titanium atom appears to form a complex with both the growing polymer chain and a molecule of monomer. The monomer attaches to the end of the chain (which remains complexed to the catalyst), leaving the titanium atom with a free site for complexation to the next molecule of monomer.

With a Ziegler–Natta catalyst, a *high-density polyethylene* (or *linear* polyethylene) can be produced with almost no chain branching and with much greater strength than common low-density polyethylene. Many other polymers are produced with improved properties using Ziegler–Natta catalysts. In 1963, Ziegler and Natta received the Nobel Prize for their work, which had revolutionized the polymer industry in only ten years.

26-5
Natural and Synthetic Rubbers

Natural rubber is isolated from a white fluid, called **latex**, that exudes from cuts in the bark of *Hevea brasiliensis*, the South American rubber tree. Many other plants secrete this polymer, as well. The name **rubber** was first used by Joseph Priestly, who used the crude material to "rub out" errors in his pencil writing. Natural rubber is soft and sticky. An enterprising Scotsman named Charles Macintosh found that rubber makes a good waterproof coating for raincoats. Natural rubber is not strong or elastic, however, so its uses were limited to waterproofing cloth and other strong materials.

Structure of Natural Rubber   Like many other plant products, natural rubber is a terpene composed of isoprene units (Section 25-8). If we imagine lining up many molecules of isoprene in the *s*-cis conformation, and moving pairs of electrons as shown below, we would produce a structure similar to natural rubber. This polymer results from 1,4-addition to each isoprene molecule, with all the double bonds in the cis configuration. Another name for natural rubber is *cis*-1,4-polyisoprene.

Imaginary polymerization of isoprene units

Natural rubber

White latex drips out of a gouge in the bark of a rubber tree in a Malaysian rubber plantation.

The cis double bonds in natural rubber force it to assume a kinky conformation that may be stretched and still return to its shorter, kinked structure when released. Unfortunately, when we pull on a mass of natural rubber, the chains slide by each other and the material pulls apart. This is why natural rubber is not suitable for uses requiring strength or durability.

Vulcanization: Cross-Linking of Rubber   In 1839, Charles Goodyear accidentally dropped a mixture of natural rubber and sulfur onto a hot stove. He was surprised to find that the rubber had become strong and elastic. This discovery led to the process that Goodyear called **vulcanization**, after the Roman god of fire and the volcano. Vulcanized rubber has much greater toughness and elasticity than natural rubber. It withstands relatively high temperatures without softening, and it remains elastic and flexible when cold.

Vulcanization also allows the casting of complicated shapes such as rubber tires. Natural rubber is putty-like, and it is easily mixed with sulfur, formed around the tire cord, and placed into a mold. The mold is closed and heated, and the gooey mass of string and rubber is vulcanized into a strong, elastic tire carcass.

On a molecular level, vulcanization causes cross-linking of the *cis*-1,4-polyisoprene chains through disulfide ($-S-S-$) bonds, similar to the cysteine bridges that link peptides (Section 24-8C). In vulcanized rubber, the polymer chains are linked together, so they can no longer slip past each other. When the material is stressed, the chains stretch, but cross-linking prevents tearing. When the stress is released, the chains return to their shortened, kinky conformations as the rubber snaps back. Figure 26-4 shows the structure of rubber before and after vulcanization.

Rubber can be prepared with a wide range of physical properties by controlling the amount of sulfur used in vulcanization. Low-sulfur rubber, made with about 1 to 3% sulfur, is soft and stretchy. It is good for rubber bands and inner tubes. Medium-sulfur rubber (about 3 to 10% sulfur) is somewhat harder, but still flexible, making good tires. High-sulfur rubber (20 to 30% sulfur) is called *hard rubber* and was once used as a hard synthetic plastic.

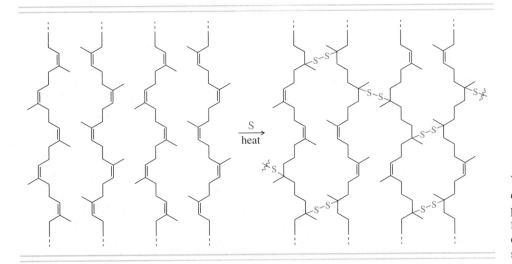

◄ Figure 26-4
Vulcanization of rubber introduces disulfide cross-links between the polyisoprene chains. Cross-linking forms a stronger, elastic material that does not pull apart when it is stretched.

Wallace Carothers, the inventor of Nylon, stretches a piece of synthetic rubber in his laboratory at the DuPont company.

PROBLEM 26-11

(a) Draw the structure of gutta-percha, a natural rubber with all its double bonds in the trans configuration.

(b) Suggest why gutta-percha is not very elastic, even after it is vulcanized.

Synthetic Rubber   There are many different formulations for **synthetic rubbers**, but the simplest is a polymer of 1,3-butadiene. Specialized Ziegler–Natta catalysts can produce 1,3-butadiene polymers where 1,4-addition has occurred on each butadiene unit, and the remaining double bonds are all cis. This polymer has properties similar to those of natural rubber, and it can be vulcanized in the same way.

1,4-polymerization of 1,3-butadiene

cis-1,4-polybutadiene

26-6
Copolymers of Two or More Monomers

All the polymers we have discussed are **homopolymers**, polymers made up of identical monomer units. Many polymeric materials are **copolymers**, made by polymerizing two or more different monomers together. In many cases, monomers are chosen so that they add selectively in an alternating manner. For example, when a mixture of vinyl chloride and vinylidene chloride (1,1-dichloroethylene) is induced to polymerize, the growing chain preferentially adds the monomer that is *not* at the end of the chain. This selective reaction gives the alternating copolymer *Saran*®, used as a film for wrapping food.

Overall reaction

vinyl chloride   vinylidene chloride   Saran®

Some polymers have three or more monomers mixed to give products with desired properties. For example, acrylonitrile, butadiene, and styrene are polymerized to give ABS plastic a strong, tough, and resilient material used for bumpers, crash helmets, and other articles that must withstand heavy impacts.

PROBLEM 26-12

Isobutylene and isoprene copolymerize to give "butyl rubber." Draw the structure of the repeating unit in butyl rubber, assuming that the two monomers alternate.

26-7
Condensation Polymers

Condensation polymers result from formation of ester or amide linkages between difunctional molecules. The reaction is called **step-growth** polymerization. Any two monomer molecules may react to form a dimer, dimers may condense to give tetramers, and so on. Each condensation is an individual *step* in the growth of the polymer, and there is no chain reaction. Many kinds of condensation polymers are known. We discuss the four most common types: polyamides, polyesters, polycarbonates, and polyurethanes.

26-7A   Polyamides: Nylon

When Wallace Carothers of DuPont discovered nylon in 1938, he opened the door to a new age of fibers and textiles. At that time, thread used for clothing was made of spun animal and plant fibers. These fibers were held together by friction or sizing, but they were weak and subject to rotting and unraveling. Silk (a protein) was the strongest fiber known at the time, and Carothers reasoned that a polymer bonded by amide linkages might approach the strength of silk. Nylon proved to be a completely new type of fiber, with remarkable strength and durability. It can be melted and extruded into a strong, continuous fiber, and it cannot rot. Thread spun from continuous nylon fibers is so much stronger than natural materials that it can be made much thinner. Availability of this strong, thin thread made possible stronger ropes, sheer fabrics, and nearly invisible women's stockings that came to be called "nylons."

Nylon is the common name for polyamides. **Polyamides** are generally made from reactions of diacids with diamines. The most common polyamide is called nylon 6,6 because it is made by reaction of a six-carbon diacid (adipic acid) with a six-carbon diamine. The six-carbon diamine, systematically named *1,6-hexanediamine*, is commonly called *hexamethylene diamine*. When adipic acid is mixed with hexamethylene diamine, a proton-transfer reaction gives a white solid called *nylon salt*. When nylon salt is heated to 250°C, water is driven off as a gas, and molten nylon results. Molten nylon is cast into a solid shape or extruded through a spinneret to produce a fiber.

Scanning electron micrograph of the material in a nylon stocking. Sheer stockings require long, continuous fibers of small diameter and enormous strength. (Magnification 150X.)

$$HO-\underset{\underset{O}{\|}}{C}-(CH_2)_4-\underset{\underset{O}{\|}}{C}-OH \; + \; H_2N-(CH_2)_6-NH_2 \longrightarrow$$

adipic acid   hexamethylene diamine

$$^-O-\underset{\overset{O}{\|}}{C}-(CH_2)_4-\underset{\overset{O}{\|}}{C}-O^-$$
$$\overset{+}{H_3N}-(CH_2)_6-\overset{+}{NH_3}$$

nylon salt

heat, $-H_2O$

$$\cdots\underset{\overset{O}{\|}}{C}-(CH_2)_4-\underset{\overset{O}{\|}}{C}\left[NH-(CH_2)_6-NH-\underset{\overset{O}{\|}}{C}-(CH_2)_4-\underset{\overset{O}{\|}}{C}\right]_n NH-(CH_2)_6-NH\cdots$$

poly(hexamethylene adipamide), called nylon 6,6

Nylon can also be made from a single monomer having an amino group at one end and an acid at the other. This reaction is similar to the polymerization of α-amino acids to give proteins. Nylon 6 is a polymer of this type, made from a six-carbon amino acid: 6-aminohexanoic acid (ε-aminocaproic acid). This synthesis starts with ε-caprolactam. When caprolactam is heated with a trace of water, some of it is hydrolyzed to the free amino acid. Continued heating gives condensation and polymerization to molten nylon 6. Nylon 6 (also called *Perlon*®) is used for making strong, flexible fibers for ropes and tire cord.

ε-caprolactam

$$\xrightleftharpoons{H_2O, \text{ heat}}$$

$$\overset{+}{H_3N}-(CH_2)_5-\underset{\overset{O}{\|}}{C}-O^-$$

ε-aminocaproic acid

heat, $-H_2O$

$$\cdots NH-(CH_2)_5-\underset{\overset{O}{\|}}{C}-NH-(CH_2)_5-\underset{\overset{O}{\|}}{C}\left[NH-(CH_2)_5-\underset{\overset{O}{\|}}{C}\right]_n NH-(CH_2)_5-\underset{\overset{O}{\|}}{C}\cdots$$

poly(6-aminohexanoic acid), called nylon 6 or Perlon®

PROBLEM 26-13

(a) *Nomex*®, a strong fire-resistant fabric, is a polyamide made from *meta*-phthalic acid and *meta*-diaminobenzene. Draw the structure of Nomex®.

(b) *Kevlar*®, made from terephthalic acid (*para*-phthalic acid) and *para*-diaminobenzene, is used in making tire cord and bulletproof vests. Draw the structure of Kevlar®.

26-7B Polyesters

The introduction of **polyester** fibers has brought about major changes in the way we care for our clothing. Nearly all modern permanent-press fabrics owe their wrinkle-free behavior to polyester, often blended with other fibers. These polyester blends have reduced or eliminated the need for starching and ironing clothes to achieve a wrinkle-free surface that holds its shape.

The most common polyester is *Dacron*®, the polymer of terephthalic acid (*para*-phthalic acid or benzene-1,4-dicarboxylic acid) with ethylene glycol. In principle, this polymer might be made by mixing the diacid with the glycol and heating the mixture to drive off water. In practice, however, a better product is obtained using a transesterification process (Section 21-5). The dimethyl ester of terephthalic acid is heated to about 150°C with ethylene glycol. Methanol is evolved as a gas, driving the reaction to completion. The molten product is spun into Dacron® fiber or cast into *Mylar*® film.

dimethyl terephthalate   ethylene glycol

poly(ethylene terephthalate) or PET, also called Dacron® polyester or Mylar® film

Dacron® fiber is used to make fabric and tire cord, and Mylar® film is used to make magnetic recording tape. Mylar® film is strong, flexible, and resistant to ultraviolet degradation. Aluminized Mylar® was used to make the Echo satellite, a huge balloon that was put into orbit around the Earth as a giant reflector. Poly(ethylene terephthalate) is also blow-molded to make plastic soft-drink bottles that are sold by the billions each year.

A polyester of ε-caprolactone is used to make bio-absorbable materials for use in the body. For example, suture thread for surgical stitches can be made from poly (ε-caprolactone).

PROBLEM 26-14

Kodel® polyester is formed by transesterification of dimethyl terephthalate with 1,4-di(hydroxymethyl)cyclohexane. Draw the structure of Kodel®.

PROBLEM 26-15

Glyptal® resin makes a strong, solid polymer matrix for electronic parts. Glyptal® is made from terephthalic acid and glycerol. Draw the structure of Glyptal®, and explain its remarkable strength and rigidity.

26-7C Polycarbonates

A *carbonate ester* is simply an ester of carbonic acid. Carbonic acid itself exists in equilibrium with carbon dioxide and water, but its esters are quite stable (Section 21-16).

$$\underset{\text{carbonic acid}}{HO-\overset{\overset{\displaystyle O}{\|}}{C}-OH} \quad \rightleftharpoons \quad CO_2 \;+\; H_2O \qquad \underset{\text{a carbonate ester}}{R-O-\overset{\overset{\displaystyle O}{\|}}{C}-O-R'}$$

Carbonic acid is a diacid; with suitable diols, it can form polyesters. For example, when phosgene (the acid chloride of carbonic acid) reacts with a diol, the product is a poly(carbonate ester). The following equation shows the synthesis of Lexan® poly-carbonate: a strong, clear, and colorless material that is used for bulletproof windows and crash helmets. The diol used to make Lexan® is a phenol called *bisphenol A*, a common intermediate in polyester and polyurethane synthesis.

Polycarbonate is a tough, clear material that withstands repeated sterilization. These properties account for its wide use in medical devices such as blood filters, surgical instruments, and intravenous line components.

Lexan® polycarbonate

PROBLEM 26-16

Propose a mechanism for the reaction of bisphenol A with phosgene.

PROBLEM 26-17

Bisphenol A is made on a large scale by a condensation of phenol with acetone. Suggest an appropriate catalyst, and propose a mechanism for this reaction. (*Hint:* This is a condensation because three molecules are joined with loss of water. The mechanism belongs to another class of reactions, though.)

26-7D   Polyurethanes

A *urethane* (Section 21-16) is an ester of a carbamic acid (R—NH—COOH), a half-amide of carbonic acid. Carbamic acids themselves are unstable, quickly decomposing to amines and CO_2. Their esters (urethanes) are quite stable, however.

$$\underset{\text{a carbamic acid}}{R-NH-\overset{\overset{\displaystyle O}{\|}}{C}-OH} \longrightarrow \underset{\text{amine}}{R-NH_2} + CO_2 \qquad \underset{\text{a urethane or carbamate ester}}{R-NH-\overset{\overset{\displaystyle O}{\|}}{C}-O-R'}$$

Because carbamic acids are unstable, normal esterification procedures cannot be used to form urethanes. Urethanes are most commonly made by treating an *isocyanate* with an alcohol or a phenol. The reaction is highly exothermic, and it gives a quantitative yield of a carbamate ester. The following reaction shows the formation of ethyl *N*-phenylcarbamate.

Latex condoms cause severe allergies for some people, and they deteriorate over time or in contact with oil-based lubricants or ointments. "Plastic condoms" made of polyurethane cause fewer allergic reactions. These polyurethane materials can be used with oil- or water-based lubricants, and they deteriorate less in storage.

$$\underset{\text{isocyanate}}{R-N=C=O} + \underset{\text{alcohol}}{HO-R'} \longrightarrow \underset{\substack{\text{carbamate ester}\\\text{(urethane)}}}{R-NH-\overset{\overset{\displaystyle O}{\|}}{C}-O-R'}$$

Example

phenyl isocyanate   ethanol   ethyl-*N*-phenylcarbamate

PROBLEM 26-18

Propose a mechanism for the reaction of phenyl isocyanate with ethanol.

A **polyurethane** results when a diol reacts with a diisocyanate, a compound with two isocyanate groups. The compound shown below, commonly called *toluene diisocyanate*, is frequently used for making polyurethanes. When ethylene glycol or another diol is added to toluene diisocyanate, a rapid condensation gives the polyurethane. Low-boiling liquids such as butane are often added to the reaction mixture. Heat evolved by the polymerization vaporizes the volatile liquid, producing bubbles that convert the viscous polymer to a frothy mass of polyurethane foam.

toluene diisocyanate   ethylene glycol

a polyurethane

PROBLEM 26-19

Explain why the addition of a small amount of glycerol to the polymerization mixture gives a stiffer urethane foam.

PROBLEM 26-20

Give the structure of the polyurethane formed by the reaction of toluene diisocyanate with bisphenol A.

26-8
Polymer Structure and Properties

Although polymers are very large molecules, we can explain their chemical and physical properties in terms of what we already know about smaller molecules. For example, when you spill a base on your polyester slacks, the fabric is weakened because the base hydrolyzes some of the ester linkages. The physical properties of polymers can also be explained using concepts we have already encountered. Although polymers do not crystallize or melt quite like smaller molecules, we can detect crystalline regions in a polymer, and we can measure the temperature at which these *crystallites* melt. In this section, we consider briefly some of the important aspects of polymer crystallinity and thermal behavior.

26-8A   Polymer Crystallinity

Polymers rarely form the large crystals characteristic of other organic compounds, but many do form microscopic crystalline regions called **crystallites**. A highly regular

crystallites

polymer that packs well into a crystal lattice will be highly crystalline, and it will generally be denser, stronger, and more rigid than a similar polymer with a lower degree of **crystallinity**. Figure 26-5 shows how the polymer chains are arranged in parallel lines in crystalline areas within a polymer.

Polyethylene shows how crystallinity affects a polymer's physical properties. Free-radical polymerization gives a highly branched, low-density polyethylene that forms very small crystallites because the random chain branching destroys the regularity of the crystallites. An unbranched, high-density polyethylene is made using a Ziegler–Natta catalyst. The *linear* structure of the high-density material packs more easily into a crystal lattice, so it forms larger and stronger crystallites. We say that high-density polyethylene has a higher degree of crystallinity, and it is therefore denser, stronger, and more rigid than low-density polyethylene.

Stereochemistry also affects the crystallinity of a polymer. Stereoregular isotactic and syndiotactic polymers are generally more crystalline than atactic polymers. By careful choice of catalysts, we can make a linear polymer with either isotactic or syndiotactic stereochemistry.

26-8B Thermal Properties

At low temperatures, long-chain polymers are *glasses*. They are solid and unyielding, and a strong impact causes them to fracture. As the temperature is raised, the polymer goes through a **glass transition temperature**, abbreviated T_g. Above T_g, a highly crystalline polymer becomes flexible and moldable. We say it is a **thermoplastic** because application of heat makes it plastic (moldable). As the temperature is raised further, the polymer reaches the **crystalline melting temperature**, abbreviated T_m. At this temperature, crystallites melt and the individual molecules can slide past one another.

Above T_m, the polymer is a viscous liquid and can be extruded through spinnerets to form fibers. The fibers are immediately cooled in water to form crystallites and then stretched (drawn) to orient the crystallites along the fiber, increasing its strength.

Long-chain polymers with low crystallinity (called **amorphous polymers**) become rubbery when heated above the glass transition temperature. Further heating causes them to grow gummier and less solid until they become viscous liquids without definite melting points. Figure 26-6 compares the thermal properties of crystalline and amorphous long-chain polymers.

These phase transitions apply only to long-chain polymers. Cross-linked polymers are more likely to stay rubbery, and they may not melt until the temperature is so high that the polymer begins to decompose.

Drugs are sometimes enclosed in water-soluble polymers to control the rate of drug release. The polymer breaks down in the body over time at a predictable rate and gradually releases the drug.

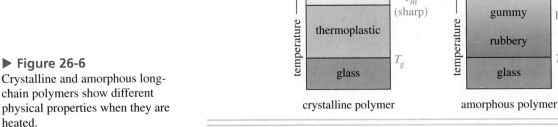

▶ **Figure 26-6**
Crystalline and amorphous long-chain polymers show different physical properties when they are heated.

26-8C   Plasticizers

In many cases, a polymer has desirable properties for a particular use, but it is too brittle—either because its glass transition temperature (T_g) is above room temperature or because the polymer is too highly crystalline. In such cases, addition of a **plasticizer** often makes the polymer more flexible. A plasticizer is a nonvolatile liquid that dissolves in the polymer, lowering the attractions between the polymer chains and allowing them to slide by one another. The overall effect of the plasticizer is to reduce the crystallinity of the polymer and lower its glass transition temperature (T_g).

A common example of a plasticized polymer is poly(vinyl chloride). The common atactic form has a T_g of about 80°C, well above room temperature. Without a plasticizer, "vinyl" is stiff and brittle. Dibutyl phthalate (see the structure in the margin) is added to the polymer to lower its glass transition temperature to about 0°C. This plasticized material is the flexible, somewhat stretchy film we think of as vinyl raincoats, shoes, and even inflatable boats. Dibutyl phthalate is slightly volatile, however, and it gradually evaporates. The soft, plasticizied vinyl gradually loses its plasticizer and becomes hard and brittle.

dibutyl phthalate

Chapter 26
Glossary

addition polymer (chain-growth polymer) A polymer that results from the rapid addition of one molecule at a time to a growing polymer chain, usually with a reactive intermediate (cation, radical, or anion) at the growing end of the chain. (p. 1183)

amorphous polymer A long-chain polymer with low crystallinity. (p. 1197)

anionic polymerization The process of forming an addition polymer by chain-growth polymerization involving an anion at the end of the growing chain. (p. 1186)

atactic polymer A polymer with the side groups on random sides of the polymer backbone. (p. 1189)

cationic polymerization The process of forming an addition polymer by chain-growth polymerization involving a cation at the end of the growing chain. (p. 1186)

chain-growth polymer See **addition polymer**.

condensation polymer (step-growth polymer) A polymer that results from condensation (bond formation with loss of a small molecule) between the monomers. In a condensation polymerization, any two molecules can condense, not necessarily at the end of a growing chain. (p. 1192)

copolymer A polymer made from two or more different monomers. (p. 1192)

crystalline melting temperature (T_m) The temperature at which melting of the crystallites in a highly crystalline polymer occurs. Above T_m, the polymer is a viscous liquid. (p. 1197)

crystallinity The relative amount of the polymer that is included in crystallites, and the relative sizes of the crystallites. (p. 1196)

crystallites Microscopic crystalline regions found within a solid polymer below the crystalline melting temperature. (p. 1196)

free-radical polymerization The process of forming an addition polymer by chain-growth polymerization involving a free radical at the end of the growing chain. (p. 1184)

glass transition temperature (T_g) The temperature above which a polymer becomes rubbery or flexible. (p. 1197)

homopolymer A polymer made from identical monomer units. (p. 1192)

isotactic polymer A polymer with all the side groups on the same side of the polymer backbone. (p. 1189)

monomer One of the small molecules that bond together to form a polymer. (p. 1182)

nylon The common name for polyamides. (p. 1193)

plasticizer A nonvolatile liquid that is added to a polymer to make it more flexible and less brittle below its glass transition temperature. In effect, a plasticizer reduces the crystallinity of a polymer and lowers T_g. (p. 1198)

polyamide (nylon) A polymer whose repeating monomer units are bonded by amide linkages, much like the peptide linkages in protein. (p. 1193)

polycarbonate A polymer whose repeating monomer units are bonded by carbonate ester linkages. (p. 1194)

polyester A polymer whose repeating monomer units are bonded by carboxylate ester linkages. (p. 1194)

polymer A large molecule composed of many smaller units (monomers) bonded together. (p. 1182)

polymerization The process of linking monomer molecules into a polymer. (p. 1182)

polyurethane A polymer whose repeating monomer units are bonded by urethane (carbamate ester) linkages. (p. 1195)

rubber A natural polymer isolated from the **latex** that exudes from cuts in the bark of the South American rubber tree. Alternatively, synthetic polymers with rubberlike properties are called **synthetic rubber**. (p. 1190)

step-growth polymer See **condensation polymer**.

syndiotactic polymer A polymer with the side groups on alternating sides of the polymer backbone. (p. 1189)

thermoplastic A polymer that becomes moldable at high temperature. (p. 1197)

vulcanization Heating of natural or synthetic rubber with sulfur to form disulfide cross-links. Cross-linking adds durability and elasticity to rubber. (p. 1191)

Ziegler–Natta catalyst Any one of a group of addition polymerization catalysts involving titanium–aluminum complexes. Ziegler–Natta catalysts produce stereoregular (either isotactic or syndiotactic) polymers in most cases. (p. 1190)

Essential Problem-Solving Skills in Chapter 26

1. Given the structure of a polymer, determine whether it is an addition or condensation polymer, and determine the structure of the monomer(s).

2. Given the structure of one or more monomers, predict whether polymerization will occur to give an addition polymer or a condensation polymer, and give the general structure of the polymer chain.

3. Use mechanisms to explain how a monomer polymerizes under acidic, basic, or free-radical conditions. For addition polymerization, consider whether the reactive end of the growing chain is more stable as a cation (acidic conditions), anion (basic conditions), or free radical (radical initiator). For condensation polymerization, consider the mechanism of the step-growth reaction.

4. Predict the general characteristics (strength, elasticity, crystallinity, chemical reactivity) of a polymer based on its structure, and explain how its physical characteristics change as it is heated past T_g and T_m.

5. Explain how chain branching, cross-linking, and plasticizers affect the properties of polymers.

6. Compare the stereochemistry of isotactic, syndiotactic, and atactic polymers. Explain how the stereochemistry can be controlled during polymerization and how it affects the physical properties of the polymer.

Study Problems

26-21   Define each term and give an example.

(a) addition polymer   (b) condensation polymer   (c) copolymer
(d) atactic polymer   (e) isotactic polymer   (f) syndiotactic polymer
(g) free-radical polymerization   (h) cationic polymerization   (i) anionic polymerization
(j) crystalline polymer   (k) amorphous polymer   (l) monomer
(m) plasticizer   (n) vulcanization   (o) Ziegler–Natta catalyst
(p) glass transition temperature   (q) polyamide   (r) polyester
(s) crystalline melting temperature   (t) polyurethane   (u) polycarbonate

26-22   Polyisobutylene is one of the components of butyl rubber used for making inner tubes.
(a) Give the structure of polyisobutylene.
(b) Is this an addition polymer or a condensation polymer?
(c) What conditions (cationic, anionic, free-radical) would be most appropriate for polymerization of isobutylene? Explain your answer.

26-23   Poly(trimethylene carbamate) is used in high-quality synthetic leather. It has the following structure.

$$\left(CH_2CH_2CH_2-\overset{\overset{\displaystyle H}{|}}{N}-\overset{\overset{\displaystyle O}{\|}}{C}-O\right)_n$$

(a) What type of polymer is poly(trimethylene carbamate)?
(b) Is this an addition polymer or a condensation polymer?
(c) Draw the products that would be formed if the polymer were completely hydrolyzed under acidic or basic conditions.

26-24   Poly(butylene terephthalate) is a hydrophobic plastic material widely used in automotive ignition systems.

$$\left(CH_2CH_2CH_2CH_2-O-\overset{\overset{\displaystyle O}{\|}}{C}-\underset{}{\bigcirc}-\overset{\overset{\displaystyle O}{\|}}{C}-O\right)_n$$

poly(butylene terephthalate)

(a) What type of polymer is poly(butylene terephthalate)?
(b) Is this an addition polymer or a condensation polymer?
(c) Suggest what monomers might be used to synthesize this polymer and how the polymerization might be accomplished.

26-25   *Urylon* fibers, used in stretchy fabrics, are composed of the following polymer.

$$\left((CH_2)_9-\overset{\overset{\displaystyle H}{|}}{N}-\overset{\overset{\displaystyle O}{\|}}{C}-\overset{\overset{\displaystyle H}{|}}{N}\right)_n$$

(a) What functional group is contained in the Urylon structure?
(b) Is Urylon an addition polymer or a condensation polymer?
(c) Draw the products that would be formed if the polymer were completely hydrolyzed under acidic or basic conditions.

26-26   Polyethylene glycol, or *Carbowax*® [($-O-CH_2-CH_2-$)$_n$], is widely used as a binder, thickening agent, and packaging additive for foods.
(a) What type of polymer is polyethylene glycol? (We have not seen this type of polymer before.)
(b) The systematic name for polyethylene glycol is poly(ethylene oxide). What monomer would you use to make polyethylene glycol?
(c) What conditions (free-radical initiator, acid catalyst, basic catalyst, etc.) would you consider using in this polymerization?
(d) Propose a polymerization mechanism as far as the tetramer.

26-27   Polychloroprene, commonly known as *neoprene*, is widely used in rubber parts that must withstand exposure to gasoline or other solvents.

$$\left(CH_2-\overset{\overset{\displaystyle H}{|}}{C}=\overset{\overset{\displaystyle Cl}{|}}{C}-CH_2\right)_n$$

polychloroprene (neoprene)

(a) What type of polymer is polychloroprene?
(b) What monomer is used to make this synthetic rubber?

26-28   Polyoxymethylene (polyformaldehyde) is the tough, self-lubricating *Delrin*® plastic used in gear wheels.
(a) Give the structure of polyformaldehyde.
(b) Formaldehyde is polymerized using an acidic catalyst. Using H⁺ as a catalyst, propose a mechanism for the polymerization as far as the trimer.
(c) Is Delrin® an addition polymer or a condensation polymer?

26-29   Acetylene can be polymerized using a Ziegler–Natta catalyst. The cis or trans stereochemistry of the products can be controlled by careful selection and preparation of the catalyst. The resulting polyacetylene is an electrical semiconductor with a metallic appearance. *cis*-Polyacetylene has a copper color, and *trans*- is silver.
(a) Draw the structures of *cis*- and *trans*-polyacetylene.
(b) Use your structures to show why these polymers conduct electricity.
(c) It is possible to prepare polyacetylene films whose electrical conductivity is *anisotropic*. That is, the conductivity is higher in some directions than in others. Explain how this unusual behavior is possible.

26-30   Use chemical equations to show how the following accidents cause injury to the clothing involved (not to mention the skin under the clothing!).
(a) An industrial chemist spills aqueous H_2SO_4 on her nylon stockings but fails to wash it off immediately.
(b) An organic laboratory student spills aqueous NaOH on his polyester slacks.

26-31   Poly(vinyl alcohol), a hydrophilic polymer used in aqueous adhesives, is made by polymerizing vinyl acetate and then hydrolyzing the ester linkages.
(a) Give the structures of poly(vinyl acetate) and poly(vinyl alcohol).
(b) Vinyl acetate is an ester. Is poly(vinyl acetate) therefore a polyester? Explain.
(c) We have seen that basic hydrolysis destroys the Dacron® polymer. Poly(vinyl acetate) is converted to poly(vinyl alcohol) by a basic hydrolysis of the ester groups. Why doesn't the hydrolysis destroy the poly(vinyl alcohol) polymer?
(d) Why is poly(vinyl alcohol) made by this circuitous route? Why not just polymerize vinyl alcohol?

26-32   In reference to cloth or fiber, the term *acetate* usually means *cellulose acetate*, a semisynthetic polymer made by treating cellulose with acetic anhydride. Cellulose acetate is spun into yarn by dissolving it in acetone or methylene chloride and forcing the solution through spinnerets into warm air, where the solvent evaporates.
(a) Draw the structure of cellulose acetate.
(b) Explain why cellulose acetate is soluble in organic solvents, even though cellulose is not.
(c) (A true story) An organic chemistry student wore a long-sleeved acetate blouse to the laboratory. She was rinsing a warm separatory funnel with acetone when the pressure rose and blew out the stopper. Her right arm was drenched with acetone, but she was unconcerned because acetone is not very toxic. About ten minutes later, the right arm of the student's blouse disintegrated into a pile of white fluff, leaving her with a ragged short sleeve and the tatters of a cuff remaining around her wrist. Explain how a substance as innocuous as acetone ruined the student's blouse.
(d) Predict what usually happens when students wear polyvinyl chloride shoes to the organic laboratory.

*****26-33**   One of the earliest commercial plastics was *Bakelite*®, formed by the reaction of phenol with a little more than one equivalent of formaldehyde under acidic or basic conditions. Baeyer first discovered this reaction in 1872, and practical methods for casting and molding Bakelite® were developed around 1909. Phenol-formaldehyde plastics and resins (also called *phenolics*) are highly cross-linked because each phenol ring has three sites (two ortho and one para) that can be linked by condensation with formaldehyde. Suggest a general structure for a phenol-formaldehyde resin, and propose a mechanism for its formation under acidic conditions. (*Hint:* Condensation of phenol with formaldehyde resembles the condensation of phenol with acetone, used in Problem 26-17, to make bisphenol A.)

*****26-34**   Plywood and particle board are often glued with cheap, waterproof urea-formaldehyde resins. Two to three moles of formaldehyde are mixed with one mole of urea and a little ammonia as a basic catalyst. The reaction is allowed to proceed until the mixture becomes syrupy, then it is applied to the wood surface. The wood surfaces are held together under heat and pressure, while polymerization continues and cross-linking takes place. Propose a mechanism for the base-catalyzed condensation of urea with formaldehyde to give a linear polymer, then show how further condensation leads to cross-linking. (*Hint:* The carbonyl group lends acidity to the N—H protons of urea. A first condensation with formaldehyde leads to an imine, which is weakly electrophilic and reacts with another deprotonated urea.)

APPENDICES

APPENDIX 1A NMR Absorption Positions of Protons in Various Structural Environments

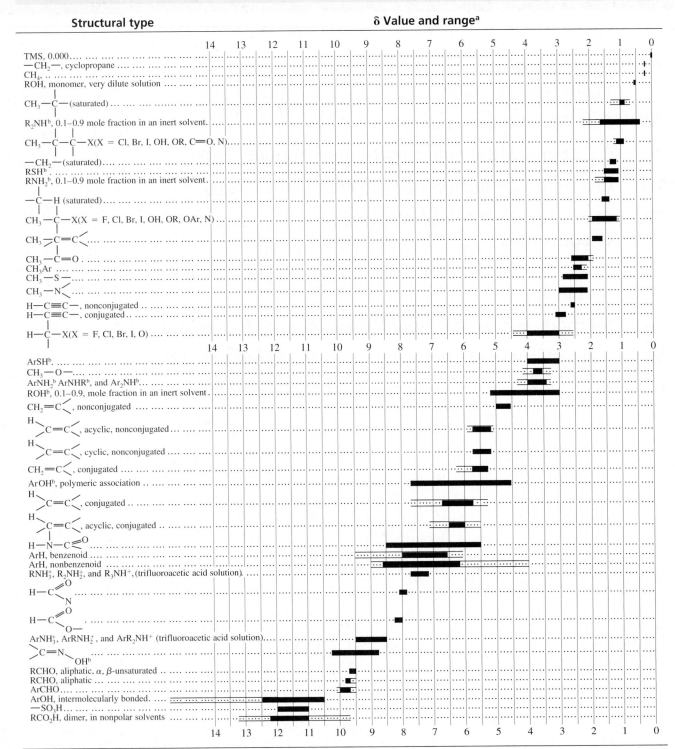

Structural type — δ Value and range[a]

[a] Normally, absorptions for the functional groups indicated will be found within the range shown in black. Occasionally, a functional group will absorb outside this range. Approximate limits are indicated by extended outlines.

[b] Absorption positions of these groups are concentration-dependent and are shifted to lower δ values in more dilute solutions.

APPENDIX 1B Spin-Spin Coupling Constants

Type	J, Hz
(geminal H-C-H)	12–15
>CH—CH<	2–9
with free rotation	~7
—C—(—C—)n—C—	~0
CH₃—CH₂—X	6.5–7.5
(CH₃)₂CH—X	5.5–7.0
H—C—C—H (X Y)	a,a 5–10 a,e 2–4 e,e 2–4
>C=C< (gem H)	0.5–3
(cis)	7–12
(trans)	13–18

Type	J, Hz
>C=C—C—H	4–10
>C=C—C<—H	0.5–2.5
H C=C C—H	~0
>C=CH—CH=C<	9–13
>CH—C≡C—H	2–3
>CH—CHO	1–3
>C=C—CHO	6–8
(benzene)	o-6–9 m-1–3 p-0–1

a = axial, e = equatorial

APPENDIX 1C ¹³C Chemical Shifts in Organic Compounds*

*Relative to internal tetramethylsilane.
Copyright 1998 by Bruker Analytik GmbH. Used by permission.

APPENDIX 2A   Characteristic Infrared Group Frequencies (s = strong, m = medium, w = weak; overtone bands are marked 2ν)

Courtesy of N. B. Colthup, Stanford Research Laboratories, American Cyanamid Company, and the editor of the *Journal of the Optical Society*.

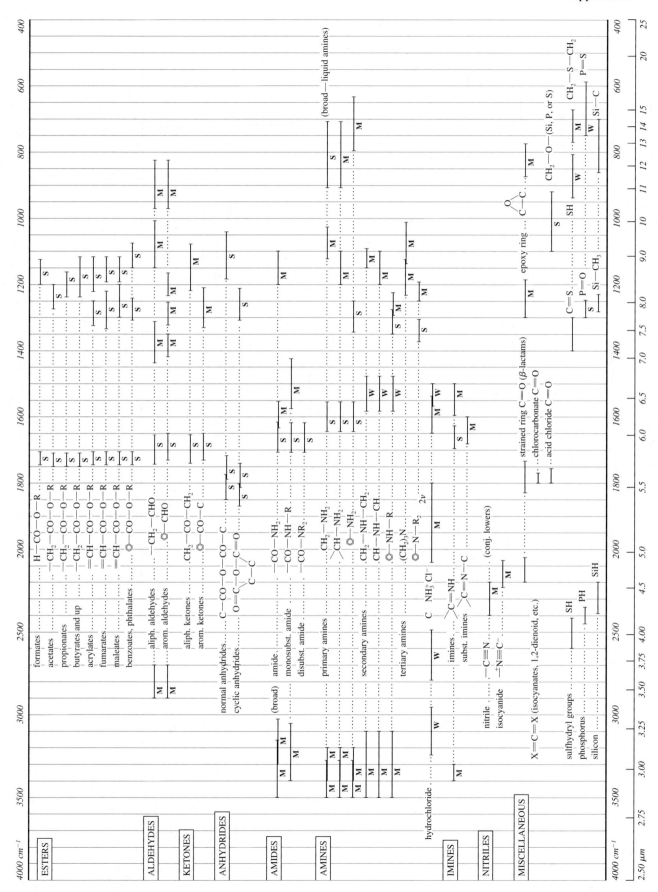

APPENDIX 2A Characteristic Infrared Group Frequencies (s = strong, m = medium, w = weak; overtone bands are marked 2ν)

INORGANIC SALTS AND DERIVED COMPOUNDS

fluorine
fluorine
fluorine
chlorine
chlorine
bromine

CF$_2$ and CF$_3$
C—F (unsat.)
C—F (sat.)
CCl$_2$ and CCl$_3$
CCl (aliph.)
CBr$_2$ and CBr$_3$
CBr (aliph.)

sulfur-oxygen compounds

ionic sulfate   (SO$_4$)$^{2-}$
ionic sulfonate   R—SO$_3^-$
sulfonic acid   R—SO$_3$H
covalent sulfate   R—O—SO$_2$—O—R
covalent sulfonate R—O—SO$_2$—R
sulfonamide   R—SO$_2$—NH$_2$
sulfone   R—SO$_2$—R
sulfoxide   R—SO—R

phosphorus-oxygen
ionic phosphate   (PO$_4$)$^{3-}$
covalent phosphate (RO)$_3$P→O
covalent phosphate CH$_2$—O—P→O
                    —O—P—O—

carbon-oxygen
ionic carbonate   (CO$_3$)$^{2-}$
covalent carbonate O=C(O—R)$_2$
imino carbonate   HN=C(O—R)$_2$

nitrogen-oxygen
ionic nitrate   (NO$_3$)$^-$
covalent nitrate   R—O—NO$_2$
nitro   R—NO$_2$
nitro   R—O—NO
covalent nitrite   R—O—NO
nitroso   R—NO

ammonium NH$_4^+$

ASSIGNMENTS

OH and NH str.
CH str.
—C≡X str.
C=O str.
C=N str.
C=C str.
NH bend
C—O str.
C—N str.
C—C str.
CH bend
OH bend
NH rock
CH rock
—C≡N str.

N. B. COLTHUP

APPENDIX 2B Characteristic Infrared Absorptions of Functional Groups

Group	Intensity[a]	Range (cm^{-1})	Group	Intensity[a]	Range (cm^{-1})
A. Hydrocarbon chromophore			d. Aromatic	v	~1600
1. C—H stretching				v	~1580
a. Alkane	m–s	2962–2853		m	~1500
b. Alkene, monosubstituted	m	3040–3010		and m	~1450
(vinyl)	and m	3095–3075	**B. Carbonyl chromophore**		
Alkene, disubstituted, cis	m	3040–3010	1. Ketone stretching vibrations		
Alkene, disubstituted, trans	m	3040–3010	a. Saturated, acyclic	s	1725–1705
Alkene, disubstituted, gem	m	3095–3075	b. Saturated, cyclic:		
Alkene, trisubstituted	m	3040–3010	6-Membered ring (and higher)	s	1725–1705
c. Alkyne	s	~3300	5-Membered ring	s	1750–1740
d. Aromatic	v	~3030	4-Membered ring	s	~1775
2. C—H bending			c. α,β-Unsaturated, acyclic	s	1685–1665
a. Alkane, C—H	w	~1340	d. α,β-Unsaturated, cyclic:		
Alkane, —CH$_2$—	m	1485–1445	6-Membered ring (and higher)	s	1685–1665
Alkane, —CH$_3$	m	1470–1430	5-Membered ring	s	1725–1708
	and s	1380–1370	e. $\alpha,\beta,\alpha',\beta'$-Unsaturated,		
Alkane, gem-dimethyl	s	1385–1380	acyclic	s	1670–1663
	and s	1370–1365	f. Aryl	s	1700–1680
Alkone, tert-butyl	m	1395–1385	g. Diaryl	s	1670–1660
	and s	~1365	h. α-Diketones	s	1730–1710
b. Alkene, monosubstituted(vinyl)	s	995–985	i. β-Diketones (enolic)	s	1640–1540
	s	915–905	j. 1,4-Quinones	s	1690–1660
	and s	1420–1410	k. Ketenes	s	~2150
Alkene, disubstituted, cis	s	~690	2. Aldehydes		
Alkene, disubstituted, trans	s	970–960	a. Carbonyl stretching vibrations:		
	and m	1310–1295	Saturated, aliphatic	s	1740–1720
Alkene, disubstituted, gem	s	895–885	α,β-Unsaturated, aliphatic	s	1705–1680
	and s	1420–1410	$\alpha,\beta,\gamma,\delta$-Unsaturated,		
Alkene, trisubstituted	s	840–790	aliphatic	s	1680–1660
c. Alkyne	s	~630	Aryl	s	1715–1695
d. Aromatic, substitution			b. C—H stretching		
type:[b] Five adjacent	v, s	~750	vibrations, two bands	w	2900–2820
hydrogen atoms	and v, s	~700		and w	2775–2700
Four adjacent hydrogen			3. Ester stretching vibrations		
atoms	v, s	~750	a. Saturated, acyclic	s	1750–1735
Three adjacent hydrogen atoms	v, m	~780	b. Saturated, cyclic:		
Two adjacent hydrogen			δ-Lactones (and larger		
atoms	v, m	~830	rings)	s	1750–1735
One hydrogen atom	v, w	~880	γ-Lactones	s	1780–1760
3. C—C multiple bond stretching			β-Lactones	s	~1820
a. Alkene, nonconjugated	v	1680–1620	c. Unsaturated:		
Alkene, monosubstituted			Vinyl ester type	s	1800–1770
(vinyl)	m	~1645	α,β-Unsaturated and aryl	s	1730–1717
Alkene, disubstituted, cis	m	~1658	α,β-Unsaturated δ-lactone	s	1730–1717
Alkene, disubstituted, trans	m	~1675	α,β-Unsaturated γ-lactone	s	1760–1740
Alkene, disubstituted, gem	m	~1653	β,γ-Unsaturated γ-lactone	s	~1800
Alkene, trisubstituted	m	~1669	d. α-Ketoesters	s	1755–1740
Alkene, tetrasubstituted	w	~1669	e. β-Ketoesters (enolic)	s	~1650
Diene	w	~1650	f. Carbonates	s	1780–1740
	and w	~1600	g. Thioesters	s	~1690
b. Alkyne, monosubstituted	m	2140–2100	4. Carboxylic acids		
Alkyne, disubstituted	v, w	2260–2190	a. Carbonyl stretching vibrations:		
c. Allene	m	~1960	Saturated aliphatic	s	1725–1700
	and m	~1060	*(continues on next page)*		

[a]Abbreviations: s = strong, m = medium, w = weak, v = variable, b = broad, sh = sharp, ~ = approximately
[b]Substituted benzenes also show weak bands in the 2000–1670 cm^{-1} region.

APPENDIX 2B Characteristic Infrared Absorptions of Functional Groups

Group	Intensity[a]	Range (cm^{-1})	Group	Intensity[a]	Range (cm^{-1})
α,β-Unsaturated aliphatic	s	1715–1690	b. N—H stretching vibrations:		
Aryl	s	1700–1680	Primary, free; two bands	m	~3500
b. Hydroxyl stretching				and m	~3400
(bonded), several bands	w	2700–2500	Primary, bonded:	m	~3350
c. Carboxylate anion	s	1610–1550	two bands	and m	~3180
stretching	and s	1400–1300	Secondary, free; one band	m	~3430
5. Anhydride stretching vibrations			Secondary, bonded;		
a. Saturated, acyclic	s	1850–1800	one band	m	3320–3140
	and s	1790–1740	c. N—H bending vibrations:		
b. α,β-Unsaturated and aryl,	s	1830–1780	Primary amides, dilute solution	s	1620–1590
acyclic	and s	1770–1720	Secondary amides	s	1550–1510
c. Saturated, 5-membered	s	1870–1820	**C. Miscellaneous chromophoric groups**		
ring	and s	1800–1750	1. Alcohols and phenols		
d. α,β-Unsaturated,	s	1850–1800	a. O—H stretching vibrations:		
5-membered ring	and s	1830–1780	Free O—H	v, sh	3650–3590
6. Acyl halide stretching vibrations			Intermolecularly hydrogen		
a. Acyl fluorides	s	~1850	bonded (change on dilution)		
b. Acyl chlorides	s	~1795	Single-bridge compounds	v, sh	3550–3450
c. Acyl bromides	s	~1810	Polymeric association	s, b	3400–3200
d. α,β-Unsaturated and aryl	s	1780–1750	Intramolecularly hydrogen		
	and m	1750–1720	bonded (no change on dilution)		
7. Amides			Single-bridge compounds	v, sh	3570–3450
a. Carbonyl stretching vibrations:			Chelate compounds	w, b	3200–2500
Primary, solid and			b. O—H bending and C—O stretching vibrations:		
concentrated solution	s	~1650	Primary alcohols	s	~1050
Primary, dilute solution	s	~1690		and s	1350–1260
Secondary, solid and			Secondary alcohols	s	~1100
concentrated solution	s	1680–1630		and s	1350–1260
Secondary, dilute solution	s	1700–1670	Tertiary alcohols	s	~1150
Tertiary, solid and all				and s	1410–1310
solutions	s	1670–1630	Phenols	s	~1200
Cyclic, δ-lactams	s	~1680		and s	1410–1310
Cyclic, γ-lactams	s	~1700	2. Amines		
Cyclic, γ-lactams, fused to			a. N—H stretching vibrations:		
another ring	s	1750–1700	Primary, free; two bands	m	~3500
Cyclic, β-lactams	s	1760–1730		and m	~3400
Cyclic, β-lactams, fused to			Secondary free; one band	m	3500–3310
another ring, dilute solution	s	1780–1770	Imines (=N—N); one band	m	3400–3300
Ureas, acyclic	s	~1660	Amine salts	m	3130–3030
Ureas, cyclic,			b. N—H bending vibrations:		
6-Membered ring	s	~1640	Primary	s-m	1650–1590
Ureas, cyclic,			Secondary	w	1650–1550
5-Membered ring	s	~1720	Amine salts	s	1600–1575
Urethanes	s	1740–1690		and s	~1500
Imides, acyclic	s	~1710	c. C—N vibrations:		
	and s	~1700	Aromatic, primary	s	1340–1250
Imides, cyclic,	s	~1710	Aromatic, secondary	s	1350–1280
6-Membered ring	and s	1700	Aromatic, tertiary	s	1360–1310
Imides, cyclic, α,β-unsaturated,	s	~1730	Aliphatic	w	1220–1020
6-Membered ring	and s	~1670		and w	~1410
Imides, cyclic, 5-membered	s	~1770	3. Unsaturated nitrogen compounds		
ring	and s	~1700	a. C≡N stretching vibrations:		
Imides, cyclic, α,β-unsaturated,	s	~1790	Alkyl nitriles	m	2260–2240
5-membered ring	and s	~1710	*(continues on next page)*		

[a] Abbreviations: s = strong, m = medium, w = weak, v = variable, b = broad, sh = sharp, ~ = approximately

APPENDIX 2B *(continued)*

Group	Intensity[a]	Range (cm^{-1})	Group	Intensity[a]	Range (cm^{-1})
α,β-Unsaturated alkyl nitriles	m	2235–2215	i. O—NO, nitrites	s	1680–1650
				and s	1625–1610
Aryl nitriles	m	2240–2220	4. Halogen compounds, C—X stretching vibrations		
Isocyanates	m	2275–2240	a. C—F	s	1400–1000
Isocyanides	m	2220–2070	b. C—Cl	s	800–600
			c. C—Br	s	600–500
b. $>$C=N— stretching vibrations (imines, oximes)			d. C—I	s	~500
			5. Sulfur compounds		
Alkyl compounds	v	1690–1640	a. S—H stretching vibrations	w	2600–2550
α,β-Unsaturated compounds	v	1660–1630	b. C=S stretching vibrations	s	1200–1050
c. —N=N— stretching vibrations, azo compounds	v	1630–1575	c. S=O stretching vibrations:		
			Sulfoxides	s	1070–1030
d. —N=C=N— stretching vibrations, diimide	s	2155–2130	Sulfones	s	1160–1140
				and s	1350–1300
e. —N$_3$ stretching vibrations, azides	s	2160–2120	Sulfites	s	1230–1150
	and w	1340–1180		and s	1430–1350
f. C—NO$_2$, nitro compounds:			Sulfonyl chlorides	s	1185–1165
Aromatic	s	1570–1500		and s	1370–1340
	and s	1370–1300	Sulfonamides	s	1180–1140
Aliphatic	s	1570–1550		and s	1350–1300
	and s	1380–1370	Sulfonic acids	s	1210–1150
g. O—NO$_2$, nitrates	s	1650–1600		s	1060–1030
	and s	1300–1250		and s	~650
h. C—NO, nitroso compound	s	1600–1500	Thioesters (C=O)S	s	~1690

[a] Abbreviations: s = strong, m = medium, w = weak, v = variable, b = broad, sh = sharp, ~ = approximately

---

Appendix 3
The Woodward–Fieser Rules for Predicting Ultraviolet–Visible Spectra

To use UV–visible spectroscopy for structure determination, we need to know what types of spectra correspond to the most common types of conjugated systems. The most useful correlations between structures and UV spectra were developed in the early 1940s by R. B. Woodward and L. F. Fieser. These correlations are called the **Woodward–Fieser rules**. The rules presented here predict only the lowest-energy $\pi \rightarrow \pi^*$ transition from the HOMO to the LUMO. Values of λ_{max} measured in different solvents can be different, so we generally assume that ethanol is the solvent.

In discussing these rules, we use the following specialized terms:

CHROMOPHORE: Any functional group (or collection of groups) responsible for absorption.

AUXOCHROME: A substituent that is not a chromophore by itself, but which alters the wavelength or the molar absorbtivity when it is attached to a chromophore.

BATHOCHROMIC SHIFT: A shift toward lower frequency and longer wavelength (longer λ_{max}).

HYPSOCHROMIC SHIFT: A shift toward higher frequency and shorter wavelength (shorter λ_{max}).

UV Spectra of Dienes and Polyenes

Bathochromic Effects of Alkyl Groups A molecule's system of conjugated double bonds (the chromophore) is the largest factor in determining its UV spectrum, but the absorption is also affected by alkyl substituents. Each alkyl group attached to the chromophore serves as an auxochrome, producing a small bathochromic shift of about 5 nm. Table A3-1 shows the effects of adding alkyl groups to 1,3-butadiene.

APPENDIX A3-1 Values of λ_{max} for Some Substituted 1,3,Butadienes

Number of Alkyl Groups	Compound	λ_{max} *(nm)*
0	$H_2C{=}CH{-}CH{=}CH_2$	217
1	$CH_3{-}CH{=}CH{-}CH{=}CH_2$	224
1	$H_2C{=}\overset{\overset{\displaystyle CH_3}{\mid}}{C}{-}CH{=}CH_2$	220
2	$H_2C{=}\overset{\overset{\displaystyle H_3C}{\mid}}{C}{-}\overset{\overset{\displaystyle CH_3}{\mid}}{C}{=}CH_2$	226
2	$CH_3{-}CH{=}CH{-}CH{=}CH{-}CH_3$	227
3	$CH_3{-}\overset{\overset{\displaystyle CH_3}{\mid}}{C}{=}CH{-}\overset{\overset{\displaystyle CH_3}{\mid}}{C}{=}CH_2$	232
4	$CH_3{-}\overset{\overset{\displaystyle CH_3}{\mid}}{C}{=}CH{-}\overset{\overset{\displaystyle CH_3}{\mid}}{C}{=}CH{-}CH_3$	241

Conformation Effects

For dienes that are predominantly in the *s*-trans conformation (either free to rotate or held in the *s*-trans conformation), Woodward and Fieser used a base value of 217 nm, the λ_{max} for unsubstituted 1,3-butadiene. To this value, add 5 nm for each alkyl substituent. For dienes that are held in the *s*-cis conformation by a six-membered ring, the base value is 253 nm for the diene, plus 5 nm for each alkyl substituent.

acyclic (*s*-trans) diene
base 217 nm

transoid cyclic diene
base 217 nm
+ 2 alkyl × (5 nm)

cisoid cyclic diene
base 253 nm
+ 2 alkyl × (5 nm)

Additional Conjugated Double Bonds

For trienes and larger conjugated systems, add 30 nm to the base value for each additional double bond. The additional double bond must be attached at the end of the conjugated system to extend the length of the polyene system in order to have this large 30-nm contribution, however.

acyclic (*s*-trans) triene
217 nm + 30 nm = base 247 nm

cisoid cyclic triene
253 nm + 30 nm = base 283 nm
+ 2 alkyl × (5 nm)

Contributions of auxochromic groups are added to the base values of the polyene chromophore. Add 5 nm for each alkyl group and 5 nm if one of the double bonds in the conjugated system is exocyclic to a ring. An exocyclic double bond is one that is attached to a ring at one end.

exocyclic double bonds

TABLE A3-2 The Woodward–Fieser Rules for Conjugated Dienes: Values for Auxochromic Groups

Grouping	Substituent Correction (nm)
another conjugated C=C	+30
alkyl group	+5
alkoxy (—OR) group	0

If one of the double bonds in the chromophore is exocyclic, add another 5 nm:

+5 (in addition to 30 nm if it lengthens the system)

exocyclic double bond

Note: These values are added to the base value for the diene system.

Table A3-2 summarizes the shifts associated with common auxochromic groups.

Examples The best way to learn to use the rules for predicting UV absorptions is to work through some examples. The following examples show several structures that follow the rules closely and one that does not.

1.

2,4-dimethyl-1,3-pentadiene

base:	217 nm
three alkyl groups:	15
predicted λ_{max}:	232 nm;

observed: 232 nm

2.

base:	217 nm
two alkyl groups:	10
exocyclic C=C:	5
predicted λ_{max}:	232 nm; observed: 230 nm

3.

base:	217 nm
two alkyl groups:	10
exocyclic C=C:	5
predicted λ_{max}:	232 nm; observed: 236 nm

4.

base:	217 nm
three alkyl groups:	15
exocyclic C=C:	5
predicted λ_{max}:	237 nm; observed: 235 nm

5.

base:	253 nm
conjugated C=C:	30
three alkyl groups:	15
exocyclic C=C:	5
predicted λ_{max}:	303 nm; observed: 304 nm

6.

1,3,5-hexatriene

base:	217 nm
conjugated C=C:	30
predicted λ_{max}:	247 nm; observed: 258 nm

TABLE A3-3 The Woodward–Fieser Rules for Conjugated Ketones and Aldehydes

$$\beta \diagdown \atop \beta \diagup C=C \diagup \atop \alpha \diagdown C \diagup R \atop \| \atop O$$

Base values: 210 nm if R = H (aldehyde)
215 nm if R = alkyl (ketone)

general structure

Grouping	Position	Correction
alkyl group, α		+10 nm
alkyl group, β		+12 nm
exocyclic position of a C=C bond		+5 nm
additional conjugated double bond		+30 nm

UV Spectra of Conjugated Ketones and Aldehydes

$\pi \rightarrow \pi^*$ **Transitions** As with dienes and polyenes, the strongest absorptions in the UV spectra of aldehydes and ketones result from $\pi \rightarrow \pi^*$ electronic transitions. These absorptions are observable ($\lambda_{max} > 200$ nm) only if the carbonyl double bond is conjugated with another double bond.

The Woodward–Fieser rules for conjugated ketones and aldehydes appear in Table A3-3. Note that bathochromic effects of alkyl groups depend on their location: 10 nm for groups α to the carbonyl and 12 nm for groups in β positions. Contributions from additional conjugated double bonds (30 nm) and exocyclic positions of double bonds (5 nm) are similar to those in dienes and polyenes.

The following examples show how the Woodward–Fieser rules predict values of λ_{max} for a variety of conjugated ketones and aldehydes. Notice that the molar absorptivities (ε) for these transitions are quite large (> 5000), as we also observed for $\pi \rightarrow \pi^*$ transitions in conjugated dienes and polyenes.

1.

Base value	210 nm
(no corrections)	
Predicted λ_{max}	210 nm

Experimental: $\lambda_{max} = 210$ nm, $\varepsilon = 11,000$

2.

Base value	215 nm
$2 \times \beta$ substituent	24 nm
Predicted λ_{max}	239 nm

Experimental: $\lambda_{max} = 237$ nm, $\varepsilon = 12,000$

3.

Base value	215 nm
α substituent	10 nm
β substituent	12 nm
Predicted λ_{max}	237 nm

Experimental: $\lambda_{max} = 233$ nm, $\varepsilon = 12,500$

4.

exocyclic
double bond

Base value	215 nm
α substituent	10 nm
β substituent	12 nm
Exocyclic double bond	5 nm
Predicted λ_{max}	242 nm

Experimental: $\lambda_{max} = 241$ nm, $\varepsilon = 5200$

$\pi \rightarrow \pi^*$ Transitions As discussed in Section 18-5E, ketones and aldehydes also show weak UV absorptions ($\varepsilon \cong 10$ to 200) from "forbidden" $n \rightarrow \pi^*$ transitions. Because the promoted electron leaves a nonbonding (n) orbital that is higher in energy than the pi bonding orbital, this transition involves a smaller amount of energy and results in a longer-wavelength (lower-frequency) absorption. $n \rightarrow \pi^*$ transitions of simple, unconjugated ketones and aldehydes give absorptions with values of λ_{max} between 280 and 300 nm. Each double bond added in conjugation with the carbonyl group increases the value of λ_{max} by about 30 nm.

In this appendix, we consider how an organic chemist systematically approaches a mechanism problem. Although there is no "formula" for solving all mechanism problems, this stepwise method should provide a starting point for you to begin building experience and confidence. Solved problems that apply this approach appear on pages 151, 303, 468, 815, 965, 1022, and 1047.

Appendix 4A
Methods and Suggestions for Proposing Mechanisms

Determining the Type of Mechanism

First, determine what conditions or catalysts are involved. In general, reactions may be classified as (a) involving strong electrophiles (includes acid-catalyzed reactions), (b) involving strong nucleophiles (includes base-catalyzed reactions), or (c) involving free radicals. These three types of mechanisms are quite distinct, and you should first try to determine which type is involved. If uncertain, you can develop more than one type of mechanism and see which one fits the facts better.

(a) In the presence of a strong acid or a reactant that can give a strong electrophile, the mechanism probably involves strong electrophiles as intermediates. Acid-catalyzed reactions and reactions involving carbocations (such as the S_N1, E1, and most alcohol dehydrations) generally fall in this category.

(b) In the presence of a strong base or strong nucleophile, the mechanism probably involves strong nucleophiles as intermediates. Base-catalyzed reactions and those whose rates depend on base strength (such as S_N2 and E2) generally fall in this category.

(c) Free-radical reactions usually require a free-radical initiator such as chlorine, bromine, NBS, AIBN, or a peroxide. In most free-radical reactions, there is no need for a strong acid or base.

Points to Watch in All Mechanisms

Once you have determined which type of mechanism to write, use the general methods discussed here to approach the problem. Regardless of the type of mechanism, however, you should follow three general rules in proposing a mechanism:

1. **Draw all bonds and all substituents of each carbon atom affected throughout the mechanism. Do not use condensed or line–angle formulas for reaction sites.** Three-bonded carbon atoms are most likely reactive intermediates: carbocations in reactions involving strong electrophiles, carbanions in reactions involving strong nucleophiles, and free radicals in radical reactions. If you draw condensed formulas or line–angle formulas, you might misplace a hydrogen atom and show a reactive species on the wrong carbon.

2. **Show only one step at a time. Do not show two or three bonds changing position in one step unless the changes really are concerted (take place simultaneously).** For example, three pairs of electrons really do move in one step in the Diels–Alder reaction; but in the dehydration of an alcohol, protonation of the hydroxyl group and loss of water are two separate steps.

3. **Use curved arrows to show *movement of electrons*, always from the nucleophile (electron donor) to the electrophile (electron acceptor).** For example, a proton has no electrons to donate, so a curved arrow should never be drawn from H^+ to anything. When an alkene is protonated, the arrow should go from the electrons of the double bond to the proton. Don't try to use curved arrows to "point out" where the proton (or other reagent) goes. In a free-radical reaction, half-headed arrows show single electrons coming together to form bonds or separating to give other radicals.

Approaches to Specific Types of Mechanisms

Reactions Involving Strong Electrophiles   General principles: When a strong acid or electrophile is present, expect intermediates that are strong acids and strong electrophiles; cationic intermediates are common. Carbocations, protonated (three-bonded) oxygen atoms, protonated (four-bonded) nitrogen atoms, and other strong acids might be involved. Any bases and nucleophiles in such a reaction are generally weak. Avoid drawing carbanions, alkoxide ions, and other strong bases. They are unlikely to coexist with strong acids and strong electrophiles.

Functional groups are often converted to carbocations or other strong electrophiles by protonation or by reaction with a strong electrophile, then the carbocation or other strong electrophile reacts with a weak nucleophile such as an alkene or the solvent.

1. Consider the carbon skeletons of the reactants and products, and identify which carbon atoms in the products are most likely derived from which carbon atoms in the reactants.

2. Consider whether any of the reactants is a sufficiently strong electrophile to react without being activated. If not, consider how one of the reactants might be converted to a strong electrophile by protonation of a Lewis basic site, complexation with a Lewis acid, or ionization.

3. Consider how a nucleophilic site on another reactant (or, in a cyclization, another part of the same molecule) can attack this strong electrophile to form a bond needed in the product. Draw the product of this bond formation.

   If the intermediate is a carbocation, consider whether it is likely to rearrange to form a bond in the product.

   If there is no possible nucleophilic attack that leads in the direction of the product, consider other ways of converting one of the reactants to a strong electrophile.

4. Consider how the product of nucleophilie attack might be converted to the final product (if it has the right carbon skeleton) or reactivated to form another bond needed in the product.

5. Draw out all the steps using curved arrows to show movement of electrons. Be careful to show only one step at a time.

Reactions Involving Strong Nucleophiles   General principles: When a strong base or nucleophile is present, expect intermediates that are strong bases and strong nucleophiles; anionic intermediates are common. Alkoxide ions, stabilized carbanions, and other strong bases might be involved. Any acids and electrophiles in such a reaction are generally weak. Avoid drawing carbocations, protonated carbonyl groups, protonated hydroxyl groups, and other strong acids. They are unlikely to coexist with strong bases and strong nucleophiles.

Functional groups are often converted to strong nucleophiles by deprotonation of the group itself; by deprotonation of the alpha position of a carbonyl group, nitro group, or nitrile; or by attack of another strong nucleophile. Then the resulting carbanion or other nucleophile reacts with a weak electrophile such as a carbonyl group, an alkyl halide, or the double bond of a Michael acceptor.

1. Consider the carbon skeletons of the reactants and products, and identify which carbon atoms in the products are most likely derived from which carbon atoms in the reactants.

2. Consider whether any of the reactants is a sufficiently strong nucleophile to react without being activated. If not, consider how one of the reactants might be converted to a strong nucleophile by deprotonation of an acidic site or by attack on an electrophilic site.

3. Consider how an electrophilic site on another reactant (or, in a cyclization, another part of the same molecule) can undergo attack by the strong nucleophile to form a bond needed in the product. Draw the product of this bond formation.

    If no appropriate electrophilic site can be found, consider another way of converting one of the reactants to a strong nucleophile.

4. Consider how the product of nucleophilic attack might be converted to the final product (if it has the right carbon skeleton) or reactivated to form another bond needed in the product.

5. Draw out all the steps, using curved arrows to show movement of electrons. Be careful to show only one step at a time.

Reactions Involving Free Radicals   General principles: Free-radical reactions generally proceed by chain-reaction mechanisms, using an initiator with an easily broken bond (such as chlorine, bromine, or a peroxide) to start the chain reaction. In drawing the mechanism, expect free-radical intermediates (especially highly substituted or resonance-stabilized intermediates). Cationic intermediates and anionic intermediates are not usually involved. Watch for the most stable free radicals, and avoid high-energy radicals such as hydrogen atoms.

Initiation

1. Draw a step involving homolytic (free-radical) cleavage of the weak bond in the initiator to give two radicals.

2. Draw a reaction of the initiator with one of the starting materials to give a free-radical version of the starting material.

    The initiator might abstract a hydrogen atom or add to a double bond, depending on what reaction leads toward the observed product. You might want to consider bond-dissociation energies to see which reaction is energetically favored.

Propagation

1. Draw a reaction of the free-radical version of the starting material with another starting material molecule to form a bond needed in the product and generate a new radical intermediate. Two or more propagation steps may be needed to give the entire chain reaction.

Termination

1. Draw termination steps showing the recombination or destruction of radicals. Termination steps are side reactions rather than part of the product-forming mechanism. Reaction of any two free radicals to give a stable molecule is a termination step, as is a collision of a free radical with the container.

Appendix 4B Suggestions for Developing Multistep Syntheses

In this appendix, we consider how an organic chemist systematically approaches a multistep synthesis problem. As with mechanism problems, there is no reliable "formula" that can be used to solve all synthesis problems, yet students need guidance in how they should begin.

In a multistep synthesis problem, the solution is rarely immediately apparent. A synthesis is best developed systematically, working backward (in the *retrosynthetic* direction) and considering alternative ways of solving each stage of the synthesis. A strict retrosynthetic approach requires considering all possibilities for the final step, evaluating each reaction, and then evaluating every way of making each of the possible precursors.

This exhaustive approach is very time-consuming. It works well on a large computer, but most organic chemists solve problems more directly by attacking the crux of the problem: steps that build the carbon skeleton. Once the carbon skeleton is assembled (with usable functionality), converting the functional groups to those required in the target molecule is relatively easy.

The following steps suggest a systematic approach to developing a multistep synthesis. These steps should help you organize your thoughts and approach syntheses like many organic chemists do: in a generally retrosynthetic direction, but with primary emphasis on the crucial steps that form the carbon skeleton of the target molecule. Solved problems that apply this approach appear on pages 357, 396, and 479.

1. Review the functional groups and carbon skeleton of the target compound, considering what kinds of reactions might be used to create them.

2. Review the functional groups and carbon skeletons of the starting materials (if specified), and see how their skeletons might fit together into the skeleton of the target compound.

3. Compare methods for assembling the carbon skeleton of the target compound. Which ones produce a key intermediate with the correct carbon skeleton and functional groups correctly positioned for conversion to the functionality in the target molecule?

   Also notice what functional groups are required in the reactants for the skeleton-forming steps and whether they are easily accessible from the specified starting materials.

4. Write down the steps involved in assembling the key intermediate with the correct carbon skeleton.

5. Compare methods for converting the key intermediate's functional groups to those in the target compound, and select reactions that are likely to give the correct product. Reactive functional groups are often added late in a synthesis, to prevent them from interfering with earlier steps.

6. Working backward through as many steps as necessary, compare methods for synthesizing the reactants needed for assembly of the key intermediate. (This process may require writing several possible reaction sequences and evaluating them, keeping in mind the specified starting materials.)

7. Summarize the complete synthesis in the forward direction, including all steps and all reagents, and check it for errors and omissions.

APPENDIX 5 pK_a Values for Representative Compounds

Compound	pK_a	Compound	pK_a	Compound	pK_a
$CH_3C{\equiv}\overset{+}{N}H$	−10.1	$O_2N{-}C_6H_4{-}\overset{+}{N}H_3$	1.0	$CH_3{-}C_6H_4{-}COH$ (O)	4.3
HI	−10	pyrimidinium	1.0	$CH_3O{-}C_6H_4{-}COH$ (O)	4.5
HBr	−9	Cl_2CHCOH (O)	1.3	$C_6H_5{-}\overset{+}{N}H_3$	4.6
CH_3CH (=$\overset{+}{O}H$)	−8	HSO_4^-	2.0	CH_3COH (O)	4.8
CH_3CCH_3 (=$\overset{+}{O}H$)	−7.3	H_3PO_4	2.1	quinolinium	4.9
HCl	−7	purine (protonated)	2.5	$CH_3{-}C_6H_4{-}\overset{+}{N}H_3$	5.1
$CH_3\overset{+}{S}H_2$	−6.8	FCH_2COH (O)	2.7	pyridinium	5.2
CH_3COCH_3 (=$\overset{+}{O}H$)	−6.5	$ClCH_2COH$ (O)	2.8	$CH_3O{-}C_6H_4{-}\overset{+}{N}H_3$	5.3
CH_3COH (=$\overset{+}{O}H$)	−6.1	$BrCH_2COH$ (O)	2.9	$CH_3C{=}\overset{+}{N}HCH_3$ (|CH_3)	5.5
H_2SO_4	−5	ICH_2COH (O)	3.2	CH_3CCH_2CH (O O)	5.9
pyrrolium	−3.8	HF	3.2	$HO\overset{+}{N}H_3$	6.0
$CH_3CH_2\overset{+}{O}CH_2CH_3$ (H)	−3.6	HNO_2	3.4	H_2CO_3	6.4
$CH_3CH_2\overset{+}{O}H$ (H)	−2.4	$O_2N{-}C_6H_4{-}COH$ (O)	3.4	imidazolium	6.8
$CH_3\overset{+}{O}H$ (H)	−2.5	$HCOH$ (O)	3.8	H_2S	7.0
H_3O^+	−1.7	$Br{-}C_6H_4{-}\overset{+}{N}H_3$	3.9	$O_2N{-}C_6H_4{-}OH$	7.1
HNO_3	−1.3	$Br{-}C_6H_4{-}COH$ (O)	4.0	$H_2PO_4^-$	7.2
CH_3SO_3H	−1.2	$C_6H_4(pyridyl){-}COH$ (O)	4.2	$C_6H_5{-}SH$	7.8
$C_6H_5{-}SO_3H$	−0.60				
CH_3CNH_2 (=$\overset{+}{O}H$)	0.0				
F_3CCOH (O)	0.2				
Cl_3CCOH (O)	0.64				
pyridine $\overset{+}{N}{-}OH$	0.79				

APPENDIX 5 pK$_a$ Values for Representative Compounds

Compound	pK$_a$	Compound	pK$_a$	Compound	pK$_a$
aziridinium (N$^+$H$_2$ ring)	8.0	cyclohexyl–$\overset{+}{N}H_3$	10.7	CH$_3$CH=O	17
H$_2$N$\overset{+}{N}H_3$	8.1	(CH$_3$)$_2\overset{+}{N}H_2$	10.7	(CH$_3$)$_3$COH	18
CH$_3$COOH	8.2	CH$_3$CH$_2\overset{+}{N}H_3$	10.7	CH$_3$CCH$_3$ (=O)	20
CH$_3$CH$_2$NO$_2$	8.6	piperidinium	11.1	CH$_3$COEt (=O)	24.5
CH$_3$CCH$_2$CCH$_3$ (2 =O)	8.9	pyrrolidinium	11.3	HC≡CH	25
purine	8.9	HPO$_4^{2-}$	12.3	CH$_3$C≡N	25
HC≡N	9.1	CF$_3$CH$_2$OH	12.4	CH$_3$CN(CH$_3$)$_2$ (=O)	30
morpholinium	9.3	EtOCCH$_2$COEt (2 =O)	13.3	NH$_3$	36
Cl–C$_6$H$_4$–OH	9.4	HC≡CCH$_2$OH	13.5	pyrrolidine (N–H)	36
$\overset{+}{N}H_4$	9.4	H$_2$NCNH$_2$ (=O)	13.7	CH$_3$NH$_2$	40
HOCH$_2$CH$_2\overset{+}{N}H_3$	9.5	CH$_3\overset{+}{N}(CH_3)CH_2CH_2OH, CH$_3$	13.9	C$_6$H$_5$–CH$_3$	41
H$_3\overset{+}{N}$CH$_2$CO$^-$	9.8	imidazole	14.4	benzene	43
C$_6$H$_5$–OH	10.0	CH$_3$OH	15.5	CH$_2$=CHCH$_3$	43
CH$_3$–C$_6$H$_4$–OH	10.2	H$_2$O	15.7	CH$_2$=CH$_2$	44
HCO$_3^-$	10.2	CH$_3$CH$_2$OH	16.0	cyclopropane	46
CH$_3$NO$_2$	10.2	CH$_3$CNH$_2$ (=O)	16	CH$_4$	50
H$_2$N–C$_6$H$_4$–OH	10.3	C$_6$H$_5$CCH$_3$ (=O)	16.0	CH$_3$CH$_3$	50
CH$_3$CH$_2$SH	10.5	pyrrole	~17		
(CH$_3$)$_3\overset{+}{N}$H	10.6				
CH$_3$CCH$_2$COEt (2 =O)	10.7				
CH$_3\overset{+}{N}H_3$	10.7				

ANSWERS TO SELECTED PROBLEMS

These short answers are sometimes incomplete, but they should put you on the right track. Complete answers to all problems are found in the *Solutions Manual*.

CHAPTER 1

1.5. (a) $\overset{\longleftrightarrow}{C-Cl}$; (b) $\overset{\longleftrightarrow}{C-O}$; (c) $\overset{\longleftrightarrow}{C-N}$; (d) $\overset{\longleftrightarrow}{C-S}$; (e) $\overset{\longleftrightarrow}{C-B}$; (f) $\overset{\longleftrightarrow}{N-Cl}$; (g) $\overset{\longleftrightarrow}{N-O}$; (h) $\overset{\longleftrightarrow}{N-S}$; (i) $\overset{\longleftrightarrow}{N-B}$; (j) $\overset{\longleftrightarrow}{B-Cl}$.
1.6. (a) +1 on O; (b) +1 on N, −1 on Cl; (c) +1 on N, −1 on Cl; (d) +1 on Na, −1 on O; (e) +1 on C; (f) −1 on C; (g) +1 on Na, −1 on B; (h) +1 on Na, −1 on B; (i) +1 on O, −1 on B; (j) +1 on N; (k) +1 on K, −1 on O; (l) +1 on O. **1.12.** (a) CH_2O, $C_3H_6O_3$; (b) $C_2H_5NO_2$, same; (c) C_2H_5Cl, same; (d) C_2H_3Cl, $C_4H_6Cl_2$. **1.13.** (a) 0.209; (b) 13.875 **1.15.** (a) favors products; (b) favors reactants; (c) favors products; (d) favors products; (e) favors products; (f) favors products. **1.16.** There is no resonance stabilization of the positive charge when the other oxygen atom is protonated. **1.17.** (a) acetic acid, ethanol, methylamine; (b) ethoxide, methylamine, ethanol. **1.21.** (a) carbon; (b) oxygen; (c) phosphorus; (d) chlorine. **1.28.** The following are condensed structures that you should convert to Lewis structures. (a) $CH_3CH_2CH_2CH_3$ and $CH_3CH(CH_3)_2$; (b) $CH_3CH_2NH_2$ and CH_3NHCH_3; (c) $CH_2(CH_2OH)_2$ and $CH_3CHOHCH_2OH$ and $CH_3OCH_2OCH_3$ and others; (d) $CH_2=CHOH$ and CH_3CHO. **1.32.** (a) C_5H_5N; (b) C_4H_9N; (c) C_4H_9NO; (d) $C_4H_9NO_2$; (e) $C_{11}H_{21}NO$; (f) $C_9H_{18}O$; (g) $C_7H_8SO_3$; (h) $C_6H_6O_3$. **1.33.** Empirical formula C_3H_6O; molecular formula $C_6H_{12}O_2$. **1.36.** (a) different compounds; (b) resonance forms; (c) resonance forms; (d) resonance forms; (e) different compounds; (f) resonance forms; (g) resonance forms; (h) different compounds; (i) resonance forms; (j) resonance forms. **1.39.** (b) The $=NH$ nitrogen atom is the most basic. **1.41.** (a) second; (b) first; (c) second; (d) first; (e) first. **1.47.** (a) $CH_3CH_2O^- Li^+ + CH_4$; (b) Methane; CH_3Li is a very strong base. **1.48.** (a) $C_9H_{12}O$; (b) $C_{18}H_{24}O_2$.

CHAPTER 2

2.2. sp^3; Two lone pairs compress the bond angle to 104.5°.
2.4. Methyl carbon; sp^3, about 109.5°. Nitrile carbon sp. 180°. Nitrile nitrogen sp, no bond angle. **2.6.** The central carbon is sp, with two unhybridized p orbitals at right angles. Each terminal $=CH_2$ group must be aligned with one of these p orbitals.
2.8. $CH_3-CH=N-CH_3$ shows cis-trans isomerism about the $C=N$ double bond, but $(CH_3)_2C=N-CH_3$ has two identical substituents on the $C=N$ carbon atom, and there are no cis-trans isomers. **2.10.** (a) cis-trans isomers; (b) constitutional isomers; (c) constitutional isomers; (d) same compound; (e) same compound; (f) same compound; (g) not isomers; (h) constitutional isomers; (i) same compound; (j) constitutional isomers; (k) constitutional isomers. **2.13.** The N—F dipole moments oppose the dipole moment of the lone pair. **2.15.** *trans* has zero dipole moment because the bond dipole moments cancel. **2.18.** (a) $CH_3CH_2OCH_2CH_3$ (b) $CH_3CH_2NHCH_3$; (c) CH_3CH_2OH; (d) CH_3COCH_3. **2.19.** (a) alkane; (b) alkene; (c) alkyne; (d) cycloalkyne; (e) cycloalkane; (f) aromatic hydrocarbon and alkene; (g) cycloalkene; (h) alkyne, alkene; (i) aromatic hydrocarbon and cycloalkene. **2.20.** (a) aldehyde; (b) alcohol; (c) ketone; (d) ether; (e) carboxylic acid; (f) ether; (g) ketone; (h) aldehyde; (i) alcohol. **2.21.** (a) amide; (b) amine; (c) ester; (d) acid chloride; (e) ether; (f) nitrile; (g) carboxylic acid; (h) cyclic ester; (i) ketone, cyclic ether; (j) cyclic amine; (k) cyclic amide; (l) amide; (m) ketone, amine; (n) cyclic ester; (o) nitrile; (p) ketone. **2.25.** No stereoisomers.

2.26. Cyclopropane has bond angles of 60°, compared with the 109.5° bond angle of an unstrained alkane. **2.29.** Formamide must have an sp^2-hybridized nitrogen atom because it is involved in pi-bonding in the other resonance form. **2.34.** Only (b) and (e). **2.35.** (a) constitutional isomers; (b) constitutional isomers; (c) cis-trans isomers; (d) constitutional isomers; (e) cis-trans isomers; (f) same compound; (g) cis-trans isomers; (h) constitutional isomers. **2.36.** CO_2 is sp-hybridized and linear; the bond dipole moments cancel. The sulfur atom in SO_2 is sp^2-hybridized and bent; the bond dipole moments do not cancel. **2.38.** Both can form H-bonds with water, but only the alcohol can form H-bonds with itself. **2.40.** (a), (c), (h), and (l) can form hydrogen bonds in the pure state. These four plus (b), (d), (g), (i), (j), and (k) can form hydrogen bonds with water. **2.42.** (a) cyclic ether; (b) cyclic alkene, carboxylic acid; (c) alkene, aldehyde; (d) aromatic, ketone; (e) alkene, cyclic ester; (f) cyclic amide; (g) aromatic nitrile, ether; (h) amine, ester.

CHAPTER 3

3.1. (a) $C_{25}H_{52}$; (b) $C_{44}H_{90}$; **3.2.** (a) 3-methylpentane; (b) 2-bromo-3-methylpentane; (c) 5-ethyl-2-methyl-4-propylheptane; (d) 4-isopropyl-2-methyldecane. **3.4.** (a) 2-methylbutane; (b) 2,2-dimethylpropane; (c) 3-ethyl-2-methylhexane; (d) 2,4-dimethylhexane; (e) 3-ethyl-2,2,4,5-tetramethylhexane; (f) 4-*t*-butyl-3-methylheptane. **3.8.** (a) $C_{10}H_{22}$; (b) $C_{15}H_{32}$. **3.9.** (a) hexane $<$ octane $<$ decane; (b) $(CH_3)_3C-C(CH_3)_3 < CH_3CH_2C(CH_3)_2CH_2CH_2CH_3 <$ octane. **3.14.** (a) 1,1-dimethyl-3-(1-methylpropyl) cyclopentane or 3-*sec*-butyl-1,1-dimethylcyclopentane; (b) 3-cyclo-propyl-1,1-dimethylcyclohexane; (c) 4-cyclobutylnonane. **3.16.** (b), (c), and (d). **3.17.** (a) *cis*-1-methyl-3-propylcyclobutane; (b) *trans*-1-*t*-butyl-3-ethylcyclohexane; (c) *trans*-1,2-dimethylcyclopropane. **3.18.** Trans is more stable. In the cis isomer the methyl groups are nearly eclipsed. **3.28.** (a) *cis*-1,3-dimethylcyclohexane, (b) *cis*-1,4-dimethylcyclohexane; (c) *trans*-1,2-dimethylcyclohexane; (d) *cis*-1,3-dimethylcyclohexane; (e) *cis*-1,3-dimethylcyclohexane; (f) *trans*-1,4-dimethylcyclohexane. **3.30.** (a) bicyclo[3.1.0]hexane; (b) bicyclo[3.3.1] nonane; (c) bicyclo[2.2.2]octane; (d) bicyclo[3.1.1]heptane. **3.33.** (a) All except the third (isobutane) are *n*-butane. (b) Top left and bottom left are *cis*-2-butene. Top center and bottom center are 1-butene. Top right is *trans*-2-butene. Lower right is 2-methylpropene. (c) The first and second are *cis*-1,2-dimethylcyclopentane. The third and fourth are *trans*-1,2-dimethylcyclopentane. The fifth is *cis*-1,3-dimethylcyclopentane. **3.37.** (a) 3-ethyl-2,2,6-trimethylheptane; (b) 3-ethyl-2,6,7-trimethyloctane; (c) 3,7-diethyl-2,2,8-trimethyldecane; (d) 2-ethyl-1,1-dimethylcyclobutane; (e) bicyclo[4.1.0]heptane; (f) *cis*-1-ethyl-3-propylcyclopentane; (g) (1,1-diethylpropyl)cyclohexane; (h) *cis*-1-ethyl-4-isopropylcyclodecane. **3.39.** (a) should be 3-methylhexane; (b) 3-ethyl-2-methyl-hexane; (c) 2-chloro-3-methylhexane; (d) 2,2-dimethylbutane; (e) *sec*-butylcyclohexane or (1 methylpropyl) cyclohexane; (f) should be *cis* or *trans*-1,2-diethylcyclopentane. **3.40.** (a) octane; (b) 2-methylnonane; (c) nonane. **3.45.** The trans isomer is more stable, because both of the bonds to the second cyclohexane ring are in equatorial positions.

CHAPTER 4

4.3. (a) One photon of light would be needed for every molecule of product formed (the quantum yield would be 1); (b) Methane does not absorb the visible light that initiates the reaction, and the quantum yield would be 1. **4.4.** (a) Hexane has three different kinds of hydrogen atoms, but

cyclohexane has only one type. (b) Large excess of cyclohexane.
4.5. (a) K_{eq} = 2.3; (b) [CH_3Br] = [H_2S] = 0.40 M,
[CH_3SH] = [HBr] = 0.60 M. **4.8.** (a) positive; (b) negative; (c) not
easy to predict. **4.10.** (a) initiation +46 kcal/mol; propagation +16
and −24 kcal/mol; (b) overall −8 kcal/mol. **4.11.** (a) first order;
(b) zeroth order; (c) first order overall. **4.13.** (a) zero, zero, zeroth
order overall; (b) rate = k_r; (c) increase the surface area of the
platinum catalyst. **4.14.** (b) +3 kcal/mol; (c) −1 kcal/mol.
4.15. (c) +27 kcal/mol. **4.17.** (a) initiation +36 kcal/mol; propagation +33 and −20 kcal/mol; (b) overall +13 kcal/mol; (c) low rate and
very unfavorable equilibrium constant. **4.18.** 1°:2° ratio of 6:2, product
ratio of 75% 1° and 25% 2°. **4.22.** (a) The combustion of isooctane
involves highly branched, more stable tertiary free radicals that react less
explosively. (b) *t*-butyl alcohol forms relatively stable alkoxy radicals that
react less explosively. **4.29.** Stability: (c) 3° > (b) 2° > (a) 1°.
4.30. Stability: (c) 3° > (b) 2° > (a) 1°.
4.38. rate = k_r[H^+][(CH_3)$_3C$—OH]; second order overall.
4.41. $PhCH_2$· > CH_2=$CHCH_2$· > (CH_3)$_3C$· > (CH_3)$_2CH$· >
CH_3CH_2· > CH_3.

CHAPTER 5

5.1. chiral: spring, desk, screw-cap bottle, rifle, knot. **5.2.** (b), (d), (e),
and (f) are chiral. **5.3.** (a) achiral, no C*; (b) achiral, no C*; (c) chiral,
one C*; (d) achiral, no C*; (e) achiral, no C*; (f) chiral, one C*;
(g) achiral, two C*; (h) chiral, two C*; (i) achiral, no C*; (j) chiral, one
C*; (k) chiral, two C*. **5.5.** (a) mirror, achiral; (b) mirror, achiral;
(c) chiral, no mirror; (d) chiral, no mirror; (e) chiral, no mirror; (f) mirror, achiral; (g) mirror, achiral; (h) chiral, no mirror. **5.6.** (a) (*R*);
(b) (*S*); (c) (*R*); (d) (*S*), (*S*); (e) (*R*), (*S*); (f) (*R*), (*S*); (g) (*R*), (*S*); (h) (*R*);
(i) (*S*). **5.8.** +8.7°. **5.10.** Dilute the sample. If clockwise, will make
less clockwise, and vice-versa. **5.12.** e.e. = 33.3%. Specific
rotation = 33.3% of +13.5° = +4.5°. **5.15.** (a), (b), (f), and (h) are
chiral; only (h) has chiral carbons. **5.16.** (a) enantiomer, enantiomer,
same; (b) same, enantiomer, enantiomer; (c) enantiomer, same, same.
5.18. (a), (d), and (f) are chiral. The others have internal mirror planes.
5.19. (from 5–17) (a) (*R*); (b) none; (c) none; (d) (2*R*), (3*R*); (e) (2*S*),
(3*R*); (f) (2*R*), (3*R*); (new ones) (g) (*R*); (h) (*S*); (i) (*S*). **5.20.** (a) enantiomers; (b) diastereomers; (c) diastereomers; (d) constitutional iso-
mers; (e) enantiomers; (f) diastereomers; (g) enantiomers; (h) same
compound; (i) diastereomers. **5.23.** (a), (b), and (d) are pairs of
diastereomers and could theoretically be separated by their physical
properties. **5.30.** (a) same compound; (b) enantiomers; (c) enantiomers; (d) enantiomers; (e) enantiomers; (f) diastereomers; (g) enantiomers; (h) same compound. **5.32.** (a) −12.5°; (b) +8.6°.
5.34. (20%e.e.) × (+12.0°) = (+)2.4°. **5.34.** (b) (−)15.90°s;
(c) 7.95°/15.90° = 50%e.e. Composition is 75% (*R*) and 25% (*S*).

CHAPTER 6

6.1. (a) vinyl halide; (b) alkyl halide; (c) aryl halide; (d) alkyl halide;
(e) vinyl halide; (f) aryl halide. **6.5.** The C—CI bond has considerably more charge separation (0.23 *e*) than the C—I bond (0.16 *e*).
6.7. Water is denser than hexane, so water forms the lower layer. Chloroform is denser than water, so chloroform forms the lower layer.
6.11. (a) substitution; (b) elimination; (c) elimination, also a reduction.
6.13. 0.02 mo/L per second. **6.14.** (a) (CH_3)$_3COCH_2CH_3$;
(b) HC≡$CCH_2CH_2CH_3$; (c) (CH_3)$_2CHCH_2NH_2$;
(d) CH_3CH_2C≡N; (e) 1-iodopentane; (f) 1-fluoropentane.
6.16. (a) (CH_3CH_2)$_2NH$, less hindered; (b) (CH_3)$_2S$, S more polarizable; (c) PH_3, P more polarizable; (d) CH_3S^-, neg. charged;
(e) (CH_3)$_3N$, N less electronegative; (f) CH_3S^-, neg. charge, more
polarizable; (g) $CH_3CH_2CH_2O^-$, less hindered; (h) I^-, more polarizable. **6.18.** methyl iodide > methyl chloride > ethyl chloride > isopropyl bromide ≫ neopentyl bromide, *t*-butyl iodide.
6.19. (a) 2-methyl-1-iodopropane; (b) cyclohexyl bromide;

(c) isopropyl bromide; (d) 2-chlorobutane; (e) isopropyl iodide.
6.24. (a) 2-iodo-2-methylbutane; (b) 2-bromo-2-methylbutane;
(c) 3-bromocyclohexene; (d) cyclohexyl bromide.
6.27. (a) (CH_3)$_2C$(OCOCH$_3$)CH_2CH_3, first order;
(b) 1-methoxy-2-methylpropane, second order; (c) 1-ethoxy-1-methyl-
cyclohexane, first order; (d) methoxycyclohexane, first order;
(e) ethoxycyclohexane, second order. **6.33.** (a) H_2C=$CHCH_2CH_3$
and CH_3CH=$CHCH_3$; (b) CH_3CH=$C(CH_2CH_3)_2$;
(c) CH_3CH=$C(CH_2CH_3)_2$ and CH_2=$CHCH(CH_2CH_3)_2$;
(d) 1-methylcyclohexene and 3-methylcyclohexene. **6.35.** 2-butanol
by the S_N2. **6.43.** (a) 2-bromo-2-methylpentane; (b) 1 chloro-1-
methylcyclohexane; (c) 1,1-dichloro-3-fluorocloheptane;
(d) 4-(2-bromoethyl)-3-(fluoromethyl)-2-methylheptane;
(e) 4,4-dichloro-5-cyclopropyl-1-iodoheptane; (f) *cis*-1,2-dichloro-
1-methylcyclohexane. **6.44.** (a) 1-chlorobutane; (b) 1-iodobutane;
(c) 4-chloro-2,2-dimethylpentane; (d) 1-bromo-2,2-dimethylpentane;
(e) chloromethylcyclohexane; (f) 3-methyl-1-bromopropane.
6.45. (a) *t*-butyl chloride; (b) 2-chlorohexane; (c) bromocyclohexane;
(d) iodocyclohexane; (e) 2-bromo-2-methylpentane; (f) 3-bromocyclo-
hexene. **6.48.** (a) rate doubles; (b) rate multiplied by six;
(c) rate increases. **6.55.** (a) (*R*)-2-butanol (inversion);
(b) (*S*)-2-iodo-3-methylpentane (inversion); (c) racemic mixture of
3-ethoxy-2,3-dimethylpentanes (racemization). **6.56.** (a) diethyl ether;
(b) $PhCH_2CH_2CN$; (c) *c*-Hx—S—CH_3; (d) 1-iododecane;
(e) N-methylpyridinium iodide; (f) (CH_3)$_3CCH_2CH_2NH_2$;
(g) tetrahydrofuran; (h) *cis*-4-methylcyclohexanol. **6.58.** (a) o.p. =
e.e. = 15.58/15.90 = 98% (99%(*S*) and 1%(*R*)); (b) The e.e. of (*S*)
decreases twice as fast as radioactive iodide substitutes, thus gives the (*R*)
enantiomer; implies the S_N2 mechanism. **6.64.** NBS provides low concd.
Br_2 for free-radical bromination. Abstraction of one of the CH_2 hydrogens
gives a resonance-stabilized free radical; product Ph-CHBr-CH_3.

CHAPTER 7

7.4. (a) one; (b) one; (c) three; (d) four; (e) five. **7.5.** (a) 4-methyl-1-
pentene; (b) 2-ethyl-1-hexene; (c) 1,4-pentadiene; (d) 1,2,4-pentatriene;
(e) 2,5-dimethyl-1,3-cyclopentadiene; (f) 4-vinylcyclohexene; (g) allylbenzene or 3-phenylpropene; (h) *trans*-3,4-dimethylcyclopentene;
(i) 7-methylene-1,3,5-cycloheptatriene. **7.6.** (1) (a), (c) and (d) show
geometric isomerism. **7.7.** (a) 2,3-dimethyl-2-pentene; (b) 3-ethyl-
1,4-hexadiene; (c) 1-methylcyclopentene; (d) give positions of double
bonds; (e) specify cis or trans; (f) (*E*) or (*Z*), not *cis*.
7.9. 2,3-dimethyl-2-butene is more stable by 1.4 kcal/mol.
7.11. (a) stable; (b) unstable; (c) stable; (d) stable; (e) unstable (maybe
stable cold); (f) stable; (g) unstable; (h) stable (i) unstable (maybe stable
cold) **7.12.** (a) *cis*-1,2-dibromoethene; (b) *cis* (*trans* has zero dipole
moment), (c) 1,2-dichlorocyclohexane. **7.17.** There is no hydrogen
trans to the bromide leaving group. **7.23.** In the first example the
bromines are axial; in the second, equatorial. **7.26.** (a) $\Delta G > 0$, disfavored. (b) $\Delta G < 0$, favored. **7.27.** (a) strong bases and nucleophiles; (b) strong acids and electrophiles; (c) free-radical chain
reaction; (d) strong acids and electrophiles. **7.32.** (a) 2-ethyl-
1-pentene; (b) 3-ethyl-2-pentene; (c) (3*E*,6*E*)-1,3,6-octatriene;
(d) (*E*)-4-ethyl-3-heptene; (e) 1-cyclohexyl-1,3-cyclo-hexadiene.
7.36. (b), (c), and (f) show geometric isomerism. **7.38.** (a) cyclopentene; (b) 2-methyl-2-butene (major) and 2-methyl-1-butene (minor);
(c) 1-methylcyclohexene (major) and methylenecyclohexane (minor);
(d) 1 methylcyclohexene (minor) and methylenecyclohexane (major).
7.42. (a) a 1-halobutane; (b) a *t*-butyl halide; (c) a 3-halopentane; (d) a
halomethylcyclohexane; (e) a 4-halocyclohexane (preferably *cis*).
7.44. (a) 2-pentene; (b) 1-methylcyclopentene; (c) 1-methylcyclohex-
ene; (d) 2-methyl-2-butene (rearrangement). **7.56.** E1 with rearrangement by an alkyl shift. The Saytzeff product violates Bredt's rule.

CHAPTER 8

8.1. (a) 2-bromopropane; (b) 2-chloro-2-methylpropane; (c) 1-iodo-1-
methylcyclohexane; (d) mixture of *cis* and *trans* 3-methyl and

4-methylcyclohexane. **8.3.** (a) 1-bromo-2-methylpropane;
(b) 1-bromo-2-methylcyclopentane; (c) 2-bromo-1-phenylpropane.
8.6. (a) 1-methylcyclopentanol; (b) 2-phenyl-2-propanol; (c) 1-phenyl-cyclohexanol. **8.10.** (b) 1-propanol; (d) 2-methyl-3-pentanol; (f) *trans*-2-methylcyclohexanol. **8.13.** (a) *trans*-2-methylcycloheptanol;
(b) mostly 4,4-dimethyl-2-pentanol; (c) — OH *exo* on the less substi-
tuted carbon. **8.16.** The carbocation can be attacked from either face.
8.22. (a) CH_2I_2 + Zn(Cu); (b) CH_2Br_2, NaOH, H_2O, PTC; (c) dehydrate (H_2SO_4), then $CHCl_3$, NaOH/H_2O, PTC. **8.28.** (a) Cl_2, H_2O;
(b) KOH/heat, then Cl_2/H_2O; (c) H_2SO_4/heat, then Cl_2/H_2O.
8.34. (a) *cis*-cyclohexane-1,2-diol; (b) *trans*-cyclohexane-1,2-diol;
(c),(f) (*R,S*)-2,3-pentanediol (+enantiomer); (d), (e) (*R,R*)-2,3-penta-
nediol (+enantiomer). **8.35.** (a) OsO_4/H_2O_2;
(b) CH_3CO_3H/H_3O^+, (c) CH_3CO_3H/H_3O^+; (d) OsO_4/H_2O_2.
8.51. H_2C=CH—$COOCH_2CH_3$. **8.55.** (a) 2-methylpropene (3°
cation); (b) 1-methylcyclohexene (3° cation); (c) 1,3-butadiene
(resonance-stabilized cation). **8.59.** (a) 1-methylcyclohexene,
RCO_3H/H_3O^+; (b) cyclooctene, OsO_4/H_2O_2; (c) *trans*-cyclodecene,
Br_2; (d) cyclohexene, Cl_2/H_2O.
8.62. $CH_3(CH_2)_{12}CH$=$CH(CH_2)_7CH_3$, cis or trans unknown.

CHAPTER 9

9.3. decomposition to its elements, C and H_2. **9.4.** Treat the mixture
with $NaNH_2$ to remove the 1-hexyne. **9.5.** (a) $Na^+{}^-C$≡CH and
NH_3; (b) $Li^+{}^-C$≡CH and CH_4; (c) no reaction; (d) no reaction;
(e) acetylene + $NaOCH_3$; (f) acetylene + NaOH; (g) no reaction;
(h) no reaction; (i) NH_3 + $NaOCH_3$. **9.7.** (a) $NaNH_2$; butyl halide;
(b) $NaNH_2$; propyl halide; $NaNH_2$; methyl halide. (c) $NaNH_2$; ethyl
halide; repeat; (d) S_N2 on *sec*-butyl halide is unfavorable; (e) $NaNH_2$;
isobutyl halide (low yield); $NaNH_2$; methyl halide; (f) $NaNH_2$ added
for second substitution on 1,8-dibromooctane might attack the halide.
9.8. (a) sodium acetylide + ethylene oxide;
(b) sodium acetylide + formaldehyde; (c) sodium
acetylide + CH_3I, then $NaNH_2$, then $CH_3CH_2CH_2CHO$;
(d) sodium acetylide + CH_3I, then $NaNH_2$, then $CH_3CH_2COCH_3$.
9.10. About 1:70. **9.13.** (a) H_2, Lindlar; (b) Na, NH_3; (c), (d) Add
halogen, dehydrohalogenate to the alkyne, reduce as needed.
9.16. (a) $CH_3CCl_2CH_2C_5H_{11}$ and $CH_3CH_2CCl_2C_5H_{11}$; (b) Lone
pairs on Cl help stabilize the carbocation. **9.18.** (a) Cl_2; (b) HBr, peroxides; (c) HBr, no peroxides; (d) excess Br_2; (e) reduce to 1-hexene,
add HBr; (f) excess HBr. **9.20.** (a) The two ends of the triple bond are
equivalent. (b) The two ends of the triple bond are not equivalent, yet not
sufficiently different for good selectivity. **9.21.** (a) 2-hexanone; hexanal;
(b) mixtures of 2-hexanone and 3-hexanone; (c) 3-hexanone for both;
(d) cyclodecanone for both. **9.24.** (a) CH_3C≡$C(CH_2)_4C$≡CCH_3
9.28. (a) ethyl-methyl-acetylene; (b) phenylacetylene; (c) *sec*-butyl-
n-propylacetylene; (d) *sec*-butyl-*t*-butylacetylene. **9.31.** Form the
heavy metal salt of the terminal alkyne. **9.38.** 1,3-cyclohexadiene with
(HC≡C—CH=CH—) at the 1 position (cis or trans).

CHAPTER 10

10.1. (a) 2-phenyl-2-propanol; (b) 5-bromo-2-heptanol; (c) 4-methyl-
3-cyclohexen-1-ol; (d) *trans*-2-methylcyclohexanol; (e) (*E*)-2-
chloro-3-methyl-2-penten-1-ol; (f) (2*R*,3*S*)-2-bromo-3-hexanol.
10.4. (a) 8,8-dimethyl-2,7-nonanediol; (b) 1,8-octanediol;
(c) *cis*-2-cyclohexene-1,4-diol; (d) 3-cyclopentyl-2,4-heptanediol.
10.5. (a) cyclohexanol; more compact; (b) 4-methylphenol; more compact,
stronger H-bonds; (c) 3-ethyl-3-hexanol; more spherical; (d) cyclooctane-
1,4-diol; more OH groups per carbon; (c) enantiomers; equal solubility.
10.7. (a) methanol; less substituted; (b) 1-chloroethanol; chlorine closer
to the OH group; (c) 2,2-dichloroethanol; two chlorines to stabilize the
alkoxide. **10.9.** The anions of 2-nitrophenol and 4-nitrophenol (but not
3-nitrophenol) are stabilized by resonance with the nitro group.
10.10. (a) The phenol (left) is deprotonated by sodium hydroxide;

it dissolves. (b) In a separatory funnel, the alcohol (right) will go into an
ether layer and the phenolic compound will go into an aqueous sodium
hydroxide layer. **10.11.** (b), (f), (g), (h). **10.15.** (a) Add
phenylmagnesium bromide to benzophenone, PhCOPh. (b) Add methylmagnesium iodide to cyclohexanone. (c) Add propylmagnesium
bromide to dicyclohexyl ketone. **10.17.** (a) 2 PhMgBr + PhCOCl;
(b) 2 CH_3CH_2MgBr + $(CH_3)_2CHCOCl$; (c) 2 *c*-HxMgBr + PhCOCl.
10.19. (a) PhMgBr + ethylene oxide;
(b) $(CH_3)_2CHCH_2MgBr$ + ethylene oxide; (c) 2-methylcyclohexyl-
magnesium bromide + ethylene oxide. **10.23.** (a) Grignard
removes NH proton; (b) Grignard attacks ester; (c) Water will
kill Grignard; (d) Grignard removes OH proton.
10.26. (a) heptanoic acid + $LiAlH_4$; or heptaldehyde + $NaBH_4$;
(b) 2-heptanone + $NaBH_4$; (c) 2-methyl-3-hexanone + $NaBH_4$;
(d) ketoester + $NaBH_4$. **10.34.** (a) 1-hexanol, larger surface area;
(b) 2-hexanol, hydrogen-bonded; (c) 1,5-hexanediol, two OH groups;
(d) 2-hexanol. **10.38.** (a) cyclohexyl methanol; (b) 2-cyclopentyl-2-pen-
tanol; (c) 2-methyl-1-phenyl-1-propanol; (d) methane +
3-hydroxycyclohexanone; (e) 5-phenyl-5-nonanol; (f) triphenylmethanol;
(g) 1,1-diphenyl-1-propanol; (h) 3-(2-hydroxyethyl)cyclohexanol;
(i) reduction of just the ketone, but not the ester; (j) isobutyl alcohol;
(k) the tertiary alcohol; (l) the secondary alcohol; (m) cyclohexane
(n) (2*S*,3*S*)-2,3-hexanediol (+enantiomer); (o) (2*S*,3*R*)-2,3-hexanediol
(+enantiomer); (p) 1,4-heptadiene. **10.39.** (a) EtMgBr; (b) Grignard
with formaldehyde; (c) *c*-HxMgBr; (d) Grignard with ethylene oxide;
(e) Grignard with formaldehyde; (f) 2 CH_3MgI; (g) cyclo-pentylmag-
nesium bromide.

CHAPTER 11

11.1. (a) oxidation, oxidation; (b) oxidation, oxidation, reduction, oxidation; (c) neither (C2 is oxidation, C3 reduction); (d) reduction;
(e) neither; (f) oxidation; (g) neither; (h) neither; (i) oxidation; (j) oxidation then neither (k) reduction then oxidation **11.6.** (a) PCC;
(b) chromic acid; (c) chromic acid or Jones reagent; (d) PCC;
(e) chromic acid; (f) Dehydrate, hydroborate, oxidize (chromic acid or
Jones reagent). **11.7.** An alcoholic has more alcohol dehydrogenase.
More ethanol is needed to tie up this larger amount of enzyme.
11.8. CH_3COCHO (pyruvaldehyde) and $CH_3COCOOH$ (pyruvic
acid). **11.10.** Treat the tosylate with (a) bromide; (b) ammonia;
(c) ethoxide; (d) cyanide. **11.14.** (a) chromic acid or Lucas reagent;
(b) chromic acid; (c) Lucas reagent; (d) Lucas reagent; allyl alcohol
forms a resonance-stabilized carbocation. (e) Lucas reagent.
11.19. (a) thionyl chloride (retention); (b) tosylate (retention), then S_N2
using chloride ion (inversion). **11.20.** resonance-delocalized cation,
positive charge spread over two carbons. **11.22.** (a) 2-methyl-2-butene
(+2-methyl-1-butene); (b) 2-pentene (+1-pentene); (c) 2-pentene
(+1-pentene); (d) *c*-Hx=$C(CH_3)_2$ (+1-isopropylcyclohexene);
(e) 1-methylcyclohexene (+3-methylcyclohexene). **11.25.** Using
R—OH and R′—OH will form R—O—R, R′—O—R′, and
R—O—R′. **11.31.** (a) $CH_3CH_2CH_2COCl$ + 1-propanol;
(b) CH_3CH_2COCl + 1-butanol;
(c) $(CH_3)_2CHCOCl$ + *p*-methylphenol; (d) benzoyl
chloride + cyclopropanol. **11.33.** An acidic solution (to protonate
the alcohol) would protonate methoxide ion. **11.34.** (a) the alkoxide of
cyclohexanol and an ethyl halide or tosylate; (b) dehydration of cyclohexanol. **11.42.** (a) Na, then ethyl bromide; (b) NaOH, then PCC to
aldehyde; Grignard, then dehydrate; (c) Mg in ether, then
$CH_3CH_2CH_2CHO$, then oxidize; (d) PCC, then EtMgBr. **11.45.** Use
CH_3SO_2Cl. **11.46.** (a) thionyl chloride; (b) tosylate, displace with
bromide. **11.52.** Compound A is 2-butanol. **11.58.** **X** is 1-buten-4-ol;
Y is tetrahydrofuran (5-membered cyclic ether).

CHAPTER 12

12.3. (a) alkene; (b) alkane; (c) terminal alkyne. **12.4.** (a) amine (primary); (b) acid; (c) alcohol. **12.5.** (a) conjugated ketone; (b) ester;

(c) primary amide. **12.6.** (a) 3070 — C — H; 1642 C $=$ C *alkene;* (b) 2712, 2814 — CHO; 1691 carbonyl-*aldehyde*; (c) over-inflated C — H region — COOH; 1703 carbonyl (maybe conjugated); 1650 C $=$ C (maybe conjugated)-*conjugated acid*; (d) 1742 ester (or strained ketone)-*ester*. **12.7.** (a) bromine (C_6H_5Br); (b) iodine (C_2H_5I); (c) chlorine (C_4H_7Cl); (d) nitrogen ($C_7H_{17}N$). **12.10.** Probably 2,6-dimethyl-3-octene or 3,7-dimethyl-3-octene. **12.11.** 126: loss of water; 111: allylic cleavage; 87: cleavage next to alcohol. **12.14.** (a) about 1660 and 1710; the carbonyl is much stronger; (b) about 1660 for both; the enol is much stronger; (c) about 1660 for both; the imine is much stronger; (d) about 1660 for both; the terminal alkene is stronger. **12.16.** (a) $CH_2=C(CH_3)COOH$; (b) $(CH_3)_2CHCOCH_3$; (c) $PhCH_2C\equiv N$; (d) $PhNHCH_2CH_3$. **12.17.** (a) 86, 71, 43; (b) 98, 69; (c) 84, 69, 87, 45.

CHAPTER 13

13.1. (a) $\delta 2.17$; (b) 0.0306 gauss; (c) $\delta 2.17$; (d) 651 Hz. **13.3.** (a) three; (b) two; (c) three; (d) five. **13.6.** (a) 2-methyl-3-butyn-2-ol; (b) *p*-dimethoxybenzene; (c) 1,2-dibromo-2-methyl-propane. **13.10.** *trans* CHCl $=$ CHCN. **13.11.** (a) 1-chloropropane; (b) methyl *p*-methylbenzoate, $CH_3C_6H_4COOCH_3$. **13.14.** (a) H^a, $\delta 9.7$ (doublet); H^b, $\delta 6.6$ (multiplet); H^c, $\delta 7.4$ (doublet); (b) $J_{ab} = 8$ Hz, $J_{bc} = 18$ Hz (approx). **13.18.** (a) Five; the two hydrogens on C3 are diastereotopic. (b) Six; all the CH_2 groups have diastereotopic hydrogens. (c) Six; three on the Ph, and the CH_2 hydrogens are diastereotopic. (d) Three; The hydrogens cis and trans to the Cl are diastereotopic. **13.21.** (a) butane-1,3-diol; (b) $H_2NCH_2CH_2OH$. **13.24.** (a) $(CH_3)_2CHCOOH$; (b) $PhCH_2CH_2CHO$; (c) $CH_3COCOCH_2CH_3$ (d) $CH_2=CHCH(OH)CH_3$; (e) $CH_3CH_2C(OH)(CH_3)CH(CH_3)_2$. **13.29.** (a) allyl alcohol, $H_2C=CHCH_2OH$. **13.30.** (a) 4-hydroxy-butanoic acid lactone. **13.31.** (a) cyclohexene. **13.32.** isobutyl bromide. **13.36.** (a) iso-propyl alcohol. **13.38.** (a) $PhCH_2CH_2OCOCH_3$. **13.42.** 1,1,2-trichloropropane. **13.45.** A is 2-methyl-2-butene (Saytzeff product); **B** is 2-methyl-1-butene. **13.47.** $PhCH_2CN$.

CHAPTER 14

14.2. $(CH_3CH_2)_2O \overset{+-}{\longrightarrow} AlCl_3$. **14.4.** (a) cyclopropyl methyl ether; methoxycyclopropane; (b) ethyl isopropyl ether; 2-ethoxypropane; (c) 2-chloroethyl methyl ether; 1-chloro-2-methoxyethane; (d) 2-methy-oxy-2,3-dimethylpentane; (e) *sec*-butyl *t*-butyl ether; 2-(1,1-dimethylethoxy) butane; (f) *trans*-2-methoxycyclohexanol (no common name). **14.6.** (a) dihydropyran; (b) 2-chloro-1,4-dioxane; (c) 3-iso-propylpyran; (d) *trans*-2,3-diethyloxirane or *trans*-3,4-epoxyhexane; (e) 3-bromo-2-ethoxyfuran; (f) 3-bromo-2,2-dimethyloxetane. **14.11.** Intermolecular dehydration of a mixture of methanol and ethanol would produce a mixture of diethyl ether, dimethyl ether, and ethyl methyl ether. **14.13.** Intermolecular dehydration might work for (a). Use the Williamson for the other two. **14.15.** (a) bromocyclohexane and ethyl bromide; (b) 1,5-diiodopentane; (c) phenol and methyl bromide; (e) phenol, ethyl bromide, and 1,4-dibromo-2-methylbutane. **14.22.** Epoxidation of ethylene gives ethylene oxide, and catalytic hydra-tion of ethylene gives ethanol. Acid-catalyzed opening of the epoxide in ethanol gives cellosolve. **14.26.** (a) $CH_3CH_2OCH_2CH_2O^-Na^+$; (b) $H_2NCH_2CH_2O^-Na^+$; (c) $Ph-OCH_2CH_2O^-Na^+$; (d) $PhNHCH_2CH_2O^-Na^+$. **14.27.** (a) 2-methyl-1,2-propanediol, ^{18}O at the C2 hydroxyl group; (b) 2-methyl-1,2-propanediol, ^{18}O at the C1 hydroxyl group; (c), (d) same products, (*S,S*) and (*R,R*). **14.28.** (a) $(CH_3)_2CH-CH_2CH_2-OH$; (b) $CH_3-CH_2C(CH_3)_2-OH$; (c) 1 cyclopentyl-1-butanol. **14.34.** (a) The old ether had autoxidized to form peroxides. On distilla-tion, the peroxides were heated and concentrated, and they detonated. (b) Discard the old ether or treat it to reduce the peroxides. **14.38.** (a) epoxide + phenylmagnesium bromide; (b) epoxide + sodium methoxide in methanol;

(c) epoxide + methanol, H^+. **14.42.** Sodium then ethyl iodide gives retention of configuration. Tosylation gives retention, then the Williamson gives inversion. Second product $(+)15.6°$. **14.46.** $(CH_3OCH_2CH_2)_2O$ **14.47.** phenyloxirane

CHAPTER 15

15.1. (a) 2,4-hexadiene < 1,3-hexadiene < 1,4-hexadiene < ; 1,5-hexadiene < 1,2-hexadiene < 1,3,5-hexatriene; (b) third < fifth < first < fourth < second. **15.6.** 3-ethoxy-1-methylcyclopentene and 3-ethoxy-3-methylcyclopentene. **15.8.** (a) **A** is 3,4-dibromo-1-butene; **B** is 1,4-dibromo-2-butene; (c) Hint: **A** is the kinetic product, **B** is the thermodynamic product; (d) Isomerization to an equilibrium mixture. 10% **A** and 90% **B**. **15.9.** (a) 1-(bromomethyl)cyclohexene and 2-bromo-1-methylenecyclo-hexane. **15.10.** (a) 3-bromocyclopentene; (b) (*cis* and *trans*) 4-bromo-2-pentene. (c) $PhCH_2Br$ **15.12.** Both generate the same allylic carbanion. **15.13.** (a) allyl bromide + *n*-butyllithium; (b) isopropyllithium + 1-bromo-2-butene. **15.20.** (b) [4 + 2] cycloaddition of one butadiene with just one of the double bonds of another butadiene. **15.21.** 800. **15.22.** (a) 353 nm; (b) 313 nm; (c) 232 nm; (d) 273 nm; (e) 237 nm. **15.24.** (a) isolated; (b) conjugat-ed; (c) cumulated; (d) conjugated and isolated; (e) conjugated. **15.25.** (a) allylcyclohexane; (b) 3-chlorocyclopentene; (c) 3-bromo-2-methyl-propene; (d) 3-bromo-1-pentene and 1-bromo-2-pentene; (e) 4-bromo-2-buten-1-ol and 1-bromo-3-buten-2-ol; (f) 5,6-dibromo-1,3-hexadiene, 1,6-dibromo-2,4-hexadiene, and 3,6-dibromo-1,4-hexa-diene (minor); (g) 1-(methoxymethyl)-2-methyl-cyclopentene and 1-methoxy-1-methyl-2-methylenecyclopentane; (h), (i) Diels–Alder adducts. **15.26.** (a) allyl bromide + isobutyl Grignard; (b) 1-bromo-3-methyl-2-butene + $CH_3CH_2C(CH_3)_2MgBr$. **15.28.** (a) 19,000; (b) second structure. **15.29.** 3-bromo-1-hexene, and (cis and trans)-1-bromo-2-hexene. **15.32.** (a) The product isomer-ized, 1630 suggests conjugated; (b) 2-propyl-1,3-cyclohexadiene.

CHAPTER 16

16.2. (a) $+7.6$ kcal/mol; (b) -21.2 kcal/mol; (c) -26.8 kcal/mol. **16.5.** Two of the eight pi electrons are unpaired in two non-bonding orbitals, an unstable configuration. **16.7.** (a) nonaromatic (internal H's prevent planarity); (b) nonaromatic (one ring atom has no *p* orbital); (c) aromatic, [14]annulene; (d) aromatic (in the outer system). **16.8.** Azulene is aromatic, but the other two are antiaromatic. **16.10.** The cation (cyclopropenium ion) is aromatic; the anion is antiaromatic. **16.12.** (a) antiaromatic if planar; (b) aromatic if planar; (c) aromatic if planar; (d) antiaromatic if planar; (e) nonaromatic; (f) aromatic if planar. **16.14.** cyclopropenium fluoroborate. **16.18.** (a) aromatic; (b) aromatic; (c) nonaromatic; (d) aromatic; (e) aro-matic; (f) nonaromatic; (g) aromatic. **16.22.** (a) fluorobenzene; (b) 4-phenyl-1-butyne; (c) 3-methyl-phenol or *m*-cresol; (d) *o*-nitrostyrene; (e) *p*-bromobenzoic acid; (f) isopropyl phenyl ether; (g) 3,4-dinitrophe-nol; (h) benzyl ethyl ether. **16.25.** 3-phenyl-2-propene-1-ol **16.28.** (a) *o*-dichlorobenzene; (b) *p*-nitroanisole; (c) 2,3-dibromoben-zoic acid; (d) 2,7-dimethoxynaphthalene; (e) *m*-chlorobenzoic acid; (f) 2,4,6-trichlorophenol; (g) 2-(1-methylpropyl)benzaldehyde; (h) cyclopropenium fluoroborate. **16.30.** The second is deprotonated to an aromatic cyclopentadienyl anion. **16.31.** (d), (e) The fourth structure, with two three-membered rings, was considered the most like-ly and was called Ladenburg benzene. **16.37.** (a) three; (b) one; (c) *meta*-dibromobenzene. **16.38.** α-chloroacetophenone **16.40.** (a) no; (b) six in each; total 12, compared with 10 in naphtha-lene; (c) $(6 \times 28.6) - 100 = 71.6$ kcal, 35.8 kcal per ring, nearly as much as benzene (36 kcal). Naphthalene has only 60 kcal, 30 kcal per ring. **16.42.** Deprotonate it to give an anion with 10 pi electrons. **16.45.** 2-isopropyl-5-methylphenol

CHAPTER 17

17.4. The sigma complex for *p*-xylene has the + charge on two 2° carbons and one 3° carbon, compared with three 2° carbons in benzene. **17.10.** Bromine *adds* to the alkene but *substitutes* on the aryl ether, evolving gaseous HBr. **17.11.** Strong acid is used for nitration, and the amino group of aniline is protonated to a deactivating — NH_3^+ group. **17.13.** 1-bromo-1-chlorocyclohexane; the intermediate cation is stabilized by a bromonium ion resonance form. **17.14.** (a) 2,4- and 2,6-dinitrotoluene; (b) 3-chloro-4-nitrotoluene and 5-chloro-2-nitrotoluene; (c) 3- and 5-nitro-2-bromobenzoic acid; (d) 4-methoxy-3-nitrobenzoic acid; (e) 5-methyl-2-nitrophenol and 3-methyl-4-nitrophenol. **17.17.** (a) phenylcyclohexane; (b) *o*- and *p*-methylanisole, with overalkylation products; (c) 1-isopropy-4-(1,1,2-trimethylpropyl)benzene. **17.18.** (a) phenylcyclohexane; (b) *t*-butylbenzene; (c) *p*-di-*t*-butylbenzene; (d) *o*- and *p*-isopropyltoluene. **17.19.** (a) *t*-butyl-benzene; (b) 2- and 4-*sec*-butyltoluene; (c) no reaction; (d) (1,1,2-trimethyl-propyl)benzene. **17.20.** (a) *sec*-butylbenzene and others; (b) OK; (c) + disub, trisub; (d) OK (some ortho); (e) OK. **17.22.** (a) $(CH_3)_2CHCH_2COCl$, benzene, $AlCl_3$; (b) $(CH_3)_3CCOCl$, benzene, $AlCl_3$; (c) PhCOCl, benzene, $AlCl_3$; (d) CO/HCl, $AlCl_3/CuCl$, anisole; (e) Clemmensen on (b); (f) $CH_3(CH_2)_2COCl$, benzene, $AlCl_3$ then Clemmensen. **17.23.** Fluoride leaves in a fast exothermic step; the C — F bond is only slightly weakened in the reactant-like transition state (Hammond postulate). **17.25.** (a) 2,4-dinitroanisole; (b) 2,4- and 3,5-dimethylphenol; (c) N-methyl-4-nitroaniline; (d) 2,4-dinitrophenylhydrazine. **17.29.** (a) (trichloromethyl) hexachlorocyclohexane; (b) 1-methyl-1,4-cyclohexadiene; (c) *cis* and *trans*-1,2-dimethylcyclohexane; (d) 1,4-dimethyl-1,4-cyclohexadiene. **17.30.** (a) benzoic acid; (b) benzoic acid; (c) *o*-phthalic acid. **17.33.** 60% beta, 40% alpha; reactivity ratio = 1.9 to 1. **17.36.** (a) 1-bromo-1-phenylpropane. **17.38.** (a) HBr, then Grignard with ethylene oxide; (b) CH_3COCl and $AlCl_3$, then Clemmensen, Br_2 and light, then $^-OCH_3$; (c) Nitrate, then Br_2 and light, then NaCN. **17.40.** (a) 3-ethoxytoluene; (b) *m*-tolyl acetate; (c) 2,4,6-tribromo-3-methylphenol; (d) 2,4,6-tribromo-3-(tribromomethyl)phenol; (e) 2-methyl-1,4-benzoquinone; (f) 2,4-di-*t*-butyl-3-methyl-phenol. **17.50.** indanone. **17.55.** The yellow species is the triphenylmethyl cation. **17.60.** kinetic control at 0°, thermodynamic control at 100°. **17.61.** Brominate, then Grignard with 2-butanone.

CHAPTER 18

18.1. (a) 5-hydroxy-3-hexanone; ethyl β-hydroxypropyl ketone; (b) 3-phenylbutanal: β-phenylbutyraldehyde; (c) *trans*-2-methoxy-cyclohexane-carbaldehyde; (d) 6,6-dimethyl-2,4-cyclohexadienone. **18.2.** (a) 2-phenylpropanal; (b) acetophenone. **18.3.** No γ-hydrogens. **18.5.** (a) <200, 280; (b) 230, 310; (c) 280, 360; (d) 270, 350. **18.8.** (a) acetophenone; (b) acetylcyclohexane; (c) 3-heptanone. **18.9.** (a) 3-heptanone; (b) phenylacetonitrile; (c) benzyl cyclohexyl ketone. **18.11.** (a) benzyl alcohol; (b) benzaldehyde; (c) hept-1-en-3-one. **18.14.** $[(CH_3)_3P — R]^+$ could lose a proton from a CH_3. **18.17.** (a) Wittig of $PhCH_2Br$ + acetone; (b) Wittig of CH_3I + $PhCOCH_3$; (c) Wittig of $PhCH_2Br$ + $PhCH=CHCHO$; (d) Wittig of CH_3I + cyclopentanone; (e) Wittig of EtBr + cyclohexanone. **18.19.** second < fourth < first < third. **18.23.** cis and trans isomers. **18.24.** (a) cyclohexanone and methylamine; (b) 2-butanone and ammonia; (c) acetaldehyde and aniline; (d) 6-amino-2-hexanone. **18.28.** (a) benzaldehyde and semicarbazide; (b) camphor and hydroxylamine; (c) tetralone and phenylhydrazine; (d) cyclohexanone and 2,4-DNP; (e) 4-(*o*-aminophenyl)-2-butanone. **18.31.** (a) tetralone and ethanol; (b) acetaldehyde and 2-propanol; (c) hexane-2,4-dione and ethanediol; (d) tetralone and 1,3-propanediol; (e) 5-hydroxypentanol and methanol; (f) $(HOCH_2CH_2CH_2)_2CHCHO$. **18.35.** (a) 4-hydroxycyclohexanecarboxylic acid; (b) 4-oxocyclohexanecarboxylic acid; (c) 3-oxocyclohexanecarboxylic acid; (d) *cis*-3,4-dihydroxycyclohexanecarboxylic

acid. **18.37.** (a) indane; (b) hexane; (c) ethyl ketal of 2-propylcyclohexanone; (d) propylcyclohexane. **18.42.** 240 nm and 300–320 nm. **18.43.** 2,5-hexanedione **18.44.** 1-phenyl-2-butanone (benzyl ethyl ketone). **18.45.** (a) 44; (b) 72; (c) 44; (d) 74. **18.47.** cyclobutanone **18.52.** (all H^+ cat.) (a) cyclobutanone and hydroxylamine; (b) benzaldehyde and cyclopentylamine; (c) benzylamine and cyclopentanone; (d) β-tetralone and ethylene glycol; (e) cyclohexylamine and acetone; (f) cyclopentanone and methanol. **18.57.** (a) $NaBD_4$, then H_2O; (b) $NaBD_4$, then D_2O; (c) $NaBH_4$, then D_2O. **18.60.** (a) $CH_3CH_2CH_2COCl$ and $AlCl_3$, then Clemmensen; (b) EtMgBr, then H_3O^+; (c) $Cl_2/FeCl_3$, then Dow process to phenol; NaOH, CH_3I, then Gatterman; (d) oxidize to the acid, $SoCl_2$, then $AlCl_3$. **18.64.** (a) 3-hexanone; (b) 2- and 3-hexanone; (c) 2-hexanone; (d) cyclodecanone; (e) 2- and 3-methyl-cyclodecanone. **18.66.** A is 2-heptanone. **18.72.** (b) The "THP ether" is an acetal, stable to base but hydrolyzed by acid. **18.73.** A is the ethylene ketal of 2-butanone; B is 2-butanone. **18.74.** *trans*-2-butenal (crotonaldehyde)

CHAPTER 19

19.1. Pyridine, 2-methylpyridine, pyrimidine, pyrrole, imidazole, indole, and purine are aromatic. **19.3.** (a) 2-pentanamine; (b) N-methyl-2-butanamine; (c) *m*-aminophenol; (d) 3-methylpyrrole; (e) *trans*-1,2-cyclopentanediamine; (f) *cis*-3-aminocyclohexanecarbaldehyde. **19.4.** (a) resolvable (chiral carbons); (b) not resolvable (N inverts); (c) symmetric; (d) not resolvable; proton on N is removable; (e) resolvable (chiral quat. salt). **19.6.** (a) aniline < ammonia < methylamine < NaOH; (b) *p*-nitroaniline < aniline < *p*-methylaniline; (c) pyrrole < aniline < pyridine; (d) 3-nitropyrrole < pyrrole < imidazole. **19.7.** (a) secondary amine; (b) primary amine; (c) alcohol. **19.8.** isobutylamine **19.9.** (a) piperidine; (b) diethylmethylamine; (c) propanal; (d) 1-propanol. **19.16.** (a) benzylamine + excess CH_3I; (b) 1-bromopentane + excess NH_3; (c) benzyl bromide + excess NH_3. **19.17.** (a) $CH_3CONHCH_2CH_3$; (b) $PhCON(CH_3)_2$; (c) N-hexanoyl piperidine. **19.23.** (a) cyclohexanediazonium chloride (then cyclohexanol and cyclohexene); (b) N-nitroso-N-ethyl-2-hexanamine; (c) N-nitrosopiperidine; (d) benzenediazonium chloride. **19.25.** (a) diazotize, then HBF_4, heat; (b) diazotize, then CuCl; (c) Protect (CH_3COCl), then 3 $CH_3I/AlCl_3$, H_3O^+, diazotize, H_3PO_2; (d) diazotize, then CuBr; (e) diazotize, then KI; (f) diazotize, then CuCN; (g) diazotize, then H_2SO_4, H_2O, heat; (h) diazotize, then couple with resorcinol. **19.26.** (a) CH_3NH_2, $Na(AcO)_3BH$; (b) H_2NOH/H^+, then $LiAlH_4$; (c) PhCHO, $Na(AcO)_3BH$; (d) aniline/H^+, then $LiAlH_4$; (e) H_2NOH/H^+, then $LiAlH_4$; (f) piperidine + cyclohexanone + $NaBH_3CN$. **19.31.** (a) nitrate, reduce; (b) brominate, then nitrate and reduce; (c) nitrate, then brominate and reduce; (d) oxidize toluene, then nitrate and reduce. **19.33.** Hofmann rearrangement goes with retention of configuration. **19.38.** only (b), (d), (f), and (h). **19.40.** (a) 2-phenylethylamine; (b) 1,4-butanediamine; (c) *trans*-2-phenylcyclopropanamine. **19.53.** (a) triethylamine; (b) An acid converts it to a solid ammonium salt. (c) Rinse the clothes with diluted vinegar (acetic acid). **19.56.** A is 2-butanamine; B is diethylamine. **19.58.** 2,2-dimethyl-1-propanamine.

CHAPTER 20

20.2. (a) 2-iodo-3-methylpentanoic acid; α-iodo-β-methylvaleric acid; (b) (*Z*)-3,4-dimethyl-3-hexenoic acid; (c) 2,3-dinitrobenzoic acid; (d) *trans*-1,3-cyclohexanedicarboxylic acid; (e) 2-chlorobenzene-1,4-dicarboxylic acid; 2-chloroterephthalic acid; (f) 3-methylhexanedioic acid; β-methyladipic acid. **20.3.** (a) first, second, third; (b) third, second, first; (c) third, second, fourth, first. **20.7.** Broad acid OH centered around 3000; conjugated carbonyl about 1690; C=C about 1650. **20.8.** (a) propanoic acid; (b) — CHO proton triplet between δ9 and

$\delta 10$. **20.11.** (a) $KMnO_4$ (b) $KMnO_4$; (c) $PhMgBr$ + ethylene oxide, oxidize; (d) PBr_3, Grignard, CO_2; (e) Conc. $KMnO_4$, heat; (f) KCN, then H_3O^+. **20.15.** (a) Methanol and salicylic acid, H^+; methanol solvent, dehydrating agent; (b) Methanol and formic acid H^+, distill product as it forms; (c) Ethanol and benzoic acid, H^+, ethanol solvent, dehydrating agent. **20.16.** (a) see Fischer esterification; (b) C—^{18}O—CH_3; (c) mass spectrometry. **20.22.** (a) phenylacetic acid and $LiAlH_4$; (b) phenylacetic acid and $LiAlH_4$, then PCC; (c) cyclopentanone-3-carboxylic acid; make ethylene acetal, then $LiAlH_4$. **20.24.** (a) benzene + CH_3CH_2COCl, $AlCl_3$; or propionic acid + 2 $PhLi$, then H_3O^+; (b) Add 2 CH_3Li, then H_3O^+. **20.36.** (a) Grignard + CO_2; or KCN, then H_3O^+; (b) concd. $KMnO_4$, heat; (c) Ag^+; (d) $SOCl_2$, then $Li(t\text{-}BuO)_3AlH$; or $LiAlH_4$, then PCC; (e) CH_3OH, H^+; or CH_2N_2; (f) $LiAlH_4$ or B_2H_6; (g) $SOCl_2$, then excess CH_3NH_2. **20.38.** diastereomers **20.40.** phenoxyacetic acid **20.45.** (a) stockroom; heptaldehyde; students; heptanoic acid; (b) air oxidation; (c) Prepare fresh samples immediately before using. **20.46.** (a) 2-phenylpropanoic acid; (b) 2-methylpropenoic acid; (c) *trans*-2-hexenoic acid.

CHAPTER 21

21.2. No aldehyde C—H at 2700 and 2800; no acid O—H centered at 3000. **21.4.** (a) acid chloride C=O at 1810; (b) primary amide C=O at 1640, two N—H around 3300; (c) anhydride C=O double absorption at 1740 and 1810. **21.5.** (a) acrylamide, H_2C—$CHCONH_2$; (b) 5-hydroxyhexanoic acid lactone. **21.8.** (a) ethanol, propionyl chloride; (b) phenol, 3-methylhexanoyl chloride; (c) benzyl alcohol, benzoyl chloride; (d) cyclopropanol, cyclohexanecarbonyl chloride. **21.9.** (a) dimethylamine, acetyl chloride; (b) aniline, acetyl chloride; (c) ammonia, cyclohexanecarbonyl chloride; (d) piperidine, benzoyl chloride. **21.10.** (i) $PhCH_2OH$; (ii) Et_2NH. **21.25.** (a) 1-butanamine; (b) cyclohexyl ethyl amine; (c) $(CH_2)_6NH$ (7 membered ring); (d) morpholine; (e) cyclohexyl methyl propyl amine. **21.30.** (a) benzene + acetyl chloride; (b) benzene + benzoyl chloride; (c) benzene + butyryl chloride, then Clemmensen. **21.32.** (a) *n*-octyl alcohol, acetic formic anhydride (formyl chloride is unavailable); (b) *n*-octyl alcohol, acetic anhydride (cheap, easy to use); (c) phthalic anhydride, ammonia (anhydride forms monoamide); (d) succinic anhydride, methanol (annhydride forms monoester). **21.34.** (a) acetic anhydride; (b) methanol, H^+; (c) diazomethane. **21.36.** (a) $SOCl_2$ $HN(CH_3)_2$, $LiAlH_4$; (b) acetic formic anhydride, then $LiAlH_4$. **21.38.** (a) $SOCl_2$, NH_3, $POCl_3$. (b) $LiAlH_4$, make tosylate, NaCN; (c) Fe/HCl, diazotize, CuCN. **21.45.** (a) ethyl benzoate; (b) acetic benzoic anhydride; (c) $PhCONHPh$; (d) 4-methoxybenzophenone; (e) Ph_3COH. **21.48.** (a) acetic formic anhydride; (b) $SOCl_2$, CH_3COONa; (c) oxalyl chloride; (d) H^+/heat, $(CH_3)_2CHOH$; (e) Ag^+, H^+. **21.51.** (after H^+) (a) $HCOOH$ + $PhOH$; (b) CH_3CH_2COOH + CH_3CH_2OH; (c) 3-(*o*-hydroxyphenyl)-propanoic acid; (d) $(CH_2OH)_2$ + $(COOH)_2$. **21.55.** (a) Ph_3COH; (b) 3 $EtMgBr$ + $EtCOOEt$, then H_3O^+. **21.58.** (a) diethyl carbonate; (b) $CH_3NHCONHCH_3$; (c) $CH_3OCONHPh$. **21.62.** Penicillin **21.65.** Acetic anhydride; add water to hydrolyze it to dilute acetic acid. **21.66.** $CH_3CH_2OCOCH_2CN$ **21.67.** δ-valerolactam **21.68.** ethyl crotonate

CHAPTER 22

22.8. (a), (b) cyclopentane carboxylate and chloroform/iodoform; (c) 2,2,6,6-tetraiodocyclohexanone; (d) $PhCOCBr_2CH_3$. **22.12.** (a) $CH_3CHBrCOOH$; (b) $PhCOOH$; (c) $HOOCCH_2CHBrCOOH$; (d) oxalic acid.

22.16. (a) $PhC(NCH_3)CH_3$; (b) CH_2=$C(h)NMe_2$; (c) cyclohexanone phenyl imine; (d) piperidine enamine of cyclohexanone. **22.17.** (a) enamine + allyl bromide; (b) enamine + $PhCH_2Br$; (c) enamine + $PhCOCl$. **22.19.** (a) 3-hydroxy-2-methylpentanal; (b) 3-hydroxy-2,4-diphenylbutanal. **22.20.** retro-aldol, reverse of aldol condensation. **22.24.** (a) 2-ethyl-2-hexenal; (b) 1,3-diphenyl-2-buten-1-one; (c) 2-cyclohexylidene-cyclohexanone. **22.26.** $PhCH$=$CHCOCH$=$CHPh$, "dibenzalacetone". **22.28.** (a) 2-methyl-3,3-diphenyl-2-propenal; (b) 4,4-dimethyl-1-phenyl-2-penten-1-one. **22.29.** benzaldehyde and acetaldehyde. **22.32.** (a) butanal and pentanal (no); (b) two $PhCOCH_2CH_3$ (yes); (c) acetone and PhCHO 9 (yes); (d) 6-oxoheptanol (yes, but also attack by enolate of aldehyde); (e) nonane-2,8-dione (yes). **22.34.** (a) transesterification to a mixture of methyl and ethyl esters; (b) saponification. **22.35.** no second alpha proton to form the final enolate to drive the reaction to completion. **22.36.** (a) methyl 2-methyl-3-ketopentanoate; (b) ethyl 2,4-diphenyl-3-ketobutyrate. **22.37.** methyl 2-benzyl-5-phenyl-3-ketopentanoate **22.38.** (a) ethyl butyrate; (b) methyl phenylacetate; (c) $c2Hx$—$CH_2CH_2COOCH_3$; (d) ethyl cyclopentanecarboxylate. **22.42.** (a) $PhCO$—$CH(Ph)COOCH_3$; (b) poor choice, four products; (c) $EtOCOCO$—CH_2COOEt; (d) $EtOCO$ $CH(CH_3)COOEt$. **22.43.** (a) $PhCOOEt$ + CH_3CH_2COOEt; (b) $PhCH_2COOMe$ + $MeOCOCOOMe$; (c) $(EtO)_2C$=O + $PhCH_2COOEt$; (d) $(CH_3)_3CCOOMe$ + $CH_3(CH_2)_3COOMe$. **22.47.** Alkylate malonic ester with; (a) $PhCH_2Br$; (b) CH_3I twice; (c) $PhCH_2CH_2Br$; (d) $Br(CH_2)_4Br$ (twice). **22.49.** (a) 4-phenyl-2-butanone; (b) cyclobutyl methyl ketone; (c) cyclopentanone. **22.50.** Alkylate acetoacetic ester with; (a) $PhCH_2Br$; (b) $Br(CH_2)_4Br$ (twice); (c) $PhCH_2Br$, then CH_2=$CHCH_2Br$ **22.53.** Alkylate the enamine of cyclohexanone with MVK. **22.56.** (a) malonic ester anion + ethyl cinnamate; (b) malonic ester anion + acrylonitrile, then H_3O^+ (c) enamine of cyclopentanone + acrylonitrile, then H_3O^+; (d) enamine of 2-methylcyclopentanone +$PhCOCH$=CH_2, then H_3O^+; (e) alkylate acetoacetic ester with CH_3I, then MVK, then H_3O^+; (f) hydrolyze the product from (a). **22.62.** (1) g < b < f < a < c < d < e; (2) a, c, d, e. **22.70.** Alkylate with: (a) $Br(CH_2)_5Br$ (twice); (b) EtBr, then CH_3CH=$CHCH_2Br$; (c) $PhCH_2Br$. **22.71.** Alkylate with: (a) CH_3I, then c-Hx—CH_2Br; (b) $Br(CH_2)_4Br$; (c) MVK (hydrolysis, decarboxylation, then Aldol gives product). **22.75.** (a) Dieckmann of dimethyl adipate, alkylation by allyl bromide, hydrolysis and decarboxylation; (b) Aldol of cyclopentanone, dehydration; (c) Robinson with CH_3CH=$CHCOCH_3$, then reduction. **22.77.** (a) $EtCOPh$ + MVK; (b) cyclohexanone and ethyl vinyl ketone; (c) cyclohexanone and $(CH_3)_2C$=$CHCOCH_3$.

CHAPTER 23

23.2. (a) two C^*, two pairs of enantiomers; (b) one C^*, one pair of enantiomers; (c) four C^*, eight pairs of enantiomers; three C^*, four pairs of enantiomers. **23.5.** (R) for D series, (S) for L series. **23.15.** 28% alpha, 72% beta. **23.19.** Galactitol is symmetrical (meso) and achiral. **23.20.** L-gulose has the same structure as D-glucose, but with the CHO and CH_2OH ends interchanged. **23.21.** (a) D-mannonic acid; (b) D-galactonic acid; (c) Br_2 does not oxidize ketoses. **23.22.** (a) D-mannaric acid; (b) D-galactaric acid. **23.23.** **A** is galactose; **B** is glucose. **23.24.** (a) non-reducing; (b) reducing; (c) reducing; (d) non-reducing; (e) reducing; (f) "sucrose" is nonreducing; should have "-oside" ending. **23.27.** glucose, benzaldehyde, and HCN (toxic). **23.28.** **A** = D-galactose; **B** = D-talose; **C** = D-lyxose; **D** = D-threose **23.39.** **E** = D-ribose; **F** = D-erythrose. **23.44.** reducing and mutarotating. **23.45.** reducing and mutarotating. **23.46.** Trehalose is α-D-glucopyranosyl-α-D-glucopyranoside.

23.47. Melibiose is 6-O-(α-D-galactopyranosyl)-D-glucopyranose.
23.58. (a) D-ribose; (b) D-altrose; (c) L-erythrose; (d) L-galactose;
(e) L-idose. **23.65.** (a) D-arabinose and D-lyxose; (b) D-threose; (c) **X** =
D-galactose; (d) No; the optically active hexose is degraded to an optically
active pentose that is oxidized to an optically active aldaric acid;
(e) D-threose gives an optically active aldaric acid. **23.68.** (a) D-tagatose
is a ketohexose, the C4 epimer of D-fructose. (b) A pyranose with the
anomeric carbon (C2) bonded to the oxygen atom of C6. **23.69.** D-altrose
23.73. (a) no; (b) yes; (c) Only applies to double stranded DNA.

CHAPTER 24

24.6. As in pyrrole, the lone pair on the indole N is part of the aromatic
sextet. One N in histidine is like that in pyridine, with the lone pair in an
sp^2 hybrid orbital. **24.9.** Reductive amination of (a) $CH_3COCOOH$;
(b) $(CH_3)_2CHCH_2COCOOH$; (c) $HOCH_2COCOOH$;
(d) $H_2NCOCH_2CH_2COCOOH$. **24.10.** Start with (a) CH_3COOH;
(b) $(CH_3)_2CHCH_2CH_2COOH$; (c) $(CH_3)_2CHCH_3COOH$;
(d) $HOOCCH_2CH_2CH_2COOH$. **24.11.** N-phthalimidomalonic ester
and (a) $(CH_3)_2CHBr$; $PhCH_2Br$; (c) $BrCH_2CH_2COO^-$;
(d) $(CH_3)_2CHCH_2Br$. **24.15.** The free amino group of the deacylated
L enantiomer should become protonated (and soluble) in dilute acid.
24.23. (a) nucleophilic aromatic substitution; (b) Edman cleaves only the
N-terminal amino acid, leaving the rest of the chain intact for further
degradation. **24.25.** Cys-Tyr-Phe-Gln-Asn-Cys-Pro-Arg-Gly $\cdot$ NH_2.
24.27. Add ethyl chloroformate, then Gly, ethyl chloroformate, then Leu.
Deprotect using H_2 and Pd. **24.30.** Add TFA (CF_3COOH), then Boc-
Gly and DCC, then TFA, then Boc-Leu and DCC, then HF.
24.34. (a) Ruhemann's purple; (b) alanine;
(c) $CH_3CONH(CH_2)_4CH(COOH)NHCOCH_3$; (d) L-proline and N-
acetyl-D-proline; (e) $CH_3CH_2CH(CH_3)CH(NH_2)CN$; (f) isoleucine;
(g) 2-bromo-3-methylbutanoic acid; (h) valine.
24.35. (a) $NH_3/H_2/Pd$; (b) Br_2/PBr_3, H_2O, excess NH_3;
(c) $NH_3/HCN/H_2O$, H_3O^+; (d) Gabriel-malonic ester synthesis.
24.37. Convert the alcohol to a tosylate and displace with excess ammonia. **24.42.** aspartylphenylalanine methyl **24.43.** Phe-Ala-Gly-Met-
Ala. **24.46.** (a) C-terminal amide ($CONH_2$), or amide (Gln) of Glu;
(b) The N-terminal Glu is a cyclic amide (a "pyroglutamyl" group) that
effectively blocks the N-terminus. The C-terminal Pro is an amide;
(c) cyclic pentapeptide. **24.49.** Ornithine is
$H_2N(CH_2)_3CH(NH_2)COOH$, a homolog of lysine, with a similar IEP.
24.51. Ala-Lys-Phe-Glu-Gly-Tyr-Arg-Ser-Leu-Ile.

CHAPTER 25

25.2. Hydrogenation of triolein (m.p. $-4°C$) gives tristearin (m.p.
$72°C$). **25.9.** Estradiol is a phenol, soluble in aqueous sodium hydroxide. **25.13.** (1) sesquiterpene; (2) monoterpene; (3) monoterpene;
(4) sesquiterpene. **25.15.** (a) a triglyceride (a fat); (b) an alkyl sulfate
detergent; (c) a wax; (d) a sesquiterpene; (e) a steroid.
25.17. (a) H_2/Ni, $LiAlH_4$; (b) H_2/Ni; (c) stearic acid from (b), add
$SOCl_2$, then l-octadecanol (a); (d) O_3, then $(CH_3)_2S$; (e) $KMnO_4$,
then H^+; (f) Br_2/PBr_3, then H_2O. **25.19.** reduce ($LiAlH_4$), esterify
with sulfuric acid. **25.21.** (a) Sodium stearate precipitates in dilute
acid or Ca^{2+}; (b) Paraffin "wax" does not saponify; (c) Myristic acid
shows acidic properties when treated with base; (d) Triolein decolorizes
Br_2 in CCl_4. **25.28.** Petroselenic acid is *cis*-6-octadecenoic acid.
25.30. The sugar-like head is polar and hydrophilic, and the alkane-like
tail is nonpolar and hydrophobic. This is a good nonionic surfactant
(that is, an uncharged detergent).

CHAPTER 26

26.1. The radical intermediates would not be benzylic if they added
with the other orientation. **26.3.** The benzylic hydrogens are more
likely to be abstracted. **26.4.** They all add to give the more highly substituted carbocation. **26.5.** (a) and (b) are possible; (c) is terrible.
26.6. The cation at the end of a chain abstracts hydride from a benzylic
position in the middle of a chain. In isobutylene, a tertiary cation would
have to abstract a hydride from a secondary position: unlikely.
26.15. The third hydroxyl group of glycerol allows for profuse cross-linking of the chains (with a terephthalic acid linking two of these hydroxyl
groups), giving a very rigid polyester. **26.19.** Glycerol allows profuse
cross-linking, as in Problem 26-15. **26.23.** (a) a polyurethane; (b) condensation polymer; (c) $HO(CH_2)_3NH_2$ and CO_2. **26.24.** (a) a polyester; (b) condensation polymer; (c) dimethyl terephthalate and
1,4-butanediol; transesterification. **26.25.** (a) a polyurea; (b) condensation polymer; (c) $H_2N(CH_2)_9NH_2$ and CO_2. **26.26.** (a) polyether
(addition polymer); (b) ethylene oxide; (c) base catalyst.
26.27. (a) addition polymer; a synthetic rubber; (b) 2-chloro-1,3-butadi-
ene ("chloroprene"). **26.28.** (a) $-CH_2-O-[CH_2-O]-$;
(c) addition polymer. **26.31.** (b) and (c) No to both. Poly(vinyl
acetate) is an addition polymer. The ester bonds are not in the main
polymer chain; (d) Vinyl alcohol (the enol form of acetaldehyde)
is not stable.

PHOTO CREDITS

INDEX

Typical Values of Proton NMR Chemical Shifts

Type of Proton			Approximate δ
alkane	$(-CH_3)$	methyl	0.9
	$(-CH_2-)$	methylene	1.3
	$(-CH-)$	methine	1.4
$-\overset{\overset{O}{\parallel}}{C}-CH_3$		methyl ketone	2.1
$-C\equiv C-H$		acetylenic	2.5
$R-CH_2-X$ (X = halogen, $-O-$)			3–4
$\overset{/}{\underset{/}{C}}=\overset{\underset{H}{/}}{C}$		vinyl	5–6
$\overset{/}{\underset{/}{C}}=\overset{\underset{CH_3}{/}}{C}$		allylic	1.7
Ph—H		aromatic	7.2
Ph—CH$_3$		benzylic	2.3
R—CHO		aldehyde	9–10
R—COOH		acid	10–12
R—OH		alcohol	variable, about 2–5
Ar—OH		phenol	variable, about 4–7
R—NH$_2$		amine	variable, about 1.5–4

These values are approximate, because all chemical shifts are affected by neighboring substituents. The numbers given here assume that alkyl groups are the only other substituents present. A more complete table of chemical shifts appears in Appendix 1.

Summary of Functional Group Nomenclature

Functional Group	Name as Main Group	Name as Substituent
Main groups in order of decreasing priority		
carboxylic acids	-oic acid	carboxy[a]
esters	-oate	alkoxycarbonyl[a]
amides	-amide	amido[a]
nitriles	-nitrile	cyano
aldehydes	-al	formyl
ketones	-one	oxo
alcohols	-ol	hydroxy
amines	-amine	amino
alkenes	-ene	alkenyl
alkynes	-yne	alkynyl
alkanes	-ane	alkyl
ethers		alkoxy
halides		halo

[a] Denotes rare usage.

Typical Values of IR Stretching Frequencies

Frequency	Functional Group	Comments
3300 cm^{-1}	alcohol O—H	Always broad.
	amine, amide N—H	May be broad, sharp, or broad with spikes.
	alkyne $\equiv$C—H	Always sharp, usually strong.
3000 cm^{-1}	alkane C—H	Alkane $\overset{\backslash}{\underset{/}{C}}$—H just below 3000 cm^{-1}.
	alkene C—H	Alkene $=\overset{\overset{H}{/}}{\underset{\backslash}{C}}$ just above 3000 cm^{-1}.
	acid O—H	Very broad, 2500–3500 cm^{-1}.
2200 cm^{-1}	alkyne —C$\equiv$C—	Alkyne C$\equiv$C just below 2200 cm^{-1}.
	nitrile —C$\equiv$N	Nitrile C$\equiv$N just above 2200 cm^{-1}.
1710 cm^{-1}	$\overset{\backslash}{\underset{/}{C}}=O$	Ketones, aldehydes, acids.
	(very strong)	Esters higher, about 1735 cm^{-1}. Conjugation lowers frequency. Amides lower, about 1650 cm^{-1}.
1660 cm^{-1}	$\overset{\backslash}{\underset{/}{C}}=\overset{/}{\underset{\backslash}{C}}$	Conjugation lowers frequency. Aromatic C=C about 1600 cm^{-1}.
	$\overset{\backslash}{\underset{/}{C}}=N$	Stronger than C=C.
	amide C=O	Stronger than C=C.

Conjugation of multiple bonds generally lowers their stretching frequencies. A more complete table of IR stretching frequencies appears in Appendix 2.